Risk Modeling, Assessment, and Management

WILEY SERIES IN SYSTEMS ENGINEERING AND MANAGEMENT

Andrew P. Sage, Editor

A complete list of the titles in this series appears at the end of this volume.

RISK MODELING, ASSESSMENT, AND MANAGEMENT

Fourth Edition

Yacov Y. Haimes
Lawrence R. Quarles Professor of Systems and Information
Engineering and Civil and Environmental Engineering

Founding Director (1987), Center for Risk Management of Engineering Systems,
University of Virginia, Charlottesville

WILEY

Copyright © 2016 by John Wiley & Sons, Inc. All rights reserved

Published by John Wiley & Sons, Inc., Hoboken, New Jersey

Published simultaneously in Canada

No part of this publication may be reproduced, stored in a retrieval system, or transmitted in any form or by any means, electronic, mechanical, photocopying, recording, scanning, or otherwise, except as permitted under Section 107 or 108 of the 1976 United States Copyright Act, without either the prior written permission of the Publisher, or authorization through payment of the appropriate per-copy fee to the Copyright Clearance Center, Inc., 222 Rosewood Drive, Danvers, MA 01923, (978) 750-8400, fax (978) 750-4470, or on the web at www.copyright.com. Requests to the Publisher for permission should be addressed to the Permissions Department, John Wiley & Sons, Inc., 111 River Street, Hoboken, NJ 07030, (201) 748-6011, fax (201) 748-6008, or online at http://www.wiley.com/go/permissions.

Limit of Liability/Disclaimer of Warranty: While the publisher and author have used their best efforts in preparing this book, they make no representations or warranties with respect to the accuracy or completeness of the contents of this book and specifically disclaim any implied warranties of merchantability or fitness for a particular purpose. No warranty may be created or extended by sales representatives or written sales materials. The advice and strategies contained herein may not be suitable for your situation. You should consult with a professional where appropriate. Neither the publisher nor author shall be liable for any loss of profit or any other commercial damages, including but not limited to special, incidental, consequential, or other damages.

For general information on our other products and services or for technical support, please contact our Customer Care Department within the United States at (800) 762-2974, outside the United States at (317) 572-3993 or fax (317) 572-4002.

Wiley also publishes its books in a variety of electronic formats. Some content that appears in print may not be available in electronic formats. For more information about Wiley products, visit our web site at www.wiley.com.

Library of Congress Cataloging-in-Publication Data:

Haimes, Yacov Y.
Risk modeling, assessment, and management / Yacov Y. Haimes. – Fourth Edition.
    pages cm. – (Wiley series in systems engineering and management)
  Includes index.
  ISBN 978-1-119-01798-1 (hardback)
1. Technology–Risk assessment. 2. Technology–Risk assessment–Case studies. 3. Risk management.
4. Risk management–Case studies. I. Title.
  T174.5.H35 2015
  658.15′5011–dc23

2015004374

Set in 10.5/13.5pt Times Ten by SPi Global, Pondicherry, India

Printed in the United States of America

Contents

Preface to the Fourth Edition   ix

The Companion Website   xv

Acknowledgments   xvii

Part I. Fundamentals of Risk Modeling, Assessment, and Management   1

1 The Art and Science of Systems and Risk Analysis   3
1.1 Introduction / 3
1.2 Systems Engineering / 4
1.3 Risk Assessment and Management / 14
1.4 Concept Road Map / 26
1.5 Epilogue / 35
References / 35

2 The Role of Modeling in the Definition and Quantification of the Risk Function   41
2.1 Introduction / 41
2.2 The Risk Assessment and Management Process: Historical Perspectives / 43
2.3 Information, Intelligence, and Models / 45
2.4 The Building Blocks of Mathematical Models / 47
2.5 On the Complex Definition of Risk, Vulnerability, and Resilience: A Systems-Based Approach / 51
2.6 On the Definition of Vulnerabilities in Measuring Risks to Systems / 56
2.7 On the Definition of Resilience in Measuring Risk to Systems / 57
2.8 On the Complex Quantification of Risk to Systems / 60
References / 65

3 Identifying Risk through Hierarchical Holographic Modeling and Its Derivatives   69
3.1 Hierarchical Aspects / 69
3.2 Hierarchical Overlapping Coordination / 70
3.3 HHM / 73
3.4 HHM and the Theory of Scenario Structuring / 76
3.5 Adaptive Multiplayer HHM Game / 79
3.6 Water Resources System / 80
3.7 Sustainable Development / 83
3.8 HHM in a System Acquisition Project / 86
3.9 Software Acquisition / 90
3.10 Hardening the Water Supply Infrastructure / 94
3.11 Risk Assessment and Management for Support of Operations other than War / 98
3.12 Automated Highway System / 103
3.13 Food-Poisoning Scenarios / 108
References / 113

4 Modeling and Decision Analysis — 115
- 4.1 Introduction / 115
- 4.2 Decision Rules Under Uncertainty / 116
- 4.3 Decision Trees / 118
- 4.4 Decision Matrix / 122
- 4.5 The Fractile Method / 124
- 4.6 Triangular Distribution / 127
- 4.7 Influence Diagrams / 128
- 4.8 Population Dynamic Models / 132
- 4.9 PSM / 139
- 4.10 Example Problems / 144
- References / 152

5 Multiobjective Trade-Off Analysis — 155
- 5.1 Introduction / 155
- 5.2 Examples of Multiple Environmental Objectives / 157
- 5.3 The Surrogate Worth Trade-Off Method / 159
- 5.4 Characterizing a Proper Noninferior Solution / 166
- 5.5 The SWT Method and the Utility Function Approach / 168
- 5.6 Example Problems / 172
- 5.7 Summary / 177
- References / 178

6 Defining Uncertainty and Sensitivity Analysis — 179
- 6.1 Introduction / 179
- 6.2 Sensitivity, Responsivity, Stability, and Irreversibility / 180
- 6.3 Uncertainties Due to Errors in Modeling / 182
- 6.4 Characterization of Modeling Errors / 183
- 6.5 Uncertainty Taxonomy / 185
- 6.6 The USIM / 196
- 6.7 Formulation of the Multiobjective Optimization Problem / 199
- 6.8 A Robust Algorithm of the USIM / 204
- 6.9 Integration of the USIM with Parameter Optimization at the Design Stage / 207
- 6.10 Conclusions / 209
- References / 209

7 Risk Filtering, Ranking, and Management — 211
- 7.1 Introduction / 211
- 7.2 Past Efforts in Risk Filtering and Ranking / 212
- 7.3 RFRM: A Methodological Framework / 213
- 7.4 Case Study: An OOTW / 220
- 7.5 Summary / 224
- References / 224

Part II. Advances in Risk Modeling, Assessment, and Management — 227

8 Risk of Extreme Events and the Fallacy of the Expected Value — 229
- 8.1 Introduction / 229
- 8.2 Risk of Extreme Events / 230
- 8.3 The Fallacy of the Expected Value / 232
- 8.4 The PMRM / 233
- 8.5 General Formulation of the PMRM / 236
- 8.6 Summary of the PMRM / 238
- 8.7 Illustrative Example / 239
- 8.8 Analysis of Dam Failure and Extreme Flood through the PMRM / 240
- 8.9 Example Problems / 243
- 8.10 Summary / 257
- References / 257

9 Multiobjective Decision-Tree Analysis — 259
- 9.1 Introduction / 259
- 9.2 Methodological Approach / 261
- 9.3 Differences between SODT and MODT / 279
- 9.4 Summary / 281
- 9.5 Example Problems / 282
- References / 293

10 Multiobjective Risk Impact Analysis Method — 295
- 10.1 Introduction / 295
- 10.2 Impact Analysis / 296
- 10.3 The Multiobjective, Multistage Impact Analysis Method: An Overview / 297
- 10.4 Combining the PMRM and the MMIAM / 298

- 10.5 Relating Multiobjective Decision Trees to the MRIAM / 304
- 10.6 Example Problems / 313
- 10.7 Epilogue / 325
- References / 326

11 Statistics of Extremes: Extension of the PMRM — 329
- 11.1 A Review of the Partitioned Multiobjective Risk Method / 329
- 11.2 Statistics of Extremes / 333
- 11.3 Incorporating the Statistics of Extremes into the PMRM / 338
- 11.4 Sensitivity Analysis of the Approximation of $f_4(\cdot)$ / 344
- 11.5 Generalized Quantification of Risk of Extreme Events / 350
- 11.6 Summary / 356
- 11.7 Example Problems / 357
- References / 368

12 Systems-Based Guiding Principles for Risk Modeling, Planning, Assessment, Management, and Communication — 371
- 12.1 Introduction / 371
- 12.2 The *Journey*: The Guiding Principles in the Broader Context of the Emerging Next Generation Developed by the Federal Aviation Administration / 372
- References / 387

13 Fault Trees — 389
- 13.1 Introduction / 389
- 13.2 Basic Fault-Tree Analysis / 391
- 13.3 Reliability and Fault-Tree Analysis / 392
- 13.4 Minimal Cut Sets / 397
- 13.5 The DARE Using Fault Trees / 400
- 13.6 Extreme Events in Fault Tree Analysis / 403
- 13.7 An Example Problem Based on a Case Study / 405
- 13.8 Failure Mode and Effects Analysis and Failure Mode, Effects, and Criticality Analysis / 409
- 13.9 Event Trees / 411
- 13.10 Example Problems / 414
- References / 420

14 Multiobjective Statistical Method — 423
- 14.1 Introduction / 423
- 14.2 Mathematical Formulation of the Interior Drainage Problem / 424
- 14.3 Formulation of the Optimization Problem / 424
- 14.4 The MSM: Step-by-Step / 425
- 14.5 The SWT Method / 427
- 14.6 Multiple Objectives / 428
- 14.7 Applying the MSM / 429
- 14.8 Example Problems / 432
- References / 438

15 Principles and Guidelines for Project Risk Management — 439
- 15.1 Introduction / 439
- 15.2 Definitions and Principles of Project Risk Management / 440
- 15.3 Project Risk Management Methods / 443
- 15.4 Aircraft Development Example / 450
- 15.5 Quantitative Risk Assessment and Management of Software Acquisition / 454
- 15.6 Critical Factors That Affect Software Nontechnical Risk / 458
- 15.7 Basis for Variances in Cost Estimation / 460
- 15.8 Discrete Dynamic Modeling / 461
- 15.9 Summary / 469
- References / 469

16 Modeling Complex Systems of Systems with Phantom System Models — 473
- 16.1 Introduction / 473
- 16.2 What Have We Learned from Other Contributors? / 474
- 16.3 The Centrality of the States of the System in Modeling and in Risk Analysis / 476
- 16.4 The Centrality of Time in Modeling Multidimensional Risk, Uncertainty, and Benefits / 477
- 16.5 Extension of HHM to PSM / 478

16.6 PSM and Meta-modeling / 480
16.7 PSM Laboratory / 486
16.8 Summary / 488
References / 489

17 Adaptive Two-Player Hierarchical Holographic Modeling Game for Counterterrorism Intelligence Analysis 493

17.1 Introduction / 493
17.2 Bayes' Theorem / 494
17.3 Modeling the Multiple Perspectives of Complex Systems / 495
17.4 Adaptive Two-Player HHM Game: Terrorist Networks versus Homeland Protection / 499
17.5 The Building Blocks of Mathematical Models and the Centrality of State Variables in Intelligence Analysis / 502
17.6 Hierarchical Adaptive Two-Player HHM Game / 504
17.7 Collaborative Computing Support for Adaptive Two-Player HHM Games / 505
17.8 Summary / 507
References / 508

18 Inoperability Input–Output Model and Its Derivatives for Interdependent Infrastructure Sectors 511

18.1 Overview / 511
18.2 Background: The Original Leontief Input–Output Model / 512
18.3 Inoperability Input–Output Model / 513
18.4 Regimes of Recovery / 516
18.5 Supporting Databases for IIM Analysis / 517
18.6 National and Regional Databases for IIM Analysis / 518
18.7 RIMS II / 522
18.8 Development of the IIM and Its Extensions / 523
18.9 The Dynamic IIM / 527
18.10 Practical Uses of the IIM / 530
18.11 Uncertainty IIM / 533
18.12 Example Problems / 536
18.13 Summary / 539
References / 540

19 Case Studies 543

19.1 A Risk-Based Input–Output Methodology for Measuring the Effects of the August 2003 Northeast Blackout / 543
19.2 Systemic Valuation of Strategic Preparedness Through Applying the IIM with Lessons Learned from Hurricane Katrina / 558
19.3 *Ex Post* Analysis Using the IIM of the September 11, 2001, Attack on the United States / 569
19.4 Risk Modeling, Assessment, and Management of Lahar Flow Threat / 575
19.5 The Statistics of Extreme Events and 6-Sigma Capability / 587
19.6 Sequential Pareto-Optimal Decisions Made During Emergent Complex Systems of Systems: An Application to the FAA NextGen / 593
References / 612

Appendix: Optimization Techniques 617

A.1 Introduction to Modeling and Optimization / 617
A.2 Bayesian Analysis and the Prediction of Chemical Carcinogenicity / 655
A.3 The Farmer's Dilemma: Linear Model and Duality / 657
A.4 Standard Normal Probability Table / 664
References / 665

Author Index 667
Subject Index 673

Preface to the Fourth Edition

Public interest in the field of risk analysis has expanded in leaps and bounds during the recent three decades. Furthermore, risk analysis has emerged as an effective and comprehensive procedure that supplements and complements the overall management of almost all aspects of our lives. Managers of health care, the environment, and physical infrastructure systems of systems (e.g., water resources, transportation, infrastructure interdependencies, homeland and cyber security, and electric power, to cite a few) all incorporate risk analysis in their decisionmaking processes. The omnipresent adaptations of risk analysis by many disciplines, along with its deployment by industry and government agencies in decisionmaking, have led to an unprecedented development of theory, methodology, and practical tools. As a fellow of seven diverse professional societies, I find technical articles on risk analysis published in all of their journals. These articles address concepts, tools, technologies, and methodologies that have been developed and practiced in such areas as planning, design, development, system integration, prototyping, and construction of physical infrastructure; in reliability, quality control, and maintenance; and in the estimation of costs and schedules and in project management.

The challenge that faces society today is that all of this knowledge has not been fully duplicated, shared, and transferred from one field of endeavor to another. This calls for a concerted effort to improve our understanding of the commonalities and differences among diverse fields for the mutual benefit of society as a whole. Such a transfer of knowledge has always been the key to advancing the natural, social, and behavioral sciences, as well as engineering. I believe that we can start meeting this challenge through our college and university classrooms and through continuing education programs in industry and government. It is essential to build bridges among the disciplines and to facilitate the process of learning from each other.

Risk, a measure of the probability and severity of adverse effects, is a concept that many find difficult to comprehend, and its quantification has challenged and confused laypersons and professionals alike. There are myriad fundamental reasons for this state of affairs. One is that risk is a complex composition and amalgamation of two components—one real (the potential damage, or unfavorable adverse effects and consequences), the other (the likelihood of projected adverse consequences), measured or estimated through an imagined mathematical human construct termed *probability*. Probability per se is intangible, yet its omnipresence in risk-based decisionmaking is indisputable. Furthermore, the measure of the probability that dominates the measure of risk is itself uncertain, especially for rare and extreme events—for example, when there exists an element of surprise.

This book seeks to balance the quantitative and empirical dimensions of risk assessment and management with the more qualitative and normative aspects of decisionmaking under risk and uncertainty. In particular, select analytical methods and tools are presented without advanced mathematics or with no mathematics at all, to enable the less math-oriented reader to benefit from them. For example, hierarchical holographic modeling (HHM) is introduced and discussed in Chapter 3 for its value as a comprehensive and systemic tool for risk identification. While all mathematical details for hierarchical coordination (within the HHM philosophy) are mostly left out of the text, they are included in my earlier book, cited in Chapter 1, *Hierarchical Multiobjective Analysis of Large-Scale Systems* [Haimes et al., 1990]. Myriad case study applications of the HHM approach for risk identification are presented here, including studies conducted for the Presidential Commission for Critical Infrastructure Protection, the US Army, General Motors, the Federal Bureau of Investigation, Virginia Department of Transportation, VA Governor's Office, Institute for Information Infrastructure Protection (I3P), US Department of Homeland Security, and the US Department of Defense, among others. The HHM philosophy is grounded on the premise that complex systems, such as air traffic control systems, should be studied and modeled in more than one way. Because such complexities cannot be adequately modeled or represented through a planar or single model or vision, overlapping of these visions is unavoidable. This can actually be helpful in providing a holistic appreciation of the interconnectedness among the various components, aspects, objectives, and decisionmakers associated with a system.

Furthermore, this holistic approach stems from the realization that the process of risk assessment and management is a blend of art and science; and although mathematical formulation and modeling of a problem are important for sound decisionmaking, they are not by themselves sufficient for that purpose. Clearly, institutional, organizational, managerial, political, and cultural considerations, among others, can be as dominant as scientific, technological, economic, or financial aspects, and must be accounted for in the decisionmaking process.

Consider, for example, the protection and management of a major water supply system. Deploying the HHM approach discussed in Chapter 3, it is possible to address the holistic nature of the system in terms of its hierarchical decisionmaking structure, which includes various time horizons, multiple decisionmakers, stakeholders, and users of the water supply system, and a host of hydrological, technological, legal, and other socioeconomic conditions and factors that require consideration. The effective identification of the myriad sources of risk to which natural, cyber, or physical system are exposed is markedly improved by considering all real, perceived, or imaginary risks from their multiple decompositions, visions, and perspectives.

The adaptive multiplayer HHM (AMP-HHM) game, introduced in Chapter 3, is an important concept with the potential to serve as a repeatable, adaptive, and systemic process that can contribute to tracking terrorism scenarios [Haimes and Horowitz, 2004]. It builds on fundamental principles of systems engineering, systems modeling, and risk analysis. The AMP-HHM game captures multiple perspectives of a system through computer-based interactions. For example, a two-player game creates two opposing views of the opportunities for carrying out acts of terrorism: one developed by a Blue team defending against terrorism, and the other by a Red team planning to carry out a terrorist act.

This book draws on my experience in the practice of risk-based decisionmaking in government and industry, and it builds on results from numerous management-based projects. It is also based on homework and exams compiled during over 40 years of teaching graduate courses in risk analysis at Case Western Reserve University and at the University of Virginia. In addition, the text incorporates the results of close to four decades of research and consulting work with industry and government that has resulted in over 80 masters and 50 doctoral theses and numerous technical papers on risk analysis.

I have also gained experience and knowledge from organizing and chairing 12 Engineering Foundation conferences on risk-based decisionmaking since 1980.

The interaction with the participants in these intensely focused meetings has markedly influenced the structure of this book. I have benefited as well from the foresight and practical orientation of hundreds of participants in numerous short courses that I taught along with colleagues from 1968 to the present. For example, for 29 consecutive years, I offered a 1-week short course titled *Hierarchical-Multiobjective Approach in Water Resources Planning and Management*. I have been offering a graduate course on risk analysis at the University of Virginia since 1987.

In preparing the first (1998), second (2004), third (2009), and fourth (2016) editions of this book, I have been guided by the following premises and needs:

1. Increasingly, international as well as US federal and state legislators and regulatory agencies have been addressing the assessment and management of risk more explicitly, whether in environmental and health protection, human safety, manufacturing, or security.
2. There is a need for a text that presents both basic and advanced methodologies in risk analysis at a sufficiently detailed level so that the reader can confidently apply specific methods to appropriate problems. To achieve this fundamental goal, risk methodologies presented in this book are supplemented with example problems and, when possible, with case study applications.
3. The modeling and assessment of risk necessarily lead to noncommensurate and conflicting objectives. Invariably, the reduction or the management of risk requires the expenditure of funds and other resources. Thus, at its simplest modeling level, at least two objectives must be considered: (i) minimizing and managing risk (e.g., environmental risk, health risk, and risk of terrorism) and (ii) minimizing the cost associated with achieving these goals. Although the concept of a multiattribute utility may be grounded on a brilliant theory, it might not be practical when applied to real-world problems and human decisionmakers. Therefore, this book emphasizes multiobjective trade-off analysis, which avoids the pre-commensuration of risks, costs, and benefits through a single utopian utility function.
4. Risk has been commonly quantified through the mathematical expectation formula. Fundamentally, the mathematical expected value concept precommensurates low-frequency events of extreme or catastrophic consequences with high-frequency events of minor impact. Although the mathematical expectation provides a valuable measure of risk, it fails to recognize or accentuate extreme event consequences. To complement the expected value of risk, this book presents a supplementary measure termed the *conditional expected value of risk* and applies it throughout the text whenever possible.
5. One of the most difficult tasks that has been least addressed in most systems analysis literature is knowing how to model a system. Most systems engineering and operations research texts offer a wealth of theories and methodologies for problem solving—that is, optimizing a pre-assumed system's model. Furthermore, most texts neglect the art and science of model building and the centrality of the state variables and other building blocks in model formulation. Given that risk cannot be managed unless it is properly assessed and that the best assessment process is realized through some form of model, the modeling process becomes an imperative step in the systemic assessment and management of risk. Consequently, this book devotes a concerted effort to the modeling task as a prelude to the ultimate assessment and management of risk.
6. Many tend to consider the field of risk analysis as a separate, independent, and well-defined discipline of its own. However, this book views the theory and methodology of risk analysis within the broader context of systems engineering (e.g., modeling and optimization), albeit with more emphasis on the stochasticity of the system and its components. This philosophical approach legitimizes the pedagogy of the separation and subsequent integration of systems modeling (risk assessment) and systems optimization and implementation (risk management). It also invites the risk

analyst to benefit fully from the utilization of the vast theories, methodologies, tools, and experience generated under the broader rubric of systems analysis and systems engineering. Indeed, imperative in any sound risk analysis is the use of such fundamental concepts as modeling, optimization, simulation, multiobjective trade-offs, regression, fault trees, fault tolerance, multiobjective decision trees, event trees, forecasting, scheduling, and numerous other tools for decisionmaking.

A book on such a broad subject as risk analysis has the potential for a significantly diverse readership. Thus, although there is a unifying theme for the theory and methodology developed for use in risk analysis, its applications can encompass every possible field and discipline. Furthermore, readers may have different levels of interest in the quantitative/empirical and the qualitative/normative aspects of risk. To at least partially meet this challenge, this book is organized in two parts.

Part I—*Fundamentals of Risk Modeling, Assessment, and Management*—which includes Chapters 1–7 and the Appendix to Part I, focuses on the more philosophical, conceptual, and decisionmaking aspects of risk analysis. It addresses fundamental concepts of modeling and optimization of systems under conditions of risk and uncertainty, articulates the intricate processes of risk assessment and management, and presents commonly known and newly developed risk analysis methodologies.

Chapter 1 provides an overview of risk analysis in the broader context of systems engineering. For example, relating Stephen Covey's book, *The Seven Habits of Highly Effective People* [1989], to systems engineering principles and from there to risk analysis is one way in which the text attempts to bridge the quantitative and qualitative dimensions of risk analysis.

Chapter 2 introduces the reader to the fundamental building blocks of mathematical models—concepts that will be understood by all who have had two courses in college calculus. The chapter has been modified and updated with a major new section on the complex definition of risk, vulnerability, and resilience: a systems-based approach. Indeed, all readers in managerial and decisionmaking positions who have a basic knowledge of college calculus and some understanding of probability can benefit from Part I of this book. To further assist the reader, the Appendix provides a review of linear and nonlinear optimization, and Bayesian analysis.

Chapter 3 (as noted earlier) addresses the HHM philosophy for risk identification and introduces the reader to the contributions made to risk management by social and behavioral scientists.

Chapter 4, as its title indicates, offers a review of fundamentals in decision analysis and the construction of evidence-based probabilities for use in decisionmaking. At various levels of the decisionmaking process, managers often encounter situations where sparse statistical data do not lend themselves to the construction of probabilities. Through illustrative examples and case studies, this chapter will make it possible for such managers to augment evidence gained through their professional experience with evidence collected through other means.

Chapter 5 introduces the uninitiated reader to the analysis of multiple objectives. One of the characteristic features of risk-based decisionmaking is the imperative need to make trade-offs among all costs, benefits, and risks. Although multiobjective analysis is the focus of this chapter, utility theory is related to this and is also briefly discussed. While the centrality of multiobjective trade-off analysis in decisionmaking is dominant in this book, and more than one chapter would be needed to adequately addresses this subject, the reader is referred to a newly republished textbook (2008) by Dover Publishing company, titled *Multiobjective Decision Making: Theory and Methodology*, by Vira Chankong and Yacov Y. Haimes.

Chapter 6 discusses sensitivity analysis and, through an uncertainty taxonomy, the broader issues that characterize uncertainty in general; also, it develops the uncertainty sensitivity index method (USIM) and its extensions. Only the extensions of the USIM component of this chapter require advance knowledge of optimization.

Chapter 7 presents a modified and improved risk filtering ranking, and management (RFRM) method. The risk ranking and filtering (RRF) method, which was developed for NASA in the early 1990s and was introduced in Chapter 4 in the first edition of this book, is only briefly discussed in this edition. The Appendix to Part I provides an overview of optimization techniques, including linear programming, Lagrange multipliers, and dynamic programming.

Part II—*Advances in Risk Modeling, Assessment, and Management*—which includes Chapters 8–19, shares with the readers the theory and ensuing methodology that define the state-of-the-art of risk analysis.

Chapter 8 covers the concept of conditional expected value of risk and discusses the partitioned multiobjective risk method (PMRM), which complements and supplements the expected (unconditional) value of risk. Several examples illustrate the erroneous analysis that is likely to result from using the conventional (unconditional) expected value as the sole measure of risk.

Chapter 9 extends the single-objective decision-tree analysis introduced in Chapter 4 to incorporate multiple objectives, and explains the multiple objective decision tree (MODT) method.

Chapter 10 extends the modeling, assessment, and management of risk from the static, time-invariant case to the dynamic case. Also, the multiobjective risk-impact analysis method (MRIAM) is described and is related to the MODT. Because the two methodologies are useful in decisionmaking at each step of the system life cycle, the theoretical and methodological relationship between MRIAM and MODT developed by Dicdican and Haimes [2005] is also presented in this chapter.

Chapter 11 incorporates the statistics of extremes with the conditional expected value of risk (developed through the PMRM), and thus it extends the theory and methodology upon which the PMRM is grounded.

Chapter 12 The old section on Bayesian analysis has been moved to the Appendix, and the remainder of the text has been replaced with systems-based guiding principles for risk modeling, planning, assessment, management, and communication.

Chapter 13 discusses the basics of fault-tree analysis, focusing on the central concept of *minimal cut sets*. It also introduces the distribution analyzer and risk evaluator (DARE) method using fault trees, and failure mode, effects, and criticality analysis (FMECA).

Chapter 14 explains the Multiobjective Statistical Method (MSM), where the symbiotic relationship between model simulation and multiobjective trade-off analysis is exploited. This chapter also focuses on modeling problems with one or more random variables, where the state variables play a central role in the modeling process.

Chapter 15 addresses principles and guidelines for project management and associated risk assessment and management issues, as well as the life cycle of software development.

Chapter 16 The old text on applying risk analysis to the space mission has been replaced with modeling complex systems of systems with phantom system models in recognition that the natural and the constructed environment are complex interdependent and interconnected systems of systems.

Chapter 17 The old text on risk modeling, assessment, and management of terrorism has been replaced with an updated text that builds on hierarchical holographic modeling (introduced in Chapter 3), with a focus on an adaptive two-player hierarchical holographic modeling game for counterterrorism intelligence analysis.

Chapter 18 is devoted in its entirety to modeling the interdependencies among infrastructures and sectors of the economy through the Leontief-based inoperability input–output model (IIM) and its derivatives: the dynamic IIM (DIIM), multiregional IIM (RIIM), and uncertainty IIM (UIIM). Detailed step-by-step derivations are presented of all the models introduced in this chapter. The chapter provides an extensive discussion on national, regional, state, and local supporting databases for the IIM and its derivatives.

Chapter 19 adds a sixth case study in this edition to further demonstrate the application of the risk-based methodologies introduced in this book. The theme of the sixth case study is on sequential Pareto-optimal decisions made within emergent complex systems of systems, with an application to the FAA NextGen.

The *Appendix* has been expanded to include Bayesian analysis for the prediction of chemical carcinogenicity (moved from old Chapter 12), and the Farmer's Dilemma, introduced in Chapter 1, has been formulated and solved using a deterministic linear model in the Appendix.

The Companion Website

This fourth edition comes with a companion website resulting from a longstanding collaboration with my colleagues and former students, Dr. Joost Santos and Dr. Zhenyu Guo. Although a large number of solved problems in risk-based decisionmaking are included in the text, the companion website contains over 200 exercises and problems that feature risk analysis theories, methodologies, and applications accompanies this Fourth Edition.

The objective of the companion website is to provide reinforced learning experiences for risk analysis scholars and practitioners through a diverse set of problems and hands-on exercises. For better tractability, these are organized similar to the chapters of this book and range from foundation topics (e.g., building blocks of modeling and structuring of risk scenarios) to relatively more complex concepts (e.g., multiobjective trade-off analysis and statistics of extremes). The problems encompass a broad spectrum of applications including disaster analysis, industrial safety, transportation security, production efficiency, and portfolio selection, among others.

The exercises and problems in the companion website are attributable to numerous students who participated in my Risk Analysis course during the last 30 years. The production of the content on the website would have not been possible without the help of the following student encoders: Dexter Galozo, Jonathan Goodnight, Miguel Guerra, Sung Nam Hwang, Jeesang Jung, Oliver Platt-Mills, Chris Story, Scott Tucker, Gen Ye, Zhenyu Guo, Joshua Bogdanor, Eva Andrijcic, and Bryan Lewis. The administrative as well as the technical support from Erika Evans and Rosemary Shaw are greatly appreciated. Last but not least, I would like to once again acknowledge Grace Zisk for her meticulous editing of the first three editions and Anne Sussman for help in editing this fourth edition.

Yacov Y. Haimes

June, 4, 2015
Charlottesville, Virginia

Acknowledgments

ACKNOWLEDGMENTS TO THE FIRST EDITION

Writing this acknowledgment is probably one of the most rewarding moments in the preparation of this book, because each of the individuals cited here played some significant role during what might be viewed as the "life cycle" of this project. Even with a careful accounting, there will likely be some who have been inadvertently missed. A great sage once said: "From all my teachers I learned and became an educated person, but my students contributed the most to my learning." This statement epitomizes the gratitude that I owe to more than 100 of my doctoral and masters students whom I have had the privilege of serving as thesis advisor and from whom I learned so much.

My long-term professional collaboration with Duan Li and the many papers that we published together during the last two decades had a major impact on the scope and contents of this book. I will always cherish his contributions to my professional growth. The painstaking and judicious technical editorial work of Grace Zisk is most appreciated and heartily acknowledged. Her personal dedication to the task of ensuring that every sentence adequately communicates its intended meaning has been invaluable. I would like to thank the undergraduate work-study students who labored long hours typing and retyping the text and drawing the figures. These students include Matthew Dombroski, Scott Gorman, Matt Heller, Matthew Kozlowski, William Martin-Gill, Luke Miller, and Elsa Olivetti. Working tirelessly, Ryan Jacoby has diligently and with great talent helped me to finally bring the text to its final version. I very much value the time and effort spent by Ganghuai Wang, who read the manuscript and made valuable comments and suggestions. My daily association and discussions with Jim Lambert have contributed to many ideas that have been formulated and finalized here. His thoughtfulness and advice have certainly improved the quality of this text. My gratitude also goes to Andrew P. Sage, the editor of the Wiley Series in Systems Engineering, to George Telecki, Executive Editor of John Wiley & Sons, and to Lisa Van Horn, Production Editor at John Wiley & Sons, for their valuable advice and for facilitating the publication of this book.

Several chapters draw from earlier articles published with my former graduate students. Prominent among them are Eric Asbeck, Vira Chankong, Steve Eisele, James Fellows, Herbert Freedman, Fernando Gomide, Lori Johnson-Payton, Per-Ola Karlsson, Peter Kuzminski, Jim Lambert, Mark Leach, Duan Li, Con Way Ling, Jim Mitsiopoulas, W. Scott Nainis, Steve Olenik, Rolf Olsen, Jerry Pannullo, Julia Pet-Edwards, Raja Petrakian, Calvin Schneiter, Richard Schooff, Takashi Shima, Kyosti Tarvainen, Vijay

Tulsiani, and Ganghuai Wang. I am indeed very grateful to all of them.

I am thankful to the following publishers who granted me and John Wiley & Sons, Inc. permission to reproduce material from published journal articles: Naval Research Logistics, The American Water Resources Association, American Geophysical Union, *Molecular Toxicology*, Taylor & Francis, *Risk Analysis*, Plenum, the Institute of Electrical and Electronics Engineers, Inc., Hemisphere Publishing Corporation, and the Defense Systems Management College Press. Specific citations are noted in the text.

The solved example problems add an important dimension to this book. They were developed as homework assignments or exams in the graduate course on risk analysis that I have taught for the last 20 years. Many of these problems were initiated and formulated by students. I am particularly indebted to the following students for their contributions: James Fellows, Hendrik Frohwein, Amy Chan Hilton, Bronwyn Jackson, Matthew M. Mehalik, Silla K. Mullei, David Romo Murillo, Michael A. Reynolds, Lauren A. Schiff, Yong Seon, Julie M. Stocker, Vijay Tuilsiani, Chenxi Wang, and Ganghuai Wang.

I would like to acknowledge the following colleagues who have either directly or indirectly influenced the outcome of this book: Clyde Chittister, Keith Hipel, Barry Johnson, Nick Matalas, David Moser, S. K. Nanda, Eugene Stakhiv, John A. Dracup, Irving Lefkowitz, Harry W. Mergler, Arden L. Bement, Jr., Betty Anderson, General Alton D. Slay, Andrew P. Sage and the late Warren A. Hall. I also would like to thank Sharon Gingras, the Center's former Office Manager, and Leslie Yowell, the Center's current Office Manager, for their assistance in bringing this book to press.

Last, but certainly not least, I thank my wife, Sonia, for her continuous support and patience. I dedicate this book to her.

Y. Y. H.

ACKNOWLEDGMENTS TO THE SECOND EDITION

To ensure that the solutions to the numerous example problems are correct and that the text is as free of typographical and other errors, I relied heavily on my graduate students for help. I am grateful to Christopher W. Anderson, Kenneth Crowther, Ruth Dicdican, Mike Diehl, Matt Dombroski, Paul Jiang, Robb Keeter, Piyush Kumar, Greg Lamm, Maria F. Leung, Chenyang Lian, Steve Liang, Marco Lopez, Alex Mosenthal, Mike Pennock, Joost R. Santos, Rod Shirbacheh, Anitha Srinivasan, Curtis Tait, Joshua Tsang, and Ganghuai Wang for their professional help and support.

Materials from papers published jointly with several of my graduate students have been incorporated into this edition of the book. These students are Matt Dombroski, Barry Ezell, Paul Jiang, Maria F. Leung, Mike Pennock, Joost R. Santos, and Rich Schooff.

I am indebted to Joost R. Santos for his invaluable help on Chapters 11 and 16, Sections A.9–A.11, and throughout this book. I also want to acknowledge the extensive help that I received from Maria F. Leung on Chapter 16 and from Steve Liang on Chapters 11 and 16.

Special appreciation is due to Christopher W. Anderson, who has masterfully typed and retyped the multiple versions of the second edition with skill, patience, and dedication beyond the call of duty; to Della Dirickson, who in addition to managing the Center for Risk Management of Engineering Systems, has put her vast experience in the production of publications to work on the production of this edition of the book; and to Grace Zisk (and to her computer helper, Burt Zisk) for her outstanding service and expert technical editing of the various drafts of the first and second editions of this book.

I also want to thank my colleagues Clyde Chittister, Barry Horowitz, Stan Kaplan, Jim Lambert, Tom Longstaff, Garrick Louis, Irv Pikus, Kent Schlussel, Paul Slovic, Mark Sulcoski, and Bill Wulf for their direct and indirect contributions to this book.

I am grateful to Andrew P. Sage, the editor of the Wiley Series in Systems Engineering, for his continued support and encouragement; George Telecki, Executive Editor at John Wiley & Sons; Cassie Craig, Wiley Editorial Assistant and Beverly Miller, the Copy Editor, and Lisa Van Horn, the Desk Editor for their culminating efforts and assistance, which have made this edition a reality.

Last, but certainly not least, I thank my wife, Sonia, for her loving patience and continuous support; I rededicate this book to her.

Y. Y. H.

ACKNOWLEDGMENTS TO THE THIRD EDITION

This third edition of *Risk Modeling, Assessment, and Management* was made possible through the generous support and technical help of many individuals to whom I owe heartfelt gratitude. As always, I relied on my graduate students who are expert at finding errors and identifying incoherent explanations of complex mathematical equations. Many thanks to Abhinav Agrawal, Kash Barker, Brett Dickey, Marcus Grimes, Ping Guan, Matt Henry, Sung Nan Hwang, Jeesang Jung, Gary Larimer, Aaron Lee, Chenyang Lian, Lindsey McGuire, Mark Orsi, Barrett Strausser, Scott Tucker, Mark Waller, and William Yan for their professional help and support.

Material from papers published jointly with several of my colleagues and graduate students have been incorporated into this edition of the book. These colleagues are Clyde Chittister, Kenneth Crowther, Barry Horowitz, Jim Lambert, and Joost Santos. The graduate students are Christopher Anderson, Kash Barker, Ruth Dicdican, Matt Henry, and Chenyang Lian.

The restructuring of Chapters 17 and 18 and the addition of Chapter 19 benefited from the help and contributions of my colleagues Kenneth Crowther and Joost Santos. I am indebted to their valuable help and contributions.

Special thanks and appreciation are due to Sung Nam Hwang, who masterfully typed and retyped the multiple versions of the third edition with patience, skill, dedication, and commitment to quality; to Erika Evans, who in addition to managing the Center for Risk Management of Engineering Systems, coordinated the many tasks that brought this edition of the book to its completion; and to our technical editor, Grace Zisk (and to her computer helper, Burt Zisk), for her friendship, commitment to quality, outstanding service, and expert technical editing of the various drafts of the first, second, and third editions of this book.

I am most appreciative and grateful to Andrew P. Sage, Editor of the Wiley Series in Systems Engineering and to George Telecki, Executive Editor at John Wiley & Sons for their continued support and encouragement. Warm thanks are due to Melissa Valentine, Editorial Assistant at Wiley, and Lisa Van Horn, Desk Editor at Wiley, for their efforts and assistance in making this edition a reality.

The frequent daily walks and stimulating colloquies with my colleague Barry Horowitz have germinated many of the ideas discussed in this edition.

I thank my wife Sonia once again for her constant encouragement and loving support. This third edition is again dedicated to her.

Y. Y. H.

ACKNOWLEDGMENTS TO THE FOURTH EDITION

This fourth edition of *Risk Modeling, Assessment, and Management* was made possible through the generous support and technical help of many individuals to whom I owe heartfelt gratitude. As always, I relied on my graduate students who are expert at finding errors and identifying incoherent explanations of complex mathematical equations. Many thanks to Zhenyu Guo, Eva Andrijcic, Evan Rust, Justin Bleistein, Joshua Bogdanor, Clay White, and Bryan Lewis.

New material from recent papers published jointly with several of my colleagues and graduate students have been incorporated into this edition of the book. These colleagues are Clyde Chittister, Barry Horowitz, Kenneth Crowther, and Andy Anderegg. The graduate students are Zhenyu Guo and Eva Andrijcic.

Special thanks and appreciation are due to Chetan Mishra and Ashutosh S. Panchang, who masterfully typed and retyped the multiple changes and additions made to the third edition with patience, skill, dedication, and commitment to quality; to Erika Evans and Rosemary Shaw, who in addition to managing the Center for Risk Management of Engineering Systems, coordinated the many tasks that brought the fourth edition of the book to its completion. I am grateful for Rosemary's keen eye for detail in the final review and to the fine contribution of our technical editor, Anne Sussman for her commitment to quality and expert technical editing of the additions to this fourth edition.

I am most appreciative and grateful to Andrew P. Sage, Editor of the Wiley Series in Systems Engineering, to Kari Capone, Editorial Program Coordinator, Global Research, Professional Practice and Learning, and Brett Kurzman, Editor, Global Research, Professional Practice and Learning at

John Wiley & Sons for their continued support and encouragement, and for their assistance to bringing this fourth edition to fruition. Warm thanks are due to George Telecki, former Executive Editor at Wiley, Melissa Valentine, Editorial Assistant at Wiley, and Lisa Van Horn, Desk Editor at Wiley, for their efforts and assistance in making the previous editions a reality.

The frequent daily walks and stimulating colloquies with my colleague Barry Horowitz have germinated many of the ideas discussed in this fourth edition.

I thank my wife Sonia once again for her constant encouragement and loving support. This fourth edition is again dedicated to her.

Y. Y. H.

Part I

Fundamentals of Risk Modeling, Assessment, and Management

1

The Art and Science of Systems and Risk Analysis

1.1 INTRODUCTION

Risk-based decisionmaking and *risk-based approaches in decisionmaking* are terms frequently used to indicate that some systemic process that deals with uncertainties is being used to formulate policy options and assess their various distributional impacts and ramifications. Today, an ever-increasing number of professionals and managers in industry, government, and academia are devoting a large portion of their time and resources to the task of improving their understanding and approach to risk-based decisionmaking. In this pursuit, they invariably rediscover (often with considerable frustration) the truism: The more you know about a complex subject, the more you realize how much still remains unknown. There are three fundamental reasons for the complexity of this subject. One is that decisionmaking under uncertainty literally encompasses every facet, dimension, and aspect of our lives. It affects us at the personal, corporate, and governmental levels, and it also affects us during the planning, development, design, operation, and management phases. Uncertainty colors the decisionmaking process regardless of whether it (i) involves one or more parties, (ii) is constrained by economic or environmental considerations, (iii) is driven by sociopolitical or geographical forces, (iv) is directed by scientific or technological know-how, or (v) is influenced by various power brokers and stakeholders. Uncertainty is inherent when the process attempts to answer the set of questions posed by William W. Lowrance: "Who should decide on the acceptability of what risk, for whom, in what terms, and why?" [Lowrance, 1976]. The second reason why risk-based decisionmaking is complex is that it is cross-disciplinary. The subject has been further complicated by the development of diverse approaches of varying reliability. Some methods, which on occasion produce fallacious results and conclusions, have become entrenched and would be hard to eradicate. The third reason is grounded on the need to make trade-offs among all relevant and important costs, benefits, and risks in a multiobjective framework, without assigning weights with which to commensurate risks, costs, and benefits.

In his book *Powershift*, Alvin Toffler [1991] states:

> As we advance into the Terra Incognito of tomorrow, it is better to have a general and incomplete map,

---

Risk Modeling, Assessment, and Management, Fourth Edition. Yacov Y. Haimes.
© 2016 John Wiley & Sons, Inc. Published 2016 by John Wiley & Sons, Inc.

subject to revision and correction, than to have no map at all.

Translating Toffler's vision into the risk assessment process implies that a limited database is no excuse for not conducting sound risk assessment. On the contrary, with less knowledge of a system, the need for risk assessment and management becomes more imperative.

Consider, for example, the risks associated with natural hazards. Causes for major natural hazards are many and diverse, and the risks associated with these natural hazards affect human lives, the environment, the economy, and the country's social wellbeing. Hurricane Katrina, which struck New Orleans in the United States on August 29, 2005, killing a thousand people and destroying properties, levees, and other physical infrastructures worth billions of dollars, is a classic example of a natural hazard with catastrophic effects [McQuaid and Schleifstein, 2006]. The medium within which many of these risks manifest themselves, however, is engineering-based physical infrastructure—dams, levees, water distribution systems, wastewater treatment plants, transportation systems (roads, bridges, freeways, and ports), communication systems, and hospitals, to cite a few. Thus, when addressing the risks associated with natural hazards, such as earthquakes and major floods, or willful hazards, that is, acts of terrorism, one must also account for the impact of these hazards on the integrity, reliability, and performance of engineering-based physical and human-based societal infrastructures. The next step is to assess the consequences—the impact on human and nonhuman populations and on the socioeconomic fabric of large and small communities.

Thus, risk assessment and management must be an integral part of the decisionmaking process, rather than a gratuitous add-on technical analysis. Figure 1.1 depicts this concept and indicates the ultimate need to balance all the uncertain benefits and costs.

For the purpose of this book, *risk* is defined as *a measure of the probability and severity of adverse effects* [Lowrance, 1976]. Lowrance also makes the distinction between risk and safety: Measuring risk is an empirical, quantitative, scientific activity (e.g., measuring the probability and severity of harm). Judging safety is judging the acceptability of risks—a

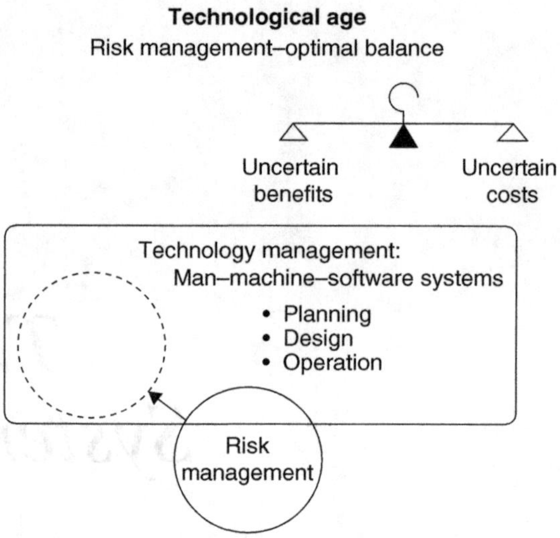

Figure 1.1 Risk management as an integral part of overall management.

normative, qualitative, political activity. Indeed, those private and public organizations that can successfully address the risks inherent in their business—whether in environmental protection, resource availability, natural forces, the reliability of man–machine systems, or future use of new technology—will dominate the technological and service-based market.

The premise that risk assessment and management must be an integral part of the overall decisionmaking process necessitates following a systemic, holistic approach to dealing with risk. Such a holistic approach builds on the principles and philosophy upon which systems analysis and systems engineering are grounded.

1.2 SYSTEMS ENGINEERING

1.2.1 What Is a System?

The human body and each organ within it, electric power grids and all large-scale physical infrastructures, educational systems from preschool to higher education, and myriad other human, organizational, hardware, and software systems are large-scale, complex, multiscale interconnected and interdependent systems with life cycles that are characterized by risk and uncertainty along with emergent behavior. But exactly what is a system? *Webster's Third New*

International Dictionary offers several insightful definitions:

> A complex unity formed of many often diverse parts subject to a common plan or serving a common purpose; an aggregation or assemblage of objects joined in regular interaction or interdependence; a set of units combined by nature or art to form an integral, organic, or organizational whole.

Almost every living entity, all infrastructures, both the natural and constructed environment, and the entire households of tools and equipment are complex systems often composed of myriad subsystems that in their essence constitute *systems of systems* (*SoS*). Each is characterized by a hierarchy of interacting and networked components with multiple functions, operations, efficiencies, and costs; the component systems are selected and coordinated according to some existing trade-offs between multiple objectives and operational perspectives. Clearly, no single model can ever attempt to capture the essence of such systems—their multiple dimensions and perspectives.

1.2.2 What Is Systems Engineering?

Even after over half a century of systems engineering as a discipline, many engineers find themselves perplexed about the following question: What is systems engineering?

Systems engineering is distinguished by its practical philosophy that advocates holism in cognition and in decisionmaking. This philosophy is grounded on the arts, natural and behavioral sciences, and engineering and is supported by a complement of modeling methodologies, state-space theory, optimization and simulation techniques, data management procedures, and decisionmaking approaches. The ultimate purpose is to (i) build an understanding of the dynamic system's nature, functional behavior, and interaction with its environment, (ii) improve the decisionmaking process (e.g., in planning, design, development, operation, management), and (iii) identify, quantify, and evaluate risks, and epistemic and aleatory uncertainties for a guided and actionable decisionmaking process.

One way of gaining greater understanding of systems engineering is to build on the well-publicized ideas of Stephen R. Covey in his best-selling book, *The Seven Habits of Highly Effective People* [Covey, 1989], and to relate these seven habits to various steps that constitute systems thinking or the systems approach to problem solving. Indeed, Covey's journey for personal development as detailed in his book has much in common with the holistic systems concept that constitutes the foundation of the field of systems engineering. Even the transformation that Covey espouses, from thinking in terms of you to me to we, is similar to moving from the perception of interactions as reactive or linear to a holistic view of connected relationships. Viewed in parallel, the two philosophies—Covey's and the systems approach—have a lot in common. The question is: How are they related, and what can they gain from each other?

Analyzing a system cannot be a selective process, subject to the single perspective of an analyst who is responsible for deciphering the maze of disparate and other knowledge. Rather, a holistic approach encompasses the multiple visions and perspectives inherent in any vast pool of data and information. Such a systemic process is imperative in order to successfully understand and address the complexity of an SoS [NRC, 2002].

1.2.3 Historical Perspectives of Systems Engineering

1.2.3.1 Classical philosophers who practiced holistic systems thinking

The *systems* concept has a long history. The art and science of systems engineering as a natural philosophy can be traced to Greek philosophers. Although the term *system* itself was not emphasized in earlier writings, the history of this concept includes many illustrious names, including *Plato* (428–348 B.C.) [Hutchins, 1952] and *Aristotle* (384–322 B.C.). The writings of *Baron von Leibniz* (1646–1716), a mathematician and philosopher, are directed by holism and systems thinking. He shares with *Isaac Newton* (1642–1727) the distinction of developing the theory of differential and integral calculus. By quantifying the causal relationships among the interplanetary SoS, Newton represents the epitome of a systems philosopher and modeler. In their seminal book, *Isaac Newton,*

The Principia, Cohen and Whitman [1999] write (p. 20):

> Newton's discovery of interplanetary forces as a special instance of universal gravity enables us to specify two goals of the Principia. The first is to show the conditions under which Kepler's laws of planetary motion are exactly or accurately true; the second is to explore how these laws must be modified in the world of observed nature by perturbations in the motions of planets and their moons.

Johann Gottlieb Fichte (1762–1814) introduced the idea of synthesis—one of the fundamental concepts of systems thinking. For example, he argued that *freedom* can never be understood unless one loses it. Thus, the *thesis* is that a man is born free, the loss of freedom is the *antithesis*, and the ability to enjoy freedom and do good works with it is the *synthesis*. In other words, to develop an understanding of a system as a whole (synthesis), one must appreciate and understand the roles and perspectives of its subsystems (thesis and antithesis). *Georg Hegel* (1770–1831), a contemporary of Fichte, was one of the most influential thinkers of his time. Like Aristotle before him, Hegel tried to develop a system of philosophy in which all the contributions of his major predecessors would be integrated. His *Encyclopedia of the Philosophical Sciences* (1817), which contains his comprehensive thoughts in a condensed form, provides important foundations for the concept of holism and the overall systems approach [Hegel, 1952].

Around 1912, *Max Wertheimer*, *Kurt Koffka*, and *Wolfgang Kohler* founded the Gestalt psychology, which emphasizes the study of experience as a *unified whole*. The German word *gestalt* means pattern, form, or shape [World Book, Inc., 1980]:

> Gestalt psychologists believe that pattern, or form, is the most important part of experience. The whole pattern gives meaning to each individual element of experience. In other words, the whole is more important than the sum of its parts. Gestalt psychology greatly influenced the study of human perception, and psychologists used Gestalt ideas in developing several principles—for example, the principle of closure (people tend to see incomplete patterns as complete or unified wholes).

1.2.3.2 Modern systems foundations

During his distinguished career, *Albert Einstein* attempted to develop a unified theory that embraces all forces of nature as a system. Feynman et al. [1963] describe a hierarchy or continuum of physical laws as distinct systems or disciplines that are cooperating and interdependent. Modern systems foundations are attributed to select scholars. Among them is *Norbert Wiener*, who in 1948 published his seminal book *Cybernetics*. Wiener's work was the outgrowth and development of computer technology, information theory, self-regulating machines, and feedback control. In the second edition of *Cybernetics* [1961], Wiener commented on the work of Leibniz:

> At this point there enters an element which occurs repeatedly in the history of cybernetics—the influence of mathematical logic. If I were to choose a patron saint for cybernetics out of the history of science, I should have to choose Leibniz. The philosophy of Leibniz centers about two closely related concepts—that of a universal symbolism and that of a calculus of reasoning. From these are descended the mathematical notation and the symbolic logic of the present day.

Ludwig von Bertalanffy coined the term *general systems theory* around 1950; it is documented in his seminal book, *General Systems Theory: Foundations, Development, Applications* [Bertalanffy, 1968/1976]. The following quotes from pages 9 to 11 are of particular interest:

> In the last two decades we have witnessed the emergence of the "system" as a key concept in scientific research. Systems, of course, have been studied for centuries, but something new has been added.... The tendency to study systems as an entity rather than as a conglomeration of parts is consistent with the tendency in contemporary science no longer to isolate phenomena in narrowly confined contexts, but rather to open interactions for examination and to examine larger and larger slices of nature. Under the banner of systems research (and its many synonyms) we have witnessed a convergence of many more specialized contemporary scientific developments. So far as can be ascertained, the idea of a "general systems theory" was first introduced by the present author prior to cybernetics, systems engineering and the emergence of related fields.

Although the term "systems" itself was not emphasized, the history of this concept includes many illustrious names.

Kenneth Boulding, an economist, published work in 1953 on *General Empirical Theory* [Boulding, 1953] and claimed that it was the same as the general systems theory advocated by Bertalanffy.

The Society for General Systems Research was organized in 1954 by the American Association for the Advancement of Science. The society's mission was to develop theoretical systems applicable to more than one traditional department of knowledge.

The major functions of the society were to (i) investigate the isomorphy of concepts, laws, and models in various fields, as well as help in useful transfers from one field to another, (ii) encourage the development of adequate theoretical models in the fields that lack them, (iii) minimize the duplication of theoretical effect in different fields, and (iv) promote the unity of science by improving communication among specialists.

Several modeling philosophies and methods have been developed over the last three decades to address the intricacy of modeling complex large-scale systems and to offer various modeling schema. They are included in the following volumes: *New Directions in General Theory of Systems* [Mesarović, 1965], *General Systems Theory* [Macko, 1967], *Systems Theory and Biology* [Mesarović, 1968], *Advances in Control Systems* [Leondes, 1969], *Theory of Hierarchical Multilevel Systems* [Mesarović et al., 1970], *Methodology for Large-Scale Systems* [Sage, 1977], *Systems Theory: Philosophical and Methodological Problems* [Blauberg et al., 1977], *Hierarchical Analyses of Water Resources Systems: Modeling and Optimization of Large-Scale Systems* [Haimes, 1977], and *Multifaceted Modeling and Discrete Event Simulation* [Zigler, 1984].

In *Synectics: The Development of Creative Capacity*, Gordon [1968] introduced an approach that uses metaphoric thinking as a means to solve complex problems. In the same era, Lowrance [1976] published an influential work considering the science of measuring the likelihood and consequence of uncertain adverse effects that emerge from complex systems. He outlined critical considerations for engineering complex systems that are characterized by uncertainty. Gheorghe [1982] presented the philosophy of systems engineering as it is applied to real-world systems. In his book *Metasystems Methodology*, Hall [1989] developed a theoretical framework to capture the multiple dimensions and perspectives of a system. Other works include Sage [1992, 1995] and Sage and Rouse [1999]. Sage and Cuppan [2001] provide a definition of emergent behavior in the context of an SoS. Slovic [2000], among his many far-reaching works, presents the capabilities of decisionmakers to understand and make *optimal* decisions in uncertain environments. Other books on systems include Fang et al. [1993], Gharajedaghi [2005], Rasmussen et al. [1994], Rouse [1991], Adelman [1991], Zeleny [2005], Blanchard and Fabrycky [1998], Kossiakoff and Sweet [2002], Maier and Rechtin [2000], Buede [1999], Blanchard [2003], Blanchard and Fabrycky [2005], Sage and Armstrong [2003], and Hatley et al. [2000].

Several modeling philosophies and methods have been developed over the years to address the complexity of modeling large-scale systems and to offer various modeling schema. In his book *Methodology for Large-Scale Systems*, Sage [1977] addressed the "need for value systems which are structurally repeatable and capable of articulation across interdisciplinary fields" with which to model the multiple dimensions of societal problems. Blauberg et al. [1977] pointed out that, for the understanding and analysis of a large-scale system, the fundamental principles of *wholeness* (representing the integrity of the system) and *hierarchy* (representing the internal structure of the system) must be supplemented by the principle of *the multiplicity of description for any system*. To capture the multiple dimensions and perspectives of a system, Haimes [1981] introduced hierarchical holographic modeling (HHM) (see Chapter 3) and asserted: "To clarify and document not only the multiple components, objectives, and constraints of a system but also its welter of societal aspects (functional, temporal, geographical, economic, political, legal, environmental, sectoral, institutional, etc.) is quite impossible with a single model analysis and interpretation." Recognizing that a system "may be subject to a multiplicity of management, control and design objectives," Zigler [1984] addressed such modeling complexity in his book *Multifaceted Modeling and Discrete Event*

Simulation. Zigler (p. 8) introduced the term *multifaceted* "to denote an approach to modeling which recognizes the existence of multiplicities of objectives and models as a fact of life." In his book *Synectics: The Development of Creative Capacity,* Gordon [1968] introduced an approach that uses metaphoric thinking as a means to solve complex problems. Hall [1989] developed a theoretical framework, which he termed *Metasystems Methodology,* to capture the multiple dimensions and perspectives of a system. Other early seminal works in this area include the book on societal systems and complexity by Warfield [1976] and the book *Systems Engineering* [Sage, 1992]. Sage identified several phases of the systems engineering life cycle; embedded in such analyses are the multiple perspectives—the structural definition, the functional definition, and the purposeful definition. Finally, the multiple volumes of the *Systems and Control Encyclopedia: Theory, Technology, Applications* [Singh, 1987] offer a plethora of theory and methodology on modeling large-scale and complex systems. Thus, multifaceted modeling, metasystems, HHM, and other contributions in the field of large-scale systems constitute the fundamental philosophy upon which systems engineering is built.

Reflecting on the origins of modern systems theory since the introduction of the Gestalt psychology in 1912, we cannot underestimate the intellectual power of the holistic philosophy that has sustained systems engineering. This multidisciplinary field transcends the arts, humanities, natural and physical sciences, engineering, medicine, and law, among others. The fact that systems engineering, systems analysis, and risk analysis have continued to grow and infiltrate other fields of study over the years can be attributed to the fundamental premise that a system can be understood only if all the intra- and interdependencies among its parts and its environment are also understood. For more than a century, mathematical models constituted the foundations upon which systems-based theory and methodologies were developed, including their use and deployment on the myriad large-scale projects in the natural and constructed environment. If we were to identify a single idea that has dominated systems thinking and modeling, it would be the state concept. Indeed, the centrality of state variables in this context is so dominant that no meaningful mathematical model of a real system can be built without identifying the critical states of that system and relating all other building blocks of the model to them (including decision, random, and exogenous variables, and inputs and outputs). In this respect, system modeling—the cornerstone of this book—has served, in many ways, as the medium with which to infuse and instill the holistic systems philosophy into the practice of risk analysis as well as of engineering and other fields.

1.2.4 Systems Engineering and Covey's Seven Habits

The concepts that Covey introduces can be compared with the systems approach as applied to the entire life cycle of a system. Through this comparison, a joint model is developed that demonstrates how the ideas from the two approaches overlap and how an understanding of this view can benefit personal development as well as systems design and development [Haimes and Schneiter, 1996].

Covey's philosophy is used in the following discussion as a vehicle with which to explain the holistic systems engineering philosophy.

1.2.4.1 Paradigm: The systems concept

From the outset, Covey stresses the understanding of paradigms—the lenses through which we see the universe. Furthermore, according to Covey, it is not what happens to us that affects our behavior; rather, it is our interpretation of what happens. Since our interpretation of the world we live in determines how we create new and innovative solutions to the problems we face, it is essential that we understand the elemental interrelationships in the world that surrounds us. Thus, both understanding the systemic nature of the universe and defining the system that we need to address are imperative requirements for our ability to solve problems.

In his book *The Fifth Discipline,* Peter Senge [1990] gives a good example of how to understand the systems concept. To illustrate the rudiments of the *new language* of systems thinking, he considers a very simple system—filling a glass of water:

From a linear viewpoint, we say, "I am filling a glass of water." But in fact, as we fill the glass, we are watching the water level rise. We monitor the gap between the level and our goal, the desired water level. As the water approaches the desired level, we adjust the faucet position to slow the flow of water, until it is turned off when the glass is full. In fact, when we fill a glass of water we operate a water-regulation system.

The routine of filling a glass of water is so basic to us that we can do it successfully without thinking about it. But when the system becomes more complex, such as building a dam across a river, it is essential to see the systemic nature of the problem to avoid adverse consequences.

Sage [1992] defines systems engineering as "the design, production, and maintenance of trustworthy systems within cost and time constraints." Sage [1990] also argues that systems engineering may be viewed as a philosophy that looks at the broader picture; it is a holistic approach to problem solving that relates interacting components to one another. Blanchard and Fabrycky [1990] define a system as all the components, attributes, and relationships needed to accomplish an objective. Understanding the systemic nature of problems is inherent in problem definition.

Understanding both the systemic nature of the world and the elements of the systems under question enables the shift to the paradigm of systems thinking. Just as the shift to Covey's Principle-Centered Paradigm [Covey, 1989] enables the adoption of his Seven Habits, the shift to systems thinking enables the successful implementation of the systems approach. This change of perspective alone, however, is not enough to make either concept or approach successful. One must carry out the steps to ensure that success.

1.2.4.2 The Seven Habits of highly effective people

The Seven Habits introduced by Covey [1989] are as follows:

Habit 1: Be proactive.
Habit 2: Begin with the end in mind.
Habit 3: Put first things first.
Habit 4: Think win–win.
Habit 5: Seek first to understand, then to be understood.
Habit 6: Synergize.
Habit 7: Sharpen the saw.

The first three of the Seven Habits are the steps toward what Covey calls *Private Victory*, and Habits 4–6 are the steps toward *Public Victory*. These habits will be examined in terms of their relationships to the systems approach as represented by its guiding universal principles and by the 13 steps that manifest it. The guiding principles are as follows:

- Adhere to the systemic philosophy of holism.
- Recognize the hierarchical decisionmaking structure (multiple decisionmakers, constituencies, power brokers, etc.).
- Appreciate the multiple-objective nature:
  - There is no single solution.
  - There are choices and trade-offs.
- Respond to the temporal domain: past, present, and future.
- Incorporate the culture, vision, mentality, and interpersonal relationships—to build an informal network of trust.
- Address the uncertain world (taxonomy of uncertainty).
- Strive for continuous improvement of quality.
- Honor the cross-disciplinary nature of quality problem solving.
- Focus on the centrality of human and interpersonal relationships.

The following is a set of 13 logical steps with which to address problems [Haimes and Schneiter, 1996]:

1. Define and generalize the client's needs. Consider the total problem environment. Clearly identify the problem.
2. Help the client determine his or her objectives, goals, performance criteria, and purpose.
3. Similar to step 1: consider the total problem's environment. Evaluate the situation, the constraints, the problem's limitations, and all available resources.
4. Study and understand the interactions among the environment, the technology, the system, and the people involved.

5. Incorporate multiple models and synthesize. Evaluate the effectiveness, and check the validity of the models.
6. Solve the models through simulation and/or optimization.
7. Evaluate various feasible solutions, options, and policies. How does the solution fulfill the client's needs? What are the costs, benefits, and risk trade-offs for each solution (policy option)?
8. Evaluate the proposed solution for the long term as well as the short term. In other words, what is the sustainability of the solution?
9. Communicate the proposed solution to the client in a convincing manner.
10. Evaluate the impact of current decisions on future options.
11. Once the client has accepted the solution, work on its implementation. If the solution is rejected, return to any of the above steps to correct it so that the client's desires are fulfilled.
12. Postaudit your study.
13. Iterate at all times.

1.2.4.3 Relating the Seven Habits to the systems approach

Covey's Seven Habits are not straightforward steps. The first three progress from dependence toward independence. Viewed in a problem-solving light, they make an essential contribution to the solution: The first habit frames the problem, the second determines the desired outcome, and the third organizes time and effort toward eventual solution. From this point, Habits 4–6 are guiding principles that enable personal growth toward interdependence. They stress communication and understanding in relationships and stress teamwork and creativity in the problem-solving process. Thus, they help *direct* the efforts mobilized in the first three habits. Habit 7 stresses constant reevaluation and improvement. This combination of elements is very similar to those necessary for successful systems engineering.

Habit 1: Be proactive

The first habit deals with how to view the problem and where to focus one's energies. Covey's primary tool for this habit is the set of concentric circles, the

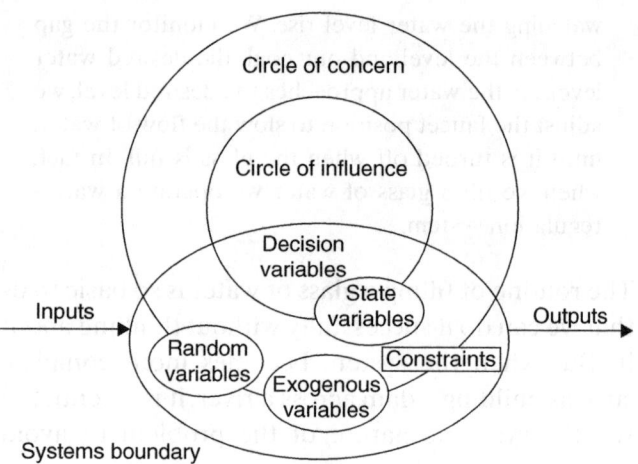

Figure 1.2 Systemic view of concentric circles. From Haimes and Schneiter [1996]; © 1996 IEEE.

circle of concern and the *circle of influence*. The circle of concern includes all things that concern us. The circle of influence includes elements that are under our control. From a systems standpoint, this perspective can relate to the definition of a system and its elements, indeed an SoS. The system's boundary defines the context within which the problem will be addressed—a subset within the circle of concern that is to be studied. (It is also possible that elements in the system lie outside the circle of concern—e.g., externalities.) The state variables, which are central to system modeling, are our primary concern; however, we do not have absolute control over them. The only variables within our circle of influence are the decision variables. Random and exogenous variables and constraints are beyond our control, although we must be cognizant of them (these terms will be defined and explained in Chapter 2).

Figure 1.2 combines Covey's key proactive circles with the elements that fully describe a system and its interrelationships.

Successful decisionmaking or problem solving requires understanding the elements within both the circle of influence and the circle of concern, that is, the elements of the SoS and its interacting environment.

Habit 2: Begin with the end in mind

In Covey's context, this habit involves mentally creating a solution to problems or developing a mission statement. Beginning with the end in mind is one of the cornerstones of systems thinking. Often referred to as the *top-down approach* to problem solving, this involves determining the overall goals for a system before beginning the design. In the filling the glass with water example, this means determining whether the goal is to fill one glass of water or many glasses or to design a useful faucet or sink. From a mathematical modeling perspective, the goal for a problem could be to minimize or maximize some function, f, of the state variables, S—for example, minimize $f(S)$. For example, we may want to minimize the distance from water level to the top of the glass, S_1, while minimizing the amount of water spilled, S_2. This can be represented as minimize $f(S_1, S_2)$.

Begin with the end in mind is also termed the leadership habit. One means of applying this is in the form of a mission statement—everything should follow from the mission statement that the leader provides. Likewise, the preliminary steps of systems engineering provide a mission for the project by determining goals, requirements, specifications, or criteria by which eventual proposed solutions will be evaluated.

In our basic example, the mental picture (goal) is a full glass of water. However, the situation is not always this simple. A more complex situation is the American effort to put a man on the moon. This is perhaps the best example of the importance of holding fast to the mental creation of an outcome. Throughout the project, the leaders kept their strong belief in this goal. This was essential because much of the necessary technology did not even exist at the outset of the project. Reliance on status quo technology or knowledge would have doomed the project—much as failure to *begin with the end in mind* would keep one from reaching personal goals.

Habit 3: Put first things first

This habit is designed to help concentrate efforts toward *more important* activities in a *less urgent* atmosphere.

Instead of trying to address the myriad problems that the first two habits may bring to the light, Covey places the emphasis on time management, leaving the eventual solution of the problem to the individual. The extensive set of actions available to help solve problems in the journey of personal growth is analogous to the array of problem-solving approaches in engineering. No specific approach is appropriate in every situation. The plethora of systems and risk-based methodologies and tools introduced in this book attest to this fact. It should be left to the individual problem solver to use the best method in a particular application. The key step is following the goal-oriented systems approach and using the most appropriate tools for the specific problem.

Time management tools commonly used in systems engineering that are analogous to Covey's time management matrix include the project evaluation and review technique (PERT) and the critical path method (CPM). Other tools such as failure mode and effects analysis (FMEA) and failure mode, effects, and criticality analysis (FMECA) are discussed in Chapter 13. In addition, Chapter 15 is devoted to project management, where time management is at the heart of project management. These help organize the order of events and assist in time management by indicating those activities whose completion times directly affect the total project time.

Habit 4: Think win–win (or no deal)

This habit illustrates the importance of the abundance mentality, a guiding principle in applying the ideas incorporated in the first three habits. Instead of focusing on outsmarting or outmaneuvering the opponent, it stresses that both parties should work together to find a mutually beneficial outcome.

This concept can come into play in the systems engineering process in several different places: in creating alternative solutions or in the working relationships of group members. Problem solving always involves trade-offs among conflicting objectives. In such situations, win–lose alternatives are abundant, but more can be gained by thinking win–win. On a more personal level, constructive cooperation between group members is essential for the eventual success of a group effort. The informal network of trust that is the foundation of successful group interaction will be eroded by win–lose thinking. A culture that embodies win–win cooperation has much greater chances for success.

Habit 5: Seek first to understand, then to be understood

This habit concerns different perspectives, implying that ordinarily adversarial roles must be overcome. This habit can be viewed on multiple levels. It is especially important in any arena where there are numerous constituencies. With the advent of cross-functional deployment, many distinct working groups are called together for a common cause. Unlike previous processes where a design group would throw plans *over the wall* to manufacturing, representatives from manufacturing are included in the design process from the start. The importance of developing a shared understanding from both perspectives is obvious.

Seek first to understand, then to be understood also highlights the importance of communication and of viewing every process from the perspective of the customer. The customer must always be satisfied, whether it is a consumer or the next workstation in an assembly process. Again, understanding the customer's perspective is essential. The application of this habit to interpersonal communication is obvious as well. Covey calls this *empathic listening*; experts in business may call this knowledge management.

Brooks [2000] offers the following succinct definition of knowledge management, which is adapted from the American Productivity and Quality Center:

> Knowledge management: Strategies and processes to create, identify, capture, organize, and leverage vital skills, information, and knowledge to enable people to best accomplish the organization mission.

In his book *Emotional Intelligence*, Goleman [1997] offers another perspective of Habit 5: "The roles for work are changing. We're being judged by a new yardstick: not just how smart we are or our expertise, but also how well we handle ourselves and each other." Relating successful individuals to personal emotional intelligence, Goleman (p. 39) quotes Gardner and Hatch [1989]: "Successful salespeople, politicians, teachers, clinicians, and religious leaders are all likely to be individuals with [a] high degree of interpersonal intelligence." Explicit in this orientation is the holistic vision that the goals of a system or a decisionmaker can be achieved by addressing and managing them as integral parts of the larger system. A central tenet of the vision of successful organizations is building and codifying trust that transcends institutions, organizations, decisionmakers, professionals, and the public at large. Their leadership has to imbue trust as the enabling landmark for knowledge management in order to lower, if not eliminate, the high *walls* and other barriers among the multiple partners of the organization. Undoubtedly, achieving this laudable goal will be a challenge in the quest to manage change.

Davenport and Prusak [1998] advocate three tenets for the establishment of trust: Trust must be visible, trust must be ubiquitous, and trustworthiness must start at the top.

Building on these three foundations of trust to realize the goals of a system means the following [Longstaff and Haimes, 2002]:

- Successful sharing of information must be built on sustained trust.
- Trust in the system is a prerequisite for its viability (e.g., a banking system that loses the trust of its customers ceases its viability).
- Trustworthiness in systems depends on their ability to be adaptable and responsive to the dynamics of people's changing expectations.
- Organizational trust cannot be achieved if the various internal and external boundaries dominate and thus stifle communication and collaboration.
- Trust in the validity of the organization's mission and agenda is a requisite for its sustained effectiveness and for the intellectual productivity of its employees; otherwise, the trust can be transient and have no problems.

Habit 6: Synergize

Habit 6 builds on the two preceding habits. With the ability to communicate openly and maturely, creative cooperation and problem solving become possible. The role of synergy in the systems approach is particularly important. According to Covey, synergy means not only that the whole is greater than the sum of the parts, but that the relationship between the parts is an element in itself. By its nature, systems engineering commonly views systems or processes as the aggregation of multiple interconnected and interdependent components. It is often helpful or instructive to understand a system by analyzing its parts, but this does not necessarily ensure a comprehensive understanding of the entire process. Only through study of the relationships among components can the true nature of the system be grasped.

Covey's discussion of synergy primarily deals with relationships among people. This, of course, is applicable to systems engineering because people with different backgrounds and positions are commonly teamed to solve a particular problem. The more successful teams will exhibit synergistic traits: They will approach the problem with open minds, they will communicate in a manner that encourages creative interaction, and they will value the differences in each other's approaches to the problem. This will enable them to recognize and assess all possible approaches as candidate solution options. Only by the inspection of all possibilities can an *optimal* solution be determined. Indeed, a basic premise of the holistic systems philosophy is that the total system is better than the sum of its parts. Chapter 3, which is devoted to modeling the multiple perspectives and dimensions of a system, highlights the imperativeness of group synergy in system modeling and thus in decisionmaking.

Habit 7: Sharpen the saw
By concluding with this habit, Covey hopes that people will continually reevaluate their personal progress, reshape their goals, and strive to improve. These issues have become quite common in engineering environment—often referred to as *kaizen*, the Japanese word for continuous improvement [Imai, 1986]. An application of this habit is also seen in the Shewhart cycle [Deming, 1986]. Iteration also plays a primary role in systems engineering. In a relationship with a client, it is necessary to receive constant feedback to ensure correct understanding, building on emotional intelligence. As our knowledge about a system develops throughout the problem-solving process, it is necessary to reevaluate the original goals. The centrality of humans in the life cycle of systems calls for individuals who can perform under pressure by continuously rejuvenating and recharging themselves.

1.2.4.4 *The Seven Habits compared to the systems approach*
The relationship between Covey's philosophy for personal change and the systems approach is further illustrated by a pairwise comparison of the two, as shown in Figure 1.3. The fact that Habit 1 corresponds to Steps 1, 3, and 4 indicates that these problem-definition steps could be grouped together. They should all be completed before the goals are determined. When these three steps are grouped together, Covey's first three habits correspond to the order of problem solving following the systems approach. First, the problem is defined, then the desired outcome is envisioned, and time and effort are organized to achieve this desired outcome. The general reference to problem solution in Habit 3, *Put first things first*, corresponds to many steps in this systems approach. Figure 1.3 indicates that these, too, could be integrated into a single category.

Habits 4–6 are more difficult to apply to specific steps. Analogous to the overriding principles enumerated in Figure 1.3, these habits are applicable throughout the problem-solving process. To the extent that these steps promote communication, the habits *think win–win* and *seek first to understand…* apply to almost every situation that involves group interaction. More specifically, *think win–win* can apply to creative problem solving and idea generation, and *seek first to understand…* directs the interaction between a systems engineer and a client. *Synergize* can also be applied on numerous levels. Finally, *sharpen the saw* directly corresponds to the constant iteration that is stressed throughout the systems engineering approach.

In sum, the side-by-side comparison of the seven habits and the steps in the systems approach serves to show how the elements of both not only correspond to, but also complement, each other. Both philosophies stress problem definition, early determination of the desired outcome, and an organized effort to determine a solution. They also promote similar overriding principles to better enable the problem-solving process. This similarity is remarkable given that the seven habits are a guide to personal development, whereas the systems approach is geared for systems design, development, and management. Most important, comparing Covey's philosophy as described earlier can help improve the understanding of systems engineering and thus better relate the process of risk assessment and management to the systems approach.

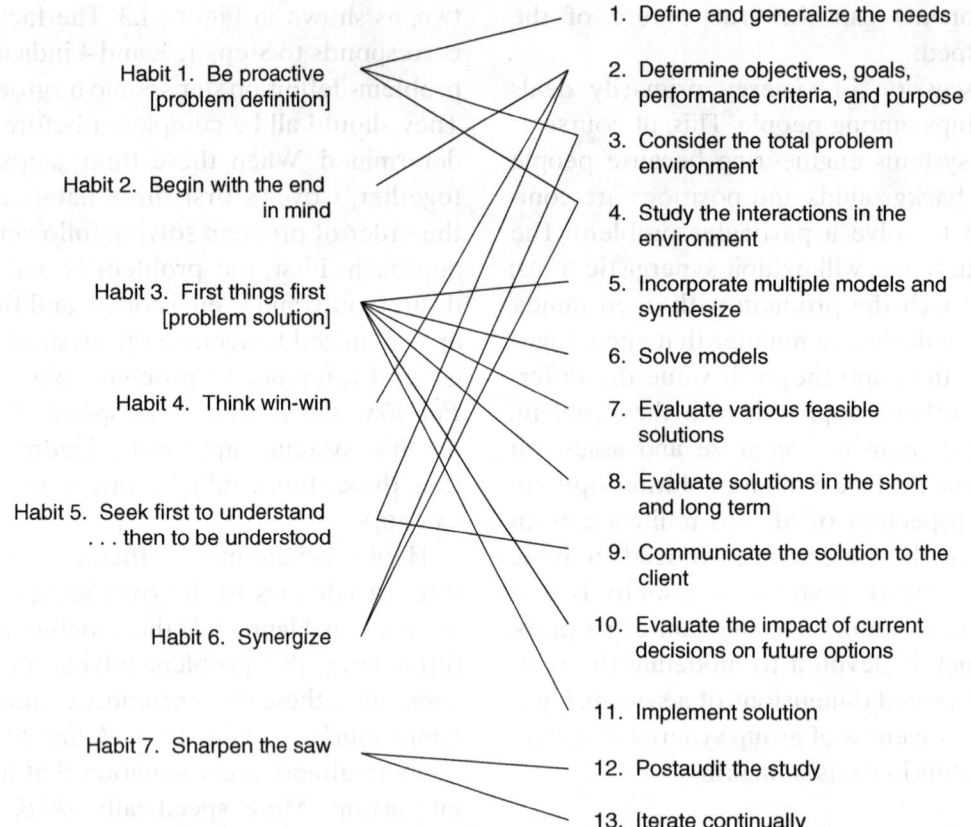

Figure 1.3 Juxtaposition of the seven habits [Covey, 1989] with the systems approach. From Haimes and Schneiter [1996]; © 1996 IEEE.

1.3 RISK ASSESSMENT AND MANAGEMENT

1.3.1 Holistic Approach

Good management of both technological and nontechnological systems must address the holistic nature of the system in terms of its hierarchical, organizational, and fundamental decisionmaking structure. Also to be considered are the multiple noncommensurate objectives, subobjectives, and sub-subobjectives, including all types of important and relevant risks; the various time horizons; the multiple decisionmakers, constituencies, power brokers, stakeholders, and users of the system; as well as a host of institutional legal and other socioeconomic conditions. Thus, risk management raises several fundamental philosophical and methodological questions [Fischhoff et al., 1983; Hall, 1989; Krimsky and Golding, 1992; Lewis, 1992; Burke et al., 1993; Wernick, 1995; Bernstein, 1996; Kunreuther and Slovic, 1996; Kaplan et al., 2001; NRC, 2002].

Engineering systems are almost always designed, constructed, integrated, and operated under unavoidable conditions of risk and uncertainty and are often expected to achieve multiple and conflicting objectives. Identifying, quantifying, evaluating, and trading off risks, benefits, and costs should constitute an integral and explicit component of the overall managerial decisionmaking process and should not be a separate, cosmetic afterthought. The body of knowledge in risk assessment and management has gained significant attention during the last three decades (and especially since the September 11, 2001, attack on the United States); it spans many disciplines and encompasses empirical and quantitative as well as normative, judgmental aspects of decisionmaking. Does this constitute a new discipline that is separate, say, from systems engineering and systems analysis? Or has systems engineering and systems analysis been too narrowly defined? When risk and uncertainty are addressed in a practical decisionmaking framework, has it been properly perceived that the body of knowledge known as risk assessment and management markedly fills a critical void that supplements and complements the theories and methodologies of systems engineering and systems analysis? Reflecting on these and other similar questions on the nature, role, and place of risk assessment and management in managing technological and nontechnological systems and in the overall managerial decisionmaking process should stem not from intellectual curiosity only. Rather, considering such questions should provide a way to bridge the gaps and remove some of the barriers that exist between the various disciplines [Haimes, 1989].

As will be discussed in more detail in this book, integrating and incorporating risk assessment and management of technological and nontechnological systems within the broader holistic approach to technology management also require the reexamination of the expected-value concept when it is used as the sole representation of risk. Many agree that in the expectation operation, commensurating high-frequency/low-damage and low-frequency/catastrophic-damage events markedly distorts their relative importance and consequences as they are viewed, perceived, assessed, evaluated, and traded off by managers, decisionmakers, and the public. Some are becoming more and more convinced of the grave limitations of the traditional and commonly used expected-value concept; and they are complementing and supplementing the concept with conditional expectation, where decisions about extreme and catastrophic events are not averaged out with more commonly occurring events. In Chapter 8 and throughout this book, risk of extreme and catastrophic events will be explicitly addressed and quantified, and the common expected-value metric for risk will be supplemented and complemented with the conditional expected value of risk.

1.3.2 The Evolution of Risk Analysis

In March 1961, Norbert Wiener, who is considered by many to be one of the fathers of what is known today as systems engineering, wrote the following in the Preface of the second edition of his book *Cybernetics* [Wiener, 1961]:

> If a new scientific subject has real vitality, the center of interest in it must and should shift in the course of years.... The role of information and the technique of measuring and transmitting information constitute a whole discipline for the engineer, for the physiologist, for the psychologist, and for the sociologist.... Thus it behooves the cyberneticist to move on to new fields and to transfer a large part of his attention to ideas which have arisen....

If one accepts the premise that good and appropriate technology management must be grounded in a holistic approach and based on Wiener's philosophical and almost prophetic statements, then it is possible that what we are witnessing today is a shift of the center of interest, an evolution toward a more holistic approach to management. Is knowledge from diverse disciplines converging into a more coherent, albeit still heterogeneous, aggregate of theory, methodologies, tools, and heuristics? To highlight this evolutionary process, let us consider Wiener's *shift* from single-objective modeling and optimization to multiple-objective modeling and optimization. The 1970s saw the emphasis shift from the dominance of single-objective modeling and optimization toward an emphasis on multiple objectives. During the past three decades, the consideration of multiple objectives in modeling and decisionmaking has grown by leaps and bounds. This has led to the emergence of a new field that has come to be known as *multiple criteria decisionmaking* (MCDM). MCDM has emerged as a philosophy that integrates common sense with empirical, quantitative, normative, descriptive, and value-judgment-based analysis. MCDM, as a subset of systems engineering, is also a philosophy that is supported by advanced systems concepts (e.g., data management procedures, modeling methodologies, optimization and simulation techniques, and decisionmaking approaches) that are grounded in both the arts and sciences for the ultimate purpose of improving the decisionmaking process. Multiple objectives are incorporated into most modeling and optimization of technological systems today.

1.3.3 Risk Communication

The risk assessment and management process is aimed at answering specific questions in order to make better decisions under uncertain conditions. In system modeling, the saying is that a model must be as simple as possible and as complex as desired and required. Similarly, the process of risk assessment and management must follow these same basic principles. These seemingly conflicting simultaneous attributes—simplicity and complexity—can be best explained and justified through effective risk communication. Invariably, the questions raised during the risk assessment and management process originate from decisionmakers at various levels of responsibilities, including managers, designers, stakeholders, journalists and other media professionals, politicians, proprietors, and government or other officials. Although the issues under consideration and their associated questions may be complex and require similarly complex sets of answers, it is imperative that their meanings and ramifications be understood by the decisionmakers. Inversely, for the risk assessment and management process to be effective and complete, decisionmakers, who originate the risk-based questions for the analysts, must be able to communicate openly, honestly, and comprehensively the multidimensional perspectives of the challenges facing them and for which they desire better understanding and possible answers. In turn, risk analysts must be able to translate complex technical analysis and results into a language to which decisionmakers can relate, understand, and incorporate into actionable decisions.

This intricate mental and intellectual dance between risk analysts and decisionmakers was comprehensively addressed in three seminal books with diverse titles: *Good to Great*, *Working with Emotional Intelligence*, and *Working Knowledge*. In his book *Good to Great*, Collins [2001] addresses the importance of the culture of discipline, transcending disciplined people, disciplined thought, and disciplined actions. He explains [p. 200]: "When you have a culture of discipline, you can give people more freedom to experiment and find their own best path to results." On the same page, Collins juxtaposes clock building with time telling: "Operating through sheer force of personality as a disciplinarian is time telling; building an enduring culture of discipline is clock building." These are important requisite traits for effective working relationships between decisionmakers and risk analysts. Goleman [1998, p. 211], in *Working with Emotional Intelligence*, identifies the following elements of competence when people collaborate and cooperate with others toward shared goals: "Balance a focus on task with attention to relationships; collaborate, sharing plans, information, and resources; promote a friendly, cooperative climate; and spot and nurture opportunities for collaboration." Goleman states on page 317 that "emotional intelligence refers to the capacity for recognizing our own feelings and those of others, for motivating ourselves, and for managing emotions well in ourselves and in our relationships." Indeed, these fundamentals are the sine qua non for effective risk communication among all parties involved in the entire process of risk assessment and management.

Invariably, complex problems cannot be solved without addressing their multiple perspectives, scales of complexity, time dependencies, and multiple interdependencies, among others. Among the many parties commonly involved in the process of risk assessment and risk management are the professionals supporting the decisionmakers, the risk analysts, and the decisionmakers themselves. Knowledge management, which builds on embracing trust, exchange of information, and collaboration within and among organization, parties, and individuals, has become essential to performing and successfully deploying the results and fruits of risk assessment and management. Moreover, knowledge management may be viewed, in many ways, as synonymous to effective risk communication. In their book *Working Knowledge*, Davenport and Prusak [1998, p. 62] identify the following five knowledge management principles that can help make the above fusion among the parties work effectively:

1. Foster awareness of the value of the knowledge sought and a willingness to invest in the process of generating it.
2. Identify key knowledge workers who can be effectively brought together in a fusion effort.
3. Emphasize the creative potential inherent in the complexity and diversity of ideas, seeing differences as positive, rather than sources of

conflict, and avoiding simple answers to complex questions.
4. Make the need for knowledge generation clear so as to encourage, reward, and direct it toward a common goal.
5. Introduce measures and milestones of success that reflect the true value of knowledge more completely than simple balance-sheet accounting.

In sum, embracing the principles advocated by these three books provides an important road map for risk communication and thus for a complete and successful risk assessment, risk management, and risk communication process (see Figure 1.5). The philosopher Peter F. Drucker [2004, p. 9] eloquently sums up his message to organizations: "Attract and hold the highest-producing knowledge workers by treating them and their knowledge as the organization's most valuable assets."

1.3.4 Sources of Failure, Risk Assessment, and Risk Management

In the management of technological systems, the failure of a system can be caused by failure of the *hardware*, the *software*, the *organization*, or the *humans* involved. Of course, the initiating events may also be natural occurrences, acts of terrorism, or other incidents.

The term *management* may vary in meaning according to the discipline involved and/or the context. *Risk* is often defined as a measure of the probability and severity of adverse effects. *Risk management* is commonly distinguished from *risk assessment*, even though some may use the term *risk management* to connote the entire process of risk assessment and management. In risk assessment, the analyst often attempts to answer the following set of triplet questions [Kaplan and Garrick, 1981]:

- What can go wrong?
- What is the likelihood that it would go wrong?
- What are the consequences?
- Here we add a fourth question: What is the time frame?

Answers to these questions help risk analysts identify, measure, quantify, and evaluate risks and their consequences and impacts. Risk management builds on the risk assessment process by seeking answers to a second set of three questions [Haimes, 1991]:

- What can be done and what options are available?
- What are the associated trade-offs in terms of all relevant costs, benefits, and risks?
- What are the impacts of current management decisions on future options?

Note that the last question is a most critical one for any managerial decisionmaking. This is so because unless the negative and positive impacts of current decisions on future options are assessed and evaluated (to the extent possible), these policy decisions cannot be deemed to be *optimal* in any sense of the word. Indeed, the assessment and management of risk is essentially a synthesis and amalgamation of the empirical and normative, the quantitative and qualitative, and the objective and subjective effort. Only when these questions are addressed in the broader context of management, where all options and their associated trade-offs are considered within the hierarchical organizational structure, can a total risk management (TRM) be realized. (The term TRM will be formally defined later.) Indeed, evaluating the total trade-offs among all important and relative system objectives in terms of costs, benefits, and risks cannot be done seriously and meaningfully in isolation from the modeling of the system and the broader resource allocation perspectives of the overall organization.

Good management must thus incorporate and address risk management within a holistic and all-encompassing framework that incorporates and addresses all relevant resource allocation and other related management issues. A TRM approach that harmonizes risk management with the overall system management must address the following four sources of failure (see Figure 1.4):

- Hardware failure
- Software failure
- Organizational failure
- Human failure

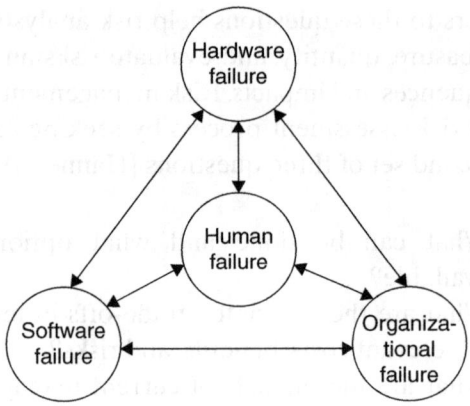

Figure 1.4 System failure.

The above set of sources of failure is intended to be internally comprehensive (i.e., comprehensive within the system's own internal environment). (External sources of failures are not discussed here because they are commonly system dependent.) These four elements are not necessarily independent of each other, however. The distinction between software and hardware is not always straightforward, and separating human and organizational failure is often not an easy task. Nevertheless, these four categories provide a meaningful foundation upon which to build a TRM framework. In his premier book on quality control, *Kaizen*, Imai [1986] states: "The three building blocks of business are hardware, software, and 'humanware.'" He further states that total quality control "means that quality control effects must involve people, organization, hardware, and software." Effective knowledge management within an organization, is instrumental in reducing the rates of these sources of failure.

Organizational errors are often at the root of failures of critical engineering systems. Yet, when searching for risk management strategies, engineers often tend to focus on technical solutions, in part because of the way risks and failures have been analyzed in the past. In her study of offshore drilling rigs, Paté-Cornell [1990] found that over 90% of the failures documented were caused by organizational errors. The following is a list of common organizational errors:

- Overlooking and/or ignoring defects
- Tardiness in correcting defects
- Breakdown in communication
- Missing signals or valuable data due to inadequate inspection or maintenance policy
- Unresolved conflict(s) between management and staff
- Covering up mistakes due to competitive pressure
- Lack of incentives to find problems
- The *kill the messenger* syndrome instead of *reward the messenger*
- Screening information, followed by denial
- Tendency to accept the most favorable hypothesis
- Ignoring long-term effects of decisions
- Loss of institutional memory
- Loss of flexibility and innovation

The importance of considering the four sources of failure is twofold. First, they are comprehensive, involving all aspects of the system's life cycle (e.g., planning, design, construction, integration, operation, and management). Second, they require the total involvement in the risk assessment and management process of everyone concerned—blue- and white-collar workers and managers at all levels of the organizational hierarchy.

1.3.5 TRM

TRM can be defined as a systematic, statistically based, holistic process that builds on quantitative risk modeling, assessment, and management. It answers the previously introduced two sets of questions for risk assessment and risk management, and it addresses the set of four sources of failures within a hierarchical–multiobjective framework. Figure 1.5 depicts the TRM paradigm (the time dimension is implicit in Figure 1.5).

The term *hierarchical–multiobjective framework* can be explained in the context of TRM. Most, if not all, organizations are hierarchical in their structure and, consequently, in the decisionmaking process that they follow. Furthermore, at each level of the organizational hierarchy, multiple, conflicting, competing, and noncommensurate objectives drive the decisionmaking process. At the heart of good management decisions is the *optimal* allocation of the organization's resources among its various hierarchical levels and subsystems. The *optimal*

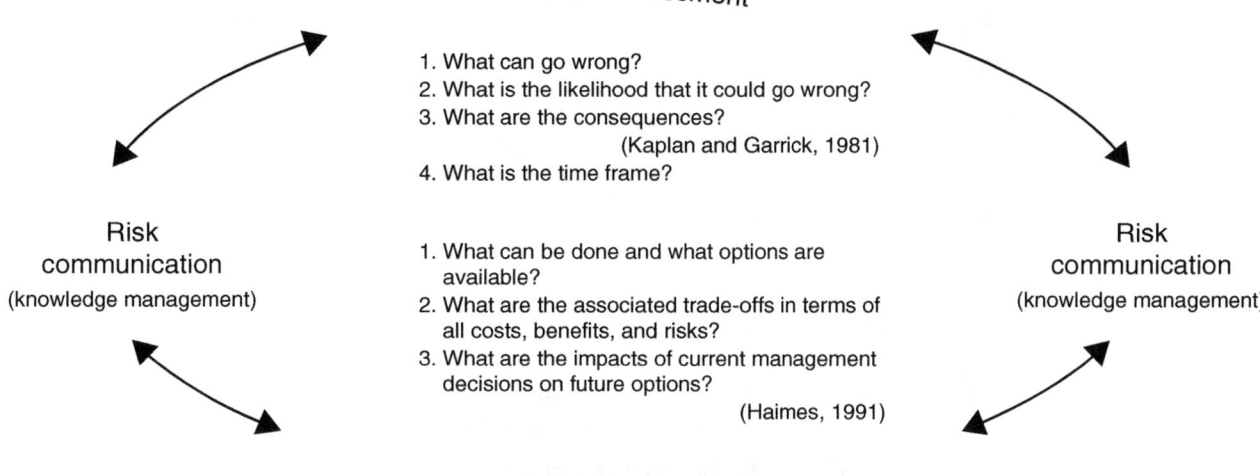

Figure 1.5 *Total risk management.*

allocation is meant in the Pareto-optimal sense, where trade-offs among all costs, benefits, and risks are evaluated in terms of hierarchical objectives (and subobjectives) and in terms of their temporal impacts on future options. Methodological approaches for such hierarchical frameworks are discussed in Haimes et al. [1990].

1.3.6 Multiple Objectives: The Student's Dilemma

The trade-offs among multiple noncommensurate and often conflicting and competing objectives are at the heart of risk management (Chapter 5 is devoted in its entirety to multiobjective analysis). Lowrance [1976] defines safety as the level of risk that is deemed acceptable, and one is invariably faced with deciding the level of safety and the acceptable cost associated with that safety [Chankong and Haimes, 1983, 2008]. The following student dilemma is used to demonstrate the fundamental concepts of Pareto-optimality and trade-offs in a multiobjective framework.

A student working part time to support her college education is faced with the following dilemma that is familiar to all of us:

$$\text{Maximize} \begin{cases} \text{income from part-time work} \\ \text{grade-point average} \\ \text{leisure time} \end{cases}$$

In order to use the two-dimensional plane for graphic purposes, we will restrict our discussion to two objectives: maximize income and maximize grade-point average (GPA). We will assume that a total of 70 h/week are allocated for studying and working. The remaining 98 h/week are available for *leisure time*, covering all other activities. Figure 1.6 depicts the income generated per week as a function of hours of work. Figure 1.7 depicts the relationship between studying and GPA. Figure 1.8 is a dual plotting of both functions (income and GPA) versus working time and studying time, respectively.

The concept of optimality in multiple objectives differs in a fundamental way from that of a single-objective optimization. *Pareto-optimality* in a multi-objective framework is that solution, policy, or option for which one objective function can be improved only at the expense of degrading another. A Pareto-optimal solution is also known as a noninferior, nondominated, or efficient solution (see Chapter 5). In Figure 1.6, for example, studying up to 60 h/week (and correspondingly working 10 h/week) is Pareto-optimal, since in this range income is sacrificed for a higher GPA. On the other hand, studying over 60 h/week (or working <10 h/week) is a non-Pareto-optimal policy, since in this range both income and GPA are diminishing. Similarly, a non-Pareto-optimal solution is also known as an inferior, dominated, or nonefficient solution. Figure 1.9 further distinguishes between Pareto- and

20 THE ART AND SCIENCE OF SYSTEMS AND RISK ANALYSIS

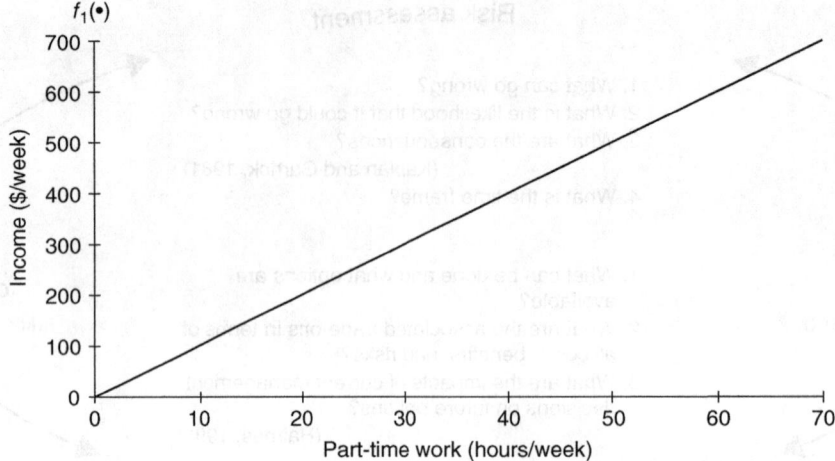

Figure 1.6 Income from part-time work.

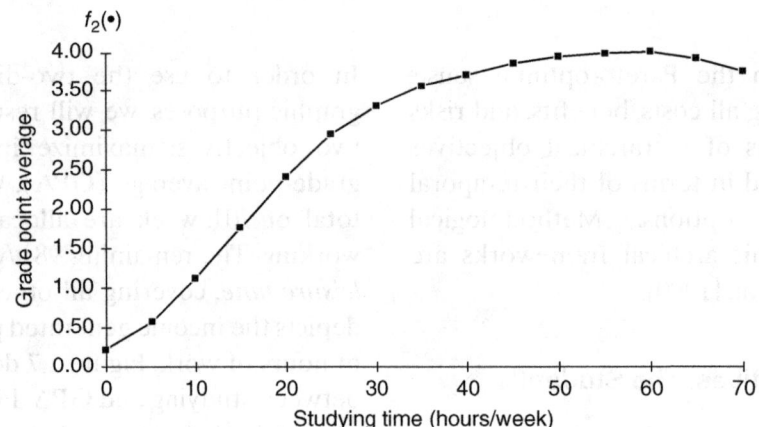

Figure 1.7 GPA as a function of studying time.

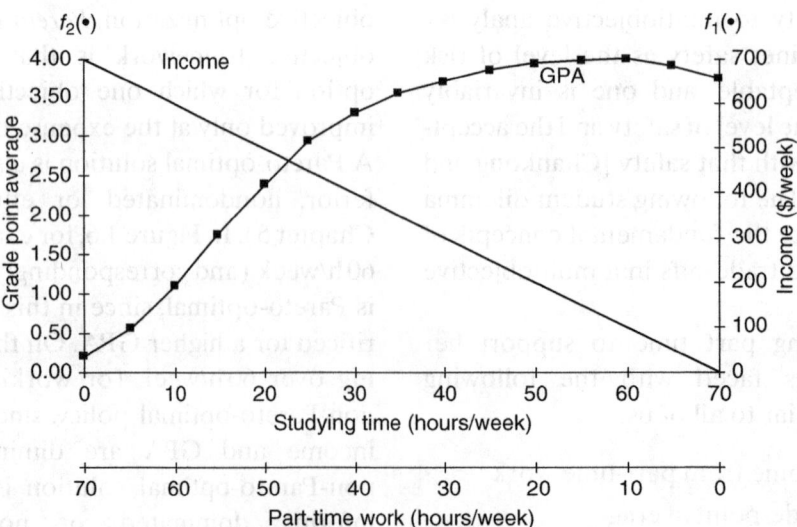

Figure 1.8 GPA versus income.

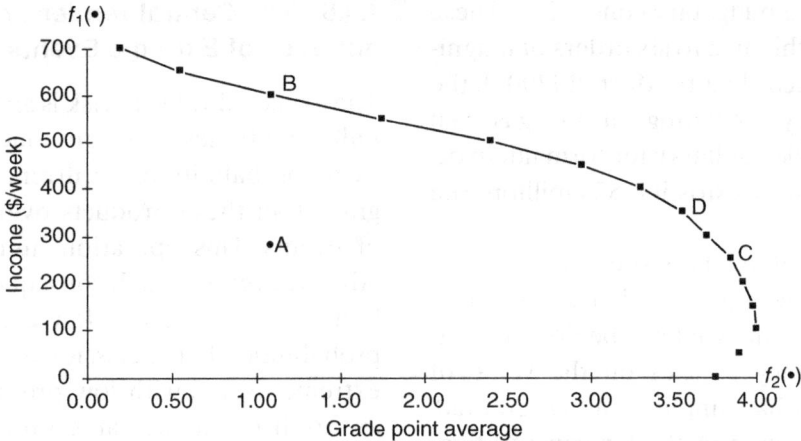

Figure 1.9 Pareto-optimal frontier.

non-Pareto-optimal solutions by plotting income versus GPA. The line connecting all the square points is called the Pareto-optimal frontier. Note that any point interior to this frontier is non-Pareto-optimal. Consider, for example, policy option A. At this point, the student makes $300 per week at a GPA of just above one, whereas at point B, she makes $600 per week at the same GPA level. One can easily show that all points (policy options) interior to the Pareto-optimal frontier are inferior points.

Consider the risk of groundwater contamination as another example. We can generate the Pareto-optimal frontier for this risk-based decisionmaking. Minimizing the cost of contamination prevention and the risk of contamination is similar in many ways to generating the Pareto-optimal frontier for the student dilemma problem. Determining the best work–study policy for the student can be compared to determining (at least implicitly) the level of safety—that is, the level of acceptability of risk of contamination and the cost associated with preventing such contamination. To arrive at this level of acceptable risk, we will again refer to the student dilemma problem illustrated in Figure 1.9. At point B, the student is making about $600 per week at a GPA of just above 1. Note that the slope at this point is about $100 per week for each 1 GPA. Thus, the student will opt to study more. At point C, the student can achieve a GPA of about 3.6 and a weekly income of about $250. The trade-off (slope) at this point is very large: By sacrificing about 0.2 GPA, the student can increase her income by about $200 per week. Obviously, the student may choose neither policy B nor C; rather she may settle for something like policy D, with an acceptable level of income and GPA. In a similar way, and short of strict regulatory requirements, a decisionmaker may determine the level of resources to allocate for preventing groundwater contamination at an acceptable level of risk of contamination.

In summary, the question is: Why should we expect environmental or other technologically based problems involving risk–cost–benefit trade-offs to be any easier than solving the student dilemma?

A single decisionmaker as in the student dilemma problem is not common, especially when dealing with public policy; rather, the existence of multiple decisionmakers is more prevalent. Indeed, policy options on important and encompassing issues are rarely formulated, traded off, evaluated, and finally decided upon at one single level in the hierarchical decisionmaking process. Rather, a hierarchy that represents various constituencies, stakeholders, power brokers, advisers, administrators, and a host of shakers and movers constitutes the true players in the complex decisionmaking process. For more on multiobjective analysis, see Chapter 5, Haimes and Hall [1974], Chankong and Haimes [2008], and Haimes et al. [1994].

1.3.7 The Perception of Risk

The enormous discrepancies and monumental gaps in the dollars spent by various federal agencies in their quest to save human lives can no longer be justified under austere budgetary constraints. These expenditures vary within five to six orders of magnitude. For example, according to Morrall [2003], the cost per life saved by regulating oil and gas well service is $100,000 (1984 dollars); for formaldehyde, it is $72 billion, and for asbestos, it is $7.4 million (see Table 1.1).

A natural and logical set of questions arises: What are the sources of these gaps and discrepancies? Why do they persist? And what can be done to synchronize federal agency policies on the value of human life? A somewhat simplistic, albeit pointed, explanation may be found in the lexicon of litigation, intimidation, fear, and public pressure in the media and by special interest groups as well as in the electoral and political processes. Larsen [2007] offers interesting views on government spending and on the perception of risk. Keeping the threat of terrorism in perspective, he writes on page 22:

> Nearly 2,000 Americans died on 9/11. It was a human tragedy on a scale that was difficult for most of us to comprehend. However, during a four-year period from January 2002 to December 31, 2005, not a single American died in our homeland from international terrorism. During the same period, 20,000 Americans died from food poisoning, 160,000 died in automobile accidents, and nearly 400,000 died from medical mistakes.

US companies have ample statistical information on the costs of improved product safety but are most careful to keep their analyses secretive and confidential [Stern and Fineberg, 1996]. Our litigious society has effectively prevented industry and government from both explicitly developing and publicly sharing such analyses [Fischhoff et al., 1983; Douglas, 1990; Sage, 1990; The Royal Society, 1992; NRC, 1996].

What is needed is at least a temporary moratorium on litigation in this area. We should extend immunity and indemnification to all analysts and public officials engaged in quantifying the cost-effectiveness of all expenditures aimed at saving human lives and/or preventing sickness or injury. In sum, we ought to generate a public atmosphere that is conducive to open dialogue and reason and to a holistic process of risk assessment and management.

1.3.8 The Central Tendency Measure of Risk and Risk of Extreme Events

The expected value of risk is an operation that essentially multiplies the consequences of each event by its probability of occurrence and sums (or integrates) all these products over the entire universe of events. This operation literally commensurates adverse events of high consequences and low probabilities with events of low consequences and high probabilities. In the classic expected-value approach, extreme events with low probability of occurrence are each given the same proportional importance regardless of their potential catastrophic and irreversible impact. This mathematical operation is similar to the precommensuration of multiple objectives through the weighting approach (see Chapter 5).

The major problem for the decisionmaker remains one of information overload: For every policy, action, or measure adopted, there will be a vast array of potential consequences as well as benefits and costs with their associated probabilities. It is at this stage that most analysts are caught in the pitfalls of the unqualified expected-value analysis. In their quest to protect the decisionmaker from information overload, analysts precommensurate catastrophic damages that have a low probability of occurrence with minor damages that have a high probability. From the perspective of public policy, it is obvious that a catastrophic dam failure or major flood that has a very low probability of happening cannot be viewed by decisionmakers in the same vein as minor flooding that has a high probability of happening. This is exactly what the expected-value function would ultimately generate. Yet, it is clear to any practitioner or public official involved in flood management that the two cases are far from being commensurate or equal. Most important, the analyst's precommensuration of these low-probability, high-damage events with high-probability, low-damage events into one expectation function (indeed some kind of a utility function) markedly distorts the relative importance of these events and consequences as they are viewed, assessed, and evaluated by the decisionmakers. This is similar to the dilemma that used to face theorists and practitioners in the field of MCDM [Haimes et al., 1990; Chankong and Haimes, 2008] (see Chapter 5 for discussion on MCDM and multiobjective analysis).

TABLE 1.1 Comparative Costs of Safety and Health Regulations

Regulation	Year	Agency	Status[a]	Initial Annual Risk Estimate[b]	Lives Saved Annually	Cost per Life Saved ($ Thousand, 1984)
Steering column protection	1967	NHTSA	F	7.7 in 10^5	1,300,000	100
Unvented space heaters	1980	CPSC	F	2.7 in 10^5	63,000	100
Oil and gas well service	1983	OSHA-S	P	1.1 in 10^3	50,000	100
Cabin fire protection	1985	FAA	F	6.5 in 10^8	15,000	200
Passive restraints/belts	1984	NHTSA	F	9.1 in 10^5	1,850,000	300
Fuel system integrity	1975	NHTSA	F	4.9 in 10^6	400,000	300
Trihalomethanes	1979	EPA	F	6.0 in 10^6	322,000	300
Underground construction	1983	OSHA-S	P	1.6 in 10^3	8,100	300
Alcohol and drug control	1985	FRA	F	1.8 in 10^6	4,200	500
Servicing wheel rims	1984	OSHA-S	F	1.4 in 10^5	2,300	500
Seat cushion flammability	1984	FAA	F	1.6 in 10^7	37,000	600
Floor emergency lighting	1984	FAA	F	2.2 in 10^8	5,000	700
Crane suspended personnel platform	1984	OSHA-S	P	1.8 in 10^3	5,000	900
Children's sleepwear flammability	1973	CPSC	F	2.4 in 10^6	106,000	1,300
Side doors	1970	NHTSA	F	3.6 in 10^5	480,000	1,300
Concrete and masonry construction	1985	OSHA-S	P	1.4 in 10^5	6,500	1,400
Hazard communication	1983	OSHA-S	F	4.0 in 10^5	200,000	1,800
Grain dust	1984	OSHA-S	P	2.1 in 10^4	4,000	2,800
Benzene/fugitive emissions	1984	EPA	F	2.1 in 10^5	0,310	2,800
Radionuclides/uranium mines	1984	EPA	F	1.4 in 10^4	1,100	6,900
Asbestos	1972	OSHA-H	F	3.9 in 10^4	396,000	7,400
Benzene	1985	OSHA-H	P	8.8 in 10^4	3,800	17,100
Arsenic/glass paint	1986	EPA	F	8.0 in 10^4	0,110	19,200
Ethylene oxide	1984	OSHA-H	F	4.4 in 10^5	2,800	25,600
Arsenic/copper smelter	1986	EPA	F	9.0 in 10^4	0,060	26,500
Uranium mill tailings/inactive	1983	EPA	F	4.3 in 10^4	2,100	27,600
Acrylonitrile	1978	OSHA-H	F	9.4 in 10^4	6,900	37,600
Uranium mill tailings/active	1983	EPA	F	4.3 in 10^4	2,100	53,000
Coke ovens	1976	OSHA-H	F	1.6 in 10^4	31,000	61,800
Asbestos	1986	OSHA-H	F	6.7 in 10^5	74,700	89,300
Arsenic	1978	OSHA-H	F	1.8 in 10^3	11,700	92,500
Asbestos	1986	EPA	P	2.9 in 10^5	10,000	104,200
DES (cattle feed)	1979	FDA	F	3.1 in 10^7	68,000	132,000
Arsenic/glass manufacturing	1986	EPA	R	3.8 in 10^5	0,250	142,000
Benzene/storage	1984	EPA	R	6.0 in 10^7	0,043	202,000
Radionuclides/DOE facilities	1984	EPA	R	4.3 in 10^6	0,001	210,000
Radionuclides/elemental phosphorus	1984	EPA	R	1.4 in 10^5	0,046	270,000
Acrylonitrile	1978	OSHA-H	R	9.4 in 10^4	0,600	308,000
Benzene/ethylbenzenol styrene	1984	EPA	R	2.0 in 10^8	0,006	483,000
Arsenic/low-arsenic copper	1986	EPA	R	2.6 in 10^4	0,090	764,000
Benzene/maleic anhydride	1984	EPA	R	1.1 in 10^6	0,029	820,000
Land disposal	1986	EPA	P	2.3 in 10^8	2,520	3,500,000
EDB	1983	OSHA-H	P	2.5 in 10^4	0,002	15,600,000
Formaldehyde	1985	OSHA-H	P	6.8 in 10^7	0,010	72,000,000

From Morrall [2003].
CPSC, Consumer Product Safety Commission; EPA, Environment Protection Agency; FAA, Federal Aviation Administration; FDA, Food and Drug Administration; NHTSA, National Highway Traffic Safety Administration; OSHA-H, Occupational Safety and Health Administration.
[a]Proposed, rejected, or final rule.
[b]Annual deaths per exposed population.

This act of commensurating the expected-value operation is analogous in some sense to the commensuration of all benefits and costs into one monetary unit. Indeed, few today would consider benefit–cost analysis, where all benefits, costs, and risks are commensurated into monetary units, as an adequate and acceptable measure for decisionmaking when it is used as the sole criterion for excellence. Close to four decades ago, multiple-objective analysis was demonstrated as a superior approach to benefit–cost analysis [Haimes, 1970; Haimes et al., 1971; Haimes and Hall, 1974]. In many respects, the expected value of risk is similar in its theoretical–mathematical construct to the commensuration of all costs, benefits, and risks into monetary units.

One of the most important steps in the risk assessment process is the quantification of risk. Yet the validity of the approach most commonly used to quantify risk—its expected value—has received neither the broad professional scrutiny it deserves nor the hoped-for wider mathematical challenge that it mandates. One of the few exceptions is the conditional expected value of the risk of extreme events (among other conditional expected values of risks) generated by the *partitioned multiobjective risk method* (PMRM) [Asbeck and Haimes, 1984] (see Chapters 8 and 11).

1.3.9 Software Risk Management

Computers have become pervasive in our society. They are integral to everything from VCRs and video games to power plants and control systems for aircraft. Computers enhance satellite communications systems that provide television nationwide; they enabled the governments (as well as CNN) to communicate during wars and other major national and international events. Computers touch the lives of most people daily.

Computers are composed of two major components. One is hardware: the power supply, printed circuit boards, and CRT screens. The other is software, sometimes thought of as the computer's intelligence.

Software engineering, unlike traditional forms of engineering, has no foundation in physical laws. The source of the structure for software engineering is in standards and policies that are defined by teams of experts. Because software is founded only in mathematics and logic and not in physical laws (except that the software logic must comply with physical laws), the risk of introducing uncertainty and other sources of failure into a software system is greater than in any other field.

Effective control of uncertainties introduced during the software development cycle should be through very stringent management. This has not been the case; to date, there has not been a well-defined process for supervising software development [Chittister and Haimes, 1994; Boehm, 2006; Jackson, 2006; Post et al., 2006]. Chapter 17 offers additional discussion on risks associated with software engineering.

The increasing dominance of computers in the design, manufacture, operation, maintenance, and management of most small- and all large-scale engineering systems has made possible the resolution of many complex technological problems. At the same time, the increased influence of software in decisionmaking has introduced a new dimension to the way business is done in engineering quarters; many former engineering decisions have been or soon will be transferred to software, albeit in a limited and controlled manner. This power shift in software functionality (from the centrality of hardware in system control and operations to software), the explicit responsibility and accountability of software engineers, and the expertise required of technical professionals on the job have interesting manifestations and implications, and they offer challenges to the professional community to adapt to new realities. All of these affect the assessment and management of risk associated with software development and use. Perhaps one of the most striking manifestations of this power shift relates to real-time control systems. Consequently, the impact of software on the reliability and performance of monitoring and warning systems for natural hazards is becoming increasingly more significant. Furthermore, the advances in hardware technology and reliability and the seemingly unlimited capabilities of computers render the reliability of most systems heavily dependent on the integrity of the software used. Thus, software failure must be scrutinized with respect to its contribution to overall system failure, along with the same diligence and tenacity that have been devoted to hardware failure.

1.3.10 Risk Characteristics of Engineering-Based Systems

In spite of some commonalities, there are inherent differences between natural systems (e.g., environmental, biological, and ecological systems) and man-made, engineering-based systems. In this section, it is constructive to focus on the characteristics of risk associated with engineering-based systems.

The following 12 risk characteristics are endemic to most engineering-based systems:

1. *Organizational failures of engineering-based systems are likely to have dire consequences.* Risk management of technological systems must be an integral part of overall systems management. Organizational failures often constitute a major source of risk of overall system failure.
2. *Risk of extreme and rare events is misrepresented when it is solely measured by the expected value of risk.* The precommensuration of rare but catastrophic events of low probability with much less adverse events of high probability in the expected-value measure of risk can lead to misrepresentation and mismanagement of catastrophic risk.
3. *Risk of project cost overrun and schedule delay.* Projects involving engineering-based systems have been experiencing major cost overruns and delays in schedule completion, particularly for software-intensive systems. The process of risk assessment and management is also the sine qua non requirement for ensuring against unwarranted delay in a project's completion schedule, cost overrun, and failure to meet performance criteria.
4. *Risk management as a requisite for engineering-based systems integration.* Effective systems integration necessitates that all functions, aspects, and components of the system must be accounted for along with an assessment of the associated risks. Furthermore, for engineering-based systems, systems integration is not only the integration of components but also an understanding of the functionality that emerges as a by-product from the integration.
5. *Rehabilitation and maintenance of physical infrastructure.* Maintaining and rehabilitating physical infrastructures, such as water distribution networks, have become an important issue as nations address the risk of their infrastructure failure. Accurate assessment of the risks of failure of deteriorating physical infrastructures is a prerequisite for the optimal allocation of limited resources.
6. *Multiple failure modes and multiple reliability measures for engineering-based systems.* Engineering-based systems often have any number of paths to failure. Evaluating the interconnected consequences of multiple modes of failure is central to risk assessment and management of engineering systems.
7. *Risk in software engineering development.* The development of software engineering—an intellectual, labor-intensive activity—has been marred by software that does not meet performance criteria while experiencing cost overruns and time and delivery delays. An integrated and holistic approach to software risk management is imperative.
8. *Risk to emergent and safety-critical systems.* Assessing and managing risk to emergent and safety-critical systems is not sufficient without building resilience in such systems. This means ensuring that even in the remote likelihood of a system failure, there will be a safe shutdown without catastrophic consequences to people or facilities. Examples of such critical systems include transportation systems, space projects, the nuclear industry, and chemical plants.
9. *Cross-disciplinary nature of engineering-based systems.* All engineering-based systems are built to serve the well-being of people. The incorporation of knowledge-based expertise from other disciplines is essential. The risk of system failures increases without incorporation of outside knowledge.

10. *Risk management: A requisite for sustainable development.* Sustainable development ensures long-term protection of the ecology and the environment, in harmony with economic development. This cannot be realized without a systemic process of risk assessment and management.
11. *Evidence-based risk assessment.* Sparse databases and limited information often characterize most large-scale engineering systems, especially during the conception, planning, design, and construction phases. The reliability of specific evidence, including the evidence upon which expert judgment is based, is essential for effective risk management of these systems.
12. *Impact analysis.* Good technology management necessarily incorporates good risk management practices. Determining the impacts of current decisions on future options is imperative in decisionmaking.

1.3.11 Guiding Principles for Risk Analysis

Numerous studies have attempted to develop criteria for what might be considered *good* risk analyses, the most prominent of which is the Oak Ridge Study [Fischhoff et al., 1980]. Good risk studies may be judged against the following list of 10 criteria. The study must be:

- Comprehensive
- Adherent to evidence
- Logically sound
- Practical, by balancing risk with opportunity
- Open to evaluation
- Based on explicit assumptions and premises
- Compatible with institutions (except when change in institutional structure is deemed necessary)
- Conducive to learning
- Attuned to risk communication
- Innovative

Chapter 12 introduces the systems-based guiding principles for risk modeling, planning, assessment, management, and communication.

1.4 CONCEPT ROAD MAP

1.4.1 Overview of the Risk Assessment and Management Process (Chapter 1)

The importance, impact on decisionmaking at all levels, and complexity of the risk assessment and management process call for iterative learning, unlearning, and relearning [Toffler, 1980]. This chapter, which provides an overview of the book, highlights the strong commonalities and interdependencies between a holistic systems engineering philosophy and a systemic quantitative risk assessment and management, where both are grounded on the arts and the sciences. Some key ideas advanced in this chapter include:

1. Risk assessment and management is a process that must answer the following set of questions [Kaplan and Garrick, 1981; Haimes, 1991]:
   What can go wrong?
   What is the likelihood?
   What are the consequences?
   (And at what time frame?)
   What can be done and what options are available?
   What are the associated trade-offs in terms of all costs, benefits, and risks?
   What are the impacts of current decisions on future options?
2. Organizational failures are major sources of risk.
3. The perception of risk and its importance in decisionmaking should not be overlooked.
4. Risk management should be an integral part of technology management, leading to multiple-objective trade-off analysis.
5. The expected value of risk leads to erroneous results when used as the sole criterion for risk measurement. Also, risk of extreme and catastrophic events should not be commensurate with high-probability/low-consequence events.

1.4.2 The Role of Modeling in the Risk Assessment Process (Chapter 2)

To provide a unified road map for this book and to relate the 19 chapters of this fourth edition to the processes of modeling, assessment, and management of risk, Chapter 2 introduces a systems-based approach to the complex definitions of risk, vulnerability, and resilience. Consider the following oversimplified farmer's dilemma that is formulated and solved in Appendix A.3.

A farmer who owns 100 acres of agricultural land is considering two crops for next season—corn and sorghum. Due to a large demand for these crops, he (the term *he* is used here generically to denote either gender) can safely assume that he can sell his entire yield. From past experience, the farmer knows that the climate in his region requires (i) an irrigation of 3.9 acre-ft of water per acre of corn and 3 acre-ft of water per acre of sorghum at a subsidized cost of $40 per acre-ft and (ii) nitrogen-based fertilizer of 200 lb/acre of corn and 150 lb/acre of sorghum at a cost of $25/100 lb of fertilizer (an acre-ft of water is a measure of one acre of area covered by one foot of water).

The farmer believes that his land will yield 125 bushels of corn per acre and 100 bushels of sorghum per acre. He expects to sell his crops at $2.80 per bushel of corn and $2.70 per bushel of sorghum.

The farmer has inherited his land and is very concerned about the loss of topsoil due to erosion resulting from flood irrigation—the method used in his farm. A local soil conservation expert has determined that the farmer's land loses about 2.2 tons of topsoil per acre of irrigated corn and about 2 tons of topsoil per acre of irrigated sorghum. The farmer is interested in limiting the total topsoil loss from his 100 acre land to no more than 210 tons per season.

The farmer has a limited allocation of 320 acre-ft of water available for the growing season, but he can draw all the credit needed for the purchasing of fertilizer. He would like to determine his optimal planting policy in order to maximize his income. He considers his labor to be equally needed for both crops, and he is not concerned about crop rotation. Note that at this stage in the case, water quality (e.g., salinity and other contamination), impact on groundwater quality and quantity, and other issues (objectives) are not addressed.

This seemingly simple farmer's dilemma includes most of the ingredients that constitute a complex, risk-based decisionmaking problem. To explore the elements of risk and uncertainty addressed in this book, in Appendix A.3, we will first model the problem with a deterministic model, focusing on the role of modeling in the risk assessment process. We will subsequently explore more realistic assumptions and situations that lend themselves to probabilistic and dynamic modeling and treatment.

Even this oversimplified version of the problem has many interesting characteristics. The following are some of the most important modeling elements:

1. There are multiple conflicting and competing objectives: Maximize crop yield and minimize soil erosion.
2. There are resource constraints: water, land, and capital.
3. These resources manifest themselves in a major modeling building block—the state variables—a concept that will be extensively explored in subsequent discussions. Examples of state variables include the state of soil erosion and soil moisture.

Note that the role of the decision variables is to bring the states of the system to the appropriate levels that ultimately optimize the objective functions. (For the farmer, it means what crops to grow, when to irrigate, etc.) To know when to irrigate and fertilize a farm, a farmer must assess the states of the soil—its moisture and level of nutrients. Although an objective function can be a state variable, the role of the decision variables is not to directly optimize the objective functions. Identifying and quantifying (to the extent possible) the building blocks of a mathematical model of any system constitutes a fundamental step in modeling, where one building block—state variables—is the sine qua non in modeling.

Although the deterministic version of the farmer's dilemma is formulated and solved in Appendix A.3, no one would expect the farmer to predict all model parameters accurately—except, of course, for the availability of 100 acres of land that he owns. All other entries are merely average estimates predicated on past experience. For example, the amount of water needed to irrigate corn and sorghum is dependent on one state variable—soil moisture, which in turn depends on the amount of irrigation or precipitation for the season. The same argument applies to prices, which fluctuate according to market supply and demand. In particular, the level of soil erosion is heavily dependent on the climate and land use. Dry seasons are likely to increase soil erosion; irrigation patterns such as flood or sprinkles irrigation combined with the type of crops being grown and climate conditions can markedly vary the rate of soil erosion.

1.4.3 Identifying Risk through HHM (Chapter 3)

To effectively model, assess, and manage risk, one must be able to identify (to the extent possible) all important and relevant sources of that risk. Clearly, the root causes of most risks are many and diverse. Farmers face numerous risks at every stage of the farming life cycle. Other examples may include the risk of project cost overrun, time delay in its completion, the risk of not meeting performance criteria, and environmental and health risks. In Chapter 3, we introduce HHM, a systemic modeling philosophy/methodology that captures the multiple aspects, dimensions, and perspectives of a system. This systemic methodology serves as an excellent medium with which to answer the first question in risk assessment (What can go wrong?) and the first question in risk management (What can be done and what options are available?). Several visions or perspectives of risk are investigated in the HHM methodology, which includes the adaptive multiplayer HHM game.

1.4.4 Decision Analysis and the Construction of Evidence-Based Probabilities (Chapter 4)

Facing numerous natural and man-made challenges, the farmer can markedly benefit from the assorted decisionmaking tools and techniques assembled under the umbrella of decision analysis. For example, the farmer may wonder whether the market for his crops will be good, fair, or poor. If he could know the market condition in advance, he would direct his crop-growing decisions accordingly. Not wanting to rely on past statistical data to make future projections, the farmer may desire to minimize his maximum loss, maximize his minimum gain, or maximize his maximum gain. Here, the minimax (or maximin) principle can be very helpful. Furthermore, the Hurwitz rule, which bridges between maximizing his maximum gain and minimizing his maximum loss, can further enhance his decisionmaking process under conditions of uncertainty.

Chapter 4 will review some of these risk-based decisionmaking tools. For example, much of the farmer's dilemma can be posed in terms of a decision tree. Although decision-tree analysis will be introduced in Chapter 4 at its rudimentary level, an extensive treatment of decision trees with multiple objectives will be presented in Chapter 9. Indeed, one may argue that since most, if not all, problems lend themselves to multiple objectives, then extending decision trees to incorporate multiple objectives is an important step forward. The reader will note that the entire concept of optimality has to be modified and extended to encompass Pareto-optimality (see Chapter 5) in multiobjective decision-tree (MODT) analysis (as discussed in Chapter 9).

Chapter 4 also will introduce two approaches for the construction of probabilities on the basis of evidence from experts, due to the lack of statistical data. These approaches are the fractile method and triangular distribution. Modeling population dynamics is important, not only to farmers (to forecast the age distribution of their livestock over time) but also for the planning of schools and hospitals, among other installations, by communities and government agencies. For this purpose, the Leslie model [Meyer, 1984] will be introduced in Chapter 4.

Finally, Chapter 4 also will introduce the Phantom System Model (PSM). This enables system modelers to effectively study, understand, and analyze major forced changes in the characteristics and performance of multiscale assured systems. One example would be the physical infrastructure of a bridge SoS and the associated major interdependent socioeconomic systems [Haimes, 2007]. (Note that the term PSM will connote the overall modeling philosophy, while PSMs will connote the modeling components.) The PSM builds on and incorporates input from HHM discussed in Chapter 3. HHM is a holistic philosophy/methodology aimed at capturing and representing the essences of the inherent diverse characteristics and attributes of a system—its multiple aspects, perspectives, facets, views, dimensions, and hierarchies.

1.4.5 Multiobjective Trade-Off Analysis (Chapter 5)

The farmer knows that the finer the soil from cultivation, the higher the expected crop yield. However, this land use management practice is likely to lead to higher soil erosion. This dilemma is at the heart of multiobjective trade-off analysis—the subject of Chapter 5. This is the expertise domain of numerous scholars around the world, most of whom have devoted their entire professional career to this subject. Indeed, the International Society on Multiple Criteria Decision Making meets about every 2 years, and experts on MCDM share their experience and knowledge.

An important component of Chapter 5 is the discussion of the surrogate worth trade-off (SWT) method [Haimes and Hall, 1974; Chankong and Haimes, 2008]. Two basic principles upon which the SWT method is grounded are as follows: (i) the premise that sound decisions cannot be made merely on the basis of the absolute values of each objective function—rather, these absolute values must be supplemented and complemented with associated trade-offs at specific levels of attainment of these objectives—and (ii) the Epsilon-constraint method [Haimes, 1970; Haimes et al., 1971; Chankong and Haimes, 2008].

In particular, multiobjective trade-off analysis (within the SWT method) avoids the need to commensurate all objectives in, say, monetary terms. The trade-offs enable the analyst and decisionmaker(s) to determine the preferred policy on the basis of the values of these objective functions and their associated trade-offs.

The farmer may make use of multiobjective trade-off analysis in many other ways. For example, he may desire to change different pieces of equipment, each with specific cost and reliability. In this case, his trade-offs are his investments in farming equipment versus reliability and performance. These types of decisions are best handled via multiobjective trade-off analysis.

Chapter 5 presents an extensive discussion on this subject with ample example problems.

1.4.6 Defining Uncertainty and Sensitivity Analysis (Chapter 6)

The farmer, having lived and worked on his farm for many years, where several past generations have passed on valuable knowledge and wisdom, is rightfully skeptical of the modeling efforts by his systems analyst. He is very well aware of the following Arabic proverb [Finkel, 1990]:

> He who knows and knows he knows,
>   He is wise—follow him;
> He who knows not and knows he knows not,
>   He is a child—teach him;
> He who knows and knows not he knows,
>   He is asleep—wake him;
> He who knows not and knows not he knows not,
>   He is a fool—shun him.

It is here that the uncertainty taxonomy presented in Chapter 6 is helpful in diffusing some of the farmer's concerns about the uncertainty and variability associated with model assumptions, databases, causal relationships, and other factors affecting his ultimate decisions. Chapter 6 is devoted to exploring and categorizing the sources of uncertainty and variability in modeling and decisionmaking under risk and uncertainty.

One of the major concerns of our farmer is the risk of bankruptcy due to one or a sequence of disastrous growing seasons. In many respects, such disasters are tantamount to a calamity with irreversible consequences. The need to assess the sensitivity, response, and stability of a system (the farm in our case) to unexpected, unplanned, or catastrophic changes is imperative for good management and prudent decisionmaking. Risk of extreme and catastrophic events is discussed in Chapters 8 and 11.

The uncertain world within which we live continuously presents surprises and unexpected events with potential dire consequences. Planning for such eventualities and assessing the impacts of current decisions on future options are at the heart of good risk assessment and management. Furthermore, the use of models in decisionmaking has markedly increased during the last four decades. Decisions involving air traffic control, nuclear reactors, petroleum refineries, manufacturing, airline reservations, and thousands of other enterprises all make extensive use of models. For example, the farm may use a simple linear programming model (see Chapter 2 and the Appendix) to determine the optimal mix of growing corn and sorghum while balancing two conflicting objectives: maximizing income from crop yields and minimizing soil erosion. Some farmers use linear models to help them determine the optimal mix of feed ingredients for their livestock as the prices fluctuate in the marketplace.

Of course, models are constructed on the basis of certain assumptions and premises, and they are composed of variables and parameters of many dimensions and characteristics (they will be discussed in detail in Chapter 2). Clearly, when making decisions on the basis of mathematical models, one must be cognizant of at least the following four eventualities:

1. Most systems are dynamic in nature, and previously assumed values for model parameters may not be representative under new conditions.
2. Model topology (e.g., its structure, dimension, and other characteristics) may not constitute a good representation of the system.
3. Model parameters may not be representative in the first place.
4. Model output may be very sensitive to certain parameters.

The uncertainty sensitivity index method (USIM) [Haimes and Hall, 1977] and its extensions [Li and Haimes, 1988] provide a methodological framework with which to evaluate the sensitivity of the model output, the objective functions, or the constraints to changes in model parameters. Furthermore, the USIM and its extension enable the analysts or decisionmaker to trade off a decrease in the sensitivity of model output with a reduction in some performance functions. (Section 18.11 presents further discussion on the USIM.)

The farmer may make use of the USIM in many ways. He may, for example, want to minimize the sensitivity of soil erosion to an assumed nominal value of the model parameter that represents soil permeability, while being willing to forgo an increased crop yield. Chapter 6 will introduce the USIM and its extensions and offer a large number of examples.

1.4.7 Risk Filtering, Ranking, and Management (RFRM) (Chapter 7)

Most people and organizations tend to rank risks by asserting that *Risk A is higher than Risk B*. Such ranking, however, is invariably made on an ad hoc basis and with no systemic or quantifiable metric. Indeed, one of the major challenges facing the risk analysis community is to develop a more universal risk-ranking method (without relying on numerical order) capable of taking into account the myriad number of attributes that deem one risk higher or lower than others.

Chapter 7 discusses one such ranking method [Haimes et al., 2002]. The farmer, for example, may desire to rank the perceived or actual risks facing his farming enterprise (to the crops, livestock, water supply, long-term investment, etc.). The application of the RFRM to a variety of studies is discussed throughout this book.

1.4.8 Risk of Extreme Events and the Fallacy of the Expected Value (Chapter 8)

Risk is a complex concept. It measures an amalgamation of two constructs: One, probability, is a mental, human-made construct that has no physical existence per se. The other is severity of adverse effects, such as contaminant concentration, loss of lives, property loss, and defects in manufactured products, among others. The correct measure of mixing probability and severity in a risk metric is the subject of Chapter 8.

The expected value (the mean or the central tendency), which does not adequately capture events of low probabilities and high consequences, is supplemented with the PMRM [Asbeck and Haimes, 1984]. In particular, risk associated with safety-critical systems cannot be assessed or managed by using the expected value as the sole metric.

The farmer, for example, may be concerned with more than one consecutive drought year. In this case, the PMRM can generate a conditional expected value of drought (e.g., rainfall of <20 in). Having this additional knowledge base, the farmer may adjust his farming policy to reduce his chance of bankruptcy. Several example problems, where extreme-event analysis is critical, are introduced and solved in this chapter.

1.4.9 MODT (Chapter 9)

Decision-tree analysis with a single-objective function was discussed in Chapter 4 as part of decision analysis. Chapter 9 extends the decision-tree methodology to incorporate multiobjective functions. Indeed, MODT [Haimes et al., 1994] adds much more realism and practicality to the power of decision trees [Raiffa, 1964].

The farmer, for example, may desire to use MODT in analyzing his policy options as to what crops to grow and at what level, what irrigation method to use and how much to irrigate, and what land use practices to follow in cultivating his land—all in order to maximize his income and reduce his soil erosion. MODT analysis is a very versatile tool in decisionmaking under risk and uncertainty. Chapter 9 is devoted in its entirety to this powerful method with many example problems.

1.4.10 Multiobjective Risk Impact Analysis Method (Chapter 10)

Chapter 10 addresses the question, *What is the impact of current decisions on future options?* This impact analysis is important whether the decisions are made under deterministic conditions or under conditions dominated by risk and uncertainty. Impact analysis is also important for emergent systems. These have features that are not designed in advance but evolve, based on sequences of events that create the motivations and responses for properties that ultimately emerge into system features. This is because our world is dynamic, and decisions thought to be optimal under current conditions may prove to be far from optimal or maybe even disastrous. In a sense, the multiobjective risk impact analysis method (MRIAM) [Leach and Haimes, 1987] combines two separately developed methodologies: the multiobjective impact analysis method (MIAM) [Gomide and Haimes, 1984] and the PMRM.

Most decisionmaking situations address systems with transitory characteristics. For example, the farmer may desire to ascertain the impact of any of the following variations on his livelihood: crop market prices over the years, water availability in future years, changes in hydrological conditions, and others.

Chapter 10 will present a section that relates the MODT introduced in Chapter 9 to the MRIAM [Dicdican and Haimes, 2005], which will also be presented with example problems.

1.4.11 Statistics of Extremes: Extension of the PMRM (Chapter 11)

Very often, historical, statistical, or experimental data are sparse, especially on extreme events (the tail of the probability distribution function). The statistics of extremes is a body of statistical theory that attempts to overcome this shortage of data by classifying most probability distributions into three families on the basis of how fast their tails decay to zero. These three families are commonly known as Gumbel type I, type II, and type III.

Chapter 11 extends Chapter 8 and builds on the body of knowledge of the statistics of extremes, incorporates the statistics of extremes with the PMRM, and extends the theory and methodology of risk of extreme events. This chapter also relates the concepts of the return period to the conditional expected value of extreme events and to the statistics of extremes.

The farmer, for example, may desire to relate the return period of a sizable flood or drought to the expected value and conditional expected value of crop yield. He can do so using parts of the methodology discussed in this chapter.

1.4.12 Systems-Based Guiding Principles for Risk Modeling, Planning, Assessment, Management, and Communication (Chapter 12)

The 10 principles set forth in this chapter are intended to provide a broad framework for understanding and practicing risk analysis—regardless of the specific domain, problem, system, or discipline. These fundamental systems-based principles build on and encapsulate the theory and methodology presented throughout this book. They are designed to guide both quantitative- and qualitative-centered risk analyses. Although these principles may be applied to a range of disciplines, to retain focus, this chapter draws from and is guided by both risk analysis and systems engineering theory, methodology, and practice.

1.4.13 Fault Trees (Chapter 13)

Assessing the reliability of an engineering system or a system component is vital to its design, development, operations, maintenance, and replacement. In particular, an analyst or a decisionmaker would invariably want to know the trade-offs among different policy options in terms of their cost and associated reliability (or unreliability). Fault trees have been developed and extensively used in myriad engineering and nonengineering applications. Most notable among them is the nuclear industry [US Nuclear Regulatory Commission, 1981].

Chapter 13 extends fault-tree analysis to incorporate a variety of probability distribution functions into a new methodology termed distribution analyzer and risk evaluator (DARE) [Tulsiani et al., 1990]. FMEA and FMECA—two important tools with extensive use in the life cycle of engineering systems—are also discussed in Chapter 13.

The farmer, for example, may desire to ascertain the reliabilities of his farm equipment or irrigation system in order to make investment decisions. He can do so using fault-tree analysis.

1.4.14 Multiobjective Statistical Method (MSM) (Chapter 14)

The MSM is grounded on adherence to the following basic premises [Haimes et al., 1980]:

1. Most, if not all, systems have a multiobjective nature.
2. State variables, which represent the essence of a system at any time period, play a dominant role in modeling.
3. Sources of risk and uncertainty can be best modeled through probabilistic modeling methods.
4. The joint use of simulation and optimization is by far more effective than the use of each one alone.
5. A good database is invaluable to good systems analysis, and the improvement of the database can be accomplished through questionnaires, expert judgment, and other mechanisms for data collection.

Our challenge in the farmer's example problem is modeling soil erosion, which is an objective function and a state variable (i.e., minimizing one objective function, which is soil erosion, is the same as minimizing the state variable soil erosion). For the purpose of this discussion, denote soil erosion by S. This state variable depends on at least three other major variables:

- Random variables ($\mathbf{r}$), such as precipitation and climate conditions (e.g., temperature, wind)
- Decision variables ($\mathbf{x}$), such as land use and irrigation patterns
- Exogenous variables ($\mathbf{e}$), such as soil characteristics (e.g., permeability and porosity and other morphological conditions)

Note that some of the variables may fall into multiple categories—this is part of the nature of the modeling process.

Through simulation, one aims at determining the causal relationships between S and the other three variables; that is, $S = S(\mathbf{r},\mathbf{x},\mathbf{e})$. Note, however, that by their nature, the random variables (precipitation and climatic conditions) are characterized by an ensemble of values over their sample space. Here, one may make use of the expected value, which is the mean or average value of the realization of each random variable. Alternatively, one may supplement and complement the expected value of the random variable with the conditional expected value as derived through the use of the PMRM [Asbeck and Haimes, 1984]. The PMRM and its extensions are extensively discussed in Chapters 8 and 11.

An analyst who is helping the farmer with crop decisions may develop a set of questionnaires to be distributed to other farmers in the region and may obtain more scientific information from the literature at agriculture experiment stations to quantify $S = S(\mathbf{r},\mathbf{x},\mathbf{e})$.

The above analyses will yield a multiobjective optimization problem where the SWT method [Haimes and Hall, 1974] can be used. The SWT method is discussed in Chapter 5.

1.4.15 Principles and Guidelines for Project Risk Management (Chapter 15)

The life cycle management of systems—small and large—is an integral part of good systems engineering and good risk management. Indeed, the increasing size and complexity of acquisition and development projects in both the public and private sectors have begun to exceed the capabilities of traditional management techniques to control them. With every new technological development or engineering feat, human endeavors inevitably increase in their complexity and ambition. This trend has led to an explosion in the size and sophistication of projects by government and private industry to develop and acquire technology-based systems. These systems are characterized by the often unpredictable interaction of people, organizations, and hardware. In particular, the acquisition of software has been marred with significant cost overruns, time delay in delivery, and the lack of meeting performance criteria.

Although the farmer has markedly increased the use of computers and, of course, the use of various software packages in his enterprise, he may not concern himself with the risk associated with software development. Nevertheless, since the software component of modern, large-scale systems continues to assume an increasingly critical role in such systems, it is imperative that software risk management be discussed in this book. Software has a major effect on any system's quality, cost, and performance. Indeed, system quality is predicated, as never before, upon the quality of its software. System risk is increasingly being defined relative to the risk associated with its software component. Acquisition officials, who previously concentrated on the hardware components of a system, instead find themselves concentrating more of their energies, concerns, and resources on the embedded hardware–software components.

Chapter 15 will address project risk management and the characteristics of software risk management and offer tools and methodologies for the management of the risk of cost overrun, the risk of time delay in software delivery, and the risk of not meeting performance criteria.

1.4.16 Modeling Complex SoS with PSM (Chapter 16)

The fact that modeling is as much an art as a science—a tedious investigative trial-and-error, learn-as-you-go process—means that an equally imaginative approach is necessary to discover the inner functionality of complex systems through modeling. In this context, Chapter 16 (i) addresses system modeling, and the inverse problem, or the system identification problem, through the PSM; (ii) analyzes the contributions of PSM as a modeling mechanism through which to experiment with creative approaches to modeling complex SoS; and (iii) relates (at the metamodeling level) the intrinsic common/shared state variables among the subsystems of the SoS, thereby offering more insight into the intra- and interdependencies among the subsystems.

1.4.17 Adaptive Two-Player HHM Game for Counterterrorism Intelligence Analysis (Chapter 17)

Intelligence gathering and analysis for countering terrorism is a vital and costly venture; therefore, approaches need to be explored that can help determine the scope of collection and improve the efficacy of analysis efforts. The Adaptive Two-Player HHM Game introduced in Chapter 3 and discussed in detail in Chapter 17 is a repeatable, adaptive, and systemic process for tracking terrorism scenarios. It builds on fundamental principles of systems engineering, system modeling, and risk analysis. The game creates two opposing views of terrorism: one is developed by a Blue Team defending against acts of terrorism, and the other by a Red Team planning to carry out a terrorist act. The HHM process identifies the vulnerabilities of potential targets that could be exploited in attack plans. These vulnerabilities can be used by the Blue Team to identify corresponding surveillance capabilities that can help to provide warning of a possible attack. Vulnerability-based scenario structuring, comprehensive risk identification and the identification of surveillance capabilities that can support preemption are all achieved through the deployment of HHM.

State variables, which represent the essence of the system, play a pivotal role in the Adaptive Two-Player HHM Game, providing an enabling road map to intelligence analysts. Indeed, vulnerabilities are defined in terms of the system's state variables: Vulnerability is the manifestation of the inherent states of a system (e.g., physical, technical, organizational, cultural) that can be exploited by an adversary to cause harm or damage. Threat is a potential adversarial intent to cause harm or damage by adversely changing the states of the system. Threat to a vulnerable system may lead to risk, which is a measure of the probability and severity of adverse effects.

1.4.18 Inoperability Input–Output Model and Its Derivatives for Interdependent Infrastructure Sectors (Chapter 18)

In assessing a system's vulnerability, it is important to analyze both the intraconnectedness of the subsystems that compose it and its interconnectedness with other external systems. This chapter develops a methodology that quantifies the dysfunctionality or *inoperability* as it propagates throughout our critical infrastructure systems or industry sectors. The inoperability that may be caused by willful attacks, accidental events, or natural causes can set off a complex chain of cascading impacts on other interconnected systems. For example, telecommunications, power, transportation, banking, and others are marked by immense complexity, characterized predominantly by strong intra- and interdependencies as well as hierarchies. The Inoperability Input–Output Model (IIM) [Haimes and Jiang, 2001; Santos, 2003; Haimes et al., 2005a, b; Lian, 2006; Crowther, 2007] and its derivatives build on the work of Wassily Leontief, who received the 1973 Nobel Prize in Economics for developing what came to be known as the Leontief Input–Output Model (I/O) of the economy [Leontief, 1951a, b, 1986]. The economy consists of a number of subsystems, or individual economic sectors or industries, which are a framework for studying its equilibrium behavior. It enables understanding and evaluating the interconnectedness among the various sectors of an economy and forecasting the effect on one segment of a change in another. The IIM is extended in Chapter 18 to model multiregional, dynamic, and uncertainty factors.

1.4.19 Case Studies (Chapter 19)

Six case studies applying risk modeling, assessment, and management to real-world problems are introduced in Chapter 19. The first case study documents the application of the IIM and its derivatives (see Chapter 18) to measure the effects of the August 2003 northeast electric power blackout in North America [Anderson et al., 2007]. Systemic valuation of strategic preparedness through applying the IIM and its derivatives with lessons learned from Hurricane Katrina is the subject of the second case study [Crowther et al., 2007]. The third case study is an ex post analysis of the September 11, 2001, attack on the United States using the IIM and its derivatives [Santos, 2003]. The focus of the fourth case study is the 5770 foot Mount Pinatubo volcano that erupted in the Philippines. We analyze the risks associated with the huge amount of volcanic materials deposited on its slopes (about $1\,mi^3$). Several concepts and methodologies introduced in this book are applied. The fifth case study provides the perspectives of the risk of extreme events when considering the six-sigma capability in quality control. The PMRM introduced in Chapter 8 and the statistics of extremes introduced in Chapter 11 are related to and compared with the six-sigma capability metric. The sixth case study provides the reader a deeper insight into the propagation of sequential Pareto-optimal decisions made within emergent complex SoS, with an application to the FAA NextGen. In particular, this case study addresses the third question in risk management—*What are the impacts of current decisions on future options?*—that is critically important for emergent complex SoS, because their conception, the evolution of their requirements and specifications, their design and development, and their ultimate operation can span several years. Furthermore, the sequential decisions made during the development of each individual subsystem of any given complex SoS will most likely affect the development of other new subsystems of the SoS in the future—with the ultimate goal that, in their totality, all of the subsystems will operate as an integrated, harmonious whole. For example, decisions to achieve specific outcomes made on subsystem **A** of an emergent SoS can change the states of subsystem **A**, but they can also affect other interconnected and interdependent subsystems that share states with subsystem **A**.

1.5 EPILOGUE

The comprehensiveness of TRM makes the systemic assessment and management of risk tractable from many perspectives. Available theories and methodologies developed and practiced by various disciplines can be adopted and modified as appropriate for TRM. Fault-tree analysis, for example, which has been developed for the assessment and management of risk associated with hardware, is being modified and applied to assess and manage all four sources of failure: hardware, software, organizational, and human. Hierarchical/multiobjective trade-off analysis is being applied to risk associated with public works and the infrastructure. As the importance of risk is better understood and its analysis is incorporated within a broader and more holistic management framework, the following progress will be likely:

1. The field of risk analysis will lose some of its current mystique, gain wider recognition, and more closely merge with the fields of systems engineering, systems analysis, and operations research.
2. The various disciplines that conduct formal risk analysis will find more common ground in their assessment and management than ever before.
3. As a by-product of 1 and 2 above, the field of risk analysis will advance by leaps and bounds as the professional community benefits from the synergistic contributions made in the area of risk assessment and management by the various disciplines: engineering, environmental science, medical health care, social and behavioral sciences, finance, economics, and others.
4. New measures of risk will likely emerge either as a substitute for, or as a supplement and complement to, the expected-value-of-risk measure.
5. Probably most important, government officials, other professionals, and the public at large will have more appreciation of, and confidence in, the process of risk assessment and management.
6. The spread of international terrorism will likely engage the attention of more and more risk analysts.

Finally, it is important to keep in mind two things: (i) Heisenberg's uncertainty principle [Feynman et al., 1963], which states that the position and velocity of a particle in motion cannot simultaneously be measured with high precision, and (ii) Einstein's statement: "So far as the theorems of mathematics are about reality, they are not certain; so far as they are certain, they are not about reality." By projecting Heisenberg's principle and Einstein's statement to the field of risk assessment and management, we assert that:

To the extent that risk assessment is precise, it is not real
To the extent that risk assessment is real, it is not precise

REFERENCES

Adelman, L., 1991, *Evaluating Decision Support and Expert Systems*, John Wiley & Sons, Inc., New York.

Anderson, C.W., J.R. Santos, and Y.Y. Haimes, 2007, A risk-based input-output methodology for measuring the effects of the August 2003 Northeast Blackout, *Economics Systems Research* **19**(2): 183–204.

Asbeck, E.L., and Y.Y. Haimes, 1984, The partitioned multiobjective risk method (PMRM), *Large Scale Systems* **6**(1): 13–38.

Bernstein, P., 1996, *Against the Gods: The Remarkable Story of Risk*, John Wiley & Sons, Inc., New York.

Bertalanffy, L., 1968/1976, *General System Theory: Foundations, Development, Applications*, George Braziller, New York.

Blanchard, B.S., 2003, *Systems Engineering Management*, Third edition, John Wiley & Sons, Inc., Hoboken, NJ.

Blanchard, B.S., and W.J. Fabrycky, 1990, *Systems Engineering and Analysis*, Prentice-Hall, Englewood Cliffs, NJ.

Blanchard, B.S., and W.J. Fabrycky, 1998, *Systems Engineering and Analysis*, Third edition, Prentice-Hall, Englewood Cliffs, NJ.

Blanchard, B.S., and W.J. Fabrycky, 2005, *Systems Engineering and Analysis*, Fourth edition, Prentice-Hall, Englewood Cliffs, NJ, 816 pp.

Blauberg, I.V., V.N. Sadovsky, and E.G. Yudin, 1977, *Systems Theory: Philosophical and Methodological Problems*, Progress Publishers, New York.

Boehm, B., 2006, Some future trends and implications for systems and software engineering processes, *Systems Engineering* 9(1): 1–19.

Boulding, K.E., 1953, *The Organizational Revolution*, Harper & Row, New York.

Brooks, C.C., 2000, Knowledge management and the intelligence community, *Defense Intelligence Journal* 9(1): 15–24.

Buede, D.M., 1999, *The Engineering Design of Systems: Models and Methods*, John Wiley & Sons, Inc., New York, 488 pp.

Burke, T., N. Tran, J. Roemer, and C. Henry, 1993, *Regulating Risk, the Science and Politics of Risk*, International Life Sciences Institute, Washington, DC.

Chankong, V., and Y.Y Haimes, 1983 *Multiobjective Decision Making: Theory and Methodology*, Elsevier-North Holland, New York.

Chankong, V., and Y.Y. Haimes, 2008, *Multiobjective Decision Making: Theory and Methodology*, Dover, Mineola, NY.

Chittister, C., and Y.Y. Haimes, 1994, Assessment and management of software technical risk, *IEEE Transactions on Systems, Man, and Cybernetics* 24(4): 187–202.

Cohen, I.B., and A. Whitman, 1999, *Isaac Newton: The Principia*, University of California Press, Berkeley, CA.

Collins, J., 2001, *Good to Great*, HarperCollins Publishers, New York

Covey, S.R., 1989, *The Seven Habits of Highly Effective People*, Simon and Schuster, New York.

Crowther, K.G., 2007, *Development of a Multiregional Preparedness Framework and Demonstration of Its Feasibility for Strategic Preparedness of Interdependent Regions*, Ph.D. dissertation, University of Virginia, Charlottesville, VA.

Crowther, K.G., Y.Y. Haimes, and G. Taub, 2007, Systemic valuation of strategic preparedness through application of inoperability input–output model with lessons learned from hurricane Katrina, *Risk Analysis* 27(5): 1345–1364.

Davenport, T.H., and L. Prusak, 1998, *Working Knowledge: How Organizations Manage What They Know*, Harvard Business Press, Boston, MA.

Deming, W.E., 1986, *Out of the Crisis, Center for Advanced Engineering Study*, Massachusetts Institute of Technology, Cambridge, MA.

Dicdican, R.Y., and Y.Y. Haimes, 2005, Relating multiobjective decision trees to the multiobjective risk impact analysis method, *Systems Engineering* 8(2): 95–108.

Douglas, M., 1990, Risk as a forensic resource, *Daedalus* 119(4): 1–16.

Drucker, P.F., 2004, *The Daily Drucker*, Harper Business, New York.

Fang, L., K.W. Hipel, and D.M. Kilgour, 1993, *Interactive Decision Making: The Graph Model for Conflict Resolution*, John Wiley & Sons, Inc., New York, 240 pp.

Feynman, R.P., R.B. Leighton, and M. Sands, 1963, *The Feynman Lectures on Physics*, Addison-Wesley, Reading, MA.

Finkel, A.M., 1990, *Confronting Uncertainty in Risk Management*, Center for Risk Management, Resources for the Future, Washington, DC.

Fischhoff, B., S. Lichtenstein, P. Slovic, R. Keeney, and S. Derby, 1980, *Approaches to Acceptable Risk: A Critical Guide*, ORNL Sub-7656, Oak Ridge National Laboratory, Eugene, OR.

Fischhoff, B., S. Lichtenstein, P. Slovic, S. Derby, and R. Keeney, 1983, *Acceptable Risk*, Cambridge University Press, Cambridge.

Gardner, H., and T. Hatch, 1989, Multiple intelligences go to school, *Educational Researcher* 18(8): 4–10.

Gharajedaghi, J., 2005, *Systems Thinking: Managing Chaos and Complexity: A Platform for Designing Business Architecture*, Second edition, Butterworth-Heinemann, Boston, MA, 368 pp.

Gheorghe, A.V., 1982, *Applied Systems Engineering*, John Wiley & Sons, Inc., New York.

Goleman, D., 1997, *Emotional Intelligence: Why It Can Matter More than IQ*, Bantam Books, New York.

Goleman, D., 1998, *Working with Emotional Intelligence*, Bantam Books, New York.

Gomide, F., and Y.Y. Haimes, 1984, The multiobjective, multistage impact analysis method: Theoretical basis, *IEEE Transactions on Systems, Man, and Cybernetics* 14: 89–98.

Gordon, W.J.J., 1968, *Synectics: The Development of Creative Capacity*, Collier Books, New York.

Haimes, Y.Y., 1970, *The Integration of System Identification and System Optimization*, Ph.D. dissertation, School of Engineering and Applied Science, University of California, Los Angeles, CA.

Haimes, Y.Y., 1977, *Hierarchical Analyses of Water Resources Systems: Modeling and Optimization of Large-Scale Systems*, McGraw-Hill, New York.

Haimes, Y.Y., 1981, Hierarchical holographic modeling, *IEEE Transactions on Systems, Man, and Cybernetics* **11**(9): 606–617.

Haimes, Y.Y., 1989, Toward a holistic approach to risk management (Guest Editorial), *Risk Analysis* **9**(2): 147–149.

Haimes, Y.Y., 1991, Total risk management, *Risk Analysis* **11**(2), 169–171.

Haimes, Y.Y., 2007, Phantom system models for emergent multiscale systems, *Journal of Infrastructure Systems* **13**(2): 81–87.

Haimes, Y.Y., and W.A. Hall, 1974, Multiobjectives in water resources systems analysis: the surrogate worth trade-off method, *Water Resources Research* **10**(4): 615–624.

Haimes, Y.Y., and W.A. Hall, 1977, Sensitivity, responsivity, stability, and irreversibility as multiobjectives in civil systems, *Advances in Water Resources* **1**: 71–81.

Haimes, Y.Y., and P. Jiang, 2001, Leontief-based model of risk in complex interconnected infrastructures, *ASCE Journal of Infrastructure Systems* **7**(1): 1–12, 111–117.

Haimes, Y.Y., and C. Schneiter, 1996, Covey's seven habits and the systems approach, *IEEE Transactions on Systems, Man, and Cybernetics* **26**(4): 483–487.

Haimes, Y.Y., L.S. Lasdon, and D.A. Wismer, 1971, On the bicriterion formulation of the integrated system identification and systems optimization, *IEEE Transactions on Systems, Man, and Cybernetics* **1**: 296–297.

Haimes, Y.Y., K.A. Loparo, S.C. Olenik, and S.K. Nanda, 1980, Multiobjective statistical method (MSM) for interior drainage systems, *Water Resources Research* **16**(3): 467–475.

Haimes, Y.Y., K. Tarvainen, T. Shima, and J. Thadathil, 1990, *Hierarchical Multiobjective Analysis of Large Scale Systems*, Hemisphere Publishing, New York.

Haimes, Y.Y., D.A. Moser, and E.Z. Stakhiv (Eds.), 1994, *Risk-Based Decision Making in Water Resources VI*, American Society of Civil Engineers, New York.

Haimes, Y.Y., S. Kaplan, and J.H. Lambert, 2002, Risk filtering, ranking, and management framework using hierarchical holographic modeling, *Risk Analysis* **22**(2): 383–397.

Haimes, Y.Y., B.M. Horowitz, J.H. Lambert, J.R. Santos, K.G. Crowther, and C. Lian, 2005a, Inoperability input–output model (IIM) for interdependent infrastructure sectors: case study, *Journal of Infrastructure Systems* **11**(2): 80–92.

Haimes, Y.Y., B.M. Horowitz, J.H. Lambert, J.R. Santos, C. Lian, K.G. Crowther, 2005b, Inoperability input–output model (IIM) for interdependent infrastructure sectors: theory and methodology, *Journal of Infrastructure Systems* **11**(2): 67–79.

Hall, A.D. III, 1989, *Metasystems Methodology: A New Synthesis and Unification*, Pergamon Press, New York.

Hatley, D.J., P. Hruschka, and I.A. Pirbhai, 2000, *Process for System Architecture and Requirements Engineering*, Dorset House Publishing Co., Inc., New York, 434 pp.

Hegel, G., 1952, *Great Books of the Western World*, Vol. **46**, Encyclopaedia Britannica, Inc., Chicago, IL.

Hutchins, R.M., Encyclopaedia Britannica, Inc., and University of Chicago, 1952, *Plato*, Vol. **7**. Great Books of the Western World. W. Benton, Chicago, IL.

Imai, M., 1986, *Kaizen: The Key to Japan's Competitive Success*, McGraw-Hill, New York.

Jackson, D., 2006, Dependable software by design, Scientific American June: 69–75.

Kaplan, S., and B.J. Garrick, 1981, On the quantitative definition of risk, *Risk Analysis* **1**(1): 11–27.

Kaplan, S., Y.Y. Haimes, and B.J. Garrick, 2001, Fitting hierarchical holographic modeling (HHM) into the theory of scenario structuring and a refinement to the quantitative definition of risk, *Risk Analysis* **21**(5): 807–819.

Kossiakoff, A., and W.N. Sweet, 2002, *Systems Engineering Principles and Practice*, John Wiley & Sons, Inc., Hoboken, NJ, 488 pp.

Krimsky, S., and D. Golding (Eds.), 1992, *Social Theories of Risk*, Praeger Publishers, Westport, CT.

Kunreuther, H., and P. Slovic (Eds.), 1996, *Challenges in Risk Assessment and Risk Management, the Annals of the American Academy of Political and Social Science*, Sage, Thousand Oaks, CA.

Larsen, R.J., 2007, *Our Own Worst Enemy: Asking the Right Questions about Security to Protect You, Your Family, and America*, Grand Central Publishing, New York.

Leach, M., and Y.Y. Haimes, 1987, Multiobjective risk-impact analysis method, *Risk Analysis* **7**: 225–241.

Leondes, C.T. (Ed.), 1969, *Advances in Control Systems*, Vol. **6**, Academic Press, New York.

Leontief, W.W., 1951a, Input/output economics, *Scientific American* **185**(4): 15–21.

Leontief, W.W., 1951b, *The Structure of the American Economy, 1919–1939*, Second edition, Oxford University Press, New York.

Leontief, W.W., 1986, Quantitative input–output relations in the economic system of the United States, *Review of Economics and Statistics* **18**(3): 105–125.

Lewis, H., 1992, *Technological Risk*, W. W. Norton, New York.

Li, D., and Y.Y. Haimes, 1988, The uncertainty sensitivity index method (USIM) and its extension, *Naval Research Logistics* **35**: 655–672.

Lian, C., 2006, *Extreme Risk Analysis of Dynamic Interdependent Infrastructures and Sectors of the Economy with Applications to Financial Modeling*, Ph.D. dissertation, University of Virginia, Charlottesville, VA.

Longstaff, T.A., and Y.Y. Haimes, 2002, A holistic roadmap for survivable infrastructure systems, *IEEE Transactions on Systems, Man, and Cybernetics* **32**(2): 260–268.

Lowrance, W.W., 1976, *Of Acceptable Risk*, William Kaufmann, Los Altos, CA.

Macko, D., 1967, *General System Theory Approach to Multilevel Systems*, Report SRC 106-A-67-44, Systems Research Center, Case Western Reserve University, Cleveland, OH.

Maier, M.W., and E. Rechtin, 2000, *The Art of Systems Architecting*, Second edition, CRC, Boca Raton, FL, 344 pp.

McQuaid, J., and M. Schleifstein, 2006, *Path of Destruction: The Devastation of New Orleans and the Coming of Age of Superstorms*, Little, Brown and Co., New York.

Mesarović, M.D., 1965, Multilevel concept of systems engineering, *Proceedings of the Systems Engineering Conference*, Chicago, IL.

Mesarović, M.D. (Ed.), 1968, *Systems Theory and Biology*, Springer-Verlag, New York.

Mesarović, M.D., D. Macko, and Y. Takahara, 1970, *Theory of Hierarchical, Multilevel Systems*, Academic Press, New York.

Meyer, W., 1984, *Concepts of Mathematical Modeling*, McGraw-Hill, New York.

Morrall, J.F. III, 2003, Saving lives: a review of the record, *Journal of Risk and Uncertainty* **27**(3): 221–237.

National Research Council (NRC), 1996, *Shopping for Safety: Providing Consumer Automotive Safety Information*, Special Report 248, Transportation Research Board, National Academy Press, Washington, DC.

National Research Council (NRC), 2002, *Making the Nation Safer: The Role of Science and Technology in Countering Terrorism*, The National Academy Press, Washington, DC.

Paté-Cornell, M.E., 1990, Organizational aspects of engineering system safety: the case of offshore platforms, *Science* **250**: 1210–1217.

Post, D.E., R.P. Kendall, and R.F. Lucas, 2006, The opportunities, challenges and risks of high performance computing in computational science and engineering, *Advances in Computers* **66**: 239–301, Marvin Zelkowitz, (Ed.), Elsevier Academic Press, New York.

Raiffa, H., 1964, *Decision Analysis: Introductory Lectures on Choices Under Uncertainty*, Random House, New York.

Rasmussen, J., A.M. Pejtersen, and L.P. Goodstein, 1994, *Cognitive Systems Engineering*, First edition, John Wiley & Sons, Inc., New York, 396 pp.

Rouse, W.B., 1991, *Design for Success: A Human-Centered Approach to Designing Successful Products and Systems*, First edition, John Wiley & Sons, Inc., New York, 304 pp.

Sage, A.P., 1977, *Methodology for Large Scale Systems*, McGraw-Hill, New York.

Sage, A.P. (Ed.), 1990, *Concise Encyclopedia on Information Processing in Systems and Organizations*, Pergamon Press, Oxford.

Sage, A.P., 1992, *Systems Engineering*, John Wiley & Sons, Inc., New York.

Sage, A.P., 1995, *Systems Management for Information Technology and Software Engineering*, John Wiley & Sons, Inc., New York.

Sage, A.P., and J.E. Armstrong, Jr., 2003, *Introduction to Systems Engineering*, John Wiley & Sons, Inc., Hoboken, NJ.

Sage, A.P., and C.D. Cuppan, 2001, On the systems engineering and management of systems of systems and federation of systems, *Information, Knowledge, Systems Management* **2**(4): 325–345.

Sage, A.P., and W.B. Rouse (Eds.), 1999, *Handbook on Systems Engineering and Management*, Second edition, John Wiley & Sons, Inc., New York.

Santos, J.R., 2003, *Interdependency Analysis: Extensions to Demand Reduction Inoperability Input–Output Modeling and Portfolio Selection*, Ph.D. dissertation, University of Virginia, Charlottesville, VA.

Senge, P.M., 1990, *The Fifth Discipline*, Doubleday, New York.

Singh, M.G., 1987, *Systems and Control Encyclopedia: Theory, Technology, Applications*, Pergamon Press, New York.

Slovic, P., 2000, *The Perception of Risk*, Earthscan Publications Ltd., Sterling, VA

Stern, P., and H. Fineberg (Eds.), 1996, *Understanding Risk: Informing Decisions in a Democratic Society*, National Research Council, National Academy Press, Washington, DC.

The Royal Society, 1992, *Risk: Analysis, Perception and Management: Report of a Royal Society Study Group*, The Royal Society, London.

Toffler, A., 1980, *The Third Wave: The Classic Study of Tomorrow*, Bantam Books, New York.

Toffler, A., 1991, *Powershift*, Bantam Books, New York.

Tulsiani, V., Y.Y. Haimes, and D. Li, 1990, Distribution analyzer and risk evaluator (DARE) using fault trees, *Risk Analysis* **10**(4): 521–538.

US Nuclear Regulatory Commission, 1981, *Fault Tree Handbook, NUREG-0492*, U.S. Government Printing Office, Washington, DC.

Warfield, J.N., 1976, *Social Systems—Planning and Complexity*, John Wiley & Sons, Inc., New York.

Wernick, I.K. (Ed.), 1995, *Community Risk Profiles*, Rockefeller University, New York.

Wiener, N., 1961, *Cybernetics*, Second edition, MIT Press, Cambridge, MA.

World Book, Inc., 1980, *The World Book Encyclopedia*, Vol. **8**, World Book—Childcraft International, Inc., Chicago, IL.

Zeleny, M., 2005, *Human Systems Management: Integrating Knowledge, Management and Systems*, First edition, World Scientific Publishing Company, Hackensack, NJ, 484 pp.

Zigler, B.P., 1984, *Multifaceted Modeling and Discrete Simulation*, Academic Press, New York.

2

The Role of Modeling in the Definition and Quantification of the Risk Function

2.1 INTRODUCTION

If the adage "To manage risk, one must measure it" constitutes the compass for risk management, then modeling constitutes the road map that guides the analyst throughout the journey of risk assessment. The process of risk assessment and management may be viewed through many lenses depending on one's perspectives, vision, and circumstances.

In this chapter, we introduce the fundamentals of systems engineering and the building blocks of mathematical models. Systems engineering provides systematic methodologies for studying and analyzing the various aspects of a system and its environment by using conceptual, mathematical, and physical models. This applies to both structural and nonstructural systems.

Systems engineering also assists in the decision-making process by selecting the best alternative policies, subject to all pertinent constraints, by using simulation and optimization techniques and other decisionmaking tools.

Figure 2.1 depicts a schematic representation of the process of system modeling and optimization, where the real system is represented by a mathematical model. The same input applied to both the real system and the mathematical model yields two different responses: the system's output and the model's output. The closeness of these responses indicates the value of the mathematical model; that is, if these two responses are consistently close (subject to a specified norm), we consider the model to be a good representation of the system. Figure 2.1 also applies solution strategies to the mathematical model or, as they are often referred to, optimization and simulation techniques. The optimal decision is considered for implementation on the physical system. One may classify mathematical models as follows:

1. Linear versus nonlinear
2. Deterministic versus probabilistic
3. Static versus dynamic
4. Distributed parameters versus lumped parameters

1. *Linear versus nonlinear.* A linear model is one that is represented by linear equations; that is, all constraints and the objective functions are linear.

Risk Modeling, Assessment, and Management, Fourth Edition. Yacov Y. Haimes.
© 2016 John Wiley & Sons, Inc. Published 2016 by John Wiley & Sons, Inc.

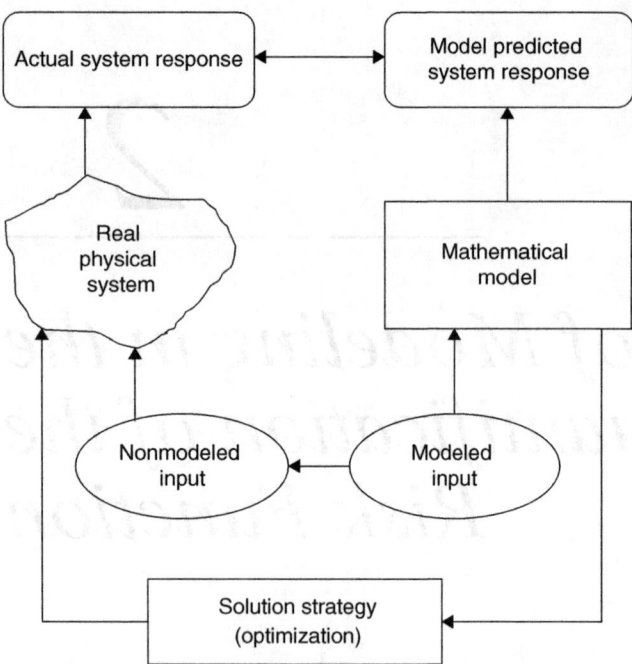

Figure 2.1 The process of system modeling and optimization [Haimes, 1977].

A nonlinear model is represented by nonlinear equations; that is, part or all of the constraints or the objective functions are nonlinear. A function $f(\cdot)$ is linear if and only if

(a) $f(ax) = af(x)$, and
(b) $f(x_1 + x_2) = f(x_1) + f(x_2)$

Examples:

Linear equations: $y = 5x_1 + 6x_2 + 7x_3$ (2.1)

Nonlinear equations: $y = 5x_1^2 + 6x_2 x_3$
$y = \log x_1$
$y = \sin x_1 + \log x_2$

2. *Deterministic versus probabilistic*. Deterministic models or elements of models are those in which each variable and parameter can be assigned a definite fixed number or a series of fixed numbers for any given set of conditions.

In probabilistic models (stochastic) models, the principle of uncertainty is introduced. Neither the variables nor the parameters used to describe the input–output relationships and the structure of the elements (and the constraints) are precisely known.

Example: "The value of x is in $(a-b, a+b)$ with 90% probability," meaning that in the long run, the value of x will be greater than $(a+b)$ or less than $(a-b)$ in 10% of the cases.

3. *Static versus dynamic*. Static models are those that do not explicitly take the variable time into account. In general, static models are of the form given by Eq. (2.1).

Dynamic models are those involving difference or differential equations, as illustrated in Eq. (2.2):

$$\min_{u_1,\ldots,u_M} \int_{t_0}^{t} F(x_1,\ldots,x_N, u_1,\ldots,u_M, t)\,dt$$

subject to the constraints

$$\frac{d}{dt} x_i = G_i(x_1,\ldots,x_N, u_1,\ldots,u_M, t), \quad i = 1, 2, \ldots, N$$
$$x_i(t_0) = x_i^0, \quad i = 1, 2, \ldots, N$$

(2.2)

Static optimization problems are often referred to as mathematical programming, while dynamic optimization problems are often referred to as optimal control problems.

4. *Distributed parameters versus lumped parameters*. A lumped parameter model ignores variations, and the various parameters and dependent variables can be considered to be homogeneous throughout the entire system.

A distributed parameter model takes into account detailed variations in behavior from point to point throughout the system. Most physical systems are distributed parameter systems. For example, the equation describing transient radial flow of a compressible fluid through a porous medium can be derived from Darcy's law:

$$T\left[\frac{1}{r}\frac{\partial}{\partial r}\left(r\frac{\partial P}{\partial r}\right)\right] = S\frac{\partial P}{\partial t} \pm Q \quad (2.3)$$

represents a distributed parameter of a groundwater system, where P is the pressure, t is the time, r is the distance along a radial coordinate, T is the transmissibility, and S is the storage.

An aggregate model of these categories is illustrated in Appendix 3, where the farmer's dilemma introduced in Chapter 1 is formulated as a linear, deterministic, static, and lumped parameter system.

Furthermore, to be realistic and representative, models may address the following five categories:

1. *Time horizon*: short, intermediate, and long term
2. *Client*: various sectors of the public
3. *Nature*: aquatic and wildlife habits
4. *Scope*: national, regional, and local needs
5. *Constraints*: legal, institutional, environmental, social, political, and economic

The Tech-Com study [The Technical Committee of the Water Resources Research Centers of the Thirteen Western States, 1974] identifies nine goals, which have been divided into two major groups:

1. *Maintenance of security*: (a) environmental security, (b) collective security, and (c) individual security
2. *Enhancement of opportunity*: (d) economic opportunity, (e) recreational opportunity, (f) aesthetic opportunity, (g) cultural and community opportunity, (h) educational opportunity, and (i) individual freedom

2.2 THE RISK ASSESSMENT AND MANAGEMENT PROCESS: HISTORICAL PERSPECTIVES

In an environmental trade-off analysis, policies should be established to promote conditions where humans and nature can exist in productive harmony. Resolution of conflicts should be achieved by balancing the advantages of development against the disadvantages to the environment. The process is one of balancing the total *benefits*, *risks*, and *costs* for both people and the environment, where the well-being of future generations is as important as that of present ones.

Fundamental to multiobjective analysis is the Pareto-optimal concept, also known as a noninferior solution. Qualitatively, a noninferior solution of a multiobjective problem is one where any improvement of one objective function can be achieved only at the expense of degrading another. The subject of multiobjective trade-off analysis and Pareto-optimality is discussed in Chapter 5.

Good systems management must address:

- The holistic nature of the system in terms of its hierarchical, organizational, and functional decisionmaking structure
- The multiple noncommensurate objectives, subobjectives, and sub-subobjectives, including all types of important and relevant risks
- The various time horizons and the multiple decisionmakers, constituencies, power brokers, stakeholders, and users of the system
- The host of institutional, legal, and other socioeconomic conditions that require consideration

Thus, risk management raises several fundamental philosophical and methodological questions, with a focus on the time frame [Hirshleifer and Riley, 1992; Kunreuther and Slovic, 1996; Slovic, 2004, 2010]. For example, Warren A. Hall's fundamental premise is as follows: Applying good systems engineering and management tools to water resources problems does not produce additional water per se; it merely ensures that water with an acceptable quality will be where, when, and in the quantity it is needed [Hall and Dracup, 1970].

Although the following discussion and definitions of technical terms are not necessarily universally acceptable, they are provided here as a common reference and to avoid ambiguities.

2.2.1 Risk and Uncertainty

Lowrance [1976] defines risk as a measure of the probability and severity of adverse effects. This definition is harmonious with the mathematical formula used to calculate the expected value of risk, to be discussed later. The Principles, Standards, and Procedures (P., S., & P.) published in 1980 by the U.S. Water Resources Council [1980] make a clear distinction between risk and uncertainty:

1. *Risk*. Situations of risk are defined as those in which the potential outcomes (i.e., consequences) can be described in reasonably well-known probability distributions. For example, if it is known that a river will flood to a specific

level on the average of once in 20 years, it is a situation of risk rather than uncertainty.
2. *Uncertainty*. In situations of uncertainty, potential outcomes cannot be described in terms of objectively known probability distributions, nor can they be estimated by subjective probabilities [Pate-Cornell, 1990, 1996].
3. *Imprecision*. In situations of imprecision, the potential outcome cannot be described in terms of objectively known probability distributions, but it can be estimated by subjective probabilities.
4. *Variability*. Variability is a result of inherent fluctuations or differences in the quantity of concern.

In addition, the P., S., & P. identifies two major sources of risk and uncertainty:

1. Risk and uncertainty arise from measurement errors and from the underlying variability of complex, natural, social, and economic situations. If the analyst is uncertain because of imperfect data or crude analytical tools, the plan is subject to measurement errors. Improved data and refined analytic techniques will obviously help minimize measurement errors. A more detailed definitions and discussion will follow in subsequent sections and chapters.
2. Some future demographic, economic, hydrologic, and meteorological events are essentially unpredictable because they are subject to random influences. The question for the analyst is whether the randomness can be described by some probability distribution. If there is a historical database that is applicable to the future, distributions can be described or approximated by objective techniques. If there is no such historical database, the probability distribution of random future events can be described subjectively, based upon the best available insight and judgment.

2.2.1.1 Risk assessment process
The risk assessment process is a set of logical, systemic, and well-defined activities that provide the decisionmaker with a sound identification, measurement, quantification, and evaluation of the risk associated with certain natural phenomena or man-made activities. The generic term *risk* will connote a multitude of risks. Some authors distinguish between risk assessment and management; others do not and incorporate risk assessment within the broader risk management label. Although we make a distinction between the two terms in this book, at the same time, we recognize that significant overlaps do exist. The following five steps constitute one vision of the entire risk assessment and management process [Haimes, 1981, 2009b]:

1. Risk identification
2. Risk modeling, quantification, and measurement
3. Risk evaluation
4. Risk acceptance and avoidance
5. Risk management
6. Risk communication

Indeed, the first three steps—risk identification; risk modeling, quantification, and measurement; and risk evaluation—relate to the following triplet of *risk assessment* questions posed by Kaplan and Garrick [1981] in Chapter 1:

1. What can go wrong?
2. What is the likelihood that it would go wrong?
3. What are the consequences?
   Here, we add a fourth question: What is the time frame?

On the other hand, the final two steps—risk acceptance and avoidance and risk management—relate to the following triplet of *risk management* questions posed by Haimes [1991, 2009b] in Chapter 1:

1. What can be done, and what options are available?
2. What are their associated trade-offs in terms of all costs, benefits, and risks?
3. What are the impacts of current management decisions on future options?

Clearly, the risk evaluation step can be associated with both assessment and management activities and is an overlapping step between the two activities. Here again is the importance of the circular–iterative process in systems engineering in general and in risk assessment and management in particular.

1. *Risk identification.* Identifying the sources and nature of risk and the uncertainty associated with the activity or phenomena under consideration is often considered the first and major step in the risk assessment process. This step calls for a complete description of the universe of risk-based events that might occur and attempts to answer the question, "What can go wrong?" The comprehensiveness of this risk identification step can be complemented by also addressing the following four sources of failure and their causes:

   - Hardware failure
   - Software failure
   - Organizational failure
   - Human failure

   Causes may include demographic, economic, hydrologic, technological, meteorological, environmental, institutional, and political elements.
2. *Risk modeling, quantification, and measurement.* This step entails (i) assessing the likelihood of what can go wrong through objective or subjective probabilities and (ii) modeling the causal relationships among the sources of risk and their impacts. Quantifying the input–output relationships with respect to the random, exogenous, and decision variables and the relations of these variables to the state variables, objective functions, and constraints is by far the most difficult step in the risk assessment process. Indeed, quantifying the probabilities and magnitude of adverse effects and their myriad consequences constitutes the heart of systems modeling.
3. *Risk evaluation.* This step constitutes the linkage or overlapping steps between the risk assessment process and risk management. Here, various policy options are formulated, developed, and optimized in a Pareto-optimal sense. Trade-offs are generated and evaluated in terms of their costs, benefits, and risks. Multiobjective analysis, which is discussed in Chapter 5, dominates the evaluation of risk.
4. *Risk acceptance and avoidance.* This is the decisionmaking step, where all costs, benefits, and risks are traded off to determine the level of acceptability of risk. Here, the decisionmakers evaluate numerous considerations that fall beyond the risk modeling and quantification process—for example, the equitable distribution of risk; potential socioeconomic, environmental, or political ramifications; and the impacts of current management decisions on future options. Indeed, it is this stage of the risk management process that answers the question, "How safe is safe enough?"
5. *Risk management.* This is the execution step of the policy options. The implementation of decisions aimed at detecting, preventing, controlling, and managing risk is not done in a vacuum. Clearly, the entire process of risk assessment and management is a circular one involving a feedback loop. At each of the five steps, the risk analyst and the decisionmaker might repeat part or all of the previous steps.

Finally, one may view the risk assessment and management process from the quantitative and empirical perspectives versus the qualitative normative perspectives. In this vision, the process constitutes three major, albeit overlapping, elements:

1. *Information measurement*—including data collection, retrieval, and processing through active public participation
2. *Model quantification and analysis*—including the quantification of risk and other objectives, the generation of Pareto-optimal policies with their associated trade-offs, and the conduct of impact and sensitivity analysis
3. *Decisionmaking*—the interaction between analysts and decisionmakers and the exercise of subjective value judgment for the selection of preferred policies.

2.3 INFORMATION, INTELLIGENCE, AND MODELS

Public officials and decisionmakers at all levels of government—local, state, regional, and national—are forced to make public policy decisions without being able to adequately and sufficiently analyze the respective risk impacts and trade-offs associated with their decisions. Thus, the need for respective data is obvious. It is wise to distinguish, however, between two kinds of data: information and intelligence.

Models, methodologies, and procedures for risk assessment (referred to, generically, as models in this section) are aimed at providing this essential service to decisionmakers—the processing of data into intelligence—so that elements of risk associated with policy decisions may be properly valued, evaluated, and considered in the decisionmaking process. For such a process to be viable, several prerequisites should be fulfilled:

1. Decisionmakers should be cognizant and appreciative of the importance of this process; they should also be capable of understanding the utility, attributes, and limitations of respective models used in risk assessment. Past experience does not provide too much encouragement in this respect.
2. Decisionmakers should also be cognizant of the affect element in their decisionmaking process.
3. Risk assessment models should be available, usable, and credible.
4. Both risk analysts and decisionmakers ought to be aware of the inherent biases that all individuals bring with them in the risk assessment and management process. Such biases are an integral part of each individual's upbringing, family tradition and culture, education, personal and professional experience, and other influences.

Evaluating the impacts and consequences of public policy decisions involving risk is an imperative step in the process of determining the acceptability of risk. Although this process is known to be complex, lengthy, and tedious (inasmuch as policy and decisionmakers must be responsive to a myriad of institutional, legal, political, historical, and other societal demands and constraints), the process must be based, to the extent possible, on firm scientific and technological foundations.

Public policies involving risks are likely to be deemed more acceptable (i) when based on credible scientific and technological information and (ii) where sound trade-off and impact analyses have been performed and made transparent.

The difficulties of dealing with the complexity of risk assessment and, particularly, the quantification of risk are familiar to all—policymakers and decisionmakers, modelers, and analysts, as well as other professionals and the public at large. This complexity is inherent in myriad considerations that transcend scientific, technological, economic, political, geographic, and legal constraints. It is not surprising, therefore, that new approaches, models, methodologies, and procedures in risk assessment have filled a real need. On the other hand, both policymakers and decisionmakers—the ultimate users of these tools—have met these relatively new approaches and risk assessment methodologies with opinions ranging from outright support to overall skepticism. One may rightfully ask why so many groups have developed not only skepticism but even antagonism toward both these analysts and their analyses. The following list summarizes some sources of skepticism to modeling, to risk analysis, and to systems analysis:

- Misuse of models and incorrect applications
- Insufficient basic scientific research for credible environmental and social aggregations
- Too much model use delegated to people who do not understand models
- Insufficient planning and resources for model maintenance and management
- Lack of incentives to document models
- Overemphasis on optional use of computers and underemphasis on efficient use of human resources
- Proliferation of models and lack of systematic inventory of available models
- Lack of proper calibration, testing, and verification of models
- Lack of communication links among modelers, users, and affected parties
- Models usable only by the developer
- Need for models to be recognized as means, not ends
- Lack of an interdisciplinary modeling team, leading to unrealistic models
- Strengths, weaknesses, and limited assumptions of models often unrecognized by the decisionmaker
- Insufficient data planning
- Lack of consideration of multiple objectives in the model

Systems analysis studies (risk assessment and management studies are no exception) have often been conducted in isolation from the decisionmakers and commissioned agencies responsible for and charged with implementing any results of these analyses.

In 1996, for example, the General Accounting Office [GAO, 1996] extensively studied ways to improve management of federally funded computerized models:

> GAO identified 519 federally funded models developed or used in the Pacific Northwest area of the United States. Development of these models cost about $39 million. Fifty-seven of the models were selected for detailed review, each costing over $100,000 to develop. They represent 55% of the $39 million of development costs in the models.
>
> Although successfully developed models can be of assistance in the management of Federal programs, GAO found that many model development efforts experienced large cost overruns, prolonged delays in completion, and total user dissatisfaction with the information obtained from the model.

The GAO study classified the problems encountered in model development into three categories: (i) 70% attributable to inadequate management planning, (ii) 15% attributed to inadequate management commitment, and (iii) 15% attributable to inadequate management coordination. Other problems stem from the fact that model credibility and reliability were either lacking or inadequately communicated to management.

2.4 THE BUILDING BLOCKS OF MATHEMATICAL MODELS

A mathematical model is a set of equations that describes and represents a real system. This set of equations uncovers the various aspects of the problem, identifies the functional relationships among all the system's components and elements and its environment, establishes measures of effectiveness and constraints, and thus indicates what data should be collected to deal with the problem quantitatively. These equations could be algebraic, differential, or other, depending on the nature of the system being modeled. Mathematical models are often solved or optimized through the use of appropriate optimization or simulation techniques.

In the following general formulation of a mathematical model, the desire is to select the set of optimal decision variables, $x_1^*, x_2^*, \ldots, x_n^*$, that maximize (minimize) the objective function, $f(x_1, x_2, \ldots, x_n)$:

$$\max f(x_1, x_2, \ldots, x_n)$$

subject to the constraints

$$\begin{aligned} g_1(x_1, x_2, \ldots, x_n) &\leq b_1 \\ g_2(x_1, x_2, \ldots, x_n) &\leq b_2 \\ &\vdots \\ g_m(x_1, x_2, \ldots, x_n) &\leq b_m \end{aligned} \quad (2.4)$$

where $f(\cdot)$ is an objective function, $x_1, x_2, \ldots, x_n$ are decision variables, $g_1(x), \ldots, g_m(x)$ are constraints, and $b_1, \ldots, b_m$ are generally known as resources.

In the formulation of mathematical models, six basic groups of variables need to be defined and will be further discussed in subsequent sections:

- Decision variables
- Input variables
- State variables
- Exogenous variables
- Random variables
- Output variables

The risk of contamination of a groundwater system with the carcinogen trichloroethylene (TCE) chemical will serve as a generic example.

Groundwater contamination is a major worldwide socioeconomic problem that has its roots in technological development. Its solution requires a scientifically sound and well-formulated public policy grounded in broad-based public participation that includes the private sector as well as the government. The lack of any one of the aforementioned elements is likely to impede viable progress toward the prevention or reduction of groundwater contamination.

To prevent groundwater contamination, one must be aware of the sources of contamination, understand the movement of contaminants through porous media, and understand the technical and socioeconomic reasons that permit, encourage, and, indeed, make groundwater contamination the widespread phenomenon that it is today.

2.4.1 Groundwater Contamination Model Building Blocks

In developing a system's model, it is essential to identify the decisionmakers and the purpose for which the model is intended. This is because the building blocks of mathematical models discussed here may be interpreted in a variety of ways depending on the context of the problem. In the groundwater model developed, the decisionmaker is the government. A completely different interpretation and representation of the building blocks would have emerged had the decisionmakers been the well owners, for example. We will return to this discussion after we define the building blocks of mathematical models.

1. *Decision variables* (**x**). These are measures controllable by the decisionmakers, such as legislation, promulgation of regulations, zoning, public education, and economic incentives and disincentives. The symbol **x** denotes a vector of such decision variables, $\mathbf{x} = (x_1, x_2, ..., x_n)$. Examples of decision variables, **x**, are as follows:

   x_1, effluent charges imposed by government agencies on polluters
   x_2, standards promulgated by the government for effluent discharges
   x_3, construction of advanced wastewater treatment plants

2. *Input variables* (**u**). These are materials discharged and/or entering the groundwater system. These input variables are not necessarily controllable by public decisionmakers; rather, they are controllable by the individual parties involved in the contamination of aquifers. Input variables include (i) the discharge of synthetic organic contaminants, such as TCE, and (ii) saltwater intrusion due to overpumping. For more refined notation and without loss of generality, the system's inputs and outputs are lumped into **u**. For example, water pumpage and artificial recharge can both be conveniently considered as part of the vector **u** in the context of modeling groundwater contamination. The symbol **u** denotes a vector of such input variables, $\mathbf{u} = (u_1, u_2, ..., u_m)$. Examples of input variables, **u**, are as follows:

   u_1, discharge of polluted effluents into the river by industry 1
   u_2, discharge of polluted effluents into the river by industry 2
   u_3, pumpage rate

   Note that if the model were developed for industry 1, for example, and not for the government, then the discharge of effluent u_1 would become a decision variable controlled by industry 1. Similarly, the effluent charges, x_1, which is a decision variable for the government, would become an input variable for the industry.

3. *Exogenous variables* (**α**). These are variables related to external factors but affect the system either directly or indirectly. Theoretically, these exogenous variables could encompass the entire universe excluding **x** and **u**. For practical purposes, however, exogenous variables such as the physical characteristics of an aquifer; water demand for industrial, urban, and agricultural development; technology assessment; and economic market forces may be considered. The symbol **α** denotes a vector of exogenous variables, $\boldsymbol{\alpha} = (\alpha_1, \alpha_2, ..., \alpha_p)$. Examples of exogenous variables, **α**, are as follows:

   α_1, water withdrawals (demand)
   α_2, nominal aquifer transmissivity coefficient
   α_3, nominal aquifer storage coefficient

   Note that water withdrawals represent an exogenous variable for the government's model, yet it would be a decision variable for an industry's model.

4. *Random variables* (**r**). A probability distribution function (PDF) may or may not be known for each random variable. For example, knowledge of probability distributions can be assumed for random processes such as precipitation and streamflow (and thus for natural recharge of aquifers). On the other hand, PDFs for random events, such as accidental spills or terrorist attacks, may not be known, and uncertainty analysis along with risk analysis might be conducted (see Chapter 6). The symbols **r** denote a vector of such random variables, events, or

processes, $\mathbf{r} = (r_1, r_2, ..., r_q)$. Examples of a random vector, $\mathbf{r} = (r_1, r_2, r_3)$, are as follows:

r_1, precipitation
r_2, streamflow
r_3, contaminant

5. *State variables* (**s**). These are variables that may represent the quantity and quality level (state) of the groundwater system at any time. Examples of such state variables include the water table level, concentration of salinity, and TCE, or biological contamination. The symbol **s** denotes a vector of such state variables, $\mathbf{s} = (s_1, s_2, ..., s_k)$. Examples of state variables, **s**, are as follows:

s_1, groundwater table
s_k, concentration of contaminant k in the groundwater

6. *Output variables* (**y**). These are variables that are closely related to the state, decision, and random variables. For linear dynamic systems, as will be discussed in Chapter 10, they are commonly represented as

$$\mathbf{y}(t) = C\mathbf{s}(t) + D\mathbf{x}(t) \quad (2.5)$$

and the state equation is written as

$$\dot{\mathbf{s}}(t) = A\mathbf{s}(t) + B\mathbf{x}(t) + \mathbf{v} \quad (2.6)$$

$$\dot{\mathbf{s}}(t_0) = \mathbf{s}_0 \quad (2.7)$$

where A, B, C, and D are exogenous variables (constants for time-invariant models).

The output variables are often represented in terms of the state variables. Examples of output variables, $\mathbf{y} = (y_1, y_2)$, are as follows:

y_1, spatial distribution of contaminants in the groundwater system
y_2, total groundwater withdrawals from the groundwater system over a period of time

The next step is to define all objective functions (including risk functions) and constraints. Here, a critical distinction must be made between the objectives of the polluter and those of the public and its representatives. The risks and costs of dumping hazardous chemical wastes, for example, are certainly different for the polluter than for the user of the contaminated groundwater.

From the aforementioned definition, it is clear that the six variables (vectors) are not all independent of each other. For example, the state of the groundwater system (**s**) depends on the quantity of contaminants (**u**) disposed of, what measures (**x**) are taken to prevent contamination, the frequency and extent at which such contamination occurs (**r**), and the physical characteristics (α) of the aquifer. Thus, $\mathbf{s} = \mathbf{s}(\mathbf{x}, \mathbf{u}, \mathbf{r}, \alpha)$ and $\mathbf{y} = \mathbf{y}(\mathbf{s})$. Figure 2.2 depicts this interdependence among the building blocks of mathematical models.

Therefore, the various objectives and constraints of the subsystems and users can be written as functions of the output vector (**y**) or state vector (**s**), whereby dependence on **x**, **u**, **r**, and α is implicit. In subsequent discussion, the objective functions will be represented in terms of the state vector (**s**).

Let $f_j(\mathbf{s})$ represent the *j*th objective function of the subsystem, $j = 1, 2, ..., J$. For example, let

$f_1(\mathbf{s}) = $ cost in dollars of contamination prevention
$f_2(\mathbf{s}) = risk$ of contamination with TCE
$f_3(\mathbf{s}) = risk$ of contamination with saltwater intrusion

The risk functions can be represented in numerous ways. For example, their representation can be in terms of probability and consequences, expected value, a utility function, or other functions. The quantification of these objective (risk) functions in terms of expected values and the conditional expected value (see Chapters 8 and 11 and throughout this book), which account for the PDFs of the random variables **r**, are also the essence of the multiobjective statistical method (MSM) to be introduced in Chapter 14 (where the building blocks

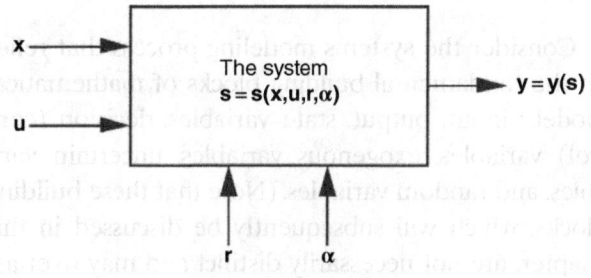

Figure 2.2 A block diagram of mathematical models.

of a mathematical model will be incorporated with risk functions).

In subsequent discussion in this book, more than one model representation will be used. Often, no knowledge of the probability density function for a specific random variable may be available, in which case one of the methodologies for uncertainty analysis, such as the uncertainty/sensitivity index method (USIM) and its extensions [Haimes and Hall, 1977; Li and Haimes, 1988], which is discussed in Chapter 6, may be used.

7. *Constraints* (**g**). Similarly, all the system's constraints (e.g., physical, economic, institutional) can be defined as

$$g_i(\mathbf{s}) \leq 0, i = 1, 2, \ldots, I$$

Examples of constraints are as follows:

$g_1(\mathbf{s})$, total budget available
$g_2(\mathbf{s})$, effluent standard limitations
$g_3(\mathbf{s})$, upper limit on pumpage rate

Thus, the set of all feasible solutions, X, that satisfy all constraints is defined as

$$X = \{\mathbf{x} \mid g_i(\mathbf{s}) \leq 0, \quad i = 1, 2, \ldots, I\} \quad (2.8)$$

The overall formulation of the groundwater problem seeks to minimize all objective functions (in a multiobjective, Pareto-optimal sense) via selection of the best feasible decision variables/measures, **x**.

Mathematically, this can be represented by

$$\min_{\mathbf{x} \in X}\{f_1(\mathbf{s}), f_2(\mathbf{s}), \ldots, f_J(\mathbf{s})\} \quad (2.9)$$

where $\mathbf{s} = \mathbf{s}(\mathbf{x}, \mathbf{u}, \mathbf{r}, \boldsymbol{\alpha})$.

The optimization of single-objective models is discussed in the Appendix, and that of multiple objectives is discussed in Chapter 5.

Consider the systems modeling process that relies on the fundamental building blocks of mathematical models: input, output, state variables, decision (control) variables, exogenous variables, uncertain variables, and random variables. (Note that these building blocks, which will subsequently be discussed in this chapter, are not necessarily distinct and may overlap; for example, input and output may be random.) All good managers desire to change the states of the systems they control in order to support better, more effective, and efficient attainment of the system objectives. Note that the role of the decision variables generated in optimizing single or multiple objectives is to bring the states of the system to levels that appropriately optimize the objective functions. Although an objective function can be a state variable, the role of the decision variables is not to directly optimize the objective functions. Identifying and quantifying (to the extent possible) the building blocks of a mathematical model of any system constitutes a fundamental step in modeling, where one building block—state variables—is the sine qua non in modeling. This is because at any instant the levels of the state variables are affected by other building blocks (e.g., decision, exogenous, and random variables, as well as inputs) and these levels determine the outputs of the system. Consider the human body and its vulnerability to infectious diseases. Different organs and parts of the body are continuously bombarded by a variety of bacteria, viruses, and other pathogens. However, only a subset of the human body is vulnerable to the threats from a subset of the would-be attackers, and due to our immune system, only a smaller subset of the human body would experience adverse effects. Thus, composites of low-level, measurable states integrate to identify higher-level fundamental state variables that define the system. Indeed, a system's vulnerability is a manifestation of the inherent *states* of that system, and each of those states is dynamic and changes in response to the inputs and other building blocks [Haimes, 2006, 2009b, 2012b].

Moreover, within any single model, it is impossible to identify and quantify the causal relationships among all relevant building blocks of models that represent the larger system of systems (S-o-S), including the state variables. There is a need to develop a body of prescriptive theory and methodology for modeling S-o-S [Haimes, 2012a]. The purpose of doing so is to enable analysts to appropriately model and understand the evolving behavior of systems due to the continued forced changes imposed on them—for example, the effects of climate variability on humans and on the natural and constructed environment. Models, laboratory experiments, and simulations are designed to answer specific questions; that is, conventional system models provide responses based on the states of a system under given conditions and assumptions. They do not, however, adequately address the behavior of unprecedented and emerging systems (e.g., the mission to Mars, the power grid for the hydrogen economy [Grant et al., 2006], or a new air traffic control system). Such systems are inherently visionary and at times elusive—they are by and large phantom entities grounded on a mix of future needs and available resources, technology, forced developments and changes, and myriad other unforeseen events [Haimes, 2007, 2009b, 2012b]. (For further discussion of this topic, see the special issue on emergent systems of the journal *Reliability Engineering and Systems Safety* [Johnson, 2006].)

2.5 ON THE COMPLEX DEFINITION OF RISK, VULNERABILITY, AND RESILIENCE: A SYSTEMS-BASED APPROACH

The risk to a system, as well as its vulnerability and resilience, can be understood, defined, and possibly quantified most effectively through a systems-based philosophical and methodological approach, and by recognizing the central role of the states of a system in this process. This is important because after the 9/11 attack on the United States, a variety of confusing and often erroneous extensions were added to the *seemingly* simple and clear definition of risk by Lowrance [1976]: "Risk is a measure of the probability and severity of adverse effects." This is not surprising, given the large set of definitions and interpretations of risk that have been accepted and published in the literature. For example, the Risk Definition Committee of the Society for Risk Analysis (SRA) printed 13 definitions of risk on the program jacket of its first meeting in 1981. The definitions of risk continue to multiply as *risk* becomes a household term.

Over time, increasing fuzziness has also led to varied conceptual and sometimes erroneous quantitative definitions and interpretations. A universally agreed-upon succinct definition of risk has been difficult to develop; one reason is that the concept is multidimensional and nuanced. It requires understanding that risk to a system is inherently and fundamentally a function of the states of the system and of its environment (and of course of the initiating event).

While risk practitioners have legitimate reasons to define risk from their perspectives, these definitions, by their very multiplicity, can lead to confusion. For the purposes of our discussion, the meaning and implications of risk are best understood on the basis of the theory, perspectives, and methodology of systems engineering/analysis. The building blocks of systems modeling and the centrality of state variables are the keys to this understanding. We will first discuss the elements of risk and then introduce the centrality of state variables in the definitions of risk to a system, as well as its vulnerability and resilience.

2.5.1 The Elements of Risk

There is ambiguity in Lowrance's [1976] definition of risk as "a measure of the probability and severity of adverse effects." What do we mean by the terms *probability* (or *likelihood*) and *adverse effects*? Consider the interpretation of the term *likelihood* in isolation (for now) of its probable consequences: Is it the likelihood of the occurrence of any kind of threat (or other initiating event), at any level or magnitude, and when, and of what duration? Or, is it the likelihood of the level and magnitude of the consequences (for every element of the vector of consequences)? Thus, the phrase *probability and severity of adverse effects* can be interpreted in two ways at the same time: (i) in terms of the probability of the *occurrence* of adverse effects and (ii) in terms of the probability of the *severity* of adverse effects, given their occurrence. Both interpretations are valid; however, each represents varied conceptual and theoretical challenges [Haimes, 2006, 2009b, 2012b].

Furthermore, when likelihood is translated into *probability*, it introduces an important conceptual and cognitive hurdle. By its very nature, probability is an abstract term often used as a model of variability and frequency or to quantify the *level of confidence that we have in the information*. In other words, probability really does not exist as a physical entity per se (although the wave functions of quantum particles may be as real as anything).

Kolmogorov builds his widely recognized definition of probability on three basic axioms. Gnedenko [1963] provides a succinct presentation of Kolmogorov's theory of probability:

> Kolmogorov starts with a set U consisting of *elementary events*. He then considers a certain family F of subsets of the set U; the elements of the family F are called *random events*. The following three conditions are imposed on the structure of the family F: (1) F is contained in the set U as one of its elements. (2) If the subsets A and B of the set U are elements of F, then the sets $A+B$, AB, $\bar{A}$ and $\bar{B}$ are also elements of F. (3) If the subsets $A_1, A_2, ..., A_n, ...$ of the set U are elements of the set F, then the sum $A_1 + A_2 + ... + A_n + ...$ of these subsets and the product $A_1 A_2 ... A_n ...$ of these subsets are also elements of F. Thus:
>
> Axiom 1: With each random event A in a field of events F, there is associated a non-negative number $\mathbf{P}(A)$, called its probability.
> Axiom 2: $\mathbf{P}(U) = 1$.
> Axiom 3: (Addition Axiom): If the events $A_1, A_2 ... A_n$ are pair-wise mutually exclusive, then $\mathbf{P}(A_1 + A_2 + ... + A_n) = \mathbf{P}(A_1) + \mathbf{P}(A_2) + ... + \mathbf{P}(A_n)$.

Kolmogorov's theory of probability has become the gold standard in the field because it is built on a holistic systems philosophy. He considers entire states of the system as subsets of the universe of event space U. In this sense, Kolmogorov's theory establishes a requirement to understand probability holistically, and since probability is a central component of risk, then risk itself must also be understood holistically.

2.5.2 The Centrality of State Variables in the Definitions of Risk, Vulnerability, and Resilience

Identifying, understanding, and quantifying the building blocks of a mathematical model of any system are fundamental steps in modeling. This is because at any instant the levels of the state variables are affected by the other building blocks, and these levels determine the outputs of the system.

All state variables are subject to continuous natural, desired, or forced changes (positive and negative). This does not mean that all models must have time-dependent state variables. This is where the art of modeling comes into play. Models are built to answer specific questions and to represent the relevant essence of the system under consideration. Thus, if small or insignificant changes in a state variable have no important effect on the answers sought from the model, then that state variable may be assumed to be static—not time dependent. Deciding whether a state variable should be modeled as static (not changing with time) or dynamic (changing with time) is similar to the modeler selecting only those state variables that represent the essence of the system. In fact, the art and science of systems modeling is characterized by a continuous process of trade-offs that modelers must make with respect to complexity and accuracy. *Any model should be as simple as possible and as complex as needed to answer the expected questions.*

2.5.3 Scenario Structuring and Systems Engineering

To perform effective risk assessment and management, the analyst must understand the system and its interactions with its environment, and this understanding is requisite to modeling the behavior of the states of the system under varied probabilistic conditions. For example, government agencies, the military, the private sector, major corporations, nongovernmental think-tank organizations, and futuristic scholars—all are interested in better understanding risk-related trends and preventing or preparing for not-unlikely sources of risk. Forced changes affect the states of the system (e.g., the 9/11 attack on the United States was a forced change that affected the states of the economy, homeland security, and international relationships, to cite a few). Thus, tracking emergent forced changes to a system through observations, precursors, intelligence and data collection, and most importantly through modeling, constitutes the essence of the risk assessment and management process. *The term emergent forced changes connotes external or internal sources of risk that may adversely affect the states of the system*, and thus the system as a whole.

The theory of scenario structuring (TSS), first presented in 1981 by Kaplan and Garrick [1981], offers an approach to understanding emergent forced changes. Indeed, the ultimate purpose and efficacy of TSS is to envision, foresee, and predict emergent forced changes based on the capability of the human imagination, the availability of evidence, and the predictions of modeling tools. Emergent forced changes, whether they originate from within or from outside the system—and when they are unanticipated, their precursors undetected or ignored, or are altogether misunderstood—are likely to affect the states of the system with adverse consequences. From a broader perspective, the risk assessment process and, to a limited extent, the risk management process are themselves supported by envisioning, discovering, and tracking emergent forced changes through the TSS.

Kaplan and Garrick [1981] introduced the following mathematical *set of triplets* definition of risk, R: $R = \{<s_i, l_i, x_i>\}$, where s_i denotes the ith *risk scenario*, l_i denotes the *likelihood* of that scenario, and x_i denotes the *damage vector* or resulting *consequences* [1981]. Since then, this definition has served the field of risk analysis well. As stated, much debate has continued to challenge the professional community about how to quantify the l_i and x_i and the meanings of *probability*, *frequency*, and *probability of frequency* in this connection [Kaplan, 1991, 1993, 1996, 1997].

In Kaplan and Garrick, the scenarios S_i were defined, somewhat informally, as answers to the question, "What can go wrong?" with the system or process being analyzed. Subsequently, Kaplan added a subscript, c, to the set of triplets: $R = \{<s_i, l_i, x_i>\}_c$ to denote that the set of scenarios, $\{s_i\}$, should be *complete*, meaning it should include "all the possible scenarios, or at least all the important ones" [Kaplan, 1991, 1993].

Kolmogorov's theory of probability can be related to the TSS. *Fundamentally, the process of risk assessment (and the TSS within it) is a systemic exercise that attempts to "discover" potential future risk-based events, to watch for precursors of forced changes, and thus to prepare for and respond to possible adverse events through risk management and risk communication* [Phimister et al., 2004]. The TSS attempts to develop risk-based scenarios systemically and methodologically. Ultimately, the completeness of the scenarios developed by the TSS, claimed in the 1981 original paper, was modified (as discussed later). In many ways, the TSS is a foundation for ensuring that we understand the scope of risk assessment with respect to the system being evaluated. It also contributes to the measurement of risk by providing a tool for interpreting the output of the analysis.

In 2001, Kaplan et al. extended the TSS by incorporating hierarchical holographic modeling (HHM) [Haimes, 1981], which will be discussed in Chapter 3. Kaplan and his colleagues refined the original *set of triplets* definition of risk so that it did not in itself assume or imply that the set of risk scenarios is finite or denumerable. Rather, this refined definition allows the set of risk scenarios to be a continuum, that is, nonenumerable. This continuous set of scenarios constitutes the *true* risk and is independent of the method used to identify them.

2.5.4 Systems Engineering and Risk

Risk analysis and systems engineering/analysis have a common philosophical approach to problem solving, but they differ in their historical evolution and technical maturity. Both aspire to the gestalt–holistic philosophy in their problem-solving methodologies. Systems modeling frameworks build on a plethora of theories, methods, tools, and techniques to provide the instruments with which problems are studied, assessed, understood, managed, and solved, to the extent possible. Thus, the previous sections of this chapter provided the systems-based foundations for our understanding of the complex definition of risk to a system as well as its measurement.

Risk analysis is similar to the systems engineering/systems analysis approach, which is predicated on the centrality of the *states of the system* and their roles in determining for each input (threat) the resulting outputs (consequences). In particular, note that:

(a) The *performance capabilities* of a system are a function of its state vector.
(b) A system's *vulnerability* and *resilience vectors* are each a function of the input, its time of occurrence, and (the vector of) the states of the system.
(c) The *consequences* are a function of the time of the event, the vector of the states, the vulnerability, and the resilience of the system.
(d) The *states* of a system are time dependent and commonly fraught with variability uncertainties and knowledge uncertainties.

(e) *Risk* is a measure of the probability and severity of adverse effects [Lowrance, 1976] (i.e., consequences).

These five premises imply that *risk is a vector of the same units (dimensions) as the consequences and is a function of:*

(i) *Time*
(ii) *The probability of the threat (initiating event) and its specificity (input)*
(iii) *The vector of the states of the system (including its performance capability, vulnerability, and resilience)*
(iv) *The vector of the resulting consequences*

Consider a sample of the multidimensional vector of consequences from Hurricane Katrina: loss of lives, displaced population, loss of property, opportunities, jobs, and the erosion of confidence in government and technology, among others. If we were to develop a risk scenario for a future hurricane with an unusually high surge of water, a similar vector of risk components would necessarily emerge from the risk assessment process. Since consequences are measured through a natural vector of noncommensurate attributes, the units of each element of the risk vector ought to correspond respectively to the same units of the vector of consequences.

The previous discussion implies that significant modeling efforts are required to first evaluate the vector of consequences for each threat scenario (as functions of the threat (initiating event), the vulnerability and resilience of the system, and the time of the event). Then each element of this vector must be paired with either (i) the probability of the scenario's occurrence or (ii) the probability of the severity of the consequences. This fundamentally complex modeling and analysis process cannot be performed correctly and effectively without relying on the states of the system being studied.

2.5.5 Relevance of the Multidimensional Risk Measure to Risk Management

Each risk scenario must address the risk assessment questions: "What can go wrong? What is the likelihood? What are the consequences?" and "What is the time frame?" The risk management process asks: "What can be done and what options are available? What are the trade-offs in terms of all relevant costs, benefits, and risks?" [Haimes, 1991].

"What are the impacts of current decisions on future options?" Risk management addresses each important element of the vector of consequences, noting that different mitigation policy options may address and impact different consequences more effectively. A one-dimensional instrument is unlikely to address the entire vector of risk elements. Rather, a portfolio of risk management policies is required so that their integrative effects are capable of addressing all important and critical risk elements. This process leads to the well-established multiple-criteria decisionmaking (MCDM) approach, where multiple competing, conflicting, and noncommensurate objectives are traded off [Keeney and Raiffa, 1976; Chankong and Haimes, 1983, 2008].

Consider a future scenario in which a hurricane similar to Katrina makes a landfall in New Orleans. The challenges associated with both the perception of risk and the quantification and management of the risk vector are loss of lives, number of people displaced, damage to infrastructures, and other varied costs. Each of these consequences is a function of the *threat* (e.g., the time of the landfall, its wind speed, and its expected water surge), the *vulnerability* of the community (e.g., those in the area below sea level, the states of the levee system, the canals, etc.), the *resilience* of the levee system (water pumps, preparedness, first responders, etc.), and the *time* of the hurricane [Slovic, 2004]. All these factors and many others must be included in any modeling efforts to assess and quantify the likely consequences associated with this scenario. Similarly, the risk management process must also address each element of the risk vector individually and the entire set as a whole. Clearly, the assessment and quantification of risk cannot be any better than the quality of the efforts devoted to quantifying the consequences.

In sum, the questions in the risk management process address the policy options, their trade-offs, the future dynamics of the system and its environment, and the emergent forced changes. These cannot be addressed correctly and effectively without adhering to and tracking the evolution of the states of the system as functions of the risk management decisions and time.

2.5.6 Risk of Low Probability with Extreme Consequences

This section discusses one of the most prevalent sources of complexity in the quantification of risk, namely, the use, and often misuse, of the expected-value-of-risk metric as the sole measurement of risk. The concerns of the public and most decisionmakers focus on events with dire consequences, even with low probabilities. In the face of calamities such as bridges falling, dams bursting, and airplanes crashing, we must acknowledge the importance of studying *extreme* events. Yet, models have been helping to *mask* the criticality of catastrophic events by adhering to the expected value of risk, which intrinsically can equate a low probability of high-consequence events with a high probability of low-consequence events. The reliance on this commonly used metric, when it is used as the *sole measure of risk*, can confuse decisionmakers and thus lead to bad choices. The problem is the expected value of risk is an operation that essentially multiplies the consequences of each event by its probability of occurrence and adds all these products over the entire probability range.

For example, imagine a catastrophic dam failure that might cause flooding of 10^6 acres of land with associated loss of human life and damage to the environment, but such a failure has a very low probability—a hurricane—for example, a probable maximum flood (10^{-6})—of happening. Obviously, this cannot be viewed by decisionmakers in the same vein as the minor flooding of 10^2 acres of land, which has a high probability of 10^{-2} of happening. Yet this is exactly what the expected-value function would ultimately generate. (Note that the products of $10^6 \times 10^{-6} = 10^2 \times 10^{-2} = 1$.) Most important, the equating of these events into one expectation function distorts the relative importance of both the events and consequences. Instead of using the traditional expected value of risk as the *sole measure of risk*, other methods, such as the partitioned multiobjective risk method (PMRM), supplement and complement the expected value of risk by generating a number of conditional expected-value-of-risk functions. (For further discussion of modeling risk of extreme and catastrophic events, including PMRM, see Chapter 8.)

Taleb [2007] ascribes three attributes to an extreme event: (i) it is an outlier, as it lies outside the realm of regular expectation; nothing in the past can convincingly point to its possibility, (ii) it carries an extreme impact, and (iii) in spite of its outlier status, human nature makes us concoct explanations for its occurrence *after the fact*, making it explainable and predictable. Risk analysts and systems engineers are challenged by the need to go beyond a simple formulation to assess the role of extreme events.

2.5.7 The Critical Role of the Time Frame in Risk Management

Risk management connotes actionable measures taken to reduce, curtail, or minimize future risks to a system at acceptable composite costs. The implicit and explicit purpose of an investment in risk management is to render the states of the system less vulnerable and more resilient. Preparedness is one important way with which risk managers change the states of natural and constructed environmental systems.

The time frame plays a central role in risk assessment, risk management, and risk communication. In *risk assessment*, the significance of the question "What can go wrong?" has everything to do with the timing of the adverse effect. When this source of risk is expected further in the future, risk managers have more flexibility. In the *risk management* process, the significance of the question "What are the impacts of current decisions on future options?" again lies in the flexibility and constraints that risk managers desire to have in the future and with emergent forced changes with which they also must cope. In *risk communication*, the time frame plays a substantial role in the perception of risk and in the acceptance of risk management policy options deployed by decisionmakers in both the private and public sectors [Slovic, 2004, 2010]. Typically, the efficacy of risk management decisions cannot be proved when they are made. There is a time lag between the decision and its impact, and public understanding of this inherent latency is paramount. Although risk management decisions are important, they may not suffice to make those at risk feel safer. The ultimate integrated steps in the risk analysis process are to explain how risk management policies and actions contribute to improved safety and to communicate and justify the residual risk to the multiple stakeholders and the affected public (when removing the entire risk is neither economically nor technically feasible) [National Research Council of the National Science Academies, 2009].

In sum, the fundamental public concerns about risk reside in the ever-lingering question: "What can go wrong in the future, and what might be the adverse consequences?" Since the present is deterministic and the future is not, there is an imperative need to assess the future states of the system as they might respond and evolve as a consequence of emergent forced changes. Thus the criticality of the time frame in risk analysis and in understanding and assessing the evolving states of the system over time.

2.5.8 Summary: A Logical Integration

The centrality of state variables in modeling, and thus in systems engineering/analysis and in risk analysis, is the sine qua non for any quest to model and understand the intricacy of any system and of complex multi- and large-scale systems in particular. The holistic nature of systems engineering/analysis, and of risk analysis, leads to systems modeling through state variables with specified system boundaries and the integration and analysis of the interdependencies and interconnectedness among all subsystems. Modeling the risk vector as an explicit function of the initiating event, its timing and level of intensity, and of the system's vulnerability and resilience (the latter two themselves being functions of the states of the system) constitutes a fundamental task in risk analysis.

The multifaceted composition of risk includes the levels of uncertainty and intensity of the initiating events or threats, the time frame, and the dynamic, probabilistic, and often nonlinear natures of the states of all natural and constructed environments on which the system's vulnerability and resilience depend. This intricacy cannot be modeled and understood on an ad hoc basis. In other words, *we must understand, model, and define the complexity of risk, vulnerability, and resilience in a systemic way and through a methodical, theoretically based systems approach, where the states of the system constitute the essence of the analysis.*

By projecting Heisenberg's uncertainty principle and Einstein's advice on the complexity of theories to the field of risk analysis, we assert:

To the extent that risk analysis is precise and simple, it is not real.
To the extent that risk analysis is real and complex, it is not precise.

2.6 ON THE DEFINITION OF VULNERABILITIES IN MEASURING RISKS TO SYSTEMS

The literature of risk analysis is replete with misleading definitions of vulnerability [Haimes, 2006, 2009a,b, 2012b]. Of particular concern is the definition of risk as the product of impact, vulnerability, and threat. Thus, in our quest to measure risks of terrorist attacks and natural disasters to critical infrastructures, we must account for the fundamental characteristics of the system. In particular, we must rely on the building blocks of mathematical models, focusing on the use of state variables, as discussed in Section 2.4.

In terms of vulnerability and criticality, infrastructures are represented (characterized) by a hierarchy of vectors of state variables. To relate the centrality of state variables in intelligence analysis to countering terrorism, it is important to define the following terms [Haimes, 2004], which broadly apply to risk analysis:

- *Vulnerability* is the manifestation of the inherent states of the system (e.g., physical, technical, organizational, cultural) that can be exploited to adversely affect (cause harm or damage to) that system.
- *Intent* is the desire or motivation to attack a target and cause adverse effects.
- *Capability* is the ability and capacity to attack a target and cause adverse effects.
- *Threat* is the *intent* and *capability* to adversely affect (cause harm or damage to) the system by adversely changing its states.
- *Risk* is the result of a threat with adverse effects to a vulnerable system.

Thus, it is clear that modeling risk as the probability and severity of adverse effects requires knowledge of the vulnerabilities, intents, capabilities, and threats to the infrastructure system.

In sum, to assess the risks to a vulnerable system, we need to (i) assess the likelihood of the threat (attack scenario), (ii) model the responses of the various interdependent state variables that characterize the system (i.e., its vulnerabilities) to the attack scenario (i.e., develop a *dose–response* function), (iii) assess the severities of consequences resulting from the dysfunctionality of the entire system or from a subset of its subsystems, and (iv) assess the likelihood of the severity of the projected consequences. *In the process of measuring risk, when the second imperative step—the critical modeling process that translates an attack scenario into consequences—is masked or skipped by simply multiplying vulnerability directly into the risk measure, then the risk measure becomes detrimentally flawed.*

2.7 ON THE DEFINITION OF RESILIENCE IN MEASURING RISK TO SYSTEMS

In the current era of worldwide terrorism, the terms *vulnerability* and *resilience* have become common in the parlance of risk analysis, and various attempts have been made to define and to quantify them. *Resilience* has been defined in the literature in many different ways. Consider, for example, the following definitions: (1) Resilience is the ability of a system to absorb external stresses [Holling, 1973]. (2) Resilience is a system capability to create foresight, to recognize, to anticipate, and to defend against the changing shape of risk before adverse consequences occur [Woods, 2005, 2006]. (3) Resilience refers to the inherent ability and adaptive responses of systems that enable them to avoid potential losses [Rose and Liao, 2005]. (4) Resilience is the result of a system (i) preventing adverse consequences, (ii) minimizing adverse consequences, and (iii) recovering quickly from adverse consequences [Westrum, 2006]. (5) Resilience engineering is a paradigm for safety management that focuses on how to help people cope with complexity under pressure to achieve success [Hollnagel et al., 2006].

The resilience of a system is a manifestation of the states of the system [Haimes, et al., 1997; Haimes, 2009a, 2012b]. Perhaps most critically, it is a vector that is time dependent. For the purposes of this chapter, resilience is defined as the ability of the system to withstand a major disruption within acceptable degradation parameters and to recover within an acceptable composite cost and time [Haimes et al., 2006, 2012b]. Moreover, resilience is similar to vulnerability in that it cannot simply be measured in a single unit metric; its importance lies in the ultimate multidimensional outputs (the consequences) of the system for any specific inputs (threats). (Note that the consequence that is considered as part of the risk metric is in fact the output of the system model, and that the input of the system's model is parallel to the concept of threat.) Indeed, the risk associated with a cyber attack on a cyber infrastructure system will depend not only on the resilience of the states of the system but also on the specific type and sophistication of the cyber attack. This is because the resilience of a system can be measured in terms of the specific threat (input), the system's recovery time, and the associated composite consequences in terms of costs and risks. Thus, different attacks would generate different consequence (output) trajectories for the same resilient system.

Consider the immunization of a population against a major strain of a flu virus termed Type B. Assume that the population develops resilience for multiple strains of viruses of Type B, except for an evolving strain of Type A. In this case, even though the population might have resilience (immunity) for Type B, the appearance of strain A into this population will likely be infectious. Here again, the risk to the population from a threat is dependent on the type of threat, the resilience of the system, and the ability of the system to withstand that specific threat.

Likewise, consider any large-scale physical infrastructure such as electric power, transportation, or telecommunication. In any such complex system, the question "What is the resilience of infrastructure x?" is unanswerable, because the question implicitly depends upon knowing whether infrastructure x would recover following any attack y within an acceptable time and composite costs and risks. Thus, the only way such a question can be answerable is when the threat (or a set of threats) is specifically identified. Indeed, the system's resilience is not merely an abstract attribute of the system; rather, it is a state of the system (composed of a vector of substates) for which any specific substate may respond differently to different inputs (threats). For example, a water distribution system may have redundancy in its electric power subsystem, and thus it may be resilient to a major storm that would shut down one of the power lines to the water distribution system, leaving the other redundant line intact. On the other hand, suppose the water distribution system depends on only one main pipe to supply water to its customers but is located in a region susceptible to earthquakes. The system is resilient only to the extent that the main pipe is functioning and can withstand an earthquake up to level 4 on the Richter scale. However, the system would likely fail during an earthquake of level 5 or 6. Here again, measuring the resilience of the water system is actually measuring the responses of the system to the specific threat, in this case the scale of the earthquake.

Furthermore, one may associate a vector of resilience with each subsystem. Thus, there can be a hierarchy of resilience attributes for any large-scale natural or constructed environment. For example, the human body as a system is made up of many subsystems (e.g., the digestive, pulmonary, and auditory systems, among others), each with a set of resilient organs and suborgans, where the level of such resilience depends on the input (physical or biological threats) and the output (a temporary or long-term loss of functionality of specific organs or suborgans). This example reinforces the thesis that system resilience can be measured in terms of the outputs for given inputs to the system. (Note that the inputs to the system, the states of the system, and the outputs are commonly time variant and probabilistic, as will be discussed subsequently.) To further appreciate the centrality of the system's input–output relationship to its resilience (states of the system), consider the fact that despite the resilience of the human body to various physical and biological attacks on it, its ultimate resilience depends upon the states of the body at the time as well as the type and strength of such attacks.

A system may also be characterized by its specific redundancy and robustness—both of which lead to a specific vector of resilience [Matalas and Fiering, 1977; Haimes et al., 1998, Haimes, 2012b]. Redundancy refers to the ability of certain components of a system to assume the functions of failed components without adversely affecting the performance of the system itself. Of course, redundancies constitute an integral part of all safety-critical systems. Robustness refers to the degree of insensitivity of a system to perturbations or to errors in the estimates of those parameters affecting the design choice.

2.7.1 On the Relationship among Preparedness, Vulnerability, and Resilience

Both vulnerability and resilience are manifestations of the states of the system. In principle, they are two sides of the same coin; vulnerability addresses only a system's protection, whereas resilience focuses also on a system's recovery following an adverse event. As states of the same system, both represent the capability of the system to withstand threats. On the one hand, vulnerability represents those states of the system that can be adversely affected by specific types and levels of magnitude of threats. On the other hand, resilience *also* represents the ability of the system to recover within an acceptable time and composite costs and risks, having been presented with a threat. That is, the vulnerability of a system does not provide information about the ability of the system to recover from a particular threat. What, then, is the relationship between preparedness and vulnerability, and what is the relationship between preparedness and resilience? If the primary objective of preparedness is reducing the vulnerability of a system to specific threats, it may (although not necessarily) also improve the resilience of the system to the same threats. For example, hardening a system against specific threat scenarios (e.g., adding more security by building fences or formulating policies and procedures that would limit access to infrastructures), but without addressing the recovery needs following a successful attack, would reduce the vulnerability to such threat scenarios; it might not, however, necessarily improve the resilience of the system in terms of its recovery time or composite costs. For example, an electric power generation unit might be hardened against terrorist attacks or major natural hazards, thus reducing its vulnerability to such events, but such hardening would not necessarily improve its resilience to an acceptable level of recovery. By the same token, improving the resilience of the electric power supply system by adding redundant power lines crossing different geographical sites could ensure an acceptable level of resilience to the same threat, but it would not lessen the vulnerability of the electric power generation unit to a physical threat.

2.7.2 On the Relationship between Resilience and Risk

Improving a system's resilience offers significant advantages in managing risk; improving the resilience of a system constitutes an integral part of the risk management process. A fundamental benefit is that an acceptable level of residual risk to the system (i.e., an acceptable level of affordable safety) can be determined for each class of threat scenario. More specifically, because of the probabilistic nature of threats, given the occurrence of a class of threat scenarios, the outputs (consequences) are best represented with PDFs. The resulting risks in terms of recovery time and composite costs can be calculated in a variety of ways, including the expected value of risks or the conditional expected value of risk of the extremes [Haimes, 2009b, 2012b]. And ultimately, the trade-offs among the various levels of risks and costs associated with each investment (e.g., through preparedness) in the system's resilience can be evaluated. The fact that the severity levels of the consequences resulting from a threat to a system are used as the metric with which the system's vector of resilience is evaluated—and given that the inputs and thus the outputs to the system are probabilistic in nature—necessarily lead us again to the triplet of risk management questions posed in Chapter 1 and at the beginning of this chapter: What can be done, and what options are available? What are the associated trade-offs in terms of all costs, benefits, and risks? What are the impacts of current management decisions on future options? [Haimes, 1991, 2009b, 2012b]. The answers to these questions, because they are so specific to each system and to each scenario, defy the ability to assign general and absolute scores to a system's resilience, as is discussed in the following section.

2.7.3 On the Challenges of Scoring a System's Resilience

The importance of resilience as a state of a system's capability to withstand forced changes to its organizational structure, functionality, and operational continuity has led to the development of capability metrics as surrogates for resilience. Indeed, several scoring systems for measuring the resilience of cyber systems are emerging. For example, Carnegie Mellon's Software Engineering Institute is developing a *Resiliency Engineering Framework* [Software Engineering Institute, 2008], which posits a vector of twenty-one capability areas that can be measured on a maturity scale that will demonstrate that cyber security and resilience practices are integrated into common and frequent management decisions. The underlying notion is that if mature business processes are in place and if they appropriately integrate common decisions with security decisions, then the organization is more likely to see, understand, and respond to a cyber threat in a way that would reduce the consequences and minimize recovery time and costs.

On the other hand, attempts to characterize the resilience of a system with a specific numerical descriptor (as a metric) and to use this metric to compare the resilience of different systems could be misleading unless we pretend to assume that these different systems will be subjected to the same exact threats and the same exact levels of such threats with the same exact probabilities. Given the diversity of the functionality and the configurations of contemporary infrastructure systems (and in particular of cyber systems), and given the immense uncertainties associated with these differences and the threats to which any such system might be subjected, it is unreasonable to assign a numerical scale or number of resilience to any of these systems.

An important conclusion that can be drawn from the previous discussion is that the resilience of a system might be measured in terms of the myriad substates that characterize that system for a specific time period and threats. It is an entirely different question, however, to project in the abstract the recovery time and costs associated with any unspecified phantom set of threat scenarios. Here again, the risk as a measure of the probability and severity of adverse affects [Lowrance, 1976] can be assessed only for a specific threat scenario at any given time.

Measuring the efficacy of system's resilience might be achieved, for example, through the unique functionality of that particular system and its responses (outputs) to specific inputs. And given that such inputs are probabilistic, so are the outputs, meaning that the system's resilience—because it is measured in terms of responses to the inputs—can be measured (quantified) only in probabilistic terms and for specific inputs. We can thus adduce the following premises for scoring resiliency [Haimes, 2012b]:

1. The probabilistic nature of threats and thus of the associated outputs necessitates a holistic, multidimensional probabilistic scoring system of resilience. Furthermore, the myriad plausible threat scenarios, each with associated magnitude and duration, necessarily limit any such method of scoring resilience to specific classes of input threats.
2. Resilience, as a vector of the states of all physical and natural systems, is time variant, and given the inherent characteristics of such systems, their resilience will deteriorate over time. Thus, even an input-limited scoring method would be further constrained by the inherent time-variant resilience of the system.

In sum, resilience—complex and composite attributes of the states of a system—which commonly constitutes a vector of substates, cannot be characterized with a single numerical descriptor. Resilience must be understood and evaluated in the context of a probabilistic and dynamic set of input threat scenarios to the system and in terms of the complex set of associated consequences attached to any such threat.

2.8 ON THE COMPLEX QUANTIFICATION OF RISK TO SYSTEMS

2.8.1 Introduction

Note that the behavior of the states of the system, as a function of time, enables modelers to determine, under certain conditions, its future behavior for any given inputs, or an initiating event including a threat.

The term *threat* connotes a malevolent attack by terrorist networks with adverse consequences to the system. As has been discussed, all systems are characterized at any moment by their respective state variables; furthermore, all state variables are under continuous natural positive or negative emergent forced changes (see Section 2.4). The decision as to whether a state variable of a system should be modeled as static (constant) or dynamic (time dependent) is one step in the modeler's determination to select only those state variables that represent the *essence* of the system. For example, risk analysts commonly update the probability of the condition (level) of the state of the system with new information, using Bayes' theorem. As is the case for most systems, where the states evolve over time, updating the conditions (levels) of the states of the system is essential for dynamic systems. (Examples include the states of all physical and cyber infrastructures, the economy, technology, health, and the states of terrorists' intent and capability [Haimes, 2012b].)

2.8.2 Quantifying the Risk Function

Multiple sources of uncertainties are associated with the risk function. Uncertainty, which will be formally discussed in Chapter 6, is commonly viewed as *the inability to determine the true state of a system*, and can be caused by *incomplete knowledge*, and/or by *stochastic variability*—two major sources of uncertainty in modeling affect risk analysis [Paté-Cornell, 1990, 1996; Apostolakis, 1999].

Knowledge (epistemic) uncertainty manifests itself in the selection of model topology (structure) and model parameters, among other sources of ignorance (e.g., lack of knowledge of important interdependencies within the states of the system and among other systems). *Variability (aleatory) uncertainty* includes all relevant and important random processes and other random events (e.g., emergent forced changes). Uncertainty dominates most decisionmaking processes and is the Achilles' heel for all deterministic and most probabilistic models. This uncertainty is commonly introduced through the selection of incorrect model's topology (structure), for example, linear for a highly nonlinear system, its parameters, data collection, and the employed processing techniques. Model uncertainties will often be introduced through human errors of both commission and omission.

2.8.2.1 Risk as a product of probability of the threat and consequences

One common method for the quantification of the risk function is presented in Eq. (2.10), where r and c are multidimensional vectors of risk and consequences, respectively, associated with a specific scenario, and $p_r(T)$ is the probability of the threat, T:

$$r = p_r(T) \cdot (c) \qquad (2.10)$$

Equation (2.10) suffers from several limitations, especially for risk associated with scenarios of extreme consequences with low probability. As noted earlier, assume that a levee may fail with a low probability of 10^{-6} during its functional time, causing the death of 100,000 people. According to Eq. (2.10), the expected value of risk to the downstream population from such a catastrophic failure would be $r_1 = 10^5 \times 10^{-6} = 10$ persons, an absurd result. Similarly, the risk to a flooded farmland, r_2, would generate a similarly meaningless result: Assuming that the total flooded land is estimated at 10^6 acres with the same probability of 10^{-6} would result in $r_2 = 10^6 \times 10^{-6} = 1$ acre.

2.8.2.2 Risk in terms of the expected value of adverse consequences

Given a threat, Eq. (2.11) represents the expected value of consequences, $E[x]$ as a surrogate function of risk, where x is a random variable representing the consequences, c, $p(x)$ is the probability density function of the consequences, and a and b are the assumed lower and upper limits of the range of consequences. Note that the probability of the scenario's occurrence is implicit ("Given the specific threat, what are the resulting consequences?"), and it does not appear in Eq. (2.11) for the continuous case or in Eq. (2.12) for the discrete case:

$$E[x] = \int_a^b x \cdot p(x) dx \qquad (2.11)$$

$$E[x] = \Sigma p_i x_i \qquad (2.12)$$

We use the probability mass function for the expected value of risk for the discrete case. Note that (i) the consequences are commonly a multidimensional vector and (ii) not all consequences resulting from a specific threat would necessarily have the same probability. In other words, since the multidimensional vector of consequences is a function of the states of the system, and a *specific threat* would not necessarily affect all states at the same level, then a *specific probability* ought to be associated with each element of the consequences. This fundamental premise has significant ramifications for the quantification of the risk function. As a result, Eqs. (2.11) and (2.12) ought to be appropriately modified to account for both the different expected value of each consequence and the correlation between and among the consequences. This important notion, which will be elaborated in the second premise (see Section 2.8.3), applies to all forms of risk function.

2.8.2.3 Modified risk in terms of the expected value of risk

Note that in the previous case, the threat was assumed as given, and thus the probability of the threat was implicit and did not appear in Eq. (2.11). In the following formulation of the expected value of risk, some analysts may incorporate the probability of the threat, $P_r(T)$, explicitly as is presented in Eq. (2.13) for the continuous case and in Eq. (2.14) for the discrete case:

$$E[x \mid T] = \left(\int_a^b x \cdot p(x) dx \right) \cdot P_r(T) \qquad (2.13)$$

$$E[x \mid T] = \left(\sum_{i=a}^b p_i x_i \right) P_r(T) \qquad (2.14)$$

This formulation is informative only for credible knowledge of the probability of threat; otherwise, the expected value of risk may be distorted either upward or downward. In such case, the risk function presented in Eq. (2.11) would be more informative.

2.8.2.4 Conditional expected value of risk for extreme events

The *expected-value-of-risk* metric, which constitutes a commensuration of low probability of high-consequence events with high probability of low-consequence events (Eqs. 2.10–2.14), has *masked* the criticality of extreme and catastrophic events in risk analysis. The PMRM, formally introduced in Chapter 8, supplements the expected-value measure of risk with a conditional expected value of risk of extreme events [Asbeck and Haimes, 1984]. A conditional expectation is defined as the expected value of a random variable, given that this value lies within some prespecified probability range. Risk perception of these different types of risk events plays an important role in the manner with which individuals, the public, and policymakers are willing to invest resources in preparedness, response, and recovery, or in the mitigation of low-probability risks of extreme events deemed far in the future [Perrow, 1984; Slovic, 2000; Bier et al., 2004].

The conditional expected value between β and ∞ can be calculated for the continuous case in Eq. (2.15) and in Eq. (2.16) for the discrete case:

$$f_4(\cdot) = E[x|x > \beta] = \frac{\int_\beta^b xp(x)dx}{\int_\beta^b p(x)dx} \quad (2.15)$$

$$f_4(\cdot) = E[x|x > \beta] = \frac{\sum_{i=\beta}^{\infty} p_i x_i}{\sum_{i=\beta}^{\infty} p_i} \quad (2.16)$$

where β is partitioning point for extreme event (consequences).

In contrast to Eq. (2.11), which provides information regarding the distribution of consequences that could result from the realization of a threat scenario, a conditional expected value of extreme events provides *additional* insight into the risk function. Moreover, although the probability of a threat may be difficult to obtain (particularly for malevolent, human-caused events), this formulation leads to a better appreciation of extreme adverse consequences that can be used more effectively for risk management.

2.8.2.5 The multidimensional probabilistic consequences and risk function

The multidimensional probabilistic consequences resulting from an initiating event yield a multi-dimensional risk function whose modeling and quantification complexity presents a considerable challenge. The selection of appropriate models to represent the essence of the system's multiple perspectives determines the effectiveness of the entire risk assessment, management, and, ultimately, communication process. In particular, the scope and effectiveness of strategic risk management options are implicitly and explicitly dependent on the perspectives of the system that are included or excluded in the ultimate modeling efforts [Haimes, 1981]. The multiple perspectives of a system may be manifested through its structure, functionality, and the services it provides, the customers it supports, the other systems on which it depends, and the time frame of all of the above.

The five premises offered in the following section augment the concepts presented in this chapter.

2.8.3 Five Premises on Terrorism for Risk Analysis

The following five premises are most relevant to the risk of terrorism against physical infrastructure and cyber infrastructure. Quantifying terrorism risks should be distinguished from risks due to natural disasters, accidents, and infrastructure failures resulting from fatigue or other reasons: This is *because there is interplay between the probability of a specific threat from terrorists and the states of the targeted system.*

First premise: There exists interdependence between a specific threat to a system by terrorist networks and the states of the targeted system, as represented through the system's vulnerability, resilience, and criticality–impact.

Intelligence gathered by terrorists on a critical targeted system can provide them with valuable information on the states of that target, which they can use to assess its vulnerability, resilience, and criticality–impact. Consider a specific threat at a given time from a would-be terrorist network that is focusing on two targeted systems, A and B, of similar functionality and importance. Target A, of high vulnerability, low resilience, and high criticality–impact, would have a higher probability of being attacked by the terrorist network than Target B (of low vulnerability, high resilience, and low criticality–impact). This is because the consequences suffered by Target A would be higher than the consequences suffered by Target B, making Target A more attractive. The fortification of critical infrastructure systems by reducing the states of their vulnerability and enhancing their resilience (for specific threats) could diminish their attractiveness as targets.

Second premise: A specific threat, its probability, its timing, the states of the targeted system, and the probability of consequences are not independent.

The second premise builds on the following three assertions: (i) the war between the terrorist networks and significant parts of the world is an asymmetric, dynamic, multiattribute, multiparty, and nonzero-sum game; (ii) the states of vulnerability, resilience, and criticality–impact of would-be targeted critical systems influence the selection of the specific threat and the *timing and probability of* being attacked; and (iii) for any specific threat, *the resulting PDF of each element of the multidimensional consequences could be different because a specific threat may impact each state of the targeted system differently.*

Terrorists require timely informative data on the states of a targeted system to launch a successful attack. On the other hand, effective risk management options employed by the defenders are aimed at reducing the vulnerability of a would-be targeted system and improving its resilience. Knowledge about any weakness of the states of critical systems (e.g., high vulnerability and low resilience for specific threats) ought to be kept guarded; masking or faking the true states of a vulnerable system is not inconsistent with the nonzero-sum game between the terrorist networks and significant parts of the world. Similarly, terrorist networks try to mask their own capability, intent, and plans regarding their most consequential attacks. The capability to mask the true states of the targeted system and of the threatening entity creates another interdependent relationship, which leads to the third premise.

Third premise: The two questions in the risk assessment process—"What is the likelihood?" and "What are the consequences?"—can be interdependent.

Based on the second premise, we make a distinction between (i) the probability of a natural event (e.g., a hurricane) and (ii) the probability of a terrorist attack on a specific target and the resulting consequences of each. The probability of a hurricane making a landfall is independent of any targeted system within the landfall location. However, the probability of a terrorist threat is undisputedly interdependent with the expected consequences affecting a specific targeted system. The consequences are related to the vulnerability and resilience to specific threats and to the criticality–impact of that system. In other words, since both the probability of the threat and the probability of the consequences depend on the vulnerability, resilience, and criticality–impact of the targeted system, any overestimate or underestimate of the states of the targeted system would introduce errors in both probabilities and thus multiply or undercount the resulting risk.

Fourth premise: Risk management can reduce both the probability of a threat to a targeted critical system and the associated probability of consequences by changing the states (e.g., vulnerability and resilience) of the system.

The ultimate goal of all decisions is to maintain or change the states of the system under consideration, to achieve specific objectives, at acceptable trade-offs, and within an acceptable time frame. Adding a parallel component to a physical system is aimed at improving its redundancy, and thus its reliability. When publicized, these risk management decisions would reduce the probability of a specific threat to that system and also reduce the probability of severe consequences. This is because the targeted system becomes less attractive to terrorists in comparison to similar targets in the same category, which may not have undergone appropriate risk management.

Fifth premise: The quantification of risk to a vulnerable system from a specific threat must be built on a systemic and repeatable modeling process, by recognizing that the states of the system are essential for the construction of quantitative metrics of the consequences based on intelligence gathering, expert evidence, and other qualitative information.

All government agencies and public and private sectors seek to understand the trends of risks associated with emergent forced changes that affect the states of their systems in order to prevent, mitigate, or prepare for undesirable future occurrences. If such trends/changes are unanticipated, undetected, misunderstood, or ignored, they affect a multitude of states of that system with potentially adverse consequences. It is imperative to be *able—through scenario structuring, modeling, and risk analysis—to* envision, discover, and track emergent forced changes and their crossovers. Emergent forced changes may be caused by natural phenomena or malevolent human actions or accidents. Precursors based on either tangible or intangible information are most effective when they are used in conjunction with other information in an effort to understand their meaning and to follow up with vigilance. Often, a crossover is ignored during the tracking of emergent forced changes and is recognized too late and only after it becomes a disaster [Perrow, 1984; Bier et al., 2004; Phimister et al., 2004; Taleb, 2007]. *Thus, in the context of modeling, crossovers connote situations where during the simulated scenarios of a dynamic system, the projected emergent forced change crosses the state of knowledge from uncertainty to a more likely eventuality of its occurrence.* Duderstadt et al. [2005] emphasize that "change is coming, and the biggest mistake could be underestimating how extensive it will be."

Effective intelligence gathering, analysis, and management should provide the knowledge required for future decisions made under uncertainty. The analysis and interpretation of such information is as much an art as a science. It is an art because the task is based on the knowledge and experience of the professionals whose task is to interpret the often incoherent, sparse, and conflicting intelligence; and it is also a scientific process because it builds on empirical data, on computer models and simulation, and on other scientific–engineering tools. *Thus, modeling cardinal consequence, supported by ordinal intelligence information using the states of the targeted system (through its vulnerability and resilience), constitutes a critical step in the quantification of risk.* The Dempster–Shafer theory [1990] is one approach to blend ordinal and cardinal information.

2.8.4 On the Complex Role of Modeling in Quantifying the Risk Function

This section focuses on the gulf between (i) modeling the complex dynamic and nonlinear relationships that exist among the elements that represent the risk function (associated with a terrorist attack on a targeted system) and (ii) the immediate need for a practical methodological framework with which to quantify the risk function, without violating any of the premises. In other words, there is an intermediate need to develop a modeling framework that recognizes the aforementioned inherent dynamic and nonlinear interdependencies and that offers a first approximation that can be extended and improved over time.

On the cause–effect relationship between (a) a specific threat to a targeted system, given its vulnerability and resilience, and (b) the probability of consequences to the targeted system.

Modeling cause–effect relationships for natural phenomena or the constructed environment has occupied and challenged scholars and modelers for generations. During the last four decades, extensive research has been performed on the impacts of chemicals and other carcinogens on the health of animals and humans and on the environment, commonly known as the study of *dose–response function*. This past history and ongoing efforts by numerous researchers from many disciplines present an important message on the enormous challenges facing risk analysts in their quest to quantify the consequences associated with terrorist threats against physical infrastructures and cyber infrastructure. For example, a cursory survey of articles on dose–response models and related studies, published in *Risk Analysis: An International Journal*, resulted in over seventy such publications during the last 10 years [Hertwich et al., 2000; Lee et al., 2006; Reiss and Griffin, 2006; McGill et al., 2007; Nauta et al., 2007]. This research shows that it is prudent to develop several probable scenarios for the specific time frame, the specific affected population, and other unique characteristics of a system threatened with potential extreme consequences. The Phantom System Models (PSM) (discussed in Chapter 16) is one approach that can assist in this task [Haimes, 2007, 2009b, 2012a].

Relating the probability of a threat to the associated probability of consequences

Based on the five premises, we draw the following conclusions:

(a) The assessed probability distribution of a threat to a targeted system and its scale is a function of (i) the time frame; (ii) the quality and level of confidence of the available intelligence and expert evidence; (iii) the states of the perceived vulnerability, resilience, and criticality–impact of the targeted system by the would-be terrorists; and (iv) the states of the would-be terrorists.

(b) The probability distribution of the consequences, given the realization of the threat, is a function of (i) the states of the would-be terrorists (e.g., their capability and intent) and thus of the magnitude and intensity of the threat, (ii) the states of the targeted system (e.g., its vulnerability, resilience, and criticality–impact), and (iii) the availability of immediate organizational and other support that may be provided to the targeted system.

Thus, for any threat scenario, the assessed probability of a terrorist attack on a specific target is a function of multiple variables associated with the *states* of the attackers and the *states* of the targeted system. On the other hand, the assessed probability of consequences is a function of the aforementioned states of the system and also a function of the states of other supporting systems (due to the interdependencies among most physical and cyber infrastructure systems).

REFERENCES

Apostolakis, G., 1999, The distinction between aleatory and epistemic uncertainties is important: an example from the inclusion of aging effects into probabilistic safety assessment, *Proceedings of PSA'99*, August 22–25, American Nuclear Society, Washington, DC.

Asbeck, E., and Y. Haimes, 1984, The partitioned multiobjective risk method (PMRM), *Large Scale Systems* **6**(1): 13–38.

Bier, V., S. Ferson, Y. Haimes, J. Lambert, and M. Small, 2004, Risk of extreme and rare events: lessons from a selection of approaches, In *Risk Analysis and Society*, T. McDaniels, and M. Small (Eds.), Cambridge University Press, Cambridge, MA.

Chankong, V. and Y.Y. Haimes, 1983, *Multiobjective Decision Making: Theory and Methodology*, North Holland, New York.

Chankong, V., and Y.Y. Haimes, 2008, *Multiobjective Decision Making: Theory and Methodology*, Dover, New York.

Duderstadt, J., W. Wulf, and R. Zemsky, 2005, Envisioning a transformed university, *Issues in Science Technology* **22**(1): 35–41.

General Accounting Office (GAO), 1996, *Ways to Improve Management of Federally Funded Computational Systems*, U.S. Government Printing Office, Washington, DC.

Gnedenko, B.V., 1963, *The Theory of Probability*, Translated from the Russian by B.D. Seckler, Chelsea Publishing Company, New York, p. 53.

Grant, P.M., C. Starr, and T.J. Overbye, 2006, A power grid for the hydrogen economy, *Scientific American*, July: 76–83.

Haimes, Y.Y., 1977, *Hierarchical Analyses of Water Resources Systems: Modeling and Optimization of Large-Scale Systems*, McGraw-Hill, New York.

Haimes, Y.Y., 1981, Hierarchical holographic modeling, *IEEE Transactions on Systems, Man, and Cybernetics* **11**(9): 606–617.

Haimes, Y.Y., 1991, Total risk management, *Risk Analysis* **11**(2): 169–171.

Haimes, Y.Y., 2004, *Risk Modeling, Assessment, and Management*, Second edition, John Wiley & Sons, Inc., Hoboken, NJ.

Haimes, Y.Y., 2006, On the definition of vulnerabilities in measuring risks to infrastructures, *Risk Analysis* **26**(2): 293–296.

Haimes, Y.Y., 2007, Phantom system models for emergent multiscale systems, *Journal of Infrastructure Systems* **13**(2): 81–87.

Haimes, Y.Y., 2009a, On the complex definition of risk: a systems-based approach, *Risk Analysis* **29**(12): 1647–1654.

Haimes, Y.Y., 2009b, *Risk Modeling, Assessment, and Management*, Third edition, John Wiley & Sons, Inc., Hoboken, NJ.

Haimes, Y.Y., 2012a, Modeling complex systems of with phantom system models, *Systems Engineering* **15**(3): 333–346.

Haimes, Y.Y., 2012b, Systems-based guiding principles for risk modeling, planning, assessment, management, and communication, *Risk Analysis*, **32**(9): 1451–1467.

Haimes, Y.Y., and W.A. Hall, 1977, Sensitivity, responsivity, stability, and irreversibility as multiobjectives in civil systems, *Advances in Water Resources* **1**: 71–81.

Haimes, Y.Y., N.C. Matalas, J.H. Lambert, B.A. Jackson, and J.F.R. Fellows, 1998, Reducing the vulnerability of water supply systems to attack, *Journal of Infrastructure Systems* **4**(4): 164–177.

Haimes, Y.Y., K.G. Crowther, and B.M. Horowitz, 2006, Homeland security preparedness: balancing protection with resilience in emergent systems, *Systems Engineering* **11**(4): 287–308.

Hall, W.A., and J.A. Dracup, 1970, *Water Resources System Engineering*, McGraw-Hill, New York.

Hertwich, E., T. McKone, W. Pease, 2000, A systematic uncertainty analysis of an evaluative fate and exposure model, *Risk Analysis* **20**: 439–454.

Hirshleifer, J., and J. Riley, 1992, *The Analytics of Uncertainty and Information*, Cambridge University Press, Cambridge.

Holling, C.S., 1973, Resilience and stability of ecological systems, *Annual Review of Ecology and Systematics* **4**(1): 1–23.

Hollnagel, E., D.D. Woods, and N. Leveson (Eds.), 2006, *Resilience Engineering: Concepts and Precepts*, Ashgate Press, Aldershot.

Johnson, C.W., 2006, What are emergent properties and how do they affect the engineering of complex systems? *Reliability Engineering and System Safety* **91**(12): 1475–1481.

Kaplan, S., 1991, The general theory of quantitative risk assessment, In *Risk-Based Decision Making in Water Resources V*, Y.Y. Haimes, D. Moser, and E. Stakhiv (Eds.), American Society of Engineers, New York, pp. 11–39.

Kaplan, S., 1993, The general theory of quantitative risk assessment—its role in the regulation of agricultural pests, *Proceedings of the APHIS/NAPPO International Workshop on the Identification, Assessment and Management of Risks due to Exotic Agricultural Pests* **11**(1): 123–126.

Kaplan, S., 1996, *An Introduction to TRIZ, the Russian Theory of Inventive Problem Solving*, Ideation International Inc., Southfield, MI.

Kaplan, S., 1997, The words of risk analysis, *Risk Analysis* **7**(4): 407–417.

Kaplan, S., and B.J. Garrick, 1981, On the quantitative definition of risk, *Risk Analysis* **1**(1): 11–27.

Kaplan, S., Y.Y. Haimes, and B.J. Garrick, 2001, Fitting hierarchical holographic modeling into the theory of scenario structuring and a resulting refinement of the quantitative definition of risk, *Risk Analysis* **21**(5): 807–815.

Keeney, R.L., and H. Raiffa, 1976, *Decisions with Multiple Objectives*, Wiley, New York.

Kunreuther, H., and P. Slovic (Eds.), 1996, *Challenges in Risk Assessment and Risk Management, Annals of the American of Political and Social Science*, Sage, Thousand Oaks, CA.

Lee, L., C. Mao, and K. Thompson, 2006, Demographic factors and their association with outcomes in pediatric submersion injury, *Academic Emergency Medicine* **13**: 308–313.

Li, D. and Y.Y. Haimes, 1988, The uncertainty sensitivity index method (USIM) and its extension, *Naval Research Logistics* **35**(6): 655–672.

Lowrance, W.W., 1976, *Of Acceptable Risk*, William Kaufmann, Los Altos, CA.

Matalas, N.C., and M.B. Fiering, 1977, Water-resource systems planning, In *Climate, Climatic Change, and Water Supply*, The National Research Council (Ed.), National Academy of Sciences, National Research Council, Washington, DC.

McGill, W., B. Ayyub, and M. Kaminsky, 2007, Risk analysis for critical asset protection, *Risk Analysis* **27**: 1265–1281.

National Research Council of the National Science Academies, 2009, *Science and Decisions: Advancing Risk Assessment*, National Academies Press, Washington, DC.

Nauta, M.J., W. Jacobs-Reitsma, and A. Havelaar, 2007, A risk assessment model for campylobacter in broiler meat, *Risk Analysis* **27**: 845–861.

Paté-Cornell, M., 1990, Organizational aspects of engineering system safety: the case of offshore platforms, *Science* **250**: 1210–1217.

Paté-Cornell, E., 1996, Uncertainties in risk analysis: six levels of treatment, *Reliability Engineering and System Safety* **54**: 95–111.

Perrow, C., 1984, *Normal Accidents: Living with High-Risk Technologies*, Princeton University Press, Princeton, NJ.

Phimister, J.R., V.M. Bier, and H.C. Kunreuther (Eds.), 2004, *Accident Precursors Analysis and Management: Reducing Technological Risk through Diligence*. National Academy of Engineering, The National Academies Press, Washington, DC.

Reiss, R., and J. Griffin, 2006, A probabilistic model for acute bystander exposure and risk assessment for soil fumigants, *Atmospheric Environment* **40**: 3548–3560.

Rose, A., and S. Liao, 2005, Modeling regional economic resilience to disasters: a computable general equilibrium analysis of water service disruptions, *Journal of Regional Science* **45**(1): 75–112.

Shafer, G., 1990, Perspectives on the theory and practice of belief functions, *International Journal of Approximate Reasoning* **3**: 1–40.

Slovic, P. 2000, *The Perception of Risk* Earthscan Publications, Ltd. Sterling, VA.

Slovic, P., 2004, *The Perception of Risk*, Earthscan Publications, Ltd, London/Sterling, VA.

Slovic, P., 2010, *The Feeling of Risk: New Perspectives on Risk Perception*, Earthscan, Abingdon, Oxon OX14.

Software Engineering Institute (SEI), 2008, *CERT Resiliency Engineering Framework*, Preview version, v0.95R. Available online: http://www.cert.org/resiliency_engineering/ (Accessed January 8, 2015).

Taleb, N.N., 2007, *The Black Swan: The Impact of the Highly Improbable*, Random House, New York.

The Technical Committee of the Water Resources Research Centers of the Thirteen Western States, 1974, *Water Resources Planning, Social Goals, and Indicators: Methodological Development and Empirical Test*, Utah Water Resources Laboratory, Utah State University, Logan, Utah.

U.S. Water Resources Council, 18 CFR Part 711, Principles and standards for water and related land resources planning-level C, Final Rule, 45, *Federal Register 64366*, (September 28, 1980).

Westrum, R., 2006, A typology of resilience situations, In *Resilience Engineering: Concepts and Precepts*, E. Hollnagel, D.D. Woods, and N. Leveson (Eds.), Ashgate Press, Aldershot, pp. 49–60.

Woods, D.D., 2005, Creating foresight: lessons for resilience from Columbia, In *Organization at the Limit: NASA and the Columbia Disaster*, M. Farjoun, and W.H. Starbuck (Eds.), Wiley-Blackwell, Malden, MA, pp. 289–308.

Woods, D.D., 2006, Essential characteristics of resilience, In *Resilience Engineering: Concepts and Precepts*, E. Hollnagel, D.D. Woods, and N. Leveson (Eds.), Ashgate Press, Aldershot, pp. 21–34.

3

Identifying Risk through Hierarchical Holographic Modeling and Its Derivatives

3.1 HIERARCHICAL ASPECTS

Most organizational as well as technology-based systems are hierarchical in nature, and thus, the risk management of such systems is driven by this hierarchical reality and must be responsive to it. The risks associated with each subsystem within the hierarchical structure contribute to and ultimately determine the risks of the overall system. The distribution of risks within the subsystems often plays a dominant role in the allocation of resources. This is manifested in the quest to achieve a level of risk that is deemed acceptable when the trade-offs among all the costs, benefits, and risks are considered.

Perhaps one of the most valuable and critical contributions of the hierarchical multiobjective framework for risk assessment and management is its ability to facilitate the evaluation of the subsystem risks and their corresponding contribution to the risks of the total system [Haimes and Tarvainen, 1981]. In particular, the ability to model the intricate relationships among the various subsystems and to account for all relevant and important elements of risk and uncertainty renders the modeling process more tractable and the risk assessment process more representative and encompassing. Consider, for example, the problem of maximizing the availability metric of an infrastructure system. A given level of availability can be achieved by many different combinations of reliability and maintainability. Reliability is defined here as the probability that the system is operational in a given time period. The system's reliability can be improved by applying a certain class of preventive maintenance policies. Maintainability is defined here as the probability that a failed system can be restored to an operational state within a specified period of time. A system's downtime may result from either scheduled or emergency shutdowns. The system's reliability or the maintainability of each of its subsystems can be independently improved if there is no budget constraint. In most real-world situations, however, a resource limitation usually acts as the driving force, and trade-offs thus exist between the reliability and the maintainability of the overall system.

Hierarchical control, when applied to risk management systems, has a harmonizing effect on the subsystems and contributes to the holistic approach within which the overall system is viewed.

Risk Modeling, Assessment, and Management, Fourth Edition. Yacov Y. Haimes.
© 2016 John Wiley & Sons, Inc. Published 2016 by John Wiley & Sons, Inc.

For example, fault-tree analysis, which is discussed in Chapter 13, a widely used analytical tool in the nuclear field as well as in others, decomposes the overall reliability problem into several levels of reliability problems. Then, it systematically calculates the failure rate of the overall (top) event from the lower level to the upper level. Studies aiming at developing risk management strategies using decomposition and higher-level coordination are currently underway. When dealing with a low-dimensional multiobjective optimization problem and identifying the impact of the subsystems' reliability on the overall system's performance, a preferred Pareto-optimal solution of a large-scale overall system can be reached by introducing coordination among the subsystems. A similar situation arises in the risk assessment and management of physical infrastructures.

Infrastructures is a general term for man-made engineered systems that include telecommunications, electric power, gas and oil, transportation, water treatment plants, water distribution networks, dams, and levees, including cyber networks. Fundamentally, such systems have a large number of components and subsystems, and therein lies their problem. Most water distribution systems, for example, must be addressed within a framework of large-scale systems, where a hierarchy of institutional and organizational decisionmaking structures (e.g., federal, state, county, and city) is often involved in determining the best replacement or repair strategy. A certain degree of coupling exists among the subsystems (e.g., the overall budget constraint imposed on the overall system), and this further complicates the management of such systems. Different replacement and repair strategies for varying subsystems often have unequal impacts on the overall system; the needs for the resources and their appropriate allocations have diverse impacts on its overall reliability.

Modeling deteriorating water distribution systems and identifying risks are focal issues in large-scale infrastructure problems [Li and Haimes, 1992a, b; Schneiter et al., 1996; ASCE, 2005; Chu and Durango-Cohen, 2007; Durango-Cohen, 2007; FHWA, 2007]. A water distribution system may consist of many subsystems. Consequently, a hierarchical approach to risk modeling, assessment, and management has proven to be an effective measure. In general, the structural nature of multilevel decomposition shows the following advantages:

1. Decomposition methods can reflect the internal hierarchical nature of large-scale multiobjective systems.
2. Trade-off analyses can be performed among subsystems and the overall system.
3. Through decomposition, the complexity of a large-scale multiobjective system can be relaxed by solving several smaller subproblems.

3.2 HIERARCHICAL OVERLAPPING COORDINATION

When modeling large-scale and complex systems, more than one mathematical or conceptual model is likely to emerge; each of these models may focus on a specific aspect of the system, yet all may be regarded as acceptable representations of it. This phenomenon is particularly common in hierarchical multilevel modeling focusing on risk and uncertainty, where more than one decomposition approach may be both feasible and desirable [Macko and Haimes, 1978]. Consequently, decomposing a system often presents a dilemma over the choice of subsystems. For example, an economic system may be decomposed into geographic regions or activity sectors. An electric power management system may be decomposed according to the various functions of the system (e.g., power generation units, energy storage units, transmission units) or along geographic or political boundaries. Another decomposition might be a time-wise decomposition into planning periods. If several aspects of the system are to be dealt with, such as the geographic regions and activity sectors of an economic system, it could be advantageous to consider several decompositions. For example, four major decomposition structures may be identified for water resources systems on the basis of political or geographical, hydrological, temporal, and functional considerations.

This section considers the decomposition and coordination problems of large-scale and complex systems that have more than one hierarchical overlapping structure. The concept and importance of hierarchical overlapping coordination (HOC) are presented through example problems [Haimes et al., 1990a, b, 2007; Yan, 2007; Yan and Haimes, 2008; Yan et al., 2008].

3.2.1 Matrix Organization

To understand HOC as a concept, consider first a very simple example. Figure 3.1 depicts a matrix organization structure of an industrial operation. For illustrative purposes, consider a decomposition of the system into a marketing division and a manufacturing division. Two sectors, which are concerned with Product A and Product B, are assumed to exist in the marketing division. Likewise, three plants, which are located in different areas, are assumed to exist in the manufacturing division. Each of the two product sectors has a manager, and each of the three plants also has a manager. Let us call the decomposition of this structure into a marketing division the *product decomposition* and call the decomposition into a manufacturing division the *plant decomposition*. Clearly, the sectors in the product decomposition overlap those in the plant decomposition. The product managers' decisions also overlap the plant managers' decisions. For example, a decision by the manager of Product Sector A overlaps the decisions of the three plant managers. The hierarchical representation of this overlapping organizational structure is depicted in the two ways shown in Figures 3.2 and 3.3. Product managers are concerned with the individual product—its development, marketing, and sales. Plant managers are concerned with the cost and efficiency of the production system. That is, these two different decompositions deal with different aspects of the system. The databases of these two decompositions differ from each other and receive different information from inside and outside the system. It is valuable to consider these different types of decompositions simultaneously. By considering different hierarchical structures together, we can expect synergistic understanding of the overall system and its corresponding sources of risk and uncertainty. The different geographical locations of the three plants, for example, may impose distinctive production constraints due to local environmental regulations. Subsequently, the manufacturing of Products A and B at the three plants may be subject to different risks of cost overrun, time delay in meeting production schedule, or not meeting performance criteria.

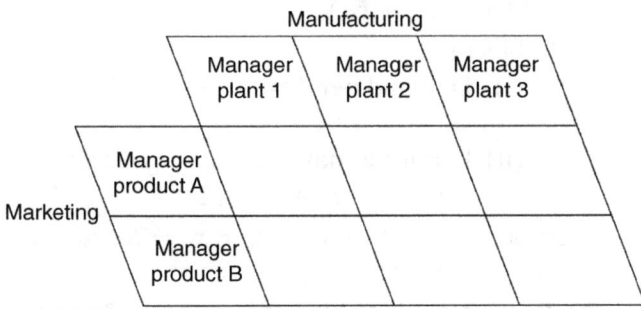

Figure 3.1 Matrix organization of a production system.

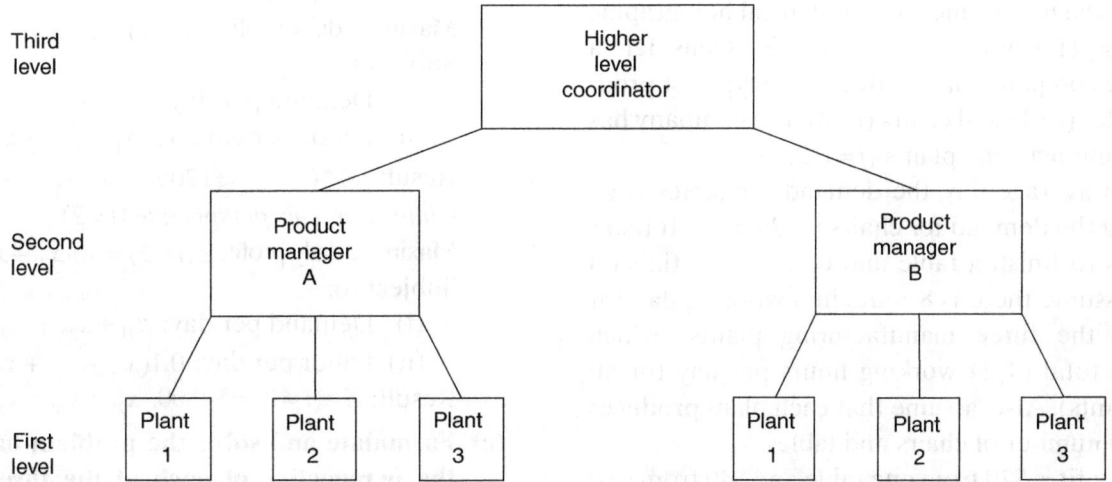

Figure 3.2 Product–plant decomposition.

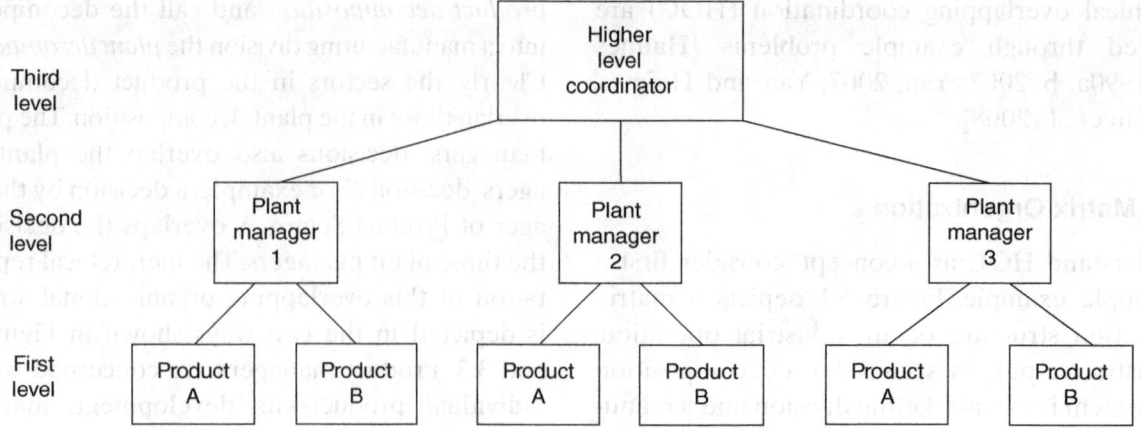

Figure 3.3 Plant–product decomposition.

Suppose that the overall objective of the matrix production system represented in Figure 3.1 is to maximize a given measure of net profit, with each manager cooperating in order to achieve it. Then, a desirable decisionmaking structure would be one in which (i) each individual manager's decisions are feasible in the overall system and (ii) the information exchange between the product managers and the plant managers leads to a sequence of decisions that produce an improved overall benefit that converges to the optimum.

So far, for simplicity, we have been discussing systems with two different hierarchical structures (i.e., decompositions). However, large-scale systems sometimes have more than two.

3.2.2 Example Problem

The following example highlights the value of HOC and thus the importance of hierarchical holographic modeling (HHM) in risk analysis. Consider a furniture company that produces two types of products: tables ($i=1$) and chairs ($i=2$). The company has three manufacturing plants ($j=1,2,3$).

On an average day, the demand for tables is 60 units and the demand for chairs is 120 units. It takes 0.2 hours to finish a table and 0.1 hours to finish a chair. Assume there is 8 hours in a working day for each of the three manufacturing plants (which means a total of 24 working hours per day for all three plants). Also, assume that each plant produces an equal number of chairs and tables.

The profit is $20 from one table and $40 from one chair. The objective is to maximize the daily profit:

(a) **Formulate and solve the problem on a company-wide level.**
Solution: Let x_{ij} be the number of units of product $i=1, 2$ to be produced per day at Plant $j=1,2,3$.
Maximize daily profit: $Z = 20(x_{11}+x_{12}+x_{13}) + 40(x_{21}+x_{22}+x_{23})$
Subject to:
  (i) Demand per day: $x_{11}+x_{12}+x_{13} \leq 60$
      $x_{21}+x_{22}+x_{23} \leq 120$
  (ii) Labor per day: $0.2(x_{11}+x_{12}+x_{13}) + 0.1(x_{21}+x_{22}+x_{23}) \leq 24$
Result: $Z^* = \$6000$; $x_{11}=x_{12}=x_{13}=20$; $x_{21}=x_{22}=x_{23}=40$

(b) **Formulate and solve the problem from the perspective of each of the two product managers.**
Table manager perspective ($i=1$)
Maximize daily profit: $Z(i=1) = 20(x_{11}+x_{12}+x_{13})$
Subject to:
  (i) Demand per day: $x_{11}+x_{12}+x_{13} \leq 60$
  (ii) Labor per day: $0.2(x_{11}+x_{12}+x_{13}) \leq 12$
Result: $Z^*(i=1) = \$1200$; $x_{11}=x_{12}=x_{13}=20$
Chair manager perspective ($i=2$)
Maximize daily profit: $Z(i=2) = 40(x_{21}+x_{22}+x_{23})$
Subject to:
  (i) Demand per day: $x_{21}+x_{22}+x_{23} \leq 120$
  (ii) Labor per day: $0.1(x_{21}+x_{22}+x_{23}) \leq 12$
Result: $Z^*(i=2) = \$4800$; $x_{21}=x_{22}=x_{23}=40$

(c) **Formulate and solve the problem based on the perspective of each of the three plant managers.**

Plant 1 manager (j=1)
Maximize daily profit: $Z(j=1) = 20x_{11} + 40x_{21}$
Subject to:
  (i) Demand per day (assume the demand for the three plants is uniformly distributed): $x_{11} \leq 60/3$; $x_{21} \leq 120/3$
  (ii) Labor per day: $0.2x_{11} + 0.1x_{21} \leq 8$
Result: $Z*(j=1) = \$2000$; $x_{11} = 20$; $x_{21} = 40$

Plant 2 manager (j=2)
Maximize daily profit: $Z(j=2) = 20x_{12} + 40x_{22}$
Subject to:
  (i) Demand per day (assume the demand for the three plants is uniformly distributed): $x_{12} \leq 20$; $x_{22} \leq 40$
  (ii) Labor per day: $0.2x_{12} + 0.1x_{22} \leq 8$
Result: $Z*(j=2) = \$2000$; $x_{12} = 20$; $x_{22} = 40$

Plant 3 manager (j=3)
Maximize daily profit: $Z(j=3) = 20x_{13} + 40x_{23}$
Subject to:
  (i) Demand per day (assume the demand for the three plants is uniformly distributed): $x_{13} \leq 20$; $x_{23} \leq 40$
  (ii) Labor per day: $0.2x_{13} + 0.1x_{23} \leq 8$
Result: $Z*(j=3) = \$2000$; $x_{13} = 20$; $x_{23} = 40$

Summary:

(a) Product decomposition yields a total profit of $6000 ($1200 for product $i=1$ and $4800 from product $i=2$).
(b) Plant decomposition yields a total profit of $6000, equally distributed among all three plants.
(c) Both decompositions also yield the same number of tables and chairs finished at each plant.
(d) Although both decompositions yield the same *optimal* solution, each provides a different perspective to the executives of the furniture company.

3.3 HHM

The fundamental attribute of large-scale systems is their inescapably multifarious nature: hierarchical noncommensurable objectives, multiple decision-makers, multiple transcending aspects, and elements of risk and uncertainty. In part, this may be a natural consequence of the fact that most large-scale systems respond to a variety of needs that are basically noncommensurable and may under some circumstances openly conflict.

It is impractical to represent within a single model all the aspects of a truly large-scale system that may be of interest at any given time (to its management, government regulators, students, or any other group). Our inability to treat the most basic attributes of large-scale systems from some relevant vantage point with some degree of commonality constitutes a remaining weakness in our theoretic modeling base.

HHM [Haimes, 1981], which forms the basis for this chapter, has emerged from a generalization of HOC. It reflects a difference in kind from previous modeling schemas. The name is suggested by holography—the technique of lensless photography. The difference between holography and conventional photography, which captures only two-dimensional planar representations of scenes, is analogous to the differences we see between conventional mathematical modeling techniques (yielding what might be termed *planar* models) and the HHM schema. In the abstract, a mathematical model may be viewed as a one-sided image of the real system that it portrays. For example, with single-model analysis and interpretation, it is quite impossible to identify and document the sources of risk associated not only with the multiple components of an infrastructure (e.g., transportation or hydroelectric power structure or food processing plants) but also with their welter of societal aspects (functional, temporal, geographical, economic, political, legal, environmental, sectoral, institutional, etc.).

Definition
HHM is a holistic philosophy/methodology aimed at capturing and representing the essence of the inherent diverse characteristics and attributes of a system—its multiple aspects, perspectives, facets, views, dimensions, and hierarchies.

Several modeling philosophies and methods have been developed over the years to address the complexity of modeling large-scale systems and to offer various modeling schema. In his book *Methodology for Large-Scale Systems*, Sage [1977] addressed the "need for value systems which are structurally repeatable and capable of articulation across interdisciplinary fields," with which to model the multiple dimensions of societal problems. Blauberg et al. [1977] pointed out that for the understanding and analysis of a large-scale system, the fundamental principles of wholeness (representing the integrity of the system) and hierarchy (representing the internal structure of the system) must be supplemented by the principle of "the multiplicity of description for any system." Recognizing that a system "may be subject to a multiplicity of management, control and design objectives," Zigler [1984] addressed such modeling complexity in his book *Multifaceted Modeling and Discrete Event Simulation*. Zigler (p. 8) introduced the term *multifaceted* "to denote an approach to modeling which recognizes the existence of multiplicities of objectives and models as a fact of life." In his book *Synectics: The Development of Creative Capacity*, Gordon [1968] introduced an approach that uses metaphoric thinking as a means to solve complex problems.

Arthur D. Hall III, whose first book on systems engineering was published in 1962, recognized the contributions of HHM in his seminal book *Metasystems Methodology* [Hall, 1989]: "In this way," he wrote, "history becomes one model needed to give a rounded view of our subject within the philosophy of Hierarchical Holographic Modeling [Haimes, 1981] being used throughout this book, defined as using a family of models at several levels to seek understanding of diverse aspects of a subject, and thus comprehend the whole." Hall developed a theoretical framework, which he termed *metasystems methodology*, with which to capture the multiple dimensions and perspectives of a system. Other early seminal works in this area include the book on societal systems and complexity by Warfield [1976] and the book *Systems Engineering* [Sage, 1992]. For example, in this book, Sage identified several phases of the systems engineering life cycle, and embedded in such analyses are the multiple perspectives—the structural definition, the functional definition, and the purposeful definition. Finally, the multiple volumes of the *Systems and Control Encyclopedia: Theory, Technology, Applications* [Singh, 1987] offer a plethora of theory and methodology on modeling large-scale and complex systems. In this sense, multifaceted modeling, metasystems, HHM, and other contributions in the field of large-scale systems constitute the fundamental philosophy upon which systems engineering and risk analysis are grounded.

3.3.1 HHM: Basic Concepts

In the abstract, a mathematical model may be viewed as a one-sided image of the real system that it portrays. With single-model analysis and interpretation, it is quite impossible to clarify and document the sources of risk associated not only with the multiple components, objectives, and constraints of a system but also with its welter of societal aspects (functional, temporal, geographical, economic, political, legal, environmental, sectoral, institutional, etc.). Given this assumption and the notion that even the integrated models we have cannot adequately cover all of a system's aspects, the concept of HHM constitutes a comprehensive theoretical framework for systems modeling and risk identification.

Central to the mathematical and systems basis of holographic modeling is the overlapping among various holographic models with respect to the objective functions, constraints, decision variables, and input–output relationships of the basic system. In this context, holographic modeling may be viewed as the generalization of HOC in the following way.

As discussed in Section 3.2, in HOC, a system's single model is divided into several decompositions in response to the various aspects of the system, and these decompositions are coordinated to yield an improved solution. Coordinating these dissociated models—that is, reassociating them via holographic modeling methodologies—can be considered a zero-order or degenerate case of holographic modeling in that, while the holographic methodology may be formally applied and even be useful, the models involved are *planar*. That is, the aggregate of all the system's objectives, constraints, and variables, as determined by the various decompositions of HOC, is identical to a system's single model.

The term *holographic* refers to the desire to have a multiview image of a system when identifying vulnerabilities (as opposed to a single view or a flat image of the system). Views of risk can include, but are not limited to, (i) economic, (ii) health, (iii) technical, (iv) political, and (v) social. In addition, risks can be geography related and time related. In order to capture a holographic outcome, the team that performs the analysis must provide a broad array of experience and knowledge.

The term *hierarchical* refers to the desire to understand what can go wrong at many different levels of the system hierarchy. HHM recognizes that for the risk assessment to be complete, one must realize that the macroscopic risks that are understood at the upper management level of an organization are very different from the microscopic risks observed at lower levels. In a particular situation, a microscopic risk can become a critical factor in making things go wrong. In order to carry out a complete HHM analysis, the team that performs it must include people who bring knowledge from up and down the hierarchy.

HHM has turned out to be particularly useful in modeling large-scale, complex, and hierarchical systems, such as defense and civilian infrastructure systems. The multiple visions and perspectives of HHM add strength to risk analysis. It has been extensively and successfully deployed to study risks for government agencies such as the President's Commission on Critical Infrastructure Protection (PCCIP), the FBI, NASA, the Virginia Department of Transportation (VDOT), and the National Ground Intelligence Center (NGIC), among others. (These cases are discussed as examples throughout this book.) The HHM methodology/philosophy is grounded on the premise that in the process of modeling large-scale and complex systems, more than one mathematical or conceptual model is likely to emerge. Each of these models may adopt a specific point of view, yet all may be regarded as acceptable representations of the infrastructure system. Through HHM, multiple models can be developed and coordinated to capture the essence of many dimensions, visions, and perspectives of infrastructure systems. One example is the study conducted for the PCCIP on the US water supply system. Sixteen different visions/perspectives (head topics) with an additional 94 subvisions (subtopics) were identified as sources of risk (see Section 3.10).

Perhaps one of the most valuable and critical aspects of HHM is its ability to facilitate the evaluation of the subsystem risks and their corresponding contributions to the risks in the total system. In the planning, design, or operational mode, the ability to model and quantify the risks contributed by each subsystem markedly facilitates identifying, quantifying, and evaluating risk. In particular, HHM has the ability to model the intricate relationships among the various subsystems and to account for all relevant and important elements of risk and uncertainty. This makes for a more tractable modeling process and results in a more representative and encompassing risk assessment process.

To present a holistic view of the elements that must be included in the model, the HHM approach involves organizing a team of experts with widely varied experience and knowledge bases (technologists, psychologists, political scientists, criminologists, and others). The broader the base of expertise that goes into identifying potential risk scenarios, the more comprehensive is the ensuing HHM. The result of the HHM process is the creation of a very large number of risk scenarios, hierarchically organized into sets and subsets. If done well, the set of scenarios at any level of the hierarchy would approach a *complete set*. The result of the HHM effort is organized into what is called the candidate *scenario model*.

The distinctive attributes of the HHM approach are summarized below:

- It provides a holographic view of a modeled system and thus is capable of identifying most, if not all, major sources of risk and uncertainty.
- It adds both robustness and resilience to modeling by capturing various system aspects and other societal elements.
- It provides more defined responsiveness in modeling development to available data so that different holographic models can make use of different databases.
- It adds more realism to the entire modeling process by recognizing that the limitations of modeling a complex system via a single model are circumvented by a model that addresses specific aspects of the system.
- It provides more responsiveness to the inherent hierarchies of multiple objectives and subobjectives and multiple decisionmakers associated with large-scale and complex systems.

The impact of HHM in the planning phase may be most profound in the way that risks and uncertainties can be integrated into the analysis. From the planning perspective, two major types of risks and uncertainties can be identified. The first type is concerned with the impact of exogenous events on the proposed plan, such as new legislation. The second is concerned with the impact of endogenous events that affect the execution of the plan, such as hardware, software, organizational, or human failures. Since the basic philosophy of HHM is to build a family of models that address different aspects of the system, this is a natural setting in which the impact of both types of risks and uncertainties can be studied in a unified way.

Several applications for HHM for risk identification are presented in subsequent sections.

3.4 HHM AND THE THEORY OF SCENARIO STRUCTURING

3.4.1 Historical Review: The Definition of Risk

In the first issue of *Risk Analysis*, Kaplan and Garrick [1981] set forth the following *set of triplets* definition of risk, R:

$$R = \{<S_i, L_i, X_i>\} \qquad (3.1)$$

where S_i, here, denotes the ith *risk scenario*, L_i denotes the likelihood of that scenario, and X_i denotes the *damage vector* or resulting consequences. This definition has served the field of risk analysis well since then, and much early debate has been thoroughly resolved about how to quantify the L_i and X_i and the meaning of *probability*, *frequency*, and *probability of frequency* in this connection [Kaplan, 1993, 1996].

In Kaplan and Garrick [1981], the S_i themselves were defined, somewhat informally, as answers to the question *What can go wrong?* with the system or process being analyzed.

Subsequently, a subscript c was added to the set of triplets by Kaplan [1991, 1993] (Eq. 3.2):

$$R = \{<S_i, L_i, X_i>\}_c \qquad (3.2)$$

to denote that the set of scenarios, $\{S_i\}$, should be *complete*, meaning it should include "all the possible scenarios, or at least all the important ones."

Also in Kaplan [1991, 1993], the idea of the *success*, or *as-planned*, scenario was introduced and denoted by S_0. The risk scenarios S_i could then be visualized as deviations from S_0. Thus, the idea began to gel that the various risk analysis methods used in different industries (e.g., failure mode and effects analysis (FMEA), fault trees, and event trees) could be viewed as just different systematic ways of identifying and categorizing these deviations, S_i. When these methods became generalized and when the Russian method of anticipatory failure determination (AFD) was added, this idea matured into what we now call the *theory of scenario structuring* (TSS) [Kaplan et al., 1999, 2001].

3.4.2 HHM and the TSS

At about the same time that the definition of risk article [Kaplan and Garrick, 1981] was published, so too was the first article on HHM [Haimes, 1981]. Central to the HHM method is a particular form of diagram, examples of which are shown in figures throughout this chapter (e.g., see Figure 3.6). This form of diagram is particularly useful for the analysis of systems with multiple interacting (perhaps overlapping) subsystems such as a regional transportation or water supply system. The different columns in the diagram reflect different *perspectives* on the overall system.

The HHM methodology recognizes that most organizational as well as technology-based systems are hierarchical in structure, and thus, the risk management of such systems must be driven by and responsive to this structure. The intent is that from this perspective, multiple methods can be compared and thus be better understood. The risk analyst then can be more confident and flexible when choosing, mixing, and designing the method applicable to a specific problem.

HHM can be seen as part of the TSS and vice versa. Under the sweeping generalization of the HHM method, the different methods of scenario structuring can lead to seemingly different sets of scenarios for the same underlying problem. This fact is a bit awkward from the standpoint of the *set of triplets* definition of risk [Kaplan and Garrick, 1981]. To eliminate this awkwardness, we refine this definition of risk to make explicit what was only implicit before: The set of risk scenarios used in a quantitative risk analysis should be (1) complete, (2) finite, and (3) disjoint. These three properties can be achieved by first noting that in realistic problems, there is always an underlying continuum of possible scenarios; we then divide this continuum into a finite set of nonoverlapping subsets. Thus, recognizing that each such subset is itself a scenario, we have our complete, finite, and disjoint set. The mathematical term for this dividing process is *partitioning*.

The HHM approach divides the continuum but does not necessarily partition it. In other words, it allows the set of subsets to be overlapping, that is, nondisjoint. It argues that disjointedness is required only when we are going to quantify the likelihood of the scenarios and, even then, only if we are going to add up these likelihoods (in which case the overlapping areas would end up counted twice.) Thus, if the risk analysis seeks mainly to identify scenarios rather than to quantify their likelihood, the disjointedness requirement can be relaxed somewhat, so that it becomes a preference rather than a necessity.

With this understanding, the risk identification and scenario structuring dimensions of HHM take their place within the TSS as an extremely general scenario identification process, alongside the other well-known but more specific processes: FMEA, hazard and operations analysis (HAZOP), fault and event trees, and AFD.

In seeing how HHM and TSS fit within each other, one key idea is to view the HHM diagram as a depiction of the success scenario S_0. Each box in the diagram may then be viewed as defining a set of actions or results required of the system, as part of the definition of *success*. Conversely then, each box also defines a set of risk scenarios, the set of scenarios in which there is failure to accomplish one or more of the actions or results defined by that box. The union of all these sets of risk scenarios is then *complete* in that it contains all possible risk scenarios.

This completeness is, of course, a very desirable feature. On the other hand, the intersection of two of our risk scenario sets, corresponding to two different HHM boxes, may not be empty. In other words, our scenario sets may not be *disjoint*. This feature of HHM is most valuable for risk-ranking purposes discussed further in Section 3.4.5 and demonstrated in Section 3.8 (also see Figure 3.10).

3.4.3 A Refinement to the Definition of Risk

In Eq. (3.1), the choice of the subscript i, on the S_i, carries with it, by conventional usage, the implicit assumption that the set of scenarios is denumerable (i.e., countable). Moreover, because Eq. (3.1) is intended to describe the result of an actual risk analysis, there is the further implicit assumption that the number of scenarios in the set $\{S_i\}$ is finite. We wish now to release both these assumptions and therefore revise Eq. (3.2) to read

$$R = \{<S_\alpha, L_\alpha, X_\alpha>\}, \quad \alpha \in A \quad (3.3)$$

where the index α now ranges over a set A, which in general is nondenumerable. The set A is therefore infinite and nondenumerable. It has the same order of infinity as the real number continuum.

From the perspective of this framework, we can now view the TSS as a study of the various techniques for achieving such a partitioning. Having defined the success scenario S_0, the process of finding the risk scenarios, S_i, consists of decomposing S_0 into *parts* or *components*. Then, putting our magnifying glass over each part in turn, we ask, *What could go wrong in this part?* In this way, we generate the S_i.

Now, we can connect Eqs. (3.2) and (3.3) by recalling the principle that every scenario, S_i, that we can describe with a finite number of words is itself a set of scenarios [Kaplan, 1991, 1993]. Thus, each S_i in Eq. (3.2) can be visualized as a subset of S_A. For practical purposes, we want the set of scenarios in our risk analysis, $\{S_i\}$, to be

1. Complete, in the sense that $U(S_i) = S_A$ where U is the set operation *union*
2. Finite
3. Disjoint, meaning that $S_i \cap S_j = \emptyset$ for all $i \neq j$, where $\cap$ is the set operation *intersection*

Such a set of subsets of S_A is termed a *partitioning*, P, of S_A. Thus, we arrive at the point of view that what we want to do in a risk analysis is to identify a partitioning of the underlying risk space S_A. The individual sets in this partitioning are the scenarios S_i, which are finite in number, disjoint, and together *cover* the underlying space S_A. We may then write

$$R_P = \{<S_i, L_i, X_i>\}_P \quad (3.4)$$

R_P is thus an approximation to R based on the partition P:

$$R_P \approx R \qquad (3.5)$$

3.4.4 Comments on the Refined Definition

Now, we observe that if S_0 is itself decomposed into a complete, finite, and disjoint set of parts, then simply defining S_i as *something goes wrong with part i* generates a complete, finite, and disjoint set of S_i. Strictly speaking, this statement holds true only insofar as *single-failure* scenarios are concerned. For true completeness, we have to add scenarios of the form *something goes wrong with parts i and j* and so forth. Pushing this idea further, if we have identified a complete, finite, and disjoint subset of risk scenarios originating in each part of S_0, then the aggregate, that is, the union of those subsets, is a complete, finite, and disjoint set of S_i for the entire problem (subject again, however, to the multiple failure comment in the preceding text).

3.4.5 The HHM Approach to Decomposing S_0

The HHM diagram may now be viewed as a portrayal of the success scenario S_0 and a decomposition of that scenario into its various parts and pieces. The decomposition strives to be complete but not necessarily disjoint. Indeed, HHM regards nondisjointness, or *overlapping* of the decomposed parts and pieces, as a useful feature, reflecting different *perspectives* on the system. Thus, HHM recognizes that most organizational as well as technology-based systems are not only hierarchical in structure but are *multiply hierarchical*, in that different overlapping hierarchical structures can be identified within the system. The risk management of such systems must then be driven by, and responsive to, this structure.

One of the valuable contributions of the HHM framework for risk assessment and management is its ability to identify risk scenarios that result from and propagate through the multiple overlapping hierarchies in real-life systems. In the planning, design, or operational mode, the ability to model and quantify the risks contributed by each subsystem markedly facilitates understanding, quantifying, and evaluating the risk from the whole system. In particular, the ability to model the intricate relationships among the various subsystems and to account for all relevant and important elements of risk and uncertainty renders the modeling process more tractable and the risk assessment process more representative and encompassing.

3.4.6 Summary

Within the subject of risk analysis, the evolving TSS and HHM aspire to be a comprehensive treatment of the process of finding, organizing, and categorizing the set of risk scenarios. As such, it should include within itself the well-known standard methods of scenario identification such as fault trees, FMEA, and failure mode, effects, and criticality analysis (FMECA) (see Chapter 13).

Along the way to showing this inclusiveness, attention is drawn to the fact that the set of risk scenarios, S_i, developed by the different methods for the same problem could well be different. This can be a bit awkward conceptually. Accordingly, Kaplan et al. [2001] found it desirable to back up and refine the original *set of triplets* definition of risk so that it did not assume or imply, as part of the definition itself, that the set of risk scenarios is finite or denumerable. Rather, this refined definition allows the set of risk scenarios to be a continuum, that is, nondenumerable. This continuous set of scenarios constitutes the *true* risk and is independent of which method is used to identify them.

For practical, computational purposes, this *true* scenario set is then partitioned into a finite, disjoint, and complete set of subsets. That is what the various risk scenario identification methods accomplish. Each such subset then *is* a risk scenario S_i, which then makes it perfectly acceptable for the different methods to arrive at different partitionings. Moreover, if the scenarios are not going to be quantified, it is also acceptable if they are not disjoint.

Thus, this refinement takes the finite set of S_i out of the definition of risk and casts it more properly as an approximation to the true, underlying, nondenumerable set of risk scenarios. Different sets of S_i, arrived at by different methods, are thus seen as just different approximations to the same underlying truth. This is a much more satisfactory viewpoint conceptually. Practically, it also suggests that the risk analyst would do well to apply more than one of the methods to a specific problem to gain more insight into and more confidence that all the important scenarios have been brought to light.

Collaborative techniques for developing HHMs for identifying threat scenarios have been a recent research development. Haimes and Horowitz [2004] discuss the adaptive two-player HHM game, a repeatable, adaptive, and systemic process for tracking terrorism scenarios, which creates opposing views of terrorism: those defending against acts of terrorism (Blue Team) and those planning terrorist acts (Red Team). This work was extended to account for multiple experts with the collaborative adaptive multiplayer HHM (CAM-HHM) [Agrawal, 2006]. Addressing the HHM building process from multiple perspectives adds richness to the resulting model.

3.5 ADAPTIVE MULTIPLAYER HHM GAME

This section introduces the adaptive multiplayer HHM (AMP-HHM) game, a new concept with the potential to serve as a repeatable, adaptive, and systemic process that can contribute to tracking terrorism scenarios [Haimes and Horowitz, 2004]. It builds on fundamental principles of systems engineering, systems modeling, and risk analysis. The AMP-HHM game captures multiple perspectives of a system through computer-based interactions. For example, for a two-player game, it creates two opposing views of the opportunities for carrying out acts of terrorism: one developed by a Blue Team defending against terrorism and the other by a Red Team planning to carry out a terrorist act. The HHM process, historically applied to system risk analysis, identifies the vulnerabilities of potential targets that could be exploited in attack plans. These vulnerabilities, separately identified by the Blue and Red Teams, can be used collectively to identify corresponding surveillance capabilities that can help to warn of a possible attack. Vulnerability-based scenario structuring, comprehensive risk identification, and the identification of surveillance capabilities that can support preemption are all achieved through the deployment of HHM.

State variables, which represent the essence of a system, play a pivotal role in the AMP-HHM game, providing an enabling road map to intelligence analysts. Indeed, vulnerabilities are defined in terms of the system's state variables: *Vulnerability* is the manifestation of the inherent states of a system (e.g., physical, technical, organizational, cultural) that can be exploited by an adversary to cause harm or damage. *Threat* is a potential adversarial intent to cause harm or damage by adversely changing the states of the system. Threat to a vulnerable system with adverse effects may lead to *risk*, which is a measure of the probability and severity of adverse effects.

The AMP-HHM game provides a methodology for intelligence collection and analysis. (The relationship between this game and classical game theory as introduced by von Neumann and Morgenstern [1972] and extended by others, e.g., Kuhn [1997], is discussed subsequently.) For pedagogical purposes, the discussion initially will be focused on intelligence analysis of terrorism.

The analysts are divided into two teams: *offense* (*Red*) and *defense* (*Blue*). The objectives of each player team are as follows:

1. For the *Blue Team—homeland defenders:* Develop a comprehensive HHM of its own system as a way of evaluating its vulnerabilities and the opportunities for adversaries to exploit such vulnerabilities. The results will be used to develop a set of surveillance efforts that could provide attack warning and assessment information to support attack preemption efforts. This team has access to all available information about the system it is defending and a set of risk specifications to consider in their analysis (e.g., level of protection against financial loss).

2. For the *Red Team—terrorist networks:* Develop a comprehensive HHM of the defender's system by collecting intelligence on potential targets and focusing on the opponent's vulnerabilities and strengths, that is, their *state variables*. This would be used as a basis for selecting possible attack scenarios.

It is imperative that two independent HHMs be developed—one from the homeland perspective and one from the terrorist perspective. Note that having the defensive Blue Team consider the opponent's HHM perspectives results in (i) the union of both, thus yielding a more complete HHM, and (ii) valuable benchmark information on the depth and breadth of the assessment. Additional benefits are greater self-understanding and knowledge of the opponent. To maximize the effectiveness of the Red Team's HHM, the inputs should represent the state variables of actual terrorist networks. These are culture, funding, sophistication, technology level, doctrinal orientation, and social levels, among others [Arquilla and Ronfeldt, 2001]. Comparing and analyzing both Red and Blue Team outputs add an important dimension to the risk filtering and management process. Clearly, the defense (Blue Team) can temper the conclusions drawn from its own HHM by relating them to the Red Team's HHM. Where they overlap, the likelihoods of an attack are higher. Where they do not, there may be a need to add elements to the Blue Team's HHM, which is easily adaptable.

In classical game theory [von Neumann and Morgenstern, 1972], the actions of the players and their consequences as well as the anticipated or perceived reactions and countermeasures are explicit in the ensuing game. The AMP-HHM game is based not only on the actions of the players and their consequences but also on an explicit understanding of the inherent characteristics of the players that necessarily lead to the observed actions and consequences. For example, the strategies and actions of the homeland Blue Team in the AMP-HHM game respond to the states of their own system as well as to those of the terrorist Red Team. Intelligence analyses for countering terrorism will be far more effective if they are driven not only by the symptoms (i.e., the actions of the terrorist networks) but also by the root causes (i.e., the states that characterize the terrorist networks). To this end, the AMP-HHM game also offers a road map for scenario tracking that accounts for the characteristics of both the root causes and the target (see Haimes, 2002; Horowitz and Haimes, 2003; Haimes et al., 2007).

While we have emphasized the two-player HHM concept, it is clear that successive games can be played involving many Red and Blue Teams. Two questions need to be addressed when conducting multiple games. First, how many game iterations involving the same situations are needed to achieve a comprehensive and relatively stable set of intelligence collection observables? The answer is that measures of convergence can potentially be developed based on the use of Bayesian and decision-tree analyses. Thus, when the observables and their corresponding probabilistic results converge using the decision trees that emerge from successive HHM analyses, the utility of the new changes to the stable HHM models have little, if any, value. Experiments involving Blue and Red Teams and using measures of convergence can establish the characteristics of the HHM convergence.

The second question is: How do results vary as the basic characteristics of the teams' players are varied? To address this, both teams need to possess a variety of skills, experience, and interests. Results can be compared, again using Bayesian and decision-tree analyses to determine the importance of the variations (see Monahan, 2000; Slovic, 2000; Monahan et al., 2001). Ultimately, the choice of Red and Blue Team participants is critical for the intelligence community. The AMP-HHM game is introduced and discussed in more detail in Chapter 17.

3.6 WATER RESOURCES SYSTEM

The Maumee River Basin (the largest subbasin of the Great Lakes Basin) spans an area of approximately 8000 sq. miles over parts of the states of Ohio, Michigan, and Indiana [Haimes, 1977]. It has been divided into five planning subareas (PSAs), each one consisting of several counties (political–geographic decomposition) as shown in Figure 3.4. The basin can also be divided into eight watersheds crossing state and county boundaries (hydrological decomposition), as shown in Figure 3.5 [Haimes and Macko, 1973]. Seven major objectives identified by the basin's Citizens' Advisory Committee have been considered in the planning process (functional decomposition). These objectives are to (i) protect agricultural land, (ii) reduce erosion and sedimentation, (iii) enhance water quality, (iv) protect fish and wildlife, (v) enhance outdoor recreational opportunities, (vi) reduce flood damage, and (vii) supply water needs. Finally, the planning time horizon

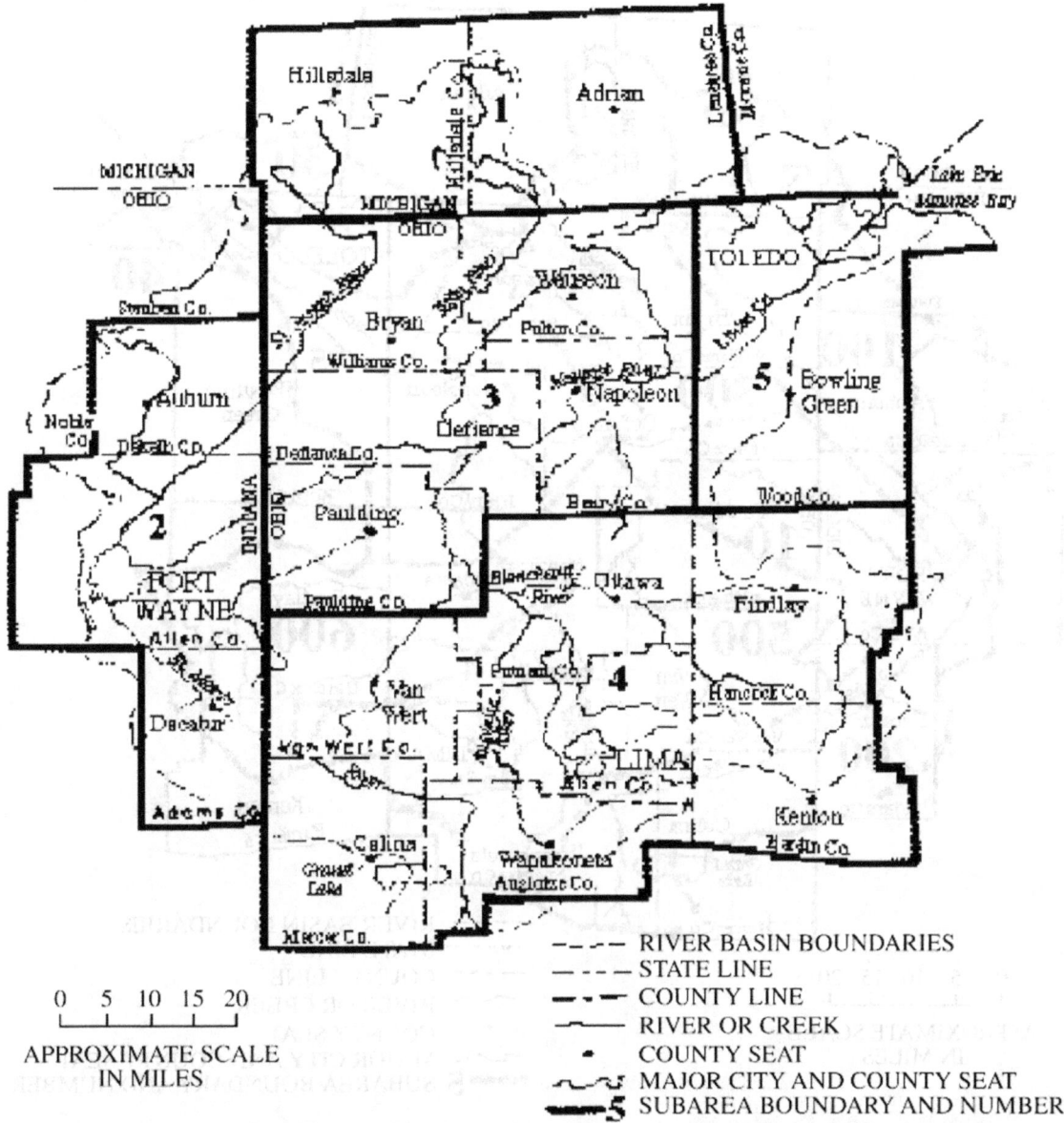

Figure 3.4 Political–geographic decomposition of the Maumee River Basin.

spans the years 1990, 2000, and 2020 (temporal decomposition).

The Maumee River Basin Planning Board, which is responsible for generating a recommended plan to the entire basin, must be responsive to the desires and needs of various groups; local, state, and federal agencies; and the environment. The board consists of seven members chaired by a study manager from the Great Lakes Basin Commission (GLBC). These members represent

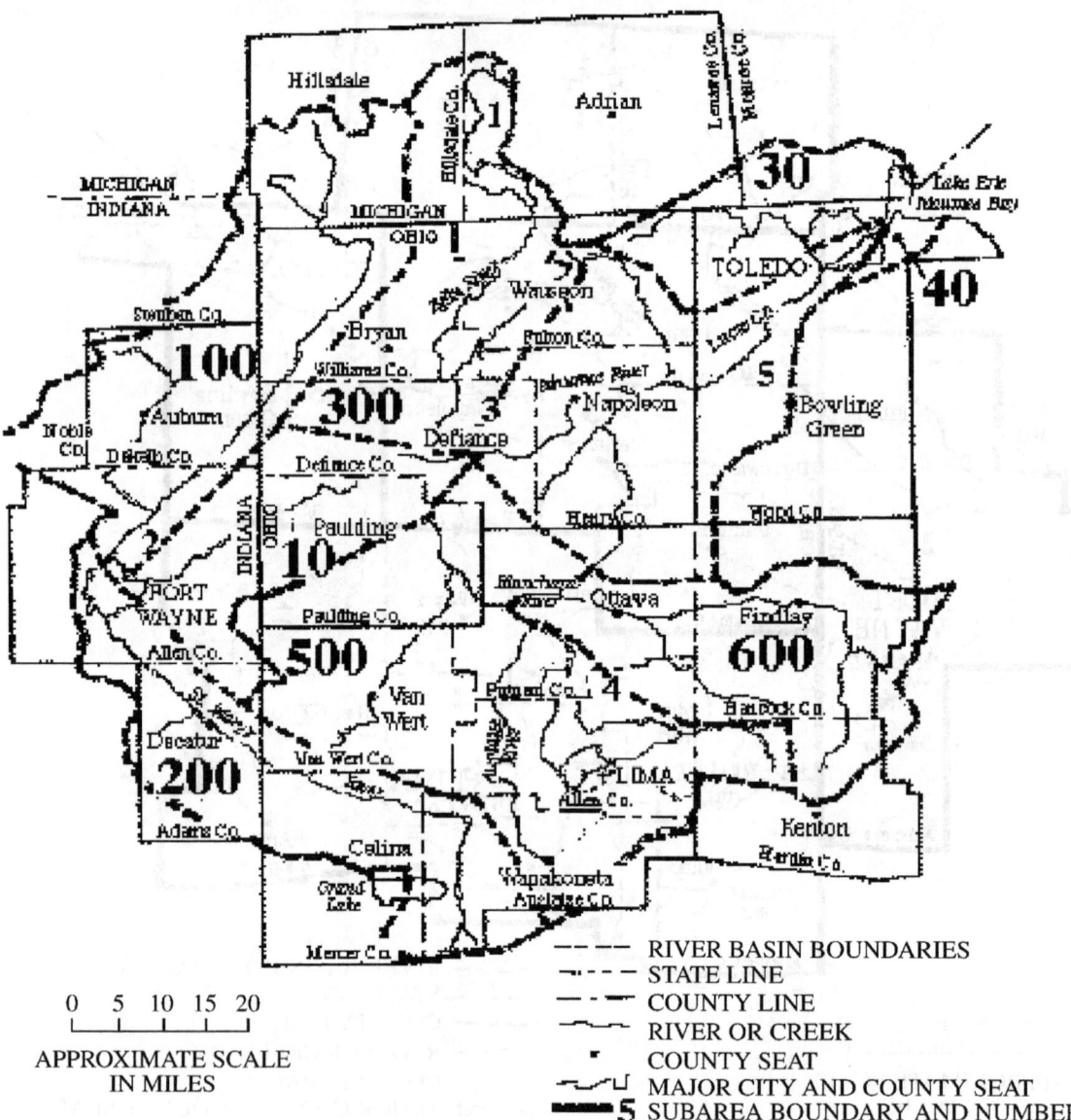

Figure 3.5 Hydrological decomposition of the Maumee River Basin.

the US Army Corps of Engineers, the Bureau of Reclamation (US Department of the Interior), the Soil Conservation Service (US Department of Agriculture), the US Environmental Protection Agency, and the states of Ohio, Michigan, and Indiana.

The planning board has one common objective: to generate for the entire basin a recommended plan that is responsive to the aforementioned seven objectives over the planning time horizon. It is evident, however, that in the planning process, each member views the planning problem differently based on the various agency responsibilities, the experience of its professional staff, the political configuration associated with tristate agencies, the information available (various types of data), and so on. A more detailed discussion of the basin's planning process and the problems and issues associated with the interagency coordination mechanism can be found in Haimes [1977] and in Haimes et al. [1979]. Each decomposition represents and uncovers important aspects not available through the other. The availability and credibility of the databases are particularly critical. Manipulating the databases to serve and suit demands and constraints that are artificially imposed through the modeling process necessitates compromise, and compromise can lead to an ultimate deterioration in model credibility. For example, data concerning streamflow, water quality, and floods are available on a hydrological basis and are collected by the US Geological Survey, the US Environmental Protection Agency, and the US Army Corps of Engineers, respectively. Data concerning population dynamics, employment, and other economic activities are available on political–geographic bases and are collected by agencies such as the US Departments of Commerce, Labor, and Treasury. HHM enables the utmost utilization of these databases with minimum manipulation or misuse. This can be achieved by resorting to two simultaneous decompositions—hydrological and political–geographic—each of which might have a number of subsystems. In general, water resources systems (as well as many other large-scale systems) lend themselves to more than one decomposition or description.

For instance, one could have a functional decomposition (the water supply and demand of various sectors—agriculture, industry, municipality, aquatic life, etc.) and a temporal decomposition (long, intermediate, and short term) as well as the hydrological and political–geographic decompositions already mentioned [Haimes et al., 1990a, b]. The HHM should facilitate coordination because each agency naturally tends to develop its own mission-oriented model using the most appropriate description or decomposition (hydrological, geographical, etc.).

Obviously, because of the multifarious aspects and needs of the basin, more than one hierarchical modeling structure may evolve. Furthermore, many possible permutations exist among the four different decompositions:

1. Five PSAs (political–geographical decompositions)
2. Eight watersheds (hydrological decomposition)
3. Seven objectives (functional decomposition)
4. Three planning periods (temporal decomposition)

3.7 SUSTAINABLE DEVELOPMENT

A worldwide environmental awakening is gathering force to save the Earth from harmful human actions that have resulted in irresponsible exploitation of our natural resources; pollution of air, water, and soil; disturbance of the delicate ecological balance in many places; catastrophic deforestation; destruction of the ozone layer; acid rain damage to freshwater lakes; and overall degradation of the environment. Mismanagement and shortsightedness are byproducts of a failure to understand the dire consequences of uncontrolled economic development; we are being forced to face what happens when little or no effort is made to consider how present policies and decisions affect the options open to future generations [Haimes, 1992].

Most people credit the term *sustainable development* to *Our Common Future*, a report by the World [Bruntland] Commission on Environment and Development (WCED) [WCED, 1987]. The WCED defines sustainable development as "development that meets the needs of the present without compromising the ability of future generations to meet their own needs."

Probably, the dominant explanation for why the holistic–systemic approach to solving worldwide environmental problems has not been adopted (or even aspired to) has been the lack, until recently, of an appropriate institutional infrastructure whose leadership has sufficient credentials in the scientific community and enough practical experience in public policy to enjoy the confidence of the political decisionmaking leadership. Although eliminating this lack is a necessary condition for the success of a holistic–systemic approach to sustainable development, also required is compliance with other operational principles. One such critical operational principle is the adherence to the process of risk assessment and management. Certainly, the trend is in the direction of a more mature, sober, and courageous approach to the spirit of sustainable development.

A systems analysis interpretation of the sustainable development paradigm necessarily leads to a vision that incorporates the following five essential operational principles of a holistic approach to economic and environmental planning, development, and management:

1. Multiobjective analysis
2. Risk analysis, including risk of extreme events
3. Impact analysis
4. The consideration of multiple decisionmakers and constituencies (e.g., regions, sectors, socioeconomic, and political subdivisions)
5. Accounting for interaction among a system's components and between the system and its environment

These five operating principles are widely addressed throughout this book in a variety of contexts.

Because of the numerous sources and causes of failure in the realization of sustainable development plans for water and related land resources, there is a need for a holistic and comprehensive analytical framework capable of identifying these myriad sources of risks. A holistic visionary quest for sustainable development can be found in the National Environmental Policy Act (NEPA) of 1969 [NEPA, 1969]. In effect, NEPA identified some major sources of risk that might stand in the way of achieving what is known today as a sustainable future:

> The Congress, recognizing the profound impact of man's activity on the interrelations of all components of the natural environment, particularly the profound influences of population growth, high-density urbanization, industrial expansion, resource exploitation, and new and expanding technological advances, and recognizing further the critical importance of restoring and maintaining environmental quality to the overall welfare and development of man, declares that it is the continuing policy of the federal government, in cooperation with state and local governments, and other concerned public and private organizations, to use all practicable means and measures, including financial and technical assistance, in a manner calculated to foster and promote the general welfare, to create and maintain conditions under which man and nature can exist in productive harmony, and fulfill the social, economic, and other requirements of present and future generations of Americans.

To capture the multivision perspectives of the multitude of sources of risks, an HHM framework is developed here. Seven decompositions, visions, considerations, or perspectives, with obvious and unavoidable overlapping among them, are introduced in Figure 3.6. These are:

1. Science and engineering (hydrological, ecological, and technological perspectives)
2. Global and geographical (international, regional, national, and local sociopolitical perspectives)
3. Institutional and organizational (governmental and nongovernmental agencies and institutions)
4. Cultural and socioeconomic (ethnicity, tradition, education, standard of living, justice, and equity)
5. Natural needs (water, land, air, forestry, food, and ecology)
6. Temporal (short, intermediate, and long term)
7. Freedom (freedom of information, religion, speech, and assembly)

Central to this HHM framework is the ability to branch out from each of the seven decompositions or considerations and explore the connectedness and ramifications within all other seven perspectives. Figures 3.7 and 3.8 present two examples of such variations in the hierarchical representation of the sources of risk. The science and engineering vision is discussed here as an example of how each of the seven visions is decomposed.

3.7.1 Science and Engineering

Science and engineering have not always served to protect the environment and the ecosystem. Indeed, in the past, technology has often been detrimental to the cause of a sustainable future. The frequent

SUSTAINABLE DEVELOPMENT   85

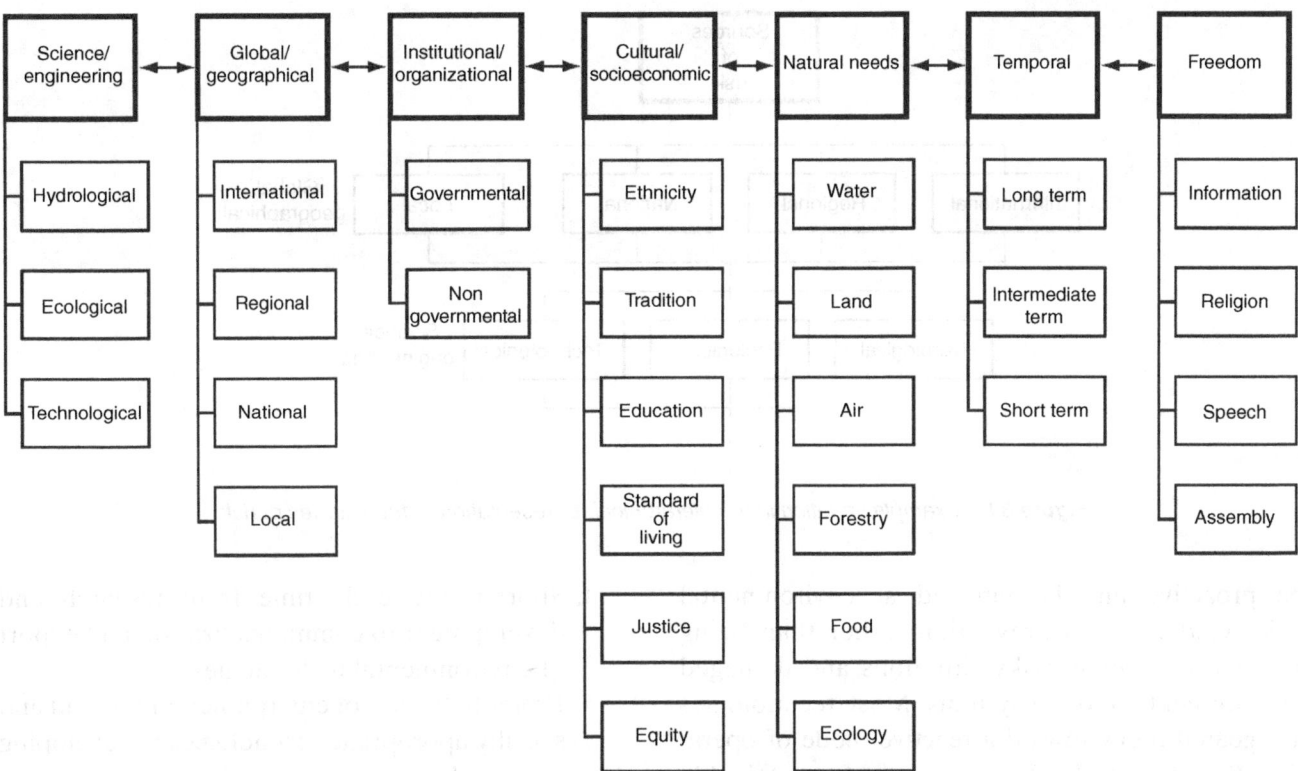

Figure 3.6 HHM framework for risk identification.

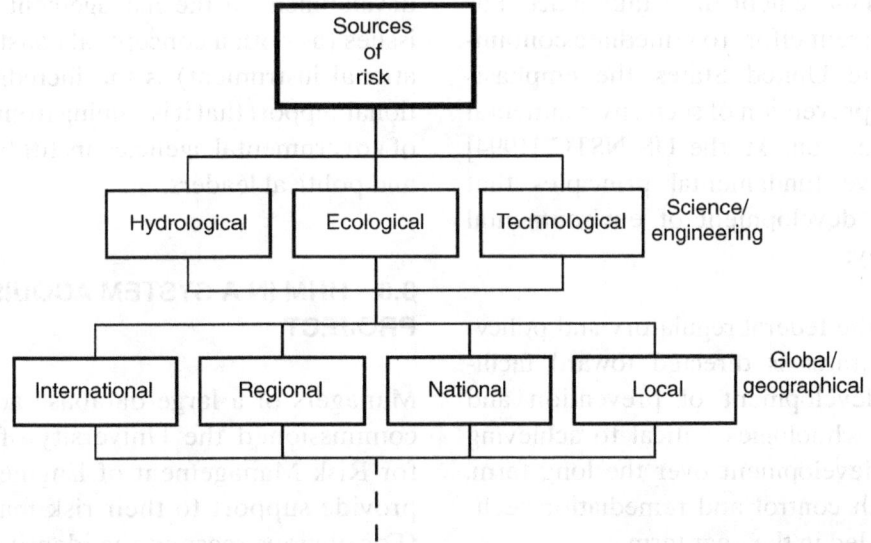

Figure 3.7 Example variation in the hierarchical representation of the sources of risk.

practice of uncontrolled large-scale cultivation and irrigation of arid lands has resulted in increased soil erosion and soil salinity. At the same time, when appropriately channeled and controlled, technology has and will continue to serve as a powerful engine toward a sustainable future. Clearly, the same know-how that has in the past exploited the Earth's natural resources without much concern for future sustainability can be a potent instrument for ensuring the future protection and viability of our natural resources, ecosystems, and economic growth [Haimes, 1992]. In particular, science and engineering should

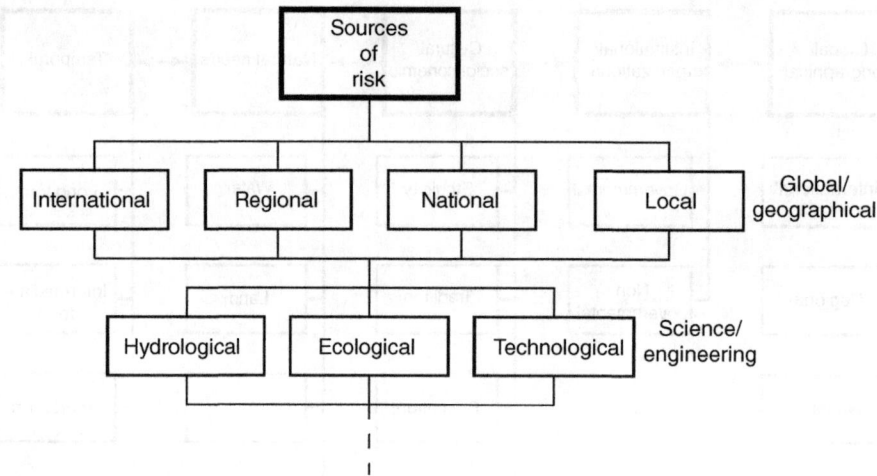

Figure 3.8 Example variation in the hierarchical representation of the sources of risk.

be proactive and be targeted at environmental risk avoidance and prevention rather than being reactive to already risky situations and damaged environments and ecosystems. Most technologies are geared today toward a reactive mode of operation. To harness technology's potential for sustainable development, however, a cultural and attitudinal paradigm shift from reactive to proactive risk assessment and management must take place. For example, in the current effort to remediate contaminated sites in the United States, the emphasis should shift to the prevention of such environmental degradation. In this context, the US NSTC [1994] has developed five fundamental principles that should guide the development of environmental technology strategy:

1. Ensure that the federal regulatory and policy-making apparatus is directed toward facilitating the development of prevention and monitoring technologies critical to achieving sustainable development over the long term, balanced with control and remediation technologies needed in the near term
2. Increase the resource efficiencies of our technological infrastructure by adopting a systems approach that employs the tenets of industrial ecology
3. Forge public–private and federal–state partnerships directed toward advancing the development, commercialization, and diffusion of environmental technologies
4. Shorten the cycle time from research and development to commercialization and export of environmental technologies
5. Promote the use of environmentally sound and socially appropriate technologies in developing nations throughout the world

In sum, what is most encouraging about sustainable development in the management of environmental issues (as both a conceptual construct and an operational instrument) is the incredibly wide international support that it is gaining from a broad spectrum of governmental agencies, institutions, and scientific and political leaders.

3.8 HHM IN A SYSTEM ACQUISITION PROJECT

Managers of a large database acquisition project commissioned the University of Virginia Center for Risk Management of Engineering Systems to provide support to their risk management effort. (For obvious reasons, the identity of the organization is kept anonymous.) The complexity of the project involved advanced hardware and software, translation of a massive database, personnel from many organizational units, transitional program phases spanning more than 5 years in implementation, and over $1.5 billion in investment. The following is a simplified and modified description of this effort.

In the earliest stage, system managers and the analysts needed to identify common program risks. Later, it was important for program managers to agree on priorities to reduce the likelihood of the program's failing to meet its schedule, cost, and performance objectives. A ranking methodology was suggested to improve the allocation of limited resources for risk mitigation. Finally, it was necessary to generate and compare alternative policies for risk management.

The analysts conducted numerous interviews with program managers and technical experts at the work site. Many oral discussions and reviews of internal documents were essential to the processes of risk identification, prioritization, and mitigation.

Information was collected by two-person teams of analysts from five major sources:

1. Interviews at the work site with approximately 20 managers
2. Reviews of the requirements documents and other program planning materials
3. Reviews of the third-party analyses of the costs and schedules for the project
4. Review of a list of risks prepared by program managers
5. Consultation with a third-party management consultant familiar with the program

Figure 3.9 depicts the multiple views of the risk identification problem for this system using the HHM approach. It consists of eight major perspectives (head topics): (1) *program consequence* (technical, cost, schedule, and the user/community), (2) *management of change* (personal trustworthiness, interpersonal trust, managerial empowerment, and institutional alignment), (3) *system acquisition* (contractor, contract management, requests for proposals and contracts, and system integration), (4) *temporal* (design and planning, transition, steady state, and system expansion), (5) *modal* (external, hardware, software, organizational, and human), (6) *information management* (process control, information storage and retrieval, information transmission, and data analysis), (7) *functional* (subsystems U, V, W, X, Y, and Z), and (8) *geographical* (primary site, secondary site, Region P, Region Q, and Region R).

The strategy for risk identification revolves around the multiple decompositions, or visions, of the HHM. After each main-level vision is introduced, a more detailed and comprehensive discussion of the entire risk assessment structure is begun. In an interview with an expert to identify new sources of risk to the large-scale technological system, an initial subset of two or more of the hierarchy's decompositions is used to formalize and structure the risk identification process. Later, inclusion of additional decompositions provides increased detail and focus to the risk identification process.

For example, one vision or decomposition of the risk associated with the database system is the functional perspective, focusing on the various services that the system will provide. From a functional view, the database system in this case was decomposed into six major subsystems. These functional areas were then evaluated for sources of risk by cross-reference to other decompositions. Another vision of the HHM relates to the acquisition process over time. Each of the overlapping stages of the system acquisition, although not sharply distinguishable, constitutes a subsystem in a temporal decomposition. Design and planning, for example, can be viewed as one frame in a fixed time in the acquisition process. For this fixed time frame, risks associated with the modal and functional decompositions are identified and articulated. The temporal domain has significance beyond the project's schedule; it articulates the change and evolution of risks over time.

The results of the identification process, which were consolidated in a master list of over 250 sources of risk, ranged in nature from technology issues through specifications documents and schedule inconsistencies to personnel and managerial leadership. There was considerable redundancy among items in the master list, which indicated the connectedness of the various levels and differing perspectives in the system. Thus, the master list gave an unfiltered impression of the perceived importance of a great number of risks to the system.

Next, each of the 250 identified sources of risk can be associated with the three most relevant HHM subtopics (areas of impact or domains designated in Figure 3.9 by the boxes under the head topics). For example, a risk item from the master list might have

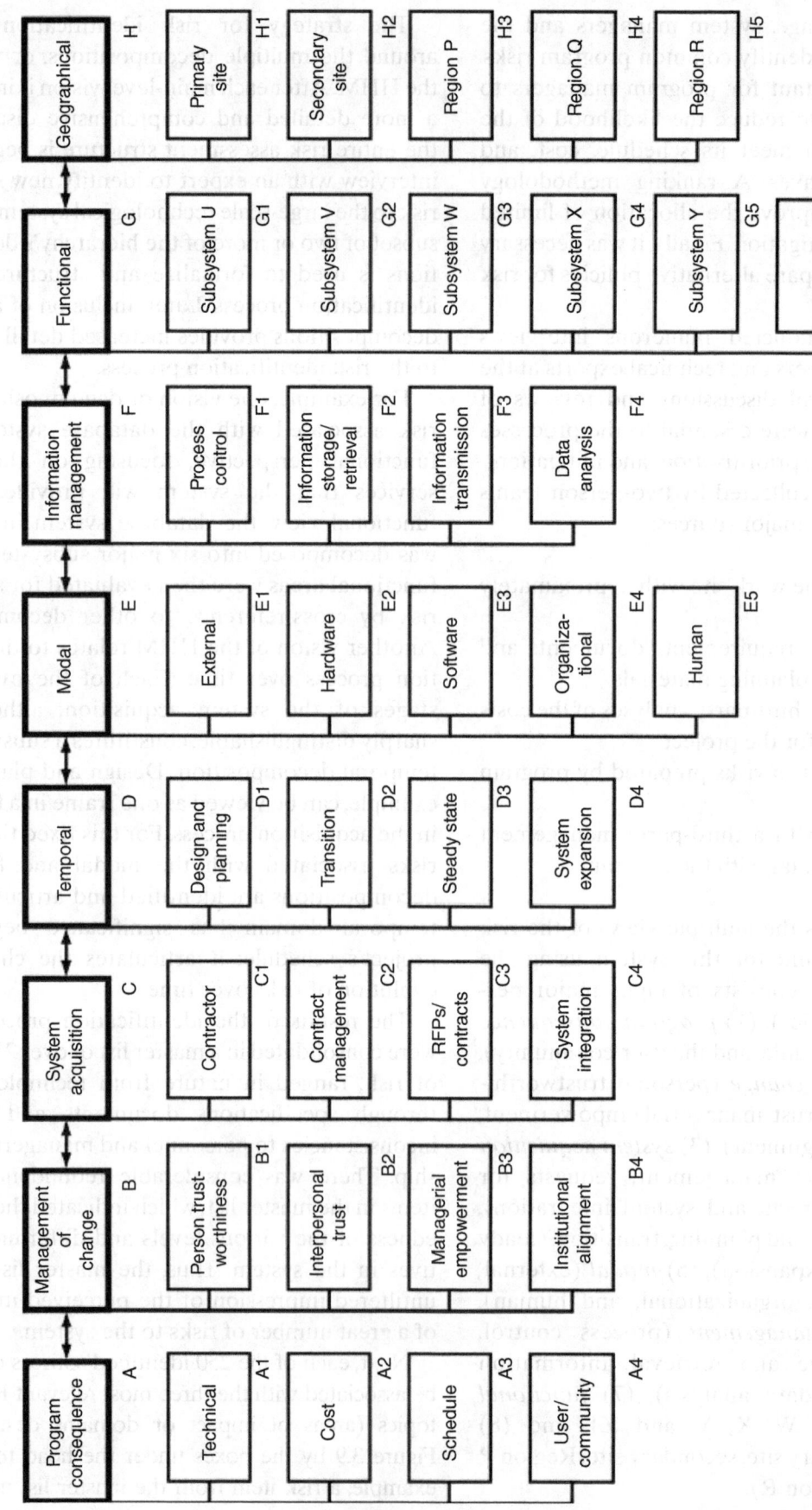

Figure 3.9 HHM framework for identification of sources of risk.

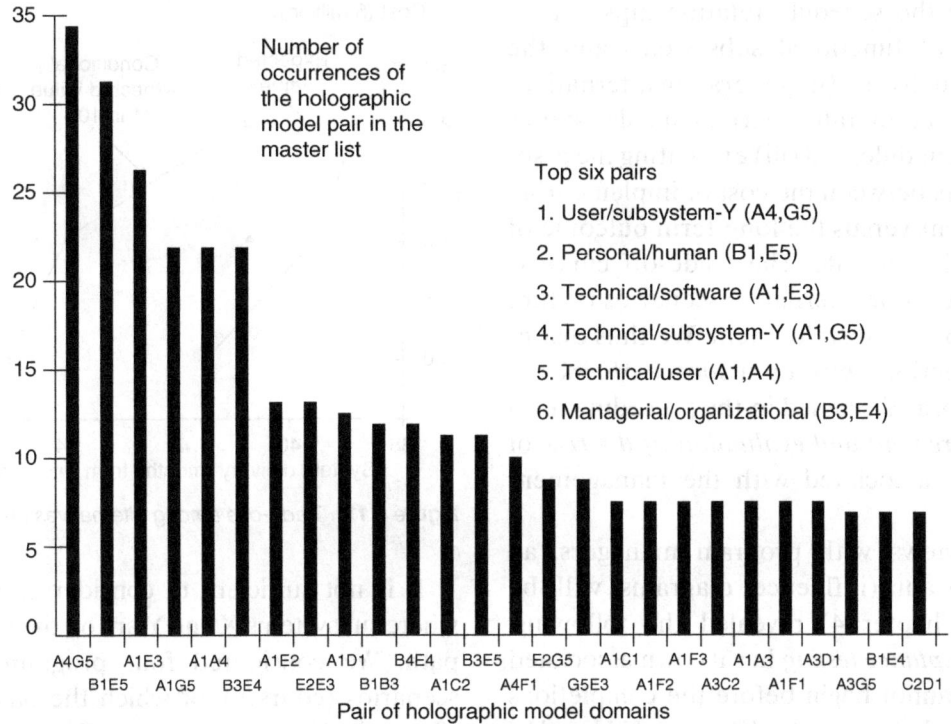

Figure 3.10 Number of matches of pairs of holographic domains with risks identified.

been a schedule risk (*program consequence*), contract management (*system acquisition*), or a *primary site* risk (*geographical*).

Counting the master list matches associated with each holographic model domain, we found that the *program consequence* and *modal* decompositions corresponded most with the master list and that the domain for *technical* risks was the greatest concern overall. In addition, we counted the domain pairs of matches. For example, *user* domains and *subsystem Y* are the pair that occurs most often on the master list. A frequently occurring pair, or intersection of two domains, is perceived to have a relatively high importance. That the intersection of *user and subsystem Y* is the first-ranked pair reflects a prevailing perception that the future uses of subsystem Y services is an important consideration for risk management (see Figure 3.10).

3.8.1 Ranking of Risks

In work with the program managers, a hierarchy of criteria was developed that would be used to prioritize the risk items in terms of their likelihoods of occurrence, the potential consequences to the program, and the efficacy and the immediacy of risk reduction efforts. Risks identified in the master list using HHM were grouped into categories of related items, reducing more than 250 items to approximately 20 broad issues.

Seven attributes developed with managers for ranking sources of risk to the system are expected impact and catastrophic impact (for program risks); service delay, error/failure, and quality degradation (for user risks); and action horizon and efficacy (for risk mitigation). The attributes and measurement scales (definitions of high, moderate, and low) were developed in consultation with the program managers.

3.8.2 Evaluating Risk Management Alternatives

Generating and evaluating risk management options is considered next. The following example considers alternatives for accelerating the system development schedule. It also studies the trade-offs between the cost of implementing risk management measures, an added expense in the short term, and the delay of system delivery, a liability in the long term. The elements of the example are

(i) quantifying the schedule relationships among the deliveries of functional subsystems and the date of system delivery, (ii) generating alternatives for managers to accelerate a particular subsystem's development schedule, and (iii) evaluating the associated trade-offs between the cost of implementing risk management versus the long-term outcome of the system delivery date. The trade-off curve is generated under four alternative scenarios (future courses) of program development. We thus distinguish in this section between a *source of risk* (or failure scenario, as discussed in the preceding text) and the *measurement and evaluation of the risk* of schedule delay associated with the management alternatives.

From interviews with program managers, an influence diagram (influence diagrams will be discussed in Chapter 4) revealed the following: *Integration acceptance testing* has its own associated duration and cannot begin before the completions of the latest developments of subsystems U, V, and W and the integration of X and Y with Z. Likewise, the integration of X and Y with Z cannot begin before the completions of the latest developments of the X, Y, and Z segments. Though a possibility, the dependence of system delivery on the completion of the subsystem Z is not modeled further in this example. From consultation with program managers, three alternatives for accelerating the development of subsystems X and Y were generated. Low, most likely, and high estimates of schedule parameters (the duration of the development period measured in months from time of contract award) and the estimated implementation costs were used for estimating the triangular probability distributions.

Figure 3.11 illustrates the trade-off between the implementation costs of the options for risk management and the completion date, both the (unconditional) expected date and the 1-in-10 worst-case date (conditional expected value) of the system delivery. With respect to the trade-off between a short-term implementation cost and the long-term issue of program delay, Alternative 1 is dominated by Alternative 2. The up-front cost of these two alternatives is the same, while the system delivery is later for the dominated alternative in terms of both the overall expected delay and the expected delay in a 1-in-10 worst case.

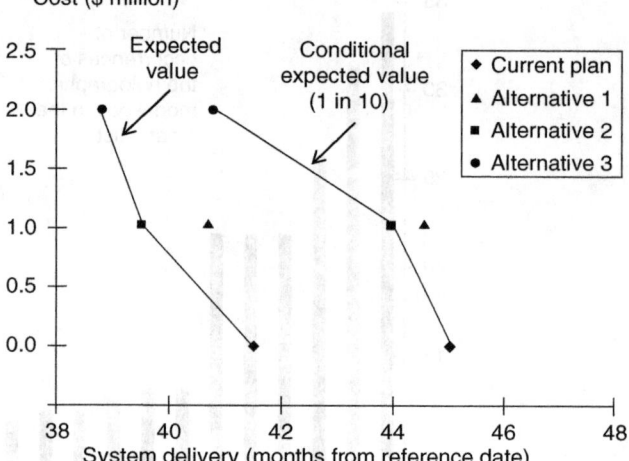

Figure 3.11 Trade-offs among alternatives and the current plan.

It is not sufficient to consider only the scenario where subsystems X and Y are on the critical schedule path. We considered four program development scenarios (courses), of which the baseline scenario (Scenario 1) is the case described in the preceding text. The three additional scenarios were specified by the program managers and accounted for possible delays of subsystems V and W so that they are potentially on the critical schedule path. It is not known which scenario (development course) was actually implemented.

3.9 SOFTWARE ACQUISITION

This section builds on the holistic representation of software acquisition through HHM [Schooff et al., 1997]. It represents software acquisition by an HHM model and enhances and extends the HHM investigative capabilities for exploring and modeling the various decompositions and submodels (see Figure 3.12) for software acquisition. Figure 3.13 depicts the six decompositions, or perspectives, indicating the multiple dimensions associated with software acquisition. The acquisition process requires the participation of numerous organizations and individuals with specific functions and responsibilities as well as requirements to coordinate their activities with the other parties. These organizations have their own goals and objectives, which are often in competition with each other. Risks and uncertainties inherent to the software

SOFTWARE ACQUISITION 91

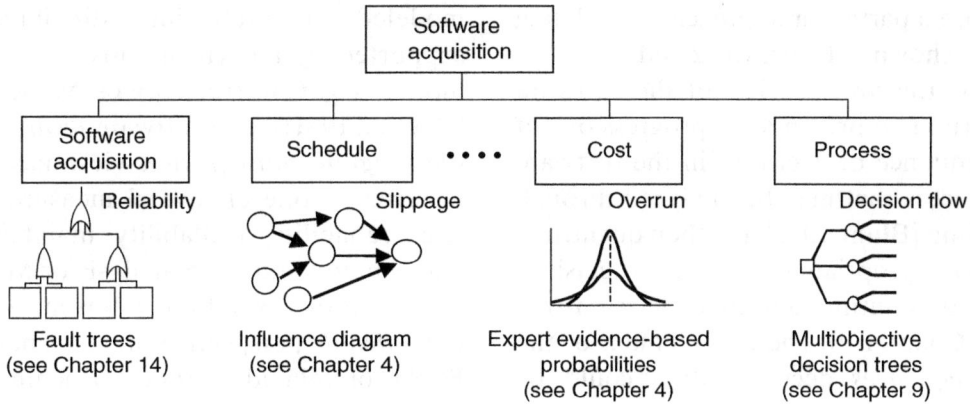

Figure 3.12 Demonstration of analytic methods for software acquisition [Schooff et al., 1997].

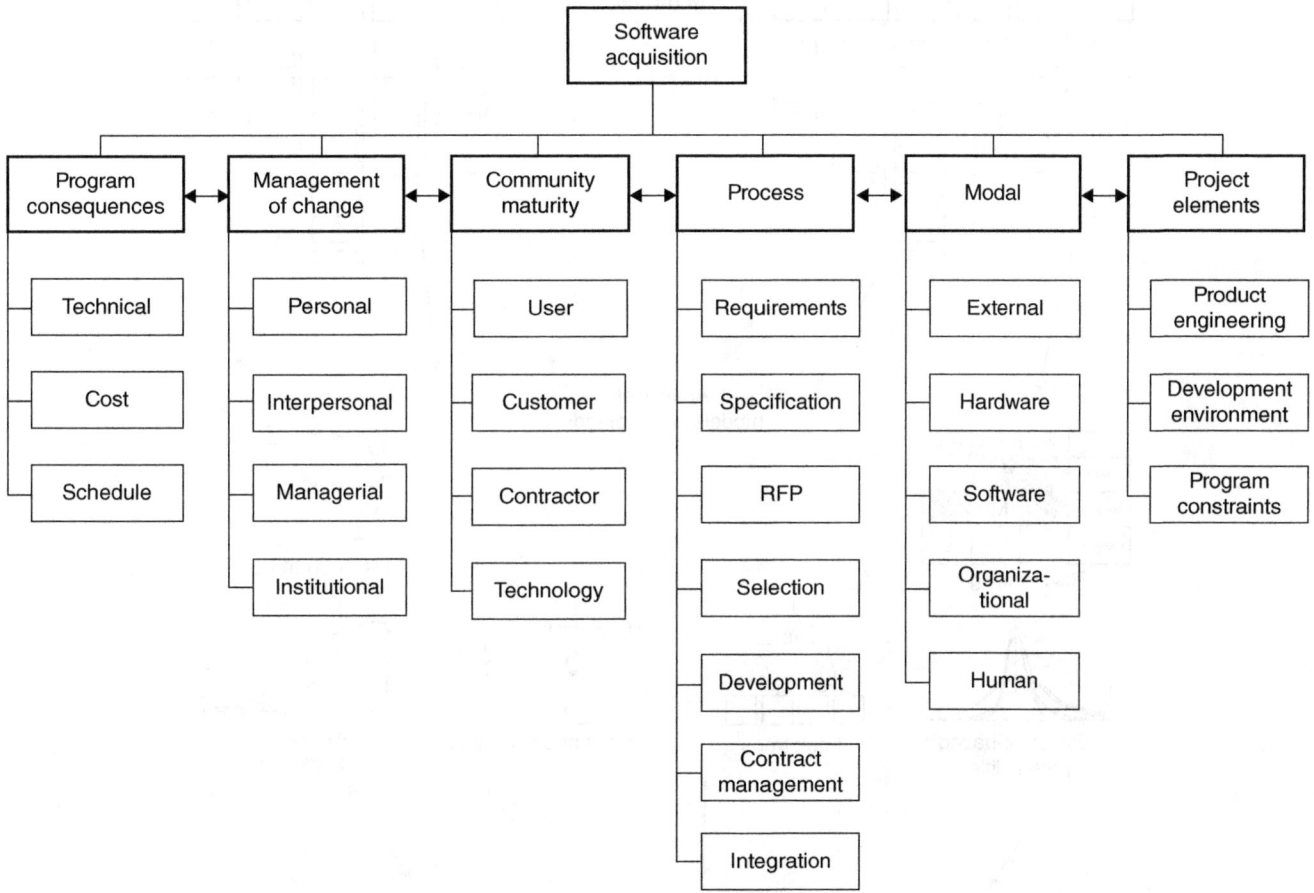

Figure 3.13 HHM for software acquisition [Schooff et al., 1997].

acquisition process complicate the several key decisions that, in turn, affect the ultimate software product. Effective management of the software acquisition process can be accomplished only by exploring the various dimensions and perspectives of the overall system's acquisition and by properly coordinating the objectives and requirements from each model perspective.

HHM provides multiple perspectives, or views, of a given problem, referred to as hierarchical holographic submodels (HHS). Each perspective has its own unique qualities, issues, limitations, and factors that may require a particular approach to modeling and analysis, as shown in Figures 3.12 and 3.14.

For instance, the *process* view of the software acquisition HHM represents a progression of events or a sequence of decisions in the software acquisition process that may be analyzed through process modeling [Blum, 1992] and then quantified by one of many appropriate tools, such as decision-tree methods or multiple-objective decision-tree methods (see Chapter 9). The *cost* element of the *program consequences* decomposition could be modeled by probability distribution analysis, supported by analytical software cost estimation models (e.g., Constructive Cost Model (COCOMO) [Boehm, 1981]). The software *technical* element of the *program consequences* view may be quantified in terms of one of several measurable objectives (e.g., reliability, availability, maintainability) and may employ fault-tree analysis or Markov process models in their solution [Johnson, 1989]. Similarly, the *schedule* perspective may be analyzed through PERT or related methods [Boehm, 1981]. While

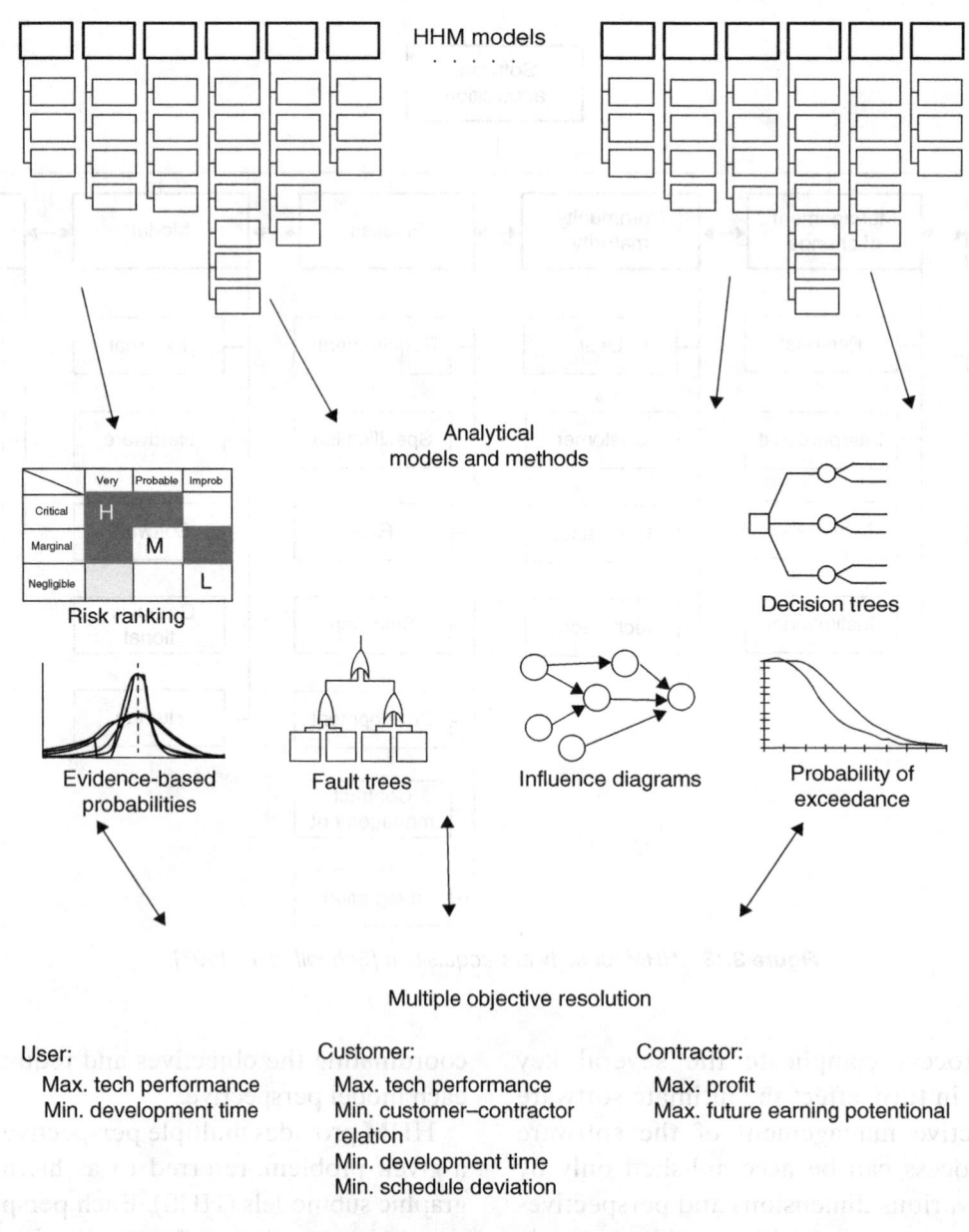

Figure 3.14 Quantitative management framework [Schooff et al., 1997].

each HHS can then be solved independently, a coordinated solution to the overall problem must be resolved at the highest level of the HHM.

3.9.1 Accepting HHM in Software Risk Management

The complexity of the software acquisition process and the multiple parties involved in that process (planning, development, delivery, and maintenance) defy the success of any attempt to represent this process by any one single model, structure, or paradigm. In fact, representation within a single model of all the aspects of software acquisition is so impracticable as never to be seriously attempted.

Many current risk identification methods, evaluation techniques, and issue investigation schemes build on the general principles embodied by HHM. For example, careful examination of the software risk taxonomy [Carr et al., 1993], its purpose and methodology, indicates a vision that is harmonious with HHM: The taxonomy is hierarchical in structure, is constituted of progressive levels of detail and abstraction, provides a way to address the multiple dimensions of a problem, and serves to identify areas of concern in a software acquisition endeavor. Recognizing the kinship of these methods to HHM strengthens the parent methodology and further demonstrates the efficacy, appropriateness, and desirability of HHM as a framework for analyzing software acquisition and other large-scale problems.

The role of models is to represent the intrinsic and indispensable properties that serve to characterize a system; that is, good models must capture the essence of the system. Clearly, the multidimensionality of the acquisition process, along with the large number of groups, organizations, and people of many disciplines that are engaged in this process, defies the capability of any single model to represent the essence of the acquisition process. To overcome the shortfalls of single planar models and to identify all sources of risk associated with the software acquisition process, an HHM framework offers a distinct answer. HHM assumes an iterative approach to provide a structure for identifying all risks. If one fails to identify a risk source with the current views of the HHM, it is possible to expand the model to include a new decomposition. This process will eventually capture all risk sources. As an example, from the *program consequences* perspective (see Figure 3.13), the software acquisition process may be decomposed into three consequence areas: technical, cost, and schedule.

1. *Technical:* In a software context, technical consequences are concerned with the quality, precision, accuracy, and performance of the software over time.
2. *Cost:* Refers to both the programmed and unexpected expenditures for procuring the software system, along with labor, capital, and other nonmonetary costs.
3. *Schedule:* Concerns the establishment of, adherence to, and changes of a temporal development plan on which systems integration schedules and operational deployment schedules are based.

For notational purposes, the model of a software acquisition subdivision will be termed the HHS. Figure 3.15 depicts one such representation from the perspective of the *program consequences* HHS, focusing on the cost risks of the software acquisition effort—in particular, the cost risks associated with each community (user, customer, contractor, and technology).

Further investigation with this HHS would focus on schedule risks and the particular schedule risks of each community (Figure 3.16). The third focus from this HHS would be to examine the technical risks associated with each community (Figure 3.17).

As depicted in Figures 3.15 through 3.17, using the *program consequences* perspective as the primary vision, one may then examine all such consequences that may be realized from the participant communities (e.g., what schedule consequences may be realized due to the customer community).

Another vision of the HHM can be obtained through the four *communities* involved in software acquisition: user, customer, contractor, and technology (Figure 3.18). Although this is a simple reversal of the decomposition, the initial focus is upon a particular program facet. Such a perspective is well suited to a manager who is focusing on one metric or performance aspect and how it can be affected. The software community maturity HHS

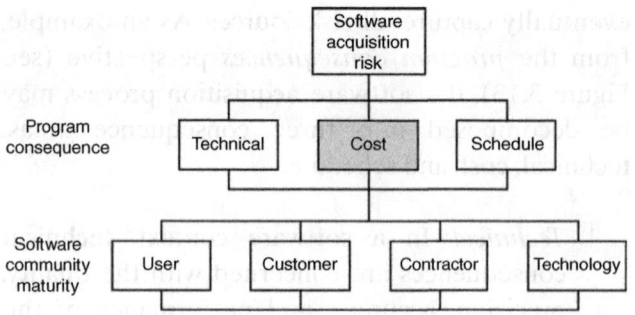

Figure 3.15 Program consequence submodel: cost focus.

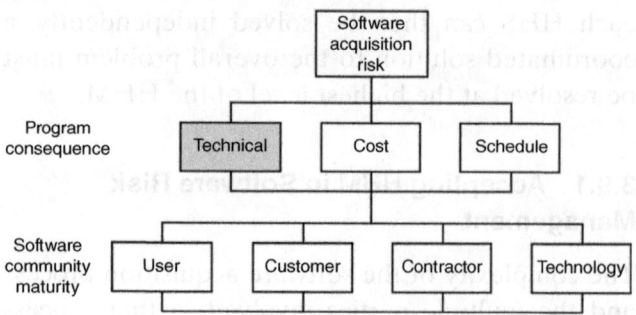

Figure 3.17 Program consequence submodel: technical focus.

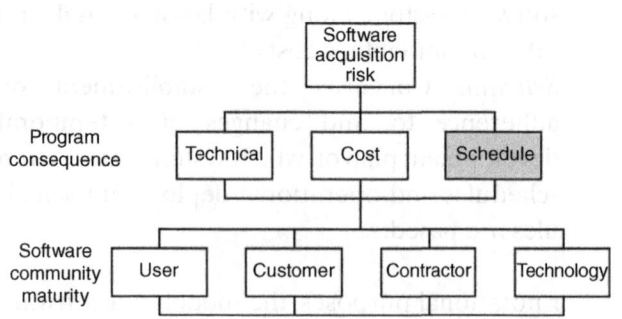

Figure 3.16 Program consequence submodel: schedule focus.

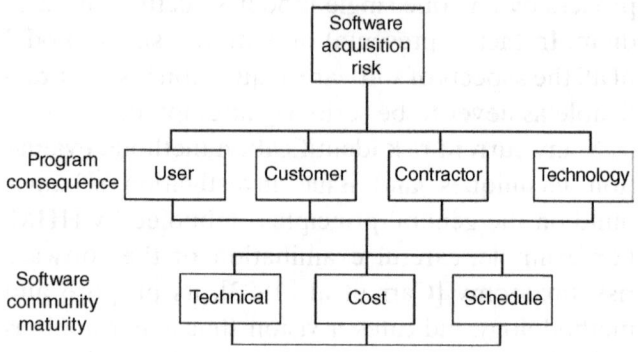

Figure 3.18 Community maturity submodel.

first emphasizes a particular community, and then it examines the impact this community may have relative to the system's performance metrics. This vision is appropriate as we examine the capability and interactions of the participant communities. Additional combinations of decompositions for each phase of the acquisition process will provide a robust scheme for risk identification.

3.10 HARDENING THE WATER SUPPLY INFRASTRUCTURE

Hardening a water supply system refers to rendering the system less vulnerable to accidents or natural hazards. The term *surety* is also commonly used to connote hardening. No system can be rendered absolutely hard. There are limits to the technology of hardening and to the public's willingness to pay for it [Haimes et al., 1998].

The HHM approach to hardening the infrastructure addresses its holistic nature in terms of its hierarchical institutional, organizational, managerial, and functional decisionmaking structure, in conjunction with factors that shape that hierarchical structure. These include the hydrologic, technologic, and legal aspects, as well as time horizons, user demands on the infrastructure, and socioeconomic conditions. Addressing the holistic nature of the water supply infrastructure by considering a large universe of real, perceived, or imagined risks from their multiple perspectives provides an effective means for identifying the myriad risks to which the infrastructure is exposed. Figure 3.19 summarizes the panoply of visions that may be useful in hardening the water supply infrastructure [Haimes et al., 1998].

In applying the HHM philosophy, the risk to a water supply infrastructure is decomposed into 16 major categories. The categories represent the risks to a water supply system from the multifaceted dimensions of each major category, including the likelihoods, root causes, consequences, and direct and indirect impacts. In general, the major categories are labeled as A, B, C,..., and their subcategories are labeled as $A_1, A_2, A_3,...$; $B_1, B_2, B_3,...$; $C_1, C_2, C_3,...$; and so on.

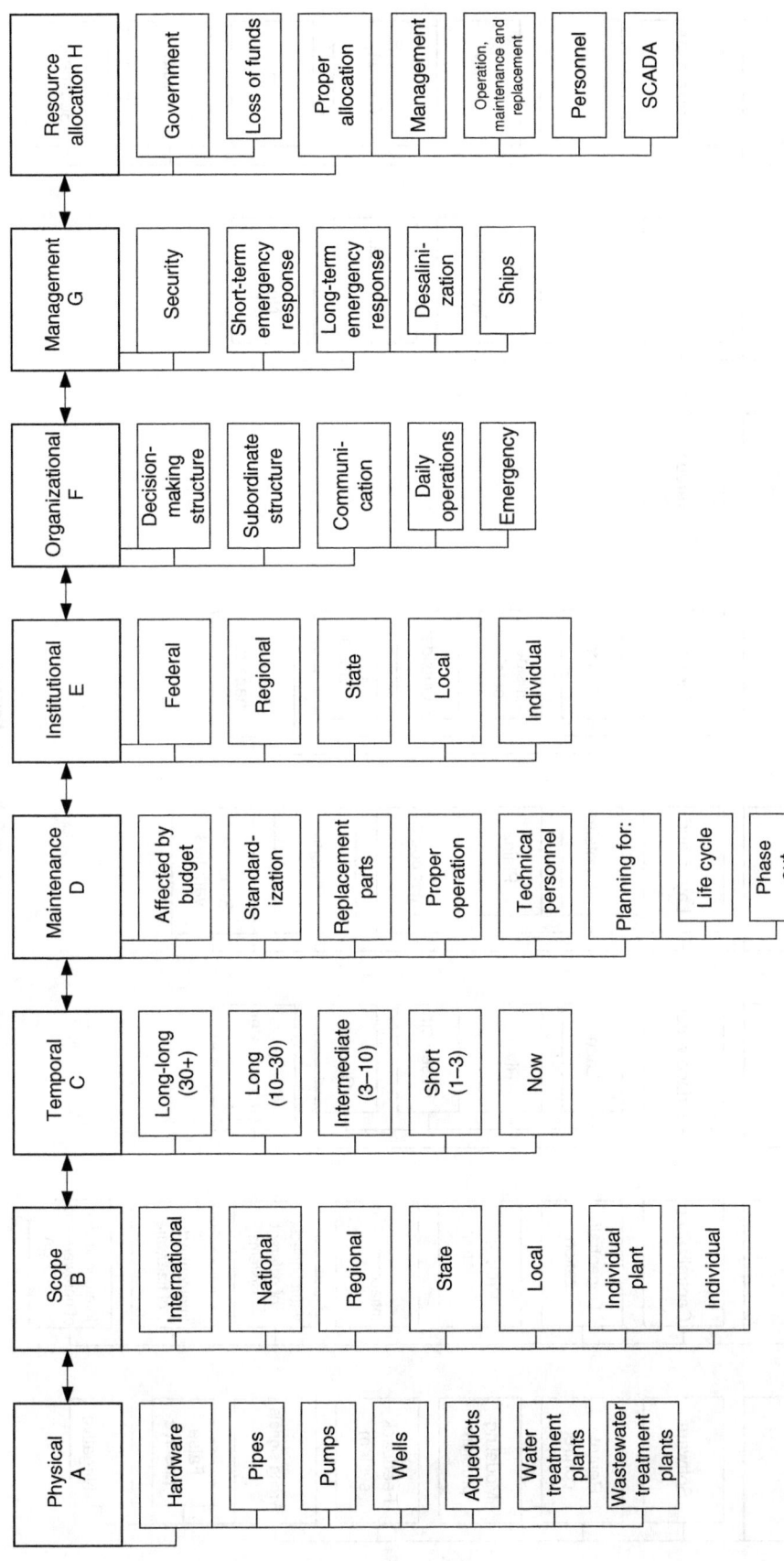

Figure 3.19 Multiple perspectives on the hardening of the water supply system.

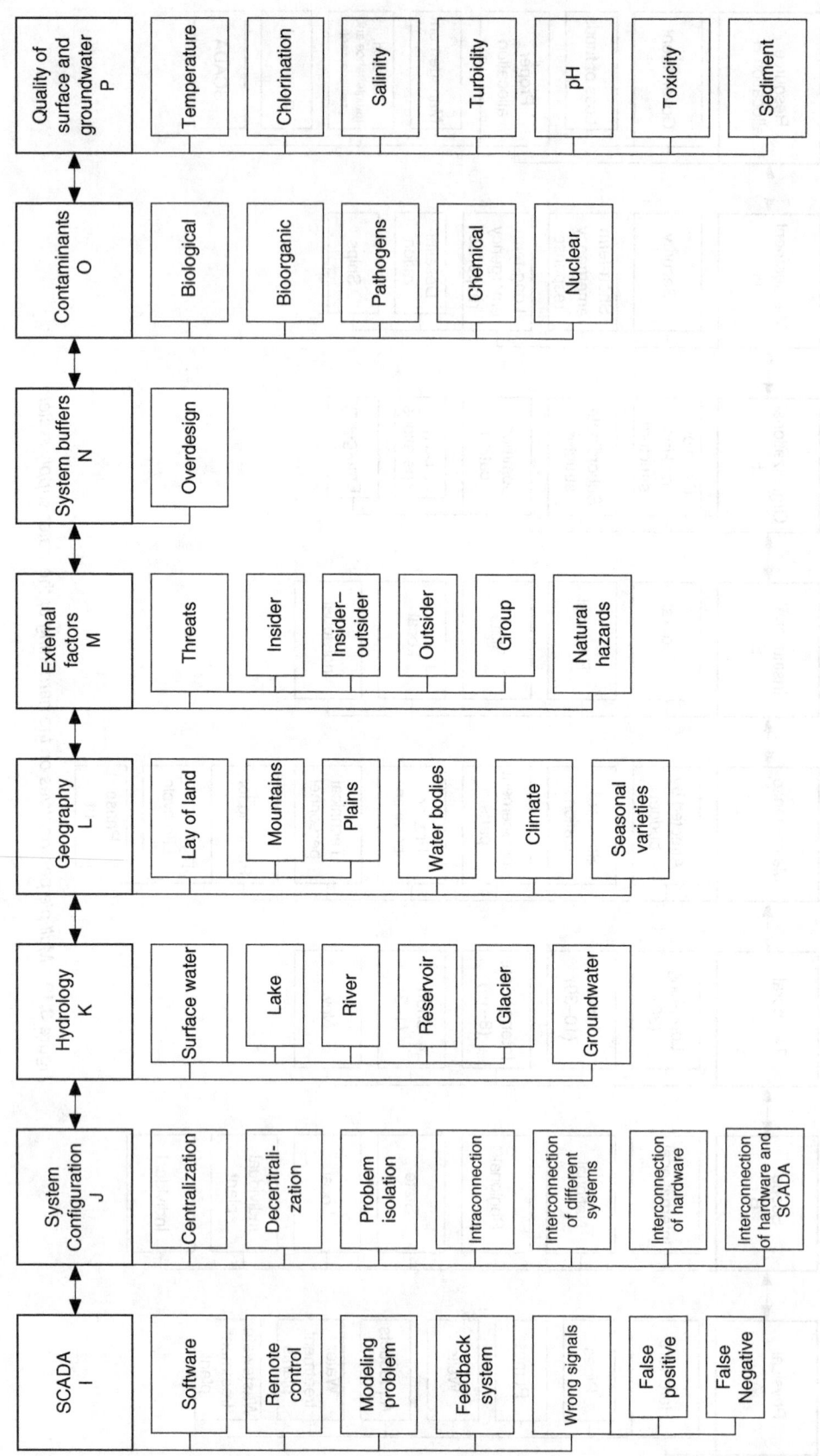

Figure 3.19 (continued)

Category A: Physical. Given the central importance of the physical components for a water supply system, the physical components are major potential targets for terrorist acts. The category is partitioned into seven subcategories or subsystems. Depending upon the scale, location, and timing, tampering with any of the subsystems could cause a major disruption in meeting the community's water demand.

Category B: Scope. This category captures the segmented target of a water supply infrastructure and its broader implications. For example, a disruption in the water supply in one community may have an impact on the nation (e.g., public policy) or the international community (e.g., international commerce). The category is partitioned into seven subcategories. The scope of the risks to water supply systems, in terms of their sources and their consequences, has implications as to how funds for hardening the systems are allocated.

Category C: Temporal. The temporal category is perhaps one of the more obscure categories of risk to water supply systems. Decisions that affect the present and future viability of a system are made continuously, involving officials at all levels of government and in the private sector. Replacing an aging component under the physical category of a system may take several years. Routine maintenance on a daily basis may enhance not only the reliability of the system but also its robustness and resilience in coping with unexpected natural or man-made disruptions. Thus, the element of time guides the decisions of water resource planners. The five-subcategory partition is somewhat arbitrary but illustrates the relevance of the temporal category in assessing the risks to water supply systems and the means of hardening them.

Category D: Maintenance. Most car owners are aware that the reliability of their cars is greatly dependent upon maintenance. The maintenance reliability attached to automobiles can be projected to large-scale, complex systems such as water supply systems with their many components distributed over several hundreds of square miles. For example, the matter of standardization in the manufacture of large-capacity pumps is an important subcategory of maintenance. Many large-capacity pumps in the world are one of a kind, and it may take several months to replace a pump. With standardization, replacement pumps could be obtained more readily and in a more timely manner, thus lessening the vulnerability of the system itself. Seven subcategories of maintenance are identified. Note that there is an overlap between the temporal category and the maintenance subcategory of planning for life cycle. It is this kind of overlap arising from different perspectives that is the strength of HHM in revealing the risks to large-scale, complex systems and the potential consequences of the risks.

Categories E, F, and G: Institutional, organizational, and management. Distinctions are made between the institutional, organizational, and management categories. The institutional infrastructure provides the basis upon which the organizational infrastructure is designed and subsequently managed. Critical policies formulated at the institutional level, such as resource allocation, can have a major impact on the wellbeing of the organization and thus on the risks to which it is exposed. Also, the culture and core values of the organization and the nature of its hierarchical decisionmaking process determine and affect the way such an organization assesses and manages its risks.

Category H: Resource allocation. Proper allocation of funds is at the heart of hardening a water supply system. Without sufficient funds to operate, maintain, expand, and protect the system, hardening cannot be achieved and maintained. The resource allocation category of risk to a water supply system pertains to (i) hardening the system by appropriating the needed funds for the system's safe and viable operation now and in the future and (ii) securing the system against unwarranted acts. Indeed, no effective risk management can be undertaken without appropriate allocation of the needed resources.

Category I: Supervisory control and data acquisition. Although not all water supply systems are operated through supervisory control and data acquisition (SCADA) electronic systems, trends suggest a rapid movement toward universal adoption of SCADA systems. There are added uncertainties and sources of risk with the use of this control system. The SCADA category addresses the opportunities and risks attendant on the control system. Studies on the protection of the Internet highlight the importance and the vulnerability of SCADA and thus the operation of water supply systems.

Category J: System configuration. Understanding the configuration of the physical infrastructure of a water supply system (including the hardware and the software) and its interconnectedness with other systems, as well as the system's institutional, organizational, and management configurations, is of paramount importance to the system's protection. Although understanding a system's configuration is important to identification of risks, an understanding of the configuration by all key system personnel is imperative for effective hardening.

Category K: Hydrology. Hydrology is the fundamental category of the design and operation of a water supply system, and therefore, this is one of the more obvious categories within the framework of HHM. Two major subcategories are identified: K_1, surface water (rivers, lakes, impoundments, and glaciers), and K_2, groundwater. Each type of water source offers unique issues within the scope of hardening a system.

Category L: Geography/physiography. Geography and physiography play important roles in the hardening of water supply systems. To some extent, geography is a determinant as to which natural hazards pose threats to a system; it is a determinant of climate and of the primary hydrologic controls on a system. The terrain dictates how conduits, pipes, canals, tunnels, and aqueducts will be laid, their configurations and depths, and the types of material used for the conduits.

Category M: External factors. Natural hazards can threaten a water supply system, as can unfriendly acts of terrorism. The lessons learned in coping with natural hazards and in responding to natural disasters such as major floods, hurricanes, and earthquakes provide guidance in coping with the consequences of unfriendly acts.

Category N: System buffers. Water resource planners have long recognized the omnipresent design uncertainties from the influence of hydrologic, economic, political, and social factors. To hedge against uncertainties, system designs are buffered through overdesign. Over the years, buffering has proven to be important in protecting systems from natural hazards. Within the context of HHM, buffering is viewed as important to hardening water supply systems.

Category O: Contaminants. Protection against water contamination, along with the recovery from such an eventuality, is an important category of risk to water supply systems. The manufacture, handling, and transport of highly toxic materials and the eventual disposal of the material residuals pose numerous hazards to the health of the nation's population. The potential contamination of the water supply due to natural hazards or accidents compounds the risk to society.

Category P: Quality of surface water and groundwater. Under normal conditions, water supply meets demand if water is delivered on schedule at the proper location, with quality meeting federal and state standards. Although there are many facets to water quality, only seven subcategories are identified.

3.11 RISK ASSESSMENT AND MANAGEMENT FOR SUPPORT OF OPERATIONS OTHER THAN WAR

The first line of defense against accidents in military operations combines good planning, intelligence, training, and ensuring adequate resources in personnel and materiel, among other factors. The following case study was performed for the US Army for operations other than war (OOTW) [Dombroski et al., 2002]. OOTW decisionmakers include all levels of the military, from strategic personnel in the Pentagon to tactical officers in the field of operations. Recent experiences of US forces involved in OOTW in Bosnia, Kosovo, Rwanda, Haiti, as well as Afghanistan, Iraq, and other nations, dramatize the need to support military planning with country information that can be clearly understood. It is necessary to carefully analyze both the geopolitical situation and the subject country to support critical initial decisions such as the nature and extent of operations and the timely marshaling of appropriate resources. Relevant details need to be screened and considered to minimize poor ad hoc decisions as well as wasted resources. Such details include information on existing roads, railways, and shipping lanes; the reliability and security of electric power; communications networks; water supply and sanitation; disease and health care; languages and cultures; police and military forces; and many others. Interagency and multinational cooperation is essential to OOTW and requires less dependence on ad hoc decisionmaking with greater attention to cultural, political, and societal concerns. An effective, holistic approach to decision support for OOTW was developed to encompass the diverse and numerous concerns affecting decisionmaking in this uncertain environment.

3.11.1 HHM for System Characterization

There are numerous ways to characterize a country as a potential theater for OOTW. Unique but important characterizations of state variables, such as its technical infrastructure, political climate, society, or environment, are essential for both risk assessment and risk management. Indeed, before US forces plan and prepare a deployment into a country for OOTW, the military needs to know practically everything important about that country. By identifying the host country's critical state variables as well as the state variables of the US forces and its allies, the military identifies (i) its own vulnerabilities (accident precursors), (ii) the threats from unfriendly elements, and (iii) the corresponding risk management options that would counter these threats. HHM served as the backbone for the risk assessment and management process in the methodology developed for the Army's NGIC and for Kosovo as a test bed.

Four HHMs were developed for OOTW: (1) The *Country HHM* identifies a broad range of criteria to characterize host countries and the demands they place on coalition forces. (2) The *US HHM* characterizes what the United States has to offer countries in need. (3) The *Alliance HHM* characterizes all forces other than US forces and organizations, such as multinational alliances and nongovernmental agencies. (4) The *Objectives HHM* recognizes the multiple and varying objectives of the many potential users of the methodology and coordinates all three HHMs.

3.11.2 Country HHM

Figure 3.20 presents a sample of a Country HHM (head topics and subtopics), which was developed using an analysis of OOTW doctrine, case studies of previous operations, and brainstorming. Analytical case study models [C520, 1995] from Operation Provide Comfort, Operation Restore Hope, Operation Joint Endeavor, and Operation Allied Force were analyzed to identify important criteria. For example, decisionmakers for a typical OOTW need to know about the culture of the people, the economic and political stability of the nation, and the strength and disposition of the country's military force. For a humanitarian relief operation, they must know about the existing health-care system, as well as food, water, and resources that the nation can provide for assistance; and for a peacekeeping mission, they are more concerned with externalities and terrorists that could potentially destabilize the existing situation. In many ways, the Country HHM constitutes a *demand* model; it represents the country's needs in terms of personnel and materiel.

3.11.3 US HHM

The US HHM addresses the supply aspect of an OOTW. The United States has a broad range of options available to address crisis situations, including diplomatic negotiations, economic assistance, and/or troops and equipment. The US HHM is separated into two major areas: (1) *Defense Decisionmaking Practice* and (2) *Defense Infrastructure*. The US HHM also provides supply-side information, helping decisionmakers to marshal supplies for an OOTW. The Defense Infrastructure subcriterion included in the US HHM documents the equipment, assets, and options that the United States can offer to an OOTW. Details of the US HHM can be found in Dombroski et al. [2002].

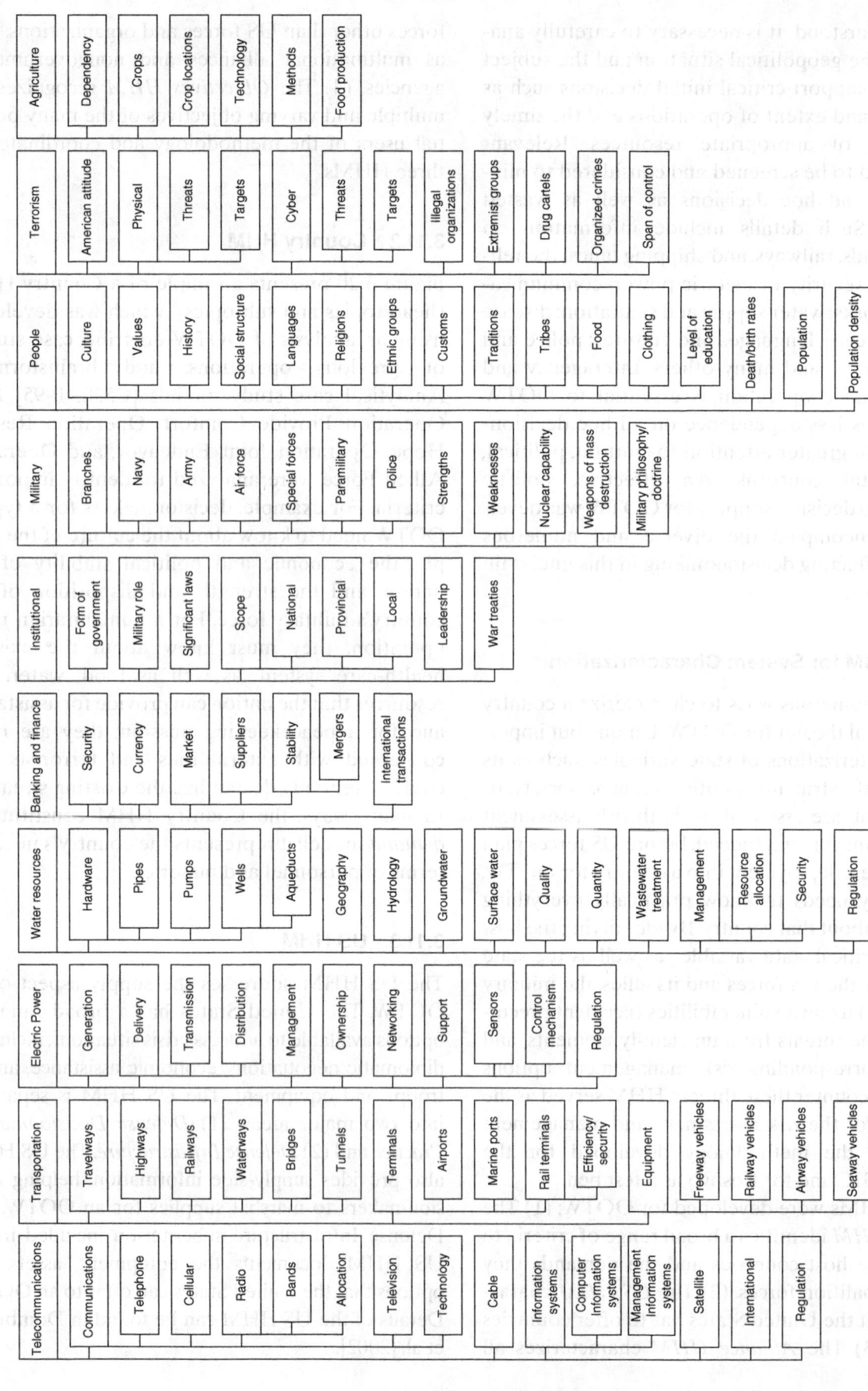

Figure 3.20 The Country HHM documenting important risks to consider about a host country for OOTW from societal, technical, political, and environmental perspectives.

Figure 3.20 (continued)

3.11.4 Alliance HHM

The Alliance HHM recognizes that the international community is more involved in maintaining international security now than it has been at any other time in world history [FM 100-8, 1997]. The Alliance HHM documents countries, multinational alliances, and permanent and temporary relief organizations involved in an OOTW. Including nongovernmental organizations (NGOs), private voluntary organizations (PVOs), and the United Nations, these stabilize the disengagement and ensure the economic, political, and social stability of a region after US military forces leave [CALL, 1993].

3.11.5 Coordination HHM

Together, the Country, US, and Alliance HHMs contain a vast amount of information pertaining to an OOTW, educating decisionmakers about the situation and helping planners and executors attain their mission goals. However, the information may not be important to all users at all times. A particular user will be concerned only with a specific subset of OOTW demands and marshal a specific subset of total characterizations for the users of the system who assist in coordinating supply and demand. The Coordination HHM identifies certain critical user-objective spaces with predictable information needs and includes the staff function, policy horizon, outcome valuation, and three decisionmaking levels: strategic, operational, and tactical. The strategic level includes *national strategic* and *theater strategic* decisionmakers.

Each decisionmaking level seeks answers to specific questions pertaining to Country HHM subtopics. These questions facilitate the identification of critical information for each decisionmaker. Strategic decisionmakers consider whether to enter into an operation. Operational decisionmakers define the operation objectives and plan missions to maintain order and prevent escalation of the situation. Tactical decisionmakers plan and execute OOTW missions to support higher objectives. Details of the Coordination HHM can be found in Dombroski et al. [2002].

3.11.6 Risk Filtering and Ranking

Due to the large number of HHM risk scenarios, decisionmakers may find it difficult to determine which kernels of information are important. Planners must focus limited resources on the most likely and uncertain sources of risk. Risk filtering, ranking, and management (RFRM) (to be presented in Chapter 7), which integrates quantitative and qualitative approaches, is used to identify these critical scenarios. Four filtering phases allow decisionmakers to sift out from 265 subtopics only the most critical 5–15.

3.11.7 Risk Management through Comparison Charts

The OOTW undertaken by the United States in the Balkans illustrates the use of comparison charts. Such charts helped determine what medical supplies were needed for the incoming refugees.

Officers viewed health-care and disease data for Serbia to understand the existing conditions in the province of Kosovo. Because the staff officers were not familiar with conditions in Serbia, they compared the data with those of the United States, China, and Croatia.

Figure 3.21 is a three-dimensional bubble chart displaying health-care metrics. Two metrics are displayed on the X and Y axes. A bubble of variable areas represents a third metric. The staff officers assume that they can draw inferences about the state of each country's health-care system by viewing Figure 3.21. It implies that Serbia's health-care system is in a state of disrepair because Serbia has fewer hospital physicians and beds per 1000 people and greater infant mortality than Croatia (the United States is used as a reference base). Even though Serbia's health-care system is not as poor as China's, staff officers infer that refugees may be in poor health, which indicates that a large variety of medical supplies might be required to conduct the operation effectively. To better understand what diseases might need treatment, the staff officers view Figure 3.22, which depicts the estimated prevalence of certain diseases in Serbia and Croatia. The metrics on the radials of Figure 3.22 indicate the percentage of population infected. The comparison

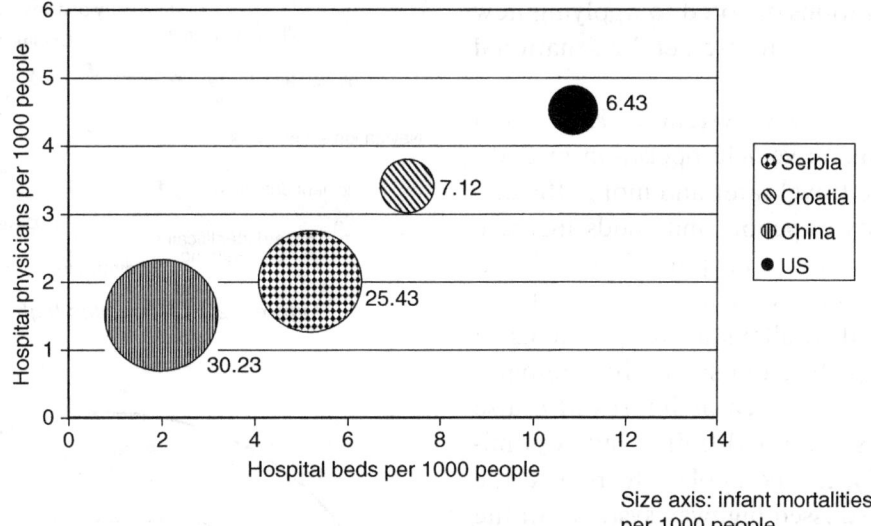

Figure 3.21 Bubble chart showing health-care metrics on each axis representing the countries of Serbia, Croatia, China, and the United States.

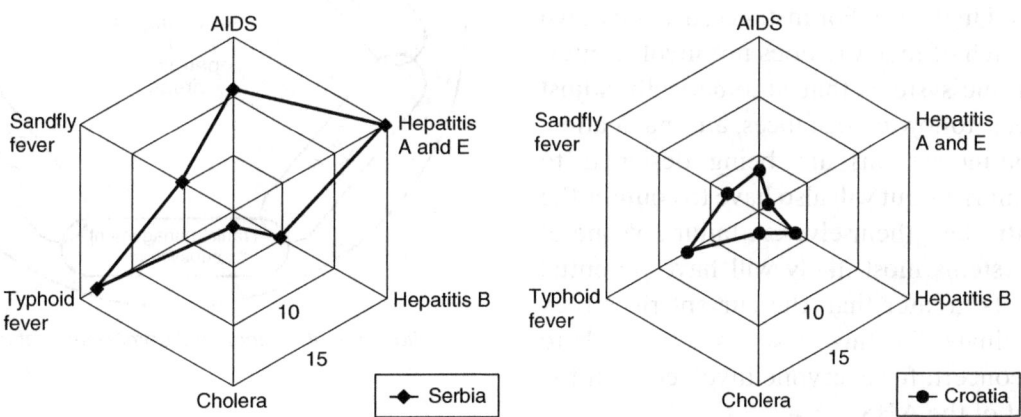

Figure 3.22 Radial chart example showing disease prevalence in Serbia and Croatia. Metrics are measured in the percentage of people infected.

shows that Serbia has more problems than Croatia with AIDS, hepatitis A and E, and typhoid fever.

3.11.8 Conclusions

The Country HHM provides nearly all information needed to correctly characterize the host country states, regardless of the type of OOTW. The US HHM provides US options to prepare for OOTW. The Alliance HHM accounts for other countries and organizations providing support to an OOTW. The Coordination HHM distinguishes users of the system and their specific needs.

3.12 AUTOMATED HIGHWAY SYSTEM

Increases in vehicular traffic are exceeding the capacity of our highway infrastructures at an alarming rate. Over *each* of the past three decades, traffic volume has increased an average of 46%. The General Accounting Office further predicts the ensuing congestion will increase by 300–400% by the year 2010. Congestion on the roadways costs the US economic system $100 billion in lost productivity annually. Statistics such as these have prompted a nationwide effort to counter the rapidly failing transportation infrastructure. Answering the call are organizations devoted to applying new technologies to solving this tremendous national problem.

An automated highway system (AHS) could provide fully automatic vehicle operation in dedicated lanes to make travel safer and more efficient, improve the mobility of people and goods, increase the productivity of surface transportation, and contribute to a better quality of life. These technologies are envisioned to alleviate the problems of highway capacity, quality, and safety. Technologies and engineered systems, such as improved cruise control, incident response, and traffic route optimization (Figure 3.23), can be applied to reduce the growing impact of stressed highway systems. In the future, an automated hands-off driving environment is anticipated [Haimes et al., 1998].

Implementing new, complex systems to our highways introduces risks not found in our current nonautomated highways. For instance, driving down a current stretch of highway does not involve interactive electronic systems that automatically adjust vehicle speeds, following distances, and navigation. The automating systems are being designed to reduce current risks but will also have to counter the risks they introduce themselves. A failure of one of these new systems most likely will have potential adverse effects greater than the current risks they serve to eliminate. For this reason, safety needs to be a major concern for everyone involved with the development of the AHS.

3.12.1 Functional Components of the AHS

Six main functional or operational areas that can benefit from automating technologies are identified here. Figure 3.24 depicts these operational areas and their interactions with the AHS. These areas and their basic descriptions are as follows:

1. Car, representing the area in which technologies can be applied to improving the capabilities of an autonomous vehicle
2. Car/car, systems that ensure vehicles communicate with one another
3. Car/road, systems where interaction and communication between vehicles and the driving surface are performed

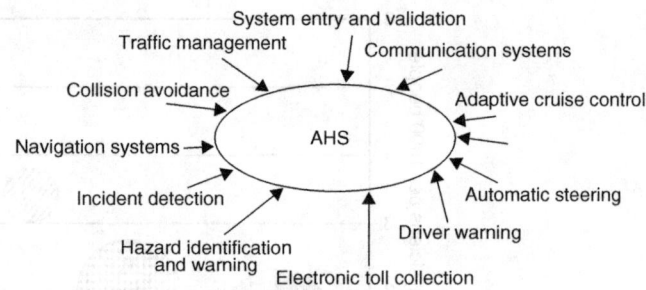

Figure 3.23 Automated highway system.

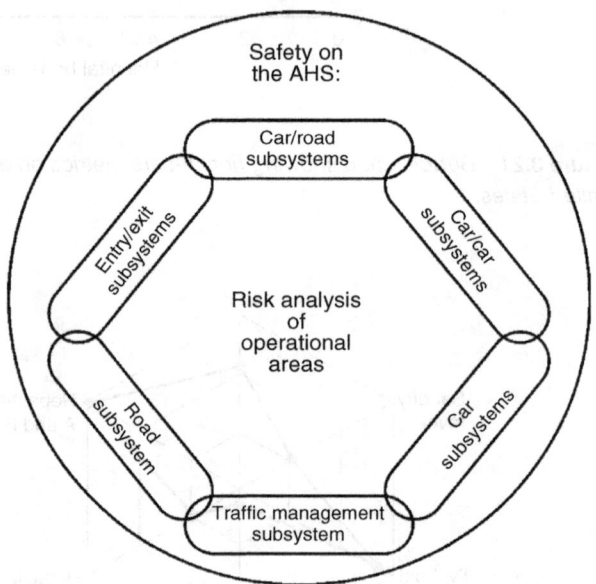

Figure 3.24 Functional components of the AHS.

4. Road, where incident detection and hazard avoidance systems can reside
5. Entry, mechanisms to ensure only safe, fully functional vehicles access the AHS
6. Traffic management, the centralized command and control for local and regional traffic decisions

The available and prospective technologies should be evaluated in terms of reliability, cost, and capacity for the six major focus areas of the AHS. Therefore, the AHS needs to be analyzed in terms of multiple failure modes.

For example, consider the two areas of the road and car/road relationship. A failure in the road subsystem could result in accidents, loss of life, limited capacity, and additional maintenance costs. However, a failure in the road surface would ultimately result in a failure in the car/road relationship due to the road's inability to communicate with the car. Hence, the additional failure modes of the car/road relationship would compound the results of the failure in the road subsystem.

Through HHM, failures are expressed in terms of hierarchy, organization, and decisionmaking structure. The structure includes time horizons, stakeholders, decisionmakers, geographical influences, technical components, and legality issues. The HHM model identifies the failure modes and their associated consequences.

3.12.2 HHM for the AHS

The AHS is a large and complex system requiring the involvement, interaction, and agreement of many stakeholders to develop technologies and guidelines for using it. Competing needs, uses, and technologies must cohere for an AHS to evolve into a new and accessible mode of transportation. Because of these divergent influences, the AHS is susceptible to a large number of failure modes. These potential failures are expected to occur not singularly or in isolation, but in aggregate.

A properly designed system anticipates failures as a way to prevent them. Failure modes are common among engineered systems, so they can often be identified in advance based on past experience. There is then a better chance of preventing these failure modes, which may or may not occur. Multiple failure modes, which often are not accounted for, can be anticipated if appropriate system views are taken.

Figure 3.25 depicts ways in which an AHS may fail at a very high level. The various perspectives are placed horizontally across the chart (head topics) and represent high-level system failure modes. These are viewed as general categories in which specific failures can be grouped. Underneath each general failure mode are listed specific failures (subtopics) that may arise in this general category. These specific failures may represent detailed individual failures or perhaps a lower-level failure mode underneath the general failure mode. For example, the infrastructure and economics subtopics are also viewed as high-level potential failures under a system source perspective. Temporal, planning period, technology, spatial/geographical, and public acceptance perspectives, among other head topics, are subdivided to their lowest-level potential failure modes as well.

3.12.2.1 System source

The system source perspective identifies the broader ways in which an AHS can fail. Societal or national influences are included in this perspective. Specific failures such as a technology failure are not included. System source is further subdivided into the following:

1. *Driver*: The operator of automated and nonautomated vehicles. Includes private and commercial interests as well as the motivations and psychology of driving
2. *Vehicle*: Motorized forms of transportation. Includes automated and nonautomated modes, gasoline and alternative fuels, and multiple- and single-occupant transport
3. *Infrastructure*: The networks of interstate highways and the public and private transportation authorities that support them
4. *Economics*: The positive and negative financial influences imposed by the use and governance of highway systems

3.12.2.2 Temporal planning

The temporal planning perspective addresses issues that are common during each of the previous planning periods. These failure modes can be identified and addressed similarly and as early as possible to resolve future problems:

1. *Excessive requirements*: System demands that exceed economic, political, or technical limits.
2. *Maintenance*: Infrastructure support becomes too costly or infeasible.
3. *Government support*: Public popularity of AHS decisions may have an impact on future political funding or decisions.
4. *Industry support*: Negative economic impact on private enterprise may deter industry support in design and planning.
5. *Cost/benefit*: Technically feasible and/or publicly acceptable consumer issues may change between current and future developments due to inflation, market economy, and so on.

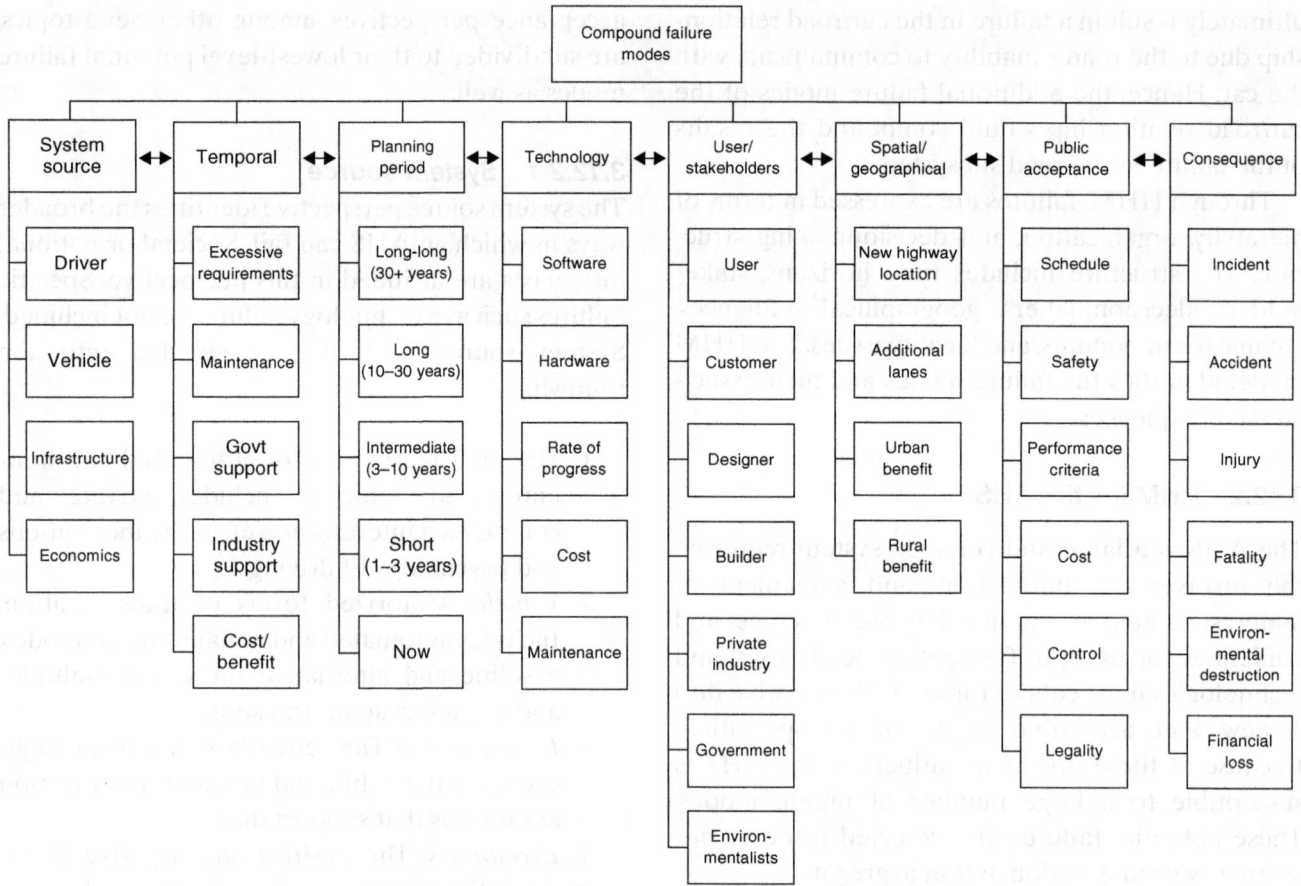

Figure 3.25 Compound failure modes for the AHS.

3.12.2.3 Planning period

The planning period addresses the varying time horizons for the design and implementation of the AHS. Each time horizon represents a different phase of the AHS, and each may or may not include failure modes present in others. Failures pertaining to the system life cycle are addressed. This perception can be used for future systems under conceptual or physical design in addition to those in use today.

3.12.2.4 Technology

The technology perspective addresses the dependence of the AHS on automating technologies. While some automating technologies are available today, significant technical hurdles must be overcome to increase the performance of mechanical and electromechanical automotive devices. Failures in the timely development of these areas will delay or prohibit evolution of the AHS as well as have a negative impact on consumer acceptance:

1. *Hardware*: Safety or performance-improving devices such as magnetic sensors, high-speed electromechanical devices, ice detection systems, and intelligent cruise control systems.
2. *Software*: Computer code used to control solid-state devices.
3. *Rate of progress*: The design and development of AHS technologies may not progress to meet expectations.
4. *Cost*: Costs for the design and development of technologies may exceed those that industry, the public, or the government may be willing to absorb.
5. *Maintenance*: Automating technologies may be feasible but have unacceptable maintenance demands.

3.12.2.5 Users/stakeholders

The users or stakeholders perspective addresses the failure modes that may arise from user interaction with the AHS. Each stakeholder has a different expectation of the system. Seven types of stakeholders are envisioned:

1. *User*: The operator of a vehicle on an automated line
2. *Customer*: A vehicle operator who must financially support AHS development
3. *Designer*: Public and private parties who influence AHS development
4. *Builder*: Public and private parties who materially construct the technologies, vehicles, and AHS transportation lanes
5. *Private industry*: Nongovernmental institutions that will use the AHS
6. *Government*: Involved through economic and developing support, collecting taxes, and providing services
7. *Environmentalists*: Function as watchdogs to ensure that technologies and infrastructure development do not have a negative impact on the environment

3.12.2.6 Spatial/geographical

The spatial/geographical perspective addresses failures specific to the geographic location of AHS lanes. This includes building totally new highways as well as adding AHS lanes to existing highways. Entry/exit ramps must be a part of these additional highways and can consume a significant quantity of land based on current estimates:

1. *New highway location*: Private landowners may object to proposed routes; available public lands may not be suitable or available.
2. *Additional lanes*: Current or future growth of existing highways will need available land. This land may not be available, or use may be undesirable.
3. *Urban benefit*: Will high-density populations benefit from the increased highway lanes considering the loss of public and private lands? Will changes to current driving behaviors be acceptable?
4. *Rural benefit*: Will low-density populations benefit from the increased highway lanes considering the loss of public and private lands? Will changes to current driving behaviors be acceptable?

3.12.2.7 Public acceptance

The public acceptance perspective addresses issues pertaining to how the public will perceive the AHS. Public acceptance affects both the acceptability as well as the evolutionary progress of the AHS. The strong support needed by the government is heavily influenced by negative public outcry:

1. *Schedule*: Will development proceed at an acceptable rate? Will the different and needed technologies be available when they are expected? Since new technologies are being developed in parallel, bottlenecks in one design may impede the development of others.
2. *Safety*: Will the public perceive the AHS as safe? Will appropriate measures be pursued to make the public aware of the safety advantages of an AHS?
3. *Performance criteria*: Will the AHS meet the expectations of public and private industry?
4. *Cost*: Will the improvements in safety be enough to justify the increased costs in the public eye?
5. *Control*: Americans enjoy the charm and freedom of automotive travel. Will they subjugate the pleasure and personal freedom derived from driving for the benefits of an AHS?
6. *Legality*: Will AHS lane access requirements be questioned? Will they be constitutional? Will AHS benefits be available to only those who can afford them?

3.12.2.8 Consequences

Identifying failure modes is not enough because it is actually their consequences that are undesirable. Different failure modes may involve more or less severe consequences, and failure modes should be addressed based on the severity. The following potential outcomes are general in nature and can be further subdivided in a more detailed analysis: *incident*, *accident*, *injury*, *fatality*, *environmental destruction*, and *financial loss*.

To identify how individual failures may interact, each failure mode perspective can be compared with the other failure modes. For example, Figure 3.26 represents possible failure mode interactions between the system source and temporal perspectives. Representing potential failures in this way not only helps to portray visually the possible interactions but also can reveal unanticipated failures. For example, it can be seen that excessive requirements must be considered for drivers, vehicles, infrastructure, and economics.

Since it is the consequences of failures that affect the system, then comparing failure modes with the general consequence category reflects additional potential consequences. Figure 3.27 graphically compares the failures of the system source perspective with the consequences listed there. For example, it illustrates how the consequence of environmental destruction can apply to drivers, vehicles, infrastructure, and economics. Enumerating each failure mode and considering how environmental damage may result ensure that the environmental destruction consequence is properly taken into consideration across all failure modes. It also demonstrates how environmental destruction can result in multiple failure modes.

3.13 FOOD-POISONING SCENARIOS

Consider the risks of meat poisoning initiated at a slaughterhouse. Figure 3.28 depicts the entire process of meat production from the farm to the consumer [Haimes and Horowitz, 2004]. Each stage of the process constitutes a subsystem that can be characterized by a number of state variables representing its essence (along with other building blocks as discussed earlier). A sample of important state variables for a slaughterhouse includes production level, employees (number, skills, types, tenure, and wages), specific equipment, and the technology used.

These important state variables would be of interest to the terrorist networks, to the manager of the slaughterhouse, and to the intelligence analyst. Hence, an AMP-HHM game can shed light on this important public health issue. The goal of the

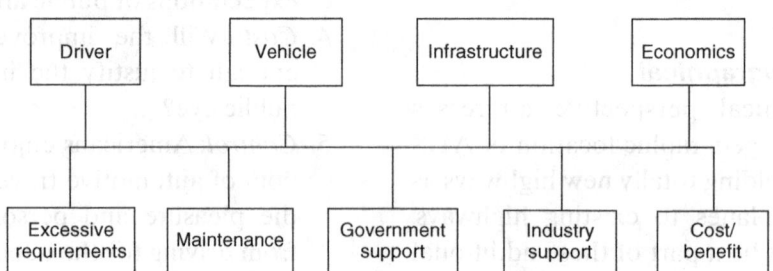

Figure 3.26 System source and temporal perspectives.

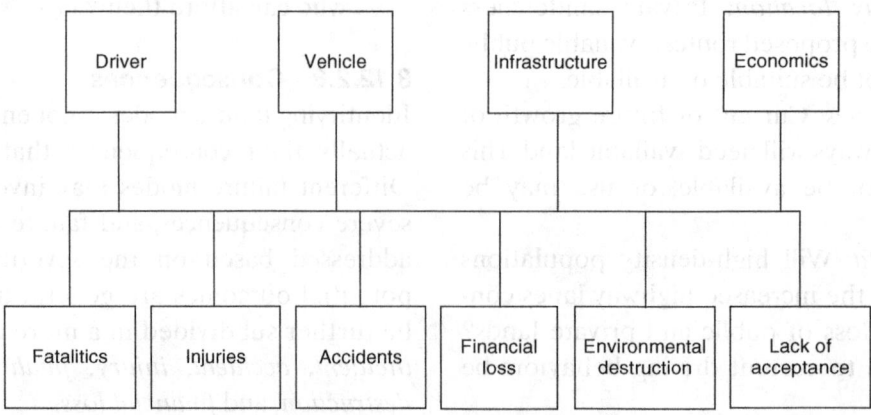

Figure 3.27 System source and consequences perspectives.

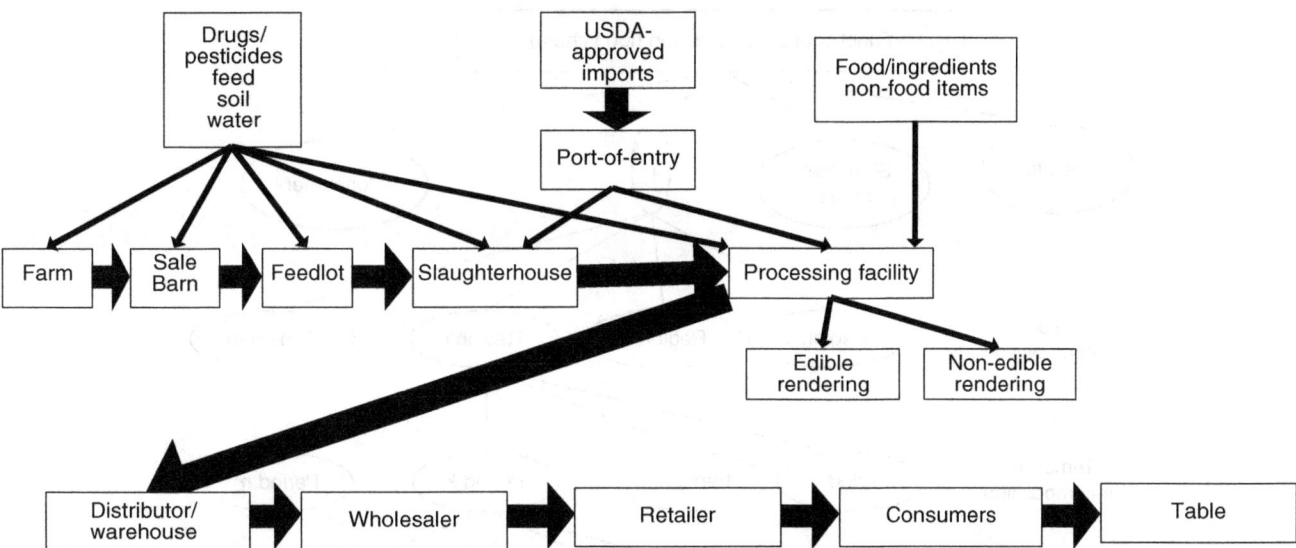

Figure 3.28 Meat from farm to table.

terrorists is to exploit or modify these state variables to their advantage; the goal of the facility manager is to operate the slaughterhouse more efficiently and effectively. Both want to alter these state variables, albeit for opposite goals. Knowledge of these state variables is also essential to the intelligence community. Perceiving the vulnerabilities of a subsystem, analysts can track, connect, and relate available intelligence to a specific scenario, assuming that such vulnerabilities are of interest to terrorists. In this sense, the states of the system (or subsystem) constitute a critically needed guide to the necessary information hidden within the chaotic databases.

The example presented in Horowitz and Haimes [2003] provides an illustrative HHM analysis that was performed by a large and capable Blue Team attempting to anticipate a food-poisoning terrorist attack to be executed at a slaughterhouse. This class of attack is a subset of Figures 3.29 and 3.30, which present the results of a broader HHM analysis for meat poisoning. In this, the slaughterhouse constitutes just one of the components in the food-poisoning problem, which also includes regional, temporal, and food product decomposition. Figure 3.31 represents a subset of an HHM for a meat-poisoning scenario at a slaughterhouse. The figure contains a variety of potential attack elements, such as avoiding the security process at the slaughterhouse, gaining employment there, and bribing the owners or key employees.

The HHM diagram in Figure 3.31 also presents the results of the Blue Team analysis focused on the slaughterhouse. For example, the head topic *Macro* provides ownership, location, slaughterhouse capacity, and customer base as critical states related to the risk of being selected as a target. For study purposes, the authors organized four undergraduate engineering students into a Red Team (offense) and four on a Blue Team (defense) to evaluate a possible slaughterhouse food-poisoning attack. This did not attempt to emulate an actual terrorist action, but was planned as a reasonable first step to illustrate the adaptive two-player HHM concept. As part of their preparation, the Red Team members looked into a number of prior terrorist attacks and open-source information about the situations leading up to each attack. The team chose its attack based on information that was available on the Internet (incidentally raising security issues about information control as well as food poisoning). Note that the Blue Team HHM never explicitly contemplated the Internet as an intelligence source for a slaughterhouse attack. In fact, for this example, the only material difference between the Red Team's analysis and the Blue Team's HHM result presented in Figure 3.31 is the addition of the subtopic *Internet* under the head topic *Macro*. The following paragraphs outline the Red Team analysis. It is based on actual data available

110 IDENTIFYING RISK THROUGH HIERARCHICAL HOLOGRAPHIC MODELING AND ITS DERIVATIVES

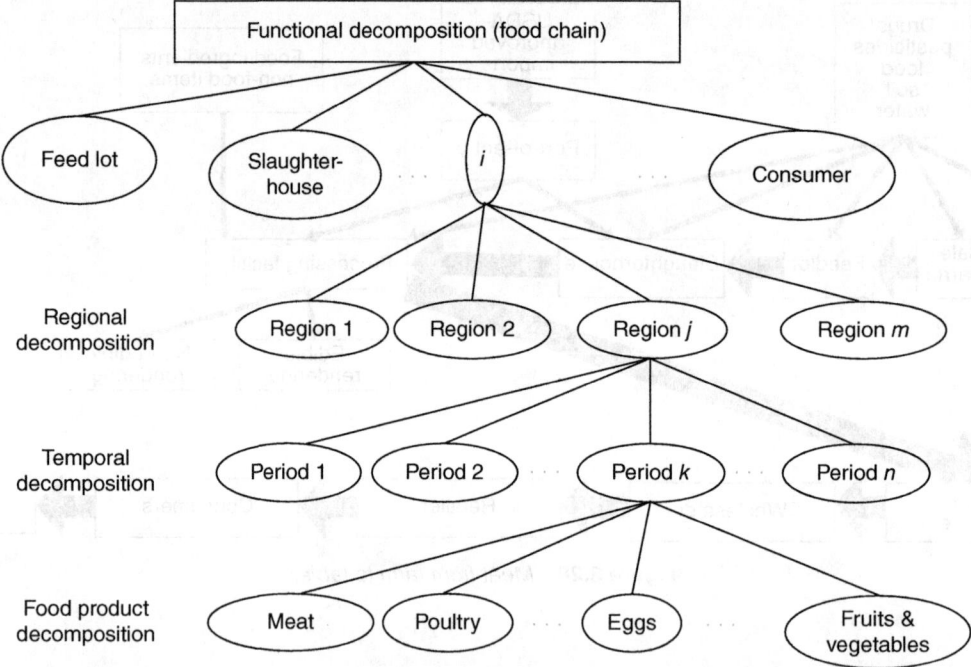

Figure 3.29 *Functional decomposition of the food chain.*

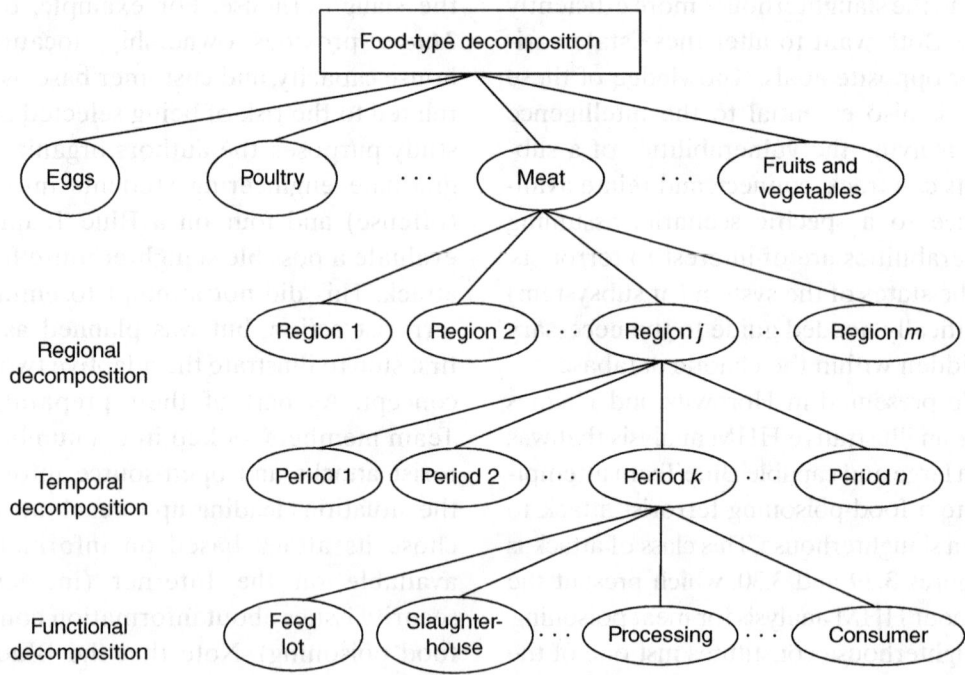

Figure 3.30 *A hierarchy of food-type decomposition.*

on the Internet. For security reasons, the websites and other specific details found on the Internet are not identified here.

The Red Team decided that a particular meat-processing center would be its target. That center provides on its website information for prospective customers about its floor plan, the transportation schedule for shipping meat to different parts of the United States, and the storage location of its packaged meat prior to shipping. All of these factors

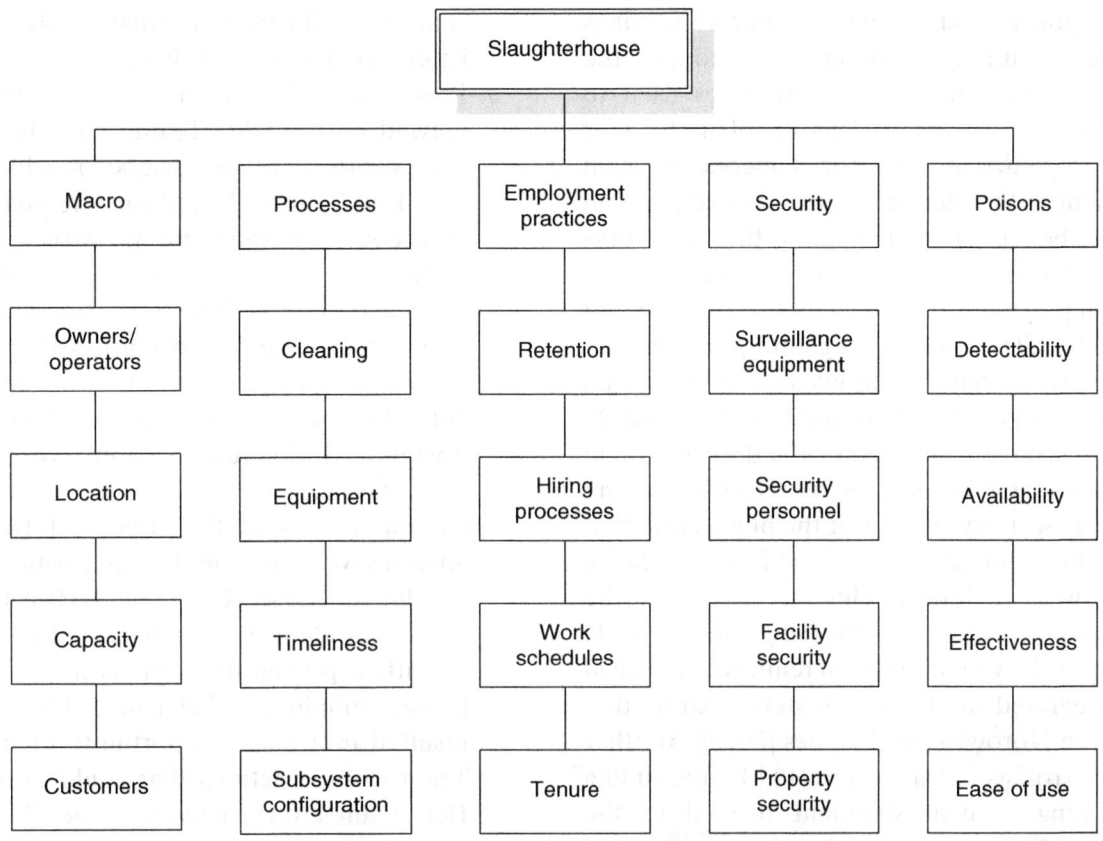

Figure 3.31 Slaughterhouse food-poisoning scenarios.

were part of the Blue Team HHM analysis, but there was no recognition that specific plant information would be available in detail on the Internet.

Next, the Red Team decided that the possible tracking of poison sales and storage locations by the US government could provide a major risk. The members of the Red Team were not experts on this subject and considered their lack of knowledge risky. They decided to search the Internet for research efforts related to poisons with the vague belief that poisons at certain research labs might not be tracked at all and might be easy to steal. In fact, on one website, the Red Team learned of a potent poison that could be delivered with a specific procedure in small quantities and yet would provide significant consequences. The Blue Team had conducted its own study of available poisons (see the head topic *Poisons* in the Blue Team HHM diagram) and had assessed the potential for many different poisons as weapons of terrorism. In their analysis, they gave too much credit to the terrorist organization in terms of their knowledge of poisons and corresponding most likely selections. In addition, as part of the set of possible choices, they never contemplated searching for poisons on the type of website found by the Red Team. The Red Team combined its information to decide that using the specific procedure could contaminate the meat that was already packaged and ready for transport. The poisoned shipment would be one scheduled for shipment to the Washington, DC, area, possibly resulting in a positive side effect of poisoning important government officials. The Red Team risk analysis concluded that the procedure used would not be noticeable enough to be detected in transport or at the retail shop. The Blue Team analysis had concluded that poisoning individual portions would be inefficient and noticeable; as a result, there was relatively low interest in that kind of scenario.

In order to poison the meat, a terrorist would have to hide in the meat storage facility. The processing center's website photos described how the meat is stored, so that simple calculations could be made about (i) the ability of a single person to inject the poison and (ii) the number of servings that could potentially be affected. In addition, the website discussed the security processes used to protect meat at the plant. This enabled the Red Team to organize a plan for someone to gain employment at the plant and use a trusted position to hide in the storage facility. Note that the workforce at such plants is known to be very transient, so gaining employment was not considered to be an unlikely event. The concepts of gaining employment and then carrying out the actions required to poison meat were included as part of the Blue Team HHM analysis, so critical strategies for the defense would be to screen new employees more carefully and improve the security systems at the plant. Note that responsibilities for employment and plant security are local, while federal intelligence would be the most likely source for identifying questionable employees. This type of situation reinforces the need for an integrated intelligence system such as that presented in Horowitz and Haimes [2003]. Another issue for terrorists in this case would be the timing for poisoning a meat shipment headed to the Washington, DC, area. From the website, the Red Team also got the daily schedule for shipments from the plant, which provided significant information for planning the timing of the attack.

While many more details could be provided for this example, the aforementioned experiment leads to several important conclusions:

1. The Red Team provided a useful set of additional considerations for the Blue Team to address. Most notable was the idea that terrorist networks could plan an attack based on Internet information. This was the only significant difference between the Blue and Red Team assessments. The exercise permitted the Blue Team to integrate Red Team results into its analysis in a straightforward fashion.
2. The Blue Team assessment of possible poisons assumed expert judgment by the terrorist team in an area where they lacked expertise. As a result, the poison selected was not considered by the Blue Team, although it included the same risks as the Red Team did in its HHM analysis. This highlights the importance of sharing intelligence information about terrorist knowledge and capabilities.
3. Poisoning individual portions of meat was viewed by the Blue Team as inefficient and discoverable and was assigned low likelihood. The Red Team selected an effective poison that could contaminate in very low doses with a specific procedure. This could help to reduce detection. The Red Team also had access to Internet information that showed the number of portions of stored meat that could be poisoned. This led to a decision that poisoning individual portions would be an effective plan.
4. Certain results of the Blue and Red Team analyses were very similar. As a critical area of overlap with the Red Team HHM, the Blue Team had identified the potential importance of either preventing employment or monitoring employee behavior. This overlap resulted in a major opportunity for the Blue Team to take actions that could prevent the Red Team attack. However, the ability of a local company to monitor employment in the suggested fashion would require intelligence collectors to transfer information for local use. Such sharing is not the practice in today's counterterrorism system.
5. The AMP-HHM game enables the Blue Team to continuously improve its HHM by incorporating missing elements gathered from the Red Team's HHM. Indeed, this is an inherent advantage of the HHM process — as additional intelligence becomes available, the HHM converges to a *complete set* of risk (or *success*) scenarios.

These conclusions point to the overall assessment that a two-player HHM analysis has the potential to help intelligence agencies deal with terrorism. The HHM approach is sufficiently flexible and adaptive to permit both defense and offense assessments that result in considerable overlap. Since both analyses start with identifying the states of the target system, comparing the analyses permits the defense to readily identify the differences and to integrate additional information into future models.

REFERENCES

Agrawal, A.B., 2006, *Integrated Risk Assessment and Management for the 2006 Virginia Gubernatorial Inauguration*, M.S. thesis, University of Virginia, Charlottesville, VA.

American Society of Civil Engineers (ASCE), 2005, *Report Card for America's Infrastructure (2005 Report Card)*, Available online: http://ascelibrary.org/doi/book/10.1061/9780784478851 (Accessed February 6, 2015).

Arquilla, J., and D. Ronfeldt, 2001, *Networks and Netwars*, National Defense Research Institute, RAND, Pittsburgh, PA.

Blauberg, I.V., V.N. Sadovsky, and E.G. Yudin, 1977, *Systems Theory: Philosophical and Methodological Problems*, Progress Publishers, New York.

Blum, B.I., 1992, *Software Engineering: A Holistic View*, Oxford University Press, New York.

Boehm, B.W., 1981, *Software Engineering Economics*, Prentice-Hall, Englewood Cliffs, NJ.

C520, 1995, *Operations Other Than War (OOTW)*, Reading material from the US Army Command and General Staff College, Fort Leavenworth, KS.

Carr, M.J., S.L. Konda, I. Monarch, F.C. Ulrich, and C.F. Walker, 1993, *Taxonomy-Based Risk Identification*, Technical Report CMU/SEI-93-TR-6, Software Engineering Institute, Carnegie Mellon University, Pittsburgh, PA.

Center for Army Lessons Learned (CALL), 1993, *Operations Other Than War Volume IV: Peace Operations*, Newsletter 93-8, US Army Combined Arms Command (CAC), Fort Leavenworth, KS.

Chu, C., and P. Durango-Cohen, 2007, Estimation of infrastructure performance models using state-space specifications of time series models, *Transportation Research Part C: Emerging Technologies* **15**(1): 1732.

Dombroski, M., Y.Y. Haimes, J.H. Lambert, K. Schlussel, and M. Sulcoski, 2002, Risk-based methodology for support of operations other than war, *Military Operations Research* **7**(1): 19–38.

Durango-Cohen, P., 2007, A time series analysis framework for transportation infrastructure management, *Transportation Research Part B: Methodological* **41**(5): 493–505.

Federal Highway Administration (FHWA), 2007, *Tables of Frequently Requested NBI Information*, Available online: http://www.fhwa.dot.gov/bridge/britab.htm (Accessed January 10, 2015).

Gordon, W.J.J., 1968, *Synectics: The Development of Creative Capacity*, Collier Books, New York.

Haimes, Y.Y., 1977, *Hierarchical Analyses of Water Resources System: Modeling and Optimization of Large-Scale System*, McGraw-Hill, New York.

Haimes, Y.Y., 1981, Hierarchical holographic modeling, *IEEE Transactions on Systems, Man, and Cybernetics* **11**(9): 606–617.

Haimes, Y.Y., 1992, Sustainable development: a holistic approach to natural resource management, *IEEE Transactions on Systems, Man, and Cybernetics* **22**(3): 413–417.

Haimes, Y.Y., 2002, Roadmap for modeling risks of terrorism to the homeland, *Journal of Infrastructure Systems* **8**(2): 35–41.

Haimes, Y.Y., and B. Horowitz, 2004, Adaptive two-player hierarchical holographic modeling game for counterterrorism intelligence analysis, *Journal of Homeland Security and Emergency Management* **1**(3): 1–21.

Haimes, Y.Y., and D. Macko, 1973, Hierarchical structures in water resources systems management, *IEEE Transactions on Systems, Man, and Cybernetics* **3**(4): 396–402.

Haimes, Y.Y., and K. Tarvainen, 1981, Hierarchical-multiobjective framework for large scale systems, In *Multicriteria Analysis in Practice*, P. Nijkamp, and J. Spronk (Eds.), Gower, London, pp. 201–232.

Haimes, Y.Y., P. Das, and K. Sung, 1979, Level-B multiobjective planning for water and land, *Journal of Water Resources Planning and Management Division* **105**(WR2): 385–401.

Haimes, Y.Y., K. Tarvainen, T. Shima, and J. Thadathil, 1990a, *Hierarchical Multiobjective Analysis of Large-Scale Systems*, Hemisphere Publications, New York.

Haimes, Y.Y., D. Li, and V. Tulsiani, 1990b, Multiobjective decision-tree analysis, *Risk Analysis* **10**(1): 111–129.

Haimes, Y.Y., N.C. Matalas, J.H. Lambert, B.A. Jackson, and J.F. Fellows, 1998, Reducing the vulnerability of water supply systems to attack, *Journal of Infrastructure Systems* **4**(4): 164–177.

Haimes, Y.Y., Z. Yan, and B.M. Horowitz, 2007, Integrating Bayes' theorem with dynamic programming for optimal intelligence collection, *Military Operations Research* **12**(4): 17–31.

Hall, A.D. III, 1989, *Metasystems Methodology: A New Synthesis and Unification*, Pergamon Press, Elmsford, NY.

Horowitz, B., and Y.Y. Haimes, 2003, Risk-based methodology for scenario tracking for terrorism: a possible new approach for intelligence collection and analysis, *Systems Engineering* **6**(3): 152–169.

Johnson, B.W., 1989, *Design and Analysis of Fault Tolerant Digital Systems*, Addison-Wesley, Reading, MA.

Kaplan, S., 1991, The general theory of quantitative risk assessment, In *Risk Based Decision Making in Water Resources V*, Y. Haimes, D. Moser, and E. Stakhiv (Eds.), American Society of Civil Engineers, New York, pp. 11–39.

Kaplan, S., 1993, The general theory of quantitative risk assessment—its role in the regulation of agricultural pests, *Proceedings of the APHIS/NAPPO International Workshop on the Identification, Assessment and Management of Risks due to Exotic Agricultural Pests* **11**(1): 123–126.

Kaplan, S., 1996, *An Introduction to TRIZ, The Russian Theory of Inventive Problem Solving*, Ideation International Inc., Southfield, MI.

Kaplan, S., and B.J. Garrick, 1981, On the quantitative definition of risk, *Risk Analysis* **1**(1): 11–27.

Kaplan, S., B. Zlotin, A. Zussman, and S. Vishnipolski, 1999, *New Tools for Failure and Risk Analysis—Anticipatory Failure Determination and the Theory of Scenario Structuring*, Ideation International Inc., Southfield, MI.

Kaplan, S., Y.Y. Haimes, and B.J. Garrick, 2001, Fitting hierarchical holographic modeling into the theory of scenario structuring and a resulting refinement of the quantitative definition of risk, *Risk Analysis* **21**(5): 807–815.

Kuhn, H.W. (Ed.), 1997, *Classics in Game Theory*, Princeton University Press, Princeton, NJ.

Li, D., and Y.Y. Haimes, 1992a, Optimal maintenance-related decision making for deteriorating water distribution systems, Part 1: Semi-Markovian model for a water main, *Water Resources Research* **28**(4): 1053–1061.

Li, D., and Y.Y. Haimes, 1992b, Optimal maintenance-related decision making for deteriorating water distribution systems, Part 2: Multilevel decomposition approach, *Water Resources Research* **28**(4): 1063–1070.

Macko, D., and Y.Y. Haimes, 1978, Overlapping coordination of hierarchical structures, *IEEE Transactions on Systems, Man, and Cybernetics* **8**(10): 745–751.

Monahan, J., 2000, The scientific status of research on clinical and actuarial predictions of violence, In *Modern Scientific Evidence: The Law and Science of Expert Testimony*, Vol. **1**, Second edition, D. Faigman, D. Kaye, M. Saks, and J. Sanders (Eds.), West Publishing Company, St. Paul, MN, pp 423–445.

Monahan, J., H. Steadman, E. Silver, P. Appelbaum, P. Robbins, E. Mulvey, L. Roth, T. Grisso, and S. Banks, 2001, *Rethinking Risk Assessment: The MacArthur Study of Mental Disorder and Violence*, Oxford University Press, New York.

National Environmental Policy Act (NEPA), P.L. 91-190, January, 1, 1969.

Sage, A.P., 1977, *Methodology for Large Scale Systems*, McGraw-Hill, New York.

Sage, A.P., 1992, *Systems Engineering*, John Wiley & Sons, New York.

Schneiter, C.R., Y.Y. Haimes, D. Li, and J.H. Lambert, 1996, Capacity reliability of water distribution networks and optimum rehabilitation decisionmaking, *Water Resources Research* **32**(7): 2271–2278.

Schooff, R.M., Y.Y. Haimes, and C. Chittister, 1997, A holistic management framework for software acquisition, *Acquisition Review Quarterly* **4**(1): 35–85.

Singh, M.G., 1987, *Systems and Control Encyclopedia: Theory, Technology, Applications*, Pergamon Press, New York.

Slovic, P, 2000, *The Perception of Risk*, Cromwell Press, Trowbridge.

United States National Science and Technology Council (US NSTC), 1994, *Technology for a Sustainable Future: A Framework for Action*, Office of Science and Technology Policy, US Government Printing Office, Washington, DC.

von Neumann, J., and O. Morgenstern, 1972, *Theory of Games and Economic Behavior*, Princeton University Press, Princeton, NJ.

Warfield, J.N., 1976, *Social Systems—Planning and Complexity*, John Wiley & Sons, New York.

World Commission on Environment and Development (WCED), 1987, *Our Common Future*, Oxford University Press, Oxford.

Yan, Z., 2007, *Risk Assessment and Management of Complex Systems with Hierarchical Analysis Methodologies*, Ph.D. dissertation, University of Virginia, Charlottesville, VA.

Yan, Z., and Y.Y. Haimes, 2008, *Hierarchical Coordinated Bayesian Modeling for Risk Analysis*, Technical Report 20–2008, Center for Risk Management of Engineering Systems, University of Virginia, Charlottesville, VA.

Yan, Z., Y.Y. Haimes, and M.G. Waller, 2008, *Modeling Sparse Data in Risk Analysis of Complex Systems with Coordinated Hierarchical Bayesian Models*, Technical Report 21–2008, Center for Risk Management of Engineering Systems, University of Virginia, Charlottesville, VA.

Zigler, B.P., 1984, *Multifaceted Modeling and Discrete Simulation*, Academic Press, New York.

4

Modeling and Decision Analysis

4.1 INTRODUCTION

The term *quantitative risk analysis* generally connotes reliance on probability and statistics. However, select quantitative risk-based decisionmaking methodologies, such as game theory, do not require knowledge of probabilities. Maximizing the minimum (maximin) gain, minimizing the maximum (minimax) loss, and maximizing the maximum (maximax) gain are but a few examples of decisionmaking criteria for handling risk and uncertainty without adhering to probabilities. The first part of this chapter will explore these decisionmaking criteria and measures.

Quantitative risk assessment builds on the existence of probabilities that describe the likelihood of outcomes, such as consequences. In general, probabilities are derived on the basis of historical records, statistical analysis, and/or systemic observations and experimentation. We commonly refer to probabilities that are derived from this process as *objective probabilities*. Often, however, situations arise where the database is so sparse and experimentation is so impractical that *objective probabilities* must be supplemented with *subjective probabilities*, or probabilities that are based on expert evidence, often referred to as *expert judgment*. In this chapter, we focus on generating probabilities on the basis of expert evidence. We will introduce two methods for generating expert evidence-based probabilities—the fractile and the triangular distribution methods.

To be responsive to the risk of extreme and catastrophic events, the expected value of risk will be supplemented in Chapter 8 with the conditional expected value. Since the concept of conditional expectation has not been discussed yet, some example problems introduced in this chapter will be revisited in Chapter 8, where the conditional expected value will be evaluated for added insight.

As a prelude to the multiobjective decision-tree analysis discussed in Chapter 9, single-objective decision-tree (SODT) analysis will be reviewed in this chapter. Finally, we will introduce influence diagrams, population dynamic models, and the Phantom

---

Risk Modeling, Assessment, and Management, Fourth Edition. Yacov Y. Haimes.
© 2016 John Wiley & Sons, Inc. Published 2016 by John Wiley & Sons, Inc.

System Model (PSM). All of the above concepts and methodologies will be illustrated with example problems.

4.2 DECISION RULES UNDER UNCERTAINTY

Most of this book is written with the assumption that the reader has the ability to generate either objective probabilities or expert evidence-based probabilities. We will use the conventional notation of $p(s_j)$ as the probability associated with scenario s_j (or the state of nature s_j). The ith decision or action adopted by the decisionmaker will be denoted by a_i, and the outcome from the combination of the scenarios and actions is the pairs (a_i, s_j). The payoff associated with the pair $(a_i, s_j), i = 1, 2, \ldots, I; j = 1, 2, \ldots, J$, will be denoted by μ_{ij}.

When $p(s_j)$ and μ_{ij} are known, the conventional criterion for decisionmaking is the expected value of gain (or loss or risk). As noted before, a supplement to the expected value of risk, termed the conditional expected value of risk, will be introduced in Chapter 8. Thus, maximizing the expected monetary value (EMV) of gain can be written as

$$\max_{1 \le i \le I} \sum_{j=1}^{J} p(s_j) \mu_{ij} \qquad (4.1)$$

In the absence of any knowledge of probabilities, it is not possible to use the expected value as a gain or risk index. The following decision rules are then common for this situation.

4.2.1 The Pessimistic Rule (Maximin or Minimax Criterion)

Following this criterion, the conservative decisionmaker seeks to maximize the minimum gain or, alternatively, minimize the maximum loss. If μ_{ij} represents a payoff, then we have

$$\max_{1 \le i \le I} \left(\min_{1 \le j \le J} \mu_{ij} \right) \qquad (4.2)$$

If μ_{ij} represents a loss or a risk, then we have

$$\min_{1 \le i \le I} \left(\max_{1 \le j \le J} \mu_{ij} \right) \qquad (4.3)$$

These criteria ensure that the decisionmakers will at least realize the minimum gain or avoid maximum loss.

4.2.2 The Optimistic Rule (Maximax Criterion)

Following this criterion, the decisionmaker is most optimistic and seeks to maximize the maximum gain. Mathematically, the maximax criterion can be represented as

$$\max_{1 \le i \le I} \left(\max_{1 \le j \le J} \mu_{ij} \right) \qquad (4.4)$$

4.2.3 The Hurwitz Rule

The Hurwitz rule offers a compromise between two extreme criteria through the use of an α-index. The decisionmaker's degree of optimism is specified through a parameter α that ranges between 0 and 1 ($0 \le \alpha \le 1$). More specifically, to apply the Hurwitz rule, one has to form a linear combination between the maximin and the maximax criteria for each alternative a_i:

$$\max_{1 \le i \le I} \mu_i(\alpha) = \max_{1 \le i \le I} \left(\alpha \min_{1 \le j \le J} \mu_{ij} + (1 - \alpha) \max_{1 \le j \le J} \mu_{ij} \right),$$
$$0 \le \alpha \le 1 \qquad (4.5)$$

Note that for $\alpha = 0$, $\max_{1 \le i \le I} \mu_i(\alpha)$ represents the maximax criterion and for $\alpha = 1$, $\max_{1 \le i \le I} \mu_i(\alpha)$ represents the maximin criterion. The following example problem should add more insight into the previous discussion.

A northern Virginia furniture corporation has excess manpower and equipment capacity. The management decided to allocate these resources to the manufacture of new products. After a detailed marketing analysis, a shortage of high-quality crutches was discovered to be prevalent in the East and Midwest. For the most effective use of resources, however, the engineering and manufacturing team recommended that the corporation manufacture crutches in only one of three possible sizes—small, regular, or large.

An engineering team was commissioned to design high-quality crutches that made use of the excess equipment capacity. The design team produced three prototypes, which were subject to elaborate testing procedures, including structural strength and reliability, cost-effectiveness, and human and aesthetic factors, among others.

Marketing analysis indicated that given the large shortage of crutches in the United States and relatively limited excess equipment capacity, factory-made crutches could be sold with the following estimated returns on investment.

Table 4.1 is commonly written in terms of a payoff matrix as in Table 4.2.

Applying the pessimistic rule (maximize the minimum gain), the maximin criterion for the sales of crutches yields the following:

For a_1 (small):   min(250, 100, −150) = −150
For a_2 (regular): min(400, 220, −30) = −30
For a_3 (large):   min(200, 100, 10) = 10

Thus, applying the maximin criterion implies a gain of at least $10,000 following a_3—that is, the manufacture of large-size crutches.

Applying the optimistic rule, that is, the maximax criterion (maximize the maximum gain from the sale of crutches), yields the following:

For a_1 (small):   max(250, 100, −150) = 250
For a_2 (regular): max(400, 220, −30) = 400
For a_3 (large):   max(200, 100, 10) = 200

Thus, the best policy following the most optimistic criterion is to manufacture regular-size crutches (a_2), yielding a return of at most $400,000.

TABLE 4.1 Profits as a Function of Sales Potential and Crutch Size

Crutch Size	Sales Potential		
	Excellent	Good	Poor
Small	$250,000	$100,000	−$150,000
Regular	$400,000	$220,000	−$30,000
Large	$200,000	$100,000	$10,000

TABLE 4.2 Payoff Matrix ($1000)

	$j=1$ (s_1)	$j=2$ (s_2)	$j=3$ (s_3)
$i=1$ (a_1)	250	100	−150
$i=2$ (a_2)	400	220	−30
$i=3$ (a_3)	200	100	10

Applying the Hurwitz rule, which compromises between two extremes through the use of the index α, yields the following:

$$\max_{a_1,a_2,a_3} \left\{ \mu_i(\alpha) = \alpha \underbrace{\min_{1\le j\le 3} \mu_{ij}}_{\text{Pessimistic}} + (1-\alpha) \underbrace{\max_{1\le j\le 3} \mu_{ij}}_{\text{Optimistic}} \right\}, \quad 0 \le \alpha \le 1 \quad (4.6)$$

For $\alpha = 1$: pessimistic
For $\alpha = 0$: optimistic

Table 4.3 summarizes the pessimistic and optimistic outcomes for each decision a_i, $i = 1, 2, 3$:

At a_1: $\mu_1(\alpha) = -150{,}000\alpha + 250{,}000(1-\alpha)$
         $= 250{,}000 - 400{,}000\alpha$ (4.7a)

At a_2: $\mu_2(\alpha) = -30{,}000\alpha + 400{,}000(1-\alpha)$
         $= 400{,}000 - 430{,}000\alpha$ (4.7b)

At a_3: $\mu_3(\alpha) = 10{,}000\alpha + 200{,}000(1-\alpha)$
         $= 200{,}000 - 190{,}000\alpha$ (4.7c)

Note that Eq. (4.7a) represents straight-line functions of the variable α, $0 \le \alpha \le 1$. Plotting each of these straight lines as a function of α is depicted in Figure 4.1.

Note that alternative a_1 is being dominated by alternative a_2 for all values of α. In other words, management should never manufacture the small-size crutches. On the other hand, for $0 \le \alpha \le 5/6$, the best policy is to manufacture regular-size crutches (a_2), and for $5/6 \le \alpha \le 1$, the best policy is to manufacture large-size crutches (a_3). This value of α can be easily determined by solving for the intersection of the two straight lines:

$$400{,}000 - 430{,}000\alpha = 200{,}000 - 190{,}000\alpha$$
$$240{,}000\alpha = 200{,}000$$
$$\alpha = 5/6$$

TABLE 4.3 Summary of Information for the Hurwitz Rule

	Sales Potential ($1000)				
	Excellent (s_1)	Good (s_2)	Poor (s_3)	Pessimistic	Optimistic
Small crutches (a_1)	250	100	−150	−150	250
Regular crutches (a_2)	400	220	−30	−30	400
Large crutches (a_3)	200	100	10	−10	200

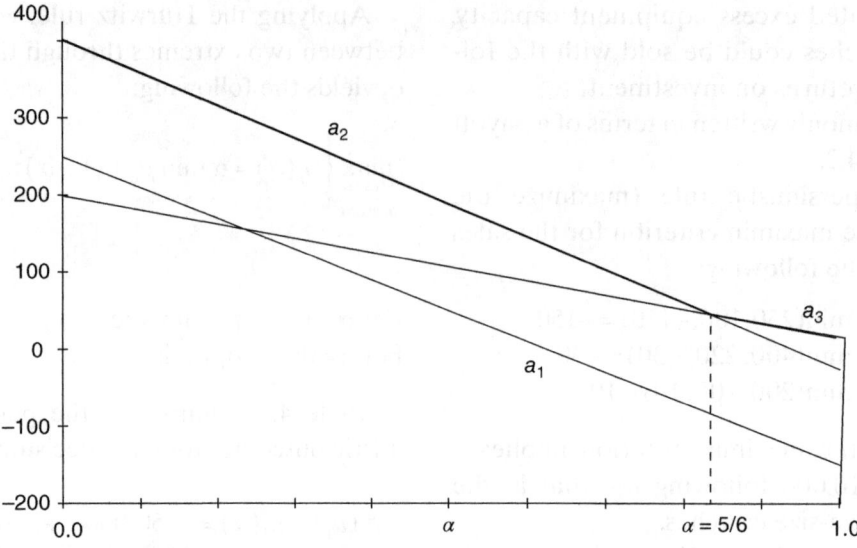

Figure 4.1 The Hurwitz rule.

Although mathematically at $\alpha = 5/6$ management is supposed to be indifferent between manufacturing regular-size and large-size crutches, other considerations are likely to dictate the ultimate choice.

4.3 DECISION TREES

Among the most commonly used tools in risk-based decisionmaking is the decision tree [Raiffa, 1968]. The popularity of the decision tree stems from its reliance on an integrative approach of graphical and analytic presentations. The graphical component is descriptive and simple to understand. The analytical component builds on Bayes' theorem. Figure 4.2 represents a generic decision tree with the following basic components:

1. *Decision node.* Decision nodes are designated by a square □. Branches emanating from a decision node represent the various decisions (actions) to be investigated. In the crutches problem, for example, there are only three options: manufacture small, regular, or large crutches. It is conventional to designate each alternative choice by a letter, for example, a_i, and identify each branch with that decision choice (i.e., a_1, a_2, and a_3 for our example problems).

2. *Chance node.* Chance nodes are designated by a circle ○. Branches emanating from a chance node represent the various states of nature with their associated probabilities. In the crutches problem, there are three states of nature:
   - Excellent potential sales, s_1, with probability $p(s_1) = 0.3$
   - Good potential sales, s_2, with probability $p(s_2) = 0.5$
   - Poor potential sales, s_3, with probability $p(s_3) = 0.2$

3. *Consequences.* The value of the consequences (outcomes) (e.g., cost, benefit, or risk) is written at the end of each branch. In Chapter 9, when we introduce multiobjective decision trees (MODTs), there will be a vector of consequences at the end of each branch. We will designate the consequence associated with the ith decision and jth state of nature by μ_{ij}. For example, in the crutches problem, the profit obtained from manufacturing small crutches (a_1) with an excellent probability of sales (s_1) is μ_{11}.

Note that one of the attractive features of decision trees is the ability to represent and analyze multiple stages in the decisionmaking process. Indeed, at each stage, new probabilities are introduced at the

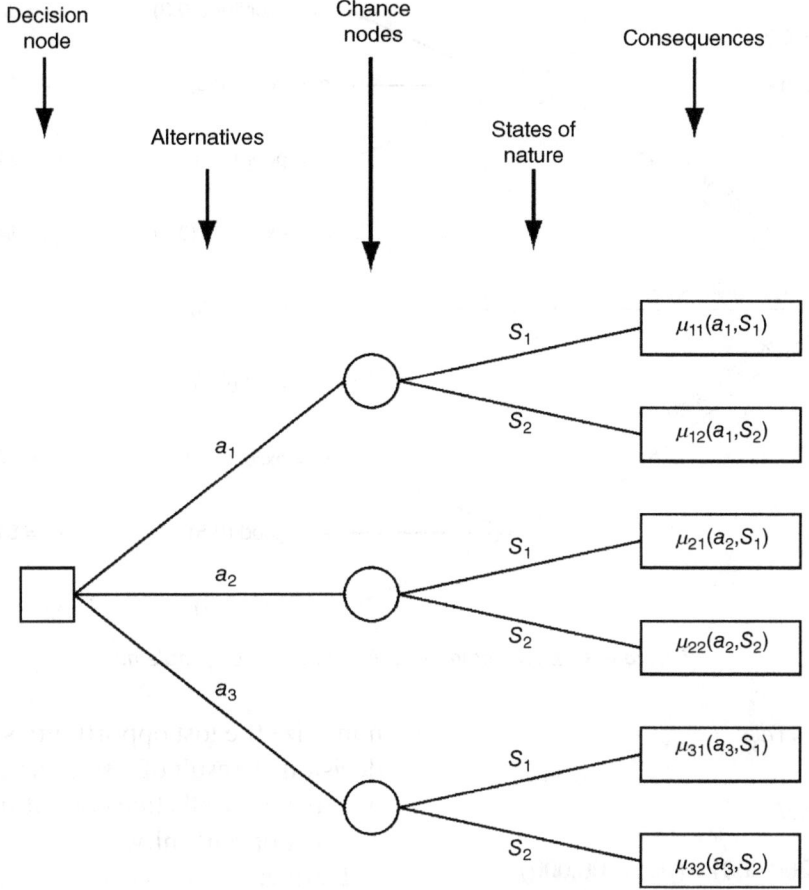

Figure 4.2 Generic decision tree.

chance nodes on the basis of new information that has been gathered over time. In this case, several sequences of *columns* of decision nodes and chance nodes will constitute the decision tree.

4.3.1 The Crutches Problem Revisited

The only modification that we are adding here to the crutches problem is our knowledge of the probabilities of potential sales. Figure 4.3 represents the decision tree for the modified crutches problem using the information presented in Table 4.2 and our knowledge of the probabilities associated with each chance node. The expected value of profits is used as the criterion with which to determine the optimal manufacturing policy. Note, however, that this measure is not necessarily the only one available to analysts, nor is it the best one under all conditions. Maximum likelihood measures, or conditional expected values, are other metrics.

4.3.1.1 Expected value of outcome

To determine the optimal manufacturing policy, we calculate the expected value of profits for each of the three alternative decision options, denoted by $E[a_i]$:

Small-size crutches (a_1):

$$E[a_1] = \sum_{j=1}^{3} p(s_j)\mu_{1j}$$
$$= (0.3)(250,000) + (0.5)(100,000) \quad (4.8)$$
$$+ (0.2)(-150,000)$$
$$= \$95,000$$

Regular-size crutches (a_2):

$$E[a_2] = \sum_{j=1}^{3} p(sj)\mu_{2j}$$
$$= (0.3)(400,000) + (0.5)(220,000) \quad (4.9)$$
$$+ (0.2)(-30,000)$$
$$= \$224,000$$

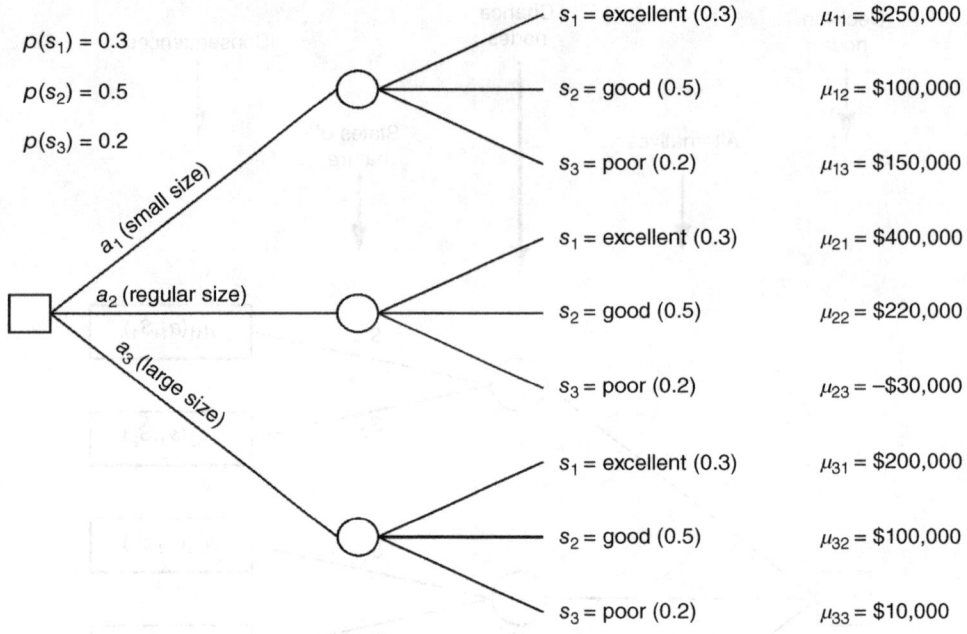

Figure 4.3 Basic information for the crutches problem.

Large-size crutches (a_3):

$$E[a_3] = \sum_{j=1}^{3} p(s_j)\mu_{3j}$$
$$= (0.3)(200,000) + (0.5)(100,000)$$
$$\quad + (0.2)(10,000) \qquad (4.10)$$
$$= \$112,000$$

The optimal manufacturing policy is determined by maximizing the expected value of profit for all policy options:

$$\max_{1\le i\le 3} \sum_{j=1}^{3} p(s_j)\mu_{ij} \qquad (4.11)$$

or

$$\max\{E[a_1], E[a_2], E[a_3]\} \qquad (4.12)$$
$$\max\{95,000, 224,000, 112,000\}$$

Clearly, the optimal policy is to manufacture regular-size crutches at an expected profit of \$224,000. Figure 4.4a depicts the expected value of profits for each of the three alternative decision options.

4.3.1.2 Expected value of opportunity loss

The expected opportunity loss (EOL) measure is essentially a modification of the expected gain metric. Instead of maximizing the net profit, the decisionmaker seeks in the EOL measure to minimize the lost opportunities associated with each decision. A result of less than the maximum possible profits under all states of nature will be considered as a lost opportunity.

Define: $M_j = \max_{1\le i\le 3}\{\mu_{ij}\}$, $j = 1, 2, 3$

For $j = 1$: $M_1 = \max_{1\le i\le 3}\{\mu_{11}, \mu_{21}, \mu_{31}\}$
$$M_1 = \max\{250,000, 400,000, 200,000\}$$
$$M_1 = 400,000$$

For $j = 2$: $M_2 = \max_{1\le i\le 3}\{\mu_{12}, \mu_{22}, \mu_{32}\}$
$$M_2 = \{100,000, 220,000, 100,000\}$$
$$M_2 = 220,000$$

For $j = 3$: $M_3 = \max_{1\le i\le 3}\{\mu_{13}, \mu_{23}, \mu_{33}\}$
$$M_3 = \{-150,000, -30,000, 10,000\}$$
$$M_3 = 10,000$$

The opportunity loss matrix (sometimes called the regret matrix) is constructed by subtracting from $M_j (j = 1, 2, 3)$ the corresponding entries in the jth column—that is, all μ_{ij} for $i = 1, 2, 3$.

Thus, the entries for $j = 1, 2, 3$, are as follows:

For $j = 1$: $\{M_1 - \mu_{i1}\}$, $i = 1, 2, 3$
For $j = 2$: $\{M_2 - \mu_{i2}\}$, $i = 1, 2, 3$
For $j = 3$: $\{M_3 - \mu_{i3}\}$, $i = 1, 2, 3$

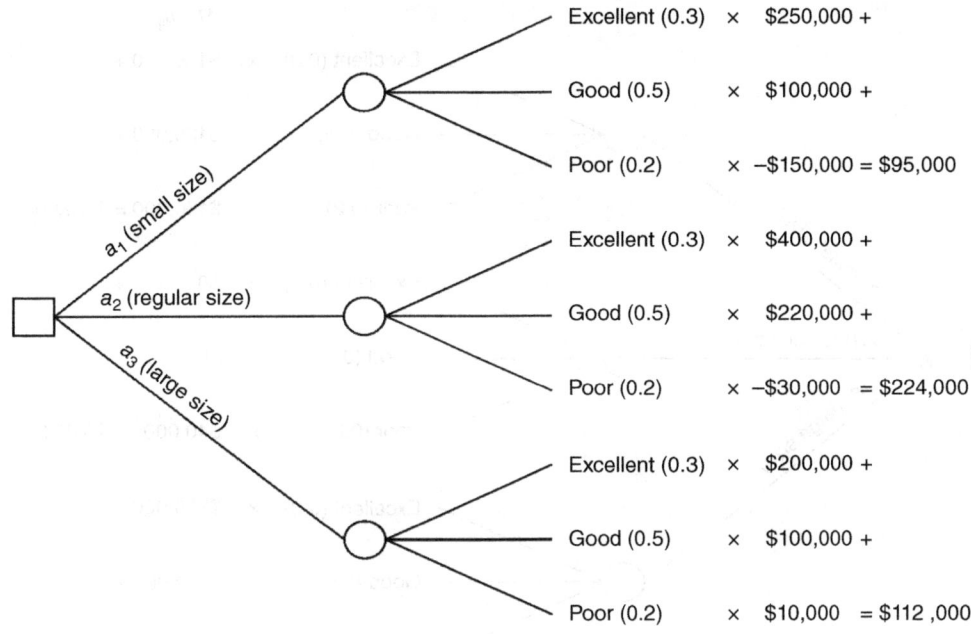

Figure 4.4a Decision tree with expected value of profits.

TABLE 4.4 Opportunity Loss Matrix

Crutch Size	Sales Potential		
	$M_1 - \mu_{j1}$ (400,000 - μ_{j1}) Excellent (j=1)	$M_2 - \mu_{j2}$ (220,000 - μ_{j2}) Good (j=2)	$M_3 - \mu_{j3}$ (10,000 - μ_{j3}) Poor (j=3)
Small (i=1)	$150,000	$120,000	$160,000
Regular (i=2)	$0	$0	$40,000
Large (i=3)	$200,000	$120,000	$0

With the information summarized in Table 4.4, we can construct a new decision tree with the expected value of the outcome minimized (since it is the EOL). Figure 4.4b depicts the EOL decision tree.

To determine the optimal policy using the EOL measure, we use the following:

$$\min_{1 \le i \le 3} \sum_{j=1}^{3} p(s_j)(M_j - \mu_{ij}) \qquad (4.13)$$

For $i = 1$: $\sum_{j=1}^{3} p(s_j)(M_j - \mu_{1j})$
$= (0.3)(400,000 - 250,000)$
$+ (0.5)(220,000 - 100,000)$
$+ (0.2)(10,000 - (-150,000))$
$= \$137,000$

For $i = 2$: $\sum_{j=1}^{3} p(s_j)(M_j - \mu_{2j})$
$= (0.3)(400,000 - 400,000)$
$+ (0.5)(220,000 - 220,000)$
$+ (0.2)(10,000 - (-30,000))$
$= \$8,000$

For $i = 3$: $\sum_{j=1}^{3} p(s_j)(M_j - \mu_{3j})$
$= (0.3)(400,000 - 200,000)$
$+ (0.5)(220,000 - 100,000)$
$+ (0.2)(10,000 - 10,000)$
$= \$120,000$

122 MODELING AND DECISION ANALYSIS

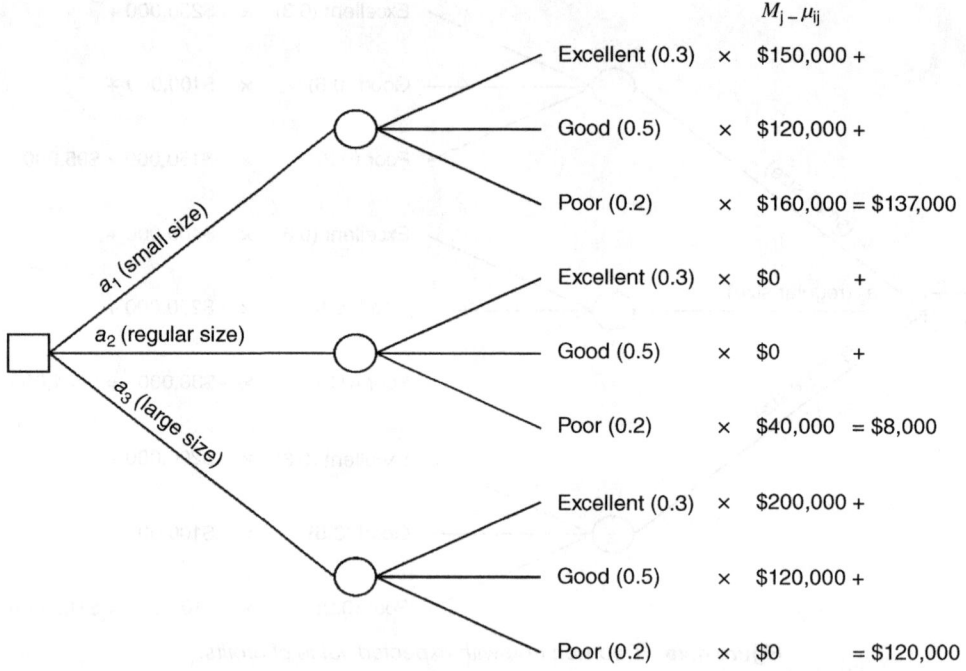

Figure 4.4b Decision tree with EOL measure.

Clearly,

$$\min\{137{,}000,\ 8{,}000,\ 120{,}000\} = 8{,}000$$

Note that both measures—maximizing the expected value of profit and minimizing the EOL—yield the same optimal policy: to manufacture regular-size crutches.

4.3.1.3 Most likely value

The most likely value (MLV) measure is not commonly used because the *optimal* results are very sensitive to the number of states of nature. In other words, the larger the number of different probabilities of outcomes (that must sum to one), the more sensitive is the optimal solution to changes in these probabilities. Figure 4.3 can still serve our purpose here. The basic difference between the expected value of outcome and the MLV measures is that in the MLV, we do not multiply the probabilities by the corresponding outcomes and sum the results. Rather, for each policy option, we select the outcome with the highest probability. The solution of the MLV measure for this example problem is simple:

For $i=1$ (small size): $\max\limits_{1 \leq j \leq 3} p(s_j) = 0.5$, corresponding to $\mu_{12} = \$100{,}000$

For $i=2$ (regular size): $\max\limits_{1 \leq j \leq 3} p(s_j) = 0.5$, corresponding to $\mu_{22} = \$220{,}000$

For $i=3$ (larger size): $\max\limits_{1 \leq j \leq 3} p(s_j) = 0.5$, corresponding to $\mu_{32} = \$100{,}000$

Thus, the optimal crutch-manufacturing policy using the MLV measure is to manufacture regular-size crutches at a most likely profit of $220,000.

4.4 DECISION MATRIX

In Chapter 5, we formally introduce the concept of multiobjective decisionmaking, and the dominant discussion will focus on objective functions that are assumed to be quantifiable. For example, cost, risk of failure, risk of time delay in meeting a project's schedule, and other factors are assumed to be expressed in terms of the state, decision, and random variables and other parameters as discussed in Chapter 2.

This section introduces a less quantitative approach for making choices among multiple objectives that are not amenable to explicit quantification in the forms discussed in Chapter 2. Choosing the *best* car (from among all possible manufacturers and models) that meets most of the customer's requirements, desires, and budget is one example. Another is to select the *best* college with the most desirable attributes for a prospective student. The common denominator of all of these decisionmaking situations is the multiple choices and the multiple attributes associated with each choice. College attributes for a prospective student may include the reputation of the university (including its specific program and faculty), tuition cost, distance from home, social life, size of student population, and others.

The decision matrix is a decisionmaking tool that can be used for these kinds of problems. It is a very simplified version of the analytic hierarchy process (AHP) [Saaty, 1980, 1988]. The following example problem explains the six-step approach of the decision matrix method.

4.4.1 Choosing a Restaurant

A group of undergraduate students at a large university applied decision matrix to select the *best* restaurant from among five candidates (policy options) here designated as A, B, C, D, and E. These establishments were subjected to the following six decision criteria (attributes): (1) taste, (2) nutrition, (3) convenience, (4) cost, (5) service, and (6) atmosphere. The following six steps summarize the decision matrix approach:

1. List all decision criteria (attributes) upon which you intend to make your choices, decisions, trade-offs, and so on.

2. Assign weights to these attributes, such that the sum of these weights is normalized to one. For example, taste = 0.25, nutrition = 0.10, convenience = 0.20, cost = 0.20, service = 0.15, and atmosphere = 0.10. Total = 1.00.

3. List all policy options (in this case, the five restaurants) from which you intend to select one option or a smaller subset of options so that you may evaluate your final decision. For each option, assign a rank from 0 to 10 for each of its attributes, where 10 is the highest rank. For example, the students ranked Restaurant A as 2 for taste, 6 for nutrition, 8 for convenience, 7 for cost, 2 for service, and 5 for atmosphere. They did the same for the four other restaurants (policy options). These rankings are summarized in Table 4.5.

4. For each policy option, multiply the rank of each attribute by the corresponding weight of that attribute, yielding a normalized weight (see Table 4.6). For example, for Restaurant A, taste, $0.25 \times 2 = 0.50$; nutrition, $0.10 \times 6 = 0.60$; convenience, $0.20 \times 8 = 1.60$; cost, $0.20 \times 7 = 1.40$; service, $0.15 \times 2 = 0.30$; atmosphere, $0.10 \times 5 = 0.50$. Total $= 0.50 + 0.60 + 1.60 + 1.40 + 0.30 + 0.50 = 4.90$ (see Table 4.6).

5. Sum these products of normalized weight for each policy option. For example, Restaurant A totals 4.90.

6. Select the best option or, better yet, the best options, and repeat the process with different weights for the attributes (sensitivity analysis).

Clearly, Restaurant D with a total score of 5.80 is the preferred choice. Of course, it is always advisable to perform sensitivity analysis. In this case, for example, Restaurant E is the closest in ranking to Restaurant D, and further analysis is warranted.

TABLE 4.5 Ranking of Restaurants according to Attributes

Restaurant	Taste	Nutrition	Convenience	Cost	Service	Atmosphere
A	2	6	8	7	2	5
B	8	7	1	1	6	3
C	4	5	4	3	4	8
D	6	2	7	4	10	4
E	9	3	3	5	3	7

TABLE 4.6 Decision Matrix for Restaurant Selection

Attribute / Alternative Restaurants	Taste 0.25	Nutrition 0.10	Convenience 0.20	Cost 0.20	Service 0.15	Atmosphere 0.10	Total Sum
			Normalized Weight				
A	0.50	0.60	1.60	1.40	0.30	0.50	4.9
B	2.00	0.70	0.20	0.20	0.90	0.30	4.3
C	1.00	0.50	0.80	0.60	0.60	0.80	4.3
D	1.50	0.20	1.40	0.80	1.50	0.40	5.8
E	2.25	0.30	0.60	1.00	0.45	0.70	5.3

4.5 THE FRACTILE METHOD

The fractile method is an effective procedure with which to construct probability distribution functions (pdfs) by soliciting expert evidence. It dissects the [0,1] probability axis into sections, termed *fractiles*, and relates each fractile to an outcome (e.g., a consequence) by soliciting evidence-based assessments from one or more experts. The cumulative distribution function (cdf) and pdf are then constructed on the basis of knowledge generated through the fractile method. As we will note in subsequent sections, the probability of exceedance, which is (1 − cdf), is used in many risk-based decisionmaking problems. The probability of exceedance will be particularly useful when we address low-probability and severe-consequence events. For completeness, we define the following.

A continuous random variable X of damages (e.g., cost overrun or time delay) has a cdf, $P(x)$, and a pdf, $p(x)$, which are defined by the relationships

$$\text{cdf:} \quad P(x) = \text{prob}[X \leq x] \quad (4.14)$$

and

$$\text{pdf:} \quad p(x) = \frac{dP(x)}{dx} \quad (4.15)$$

The cdf represents the nonexceedance probability of x. The exceedance probability of x is defined as the probability that X is observed to be greater than x and is equal to one minus the cdf evaluated at x.

The expected value, average, or mean value of the random variable X is defined as

$$E[X] = \int_{-\infty}^{\infty} x p(x) \, dx \quad (4.16)$$

For the discrete case, Eq. (4.16) takes the form of Eq. (4.17). In this case, the pdf is divided into n segments of consequences x_i, each with a corresponding probability of p_i, $i = 1, 2, \ldots, n$:

$$E[X] = \sum_{i=1}^{n} p_i x_i \quad (4.17)$$

where

$$\sum_{i=1}^{n} p_i = 1, \quad p_i \geq 0, \quad i = 1, 2, \ldots, n \quad (4.18)$$

4.5.1 Example Problem 1: Airplane Acquisition

Assume that the US Department of Defense (DoD) is considering a new strategic airplane that will constitute the flagship of the Air Force. Aware of the power shift from hardware to software in technology and the emerging centrality of software as the overall system integrator and coordinator, the DoD considers the software development for this airplane to be of paramount importance. The Air Force commissions the assistance of a support organization to develop requirements for the software-intensive system, and it also makes a request for proposal (RFP) for designing, prototyping, and developing the software needed for the flagship airplane. Following a detailed and tedious process of qualifying prospective bidders, the Air Force issues an RFP for the development of the required software engineering. This time, however, the RFP includes contractual requirements that had not been requested previously. For example, the RFP requires that each contractor provide variances along with the estimated project's cost and completion schedule, instead of the commonly practiced requirement of single deterministic values. The RFP leaves it up to the contractors to determine the form that these variances take, including, if the contractor so desires, the type of pdf selected for each estimate. The Air Force and its support team, planning to use the same approach themselves in evaluating the various proposals, recommend in the RFP the optional use of the fractile method or the triangular distribution when appropriate statistical information is not readily available.

To capture the mathematical details entailed in the process of developing representative pdfs for cost and completion time, a step-by-step procedure using the fractile method (adopted by Contractors A and B) is presented here. A detailed analysis is presented for Contractor A only, and results for Contractor B and the customer are shown in Figure 4.7. The team from Contractor A estimates a most likely cost of $150 million. Using the fractile method, along with brainstorming sessions with experts, the following evidence-based information emerges:

- Best-case project cost increase = 0% (i.e., project cost is $150 million)
- Worst-case project cost increase = 50% (i.e., project cost increase is $75 million, for a total of $225 million)
- Median value of project cost increase (equal likelihood of being greater or less than this value) = 15% (i.e., project cost increase is $22.5 million, for a total of $172.5 million)
- A 50–50 chance that the actual project cost would be within 5% of the 15% median estimate (i.e., project cost increase is $(15 \pm 5)\%$)

From the above information, the following fractiles (percentiles) are readily determined:

- The best scenario of no cost overrun (0% cost increase, i.e., a total cost of $150 million) represents the 0.00 fractile (0 percentile).
- The worst scenario of 50% cost overrun (a total cost of $225 million) represents the 1.00 fractile (100th percentile).
- The median value of 15% cost overrun (a total cost of $172.5 million) represents the 0.50 fractile (50th percentile).

TABLE 4.7 Comparative CDFs

	Project Cost Increase (%)		
Fractile	Customer	Contractor A	Contractor B
0.00	0	0	0
0.25	5	10	15
0.50	10	15	20
0.75	15	20	25
1.00	30	50	40

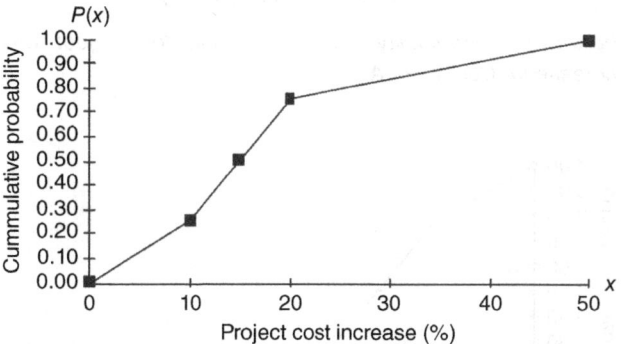

Figure 4.5 Graphical cdf for project cost increase for Contractor A.

- The 0.25 fractile (25th percentile) is $(15-5)\% = 10\%$ increase over $150 million (a total cost of $165 million).
- The 0.75 fractile (75th percentile) is $(15+5)\% = 20\%$ increase over $150 million (a total cost of $180 million).

The above assessment of project cost for Contractor A (and similar hypothetical costs for Contractor B and for the customer) is summarized in Table 4.7 and is used as a basis for constructing the corresponding cdf for Contractor A (see Figure 4.5).

The cdf (Figure 4.5) can now be represented in terms of a pdf (Figure 4.6). To construct the pdf, one must be guided by the following principles: (i) the sum of the shaded area (the pdf) must be equal to 1; and (ii) the first quartile in Figure 4.6 (representing 25% of the probabilities) spans a cost overrun from 0 to 10%. Thus, the corresponding area of the pdf (Figure 4.6) must be equal to one-fourth of the total area, that is, 0.25. Dividing 0.25 by 10 yields a height of 0.025 for the first rectangle in Figure 4.6. Similarly, each of the second and third quartiles spans 5% of the project cost increase. Thus, the area of each of

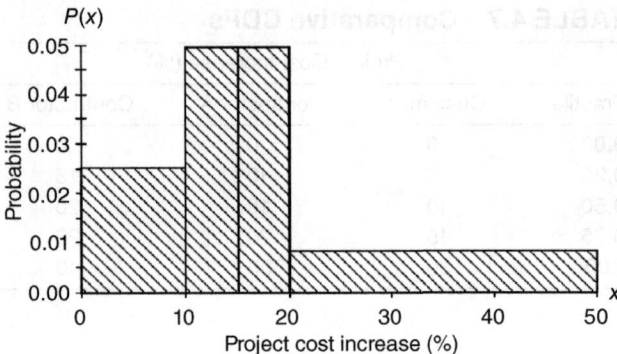

Figure 4.6 Probability density function for project cost increase for Contractor A.

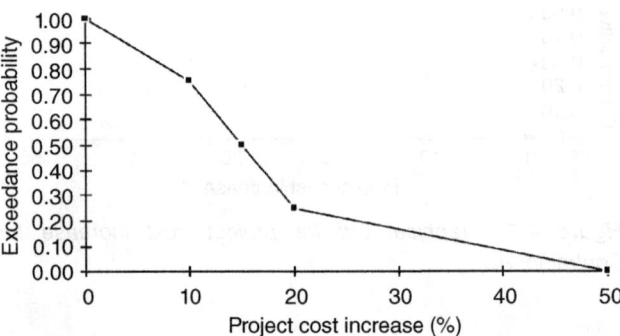

Figure 4.7 Exceedance probability for project cost increase for Contractor A.

the second and third rectangles of the pdf (Figure 4.6) is 0.25 and, when divided by 5, yields a height of 0.05 on the probability axis. Finally, the last quartile spans a cost overrun of 30% (from 20 to 50%). The area of the rectangle is 0.25 and, when divided by 30, yields a height of 0.0083 on the probability axis. Figure 4.7 depicts the exceedance probability (1 − cdf) versus project cost increase.

The expected value of the percentage of project cost increase can be determined graphically (Figure 4.6) and by using Eq. (4.17):

$$E[X] = \sum_{i=1}^{n} p_i x_i$$

$$E[X] = 0.25\left[0 + \frac{(10-0)}{2}\right] + 0.25\left[10 + \frac{(15-10)}{2}\right]$$
$$+ 0.25\left[15 + \frac{(20-15)}{2}\right] + 0.25\left[20 + \frac{(50-20)}{2}\right]$$
$$= 0.25(5) + 0.25(12.5) + 0.25(17.5) + 0.25(35)$$
$$= 0.25(70) = 17.5\%, \text{ or } 26.25 \text{ million.}$$

In other words, the expected value of the total cost of the project is $176.25 (150+26.25) million.

The expected value of the percentage of project cost increase may also be calculated using Eq. (4.16):

$$E[X] = \int_0^\infty x p(x)\, dx$$

$$E[X] = \int_0^{10} x p(x)\, dx + \int_{10}^{15} x p(x)\, dx + \int_{15}^{20} x p(x)\, dx$$
$$+ \int_{20}^{50} x p(x)\, dx$$

$$= \int_0^{10} \left(\frac{0.25}{10}\right) x\, dx + \int_{10}^{15} \left(\frac{0.25}{5}\right) x\, dx$$
$$+ \int_{15}^{20} \left(\frac{0.25}{5}\right) x\, dx + \int_{20}^{50} \left(\frac{0.25}{30}\right) x\, dx$$

$$= 0.025 \frac{x^2}{2}\bigg|_0^{10} + 0.05 \frac{x^2}{2}\bigg|_{10}^{15} + 0.05 \frac{x^2}{2}\bigg|_{15}^{20}$$
$$+ \left(\frac{0.25}{30}\right)\frac{x^2}{2}\bigg|_{20}^{50}$$

$$= 0.025\left(\frac{100-0}{2}\right) + 0.05\left(\frac{225-100}{2}\right)$$
$$+ 0.05\left(\frac{400-225}{2}\right)$$
$$+ \left(\frac{0.25}{30}\right)\left(\frac{2500-400}{2}\right)$$
$$= 1.25 + 3.125 + 4.375 + 8.75$$
$$E[X] = 17.50\%$$

Note that the expected value of cost overrun of $26.25 million (i.e., total cost of $176.25 million) for Contractor A does not provide any vital information on the probable extreme behavior of the project cost. Also, note that there is a one-to-one functional relationship between the probability axis and the percentage of project cost increase as is depicted in Figure 4.7. For example, there is a 0.1 probability (one chance in 10) that the project cost increase will be equal to or above 38%. This result is generated as follows (here, we are interested in the probability of exceedance 0.1—i.e., $\alpha = 0.90$, or $(1-\alpha) = 0.10$):

$$\frac{x-20}{50-20} = \frac{a-b}{a-c} = \frac{0.25-(1-\alpha)}{0.25}$$

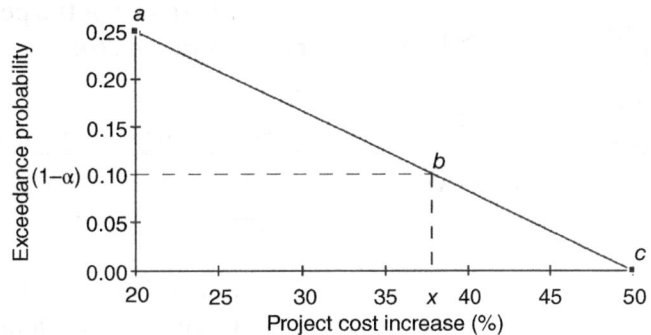

Figure 4.8 Computing the partition point on the damage axis for Contractor A.

Thus,

$$x = 30 - \frac{30(1-\alpha)}{0.25} + 20 = 38\% \quad \text{for } \alpha = 0.9$$

In other words, $\alpha = 0.9$ means that

$$\Pr[X \leq 38] = 0.9$$

and $1 - \alpha = 0.1$ means that

$$\Pr[X > 38] = 0.1$$

Alternatively, we can compute from Figure 4.7 the partition point x (the percentage of increase in cost) that corresponds to a probability of 0.1 as shown in Figure 4.8.

The height of the probability axis, h, is derived from Figure 4.6 as

$$h = \frac{0.25}{50 - 20} = 0.0083$$

Note that x on the damage axis (see Figure 4.7) corresponds to $(1-\alpha)$ on the probability axis. The area $(50-x)h$ must correspond to the probability $(1-\alpha)$. Thus, $(1-\alpha) = (50-x)h$ or

$$x = 50 - \left(\frac{1-\alpha}{h}\right) = 50 - \frac{(1-0.9)}{0.0083} = 38\% \quad \text{for } \alpha = 0.9$$

As we noted earlier, this example problem will be revisited in Chapter 8.

4.6 TRIANGULAR DISTRIBUTION

When constructed on the basis of expert evidence-based knowledge, the triangular distribution follows a path similar to the one discussed in the fractile method. Here, the expert is not asked to assess probabilities. Rather, only three assessments of outcomes are solicited from the expert: lowest value (a), highest value (b), and most likely value (c). Figure 4.9 depicts a triangular distribution. Equations (4.21) and (4.22) [Law and Kelton, 1991] present the functional relationships for the triangular distribution.

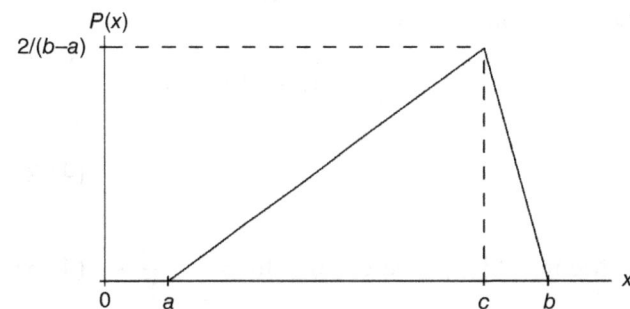

Figure 4.9 Triangular distribution.

In many respects, the triangular distribution is an ideal approach for soliciting expert evidence when the expert is not comfortable with probabilities, as is required in the fractile method. Note that the area of the triangle in Figure 4.9 must be equal to 1 for the triangle to qualify as a pdf.

From this fact, the frequency of the most likely value of the outcome (point c in Figure 4.9) can be readily calculated using Eq. (4.19) for the area of the triangle — area = [(base)(height)]/2:

$$(b-a)p(c)/2 = 1 \qquad (4.19)$$

where $p(c)$ is the height of the triangle. Thus,

$$p(c) = 2/(b-a) \qquad (4.20)$$

$$\text{Density}\,[p(x)] = \begin{cases} \dfrac{2(x-a)}{(b-a)(c-a)} & \text{if } a \leq x \leq c \\ \dfrac{2(b-x)}{(b-a)(b-c)} & \text{if } c < x \leq b \\ 0 & \text{otherwise} \end{cases} \quad (4.21)$$

$$\text{Distribution}\,[P(x)] = \begin{cases} 0 & \text{if } x < a \\ \dfrac{(x-a)^2}{(b-a)(c-a)} & \text{if } a \leq x \leq c \\ 1 - \dfrac{(b-x)^2}{(b-a)(b-c)} & \text{if } c < x \leq b \\ 1 & \text{if } x > b \end{cases} \quad (4.22)$$

$$\text{Mean} = E[X] = \text{expected value} = \frac{a+b+c}{3} \quad (4.23)$$

$$\text{Variance} = \frac{a^2 + b^2 + c^2 - ab - ac - bc}{18} \quad (4.24)$$

4.6.1 Example Problem 2: Performance Assessment

Let us reconsider the airplane acquisition problem discussed in Section 4.5.1, focusing on the expected performance of the aircraft. Three values are solicited from the expert:

Worst-case performance: $a = 50\%$
Best-case performance: $b = 110\%$
Most likely performance: $c = 100\%$

A 100% performance means that the aircraft meets its designed performance criteria; a 110% performance indicates a better performance—for example, higher speed or higher load capability; a 50% performance is meeting only one-half of its performance criteria.

The expected value of the aircraft's performance (see Eq. 4.23) based on the expert's evidence-based knowledge is given in Eq. (4.25):

$$E(x) = \frac{a+b+c}{3} = \frac{50+110+100}{3} = 86.7\% \quad (4.25)$$

The variance of the performance (see Eq. 4.22) is given in Eq. (4.26):

Variance
$$= \frac{(50)^2 + (110)^2 + (100)^2 - (50)(110) - (50)(100) - (110)(100)}{18}$$
$$= \frac{3100}{18} \quad (4.26)$$

The standard deviation is 13.12%.

This very high standard deviation indicates a major variability in the ultimate performance of the designed aircraft.

For normal distributions, about 68% of the distribution lies in an interval extending from one standard deviation to the left of the mean to one standard deviation to the right of the mean. We will revisit this example problem in Chapter 8 after we introduce the conditional expected value concept, focusing on extreme events.

4.7 INFLUENCE DIAGRAMS

The art and science of systems modeling builds on diverse philosophies, theories, tools, and methodologies. Probably the most basic, logical, and intuitive of all are influence diagrams [Oliver and Smith, 1990]. They are effective because they enable the systems analyst and decisionmaker alike to represent the causal relationships among the very large number of variables affecting and characterizing the system. Furthermore, through the use of conventional symbols, such as decision nodes and chance nodes, influence diagrams capture the probabilistic nature of the randomness associated with the system. (See Section 4.3 on decision trees and Chapter 9 on MODTs.) Consequently, the quantification of risk, which is a measure of the probability and severity of adverse effects, can be performed on sound foundations.

The most effective deployment of influence diagrams is through brainstorming sessions with all principal parties involved with the system. In this setting, the varied expertise of the study team members produces a deeper understanding of the interactions among the important and critical variables of the system. Similar to an engineering design project, the initial phase of constructing an influence diagram may result in an unwieldy *mess chart* that includes trivial, as well as critical, components. Through an open and constructive dialogue among the systems analyst(s) and decisionmaker(s), the *mess chart* becomes more coherent and includes what is deemed to be only essential variables and building blocks of the system's model.

To avoid further generalities, the deployment of an influence diagram in a study for the US Army Corps of Engineers is presented.

4.7.1 Channel Reliability of the Upper Mississippi River

With its ability to transport easily and cheaply billions of dollars in bulk commodities such as grain, coal, and petroleum, the Upper Mississippi River navigation system is a major contributor to the economic prosperity of Middle America. Over the almost 60 years of the navigation system's operation under its present dimensions, commercial traffic on the river increased by several orders of magnitude to over 100 million tons of cargo per year [Tulsiani, 1996].

For more than a century, the US Army Corps of Engineers (Corps) has been responsible for the construction, operation, and maintenance of the Upper Mississippi River navigation system. The required navigation standard is maintained through the use of structural measures, such as wing dams and closing dams, as well as through maintenance dredging. There are various costs associated with this function, such as for dredging and structural dredge material. Furthermore, deterioration of the various structures, including wing dams and closing dams, has an impact on the navigability of the channel. Due to these costs and concerns, as well as the fact that channel closure conditions can occur in a short period of time, maintaining the navigation system is a complex process.

The objective of the modeling effort is to develop a reliability model for the navigation channel to be used by the Corps in a planning and management framework of the river navigation system. This includes examining the trade-offs among costs, benefits, and reliability in making rehabilitation and maintenance decisions for operating the system. To do so, we identify the basic building blocks of the mathematical model using influence diagrams [Tulsiani, 1996].

4.7.1.1 The process of channel failure

Alluvial channels continuously undergo self-adjustment in their slope, width, depth, and velocity. These changes depend on the magnitude of water and sediment discharges in the channel. Due to these changes, some portions of the river undergo erosion (removal of sediment), while other portions may undergo deposition (addition of sediment). These changes in the river channel, creating shallow reaches known as crossings and deep reaches known as pools, have an impact on the navigability of the river.

In addition to the natural effects, the construction of locks and dams has affected the deposition and erosion patterns in the Upper Mississippi River. For low and intermediate flows, (i) the water surface profile is flatter close to the dams, and (ii) velocities—and therefore the sediment transport rates—are higher in the upper reaches farther away from the dam. Erosion thus occurs in the upper reaches, and deposition occurs in the lower reaches closer to the dams. During higher flows, some of the sediment deposited at the lower reaches is eroded and transported downstream. The results of this yearly cycle are a net erosion at the upper reaches and a net deposition in the lower reaches.

Combined with the effect of the dams is the impact of these flow variations on the river crossings. During high flows, sediment is deposited on the crossings. When the flows return to lower levels, these deposits are eroded. However, the rate of erosion depends upon the time period at the intermediate flows. If the fall in stage is rapid, there is insufficient time for the deposits to erode away. At lower stages, there is a net deposition from the corresponding low stage in the previous cycle, thus reducing the depth available for navigation.

One possible measure of the river navigability is the reliability of the navigation channel—that is, the probability that the channel cross section (depth and width) meets the minimum requirements. This reliability is affected by a large number of variables. Figure 4.10 shows an influence diagram that illustrates some of the interactions between the decision, exogenous, and random variables that have an impact on the channel depth and width. Identifying

130 MODELING AND DECISION ANALYSIS

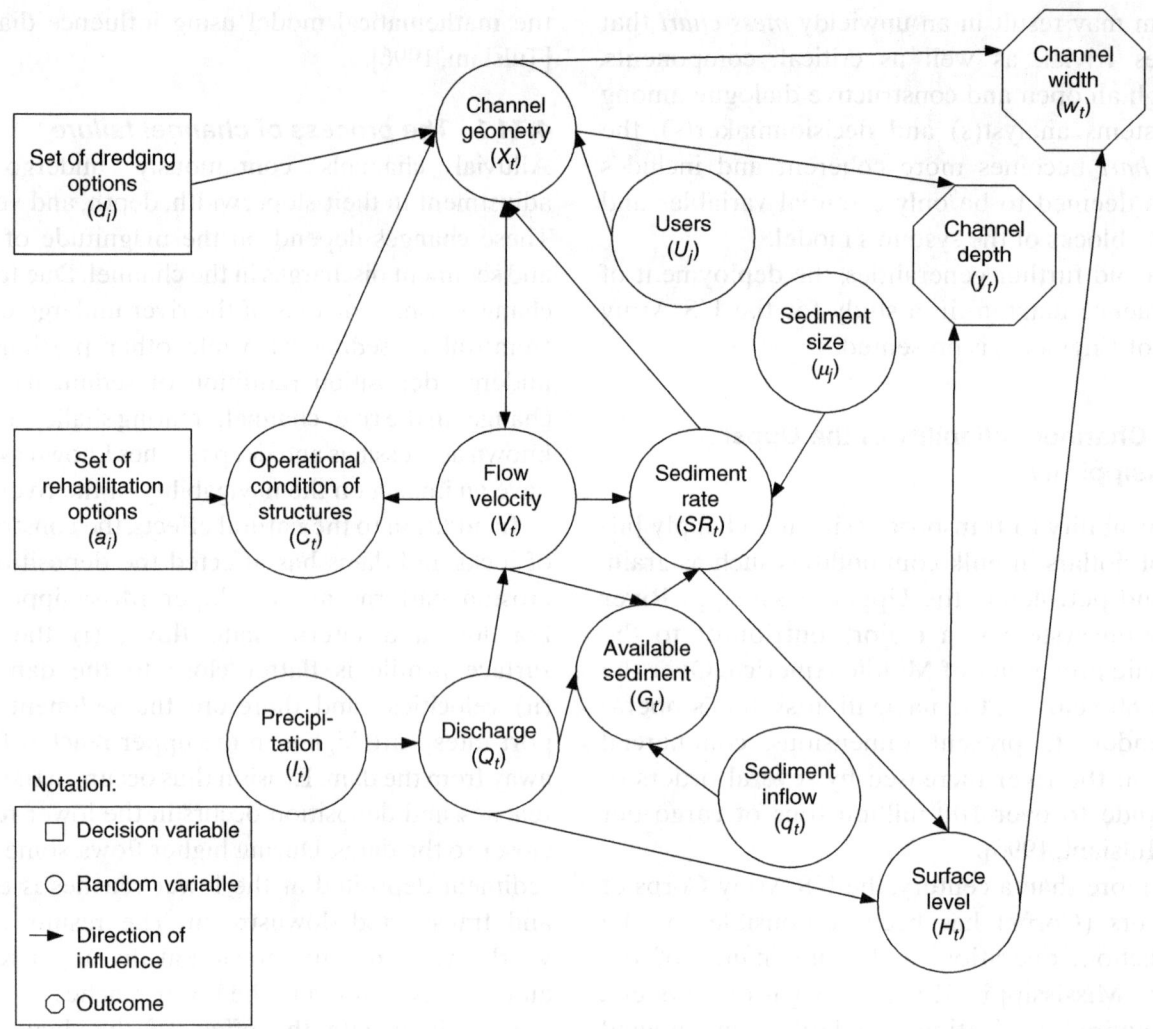

Figure 4.10 Influence diagram for variables that affect channel reliability.

these variables and their impacts plays an important part when developing the models of navigation channel reliability.

4.7.1.2 Variables affecting the channel reliability

Through intensive discussions among the University of Virginia research team and dozens of engineers and economists from three districts of the US Army Corps of Engineers and through the use of influence diagrams (see Figure 4.10), the following building blocks of the reliability model were developed:

Set of dredging options (d_i). The set of dredging sites selected and the volume of material dredged at each site have an impact on future as well as on current options. In the current scenario, selection of a particular site reduces the dredging capacity available elsewhere in the system. In future scenarios, the volume of dredge material from a particular site may influence the need for subsequent dredging at that site.

Set of rehabilitation options (a_i). The rehabilitation options selected also affect future as well as current options available. In the current scenario, they reduce the funds available for rehabilitation elsewhere in the system. In future scenarios, they might reduce the volume of material to be dredged from that site. However, they also may lead to an increase in the dredge material from the downstream sites.

Operational condition of structures (C_t). The operational condition of structures at a particular site affects the volume of material that may require dredging. Structures in good condition may work effectively in channeling the flow so that little or no dredging is required, while structures in a degraded condition may allow the water velocity to slow down, thus causing sedimentation.

Channel geometry (X_t). Channel geometry is the primary variable that, in combination with the water level, determines channel reliability. It can be affected by the dredging options, the sedimentation, and water velocity, among others.

Flow velocity (V_t). Flow velocity is one of the primary variables that affect the sedimentation rate. Increased flow velocity gives the river the additional power required to move the sediment downstream. Thus, flow velocity affects and is affected by channel geometry.

Discharge (Q_t). Water discharge is a function of precipitation in the watershed and the inflow from upstream. The discharge is the primary variable affecting the flow velocity and surface level. The stage–discharge (rating curve) is a primary means of determining water surface levels for varying levels of discharge.

Surface level (H_t). Water surface level and channel geometry are the primary variables affecting the navigability of the river channel. The surface level (stage) can usually be determined by the discharge.

Precipitation (I_t). Watershed precipitation and the upstream inflow determine the river discharge at a particular site. High precipitation in the watershed can cause a rapid increase in the river stage as well as an increase in the sediment inflow.

Sediment inflow (q_t). Sediment inflow in a region can determine the extent of the navigation problems in a channel. The sediment inflow is typically influenced by the topography of a region, such as the amount of forest cover, land use, vegetation, and so on.

Available sediment (G_t). The available sediment at a particular site is a function of the sediment inflow from the watershed, the discharge, and the flow velocity. The available sediment determines the sedimentation rate. There is usually a sediment deficiency downstream of locks and dams due to sedimentation in the pool just upstream of the dam.

Sedimentation rate (SR_t). The sedimentation rate at a particular site is a function of the sediment inflow from the watershed, the sediment grain size, and the flow velocity. A large grain size and a slow water velocity lead to an increase in the sedimentation rate.

Sediment size (μ_j). Sediment grain size in a particular region can affect the sedimentation rate within the region and downstream. Large sediment grains can armor the riverbed, leading to sediment deficiency downstream. Fine sediment grain can increase the sediment carrying capacity of the river and cause sedimentation problems downstream.

Users (U_j). The number of river barges and other traffic can influence the channel geometry by increasing shore and bank erosion due to wave motion.

Channel width (w_t). Channel width is one of the two primary outcomes of interest. It is determined by the channel geometry and the river stage.

Channel depth (y_t). Channel depth is the other primary outcome of interest. It is also determined by the channel geometry and the river stage.

4.7.1.3 Variable impact

The joint impact of these effects yields a net deposition in certain reaches of the river. When this deposition is large enough to endanger normal navigation through the reach, dredging is used to correct the problem. Channel failure occurs when the deposition causes navigation to be considered unsafe. Since we assume that this channel failure is caused by sedimentation, the hydraulics behind the sedimentation process become an important topic of investigation.

The primary variables of interest are the channel width and depth, since the navigation channel reliability is dependent on these two variables. Their values are dependent upon the channel geometry and the water surface level in the river at any particular time. The surface level is dependent upon the magnitude of the water discharge and the amount of sedimentation in the river. Similarly, the channel geometry is dependent upon the flow velocity, the operational condition of the navigation structures, the sedimentation rate, and the magnitude and location of the dredging.

4.8 POPULATION DYNAMIC MODELS

4.8.1 Macropopulation Model

Recall that risk management builds on the risk assessment process by seeking answers to the following set of three questions: What can be done and what options are available? What are their associated trade-offs in terms of all costs, benefits, and risks? And what are the impacts of current management decisions on future options? (See Section 1.3.4.) Any attempt to address the third question quantitatively necessarily lends itself to a dynamic modeling effort. Although risk assessment and management of dynamic models are discussed in Chapter 10, an exposure to discrete dynamic models seems appropriate here. To avoid abstract theoretical discussion, we introduce the formulation of discrete dynamic models through a specific population dynamic model. Indeed, to address the impacts of current decisions on future options, one must be able to project the consequences of current decisions into the future.

In the following population dynamic model, we focus on one state variable, $p(t)$, the level of the total population at time t. Other micromodels such as the Leslie model [Meyer, 1984] divide the reproductive portion of the population into n segments on the basis of age categories. Such models yield more than one state variable. For simplicity, we assume that the numbers of births and deaths in any one year are exogenous variables (i.e., uncontrollable) and that they do not change significantly over time.

DEFINITIONS

$p(t)$: the level of population at time t
B: the number of births in any one year
D: the number of deaths in any one year
$b(t)$: birth rate for the time between t and $t+1$
$d(t)$: death rate for the time between t and $t+1$

$$b(t) = B/p(t) \qquad (4.27)$$
$$d(t) = D/p(t) \qquad (4.28)$$

We assume that the population level at time $t=0$ is known; that is, $p(0)$ is known. In this discussion, we further assume that the birth and death rates do not change significantly with time over the planning time horizon, namely, $b(t) = b$ and $d(t) = d$. The balance of population growth from time t to $t+1$ yields

$$\begin{aligned} p(t+1) &= p(t) + B - D \\ &= p(t) + bp(t) - dp(t) \\ &= p(t)[1 + b - d] \end{aligned} \qquad (4.29)$$

Let $r = [1+b-d]$ denote the overall growth rate; the growth rate is also known as the Malthusian parameter [Meyer, 1984]. Then

$$p(t+1) = p(t)r \qquad (4.30)$$

For $t=0$, $p(t)$ is known; then

$$p(1) = p(0)r \qquad (4.31)$$

For $t=1$, Eq. (4.30) becomes

$$p(2) = p(1)r \qquad (4.32)$$

Substituting Eq. (4.31) into Eq. (4.32) yields

$$p(2) = [p(0)r]r = p(0)r^2 \qquad (4.33)$$

and for any period t,

$$p(t) = p(0)r^t, \quad t = 0,1,2,\ldots \qquad (4.34)$$

This macropopulation model is also called an exponential model because the growth rate is in the form of an exponential function.

4.8.2 Example Problems

4.8.2.1 Example problem 1

Assume that the current population of Country A is 1,000,000,000 people and that of Country B is 900,000,000. Assuming that the birth rates in Countries A and B are 0.015 and 0.025, respectively, and that the death rates in Countries A and B are

0.010 and 0.012, respectively, how long would it take for the two populations to be the same? Let

$p_a(0)$ = initial population of Country A
$p_b(0)$ = initial population of Country B
$p_a(x)$ = population of Country A in year x
$p_b(x)$ = population of Country B in year x
r_a = growth rate of Country A
r_b = growth rate of Country B

Then

$$p_a(x) = p_a(0)r_a^x \qquad (4.35)$$

and

$$p_b(x) = p_b(0)r_b^x$$
$$r_a = 1 + 0.015 - 0.010 = 1.005$$
$$r_b = 1 + 0.025 - 0.012 = 1.013 \qquad (4.36)$$

Let the two populations be the same in year x; then

$$p_a(x) = p_b(x), \quad \text{or}$$
$$p_a(0)r_a^x = p_b(0)r_b^x$$
$$(1{,}000{,}000{,}000)(1.005)^x = (900{,}000{,}000)(1.013)^x$$
$$x = 13.29 \cong 13 \text{ years}$$

In other words, it would take about 13 years for the populations of the two countries to be the same.

4.8.2.2 Example problem 2

Assume that the current faculty population at a major university is 1500 professors. The rate of increase due to new hiring has been 0.03, and the rate of faculty leaving the university (including retirement) has been 0.01.

(a) How many faculty will be at that university in 10 years?
(b) How many years will it take for the faculty to double?
(c) How many new faculty will join the university between years 8 and 9?

Solutions

(a) $p(0) = 1500, \ b = 0.03, \ d = 0.01$
$$r = 1 + b - d = 1.02$$
$$p(k) = p(0)r^k$$
$$p(10) = (1500)(1.02)^{10} \cong 1828 \text{ professors}$$

Thus, the number of faculty is expected to be about 1828 in 10 years:

(b) $p(x) = 2p(0) = P(0)r^x$,
where x = number of years of doubling

or

$$2 = (1.02)^x$$
$$x = \log 2 / \log 1.02 \cong 35 \text{ years}$$

Thus, it would take 35 years for the faculty to double.

(c) Let $q(8)$ = number of new faculty between years 8 and 9. Therefore,

$$q(8) = bp(8) = bp(0)r^8 = (0.03)(1500)(1.02)^8$$
$$\cong 53 \text{ professors}$$

Thus, about 53 new faculty would join the university between years 8 and 9.

4.8.2.3 Example problem 3

The word *planning* in river basin planning connotes a time horizon beyond the present. Therefore, the models that are built for such a planning activity must be able to accommodate the changes that take place over time. Discrete dynamic models can be very helpful in this regard, and the objective of this example problem is to extend such models beyond one state variable.

In regions with a limited water supply, water demand for a major livestock industry may be of a special concern. The competition for a limited water supply that exists between urban and rural populations and between large and small livestock is the subject of this example. In an attempt to model the dynamics of water usage, a four-state discrete dynamic model is presented with the following assumptions:

1. There are no uncertainties (no random variables, i.e., deterministic model).

2. Decisions are made only once (at time $t=0$), and their consequences are evaluated in subsequent years.
3. Urban and rural demands for water and livestock are always met.
4. The birth rates of large and small livestock are controllable.
5. The migration rate of rural population to urban areas is controllable (by providing employment, subsidies, or other incentives).

State variables

$s_1(t)$ = urban population at time t
$s_2(t)$ = rural population at time t
$s_3(t)$ = number of large livestock at time t
$s_4(t)$ = number of small livestock at time t

To construct the discrete dynamic model, we introduce the following building blocks:

Decision variables

x_1 = fraction of investment in large livestock

Output variables

$y_1(t)$ = urban water consumption at time t
$y_2(t)$ = total rural water consumption at time t
$y_3(t)$ = number of large livestock (including purchases) at time t
$y_4(t)$ = number of small livestock (including purchases) at time t

Input variables

$u_1(t)$ = government investment in the region at time t
$u_2(t)$ = other investments in the region at time t

Exogenous variables

a = rate at which investment in rural area influences changes in migration
b_1 = birth rate of urban population
d_1 = death rate of urban population
b_2 = birth rate of rural population
d_2 = death rate of rural population
c_1 = percent increase per capita in urban water consumption
c_2 = percent increase per capita in rural water consumption
d_3 = death rate of large livestock
d_4 = death rate of small livestock
e_1 = cost of one head of large livestock
e_2 = cost of one head of small livestock
f = base rate of rural migration per year
g_1 = urban per capita water consumption (in l/day)
g_2 = rural per capita water consumption (in l/day)
g_3 = large livestock water use (in l/day)
g_4 = small livestock water use (in l/day)

For pedagogical purposes, we will construct the discrete dynamic model in stages, accounting only for one aspect at each stage:

1. Population growth—Scenario 1

$$s_1(t+1) = (1+b_1-d_1)s_1(t) \\ + \{f - a[u_1(t)+u_2(t)]\}s_2(t) \quad (4.37)$$

$$s_2(t+1) = (1+b_2-d_2)s_2(t) \\ + \{f - a[u_1(t)+u_2(t)]\}s_2(t) \quad (4.38)$$

$$s_3(t+1) = (1-d_3)s_3(t) \\ + [1/e_1][u_1(t)+u_2(t)]x_1 \quad (4.39)$$

$$s_4(t+1) = (1-d_4)s_4(t) \\ + [1/e_2][u_1(t)+u_2(t)][1-x_1] \quad (4.40)$$

2. Water consumption—Scenario 2

$$y_1(t+1) = g_1(1+c_1/100)^t s_1(t+1) \quad (4.41)$$

$$y_2(t+1) = g_2(1+c_2/100)^t s_2(t+1) \\ + g_3 s_3(t+1) + g_4 s_4(t+1) \quad (4.42)$$

3. New Livestock—Scenario 3

$$y_3(t+1) = [1/e_1][u_1(t)+u_2(t)]x_1 \quad (4.43)$$

$$y_4(t+1) = [1/e_2][u_1(t)+u_2(t)][1-x_1] \quad (4.44)$$

The database for exogenous variables is presented in Table 4.8. Several scenarios are developed and the dynamic model is solved for five periods.

Scenario 1
Decision variable: $x_1 = 0.95$

Scenario 2
Decision variable: $x_1 = 0.95$

TABLE 4.8 Database for Exogenous Variables

a	b_1	b_2	c_1	c_2	d_1	d_2	d_3	d_4	e_1	e_2	f	g_1	g_2	g_3	g_4
0.00001	0.0274	0.0317	0.5	0.5	0.006	0.0116	0.1	0.2	5	0.2	0.001	80	25	45	7

TABLE 4.9 Scenario 1 Results (Population Growth)

t	0	1	2	3	4	5
u_1	1,000	1,100	1,200	1,300	1,400	1,500
u_2	1,000	1,100	1,200	1,300	1,400	1,500
s_1	1,400	1,424	1,446	1,468	1,489	1,508
s_2	6,300	6,433	6,570	6,711	6,857	7,007
s_3	3,400	3,440	3,514	3,619	3,751	3,908
s_4	2,100	2,180	2,294	2,435	2,598	2,779
y_1	11,200	114,462	116,873	119,224	121,507	123,716
y_2	325,200	331,687	340,083	350,193	361,846	374,893
y_3		380	418	456	494	532
y_4		500	550	600	650	700

TABLE 4.10 Scenario 2 Results (Water Consumption)

t	0	1	2	3	4	5
u_1	1,000	1,000	1,000	1,000	1,000	1,000
u_2	1,000	1,000	1,000	1,000	1,000	1,000
s_1	1,400	1,424	1,448	1,472	1,497	1,522
s_2	6,300	6,433	6,569	6,707	6,849	6,993
s_3	3,400	3,440	3,447	3,508	3,538	3,564
s_4	2,100	2,180	2,244	2,295	2,336	2,369
y_1	11,200	114,462	116,977	119,544	122,165	124,842
y_2	325,200	331,687	337,991	344,154	350,213	356,201
y_3		380	380	380	380	380
y_4		500	500	500	500	500

TABLE 4.11 Scenario 3 Results (New Livestock)

t	0	1	2	3	4	5
u_1	1,000	1,100	1,200	1,300	1,400	1,500
u_2	1,000	1,100	1,200	1,300	1,400	1,500
s_1	1,400	1,424	1,448	1,472	1,497	1,522
s_2	6,300	6,433	6,569	6,707	6,849	6,993
s_3	3,400	3,420	3,438	3,454	3,469	3,482
s_4	2,100	2,680	3,144	3,515	3,812	4,050
y_1	11,200	114,462	116,977	119,544	122,165	124,842
y_2	325,200	334,287	342,581	350,255	357,450	364,281
y_3		360	360	360	360	360
y_4		1,000	1,000	1,000	1,000	1,000

Scenario 3
Decision variable: $x_1 = 0.90$

The results for Scenarios 1, 2, and 3 are set out in Tables 4.9, 4.10, and 4.11, respectively. Figures 4.11, 4.12, and 4.13 depict the dynamics for Scenarios 1–3 over 5 years for population growth, water consumption, and new livestock, respectively.

4.8.3 Micropopulation Model: The Leslie Model

4.8.3.1 Model overview

Demographic changes in communities, large and small, are the driving force in resource allocation for schools, housing, transportation systems, hospitals

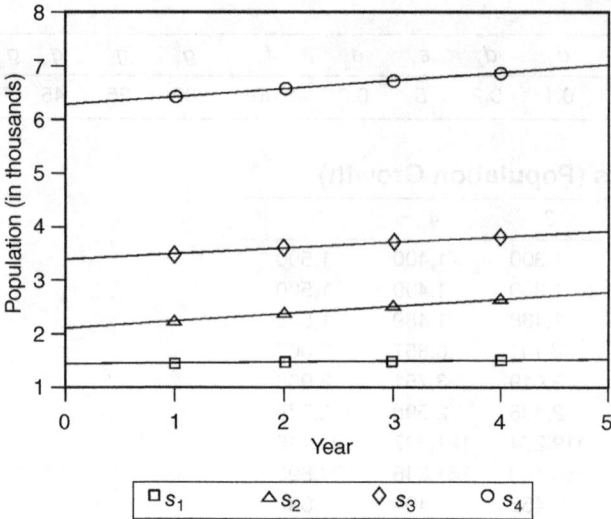

Figure 4.11 Scenario 1: population growth.

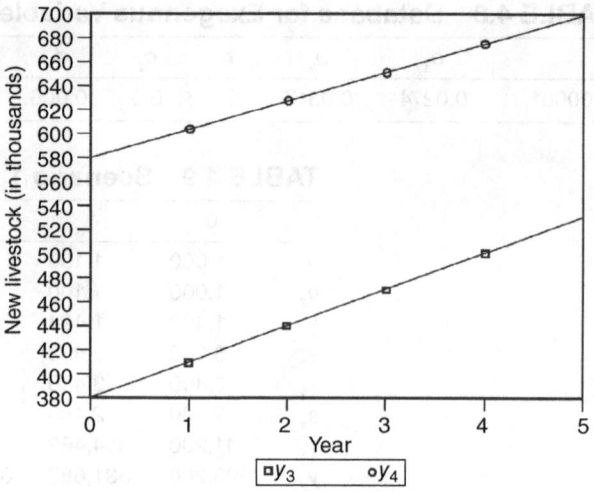

Figure 4.13 Scenario 3: new livestock.

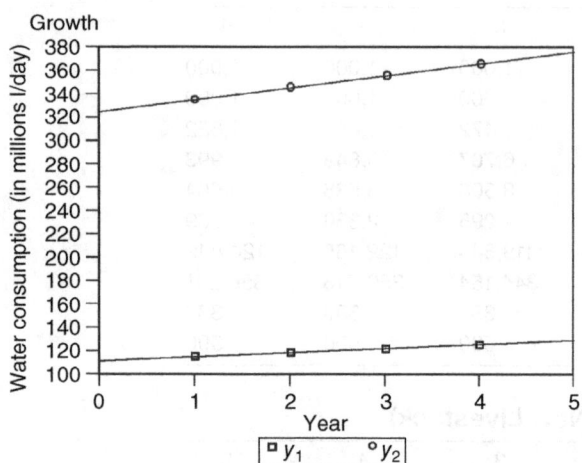

Figure 4.12 Scenario 2: water consumption.

and healthcare delivery, water, electric power and other utilities, and social security, among others. To model these inevitable and critical changes in communities around the world, the Leslie model is often used for the projection of population growth. Consider, for example, the challenge facing a school planning board in a large metropolitan area. The present capacity of classrooms for the elementary, middle, and high schools is already at peak capacity. No one questions the need to build new schools; however, deciding on the size of each of the three school levels must be based on sound analysis. The Leslie model is very effective for this analysis. Given the database on growth projections available to the planning board, along with other more recent information on the demographic composition of the pupils in the metropolitan area, it is possible to make credible projections on the future demand for elementary, middle, and high school buildings and classrooms. This analysis serves multiple purposes: It is cost-effective, and it avoids unnecessary expansion as well as overcrowded classrooms. So far, the focus has been on building space. The same analysis applies to estimating the number of future teachers needed for each class or age category, the associated administrative and maintenance staff, and budgetary and other resource allocations needed to accommodate the projected growth. The following simplified version of the Leslie model is adapted from Meyer [1984].

Assumptions

1. Only the female population will be considered.
2. The female population is divided into n age categories:
   $[0, \Delta), [\Delta, 2\Delta), \ldots, [(n-1)\Delta, n\Delta)$, where, Δ is the width of each age interval of the population. For example, the age interval, Δ, can be 1 year, 5 years, or longer, depending on the planning needs.

Definitions

1. $F_i(t)$ = number of females in the ith age group at time t, namely, the number of females in the interval age group $[i\Delta, (i+1)\Delta)$ at time t.
2. $\mathbf{F}(t)$ is called the age distribution vector at time t.

3. **F**(0) is the age distribution vector at time 0 (or the current age distribution).
4. d_i is the graduation (or withdrawal) rate of the ith age group.
5. $p_i = 1 - d_i$ is the survival rate of the ith age group.
6. m_i is the Δ-year maternity rate for the ith age group, and it is assumed in this simplified model to be invariant over time. This maternity rate implies that at time t, the average female in the ith age group (i.e., in the interval age group $[i\Delta, (i+1)\Delta)$ will contribute m_i children to the lowest age group at time $t+1$).
7. No immigration is incorporated into this simplified model.

4.8.3.2 Model formulation

The female population of the ith age group at the next $(t+\Delta)$ period is given in Eq. (4.45):

$$F_{i+1}(t+\Delta) = (1-d_i)F_i(t) = P_i F_i(t), \quad t = 0, \Delta, 2\Delta, \ldots$$

(4.45)

The number of newborns at the lowest age group (age zero) at time $(t+\Delta)$ is given in Eq. (4.46):

$$F_0(t+\Delta) = \sum_{i=0}^{n-1} m_i F_i(t) \quad (4.46)$$

Equations (4.45) and (4.46) represent the population pyramids of the Leslie model. Figure 4.14, which is adopted from Meyer [1984], depicts these two equations graphically.

Combining Eqs. (4.45) and (4.46) yields

$$\begin{bmatrix} F_0(t+\Delta) \\ F_1(t+\Delta) \\ \vdots \\ F_{n-1}(t+\Delta) \end{bmatrix} = \begin{bmatrix} m_0 & m_1 & m_2 & \ldots & \ldots & m_{n-1} \\ P_0 & 0 & 0 & \ldots & \ldots & 0 \\ 0 & P_1 & 0 & \ldots & \ldots & 0 \\ \ldots & \ldots & \ldots & \ldots & \ldots & \ldots \\ 0 & 0 & 0 & \ldots & P_{n-2} & 0 \end{bmatrix} \begin{bmatrix} F_0(t) \\ F_1(t) \\ \vdots \\ F_{n-1}(t) \end{bmatrix}$$

(4.47)

$$t = 0, \Delta, 2\Delta, \ldots$$

The $(n \times n)$ matrix in Eq. (4.47), denoted by M, is called the Leslie matrix. Equations (4.48) and (4.49) capture the dynamics depicted in Figure 4.14:

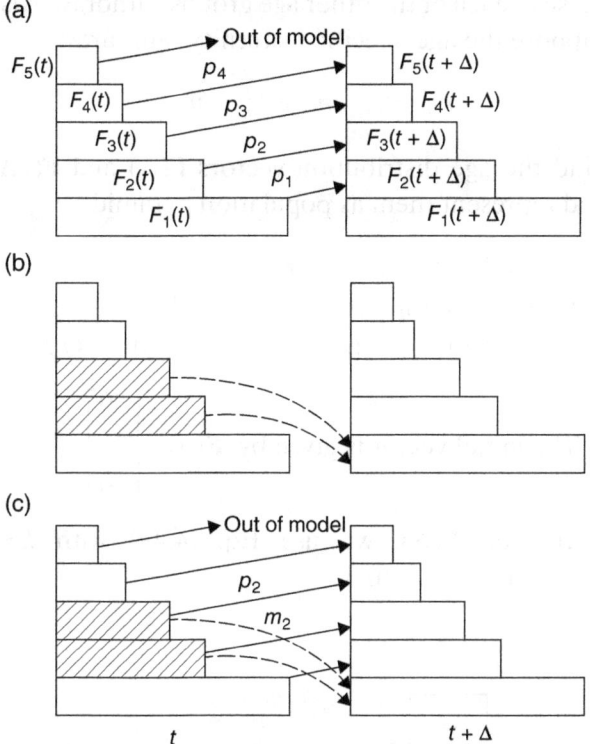

Figure 4.14 The pyramids of the Leslie model [Meyer, 1984]. (a) People move up because they die or grow up. (b) Maternity giving birth. (c) Combination of (a) and (b).

$$\mathbf{F}(t+\Delta) = M\mathbf{F}(t) \quad t = 0, 1, 2, \ldots \quad (4.48)$$

for $t = 0$

$$\mathbf{F}(\Delta) = M\mathbf{F}(0)$$

for $t = \Delta$

$$\mathbf{F}(2\Delta) = M\mathbf{F}(\Delta) = MM\mathbf{F}(0) = M^2\mathbf{F}(0)$$

Likewise,

$$\mathbf{F}(3\Delta) = MM^2\mathbf{F}(0) = M^3\mathbf{F}(0)$$

$$\mathbf{F}(k\Delta) = M^k\mathbf{F}(0) \quad k = 0, 1, \ldots \quad (4.49)$$

4.8.3.3 Example problem

You are given a population that is divided into three age groups at time $t = 0$ as depicted in Figure 4.15.

As one time unit passes, everyone in the oldest group leaves the school district, and one-fourth of those in each of the other age groups withdraw. Also, suppose the age-specific maternity rates are:

$$m_0 = 0; \quad m_1 = 2; \quad m_2 = 3$$

Find the age distribution vectors $\mathbf{F}(\Delta)$ and $\mathbf{F}(2\Delta)$, and represent them as population pyramids:

If $\begin{matrix} d_i = 1/4 \\ P_0 = 3/4 \\ P_1 = 3/4 \end{matrix}$ for $\begin{bmatrix} m_0 & m_1 & m_2 \\ P_0 & 0 & 0 \\ 0 & P_1 & 0 \end{bmatrix} = \begin{bmatrix} 0 & 2 & 3 \\ 3/4 & 0 & 0 \\ 0 & 3/4 & 0 \end{bmatrix}$

The initial vector is given by $\mathbf{F}(0) = \begin{bmatrix} 50 \\ 30 \\ 10 \end{bmatrix}$

To find $\mathbf{F}(\Delta)$, we use Eq. (4.49) with $k=1$ $(\mathbf{F}(\Delta) = M^1 \mathbf{F}(0))$.

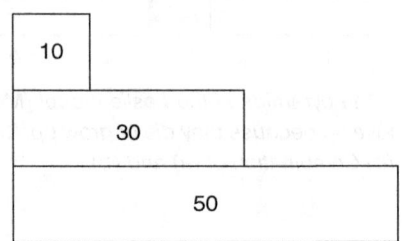

Figure 4.15 *The initial population breakdown.*

(Age Distribution Vector)

$$\mathbf{F}(\Delta) = \begin{bmatrix} F_0(\Delta) \\ F_1(\Delta) \\ F_2(\Delta) \end{bmatrix} = \begin{bmatrix} 0 & 2 & 3 \\ 3/4 & 0 & 0 \\ 0 & 3/4 & 0 \end{bmatrix} \begin{bmatrix} 50 \\ 30 \\ 10 \end{bmatrix} = \begin{bmatrix} 90 \\ 37.5 \\ 22.5 \end{bmatrix}$$

To calculate change in population over two periods,

$$\mathbf{F}(2\Delta) = \begin{bmatrix} F_0(\Delta) \\ F_1(\Delta) \\ F_2(\Delta) \end{bmatrix} = \begin{bmatrix} 0 & 2 & 3 \\ 3/4 & 0 & 0 \\ 0 & 3/4 & 0 \end{bmatrix} \begin{bmatrix} 90 \\ 37.5 \\ 22.5 \end{bmatrix} = \begin{bmatrix} 142.5 \\ 67.5 \\ 28.125 \end{bmatrix}$$

or

$$\mathbf{F}(2\Delta) = \begin{bmatrix} F_0(\Delta) \\ F_1(\Delta) \\ F_2(\Delta) \end{bmatrix}$$

$$= \begin{bmatrix} 0 & 2 & 3 \\ 3/4 & 0 & 0 \\ 0 & 3/4 & 0 \end{bmatrix} \begin{bmatrix} 0 & 2 & 3 \\ 3/4 & 0 & 0 \\ 0 & 3/4 & 0 \end{bmatrix} \begin{bmatrix} 50 \\ 30 \\ 10 \end{bmatrix}$$

$$= \begin{bmatrix} 3/2 & 9/4 & 0 \\ 0 & 3/2 & 9/4 \\ 9/16 & 0 & 0 \end{bmatrix} \begin{bmatrix} 50 \\ 30 \\ 10 \end{bmatrix} = \begin{bmatrix} 142.5 \\ 67.5 \\ 28.125 \end{bmatrix}$$

These results are presented in Figure 4.16.

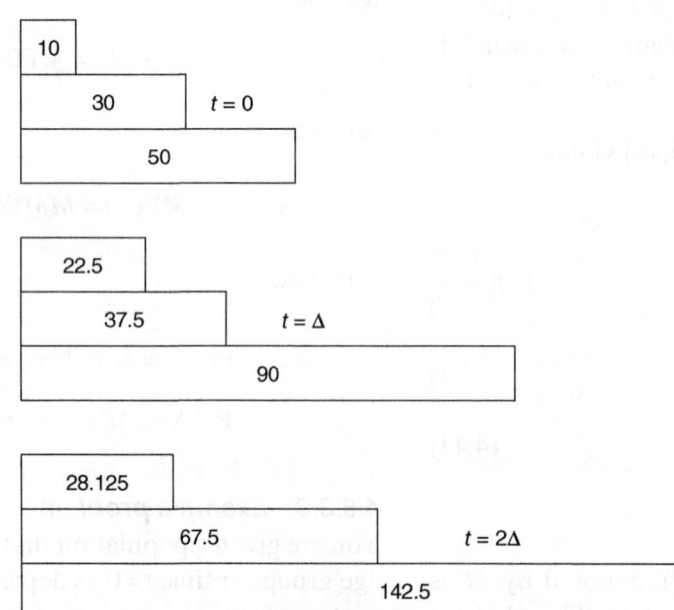

Figure 4.16 *The population breakdown at $t = 0$, Δ, and 2Δ.*

4.9 PSM*

4.9.1 Introduction

Since the 1997 report by the President's Commission on Critical Infrastructure Protection [PCCIP, 1997], billions of public and private dollars have been spent in the United States to assess and manage risks to the homeland from multiscale natural, technological, and human-generated hazards. Unfortunately, we still do not have adequate and appropriate metrics, models, and evaluation procedures with which to measure the costs, benefits, and remaining risks associated with preparedness and security expenditures. In other words, we must be able to measure the efficacy of risk assessment and management against catastrophic and contextual risks (i.e., risks to system performance resulting from external changes in an interdependent socioeconomic landscape). Such measures have been called for over three decades [White and Haas, 1975] and remain urgently necessary if disaster research is to have an appropriate impact on national and regional preparedness policies. Similarly, billions of dollars are spent on education and other economic initiatives. Yet we are losing the global edge of economic competitiveness, according to the report of the New Commission on the Skills of the American Workforce [NCEE, 2007]. One reason is the inability to appropriately rationalize investments in risk management against the background of emergent economies and associated contextual risks.

As discussed by Haimes [2007, 2008, 2012], these two dissimilar and seemingly unrelated national concerns—the economy and homeland security—share inherent characteristics, namely, they are multiscale systems characterized by emergent risks with potentially significant national economic and security ramifications. There is a need to (i) have the ability to model and assess the costs, benefits, and remaining risks associated with each viable risk management policy option and (ii) produce methods that support continued, measured learning that can feed an adaptive resource allocation process.

No single model can capture all the dimensions necessary to adequately evaluate the efficacy of risk assessment and management activities. This is because it is impossible to identify all relevant state variables and their substates that adequately represent large and multiscale systems [Haimes, 1977, 1981, 2004, 2007]. There is a need for theory and methodology that will enable regions to appropriately rationalize risk management decisions through a process that:

(a) Identifies existing and potential emergent risks systemically.
(b) Evaluates, prioritizes, and filters these risks based on justifiable selection criteria.
(c) Collects, integrates, and develops appropriate metrics and a collection of models to understand the critical aspects of regions.
(d) Recognizes emergent risks that produce large impacts and risk management strategies that potentially reduce those impacts for various time frames.
(e) Optimally learns from implementing risk management strategies.
(f) Adheres to an adaptive risk management process that is responsive to dynamic, internal, and external forced changes. *To do so effectively, models must be developed to periodically quantify, to the extent possible, the efficacy of risk management options in terms of their costs, benefits, and remaining risks.*

A risk-based, multimodel, systems-driven approach can effectively address these emergent challenges at both the national and regional levels. Such an approach must be capable of maximally utilizing what is known now and optimally learn, update, and adapt through time as decisions are made and more information becomes available at various regional levels. The methodology must quantify risks as well as measure the extent of learning to quantify adaptability. This learn-as-you-go tactic will result in reevaluation and evolving/learning risk management over time.

4.9.2 Risk Modeling, Assessment, and Management

In Chapter 1, we cited Lowrance [1976] who described *risk* as "a measure of the *probability* and *severity* of *adverse effects.*" In Chapter 3, we cited

---

*This chapter is based on Haimes [2007]. Chapter 16 is devoted in its entirety to the PSM and to modeling systems of systems.

Kaplan and Garrick [1981] who were the first to formalize a theory of quantitative risk assessment with the triplet {S, L, C} questions, where S is the set of risk scenarios or adverse events, L is the set of likelihoods or probabilities, and C is the associated set of consequences describing severity of impacts from risk scenarios. This definition, although very descriptive of risk, has resulted in some operational challenges in its implementation. Consider, for example, the challenges faced by modelers of dose–response functions stemming from the exposure of humans and animals to chemicals and other dangerous agents. (See, e.g., early work by Lamanna [1959] or Lowrance [1976].) The professional community has been hard at work relating human actions to effects on human health, the environment, and the ecology; their achievements have not been gained overnight and without significant and concerted efforts. Decades of research have resulted in the development of cause–effect relationships that served as the foundation of risk–cost–benefit analyses and strategic decisions related to food, air quality, water quality, pollution, and many other risk-based decisions. Deconstructing the quantitative dose–response-type risk assessment has illuminated a strong need to focus modeling efforts on identifying and quantifying the *state of the system* (see Chapter 2).

Any risk modeling exercise must consider the state of the system, X. We define the *state of the system* as *those characteristics and parameters that fundamentally represent the system and provide insight into the relationships between scenarios, likelihoods, and their consequences.* Vulnerability and threat are both manifestations of inherent states of systems. *Vulnerability* is a manifestation of the states of the system, and it refers to the system's performance objectives that we are trying to secure or control [Haimes, 2005, 2006, 2007]. (Let X_1 be the set of states of the assured system.) *Threat* is a manifestation of the inherent capabilities and intents (in the case of human adversaries) of potential antagonistic systems such as attacks, accidents, or natural disasters. (Let X_2 be the set of states of antagonistic systems.) Chapter 17 presents a more detailed discussion on the relationship between vulnerability and the states of the system.

To illustrate, consider national competitiveness. It can be measured by national productivity [Porter, 1998, 2003; Li and Xu, 2004], which would translate into wage/income levels. However, state variables, X_1, measure education and skills as well as the production of specific industries and assets. These state variables (and their substates) are not static in their levels of operation and functionality and form the foundation for any models that support risk-based decisionmaking. Correspondingly, the level of vulnerability fluctuates with the state of the system under examination. With this understanding, the set of risk scenarios can now be considered potential threats to system vulnerabilities that can result in adverse effects at specific times.

Infrastructures, for example, the educational system or homeland critical facilities, commonly incorporate myriad components, such as cyber, physical, and organizational. These can be modeled by dynamic hierarchies of interconnected and interdependent subsystems that are threatened by natural hazards, evolving terrorist networks, and emerging global economies. Indeed, models of emergent multiscale systems may be represented by one or more of the following characteristics and attributes:

- Micro- or macroperspectives
- Dynamic or static conditions
- Linear or nonlinear relationships
- Lumped or spatially distributed elements
- Deterministic or stochastic levels of uncertainty
- Acceptable levels of risk or risks of extreme and catastrophic events
- Single or multiple conflicting and competing goals and objectives
- Hardware, software, human, organizational, or political dimensions
- Short-, intermediate-, or long-term temporal domains
- Single or multiple agencies with different missions, resources, timetables, and agendas
- Single or multiple decisionmakers and stakeholders
- Local, regional, national, or international relationships

Thus, it is a major challenge to understand and then model the intra- and interrelationships among these multidimensional and multiperspective subsystems and their ultimate integration into a coherent homeland security-based system. Multiple models and submodels built for these purposes are inherently different in their structures and roles. *Therefore, we must match a flexible, agile, and responsive modeling schema to the plethora of characteristics and attributes of these complex multiscale systems.*

A major deficiency remains in our ability to model emergent multiscale systems—that is, to develop appropriate modeling capabilities. To do so constitutes the theme of this section.

4.9.3 The PSM

According to *Webster's New International Dictionary*, a phantom is "Something that is apparent to the sight or other senses but has no actual substantial existence; something elusive or visionary." The PSM [Haimes, 2007] enables research teams to effectively analyze major forced changes in the characteristics and performance of multiscale assured systems such as cyber and physical infrastructure systems or major socioeconomic systems. (*Note that the term PSM will connote the overall modeling philosophy, while PSMs will connote the modeling components.*) Forced changes are manifestations of the states of antagonistic systems, X_2, that have a direct impact on the states of the assured system, X_1. Thus, we consider as forced changes both the risks of weapons of mass destruction (WMD) and the risks of losing American global competitiveness to foreign economies. The PSM introduced in this paper builds and expands on hierarchical holographic modeling (HHM) [Haimes, 1981, 2004], various analytical modeling methods, and simulation tools to present comprehensive views and perspectives on unknowable emergent systems. (See Chapter 3 for a more elaborate discussion of HHM.) By building on and incorporating input from HHM, the PSM seeks to develop causal relationships through various modeling and simulation tools. In doing so, the PSM imbues life and realism into visionary ideas for emergent multiscale systems—ideas that otherwise would never be realized. In other words, with different modeling and simulation tools, PSM legitimizes exploring and experimenting with out-of-the-box and seemingly *crazy* ideas.

Ultimately, it discovers insightful implications that otherwise would have been completely missed and dismissed. In this sense, it allows for *nonconsensus* ideas or an *agree-to-disagree* process for further exploration and study.

The output of the HHM is a taxonomy of identified risk scenarios or multiple perspectives of a system for modeling. Alternatively, the output of the PSM is a justification or rationalization of investment in preparedness or learning activities to protect against critical forced changes or emergent risks—investment that might not otherwise have been approved. *Through logically organized and systemically executed models, the PSM provides a reasoned experimental modeling framework with which to explore and thus understand the intricate relationships that characterize the nature of multiscale emergent systems.* The PSM philosophy rejects dogmatic problem solving that relies on a single modeling approach structured on one school of thinking [Bertalanffy 1968/1976]. Rather, its modeling schema builds on the multiple perspectives gained through generating multiple scenarios. This leads to the construction of appropriate models to deduce tipping points as well as meaningful information for logical conclusions and future actions. Currently, models assess what is optimal, given what we know, or what we think we know. We want to extend these models to answer the following questions:

1. What do we need to know?
2. What value might appear from risk reduction results producing more precise and updated knowledge about complex systems?
3. Where is that knowledge needed for acceptable risk management and decisionmaking?

Models, experiments, and simulations are conceived and built to answer specific questions. Conventional system models attempt to provide answers based on the responses on the states of a system under given conditions and assumptions. For example, the Leontief input–output economic model [Leontief, 1951a, b, 1966], discussed in Chapter 18, enables analysts to ask: What are the relationships between production and consumption among the interdependent sectors of the economy? For emergent multiscale systems, analysts may ask an entirely different type of question through the PSM: What kind of a multiscale system and its influencing environment may emerge in the future, where today's known relationship between production and consumption may or may not hold or be applicable? Answering this mandates seeking the *truth* about the unknowable complex nature of emergent systems; it requires intellectually bias-free modelers and thinkers who are empowered to experiment with a multitude of modeling and simulation approaches and to collaborate for appropriate solutions. PSM users will be expected to build on the knowledge generated through the diverse models employed and on the contributions made by analysts of diverse disciplines and expertise.

An artist's first painting is usually not a masterpiece. To achieve this, the artist must usually select and explore various themes to develop knowledge and understanding. The final product can then be carefully designed based upon what is learned through experience. The PSM is a modeling paradigm that is congruent with and responsive to the uncertain and ever-evolving world of emergent systems. In this sense, it serves as an adaptive process, *a learn-as-you-go modeling laboratory*, where different scenarios of needs and developments for emergent systems can be explored and tested. (These scenarios are generated through the collaborative adaptive multiplayer HHM (CAM-HHM) game, introduced in Chapter 3.) In other words, to represent and understand the uncertain and imaginary evolution of a future emergent system, we need to deploy an appropriate modeling technology that is equally agile and adaptive. One may view the PSM as matching methodology and technology; emergent systems are studied through this model similar to the way other appropriate models are constructed for systems with different characteristics. (Examples are difference equations and differential equations for dynamic systems, algebraic equations for static systems, and probabilities for systems that are driven by random events and processes.) The PSM can be continually manipulated and reconfigured in our attempts to answer difficult emergent questions and challenges.

FEMA's HAZUS-MH for Hurricanes [FEMA, 2006] is an example of the type of tool that might emerge from a PSM process. Although hurricanes are not necessarily emergent, the construction of the HAZUS-MH has resulted from integrating databases, models, and simulation tools that have been developed across many disciplines over the last several decades; as an integrated tool, it can be used to study the impacts of various hurricane scenarios on regions and their system states. At the basic modeling level, there are databases of buildings, businesses, essential facilities, and other fundamental structural and regional facts that characterize the state of the region under study (i.e., X_1). These databases are editable to enable exploring agile properties of structures that may change the impact of hurricanes. Scientific models from decades of research estimate probabilistic structural damage from wind gusts striking various structural vulnerabilities. Finally, there is a hazard model to estimate peak wind gust given historical or user-defined catastrophes, that is, user-defined/user-imagined states of the antagonistic system, X_2. Integrating databases, causal damage models, and flexible hazard simulations results in a tool that enables regions to fully explore ranges of *phantom* situations. (This includes both uncertain/emergent changes in the threats by changes in X_2 and controllable mitigation actions represented by changes in X_1.) In this context, PSM also can be viewed as a methodological process for developing tools that will have the flexibility to capture emergent behavior of both regional vulnerabilities and threats. Moreover, these solutions to a PSM process result in a method to trace changes in problem definitions, critical variables, critical metrics, available data, and others, in a way that enables us to measure learning, changing, and improvement in risk management activities over time.

An example application resulting from PSM might integrate databases of students, training programs, and part-time jobs with probabilistic learning models and simulations of part-time job growth and student success. Such a tool could engender proposals that adolescents fill a stronger role in skilled labor through vocational training and part-time work during high school while simultaneously preparing for college. PSM can provide a formal framework in which such ideas can be imagined and then realized through modeling and simulation suites that act as large-scale experimental laboratories. Thus, researchers gain added knowledge of the systems they are discovering. The results of such activities simultaneously support effective resource allocation for risk management. In other words, *PSM is the process by which identified emergent risk scenarios can guide the creation of modeling and simulation suites to cost-effectively explore and rationalize preparedness against a host of emergent threats that are unpredictable.*

4.9.4 Modeling Engines That Drive the PSM

A plethora of models, methodologies, and tools that have been developed over several decades by different disciplines are marshaled by the PSM to shed light on and provide answers to several modeling questions that constitute the essence of the risk assessment and management process (see Chapter 1). Two major modeling groups are explored and briefly developed in this section as a part of the PSM framework for evaluating the efficacy of risk assessment and adaptive risk management.

4.9.4.1 Decision-based modeling and simulation

The contributions of PSM are even more specific and significant when various decisions and policy analyses can be made by experimenting with multiple models and systems-based methodologies. Two major groups of models are required for the success of the PSM framework: decision-based models and domain-specific models. *Decision-based* models are extremely flexible and provide outputs such as optimums, trade-offs, tipping points, and others that are useful and supportive of specific decision questions. Their flexibility enables a wide variety of applications to answer questions on various geographic and temporal scales. *Domain-specific* models are those that are developed around a phenomenon or behavior. They are built on a fundamental understanding of scientific principles, and they are traditionally more narrowly applicable and can be distilled into sophisticated computer applications.

Examples of decision-based models include decision trees, dynamic programming, and adaptive management. Also, Bellman's principle of optimality may be suitable to address the sequential feature of decisionmaking in resource allocation. In multiple stages of resource allocation across multiple objectives, MODTs (see Chapter 9) can be used to model the impact of the agency's current decisions on future options. For example, risk-based adaptive management using MODT would add measurable assurance for decisions made on resource allocations and on the impacts such current allocations might have on future scenarios and needs. Fitting MODT into the PSM framework supports validating information operations in regional and national strategies that would support adaptive risk management. The result of such an effort may modify the MODT solution paradigm: from proving the *best* or *correct* policies to developing the capacity to improve learning, adaptation, and communication against emergent risks. Valid information operations would result from viewing impacted solutions as somehow *getting better* in response to changing risks.

4.9.4.2 Graphical depiction of the methodological framework

Through the PSM, changes that result in emergent risks are modeled and analyzed from their numerous perspectives through multiple models of varied structure, mathematical rigor, analytical level, or heuristics. This is a dynamic and ever-evolving process in response to forced events that are inherently dynamic and unpredictable. It also captures diverse perspectives of the system and its risks and opportunities. For example, this process may result in an HHM that is filtered through a modified risk filtering, ranking, and management (RFRM) procedure (discussed in Chapter 7). The state variables and emergent risk scenarios become the foundation for laboratory-like experimentation through the strategic development of a modeling suite that includes both domain-specific and decision-based models and simulations. The final outputs of a single PSM iteration are trade-offs, optimums, tipping points, and support for resource allocation. These will support future PSM exercises for adaptive learning.

4.9.5 Summary

Unprecedented and emerging multiscale systems are inherently elusive and visionary—they are by and large phantom entities grounded on a mix of future needs and available resources, technology, forced developments and changes, and myriad other unforeseen events. From the systems engineering perspective, understanding and effectively responding to and managing these evolving forced changes require an equally agile and flexible multiplicity of models. Both models and modelers must represent broad perspectives and possess matching capabilities, wisdom, and foresight for futuristic and out-of-the-box thinking. These three components—the emergent systems, the agile and flexible multiplicity of models, and the human systems engineering experience, expertise, and capabilities—together constitute the PSM. In this sense, the PSM is a real-to-virtual laboratory for experimentation, a learn-as-you-go facility, and a process for emergent systems that are not yet completely designed and developed. The Human Genome Project may be considered another multiscale audacious emergent system, fraught with uncertainties and involving participants from multiple disciplines with varied perspectives, experience, skills, and backgrounds. In an October 30, 2006, interview in US News & World Report [Hobson, 2006], Eric Lander, genetic researcher and a leader in the Human Genome Project, was asked, "The right way to decipher the genome wasn't at all clear. How did you lead in that environment?" He answered:

> A lot of it is managing in the face of tremendous uncertainty. You have to be willing to rethink the plan at least every six months. It was destabilizing—but really important—that we were prepared to put on the table every three to four months whether we were doing the right thing.... We made many, many midcourse corrections.

Finally, it is not too unrealistic to compare the evolving process of the PSM to the *modeling* experience of children at play. They experiment and explore their uncorrupted, imaginative emergent world with Play-Doh and Legos while patiently embracing construction and reconstruction in an endless trial-and-error process with great enjoyment and some success.

4.10 EXAMPLE PROBLEMS

4.10.1 Testing Problem

Payton Products is currently manufacturing gas chromatographs. These are commonly used for large amounts of drug testing, such as drug testing of

TABLE 4.12 Cost of Defective Part Returns

Costs	Excellent Quality	Good Quality	Poor Quality
No testing	$1000	$2500	$5000
Moderate testing	$100	$250	$500
Extensive testing	$25	$62	$125

TABLE 4.13 Testing Costs

Costs	Excellent Quality	Good Quality	Poor Quality
No testing	$0	$0	$0
Moderate testing	$2000	$2000	$2000
Extensive testing	$4000	$4000	$4000

athletes. The company is trying to decide how much testing of the product should be done in order to maximize quality and increase customer satisfaction. It is trying to determine whether to (i) do no testing, (ii) do moderate testing, or (iii) do extensive testing. No testing is identified as 0 days of product testing, moderate testing is defined as 4 days, and extensive testing is identified as 8 days. Three quality-defect levels have been established: 1% and lower would be excellent quality, 1.1–4.9% would be good, and a defect level of 5% and greater would be considered poor quality. The anticipated costs are a function of the amount of testing needed and the quality level of the gas chromatographs.

It costs Payton Products $100 for any defective part that is sent back by a customer, and the company assumes that any defective product will be returned. Thus, the total cost is $100 multiplied by the number of defective parts that do not pass inspection. The total cost for the number of returned defective parts for each testing and quality level is summarized in Table 4.12.

The testing costs are based on a set fee of $500 times the number of days the product is being tested. A summary of the testing costs is shown in Table 4.13.

Thus, the testing costs plus the cost of defective parts returned constitute the total cost. A summary of the total costs is shown in Table 4.14 as a combined function of the testing and quality levels.

It will be necessary to apply the Hurwitz rule to determine the company's best policy for reducing its cost and also improving customer satisfaction.

Definition of problem
- Actions
  1. No testing (a_1)
  2. Moderate testing (a_2)
  3. Extensive testing (a_3)
- Quality levels
  1. Excellent (s_1)
  2. Good (s_2)
  3. Poor (s_3)

The payoff matrix (presented in terms of negative profits) is given in Table 4.15.

Analysis

The Hurwitz rule, which is defined as

$$\max_{1 \le i \le 3} \left\{ \mu_i(\alpha) = \alpha \min_{1 \le j \le J} \mu_{ij} + (1-\alpha) \max_{1 \le j \le J} \mu_{ij} \right\},$$

compromises between the two extremes through the use of the index α, where $0 \le \alpha \le 1$, where $\alpha = 1$ implies a pessimistic criterion and $\alpha = 0$ implies an optimistic criterion. The pessimistic and optimistic outcomes for each action are shown in Table 4.16:

At a_1: $\alpha(-5000) + (1-\alpha)(-1000) = -1000 - 4000\alpha$
At a_2: $\alpha(-2500) + (1-\alpha)(-2100) = -2100 - 400\alpha$
At a_3: $\alpha(-4125) + (1-\alpha)(-4025) = -4025 - 100\alpha$

TABLE 4.14 Costs as Function of Amount of Testing and Quality Levels

Costs	Excellent Quality	Good Quality	Poor Quality
No testing	−$1000	−$2500	−$5000
Moderate testing	−$2100	−$2250	−$2500
Extensive testing	−$4025	−$4062	−$4125

TABLE 4.15 Payoff Matrix

	$j=1$ (s_1)	$j=2$ (s_2)	$j=3$ (s_3)
No testing $i=1$ (a_1)	−$1000	−$2500	−$5000
Moderate testing $i=2$ (a_2)	−$2100	−$2250	−$2500
Extensive testing $i=3$ (a_3)	−$4025	−$4062	−$4125

These actions are displayed graphically in Figure 4.17. Now, it is necessary to solve for the value of α based on the decisionmaker's degree of optimism. Based on the graph of the functions, it is easy to determine the value of α that will help the decisionmaker choose the best action according to his or her level of optimism. The calculations for determining α follow:

Calculations for α:

$$-1000 - 4000\alpha = -2100 - 400\alpha$$
$$1100 = 3600\alpha$$
$$\alpha = 11/36 = 0.306$$

Therefore, for $\alpha < 0.306$, Action 1 should be taken—that is, no testing of gas chromatographs. For $\alpha > 0.306$, Action 2 should be taken—that is, moderate testing of the gas chromatographs. Clearly, Action 3 is dominated by the other actions for all values of α.

4.10.1.1 Expected Monetary Value (EMV)

From past experience regarding the quality of the gas chromatographs, Payton Products knows that quality is excellent 50% of the time, good 25% of the time, and poor 25% of the time. The EMV of the profit is defined as follows:

$$\text{EMV} = \max_{1 \le i \le 3} \sum_{j=1}^{3} P(s_j) \mu_{ij}$$

Figure 4.18 shows a graphical representation of the problem through a decision tree.

The EMV for all actions is calculated below:

At a_1: $\sum_{j=1}^{3} P(s_j) \mu_{ij} = P(s_1)\mu_{11} + P(s_2)\mu_{12} + P(s_3)\mu_{13}$
$= 0.5(-1000) + 0.25(-2500) + 0.25(-5000)$
$= -\$2375$

At a_2: $\sum_{j=1}^{3} P(s_j) \mu_{ij} = P(s_1)\mu_{21} + P(s_2)\mu_{22} + P(s_3)\mu_{23}$
$= 0.5(-2100) + 0.25(-2250) + 0.25(-2500)$
$= -\$2237$

TABLE 4.16 Pessimistic and Optimistic Outcomes

	Excellent (s_1)	Good (s_2)	Poor (s_3)	Optimistic	Pessimistic
No testing (a_1)	−$1000	−$2500	−$5000	−$1000	−$5000
Moderate testing (a_2)	−$2100	−$2250	−$2500	−$2100	−$2500
Extensive testing (a_3)	−$4025	−$4062	−$4125	−$4025	−$4125

146 MODELING AND DECISION ANALYSIS

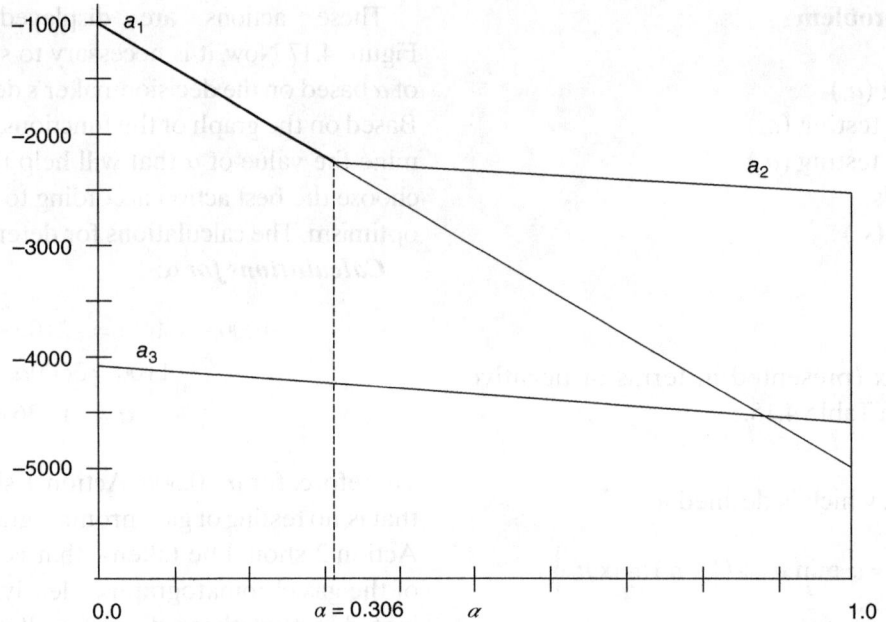

Figure 4.17 Hurwitz rule applied to the testing problem.

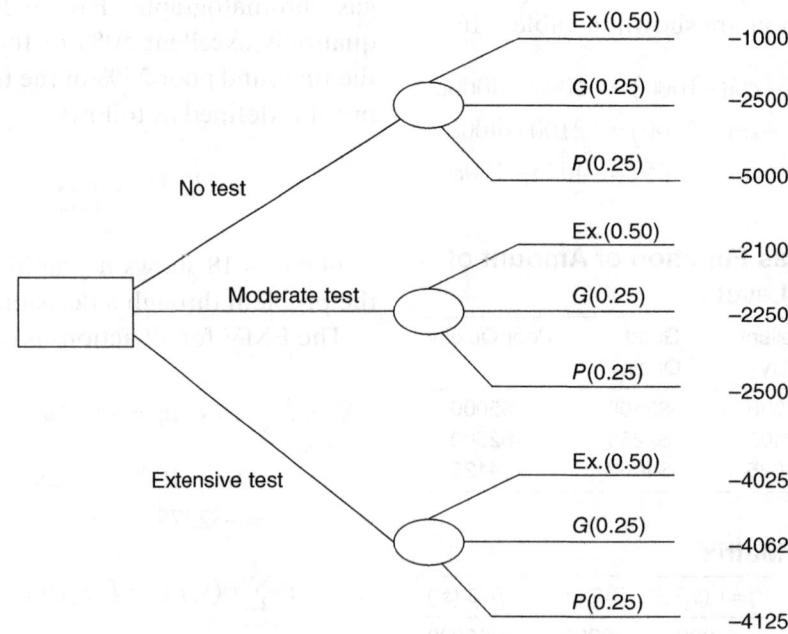

Figure 4.18 Decision-tree representation.

At a_3: $\sum_{j=1}^{3} P(s_j)\mu_{ij} = P(s_1)\mu_{31} + P(s_2)\mu_{32} + P(s_3)\mu_{33}$

$= 0.5(-4025) + 0.25(-4062) + 0.25(-4125)$

$= -\$4059$

$\text{EMV} = \max(-2375, -2237, -4059)$

$= -2237$

Thus, since the EMV should be maximized, the action that should be chosen is action 2, moderate testing, since the EMV is −$2237, which is the least cost for testing the gas chromatograph.

The fact that both methods give the same result, moderate testing, shows that Payton Products should try to do moderate testing of the gas chromatograph to improve customer satisfaction. However, analysts must also keep in mind that different methods often give different solutions. (Using the Hurwitz rule, *no testing* was also found to be a solution.) Therefore, the final conclusion is up to the decisionmaker as to which action is the best.

4.10.2 A Deicing Problem

Although multiobjective analysis is discussed in the following chapter and multiobjective decision-tree analysis is discussed in Chapter 9, considering two objectives in this example problem should be easy to comprehend. The same analysis follows single and multiobjective decision-tree analyses with a minor modification in the final analysis.

The County Board of Supervisors must decide if and when to send deicing crews on county roads when there is precipitation. Icing occurs when the temperature is under 32°F. These decisions are made in 12 h periods. Thus, in bad weather, two decisions must be made each day. The county wishes to minimize the cost (C) of deicing and also minimize residents' property damage (PD) due to accidents. It is assumed that C and PD are noncommensurate, that is, they cannot be added up. For example, PD may just denote the number of accidents. Below are the associated costs for deicing:

- DI_1 : Deice in Stage 1 : $5,000
- DI_2 : Deice in Stage 2 : $3,000, if no ice in Stage 1
  $4,000, if ice in Stage 1
- DN_1 : Do nothing in Stage 1 : $0
- DN_2 : Do nothing in Stage 2 : $0

Further, the following assumptions are made:

- Deicing in Stage 1 also avoids ice problems in Stage 2.
- Deicing leads to PD = 0 in the deiced stage.
- If icing occurs only in Stage 1, then PD = 40.
- If icing occurs only in Stage 2, then PD = 60, due to higher traffic volume.
- If icing occurs in both periods, then PD = 100.

- If icing occurs in Stage 1 (without deicing) and a deicing decision is made for Stage 2, then PD = 50 in Stage 2, if the temperature does indeed fall below 32°F in Stage 2, because of some residual ice from Stage 1.

Two equally likely lognormal pdfs represent the air temperature in the winter: $T_1 = $ LN (3.9,1) and $T_2 = $ LN (3.4,1).

There are two possible events at the end of the first period:

1. There is ice ($T \leq 32°$).
2. There is no ice ($T > 32°$).

The property damage (PD) and cost (C) associated with each incident are depicted in the decision tree shown in Figure 4.19:

Calculations

$$\Pr(\text{ice}) = \Pr(\text{ice}|T_1)\Pr(T_1) + \Pr(\text{ice}|T_2)\Pr(T_2)$$

$$\Pr(\text{ice}) = \Pr(T \leq 32|T_1)\Pr(T_1) + \Pr(T \leq 32|T_2)\Pr(T_2)$$

$$\Pr(T \leq 32|T_1) = \Pr\left(z \leq \left(\frac{\ln(32) - 3.9}{1}\right)\right)$$

$$\Pr(z \leq -0.434 = 1 - \phi(0.434) = 0.3318)$$

$$\Pr(T \leq 32|T_2) = \Pr\left(z \leq \left(\frac{\ln(32) - 3.4}{1}\right)\right)$$

$$\Pr(z \leq 0.065) = 0.5260$$

$$\Pr(\text{ice}) = \frac{1}{2}(0.3318 + 0.5260) = 0.4289$$

$$\Pr(\text{no ice}) = \Pr(\text{no ice}|T_1)\Pr(T_1) + \Pr(\text{no ice}|T_2)\Pr(T_2)$$

$$= 1 - \Pr(\text{ice}) = 1 - 0.4289 = 0.5711$$

At the beginning of Stage 2, we can compute the posterior probabilities:

$$\Pr(T_1|\text{ice}) = \frac{\Pr(\text{ice}|T_1)\Pr(T_1)}{\Pr(\text{ice}|T_1)\Pr(T_1) + \Pr(\text{ice}|T_2)\Pr(T_2)}$$

$$= \frac{\frac{1}{2}(0.3318)}{\frac{1}{2}(0.3318) + \frac{1}{2}(0.5260)} = 0.3868$$

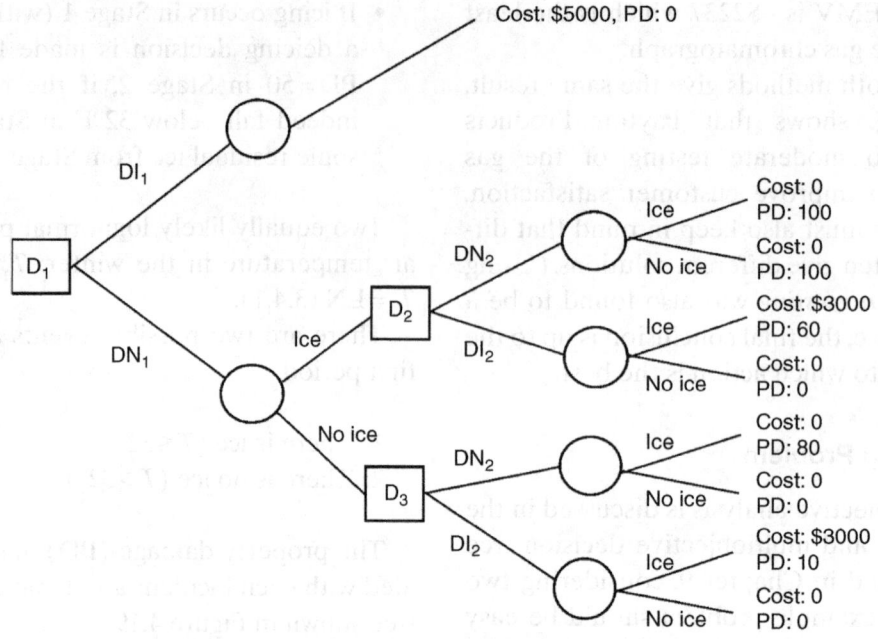

Figure 4.19 Property damage caused by each incident.

$$\Pr(T_1 \mid \text{no ice})$$

$$= \frac{\Pr(\text{no ice} \mid T_1)\Pr(T_1)}{\Pr(\text{no ice} \mid T_1)\Pr(T_1) + \Pr(\text{no ice} \mid T_2)\Pr(T_2)}$$

$$= \frac{\frac{1}{2}(0.6682)}{\frac{1}{2}(0.6682) + \frac{1}{2}(0.4740)} = 0.5850$$

Similarly,

$$\Pr(T_2 \mid \text{ice}) = \frac{0.5260}{0.3318 + 0.5260} = 0.6132$$

$$\Pr(T_2 \mid \text{no ice}) = \frac{0.4740}{0.6682 + 0.4740} = 0.4150$$

$$\Pr(\text{ice} \mid \text{ice in stage 1}) = \Pr(\text{ice} \mid T_1)\Pr(T_1 \mid \text{ice})$$
$$+ \Pr(\text{ice} \mid T_2)\Pr(T_2 \mid \text{ice})$$
$$= (0.3318)(0.3868)$$
$$+ (0.5260)(0.6312) = 0.4509$$

$$\Pr(\text{ice} \mid \text{no ice in stage 1}) = \Pr(\text{ice} \mid T_1)$$
$$\Pr(T_1 \mid \text{no ice in stage 1})$$
$$+ \Pr(\text{ice} \mid T_2)$$
$$\Pr(T_2 \mid \text{no ice in stage 1})$$
$$= (0.3318)(0.5850)$$
$$+ (0.5260)(0.4150) = 0.4124$$

The expected [C; PD] vectors associated with C_1 through C_4 can then be calculated as follows:

C_1: [0; (0.4509)(100) + (0.5491)(40)] = [0; 67.05]
C_2: [4000; (0.4509)(50) + (0.5491)(40)] = [4000; 44.51]
C_3: [0; (0.4124)(60) + (0.5876)(0)] = [0; 24.74]
C_4: [3000; 0] = [3000; 0]

At D_2, we have to decide between the solutions associated with C_1 and C_2. However, neither one dominates the other, so we fold [0; 67.05] and [4000; 44.51] back to D_2. Similarly, we fold [0; 24.74] and [3000; 0] from C_3 and C_4 back to D_3.

At C_0, we have to average out, individually, each of the two vectors associated with D_2 with those two vectors associated with D_3. We obtain:

(a) Pr(ice) [C_1] + Pr(no ice) [C_3]
    0.4289 [0; 67.05] + 0.5711 [0; 24.74] = [0; 42.89]
(b) Pr(ice) [C_1] + Pr(no ice) [C_4]
    0.4289 [0; 67.05] + 0.5711 [3000; 0] = [1713.3; 28.76]
(c) Pr(ice) [C_2] + Pr(no ice) [C_3]
    0.4289 [4000; 44.51] + 0.5711 [0; 24.74] = [1715.6; 33.22]
(d) Pr(ice) [C_2] + Pr(no ice) [C_4]
    0.4289 [4000; 44.51] + 0.5711 [3000; 0] = [3412.7; 19.09]

Clearly, (c) is dominated by (b), and at this point, we can delete the option "if *ice* in Stage 1, then *deice* in Stage 2, and if *no ice* in Stage 1, then *do nothing* in Stage 2" from further consideration. We fold (a), (b), and (d) back to D_1 and compare them with the alternative DI_1 ([5000; 0]). This alternative neither dominates nor is dominated by one of the other three remaining alternatives.

In conclusion, four out of five possible strategies are nondominated. A selection will have to be made based on the decisionmaker's preferences concerning cost and property damage.

4.10.3 Computer Manufacturing Decision Analysis

A small computer company wishes to come out with a new line of computers. They decide they can make a high-performance, medium-performance, or economic (low-performance) model. It is assumed the company knows the sales potential for each of the computer lines as either excellent, good, or poor (see Table 4.17). The probability of excellent sales is 0.25, good is 0.6, and poor is 0.15. The company wishes to decide on the best development plan based upon minimizing risk of financial loss (EOL) and/or maximizing the expected profit.

Building blocks of the mathematical model
  Objectives
  • Minimize risk of financial loss or EOL.
  • Maximize the expected profit.
  Assumptions
  • The company will produce only one type of computer system.
  • The net return as a function of the sales is given (see Table 4.17).
  • Probability of excellent sales is 0.25.
  • Probability of good sales is 0.6.
  • Probability of poor sales is 0.15.

TABLE 4.17 Sales Potential

Computer System	Excellent	Good	Poor
Economical	$150,000	$50,000	$20,000
Medium	$300,000	$175,000	–$100,000
High	$450,000	$150,000	–$150,000

Decision variables
  • Which computer system to develop and sell on the open market
Input variables
  • State and federal support for small businesses
  • Federal regulation of the open market to maintain prices and/or stimulate the market with incentives
Exogenous variables
  • Cost of manufacture for each of the systems
  • Financial return for each of the systems as a function of the sales potential
  • Cost for advertising new system
  • Probability of sales potential assumed exogenous variable
    • Probability of excellent sales potential is 0.25.
    • Probability of good sales potential is 0.6.
    • Probability of poor sales potential is 0.15.
Random variables
  • Periodic fluctuation of the market
  • Operations, maintenance, and replacement fees for maintaining the production facility
State variables
  • Number of each type of computer system produced
  • Financial return
Output variables
  • Total number of each type of computer system produced
  • Net profit
Constraints
  • Regulatory laws regarding investment in the production and sales of computer systems
  • Resources available to manufacture computers

Hurwitz model
  Objective

$$\max_{\substack{1 \leq i \leq 3 \\ a_1, a_2, a_3}} \left\{ \mu_i(\alpha) = \alpha \min_{1 \leq j \leq 3} \mu_{ij} + (1-\alpha) \max_{1 \leq j \leq 3} \mu_{ij} \right\}, \quad 0 \leq \alpha \leq 1$$

For $\alpha = 1$: pessimistic
For $\alpha = 0$: optimistic

Table 4.18 presents a summary of the problem's assumptions.

4.10.3.1 Solution

Table 4.19 and Figure. 4.20 present a summary of the solution.

Therefore,

- At a_1: $u_1(\alpha) = 20{,}000\alpha + 150{,}000(1-\alpha) = 150{,}000 - 130{,}000\alpha$
- At a_2: $u_2(\alpha) = -100{,}000\alpha + 300{,}000(1-\alpha) = 300{,}000 - 400{,}000\alpha$
- At a_3: $u_3(\alpha) = -150{,}000\alpha + 450{,}000(1-\alpha) = 450{,}000 - 600{,}000\alpha$

Intersection occurs at $450{,}000 - 600{,}000\alpha = 150{,}000 - 170{,}000\alpha$.

TABLE 4.18 Hurwitz Data

i	1 (s_1)	2 (s_2)	3 (s_3)
1 (a_1)	150	50	20
2 (a_2)	300	175	−100
3 (a_3)	450	150	−150

Therefore, $\alpha = 30/43 \approx 0.698$

Thus, an analysis of the Hurwitz rule model indicates that for relatively optimistic decision levels, $\alpha \leq 0.698$, the high-performance computer should be produced and sold. For less optimistic decision levels of $\alpha \geq 0.698$, however, the economical computer should be considered.

4.10.4 Dingo Population Example

An Australian biologist wishes to model the population dynamics of the endangered dingo population on the island of Tasmania in order to assess possible conservation policies. The biologist has divided the population into four groups based on age. In the study, she selectively studied the females. Based on observations, the dingo population has a constant maternity rate based on the age category as follows: $m_0 = 0$, $m_1 = 1$, $m_2 = 3$, $m_3 = 1$. It has also been shown that the survival rate for the individual population cohorts is $p_0 = 1/2$, $p_1 = 3/4$, $p_2 = 3/4$, $p_3 = 0$. The population of dingoes on

TABLE 4.19 Pessimistic and Optimistic Outcomes

	Excellent (s_1)	Good (s_2)	Poor (s_3)	Optimistic	Pessimistic
Economical (a_1)	150	50	20	20	150
Medium (a_2)	300	175	−100	−100	300
High (a_3)	450	150	−150	−150	450

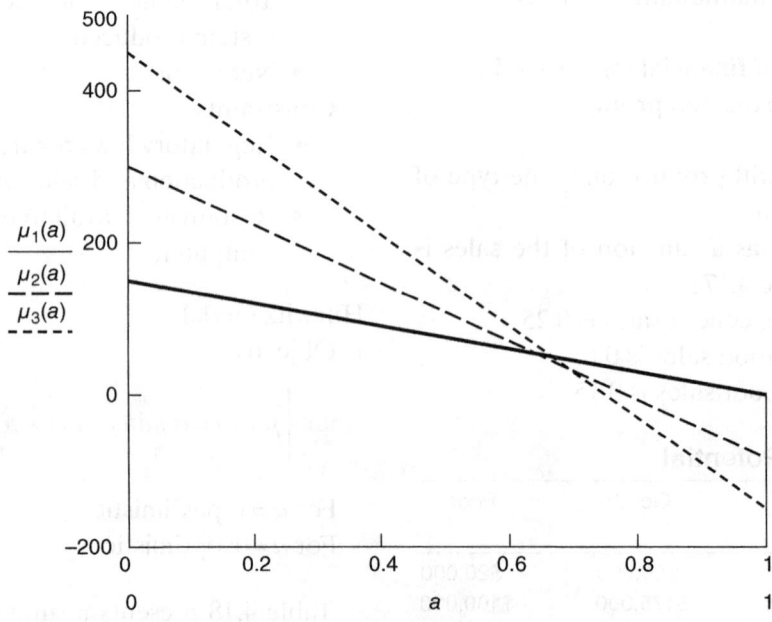

Figure 4.20 Results for the Hurwitz model.

Tasmania (at the time of the study) are (in hundreds) $F_0(0)=100$, $F_1(0)=80$, $F_2(0)=60$, $F_3(0)=40$. The biologist wishes to project the population dynamics for two terms.

4.10.4.1 Building blocks of the Leslie matrix

Objectives
- To model the population dynamics of a particular group of dingoes on the island of Tasmania

Assumptions
- Average birthing and survival rate is constant from one time interval to the next.
- Study involved only the female population of dingoes.
- Average birthing rate for a healthy dingo based on age classification is $m_0=0$, $m_1=1$, $m_2=3$, $m_3=1$.
- Average survival rate for a healthy dingo based on age classification is $p_0=1/2$, $p_1=3/4$, $p_2=3/4$, $p_3=0$.
- Initial population of dingoes based on age classification is $F_0(0)=100$, $F_1(0)=80$, $F_2(0)=60$, $F_3(0)=40$.
- Analysis is carried out for only two terms.

Decision variables
- Population conservation methodology used
- Extent of population conservation

Input variables
- Federal (Australian) and local support for population conservation
- Federal (Australian) and local funding for population conservation

Exogenous variables
- Costs of associated population conservation methodologies
- Birth rate of healthy dingo population for each of the population cohorts (this may also be viewed as a random variable)
- Survival rate of healthy dingo population for each of the population cohorts (may also be viewed as a random variable)

Random variables
- Death due to unnatural causes for healthy dingo population
- Death due to infection of dingo population
- Death due to natural conditions and disasters such as drought and decreasing food supply

State variables
- $F_0(t)$, population of dingoes in Cohort 0 at time t
- $F_1(t)$, population of dingoes in Cohort 1 at time t
- $F_2(t)$, population of dingoes in Cohort 2 at time t
- $F_3(t)$, population of dingoes in Cohort 3 at time t

Output variables
- Total population of dingoes
- Population of healthy male dingoes
- Population of healthy female dingoes

Constraints
- Project funding limitations
- Regulatory requirements regarding the dingo population
- Resources available for population analysis and conservation project

4.10.4.2 Leslie matrix

m_0	m_1	m_2	m_3	p_0	p_1	p_2	p_3	$F_0(0)$	$F_1(0)$	$F_2(0)$	$F_3(0)$
0	1	3	1	1/2	3/4	3/4	0	100	80	60	40

$$F(\Delta) = \begin{pmatrix} 0 & 1 & 3 & 1 \\ \frac{1}{2} & 0 & 0 & 0 \\ 0 & \frac{3}{4} & 0 & 0 \\ 0 & 0 & \frac{3}{4} & 0 \end{pmatrix} \cdot \begin{pmatrix} 100 \\ 80 \\ 60 \\ 40 \end{pmatrix} = \begin{pmatrix} 300 \\ 50 \\ 60 \\ 45 \end{pmatrix}$$

$$F(2\Delta) = \begin{pmatrix} 0 & 1 & 3 & 1 \\ \frac{1}{2} & 0 & 0 & 0 \\ 0 & \frac{3}{4} & 0 & 0 \\ 0 & 0 & \frac{3}{4} & 0 \end{pmatrix}^2 \cdot \begin{pmatrix} 100 \\ 80 \\ 60 \\ 40 \end{pmatrix} = \begin{pmatrix} 275 \\ 150 \\ 37.5 \\ 45 \end{pmatrix}$$

4.10.4.3 Solution

$F_0(1\Delta)$	$F_1(1\Delta)$	$F_2(1\Delta)$	$F_3(1\Delta)$	$F_0(2\Delta)$	$F_1(2\Delta)$	$F_2(2\Delta)$	$F_3(2\Delta)$
300	50	60	45	275	150	37.5	45

4.10.4.4 Comments

In this model, the large disparity between the first cohort (newborn dingoes) and the later cohorts can be attributed to the relatively small survival rates of the older fertile dingoes. The first survival rate, p_0, has an especially large impact on the system because it eliminates half of the growing newborn population before that population can reproduce. Because of these low survival rates, the overall number of fertile dingoes drops between $F(1\Delta)$ and $F(2\Delta)$. A result of this drop is a decrease in the newborn generation F_0 between $F(1\Delta)$ and $F(2\Delta)$. Mathematically, we can express this as

$$1(80) + 3(60) + 1(40) < 1(50) + 3(60) + 1(45)$$

The left-hand side of this equation represents the drawn-out matrix equation for $F_0(\Delta)$, and the right-hand side represents the same for $F_0(2\Delta)$.

Although the newborn generation drops in the second time period, we can predict that it will rise again in future periods as a larger number of newborns move into older age groups and become fertile. For example, in $F(2\Delta)$, the second age group jumps up to 150, which greatly increases the overall number of fertile dingoes. A reproducing population of 180 dingoes resulted in a 29% increase (180–232) in their population group over one time period. We can use this information to predict future increases within that group and the newborn group while holding the maternal and survival rates constant.

REFERENCES

Bertalanffy, L., 1968/1976, *General System Theory: Foundations, Development, Applications*, Revised edition, George Braziller, New York.

FEMA, 2006, *Multi-Hazard Loss Estimation Methodology for Hurricanes: HAZUS-MH Technical Manual*, Department of Homeland Security, Federal Emergency Management Agency, Washington, DC.

Haimes, Y.Y., 1977, *Hierarchical Analyses of Water Resources Systems: Modeling and Optimization of Large-Scale Systems*, McGraw-Hill, New York.

Haimes, Y.Y. (Ed.), 1981, *Risk/Benefit Analysis in Water Resources Planning and Management*, Plenum Publishing Company, New York.

Haimes, Y.Y., 2004, *Risk Modeling, Assessment, and Management*, John Wiley & Sons, Inc. Hoboken, NJ.

Haimes, Y.Y., 2005, Managing risks of catastrophic and extreme events, *Risk Analysis* **25**(4): 1083.

Haimes, Y.Y., 2006, On the definition of vulnerabilities in measuring risks to infrastructures, *Risk Analysis* **26**(2): 293–296.

Haimes, Y.Y., 2007, Phantom system models for emergent multiscale systems, Journal of Infrastructure Systems June: 81–87.

Haimes, Y.Y., 2008, Models for risk management of systems of systems, *International Journal of Systems of Systems Engineering* **1**(1/2): 222–236.

Haimes, Y.Y., 2012, Modeling complex systems of systems with phantom system models, *Systems Engineering* **15**(3): 333–346.

Haimes, Y.Y., and B.M. Horowitz, 2004, Adaptive two-player hierarchical holographic two-player game for counterterrorism intelligence analysis, *Journal for Homeland Security and Emergency Management* **1**(3): 1–12. Article 302.

Hobson, K., November 22, 2006, *Science Across the Borders (Interview with Eric Lander)*, US News and World Reports, Money and Business Section. Available online: http://connection.ebscohost.com/c/interviews/22831945/science-across-borders (Accessed February 6, 2015).

Kaplan, S., and B.J. Garrick, 1981, On the quantitative definition of risk, *Risk Analysis* **1**(1): 11–27.

Lamanna, C., 1959, The most poisonous poison, *Science* **130**: 763–765.

Law, A.M., and W.D. Kelton, 1991, *Simulation Modeling and Analysis*, McGraw-Hill, New York.

Leontief, W.W., 1951a, Input/output economics, *Scientific American* **185**(4): 15–21.

Leontief, W.W., 1951b, *The Structure of the American Economy, 1919–1939*, Second edition, Oxford University Press, New York.

Leontief, W.W., 1966, *Input–Output Economics*, Oxford University Press, New York.

Li, W., and L.C. Xu, 2004, The impact of privatization and competition in the telecommunications sector around the world, *Journal of Law and Economics* **47**(2): 395–430.

Lowrance, W.W., 1976, *Of Acceptable Risk: Science and the Determination of Safety*, William Kaufmann, Inc., Los Altos, CA.

Meyer, W.J., 1984, *Concepts of Mathematical Modeling*, McGraw-Hill, New York.

National Center on Education and the Economy (NCEE), 2007, *Tough Choices or Tough Times: The Report of the*

New Commission on the Skills of the American Workforce, Jossey-Bass, Hoboken, NJ.

Oliver, R.M., and J.Q. Smith (Eds.), 1990, *Influence Diagrams, Belief Nets and Decision Analysis*, John Wiley & Sons, New York.

PCCIP, 1997, *Critical Foundations: Protecting Americas Infrastructures*, The Report of the President's Commission on Critical Infrastructure Protection (PCCIP), Washington, DC, October.

Porter, M.E., 1998, *The Competitive Advantage of Nations: With a New Introduction*, The Free Press, New York.

Porter, M.E., 2003, The economic performance of regions, *Regional Studies* **37**(6&7): 549–678.

Raiffa, H., 1968, *Decision Analysis: Introductory Letters on Choices Under Uncertainty*, Addison-Wesley, Menlo Park, CA.

Saaty, T.L., 1980, *The Analytic Hierarchy Process*, McGraw-Hill, New York.

Saaty, T.L., 1988, *Mathematical Methods of Operations Research*, Dover Publications, New York.

Tulsiani, V., 1996, *Reliability-Based Management of River Navigation Systems*, Ph.D. dissertation, Systems Engineering Department, University of Virginia, Charlottesville, VA.

White, G.F., and J.E. Haas, 1975, *Assessment of Research on Natural Hazards*, The Massachusetts Institute of Technology Press, Cambridge, MA.

5

Multiobjective Trade-Off Analysis*

5.1 INTRODUCTION

During the past three decades, the consideration of multiple objectives in modeling and decision-making has grown by leaps and bounds. The 1980s in particular saw the emphasis shift from the dominance of single-objective modeling and optimization toward an emphasis on multiple objectives. This has led to the emergence of the new field of multiple-criteria decisionmaking (MCDM).

Most (if not all) real-world decisionmaking problems are characterized by multiple, noncommensurate, and often conflicting objectives. For most such problems, there exists a hierarchy of objectives, subobjectives, sub-subobjectives, and so on. In modeling, it is important to identify this hierarchy of objectives and avoid comparing and trading off objectives that belong to different levels.

5.1.1 MCDM as a Philosophy and the Fallacy of Optimality

MCDM has emerged as a philosophy that integrates common sense with empirical, quantitative, normative, and descriptive analysis. It is a philosophy supported by advanced systems concepts (e.g., data management procedures, modeling methodologies, optimization and simulation techniques, and decisionmaking approaches) that are grounded in both the arts and the sciences for the ultimate purpose of improving the decisionmaking process.

An optimum *does not exist in an objective sense per se*. An *optimum* solution exists for a model; however, to a real-life problem, it depends on myriad factors, which include the identity of the decisionmakers (DMs), their perspectives, the biases of the modeler, the credibility of the database, and others. Therefore, a mathematical optimum for a model does not necessarily correspond to the optimum for the real-life problem.

In general, multiple decisionmakers (MDMs) are associated with any single real-world decisionmaking problem. These MDMs may represent different constituencies, preferences, and perspectives; they may be elected, appointed, or commissioned and may be public servants, professionals, proprietors, laypersons, and so on; also, they are often associated

---

* This chapter is based on Chapter 7 of Haimes [1977] and on Chankong and Haimes [2008].

Risk Modeling, Assessment, and Management, Fourth Edition. Yacov Y. Haimes.
© 2016 John Wiley & Sons, Inc. Published 2016 by John Wiley & Sons, Inc.

or connected with a specific level of the various hierarchies of objectives mentioned earlier.

Solutions to a multiobjective optimization problem (MOP) with MDMs are often reached through negotiation, either through the use of group techniques of MCDM or on an ad hoc basis. Such solutions are often referred to as compromise solutions. Beware, however, of a nonwin–win compromise solution that is reached among MDMs where one or more DMs lose in the voting or negotiation process, even though the rules of the game have not been violated. A DM in a losing group may be influential enough to sabotage the compromise solution and prevent its implementation. Behind-the-scenes horse trading is a reality that must be accepted as part of human behavior. If a stalemate arises and a compromise solution is not achievable (e.g., if a consensus rule is followed and one or more DMs object to a noninferior solution that is preferred by all others), the set of objectives may be enlarged or the scope of the problem may be broadened. Finally, it is imperative that decisions be made on a timely basis—a *no-decision* stance could be costly.

5.1.2 Risk Assessment and Risk Management in Relation to MCDM

Risk assessment should be an integral part of the multiple-objective modeling effort, and risk management should be an imperative part of the multiple-objective decisionmaking process—not an after-the-fact vacuous exercise. Risk assessment, as discussed in Chapter 1, is defined here as a process that encompasses all the following four elements or steps: risk identification, risk quantification, risk evaluation, and risk management. Risk management is defined as the formulation of policies and the development of risk control options (i.e., measures to reduce or prevent risk). The obvious and inevitable overlapping of risk assessment and risk management has led many to consider the former as part of the latter.

5.1.3 Modeling and Decisionmaking versus Optimization

Most of the effort in MCDM should be devoted to the modeling activity. This should include the interaction between the DM(s) and the modeler, which has as its purpose (i) developing a causal relationship among the various systems' inputs and outputs and (ii) determining the preferences of each DM in order to arrive at his or her indifference band and preferred solution. Generally, much less effort is needed for optimization, namely, generating an appropriate set of noninferior (Pareto-optimal) solutions and their associated trade-offs, than for modeling.

In determining a preferred solution or policy in an MCDM framework, it is not sufficient to provide the DM with only the values of the objective functions at each alternative policy option (on the noninferior frontier set). A solution to an MOP is termed noninferior, or Pareto-optimal, if improving one objective function can be achieved only at the expense of degrading another one. A formal, mathematical definition will be introduced in a subsequent section. For sound and informative decisionmaking, it is imperative that the DM also be provided with the trade-off values associated with the respective objectives.

5.1.4 The Fine Line between an Inferior and a Noninferior Solution

Modelers and systems analysts place great emphasis on generating only noninferior solutions (i.e., discarding inferior solutions). This emphasis, though justifiable, should be moderately balanced by the fact that a noninferior solution to, for example, a three-objective function could become an inferior solution if one of the three objectives is ignored. Similarly, an inferior solution could become noninferior if the number of objectives is increased while making no changes in the meaning or definition of any objective. This observation is further supported by the fact that the number of objectives that are formally considered in the MCDM process in the first place is subject to value-judgment-based decisions. This cautious remark is not unrelated to the overconfidence and reverence that systems analysts place in the optimality of a single-objective model.

5.1.5 Decision Support Systems and MCDM

Decision support systems (DSS) are interactive computer-based systems that help DMs utilize data, mathematical models, and simulation and optimization methodologies to generate alternative policy options and solve both structured and unstructured problems. True DSS must be grounded on the same premises as MCDM. From a practical standpoint, DSS and MCDM should be supplementary and complementary to each other (and both should, of course, include the consideration of risk assessment and management), and ultimately, they should aim at the same goal. The goals of MCDM and DSS are the same—to improve decisionmaking—albeit the emphasis in each and the ways and means for achieving these goals may be different. A similar argument can be made about how MCDM and DSS are related to artificial intelligence (AI), which is the study of ideas that enable computers to be intelligent. The fundamental principle underlying AI is the use of information for learning purposes. Thus, for a DM, a DSS will be effective if it incorporates multiple objectives and, at the same time, has the capability of self-learning and model updating.

5.1.6 Sensitivity within the MCDM Process

One should take into account the multiplicity of errors and uncertainties associated with the MCDM process, including errors associated with (i) the database; (ii) the modeling effort; (iii) the optimization; (iv) the DM's perception of his or her values, needs, and preferences; and (v) the decisionmaking process itself. The diversity of errors associated with the MCDM process is likely to add instability to the preferred solution. For example, the values of certain exogenous variables may, in reality, deviate from their assumed nominal values. Constructing and adding one or more new sensitivity functions that are minimized along with the other original multiobjective functions (as done in Chapter 6) could add some of the needed stability to the resulting preferred solution or selected policy.

5.1.7 Optimizing the Objectives Correctly

It is a mistake to try to optimize a set of objectives that are limited to present aspirations or are not responsive to future needs. The future impacts of present decisions and policies must be accounted for. Therefore, impact analysis should be incorporated into the MCDM process so that (i) the attainment of present objectives can be juxtaposed against potential or perceived objectives (e.g., maximizing present profit vs. maximizing future technological and economic competitiveness through an investment in research and development) and (ii) more flexibility may be added to ensure against adverse irreversible consequences. For example, evaluating the consequences and future flexibility of two preferred noninferior solutions could dictate a distinct choice between two seemingly equivalent options. The value and importance of impact analysis are even more critical for multistage problems, which are characterized by multiple objectives at each stage of the decisionmaking process (as discussed in Chapter 10). In other words, a trade-off between the attainment of present objectives and future flexibility can be incorporated within the MCDM process.

5.1.8 Importance of Modeling Multiple Perspectives into MCDM

Should the system modeler or the DM always be satisfied with a single-perspective model of the system under study? The answer is no. We emphasized in Chapter 3 that invariably single models cannot adequately capture the multifarious nature of large-scale systems, their bewildering variety of resources and capabilities, their multiple noncommensurable objectives, and their diverse users, constituencies, and DMs. When concepts from hierarchical holographic modeling (HHM) are incorporated into MCDM, the modeling base is broadened, and an opportunity is provided for a modeling and decisionmaking framework that is more responsive to users and DMs. Approaches that allow this incorporation seem especially worthwhile for group decisionmaking situations.

5.2 EXAMPLES OF MULTIPLE ENVIRONMENTAL OBJECTIVES

The planning of water and related land resources in a river basin (or a region) is a vital element in the formulation of public policy on this critical resource. Such planning should be responsive to the inherent multiple objectives and goals and should account for the trade-offs among these objectives with respect to myriad objectives, including the following five categories of concern [Haimes, 1977]:

1. Time horizon: short, intermediate, and long term
2. Client: various sectors of the public
3. Nature: aquatic and wildlife habitats
4. Scope: national, regional, and local needs
5. Constraints: legal, institutional, environmental, social, political, and economic

There are many ways and means of identifying and classifying objectives and goals for such a planning effort. The US Water Resource Council advocated the enhancement of four major objectives: (i) national economic development, (ii) regional economic development, (iii) environmental quality, and (iv) social well-being.

The Technical Committee study [Peterson, 1974] identifies nine goals, which have been divided into two major groups:

1. Maintenance of security: (a) environmental security, (b) collective security, and (c) individual security
2. Enhancement of opportunity: (d) economic opportunity, (e) recreational opportunity, (f) aesthetic opportunity, (g) cultural and community opportunity, (h) educational opportunity, and (i) individual freedom

In an environmental trade-off analysis, policies should be established to promote conditions where human and nature can exist in harmony. Resolution of conflicts should be achieved by balancing the advantages of development against the disadvantages to the environment and the aquatic system. The process is one of balancing the total *benefits*, *risks*, and *costs* for both people and the environment, where the well-being of future generations is as important as that of present ones. Fundamental to multiobjective analysis is the Pareto-optimum concept.

5.2.1 Flood Control versus Hydropower Generation

Consider two major objectives in the operation of reservoir systems [Haimes et al., 1990]:

1. Minimize hydroelectric power generation losses from the reservoir.
2. Minimize flood damages.

Obviously, these two objectives are in conflict and competition (see Figures 5.1 and 5.2). The higher the level of the reservoir, the more electric

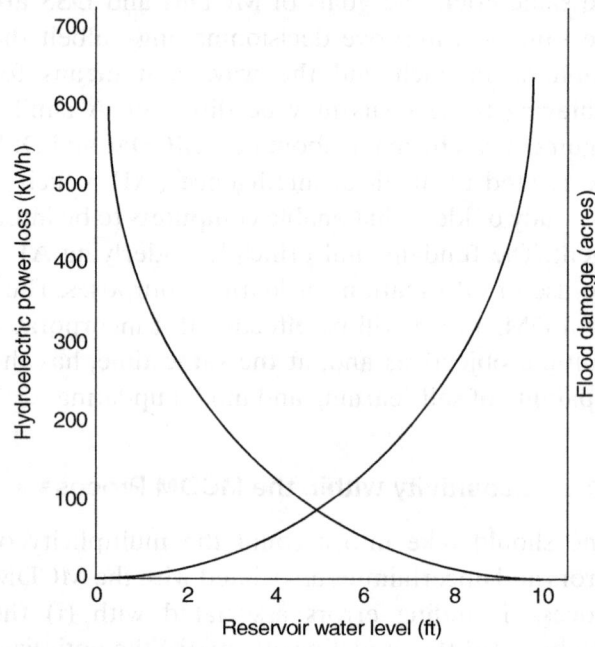

Figure 5.1 *Flood damage and hydroelectric power loss in the decision space.*

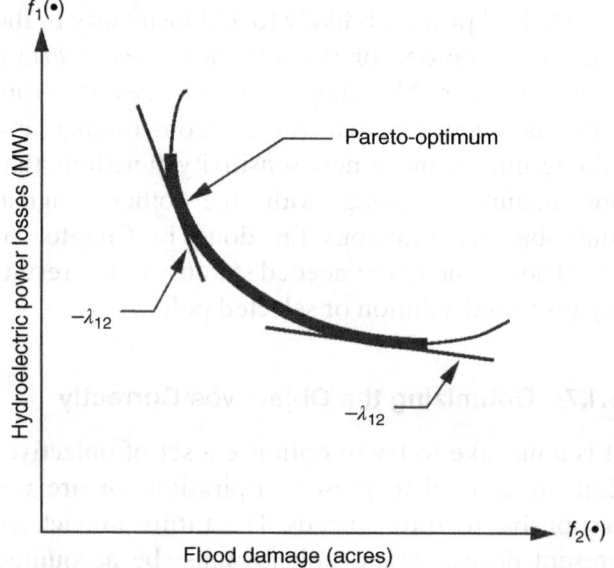

Figure 5.2 *Flood damage versus hydroelectric power loss in the functional space.*

power generation is possible because of the high waterhead, yet less water storage is available for flood control purposes. Clearly, one can identify, within the active storage capacity of that reservoir, a Pareto-optimum region whereby the enhancement of the first objective can be achieved only at the expense or degrading of the second, namely, flood control.

Also note that the units of these two objectives are noncommensurable. The first objective, which minimizes the hydropower losses, may be measured in units of energy and not necessarily in monetary units, where the second objective can be measured in terms of acres of land, livestock, or human life lost.

The function, f_1, represents the hydropower output lost (in kWh), while f_2 represents the expected damage (in acres flooded). The maximum water level possible for the reservoir is 10, where

$$f_1(x) = 1000\, e^{-x}$$

$$f_2(x) = e^{0.65x}$$

and where x denotes the water level at the reservoir.

The MOP is

$$\underset{x}{\text{minimize}} \begin{cases} f_1(x) = 1000 e^{-x} \\ f_2(x) = e^{0.65x} \end{cases}$$

subject to the constraint

$$0 < x \le 10$$

Figures 5.1 and 5.2, which are generic graphs typical for these objective functions, show the trade-offs between flood damage and kilowatt hours lost in the decision space and functional space, respectively.

The water level can be set at a number of levels, all of which are technically Pareto-optimal, because each change in x degrades one objective function while improving the other. Note, however, that at roughly $x = 2$ and $x = 8$, one of the two objective functions stays virtually constant (thus not degraded), while the other objective is improved. Therefore, this range of water levels was chosen for the sample Pareto-optimal solutions shown in Table 5.1.

TABLE 5.1   Pareto-Optimal Solutions

Water Reservoir Level (x)	Flood Damage (Acres)	Hydropower Loss (kWh)	Trade-Off (Slope)
2.0	3.7	135.3	−37.8
2.5	5.1	82.0	−16.6
3.0	7.0	49.8	−7.3
3.5	9.7	30.2	−3.2
4.0	13.5	18.3	−1.4
4.5	18.6	11.1	−0.6
5.0	25.8	6.7	−0.3
5.5	35.7	4.1	−0.1
6.0	49.4	2.5	−0.1

Table 5.1 presents a set of Pareto-optimal solutions with their associated trade-off values. Note that these trade-offs are calculated using the relationship

$$\lambda_{12} = -\frac{\Delta f_1}{\Delta f_2}$$

Figure 5.2 is a representation of the trade-offs in the functional space. Note that $\lambda_{12} > 0$ is a necessary condition for Pareto-optimality, and thus, the slope $\Delta f_1 / \Delta f_2$ must be negative.

5.3   THE SURROGATE WORTH TRADE-OFF METHOD

5.3.1   Formulation of MOPs

To define a noninferior solution mathematically, consider the following multiobjective function problem, also known as an MOP:

$$\text{MOP}: \quad \begin{aligned} & \min_{\mathbf{x} \in X} \{f_1(\mathbf{x}), f_2(\mathbf{x}), \ldots f_n(\mathbf{x})\} \\ & X = \{\mathbf{x} \mid g_i(\mathbf{x}) \le 0,\ i = 1, 2, \ldots, m\} \end{aligned} \quad (5.1)$$

where $\mathbf{x}$ is an N-dimensional vector of decision variables, X is the set of all feasible solutions, and $g_i(\mathbf{x})$ is the ith constraint.

Definition

A decision $\mathbf{x}^*$ is said to be a noninferior solution to the system posed by the MOP (5.1) if and only if there does not exist another $\bar{\mathbf{x}}$ so that $f_j(\bar{\mathbf{x}}) \le f_j(\mathbf{x}^*)$, $j = 1, 2, \ldots, n$, with strict inequality holding for at least one j.

Clearly, the solution to the multiobjective problem posed by Eq. (5.1) is not unique, and some kind of subjective judgment by the DM(s) should be added to the quantitative analysis. Although more than one DM may be involved in the selection of an acceptable and preferred solution, to avoid complexity in notation, a reference to a DM will denote MDMs unless it is specified otherwise.

The various available methodologies for solving Eq. (5.1) differ in two major ways: (i) the procedures used to generate noninferior solutions and (ii) the timing and the ways and means used to interact with the DMs and the type of information made available to them in the process (such as trade-offs). The weighting method, also known as the parametric approach, was the most common method used for solving multiobjective problems until recently. The MOP (5.1) is essentially converted in the weighting method into a scalar optimization $p(w)$ as given below:

$$p(w): \begin{array}{c} \min_{x \in X} \sum_{i=1}^{n} w_i f_i(\mathbf{x}) \\ \\ \sum_{i=1}^{n} w_i = 1, \quad w_i \geq 0 \end{array} \quad (5.2)$$

A subjective determination of the levels of the weighting coefficients, w_i, is necessary. Subsequently, this parametric approach may yield meaningful results to the DM only when solved (parametrically) many times for different values of $w_i, i = 1, 2, \ldots, n$. The potential existence of a duality gap is an additional important drawback to this method (see Section A.7).

There exist numerous methods for solving multiobjective problems, such as utility functions; indifference functions; the lexicographic, parametric, and ε-constraint approaches; goal programming; the goal attainment method; the adaptive search approach; interactive approaches; the ELECTRE method; the surrogate worth trade-off (SWT) method; and others [Chankong and Haimes, 1983a, 2008]. Several recent volumes discussing multiobjective decisionmaking include the works by Belton and Stewart [2002], Collette and Siarry [2004], and Ehrgott [2005]. This chapter will review the SWT method and its extensions [Haimes and Hall, 1974].

5.3.2 The ε-Constraint Method

The SWT method recognizes that the optimization theory is usually much more concerned with the relative value of additional increments of the various noncommensurable objectives, at a given value of each objective function, than it is with their absolute values. Furthermore, given any current set of objective levels attained, it is much easier to turn to the DMs to assess the relative value of the trade-off of marginal increases and decreases between any two objectives than it is to assess their absolute average values. In addition, the optimization procedure can be developed so that it assesses whether one more quantity of one objective is worth more or less than that lost from another at any given level. Ordinal scale can then be used with much less concern for the distortions that relative evaluation introduces into attempts to commensurate the total value of all objectives.

Since the dimension of the decision space N for most real-world problems is generally higher than the dimension of the functional space n (N decisions and n objectives, $N \gg n$), as a further simplification one should make decisions in the functional space and only later transfer the information to the decision space.

A basic approach to treating noncommensurable objectives is selecting a primary or dominating objective to be optimized while constraining the decisions considered to ensure that some minimum level for all others is attained in the process. If all objectives are equal to or better than this minimum level of attainment with some proposed decision, such a decision can be termed satisfactory. So long as any decision set exists that is satisfactory, it is unnecessary to consider any decision that results in a poorer condition in any objective. Hence, this approach will also help reduce the field of decisions to explore.

Let

$$\underline{f_j} = \min_x f_j(\mathbf{x}), \quad \mathbf{x} \in X; j = 1, 2, \ldots, n \quad (5.3)$$

The ε-constraint approach replaces $(n-1)$ objective functions by $(n-1)$ constraints as given by $P_k(\varepsilon)$ in Eq. (5.4):

$$P_k(\varepsilon): \min_x f_i(x) \quad \text{subject to} \quad f_j(x) \leq \varepsilon_j,$$
$$j \neq i; j = 1, 2, \ldots, n; x \in X \quad (5.4)$$

where $\varepsilon_j, j \neq i, j = 1, 2, \ldots, n$, are variables ($\varepsilon_j = f_j + \bar{\varepsilon}_j$), because $\bar{\varepsilon}_j > 0$ are variables.

The levels of satisfactory ε_j can be varied parametrically to evaluate the impact on the single-objective function $f_i(\mathbf{x})$. Of course, the ith objective, $f_i(\mathbf{x})$, can be replaced by the jth objective, $f_j(\mathbf{x})$, and the solution procedure repeated. The equivalence between Eqs. (5.1) and (5.4) is well documented in the literature [Haimes et al., 1971]. The ε-constraint approach facilitates the generation of noninferior solutions as well as trade-off functions, as will be discussed later.

By considering one objective function as primary and all others at minimum satisfying levels as constraints, the Lagrange multipliers related to the $(n-1)$ objectives as constraints will be zero or nonzero. (Lagrange multipliers are discussed in the Appendix.) If nonzero, that particular constraint does limit the optimum. It will be shown that positive Lagrange multipliers correspond to the noninferior set of solutions. Furthermore, the set of nonzero Lagrange multipliers represents the set of trade-off ratios between the principal objective and each of the constraining objectives, respectively. Clearly, these Lagrange multipliers are functions of the optimal level attained by the principal objective function, as well as the level of all other objectives satisfied as equality (binding) constraints. Consequently, these Lagrange multipliers form a matrix of trade-off functions.

The question of the worth ratios still remains after the matrix of trade-off functions has been computed. The worth ratios are essentially achieved through an interaction with the DM. However, since the worth ratio need only represent relative worth of the objectives, not the absolute level of worth, any surrogate ratio that varies monotonically with the correct one will suffice.

5.3.3 The Trade-Off Function

The following development shows that the trade-off functions can be found from the values of the dual variables associated with the constraints in a reformulated problem. Reformulate the system MOP (5.1) with the $P_k(\varepsilon)$ (5.4), where $\varepsilon_j = f_j + \bar{\varepsilon}_j, \bar{\varepsilon}_j > 0, j = 2, 3, \ldots, n$, and f_j were defined in Eq. (5.3), and $\bar{\varepsilon}_j$ will be varied parametrically in the process of constructing the trade-off function.

Form the generalized Lagrangian, L, to the system:

$$L = f_1(\mathbf{x}) + \sum_{j=2}^{n} \lambda_{1j} \left[f_j(\mathbf{x}) - \varepsilon_j \right] \quad (5.5)$$

where $\lambda_{1j}, j = 2, 3, \ldots, n$, are generalized Lagrange multipliers. The subscript $1j$ in λ denotes that λ is the Lagrange multiplier associated (in the ε-constraint vector optimization problem) with the jth constraint, where the objective function is $f_1(\mathbf{x})$. Subsequently, λ_{1j} will be generalized to associate with the ith objective function and the jth constraint, λ_{ij}. Denote by $\hat{X}$ the set of all $x_i, i = 1, 2, \ldots, N$, and by Ω the set of all $\lambda_{ij}, j = 2, 3, \ldots, n$, that satisfy the Kuhn–Tucker condition for Eq. (5.5) (see the Appendix). The conditions of interest to our analysis are

$$\lambda_{1j} \left[f_j(\mathbf{x}) - \varepsilon_j \right] = 0, \quad \lambda_{1j} \geq 0; j = 2, 3, \ldots, n \quad (5.6)$$

Note that if $f_j(\mathbf{x}) < \varepsilon_j$ for any $j = 2, 3, \ldots, n$ (i.e., the constraint is not binding), then the corresponding Lagrange multiplier λ_{1j} equals 0.

The value of $\lambda_{1j}, j = 2, 3, \ldots, n$, corresponding to a binding constraint, is of special interest since it indicates the marginal benefit (cost) of the objective function $f_1(\mathbf{x})$ due to an additional unit of ε_j. From Eq. (5.5), assuming that the solution is global, the following results can be derived:

$$\lambda_{1j}(\varepsilon_j) = -\frac{\partial L}{\partial \varepsilon_j}, \quad j = 2, 3, \ldots, n \quad (5.7)$$

Note, however, that for $x \in \hat{X}$, $\lambda_{ij} \in \Omega$ for all j, we obtain

$$f_1(\mathbf{x}) = L \quad (5.8)$$

Thus,

$$\lambda_{1j}(\varepsilon_j) = -\frac{\partial f_1(\cdot)}{\partial \varepsilon_j}, \quad j = 2, 3, \ldots, n \quad (5.9)$$

In the derivation of the trade-off functions in the SWT method, only those $\lambda_{ij} > 0$ corresponding to $f_j(\mathbf{x}) = \varepsilon_j$ are of interest (since they correspond to the noninferior solution). Thus, for $f_j(\mathbf{x}) = \varepsilon_j$, Eq. (5.9) can be replaced by Eq. (5.10):

$$\lambda_{1j}(\varepsilon_j) = -\frac{\partial f_1(\cdot)}{\partial f_j(\cdot)}, \quad j = 2, 3, \ldots, n \quad (5.10)$$

Clearly, Eq. (5.10) can be generalized where the index of performance is the ith objective function of the system (5.1) rather than objective function $f_1(\cdot)$. In this case, the index i should replace the index 1 in λ_{1j} yielding λ_{ij} Accordingly,

$$\lambda_{ij}(\varepsilon_j) = -\frac{\partial f_i(\cdot)}{\partial f_j(\cdot)}, \qquad i \neq j; i,j = 1,2,3,\ldots n \quad (5.11)$$

For the rest of this section, only $\lambda_{ij}(\varepsilon_j) > 0$ (which correspond to binding constraints) are considered, since there exists a direct correspondence between λ_{ij} associated with the binding constraints and the noninferior set in Eq. (5.1).

The possible existence of a duality gap and its effect on the SWT method is discussed in detail elsewhere (see Chankong and Haimes [1983a, b, 2008]). A duality gap occurs when the minimum of the primal problem is not equal to the maximum of the dual problem. This is the same situation when a saddle point does not exist for the Lagrangian function (see the Appendix). Note that if a duality gap does exist, the ε-constraint method still generates all needed noninferior solutions. However, a given value of the trade-off function λ_{ij} may correspond to more than one noninferior solution. On the other hand, if a duality gap does exist, then not all Pareto-optimal solutions can be generated for the weighting problem $p(w)$ posed in Eq. (5.2).

Definition

The *indifference band* is defined to be a subset of the noninferior set where the improvement of one objective function is equivalent (in the mind of the DM) to the degradation of another.

Definition

An *optimum solution* (or *preferred solution*) is defined to be any noninferior feasible solution that belongs to the indifference band.

The computational derivation of the trade-off function λ_{ij} will be demonstrated through the derivation of λ_{ij} as follows:

The system given by Eq. (5.5) is solved for K values of ε_2, say, $\varepsilon_2^1, \varepsilon_2^2, \ldots, \varepsilon_2^K$, where all other $\varepsilon_j, j = 3,4,\ldots,n$, are held fixed at some level ε_j^0. Only those $\lambda_{12}^k > 0$ that correspond to the binding constraints $f_2^k(x) = \varepsilon_2^k$ $k = 1,2,\ldots,K$, are of interest, since they belong to the noninferior solution.

Assume that for $\varepsilon_2^1, \lambda_{12}^1 > 0$ with the corresponding solution $\mathbf{x}^1$. Then $f_2(\mathbf{x}^1) = \varepsilon_2^1$. Clearly, not all other $\lambda_{ij}, j = 3,4,\ldots,n$, corresponding to this solution $(\mathbf{x}^1)$ are positive. Thus, the following equation is solved:

$$\min_x f_1(\mathbf{x}); \mathbf{x} \in X \quad \text{so that} \quad f_j(\mathbf{x}) < f_j(\mathbf{x}^1), j = 2,3,\ldots,n \quad (5.12)$$

where ε_j^0 were replaced by $f_j(\mathbf{x}^1), j = 3,4,\ldots,n$. A small variation δ_j may be needed to ensure positive $\lambda_{1j}, j = 3,4,\ldots,n$, in the computational procedure. The trade-off λ_{12} is a function of all $\varepsilon_j, j = 2,3,\ldots,n$ (i.e., $\lambda_{12} = \lambda_{12}(\varepsilon_2,\ldots,\varepsilon_n)$). It will be shown in subsequent discussions that the trade-off function $\lambda_{ij}(\cdot)$ may be constructed (via multiple regression) in the vicinity of the indifference band.

Similarly, the trade-off function $\lambda_{13}\sigma$ can be generated, where again the prime objective function is $f_1(\mathbf{x})$, and the system (5.5) is solved for K' different values of $\varepsilon_3^k, k = 1,2,\ldots,K'$, with a fixed level of $\varepsilon_2^0, \varepsilon_4^0, \ldots, \varepsilon_n^0$. Similarly, the trade-off functions λ_{1j} can be generated for $j = 4,5,\ldots,n$. Once all trade-off functions $\lambda_{1j}, j = 1,2,3,\ldots,n$, have been generated, the prime objective may be changed to the ith, and thus, all trade-off functions $\lambda_{ij}, i \neq j; i, j = 1,2,3,\ldots,n$, can be generated. It can be shown, however, that not all λ_{ij} need to be generated computationally since the following relationships hold:

$$\lambda_{ij} = \lambda_{ik}\lambda_{kj} \text{ for } \lambda_{ij} > 0; i \neq j; i,j = 1,2,\ldots,n \quad (5.13)$$

In addition, the relationship $\lambda_{ij} = 1/\lambda_{ji}$ for $\lambda_{ji} \neq 0$ can also be used.

5.3.4 The Surrogate Worth Function

The surrogate worth function provides the interface between the DM and the mathematical model. The value of the surrogate worth function W_{ij} is an assessment by the DM as to how much (on an ordinal scale, say, from −10 to +10, with zero signifying equal preference) he or she prefers trading λ_{ij} marginal units of f_i for one marginal unit of f_j, given the values of all the objectives $f_i,\ldots,f_n$ corresponding to λ_{ij}. Note that $W_{ij} > 0$ means the DM does prefer making such a trade, $W_{ij} < 0$ means he or she does not, and $W_{ij} = 0$ implies indifference. A formal definition of W_{ij} is given below:

$$W_{ij} = \begin{cases} >0 & \text{when } \lambda_{ij} \text{ marginal units of } f_i(x) \\ & \text{are preferred over one marginal} \\ & \text{unit of } f_j(x), \text{ given the satisfaction} \\ & \text{of all objectives at level } \varepsilon_k, k=1,2,\ldots,n \\ =0 & \text{when } \lambda_{ij} \text{ marginal units of } f_i(x) \\ & \text{are equivalent to one marginal unit} \\ & \text{of } f_j(x), \text{ given the satisfaction} \\ & \text{of all objectives at level } \varepsilon_k, k=1,2,\ldots,n \\ <0 & \text{when } \lambda_{ij} \text{ marginal units of } f_i(x) \\ & \text{are not preferred to one marginal} \\ & \text{unit of } f_j(x), \text{ given the satisfaction} \\ & \text{of all objectives at level } \varepsilon_k, k=1,2,\ldots,n \end{cases}$$

It is important to note here that the DM is provided with the trade-off value (via the trade-off function) of any two objective functions at a given level of attainment of the other objective functions. Furthermore, all trade-off values generated from the trade-off function are associated with the noninferior set. Thus, any procedure that can generate a surrogate worth function, which in turn can provide the indifference band of λ_{ij}, $i \neq j$, $i,j = 1,2,3,\ldots,n$, will solve the multiobjective problem. In this respect, much of the experience developed and gained in the fields of decision theory and team theory can be utilized in the SWT method.

The band of indifference can be determined as follows: The DM is asked whether λ_{ij} units of $f_i(\mathbf{x})$ is {<} one unit of $f_j(\mathbf{x})$ for two distinct values of λ_{ij}. A linear interpolation of the corresponding two answers $W_{ij}(\lambda_{ij})$ obtained from the DM in ordinal scale can be made (see Figure 5.3). Then the value of $\lambda_{ij} = \lambda_{ij}^*$ is chosen so that $W_{ij}(\lambda_{ij}^*) = 0$ on the line segment fitting the two values of λ_{ij}. With λ_{ij}^* determined, the indifference band is assumed to exist within the neighborhood of λ_{ij}^*. Additional questions to the DM can be asked in the neighborhood of λ_{ij}^* to improve the accuracy of λ_{ij}^* and the band of indifference. The surrogate worth function assigns a scalar value (on an ordinal scale) to any given noninferior (efficient, Pareto-optimal) solution.

There are three ways of specifying a noninferior solution:

1. By the values of its decision variables, $x_1, \ldots, x_N$
2. By the trade-off functions $\lambda_{i1}, \ldots, \lambda_{in}$
3. By its objective function values $f_1, \ldots, f_n$

Hence, we can have $W_{ij}(x_1, \ldots, x_N)$ or $W_{ij}(\lambda_{i1}, \ldots, \lambda_{in})$ or $W_{ij}(f_1, \ldots, f_n)$. The first is generally ruled out by the inefficiencies of decision space manipulations. The second may suffer from problems when discontinuities or nonconvexities occur in the functional space but can be used in other problems. The third approach, using objective function space values, appears to be best.

As an example of how the method works, consider a three-objective problem. Several noninferior points, $(f_2, f_3)_0, \ldots, (f_2, f_3)_k$, and their trade-offs, $(\lambda_{12}, \lambda_{13})_0, \ldots, (\lambda_{12}, \lambda_{13})_k$ are determined, for example, via the ε-constraint method. The DM is then questioned to get values

$$W_{12}(f_2, f_3)_0, \ldots, W_{12}(f_2, f_3)_k$$

and

$$W_{13}(f_2, f_3)_0, \ldots, W_{13}(f_2, f_3)_k$$

(It can be shown that the other W_{ij} need not be determined.) Now, since generally none of these will be zero, we must determine more noninferior solutions and their trade-offs than before, and we must ask more questions of the DM until we find an $(f_2, f_3)^*$ so that $W_{12}(f_2, f_3)^*$ and $W_{13}(f_2, f_3)^*$ both equal to zero.

The use of a functional relation (via regression or interpolation) for $W_{12}(f_2, f_3)$ and $W_{13}(f_2, f_3)$ can be used as an approximation when setting new constraint levels in determining new noninferior solutions.

Since the worth is evaluated only at known noninferior points, it is guaranteed that $(f_2, f_3)^*$ will give rise to a feasible solution when put into the overall mathematical model. The same guarantee holds when $W_{ij}(\lambda_{i1}, \ldots, \lambda_{in})$ is used.

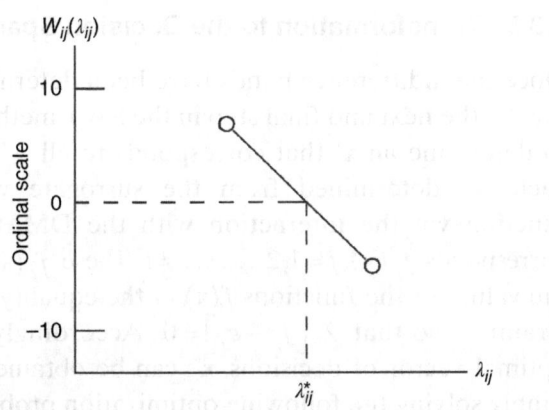

Figure 5.3 Determination of the indifference band at λ_{ij}^*.

What happens if there cannot be found a pair of $(f_2, f_3)^*$ whose worth functions are both zero? In that case, we can take the one whose worth functions are closest to zero as an approximate preferred solution. Note that the noninferior solutions whose surrogate worth functions are all zero correspond to the maximum utility solutions. The noninferior solution whose worth functions are closest to zero will be the one closest to the maximum utility solution.

There is a close relation between the surrogate worth function, W_{ij}, and the partial derivatives of the utility function.

In multiobjective analysis, it is assumed implicitly that the DM maximizes his utility, which is a function of the various objective functions. Given a decision $\mathbf{x}$ and the associated consequences $f(\mathbf{x})$, the utility is given by

$$U = U[f_1(\mathbf{x}), \ldots, f_n(\mathbf{x})] \quad (5.14)$$

For a small change in f_i, one can linearize Eq. (5.14):

$$\Delta U = \frac{\partial U}{\partial f_1} \Delta f_1 + \frac{\partial U}{\partial f_2} \Delta f_2 + \cdots + \frac{\partial U}{\partial f_n} \Delta f_n \quad (5.15)$$

However, for noninferior points, we obtain

$$\Delta f_1 = \sum_{i=2}^{n} \frac{\partial f_1}{\partial f_i} \Delta f_i = -\sum_{i=2}^{n} \lambda_{1i} \Delta f_i \quad (5.16)$$

Eliminating Δf_1 (Eq. (5.16)) from Eq. (5.15) yields

$$\Delta U = \sum_{i=2}^{n} \left(\frac{\partial U}{\partial f_i} - \frac{\partial U}{\partial f_1} \lambda_{1i} \right) \Delta f_i = \sum_{i=2}^{n} \Delta U_{1i} \Delta f_i$$

where

$$\Delta U_{1i} = \left(\frac{\partial U}{\partial f_i} - \frac{\partial U}{\partial f_1} \lambda_{1i} \right)$$

Let

$$a_i = \frac{\partial U}{\partial f_i}$$

then,

$$\Delta U_{1i} = (a_i - a_1 \lambda_{1i})$$

The surrogate worth function W_{1i} is a monotonic function of ΔU_{1i} with the property that $W_{1i} = 0 \leftrightarrow \Delta U_{1i} = 0$ and can therefore be written as

$$W_{1i} = h_i (a_i - a_1 \lambda_{1i})$$

where h_i is some monotonic increasing function of its argument, with a range of -10 to $+10$ and with the property that $h_i(0) = 0$. If a_i is considered constant or varies only slightly with f_i, $i = 1, \ldots, n$, then it is possible to assume that W_{1i} depends only on λ_{1i}.

Finally, one may question whether an interaction with the DM in the function space should always yield a $W_{12}(\lambda_{12}) = 0$—that is, an indifference solution. Two cases may be identified here:

1. The DM's response is always on one side of the W_{12} scale for all λ_{12} corresponding to the Pareto-optimal solutions. That is to say, the DM's answers are either all on the positive or all on the negative scale of W_{12}. This really means that the DM is always willing to improve objective 1, for example, at the expense of degrading objective 2 in the entire Pareto-optimal space. This case, while it may actually happen, is of no particular interest here, since it reduces the multiobjective problem to a single-objective optimization problem.

2. Should the DM's response in the function space be on the positive scale of W_{12} for some values of λ_{12} and negative for other sets of values of λ_{12}, then (assuming consistency in the DM's response and continuity in λ_{12}) it can be guaranteed that a value of $W_{12} = 0$ exists, which corresponds to an indifference solution with λ_{12}^*, that is, $W_{12}(\lambda_{12}^*) = 0$.

5.3.5 Transformation to the Decision Space

Once the indifference bands have been determined for λ_{ij}^*, the next and final step in the SWT method is to determine an $\mathbf{x}^*$ that corresponds to all λ_{ij}^*. For each λ_{ij}^* determined from the surrogate worth function via the interaction with the DM, there corresponds $f_j^*(\mathbf{x}), j = 1, 2, \ldots, n, j \neq i$. These $f_j^*(\mathbf{x})$ are the values of the functions $f_j(\mathbf{x})$ at the equality constraints ε_j so that $\lambda_{ij}^* \left[f_j^* - \varepsilon_j \right] = 0$. Accordingly, the optimal vector of decisions, $\mathbf{x}^*$, can be obtained by simply solving the following optimization problem:

$$\min_{\mathbf{x} \in X} f_i(\mathbf{x}) \quad \text{subject to} \quad f_j(\mathbf{x}) \le f_j^*(\mathbf{x}),$$
$$j = 1, 2, \ldots, n, j \ne i \tag{5.17}$$

Equation (5.17) is a common optimization problem with a single-objective function. The solution of Eq. (5.17) yields the desired $\mathbf{x}^*$ for the total vector optimization problem posed by Eq. (5.1).

The consistency of the DM should not always be assumed. The DM may show nonrational behavior or provide conflicting information at times. The SWT method safeguards against this by cross-checking the resulting λ_{ij}^*. It has been shown elsewhere that one set of $\lambda_{1i}, \ldots, \lambda_{1n}$ will suffice for solving the multiobjective problem posed previously. It is always possible, however, to generate, for example, $\lambda_{12}^*, \ldots, \lambda_{23}^*$, and λ_{13}^* (via an interaction with the DM) and to check that indeed the following relation holds: $\lambda_{13}^* = \lambda_{12}^* \lambda_{23}^*$ (i.e., satisfies the general relationship $\lambda_{ij} = \lambda_{ik} \lambda_{kj}$ for $\lambda_{ij} > 0; i, j, = 1, 2, \ldots, n$).

Theorem 5.1 For every feasible set of λ_{ij}^* associated with the multiobjective problem given in Eq. (5.1), there exists a corresponding feasible set of decisions $\mathbf{x}^*$.
Proof. Rewrite Eq. (5.1) as follows:

$$\min_{\mathbf{x} \in X} \left\{ f_i(\mathbf{x}) + \sum_{j \ne i} \lambda_{ij}^* f_j(\mathbf{x}) \right\}$$

If all $f_k(\mathbf{x})$, $k = 1, 2, \ldots, n$, are continuous and the solution set X is compact (a set X is said to be compact if it is both closed and bounded—that is, if it is closed and is contained within some sphere of finite radius), then this problem must have a solution (by Weierstrass's theorem).

These assumptions are very mild. Compactness of X can be guaranteed by imposing finite upper and lower bounds on each component of the decision vector $\mathbf{x}$, assuming the constraint functions $g_i(\mathbf{x})$ are continuous. A continuity assumption of all $f_j(\mathbf{x})$ and $g_i(\mathbf{x})$ (as defined in Eq. (5.1)) is common in mathematical programming.

Let $\mathbf{x}^*$ be a solution for a given λ_{ij}^*. Then λ_{ij}^*'s are the optimal trade-off values (Lagrange multipliers) for the problem. Thus, $\mathbf{x}^*$ is in X and λ_{ij}^*'s are the desired Lagrange multipliers.

The feasibility of a solution $\mathbf{x}^*$ corresponding to λ_{ij}^* can also be shown on the basis of the Lambda theorem by Everett [1963]. It is helpful to summarize the three major steps in the SWT method. These are:

Step 1. Identify and generate noninferior (Pareto-optimal) solutions, along with the trade-off functions, λ_{ij}^*, between any two objective functions $f_i(\mathbf{x})$ and $f_j(\mathbf{x})$, $i \ne j$. It can be shown that under certain mild conditions, one set of n trade-off functions, $\lambda_{11}, \ldots, \lambda_{1n}$, will suffice to generate all other $\lambda_{ij}, i \ne j, i, j = 1, 2, \ldots, n$.

Step 2. Interact with the DM to assess the indifference band where the surrogate worth function $W_{ij}(\lambda_{ij}^*) = 0$. It was shown that under certain mild conditions, W_{ij} depends only on λ_{ij}.

Step 3. Determine the optimal decision set, $\mathbf{x}^*$, using the optimal trade-off values λ_{ij}^*.

5.3.6 The SWT Method with MDMs

Water resource systems, like most other civil systems, are characterized by MDMs at the various levels of the decisionmaking process. This is true for both planning and management purposes. In the case study discussed in Chapter 3, for example, the Planning Board of the Maumee River Basin consists of eight members from federal, state, and regional agencies. The board is in charge of developing a basin-wide comprehensive plan that is responsive to environmental, economic, social, legal, political, and institutional needs. However, members of the board, as DMs, exercise their mandate to be responsive along with their professional judgment, the agency's stand, and the public preferences as voiced by various public hearings and other media. Clearly, in applying the SWT method, different indifference bands may result by interacting with each Planning Board member. The key question is how to modify the SWT method to handle this situation.

Three major cases of MOPs with MDMs are commonly discussed in the literature: direct group decisionmaking systems, representative decisionmaking systems, and political decisionmaking. For simplicity, a more general case will be assumed here.

Consider the MOP posed by Eq. (5.1), where an interaction with the DMs takes place for assessing the corresponding trade-offs and preferences that lead to $W_{ij} = 0$. Two cases will be identified here: the ideal and the probable.

The ideal case. In assessing trade-offs and preferences with the DMs, it is assumed in the ideal case that the indifference bands generated by all the DMs for all W_{ij}, $i \neq j, j = 1, 2, \ldots, n$, have a common indifference band, Δ, as depicted in Figure 5.4. This situation is unlikely to happen; however, it provides a medium for understanding the probable case. All the indifference bands in Figure 5.4 correspond, of course, to $W_{ij} = 0$; however, they are plotted at different levels on the W_{ij} scale in order to distinguish among the indifference bands of the various DMs.

The probable case. In the probable case, no common indifference band can be found for all the DMs. This case is depicted in Figure 5.5. The SWT method provides an explicit and quantitative mechanism for simulating the DMs' preferences with respect to the trade-offs between any two objective functions. Identifying the differences in the DMs' preferences is a first step in closing these gaps through the inevitable process of negotiation and compromise. These negotiations may take different forms and are expected to lead to an agreeable decision (depending on whether a simple majority, absolute majority, consensus, or other guideline is needed for an agreed-upon decision).

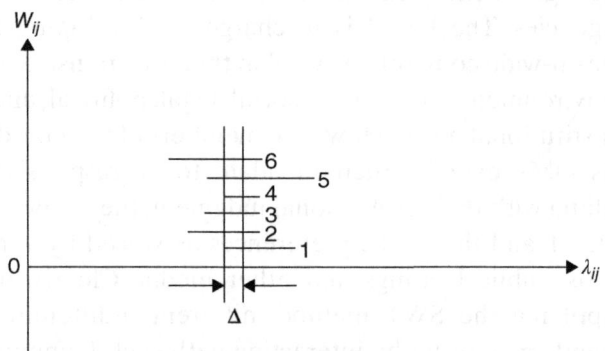

Figure 5.4 Common indifference band in the ideal case.

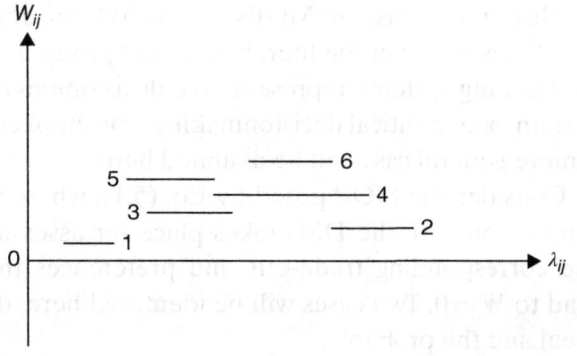

Figure 5.5 Indifference bands in the probable case.

5.3.7 Summary

The SWT method can be used to analyze and optimize MOPs. The following is a brief summary of this method:

1. It is capable of generating all needed noninferior solutions to a vector optimization problem.
2. The method generates the trade-offs between any two objective functions on the basis of duality theory in nonlinear programming. The trade-off function between the ith and jth objective functions, λ_{ij}, is explicitly evaluated and is equivalent to $-\partial f_i / \partial f_j$.
3. The DM interacts with the systems analyst and the mathematical model at a general and very moderate level. This is done via the generation of the surrogate worth functions, which relate the DMs' preferences to the noninferior solutions through the trade-off functions. These preferences are constructed in the objective function space (more familiar and meaningful to DMs) and only then transferred to the decision space. This is particularly important, since the dimensionality of the objective function space is often smaller than that of the decision space. These preferences yield an indifference band where the DM is indifferent to any further trade-off among the objectives.
4. The SWT method provides for the quantitative and qualitative analysis of noncommensurable objective functions.
5. The method is very well suited to the analysis and optimization of multiobjective functions with multiple DMs.
6. The method has an appreciable computational advantage over all other existing methods when the number of objective functions is three or more.

5.4 CHARACTERIZING A PROPER NONINFERIOR SOLUTION

The concept of a proper noninferior solution was first introduced by Kuhn and Tucker [1951], and it was later modified by Geoffrion [1968]. A feasible solution $\mathbf{x}^*$ is a proper noninferior solution if there exists at least a pair of objectives, say, f_i and f_j, for which a finite improvement of one objective is possible only at the expense of some reasonable degradation of the other. More precisely, a proper noninferiority of **x*** implies the existence of a constant $M>0$ such that for each i, $i=1, \ldots, n$, and each $\mathbf{x} \in X$ satisfying $f_i(\mathbf{x}) < f_i(\mathbf{x}^*)$, there exists at least one $j \neq i$ with $f_j(\mathbf{x}) > f_j(\mathbf{x}^*)$, and $\left[f_i(\mathbf{x}) - f_i(\mathbf{x}^*) \right]\left[f_j(\mathbf{x}^*) - f_j(\mathbf{x}) \right] \leq M$. Naturally, one should only seek, as candidates for the best compromise solution, proper noninferior solutions. A noninferior solution that is not proper is an *improper noninferior solution*.

Geoffrion [1968] characterizes proper noninferior solutions by showing the following. A sufficient condition for **x*** to be proper and noninferior is that it solves a weighting problem $P(w)$, with w being a vector of strictly positive weights. The condition becomes necessary if convexity for all functions is also assumed. This implies that a necessary and sufficient condition for **x*** to be a proper noninferior solution for a linear MOP is that it solves $P(w)$ with strictly positive weights w.

Chankong [1977] and Chankong and Haimes [1983a, b, 2008] then characterize proper noninferiority by means of the ε-constraint problem discussed in Section 5.3.2. Assuming continuous differentiability of all functions and the regularity of the point **x*** of the binding constraints of $P_k(\varepsilon^k)$, a necessary condition for **x*** to be properly noninferior is that **x*** solves $P_k(\varepsilon^k)$, with all the Kuhn–Tucker multipliers associated with the constraints $f_j(\mathbf{x}) \leq \varepsilon_j, j \neq k$, being strictly positive. The condition becomes sufficient if convexity for all functions is further assumed. This condition, as depicted in Figure 5.6, is often easy to verify when the ε-constraint approach is used as a means for generating noninferior solutions. Relationships between improper noninferiority and positivity of the Kuhn–Tucker multipliers can also be established, as displayed in Figure 5.6. Figure 5.7 illustrates a potential use of results depicted in Figure 5.6.

Consider the following vector minimization problem:

$$f_1(\mathbf{x}) = (x_1 - 1)^2 + (x_2 - 1)^2$$

$$f_2(\mathbf{x}) = (x_1 - 6)^2 + (x_2 - 2)^2$$

$$f_3(\mathbf{x}) = (x_1 - 2)^2 + (x_2 - 5)^2$$

and

$$X = \{ \mathbf{x} \mid x \in R^2, x_1 \geq 0, x_2 \geq 0 \}$$

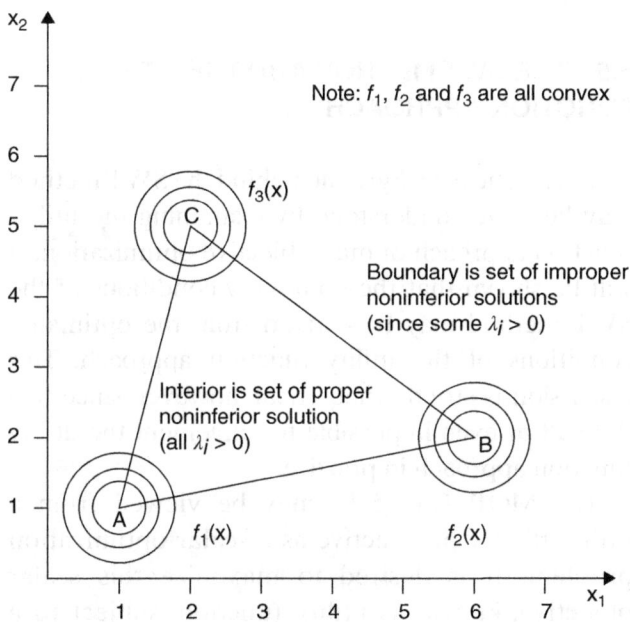

Figure 5.7 *Graphical illustration of relationships between positivity of λ's and proper noninferiority.*

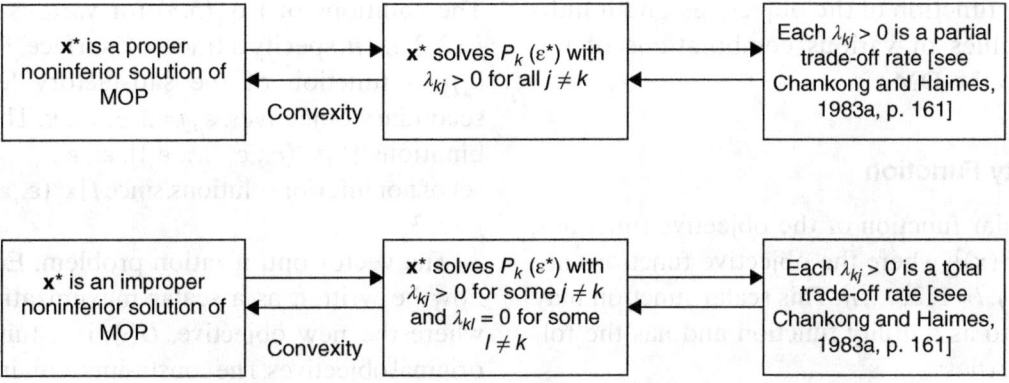

Figure 5.6 *Relationships between proper noninferiority and Kuhn–Tucker multipliers.*

It can be shown (see Chankong [1977]; Chankong and Haimes [1983a, b, 2008]) through the use of either the weighting problem (note that all objective functions are and must be convex) or the ε-constraint problems that the set of all noninferior solutions consists of all points within and on the boundary of the triangle *ABC* in Figure 5.7. If f_1 is taken to be the primary objective in the ε-constraint formulation, then it can be shown that the Kuhn–Tucker multipliers ($\lambda_{12}, \lambda_{13}$), corresponding to each point within the triangle, are strictly positive, while at least one λ_{ij} corresponding to points on the boundary of the triangle is zero. Consequently, each interior point of the triangle is a proper noninferior solution, whereas each boundary point of the triangle is an improper noninferior solution.

5.5 THE SWT METHOD AND THE UTILITY FUNCTION APPROACH

The theoretical background behind the SWT method may be better understood by examining the utility function approach of multiobjective optimization. It will be shown that the optimality conditions of the SWT method may be derived from the optimality conditions of the utility function approach. This discussion is strictly theoretical, however, since it is difficult or even impossible to implement the utility function approach in practice.

The MOP, Eq. (5.1), may be viewed from a utility theory perspective as a scalar optimization problem. It is desired to maximize this scalar objective, known as utility function, subject to a number of constraints. The DM may represent a consumer, while the various objectives represent goods that the DM desires. The utility function is thus a scalar function of the objectives, and it indicates the values of various combinations of the objectives to the DM.

5.5.1 Utility Function

Define a scalar function of the objective functions, $U[f_1(\mathbf{x}), \ldots, f_n(\mathbf{x})]$, where the objective functions are given by $f_i(\mathbf{x}), i = 1, 2, \ldots, n$. This scalar function may be referred to as a utility function and has the following properties:

1. If $f_i(\mathbf{x}^1) \leq f_i(\mathbf{x}^2)$, for $i = 1, 2, \ldots, n$, then $U[f_1(\mathbf{x}^1), f_2(\mathbf{x}^1), \ldots, f_n(\mathbf{x}^1)] \geq U[f_1(\mathbf{x}^2), f_2(\mathbf{x}^2), \ldots, f_n(\mathbf{x}^2)]$.
2. $U[f_1(\mathbf{x}^1), f_2(\mathbf{x}^1), \ldots, f_n(\mathbf{x}^1)] \geq U[f_1(\mathbf{x}^2), f_2(\mathbf{x}^2), \ldots, f_n(\mathbf{x}^2)]$ implies that the combination of objectives $[f_1(\mathbf{x}^1), f_2(\mathbf{x}^1), \ldots, f_n(\mathbf{x}^1)]$ is preferred to the combination of objectives $[f_1(\mathbf{x}^2), f_2(\mathbf{x}^2), \ldots, f_n(\mathbf{x}^2)]$.
3. If $U[f_1(\mathbf{x}^1), f_2(\mathbf{x}^1), \ldots, f_n(\mathbf{x}^1)] = U[f_1(\mathbf{x}^2), f_2(\mathbf{x}^2), \ldots, f_n(\mathbf{x}^2)]$, then the DM is indifferent to the combinations $[f_1(\mathbf{x}^1), f_2(\mathbf{x}^1), \ldots, f_n(\mathbf{x}^1)]$ and $[f_1(\mathbf{x}^2), f_2(\mathbf{x}^2), \ldots, f_n(\mathbf{x}^2)]$; in other words, given the choice, the DM would not have a preference or be able to choose between the two combinations.

While it is extremely difficult or impossible to actually determine the DM's utility function—that is, to assign numerical utilities to the various combinations of the objectives—the following theoretical development will be useful in developing the optimality conditions of the SWT method. This discussion should serve to motivate further development of the SWT method.

The contours of the utility function, $U[f_1(\mathbf{x}), f_2(\mathbf{x}), \ldots, f_n(\mathbf{x})] = c$, are called indifference curves because the DM is indifferent to any pair of combinations along a given curve. However, if $c^1 > c^2$, then all combinations along the curve $U[f_1(\mathbf{x}^1), f_2(\mathbf{x}^1), \ldots, f_n(\mathbf{x}^1)] = c^1$ are preferred to any combination along the curve $U[f_1(\mathbf{x}^2), f_2(\mathbf{x}^2), \ldots, f_n(\mathbf{x}^2)] = c^2$. Again, it may be difficult or impossible to determine these curves.

Let the solution of the ε-constraint problem, Eq. (5.3), where $i = 1$, be represented by the vector $\mathbf{x}^*(\varepsilon_2, \varepsilon_3, \ldots, \varepsilon_n)$ where ε_j are specified for $j = 2, 3, \ldots, n$. The value of the primary objective attained, given ε_j, $j = 2, 3, \ldots, n$, is then $f_1[\mathbf{x}^*(\varepsilon_2, \varepsilon_3, \ldots, \varepsilon_n)]$ if we assume that all constraints are binding at $\mathbf{x}^*(\varepsilon_2, \varepsilon_3, \ldots, \varepsilon_n)$. The solutions of Eq. (5.3) for various values of ε_j, $j = 2, 3, \ldots, n$, specify a trade-off surface, $f_1[\mathbf{x}^*(\varepsilon_2, \varepsilon_3, \ldots, \varepsilon_n)]$, a function of the satisfactory levels of the secondary objectives, $\varepsilon_j, j = 2, 3, \ldots, n$. Thus, the combinations $\{f_1[\mathbf{x}^*(\varepsilon_2, \varepsilon_3, \ldots, \varepsilon_n)], \varepsilon_2, \varepsilon_3, \ldots, \varepsilon_n\}$ form the set of noninferior solutions, since $f_j[\mathbf{x}^*(\varepsilon_2, \varepsilon_3, \ldots, \varepsilon_n)] = \varepsilon_j$, $j = 2, 3, \ldots, n$.

The vector optimization problem, Eq. (5.1), may now be written as a scalar maximization problem, where the new objective, $U(\cdot)$, is a function of the original objectives. The constraints remain unchanged.

Therefore, we have

$$\max_{\mathbf{x} \in X} U[f_1(\mathbf{x}), f_2(\mathbf{x}), \ldots, f_n(\mathbf{x})] \quad (5.18)$$

where

$$X = \{\mathbf{x} \mid g_i(\mathbf{x}) \leq 0, i = 1, 2, \ldots, m\}$$

The utility function is constructed, however, so that noninferior solutions are preferred to inferior ones. Therefore, only noninferior solutions must be examined in Eq. (5.18). The solution of the ε-constraint problem, Eq. (5.3), generates the noninferior region. Restricting the utility maximization problem to the noninferior region simplifies the approach considerably. The decision variables are now the desired levels of the objectives, $\varepsilon_j, j=2, 3, \ldots, n$, rather than the original decision variables, $\mathbf{x}$. The optimization is carried on in the objective function space, E^{n-1}, not in the decision variable space, E^N. As mentioned before, in most realistic problems, $N \gg n$. Only $n-1$ objective values must be specified, since the primary objective, $f_1[\mathbf{x}^*(\varepsilon_2, \varepsilon_3, \ldots, \varepsilon_n)]$, is specified by the solution of Eq. (5.3).

Substituting the optimal values of the original decision variables, $\mathbf{x}^*(\varepsilon_2, \varepsilon_3, \ldots, \varepsilon_n)$, given desired levels of the secondary objectives, $\varepsilon_j, j=2, 3, \ldots, n$, the utility maximization problem may be restated as follows:

$$\max U\{f_1[\mathbf{x}^*(\varepsilon_2, \varepsilon_3, \ldots, \varepsilon_n)], \varepsilon_2, \varepsilon_3, \ldots, \varepsilon_n\} \quad (5.19)$$

No constraints are involved in Eq. (5.19) since all constraints were considered in the solution of the ε-constraint problem (Eq. (5.3)). Again, the decision variables in the utility maximization problem are now the desired levels of the secondary objectives, $\varepsilon_j, j=2, 3, \ldots, n$. The original decision variables, $\mathbf{x}$, are ignored at this stage, having been employed to determine the trade-off surface, $f_1[\mathbf{x}^*(\varepsilon_2, \varepsilon_3, \ldots, \varepsilon_n)]$, by repeated solution of Eq. (5.3), with various values of the secondary objectives, $\varepsilon_j, j=2, 3, \ldots, n$. Once the optimal values of the objectives are determined, ε_j^*, $j=2, 3, \ldots, n$, Eq. (5.3) will be solved once more, to find the optimal values of the decision variables, $x^*(\varepsilon_2^*, \varepsilon_3^*, \ldots, \varepsilon_n^*)$. The optimal values of the objectives are found by solving Eq. (5.19).

Since Eq. (5.19) involves unconstrained optimization, the necessary first-order conditions for a stationary point, $(\varepsilon_2^*, \varepsilon_3^*, \ldots, \varepsilon_n^*)$, are as follows:

$$\frac{\partial U(\cdot)}{\partial \varepsilon_j} = 0, j = 2, 3, \ldots, n \quad (5.20)$$

Applying the chain rule on Eq. (5.20) yields

$$\frac{\partial U(\cdot)}{\partial f_1(\cdot)} \frac{\partial f_1(\cdot)}{\partial \varepsilon_2} + \frac{\partial U(\cdot)}{\partial \varepsilon_2} = 0 \quad (5.21a)$$

$$\frac{\partial U(\cdot)}{\partial f_1(\cdot)} \frac{\partial f_1(\cdot)}{\partial \varepsilon_3} + \frac{\partial U(\cdot)}{\partial \varepsilon_3} = 0 \quad (5.21b)$$

$$\frac{\partial U(\cdot)}{\partial f_1(\cdot)} \frac{\partial f_1(\cdot)}{\partial \varepsilon_n} + \frac{\partial U(\cdot)}{\partial \varepsilon_n} = 0 \quad (5.21c)$$

Equations (5.21a)–(5.21c) yield

$$\frac{\partial U(\cdot)}{\partial f_1(\cdot)} = -\frac{\partial U(\cdot)}{\partial \varepsilon_2} \bigg/ \frac{\partial f_1(\cdot)}{\partial \varepsilon_2} \quad (5.22a)$$

$$\frac{\partial U(\cdot)}{\partial f_1(\cdot)} = -\frac{\partial U(\cdot)}{\partial \varepsilon_3} \bigg/ \frac{\partial f_1(\cdot)}{\partial \varepsilon_3} \quad (5.22b)$$

$$\frac{\partial U(\cdot)}{\partial f_1(\cdot)} = -\frac{\partial U(\cdot)}{\partial \varepsilon_n} \bigg/ \frac{\partial f_1(\cdot)}{\partial \varepsilon_n} \quad (5.22c)$$

Equating (5.22a), (5.22b), and (5.22c) yields

$$-\frac{\partial U(\cdot)}{\partial \varepsilon_2} \bigg/ \frac{\partial f_1(\cdot)}{\partial \varepsilon_2} = -\frac{\partial U(\cdot)}{\partial \varepsilon_3} \bigg/ \frac{\partial f_1(\cdot)}{\partial \varepsilon_3}$$
$$= \cdots = -\frac{\partial U(\cdot)}{\partial \varepsilon_n} \bigg/ \frac{\partial f_1(\cdot)}{\partial \varepsilon_n} = \frac{\partial U(\cdot)}{\partial f_1(\cdot)} \quad (5.23)$$

Since utility is measured in arbitrary units, we may assign

$$\frac{\partial U(\cdot)}{\partial f_1(\cdot)} = 1 \quad (5.24)$$

In this way, utility is now measured in units commensurable with the units of the objective $f_1(\mathbf{x})$.

5.5.2 Trade-Offs and Marginal Rate of Substitution

Since the trade-off surface, $f_1[\mathbf{x}^*(\varepsilon_2, \varepsilon_3, \ldots, \varepsilon_n)]$, was determined by repeated solution of Eq. (5.3) for various values $\varepsilon_j, j=2, 3, \ldots, n$, the trade-offs, $\lambda_{1j}, j=2, 3, \ldots, n$, are known as well. Rewriting Eq. (5.9) as

$$\lambda_{1j}(\varepsilon_2, \varepsilon_3, \ldots, \varepsilon_n) = \frac{\partial f_1(\cdot)}{\partial \varepsilon_j}, j = 2, 3, \ldots, n \quad (5.25)$$

and combining it with Eqs. (5.23) and (5.24) yields

$$\frac{\partial U(\cdot)}{\partial \varepsilon_2}\bigg/\lambda_{12}(\varepsilon_2,\varepsilon_3,\ldots,\varepsilon_n) = \frac{\partial U(\cdot)}{\partial \varepsilon_3}\bigg/\lambda_{13}(\varepsilon_2,\varepsilon_3,\ldots,\varepsilon_n) =$$
$$= \frac{\partial U(\cdot)}{\partial \varepsilon_n}\bigg/\lambda_{1n}(\varepsilon_2,\varepsilon_3,\ldots,\varepsilon_n) = 1 \quad (5.26)$$

Therefore, at the optimum, we obtain

$$\lambda_{12}(\varepsilon_2^*,\varepsilon_3^*,\ldots,\varepsilon_n^*) = \frac{\partial U(\cdot)}{\partial \varepsilon_2}$$
$$\lambda_{13}(\varepsilon_2^*,\varepsilon_3^*,\ldots,\varepsilon_n^*) = \frac{\partial U(\cdot)}{\partial \varepsilon_3} \quad (5.27)$$
$$\lambda_n(\varepsilon_2^*,\varepsilon_3^*,\ldots,\varepsilon_n^*) = \frac{\partial U(\cdot)}{\partial \varepsilon_n}$$

where $\varepsilon_2^*, \varepsilon_3^*, \ldots, \varepsilon_n^*$ represent the optimal values of the secondary objectives, $\varepsilon_j, j = 2, 3, \ldots, n$.

Define a new function

$$m_{1j}(\varepsilon_2,\varepsilon_3,\ldots,\varepsilon_n) = \frac{\partial U(\cdot)}{\partial \varepsilon_j}, j = 2,3,\ldots,n \quad (5.28)$$

It is instructive to define by Eq. (5.28) a new function, $m_{1j}(\varepsilon_2, \varepsilon_3, \ldots, \varepsilon_n)$, which represents the marginal rate of substitution of objective $f_1(\mathbf{x})$ with respect to objective $f_j(\mathbf{x})$, given levels of the remaining objectives, $f_i(\mathbf{x}), i = 2, 3, \ldots, n$. That is, these rates indicate the trade-offs that the DM is willing to make. The trade-off that must be made to remain on the trade-off surface is given by $\lambda_{1j}(\varepsilon_2, \varepsilon_3, \ldots, \varepsilon_n)$. Given $\{f_1[\mathbf{x}^*(\varepsilon_2, \varepsilon_3, \ldots, \varepsilon_n)], \varepsilon_2, \varepsilon_3, \ldots, \varepsilon_n\}$, it would cost the DM $\lambda_{1j}(\varepsilon_2, \varepsilon_3, \ldots, \varepsilon_n)$ units of $f_1(\cdot)$ to reduce ε_j by one unit, while he or she would be willing to spend $m_{1j}(\varepsilon_2, \varepsilon_3, \ldots, \varepsilon_n)$ units of $f_1(\cdot)$ to make the same reduction in ε_j.

At the optimum, therefore, the marginal rates of substitution must be equal to the corresponding trade-offs for $f_1(\cdot)$ with respect to $\varepsilon_j, j = 2, 3, \ldots, n$. The utility function is then tangent to the trade-off surface. Analogously, this procedure (utility maximization) finds the highest indifference curve tangent to the trade-off surface.

The optimality condition for Eq. (5.19) is then

$$m_{1j}(\varepsilon_2,\varepsilon_3,\ldots,\varepsilon_n) = \lambda_{1j}(\varepsilon_2,\varepsilon_3,\ldots,\varepsilon_n), j = 2,3,\ldots,n \quad (5.29)$$

where ε_j^* represents the optimal desired value of the jth objective, $j = 2, 3, \ldots, n$. Of course, it may be possible to satisfy these conditions exactly. However, it is sufficient to determine the range of values for which the optimality conditions are approximately satisfied (within some specified tolerance). This range indicates the indifference band, and any solution within the indifference band will be satisfactory. The λ_{1j}'s are determined by the solution of the ε-constraint problem, Eq. (5.3), for various values of $\varepsilon_j, j = 2, 3, \ldots, n$. The m_{1j}'s, however, are specified by the DM, through his or her objective interpretation of the utility function, relating the DM's preferences among the competing, multiple objectives.

The first phase in solving a multiobjective problem is to determine the trade-offs, $\lambda_{1j}(\varepsilon_2, \varepsilon_3, \ldots, \varepsilon_n), j = 2, 3, \ldots, n$, and the trade-off surface, $f_1[\mathbf{x}^*(\varepsilon_2, \varepsilon_3, \ldots, \varepsilon_n)]$ for various levels of the objectives satisfied as constraints, $\varepsilon_j, j = 2, 3, \ldots, n$. This phase is concerned with optimization in the decision variable space, E^N, choosing optimal values of the decision variables, $\mathbf{x}^*(\varepsilon_2, \varepsilon_3, \ldots, \varepsilon_n)$. The entire noninferior region may be found in this phase. Several approaches are available for this phase, including the ε-constraint method and the weighting approach.

The second phase involves interaction with the DM to determine the desired levels of the multiple objectives, $\varepsilon_j^*, j = 2, 3, \ldots, n$. While the DM may not actually know his or her utility function, the development of the utility maximization problem provides a worthwhile motivation for formulating the SWT method. This phase requires the satisfaction of the optimality conditions, Eq. (5.29), for the utility maximization problem; however, the DM may not have enough information available to determine the m_{1j}'s. Again, several alternative approaches are available. The DM is questioned about his or her preferences among the multiple objectives. From the noninferior solutions, the indifference band is determined. An optimal solution is then chosen from the indifference band.

5.5.3 Interactive Procedures

Several types of interaction with the DM may be possible, depending on the complexity of the information required. Of the proposed schemes, the SWT method requires the least information. While the DM may not actually know the utility function, it may be possible to infer information concerning its shape, through the interaction process. The DM is asked the question, At what point(s) along the trade-off surface would you be indifferent to changes in either direction of ε_j, given levels of ε_i, $i = 2, 3, ..., n, i \neq j$? This questioning would be repeated for all ε_j, $j = 2, 3, ..., n$. The optimal solution would occur at that point where the DM is simultaneously indifferent to move in any direction.

Another interactive scheme involves asking the DM the following question: Given levels of objectives $f_1(\mathbf{x}), f_2(\mathbf{x}), ..., f_n(\mathbf{x})$ satisfied as constraints, $\varepsilon_2, \varepsilon_3, ..., \varepsilon_n$, how much would you be willing to spend to reduce ε_j by one unit? This scheme attempts to determine the marginal rates of substitution. If no point is found at which the trade-offs exactly match the marginal rates of substitution, a linear multiple regression analysis would be required of the differences between the trade-offs and the corresponding marginal rates of substitution versus the various levels of the multiple objectives. These linear equations could then be solved for the point at which all the differences simultaneously equal zero.

The scheme proposed by the SWT method involves an ordinal ranking of the trade-offs, as compared with the marginal rates of substitution. The DM would be asked: Given levels of objectives $f_1(\cdot)$, $f_2(\cdot), ..., f_n(\cdot)$, would you be willing to spend (i) much more, (ii) more, (iii) about the same, (iv) less, or (v) much less than $\lambda_{1j}(\varepsilon_2, \varepsilon_3, ..., \varepsilon_n)$ units of $f_1(\cdot)$ to reduce ε_j by one unit? The surrogate worth function, $W_{1j}(\varepsilon_2, \varepsilon_3, ..., \varepsilon_n)$, would be based on an ordinal scale, where -10 might indicate that λ_{1j} units of $f_1(\cdot)$ are very much less worthwhile than one marginal unit of ε_j and $+10$ would indicate the opposite extreme, while 0 would signify that the exchange is an even trade; that is, the solution belongs to the indifference band. The optimum is found at the point where all surrogate worth functions are simultaneously equal to zero. By questioning the DM and determining the surrogate worth function, the shape of the utility function may be inferred. The surrogate worth function tends to make the objectives commensurable.

Several algorithms for computational implementation of the SWT method are available. In general, the surrogate worth function may take any form, so that $W_{1j}(\varepsilon_2^*, \varepsilon_3^*, ..., \varepsilon_n^*) = 0$, $j = 2, 3, ..., n$, implies that the optimality conditions for Eq. (5.29) are satisfied. For example, alternative forms of $W_{1j}(\cdot)$ may be

$$W_{1j}(\varepsilon_2, \varepsilon_3, ..., \varepsilon_n) = m_{1j}(\varepsilon_2, \varepsilon_3, ..., \varepsilon_n) - \lambda_{1j}(\varepsilon_2, \varepsilon_3, ..., \varepsilon_n) \quad (5.30)$$

or

$$W_{1j}(\varepsilon_2, \varepsilon_3, ..., \varepsilon_n) = \log \frac{m_{1j}(\varepsilon_2, \varepsilon_3, ..., \varepsilon_n)}{\lambda_{1j}(\varepsilon_2, \varepsilon_3, ..., \varepsilon_n)} \quad (5.31)$$

Obviously, if either of the above forms of the surrogate worth function is employed, the conditions $W_{1j}(\varepsilon_2^*, \varepsilon_3^*, ..., \varepsilon_n^*) = 0$ do indeed imply that the objective values $\varepsilon_2^*, \varepsilon_3^*, ..., \varepsilon_n^*$ are optimal.

Any surrogate worth function may be used as long as:

1. $W_{1j}(\varepsilon_2, \varepsilon_3, ..., \varepsilon_n) > 0$ implies $m_{1j}(\varepsilon_2, \varepsilon_3, ..., \varepsilon_n) > \lambda_{1j}(\varepsilon_2, \varepsilon_3, ..., \varepsilon_n)$.
2. $W_{1j}(\varepsilon_2, \varepsilon_3, ..., \varepsilon_n) < 0$ implies $m_{1j}(\varepsilon_2, \varepsilon_3, ..., \varepsilon_n) < \lambda_{1j}(\varepsilon_2, \varepsilon_3, ..., \varepsilon_n)$.
3. $W_{1j}(\varepsilon_2, \varepsilon_3, ..., \varepsilon_n) = 0$ implies $m_{1j}(\varepsilon_2, \varepsilon_3, ..., \varepsilon_n) = \lambda_{1j}(\varepsilon_2, \varepsilon_3, ..., \varepsilon_n)$.

An ordinal ranking will suffice if enough information is not available to actually assign numerical values to the m_{1j}'s.

The SWT method does not depend on the utility function, but only upon an ordinal ranking of trade-offs and marginal rates of substitution along the trade-off surface—that is, in the noninferior region. The DM must compare the trade-offs on the trade-off surface with the trade-offs that he or she is actually willing to make. The optimality conditions of the utility maximization problem are employed to formulate a surrogate worth function, which may be determined with less information than is required for the utility maximization approach.

Once the optimal values of the multiple objectives are determined, $\varepsilon_j^*, j = 2, 3, ..., n$, the final phase of the decisionmaking process involves solving the ε-constraint problem (Eq. (5.4)) with the optimal objective values at the right-hand side. The optimal decision variables are given by $\mathbf{x}^*(\varepsilon_2^*, \varepsilon_3^*, ..., \varepsilon_n^*)$, and the solution of the MOP is complete. For a more extensive discussion on multiobjective optimization and on the SWT method, the reader is referred to Chankong and Haimes [1983a, 2008].

In summary, then, the first phase of solving Eq. (5.1) involves the generation of the noninferior solutions by solving the ε-constraint problem with a number of different right-hand sides. The results of this phase include the trade-offs and the trade-off surface. Next, interaction with the DM is employed to determine the surrogate worth functions. The optimality conditions are then satisfied, yielding the optimal values of the objective functions. Finally, the optimal objective values are substituted in the ε-constraint problem, resulting in the optimal decision variables. Example problems are solved in the following section.

5.6 EXAMPLE PROBLEMS

Two example problems are presented here mainly for pedagogical purposes. There are two objective functions and two decision variables in Example 1, and there are three objective functions and two decision variables in Example 2. The corresponding solutions are relatively simple; therefore, they do not necessarily demonstrate the actual computational procedures involved in large-scale problems.

5.6.1 Example Problem 1

Solve the following MOP via the SWT method:

$$\min \begin{Bmatrix} f_1(x_1,x_2) = (x_1-2)^2 + (x_2-4)^2 + 5 \\ f_2(x_1,x_2) = (x_1-6)^2 + (x_2-10)^2 + 6 \end{Bmatrix} \mathbf{x}_1 \geq 0, \mathbf{x}_2 \geq 0 \quad (5.32)$$

A solution to Eq. (5.32) necessitates the existence of a DM who selects a preferred solution from the noninferior solutions. For simplicity, no constraints are introduced in this example problem.

Solution. The first phase in applying the SWT method is converting Eq. (5.32) into the ε-constraint form presented by Eq. (5.33):

$$\text{Subject to } \begin{Bmatrix} \min f_1(x_1,x_2) \\ f_2(x_1,x_2) \leq \varepsilon_2 \end{Bmatrix} \quad (5.33)$$

Form the Lagrangian function, $L(x_1, x_2, \lambda_{12})$:

$$L(x_1, x_2, \lambda_{12}) = f_1(x_1, x_2) + \lambda_{12}[f_2(x_1, x_2) - \varepsilon_2] \quad (5.34)$$

Substituting Eq. (5.32) into Eq. (5.34) yields

$$\begin{aligned} L(x_1,x_2,\lambda_{12}) &= (x_1-2)^2 + (x_2-4)^2 + 5 \\ &+ \lambda_{12}\left[(x_1-6)^2 + (x_2-10)^2 + 6 - \varepsilon_2\right] \end{aligned} \quad (5.35)$$

Note that the Kuhn–Tucker [1951] necessary conditions for stationarity (see Appendix) are simplified here, since there are no constraints on $\mathbf{x}_1$ and $\mathbf{x}_2$. These conditions are reduced to Eqs. (5.36)–(5.40):

$$\frac{\partial L(\cdot)}{\partial x_1} = 2(x_1-2) + 2\lambda_{12}(x_1-6) = 0 \quad (5.36)$$

$$\frac{\partial L(\cdot)}{\partial x_2} = 2(x_2-4) + 2\lambda_{12}(x_2-10) = 0 \quad (5.37)$$

$$\frac{\partial L(\cdot)}{\partial \lambda_{12}} = \left[(x_1-6)^2 + (x_2-10)^2 + 6 - \varepsilon_2\right] \leq 0 \quad (5.38)$$

$$\lambda_{12}[(x_1-6)^2 + (x_2-10)^2 + 6 - \varepsilon_2] = 0 \quad (5.39)$$

$$\lambda_{12} \geq 0 \quad (5.40)$$

Equation (5.36) yields

$$\lambda_{12} = \frac{x_1-2}{6-x_1} \quad (5.41)$$

Equation (5.37) yields

$$\lambda_{12} = \frac{x_2-4}{10-x_2} \quad (5.42)$$

Since $\lambda_{12} > 0$ guarantees a noninferior solution, Eqs. (5.38)–(5.40) are reduced to Eqs. (5.43) and (5.44):

$$(x_1-6)^2 + (x_2-10)^2 + 6 - \varepsilon_2 = 0 \quad (5.43)$$

$$\lambda_{12} > 0 \quad (5.44)$$

Note that both Eqs. (5.41) and (5.42) should be satisfied. Therefore, these equations yield Eq. (5.45):

$$\lambda_{12} = \frac{x_1-2}{6-x_1} = \frac{x_2-4}{10-x_2} \quad (5.45)$$

Upper and lower limits on x_1 and x_2 may easily be derived by satisfying Eqs. (5.41), (5.42), and (5.44):

$$2 < x_1 < 6 \quad (5.46)$$

$$4 < x_2 < 10 \quad (5.47)$$

The boundary points 2 and 6 for $\mathbf{x}_1$, and 4 and 10 for $\mathbf{x}_2$, result in either $\lambda_{12} = 0$ or $\lambda_{12} = \infty$.

Solving Eq. (5.45) simplifies the generation of noninferior points as is presented in Table 5.2:

$$x_2 = 1.5x_1 + 1 \quad (5.48)$$

Figures 5.8, 5.9, and 5.10 depict the noninferior solution in the functional space $f_1(x_1, x_2)$ and $f_2(x_1, x_2)$, the noninferior solution in the decision space $\mathbf{x}_1$ and $\mathbf{x}_2$, and trade-off function $\lambda_{12}(f_2)$ versus $f_2(x_1, x_2)$, respectively. Assuming that an interaction with a DM does take place resulting in a selection of an indifference level of trade-off, λ_{12}^*, then the corresponding preferred solution $\mathbf{x}_1^*$ and $\mathbf{x}_2^*$ can be obtained either directly from Table 5.2 or by solving Eq. (5.45) with $\lambda_{12} = \lambda_{12}^*$.

The reader should note that noninferior solutions and their corresponding trade-off values were not generated by varying ε_2, as is suggested by the

TABLE 5.2 Noninferior Solutions and Trade-Off Values for Example Problem 5.2

x_1	x_2	$f_1(x_1, x_2)$	$f_2(x_1, x_2)$	λ_{12}
2.00	4.00	5.00	58.00	0
2.50	4.75	5.81	45.81	0.14
3.00	5.50	8.25	35.25	0.33
3.50	6.25	12.31	26.31	0.60
4.00	7.00	18.00	19.00	1.00
4.50	7.75	25.31	13.31	1.67
5.00	8.50	34.25	9.25	3.00
5.50	9.25	44.81	6.81	7.00
6.00	10.00	57.00	6.00	∞

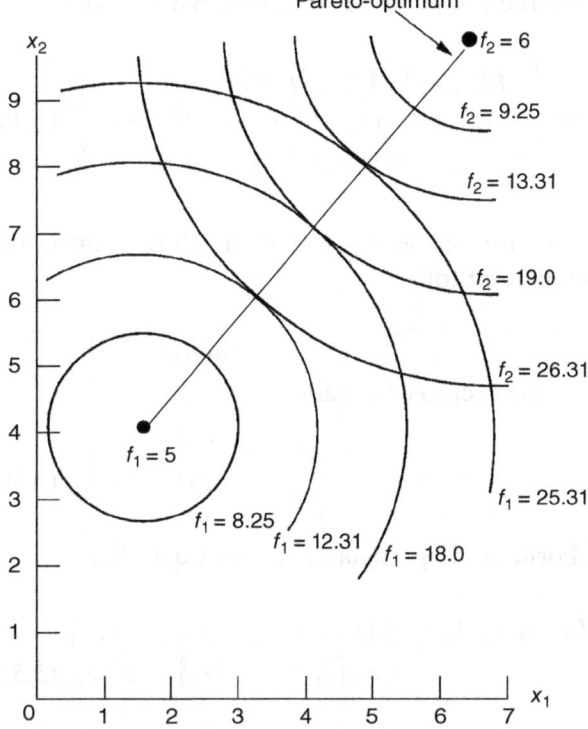

Figure 5.9 Noninferior solution in the decision space.

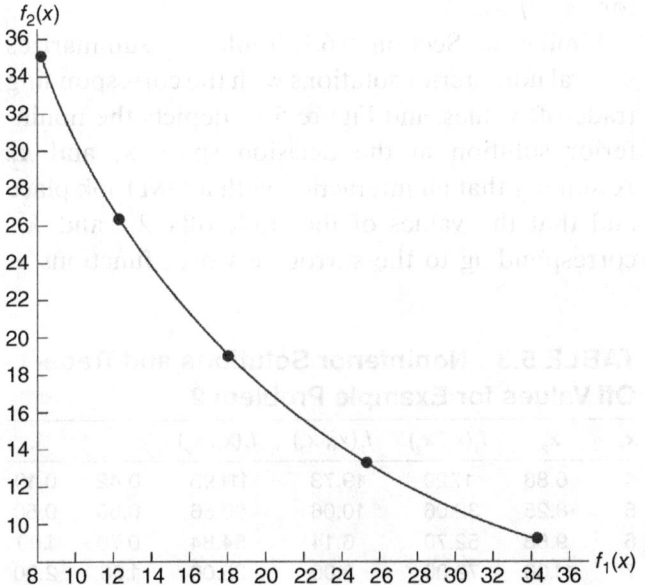

Figure 5.8 Noninferior solution in the functional space.

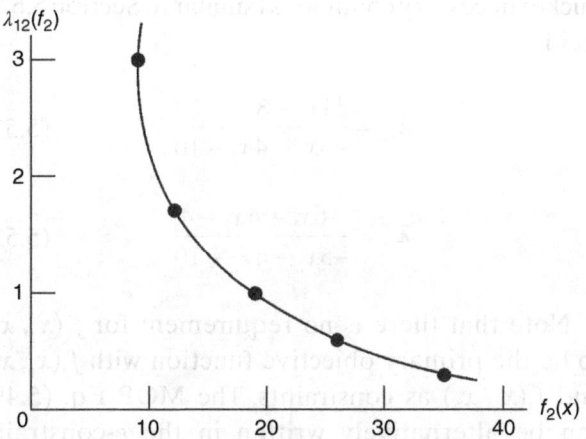

Figure 5.10 Trade-off function $\lambda_{12}(f_2)$ versus $f_2(x)$.

SWT method, because a closed-form and direct solution was obtained instead. In larger-scale problems with decision variables exceeding even 4 or 5, the aforementioned closed form will not be computationally tractable, and noninferior solutions would be generated by varying the ε's. This explanation also applies to Example Problem 2, discussed in the next section.

5.6.2 Example Problem 2

Solve the following MOP via the SWT method:

$$\min \begin{cases} f_1(x_1,x_2) = (x_1-2)^2 + (x_2-4)^2 + 5 \\ f_2(x_1,x_2) = (x_1-6)^2 + (x_2-10)^2 + 6 \\ f_3(x_1,x_2) = (x_1-10)^2 + (x_2-15)^2 + 10 \end{cases} \quad (5.49)$$

Solution. Rewrite problem (5.49) into the ε-constraint form:

$$\begin{aligned} & \min f_1(x_1,x_2) \\ & \text{Subject to constraints} \\ & \quad f_2(x_1,x_2) \le \varepsilon_2 \\ & \quad f_3(x_1,x_2) \le \varepsilon_3 \end{aligned} \quad (5.50)$$

Form the Lagrangian $L_1(\cdot)$ for Eq. (5.50):

$$L(x_1,x_2,\lambda_{12},\lambda_{13}) = f_1(x_1,x_2) + \lambda_{12}[f_2(x_1,x_2) - \varepsilon_2] + \lambda_{13}[f_3(x_1,x_2) - \varepsilon_3] \quad (5.51)$$

Substituting the values of $f_1(\cdot), f_2(\cdot)$, and $f_3(\cdot)$ from Eq. (5.49) into Eq. (5.51) and solving the Kuhn–Tucker necessary conditions (similar to Section 5.6.1) yield

$$\lambda_{12} = \frac{11x_1 - 8x_2 + 10}{-5x_1 - 4x_2 - 10} \quad (5.52)$$

$$\lambda_{13} = \frac{-6x_1 + 4x_2 - 4}{-5x_1 - 4x_2 - 10} \quad (5.53)$$

Note that there is no requirement for $f_1(x_1, x_2)$ to be the primary objective function with $f_2(x_1, x_2)$ and $f_3(x_1, x_2)$ as constraints. The MOP Eq. (5.49) can be alternatively written in the ε-constraint form as follows:

$$\begin{aligned} & \min f_2(x_1,x_2) \\ & \text{Subject to constraints} \\ & \quad f_1(x_1,x_2) \le \varepsilon_1 \\ & \quad f_3(x_1,x_2) \le \varepsilon_3 \end{aligned} \quad (5.54)$$

Form the Lagrangian $L_2(\cdot)$ for Eq. (5.54):

$$j = 1,2 \quad (5.55)$$

Again, substituting the values of $f_1(\cdot), f_2(\cdot)$, and $f_3(\cdot)$ from Eq. (5.49) into Eq. (5.55) and solving the Kuhn–Tucker necessary conditions yield

$$\lambda_{21} = \frac{-5x_1 - 4x_2 - 10}{11x_1 - 8x_2 + 10} \quad (5.56)$$

$$\lambda_{23} = \frac{-6x_1 + 4x_2 - 4}{11x_1 - 8x_2 + 10} \quad (5.57)$$

Note that Eqs. (5.52), (5.53), (5.56), and (5.57) satisfy Eq. (5.13), which is rewritten here for convenience:

$$\lambda_{ij} = \lambda_{ik}\lambda_{kj} \quad (5.58)$$

for positive λ's and $i \ne j \ne k$, and

$$\lambda_{ij} = \frac{1}{\lambda_{ji}} \quad (5.59)$$

for $i \ne j$, $\lambda j_i > 0$.

Similar to Section 5.6.1, Table 5.3 summarizes several noninferior solutions with the corresponding trade-off values, and Figure 5.11 depicts the noninferior solution in the decision space $\mathbf{x}_1$ and $\mathbf{x}_2$. Assuming that an interaction with a DM took place and that the values of the trade-offs λ_{12}^* and λ_{13}^* corresponding to the surrogate worth functions at

TABLE 5.3 Noninferior Solutions and Trade-Off Values for Example Problem 2

x_1	x_2	$f_1(x_1, x_2)$	$f_2(x_1, x_2)$	$f_3(x_1, x_2)$	λ_{12}	λ_{13}
4	6.88	17.29	19.73	111.93	0.42	0.19
5	8.25	32.06	10.06	80.56	0.50	0.50
6	9.63	52.70	6.14	54.84	0.70	1.00
7	11.00	79.00	8.00	35.00	1.00	2.00
8	12.38	111.22	15.66	20.86	2.17	5.17

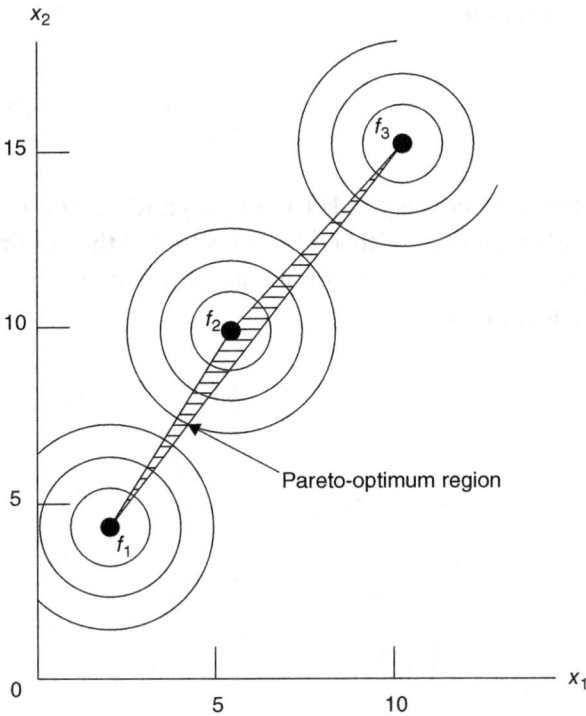

Figure 5.11 Noninferior solution in the decision space.

$W_{12} = 0$ and $W_{13} = 0$, respectively, were obtained, then the preferred solution can be generated by substituting the values of λ_{12}^* and λ_{13}^* into Eqs. (5.52) and (5.53) and solving x_1^* and x_2^*.

The reader is again reminded that for larger problems a closed-form solution may not be obtained, as is the case in this example, and the generation of noninferior solutions and their corresponding trade-off values then should be obtained by varying the ε's.

5.6.3 The Limitation of Pareto-Optimal Solutions

Example Problems 1 and 2 have two common objective functions; however, a third objective function in Example Problem 2 has been added to demonstrate an important attribute that characterizes all multiple-objective optimization problems. Namely, the set of Pareto-optimal solutions is critically dependent not only on the form but also on the number of objective functions that constitute the system's model. Note, for example, that the Pareto-optimal set in the decision space for the two-objective optimization problem (Section 5.6.1) lies on a straight line (see Figure 5.8). Yet, by adding a third objective function, the Pareto-optimal set in the decision space now constitutes an entire plane (see the shaded triangle in Figure 5.11). This means that a large number of Pareto-optimal solutions have been added. Conversely, by deleting one or more objective functions, the Pareto-optimal frontier will be reduced markedly.

The direct and sobering conclusion is that a large set of what were previously considered optimal solutions (in the Pareto sense) have suddenly become inferior, non-Pareto-optimal solutions. This is indeed a humbling experience for all modelers who consider any Pareto-optimal set to an MOP as a *sacred* and undisputed *optimal set of solutions*. In particular, remember that commonly DMs have a number of objectives that they desire to optimize, and thus, adding or deleting a secondary or a tertiary set of objectives is not only plausible but most probable.

5.6.4 The Reid–Vemuri Example Problem

Reid and Vemuri [1971] and Haimes and Hall [1974] introduced the following multiobjective function problem in water resource planning:

> A dam of finite height impounds water in the reservoir and that water is required to be released for various purposes such as flood control, irrigation, industrial and urban use, and power generation. The reservoir may also be used for fish and wildlife enhancement, recreation, salinity and pollution control, mandatory releases to satisfy riparian rights of downstream users, and so forth. The problem is essentially one of determining the storage capacity of the reservoir so as to maintain the net benefits accrued.

There are two decision variables: x_1, the total man-hours devoted to building the dam, and x_2, the mean radius of the lake impounded in some fashion. There are three objective functions: $f_1(x_1, x_2)$, the capital cost of the project; $f_2(x_2)$, the water loss (volume/year) due to evaporation; and $\hat{f}_3(x_1, x_2)$, the total volume capacity of the reservoir. In order to change the volume objective to a minimization problem, the reciprocal function $f_3(x_1, x_2)$ was formed, namely,

$$f_3(x_1, x_2) = 1/\hat{f}_3(x_1, x_2) \qquad (5.60)$$

where

$$f_1(x_1, x_2) = \exp(0.01x_1)(x_1)^{0.02}(x_2)^2 \quad (5.61)$$

$$f_2(x_2) = \frac{1}{2}(x_2)^2 \quad (5.62)$$

$$f_3(x_1, x_2) = \exp(-0.005x_1)(x_1)^{-0.01}(x_2)^{-2} \quad (5.63)$$

All decisions and objectives are constrained to be nonnegative. Although this problem is far from representing a realistic decisionmaking water resource problem (there are only two decision variables), it was chosen because of the general interest that Reid and Vemuri had generated by their paper.

The first step of the SWT method is to find the minimum values for each objective function. Clearly, $\bar{f}_1 = 0, \bar{f}_2 = 0$ at $x_2 = 0$, and $\bar{f}_3 = 0$ at $x_1 = \infty$. The constraint formulation is now adopted to generate λ_{12} and λ_{13}:

$$\min\left\{\exp(0.01x_1)(x_1)^{0.02} x_2^2\right\} \quad (5.64)$$

subject to

$$\frac{1}{2}x_2^2 \leq \varepsilon_2 \quad (5.65)$$

$$\exp(-0.005x_1)(x_1)^{-0.01} x_2^{-2} \leq \varepsilon_3 \quad (5.66)$$

$$x_1 \geq 0 \quad x_2 \geq 0 \quad (5.67)$$

From the Lagrangian, $L(\cdot)$:

$$L(\cdot) = \exp(0.01x_1)(x_1)^{0.02} x_2^2 + \lambda_{12}\left(0.5x_2^2 - \varepsilon_2\right)$$
$$+ \lambda_{13}\left[\exp(-0.005x_1)(x_1)^{-0.01} x_2^{-2} - \varepsilon_3\right] \quad (5.68)$$

The Kuhn–Tucker necessary conditions for a minimum are

$$x_i \frac{\partial L}{\partial x_i} = 0; \frac{\partial L}{\partial x_i} \geq 0; x_i \geq 0; i = 1, 2 \quad (5.69)$$

$$\lambda_{ij} \frac{\partial L}{\partial \lambda_{ij}} = 0; \frac{\partial L}{\partial \lambda_{ij}} \leq 0; \lambda_{ij} \geq 0; j = 1, 2 \quad (5.70)$$

The above conditions were solved for various values of ε_2 and ε_3 (see Table 5.4).

Note that

$$f_3(x_1, x_2) = \frac{1}{\hat{f}_3(x_1, x_2)} \quad (5.71)$$

Since λ_{13} corresponds to $f_3(x_1, x_2)$ and yet the DM is rather familiar with $\hat{f}_3(x_1, x_2)$, which is the volume capacity of the reservoir, a trade-off function $\hat{\lambda}_{13}$ is needed, that is,

Given: $f_3 = \dfrac{1}{\hat{f}_3}$; $\lambda_{13} = -\dfrac{\partial f_1}{\partial f_3}$; $\hat{\lambda}_{13} = -\dfrac{\partial f_1}{\partial \hat{f}_3}$

$$\hat{\lambda}_{13} = -\frac{\partial f_1}{\partial \hat{f}_3} = -\frac{\partial f_1}{\partial f_3}\frac{\partial f_3}{\partial \hat{f}_3} = \lambda_{13} \frac{\partial f_3}{\partial \hat{f}_3} = \lambda_{13} \frac{\partial f_3}{\partial\left(\dfrac{1}{f_3}\right)}$$

$$= \lambda_{13} \frac{\partial f_3 / \partial x}{\partial\left(\dfrac{1}{f_3}\right)/\partial x} = \lambda_{13} \frac{\partial f_3 / \partial x}{-(f_3)^{-2}\dfrac{\partial f_3}{\partial x}} = -\frac{\lambda_{13}}{(f_3)^{-2}}$$

$$= -\lambda_{13}(f_3)^2$$

Thus,

$$\hat{\lambda}_{13} = -\frac{\lambda_{13}}{(\hat{f}_3)^2} \quad (5.72)$$

A multiple regression analysis for the construction of λ_{12} and λ_{13} as functions of f_2 and f_3 by using the wide band of noninferior points (Table 5.4) resulted in a correlation coefficient of only 0.80. This is attributed to the exponential nature of the objective functions. Consequently, the second approach was adopted (as is explained in the section on computational procedure for constructing the trade-off function), where the DM provided the surrogate worth values W_{12} and W_{13} for those values of λ_{12} and λ_{13} given in Table 5.4. Clearly, for each λ_{12} and λ_{13}, the corresponding f_1, f_2, and f_3 can also be found in Table 5.4. Should the DM need additional information in the neighborhood of λ_{12}^* and λ_{13}^*, then a multiple regression analysis can be conducted to yield the needed information.

The values of surrogate worth functions generated with a DM are tabulated as W_{12} and W_{13} in Table 5.4. More than one set of trade-offs resulted in an indifference band, namely $W_{ij} = 0$. The corresponding values of λ_{12}, λ_{13}, f_1, f_2, and f_3 can be read directly from Table 5.4, rows 9, 25, 30, and 32.

TABLE 5.4 Noninferior Points and Decisionmaker Responses

	x_1	x_2	f_1	f_2	f_3	λ_{12}	$\hat{\lambda}_{13}$	W_{12}	W_{13}
1	0.70	22.36	499.95	250.00	500.00	2.00	−2.00	+8	+6
2	128.91	22.36	2000.00	250.00	1000.00	8.00	−4.00	+2	+2
3	239.59	22.36	6124.45	250.00	1750.00	24.50	−7.00	−2	−2
4	310.41	22.36	12,499.99	250.00	2500.00	50.00	−10.00	−5	−5
5	391.04	22.36	28,124.09	250.00	3750.00	112.49	−15.00	−10	−10
6	448.28	22.36	49,984.46	250.00	5000.00	199.88	−19.99	−10	−10
7	24.43	38.73	2041.46	750.00	1750.00	2.72	−2.33	+7	+5
8	93.09	38.73	4166.41	750.00	2500.00	5.55	−3.33	+4	+3
9	172.95	38.73	9374.98	750.00	3750.00	12.50	−5.00	0	0
10	229.91	38.73	16,665.71	750.00	5000.00	22.22	−6.67	−2	−2
11	421.71	14.14	15,310.72	100.00	1750.00	153.09	−17.50	−10	−10
12	102.65	31.62	3062.14	500.00	1750.00	6.12	−3.50	+4	+3
13	573.53	14.14	70,310.77	100.00	3750.00	703.09	−37.50	−10	−10
14	253.27	31.62	14,060.19	500.00	3750.00	28.12	−7.50	−3	−3
15	116.19	44.72	7029.45	1000.00	3750.00	7.03	−3.75	+3	+2
16	150.47	14.56	1055.33	106.00	473.00	9.96	−4.46	0	+1
17	151.74	8.17	336.83	33.40	150.00	10.08	−4.49	0	+1
18	151.74	25.85	3368.26	334.00	1500.00	10.08	−4.49	0	+1
19	150.47	46.04	10,553.25	1060.00	4730.00	9.96	−4.46	0	+1
20	310.41	7.95	1580.00	31.60	316.00	50.00	−10.00	−5	−5
21	609.47	2.58	3367.91	3.34	150.00	1008.25	−44.90	−10	−10
22	379.22	10.91	5943.42	59.50	841.00	99.89	−14.13	−10	−9
23	219.38	13.33	1776.34	88.90	562.00	19.98	−6.32	−2	−1
24	609.47	8.17	33,679.12	33.40	1500.00	1008.25	−44.90	−10	−10
25	172.95	14.14	1250.00	100.00	500.00	12.50	−5.00	0	0
26	310.41	14.14	5000.00	100.00	1000.00	50.00	−10.00	−5	−5
27	630.86	14.14	124,971.65	100.00	5000.00	1249.43	−49.98	−10	−10
28	0.70	31.62	999.89	500.00	1000.00	2.00	−2.00	+8	+6
29	310.41	31.62	24,999.99	500.00	5000.00	50.00	−10.00	−5	−5
30	172.95	44.72	12,499.97	1000.00	5000.00	12.50	−5.00	0	0
31	492.62	14.14	31,209.63	100.00	2500.00	311.69	−24.95	−10	−10
32	172.95	31.62	6249.99	500.00	2500.00	12.50	−5.00	0	0
33	37.39	44.72	3125.00	1000.00	2500.00	3.12	−2.50	+7	+5

All solutions corresponding to these rows are optimal in the sense defined in Section 5.3.3 on the derivation of the trade-off function: they are noninferior solutions that belong to the indifference band.

The decision variables corresponding to the aforementioned optimal solutions can be obtained in several ways. The simplest way in this example is the use of Table 5.4. Thus, for example, row 9 provides the following optimal decisions and values of the objective functions: $x_1 = 172.95$, $x_2 = 38.73$, $f_1 = 9374.98$, $f_2 = 750.00$, $f_3 = 3750.00$. By using Table 5.4 to generate the optimal decisions x_1 and x_2, one may need to make an additional analysis in the case where there is no row with both W_{12} and W_{13} equal to zero. It is also possible to solve Eqs. (5.63–5.66) for $\varepsilon_2 = f_2(\lambda_{12}{}^*, \lambda_{13}{}^*)$ and $\varepsilon_3 = f_3(\lambda_{12}{}^*, \lambda_{13}{}^*)$, as was described in the second approach in the preceding section. Since Table 5.4 was used in deriving the optimal solution without the need for a further multiple regression analysis, the trade-off functions λ_{23}, λ_{21}, λ_{31}, and λ_{32} were not needed and thus not derived.

5.7 SUMMARY

The major characteristics and advantages of the SWT method are as follows:

1. Noncommensurable objective functions can be handled quantitatively.
2. The surrogate worth functions, which relate the DM's preferences to the noninferior solutions

through the trade-off functions, are constructed in the functional space and only then are transformed into the decision space.
3. The DM interacts with the mathematical model at a general and a very moderate level. The DM makes decisions on his or her subjective preference in the functional space (more familiar and meaningful to the DM) rather than in the decision space. This is particularly important, since the dimensionality of the decision space N is generally much larger than the dimensionality of the functional space n.

REFERENCES

Belton, V. and T.J. Stewart, 2002, *Multiple Criteria Decision Analysis: An Integrated Approach*, Kluwer Academic Publishers, Norwell, MA.

Chankong, V., 1977, Multiobjective Decision Making Analysis: The Interactive Surrogate Worth trade-off method, Ph.D. dissertation, Systems Engineering Department, Case Western Reserve, Cleveland, OH.

Chankong, V. and Y.Y. Haimes, 1983a, *Multiobjective Decision Making: Theory and Methodology*, Elsevier, New York.

Chankong, V. and Y.Y. Haimes, 1983b, Optimization-based methods for multiobjective decision-making: An overview, *Large Scale Systems* **5**: 1–33.

Chankong, V. and Y.Y. Haimes, 2008, *Multiobjective Decision Making: Theory and Methodology*, Dover, New York.

Collette, Y. and P. Siarry, 2004, *Multiobjective Optimization: Principles and Case Studies*, Springer, New York.

Ehrgott, M., 2005, *Multicriteria Optimization*, second edition, Springer, New York.

Everett, H., III, 1963, Generalized lagrange multiplier method for solving problems of optimum allocation of resources, *Operations Research II* **399**: 418.

Geoffrion, A.M., 1968, Proper efficiency and theory of vector maximization, *Journal of Mathematical Analysis and Applications* **22**: 618–630.

Haimes, Y.Y., 1977, *Hierarchical Analyses of Water Resource Systems: Modeling and Optimization of Large-Scale Systems*, McGraw-Hill, New York.

Haimes, Y.Y. and W.A. Hall, 1974, Multiobjectives in water resources systems analysis: The surrogate worth trade-off method, *Water Resources Research* **10**(4): 615–624.

Haimes, Y.Y., L.S. Lasdon, and D.A. Wismer, 1971, On the bicriterion formulation of the integrated system identification and systems optimization, *IEEE Transactions on Systems, Man, and Cybernetics* **1**: 296–297.

Haimes, Y.Y., K. Tarvainen, T. Shima, and J. Thadathil, 1990, *Hierarchical Multiobjective Analysis of Large-Scale Systems*, Hemisphere Publishing Corporation, New York.

Kuhn, H.W. and A.W. Tucker, 1951, Nonlinear programming, *Proceedings, 2nd Berkeley Symposium on Mathematical Statistics and Probability*, University of California Press, Berkeley, CA, pp. 481–492.

Peterson, D.F., chair, Technical Committee (TECHCOM) 1974, Water resources planning, social goals, and indicators: Methodological development and empirical tests, PRWG 131-1, Utah Water Research Laboratory, Utah State University, Logan, UT.

Reid, R.W. and V. Vemuri, 1971, On the noninferior index approach to large-scale multi-criteria systems, *Journal of The Franklin Institute* **291**(4): 241–254.

6

Defining Uncertainty and Sensitivity Analysis

6.1 INTRODUCTION

Most mathematical models treat important system characteristics such as risk, uncertainty, sensitivity, stability, responsivity, and irreversibility either by means of system constraints or by artificially embedding them in the overall index of performance. The systems analyst (the modeler) assumes the roles of both professional analyst and decisionmaker by explicitly or implicitly assigning weights to these and other noncommensurate system characteristics, thus commensurating them into the performance index (the mathematical model's function). Obviously, this process deserves further scrutiny, even where the analyst is the decisionmaker.

The above system characteristics should be quantified to the extent possible, and they may even be included in the mathematical models as separate objective functions. These should then be optimized along with the original model's objective function (index of performance), to allow the decisionmaker(s) to select a preferred policy (solution) from within the Pareto-optimal set.

Decisionmaking problems with uncertain parameters have generated increasing concern in recent years. In many cases, uncertainties prevent the formulation of deterministic models. Moreover, in formulating viable and *best* policies, it is often necessary to assess the behavior of a system under varying conditions. The literature offers some confusion about the terms *risk* and *uncertainty*, and this necessitates a restatement here of their conventional definitions: The term *risk* refers to a situation in which the potential outcomes can be described in objectively known probability distributions. Risk is a measure of the probability and severity of adverse effect. The term *uncertainty* refers to a situation in which no reasonable probabilities can be assigned to the potential outcomes. Uncertainty is the inability to determine the true state of affairs of a system.

As far back as April 1971, the Committee on Public Engineering Policy (COPEP) [1972] of the National Academy of Engineering organized a colloquium on *perspectives on benefit–risk decisionmaking*. It primarily addressed risks to life, health,

---

Risk Modeling, Assessment, and Management, Fourth Edition. Yacov Y. Haimes.
© 2016 John Wiley & Sons, Inc. Published 2016 by John Wiley & Sons, Inc.

or safety, and it focused on the following major categories of decisionmaking:

1. Individual or voluntary risks (e.g., sports, smoking)
2. Risks where the individual's options are somewhat limited by regulations
3. Risks in which voluntary individual decisionmaking is preempted (e.g., air pollution, nuclear energy, and public health)

In that colloquium, COPEP extended the benefit–cost concept to include the evaluation of all the benefits and costs of a proposed action. It also identified *the necessary ingredients of a process of rational analysis* when addressing the benefit–risk subject. These are:

1. The explicit recognition of uncertainty
2. Consistency in assessment of values
3. Distinguishing between decisions and outcomes (i.e., because of bad luck or unforeseeable events, a good decision could lead to an undesirable outcome)
4. Consideration of time preferences (i.e., giving proper weighting to short-term and long-term benefits and risks)

More than 35 years ago, Starr [1969, 1972] recognized the importance of trade-off analysis. Once systems characteristics such as risk, sensitivity, responsivity, irreversibility, and others are quantified, trade-offs among all benefits and costs can be generated via multiobjective optimization analyses. In his paper, Starr concluded:

> It is evident that we need much more study of the methodology for evaluating social benefits and costs. The fatality measure of public risk is perhaps more advanced than most because of decades of data collection. Nevertheless, even the use of crude measures of both benefits and costs would assist in the development of the insight needed for national policy purposes. We should not be discouraged by the complexity of this problem—the answers are too important, if we want a rational society.

Uncertainty dominates most decisionmaking processes and is the Achilles' heel for all deterministic and for some probabilistic models. Sensitivity, responsivity, and irreversibility are introduced as important factors in modeling and decisionmaking. Modeling, which constitutes the basis for most, if not all, decisionmaking processes that rely on quantitative or other formal analyses, is particularly prone to errors that originate from uncertainty. Sections 6.3 and 6.4 address these concerns. An uncertainty taxonomy is subsequently presented to provide the readers with a road map in this rugged terrain of uncertainty. The uncertainty sensitivity index method (USIM) and its extension are then developed and explained through an example problem [Haimes and Hall, 1977; Li and Haimes, 1988]. The USIM and its extension are grounded on the premise that systems characteristics, such as sensitivity, should be quantified and should be included in the system's model as separate objective functions. The new objective functions should then be optimized along with the system's original objective functions to allow the decisionmaker to select a preferred solution. Sections 6.2–6.7 are based on Haimes and Hall [1977].

The USIM and its extension are applied to three cases: (i) optimization problems with more than one uncertain parameter, (ii) dynamic optimization problems under uncertainty, and (iii) optimization problems with equality constraints having uncertain parameters. Section 6.8 investigates the case where the nominal value of the uncertain parameter is itself an uncertain variable. A robust algorithm is developed to guarantee an ideal solution for this problem. Section 6.9 addresses a design problem and suggests a method to identify the best compromise nominal values of certain parameters by integrating the USIM and the envelope approach.

6.2 SENSITIVITY, RESPONSIVITY, STABILITY, AND IRREVERSIBILITY

Thinking of risk as of an objective to be minimized appears deceptively simple but is in fact extremely complex. The question is, Risk of what? The answer to this question is usually a long list of undesirable outcomes and combinations of outcomes, each with a nonnegligible probability of occurring.

While in some cases a specific quantitative risk index can be defined and used as the objective, more often, there will be an excessive number of such indices. In such cases, it is possible that certain risk-related characteristics of the system can be identified, quantified, and used to serve as a single measure of many of those individual risk objectives. Among these characteristics, sensitivity, responsivity, stability, and irreversibility appear to be particularly important.

Although we recognize that the current state of the art in risk analysis is not yet fully capable of quantitatively treating all of these characteristics, it is essential that they be considered as thoroughly as possible. They are defined as follows [Haimes et al., 1975]:

- *Sensitivity* relates changes in the system's performance index (or output) to possible variations in the decision variables, constraint levels, and uncontrolled parameters (model coefficients).
- *Responsivity* represents the ability of the system to be dynamically responsive to changes (including random variations) in decisions over a period of time.
- *Stability* relates to the degree of variation of the mean system to fixed decisions. A stable system yields an invariant mean response to fixed decisions. In other words, a stable system yields an invariant mean response to the mean value of a decision set. A system may be stable and still have an important random component.
- *Irreversibility* measures the degree of difficulty involved in restoring previous states or conditions once the system has been altered by a decision (including the decision to do nothing).

6.2.1 Sensitivity

One can construct hypothetical situations in which the deterministic mathematical optimum decision would be the worst possible unless the decision variable could be very precisely controlled. Figure 6.1 shows such a situation in which it is presumed that the decision variable can be controlled only within limits, x_c, and that with equal likelihood x may take on any value within these limits. The deterministic mathematical maximum is far from being the practical optimum decision. In this contrived example, x_2^* is clearly a *better* decision than x_1^* (because x_1^* in

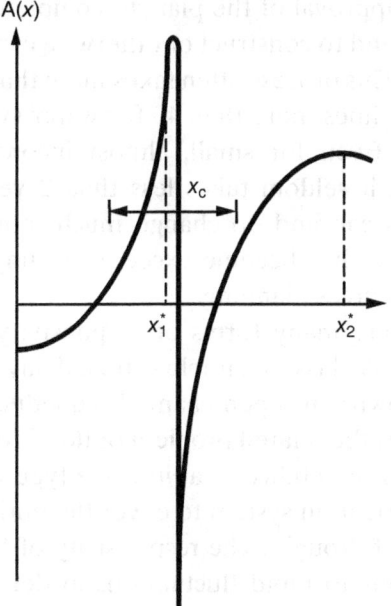

Figure 6.1 Sensitivity band, x_c.

Figure 6.1 could exceed the upper limit of x_c), unless the decisionmaker is more interested in gambling than avoiding risk.

Even if the example is treated by maximizing the mathematical expectation of $f(\mathbf{x})$, it does not follow that a resulting *optimum* at x_1 is superior to x_2. For this to be true, the appropriate objective must indeed be to maximize or minimize the expected value of $f(\mathbf{x})$. This is seldom true where risk is a major consideration. The *gambler's ruin* problem is the classic example where this is clearly not the objective.

Note, for example, that the decision that maximizes the expected value of the return to the gambler may also correspond to a maximization of the risk of getting little or nothing. In reality, there are at least two noncommensurable objectives in this case: avoiding risk and gaining economic return.

6.2.2 Responsivity

This is the capability of the system to respond in a reasonable time to a variable (changing) decision. It is generally related to *frictions* in the system and delayed response. One of the most important responsivity characteristics of many civil systems is the long lead time usually required to observe a need, to conceive a possible means of meeting that need, to develop a preliminary plan, to get basic political approval of the plan, to complete the final decision, and to construct or otherwise carry out the decision. This process often takes more than 20 years and sometimes more than 40 for water systems, for example. Even for small, almost inconsequential problems, it seldom takes less than 2 years. Since objectives can and do change much more rapidly, responsivity has become exceedingly important in water resources planning.

There are many forms of responsivity in water resources. A classic example is time delay in routing water down an open-channel aqueduct system. Another is the related problem of flood routing. Yet another is the ability of a *movable* type of supplemental irrigation system to cover the entire field in the face of drought. The responsivity of hydroelectric systems to rapid fluctuations in demand is an economically useful element of these systems.

The responsivity of water use to price, for example, is also very important in water resources systems. In many instances, costs, which vary with the amount of water used, are quite small relative to costs, which are insensitive to the amount used (largely irreversible capital investment). This may result in a response delay that severely affects the investments involved.

6.2.3 Stability

Stability measures the resistance to nondecision modification of the mean response of the system. For most environmental systems, frequently, the response of the system will vary appreciably even for a fixed decision. If the effect of the variation is to return the system automatically to the *output* or objective value represented by the decision, the decision system is stable. On the other hand, if autocatalytic effects cause the response to move away from that intended by the decisions, the decision system is unstable. Many environmental and other civil systems have highly unstable decision systems. One obvious example is the flood control decision system. It has been asserted that providing partial flood control, commensurate with one set of predicted future conditions, has resulted in attracting more economic activity into the *protected* area—making the original decision for partial control quite improper for the new situation. Transportation routing is another classic example of stability problems.

6.2.4 Irreversibility

This is a measure of the difficulty in returning a system to its original state once a decision change has been made. Suicide is an extreme example of an irreversible decision. In other cases, a decision can be reversed, but only at great social or economic cost. Humpty Dumpty is the literary personification of this important objective of water resources and many other civil systems.

Some decisions are completely irreversible but can be somewhat changed in time. That is, the state of the system, s, can be changed by arbitrary small modifications over time, t (or space), in one direction, but it cannot be reversed. Mathematically, this form can be represented by $\partial s / \partial t \geq 0$. We can burn fossil fuel but we cannot unburn it. In other decisions, complete irreversibility is a matter of degree, either continuous or discontinuous. A highway is an excellent example of a variable *irreversibility*, since it can be removed or expanded only at considerably greater cost than if the proper decision had been made originally.

6.3 UNCERTAINTIES DUE TO ERRORS IN MODELING

Not all of the uncertainties in civil or military systems have to do with the actual system itself. A significant uncertainty, all too often ignored in the quest for quantitative predictive models, is how well the models used actually represent the real system's significant behavior. This uncertainty can be introduced through the model's topology, its parameters, and the data collection and processing techniques. Model uncertainties will often be introduced through human errors of both commission and omission. An *optimized* decision set is truly optimal only if the mathematical model used to generate it closely represents the significant behavior of the actual system over time and space. The fact that some socioeconomic elements of the real system can react competitively or in complement to the chosen decision set only emphasizes this shortcoming of most mathematical models. In fact, there are actually no civil systems with a single decisionmaker, despite this customary assumption in optimal decision modeling.

The necessary condition for reasonable use of any decision set obtained through model optimization is that the important responses of the real system to those decisions are the same, within a tolerable limit for error. Since, for example, water resource decisions are often made only once (e.g., building a dam on a site), it may be difficult to evaluate modeling errors, let alone reduce them to quantitative probability measures. This significant source of uncertainty is probably one of the major reasons for the slow, cautious adoption in civil systems of the products of research, particularly of systems modeling [IAEA, 1989].

There is extensive literature about the sensitivity of optimizing solutions to variations in the parameters. In general, these evaluations are based on the properties of the first partial derivatives and higher-order partial derivatives of the objective function with respect to the constraints or other modeling parameters at the optimized values of the objective. To the extent that point properties reflect the risk concerns, any of these possibilities can be utilized as objective functions for multiple objective optimization. They are limited only by the degree to which the decisionmaker can understand their significance in context with his or her often-qualitative version of the risk problem. Obviously, one can create a situation where *point* properties evaluated at the optimum are poor indicators of the risk impacts at other points removed even a relatively small distance from the analytical optimum. Thus, for some problems where control is imprecise or indirect, a spatially distributed index may be preferable over a point index.

Little or no work has been accomplished with respect to the quantification of indices for specific systems and modeling characteristics. This must be done if they are to be useful in practical application.

First, the index should measure the pertinent characteristics of the problem. In particular, if a problem contains a large number of parameters, one must decide whether to use an index that measures the sensitivity of each individual parameter. Furthermore, only those parameters with deviations having the greatest effect on the optimal solution should be considered. This will avoid excessive computation and the generation of irrelevant information.

Second, information conveyed by the index should be clearly understood. The conceptual basis underlying the sensitivity measure must be easy to grasp, because it may be that the decisionmakers analyzing the problem have little technical understanding. This is often the case when solving large-scale multiobjective problems involving public investment.

Third, the index should not be difficult to calculate. When making a multiobjective analysis, it is often necessary to generate many noninferior points before a preferred solution can be found. Evaluating the sensitivity at each noninferior point may entail a heavy computational burden if the calculations used in determining the index are complex. Accordingly, it is desirable to have an index that utilizes information calculated by the particular optimization algorithm used in solving the problem.

6.4 CHARACTERIZATION OF MODELING ERRORS

The validity of the optimal solution x^* to any maximization or minimization problem depends (among other things) on the accuracy with which the mathematical model represents the real system. In turn, this accuracy depends on the closeness of the real system to the model's input–output relationships. The sources of uncertainties and errors can be associated with at least six major characteristics, which are discussed on the following pages: model topology (α_1), model parameters (α_2), model scope or focus (α_3), data (α_4), optimization technique (α_5), and human subjectivity (α_6).

6.4.1 Model Topology (α_1)

Model topology refers to the order, degree, and form of the equations that represent the real system. For example, a dynamic water system might be represented by a system of differential equations (ordinary or partial), and a static system might be represented by sets of algebraic equations such as polynomials.

For example, consider a groundwater system of both confined (bounded by impermeable rock) and unconfined (bounded by permeable rock) aquifers. To model the dynamic response of the aquifer's hydraulic head to any future demands (withdrawals or recharges) on the groundwater system, one may use a system of differential equations. Linear, second-order partial differential equations (PDEs) may be adequate for modeling the confined aquifer, whereas nonlinear, second-order PDEs might be needed for the unconfined aquifer. Furthermore, a homogeneous aquifer may be adequately modeled by a two-dimensional system, but a stratified, nonhomogeneous one ought to be modeled by a three-dimensional PDE. Clearly, selecting one model topology over another introduces uncertainties and errors into the accuracy of the model.

Model topology is particularly important in decisionmaking and optimization. Almost any function form can be used to approximate the absolute value of any cause–effect relationship. However, optimal decisions are usually not as concerned with the magnitude of these functions as with their derivatives (or incremental ratios). Thus, because of the characteristics of linear system optimization, a linear least-squares regression model of a nonlinear system is likely to select *decisions* at points that have the greatest errors in the representation of the true derivative.

6.4.2 Model Parameters (α_2)

Once the model topology has been selected, the choice of model parameters (often called parameter identification, parameter estimation, system identification, model calibration, etc.) determines the accuracy with which the model represents the real system. Consider the groundwater system discussed earlier. Once the customary system of parabolic PDEs is selected, the proper values of the coefficients need to be determined (e.g., storage capability and transmissivity as functions of the spatial coordinates). This parameter estimation process introduces uncertainties and errors that affect the accuracy of the calculated values of the parameters and in turn of the model itself.

6.4.3 Model Scope (α_3)

Model scope refers to the type and level of resolution used in the model for the description of the real system. Common descriptions of, for example, water resources systems include temporal, physical–hydrological, political–geographical, and goal or functional descriptions. Chapter 3 presents a more detailed discussion on the multiple descriptions of a system through the hierarchical holographic modeling concept. The characteristic parameters of uncertainty and error associated with the selection of the model scope are denoted by the set α_3.

In referring again to the groundwater system, one may wish to study the behavior (response) of the system under planned development for short-, intermediate-, and long-term planning. The groundwater system itself, which may consist of several aquifers, may be described on the basis of physical–hydrological characteristics or political–geographical boundaries. Finally, if the groundwater system is to be managed as part of a larger water resources system with concern for water quality, storage, recharge, and so on, then different approaches may be more advantageous, such as goal description. Clearly, while these four descriptions have individual merits, each portrays the system from a narrow point of view. The system in totality may never be well represented by any one description, and thus, the selection of a model's scope introduces yet another source of uncertainty and error into the system's representation. Scope is particularly important where the system is controlled by many relatively independent decisionmakers, each with somewhat different objectives. Even so, such systems are often modeled as though a single *rational* decisionmaker were at the helm—that is, as though a single point of view could be asserted.

6.4.4 Data (α_4)

Access to enough representative data for model construction, calibration, identification, testing, validation, and, hopefully, implementation is obviously very important in risk and in systems analysis. Clearly a lack of either accurate or sufficient data due to such problems in collecting, acquiring, processing, and analyzing it may cause substantial errors. Consider again the above groundwater system: The value of the model parameters identified is likely to depend on the available data. An insufficient number of sampling sites, the number of samples, and sampling accuracy (within each location) may introduce significant uncertainties and errors into the system model.

6.4.5 Optimization Techniques (α_5)

Once the mathematical model has been constructed and its parameters identified, selecting and applying suitable optimization methodologies (solution strategies) introduce another source of uncertainty and error into the system model. In the groundwater system discussed earlier, potential sources of uncertainty and error in the solution include selecting the method of numerical integration of PDEs with the associated grid size, boundaries, and initial conditions and computer storage capacity and accuracy. As another example, consider a nonlinear objective function with a nonlinear system of inequality constraints representing a power and water supply system. If the optimization method for solving this system is the simplex method (via linearization of the system model), then the accuracy of the solution obtained may be questionable. This is particularly true for highly nonlinear systems.

It is important to note that selecting the optimization technique generally coincides (or should coincide) with the model's construction. Consequently, any trade-offs between the sophistication (or simplification) of the model and the accuracy (or approximation) of the solution should be made during model development.

6.4.6 Human Subjectivity (α_6)

Human subjectivity strongly influences the outcome of the systems analysis and thus the risk assessment and management process. This factor includes the background, training, and experience of the analyst(s), personal preference, self-interest, and proficiency. Clearly, human subjectivity can influence all of the other five major categories of model characteristics.

A civil engineer, a hydrologist, or a systems engineer, for example, all involved in planning the development of the above groundwater system, may each conceive a different approach or methodology. While human subjectivity plays a very important role in the selection of all major model characteristics, each of which could introduce uncertainties and errors into the system model, there is no way to analyze to what extent this could happen. Rather than try to quantify such cause-and-effect relationships here, the importance of each characteristic is indicated and a framework for its analysis is suggested.

In analyzing the sources of uncertainty and error as they affect sensitivity, responsivity, stability, irreversibility, and, ultimately, optimality, the systems analyst may encounter any of the three conditions: (i) A complete knowledge of α is available, that is, α is a deterministic variable; (ii) alternatively, the vector α could be a stochastic variable, but an estimate of its probability distribution function is available; or (iii) the vector α could also be a stochastic variable where no knowledge is available on the probability distribution function.

It is assumed that for any given system, some analytical functions can be constructed relating sensitivity, stability, and irreversibility to α. Furthermore, depending on which element of α is under consideration, the knowledge of its mean and variance can vary between full knowledge and no knowledge. In any event, noncommensurable objective functions will result regardless of the degree of knowledge of α.

6.5 UNCERTAINTY TAXONOMY

Uncertainty is the inability to determine the true state of affairs of a system. It can be caused by incomplete knowledge or stochastic variability and surrounds all aspects of decisionmaking, encompassing many of the concepts integral to effective policy analysis. Uncertainty can arise from the inability to predict future events; for example, what will the prime rate be at the end of the decade? Or uncertainty can come from a limited understanding of a true process, for example, How does a virus weaken the central nervous system? Uncertainty can be caused by inaccurate communication of information: Does the phrase *flying is dangerous* mean that one should never fly or that one should fly only with caution? Even when there is complete understanding, there is still uncertainty in personal preferences and values: Should location, job, and/or salary be the main criterion in a job search? Sometimes, the value of interest is inherently uncertain: The moon's elliptical orbit means that the distance between the earth and moon is variable. This inherent uncertainty leads to an even more confusing concept: There are occasions when there is uncertainty concerning the variability of a value. If we did not understand the moon's orbit, there would be uncertainty about the representation of variation in distance between the moon and earth. The numerous types, sources, and terminologies concerning uncertainty generate confusion, which ultimately hampers the decisionmaking process [Ling, 1993]. The ability to identify and understand the different types and sources of uncertainty, as presented in Chapter 3, can facilitate its representation, which in turn can improve the decisionmaking process [Haimes et al., 1994].

The type and source of uncertainty can have an impact on the effectiveness of an uncertainty analysis and can dictate the methods used to characterize uncertainty [Hoffman and Hammonds, 1994]. In addition to affecting methodology, understanding the possible types or sources can improve the communication and interpretation of statements of uncertainty [Teigen, 1988]. The influence of uncertainty on methodology and perception emphasizes the importance of identifying uncertainty types and sources [Hirshleifer and Riley, 1992].

Several groups have addressed individual types and sources of uncertainty. An International Atomic Energy Agency report [IAEA, 1989] discussed the basic differences between a deterministic and probabilistic result and how these differences affect uncertainty. Other works have addressed the differences in uncertainty caused by stochastic variance versus incomplete knowledge. Some works have focused on uncertainty sources related to measurable properties [Morgan and Henrion, 1990]. Still others have provided general frameworks for the sources of uncertainties found in the basic components of a decision process [Finkel, 1990; Rowe, 1994]. The combined result of all of these works provides an adequate but often confusing picture of uncertainty.

This confusion is caused by overlapping ideas expressed by differing terminology and viewpoints. The current works tend to focus on individual areas of uncertainty. Although this focus results in an understanding of the specific areas, it is often difficult to assimilate each area into an overall picture of uncertainty. This chapter strives to develop a taxonomy of uncertainty by combining existing works and filling gaps. An overview of uncertainty should improve the understanding and communication of uncertainty types and sources [Ling, 1993].

6.5.1 Terminology

The first area of confusion arises in the terminology used to describe uncertainty and variability. This terminology is aimed at distinguishing between two types of value ambiguity: (i) that caused by incomplete knowledge and (ii) that caused by stochastic variability. The literature is basically divided into two groups. The first group considers both forms of ambiguity as a type of uncertainty. The second group takes a semantically different approach, labeling incomplete knowledge as uncertainty and stochastic variability as variability. Although this confusion is a matter of semantics, this text follows the first group and views the two forms of ambiguity as two types of uncertainty, addressing incomplete knowledge as knowledge uncertainty and stochastic variability as variability uncertainty.

This view is taken because both incomplete knowledge and stochastic variability affect one's ability to determine or state the true value of a quantity of concern. This means that both types fall into our earlier definition of uncertainty. In addition, from a practical viewpoint, it is rare to encounter one type without the other. Viewing incomplete knowledge and stochastic variability as types of uncertainty can clarify our understanding and communication of their relationship. Thus, to build our taxonomy of uncertainty, we classify uncertainty into two types: variability and knowledge. The characteristics and sources of the two types of uncertainty are discussed in the following sections.

6.5.2 Variability

Uncertainty caused by variability is a result of inherent fluctuations or differences in the quantity of concern. More precisely, variability occurs when the quantity of concern is not a specific value but rather a population of values. The three major sources of variability are [Taylor, 1993] (see Figure 6.2):

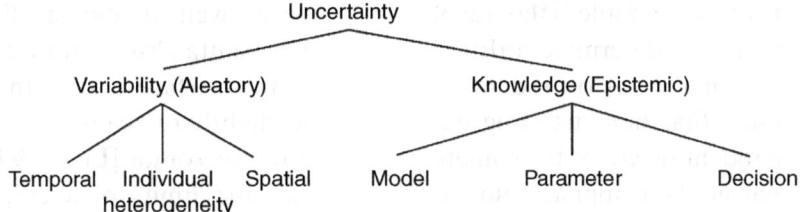

Figure 6.2 Major sources of uncertainty.

1. Temporal
2. Spatial
3. Individual heterogeneous

Temporal variability occurs when values fluctuate according to time. For example, the pollen count in the atmosphere varies with the seasons. Spatial variability affects values, which depends upon location or area. For example, the average rainfall in April varies according to geographical location, or the amount of fish eaten in a diet may depend on the proximity to waterways. The final category, individually heterogeneous, effectively covers all other sources of variability. Many quantities vary according to characteristics unique to their group or subgroup. For example, resistance to pesticides may vary according to the species of insect. The distinction between sources of variability is not mutually exclusive. Sources can and do overlap. For example, the pollen count in the atmosphere may depend on the seasons, but it also depends on the geographical location.

Pure variability contains no uncertainty that is due to a lack of knowledge. This means that all relationships are known, and if the source of variability is taken into account, a quantity can be calculated. A pure situation is rare, however; variability is usually complicated with uncertainty due to a lack of knowledge.

6.5.3 Knowledge

The second type of uncertainty is due to incomplete knowledge. It arises when the particular value or population of values of concern cannot be presented with complete confidence because of a lack of understanding or limitation of knowledge. The ease of identifying these sources ranges from simply remembering the source for some types, to considerable creative delving for the others. The impact of these sources on overall uncertainty also varies; some are almost insignificant, while others can change the uncertainty picture altogether. The main sources for uncertainty due to knowledge are depicted in Figure 6.2 on the right side of the taxonomy tree. These sources are explained in the following.

6.5.3.1 Model uncertainty

Model, or structural, uncertainty refers to uncertainties in the general knowledge of a process. Models are simplified representations of real-world processes; as such, they must make certain assumptions concerning the true state of nature. Model uncertainty can arise from oversimplification or from the failure to capture important characteristics of the process under investigation [Finkel, 1990]. If this uncertainty is improperly understood, it can be potentially the largest contributor of error, leading to significant misrepresentations of processes. Addressing this type of uncertainty is part of the art component of the art and science of modeling and constitutes the coarse-tuning function of the analysis; it is better understood by studying its major sources. These are discussed by Finkel [1990] and Morgan and Henrion [1990]. Paté-Cornell [1996] and Apostolakis [1999] characterized uncertainty within two categories: Aleatory (Variability) and Epistemic (Knowledge) uncertainty.

1. *Surrogate variables* are those quantities that are used in place of the actual quantity of concern. They are used when the quantity of concern is too difficult or too expensive to assess, and the surrogate variable is assumed to be a close substitute that can be dealt with more easily. The surrogate variable is an approximation of the real value; when used, the benefit of using a more accessible variable must be weighted against the disadvantages of using an estimate.

An example of a surrogate variable is the use of drug testing on rodents to determine a drug's effect on humans. Testing drugs on rodents is obviously more feasible than human testing, but the impact of using rodent reactions to estimate human reactions may not be completely understood. Thus, although surrogate variables are very appealing because of resource savings, they should be used with caution because they may increase the uncertainty of the results if the relationship between the surrogate estimate and the real value is not completely understood.

2. The second source of model uncertainty stems from *excluded variables* [Finkel, 1990]. Excluded variables are those deemed insignificant in a model of the process under investigation. The removal of certain variables or factors may introduce large uncertainties into the model. For example, many environmental risk assessment methods do not consider the propagating effects of hazardous chemicals through vegetation [R.C. Johnson, personal communication, 1992]. The effect of contaminated soil on the human consumption of vegetation may not be included in these models because it is not well understood. The exclusion of this variable may be significant if future research finds that vegetable consumption plays a significant role in the propagation of contaminants into the human body. Attempting to address excluded variables raises a natural paradox: We may not know that something has been overlooked until it is too late [Finkel, 1990]. This makes it very difficult to account for excluded variables. Unfortunately, as illustrated earlier, inattention to this source can lead to serious misrepresentations.

3. The impact of *abnormal situations* on models is the third source of model uncertainty. The very nature of a model requires that it simplifies real processes by aggregating numerous circumstances into a few broad categories. Problems arise when a model is used to represent a situation outside of its design. For example, a carpenter's level models a horizontal line using an air bubble inside a tube of fluid and the assumption that gravity is perpendicular to horizontal. This type of level works well at almost all locations; however, near Santa Cruz, California, an anomaly in the earth's surface causes the force of gravity to be slightly off, causing a level to misrepresent true horizontal [Ling, 1993]. Failure to recognize the limits of a carpenter's level causes people to draw erroneous conclusions in this area. The potential for unforeseen abnormal situations increases the uncertainty in the use of models to represent real-world situations.

4. *Approximation uncertainty* is the fourth source of model uncertainty [Morgan and Henrion, 1990]. This source covers the remaining types of uncertainty due to model generalization. An example of approximation of uncertainty can be found in the use of discrete probability distributions to represent a continuous real-world process, or in the limitation of finite runs used in a Monte Carlo analysis [Morgan and Henrion, 1990].

5. The fifth type of model uncertainty, *incorrect form*, is initially the most obvious but can easily be overlooked once an analysis has been started. This uncertainty concerns the validity, or accuracy, of the basic model being used to represent the real world. The impact of this uncertainty can potentially wipe out the significance of any other type of uncertainty. Haimes and Hall [1977] provide the following example to illustrate this type of uncertainty. A flood control system recommends the use of partial flood controls based on the current characteristics of a designated area. Unfortunately, the development of partial flood controls results in the attraction of more economic activity to this protected area, thereby making the original decision for partial control quite inappropriate for the new situation. In other words, the original model was designed for a static situation but was inappropriately applied to a dynamic scenario. To properly address this source, decisionmakers must remember that all results are directly dependent on the validity of the assumed model's representation of the true process being modeled.

6. The final source of model uncertainty is derived from *disagreement* [Morgan and Henrion, 1990]. Conflicting expert opinion or

data interpretation can cause differences in beliefs concerning the fundamental processes. Conflicting opinion may be due to hidden agendas or differing viewpoints. Sometimes, disagreement may occur because experts have a personal stake in the realization of a certain outcome. Conflicting experiments can also cause uncertainty as to the true value of concern. This source of uncertainty can sometimes be reduced over time as more information and research are available.

Model uncertainty can potentially contribute the most uncertainty to an analysis. However, its reduction is not straightforward or simple. It requires research into the understanding of the process under investigation and an effective balance between the cost of research and the cost of model errors. Proper identification and representation of model uncertainty can aid in understanding the overall level of uncertainty in an analysis.

6.5.3.2 Parameter uncertainty

The next general category of uncertainty due to a lack of knowledge is parameter uncertainty. This is found in the process of developing a specific value or population of values for the quantity of concern and can be thought of as fine-tuning the model. On the average, parameter uncertainty does not cause the large variations found in model uncertainty; but in total, it does represent a large portion of the uncertainty found in an analysis.

1. Probably the most common and best understood parameter uncertainty is *random error in direct measurements*. This source has been referred to as metrical error [Rowe, 1994], measurement error [Finkel, 1990], random error [Morgan and Henrion, 1990], and statistical variation [Morgan and Henrion, 1990]. The term *statistical variation* should not be confused with *uncertainty* due to inherent variations, which was discussed earlier. Statistical variation refers to the inability to provide an exact answer to a deterministic question because of knowledge limitations. Inherent variations refer to the need to use a population of values to answer a probabilistic question.

   Measurement error describes error caused by knowledge and technical limitations, not variability. It occurs because no measurement of a quantity can be exact. Imperfections in the measuring instrument and in observational techniques lead to imprecision and inaccuracies in measurements. For example, a yardstick may only be accurate to within an eighth of an inch. Fortunately, these variations in measurements can usually be reduced and quantified by repeating the procedure many times and developing summary statistics.

   Measurement errors do not always involve analytical hardware. For example, the conclusions drawn by many sociological studies are highly dependent on the accuracy and integrity of responses to survey questions. The potential for responders to answer survey questions inaccurately or untruthfully creates uncertainty in measuring society's true values or beliefs. Although the measurement error associated with each individual event in a model may appear minimal, typically the sheer number of events that are measured propagates measurement error into a significant factor of uncertainty. The effects of measurement error should not be underestimated.

2. The second and possibly largest source of parameter uncertainty is *systematic error*. Both Finkel [1990] and Morgan and Henrion [1990] address this source. Systematic error is sometimes called *error due to subjective judgment*, and it is defined as the difference between the true value and the mean of the value to which measurements converge [Morgan and Henrion, 1990]. This means that systematic error does not decrease with a larger sample size as does random error. For example, consider the situation where a lobby wants to determine public opinion of a new Republican tax law. An exit poll completed in an area known to be popular among Democrats will most likely misrepresent the true opinion of the general population. Polling a larger sample of Democrats will not reduce this error; if anything, it may obscure the systematic uncertainty because a larger sample size when using standard techniques to measure random error may lead to

overconfidence. The misrepresentation in this example is rather obvious because of our knowledge about the negative correlation between Democratic and Republican values. In many situations, the correlation between measurements and the environment is not as well known. In these cases, it is much more difficult to identify and reduce potential sources of systematic error. Reduction of systematic uncertainty can be accomplished by modifying the sampling technique or compensating for the error.

3. The third type of error is caused by *sampling*. This source has been termed both *random error* and *sampling error*. Even though the nomenclature overlaps with the measurement error described earlier, there is a difference between them. As discussed, measurement error arises due to the imprecision of measuring techniques. Sampling error appears when one draws inferences about a population from a limited representation. Sampling is conducted when it is too expensive or too impractical to analyze an entire population; instead, a small portion is studied and assumed to represent the whole. For example, consider the situation where a factory manager wants to estimate the quality of 400 electronic parts. Instead of testing each one, she may choose to test 40 of the devices and, based on these results, make a decision concerning the entire shipment. Sampling causes uncertainty in the degree to which the sample represents the whole. Well-developed statistical techniques such as confidence intervals, variation, and sample size are rich in this area and help to quantify this type of uncertainty (see Hogg and Tanis [1988]). The impact of this source is relatively simple to quantify.

4. The fourth type of parameter uncertainty is caused by *unpredictability*. Morgan and Henrion [1990] also call this source as *randomness*. It refers to the uncertainty that the extreme sensitivity of nonlinear systems exhibit to initial conditions. For example, consider the scenario in which a county government is concerned about the atmospheric effects of a possible release of gas from a chemical plant. Limitations in knowledge and the inherent unpredictability of the process make it impossible to predict the wind direction and velocity at a future date. The best that can be done is to assume knowledge based on current information and assume the value is variable. The distinction between unpredictability and uncertainty due to stochastic variance is subtle and for practical purposes can be considered similar. Unpredictability is presented here for completeness.

5. The fifth source of uncertainty is caused by *linguistic imprecision*. Everyday language and communication is rather imprecise. For example, the statement *Mary Anne is tall* is relative to a person's point of view. Is the statement true if she is five foot ten or five foot eight? Spedden and Ryan [1992] describe another example where people place varying numerical probabilities to the terms *probable* and *possible* in carcinogenic risk assessments. Tversky and Kahneman [1974] provide other examples of biases that may affect interpretations of uncertainty. Imprecise statements such as these create uncertainty as to the quantity of concern. Howard [1988] provides a clarity test to control this source of uncertainty. The test asks whether a clairvoyant would be able to determine the value of concern in question. If the clairvoyant can respond, then the question is precisely phrased; if not, the question should be modified. Other methods attempt to account for linguistic imprecision in the belief that it is unavoidable in human communication. For example, fuzzy set theory [Zadeh, 1984] classifies statements like *Mary Anne is tall* into sets defined by a fuzzy membership function. Another work has been completed in studying the relationship between verbal phrases such as *very possible* and the actual quantitative interpretation [Morgan and Henrion, 1990]. The ramifications of these methods for handling linguistic imprecision are not yet clear, and at this stage, it seems wiser to reduce linguistic uncertainty through clear specifications of events and values [Morgan and Henrion, 1990]. This source of uncertainty is relatively easy to remove compared to other sources that require large increases in research and resources.

This completes the explanation of the major sources of parameter uncertainty. The magnitude of uncertainty from each individual source may be rather small, but the great number of occurrences from each source makes parameter uncertainties a major contributor to overall uncertainty.

Both model uncertainty and parameter uncertainty pertain to uncertainty in the development and measurement of information. These two forms do not address the uncertainty surrounding questions of how to use the information in the implementation of a policy.

6.5.3.3 Decision uncertainty

Decision uncertainty arises when there is controversy or ambiguity concerning how to compare and weigh social objectives [Finkel, 1990]. Unfortunately, this source is often overlooked as part of the total assessment of uncertainty. In many circumstances, knowing how to implement the results of an analysis is just as problematic as completing the analysis. In practice, this source of ambiguity is more the rule than the exception and is often responsible for the inability of decisionmakers to take effective action. It is important to understand the overall impact of decision uncertainty on decisionmaking. The three major sources of decision uncertainty are illustrated in Figure 6.3:

1. *Risk measurement*, the first source of decision uncertainty, occurs during the selection of an index to determine the level of risk. This is required; with no risk measure, it is impossible to determine where one is in the process. Is there improvement? Are the objectives being met? The selection of a risk measure is both an art and a science because the measure must be as technically correct as possible while still being both valid and meaningful. Examples of risk measures may be the average life expectancy of a person exposed to radiation or the number of deaths reduced by a new braking system. Uncertainty arises in choosing a risk measure because there may be ambiguity about which measures portray the true situation better. A decisionmaker can rarely be completely sure that the risk measures chosen are the most representative of the real situation.

2. The second source of decision uncertainty lies in deciding the *social cost of risk*. To make differing risk measures commensurate, the decisionmaker is often forced to quantify different risks into comparable quantities. The difficulties in this process are clearly illustrated in the concept of developing a monetary equivalent for the value of life. Calculations for the value of life may be completed but are always open to heated debate over the derived value and the ethical implications of attaching a monetary value to life. In some scenarios, decisionmakers may be able to bypass the process of transforming all risk measures into comparable quantities, but in these situations, there still remains ambiguity in the evaluation process. Evaluating the social cost of risk creates uncertainty because there is rarely a clear, objective relationship between risk and social cost.

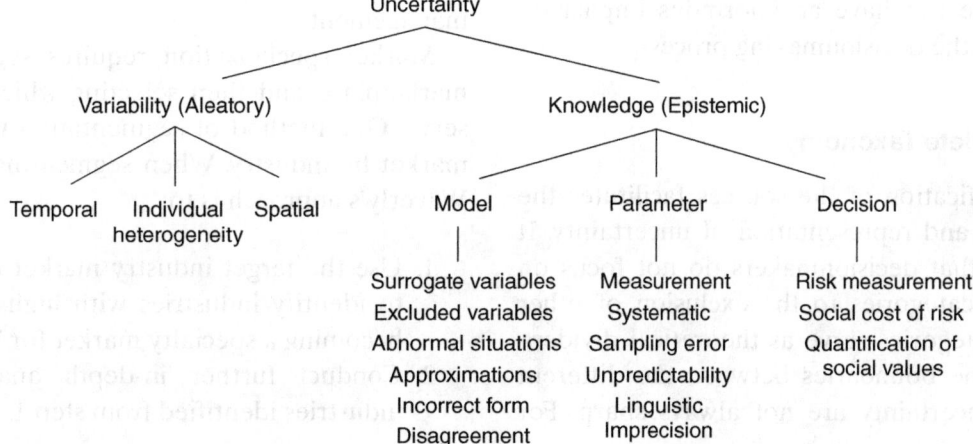

Figure 6.3 Component sources of knowledge uncertainty.

3. The *quantification of social values* is the third source of decision uncertainty. Once a risk measure and the cost of risk are generated, controversy still remains over what level of risk is acceptable [Lowrance, 1976]. This level is dependent upon determining society's risk attitude, but this brings more ambiguity and uncertainty into the process. Questions such as the following need to be addressed: How does one aggregate individual risk preferences to form a risk attitude for society? Should the risk be equally distributed, or should some suffer more to reduce risk for the majority? Another aspect is the concept of time. A decisionmaker must assess society's views on which is more preferred: risk today or risk tomorrow. These concepts are quite ambiguous and add significant uncertainty into a decision process.

Decision uncertainty can be difficult to address. The issues raised here are cursory in nature and only touch the surface of the numerous other issues that may be considered. The purpose of this section is to inform the reader that many of the assumptions made in a risk-based decision process are not as clear as they seem. For example, the assumed goal of minimizing risk does not have the same meaning to everyone; it is based on one's values, the measure of risk, and the comparison of risk values. These three sources of decision uncertainty contain numerous uncertainties and ambiguities. In the end, the perfect handling of model and parameter uncertainty can be insignificant if the information provided is implemented incorrectly in the decision phase. Recognizing the uncertainty present during the decision phase can have an enormous impact on the success of the decisionmaking process.

6.5.4 Complete Taxonomy

Proper identification of the sources facilitates the identification and representation of uncertainty. It is important that decisionmakers do not focus on the separate categories to the exclusion of other issues. The categories serve as theoretical dividers. In practice, the boundaries between the different sources of uncertainty are not always sharp. For example, there are similarities between systematic error and excluded variable uncertainty. Many of the distinctions among different sources of uncertainties are subtle. Instead of worrying about *distinguishing* among sources, the risk manager should be concerned with the *identification* of sources.

In conclusion, the taxonomy of uncertainty sources seeks to improve the information provided in a risk assessment. A decisionmaker should use the taxonomy to aid in the identification of uncertainty; this in turn can facilitate the process of managing uncertainty and lead to improved decisionmaking.

6.5.5 Application of the Taxonomy to the Selection of Target Markets

The previous section of this chapter presented a taxonomy for the types and sources of uncertainty. In each case, examples of the different sources were provided. To develop a more complete understanding, we can examine how Waverly Banking Corporation used the taxonomy to identify potential target markets.

6.5.5.1 Overview of the market selection process

To better meet the needs of its customers and to consistently improve the performance of its portfolio, Waverly Bank's Commercial Line of Business is implementing strategies that target specific markets. The Commercial Line of Business is the unit within Waverly that provides loans and services to middle-market businesses. This department believes market specialization increases Waverly's ability to develop in-depth market expertise and knowledge, thereby improving customer service and portfolio management.

Market specialization requires segmenting the marketplace and then selecting which markets to serve. One method of segmentation is to view the market by industry. When segmenting by industry, Waverly's approach is to:

1. Use the target industry market (TIM) model to identify industries with high potential for becoming a specialty market for Waverly.
2. Conduct further in-depth analysis of the industries identified from step 1.

3. Select an industry.
4. Test-market the industry.
5. If results from step 4 are favorable, roll out full specialization. Otherwise, return to step 1 or 2 with new information gained from step 4.

The true facilitator of this process is the TIM model, which reduces the time and costs associated with identifying industries by allowing Waverly to focus its resources immediately on a select number of industries.

The TIM model ranks industries based on four criteria:

1. Industry's market potential
2. Industry's risk and consistency
3. Waverly's industry expertise
4. Waverly's performance in the industry

The industry's market potential measures the amount of business Waverly may expect from an industry. The industry's risk and consistency addresses growth, cyclicality, and overall economic performance, helping to predict corporate bankruptcy and subsequently loan default levels. Waverly's industry expertise, the third criterion, gauges Waverly's current level of knowledge for the industry. The final criterion, Waverly's performance in the industry, addresses Waverly's past and present performance in the industry. Thus, the first two criteria evaluate the intrinsic attractiveness of a particular industry, and the latter two represent Waverly's prior experience in that industry. The model uses the two categories of criteria to represent the balance between the appeal of highly attractive industries and the value of proven industry experience.

The balance between industry attractiveness and Waverly experience is managed by eliciting preferences from the decisionmakers, who are able to identify their values of importance for each criterion. The model then uses these inputs to develop a final industry score. This represents a combination of how the industry scored within each criterion and the relative importance of each criterion according to the decisionmaker.

6.5.5.2 Waverly's application of the taxonomy

As mentioned earlier, the TIM model is used by Waverly as an initial screening tool. After the TIM model selects an industry, much time and effort is required before the industry can be developed into a specialty. For example, resources are required to conduct industry research, to complete company interviews, to analyze market tests, and to develop appropriate products. Because these resources are limited, the TIM model must be able to reduce the number of prospective industries from approximately 1000 to 5. The TIM model accomplishes this rapid reduction by using assumptions that in turn create uncertainties.

These uncertainties are identified by Waverly for the final five candidate industries through the use of the taxonomy. The identified uncertainties are then prioritized based on the perceived risk relative to the perceived success of developing a specialty in the candidate industry. The prioritized list of uncertainties is then used as the basis for the design of subsequent research, interviews, and tests. Research is continued until uncertainties are reduced to a level at which the expected cost of further study outweighs the perceived benefit of continued analysis. At this point, Waverly uses the current information to decide if the candidate industry should be developed into a specialty. In this manner, the taxonomy facilitates the identification and reduction of uncertainties during Waverly's selection of future markets.

To more clearly illustrate how Waverly uses the taxonomy, the following pages provide examples of the sources of uncertainty identified by Waverly in its TIM model. After a brief discussion of uncertainty due to stochastic variability, this section illustrates sources of uncertainty due to incomplete knowledge.

6.5.5.3 Stochastic variability: Temporal, individual, and spatial

Variability is encountered when the TIM model estimates industry market potential using an industry's sales in dollars. The level of sales may vary according to the seasons (temporal variability). On the other hand, if a time reference unit is specifically stated, then the actual dollar value of sales can be calculated.

For example, the question *What was the dollar value of sales for the media and publishing industry at the end of the first quarter of 1998?* can be answered with a specific dollar figure. The ability to reduce variability and knowledge uncertainty requires the analyst to estimate the feasibility and benefit of running the TIM model for a specific quarter, removing sales variability, versus the impact of applying the model for a year, keeping sales variability. In this situation, reducing variability may be undesirable: Running the model for a specific quarter may reduce its applicability and accuracy for more general situations.

Other examples of variability arise within the TIM model. The industry risk score varies according to a credit score calculated for specific companies within the industry (individual variability). Sales numbers for different industries vary according to geography (spatial variability). As with the earlier example, these two variability sources can be reduced if the TIM model is run for specific companies and geographical locations, respectively.

6.5.5.4 Knowledge

The following is an example of uncertainty in Waverly's TIM model caused by incomplete knowledge.

Model uncertainty

Surrogate variables. Surrogate variable uncertainty occurs when the TIM model uses the characteristics of one quantity to represent the characteristics of another. For example, Waverly prefers sales figures provided directly by companies when calculating the market potential score. When sales data are not available, the market potential score uses size of workforce, geographical location, and industry averages as surrogate variables for sales data. This methodology may lead to a misrepresentation of the true sales data because factors such as management style, company culture, and organizational structure also affect a company's sales.

Excluded variables. In the TIM model, excluded variable uncertainty can occur in the development of the industry market potential score. This score is based on the number and size of firms in Waverly's current market and in the larger national market. The model does not include the impact of specialties developed by competing banks. This excluded variable that may affect an industry's market potential. For example, the insurance industry with 100 firms may appear more appealing than the apparel industry with 50 firms when Waverly's competitors are not considered. But when competitors are included and it is disclosed that one bank services 95% of the insurance firms and no bank services more than 1% of the apparel industry, the apparel industry may actually be more attractive. Excluding competitor information may affect the accuracy of the TIM model's ability to measure market potential.

Abnormal situations. Another source of uncertainty is caused by the failure of models to account for abnormal situations. For example, the TIM model is designed for modern economic conditions. It is questionable whether the assumptions applied to the model would result in proper recommendations in times of severe depression, extreme growth, or political instability. Underlying assumptions and abnormal situations are often not accounted for in this model and can cause uncertainty.

Approximations. Approximation uncertainty arises because models are only simplified representations of the real world. These simplifications may add to the computational ease of a model but often reduce the correlation between the real-world results and the model results. For example, the TIM model assumes the performance of different industries to be independent. This independence assumption may make the analysis more tractable and easier to complete, but it also creates uncertainty. For example, the performance of the steel industry may be influenced by the performance of the auto industry. The TIM model views an exposure to the auto and steel industries as a diversified situation with decreased risk; but due to their interdependence, this exposure may actually increase the risk of Waverly's portfolio. Thus, approximations add uncertainty through simplifications within the model.

Incorrect form. Incorrect form uncertainty also affects the validity of the model: Does the model represent the real world? For example, the TIM model assumes that the final industry score should be linearly correlated to changes in market potential, industry risk, and so on. Uncertainty arises because the correct form of the model may be a nonlinear representation, although currently there are not enough data to properly justify another form.

Disagreement. Disagreement is the final source of model uncertainty. An example of disagreement uncertainty in the TIM model occurs in the industry risk score. The industry risk score includes economic predictions for an industry's growth or cyclicality. These predictions are based on the expert opinions of different economists. These economists may interpret economic data and indicators differently based on their past experiences and biases. This leads to differing conclusions, which creates uncertainty as to whose industry risk score should be used in the TIM model.

In conclusion, the actual identification of model uncertainties is often difficult to implement a priori but can be facilitated through an understanding of the sources depicted in the taxonomy.

Parameter uncertainty

Measurement. Measurement uncertainty is prevalent throughout the TIM model because of its large dependence on data collection. For example, there is uncertainty in the processes used to estimate the number of companies, the dollar amount of loans held, and the percentage of growth in an industry. Uncertainty arises because Waverly's market is composed of privately held companies; very little information on these companies is publicly available. The number of companies in a region is sometimes measured by counting those listed in the Yellow Pages. The Yellow Pages, however, does not properly represent new companies, companies no longer in existence, or companies choosing not to be listed. The potential inaccuracies in consulting the Yellow Pages can therefore lead to measurement uncertainty in the model.

Systematic. Systematic uncertainty is often more difficult to identify than measurement uncertainty. Systematic uncertainty is caused by a fundamental bias in procedures. For example, in some instances, information on companies is retrieved through the Securities and Exchange Commission (SEC) filings. This means that the model's information on industries is systematically biased toward larger, public companies that are required to file with the SEC. Smaller, privately held companies are not represented in SEC filings, and therefore, they are not directly represented in some of the model's conclusions.

Sampling error. Sampling error occurs when a limited number of events are used to draw inferences about a parent population. The TIM model's inference of Waverly's performance in an industry is an example of sampling error. The model assumes that Waverly's performance with prospective companies in an industry will be similar to its known performance with a few companies in the same industry. For example, if Waverly's current performance with tobacco companies provides a return of equity of 30%, the model assumes that all future relationships with tobacco companies will yield a 30% return. This assumption overstates the industry's profitability if Waverly's current relationships represent the most profitable tobacco companies in Waverly's region. Uncertainty arises because the current sample of tobacco companies may not be representative of the industry as a whole.

Unpredictability. Unpredictability of business events also causes uncertainty in the TIM model. For example, the model assumes the companies in its model will remain independent business entities. Although it may be possible to foresee potential industry consolidations, from an outsider's viewpoint, it is arguably impossible to predict the potential merger or acquisition of a company. This unpredictability is attributed to the difficulty an outsider faces when trying to estimate the behavior of a board of directors or the rationality of shareholders. The inability to predict future actions by directors or shareholders creates uncertainties in the model.

Linguistic imprecision. The wide array of services and products in today's larger companies provides room for linguistic uncertainty when classifying companies by industry. For example, would a representative of Honda Motor Company classify the company in the automobile, motorcycle, or small engine industry? This uncertainty may cause the TIM model to misrepresent the number and sizes of companies within different industries.

Parameter uncertainty and model uncertainty both address uncertainty within the analysis process. The next type of uncertainty occurs when implementing results of the analysis into actionable decisions.

Decision uncertainty
Decision uncertainty surrounds the implementation of analytical results into actual decisions and policy. In regard to the TIM model, this component of uncertainty can be the major roadblock in developing an industry specialization.

Risk measurement. The first source of decision uncertainty is caused by the ambiguity surrounding the development of a risk measure. For example, the TIM model represents the risk to shareholders of a poorly performing loan portfolio by estimating the chance that a company will go bankrupt or encounter economic difficulties. This source of uncertainty is created by transforming a qualitative idea, such as shareholder preference, into a quantitative measure, such as the probability that a company will go bankrupt. The uncertainty arises due to the question of how well this measure represents the real-world risk of concern.

Social cost of risk. The second source of uncertainty is equating the risk measure to a social cost of risk. In this example, uncertainty arises when trying to tie the risk measure to Waverly's shareholder cost of risk. Waverly may decide to evaluate the social cost of risk in terms of the dollar change in stock price or the dollar change in dividend. Difficulties and ambiguities occur when trying to make different units commensurate with a single measure of shareholder cost. For example, how should the model convert an increase in the probability of a company's going bankrupt to a shareholder's cost of a change in the stock price? It is apparent that much uncertainty surrounds this process of developing a social cost of risk.

Quantification of social values. The quantification of social values is the final source of decision uncertainty. This addresses the ambiguity in the estimate of society's preferences between realizations of the social cost of risk. In Waverly's case, ambiguity occurs in trying to represent a shareholder's preference between changes in stock price. For example, will shareholders be just as willing to approve a loan to a company that may change the stock price up or down 1/8 of a point when the stock is trading at $4 as they would when the stock is trading at $43? This type of uncertainty is difficult to address and can contribute greatly to the overall uncertainties that may prevent effective decisionmaking.

After Waverly's use of the taxonomy to identify sources of uncertainty in its TIM model, the identified uncertainties would now form the basis for the design of subsequent analyses. The use of the taxonomy provides Waverly with a systematic method for addressing and accounting for uncertainties in its decisions to enter new markets.

6.6 THE USIM

Some major tasks yet to be accomplished through research are (i) the quantification of the concepts of sensitivity, responsivity, stability, irreversibility, risk, and uncertainty and (ii) the construction of the associated indices so that they can be considered as objective functions in a multiobjective optimization framework. Examples of such indices were introduced in the previous section.

To provide proper motivation for, and better understanding of, the USIM, we first consider the following mathematical model:

$$y(x,\alpha) = 2x^2 - 2x(\alpha - 1) - \alpha^2 \quad (6.1)$$

where

$y(x, \alpha)$ denotes the system's output response
x denotes the model's decision variable
α denotes the model's parameter

Let $\hat{\alpha}$ denote the nominal value of α, which may be determined using any systems identification procedure. This is the value actually used in the optimization process.

Let the model's function be $f_1(x, \alpha)$. For simplicity, the objective is to minimize the output (e.g., cost):

$$f_1(x, \alpha) = y(x, \alpha) \quad (6.2)$$

or

$$f_1(x, \alpha) = 2x^2 - 2x(\alpha - 1) - \alpha^2 \quad (6.3)$$

Both $y(\cdot)$ and $f_1(\cdot)$ are written as functions of both x and α to emphasize this dependency not only on x alone but also on α. Let

$$\hat{\alpha} = 2 \quad (6.4)$$

the corresponding nominal output response is given by Eq. (6.5):

$$y(x,\hat{\alpha}) = 2x^2 - 2x - 4 \quad (6.5)$$

Define a sensitivity index, $f_2(\cdot)$, which measures the changes in the model's response to changes in α as follows:

$$f_2(x,\hat{\alpha}) = \left[\frac{\partial y(x,\alpha)}{\partial \alpha}\bigg|_{\alpha=\hat{\alpha}}\right]^2 \quad (6.6a)$$

where

$$\frac{\partial y(x,\alpha)}{\partial \alpha} = -2x - 2\alpha \quad (6.6b)$$

Thus,

$$f_2(x,\alpha) = 4x^2 + 8\alpha x + 4\alpha^2 \quad (6.7)$$

The joint *optimality* and *sensitivity* problem can be written in a multiobjective framework as follows:

$$\min \begin{bmatrix} f_1(x,\hat{\alpha}) \\ f_2(x,\hat{\alpha}) \end{bmatrix} \quad (6.8)$$

There are no constraints on x. Substituting Eqs. (6.3)–(6.7) into Eq. (6.8) yields

$$\min \begin{bmatrix} f_1(x,\hat{\alpha}) = 2x^2 - 2x - 4 \\ f_2(x,\hat{\alpha}) = 4x^2 + 16x + 16 \end{bmatrix} \quad (6.9)$$

Problem (6.9) can now be solved using the same procedures discussed earlier in Chapter 5. Transfer Eq. (6.9) into the ε-constraint form:

$$\min[2x^2 - 2x - 4] \quad (6.10)$$

subject to the constraint

$$4x^2 + 16x + 16 \leq \varepsilon_2 \quad (6.11)$$

Form the Lagrangian function for Eqs. (6.10) and (6.11):

$$L(x,\hat{\alpha},\lambda_{12}) = 2x^2 - 2x - 4 + \lambda_{12}[4x^2 + 16x + 16 - \varepsilon_2] \quad (6.12)$$

The Kuhn and Tucker [1950] necessary conditions for stationarity corresponding to Eqs. (6.10) and (6.11) (see the Appendix) yield

$$\frac{\partial L}{\partial x} = 4x - 2 + (8x + 16)\lambda_{12} = 0 \quad (6.13a)$$

$$x\frac{\partial L}{\partial x} = x(4v - 2 + (8x + 16)\lambda_{12}) = 0 \quad (6.13b)$$

$$\frac{\partial L}{\partial \lambda_{12}} = 4x^2 + 16x + 16 - \varepsilon_2 \leq 0 \quad (6.14)$$

$$\lambda_{12}[4x^2 + 16x + 16 - \varepsilon_2] = 0 \quad (6.15)$$

$$\lambda_{12} \geq 0 \quad (6.16)$$

Solving Eq. (6.13) yields

$$\lambda_{12} = \frac{2 - 4x}{16 + 8x} \quad (6.17)$$

Table 6.1 lists several noninferior solutions with the corresponding trade-off values. Figure 6.4 depicts the noninferior solution in the functional space $f_1(\cdot)$ and $f_2(\cdot)$. Note that $\lambda_{12} > 0$ guarantees Pareto-optimality.

Let x^* and $\hat{x}$ denote the decision variables that minimize $f_1(x,\hat{\alpha})$ and $f_2(x,\hat{\alpha})$, respectively:

$$\min f_1(x,\hat{\alpha}) = f_1(x^*,\hat{\alpha}) \quad (6.18)$$

$$\min f_2(x,\hat{\alpha}) = f_2(\hat{x},\hat{\alpha}) \quad (6.19)$$

Both x^* (business-as-usual policy) and $\hat{x}$ (most conservative policy) can be easily obtained (e.g., from Eqs. (6.18) and (6.19)), resulting in

$$x^* = 0.5 \quad (6.20)$$

$$\hat{x} = -2 \quad (6.21)$$

TABLE 6.1 Noninferior Solutions and Trade-Off Values for the Example Problem

x	$f_1(x,\hat{\alpha})$	$f_2(x,\hat{\alpha})$	λ_{12}
0	−4.00	16.00	0.13
−0.20	−3.52	12.96	0.19
−0.50	−2.50	9.00	0.33
−1.00	0	4.00	0.75
−1.50	3.50	1.00	2.00
−1.60	4.32	0.64	2.63
−1.75	5.63	0.25	4.50
−1.80	6.08	0.16	5.75
−1.90	7.02	0.04	12.00

To dramatize the trade-offs between the sensitivity objective function $f_2(\cdot)$ and the optimality objective function $f_1(\cdot)$, the latter is evaluated at x^* and $\hat{x}$ as a function of α. The resulting functions $f_1(x^*, \alpha)$ and $f_1(\hat{x}, \alpha)$ are plotted in Figure 6.5 as functions of α. These functions are given by Eqs. (6.22) and (6.23), respectively:

$$f_1(x^*, \alpha) = 0.5 - (\alpha - 1) - \alpha^2 \qquad (6.22)$$

$$f_1(\hat{x}, \alpha) = 8 + 4(\alpha - 1) - \alpha^2 \qquad (6.23)$$

Note that at the nominal value of α (i.e., $\hat{\alpha} = 2$), $f_1(x^*, \hat{\alpha})$ changes rapidly with a slope equal to -5, where at the same point ($\hat{\alpha} = 2$), $f_1(\hat{x}, \hat{\alpha})$ is stable with a slope equal to zero as given by Eqs. (6.24) and (6.25), respectively:

$$\left.\frac{\partial f_1(x^*, \alpha)}{\partial \alpha}\right|_{\alpha = \hat{\alpha}} = -5 \qquad (6.24)$$

$$\left.\frac{\partial f_1(\hat{x}, \alpha)}{\partial \alpha}\right|_{\alpha = \hat{\alpha}} = 0 \qquad (6.25)$$

Furthermore, Figure 6.6 depicts the changes that take place in $f_1(x^*, \alpha)$ and $f_1(\hat{x}, \alpha)$ when the nominal value $\hat{\alpha}$ is perturbed by $\Delta \alpha = -0.5$. The corresponding variations are given below:

$$f_1(x^*, \hat{\alpha}) = -4.5 \qquad (6.26)$$

$$f_1(x^*, \hat{\alpha} - 0.5) = -2.25 \qquad (6.27)$$

$$\left| f_1(x^*, \hat{\alpha}) - f_1(x^*, \hat{\alpha} - 0.5) \right| = 2.25 \qquad (6.28)$$

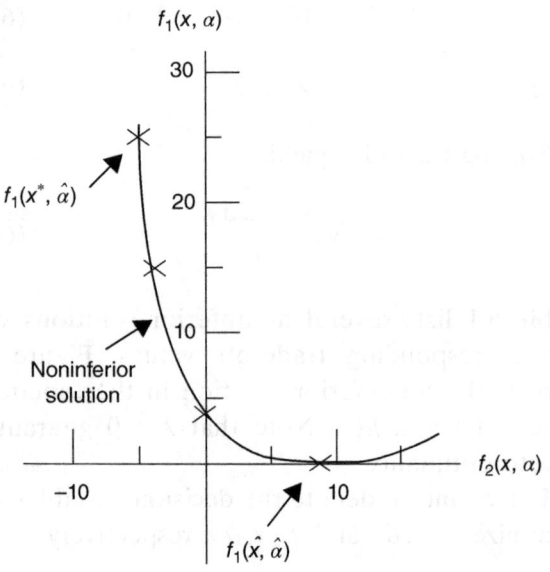

Figure 6.4 Noninferior solution in the functional space.

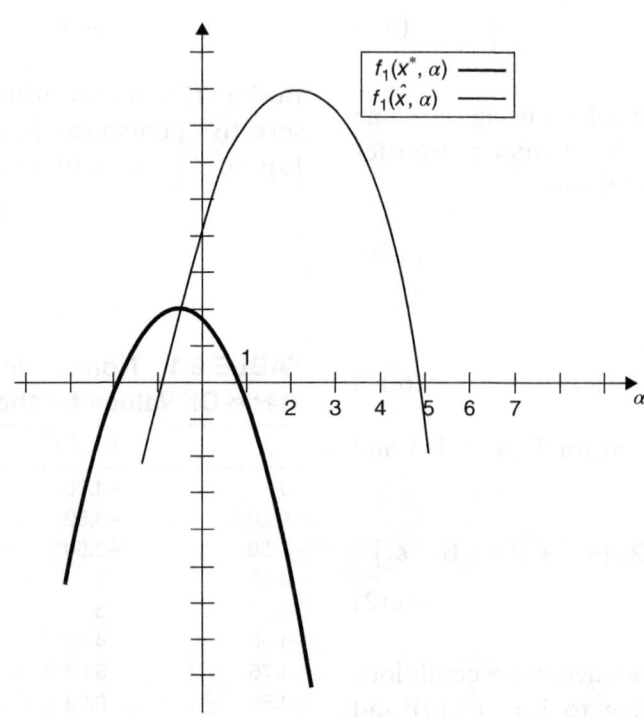

Figure 6.5 The functions $f_1(x^*, \alpha)$ and $f_1(\hat{x}, \alpha)$ versus α (in the neighborhood of $\hat{\alpha}$).

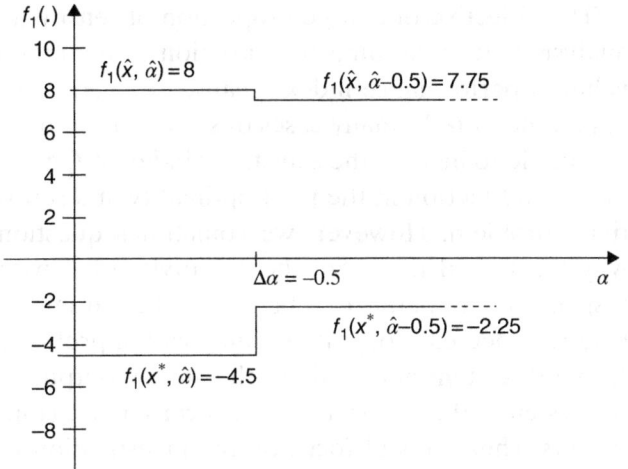

Figure 6.6 The functions $f_1(x^*, \alpha)$ and $f_1(\hat{x}, \alpha)$ versus perturbation in α.

Let $\eta(x^*, 0.75\hat{\alpha})$ denote the percentage of change in $f_1(x^*, \hat{\alpha})$ with a perturbation of 25% in $\hat{\alpha}$. Then

$$\eta(x^*, 0.75\hat{\alpha}) = 50\%$$

Similarly,

$$f_1(\hat{x}, \hat{\alpha}) = 8 \quad (6.29)$$

$$f_1(\hat{x}, \hat{\alpha} - 0.5) = 7.75 \quad (6.30)$$

$$|f_1(\hat{x}, \hat{\alpha}) - f_1(\hat{x}, \hat{\alpha} - 0.5)| = 0.25 \quad (6.31)$$

Let $\eta(\hat{x}, 0.75\hat{\alpha})$ denote the percentage of change in $f_1(\hat{x}, \hat{\alpha})$ with a perturbation of 25% in $\hat{\alpha}$. Then

$$\eta(\hat{x}, 0.75\hat{\alpha}) = 3\%$$

The results given in Figure 6.6 indicate that following a conservative policy that trades optimality (cost objective) for a less sensitive outcome provides a very stable solution (3 vs. 50% changes in $f_1(\cdot)$ with a deviation of 25% from the nominal value $\hat{\alpha}$). Clearly, neither the solution x^* nor $\hat{x}$ is likely to be recommended. From the use of Table 6.1 and the SWT method, with an interaction with a decision-maker the selection of a preferred level of x should evolve, where

$$\hat{x} \leq x \leq x^*$$

6.7 FORMULATION OF THE MULTIOBJECTIVE OPTIMIZATION PROBLEM

The general form of sensitivity indices can be presented as follows:

$$\psi_1(\mathbf{x}, \alpha), \ldots, \psi_J(\mathbf{x}, \alpha)$$

where $\mathbf{x}$ is a vector of decision (control) variables, α is a vector of model parameters, and $\psi_1(\cdot), \ldots, \psi_J(\cdot)$ are functions representing sensitivity, responsivity, and so on. Sections 6.7–6.9 are based on Li and Haimes [1988].

Consider the following classical formulation of an optimization problem:

$$\min_{\mathbf{x} \in X} f_1(\mathbf{x}, \alpha) \quad (6.32)$$

where X is the set of all feasible solutions and $g_k(\mathbf{x}, \alpha)$ are constraints. Specifically,

$$X = \{\mathbf{x} \mid g_k(\mathbf{x}, \alpha) \leq 0, k = 1, 2 \ldots, K\}$$

Problem (6.32) can be modified to include one or more of the earlier indices $\psi_j(\mathbf{x}, \alpha)$, $j = 1, 2, \ldots, J$, such as

$$\min_{\mathbf{x} \in X} \begin{bmatrix} f_1(\mathbf{x}, \alpha) \\ \Psi_1(\mathbf{x}, \alpha) \end{bmatrix} \quad (6.33)$$

Problem (6.33) is a multiobjective optimization problem. It is possible, of course, that the original problem itself is given in a multiobjective optimization form; and with the addition of sensitivity and other indices, the new problem may have the following form:

$$\min_{\mathbf{x} \in X}[f_1(\mathbf{x}, \alpha), \ldots, f_n(\mathbf{x}, \alpha), \Psi_1(\mathbf{x}, \alpha), \ldots, \Psi_J(\mathbf{x}, \alpha)] \quad (6.34)$$

It is assumed that all functions $f_i(\mathbf{x}, \alpha)$, $\psi_j(\mathbf{x}, \alpha)$, $g_k(\mathbf{x}, \alpha)$ are properly defined and continuous. The surrogate worth trade-off (SWT) method and its extensions, discussed in Chapter 5, can then be used to solve problems (6.32) and (6.34).

6.7.1 General Formulation of the USIM

Consider the following optimization problem:

$$\min f_1(\mathbf{x}, y; \hat{\alpha}) \quad (6.35)$$

subject to
$$y = h(\mathbf{x}; \hat{\alpha}) \quad (6.36)$$

where $\mathbf{x}$ denotes an n-dimensional vector of decision variables; α denotes a random systems parameter with an unknown probability distribution function; $\hat{\alpha}$ denotes the nominal value of α; $f_1(\mathbf{x}, y; \hat{\alpha})$ denotes the system's objective function, which may itself consist of multiple objectives; and $y \in R$ denotes the system's output, which is differentiable with respect to $\mathbf{x}$ and α. At this stage, we assume that the value of $\hat{\alpha}$ is available and α varies in the neighborhood of $\hat{\alpha}$.

The USIM represents the uncertainty associated with potential variations of the system's parameter by defining a sensitivity function of the system's output in the following way:

$$f_2(\mathbf{x}; \hat{\alpha}) = [\partial y(\mathbf{x}; \alpha)/\partial \alpha]^2 \big|_{\alpha = \hat{\alpha}} \quad (6.37)$$

Note that to reduce the system's sensitivity, the quadratic form given in Eq. (6.37) is the most common, since it is mathematically tractable. However, some other forms of sensitivity functions based on different considerations can also be chosen.

The overall joint optimality and sensitivity problem in which uncertainty is intrinsically considered in the decisionmaking process is now expressed by

$$\min \begin{bmatrix} f_1(\mathbf{x}, y; \hat{\alpha}) \\ f_2(\mathbf{x}, \hat{\alpha}) \end{bmatrix} \quad (6.38a)$$

subject to
$$y = h(\mathbf{x}; \hat{\alpha}) \quad (6.38b)$$

The best compromise solution of this multiobjective optimization problem is a policy that reflects the decisionmaker's preference in terms of how much reduction in the original objective function of the system should be traded off for a reduction in the system's sensitivity.

The best compromise solution sought here is one that allows the system to react weakly to parameter fluctuations. This solution is nonadaptive. After the solution of the joint optimality and sensitivity model is generated, based on the nominal value $\hat{\alpha}$, it is implemented in a real process whose parameter, α, varies in the neighborhood of $\hat{\alpha}$. The variation of the uncertainty parameter is unknown during the process.

The objective of a large proportion of sensitivity analysis is to determine the variation in minimum value of performance index f_1 caused by variations in parameter α. In many cases of sensitivity study, it is suitable to include the sensitivity index of f_1 as an objective function in the joint optimality and sensitivity problem. However, we sometimes question why we should minimize the sensitivity of f_1 with respect to parameter α if changes in the parameter lead to a decrease of f_1 in a minimization problem. In many system applications, the steady output of the system is the major stability concern of decisionmakers. Thus, we will focus on the investigation of the joint optimality and sensitivity problem that was posed in Eq. (6.38).

The principle of the USIM can be easily extended to the three classes of problems discussed in the following sections.

6.7.2 The USIM with Multiple Uncertain Parameters

Assume that in the problem presented in Eq. (6.38), there are n uncertain parameters that vary in the neighborhood of their nominal value $(\hat{\alpha}_1, \hat{\alpha}_2, \ldots, \hat{\alpha}_n)$. Using the Taylor-series expansion, we have

$$y(\mathbf{x}; \hat{\alpha}_1 + \Delta \hat{\alpha}_1, \ldots, \hat{\alpha}_n + \Delta \tilde{\alpha}) \cong y(\mathbf{x}; \hat{\alpha}_1, \ldots, \hat{\alpha}_n) + \sum_{i=1}^{n} \{[\partial y(x; \hat{\alpha}, \ldots, \hat{\alpha}_n / \partial \hat{\alpha}_i)] \Delta \hat{\alpha}_i\}$$
$$(6.39)$$

where $\Delta \hat{\alpha}_i$ is very small, $\forall i = 1, 2, \ldots, n$.

It follows that due to variations in the parameters, the variation of the system's output y is approximately equal to the second term of the right-hand side of Eq. (6.39). Using the Cauchy–Schwarz inequality, we have

$$\sum_{i=1}^{n} \left[\frac{\partial}{\partial \alpha_i} y(\mathbf{x}; \hat{\alpha}_1, \ldots, \hat{\alpha}_n) \Delta \hat{\alpha}_i \right]$$
$$\leq \left(\sum_{i=1}^{n} \left[\frac{\partial}{\partial \alpha_i} y(x; \hat{\alpha}_1, \ldots, \hat{\alpha}_n) \right]^2 \right)^{1/2} \left(\sum_{i=1}^{n} (\Delta \hat{\alpha}_i)^2 \right)^{1/2}$$
$$(6.40)$$

Thus, in order to reduce the variation of the system's output associated with variations in the parameters, we can choose a control policy **x** that makes $\sum_{i=1}^{n}[(\partial/\partial\alpha_i)y(\mathrm{x};\hat{\alpha}_1,....\hat{\alpha}_n)]^2$ attain its minimum. Based on this recognition, the sensitivity function f_2 is defined as follows:

$$f_2(x;\hat{\alpha}_1,...,\hat{\alpha}_n) = \sum_{i=1}^{n}\left[\frac{\partial}{\partial\alpha_i}y(x;\hat{\alpha}_1,...,\hat{\alpha}_n)\right]^2 \quad (6.41)$$

and one can deal with the joint optimality and sensitivity problem in the multiobjective framework given in Eq. (6.38), except that $\hat{\alpha}$ now is a vector.

Example 6.1 Consider a system that has the following output and objective function:

$$y(x;\alpha_1,\alpha_2) = (\alpha_1^2 + \alpha_2^2)x,$$
$$f_1(x,y;\alpha_1,\alpha_2) = x^2 - 2\alpha_1\alpha_2^2 x + \alpha_1^2\alpha_2$$

From Eq. (6.41), the system's sensitivity function is

$$f_2(x;\alpha_1;\alpha_2) = 4(\alpha_1^2 + \alpha_2^2)x^2$$

Assume that the nominal values of α_1 and α_2 are $\hat{\alpha}_1 = 1$ and $\hat{\alpha}_2 = 2$. This will yield the following joint optimality and sensitivity problem:

$$\min\begin{bmatrix} f_1 = x^2 - 8x + 2 \\ f_2 = 20x^2 \end{bmatrix}$$

Using the SWT method discussed in Chapter 5 and the ε-constraint method [Haimes et al., 1971] (or any other multiobjective generating method), we can determine that the set of noninferior solutions is $\{0 \le x \le 4\}$. Table 6.2 presents a sample of noninferior solutions and their associated trade-off values between the system's original objective function f_1 and the sensitivity index f_2. Table 6.2 also presents the values of variation of the system's output y when the nominal values $\hat{\alpha}_1$ and $\hat{\alpha}_2$ are perturbed by $\Delta\alpha_1 = 0.1$ and $\Delta\alpha_2 = 0.1$ [Li and Haimes, 1988].
Define:

- x^* as the optimal decision for the business-as-usual policy
- $\hat{x}$ as the optimal decision for the most conservative policy

TABLE 6.2 A Sample of Noninferior Solutions for Example 6.1 and Corresponding System Outcomes

x	0	1	2	3	4
f_1	2	−5	−10	−13	−14
f_2	0	20	80	180	320
λ_{12}	∞	3/20	1/20	1/60	0
y	0	5	10	15	20
Δy	0	0.62	1.24	1.86	2.48

$$\min f_1(x;\hat{\alpha}_1,\hat{\alpha}_2) = f_1(x^*;\hat{\alpha}_1,\hat{\alpha}_2)$$
$$\min f_2(x;\hat{\alpha}_1,\hat{\alpha}_2) = f_2(\hat{x};\hat{\alpha}_1,\hat{\alpha}_2)$$

By construction, the most conservative policy, $\hat{x}$, which corresponds to $x = 0$, provides a very stable solution, while the conventional business-as-usual solution, which corresponds to $x^* = 4$, suffers the highest deviation from its nominal value in comparison with the set of all noninferior solutions. Based on the preference of the decisionmaker, the best compromise solution may be selected from among the solution set $\{0 \le x \le 4\}$ by using the SWT or some other multiobjective optimization method.

Note that probability distributions derived from expert elicitation, such as the ones derived using the fractile method or the triangular distribution (see Chapter 4), can suffer from a number of errors [Bier, 2004; Taleb, 2007; Lin and Bier, 2008]. To address these potential errors, the USIM was extended to assess the sensitivity of model outcomes to expert-driven probabilistic model inputs [Barker, 2008; Barker and Haimes, 2008a]. Furthermore, the USIM was applied to the study of uncertainties in the analysis of interdependent infrastructure sectors [Barker and Haimes, 2008b], a summary of which is provided in Section 18.11.

6.7.3 Application of the USIM to Dynamic Systems

Consider the following primal control problem [Li and Haimes, 1988]:

$$\min J_1 = \int_0^T g(\mathbf{x},\mathbf{u},t;\alpha)dt \quad (6.42a)$$

subject to

$$dx(t;\alpha)/dt = f(x, u, t; \alpha) \quad (6.42b)$$

$$y(t;\alpha) = h(x, u, t; \alpha) \quad (6.42c)$$

where $x \in R^n$ is the state vector, $u \in R^m$ is the control vector, $y \in R^p$ is the output vector, α is the uncertain parameter, T is the final time, and t is time.

In order to consider the system's sensitivity along with its primal performance index, the state trajectory sensitivity vector is defined as follows:

$$\lambda(t) = \partial x(t;\alpha)/\partial \alpha \quad (6.43)$$

Differentiating $\lambda(t)$ with respect to t, we obtain

$$d\lambda(t)/dt = [\partial f(x, u, t; \alpha)/\partial x]\lambda(t) + \partial f(x, u, t; \alpha)/\partial \alpha \quad (6.44)$$

The system output sensitivity vector $\eta(t)$ is defined in a similar manner:

$$\eta(t) = \partial y(t;\alpha)/\partial \alpha = [\partial h(x, u, t; \alpha)/\partial x]\lambda(t) + \partial h(x, u, t; \alpha)/\partial \alpha \quad (6.45)$$

Equations (6.44) and (6.45) define the sensitivity model. For the assumed nominal value $\hat{\alpha}$, the nominal solution $x(t;\hat{\alpha})$ can be calculated by solving the problem given in Eq. (6.42). The variation of the output y due to the perturbation of the uncertain parameter α can be expressed approximately as follows:

$$\delta y(t;\hat{\alpha}) \cong \left[\frac{\partial y(t;\hat{\alpha})}{\partial \alpha}\right]\delta \alpha = \eta(t)\delta \alpha \quad (6.46)$$

In order to get a solution with low sensitivity, we introduce the system sensitivity index:

$$J_2 = \int_0^T \eta'(t)S(t)\eta(t)dt \quad (6.47)$$

where $S(t)$ is an assigned weighting matrix.

The joint optimality and sensitivity problem can now be posed as follows:

$$\min \begin{bmatrix} J_1 = \int_0^T g(x, u, t; \hat{\alpha})dt \\ J_2 = \int_0^T \eta'(t)S(t)\eta(t)dt \end{bmatrix} \quad (6.48a)$$

subject to the augmented system's state:

$$dx(t;\hat{\alpha})/dt = f(x, u, t; \hat{\alpha}) \quad x(0) \text{ is given} \quad (6.48b)$$

$$d\lambda(t)/dt = [\partial f(x, u, t; \hat{\alpha})/\partial x]\lambda(t) + \partial f(x, u, t; \hat{\alpha})/\partial \alpha \quad (6.48c)$$

$$\lambda(0) = \partial x(u, t; \hat{\alpha})/\partial \alpha \big|_{t=0}$$

and the augmented system's output:

$$y(t;\hat{\alpha}) = h(x, u, t; \hat{\alpha}) \quad (6.48d)$$

$$\eta(t) = [\partial h(x, u, t; \hat{\alpha})/\alpha x]\lambda(t) + \partial h(x, u, t; \hat{\alpha})/\partial \alpha \quad (6.48e)$$

The problem given in Eq. (6.48) can be solved either by the weighting method [Gass and Saaty, 1955; Zadeh, 1963] if the problem is convex or by the ε-constraint method [Haimes et al., 1971] as discussed in Chapter 5. The best compromise solution of this multiobjective optimization problem is a policy reflecting the decisionmaker's preference as to how much reduction in the optimality function he or she is willing to trade for a reduction in the system's sensitivity.

Note that the noninferior control u generated by the earlier joint optimality and sensitivity problem is an open-loop control. If we want to have feedback control $u = \psi(x;\hat{\alpha})$, the differential equation for the trajectory sensitivity function should be modified as follows:

$$d\lambda(t)/dt = [\partial f/\partial x + (\partial f/\partial u)(\partial \psi/\partial x)]\lambda(t) + \partial f/\partial \alpha \quad (6.49)$$

and the equation for the output sensitivity also needs to be modified as

$$\eta(t) = [\partial h/\partial x + (\partial h/\partial u)(\partial \psi/\partial x)]\lambda(t) + \partial h/\partial x \quad (6.50)$$

For a problem under uncertainty, $d\psi(x;\hat{\alpha})/dx = [\partial \psi(x;\hat{\alpha})/\partial x][\partial x/\partial \alpha]$. This does not include the term $\partial \psi/\partial \alpha$ since the calculation of $u = \psi(x;\hat{\alpha})$ is only based on the nominal value $\hat{\alpha}$.

6.7.4 Extension of the USIM to Problems with Equality Constraints

Assume that there exist some equality constraints on the decision vector **x** in the system's model given in Eq. (6.35). Thus, the primal problem becomes

$$\min f_1(\mathbf{x}, \mathbf{y}; \alpha, \beta) \tag{6.51}$$

subject to

$$\mathbf{y} = \mathbf{h}(\mathbf{x}; \alpha) \tag{6.52}$$

$$\phi(\mathbf{x}, \beta) = 0 \tag{6.53}$$

where $\mathbf{x} \in R^n$ is the decision vector, $\mathbf{y} \in R^p$ is the system's output vector, f_1 is the system's objective function, ϕ is an m-dimensional system constraint vector, and $\alpha \in R$ and $\beta \in R$ are two uncertain parameters [Li and Haimes, 1988].

We specify the constraint ϕ in Eq. (6.53) as the external constraint [Wierzbicki, 1984]. The external constraints represent the control goals of the system, such as some of its desired economic bounds. In contrast with the system's internal physical constraints, which must be satisfied all the time, the external constraints generally will not be satisfied, because of the differences between the system model and the real-world process, which is uncertain. Wierzbicki proposes that a penalty term for deviations from the control goal be introduced into the performance index. A multiobjective approach can help the decisionmaker to understand better the uncertainty system and provide trade-offs between reducing the impact of the system's uncertainty and degrading the system's performance index.

In order to minimize the level of constraint violation due to variations in the uncertain parameter β, we introduce the following pairs of indexes: the system's output sensitivity index:

$$f_2(\mathbf{x}; \alpha) = [\partial \mathbf{y}(\mathbf{x}; \alpha)/\partial x]'[\partial \mathbf{y}(\mathbf{x}; \alpha)/\partial \alpha] \tag{6.54a}$$

and the constraint sensitivity index:

$$f_3(\mathbf{x}; \beta) = [\partial \phi(\mathbf{x}; \beta)/\partial \beta]'[\partial \phi(\mathbf{x}; \beta)/\partial \beta] \tag{6.54b}$$

The overall joint optimality and sensitivity problem can now be stated as follows:

$$\min \begin{bmatrix} f_1(\mathbf{x}; \mathbf{y}; \hat{\alpha}, \hat{\beta}) \\ f_2(\mathbf{x}; \hat{\alpha}) \\ f_3(\mathbf{x}; \hat{\beta}) \end{bmatrix} \tag{6.55a}$$

subject to

$$\mathbf{y} = \mathbf{h}(\mathbf{x}; \hat{\alpha}), \tag{6.55b}$$

$$\phi(\mathbf{x}, \hat{\beta}) = 0 \tag{6.55c}$$

where $\hat{\alpha}$ and $\hat{\beta}$ are the nominal values of α and β, respectively.

We can think of f_3 as the degree of the feasibility of the decision vector **x**. Based on this consideration, the new objective function f_3 is given the highest priority within the problem expressed in Eq. (6.55).

Definition

$\mathbf{x}^*$ is an optimal solution to the problem in Eq. (6.55) if $\mathbf{x}^*$ is a noninferior solution of Eq. (6.55) and $\mathbf{x}^*$ minimizes f_3.

Note that if the minimization of f_3 leads to a unique solution $\mathbf{x}^*$, then $\mathbf{x}^*$ is the unique solution of Eq. (6.55). On the other hand, if the minimization of f_3 has multiple solutions, then at least one of them is a noninferior solution of Eq. (6.55).

According to this definition, the optimization of the problem posed by Eq. (6.55) can be decomposed into two steps:

Step 1: Solve the single-objective optimization problem:

$$\min f_3(\mathbf{x}; \hat{\beta}) \tag{6.56a}$$

subject to

$$\phi(\mathbf{x}, \hat{\beta}) = \mathbf{0} \tag{6.56b}$$

and obtain the solution set S. If there is only one element in S, then the optimization problem is completed; otherwise, go to step 2.

Step 2: Solve the multiobjective optimization problem:

$$\min \begin{bmatrix} f_1(\mathbf{x}; \mathbf{y}; \hat{\alpha}, \hat{\beta}) \\ f_2(\mathbf{x}; \hat{\alpha}) \end{bmatrix} \tag{6.57a}$$

subject to

$$\mathbf{y} = \mathbf{h}(\mathbf{x}; \hat{\alpha}) \tag{6.57b}$$

$$\mathbf{x} \in S. \tag{6.57c}$$

Example 6.2 Consider the following primal problem [Li and Haimes, 1988]:

$$\min f_1 = y$$
$$\text{subject to} \quad y = -x_1\alpha + x_2 + x_3^2$$
$$\phi(x, \beta) = x_1^2 e^\beta - x_2\beta + x_3 = 0$$

where both $\hat{\alpha}$ and $\hat{\beta}$ are assumed to be equal to 1.

From Eq. (6.54), the system's output sensitivity index and the constraint sensitivity index are, respectively,

$$f_2(x;\alpha) = [\partial y(x;\alpha)/\partial \alpha]^2 = x_1^2$$
$$f_3(x;\beta) = [\partial \phi(x,\beta)/\partial \beta]^2 = \left(x_1^2 e^\beta - x_2\right)^2$$

The overall joint optimality and sensitivity problem is

$$\min \begin{bmatrix} f_1 = -x_1 + x_2 + x_3^2 \\ f_2 = x_1^2 \\ f_3 = (x_1^2 e - x_2)^2 \end{bmatrix}$$
$$\text{subject to} \quad x_1^2 e - x_2 + x_3 = 0$$

Step 1:
$$\min \left(x_1^2 e - x_2\right)^2$$
$$\text{subject to} \quad x_1^2 e - x_2 + x_3 = 0$$

The solution set is expressed as

$$S = \{(x_1, x_2, x_3) | x_2 = x_1^2 e, x_3 = 0\}$$

Step 2:
$$\min \begin{bmatrix} f_1 = -x_1 + x_2 + x_3^2 \\ f_2 = x_1^2 \end{bmatrix}$$
$$\text{subject to} \quad x_2 = x_1^2 e; \quad x_3 = 0$$

Using the ε-constraint method, we can find the set of noninferior solutions and the trade-off, λ_{12}, between f_1 and f_2, where $\lambda_{12} = -\partial f_1/\partial f_2$:

$$x_1 \in [0, 1/(2e)]; \quad x_2 = ex_1^2; \quad x_3 = 0$$
$$\lambda_{12} = (1 - 2ex_1)/(2x_1)$$
$$f_1 = y = -x_1 + ex_1^2; f_2 = x_1^2; f_3 = 0$$

TABLE 6.3 A Sample of Noninferior Solutions for Example 6.2 and Corresponding System Outputs

x_1	0	1/(4e)	1/(2e)
x_2	0	1/(16e)	1/(4e)
x_3	0	0	0
f_1	0	−3/(16e)	−1/(4e)
f_2	0	1/(16e²)	1/(4e²)
f_3	0	0	0
Δy	0	−0.5/(40e)	−1/(40e)
$\Delta \phi$	0	0.00003	0.0001

Table 6.3 presents a sample of noninferior solutions along with the values of variations of the system's output and constraints when the nominal values of α and β are perturbed by $\Delta \alpha = 0.05$ and $\Delta \beta = 0.05$.

Note that solving the primal problem of Example 6.2 without consideration of the sensitivity indices yields $x_1 = 1/(2e)$, $x_2 = 1/(4e) - 1/2$, $x_3 = -1/2$, $f_1 = -1/(4e) - 1/4$, $\Delta y = -1/(40e)$, and $\Delta \phi = 0.025$. The results show that the extension of the USIM has a better performance than the conventional solution in both the system's output sensitivity index and in the constraint sensitivity index.

6.8 A ROBUST ALGORITHM OF THE USIM

It is important to keep in mind that the above results are meaningful in the neighborhood of the nominal value of α. This is, however, only a point property. Consider the situation depicted in Figure 6.7. When α is equal to the assumed nominal value $\hat{\alpha}$, the control $\hat{x}$, which represents the most conservative policy, yields the least sensitivity of the system to a variation of the uncertain parameter α. However, if the actual nominal value is $\tilde{\alpha}$ instead of $\hat{\alpha}$, we can see from Figure 6.7 that the control $\hat{x}$ is worse than $\tilde{x}$ in both senses of the system's original performance index f_1 and the system's sensitivity index [Li and Haimes, 1988]. The evident conclusion is that in many situations, point properties evaluated at the nominal value may be poor indicators for that property when a relatively small perturbation from the nominal value is introduced. Thus, for some problems under uncertainty, a spatially distributed index is preferable over a point index. There is a need to investigate the impact that the nominal value of α

has on the family of noninferior frontiers, which is shown in Figure 6.8.

In this section, we consider the case where the nominal value is itself an uncertain parameter. This is similar to a risk case where the random variable **x** follows a normal probability density function and the mean value of this probability density function is also a random variable with normal distribution.

The modified version of the problem in Eq. (6.38) is given as follows:

$$\min \begin{bmatrix} f_1(\mathbf{x}, \mathbf{y}; \hat{\alpha}_j) \\ f_2(x; \hat{\alpha}_j) \end{bmatrix} \quad (6.58a)$$

Subject to

$$\mathbf{y} = \mathbf{h}(\mathbf{x}; \hat{\alpha}_j) \quad (6.58b)$$

where $\hat{\alpha}_j$ is a random variable with an unknown probability distribution function and j takes values in the set $\{1, 2, ..., N\}$. The indicant parameter j serves to index the nominal parameter $\hat{\alpha}$. Therefore, the N modes of the joint optimality and sensitivity problem are characterized by the value of $j \in \{1, 2, ..., N\}$.

Denote by P_j the joint optimality and sensitivity multiobjective problem corresponding to $\hat{\alpha}_j$, and define the set of noninferior solutions of problem P_j as

$$X_j^* = \{\mathbf{x} | \mathbf{x} \text{ is a noninferior solution of problem } P_j\}$$

The same control **x** will yield different points in the functional space for different values of $\hat{\alpha}_j$. Note that the following situations are always realized for some **x** such that $\mathbf{x} \in X_i^*$ and $\mathbf{x} \notin X_k^*$, $i \ne k$, $k \in \{1, 2, ..., N\}$. For some control **x**, which is a noninferior solution for all problems $P_j(j = 1, 2, ..., N)$, it must belong to the following set:

$$X^* = \bigcap_{j=1}^{N} X_j^* \quad (6.59)$$

What follows is a robust algorithm capable of generating a best compromise solution for the joint optimality and sensitivity problem with N modes of nominal values of the uncertain parameter. We distinguish between two cases: Case 6.1, in which X^* is not an empty set, and Case 6.2, in which X^* is an empty set.

Case 6.1 X^* is not an empty set. In this case, our best compromise solution must belong to X^*, since only in this way can we guarantee that the solution we choose will be noninferior for any nominal value $\hat{\alpha}_j$, $j \in \{1, 2, ..., N\}$. The best compromise solution is selected from X^* according to the minimax criterion, as will be discussed later.

As we can see from Figure 6.9, each control **x**, which belongs to the set X^*, yields a curve $S_\mathbf{x}$ in the functional space. For each $S_\mathbf{x}$, $\forall \mathbf{x} \in X^*$ (in most cases, discretization of the decision space is necessary), the analyst interacts with the decisionmaker to determine the most unfavorable point, $f_\mathbf{x}^w$, on the curve $S_\mathbf{x}$. After the family of $f_\mathbf{x}^w$, $\forall \mathbf{x} \in X^*$, is obtained, the point that the decisionmaker most favors among this family will be selected. The corresponding control, which is denoted by $\tilde{\mathbf{x}}$, is the best compromise solution. Note the following:

1. By adopting this procedure, the best compromise solution is always noninferior for any nominal value $\hat{\alpha}_j$, $\forall j = 1, 2, ..., N$.

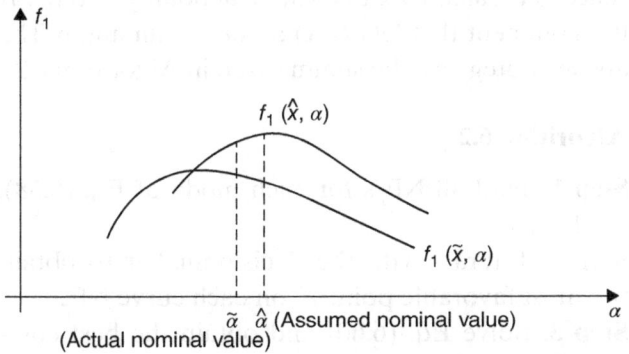

Figure 6.7 One possible solution when the actual nominal value is different from the assumed one.

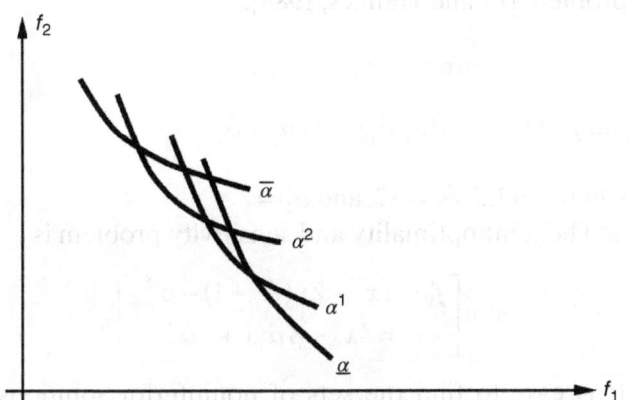

Figure 6.8 Family of efficient frontiers for different nominal values of α.

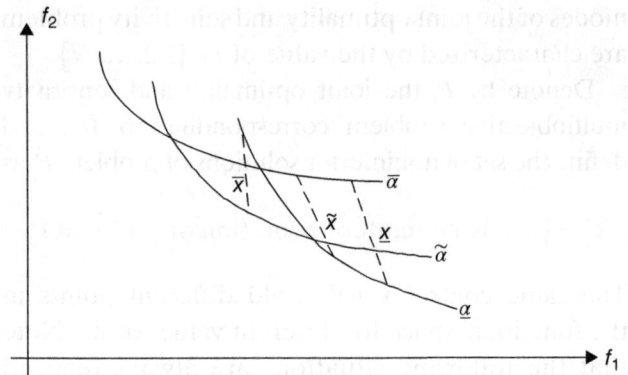

Figure 6.9 Trajectories of objective functions corresponding to different decision variables x. (– – –) Family of trajectories of objective functions f_1 and f_2 corresponding to different control policies x. (—) Family of efficient frontiers corresponding to different nominal values of parameter α.

2. The *minimax* criterion is used to select the best compromise $\tilde{x}$ solution from the set X^*. Because the decisionmaker lacks knowledge of what the actual $\hat{\alpha}_j$ is, he or she behaves in this scheme in a pessimistic way when considering a multiobjective optimization problem with the sensitivity index.

The above strategy can be summarized in Algorithm 6.1.

Algorithm 6.1

Step 1: Find all the sets X_j^*, $\forall j = 1, 2, \ldots, N$.

Step 2: Form the set $X^* = \bigcap_{j=1}^{N} X_j^*$.

Step 3: Interact with the decisionmaker to assess f_x^w on S_x for each $x \in X^*$ (in most cases, discretization of the decision space is necessary).

Step 4: Interact with the decisionmaker to identify $f_{\tilde{x}}^w$ —the most favorable point among the family of f_x^w. Select $\bar{f}$ as the best compromise control.

Case 6.2 X^* is an empty set. In this case, there does not exist any control x that is noninferior for all $\hat{\alpha}_j$'s. The problem consists of choosing (on the basis of a criterion such as Eq. (6.60)) a control x that approximates the ideal case.

For each $\hat{\alpha}_j$, the problem in Eq. (6.58) is solved and the noninferior frontier in the functional space, which is denoted by NF_j, is obtained. Using the SWT method or some other multiobjective method, the most favorable point on NF_j can be identified based on the preference of the decisionmaker. We call this point f_j^b. The values of f_j^b, $j = 1, 2, \ldots, N$, yield the ideal points that the decisionmaker favors the most for different modes of the joint optimality and sensitivity problem in Eq. (6.58).

Given f_j^b, $j = 1, 2, \ldots, N$, the ideal control $\bar{x}$, which approximates the ideal case, is the control that minimizes the following function:

$$D(\mathbf{x}) = \sum_{j=1}^{N} \left\| f_j(\mathbf{x}, \mathbf{y}; \hat{\alpha}_j) - f_f^b \right\|^2$$
$$= \sum_{j=1}^{N} \left\{ \left(f_1(\mathbf{x}, \mathbf{y}; \hat{\alpha}_j) - f_{j1}^b \right)^2 + \left(f_2(\mathbf{x}; \hat{\alpha}_j) - f_{j2}^b \right)^2 \right\}$$
(6.60)

where f_{jk}^b, $k = 1, 2$, are the first and second components of f_j^b, respectively.

This strategy has a very clear geometrical interpretation. The value of $D(\mathbf{x})$ is the summation of the square of the distances from the point in the functional space generated by $\mathbf{x}$ to each ideal point f_j^b. And $\bar{x}$ is the argument that lets $D(\mathbf{x})$ attain its minimum. The above strategy can be summarized in Algorithm 6.2.

Algorithm 6.2

Step 1: Find all NF_j's for each mode of Eq. (6.58), $j = 1, \ldots, N$.

Step 2: Interact with the decisionmaker to obtain the most favorable point f_j^b on each curve NF_j.

Step 3: Solve Eq. (6.60) and obtain the best compromise solution $\bar{x}$.

Example 6.3 Consider the following primal problem [Li and Haimes, 1988]:

$$\min f_1(x, y, \hat{\alpha}_j) = 2x^2 + 2x - y$$

subject to $\quad y(x; \hat{\alpha}_j) = 2x\hat{\alpha}_j + \hat{\alpha}_j^2$

where $j = 1, 2, \hat{\alpha}_1 = -2$, and $\hat{\alpha}_2 = 2$.

The joint optimality and sensitivity problem is

$$\min \begin{bmatrix} f_1 = 2x^2 - 2x(\hat{\alpha}_j - 1) - \hat{\alpha}_j^2 \\ f_2 = 4x^2 + 8\hat{\alpha}_j x + 4\hat{\alpha}_j^2 \end{bmatrix}$$

It is easy to find the sets of noninferior solutions for P_1 and P_2, which are expressed, respectively, as follows:

$$X_1^* = \{x | -3/2 \le x \le 2\}$$
$$X_2^* = \{x | -2 \le x \le 1/2\}$$

The intersection of x_1^* and x_2^* is

$$X^* = \{x | -3/2 \le x \le 1/2\}$$

Table 6.4 presents the hypothetical decisionmaking process for determining the best compromise solution $\bar{x}$, where $f_x(\alpha)$ represents the values of the two objective functions on the trajectory S_x.

Example 6.4 Consider the following primal problem:

$$\min f_1 = x_1^2 + (x_2 - \hat{\alpha}_j)^2$$

subject to $\quad y = \hat{\alpha}_j x_1 + (1/2) \hat{\alpha}_j^2 x_2$

where $j = 1, 2$; $\hat{\alpha}_1 = 1$, $\hat{\alpha}_2 = -1$.

The joint optimality and sensitivity problem is

$$\min \begin{bmatrix} f_1 = x_1^2 + (x_2 - \hat{\alpha}_j)^2 \\ f_2 = (x_1 + \hat{\alpha}_j x_2)^2 \end{bmatrix}$$

Using the weighting method to minimize $\theta f_1 + (1-\theta) f_2$, the set of noninferior solutions for $\hat{\alpha}_1$ and $\hat{\alpha}_2$ can be obtained as follows:

$$\text{For } \hat{\alpha} = 1: x_1 = \frac{-\theta}{1+\theta}, \quad x_2 = \frac{1}{1+\theta},$$
$$f_1 = \frac{2(1-\theta)^2}{(\theta-2)^2}, \quad f_2 = \frac{(1-\theta)^2}{(1+\theta)^2}$$

$$\text{For } \hat{\alpha} = 2: x_1 = \frac{1-\theta}{\theta-2}, \quad x_2 = \frac{1}{\theta-2},$$
$$f_1 = \frac{2(1-\theta)^2}{(\theta-2)^2}, \quad f_2 = \frac{\theta^2}{(\theta-2)^2}$$

where $0 \le \theta \le 1$ is a weighting coefficient.

TABLE 6.4 The Generation Process of the Best Compromise Solution for Example 6.3

x	$f_x(\hat{\alpha}_1)$	$f_x(\hat{\alpha}_2)$	f_x^w	$\bar{x}$
−1.5	(−8.5, 49)	(3.5, 1)	(−8.5, 49)	…
−1	(−8, 36)	(0, 4)	(−8, 36)	…
−0.5	(−6.5, 25)	(−2.5, 9)	(−6.5, 25)	…
0	(−4, 16)	(−4, 16)	(−4, 16)	0
0.5	(−0.5, 9)	(−4.5, 25)	(−4.5, 25)	…

It is easy to verify that X^* is an empty set. Assume that the decisionmaker's favorite solution for $\hat{\alpha}_1$ is $\{\theta = 0.5, \ x_1 = -1/3, \ x_2 = 2/3, \ f_{11}^b = 2/9, \ f_{12}^b = 1/9\}$ and the favorite noninferior solution for $\hat{\alpha}_2$ is $\{\theta = 0.5, \ x_1 = -1/3, \ x_2 = 2/3, \ f_{21}^b = 2/9, \ f_{22}^b = 1/9\}$. Then, according to Eq. (6.60), the ideal solution can be found by solving the following problem:

$$\min D(\mathbf{x}) = [f_1(\mathbf{x}, \mathbf{y}; \hat{\alpha}_1 - 2/9]^2 + [f_2(\mathbf{x}; \hat{\alpha}_1) - 1/9]^2$$
$$+ [f_1(\mathbf{x}, \mathbf{y}; \hat{\alpha}_2 - 2/9)]^2 + [f_2(\mathbf{x}; \hat{\alpha}_2) - 1/9]^2$$

The ideal solution $[\bar{x}_1, \bar{x}_2]$ is $[0,0]$.

6.9 INTEGRATION OF THE USIM WITH PARAMETER OPTIMIZATION AT THE DESIGN STAGE

So far, we have investigated the optimality and sensitivity of a system in the neighborhood of a nominal point of an uncertain parameter. We have also observed that the system's performance (in both the optimality and sensitivity aspects) is determined not only by the decision variable x but also by the assumed nominal value of the uncertain parameter [Li and Haimes, 1988].

In this section, we will consider a parameter optimization problem in the design stage. That is, we will learn how to select the value of a parameter such that the system can have satisfactory performance in both optimality and sensitivity (subject to an acceptable trade-off). We assume that a system's uncertain parameter is partially controllable. That is, once a nominal value has been assigned to the uncertain parameter, the parameter may take a value, at random, near its nominal value.

Consider the following design problem:

$$\min \begin{bmatrix} f_1(\mathbf{x}, \mathbf{y}; \hat{\boldsymbol{\alpha}}) \\ f_2(\mathbf{x}, \mathbf{y}; \hat{\boldsymbol{\alpha}}) \end{bmatrix} \quad (6.61a)$$

subject to

$$\mathbf{y} = \mathbf{h}(\mathbf{x}; \hat{\boldsymbol{\alpha}}) \quad (6.61b)$$

where $\mathbf{x} \in R^n$ is the control, $\mathbf{y} \in R^p$ is the output, $\hat{\boldsymbol{\alpha}} \in R^m$ is the nominal value of an uncertain vector of parameters to be optimally selected at the design stage, f_1 is the system's primal performance index, and f_2 is the system's sensitivity index.

If the nominal value $\hat{\alpha}$ is fixed, we can solve Eq. (6.61) and generate a noninferior frontier in the functional space. As we vary the value of $\hat{\alpha}$, we will generate a family of noninferior frontiers in the functional space. Under certain conditions, it can be proved that all noninferior solutions of Eq. (6.61) lie on the envelope of this family of Pareto-optimal solutions [Li and Haimes, 1987, 1988]. In other words, all the best possible solutions of the joint optimality and sensitivity problem can be estimated by using the envelope approach in the design stage.

Assume that for each given value of $\hat{\alpha}$, the noninferior frontier for Eq. (6.61) is expressed in a parametric form as follows:

$$f_1^* = f_1^*(\theta; \hat{\alpha}) \qquad (6.62a)$$

$$f_2^* = f_2^*(\theta; \hat{\alpha}) \qquad (6.62b)$$

where $\theta \in R$ is the parameter of the noninferior frontier. The parameter θ may be the weighting coefficient or the ε value used in the ε-constraint method.

The envelope of the family of curves given in Eq. (6.62) can be obtained by the following formulas [Li and Haimes, 1987, 1988]:

$$f_1^* = f_1^*(\theta; \hat{\alpha}) \qquad (6.63a)$$

$$f_2^* = f_2^*(\theta; \hat{\alpha}) \qquad (6.63b)$$

$$\frac{\partial f_2^*}{\partial \theta} \frac{\partial f_1^*}{\partial \hat{\alpha}} - \frac{\partial f_1^*}{\partial \theta} \frac{\partial f_2^*}{\partial \hat{\alpha}} = 0 \qquad (6.63c)$$

Once the envelope curve is generated, the analyst can interact with the decisionmaker to identify the most favored point on the envelope. The value of $\hat{\alpha}$ corresponding to this point is thus selected as the nominal value of the uncertain vector of parameters, α, in the design stage.

Example 6.5 Consider the following system:

$$\min f_1(x, y; \hat{\alpha}) = 2x^2 + 2x - y$$

subject to

$$y(x; \hat{\alpha}) = 2x\hat{\alpha} + \hat{\alpha}^2$$

The aim in the design stage is to select a nominal value $\hat{\alpha}$ between 1 and 3 that gives the system satisfactory properties in both optimality and sensitivity. The system's sensitivity index can be derived by™

$$f_2(x, y; \hat{\alpha}) = (\partial y / \partial \hat{\alpha})^2 = 4x^2 + 8\hat{\alpha}x + 4\hat{\alpha}^2$$

Thus, the joint optimality and sensitivity design problem is given as follows:

$$\min \begin{bmatrix} f_1 = 2x^2 + 2x - y \\ f_2 = 4x^2 + 8\hat{\alpha}x + 4\hat{\alpha}^2 \end{bmatrix}$$

subject to $y = 2x\hat{\alpha} + \hat{\alpha}^2$. For each given value of $\hat{\alpha}$, the above problem can be solved by the ε-constraint method, and the set of noninferior solutions can be expressed as follows:

$$x = \sqrt{\varepsilon}/2 - \hat{\alpha}, \quad -\hat{\alpha} \le x \le (\hat{\alpha} - 1)/2$$
$$\lambda_{12} = (\hat{\alpha} - 1 - 2x)/[4(x + \hat{\alpha})]$$
$$f_1 = \varepsilon/2 + (1 - 3\hat{\alpha})\sqrt{\varepsilon} + 3\hat{\alpha}^2 - 2\hat{\alpha}$$
$$f_2 = \varepsilon$$

where $\varepsilon \ge 0$ is the value of the second ε-constraint objective and λ_{12} is the trade-off value between the first and second objectives.

We generate a family of curves $\{f_1 = f_1(\varepsilon; \hat{\alpha}), f_2 = f_2(\varepsilon; \hat{\alpha})\}$ for different values of $\hat{\alpha}$. The envelope of this family can be calculated by using Eq. (6.63):

$$f_1 = \varepsilon/2 + (1 - 3\hat{\alpha})\sqrt{\varepsilon} + 3\hat{\alpha}^2 - 2\hat{\alpha}$$
$$f_2 = \varepsilon$$
$$\frac{\partial f_2^*}{\partial \varepsilon} \frac{\partial f_1^*}{\partial \varepsilon} - \frac{\partial f_1^*}{\partial \varepsilon} \frac{\partial f_2^*}{\partial \hat{\alpha}} = -3\sqrt{\varepsilon} + 6\hat{\alpha} - 2 = 0$$

It can be shown that after some mathematical manipulation and simplification, the following relationships hold on the envelope of the noninferior frontiers:

$$\hat{\alpha} = \sqrt{\varepsilon}/2 + 1/3$$
$$x = -1/3$$
$$f_1 = -(1/4)\varepsilon - 1/3$$
$$f_2 = \varepsilon$$

Figure 6.10 depicts the envelope associated with this example problem, and Table 6.5 lists several points on the envelope. Assume the decisionmaker's most preferred point on the envelope is

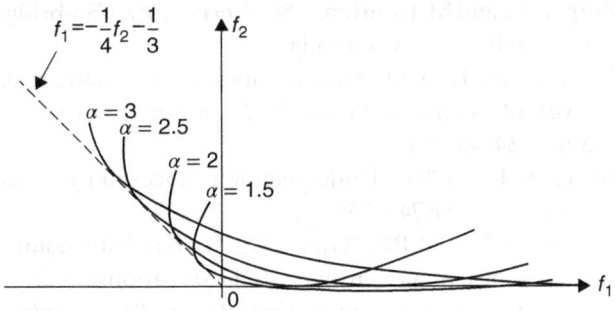

Figure 6.10 The envelope of the family of efficient frontiers.

TABLE 6.5 A Sample of Points on the Envelope and Their Corresponding $\hat{\alpha}$

$\hat{\alpha}$	f_1	f_2
1.33	−1.33	4
1.83	−2.58	9
2.33	−4.33	16
2.83	−6.58	25

$[f_1 = -4.33, f_2 = 16]$. Therefore, the nominal value of 2.33 will be chosen.

6.10 CONCLUSIONS

This chapter introduced an uncertainty taxonomy and established common analytic characteristics for the joint optimality and sensitivity analysis of decisionmaking problems under uncertainty. The consideration of the system's sensitivity in a multiobjective framework possesses several advantages. It can (i) help the analyst and the decisionmaker to understand better the problem under study, (ii) handle the optimality and sensitivity systematically and simultaneously, and (iii) display the trade-offs between reducing the system's uncertainty and degrading the original system's performance index. For additional deployment of the USIM and its extensions, the reader is referred to Section 18.11.

It is important to note that in the joint optimality and sensitivity analysis, we have made use of a first-order approximation. Neglecting higher-order terms makes the results formally correct only for small distributions.

REFERENCES

Apostolakis, G., 1999, The distinction between aleatory and epistemic uncertainties is important: an example from the inclusion of aging effects into probabilistic safety assessment. *Proceedings of PSA'99*, August 22–25, 1999, American Nuclear Society, Washington, DC.

Barker, K., 2008, *Extensions of Inoperability Input-Output Modeling for Preparedness Decisionmaking: Uncertainty and Inventory*, Ph.D. dissertation, University of Virginia, Charlottesville, VA.

Barker, K. and Y.Y. Haimes, 2008a, *Uncertainty Analysis of Interdependencies in Dynamic Infrastructure Recovery: Applications in Risk-Based Decisionmaking*, Technical Report 20-08, Center for Risk Management of Engineering Systems, University of Virginia, Charlottesville, VA.

Barker, K. and Y.Y. Haimes, 2008b, *Assessing Uncertainty in Extreme Events: Applications to Risk-Based Decisionmaking in Interdependent Infrastructure Sectors*, Technical Report 21-08, Center for Risk Management of Engineering Systems, University of Virginia, Charlottesville, VA.

Bier, V.M., 2004, Implications of the research on expert overconfidence and dependence, *Reliability Engineering and System Safety*, **85**(1–3): 321–329.

Committee on Public Engineering Policy (COPEP), 1972, *Perspectives on Benefit–Risk Decision Making*, National Academy of Engineering, Washington, DC.

Finkel, A., 1990, *Confronting Uncertainty in Risk Management: A Guide for Decision-Makers*, Resources for the Future, Center for Risk Management, Washington, DC.

Gass, S. and T. Saaty, 1955, The computational algorithm for the parametric objective function, *Naval Research Logistics Quarterly* **2**: 39–45.

Haimes, Y.Y. and W.A. Hall, 1977, Sensitivity, responsivity, stability, and irreversibility as multiobjectives in civil systems, *Advances in Water Resources* **1**: 71–81.

Haimes, Y.Y., L.S. Lasdon, and D.A. Wismer, 1971, On the bicriterion formulation of the integrated system identification and systems optimization, *IEEE Transactions on Systems, Man, and Cybernetics* **1**: 296–297.

Haimes, Y.Y., W.A. Hall, and H.T. Freedman, 1975, *Multiobjective Optimization in Water Resource Systems: The Surrogate Worth Trade-off Method*, Elsevier, Amsterdam.

Haimes, Y.Y., T. Barry, and J.H. Lambert (Eds.), 1994, When and how can you specify probability distribution when you don't know much? Editorial and workshop proceedings, *Risk Analysis* **14**(4): 661–706.

Hirshleifer, J. and J. Riley, 1992, *The Analytics of Uncertainty and Information*, Cambridge University Press, Cambridge.

Hoffman, F.O. and J.S. Hammonds, 1994, Propagation of uncertainty in risk assessments: the need to distinguish between uncertainty due to lack of knowledge and uncertainty due to variability, *Risk Analysis* **14**(5), 707–712.

Hogg, R.V. and E.A. Tanis, 1988, *Probability and Statistical Inference*, Third edition, Macmillan, New York.

Howard, R., 1988, Decision analysis: practice and promise, *Management Science* **34**(6): 693–697.

International Atomic Energy Agency (IAEA), 1989, Evaluating the reliability of predictions made using environmental transfer models, *Safety Practice Publications of the IAEA*, IAEA Safety Series No. 100, pp. 1–106 (STI/PUB/835, IAEA, Vienna, Austria).

Kuhn, H.W. and A.W. Tucker, 1950, Nonlinear programming, *Proceedings of the 2nd Berkeley Symposium on Mathematical Statistics and Probability*, University of California Press, Berkeley, CA, pp. 481–492.

Li, D. and Y.Y. Haimes, 1987, The envelope approach for multiobjective optimization problems, *IEEE Transactions on Systems, Man, and Cybernetics* **SMC-17**: 1026–1038.

Li, D. and Y.Y. Haimes, 1988, The uncertainty sensitivity index method (USIM) and its extension, *Naval Research Logistics* **35**: 655–672.

Lin, S.W. and V.M. Bier, 2008, A study of expert overconfidence. *Reliability Engineering and System Safety* **93**(5): 711–721.

Ling, C.W., 1993, *Characterizing Uncertainty: A Taxonomy and an Analysis of Extreme Events*, M.S. thesis, Systems Engineering Department, University of Virginia, Charlottesville, VA.

Lowrance, W.W., 1976, *Of Acceptable Risk*, William Kaufmann, Los Altos, CA.

Morgan, G. and M. Henrion, 1990, *Uncertainty*, Cambridge University Press, Cambridge.

Paté-Cornell, E., 1996, Uncertainties in risk analysis: six levels of treatment, *Reliability Engineering and System Safety* **54**: 95–111.

Rowe, W.D., 1994, Understanding uncertainty, *Risk Analysis* **14**(5): 743–750.

Spedden, S.E. and P.B. Ryan, 1992, Probabilistic connotations of carcinogen hazard classifications: analysis of survey data for anchoring effects, *Risk Analysis* **12**(4): 535–542.

Starr, C., 1969, Social benefit versus technological risk, *Science* **166**: 1232.

Starr, C., 1972, *Benefit–Cost Studies in Socio-Technical Systems*, Report of the Colloquium by the Committee on Public Engineering Policy, National Academy of Engineering, Washington, DC, pp. 17–24.

Taleb, N.N., 2007, *The Black Swan: The Impact of the Highly Improbable*, Random House, New York.

Taylor, A.C., 1993, Using objective and subjective information to develop distributions for probabilistic exposure assessment, *Journal of Exposure Analysis and Environmental Epidemiology* **3**: 285–298.

Teigen, K.H., 1988, The language of uncertainty, *Acta Psychology* **68**: 27–38.

Tversky, A. and D. Kahneman, 1974, Judgment under uncertainty: heuristics and biases, *Science* **85**: 1124–1131.

Wierzbicki, A., 1984, *Models and Sensitivity of Control Systems*, Elsevier, Amsterdam.

Zadeh, L.A., 1963, Optimality and nonscalar-values performance criteria, *IEEE Transactions on Automatic Control*, **8**(1): 59–60.

Zadeh, L.A., 1984, Making computers think like people, *IEEE Spectrum*, **21**(8): 26–32.

7

Risk Filtering, Ranking, and Management*

7.1 INTRODUCTION

The need for ranking risks arises in a variety of situations. The following are a few examples where risk ranking is not only desirable but essential: thousands of military and civilian sites have been contaminated with toxic substances; myriad sources of risk are commonly identified during the development of software-intensive engineering systems; and each year, thousands of the space shuttle's mechanical and electronic components are placed on a critical item list (CIL) to identify items that contribute significantly to program risk. The common element in such risk identification procedures is the need to establish priorities among a large number of individual contributions to the overall system risk. A dependable and efficient ranking of identified risk elements can be a step toward systemic risk reduction.

Infrastructure operation and protection highlights the challenges to risk filtering, ranking, and management (RFRM) in large-scale systems. Our man-made engineered infrastructures are becoming increasingly vulnerable to natural and willful hazards; these systems include telecommunications, electric power, gas and oil, transportation, water treatment plants, water distribution networks, dams, and levees (see Chapter 17). Fundamentally, such systems have a large number of components and subsystems. Most water distribution systems, for example, fall within a framework of large-scale systems, where a hierarchy of institutional and organizational decisionmaking structures (e.g., federal, state, county, and city) is often involved in their management. Coupling exists among the subsystems (e.g., the overall budget constraint is one factor), and this further complicates their management. A better understanding of the interrelationship among natural, willful, and accidental hazards is a logical step in improving the protection of critical national infrastructures. Such efforts should build on the experience gained over the years from the recovery and survival of infrastructures assailed by natural and human hazards. Furthermore, it is imperative to model critical infrastructures as dynamic systems in which current decisions have impacts on future consequences and options.

---

*This chapter is based on Haimes et al. [2002].

Risk Modeling, Assessment, and Management, Fourth Edition. Yacov Y. Haimes.
© 2016 John Wiley & Sons, Inc. Published 2016 by John Wiley & Sons, Inc.

In total risk management, identifying what can go wrong and the associated consequences and likelihoods (risk assessment) helps generate mitigation options with their trade-offs and impacts on future decisions. Ranking critical elements contributes to the analysis of options by forcing a seemingly intractable decision problem to focus on the most important contributors to the risk.

This chapter presents a methodological framework to identify, prioritize, assess, and manage scenarios of risk to a large-scale system from multiple overlapping perspectives. After reviewing earlier efforts in risk filtering and ranking, we describe the guiding principles and the eight phases of the RFRM methodology. This is followed by several examples, including applying the framework to a mission in support of an operation other than war (OOTW).

7.2 PAST EFFORTS IN RISK FILTERING AND RANKING

The RFRM methodology is a modified and much-improved version of risk ranking and filtering (RRF), which was developed a decade ago for NASA for the space shuttle [CRMES, 1991; Haimes et al., 1992]. It was introduced and discussed in Section 4.7 of the first edition of this book. In RRF, the risk prioritization task considers both multiple quantitative factors (such as reliability estimates) and qualitative factors (such as expert rankings of component criticality). Measurement theory was used in the development of RRF; this can ensure that engineering judgments represent both preferences and available information.

The key aspects of the RRF method are (i) a hierarchy of five major contributors to program risk, which constitute the criteria of the ranking; (ii) a quantification of program risk by measurable attributes; (iii) a graphical risk *fingerprint* to distinguish among critical items; (iv) a telescoping filter approach to reducing the CIL to the most critical number of sources of risk, often referred to as the top n; and (v) a weighted-score method, adapted from the analytic hierarchy process (AHP) [Saaty, 1988], augmenting the criteria hierarchy and risk fingerprint to support interactive prioritization of the top n. Eliciting engineering judgment is minimal until the list has been reduced to the top n, at which point the AHP, hierarchy, and fingerprint comprise a decision-support environment for the ultimate prioritization.

Within a program risk hierarchy of the RFF, the following four elements (criteria) of program risk are considered: (1) prior risk information, (2) moderate-event risk, (3) extreme-event risk, and (4) fault tolerance. A fifth element (criterion)—risk reduction potential—may also be considered.

Several scholars have addressed in the literature ranking of attributes. Sokal [1974] discusses classification principles and procedures that create a distinction between two methods: monothetic and polythetic. The *monothetic* category establishes classes that differ by at least one property that is uniform among members of each class, whereas the *polythetic* classification groups individuals or objects that share a large number of traits but do not agree necessarily on any one trait. Webler et al. [1995] outline a risk ranking methodology through an extensive survey example dealing with an application of sewage sludge on a New Jersey farmland. Working with expert and lay community groups, two perceptions of risk are developed and categorized, and weights are used to balance the concerns of the two groups. They demonstrate how discussion-oriented approaches to risk ranking can supplement current methodological approaches, and they present a taxonomy that addresses the substantive need for public discussion about risk.

Morgan et al. [1999, 2000] propose a ranking methodology designed for use by federal risk management agencies, calling for interagency task forces to define and categorize the risks. The task forces would identify the criteria that all agencies should use in their evaluations. The ranking would be done by four groups: (1) federal risk managers drawn from inside and outside the concerned agency, (2) laypeople selected somewhat randomly, (3) a group of state risk managers, and (4) a group of local risk managers. Each ranking group would follow two different procedures: (1) a reductionist–analytic approach and (2) a holistic–impressionistic approach. The results would then be combined to refine a better ranking, and the four groups would meet together to discuss their findings. In a most recent contribution in this area, *Categorizing Risks for Risk Ranking*, Morgan et al. [2000] discuss the problems inherent in grouping a large number of risk scenarios into easily managed categories and argue that such risk categories must be evaluated with respect to a set of criteria. This is particularly important when hard choices must be made in comparing and ranking thousands of specific risks. The ultimate risk characterization should be logically consistent, administratively compatible, equitable, and compatible with cognitive constraints and biases. Baron et al. [2000] conducted several extensive surveys of experts and nonexperts in risk analysis to ascertain their priorities as to personal and government action for risk reduction, taking into account the severity of the risk, the number of people affected, worry, and probabilities for hazards to self and others. A major finding of these surveys is that *concern for action, both personal and government, is strongly related to worry. Worry, in turn, is affected mainly by beliefs about probability.*

7.3 RFRM: A METHODOLOGICAL FRAMEWORK

7.3.1 Guiding Principles

It is constructive to identify again the two basic structural components of hierarchical holographic modeling (HHM). First are the *head topics*, which constitute the major visions, concepts, and perspectives of success. Second are the *subtopics*, which provide a more detailed classification of requirements for the success scenarios or sources of risk for the risk scenarios. Each such requirement class corresponds to a class of risk scenarios, namely, those that have an impact on that requirement. In this sense, each requirement is also considered a *source of risk*.

Thus, by its nature and construction, the HHM methodology generates a comprehensive set of sources of risk, that is, categories of risk scenarios, commonly in the order of hundreds of entries. Consequently, there is a need to discriminate among these sources as to the likelihood and severity of their consequences and to do so systematically on the basis of principled criteria and sound premises. For this purpose, the proposed framework for risk filtering and ranking is based upon the following major considerations:

- It is often impractical (e.g., due to time and resource constraints) to apply quantitative risk analysis to hundreds of sources of risk. In such cases, qualitative risk analysis may be adequate for decision purposes under certain conditions.
- All sources of evidence should be harnessed in the filtering and ranking process to assess the significance of the risk sources. Such evidence items include professional experience, expert knowledge, statistical data, and common sense.
- Six basic questions characterize the process of risk assessment and management (see Chapter 1) and serve as the compass for the RFRM approach. For the risk assessment process, there are three questions [Kaplan and Garrick, 1981]: What can go wrong? What is the likelihood of that happening? What are the consequences? There are also three questions for the risk management process [Haimes, 1991]: What can be done and what are the available options? What are the associated trade-offs in terms of costs, benefits, and risks? What are the impacts of current decisions on future options?

To deploy the RFRM methodology effectively, we must consider the variety of sources of risks, including those representing hardware, software, organizational, and human failures. Risks that also must be addressed include programmatic risks (such as project cost overrun and time delay in meeting completion schedules) and technical risks (such as not meeting performance criteria).

An integration of empirical and conceptual, descriptive and normative, quantitative and qualitative methods and approaches is always superior to the *either-or* choice. For example, relying on a mix of simulation *and* analytically based risk methodologies is superior to either one alone. The trade-offs that are inherent in the risk management process manifest themselves in the RFRM methodology as well. The multiple noncommensurate and often conflicting objectives that characterize most real systems guide the entire process of risk filtering and ranking.

The risk filtering and ranking process is aimed at providing priorities in the scenario analysis. This does not imply ignoring the sources of risks that have been filtered out earlier; it just means exploring the more urgent sources of risks or scenarios first.

7.3.2 RFRM Phases

Eight major phases constitute the RFRM method:

Phase I: Scenario identification. An HHM is developed to describe the system's *as-planned* or *success* scenario.

Phase II: Scenario filtering. The risk scenarios identified in Phase I are filtered according to the responsibilities and interests of the current system user.

Phase III: Bicriteria filtering and ranking. The remaining risk scenarios are further filtered using qualitative likelihoods and consequences.

Phase IV: Multicriteria evaluation. Eleven criteria are developed that relate the ability of a risk scenario to defeat the defenses of the system.

Phase V: Quantitative ranking. Filtering and ranking of scenarios continue based on quantitative and qualitative matrix scales of likelihood and consequence.

Phase VI: Risk management. Identifying risk management options for dealing with the filtered scenarios and estimating the cost, performance benefits, and risk reduction of each.

Phase VII: Safeguarding against missing critical items. Evaluating the performance of the options selected in Phase VI against the scenarios previously filtered out during Phases II–V.

Phase VIII: Operational feedback. Using the experience and information gained during application to refine the scenario filtering and decision processes of earlier phases.

These eight phases reflect a philosophical approach rather than a mechanical methodology. In this philosophy, the filtering and ranking of discrete scenarios are viewed as a precursor to, rather than a substitute for, considering all risk scenarios.

7.3.2.1 Phase I: Identifying risk scenarios through HHM

Most, if not all, sources of risk are identified through the HHM methodology, as discussed earlier. In their totality, these sources of risk describe *what can go wrong* in the *as-planned* or success scenario. Included are acts of terrorism, accidents, and natural hazards. Therefore, each subtopic represents a category of risk scenarios, that is, descriptions of what can go wrong. Thus, through the HHM, we generate a diagram that organizes and displays the complete set of system success criteria from multiple overlapping perspectives. Each box in the diagram represents a set of sources of risk or requirements for the successful operation of the system. Note that the head topics and the subtopics in the HHM may be viewed in two different, albeit complementary, ways: (1) as sources of risk scenarios and (2) as requirements for success scenarios. At the same time, any failure will show up as a deficiency in one or more of the boxes. To demonstrate the applications of the RFRM to a real-world problem, we revisit here the HHM developed in support for OOTW. Figure 7.1 is an excerpt from the HHM introduced in Section 3.11. It is important to note the trade-off inherent in the construction of the HHM: A more detailed HHM yields a more accurate picture of the success scenario and consequently leads to a better assessment of the risk situation. In other words, having more levels in the hierarchy describes the system structure in greater detail and facilitates identifying the various failure modes. A less detailed HHM, however, encapsulates a larger number of possible failure scenarios within each subtopic. This leads to less specificity in identifying failure scenarios. Of course, a more detailed HHM is more expensive to construct in terms of time and resources. Therefore, as in all modeling efforts, there is a trade-off: detail and accuracy versus time and resources. Consequently, the appropriate level of detail for an HHM is a matter of judgment dependent upon the resources available for risk management and the nature of the situation to which it is applied.

7.3.2.2 Phase II: Scenario filtering based on scope, temporal domain, and level of decisionmaking

In Phase II, filtering is done at the level of *subtopics* or *sources of risk*. As mentioned earlier, the plethora of sources of risk identified in Phase I can be overwhelming. The number of subtopics in the HHM may easily be in the hundreds (see Chapter 3).

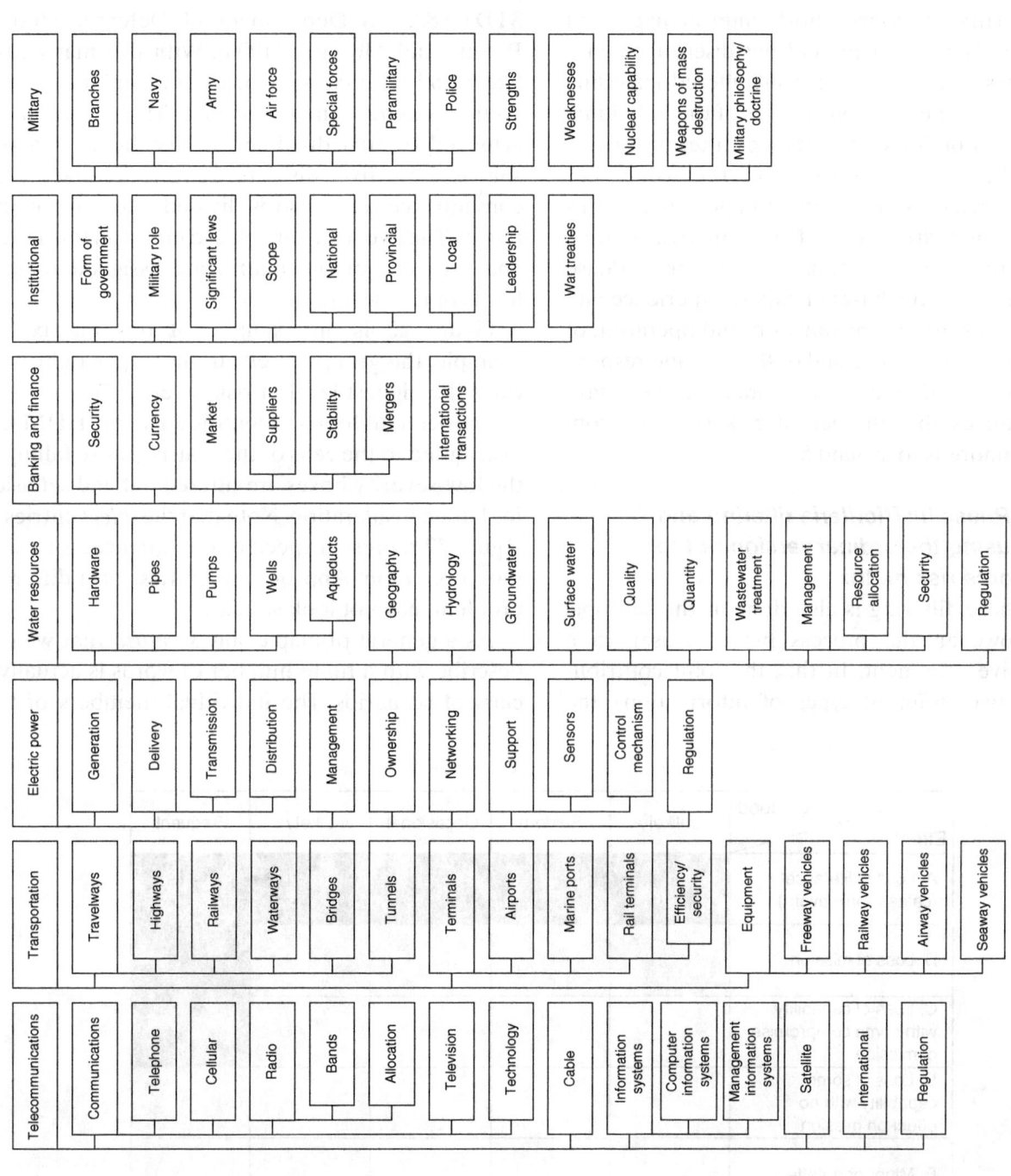

Figure 7.1 Excerpt from a hierarchical holographic model developed to identify sources of risk to operations other than war [Dombroski et al., 2002].

Clearly, not all of these subtopics can be of immediate concern to all levels of decisionmaking and at all times. For example, in OOTW, one may consider at least three decisionmaking levels (strategic, planning, and operational) and several temporal domains (first 48 hours; short, intermediate, and long-term disengagement; and postdisengagement). At this phase, the sources of risk are filtered according to the interests and responsibilities of the individual risk manager or decisionmaker. The filtering criteria include the decisionmaking level, the scope (i.e., what risk scenarios are of prime importance to this manager), and the temporal domain (which time periods are important). Thus, the filtering in Phase II is achieved on the bases of expert experience and knowledge of the nature, function, and operation of the system being studied and of the role and responsibility of the individual decisionmaker. This phase often reduces the number of risk sources from several hundreds to around 50.

7.3.2.3 Phase III: Bicriteria filtering and ranking using the ordinal version of the US Air Force risk matrix

In this phase, filtering is also done at the subtopic level. However, the process moves closer to a quantitative treatment. In this, the joint contributions of two different types of information—the likelihood of what can go wrong and the associated consequences—are estimated on the basis of the available evidence. This phase is accomplished in the RFRM by using the ordinal version of the matrix procedure adapted from Military Standard (MIL-STD) 882, US Department of Defense, cited in Roland and Moriarty [1990]. With this matrix, the likelihoods and consequences are combined into a joint concept called *severity*. The mapping is achieved by first dividing the likelihood of a risk source into five discrete ranges. Similarly, the consequence scale also is divided into four or five ranges. The two scales are placed in matrix formation, and the cells of the matrix are assigned relative levels of risk severity.

Figure 7.2 is an example of this matrix, for example, the group of cells in the upper right indicates the highest level of risk severity. The scenario categories (subtopics) identified by the HHM are distributed to the cells of the matrix. Those falling in the low-severity boxes are filtered out and set aside for later consideration. Note that the *effect* entries in Figure 7.2 represent specific consequences for a military operation. Appropriate entries for a different problem may not look similar.

As a general principle, any *scenario* that we can describe with a finite number of words is actually a class of scenarios. The individual members of this

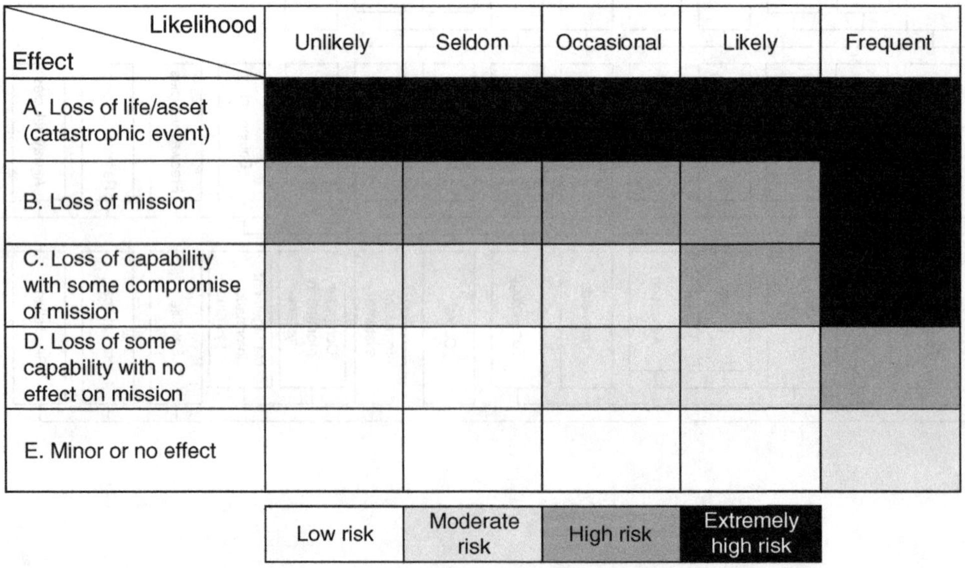

Effect \ Likelihood	Unlikely	Seldom	Occasional	Likely	Frequent
A. Loss of life/asset (catastrophic event)					
B. Loss of mission					
C. Loss of capability with some compromise of mission					
D. Loss of some capability with no effect on mission					
E. Minor or no effect					

Low risk | Moderate risk | High risk | Extremely high risk

Figure 7.2 *Example risk matrix for Phase III.*

class are subscenarios of the original scenario. Similarly, any subtopic from the HHM diagram placed into the matrix represents a class of failure scenarios. Each member of the class has its own combination of likelihood and consequence. There may be failure scenarios that are of low probability and high consequence and scenarios that are of high probability and low consequence. In placing the subtopic into the matrix, the analyst must judge the likelihood and consequence range that characterizes the subtopic as a whole. This judgment must avoid overlooking potentially critical failure scenarios and also avoid overstating the likelihood of such scenarios.

7.3.2.4 Phase IV: Multicriteria evaluation

In Phase III, we distributed the individual risk sources, by judgment, into the boxes defined in Figure 7.2 by the consequence and likelihood categories. Those sources falling in the upper right boxes of the risk matrix were then judged to be those requiring priority attention.

In Phase IV, we take the process one step further by reflecting on the ability of each scenario to defeat three defensive properties of the underlying system: *resilience, robustness,* and *redundancy.* Classifying the defenses of the system as resilience, robustness, and redundancy (3 Rs) is based, in part, on an earlier and related categorization of water resources systems by Matalas and Fiering [1977], updated by Haimes et al. [1997]. *Redundancy* refers to the ability of extra components of a system to assume the functions of failed components. *Robustness* refers to the insensitivity of system performance to external stresses. *Resilience* is the ability of a system to recover following an emergency. Scenarios able to defeat these properties are of greater concern and thus are scored as more severe. As an aid to this reflection, we present a set of 11 *criteria* defined in Table 7.1. (These criteria are intended to be generally applicable, but of course, the user may modify them to suit the specific system under study.)

As a further aid to this reflection, it may be helpful to rate the scenario of interest as *high, medium,* or *low* against each criterion (using Table 7.2 for guidance) and then to use this combination of ratings to judge the ability of the scenario to defeat the system.

The criteria of risk scenarios related to the three major defensive properties of most systems are presented in Table 7.1. These (example) criteria are intended to be used as a base for Phase V.

After the completion of Phase IV, Phase V ranks the remaining scenarios with quantitative assessments of likelihood and consequence. Scenarios that are judged to be less urgent (based on Phase IV) can be returned to for later study.

TABLE 7.1 Eleven Criteria Relating the Ability of a Risk Scenario to Defeat the Defenses of the System

Undetectability refers to the absence of modes by which the initial events of a scenario can be discovered before harm occurs
Uncontrollability refers to the absence of control modes that make it possible to take action or make an adjustment to prevent harm
Multiple paths to failure indicates that there are multiple and possibly unknown ways for the events of a scenario to harm the system, such as circumventing safety devices
Irreversibility indicates a scenario in which the adverse condition cannot be returned to the initial, operational (pre-event) condition
Duration of effects indicates a scenario that would have a long duration of adverse consequences
Cascading effects indicates a scenario where the effects of an adverse condition readily propagate to other systems or subsystems, that is, cannot be contained
Operating environment indicates a scenario that results from external stressors
Wear and tear indicates a scenario that results from use, leading to degraded performance
Hardware, software, human, and organizational (HW/SW/HU/OR) interfaces indicates a scenario in which the adverse outcome is magnified by interfaces among diverse subsystems (e.g., human and hardware)
Complexity/emergent behaviors indicates a scenario in which there is a potential for system-level behaviors that are not anticipated even with knowledge of the components and the laws of their interactions
Design immaturity indicates a scenario in which the adverse consequences are related to the newness of the system design or other lack of a proven concept

TABLE 7.2 Rating Risk Scenarios in Phase IV against the 11 Criteria

Criterion	High	Medium	Low	Not Applicable
Undetectability	Unknown or undetectable	Late detection	Early detection	Not applicable
Uncontrollability	Unknown or uncontrollable	Imperfect control	Easily controlled	Not applicable
Multiple paths to failure	Unknown or many paths to failure	Few paths to failure	Single path to failure	Not applicable
Irreversibility	Unknown or no reversibility	Partial reversibility	Reversible	Not applicable
Duration of effects	Unknown or long duration	Medium duration	Short duration	Not applicable
Cascading effects	Unknown or many cascading effects	Few cascading effects	No cascading effects	Not applicable
Operating environment	Unknown sensitivity or very sensitive to operating environment	Sensitive to operating environment	Not sensitive to operating environment	Not applicable
Wear and tear	Unknown or much wear and tear	Some wear and tear	No wear and tear	Not applicable
Hardware/software/ human/organizational	Unknown sensitivity or very sensitive to interfaces	Sensitive to interfaces	No sensitivity to interfaces	Not applicable
Complexity and emergent behaviors	Unknown or high degree of complexity	Medium complexity	Low complexity	Not applicable
Design immaturity	Unknown or highly immature design	Immature design	Mature design	Not applicable

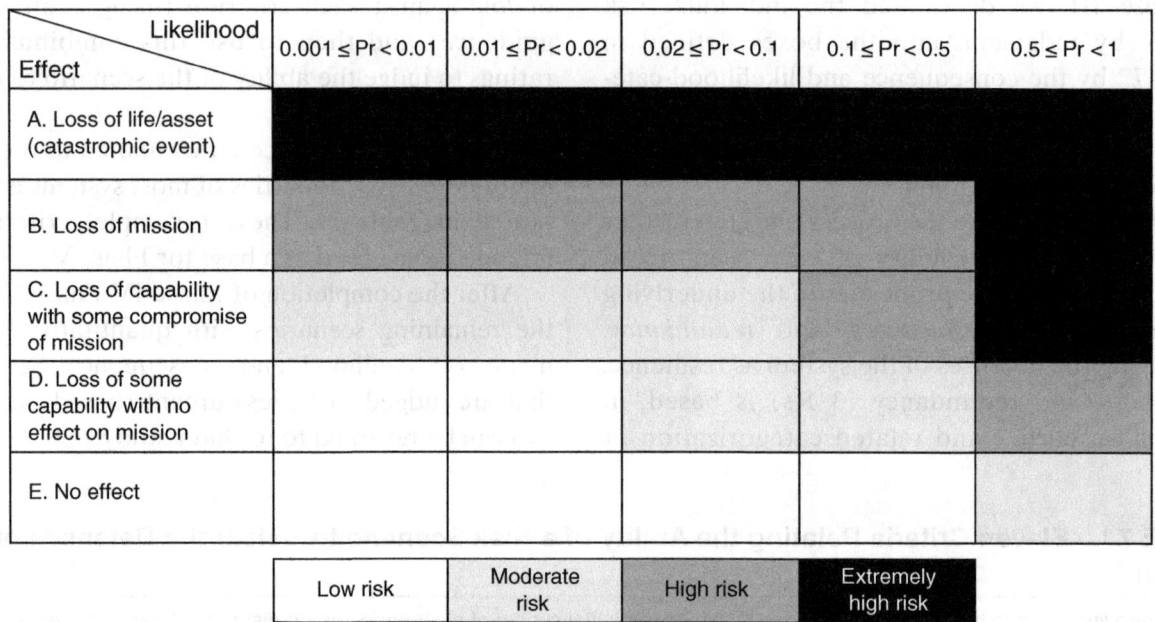

Figure 7.3 Risk matrix with numerical values for use in Phase V.

7.3.2.5 Phase V: Quantitative ranking using the cardinal version of the MIL-STD 882 risk matrix

In Phase V, we quantify the likelihood of each scenario using Bayes' theorem and all the relevant evidence available. The quantification of likelihood should, of course, be based on the totality of relevant evidence available and should be done by processing the evidence items through Bayes' theorem. The value of quantification is that it clarifies the results, disciplines the thought process, and replaces opinion with evidence. See Chapter 12 for more on the use of Bayes' theorem.

Calculating the likelihoods of scenarios avoids possible miscommunication when interpreting verbal expressions such as *high*, *low*, and *very high*. This approach yields a matrix with ranges of probability on the horizontal axis, as shown in Figure 7.3. This is the *cardinal* version of the *ordinal* risk matrix first deployed in Phase III. Filtering and ranking the risk scenarios through this matrix typically reduce the number of scenarios from about 20 to about 10.

7.3.2.6 Phase VI: Risk management

Having quantified the likelihood of the scenarios in Phase V and having filtered the scenarios by likelihood and consequence in the manner of Figure 7.3, we have now identified a number of scenarios, presumably small, constituting most of the risk for our subject system. (Note that the *effect* and *likelihood* entries in Figure 7.3 represent specific sets of consequences and likelihood for a military operation. Appropriate entries for a different problem may not look similar.) We therefore now turn our attention to risk management and ask, *What can be done and what options are available?* and *What are the associated trade-offs in terms of costs, benefits, and risks?* The first of these questions puts us into a creative mode. Knowing the system and the major risk scenarios, we create options for actions, asking, *What design modifications or operational changes could we make that would reduce the risk from these scenarios?* Having set forth these options, we then shift back to an analytical and quantitative thought mode: *How much would it cost to implement (one or more of) these options? How much would we reduce the risk from the identified scenarios? Would these options create new risk scenarios?*

Moving back and forth between these modes of thought, we arrive at a set of acceptable options (in terms of the associated trade-offs) that we now would like to recommend for implementation. However, we must remember that we have evaluated these options against the filtered set of scenarios remaining at the end of Phase V. Thus, in Phase VII, we look at the effect these options might have on the risk scenarios previously filtered out.

7.3.2.7 Phase VII: Safeguarding against missing critical items

Reducing the initial risk scenarios to a much smaller number at the completion of Phase V may inadvertently filter out scenarios that originally seemed minor but could become important if the proposed options were actually implemented. Also, in a dynamic world, early indicators of newly emerging critical threats and other sources of risk should not be overlooked. Following the completion of Phase VI, which generates and selects risk management policy options and their associated trade-offs, we ask the question, *How robust is the policy selection and risk filtering and ranking process?* Phase VII, then, is aimed at providing added assurance that the proposed RFRM methodology creates flexible reaction plans if indicators signal the emergence of new or heretofore undetected critical items. In particular, in Phase VII of the analysis, we:

1. Ascertain the extent to which the risk management options developed in Phase VI affect or are affected by any of the risk scenarios discarded in Phases II–V. That is, in the light of the interdependencies within the success scenario, we evaluate the proposed management policy options against the risk scenarios previously filtered out.
2. From what was learned in Step 1, make appropriate revisions to the risk management options developed in Phase VI.

Thus, in Phase VII, we refine the risk management options in the light of previously screened-out scenarios.

The detailed deployment of Phase VII is mostly driven by the specific characteristics of the system. The main guiding principle in this phase focuses on cascading effects due to the system's intra- and interdependencies that may have been overlooked during the filtering processes in Phases I–V. The defensive properties that are addressed in Phase IV may be revisited as well to ensure that the system's redundancy, resilience, and robustness remain secure by the end of Phase VII.

7.3.2.8 Phase VIII: Operational feedback

As with other methodologies, the RFRM can be improved on the basis of the feedback accumulated during its deployment. The following are guiding principles for the feedback data collection process:

- The HHM is never considered finished; new sources of risk should be added as additional categories or new topics.
- Be cognizant of all benefits, costs, and risks to human health and the environment.

Remember that no single methodology or tool can fit all cases and circumstances. However, the viability and effectiveness of the risk filtering and ranking methodology can be maintained by a systematic data collection process that is cognizant of the dynamic nature of the evolving sources of risk and their criticalities.

7.4 CASE STUDY: AN OOTW

To demonstrate the RFRM methodology, we use a case study of OOTW [Dombroski et al., 2002]. This was conducted with the National Ground Intelligence Center, US Department of Defense, and the US Military Academy at West Point and focused on the US and allied operations in the Balkans in the late 1990s. The overall aim of the study was to ensure that the deployment of US forces abroad for an OOTW would be effective and successful, with minimal casualties, losses, or surprises.

This case study focuses on the following mission: US and allied forces deployed in the Balkans are asked to establish and maintain security for 72 hours at a bridge crossing the Tirana River in Bosnia. The purpose is to support the exchange, using the bridge, of humanitarian medical and other supplies among several nongovernmental organizations and public agencies. These entities and the allied forces must communicate in part over public telecommunications networks and the Internet regarding the security status of the bridge. The public also will need to be informed about the status of the bridge using radio, television, and the Internet. RFRM will be used to identify, filter, and rank scenarios of risk for the mission.

7.4.1 Phase I: Developing the HHM

To identify risk scenarios that allied forces might encounter, the following four HHMs were developed:

1. Country HHM
2. US HHM
3. Alliance HHM
4. Coordination HHM

To limit the size of the example, our demonstration focuses only on the *Telecommunications* head topic of the Country HHM (see Figure 7.1).

From the *Telecommunications* head topic, we choose the 11 subtopics (risk scenarios) for input to the Phase II filtering, as follows: telephone, cellular, radio, television, technology, cable, computer information systems (CIS), management information systems (MIS), satellite, international, and regulation.

7.4.2 Phase II: Scenario Filtering by Domain of Interest

In Phase II, we filter out all scenarios except those in the decisionmaker's domain of interest and responsibilities. In OOTW, one may identify three levels of decisionmakers: *strategic* (e.g., chiefs of staff), *operational* (e.g., generals and colonels), and *tactical* (e.g., captains and majors). The concerns and interest relevant to a specific subset of the risk scenarios will depend on the decisionmaking level and on the temporal domain under consideration. At the strategic level, generals may not be concerned with the specific location of a company's base and the risks associated with it, while the company's commander would be. For this example, we assume that the risk scenarios *technology* and *regulation* were filtered out based on the decisionmaker's responsibilities. The following surviving set of nine risk scenarios becomes the input to Phase III: telephone, cellular, radio, television, cable, CIS, MIS, satellite, and international.

7.4.3 Phase III: Bicriteria Filtering

To further reduce the number of risk scenarios, in Phase III, we subject the remaining nine subtopics (risk scenarios) to the qualitative severity scale matrix as shown in Figure 7.4. We have assumed that evidence for the evaluations shown in Figure 7.4 came from reliable intelligence sources providing knowledge about the telecommunications infrastructure in Bosnia. Also, for the purpose of this example, we further assume that the decisionmaker's analysis of the subtopics (risk scenarios) results in removing the risk scenarios that received a moderate- or low-risk valuation from the subtopic set. Based on the decisionmaker's preferences, the subtopics *radio*, *television*, and *MIS*, which attained a moderate valuation, are removed. The remaining set of six risk scenarios follows: (i) telephone, (ii) cellular, (iii) cable, (iv) CIS, (v) satellite, and (vi) international.

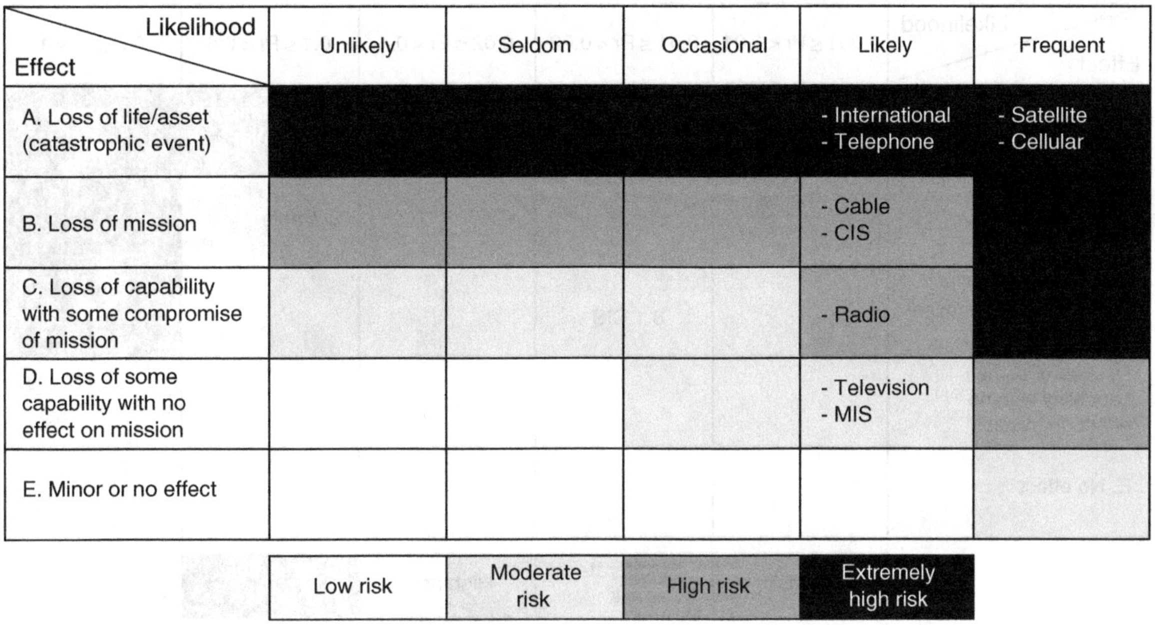

Figure 7.4 Qualitative severity scale matrix.

Effect \ Likelihood	Unlikely	Seldom	Occasional	Likely	Frequent
A. Loss of life/asset (catastrophic event)				- International - Telephone	- Satellite - Cellular
B. Loss of mission				- Cable - CIS	
C. Loss of capability with some compromise of mission				- Radio	
D. Loss of some capability with no effect on mission				- Television - MIS	
E. Minor or no effect					

Low risk | Moderate risk | High risk | Extremely high risk

TABLE 7.3 Risk Scenarios for Six Remaining Subtopics

Subtopic	Risk Scenario
Telephone	Failure of any portion of the telephone network for more than 48 hours
Cellular	Failure of any portion of the cellular network for more than 24 hours
Cable	Failure of any portion of the coaxial and/or fiber-optic cable networks for more than 12 hours
CIS	Loss of access to the Internet throughout the entire country for more than 48 hours
Satellite	Failure of the satellite network for more than 12 hours throughout the region
International	Failure of international communications network for more than 6 hours

TABLE 7.4 Scoring of Subtopics for OOTW Using the Criteria Hierarchy

Criteria	Telephone	Cellular	Cable	CIS	Satellite	International
Undetectability	Low	Low	Med	High	Low	High
Uncontrollability	Med	Med	High	High	Med	High
Multiple paths to failure	High	Med	High	High	Med	High
Irreversibility	Med	High	Med	High	High	Low
Duration of effects	High	High	High	High	High	High
Cascading effects	Med	Med	Low	Low	High	High
Operating environment	High	High	High	High	Med	High
Wear and tear	Med	High	Low	High	Med	High
Hardware/software/human/organizational	High	High	Med	High	High	High
Complexity and emergent behaviors	Med	High	Low	High	High	High
Design immaturity	Med	High	Med	High	High	Med

7.4.4 Phase IV: Multicriteria Filtering

Now that the risk scenarios have been narrowed down to a more manageable set, the decisionmaker can perform a more thorough analysis on each subtopic. Table 7.3 lists the remaining six subtopics (risk scenarios) and gives each a more specific definition. In Phase IV, the decisionmaker assesses each of these remaining subtopics in terms of the 11 criteria identified in Table 7.1. Table 7.4 summarizes these assessments. As part of our example, we assume that

222 RISK FILTERING, RANKING, AND MANAGEMENT

Effect \ Likelihood	$0.001 \leq Pr < 0.01$	$0.01 \leq Pr < 0.02$	$0.02 \leq Pr < 0.1$	$0.1 \leq Pr < 0.5$	$0.5 \leq Pr < 1$
A. Loss of life/asset (catastrophic event)			1.1 Telephone	5. International 1.2 Cellular	4. Satellite
B. Loss of mission				2. Cable	
C. Loss of capability with some compromise of mission		3.1 CIS			
D. Loss of some capability with no effect on mission					
E. No effect					

Low risk | Moderate risk | High risk | Extremely high risk

Figure 7.5 Quantitative severity scale matrix.

these assessments result from analyzing each of the subtopics (risk scenarios) against the criteria, using intelligence data and expert analysis.

7.4.5 Phase V: Quantitative Ranking

Thus far, the important scenario list has been reduced from 11 to 6. Employing the quantitative severity scale matrix and the criteria assessments in Phase IV, the decisionmaker will now reduce the set further. In Phase V, the same severity scale index introduced in Phase III is used, except that the likelihood is now expressed quantitatively as shown in Figure 7.5.

7.4.5.1 Telephone

Likelihood of failure = 0.05; effect = A (loss of life); risk = extremely high.

This failure will cause loss of life and incapacitate the mission. Based on intelligence reports, however, enemy forces operating in Bosnia do not appear to be preparing for an attack against the telephone network. Therefore, we assign only 5% probability to this scenario. Should such an attack occur, a failure would be detectable.

The Bayesian reasoning behind this assignment is as follows: Let A denote an enemy attack against the phone network. Let E denote the relevant evidence—that the intelligence reports no preparations for an attack.

By Bayes' theorem, then

$$\Pr(A|E) = \Pr(A)\Pr(E|A)/\Pr(E)$$

$$\Pr(E) = \Pr(E|A)\Pr(A) + \Pr(E|\text{not } A)\Pr(\text{not } A)$$

Our prior state of knowledge about A, before receiving the evidence, is $P_0(A) = 0.5 = P(\text{not}A)$.

The probability of intelligence seeing evidence E, that is, no preparations for an enemy attack, is small. We take it as $P(E|A) = 0.05$. (This is our appraisal of the effectiveness of our intelligence.)

The probability of intelligence not seeing preparations given that the enemy is not going to attack is high $P\{E|\text{not}A) = 0.99$. (This expresses our confidence that the enemy would not make preparations as a deceptive maneuver.)

Therefore,

$$\Pr(E) = (0.05)(0.5) + (0.99)(0.5) = 0.025 + 0.495 = 0.52$$

$$\Pr(A|E) = (0.5)(0.05)/(0.52) = 0.05$$

7.4.5.2 Cellular

Likelihood of failure = 0.45; effect = A (loss of life); risk = extremely high.

US forces will be dependent on cellular communications; thus, this failure could cause loss of mission and loss of life. Intelligence reports and expert analysis show that insurgent forces may be preparing for an attack on the cellular network, knowing that coalition forces are utilizing it. Therefore, we assign a 45% likelihood that the risk scenario will occur during the operation as assessed by this intelligence. Analysis also shows that an attack's effects will be difficult to reverse.

CIS: Likelihood of failure = 0.015; effect = C (loss of some capability with compromise of some mission objectives); risk = moderate.

US forces would not be immediately dependent upon the CIS network, so this may cause some loss of capability but should not cause the mission to fail. Detailed analysis of the CIS network shows that if an attack occurs against the existing Bosnian network, its effects may be severe with a low likelihood (about 0.015).

7.4.5.3 Cable

Likelihood of failure = 0.3; effect = B (loss of mission); risk = high.

US forces utilize existing fiber-optic and coaxial cable networks to communicate over the region. Intelligence of insurgent and enemy activity shows that forces are preparing for an attack on the cable network due to its vulnerability across the country. However, the network is not a primary communications platform. Therefore, we assign a likelihood of 0.3 for this risk scenario, given the current security over the network.

7.4.5.4 Satellite

Likelihood of failure = 0.55; effect = A (loss of life); risk = extremely high.

Because US forces are strongly dependent on satellite communications, any loss for 12 hours or more can result in a loss of life and mission. An intelligence analysis of the satellite network shows that it is protected throughout Bosnia, but not enough to ensure that forces opposing the operation will fail when attacking it. Due to the criticality of the network, enemy forces will likely target the network. Based on this assessment, the likelihood of the failure scenario occurring is high (0.55).

7.4.5.5 International

Likelihood of failure = 0.15; effect = A (loss of life); risk = extremely high.

Here, we assume that any loss of international communications for 6 hours or longer throughout the region would cut off US forces from their strategic decisionmakers and from other countries. Therefore, this is a very high-risk failure. Due to expert analysis of forces opposing the operation, an attack against international communications would be difficult but fairly likely. Therefore, we assign the likelihood of 0.15 to this scenario. If it did occur, however, its effects might be somewhat reversible within 6 hours.

Assuming that we filter out all subtopics (risk scenarios) attaining a risk valuation of moderate or low risk, CIS is filtered out. Therefore, based on the assessments shown previously and in Figure 7.5, planners of the operation would surely want to concentrate resources and personnel on protecting the remaining five critical risk scenarios—cellular, cable, satellite, telephone, and international communications networks.

7.4.6 Phase VI: Risk Management

In Phase VI, a complete quantitative decision analysis is performed, involving estimates of cost, performance benefits, and risk reduction and of management options for dealing with the most urgent remaining scenarios.

Examples for Phases VI–VIII are beyond the scope of the risk filtering and ranking aspects of this chapter. Further information on the deployment of these phases may be found in Dombroski [2001], Lamm [2001], and Mahoney [2001].

7.4.7 Phase VII: Safeguarding against Missing Critical Items

In Phase VII, we examine the performance of the options selected in Phase VI against the scenarios that have been filtered out during Phases II–V.

7.4.8 Phase VIII: Operational Feedback

Phase VIII represents the operational phase of the underlying system, during which the experience and information gained are used to continually update the scenario filtering and decision processes, Phases II–VII.

7.5 SUMMARY

Most safety critical systems, including military OOTW, require serious analysis. Risk analysts must identify all conceivable sources of risk, impose priorities, and take appropriate actions to minimize these risks. The RFRM methodological framework presented here addresses this process. The eight phases of the methodology reflect a philosophical approach rather than a mechanical process. The philosophy can be specialized to particular contexts, for example, OOTW, an aerospace system, contamination of drinking water, or the physical security of an embassy. In this philosophy, filtering and ranking discrete classes of scenarios are viewed as a precursor to, rather than a substitute for, analysis of the totality of all risk scenarios. The RFRM has been used in the following studies: Leung et al. [2004] applied the RFRM to prioritize transportation assets for protection against terrorist events. The RFRM was combined with the balanced scorecard [Kaplan and Norton, 1992, 1996], a strategy management approach, for the identification and prioritization of the US Army's critical assets [Anderson et al., 2008].

REFERENCES

Anderson, C.W., K. Barker, and Y.Y. Haimes, 2008, Assessing and prioritizing critical assets for the United States Army with a modified RFRM methodology, *Journal of Homeland Security and Emergency Management* **5**(1): 1–24. Article 5.

Baron, J., J.C. Hershey, and H. Kunreuther, 2000, Determinants of priority for risk reduction: the role of worry, *Risk Analysis* **20**(4): 413–427.

CRMES, 1991, *Ranking of Space Shuttle FMEA/CIL Items: The Risk Ranking and Filtering (RRF) Method*, Center for Risk Management of Engineering Systems, University of Virginia, Charlottesville, VA.

Dombroski, M., 2001, A Risk-Based Decision Support Methodology for Operations Other Than War, M.S. thesis, Department of Systems and Information Engineering, University of Virginia, Charlottesville, VA.

Dombroski, M., Y.Y. Haimes, J.H. Lambert, K. Schlussel, and M. Sulcoski, 2002, Risk-based methodology for support of operations other than war, *Military Operations Research* **7**(1): 19–38.

Haimes, Y.Y., 1991, Total risk management, *Risk Analysis* **11**(2): 169–171.

Haimes, Y.Y., J. Lambert, D. Li, J. Pet-Edwards, V. Tulsiani, and D. Tynes, 1992, Risk Ranking, and Filtering Method, Technical Report, Center for Risk Management of Engineering Systems, University of Virginia, Charlottesville, VA.

Haimes, Y.Y., N.C. Matalas, J.H. Lambert, B.A. Jackson, and J.F.R. Fellows, 1997, Reducing the vulnerability of water supply systems to attack, *Journal of Infrastructure Systems* **4**(4): 164–177.

Haimes, Y.Y., S. Kaplan, and J.H. Lambert, 2002, Risk filtering, ranking, and management framework using hierarchical holographic modeling, *Risk Analysis* **22**(2): 383–397.

Kaplan, S., and B.J. Garrick, 1981, On the quantitative definition of risk, *Risk Analysis* **1**(1): 11–27.

Kaplan, R.S., and D.P. Norton, 1992, The balanced scorecard—measures that drive performance, Harvard Business Review **January/February**: 71–79.

Kaplan, R.S., and D.P. Norton, 1996, *The Balanced Scorecard: Translating Strategy into Action*, Harvard Business School Press, Boston, MA.

Lamm, G., 2001, Assessing and Managing Risks to Information Assurance: A Methodological Approach, M.S. thesis, Department of Systems and Information Engineering, University of Virginia, Charlottesville, VA.

Leung, M.F., J.H. Lambert, and A. Mosenthal, 2004, A risk-based approach to setting priorities in protecting bridges against terrorist attacks, *Risk Analysis* **24**(4): 963–984.

Mahoney, B., 2001, Quantitative Risk Analysis of GPS as a Critical Infrastructure for Civilian Transportation Applications, M.S. thesis, Department of Systems and Information Engineering, University of Virginia, Charlottesville, VA.

Matalas, N.C., and M.B. Fiering, 1977, Water-resource system planning, in climate, climate change and water supply, In *Studies in Geophysics*, National Research Council, National Academy of Sciences, Washington, DC, pp. 99–109.

Morgan, M.G., B. Fischhoff, L. Lave, and P. Fischbeck, 1999, A proposal for risk ranking within federal agencies, In *Comparing Environmental Risks: Tools for Setting Government Priorities*, J. Clarence Davies (ed.), Resources for the Future, Washington, DC.

Morgan, M.G., H.K. Florig, M.L. DeKay, and P. Fischbeck, 2000, Categorizing risks for risk ranking, *Risk Analysis* **20**(1): 49.

Roland, H.E., and B. Moriarty, 1990, *System Safety Engineering and Management*, Second edition, John Wiley & Sons, New York.

Saaty, T.L., 1988, *Mathematical Methods of Operations Research*, Dover Publications, New York.

Sokal, R.R., 1974, Classification: purposes, principles, progress, prospects, *Science* **185**(4157): 1115–1123.

Webler, T., H. Rakel, O. Renn, and B. Johnson, 1995, Eliciting and classifying concerns: a methodological critique, *Risk Analysis* **15**(3): 421–436.

Part II

Advances in Risk Modeling, Assessment, and Management

8

Risk of Extreme Events and the Fallacy of the Expected Value

8.1 INTRODUCTION

With the increase of public interest in risk-based decisionmaking and the involvement of a growing number of professionals in the field, a relatively new professional niche of risk analysts has gained maturity. The professionals involved in risk-based decisionmaking are experiencing the same evolutionary process that systems analysts and systems engineers went through a few decades ago. That is, risk analysts are realizing and appreciating the efficacy as well as the limitations of mathematical tools and systematic analysis. In fact, there are many who simply see risk analysis as a specialized extension of the body of knowledge and evaluation perspectives that have come to be associated with systems analysis. Professionals from diverse disciplines are responding much more forcefully and knowledgeably to risks of all kinds as well, and in many instances, they are leading what has ultimately come to be a political debate. This professional community is more willing to accept the premise that a truly effective risk analysis study must, in most cases, be cross-disciplinary, relying on social and behavioral scientists, engineers, regulators, and lawyers. At the same time, this professional community has become more critical of the tools that it has developed because it recognizes their ultimate importance and usefulness in the resolution of critical societal problems. For risk methodologies and tools to be useful and effective, they must be representative; that is, they must capture not only the average risks but also the extreme and catastrophic ones.

The ultimate utility of decision analysis, including risk-based decisionmaking, is not necessarily to articulate the best policy option, but rather to avoid the extreme, the worst, and the most disastrous policies—those actions in which the cure is worse than the disease.

In his book *The Black Swan: The Impact of the Highly Improbable*, Taleb [2007] explains that hindsight—understanding produced by assessing signals (precursors) after an event—is usually not general enough to yield insight, misses causal understanding, and often lacks meaningful impact on decisions. It is the improbable events that result in the greatest impact (e.g., the unimagined terrorist attack on passenger planes, the implausible combination of a hurricane and a failure of critical infrastructure). By their definition, disasters constitute extreme and

---

Risk Modeling, Assessment, and Management, Fourth Edition. Yacov Y. Haimes.
© 2016 John Wiley & Sons, Inc. Published 2016 by John Wiley & Sons, Inc.

catastrophic events; thus, their probabilities and associated consequences defy any common expected-value representation of risk. Taleb [2007] ascribes three attributes to an extreme event: (i) It is an outlier, as it lies outside the realm of regular expectation (nothing in the past can convincingly point to its possibility), (ii) it carries an extreme impact, and (iii) in spite of its outlier status, human nature makes us concoct explanations for its occurrence *after the fact*, making it explainable and predictable. The prolific literature on risk of extreme and catastrophic events spans social and behavioral scientists, natural scientists and engineers, and economists, to cite a few, and defies the reliance on the expected value of risk. This is due to the mathematical fact that the expected value of risk (i.e., the mean) commensurates events of high probabilities and low consequences with events of low probabilities and high consequences.

Risk is commonly defined as a measure of the probability and severity of adverse effects [Lowrance, 1976]. With this definition of risk widely adopted by many disciplines, its translation into quantitative terms has been a major source of misunderstanding and misguided use and has often led to erroneous results and conclusions. The most common quantification of risk—the use of the mathematical construct known as the expected value—is probably the dominant reason for this chaotic situation in the quantification of risk. Whether the probabilities associated with the universe of events are viewed by the analyst as discrete or continuous, the expected value of risk is an operation that essentially multiplies each event by its probability of occurrence and sums (or integrates) all these products over the entire universe of events. This operation literally commensurates adverse events of high consequences and low probabilities of exceedance with events of low consequences and high probabilities of exceedance. (Recall that probability of exceedance is one minus the cumulative distribution functions, i.e., 1 − cdf.) This chapter addresses the misuse, misinterpretation, and fallacy of the expected value *when it is used as the sole criterion for risk in decisionmaking*. Many experts who are becoming more and more convinced of the grave limitations of the traditional and commonly used expected-value concept are augmenting this concept with a supplementary measure to the expected value of risk—the conditional expectation. In this, decisions about extreme and catastrophic events are not averaged with more commonly occurring high-frequency/low-consequence events.

8.2 RISK OF EXTREME EVENTS

Most analysis and decision theorists are beginning to recognize a simple yet fundamental philosophical truth. In the face of such unforeseen calamities as bridges falling, dams bursting, and airplanes crashing, we must acknowledge the importance of studying *extreme* events. Modern decision analysts are no longer asking questions about expected risk; instead, they are asking questions about expected catastrophic or unacceptable risk. These analysts are focusing their efforts on forming a more robust treatment of extreme events, in both a theoretical and a practical sense. Furthermore, managers and decisionmakers are most concerned with the risk associated with a specific case under consideration and not with the likelihood of the average adverse outcomes that may result from various risk situations. In this sense, the expected value of risk, which until recently has dominated most risk analysis in the field, is not only inadequate, but can lead to fallacious results and interpretations. Indeed, people in general are not risk neutral. They are often more concerned with low-probability catastrophic events than with more frequently occurring but less severe accidents. In some cases, a slight increase in the cost of modifying a structure might have a very small effect on the unconditional expected risk (the commonly used business-as-usual measure of risk) but would make a significant difference to the conditional expected catastrophic risk. Consequently, the conditional expected catastrophic risk can be of a significant value in many multiobjective risk problems.

Two difficult questions—How safe is safe enough, and what is an acceptable risk?—underline the normative, value-judgment perspectives in risk-based decisionmaking. No mathematical, empirical knowledge base today can adequately model the perception of risks in the mind of decisionmakers. In the study of multiple-criteria decisionmaking (MCDM), we clearly distinguish between the quantitative element in the decisionmaking process, where efficient (Pareto-optimal) solutions and their corresponding trade-off values are generated, and the normative value-judgment element, where the decisionmakers make use of these efficient solutions and trade-off values to determine their preferred (compromise) solution [Chankong and Haimes, 1983]. In many ways, risk-based decisionmaking can and should be viewed as a type of stochastic MCDM in which some of the objective functions represent risk functions. This analogy can be most helpful in making use of the extensive knowledge already generated by MCDM (witness the welter of publications and conferences on the subject).

It is worth noting that there are two modalities to the considerations of risk-based decisionmaking in a multiobjective framework. One is viewing risk (e.g., the risk of dam failure) as an objective function to be traded off with cost and benefit functions. The second modality concerns the treatment of damages of different magnitudes and different probabilities of occurrence as noncommensurate objectives, which thus must be augmented by a finite, but small, number of risk functions (e.g., a conditional expected-value function as will be formally introduced in subsequent discussion). Probably the most important aspect of considering risk-based decisionmaking within a stochastic MCDM framework is the handling of extreme events.

To dramatize the importance of understanding and adequately quantifying the risk of extreme events, the following statements are adopted from Runyon [1977]:

Imagine what life would be like if

- Our highways were constructed to accommodate the average traffic load of vehicles of average weight.
- Mass transit systems were designed to move only the average number of passengers (i.e., total passengers per day divided by 24 hours) during each hour of the day.
- Bridges, homes, and industrial and commercial buildings were constructed to withstand the average wind or the average earthquake.
- Telephone lines and switchboards were sufficient in number to accommodate only the average number of phone calls per hour.
- Your friendly local electric utility calculated the year-round average electrical demand and constructed facilities to provide only this average demand.
- Emergency services provided only the average number of personnel and facilities during all hours of the day and all seasons of the year.
- Our space program provided emergency procedures for only the average type of failure.
- Chaos is the word for it. Utter chaos.

Lowrance [1976] makes an important observation on the imperative distinction between the quantification of risk, which is an empirical process, and the determination of safety, which is a normative process. In both of these processes, which are seemingly dichotomous, the influence and imprint of the analyst cannot and should not be overlooked. The essential role of the analyst, sometimes hidden but often explicit, is not unique to risk assessment and management; rather, it is indigenous to the process of modeling and decisionmaking [Kunreuther and Slovic, 1996].

The major problem for the decisionmaker remains one of information overload: For every policy (action or measure) adopted, there will be a vast array of potential damages as well as benefits and costs with their associated probabilities. It is at this stage that most analysts are caught in the pitfalls of the unqualified expected-value analysis. In their quest to protect the decisionmaker from information overload, analysts precommensurate catastrophic damages that have a low probability of occurrence with minor damages that have a high probability. From the perspective of public policy, it is obvious that a catastrophic dam failure, which might cause flooding of, say, 10^6 acres of land with associated damage to human life and the environment but which has a very low probability (say, 10^{-6}) of happening, cannot be viewed by decisionmakers in the same vein as minor flooding of, say, 10^2 acres of land, which has a high probability of 10^{-2} of happening. Yet this is exactly what the expected-value function would ultimately generate. Most importantly, the analyst's precommensuration of these low-probability occurrence/high-damage events with high-probability, low-damage events into one expectation function (indeed some kind of a utility function) markedly distorts the relative importance of these events and consequences as they are viewed, assessed, and evaluated by the decisionmakers.

8.3 THE FALLACY OF THE EXPECTED VALUE

One of the most dominant steps in the risk assessment process is the quantification of risk, yet the validity of the approach most commonly used to quantify risk—its expected value—has received neither the broad professional scrutiny it deserves nor the hoped-for wider mathematical challenge that it mandates. The conditional expected value of the risk of extreme events (among other conditional expected values of risks) generated by the partitioned multi-objective risk method (PMRM) [Asbeck and Haimes, 1984] is one of the few exceptions.

Let $p_x(x)$ denote the probability density function (pdf) of the random variable X, where X is, for example, the concentration of the contaminant trichloroethylene (TCE) in a groundwater system, measured in parts per billion (ppb). The expected value of the contaminant concentration (the risk of the groundwater being contaminated by TCE at an average concentration of TCE) is $E(X)$ ppb. If the pdf is discretized to n regions over the entire universe of contaminant concentrations, then $E(X)$ equals the sum of the product of p_i and x_i, where p_i is the probability that the ith segment of the probability regime has a TCE concentration of x_i. Integration (instead of summation) can be used for the continuous case. Note, however, that the expected-value operation commensurates contaminations (events) of low concentration and high frequency with contaminations of high concentration and low frequency. For example, events $x_1 = 2$ pbb and $x_2 = 20,000$ ppb that have the probabilities $p_1 = 0.1$ and $p_2 = 0.00001$, respectively, yield the same contribution to the overall expected value: $(0.1)(2) + (0.00001)(20,000) = 0.2 + 0.2$. However, to the decisionmaker in charge, the relatively low likelihood of a disastrous contamination of the groundwater system with 20,000 ppb of TCE cannot be equivalent to the contamination at a low concentration of 0.2 ppb, even with a very high likelihood of such contamination. Due to the nature of mathematical smoothing, the averaging function of the contaminant concentration in this example does not lend itself to prudent management decisions. This is because the expected value of risk does not accentuate the catastrophic events and their consequences, thus misrepresenting what would be perceived as an unacceptable risk.

It is worth noting that the number of *good* decisions that managers make during their tenure is not the only basis for rewards, promotion, and advancement; rather, they are likely to be penalized for any disastrous decisions, no matter how few, made during their career. The notion of *not on my watch* clearly emphasizes the point. In this and other senses, the expected value of risk fails to represent a measure that truly communicates the manager's or the decisionmaker's intentions and perceptions. The conditional expected value of the risk of extreme events generated by the PMRM, when used in conjunction with the (unconditional) expected value, can markedly contribute to the total risk management approach. In this case, the manager must make trade-offs not only between the cost of preventing contamination by TCE and the expected value of such risk of contamination but also between the cost of prevention and the conditional expected value of extreme contamination by TCE. Such a dual multiobjective analysis provides the manager with more complete, more factual, and less aggregated information about all viable policy options and their associated trade-offs [Haimes, 1991].

This act of commensurating the expected-value operation is analogous in some sense to the commensuration of all benefits and costs into one monetary unit. Indeed, few today would consider benefit–cost analysis, where all benefits, costs, and risks are commensurated into monetary units, as an adequate and acceptable measure for decisionmaking when it is used as the sole criterion for excellence. Multiobjective analysis has been demonstrated as a superior approach to benefit–cost analysis [Haimes and Hall, 1974].

To demonstrate the limitation of the expected-value approach, consider a design problem where four design options are being considered. Associated with each option are the cost, the mean of a failure rate (i.e., the expected value of failures for a normally distributed pdf of a failure rate), and the standard deviation (see Table 8.1). Figure 8.1 depicts the normally distributed pdf's of failure rates for each of the four designs. Clearly on the basis of the expected value alone, the least-cost design (Option 4) seems to be preferred, at a cost of $40,000. However, consulting the variances, which provide an indication of extreme failures, reveals that this choice might not be the best after all, and it calls for a more in-depth trade-off analysis.

8.4 THE PMRM

Before the PMRM was developed, problems with at least one random variable were solved by computing and minimizing the unconditional expectation of the random variable representing damage. In contrast, the PMRM isolates a number of damage ranges (by specifying so-called partitioning probabilities) and generates conditional expectations of damage, given that the damage falls within a particular range. A *conditional expectation* is defined as the expected value of a random variable, given that this value lies within some prespecified probability range. Clearly, the values of conditional expectations depend on where the probability axis is partitioned. The analyst subjectively chooses where to partition in response to the extremal characteristics of the decisionmaking problem. For example, if the decisionmaker is concerned about the once-in-a-million-years catastrophe, the partitioning should be such that the expected catastrophic risk is emphasized.

The ultimate aim of good risk assessment and management is to suggest some theoretically sound and defensible foundations for regulatory agency guidelines for the selection of probability distributions. Guidelines for the selection of probability distributions should help incorporate meaningful decision criteria, accurate assessments of risk in regulatory problems, and reproducible and persuasive analyses. Since these risk evaluations are often tied to highly infrequent or low-probability catastrophic events, it is imperative that the guidelines consider and build on the statistics of extreme events in the selection of probability distributions. Selecting probability distributions to characterize the risk of extreme events is a subject of emerging studies in risk management [Haimes et al., 1992; Lambert et al., 1994; Leemis, 1995; Bier et al., 2004].

There is abundant literature that reviews the methods of approximating probability distributions from empirical data. Goodness-of-fit tests determine whether hypothesized distributions should be

TABLE 8.1 Design Options Data and Results

Option Number	Cost ($)	Mean (*m*) Expected Value	Standard Deviation (*s*)
1	100,000	5	1
2	80,000	5	2
3	60,000	5	3
4	40,000	5	4

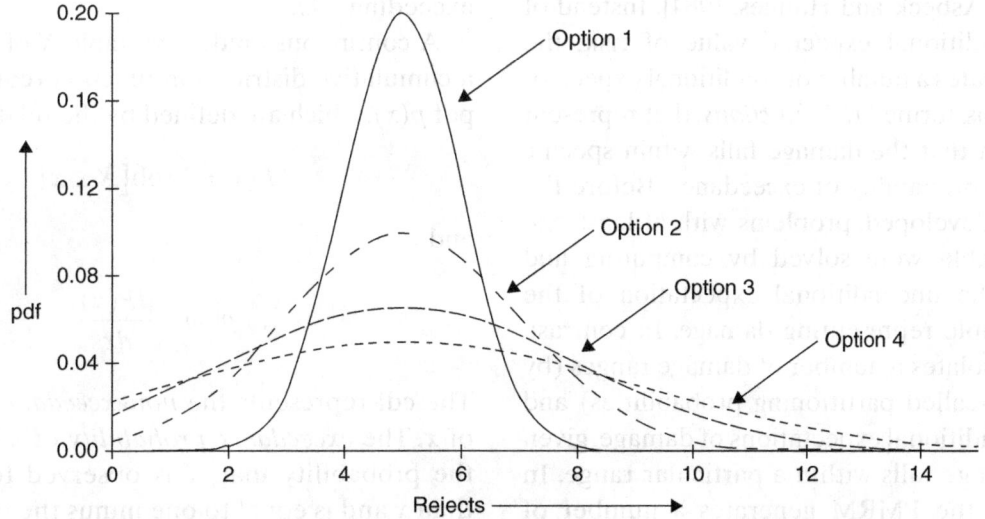

Figure 8.1 Mapping of the probability partitioning onto the damage axis.

rejected as representations of empirical data. Approaches such as the method of moments and maximum likelihood are used to estimate distribution parameters. The caveat in directly applying accepted methods to natural hazards and environmental scenarios is that most deal with selecting the best matches for the *entire* distribution. The problem is that natural hazards and environmental assessments and decisions typically address worst-case scenarios on the tails of distributions. The differences in distribution tails can be very significant even if the parameters that characterize the central tendency of the distribution are similar. A normal distribution and a uniform distribution that have similar expected values can markedly differ on the tails. The possibility of significantly misrepresenting potentially the most relevant portion of the distribution, the tails, highlights the importance of bringing the consideration of extreme events into the selection of probability distributions.

More time and effort should be spent to characterize the tails of distributions along with modeling the entire distribution. Improved matching between extreme events and distribution tails provides policymakers with more accurate and relevant information. Major factors to consider when developing distributions that account for tail behaviors include (i) availability of data; (ii) characteristics of the distribution tail, such as shape and rate of decay; and (iii) value of additional information in assessment.

The PMRM is a risk analysis method developed for solving multiobjective problems of a probabilistic nature [Asbeck and Haimes, 1984]. Instead of using the traditional expected value of risk, the PMRM generates a number of conditional expected-value functions, termed *risk functions*, that represent the risk given that the damage falls within specific ranges of the probability of exceedance. Before the PMRM was developed, problems with at least one random variable were solved by computing and minimizing the unconditional expectation of the random variable representing damage. In contrast, the PMRM isolates a number of damage ranges (by specifying so-called partitioning probabilities) and generates conditional expectations of damage, given that the damage falls within a particular range. In this manner, the PMRM generates a number of risk functions, one for each range, which are then augmented with the original optimization problem as new objective functions.

The conditional expectations of a problem are found by partitioning the problem's probability axis and mapping these partitions onto the damage axis. Consequently, the damage axis is partitioned into corresponding ranges. *A conditional expectation is defined as the expected value of a random variable given that this value lies within some prespecified probability range.* Clearly, the values of conditional expectations are dependent on where the probability axis is partitioned. The choice of where to partition is made subjectively by the analyst in response to the extreme characteristics of the problem. If, for example, the analyst is concerned about the once-in-a-million-years catastrophe, the partitioning should be such that the expected catastrophic risk is emphasized. Although no general rule exists to guide the partitioning, Asbeck and Haimes [1984] suggest that if three damage ranges are considered for a normal distribution, then the $+1s$ and $+4s$ partitioning values provide an effective rule of thumb. These values correspond to partitioning the probability axis at 0.84 and 0.99968; that is, the low-damage range would contain 84% of the damage events, the intermediate range would contain just under 16%, and the catastrophic range would contain about 0.032% (probability of 0.00032). In the literature, catastrophic events are generally said to be events with a probability of exceedance of 10^{-5} (see, for instance, the NRC report on dam safety [National Research Council, 1985]). This probability corresponds to events exceeding $+4s$.

A continuous random variable X of damages has a cumulative distribution function (cdf) $P(x)$ and a pdf $p(x)$, which are defined by the relationships

$$P(x) = \text{Prob}[X \leq x] \tag{8.1}$$

and

$$p(x) = \frac{dP(x)}{dx} \tag{8.2}$$

The cdf represents the *nonexceedance probability* of x. The *exceedance probability* of x is defined as the probability that X is observed to be greater than x and is equal to one minus the cdf evaluated at x.

The expected value, average, or mean value of the random variable X is defined as

$$E[X] = \int_0^\infty x p(x)\, dx \quad (8.3)$$

For the discrete case, where the universe of events (sample space) of the random variable X is discretized into I segments, the expected value of damage $E[X]$ can be written as

$$E[X] = \sum_{i=1}^{I} p_i x_i \quad (8.4)$$

$$p_i \geq 0 \quad (8.5)$$

$$\sum p_i = 1 \quad (8.6)$$

where x_i is the ith segment of the damage.

In the PMRM, the concept of the expected value of damage is extended to generate multiple *conditional expected-value functions*, each associated with a particular range of exceedance probabilities or their corresponding range of damage severities. The resulting conditional expected-value functions, in conjunction with the traditional expected value, provide a family of risk measures associated with a particular policy.

Let $1-\alpha_1$ and $1-\alpha_2$, where $0 < \alpha_1 < \alpha_2 < 1$, denote exceedance probabilities that partition the domain of X into three ranges as follows. On a plot of exceedance probability, there is a unique damage β_1 on the damage axis that corresponds to the exceedance probability $1-\alpha_1$ on the probability axis. Similarly, there is a unique damage β_2 that corresponds to the exceedance probability $1-\alpha_2$. Damages less than β_1 are considered to be of low severity, and damages greater than β_2 are of high severity. Similarly, damages of a magnitude between β_1 and β_2 are considered to be of moderate severity. The partitioning of risk into three severity ranges is illustrated in Figure 8.2. If the partitioning probability α_1 is specified, for example, to be 0.05, then β_1 is the fifth exceedance percentile. Similarly, if α_2 is 0.95 (i.e., $1-\alpha_2$ is equal to 0.05), then β_2 is the 95th exceedance percentile.

For each of the three ranges, the conditional expected damage (given that the damage is within that particular range) provides a measure of the risk associated with the range. These measures are obtained through the definition of the *conditional expected value*. Consequently, the new measures of risk are $f_2(\cdot)$, of high exceedance probability and low severity; $f_3(\cdot)$, of medium exceedance probability and moderate severity; and $f_4(\cdot)$, of low exceedance probability and high severity. The function $f_2(\cdot)$ is the conditional expected value of X, given that x is less than or equal to β_1:

$$f_2(\cdot) = E[X \mid X \leq \beta_1]$$

$$f_2(\cdot) = \dfrac{\int_0^{\beta_1} x p(x)\, dx}{\int_0^{\beta_1} p(x)\, dx} \quad (8.7)$$

Similarly, for the other two risk functions, $f_3(\cdot)$ and $f_4(\cdot)$

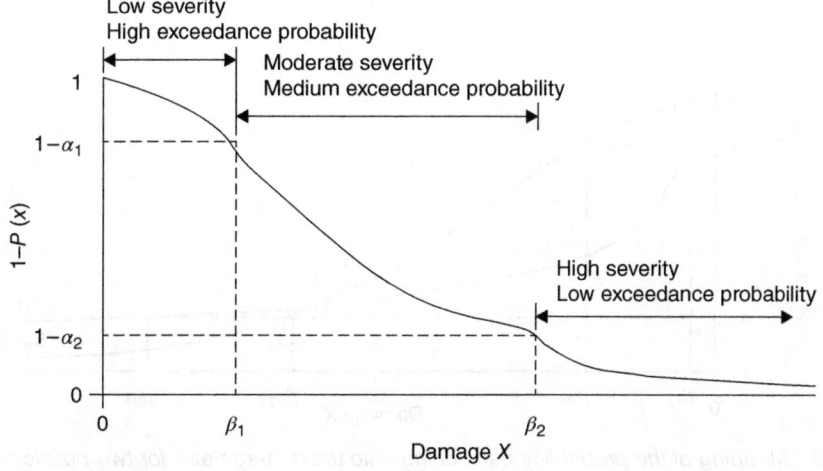

Figure 8.2 *pdf of failure rate distributions for four designs.*

$$f_3(\cdot) = E[X|\beta_1 \leq X \leq \beta_2]$$

$$f_3(\cdot) = \frac{\int_{\beta_1}^{\beta_2} xp(x)\,dx}{\int_{\beta_1}^{\beta_2} p(x)\,dx} \quad (8.8)$$

and

$$f_4(\cdot) = E[X|X > \beta_2]$$

$$f_4(\cdot) = \frac{\int_{\beta_2}^{\infty} xp(x)\,dx}{\int_{\beta_2}^{\infty} p(x)\,dx} \quad (8.9)$$

Thus, for a particular policy option, there are three measures of risk, $f_2(\cdot)$, $f_3(\cdot)$, and $f_4(\cdot)$, in addition to the traditional expected value denoted by $f_5(\cdot)$. The function $f_1(\cdot)$ is reserved for the cost associated with the management of risk. Note that

$$f_5(\cdot) = \frac{\int_0^{\infty} xp(x)\,dx}{\int_0^{\infty} p(x)\,dx} = \int_0^{\infty} xp(x)\,dx \quad (8.10)$$

since the total probability of the sample space of X is necessarily equal to one. In the PMRM, all or some subset of these five measures are balanced in a multiobjective formulation. The details are made more explicit in the next two sections.

8.5 GENERAL FORMULATION OF THE PMRM

Assume that the damage severity associated with the particular policy s_j, $j \in \{1,\ldots,q\}$ can be represented by a continuous random variable X, where $p_X(x;s_j)$ and $P_X(x;s_j)$ denote the pdf and the cdf of damage, respectively. Two partitioning probabilities, α_i, $i=1,2$, are preset for the analysis and determine three ranges of damage severity for each policy s_j. The damage, β_{ij}, corresponding to the exceedance probability $(1-\alpha_i)$, can be found due to the monotonicity of $P_X(x;s_j)$. The policies s_j, the partitions α_i, and the bounds β_{ij} of damage ranges are related by the expression

$$P_X(\beta_{ij};s_j) = \alpha_i, \quad i=1,2, \; \forall j \quad (8.11)$$

This partitioning scheme is illustrated in Figure 8.3 for two hypothetical policies s_1 and s_2. The ranges of damage severity include high exceedance probability and low damage, $\{x : x \in [\beta_{0j}, \beta_{1j}]\}$, the set of possible realizations of X for which it is true that $x \in [\beta_{0j}, \beta_{1j}]$; medium exceedance probability and medium damage, $\{x \in [\beta_{1j}, \beta_{2j}]\}$; and low exceedance probability and high damage (extreme event), $\{x : x \in [\beta_{2j}, \beta_{3j}]\}$, where β_{0j} and β_{3j} are the lower and upper bounds of damage X.

The conditional expected-value risk functions f_i, $i=2,3,4$, are given by

$$f_i(s_j) = E[X|p_x(x;s_j), x \in [\beta_{i-2,j}, \beta_{i-1,j}]],$$
$$i=2,3,4; \; j=1,\ldots,q \quad (8.12)$$

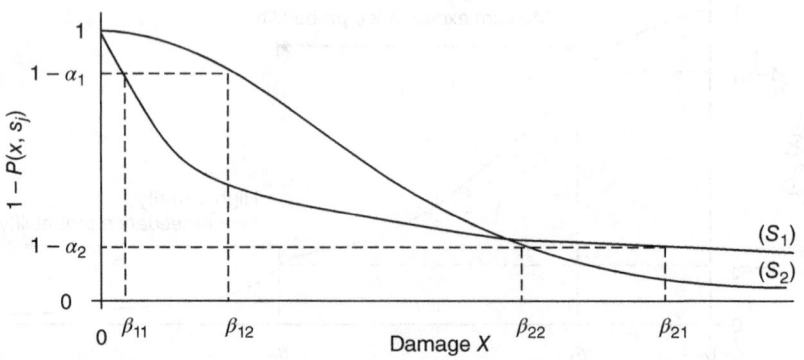

Figure 8.3 Mapping of the probability partitioning onto the damage axis for two policies s_1 and s_2.

and, equivalently,

$$f_i(s_j) = \frac{\int_{\beta_{i-2,j}}^{\beta_{i-1,j}} x p_x(x; s_j)\, dx}{\int_{\beta_{i-2,j}}^{\beta_{i-1,j}} p_x(x; s_j)\, dx}, \quad i = 2, 3, 4;\ j = 1, \ldots, q \tag{8.13}$$

The denominator of Eq. (8.13) is defined to be $q_i, i = 2, 3, 4$, as follows:

$$q_2 = \int_0^{\beta_{1j}} p_x(x; s_j)\, dx \tag{8.14}$$

$$q_3 = \int_{\beta_{1j}}^{\beta_{2j}} p_x(x; s_j)\, dx \tag{8.15}$$

$$q_4 = \int_{\beta_{2j}}^{\infty} p_x(x; s_j)\, dx \tag{8.16}$$

If the unconditional expected value of the damage from policy s_j is defined to be $f_5(s_j)$, then the following relationship holds:

$$f_5(s_j) = q_2 f_2(s_j) + q_3 f_3(s_j) + q_4 f_4(s_j) \tag{8.17}$$

with $q_i > 0$ and $q_2 + q_3 + q_4 = 1$. The q_i are the probabilities that X is realized in each of the three damage ranges and are independent of the policies s_j.

The preceding discussion has described the partitioning of three damage ranges by fixed exceedance probabilities $\alpha_i, i = 1, 2$. Alternatively, the PMRM provides for the partitioning of damage ranges by preset thresholds of damage. For example, the meaning of $f_4(s_j)$ in partitioning by a fixed damage becomes the expected damage resulting from policy j given that the damage exceeds a fixed magnitude. For further details on the partitioning of damage ranges, see Asbeck and Haimes [1984], Karlsson and Haimes [1988a, b], and Haimes et al. [1992].

In sum, the conditional expected-value functions in the PMRM are multiple, noncommensurate measures of risk, each associated with a particular range of damage severity. In contrast, the traditional expected-value commensurate risks from all ranges of damage severity represent only the central tendency of the damage.

Combining any one of the generated conditional expected risk functions or the unconditional expected risk function with the cost objective function f_1 creates a set of multiobjective optimization problems:

$$\min[f_1, f_i], \quad i = 2, 3, 4, 5 \tag{8.18}$$

This formulation offers more information about the probabilistic behavior of the problem than the single formulation $\min[f_1, f_5]$. The trade-offs between the cost function f_1 and any risk function $f_i, i \in \{2, 3, 4, 5\}$ allow decisionmakers to consider the marginal cost of a small reduction in the risk objective, given a particular risk assurance for each of the partitioned risk regions and given the unconditional risk functions f_5. The relationship of the trade-offs between the cost function and the various risk functions is given by

$$\frac{1}{\lambda_{15}} = \frac{q_2}{\lambda_{12}} + \frac{q_2}{\lambda_{13}} + \frac{q_4}{\lambda_{14}} \tag{8.19}$$

where

$$\lambda_{1i} = -\frac{\partial f_1}{\partial f_i} \tag{8.20}$$

with q_2, q_3, and q_4 as defined earlier. A knowledge of this relationship among the marginal costs provides decisionmakers with insights that are useful for determining an acceptable level of risk. Any multiobjective optimization method can be applied at this stage—for example, the surrogate worth trade-off (SWT) method discussed in Chapter 5.

It has often been observed that expected catastrophic risk is very sensitive to the partitioning policy. This sensitivity may be quantified using the statistics of extremes approach suggested by Karlsson and Haimes [1988a, b] and Haimes et al. [1990] and discussed in Chapters 11 and 12. In many applications, if given a database representing a random process (e.g., hydrological data related to flooding), it is very difficult to find a specific distribution that represents this database. In some cases, one can exclude some pdf's or guess that some are more representative than others. Quite often, one is given a very limited database that does not contain information about the extreme events. In particular, nothing can be said with certainty about the probable maximum flood (PMF), which corresponds to a flood with a return period between 10^4 and 10^6 years. Events of a more extreme character are very important because they determine the expected catastrophic risk. The conditional expectations in the PMRM are dependent on the probability partitions and on the choice of the pdf representing the probabilistic behavior of the data [Karlsson and Haimes, 1988a, b].

8.6 SUMMARY OF THE PMRM

This section compares partitioning on the damage axis with partitioning on the probability axis. Equations (8.21)–(8.23) are measures of the conditional expected values—$f_2(\cdot), f_3(\cdot)$, and $f_4(\cdot)$—of the random variable that represents damage. Equation (8.24) represents the unconditional expected-value function $f_5(\cdot)$. Figure 8.4 depicts the partitioning on the damage axis.

Risk functions:

$$f_2(\cdot) = E[X \mid X \le \beta_1] \qquad (8.21)$$
$$f_3(\cdot) = E[X \mid \beta_1 < X \le \beta_2] \qquad (8.22)$$
$$f_4(\cdot) = E[X \mid X > \beta_2] \qquad (8.23)$$
$$f_5(\cdot) = E[X] \qquad (8.24)$$

In parallel with partitioning on the damage axis, Eqs. (8.25)–(8.27) are measures of the same conditional expected values with partitioning on the probability axis. Similar to Eq. (8.24), Eq. (8.28) represents the unconditional expected-value function $f_5(\cdot)$. Figure 8.5 depicts the partitioning on the probability axis.

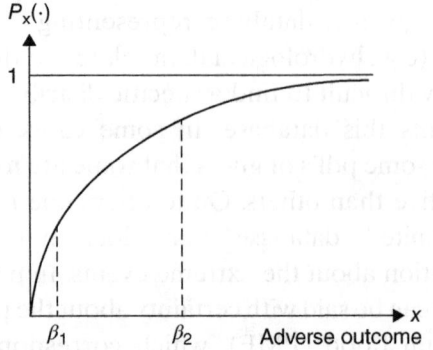

Figure 8.4 Partitioning on the damage axis.

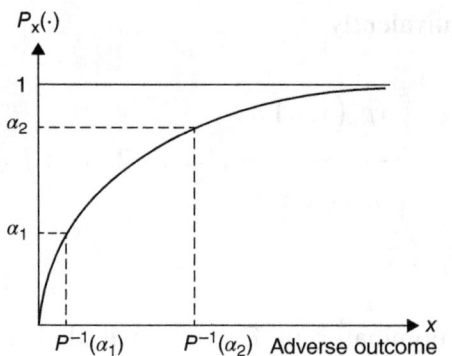

Figure 8.5 Partitioning on the probability axis.

Risk functions:

$$f_2(\cdot) = E\left[X \mid X \le P^{-1}(\alpha_1)\right] \qquad (8.25)$$
$$f_3(\cdot) = E\left[X \mid P^{-1}(\alpha_1) < X \le P^{-1}(\alpha_2)\right] \qquad (8.26)$$
$$f_4(\cdot) = E\left[X \mid X > P^{-1}(\alpha_2)\right] \qquad (8.27)$$
$$f_5(\cdot) = E[X] \qquad (8.28)$$

To gain further insight into the two partitioning schemes and their implications, Table 8.2 juxtaposes them.

Figure 8.6 depicts the partitioning of the exceedance probability $[1-P_x(X)]$ on the probability axis. Note that the denominator of the conditional expected-value functions (see Step 4 in Table 8.2) remains constant for different policies (scenarios) s_j (see Eq. 8.29):

$$\int_{\beta_{1,j}}^{\beta_{2,j}} p_x(x;s_j)\,dx = (1-\alpha_1)-(1-\alpha_2) = \alpha_2 - \alpha_1 \qquad (8.29)$$

We further note from Figure 8.6 that the projections of the partitioning probabilities on the damage axis are not the same; namely, $[\beta_{11}, \beta_{12}]$ and $[\beta_{22}, \beta_{21}]$ are not the same.

Similarly, Figure 8.7 depicts the mapping of the partitioning of the exceedance probability $(1-P_x(X))$ on the damage axis. Note that the denominators of the conditional expected-value functions (see Step 4 in Table 8.2) are different for different policies (scenarios) s_j. We further note from Figure 8.7 that the projections of these damage partitionings on the probability axis are not the same. In sum, in partitioning the exceedance probability on the damage axis, different weighting coefficients in the denominator are experienced for different scenarios s_j, while the same damage regions remain.

TABLE 8.2 Comparison of the PMRM with Partitioning on the Damage Axis and on the Probability Axis

Partitioning the Probability Axis and Projecting onto the Damage Axis (Figure 8.6)	Partitioning the Damage Axis and Projecting onto the Probability Axis (Figure 8.7)
Step	Step
1. Generate the probability of exceedance: $1 - P_x(\cdot)$	1. Generate the probability of exceedance: $1 - P_x(\cdot)$
2. Partition on the damage axis: $[\beta_i, \beta_{i+1}] \quad i = 1, 2, \ldots, N$	2. Partition on the probability axis: $[1-\alpha_i, 1-\alpha_{i+1}] \quad i = 1, 2, \ldots, N$
3. Not applicable	3. Map the partitioning of the probability axis to the damage axis for each scenario (policy) s_j: $[\beta_{i,j}, \beta_{i+1,j}], \quad i = 1, 2, \ldots, N, \quad \forall j$ where $\beta_{i,j} = P_x^{-1}(1-\alpha_i) \forall j$ $\beta_{i+1,j} = P_x^{-1}(1-\alpha_{i+1}) \forall j$
4. Calculate conditional expectations: $E[X \mid \beta_i < X \leq \beta_{i+1}] = \dfrac{\int_{\beta_i}^{\beta_{i+1}} x p_x(x; s_j) dx}{\int_{\beta_i}^{\beta_{i+1}} p_x(x; s_j) dx}$	4. Calculate conditional expectations: $E[X \mid \beta_{i,j} < X \leq \beta_{i+1,j}] = \dfrac{\int_{\beta_{i,j}}^{\beta_{i+1,j}} x p_x(x; s_j) dx}{\int_{\beta_{i,j}}^{\beta_{i+1,j}} p_x(x; s_j) dx}$

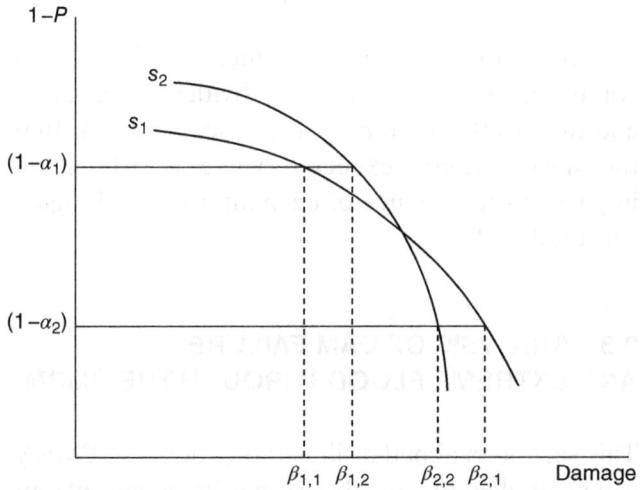

Figure 8.6 Mapping of the partitioning of the probability exceedance axis onto the damage axis for two policies s_1 and s_2.

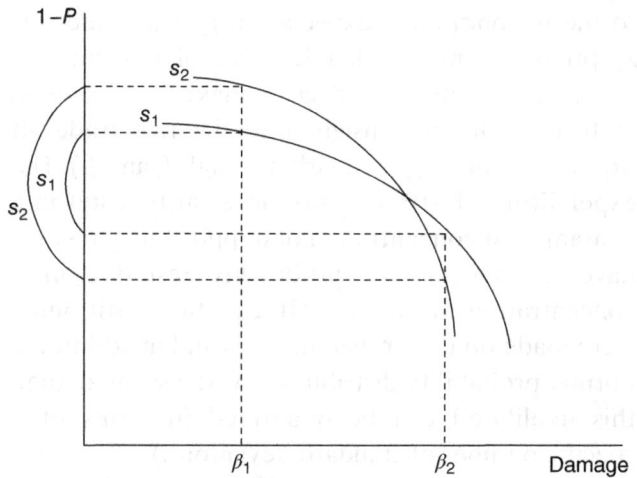

Figure 8.7 Mapping of the partitioning of the damage axis onto the probability axis for two policies s_1 and s_2.

8.7 ILLUSTRATIVE EXAMPLE

To illustrate the usefulness of the additional information provided by the PMRM, consider Figure 8.8 where the cost of prevention of groundwater contamination f_1 is plotted against (1) the conditional expected value of contaminant concentration at the low probability of exceedance/high-concentration range f_4 and (2) the unconditional expected value of contaminant concentration f_5. Note that with Policy A, an investment of $\$2 \times 10^6$ in the prevention of groundwater contamination results in an expected value of contaminant concentration of 30 ppb; however, under the more conservative view (as presented by f_4), the conditional expected value of contaminant concentration (given that the state of nature will be in a low probability of exceedance/high-concentration region) is twice as high (60 ppb). Policy B, $\$10^6$ of expenditure, reveals similar results: 60 ppb for the unconditional expectation f_5 but 110 ppb for the conditional

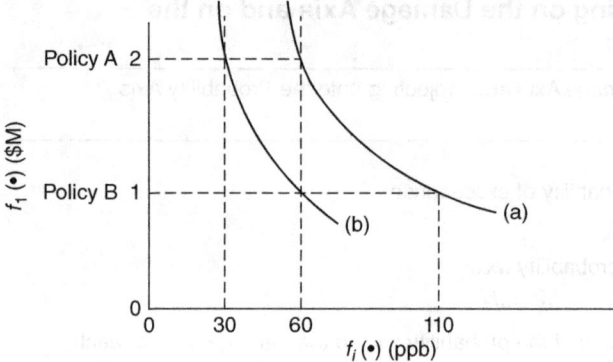

Figure 8.8 (a) Cost functions versus conditional expected value of contaminant concentration $f_4(\cdot)$ and (b) cost function versus expected value of contaminant concentration $f_5(\cdot)$.

TABLE 8.3 Values of Conditional Expected Values of Extreme Failure, $f_4(\cdot)$

Option Number	Cost ($)	Mean (m) Expected Value	Standard Deviation (s)	Conditional Expected Value, $f_4(\cdot)$
1	100,000	5	1	8.37
2	80,000	5	2	11.73
3	60,000	5	3	15.10
4	40,000	5	4	18.47

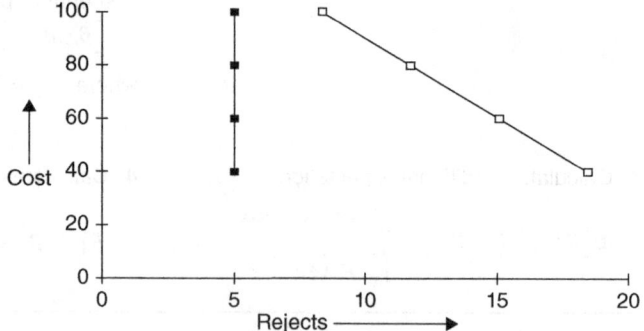

Figure 8.9 Pareto-optimal frontier.

expectation f_4. Also, note that the slopes of the noninferior frontiers with policies A and B are not the same. The slope of f_5 between policies A and B is smaller than that of f_4, indicating that a further investment beyond 10^6 would contribute more to a reduction of the extreme-event risk f_4 than it would to the unconditional expectation f_5. The trade-offs λ_{1i} provide a most valuable piece of information. More specifically, the decisionmaker is provided with an additional insight into the risk trade-off problem through f_4 (similarly through f_2 and f_3). The expenditure of 10^6 may not necessarily result in a contaminant concentration of 60 ppb; it may instead have a nonnegligible probability resulting in a concentration of 100 ppb. (If, e.g., the partitioning were made on the probability axis and in addition a normal probability distribution were assumed, then this likelihood can be quantified in terms of a specific number of standard deviations.)

Furthermore, with an additional expenditure of 10^6 (Policy A), even the extreme event of likely concentration is 60 ppb—closer to the range of acceptable standards. It is worth remembering that the additional conditional risk functions provided by the PMRM do not invalidate the traditional expected-value analysis per se—they improve on it by providing additional insight into the nature of risk to a system.

Let us revisit the design problem with its four alternatives. Table 8.3 summarizes the values of the conditional expected value of extreme failure, f_4. Figure 8.9 depicts the cost of each design versus the unconditional expected value, f_5, and the cost versus the conditional expected value, f_4. Clearly, the conditional expected value f_4 provides much valued additional information on the associated risk than the unconditional expected value f_5, where the impact of the variance of each alternative design is captured by f_4.

8.8 ANALYSIS OF DAM FAILURE AND EXTREME FLOOD THROUGH THE PMRM

This section is aimed at illustrating how the PMRM can be applied to a real but somewhat idealized dam safety case study [Petrakian et al., 1989]. During the course of the analysis, useful relationships are derived that greatly facilitate the application of the PMRM method, not only to dam safety but also to a variety of other risk-related problems. Apart from theoretical investigations, the practical usefulness of the PMRM is examined in detail through its use in the evaluation of various dam safety remedial actions.

Dams are designed, in part, to control the extreme variability in natural hazards (floods and drought), but they simultaneously impose an even larger, though much less frequent, technological hazard: potential dam failure [Stedinger and Grygier, 1985].

Therefore, a low-probability/high-consequence (LP/HC) risk analysis of dams is the most appropriate approach to tackle the issue of dam safety.

The main function of a dam's spillway is to protect the dam itself during extreme floods. Spillways help to avoid dam failure by passing excess water—that is, water beyond the design flood volume—that might otherwise cause the dam to be overtopped or breached. The hazards posed by inadequate spillways might approach or even exceed damages that would have occurred under natural flood conditions without the existence of the dam.

Two preventative remedial actions are of interest: widening the spillway and raising the dam's height. Inherent in each of these actions is a trade-off between two situations. For example, widening the spillway reduces the chances of a failure caused by rare floods with high magnitudes that overtop the dam; but greater damage is incurred downstream by medium-sized floods that pass through the spillway. Similarly, augmenting the dam's height reduces the likelihood of a dam failure but increases the severity of downstream damages in the event of a failure. This reflects an incommensurable trade-off in risk reduction. Each alternative can meet a stated design objective, but the damages occur in different parts of the frequency spectrum. The expected-value approach cannot capture this risk reduction. Sixteen remedial actions, which variously combine changes in the spillway's width and the dam's height, will be considered here.

8.8.1 Flood-Frequency Distribution for Rare Floods

The lognormal distribution has been widely used as a flood-frequency distribution, in particular for floods with moderate return periods. The Pareto distribution (Pearson type IV), which has a tail similar to that of the Gumbel, is often used by the Bureau of Reclamation as a flood-frequency distribution. The Weibull distribution is widely employed in reliability models; it takes on shapes similar to the gamma distribution. The Weibull distribution is also known as the extreme value type III distribution of the smallest value. The Gumbel distribution might be proper for representing maximum yearly floods, which can be considered the extreme values of daily floods.

The cumulative distribution derived from the assumed flood-frequency distribution between the PMF and the 100-year flood will be interpolated, but first, it will be necessary to estimate T, the return period of the PMF. This task involves many uncertainties and in general yields inaccurate estimates. The return period of the PMF is sometimes estimated to be as low as 10^4, but the American Nuclear Society [1981], for example, has estimated it to be a larger than 10^7. Therefore, it was decided to perform a sensitivity analysis on the value of the return period of the PMF; the values 10^4, 10^5, 10^6, and 10^7 were examined. The following notation will be used: $T_4 = 10^4$, $T_5 = 10^5$, $T_6 = 10^6$, and $T_7 = 10^7$.

8.8.2 Computational Results

Sixteen alternatives were considered for combining the remedial actions of raising the dam's height and increasing the spillway's width. They are described in detail in Table 8.4 [US Army Corps of Engineers, 1988].

If the dam's height is raised by 10 ft to an elevation of 920 ft above sea level and if the present spillway width is maintained, the dam will safely pass the PMF. Similarly, if the present dam's height is kept and if the spillway is widened to 2.4 times the current size, the dam will also safely pass the PMF. Alternatives such as increasing the spillway's width by more than 2.4 times or raising the dam's height by more than 10 ft were disregarded, since corresponding added construction costs only ensure that the dam would pass floods larger than the PMF. Floods of such large magnitude are considered to be very unlikely, however, and have generally been ignored by analysts in the field of dam safety.

Cost estimates of remedial actions for the Tomahawk Dam were derived by the US Army Corps of Engineers [1983, 1985]. These values have

TABLE 8.4 Description of the Alternatives s_j ($j = 1, 2, \ldots, 16$)

Increase in Dam Height (ft)	Spillway Width (1 Unit = 620 ft)			
	1	1.5	2	2.4
0	s_1	s_5	s_9	s_{13}
3	s_2	s_6	s_{10}	s_{14}
6	s_3	s_7	s_{11}	s_{15}
10	s_4	s_8	s_{12}	s_{16}

been used to obtain the cost estimates for the 16 alternative actions (see Table 8.4).

Consider the following results obtained by Petrakian et al. [1989] on the Shoohawk Dam study. Two decision variables are considered: (1) raising the dam's height and (2) increasing the dam's spillway capacity. Although Petrakian et al. considered several policy options or scenarios, only a few are discussed here. Table 8.5 presents the values of $f_1(x)$ (the cost associated with increasing the dam's height and the spillway capacity) and of $f_4(x)$ and $f_5(x)$ (the conditional and unconditional expected value of damages, respectively). The conditional expected-value function $f_4(x)$ is evaluated for a partitioning of the probability axis at $\alpha = 0.999$. These values are listed for each of the selected scenarios. Note that the range of the unconditional expected value of the damage, $f_5(x)$, is $161.5–161.7 million for the various scenarios. The range of the low-frequency, high-damage conditional expected value, $f_4(x)$, varies between $719 and $1260 million—a marked difference. Thus, while an investment in the safety of the dam at a cost, $f_1(x)$, ranging from $0 to $46 million, does not appreciably reduce the unconditional expected value of damages, such an investment markedly reduces the conditional expected value of extreme damage from about $1260 to $720 million.

This significant insight into the probable effect of different policy options on the safety of the Shoohawk Dam would have been completely lost without the consideration of the conditional expected value derived by the PMRM. Figure 8.10 depicts the plotting of $f_1(x)$ versus $f_4(x)$ and $f_5(x)$. Note that the unusually high values of $f_1(x)$, on the order of $160 million, are attributed to the assumptions concerning antecedent flood conditions (in compliance with the guidelines and recommendations established by the

TABLE 8.5 Cost of Improving the Dam's Safety and Corresponding Conditional and Unconditional Expected Damages

Scenarios	$f_1(x)$ $10^6	$f_4(x)$ $10^6	$f_5(x)$ $10^6
1	0	1260	161.7
2	20	835	161.6
3	26	746	161.6
4	36	719	161.5
5	46	793	160.5

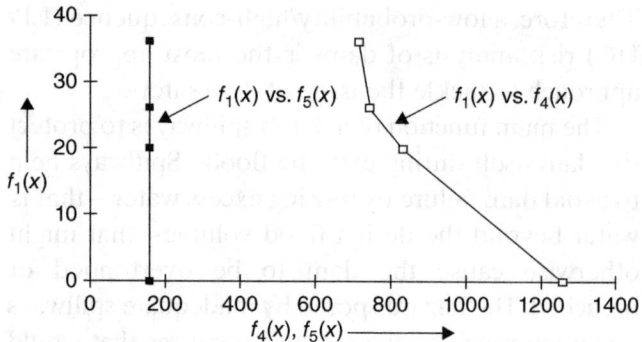

Figure 8.10 Pareto-optimal frontiers of $f_1(x)$ versus $f_4(x)$ and $f_1(x)$ versus $f_5(x)$.

US Army Corps of Engineers). This dam safety problem is discussed further in more detail in the next section.

In sum, new metrics to represent and measure the risk of extreme events are needed to supplement and complement the expected-value measure of risk, which represents the central tendency of events. There is much work to be done in this area, including the extension of the PMRM. Research efforts directed at using results from the area of statistics of extremes in representing risk of extreme events have been proven very promising and should be continued. Chapter 11 introduces the statistics of extremes as they support the formulation and presentation of the PMRM.

8.8.3 Analysis of Results

This section contains a discussion of the results obtained by applying the PMRM to the dam safety problem. In particular, a sensitivity analysis is performed on the distribution used to extrapolate the frequency curve to the PMF, the return period of the PMF, and the partitioning points.

Traditionally, risk analysis has relied heavily on the concept of the yearly expected value. Note that f_5, the yearly expected damage, takes unusually high values (on the order of 161×10^6 to 162×16^6). This is due to the assumptions concerning antecedent floods—in particular the assumptions that the reservoir is filled to the spillway's crest and that the outlet is open to 75% of its capacity. Therefore, any small inflow into the reservoir will cause large damages, on the order of 160×10^6. These two assumptions were made to comply with the guidelines and recommendations established by the US Army Corps of Engineers.

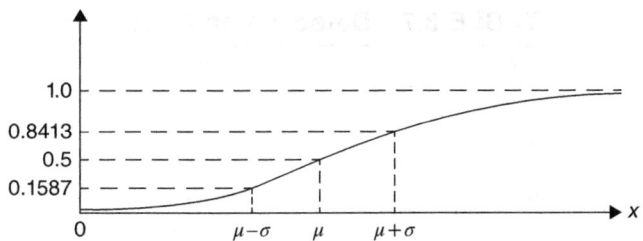

Figure 8.11 Cumulative probability distribution function.

It is also apparent from Table 8.5 that when the dam's height is increased, f_5 decreases, but by less than 0.3% (see Figure 8.11). Furthermore, when the spillway's width is increased, f_5 increases in general; and when it decreases, it does so by less than 0.02%. These observations could lead the decisionmaker to conclude that increasing the spillway's width is not an attractive solution because any investment in such an action will mainly increase the risks. By looking at the trade-offs, the decisionmaker could also find incentives not to invest money to raise the dam, since under alternative s_2 an investment of 10^6 US$ will not reduce the expected yearly damages by more than $25,386.

But if the decisionmaker takes into consideration the rest of the risk objective functions, in particular $f_4(s_j)$, then the picture of the problem might radically change. First, notice that f_4 decreases greatly when the spillway's width is increased but that f_3 increases. In other words, the decisionmaker will be able to see that by increasing the spillway's width, the risks in the LP/HC domain are decreasing, because spillway widening reduces both the probability of dam failure and the damages in case of failure. The decisionmaker will also note that the risks associated with less extreme events are increasing, because floods that are relatively frequent will cause more downstream damage. Moreover, even when compared to increasing the dam's height, spillway widening could still be an attractive solution. For example, s_6, which would have been disregarded if traditional risk analysis methods were used, becomes a noninferior solution if the risk objective f_4 is considered. Thus, by using the PMRM, the decisionmaker can better understand the trade-offs among risks that correspond to the various risk domains.

Moreover, regarding the alternative of increasing the dam's height, the use of f_4 allows explicit quantification of risks in the LP/HC risk domain, and this might induce the decisionmaker to invest money in some situations where such an investment might not have been made had only f_5 been considered. Using the same example as in the preceding text, investing $1 million under alternative s_2 only reduces the expected yearly damages by $25,386. It is apparent that if f_4 is included, then, in the case of an extreme event, up to $31,924,280 in yearly damages might be saved with a probability of 7.371×10^{-4}.

Notice that for this problem, because smaller inflows caused the same amount of damages for all alternatives, $f_2(s_j)$ is constant for all alternatives and therefore is of no interest to the decisionmaker. This can be interpreted to mean that the HP/LC risk domain provides no additional information and for this reason will be disregarded. By using the PMRM, the decisionmaker is able to grasp certain aspects of the problem that would have been completely ignored had he or she simply used the yearly expected value of damages. These aspects were mainly associated with LP/HC risks in this case, but this is not a general restriction.

8.9 EXAMPLE PROBLEMS

8.9.1 Groundwater Contamination

There are several processes available today to clean up contaminated groundwater, including air stripping (aeration) and the use of granular activated carbon (GAC). Each of these two processes can be used at different levels and in combination with each other. As one might expect, the more intensive the cleanup process, the better its performance and the higher its cost.

One of the major chemical companies has recently completed a study that provides the relationship between the level of concentration reduction of the contaminant and the probability of achieving that level. Table 8.6 provides the cumulative probability associated with each level of resultant concentration for six different cleanup policies. The ith cleanup policy, which is designated by the notation u_i, denotes the cost in millions of dollars associated with that policy.

Because of the limited available information, it is assumed that the cleanup process follows a normal

TABLE 8.6 Database on Contaminant Concentration

Cumulative Probability	Resultant Concentration of Contaminant, x_i (ppb)					
	u_1	u_2	u_3	u_4	u_5	u_6
0.1234	10	13	20	40	70	100
0.2572	11	15	25	50	80	120
0.3577	12	20	27	70	100	150
0.4321	14	25	35	90	120	180
0.5123	16	30	45	110	150	200
0.6915	20	40	60	130	180	250
0.8413	22	50	80	150	200	300
0.9938	23	55	100	180	220	320
0.9981	24	57	105	190	250	340
0.9999	25	59	110	200	280	350

TABLE 8.7 Database on Cost

Policy	Cost ($M)
u_1	25
u_2	20
u_3	15
u_4	10
u_5	5
u_6	0

distribution. Using the PMRM, the probability axis is partitioned into three segments:

(a) $\quad -\infty < P \leq \mu - \sigma$
(b) $\quad \mu - \sigma \leq P \leq \mu + \sigma$ $\qquad$ (8.30)
(c) $\quad P > \mu + \sigma$

$$P[x < \mu - \sigma] = P\left[\frac{x - \mu}{\sigma} < -1\right] = P[z < -1] = \Phi(-1)$$

$$\Phi(-1) = 1 - \Phi(1) = 1 - 0.8413 = 0.1587$$

(8.31)

$$P[x < \mu + \sigma] = P\left[\frac{x - \mu}{\sigma} < 1\right] = \Phi(1) = 0.8413$$

(8.32)

where $\Phi(\cdot)$ is the standard normal probability, and the above probability partition points are true for any values of μ and σ (provided the distribution is normal).

Tables 8.6 and 8.7 summarize the database. For example, under policy u_4, the probability of achieving 150 ppb or less is 0.8413, and the cost of implementing policy u_4 is $10M. Figure 8.11 depicts the cdf.

The following relationships, which are used in this example problem, are derived in Chapter 10 for normal distributions:

$$\begin{aligned} f_2(u) &= \mu(u) + \beta_2 \sigma \\ f_4(u) &= \mu(u) + \beta_4 \sigma \\ f_3(u) &= \mu(u) + \beta_3 \sigma \\ f_5(u) &= \mu(u) \end{aligned}$$

(8.33)

The following values of β_2, β_3, and β_4 (associated with the partitioning given in Eqs. 8.34–8.36) are based on derivations presented in Chapter 10:

$$\beta_2 = \frac{\int_{-\infty}^{\mu-\sigma} \frac{\tau}{(2\pi)^{1/2}} e^{-\tau^2/2} d\tau}{\int_{-\infty}^{\mu-\sigma} \frac{1}{(2\pi)^{1/2}} e^{-\tau^2/2} d\tau} = -1.5247, \quad 0 \leq P < \mu - \sigma$$

(8.34)

$$\beta_3 = \frac{\int_{\mu-\sigma}^{\mu+\sigma} \frac{\tau}{(2\pi)^{1/2}} e^{-\tau^2/2} d\tau}{\int_{\mu-\sigma}^{\mu+\sigma} \frac{1}{(2\pi)^{1/2}} e^{-\tau^2/2} d\tau} = 0, \quad \mu - \sigma \leq P \leq \mu + \sigma$$

(8.35)

$$\beta_4 = \frac{\int_{\mu+\sigma}^{\infty} \frac{\tau}{(2\pi)^{1/2}} e^{-\tau^2/2} d\tau}{\int_{\mu+\sigma}^{\infty} \frac{1}{(2\pi)^{1/2}} e^{-\tau^2/2} d\tau} = 1.5247, \quad P > \mu + \sigma$$

(8.36)

Solution: Since we have an expression for $f_4(u)$ in terms of μ, σ, and a constant, we need only approximate μ and σ for each (u). The process is normal; therefore, the best estimates for μ and σ are the maximum likelihood indicators. Let x_i denote the contaminant concentration for the ith probability:

$$\mu(u) \cong \bar{x}(u) = \frac{1}{n} \sum_{i=1}^{n} x_i \qquad (8.37)$$

$$\sigma^2(u) \cong s^2(u) = \frac{1}{n-1} \sum_{i=1}^{n} (x_i - \bar{x}(u))^2 \qquad (8.38)$$

Thus,

for u_1, $\bar{x}(u_1) = 17.1$; $s^2(u_1) = 33.12$, $s(u_1) = 5.755$
for u_2, $\bar{x}(u_2) = 36.4$; $s(u_2) = 18.026$
for u_3, $\bar{x}(u_3) = 60.7$; $s(u_3) = 35.409$
for u_4, $\bar{x}(u_4) = 121$; $s(u_4) = 58.395$
for u_5, $\bar{x}(u_5) = 165$; $s(u_5) = 72.763$
for u_6, $\bar{x}(u_6) = 231$; $s(u_6) = 93.506$

Thus, using the expression for $f_4(u)$, we obtain

$$f_4(u) = \bar{x}(u) + (1.525)s(u)$$
$$f_5(u) = \bar{x}(u) \quad \{\text{unconditional expectation}\}$$

Using the values of $\bar{x}(u)$ and $s(u)$ computed earlier, we can derive Table 8.8 and Figure 8.12.

Note that the conditional expected value of contaminant concentration is higher than the unconditional expected value for all corresponding policies. For example, for policy u_4 and at a cost of $10 million, the expected value of contaminant concentration is 121 ppb, while the conditional expected value is close to double that value (210 ppb). This example highlights the distortion of the averaging effect of the unconditional expected value of risk.

8.9.2 Environmental Health and Safety

A chemical facility is leaking and the chemical waste is ending up, in part, in a nearby well currently used for drinking water. The objective is to find a cost-effective way of minimizing the potential for groundwater contamination.

Solution: Three methods for cleanup are considered:

1. Use neutralizing chemical 1—effective but expensive.
2. Use neutralizing chemical 2—less expensive but less effective.
3. Do nothing.

Two methods for future storage are considered:

1. Storage in open holding pond
2. Storage in steel drums

Six alternative management options are considered:

A: Use neutralizing chemical 1 to clean up and store in holding pond.
B: Use neutralizing chemical 2 to clean up and store contaminant in holding pond.
C: Do nothing to clean up; store contaminant in holding pond.
D: Use neutralizing chemical 1 to clean up and store contaminant in steel drums.
E: Use neutralizing chemical 2 to clean up and store contaminant in steel drums.
F: Do nothing to clean up and store contaminant in steel drums.

The six alternative policy options and their corresponding cost, mean (in ppb), and standard deviation are summarized in Table 8.9.

Assuming that the six policy options are governed by lognormal distributions, we calculate the (unconditional) expected value, $f_5(\cdot)$ (using Eq. 8.39), and the conditional expected value, $f_4(\cdot)$

TABLE 8.8 Summary of Results (Groundwater Contamination)

u	$f_1(u)$ ($M10)	$f_5(u)$ (ppb)	$f_4(u) = \bar{x} + \beta_4 s$ (ppb)
u_1	25	17.7	26.475
u_2	20	36.4	63.884
u_3	15	60.7	114.69
u_4	10	121	210.03
u_5	5	165	275.942
u_6	0	231	373.57

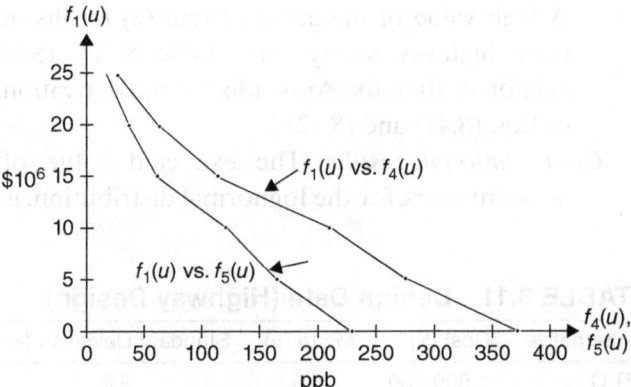

Figure 8.12 Cost versus unconditional $f_5(u)$ and conditional $f_4(u)$ expected value of risk.

246 RISK OF EXTREME EVENTS AND THE FALLACY OF THE EXPECTED VALUE

(using Eq. 8.40), at a partitioning on the probability axis for $\alpha=0.999$.

Expected value of accident rate:

$$f_5(\cdot) = \exp\left[\mu + \frac{\sigma^2}{2}\right] \quad (8.39)$$

Conditional expected value of accident rate:

$$f_4 = \frac{\exp\left[\mu + (\sigma^2/2)\right]}{(1-\alpha)}\left\{1 - \Phi\left[\Phi^{-1}(\alpha) - \sigma\right]\right\} \quad (8.40)$$

TABLE 8.9 Design Data (Environmental Cleanup)

Alternative	Cost ($)	Mean (μ)	Standard Deviation (σ)
A	400,000	0.5	0.5
B	300,000	1.0	0.5
C	100,000	1.5	0.5
D	600,000	0.5	0.1
E	500,000	1.0	0.1
F	300,000	1.5	0.1

TABLE 8.10 Compiled Results (Environmental Cleanup)

Alternative	Cost ($)	$f_5(\cdot)$	$f_4(\cdot)$
A	400,000	1.868246	8.965713
B	300,000	3.080217	14.78196
C	100,000	5.078419	24.37133
D	600,000	1.656986	2.311495
E	500,000	2.731907	3.811011
F	300,000	4.504154	6.283294

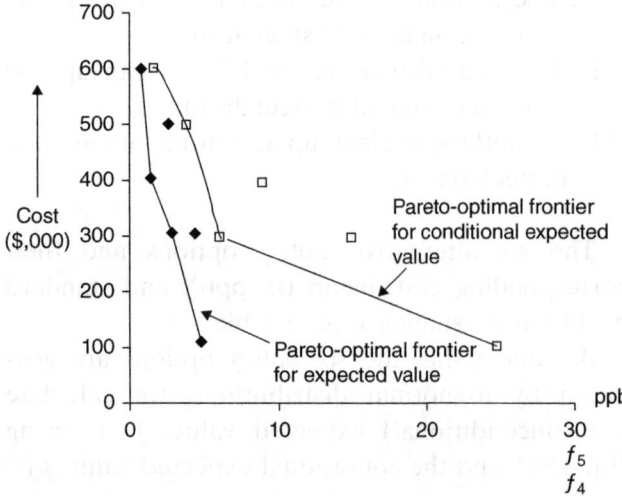

Figure 8.13 *Cost versus expected value and conditional expected value of contaminant concentration.*

where $\alpha=0.999$ is the partition point on the probability axis. Table 8.10 summarizes these results. [See Section A.10 in the Appendix for the derivation of Eqs. (8.39) and (8.40).]

The cost of risk management versus the expected value and the conditional expected value of contaminant concentration are depicted in Figure 8.13.

8.9.3 Highway Design

Design a new highway taking into consideration the various environment-related design factors affecting accident rate (number of accidents per week). There are two objectives: minimize accident rate and minimize construction cost.

Solution: Environmental design factors affecting accident rate include roadway type, type of intersection and interchange, grades, curves, roadside hazards, speed differentials, and stopping sight distance. The following are the design factors considered:

Roadway type: four-lane divided highway (R_1), four-lane undivided highway (R_2), and two-lane undivided highway (R_3)
Curves: gradual curves (C_1) and sharp curves (C_2)

Thus, the total number of combinations considered is $2 \times 3 = 6$.

Design data: The accident rate (λ) for the highway is assumed to be of a lognormal distribution $LN(\mu, \sigma)$, where the parameters μ and σ are determined by the different design options. A high value of the accident rate (λ) results in poor highway safety (see Table 8.11). (See Section A.10 in the Appendix for the derivation of Eqs. (8.41) and (8.42).)

Computational results: The expected value of accident rates, for the lognormal distribution, is

TABLE 8.11 Design Data (Highway Design)

Alternative	Cost ($)	Mean (μ)	Standard Deviation (σ)
R_1C_1	1,000,000	0.3	0.3
R_2C_1	500,000	1.0	0.5
R_3C_1	200,000	1.5	1.0
R_1C_2	800,000	1.2	0.4
R_2C_2	700,000	1.8	0.6
R_3C_2	400,000	2.0	0.6

$$f_5(\cdot) = \exp\left[\mu + \frac{\sigma^2}{2}\right] \quad (8.41)$$

The conditional expected value of accident rate is

$$f_4(\cdot) = \frac{\exp\left[\mu + (\sigma^2/2)\right]}{(1-\alpha)}\left\{1 - \Phi\left[\Phi^{-1}(\alpha) - \sigma\right]\right\} \quad (8.42)$$

where α is the partition point on the probability axis ($\alpha = 0.99$). Table 8.12 summarizes the values of $f_5(\cdot)$ and $f_4(\cdot)$ associated with the six design options. Figure 8.14 depicts the cost versus $f_4(\cdot)$ and $f_5(\cdot)$ for all six design options.

TABLE 8.12 Compiled Results (Highway Design)

Alternative	Cost ($)	$f_5(\cdot)$	$f_4(\cdot)$
R_1C_1	1,000,000	1.412	2.993
R_2C_1	500,000	3.080	10.349
R_3C_1	200,000	7.389	67.831
R_1C_2	800,000	3.896	10.441
R_2C_2	700,000	7.243	30.276
R_3C_2	400,000	8.846	36.976

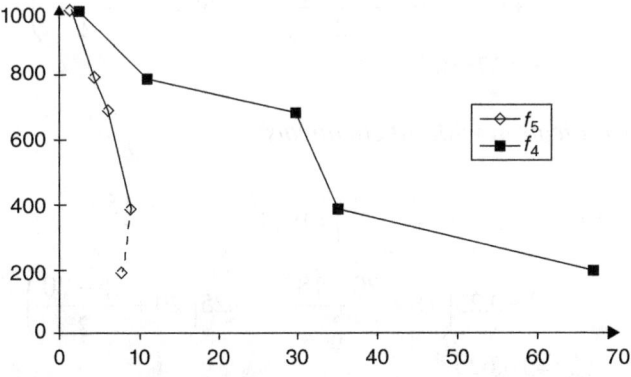

Figure 8.14 Cost versus expected value (f_5) and conditional expected value (f_4) of risk.

8.9.4 The Medfly Problem

The Mediterranean fruit fly is a major concern of agriculture throughout the world and has recently become a threat to US agriculture. The USDA [1994] has evaluated five policy options in the event that the United States becomes infested with the *medfly*. These include:

1. No action
2. Suppression with chemicals
3. Suppression without chemicals
4. Eradication with chemicals
5. Eradication without chemicals

In order to better assess these policy options, extreme-event analysis is used.

8.9.4.1 Fractile method

Probability distributions in this example problem are determined based upon expert evidence and scientific studies. In this case, data were derived using the fractile method. For pedagogical purposes, we start the solution with the suppression with chemicals option:

- Worst case of agriculture loss = 40%.
- Best case of agricultural loss = 2%.
- Median value (equal likelihood of being greater than or less than this value) = 20%.
- 25th percentile is 20 − 10 = 10%.
- 75th percentile is 20 + 10 = 30%.

Table 8.13 summarizes the above expert evidence information for all five options. Figures 8.15 and 8.16 depict the cdf and pdf for the suppression with chemicals option.

To compute the height (frequency) of the bars for Figure 8.16, we apply simple geometry. Since the total shaded area of the pdf is equal to one, then the area of each one of the four blocks is 0.25. Accordingly, the height of the first block is equal to its area (0.25) divided by its base (10 − 2), that is, 0.25/8 = 0.031.

The expected value of the percentage of agricultural loss can be calculated geometrically:

$$E[x] = f_5(\cdot) = p_1x_1 + p_2x_2 + p_3x_3 + p_4x_4$$

TABLE 8.13 Agricultural Percentage Loss for Each Option

	Best (0)	25th	Median 50th	75th	Worst (100)
No action	2	50	60	90	100
Suppression with chemicals	2	10	20	30	40
Suppression—no chemicals	2	12	22	35	42
Eradication with chemicals	0	8	10	12	15
Eradication—no chemicals	2	15	18	20	25

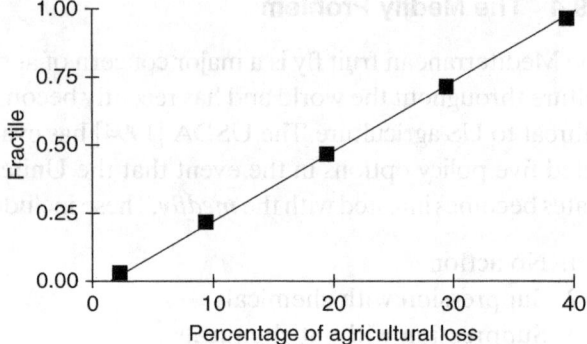

Figure 8.15 Suppression with chemicals option: cumulative distribution function (cdf).

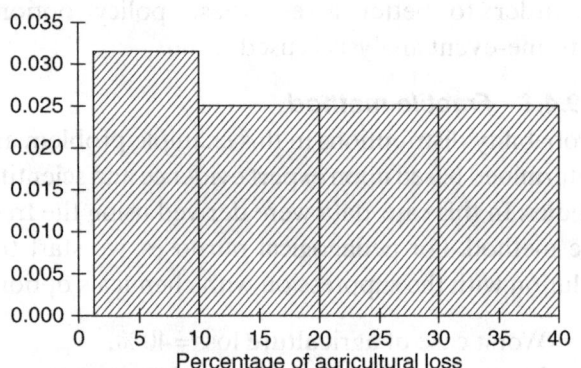

Figure 8.16 Suppression with chemicals: probability density function (pdf).

No action:

$$f_5 = 0.25\left(2+\frac{50-2}{2}\right) + 0.25\left(50+\frac{60-50}{2}\right)$$
$$+ 0.25\left(60+\frac{90-60}{2}\right) + 0.25\left(90+\frac{100-90}{2}\right)$$
$$= 6.5 + 13.75 + 18.75 + 23.75 = 62.75\%$$

Suppression with chemicals:

$$f_5 = 0.25\left(2+\frac{10-2}{2}\right) + 0.25\left(10+\frac{20-10}{2}\right)$$
$$+ 0.25\left(20+\frac{30-20}{2}\right) + 0.25\left(30+\frac{40-30}{2}\right)$$
$$= 20.25\%$$

Suppression without chemicals:

$$f_5(\cdot) = 0.25\left(2+\frac{12-2}{2}\right) + 0.25\left(12+\frac{22-12}{2}\right)$$
$$+ 0.25\left(22+\frac{35-22}{2}\right) + 0.25\left(35+\frac{42-35}{2}\right)$$
$$= 22.75\%$$

TABLE 8.14 Summary: Fractile Method

Policy	Estimated Cost ($M)	$f_5(\cdot)$ %
No action	0	62.75
Suppression with chemicals	50	20.25
Suppression without chemicals	30	22.75
Eradication with chemicals	100	9.38
Eradication without chemicals	75	16.63

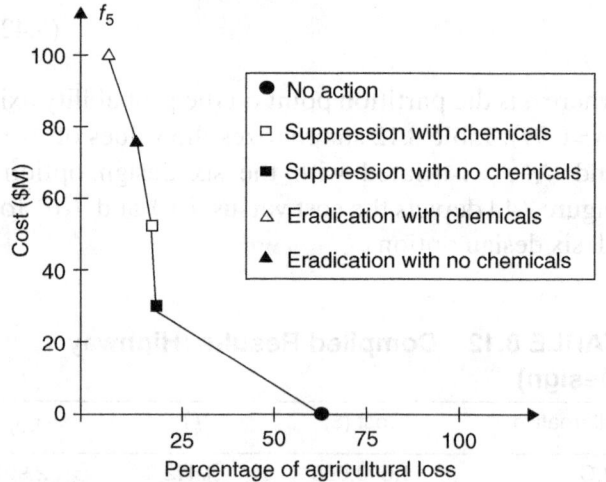

Figure 8.17 Cost versus percentage of agricultural loss.

Eradication with chemicals:

$$f_5(\cdot) = 0.25\left(0+\frac{8-0}{2}\right) + 0.25\left(8+\frac{10-8}{2}\right)$$
$$+ 0.25\left(10+\frac{12-10}{2}\right) + 0.25\left(12+\frac{15-12}{2}\right)$$
$$= 9.375\%$$

Eradication without chemicals:

$$f_5(\cdot) = 0.25\left(2+\frac{15-2}{2}\right) + 0.25\left(15+\frac{18-15}{2}\right)$$
$$+ 0.25\left(18+\frac{20-18}{2}\right) + 0.25\left(20+\frac{25-20}{2}\right)$$
$$= 16.625\%$$

A graphical representation of the cost associated with each policy (see Table 8.14) versus the expected value of the percentage of agricultural loss is depicted in Figure 8.17. The USDA is also interested in the worst 10% scenario, that is, the conditional expected value of percentage of agricultural loss, given that the loss occurs with a probability of

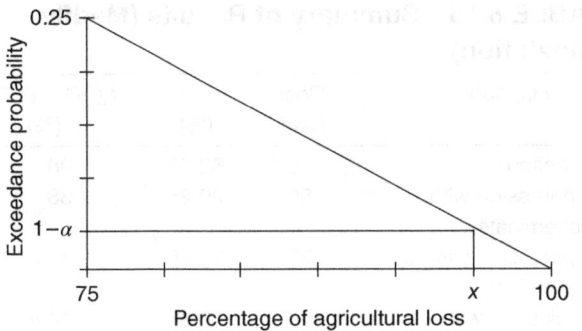

Figure 8.18 Exceedance probability versus percentage agricultural loss.

0.10 or lower. Therefore, the partition point on the damage axis corresponding to $(1-\alpha)=0.1$ is computed. In other words, to compute the conditional expected value of agricultural loss, we need to project the partitioning of the probability axis at $\alpha = 0.9$ to the damage axis (i.e., agricultural loss). Figure 8.18 depicts the probability of exceedance versus the percentage of agricultural loss. Note that $\alpha = 0.9$ translates into what the USDA considers, in this case, as the worst 10% scenario. Using simple geometry, we calculate the percentage of agricultural loss that corresponds to a probability of exceedance of 0.1 for each policy option.

No action:

$$\frac{x-90}{100-90} = \frac{0.25-(1-\alpha)}{0.25};$$

$$\frac{x-90}{10} = \left(\frac{0.25-0.1}{0.25}\right); \quad x = 96\%$$

This means that the probability of exceeding 96% of agricultural loss is 0.10. The partition points on the damage axis are computed for the other four policy options:

Suppression with chemicals:

$$\frac{x-30}{40-30} = \frac{0.25-0.1}{0.25}; \quad x = 36\%$$

Suppression with no chemicals:

$$\frac{x-35}{42-35} = \frac{0.25-0.1}{0.25}; \quad x = 39.2\%$$

Eradication with chemicals:

$$\frac{x-12}{15-12} = \frac{0.25-0.1}{0.25}; \quad x = 13.8\%$$

Eradication with no chemicals:

$$\frac{x-20}{25-20} = \frac{0.25-0.1}{0.25}; \quad x = 23\%$$

The conditional expected values, $f_4(\cdot)$, are then computed with these partition points. Note that the straight line of the exceedance probability (see Figure 8.18) means that the cdf is also a straight line, representing a pdf of a uniform distribution. Thus, the conditional expected value of a uniform distribution is the average between the lowest and highest values.

No action:

$$f_4(\cdot) = \frac{96+100}{2} = 98\%$$

This geometry-based calculation can be also computed using integration:

$$f_4(\cdot) = \frac{\int_{96}^{100} x p(x)\, dx}{\int_{96}^{100} p(x)\, dx} = \frac{\int_{96}^{100} xK\, dx}{\int_{96}^{100} K\, dx} = \frac{\left.\frac{x^2}{2}\right|_{96}^{100}}{\left.x\right|_{96}^{100}} \quad (8.43)$$

$$= \frac{(10{,}000 - 9{,}216)}{2(100-96)}$$

$$f_4(\cdot) = 98\% \text{ of agricultural loss}$$

Suppression with chemicals:

$$f_4(\cdot) = \frac{36+40}{2} = 38\%$$

Using integration, we obtain

$$f_4(\cdot) = \frac{\int_{36}^{40} x p(x)\, dx}{\int_{36}^{40} p(x)\, dx} = \frac{\int_{36}^{40} xK\, dx}{\int_{36}^{40} K\, dx} = \frac{\left.\frac{x^2}{2}\right|_{36}^{40}}{\left.x\right|_{36}^{40}}$$

$$= \frac{(1600 - 1296)}{2(40-36)} \quad (8.44)$$

$$f_4(\cdot) = 38\% \text{ of agricultural loss}$$

Suppression without chemicals:

$$f_4(\cdot) = \frac{39.2 + 42}{2} = 40.6\%$$

Using integration, we obtain

$$f_4(\cdot) = \frac{\int_{39.2}^{42} xp(x)\,dx}{\int_{39.2}^{42} p(x)\,dx} = \frac{\int_{39.2}^{42} xK\,dx}{\int_{39.2}^{42} K\,dx} = \frac{\left.\frac{x^2}{2}\right|_{39.2}^{42}}{x\big|_{39.2}^{42}} \quad (8.45)$$

$$= \frac{1764 - 1536.64}{2(42 - 39.2)}$$

$$f_4(\cdot) = 40.6\% \text{ of agricultural loss}$$

Eradication with chemicals:

$$f_4(\cdot) = \frac{13.8 + 15}{2} = 14.4\%$$

Using integration, we obtain

$$f_4(\cdot) = \frac{\int_{13.8}^{15} xp(x)\,dx}{\int_{13.8}^{15} p(x)\,dx} = \frac{\int_{13.8}^{15} xK\,dx}{\int_{13.8}^{15} K\,dx} = \frac{\left.\frac{x^2}{2}\right|_{13.8}^{15}}{x\big|_{13.8}^{15}}$$

$$= \frac{225 - 190.44}{2(15 - 13.8)} \quad (8.46)$$

$$f_4(\cdot) = 14.4\% \text{ of agricultural loss}$$

Eradication without chemicals:

$$f_4(\cdot) = \frac{23 + 25}{2} = 24\%$$

Using integration, we obtain

$$f_4(\cdot) = \frac{\int_{23}^{25} xp(x)\,dx}{\int_{23}^{25} p(x)\,dx} = \frac{\int_{23}^{25} xK\,dx}{\int_{23}^{25} K\,dx} = \frac{\left.\frac{x^2}{2}\right|_{23}^{25}}{x\big|_{23}^{25}}$$

$$= \frac{625 - 529}{2(25 - 23)} \quad (8.47)$$

$$f_4(\cdot) = 24\% \text{ of agricultural loss}$$

The above results are summarized in Table 8.15, and the conditional expected value, $f_4(\cdot)$, is plotted in Figure 8.19 alongside the unconditional (traditional) expected value for further insight.

This analysis shows that suppression with no chemicals and no action policies have a larger risk of extreme events. Furthermore, using only traditional expected value, it appears that suppression with no chemicals is about on par with eradication with no chemicals in terms of % agricultural loss. Should prices change, either option may appear to be favorable. However, when we analyze the conditional expected values, eradication with no chemicals is much more stable and appears to be a more favorable policy than suppression with no chemicals.

TABLE 8.15 Summary of Results (Medfly Infestation)

Policy Options	Cost ($M)	$f_5(\cdot)$ (%)	$f_4(\cdot)(1-\alpha=0.1)$ (%)
No action	0	62.75	98
Suppression with chemicals	50	20.25	38
Suppression without chemicals	30	22.75	40.6
Eradication with chemicals	100	9.38	14.4
Eradication without chemicals	75	16.63	24

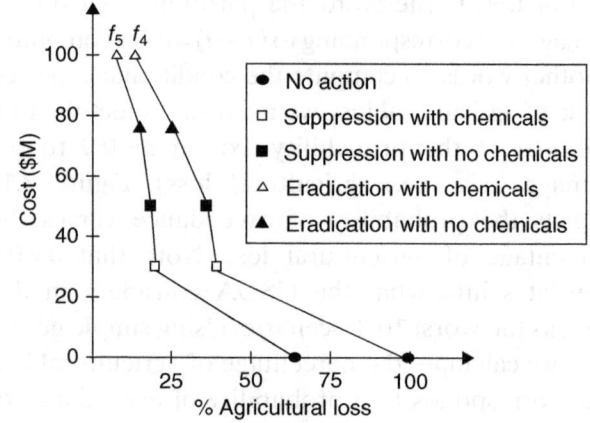

Figure 8.19 A comparison of the conditional and traditional expected value.

8.9.5 Airplane Acquisition Revisited

In Section 4.5.1, we introduced the airplane acquisition problem. Here, we add the conditional expected value to the analysis. Recall that based on the geometry presented in Figure 4.8, we found a 38% increase in project cost corresponding to an exceedance probability of 0.1 $(1-\alpha=0.1)$.

The conditional expected value of project cost can be calculated for several scenarios to shed light on the behavior of the tail of the pdf. For example, from Figures 4.7 and 4.8, given that there is 0.1 probability of project cost overrun that would be equal to or exceed 38% of its original scheduled budget, management might be interested in answering the following question: What is the conditional expected value of extreme cost overrun beyond the 38% (or extreme cost overrun with exceedance probability that is below 0.1)? Or posed differently, within the range of exceedance probabilities between 0.1 and 0.0 and range of cost overruns between 38 and 50%, what is the expected value of project cost overrun? Note that (i) the maximum cost overrun was predicted not to exceed 50%, (ii) the conditional expected value is the common expected value limited between specific levels of cost overruns instead of the entire range of possible cost overruns, and (iii) the expected value is a weighted average of possible cost overruns multiplied by their corresponding probabilities of occurrence and summed over that entire range.

Using Eq. (4.17), the common, unconditional expected value of cost overrun, $f_5(\cdot)$, was calculated earlier to be 17.5%:

$$f_5(\cdot) = \int_0^{10} xp_x(x)\,dx + \int_{10}^{20} xp_x(x)\,dx + \int_{20}^{50} xp_x(x)\,dx$$

$$f_5(\cdot) = \int_0^{10} 0.025x\,dx + \int_{10}^{20} 0.05x\,dx + \int_{20}^{50} 0.00833x\,dx$$

$$= 0.025 \frac{x^2}{2}\bigg|_0^{10} + 0.05\frac{x^2}{2}\bigg|_{10}^{20} + 0.00833\frac{x^2}{2}\bigg|_{20}^{50}$$

$$= 0.025(50) + 0.05(200-50) + 0.00833(1250-200)$$

$$= 1.25 + 7.50 + 8.75$$

$$= 17.50\% \text{ (i.e., total cost of } \$(150+26.25) \text{ million)}$$

Note that the expected value of cost overrun of $26.25 million (i.e., total cost of $176.25 million) does not provide any vital information on the probable extreme behavior of project cost. Also, note that there is a one-to-one functional relationship between 0.1 probability of exceedance and 38% cost overrun; this relationship is depicted in Figure 4.8 in Chapter 4 and is generated as follows (here, we are interested in the probability of exceedance of 0.1 — i.e., $a = 0.90$, or $(1-a) = 0.10$):

$$\frac{x-20}{50-20} = \frac{0.25-(1-\alpha)}{0.25}$$

Thus,

$$x = 30 - \frac{30(1-\alpha)}{0.25} + 20 = 30\%; \quad \text{for } \alpha = 0.9$$

Alternatively, we can compute from Figure 4.8 the partition point x (the percentage of increase in cost) that corresponds to a probability of 0.1 as shown below:

$$(1-\alpha) = (50-x)/h$$

where h is the height in the probability axis:

$$h = \frac{0.25}{50-20} = 0.0083$$

$$x = 50 - \left(\frac{1-\alpha}{h}\right) = 50 - \frac{(1-0.9)}{0.0083} = 38\%; \quad \text{for } \alpha = 0.9$$

Similarly, the conditional expected value of cost overrun under the scenario of 0.1 probability of exceeding the original cost estimate (by 38% or by $57 million), computed using Eq. (8.49), yields $f_4(\cdot) = 44\%$:

$$f_4(\cdot) = \frac{\int_{38}^{50} xp(x)\,dx}{\int_{38}^{50} p(x)\,dx} = \frac{\int_{38}^{50} xK\,dx}{\int_{38}^{50} K\,dx} = \frac{\frac{x^2}{2}\big|_{38}^{50}}{x\big|_{38}^{50}} \quad (8.48)$$

$$f_4(\cdot) = 44\% \text{ (i.e., } \$(150+66) = \$216 \text{ million)}$$

Note that the pdf of the cost overrun portion from 20% and beyond is a linear function (kx). Alternatively, the conditional expected value can be computed (on the basis of the geometry of the pdf) as the mean of the shaded area in Figure 8.20, yielding, of course, the same result:

$$f_4(\cdot) = 38 + \frac{50-38}{2} = 44\% \quad (8.49)$$

In other words, the adjusted (conditional) expected value of cost overrun, when it is in the range of 38–50% of the original scheduled cost, is 44%.

Even if the project is a cost-plus contract, the interpretation of these results should alarm the top management of contractor A: Although the expected cost overrun of the proposed budget is 17.50% above the budgeted cost of $150 million, there is a 10%

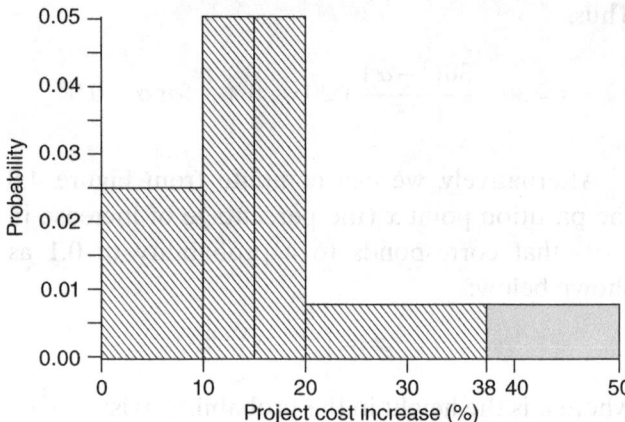

Figure 8.20 Computing the conditional expected value (f_4) for contractor A.

chance (0.1 probability) that the cost overrun will exceed 38% of the budgeted cost! Furthermore, at a 10% chance of cost overrun, the conditional expected value of cost overrun that exceeds 38% is 44% above the original budget—that is, an exceedance of $66 million. In other words, under these conditions, the conditional expected value of the total cost will be ($150 + $66) million = $216 million.

It is worthwhile to clarify at this point the meaning of the two distinct terms of cost overrun: 38 and 44%. The term *38% cost overrun* corresponds to a single probability point and is derived directly from Figure 4.8. The term *44%* represents the conditional expected value, the averaging of all the probabilities from 0.10 to 0 multiplied by the corresponding cost overruns from 38 to 50%, summed as appropriate and scaled. Thus,

$$f_4(\cdot) = E[X | > 38\% \text{ cost overrun}] = 44\%$$

or, equivalently,

$$f_4(\cdot) = E[X | > \$207 \text{ million}] = \$216 \text{ million}$$

It is constructive to further clarify the information summarized in Table 8.16. Consider the customer's column. According to the customer's estimates, the common, unconditional expected value of cost overrun is 11.25%. Through mathematical calculations based on the information provided by the customer (as shown in Tables 8.16 and 8.17), it can be determined that there is a 0.1 probability of project cost overrun that would exceed 24% of its original scheduled cost (see Haimes and Chittister, 1995). Thus, the

TABLE 8.16 Comparative Tabular cdf

Fractile	Project Cost Increase (%)		
	Customer	Contractor A	Contractor B
0.00	0	0	0
0.25	5	10	15
0.50	10	15	20
0.75	15	20	25
1.00	30	50	40

TABLE 8.17 Summary of Results (Airplane Acquisition)

	Customer	Contractor A	Contractor B
Unconditional expected value, $f_5(\cdot)$	11.25%	17.50%	20.00%
Partitioning point	$\alpha = 0.90$	$\alpha = 0.90$	$\alpha = 0.90$
Corresponding percentage of cost increase	$x = 24\%$	$x = 38\%$	$x = 34\%$
Conditional expected value, $f_4(\cdot)$	27%	44%	37%

conditional expected value of extreme cost overrun between 24 and 30% (or extreme cost overrun with exceedance probability below 0.1) is 27%.

8.9.6 Risks of Cyber Attack to a Water Utility: Supervisory Control and Data Acquisition Systems*

Water systems are increasingly monitored, controlled, and operated remotely through supervisory control and data acquisition (SCADA) systems. The vulnerability of the telecommunications system renders the SCADA system vulnerable to intrusion by terrorist networks or by other threats. This case study addresses the risks of willful threats to water utility SCADA systems. As a surrogate for terrorist networks, the focus in this case study is on a disgruntled employee's attempt to reduce or eliminate the water flow in a city we'll call XYZ. The data are based on actual survey results that revealed that the primary concern of US water utility managers in City XYZ was the disgruntled employees [Ezell, 1998].

___

*This example is adopted from Ezell et al. [2001].

8.9.6.1 Identifying risks through system decomposition

Using hierarchical holographic modeling (HHM), the following major head topics and subtopics were identified [Ezell et al., 2001] (see Figure 8.21).

Head topic A: *Function.* Given the importance of the water distribution system, its function is a major source of risk from cyber intrusion. This category may be partitioned into three subtopics.

Head topic B: *Hardware.* The hardware of SCADA is vulnerable to tampering in a variety of configurations. Depending on the tools and skill of an attacker, these subtopics could have a significant impact on water flow for a community.

Head topic C: *Software.* Perhaps the most complex, this head topic also represents the most dynamic aspects of changes in water utilities. Software has many components that are sources of risk—among them are C_1 *controlling* and C_2 *communication*.

Head topic D: *Human.* There are two major subtopics: D_1 *employees* and D_2 *attackers*. This head topic addresses a decomposition of those capable of tampering with a system.

Head topic E: *Tools.* A distinction is made between the various types of tools an intruder may use to tamper with a system. There are six subtopics.

Head topic F: *Access.* There are many paths into a system. An intruder can exploit these vulnerabilities and pose a severe risk. There are five subtopics. A system may be designed to be safe, yet its installation and use may lead to multiple sources of risk.

Head topic G: *Geographic.* Location is not relevant for many risks of cyber intrusion, as tampering with a SCADA system can have global sources. International borders are virtually nonexistent because of the Internet. Four subtopics are identified.

Head topic H: *Temporal.* The temporal category seeks to show how present or future decisions affect the system. The decision to replace a legacy SCADA system in 10 years may have to be made today. Therefore, this head topic addresses the life cycle of the system. There are four partitions.

8.9.6.2 City XYZ

City XYZ is relatively small with a population of 10,000 [Wiese et al., 1997]. It has a water distribution system that accepts processed and treated water *as is* from an adjacent city. The water utility of XYZ is primarily responsible for an uninterrupted flow of water to its customers. The SCADA system uses a master–slave relationship, relying on the total control of the SCADA master; the remote terminal units are dumb. There are two tanks and two pumping stations as shown in Figure 8.22. The first tank serves the majority of customers; the second tank serves relatively fewer customers in a topographically high-level area. Tank II is at a point higher than the highest customer served. The function of the tanks is to provide a buffer and to allow the pumps to be sized lower than peak instantaneous demand.

The tank capacity has two component segments: One is a reserve storage that allows the tank to operate over a peak week when demand exceeds pumping capacity. The other component is control storage; this is the portion of the tank between the pump cutout and cut-in levels. Visually, the control storage is the top portion of the tank. If demand is less than the pump rate (low-demand periods), the level rises until it reaches the pump cutout level. When the water falls to the tank cut-in level, it triggers the pump to start operating. If the demand is greater than the pump rate, the level will continue to fall until it reaches reserve storage. The water level will stay in this area until the demand has fallen for a sufficient time to allow it to recover. The reserve storage is sized according to demand (e.g., Tank I with its larger reserve storage serves more customers).

The SCADA master communicates directly with Pumping Stations I and II and signals the unit when to start and stop. The operating levels are kept in the SCADA master. Pumping Station I boosts the flow of water beyond the rate that can be supplied by gravity. The function of Pumping Station II is to pump water off-peak from Tank I to Tank II. The primary operational goal of both stations is to maximize gravity flow and, as necessary, to pump off-peak as much as possible. The pumping stations receive a start command from the SCADA master via the master terminal unit (MTU) and attempt to start the duty pump. At each tank, there are separate inlets (from the source) and outlets (to the customer).

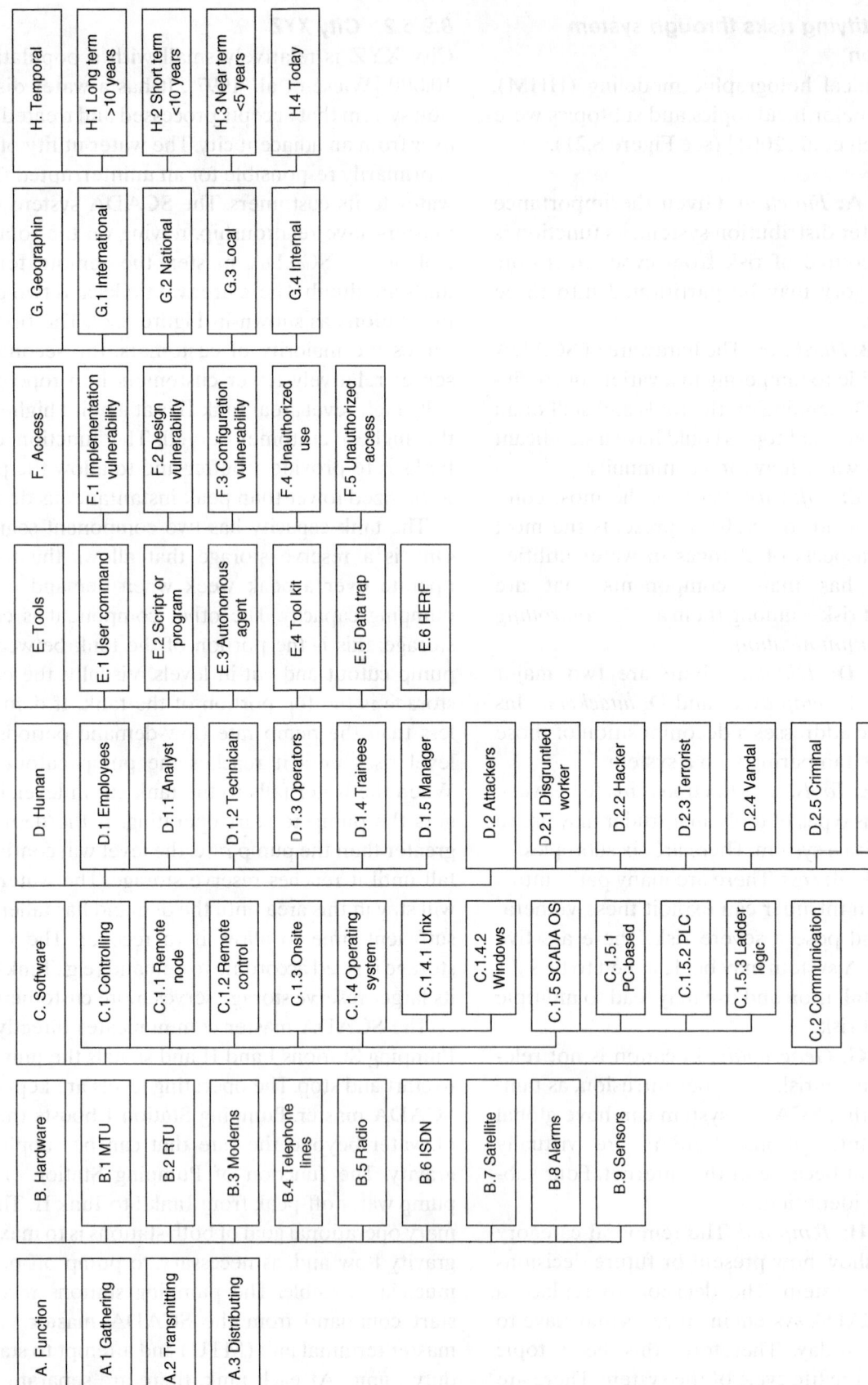

Figure 8.21 A framework for system decomposition that can be used to identify sources of risk to a water utility.

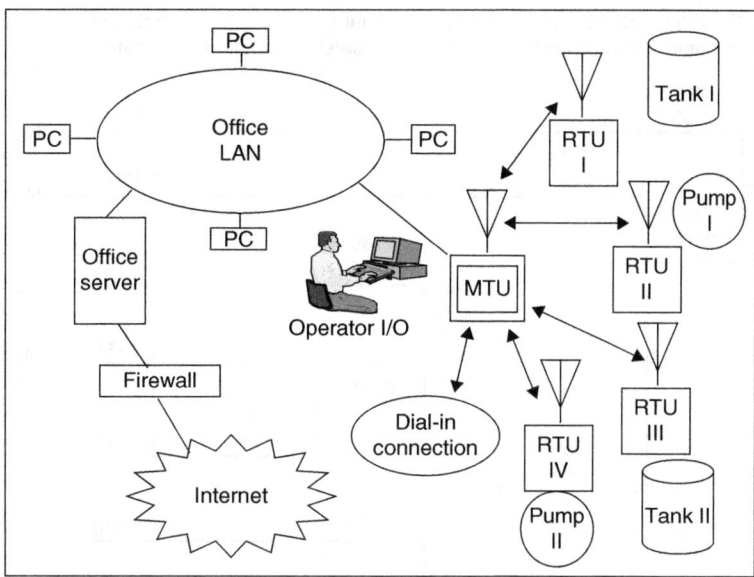

Figure 8.22 Interconnectedness of the SCADA system, local area network, and the Internet.

Water level and flows in and out are measured at each. An altitude control valve shuts the inlet when the tank is full. The tank's *full* position is defined above the pump cutout level, so there is no danger of shutting the valve while pumping. If something goes wrong and the pump does not shut off, the altitude valve will close and the pump will stop delivery on overpressure to prevent the main from bursting.

The SCADA system is always dependent on the communications network of the MTU and the SCADA master, who regularly polls all remote sites. Remote terminal units respond only when polled to ensure no contention on the communications network. The system operates automatically; the decision to start and stop pumps is made by the SCADA master and not by an operator sitting at the terminal. The system has the capability to contact operations staff after hours through a paging system in the event of an alarm.

In the example, the staff has dial-in access. If contacted, they can dial in from home and diagnose the extent of the problem. The dial-in system has a dedicated PC that is connected to the Internet and the office's local area network (LAN). A packet filter firewall protects the LAN and the SCADA. The SCADA master commands and controls the entire system. The communications protocols in use for the SCADA communications are proprietary. The LAN, the connection to the Internet, and the dial-in connection all use transmission control protocol and Internet protocol (TCP/IP). Instructions to the SCADA system are encapsulated with TCP/IP as well. Once the instructions are received by the LAN, the SCADA master de-encapsulates TCP/IP, leaving the proprietary terminal emulation protocols for the SCADA system. The central facility is organized into different access levels for the system, and an operator or technician has a level of access, depending on the need.

8.9.6.3 Identifying risks through system decomposition

The head topics A–H identified earlier through HHM, and the corresponding subtopics, identify 60 sources of risk for the centrally controlling SCADA system of City XYZ. The access points for the system are the dial-in connection points and the firewall that connect the utility to the Internet. For this example, the intruder might use the dial-in connection to gain access to and control of the system.

An intruder's most likely course of action is to use a password to access the system and its control devices. Since physical damage to equipment from dial-in access is inherently due to analog fail-safes, managers conclude that the intruder's probable goal is to manipulate the system to adversely affect the flow of water to the city. For example, creating water hammers may burst mains and damage customers' pipes. Or an intruder could shut off valves and pumps to reduce water flow. After discussing the potential threats, the managers may conclude that their greatest

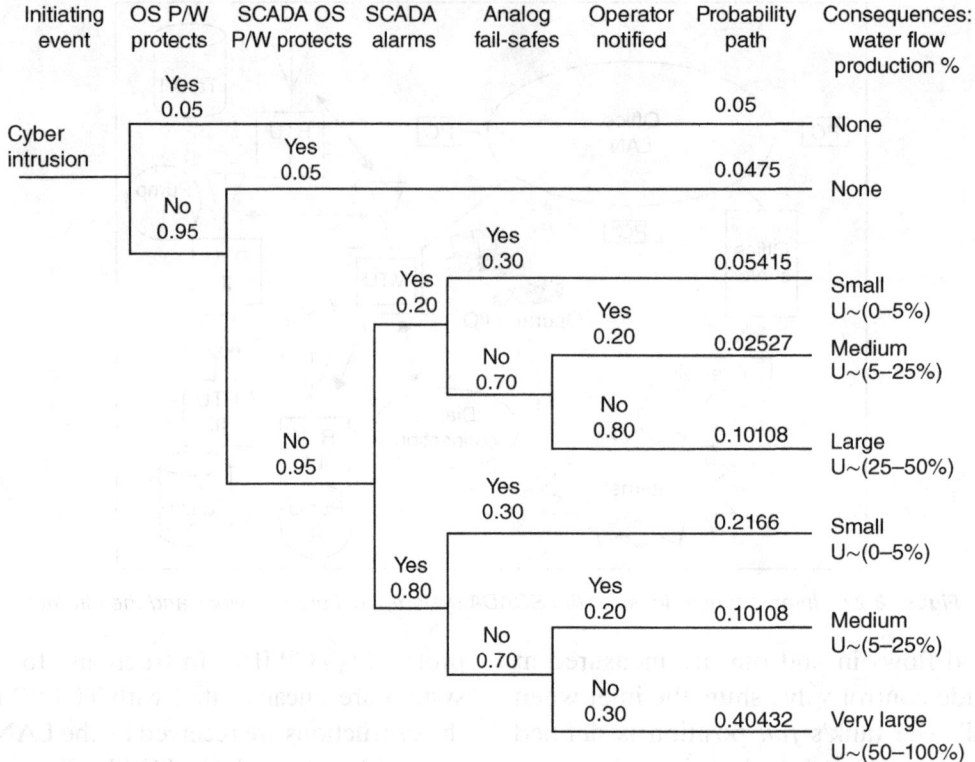

Figure 8.23 Event tree modeling the mitigating events in place to protect the system.

concern is the prospect of a disgruntled employee tampering with the SCADA system in such ways.

8.9.6.4 Risk management using PMRM

For each alternative, the managers would benefit from knowing both the expected percentage of water flow reduction and the conditional expected extreme percentage reduction in 1-in-100 outcomes (corresponding to β). Hence, the PMRM will partition the framework $s_1, s_2, s_3, \ldots, s_n$ on the consequence (damage) axis at β for all alternative risk management policies. For this presentation, we used the assessment of Expert A[†] for the event tree in Figure 8.23. This represents the current system's state of performance given an expert's assessment of an intruder's ability to transition through the mitigating events of the event tree. The initiating event, cyber intrusion, engenders each event and culminates with consequences at the end of each path through the event tree.

Assuming a uniform distribution, U, of damage for each path through the tree, a composite, or mixed, probability density is generated. The uniform distribution is appropriate because the managers were indifferent beyond the upper and lower bounds (see Figure 8.23).

The conditional expected value of water flow reduction for the current system at the partitioning of the worst-case probability axis at 1 in 100 corresponds to $\beta=98.7\%$. Thus, the conditional expected value for this new region is 99.5%. Using Eqs. (8.9) and (8.10), five expected values of risk $E(x)$ and several conditional expected values of risk, $f_4(\beta)$, can be generated.

8.9.6.5 Assessing risk using multiobjective trade-off analysis

Figure 8.23 depicts the plot of each alternative's cost on the vertical axis and the consequences on the horizontal axis. In the unconditional expected-value-of-risk region, alternatives 5 and 6 are efficient. For

---

[†]Nelson, A., Expert A, Personal Email on scenario and estimates, 1998, Nelson works with a wide range of applications from business and government accounting through technical applications such as Electronic and Mechanical Computer Assisted Drafting. He has worked on a variety of standard and proprietary platforms (i.e. UNIX, PC, DOS, and networks). He implements security measures at the computer hardware level, the operating systems level, and the applications level.

example, alternative 5 outperforms alternative 3 and costs $56,600 less. In the conditional expected value of risk (worst-case region), only alternatives 5 and 6 are efficient (Pareto-optimal policies). Alternative 5 reduces the expected value of water flow reduction by 57% for the 1-in-100 worst case. Note, for example, that while alternative Policy 5 yields a relatively low expected value of risk, at the partitioning β, the conditional expected value of risk is markedly higher (over 40%). To supplement the information from our analysis, the managers apply judgment to arrive at an acceptable risk management policy.

8.9.6.6 Conclusions

This case study illustrates how risk assessment and management was used to help decisionmakers determine preferred solutions to cyberintruder threats. The approach was applied to a small city using information learned from experts' input. The limitations of this approach are the following: (i) currently, it relies on expert opinion to estimate probabilities for the event tree; (ii) the model is not dynamic, so it does not completely represent the changes in the system during a cyber attack; and (iii) the event tree produces a probability mass function that must be converted to a density function in order for the exceedance probability to be partitioned. The underlying assumption that damages are uniformly distributed must be further explored.

8.10 SUMMARY

There is considerable literature on risks of extreme events, which by their nature and definition connote phenomena with dire and possibly catastrophic consequences but with low probabilities of occurrence. In terms of their representation in a histogram or a pdf, the data points on the tails of extreme-event distributions are sparse. Theory and methodology developed in this area have been driven by critical natural hazards, such as earthquakes, hurricanes, tornadoes, volcanoes, or severe droughts. In this connection, terrorism, although not a new phenomenon in world history, is now being studied analytically and with rigor (see Chapter 17).

The expected-value metric for risk evaluation falls very short in representing the true risk of safety-critical systems for which the consequences may be catastrophic, even though the probability of such an event is very low. Therefore, the risk of such systems should *not* be measured solely by the expected-value metric, especially when the consequences are unacceptable.

This chapter focuses on the development of the PMRM, a metric to complement the expected value of risk for extreme and catastrophic events. In Chapter 11, we will expand on this metric by incorporating the theory of statistics of extremes into the PMRM. Indeed, the theory of the statistics of extremes has been one of the more useful approaches for analyzing and understanding the behavior of the tails of extreme events.

REFERENCES

American Nuclear Society (ANS), 1981, *American National Standard for Determining Design Basis Flooding at Power Reactor Sites*, Report ANSI/ANS-2.8-1981. ANS, La Grange Park, IL.

Asbeck, E., and Y.Y. Haimes, 1984, The partitioned multiobjective risk method, *Large Scale Systems* **6**(1): 13–38.

Bier, V.M., S. Ferson, Y.Y. Haimes, J.H. Lambert, and M.J. Small, 2004, Risk of extreme and rare events: lessons from a selection of approaches, In *Risk Analysis and Society*, T. McDaniels, and M. Small (Eds.), Cambridge University Press, Cambridge/New York, pp. 74–118.

Chankong, V., and Y.Y. Haimes, 1983, *Multiobjective Decision-making: Theory and Methodology*, North-Holland, New York.

Ezell, B.C., 1998, *Risk of Cyber Attacks to Supervisory Control and Data Acquisition for Water Supply*, M.S. thesis, Systems Engineering Department, University of Virginia, Charlottesville, VA.

Ezell, B., Y.Y. Haimes, and J.H. Lambert, 2001, Risk of cyber attack to water utility supervisory control and data acquisition systems, *Military Operations Research* **6**(2): 23–33.

Haimes, Y.Y., 1991, Total risk management, *Risk Analysis* **11**(2): 169–171.

Haimes, Y.Y., and C. Chittister, 1995, An acquisition process of the management of non-technical risks associated with software development, *Acquisition Review Quarterly* **11**(2): 121–154.

Haimes, Y.Y., and W.A. Hall, 1974, Multiobjectives in water resources systems analysis: the surrogate worth trade-off method, *Water Resources Research* **10**(4): 615–624.

Haimes, Y.Y., D. Li, P. Karlsson, and J. Mitsiopoulos, 1990, Extreme events: risk, management, In *System and Control Encyclopedia*, Supplementary Vol. **1**, M.G. Singh (Ed.), Pergamon Press, Oxford.

Haimes, Y.Y., J.H. Lambert, and D. Li, 1992, Risk of extreme events in a multiobjective framework, *Water Resources Bulletin* **28**(1), 201–209.

Karlsson, P.O., and Y.Y. Haimes, 1988a, Probability distributions and their partitioning, *Water Resources Research* **24**(1): 21–29.

Karlsson, P.O., and Y.Y. Haimes, 1988b, Risk-based analysis of extreme events, *Water Resources Research* **24**(1): 9–20.

Kunreuther, H., and P. Slovic (Eds.), 1996, *Challenges in Risk Assessment and Risk Management*, The Annals of the American Academy of Political and Social Science, Sage, Thousand Oaks, CA.

Lambert, J.H., N.C. Matalas, C.W. Ling, Y.Y. Haimes, and D. Li, 1994, Selection of probability distributions in characterizing risk of extreme events, *Risk Analysis* **149**(5): 731–742.

Leemis, M.L., 1995, *Reliability; Probabilistic Models and Statistical Methods*, Prentice Hall, Englewood Cliffs, NJ.

Lowrance, W., 1976, *Of Acceptable Risk*, William Kaufmann, Los Altos, CA.

National Research Council (NRC), 1985, *Safety of Dams—Flood and Earthquake Criteria*, Committee on Safety Criteria for Dams, National Academy Press, Washington, DC.

Petrakian, R., Y.Y. Haimes, E.Z. Stakhiv, and D.A. Moser, 1989, Risk analysis of dam failure and extreme floods, In *Risk Analysis and Management of Natural and Man-Made Hazards*, Y.Y. Haimes, and E.Z. Stahkiv (Eds.), American Society of Civil Engineers, New York.

Runyon, R.P., 1977, *Winning the Statistics*, Addison-Wesley, Reading, MA.

Stedinger, J., and J. Grygier, 1985, Risk–cost analysis and spillway design, In *Computer Applications in Water Resources*, H. Torno (Ed.), American Society of Civil Engineers, New York.

Taleb, N.N., 2007, *The Black Swan: The Impact of the Highly Improbable*, Random House, New York.

U.S. Army Corps of Engineers, 1983, *Design Memorandum for Correction of Spillway Deficiency for Mohawk Dam*, Prepared by Huntington District, Corps of Engineers, Huntington, WV.

U.S. Army Corps of Engineers, 1985, *Justification for Correction of Spillway Deficiency for Mohawk Dam*, Prepared by Huntington District, Corps of Engineers, Huntington, WV.

U.S. Army Corps of Engineers, 1988, *Multiobjective Risk Partitioning: An Application to Dam Safety Risk Analysis*, Prepared by the Institute for Water Resources, Ft. Belvoir, VA.

USDA Medfly Study, 1994, *Center for Risk Management of Engineering Systems*, University of Virginia, Charlottesville, VA.

Wiese, I., C. Hillebrand, and B. Ezell, 1997, *Scenarios One and Two: Source to No 1 PS to No 1 Tank to No 2 PS to No 2 Tank (High Level) for a Master-Slave SCADA Systems*, SCADA Consultants, SCADA Mail List, scada@gospel.iinet.au.

9

Multiobjective Decision-Tree Analysis

9.1 INTRODUCTION

Decision-tree analysis (introduced in Chapter 4) has emerged over the years as an effective and useful tool in decisionmaking. Three decades ago, Howard Raiffa [1968] published the first comprehensive and authoritative book on decision-tree analysis. Ever since, its application to a variety of problems from numerous disciplines has grown by leaps and bounds [Pratt et al., 1995]. Advances in science and in scientific approaches to problem solving are often made on the basis of earlier works of others. In this case, the foundation for Raiffa's contributions to decision-tree analysis can be traced to the works of Bernoulli on utility theory [see von Neumann and Morgenstern, 1953; Edwards, 1954; Savage, 1954; Adams, 1960; Arrow, 1963; Shubik, 1964; Luce and Suppes, 1965; Schlaifer, 1969]. This chapter, in an attempt to build on the above seminal works, extends and broadens the concept of decision-tree analysis to incorporate (i) multiple, noncommensurate, and conflicting objectives (see Chapter 5); (ii) impact analysis (see Chapter 10); and (iii) the risk of extreme and catastrophic events (see Chapter 8). Indeed, the current practice often involves one-sided use of decision trees—optimizing a single-objective function and commensurating infrequent catastrophic events with more frequent noncatastrophic events using the common unconditional mathematical expectation [see Haimes et al., 1990].

9.1.1 Multiple Objectives

The single-objective models that were advanced in the 1950s, 1960s, 1970s, and 1980s are today considered by many to be unrealistic, too restrictive, and often inadequate for most real-world problems. The proliferation of books, articles, conferences, and courses during the last decade or two on what has come to be known as multiple-criteria decisionmaking (MCDM) is a vivid indication of this somber realization and of the maturation of the field of decisionmaking (see Chapter 5 and Chankong and Haimes [1983]). In particular, an optimum derived from a single-objective mathematical model, including that which is derived from a decision tree, often may be far from representing reality, thereby misleading the analysts as well as the decisionmakers. Fundamentally, most complex problems involve,

Risk Modeling, Assessment, and Management, Fourth Edition. Yacov Y. Haimes.
© 2016 John Wiley & Sons, Inc. Published 2016 by John Wiley & Sons, Inc.

among other things, minimizing costs, maximizing benefits (not necessarily in monetary values), and minimizing risks of various kinds. Decision trees can better serve both analysts and decisionmakers when they are extended to deal with the above multiple objectives. They are a powerful mechanism for analyzing complex problems.

9.1.2 Impact Analysis

On a long-term basis, managers and other decisionmakers are often rewarded not because they have made many optimal decisions in their tenure but because they avoided adverse and catastrophic consequences. If one accepts this premise, then the role of impact analysis—studying and investigating the consequences of present decisions on future policy options—might be even more important than generating an optimum for a single-objective model or identifying a Pareto-optimal set for a multiobjective model. Certainly, when the ability to generate both is present, having an appropriate Pareto-optimal set and knowing the impact of each Pareto optimum on future policy options should enhance the overall decisionmaking process within the decision-tree framework.

9.1.3 Review of Risk of Extreme and Catastrophic Events

To streamline the incorporation of risk of extreme and catastrophic events into multiobjective decision-tree (MODT) analysis, the following is a brief summary of the partitioned multiobjective risk method (PMRM) discussed in Chapter 8 and some of the results derived there. The PMRM separates extreme events from other noncatastrophic events, thereby providing decisionmakers with additional valuable and useful information. In addition to using the traditional expected value, the PMRM generates a number of conditional expected-value functions, termed here *risk functions*, which represent the risk, given that the damage falls within specific probability ranges (or damage ranges). Assume that the risk can be represented by a continuous random variable X with a known probability density function (pdf) $p_x(x; s_j)$, where s_j ($j = 1, \ldots, q$) is a control policy. The PMRM partitions the probability axis into three ranges. Denote the partitioned points on the probability axis by α_i ($i = 1, 2$). For each α_i and each policy s_j, it is assumed that there exists a unique damage β_{ij} such that

$$P_x(\beta_{ij}; s_j) = \alpha_i \qquad (9.1)$$

where P_x is the cumulative distribution function of X. These β_{ij} (with β_{0j} and β_{3j} representing, respectively, the lower bound and upper bound of the damage) define the conditional expectation as follows:

$$f_i(s_j) = E\left[X \mid p_x(x; s_j), X \in \left[\beta_{i-2,j}, \beta_{i-1,j}\right]\right], \\ i = 2, 3, 4; \ j = 1, \ldots, q \qquad (9.2)$$

or

$$f_i(s_j) = \frac{\int_{\beta_{i-2,j}}^{\beta_{i-1,j}} x p_x(x; s_j)\, dx}{\int_{\beta_{i-2,j}}^{\beta_{i-1,j}} p_x(x; s_j)\, dx}, \ i = 2, 3, 4; \ j = 1, \ldots, q \qquad (9.3)$$

where f_2, f_3, and f_4 represent the risk with high probability of exceedance and low damage, the risk with medium probability of exceedance and medium damage, and the risk with low probability of exceedance and high damage, respectively. The unconditional (conventional) expected value of X is denoted by $f_5(s_j)$. The relationship between the conditional expected values (f_2, f_3, f_4) and the unconditional expected value (f_5) is given by

$$f_5(s_j) = \theta_2 f_2(s_j) + \theta_3 f_3(s_j) + \theta_4 f_4(s_j) \qquad (9.4)$$

where θ_i ($i = 2, 3, 4$) is the denominator of Eq. (9.3). From the definition of β_{ij}, it can be seen that $\theta_i \geq 0$ is a constant, and $\theta_2 + \theta_3 + \theta_4 = 1$.

Combining either the generated conditional expected risk function or the unconditional expected risk function with the cost objective function f_1 creates a set of multiobjective optimization problems:

$$\min[f_1, f_i]^t, \quad i = 2, 3, 4, 5 \qquad (9.5)$$

where the superscript t denotes the transpose operator. This formulation offers more information about the probabilistic behavior of the problem than the single multiobjective formulation $\min[f_1, f_5]^t$. The trade-offs between the cost function f_1 and any risk function f_i ($i \in \{2, 3, 4, 5\}$) allow decisionmakers to consider the marginal cost of a small reduction in the risk objective, given a particular level of risk assurance for each of the partitioned risk regions and given the unconditional risk function f_5. The relationship of the trade-offs between the cost function and the various risk functions is given by

$$1/\lambda_{15} = \theta_2/\lambda_{12} + \theta_3/\lambda_{13} + \theta_4/\lambda_{14} \quad (9.6)$$

where

$$\lambda_{1i} = -\partial f_1/\partial f_i, \quad i = 2,3,4,5 \quad (9.7)$$

and θ_2, θ_3, and θ_4 are as defined earlier. A knowledge of this relationship among the marginal costs provides the decisionmakers with insights that are useful for determining an acceptable level of risk.

9.2 METHODOLOGICAL APPROACH

9.2.1 Extension to Multiple Objectives

Similar to the decision tree in conventional single-objective analysis (see Chapter 4), an MODT (Figure 9.1) is composed of decision nodes and chance nodes [Haimes et al., 1990]. Each pairing of an alternative and a state of nature, however, is now characterized by a vector-valued performance measure.

At a decision node, usually designated by a square, the decisionmaker selects one course of action from the feasible set of alternatives. We assume that there are only a finite number of alternatives at each decision node. These alternatives are shown as branches emerging to the right side of the decision node. The performance vector associated with each alternative is written along the corresponding branch. Each alternative branch may lead to another decision node, a chance node, or a terminal point.

A chance node, designated by a circle on the tree, indicates that a chance event is expected at this point; that is, one of the states of nature may occur. We consider two cases in this chapter: (1) a discrete case, where the number of states of nature is assumed finite, and (2) a continuous case, where the possible states of nature are assumed continuous. The states of nature are shown on the tree as branches to the right of the chance nodes, and their known probabilities are written above the branches. The states of nature may be followed by another chance node, a decision node, or a terminal point.

Allowing for the evaluation of the multiple objectives at each decision node constitutes an important

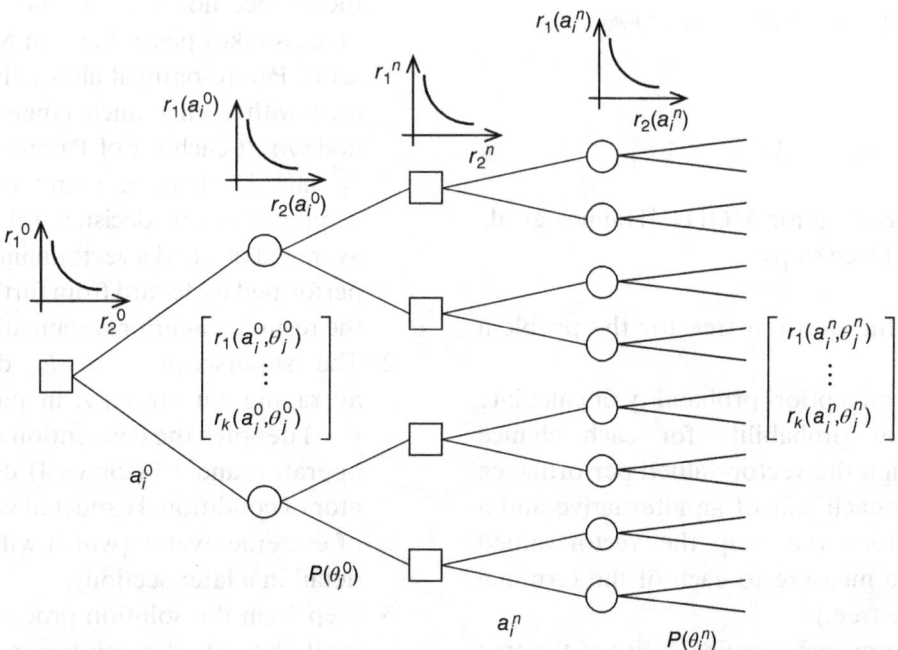

Figure 9.1 Structure of multiobjective decision trees. From Haimes et al. [1990]; © 1990 Wiley.

feature of this approach. It is a significant extension of the average-out-and-fold-back strategy used in conventional single-objective decision-tree (SODT) methods.

To allow for this extension, we first define a k-dimensional vector-valued performance measure associated with an action a_n and a state of nature θ_n as follows:

$$r(a_n, \theta_n) = [r_1(a_n, \theta_n), r_2(a_n, \theta_n), \ldots, r_k(a_n, \theta_n)]^t \quad (9.8)$$

A point $r = [r_1, r_2, \ldots, r_k]^t$ in the objective function space is said to be noninferior (for a vector minimization) if there does not exist another feasible point $r' = [r_1', r_2', \ldots, r_k']^t$ such that

$$r_i' \leq r_i, \quad i = 1, 2, \ldots, k \quad (9.9)$$

with at least one strict inequality holding for $i = 1, 2, \ldots, k$.

The sequential structure of MODT necessitates introducing a vector of operators that combine the vectors of performance measures of successive decision nodes. Let o denote a k-dimensional vector of binary operators, which are to be applied to elements corresponding to the same components or any two vectors of a performance measure. For example, if

$$r_1 = [2,3], \quad r_2 = [-3,2], \quad o = (+, \bullet)$$

then

$$r_1 \circ r_2 = [2-3, 3 \bullet 2] = [-1, 6]$$

The solution procedure for MODT [Haimes et al., 1990] is stated in three steps:

Step 1. Chart the decision tree for the problem under study.
Step 2. Assign an a priori probability or calculate the posterior probability for each chance branch. Assign the vector-valued performance measure for each pair of an alternative and a state of nature. (Or map the vector-valued performance measure to each of the terminal points of the tree.)
Step 3. Start from each terminal point of the tree and fold backward on the tree.

At each decision node n and at each branch emerging to the right side of the decision node, find the corresponding set of vector-valued performance measures, $r(a_i^n)$, for each alternative a_i and identify the set of noninferior solutions by solving

$$r^n = \min \bigcup_i r(a_i^n) \quad (9.10)$$

where $\bigcup$ is the union operator on sets $r(a_i^n)$.

Note: In MODT analysis, instead of having a single optimal value associated with an SODT, we have r^n, a set of vector-valued performance measures of noninferior alternatives at decision node n.

At each chance node m and at branches emerging to the right side of the chance node, find the corresponding set of vector-valued performance measures r_j^m, for each state of nature θ_j^m, and then calculate the vector-valued expected performance measure or other specified vector-valued *risk* performance measure, which is denoted by r^m:

$$r^m = \min_j E_j^s \{r_j^m\} \quad (9.11)$$

Note that:

1. In SODT analysis, there is no choice process at the chance nodes, since only an averaging-out process takes place there. In MODT analysis, a set of Pareto-optimal alternatives, r_j^m, is associated with each branch emerging from chance node m. If each set of Pareto-optimal solutions r_j^m has d_j^m elements, then there exist $\prod_j \{d_j^m\}$ combinations of decision rules needing to be averaged out, and a vector minimization must be performed to discard from further consideration the resulting inferior combinations.
2. The superscript s in E^s denotes the sth averaging-out strategy; in particular, E^5 (for $s=5$) denotes the conventional expected-value operator, and E^4 (for $s=4$) denotes the operator of conditional expected value in the region of extreme events (which will be discussed in detail in a later section).
3. Step 3 (in the solution procedure) is repeated until the set of noninferior solutions at the starting point of the tree is obtained.

9.2.2 Impact of Experimentation

The impact of an added piece of information (obtained, e.g., through experimentation) on different objectives is now addressed, and the value of the information is quantified by a vector-valued measure. In conventional decision-tree analysis, whether an experiment should be performed depends on an assessment of the expected value of experimentation (EVE), which is the difference between the expected loss without experimentation and the expected loss with experimentation. If the EVE is negative, experimentation is deemed unwarranted; otherwise, the experiment that yields the lowest loss is selected. In MODT analysis, the monetary index does not constitute the sole consideration; rather, the value of experimentation is judged in a multiobjective way where, in many cases, the noninferior frontiers generated with and without experimentation do not dominate each other. The added experimentation in these cases reshapes the feasible region (and thus the noninferior frontier) and generates new and better options for the decisionmakers (Figure 9.2). MODT analysis involves extensive mathematical manipulations. The following MODT analysis of a flood warning and evacuation system developed for the Institute for Water Resources, US Army Corps of Engineers, provides an example illustration [see Haimes et al., 1990, 1996].

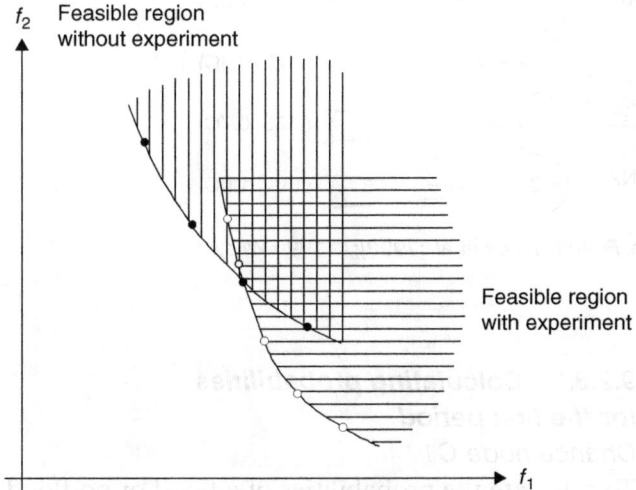

Figure 9.2 Reshape of the feasible region by experimentation. From Haimes et al. [1990]; © 1990 Wiley.

9.2.3 Example for the Discrete Case

9.2.3.1 Problem definition

The example problem discussed here concerns a simplified flood warning and evacuation system. Three possible actions—evacuation, issuing a flood watch, and doing nothing—are under consideration. There are cost factors associated with the first two options. The decision tree covers two time periods, and the cost associated with each option is a function of the period in which the action is taken. The complete decision tree for the problem is shown in Figure 9.3. The following assumptions are made:

1. There are three possible actions with associated costs for the first period:
   a. Issuing an evacuation order at a cost of $5 million [EV1]
   b. Issuing a flood watch at a cost of $1 million [WA1]
   c. Doing nothing at no cost [DN1]
2. For the second period, the actions and the corresponding costs are:
   a. Issuing an evacuation order at a cost of $3 million [EV2]
   b. Issuing a flood watch at a cost of $0.5 million [WA2]
   c. Doing nothing at no cost [DN2]
3. The flood stage is reached at water flow $(W) = 50,000$ cfs.
4. There are three underlying pdf's for the water flow:
   a. $W \sim$ lognormal (10.4,1), represented as LN_1
   b. $W \sim$ lognormal (9.1,1), represented as LN_2
   c. $W \sim$ lognormal (7.8,1), represented as LN_3
   The prior possibilities that any of these pdf's is the actual pdf are equal.
5. There are four possible events at the end of the first period given that the current water flow is $5,000 \text{ cfs} \leq W \leq 15,000 \text{ cfs}$:
   a. A flood $(W \geq 50,000 \text{ cfs})$ occurs.
   b. The water flow is greater than in the previous period $(15,000 \text{ cfs} \leq W \leq 50,000 \text{ cfs})$, represented as $W1$.
   c. The water flow is in the same range as in the previous period $(5,000 \text{ cfs} \leq W \leq 15,000 \text{ cfs})$, represented as $W2$.
   d. The water flow is lower than in the previous period $(W \leq 5000 \text{ cfs})$, represented as $W3$.

264 MULTIOBJECTIVE DECISION-TREE ANALYSIS

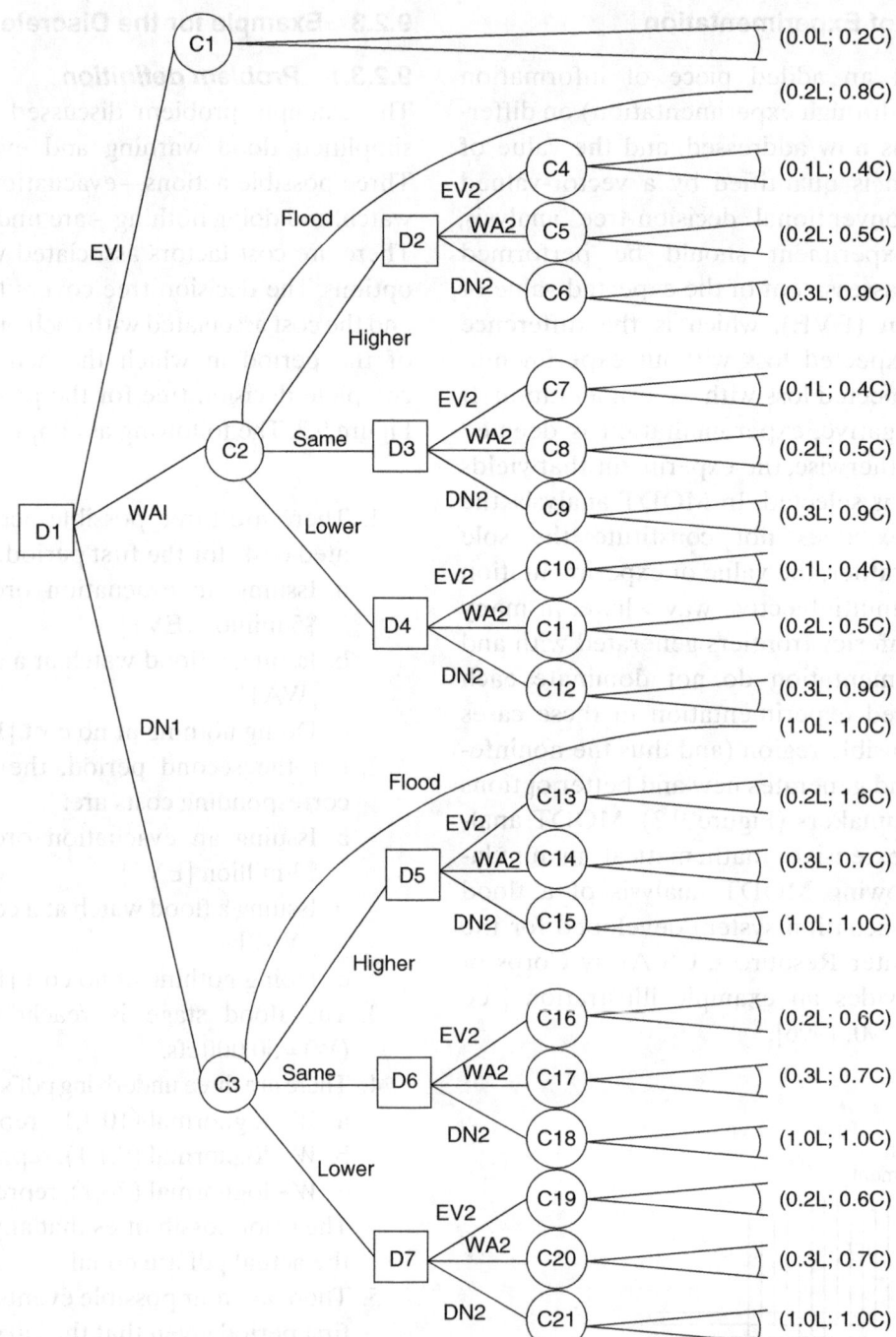

Figure 9.3 Decision tree for the discrete case. From Haimes et al. [1990]; © 1990 Wiley.

6. $L = 7$ and $C = \$7,000,000$ are, respectively, the maximum possible loss of lives and property values, given no flood warning. All costs and loss of lives at the end of the second-period chance nodes shown in Figure 9.3 are given by the US Army Corps of Engineers.

9.2.3.2 Calculating probabilities for the first period
Chance node C1
To calculate the probabilities of a flood or no-flood event at the end of the second period (see Figure 9.4), we use the facts that the possible pdf of the water

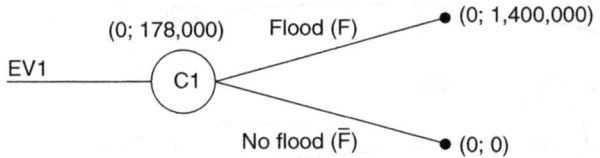

Figure 9.4 Averaging out at chance node C1 (discrete case).

flow (W) is LN_i with probability 1/3, $i=1, 2, 3$, and that the flood stage is at $W=50,000$ cfs. The probability of a flood event can be calculated as follows:

$$\Pr(\text{flood}) = \sum_{i=1}^{3} \Pr(\text{flood} \mid LN_i) \Pr(LN_i) \quad (9.12a)$$
$$= \sum_{i=1}^{3} (1/3) \Pr(X \geq 50,000 \text{ cfs} \mid LN_i)$$

where

$$\Pr(X \geq 50,000 \text{ cfs} \mid LN_i)$$
$$= \int_{50,000}^{\infty} \frac{\exp\left[-\{\ln(x) - \mu_i\}^2 / 2\sigma_i^2\right] dx}{\sqrt{2\pi} x \sigma_i} \quad (9.12b)$$

Equation (9.12b) is converted into a standard normal distribution by using

$$z = [\ln(x) - \mu_i]/\sigma_i \quad (9.13)$$

$dz = dx/x\sigma_i$ yielding

$$\Pr(X \geq 50,000 \text{ cfs} \mid LN_i) = \int_{(\ln 50,000 - \mu_i)/\sigma_i}^{\infty} \frac{\exp(-z^2/2)}{\sqrt{2\pi}} dz \quad (9.14)$$

Equation (9.14) is evaluated using standard normal distribution tables. For a more detailed calculation, see Section A.8 of the Appendix. This yields

$$\Pr(\text{flood}) = \Pr(X \geq 50,000) = 0.1271$$

Chance nodes C2 and C3
Nodes C2 and C3 each present four possible events at the beginning of the second period: a flood event, a higher water flow, the same water flow, and a lower water flow than in the previous period (see Figure 9.5). The distribution of water flow at the end of the first period is given by assumption 4 (Section 9.2.3.1). The probability of each event is calculated using Eqs. (9.12a), (9.13), and (9.14) with modified integral intervals:

$$\Pr(\text{flood}) = \Pr(50,000 \leq X \leq \infty) = 0.1271$$
$$\Pr(\text{higher}) = \Pr(15,000 \leq X \leq 50,000) = 0.2466$$
$$\Pr(\text{same}) = \Pr(5,000 \leq X \leq 15,000) = 0.2685$$
$$\Pr(\text{lower}) = \Pr(0 \leq X \leq 5,000) = 0.3577$$

9.2.3.3 Calculating probabilities for the second period

Regardless of whether a watch action (WA1) or a do-nothing action (DN1) was taken at the first period, three possible actions must be considered at the second period—evacuate, issue another flood watch, or do nothing. Depending on the actions taken in the first and the second periods and on the water flow at the second period, different values of the expected losses for each of the terminal chance nodes are calculated. Three equally probable underlying pdf's for the water flow prevail in the first period. At the end of the first period, after measuring the water flow W_j, the posterior probabilities for each of these pdf's are calculated using Bayes' formula:

$$\Pr(LN_i \mid W_j) = \frac{\Pr(W_j \mid LN_i) \Pr(LN_i)}{\sum_{i=1}^{3} \Pr(W_j \mid LN_i) \Pr(LN_i)} \quad (9.15)$$

where $\Pr(LN_i) = 1/3$ and W_j is given in assumption 5 (Section 9.2.3.1) and $\Pr(W_j \mid LN_i)$ is calculated using Eqs. (9.12a), (9.13), and (9.14). Then, the probability of a flood event at any chance node is calculated as

$$\Pr(\text{flood} \mid W_j) = \sum_{i=1}^{3} \Pr(\text{flood} \mid LN_i) \Pr(LN_i \mid W_j) \quad (9.16)$$

For example,

$$\Pr(\text{flood} \mid \text{higher}) = \Pr(\text{flood} \mid LN_1) \Pr(LN_1 \mid \text{higher})$$
$$+ \Pr(\text{flood} \mid LN_2) \Pr(LN_2 \mid \text{higher})$$
$$+ \Pr(\text{flood} \mid LN_3) \Pr(LN_3 \mid \text{higher})$$

The values of $\Pr(\text{flood} \mid LN_i)$ ($i=1, 2, 3$) are calculated using Eqs. (9.12a), (9.13), and (9.14), and the values of $\Pr(LN_i \mid \text{higher})$ ($i=1, 2, 3$) are calculated using Eq. (9.15). Therefore, from Eq. (9.16),

266   MULTIOBJECTIVE DECISION-TREE ANALYSIS

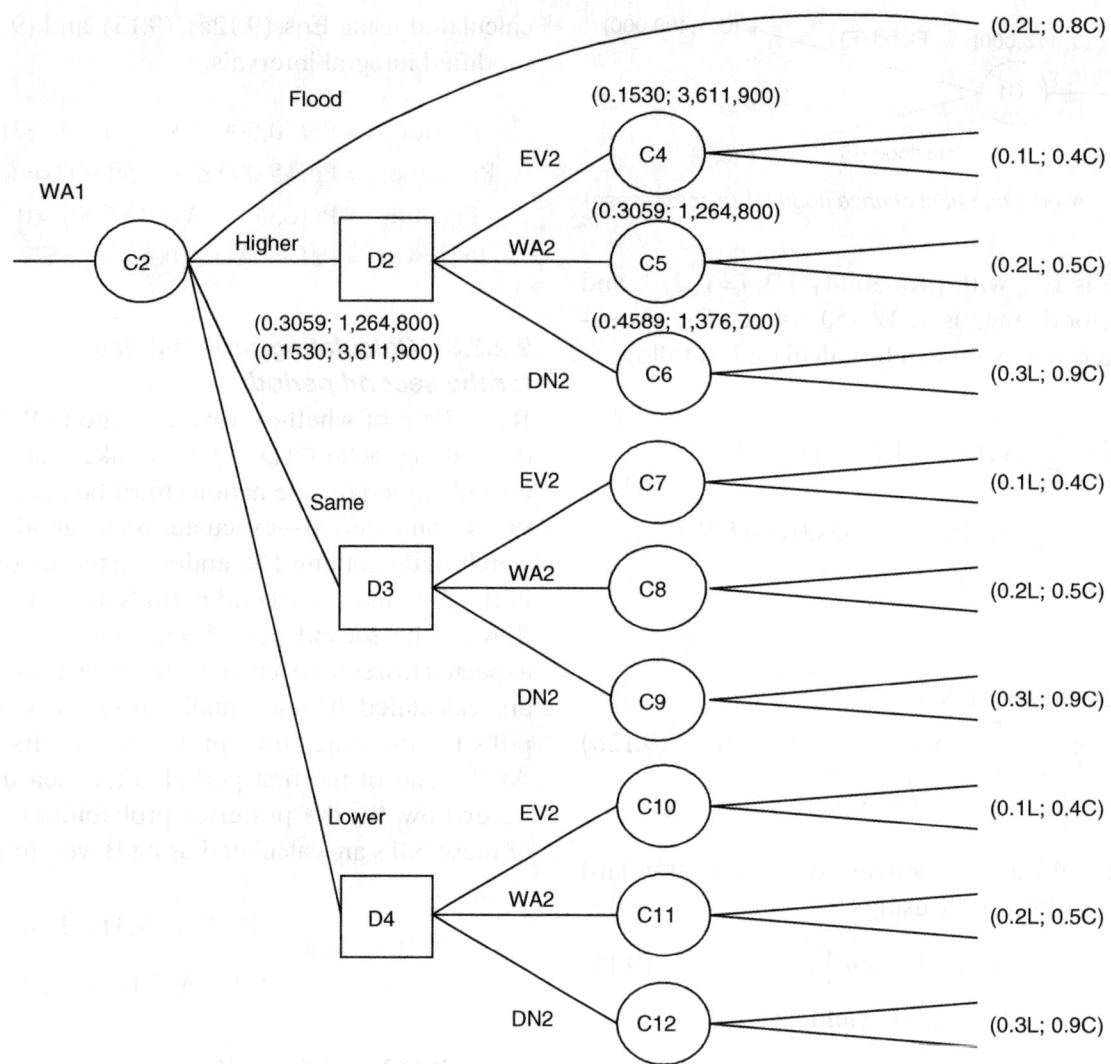

Figure 9.5 *Second-stage tree corresponding to chance node C2 (discrete case). Note that L=7 lives and C=$7,000,000. From Haimes et al. [1990]; © 1990 Wiley.*

$$\begin{aligned}\Pr(\text{flood}\,|\,\text{higher}) &= (0.3372)(0.6031) \\ &\quad + (0.0427)(0.3517) \\ &\quad + (0.0013)(0.0452) \\ &= 0.2185\end{aligned}$$

Similarly,

$$\Pr(\text{flood}\,|\,\text{same}) \cong 0.1006$$
$$\Pr(\text{flood}\,|\,\text{lower}) \cong 0.0214$$

(Note that the values for loss of life and cost are rounded off throughout this example problem.)

The required value of the loss vector-valued functions is then computed by multiplying the flood probability by the damage vector. Consider, for example, arc EV2 corresponding to decision node D2 in Figure 9.5:

$$L_{\text{EV2}|\text{D2}} = (0.2185)(0.7) \cong 0.1530$$
$$C_{\text{EV2}|\text{D2}} = (0.2185)(2,800,000) + 3,000,000 \cong 3,611,900$$

Table 9.1 presents the values of the loss vectors for the second-period decision arcs. Folding back at each decision node, the vector-valued functions are compared, and all dominated (inferior) solutions are eliminated. Consider, for example, decision node D2. The vector corresponding to the decision DN2 is inferior to the vector corresponding to the decision WA2:

$$\begin{bmatrix} 0.3059 \\ 1,264,800 \end{bmatrix}_{WA2} < \begin{bmatrix} 0.4589 \\ 1,376,700 \end{bmatrix}_{DN2}$$

Table 9.2 presents the noninferior decisions for the second-period decision arcs. Averaging out at the chance nodes for the first period, each noninferior decision corresponding to each arc is multiplied by the probability for that arc, yielding a single decision rule for the first-period decision node. For example, we have 18 different combinations at WA1, one of which is (EV2|higher, EV2|same, EV2|lower). The value of the loss vector for this combination is

$$\begin{bmatrix} 0.1780 + (0.1530)(0.2466) + (0.0704)(0.2685) \\ +(0.0150)(0.3577) \\ 711,800 + (3,611,900)(0.2466) \\ +(3,281,600)(0.2685) + (3,060,000)(0.3577) \\ +1,000,000 \end{bmatrix}$$
$$\cong \begin{bmatrix} 0.2400 \\ 4,578,500 \end{bmatrix}$$

where the first and second elements represent a loss of lives of 0.2400 and a cost of $4,578,500, respectively. Table 9.3 presents the values of the vector of objectives for the first-period decision node. A total of nine noninferior decisions are generated for action WA1. Similarly, there are eight noninferior solutions by self-comparison of all vectors for action DN1 and only five after comparison of all decisions. There are a total of 15 noninferior solutions for decision node D1 (see Figure 9.6). Figure 9.7 depicts the graph of all noninferior solutions.

9.2.3.4 Summary

The following is a summary of the MODT steps. Note that the probability values here (and in the text) differ from those in Appendix A.8 due to rounding errors used in the standard normal tables in the Appendix:

TABLE 9.1 Expected Value of Loss Vectors for the Second-Period Decision Arcs

Node	Arc	Lives (L)	Cost (C)($)
D2	EV2	0.1530	3,611,900
	WA2	0.3059	1,264,800
	DN2	0.4589	1,376,700
D3	EV2	0.0704	3,281,600
	WA2	0.1408	852,000
	DN2	0.2112	633,700
D4	EV2	0.0150	3,060,000
	WA2	0.0300	575,000
	DN2	0.0450	135,000
D5	EV2	0.3059	3,917,800
	WA2	0.4589	1,570,700
	DN2	1.5296	1,529,600
D6	EV2	0.1408	3,422,400
	WA2	0.2112	992,900
	DN2	0.7041	704,100
D7	EV2	0.0300	3,090,000
	WA2	0.0450	605,000
	DN2	0.1500	150,000
C2	F	0.1780	711,800
C3	F	0.8898	889,800

C, cost ($); L, loss of lives.

TABLE 9.2 Noninferior Decisions for the Second-Period Decision Nodes (Discrete Case)

Node	Noninferior Decisions
D2	EV2, WA2
D3	EV2, WA2, DN2
D4	EV2, WA2, DN2
D5	EV2, WA2, DN2
D6	EV2, WA2, DN2
D7	EV2, WA2, DN2

Step 1. Generate the following probabilities: Pr(flood), Pr(higher), Pr(same), and Pr(lower).

1a. Pr(flood):

$$\Pr(\text{flood}) = \sum_{i=1}^{3} \Pr(\text{flood}|LN_i) \Pr(LN_i)$$

Condition of flood:
$\Pr(\text{flood}) = \Pr(X \geq 50,000 \text{ cfs})$

Get the probability of flood event given LN_i using the formula

TABLE 9.3 Decisions for the First-Period Node (Discrete Case)

First-Period Decision	Second-Period Decision Higher	Same	Lower	Loss Vector Lives	Cost ($)
EV1[a]	—	—	—	0.0000	5,178,000
WA1[a]	EV2	EV2	EV2	0.2400	4,578,500
WA1[a]	EV2	EV2	WA2	0.2453	3,689,500
WA1[a]	EV2	EV2	DN2	0.2507	3,532,100
WA1	EV2	WA2	EV2	0.2589	3,926,100
WA1[a]	EV2	WA2	WA2	0.2642	3,037,100
WA1[a]	EV2	WA2	DN2	0.2696	2,879,700
WA1	EV2	DN2	EV2	0.2778	3,867,500
WA1	EV2	DN2	WA2	0.2831	2,978,500
WA1[a]	EV2	DN2	DN2	0.2885	2,821,100
WA1	WA2	EV2	EV2	0.2777	3,999,600
WA1	WA2	EV2	WA2	0.2830	3,110,600
WA1	WA2	EV2	DN2	0.2884	2,953,200
WA1	WA2	WA2	EV2	0.2966	3,347,300
WA1[a]	WA2	WA2	WA2	0.3019	2,458,200
WA1[a]	WA2	WA2	DN2	0.3073	2,300,800
WA1	WA2	DN2	EV2	0.3155	3,288,600
WA1	WA2	DN2	WA2	0.3209	2,399,600
WA1[a]	WA2	DN2	DN2	0.3262	2,242,200
DN1	EV2	EV2	EV2	1.0138	3,880,400
DN1	EV2	EV2	WA2	1.0191	2,991,400
DN1	EV2	EV2	DN2	1.0567	2,828,700
DN1	EV2	WA2	EV2	1.0327	3,228,100
DN1	EV2	WA2	WA2	1.0380	2,339,100
DN1[a]	EV2	WA2	DN2	1.0756	2,176,300
DN1	EV2	DN2	EV2	1.1650	3,150,500
DN1	EV2	DN2	WA2	1.1704	2,261,500
DN1	EV2	DN2	DN2	1.2079	2,098,800
DN1	WA2	EV2	EV2	1.0515	3,301,600
DN1	WA2	EV2	WA2	1.0568	2,412,600
DN1	WA2	EV2	DN2	1.0944	2,249,800
DN1	WA2	WA2	EV2	1.0704	2,649,200
DN1[a]	WA2	WA2	WA2	1.0758	1,760,200
DN1[a]	WA2	WA2	DN2	1.1133	1,597,400
DN1	WA2	DN2	EV2	1.2027	2,571,700
DN1	WA2	DN2	WA2	1.2081	1,682,700
DN1[a]	WA2	DN2	DN2	1.2457	1,519,900
DN1	DN2	EV2	EV2	1.3156	3,291,400
DN1	DN2	EV2	WA2	1.3209	2,402,400
DN1	DN2	EV2	DN2	1.3585	2,239,600
DN1	DN2	WA2	EV2	1.3345	2,639,100
DN1	DN2	WA2	WA2	1.3398	1,750,100
DN1	DN2	WA2	DN2	1.3774	1,587,300
DN1	DN2	DN2	EV2	1.4668	2,561,500
DN1	DN2	DN2	WA2	1.4722	1,672,500
DN1[a]	DN2	DN2	DN2	1.5097	1,509,700

[a]Noninferior decision.

METHODOLOGICAL APPROACH   269

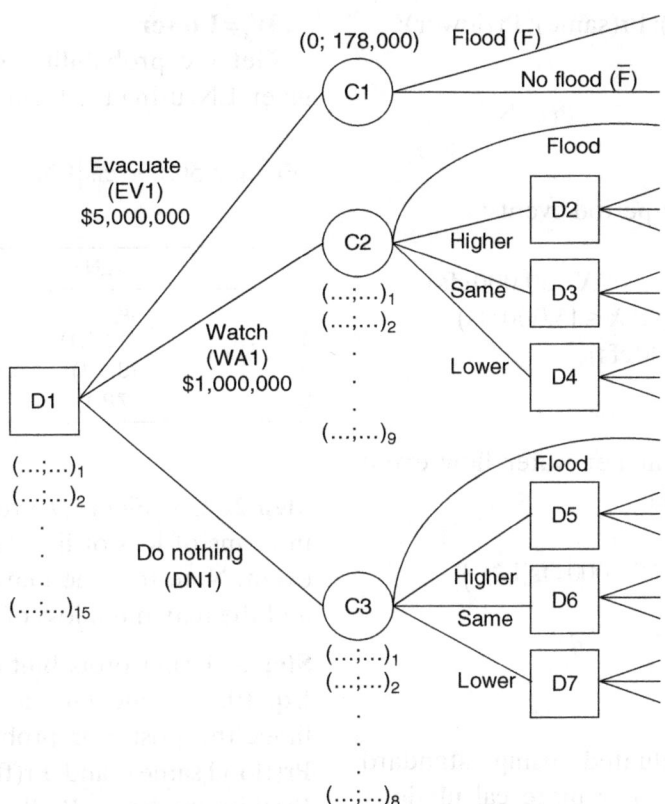

Figure 9.6 Decision tree for the first stage (discrete case). From Haimes et al. [1990]; © 1990 Wiley.

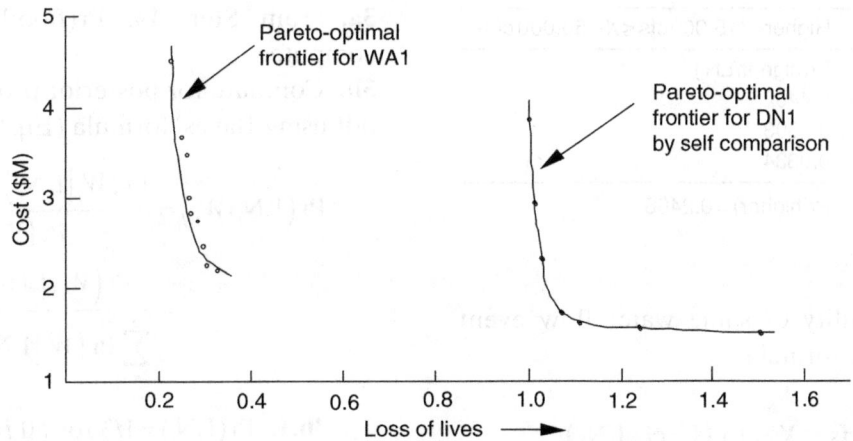

Figure 9.7 Pareto-optimal frontier (discrete case). From Haimes et al. [1990]; © 1990 Wiley.

$$\Pr(X \geq 50{,}000 \text{ cfs} | LN_i) = \int_{(\ln 50{,}000 - \mu_i)/\sigma_i}^{\infty} \frac{\exp(-z^2/2)}{\sqrt{2\pi}} \, dz$$

i	LN_i (μ_i, σ_i)	Flood = $(X \geq 50{,}000\text{ cfs})$ $\Pr(\text{flood}|LN_i)$
1	(10.4, 1)	0.3373
2	(9.1, 1)	0.0427
3	(7.8, 1)	0.0013

$$\begin{aligned}
\Pr(\text{flood}) &= \Pr(\text{flood}|LN_1)\Pr(LN_1) + \\
&\quad \Pr(\text{flood}|LN_2)\Pr(LN_2) + \\
&\quad \Pr(\text{flood}|LN_3)\Pr(LN_3) \\
&= (0.3373)(1/3) + (0.0427)(1/3) \\
&\quad + (0.0013)(1/3) \\
&= 0.1271
\end{aligned}$$

1b. $\Pr(W_j) \to \Pr(\text{higher}); \Pr(\text{same}); \Pr(\text{lower})$:

$$\Pr(W_j) = \sum_{i=1}^{3} \Pr(W_j|LN_i)\Pr(LN_i)$$

Condition of the second-period events:

- $\Pr(\text{higher}) = \Pr(15{,}000\,\text{cfs} \le X \le 50{,}000\,\text{cfs})$
- $\Pr(\text{same}) = \Pr(5{,}000\,\text{cfs} \le X \le 15{,}000\,\text{cfs})$
- $\Pr(\text{lower}) = \Pr(X \le 5{,}000\,\text{cfs})$

$W_1 = $ Higher

Get the probability of higher water flow event given LN_i using the formula

$$\Pr(15{,}000\,\text{cfs} \le X \le 50{,}000\,\text{cfs}|LN_i)$$
$$= \int_{(\ln 15{,}000 - \mu_i)/\sigma_i}^{(\ln 50{,}000 - \mu_i)/\sigma_i} \frac{\exp(-z^2/2)}{\sqrt{2\pi}}\,dz$$

The equation is evaluated using standard normal distribution tables. For more calculations, see Section A.9 of the Appendix.

	LN_i	Higher $= (15{,}000\,\text{cfs} \le X \le 50{,}000\,\text{cfs})$
i	(μ_i, σ_i)	$\Pr(\text{higher}/LN_i)$
1	(10.4, 1)	0.4462
2	(9.1, 1)	0.2603
3	(7.8, 1)	0.0334

$\Pr(\text{higher}) = 0.2466$

$W_2 = $ Same

Get the probability of same water flow event given LN_i using the formula

$$\Pr(5{,}000\,\text{cfs} \le X \le 15{,}000\,\text{cfs}|LN_i)$$
$$= \int_{(\ln 5{,}000 - \mu_i)/\sigma_i}^{(\ln 15{,}000 - \mu_i)/\sigma_i} \frac{\exp(-z^2/2)}{\sqrt{2\pi}}\,dz$$

	LN_i	Same $= (5{,}000\,\text{cfs} \le X \le 15{,}000\,\text{cfs})$
i	(μ_i, σ_i)	$\Pr(\text{same}/LN_i)$
1	(10.4, 1)	0.1866
2	(9.1, 1)	0.4170
3	(7.8, 1)	0.2019

$\Pr(\text{same}) = 0.2685$

$W_3 = $ Lower

Get the probability of lower water flow event given LN_i using the formula

$$\Pr(X \ge 50{,}000\,\text{cfs}|LN_i) = \int_{-\infty}^{(\ln 5000 - \mu_i)/\sigma_i} \frac{\exp(-z^2/2)}{\sqrt{2\pi}}\,dz$$

	LN_i	Lower $= (X \le 5000\,\text{cfs})$
i	(μ_i, σ_i)	$\Pr(\text{lower}|LN_i)$
1	(10.4, 1)	0.0299
2	(9.1, 1)	0.2800
3	(7.8, 1)	0.7634

$\Pr(\text{lower}) = 0.3577$

Step 2. Use Figure 9.3 to obtain the consequences in terms of loss of lives and property lost for each event. Note that the maximum loss of lives is $L = 7$ and the maximum loss of property is $C = \$7$ million.

Step 3. Using probabilities obtained in step 1 and Eq. (9.16), calculate for second-period probabilities, the posterior probabilities: $\Pr(\text{flood}|\text{higher})$, $\Pr(\text{flood}|\text{same})$, and $\Pr(\text{flood}|\text{lower})$. For example, $\Pr(\text{flood}|\text{higher}) = \Pr(\text{flood}|LN_1)\Pr(LN_1|\text{higher}) + \Pr(\text{flood}|LN_2)\Pr(LN_2|\text{higher}) + \Pr(\text{flood}|LN_3)\Pr(LN_3|\text{higher})$.

3a. From Step **1a**, $\Pr(\text{flood}|LN_i)$ have been computed.

3b. Compute for posterior probabilities for each pdf using Bayes' formula (Eq. 9.15):

$$\Pr(LN_i|W_j) = \frac{\Pr(W_j|LN_i)\Pr(LN_i)}{\Pr(W_j)}$$
$$= \frac{\Pr(W_j|LN_i)\Pr(LN_i)}{\sum_{i=3}^{3}\Pr(W_j|LN_i)\Pr(LN_i)}$$

3b.1. $\Pr(LN_i) = 1/3$ for all $i = 1, 2, 3$ (*given*).

3b.2. From step **1b**, $\Pr(W_j|LN_i)$ for all $i = 1, 2, 3$ and $j = 1$ (higher), 2 (same), 3 (lower) have been computed.

3b.3. $\Pr(LN_i|W_j)$ for all $i = 1, 2, 3$:

$$\Pr(LN_i|W_j) = \frac{\Pr(W_j|LN_i)\Pr(LN_i)}{\Pr(W_j)}$$
$$= \frac{\Pr(W_j|LN_i)\Pr(LN_i)}{\sum_{i=3}^{3}\Pr(W_j|LN_i)\Pr(LN_i)}$$

W_1 = Higher

LN_i	A Pr(higher|LN_i) (from step **1b**)	B Pr(LN_i) (given)	C = A × B Pr(higher|LN_i) × Pr(LN_i)	D = C/Total C Pr(LN_i|higher)
1	0.4462	1/3	0.1487	0.6031
2	0.2603	1/3	0.0868	0.3517
3	0.0334	1/3	0.0111	0.0452
			0.2466	

W_2 = Same

LN_i	Pr(same|LN_i)	Pr(LN_i)	Pr(same|LN_i) × Pr(LN_i)	Pr(LN_i|same)
1	0.1866	1/3	0.0622	0.2317
2	0.4170	1/3	0.1390	0.5177
3	0.2019	1/3	0.0673	0.2507
			0.2685	

W_3 = Lower

LN_i	Pr(Lower|LN_i)	Pr(LN_i)	Pr(lower|LN_i) × Pr(LN_i)	Pr(LN_i|lower)
1	0.0299	1/3	0.0100	0.0278
2	0.2800	1/3	0.0933	0.2609
3	0.7634	1/3	0.2545	0.7113
			0.3577	

3c. Computing for second-period probabilities, Pr(flood\|W_j):

$$\Pr(\text{flood}|W_j) = \sum_{i=1}^{3} \Pr(\text{flood}|LN_i)\Pr(LN_i|W_j)$$

W_1 = Higher

LN_i	A Pr(flood|LN_i)	B Pr(LN_i|higher)	C = A × B Pr(flood|higher)
1	0.3373	0.6031	0.2034
2	0.0427	0.3517	0.0150
3	0.0013	0.0452	0.0001
			0.2185

W_2 = Same

LN_i	Pr(flood|LN_i)	Pr(LN_i|same)	Pr(flood|same)
1	0.3373	0.2317	0.0781
2	0.0427	0.5177	0.0221
3	0.0013	0.2507	0.0003
			0.1006

W_3 = Lower

LN_i	Pr(flood|LN_i)	Pr(LN_i|lower)	Pr(flood|lower)
1	0.3373	0.0278	0.0094
2	0.0427	0.2609	0.0111
3	0.0013	0.7113	0.0009
			0.0214

Summary of Second-Period Probabilities

	Second-Period Events W_j		
	Higher	Same	Lower
Pr(flood/W_j)	0.2185	0.1006	0.0214

Step 4. Calculate the expected value of the loss of lives and total cost (property loss and other costs) for each EV2, WA2, and DN2 branch (see Figure 9.5). These values are summarized in Table 9.1.

Consider the D2 node (higher water flow):

For EV2 arc:
Lives lost:

$$(0.1\,L)\Pr(\text{flood}\,|\,\text{higher}) = (0.1)(7)(0.2185)$$
$$\cong 0.1530$$

Total cost:

$$(0.4C)\Pr(\text{flood}\,|\,\text{higher}) + \text{cost}(EV2)$$
$$= (0.4)(7,000,000)(0.2185) + 3,000,000$$
$$\cong \$3,611,900$$

Similarly, for WA2:
Lives lost:

$$(0.2\,L)\Pr(\text{flood}\,|\,\text{higher}) = (0.2)(7)(0.2185)$$
$$\cong 0.3059$$

Total cost:

$$(0.5C)\Pr(\text{flood}\,|\,\text{higher}) + \text{cost}(WA2)$$
$$= (0.5)(7,000,000)(0.2185) + 500,000$$
$$\cong \$1,264,800$$

Finally, for DN2:
Lives lost:

$$(0.3\,L)\Pr(\text{flood}\,|\,\text{higher}) = (0.3)(7)(0.2185)$$
$$\cong 0.4589$$

Total cost:

$$(0.9C)\Pr(\text{flood}\,|\,\text{higher}) + \text{cost}(DN2)$$
$$= (0.9)(7,000,000)(0.2185) + 0$$
$$\cong \$1,376,700$$

Consider the D3 node (same water flow):

For EV2 arc:
Lives lost:
$$(0.1\,L)\,\Pr(\text{flood}\,|\,\text{same}) = (0.1)(7)(0.1006)$$
$$\cong 0.0704$$

Total cost:
$$(0.4C)\,\Pr(\text{flood}\,|\,\text{same}) + \text{cost}(\text{EV2})$$
$$= (0.4)(7{,}000{,}000)(0.1006) + 3{,}000{,}000$$
$$\cong \$3{,}281{,}600$$

Similarly, for WA2:
Lives lost:
$$(0.2\,L)\,\Pr(\text{flood}\,|\,\text{same}) = (0.2)(7)(0.1006)$$
$$\cong 0.1408$$

Total cost:
$$(0.5C)\,\Pr(\text{flood}\,|\,\text{same}) + \text{cost}(\text{WA2})$$
$$= (0.5)(7{,}000{,}000)(0.1006) + 500{,}000$$
$$\cong \$852{,}100$$

Finally, for DN2:
Lives lost:
$$(0.3\,L)\,\Pr(\text{flood}\,|\,\text{same}) = (0.3)(7)(0.1006)$$
$$\cong 0.2112$$

Total cost:
$$(0.9C)\,\Pr(\text{flood}\,|\,\text{same}) + \text{cost}(\text{DN2})$$
$$= (0.9)(7{,}000{,}000)(0.1006) + 0$$
$$\cong \$633{,}700$$

Consider the D4 node (lower water flow):

For EV2 arc:
Lives lost:
$$(0.1\,L)\,\Pr(\text{flood}\,|\,\text{lower}) = (0.1)(7)(0.0214)$$
$$\cong 0.0150$$

Total cost:
$$(0.4C)\,\Pr(\text{flood}\,|\,\text{lower}) + \text{cost}(\text{EV2})$$
$$= (0.4)(7{,}000{,}000)(0.0214) + 3{,}000{,}000$$
$$\cong \$3{,}060{,}000$$

Similarly, for WA2:
Lives lost:
$$(0.2\,L)\,\Pr(\text{flood}\,|\,\text{lower}) = (0.2)(7)(0.0214)$$
$$\cong 0.0300$$

Total cost:
$$(0.5C)\,\Pr(\text{flood}\,|\,\text{lower}) + \text{cost}(\text{WA2})$$
$$= (0.5)(7{,}000{,}000)(0.0214) + 500{,}000$$
$$\cong \$575{,}000$$

Finally, for DN2:
Lives lost:
$$(0.3\,L)\,\Pr(\text{flood}\,|\,\text{lower}) = (0.3)(7)(0.0214)$$
$$\cong 0.0450$$

Total cost:
$$(0.9C)\,\Pr(\text{flood}\,|\,\text{lower}) + \text{cost}(\text{DN2})$$
$$= (0.9)(7{,}000{,}000)(0.0214) + 0$$
$$\cong \$135{,}000$$

Consider flood arc from C2 node:

Lives lost:
$$(0.2\,L)\,\Pr(\text{flood}) = (0.2)(7)(0.1271) \cong 0.1780$$

Total cost:
$$(0.8C)\,\Pr(\text{flood}) = (0.8)(7{,}000{,}000)(0.1271)$$
$$\cong \$711{,}800$$

The results are summarized in Table 9.1.

Step 5. Folding back at each decision node, the vector-valued functions are compared, and dominated (inferior) solutions are eliminated. Summarize noninferior solutions for the second-period decision nodes. Table 9.2 summarizes the noninferior decisions for the second-period decision nodes.

Consider, for example, decision node D2.

The vector corresponding to the decision DN2 is inferior to WA2. Table 9.2 presents the noninferior decisions for the second period.

Node	Arc	Loss of Lives	Total Cost
D2	EV2	0.1530	3,611,900
	WA2	0.3059	1,264,800
	DN2	0.4589	1,376,700

The vector corresponding to DN2 is inferior to the vector corresponding to WA2:

$$\begin{bmatrix} 0.3059 \\ 1,264,800 \end{bmatrix}_{WA2} < \begin{bmatrix} 0.4589 \\ 1,376,700 \end{bmatrix}_{DN2}$$

Follow the same procedure for the other decision nodes (see Table 9.2).

Step 6. Averaging out at the chance nodes for the first period, each noninferior decision corresponding to each arc is multiplied by the probability for that arc, yielding a single decision value for the first-period decision node. Calculate the expected-value vector for all permutations of the noninferior solutions for the first period. These values are summarized in Table 9.3.

6a. Calculate for the expected value for all permutations (see Table 9.3).

6b. Complete generation of Table 9.3 and summarize decisions for the first-period node.

6c. Folding back at each decision node, the vector functions are compared, and inferior (dominated) solutions are eliminated, for example:

WA1	EV2	EV2	WA2	0.2453	3,689,500	
WA1	EV2	WA2	EV2	0.2589	3,926,100	➡ Inferior solution

Table 9.3 lists 18 different combinations for WA1, nine of which are noninferior (designated by the superscript a). Similarly, 27 different combinations for DN1 are listed. There are eight noninferior decisions for DN1. Thus, there are a total of 17 noninferior solutions for both WA1 and DN1 (see Figure 9.7).

First Period	Second Period
WA1[a] Lives	Higher EV2 Same EV2 Lower EV2 = (Lives lost for EV2|higher)Pr(higher) + (lives lost for EV2|same)Pr(same) + (lives lost for EV2|lower) Pr(lower) + lives lost with floods = (0.1530)(0.2466) + (0.0704)(0.2685) + (0.0150)(0.3577) + 0.1780 ≅ 0.2400
Cost	= (Property lost for EV2|higher)Pr(higher) + (property lost for EV2|same)Pr(same) + (property lost for EV2|lower)Pr(lower) + property lost with floods + cost(WA1) = (3,611,900)(0.2466) + (3,281,600)(0.2685) + (3,060,000)(0.3577) + 711,800 + 1,000,000 ≅ 4,578,500
WA1[a] Lives	Higher EV2 Same EV2 Lower WA2 = (Lives lost for EV2|higher)Pr(higher) + (lives lost for EV2|same)Pr(same) + (lives lost for WA2|lower)Pr(lower) + lives lost with floods = (0.1530)(0.2466) + (0.0704)(0.2685) + (0.0300)(0.3577) + 0.1780 ≅ 0.2453
Cost	= (Property lost for EV2|higher)Pr(higher) + (property lost for EV2|same)Pr(same) + (property lost for A2|lower)Pr(lower) + property lost with floods + cost(WA1) = (3,611,900)(0.2466) + (3,281,600)(0.2685) + (575,000)(0.3577) + 711,800 + 1,000,000 ≅ 3,689,500
WA1[a] Lives	Higher WA2 Same DN2 Lower DN2 = (Lives lost for WA2|higher)Pr(higher) + (lives lost for DN2|same)Pr(same) + (lives lost for DN2|lower)Pr(lower) + lives lost with floods = (0.3059)(0.2466) + (0.2112)(0.2685) + (0.0450)(0.3577) + 0.1780 ≅ 0.3262
Cost	= (Property lost for WA2|higher)Pr(higher) + (property lost for DN2|same)Pr(same) + (property lost for DN2|lower)Pr(lower) + property lost with floods + cost(WA1) = (1,264,800)(0.2466) + (633,700)(0.2685) + (135,000)(0.3577) + 711,800 + 1,000,000 ≅ 2,242,200

A large number of significant digits are kept for the benefit of the reader who would generate these values; however, for practical purposes, the final result is truncated.

9.2.4 Extension to Multiple Risk Measures

Determining the fold-back strategy associated with conditional expected values is substantially different from such an operation using the conventional expected value. Unlike the latter, which is a linear operation, the conditional expected-value operator is nonlinear. This nonlinearity represents an obstacle in decomposing the overall value of the conditional expected value and in calculating it at different decision nodes. Thus, in calculating conditional risk functions f_4, all performance measures at the different branches are mapped to the terminal points where the partitioning is performed.

In order to develop a fold-back strategy for the conditional expected value f_4 (the schemes for f_2 and f_3 are similar and thus are omitted here), some properties in a sequential calculation of f_4 will first be discussed.

Consider a two-stage decision-tree problem with a damage function $f(a_1, \theta_1, a_2, \theta_2)$, where a_j is the action at stage j and θ_j is the state of nature at stage j ($j = 1$ and 2). The optimal value of f_4 is given by

$$f_4^* = \min_{a_1, a_2} \frac{\iint_{f(a_1,\theta_1,a_2,\theta_2) \geq P^{-1}(\alpha)} f(a_1, \theta_1, a_2, \theta_2) p(\theta_1, \theta_2 | a_1, a_2)\, d\theta_1 d\theta_2}{\iint_{f(a_1,\theta_1,a_2,\theta_2) \geq P^{-1}(\alpha)} p(\theta_1, \theta_2 | a_1, a_2)\, d\theta_1 d\theta_2}$$

(9.17)

where α is the partitioning point on the probability axis. Rewrite

$$p(\theta_1, \theta_2 | a_1, a_2) = p(\theta_2 | \theta_1, a_1, a_2) p(\theta_1 | a_1) \quad (9.18)$$

The fact that an action at a subsequent stage does not affect the state of nature at a previous stage is seen in Eq. (9.18). Consequently, the optimization problem in Eq. (9.17) can be evaluated in a two-stage form:

$$f_4^* = \min_{a_1, a_2} \frac{\int_{f(a_1,\beta_1,a_2,\beta_2) \geq P^{-1}(\alpha)} \left[\int f(a_1,\theta_1,a_2,\theta_2) p(\theta_2|\theta_1,a_1,a_2) d\theta_2\right] p(\theta_1|a_1) d\theta_1}{\int_{f(a_1,\beta_1,a_2,\beta_2) \geq P^{-1}(\alpha)} \left[\int p(\theta_2|\theta_1,a_1,a_2) d\theta_2\right] p(\theta_1|a_1) d\theta_1}$$

(9.19)

The optimization problem in Eq. (9.19) is nonseparable. To separate the objective function with respect to stages, it is thus necessary to record two numbers at each stage: the values of the numerator and the denominator for each optimal conditional expected value. A more serious problem related to the decomposition of Eq. (9.19) is its nonmonotonicity. This can be easily observed in the fact that minimization of $a(\cdot)/b(\cdot)$ does not necessarily lead to the solution of minimization of $[c+a(\cdot)]/[d+b(\cdot)]$, where c and d are two constants and b and d are positive. The only exception to the above is the case where b remains a constant. The following simplification will be introduced to make possible the stagewise calculation of the value of the conditional expectation f_4. From the definition, we have $P[f(\theta_1, \theta_2) \geq P^{-1}(\alpha)] = 1 - \alpha$. When the value of θ_1 is fixed, $P[f(\theta_1, \theta_2) \geq P^{-1}(\alpha)|\theta_1]$ is not necessarily equal to $1-\alpha$. In order to have a common denominator, we introduce a set of $P^{-1}(\alpha)$ to keep $P\left[f(\theta_1, \theta_2) \geq P_1^{-1}(\alpha) | \theta_1\right] = 1 - \alpha$, where P_1 is the conditional cumulative distribution function of θ_2, given the value of θ_1. When we fold back, this simplification yields

$$\int_{\theta_1} P\left[f(\theta_1, \theta_2) \geq P_1^{-1}(\alpha) | \theta_1\right] p(\theta_1) d\theta_1 = 1 - \alpha \quad (9.20)$$

In summary, we should adhere to the following rules when calculating the conditional expected value in the fold-back step of decision trees:

1. Partition and calculate f_4 at terminal points according to the conditional pdf.
2. Fold back and perform at each chance node the operation of the expected value.

Note that although reducing the variance (uncertainty) of the risk may not contribute much to reducing the expected value f_5, it often markedly reduces the conditional expected value f_4 associated with extreme events (see Figure 9.8). Two benefits that result from additional experimentation include reducing the expected loss and reducing the uncertainty associated with decisionmaking under risk. However, in most cases, these two aspects of experimentation conflict with each other. The general framework of MODT analysis proposed here provides a medium with which these dual aspects can be captured by investigating the multiple impacts of experimentation.

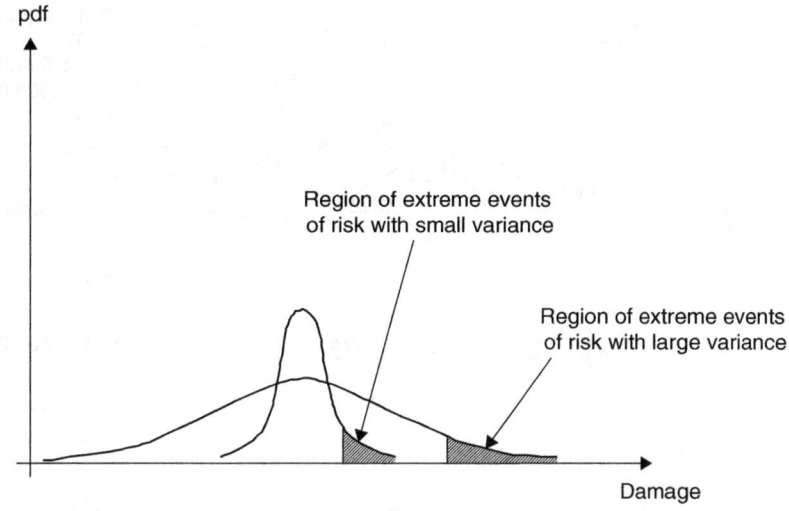

Figure 9.8 Variances and regions of extreme events. From Haimes et al. [1990]; © 1990 Wiley.

9.2.5 Example Problem for the Continuous Case

9.2.5.1 Problem definition

The flood warning problem developed in the previous example (Section 9.2.3) for the discrete case is modified here to handle continuous loss functions and extreme random events. The main difference between the discrete and the continuous cases lies in calculating the damage vector for the terminal nodes, which can be determined using the expected value $f_5(\cdot)$ and/or the conditional expected value $f_4(\cdot)$. The subsequent computations are similar to those carried out for the discrete case. Consequently, assumption 6 (Section 9.2.3.1) for the discrete case is modified as follows:

6. L and C are, respectively, the possible loss of lives and the cost, given that no flood warning is issued; they are linear functions of the water flow W. All other costs (as shown in Figure 9.9) are given in terms of the loss functions L and C, where $L = WL_F$, $L_F = 0.0001$, $C = WC_F$, and $C_F = 100$. The complete decision tree for this case is shown in Figure 9.9.

The loss functions L and C are calculated using the unconditional expected-value function $f_5(\cdot)$ and/or the conditional expected-value function $f_4(\cdot)$. The unconditional expected loss $f_5(\cdot)$ is given by

$$f_5(\cdot) = P_f \int_{50{,}000}^{\infty} \frac{W}{\sqrt{2\pi}\sigma} \frac{1}{W} \exp\left[-\frac{1}{2}\left[\frac{\ln(W)-\mu}{\sigma}\right]^2\right] dW$$

$$= P_f \left[1 - \Phi\left[\frac{10.82-\mu}{\sigma}\right]\right] \exp(\mu + \sigma^2/2)$$

$$\left[1 - \Phi\left[\frac{10.82-\mu}{\sigma} - \sigma\right]\right] \quad (9.21)$$

where P_f is equal to L_f or C_f when Eq. (9.21) is used to calculate f_5 for loss of lives or monetary costs, respectively. The conditional expected loss $f_4(\cdot)$ is given by

$$f_4(\cdot) = P_f \left[1 - \Phi\left[\Phi^{-1}(\alpha) - \sigma\right]\right]$$
$$\exp(\mu + \sigma^2/2)/(1-\alpha) \quad (9.22)$$

where P_f is equal to L_f or C_f when Eq. (9.22) is used to calculate f_4 for loss of lives or monetary costs, respectively, and α is the partitioning point on the probability axis, which is 0.99 in this case. With the use of Eqs. (9.21) and (9.22), the cost (C) and the loss of lives (L) are calculated using $f_4(\cdot)$ and $f_5(\cdot)$ at all the terminal nodes for each of the decision arcs. Note that each of the risk functions $f_4(\cdot)$ and $f_5(\cdot)$ is composed of two components: cost and loss of lives.

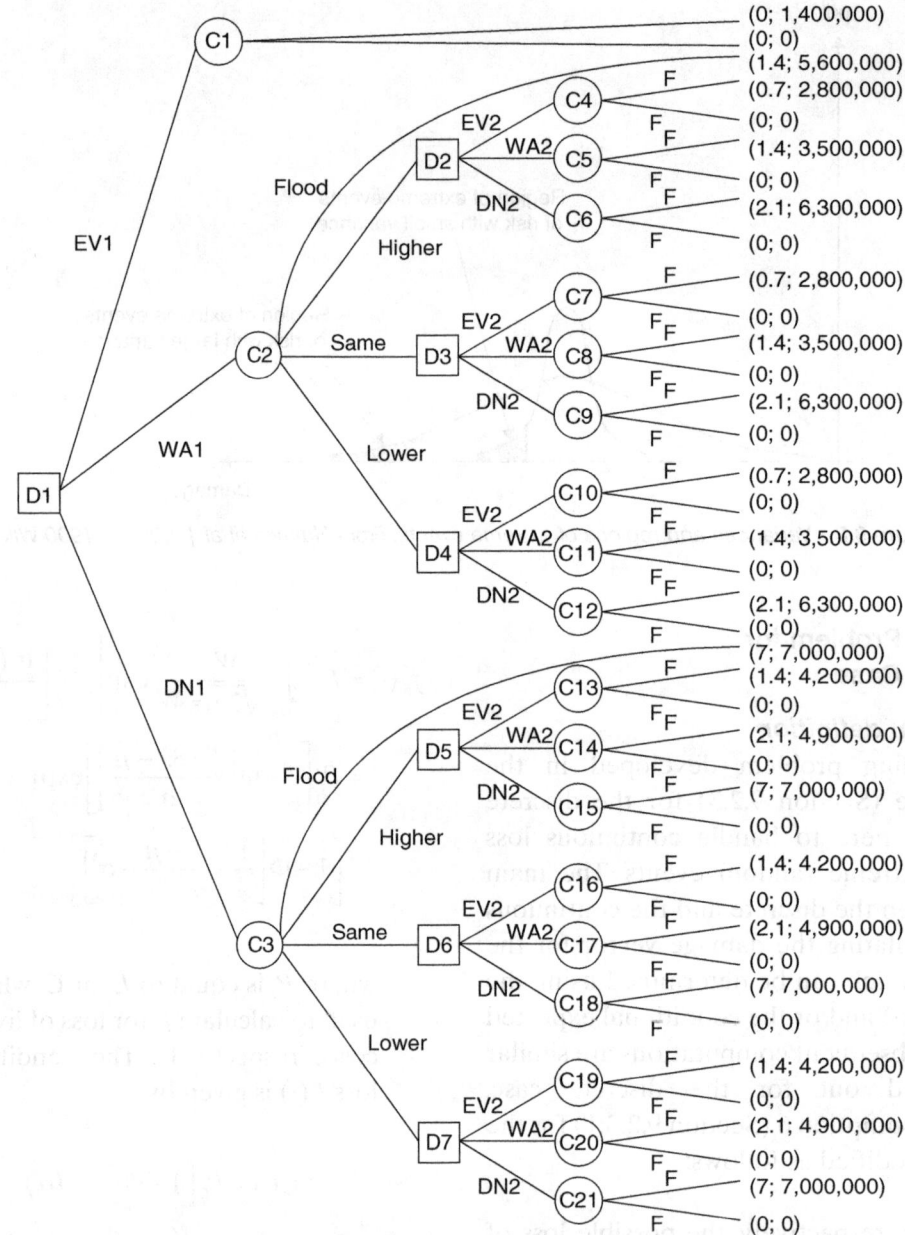

Figure 9.9 Decision tree for the continuous case. From Haimes et al. [1990]; © 1990 Wiley.

9.2.5.2 Calculating the loss vectors for the first period

Chance node C1
Assuming that the possible pdf of the water flow (W) is LN_i with probability $1/3$, $i = 1, 2, 3$, and that the flood stage is at $W = 50,000$ cfs, two outcomes are considered at the end of the second period: a flood or a no-flood event (see Figure 9.10). The values of the components of $f_4(\cdot)$ and $f_5(\cdot)$ for node C1 are calculated using Eqs. (9.21) and (9.22), respectively.

The value of the loss vector for C1 using $f_5(\cdot)$ is shown in Figure 9.10.

Chance nodes C2 and C3
Four possible outcomes at the beginning of the second period are investigated at nodes C2 and C3: a flood event, a higher water flow, the same water flow, and a lower water flow (see Figure 9.11). Similar to the discrete case, the probabilities of these outcomes are calculated using Eqs. (9.12a), (9.13), and (9.14).

METHODOLOGICAL APPROACH 277

9.2.5.3 Calculating the loss vectors for the second period

Regardless of whether a watch (WA1) was ordered or a do-nothing (DN2) action was followed at the first period, the same three possible actions are evaluated at the second period: evacuate, order another flood watch, or do nothing. Depending on the actions taken at the first and second periods and the water flow level at the second period, different values of losses are generated for each terminal chance node. There are three equally probable underlying pdf's for the water flow for the first period. After measuring the water flow W_j at the end of the first period, the posterior probabilities are calculated using Eq. (9.15). The required value of the loss vector [of $f_4(\cdot)$ and $f_5(\cdot)$] is then calculated using Eqs. (9.21) and (9.23) for $f_5(\cdot)$ and Eqs. (9.22) and (9.24) for $f_4(\cdot)$:

$$f_5(\cdot|W_j) = \sum_{i=1}^{3}[f_5(\cdot)|LN_i]\Pr(LN_i|W_j) \quad (9.23)$$

$$f_4(\cdot|W_j) = \sum_{i=1}^{3}[f_4(\cdot)|LN_i]\Pr(LN_i|W_j) \quad (9.24)$$

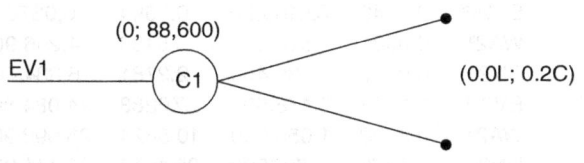

Figure 9.10 Averaging out chance node C1 using f_5 (continuous case). From Haimes et al. [1990]; © 1990 Wiley.

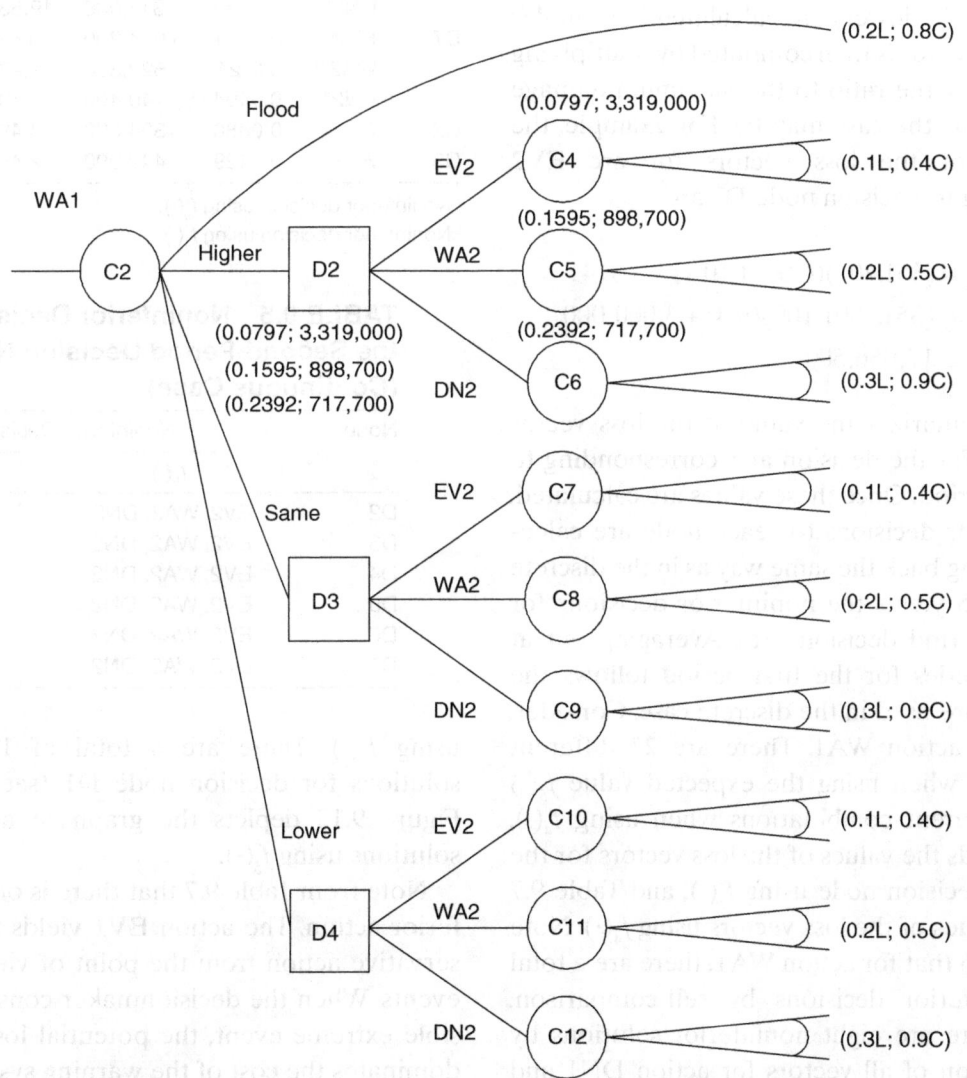

Figure 9.11 Second-stage tree corresponding to chance node C2 using f_5. From Haimes et al. [1990]; © 1990 Wiley.

For example,

$$f_4(\cdot|\text{higher}) = f_4(\cdot|\text{LN}_1)\Pr(\text{LN}_1|\text{higher})$$
$$+ f_4(\cdot|\text{LN}_2)\Pr(\text{LN}_2|\text{higher})$$
$$+ f_4(\cdot|\text{LN}_3)\Pr(\text{LN}_3|\text{higher})$$

The values of $\Pr(\text{LN}_i|\text{higher})$ ($i = 1, 2, 3$) are calculated using Eq. (9.15), and the values of $f_4(\cdot|\text{LN}_i)$ are calculated using Eq. (9.22). Therefore, Eq. (9.24) yields

$$f_4(\cdot|\text{higher}) = (500,400)(0.6031) + (136,400)(0.3517)$$
$$+ (37,200)(0.0452)$$
$$\cong 351,400$$

The values for $f_4(\cdot|\text{same})$, $f_4(\cdot|\text{lower})$, $f_5(\cdot|\text{higher})$, $f_5(\cdot|\text{same})$, and $f_5(\cdot|\text{lower})$ are calculated in a similar way. The loss vector is then computed by multiplying these results by the ratio to the maximum damage and L_f or C_f, as the case may be. For example, the components of the loss vectors for arc EV2 corresponding to decision node D2 are

$$L_{\text{EV2}|\text{D2},f_4(\cdot)} = (351,400)(0.0001)(0.1) \cong 3.5141$$
$$C_{\text{EV2}|\text{D2},f_4(\cdot)} = (351,400)(100)(0.4) + 3,000,000$$
$$\cong 17,056,500$$

Table 9.4 summarizes the value of the loss vector $f_5(\cdot)$ and $f_4(\cdot)$ for the decision arcs corresponding to the second period. Once these values are calculated, the noninferior decisions for each node are calculated by folding back the same way as in the discrete case. Table 9.5 yields the noninferior decisions for the second-period decision arcs. Averaging out at the chance nodes for the first period follows the same procedure used in the discrete case. Consider, for example, action WA1. There are 27 different combinations when using the expected value $f_5(\cdot)$ and four different combinations when using $f_4(\cdot)$. Table 9.6 yields the values of the loss vectors for the first-period decision node using $f_5(\cdot)$, and Table 9.7 yields the values of the loss vectors using $f_4(\cdot)$. Note from Table 9.6 that for action WA1, there are a total of 10 noninferior decisions by self-comparison. Similarly, there are eight noninferior solutions by self-comparison of all vectors for action DN1 and six after comparison of all decisions for all actions

TABLE 9.4 Loss Vectors for the Second-Period Decision Arcs (Continuous Case)

Node	Arc	$f_5(\cdot)$ L	$f_5(\cdot)$ C($)	$f_4(\cdot)$ L	$f_4(\cdot)$ C($)
D2	EV2[a,b]	0.0797	3,319,000	3.5141	17,056,500
	WA2[a]	0.1595	898,700	7.0283	18,070,700
	DN2[a]	0.2392	717,700	10.5424	31,627,200
D3	EV2[a,b]	0.0312	3,124,800	1.9583	10,833,100
	WA2[a,b]	0.0624	656,000	3.9165	10,291,300
	DN2[a]	0.0936	280,800	5.8748	17,624,400
D4	EV2[a,b]	0.0040	3,016,200	0.7594	6,037,500
	WA2[a,b]	0.0081	520,200	1.5188	4,296,900
	DN2[a]	0.0121	36,400	2.2781	6,834,400
D5	EV2[a,b]	0.1595	3,478,500	7.0283	24,084,800
	WA2[a]	0.2392	1,058,200	10.5424	25,098,900
	DN2[a]	0.7975	797,500	35.1413	35,141,300
D6	EV2[a,b]	0.0624	3,187,200	3.9165	14,749,600
	WA2[a,b]	0.0936	718,400	5.8748	14,207,900
	DN2[a]	0.3120	312,000	19.5827	19,582,700
D7	EV2[a,b]	0.0081	3,024,300	1.5188	7,556,300
	WA2[a,b]	0.0121	528,300	2.2781	5,815,600
	DN2[a]	0.0404	40,400	7.5938	7,593,800
C2	F	0.0886	354,300	4.4928	17,971,300
C3	F	0.4429	442,900	22.4641	22,464,100

[a] Noninferior decision using $f_5(\cdot)$.
[b] Noninferior decision using $f_4(\cdot)$.

TABLE 9.5 Noninferior Decisions for the Second-Period Decision Nodes (Continuous Case)

Node	Noninferior Decision $f_5(\cdot)$	Noninferior Decision $f_4(\cdot)$
D2	EV2, WA2, DN2	EV2
D3	EV2, WA2, DN2	EV2, WA2
D4	EV2, WA2, DN2	EV2, WA2
D5	EV2, WA2, DN2	EV2
D6	EV2, WA2, DN2	EV2, WA2
D7	EV2, WA2, DN2	EV2, WA2

using $f_5(\cdot)$. There are a total of 17 noninferior solutions for decision node D1 (see Figure 9.12). Figure 9.13 depicts the graph of all noninferior solutions using $f_5(\cdot)$.

Note from Table 9.7 that there is only one noninferior action. The action EV1 yields the most conservative action from the point of view of extreme events. When the decisionmaker considers the possible extreme event, the potential loss of property dominates the cost of the warning system. Thus, the two objective functions do not conflict in this case.

TABLE 9.6 Decisions for the First-Period Node Using f_5 (Continuous Case)

First-Period Decision	Second-Period Decision			Loss Vector	
	Higher	Same	Lower	L	C($)
EV1[a]	—	—	—	0.0000	5,088,600
WA1[a]	EV2	EV2	EV2	0.0408	3,781,700
WA1[a]	EV2	EV2	WA2	0.0422	2,888,800
WA1[a]	EV2	EV2	DN2	0.0436	2,715,700
WA1	EV2	WA2	EV2	0.0491	3,118,800
WA1[a]	EV2	WA2	WA2	0.0506	2,225,900
WA1[a]	EV2	WA2	DN2	0.0520	2,052,800
WA1	EV2	DN2	EV2	0.0575	3,018,100
WA1	EV2	DN2	WA2	0.0590	2,125,100
WA1[a]	EV2	DN2	DN2	0.0604	1,952,000
WA1	WA2	EV2	EV2	0.0604	3,184,800
WA1	WA2	EV2	WA2	0.0619	2,291,800
WA1	WA2	EV2	DN2	0.0633	2,118,800
WA1	WA2	WA2	EV2	0.0688	2,521,900
WA1[a]	WA2	WA2	WA2	0.0702	1,629,000
WA1[a]	WA2	WA2	DN2	0.0717	1,455,900
WA1	WA2	DN2	EV2	0.0772	2,421,100
WA1	WA2	DN2	WA2	0.0786	1,528,200
WA1[a]	WA2	DN2	DN2	0.0801	1,355,100
WA1	DN2	EV2	EV2	0.0801	3,140,100
WA1	DN2	EV2	WA2	0.0815	2,247,200
WA1	DN2	EV2	DN2	0.0830	2,074,100
WA1	DN2	WA2	EV2	0.0885	2,477,200
WA1	DN2	WA2	WA2	0.0899	1,584,300
WA1	DN2	WA2	DN2	0.0914	1,411,200
WA1	DN2	DN2	EV2	0.0968	2,376,500
WA1	DN2	DN2	WA2	0.0983	1,483,600
WA1[a]	DN2	DN2	DN2	0.0997	1,310,500
DN1	EV2	EV2	EV2	0.1153	2,851,900
DN1	EV2	EV2	WA2	0.1167	1,959,000
DN1	EV2	EV2	DN2	0.1269	1,784,500
DN1	EV2	WA2	EV2	0.1237	2,189,000
DN1[a]	EV2	WA2	WA2	0.1251	1,296,100
DN1[a]	EV2	WA2	DN2	0.1352	1,121,600
DN1	EV2	DN2	EV2	0.1823	2,079,900
DN1	EV2	DN2	WA2	0.1838	1,187,000
DN1	EV2	DN2	DN2	0.1939	1,012,500
DN1	WA2	EV2	EV2	0.1350	2,255,000
DN1	WA2	EV2	WA2	0.1364	1,362,100
DN1	WA2	EV2	DN2	0.1465	1,187,600
DN1	WA2	WA2	EV2	0.1433	1,592,100
DN1[a]	WA2	WA2	WA2	0.1448	699,200
DN1[a]	WA2	WA2	DN2	0.1549	524,700
DN1	WA2	DN2	EV2	0.2020	1,483,000
DN1	WA2	DN2	WA2	0.2034	590,100
DN1[a]	WA2	DN2	DN2	0.2135	415,500
DN1	DN2	EV2	EV2	0.2726	2,190,700
DN1	DN2	EV2	WA2	0.2741	1,297,800
DN1	DN2	EV2	DN2	0.2842	1,123,200
DN1	DN2	WA2	EV2	0.2810	1,527,800
DN1	DN2	WA2	WA2	0.2825	634,900
DN1	DN2	WA2	DN2	0.2926	460,400
DN1	DN2	DN2	EV2	0.3397	1,418,700
DN1	DN2	DN2	WA2	0.3411	525,800
DN1[a]	DN2	DN2	DN2	0.3512	351,200

[a] Noninferior decisions.

TABLE 9.7 Decisions for the First-Period Node Using f_4 (Continuous Case)

First-Period Decision	Second-Period Decision			Loss Vector	
	Higher	Same	Lower	L	C($)
EV1[a]	—	—	—	0.0000	9,492,800
WA1	EV2	EV2	EV2	2.2353	12,559,700
WA1	EV2	EV2	WA2	2.5069	11,937,000
WA1	EV2	WA2	EV2	2.7611	12,414,300
WA1	EV2	WA2	WA2	3.0327	11,791,600
DN1	EV2	EV2	EV2	6.1837	15,459,200
DN1	EV2	EV2	WA2	6.4554	14,836,500
DN1	EV2	WA2	EV2	6.7096	15,313,700
DN1	EV2	WA2	WA2	6.9812	14,691,000

[a] Noninferior decisions.

9.3 DIFFERENCES BETWEEN SODT AND MODT

It is worthwhile to summarize the basic differences between an SODT and an MODT:

1. Since most, if not all, real-world systems and problems are characterized by multiple noncommensurate and competing objectives, the first difference is a better and more realistic representation of the essence of the system through MODT. Indeed, in MODT, there is no compulsive need to force all attributes and objectives into a simple metric.
2. The end nodes of SODT consist of single values (the outcomes of a given course of action/ strategy with respect to the single objective). In MODT, on the other hand, the end nodes comprise a vector of values, reflecting the value of each objective function associated with a given action.

280 MULTIOBJECTIVE DECISION-TREE ANALYSIS

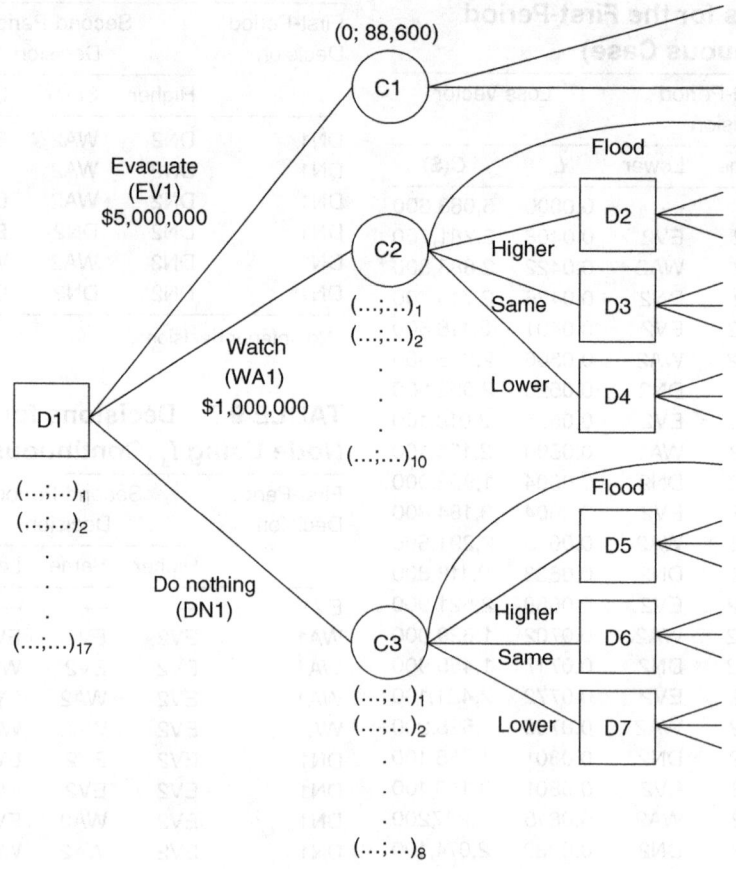

Figure 9.12 Decision tree for the second stage using f_5 (continuous case). From Haimes et al. [1990]; © 1990 Wiley.

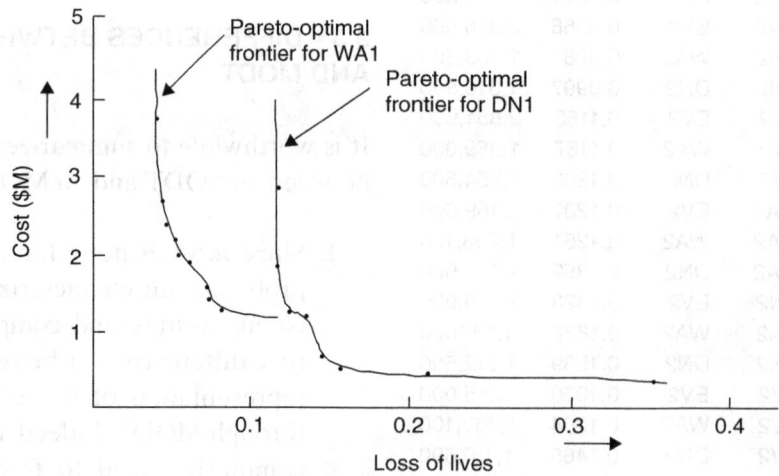

Figure 9.13 Pareto-optimal frontier using f_5 (continuous case). From Haimes et al. [1990]; © 1990 Wiley.

3. The outcomes at a chance node just prior to an end node are *averaged out* according to the probabilities associated with the chance node's branches. The SODT results in only one number, which is the expected value of the associated outcome. In an MODT, on the other hand, there will be a vector of expected values of outcomes (objective functions).

4. Consider a decision node that is the first node prior to an end node. In SODT, we select only one optimal alternative action—the one that maximizes (or minimizes, as appropriate) the

objective functions. All other alternative options are then discarded. In MODT, however, it is common to have more than one noninferior solution (alternative option) for that node. This means that we roll back all noninferior solutions to the decision node.

5. Consider a chance node that is somewhere in the middle of an SODT and an MODT. In an SODT, there is a single scalar associated with all chance or decision nodes. The value that will be rolled back to that chance or decision node will be just one scalar: the expected value of all attached nodes. In MODT, one or more of the attached nodes (to the right) may have associated with them more than one noninferior vector.

Suppose there are N nodes attached to the chance node and node j ($j=1, ..., N$) has M_j noninferior vectors associated with it. Then for our current chance node, we need to consider $M_1 \times M_2 \times \cdots \times M_N$ possible vectors. (*Each* vector associated with node 1 has to be averaged out, as described earlier, with *each* vector associated with nodes 2 and 3 ... and node N.) From these $M_1 \times \cdots \times M_N$ vectors, the noninferior ones have to be identified and rolled back, that is, associated with the current chance node. In the flood warning and evacuation system discussed earlier (see Tables 9.4, 9.5, and 9.6), D2, D3, and D4 each have three associated noninferior vectors (see Tables 9.4 and 9.5). Thus, for C2, we have to consider $3 \times 3 \times 3 = 27$ vectors; however, only 10 out of these turn out to be noninferior when we combine the three sets (see Table 9.6). Consequently, these 10 vectors are rolled back to C2.

6. For a decision node that is somewhere in the tree (and could also be initial decision node), the following procedure applies: In SODT, all attached nodes (to the right) will have just one value associated with them; we only need to select the optimal one and roll it back to the current decision node. In MODT, any of the attached nodes could have more than one noninferior vector attached to them. We need to consider (i) the totality of *noninferior* vectors associated with those attached nodes, and (ii) if a *noninferior* vector associated with one node is possibly inferior to a noninferior vector associated with another node, we only keep those vectors that are *truly* noninferior (i.e., noninferior in the combined set of *noninferior* vectors) and roll these back to the current decision node.

Example: D1 has one noninferior vector attached to it via its EV1 branch, 10 noninferior vectors via its WA1 branch, and 8 noninferior vectors via its DN1 branch. However, only 17 out of this set of 19 vectors are truly noninferior when we consider the totality of these 19 vectors (i.e., two vectors associated with DN1 are eliminated).

7. Considering the above, the result is that in an SODT, only one scalar is associated with (rolled back to) the initial node (the root of the tree). It is the expected value of the optimal strategy, that is, the optimal *path* through the tree. In an MODT, we can have one or (more likely) more than one noninferior vector associated with (rolled back to) the initial node. These reflect the Pareto-optimal set of strategies for the given problem. This is due to the fact that without information on the decisionmaker's preference, there usually does not exist a single optimal strategy under multiple objectives. One strategy may be best for one objective, and another strategy may be advantageous for another objective.

9.4 SUMMARY

MODT analysis is an extension of the single-objective-based decision-tree analysis discussed in Chapter 4 and formally introduced three decades ago by Howard Raiffa [1968]. This extension is made possible by synthesizing the traditional method with the more recently developed approaches used for multiobjective analysis and for the risk of extreme and catastrophic events. Successful applications of SODT analysis to numerous business, engineering, and governmental decisionmaking problems over the years have made the methodology into an important and valuable tool in systems analysis. Its extension—incorporating multiple noncommensurate objectives, impact analysis, and the conditional expected value for extreme and catastrophic events—might be viewed as an indicator of growth in the broader field of systems analysis and in decisionmaking under risk and uncertainty. Undoubtedly, there remain several theoretical challenges that must be addressed to fully realize the strengths and usefulness of the MODT. Additional studies on MODT can be found in Frohwein et al. [1999, 2000] and Frohwein and Lambert [2000]. This work involves the calculation of the PMRM metrics f_5 and f_4 developed in Chapter 8. MODT was also extended to include sequential decisionmaking involving multiple, interdependent infrastructure sectors by Santos et al. [2008], whose work is referred to as multiobjective inoperability decision trees (MOIDT).

9.5 EXAMPLE PROBLEMS

9.5.1 Interstate Transportation Problem

The consulting firm Better Decisions, Inc., was commissioned by a state agency to model and analyze the maintenance policy for a bridge on Interstate 64 in the Hampton Roads area. Three policy options were considered: replace the bridge, repair it, or do nothing. The problem is modeled using MODT. Two objectives are considered: the cost associated with each policy option and the mean time to failure (MTTF) of the bridge.

The following assumptions are made for this problem:

1. The cost of a new bridge is $1 million.
2. The condition of the bridge can be judged by a parameter s, which represents a declining factor of the age of the bridge.
3. The cost of repair depends upon the parameter s and is given by

$$C_{\text{REPAIR}} = 200{,}000 + 4{,}000{,}000(s - 0.05)$$

4. The parameter s is uncertain in nature and can take the following values:

$$s = s_1 = 0.050$$
$$s = s_2 = 0.075$$
$$s = s_3 = 0.100$$

5. The prior probability distribution of s is

$$p(s_1) = 0.25$$
$$p(s_2) = 0.50$$
$$p(s_3) = 0.25$$

6. A test to reduce the uncertainty in s can be performed at a cost of $50,000.
7. The test to reduce the uncertainty in s can have three possible outcomes:

$$T_1 = \text{higher uncertainty}$$
$$T_2 = \text{same uncertainty}$$
$$T_3 = \text{lower uncertainty}$$

8. The conditional probabilities of the test results (T_1, T_2, T_3) are as shown:

$$p(T_1|s_1) = 0.50, \quad p(T_2|s_1) = 0.25, \quad p(T_3|s_1) = 0.25$$
$$p(T_1|s_2) = 0.25, \quad p(T_2|s_2) = 0.50, \quad p(T_3|s_2) = 0.25$$
$$p(T_1|s_3) = 0.25, \quad p(T_2|s_3) = 0.25, \quad p(T_3|s_3) = 0.50$$

9. The value of λ for the exponential distribution of time to failure of a new bridge is 0.1.
10. The value of λ for the exponential distribution of time to failure of a repaired bridge is 0.15.

9.5.1.1 Solving the problem

To solve the problem, we construct a decision tree and compute the set of Pareto-optimal decisions for one branch of the tree corresponding to decision node D2. Note that:

1. Bridge failure is defined as any event that causes the closure of the bridge.
2. MTTF is defined as the amount of time that can be expected to pass before a bridge failure occurs.
3. For the exponential distribution with parameter λ, the mean time to failure is MTTF $= 1/\lambda$.
4. The pdf of time to failure of a new bridge is given by an exponential distribution with mean $1/\lambda$.
5. The pdf of time to failure of an old bridge is given by an exponential distribution with mean $1/(0.1 + s)$.
6. If repair is done immediately, the pdf of time to failure is given by an exponential distribution with mean $1/(0.15)$.

The decision tree for the problem is given in Figure 9.14. The two objective functions are to maximize MTTF and minimize cost.

Computing the MTTF
For an exponential distribution, the MTTF is given by $1/\lambda$, where λ is the parameter of the exponential distribution.

For a new bridge, $\lambda = 0.1$:

$$\Rightarrow \text{MTTF} \mid \text{replace} = 1/\lambda = 10 \text{ years}$$

For a repaired bridge, $\lambda = 0.15$:

$$\Rightarrow \text{MTTF} \mid \text{repair} = 1/\lambda = 6.6667 \text{ years}$$

For the do-nothing option, the MTTF is a function of the value of s.

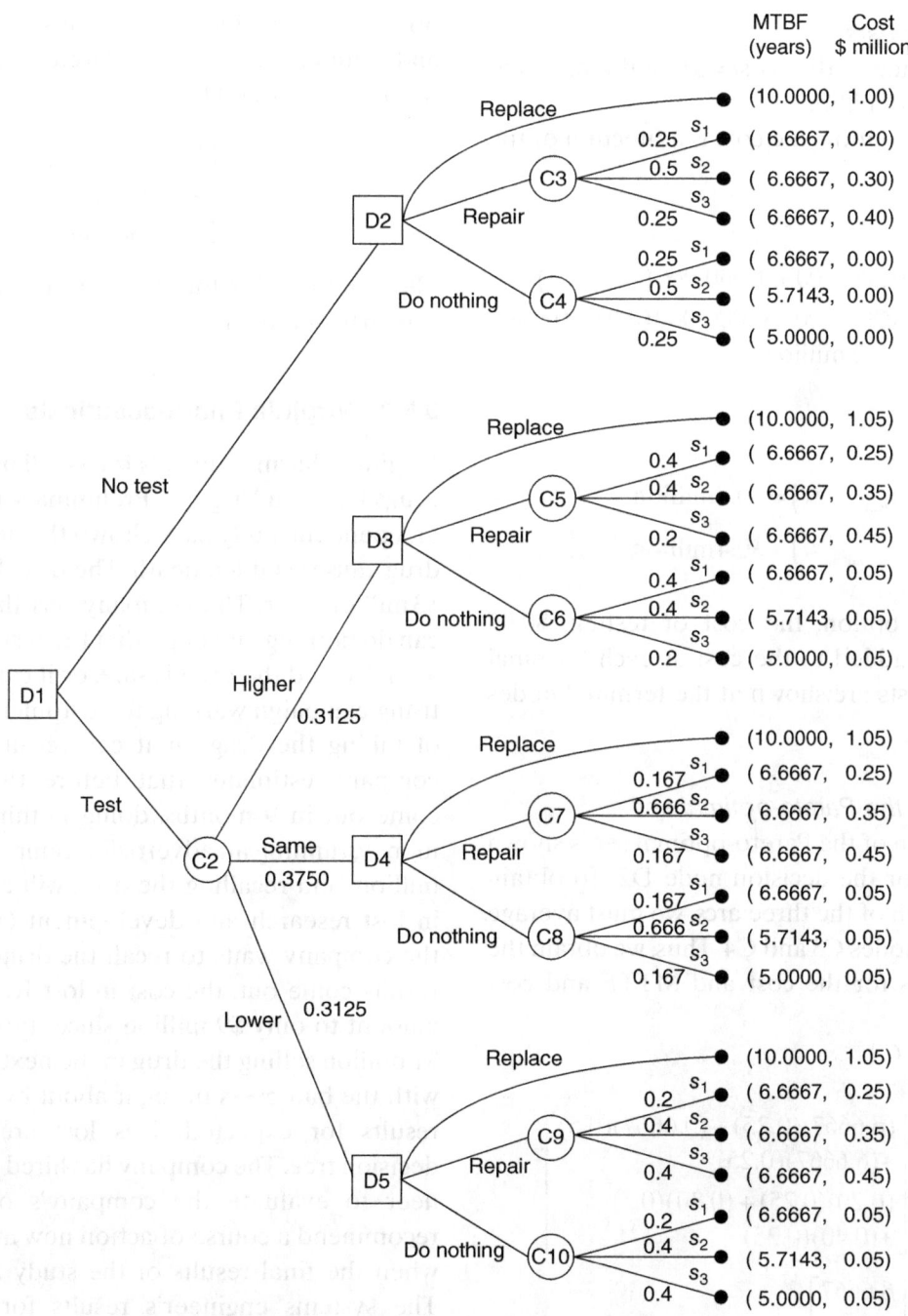

Figure 9.14 Decision tree for the bridge maintenance problem.

For $s=s_1$, $\lambda=0.1+0.05$:

$$\Rightarrow \text{MTTF}|s_1 = 1/0.15 = 6.6667 \text{ years}$$

For $s=s_2$, $\lambda=0.1+0.075$:

$$\Rightarrow \text{MTTF}|s_2 = 1/0.175 = 5.7143 \text{ years}$$

For $s=s_3$, $\lambda=0.1+0.1$:

$$\Rightarrow \text{MTTF}|s_3 = 1/0.2 = 5 \text{ years}$$

Computing the costs

For a new bridge, the cost = $1 million. Thus, (cost|replace) = $1 million.

For the repair option, the cost is a function of the value of s.

For $s=s_1$,

$$\begin{aligned}(\text{Cost}_{\text{Repair}}|s_1) &= 200{,}000 + 4{,}000{,}000(s_1 - 0.05) \\ &= 200{,}000 + 4{,}000{,}000(0.05 - 0.05) \\ &= 0.2 \text{ million}\end{aligned}$$

Similarly,

$$(\text{Cost}_{\text{Repair}}|s_2) = \$0.3 \text{ million}$$
$$(\text{Cost}_{\text{Repair}}|s_3) = \$0.4 \text{ million}$$

For the test option, the cost of testing, $0.05 million, will be added to the cost at each terminal node. All the costs are shown at the terminal nodes in Figure 9.14.

Computation of the Pareto-optimal set

The computation of the Pareto-optimal set is shown in Figure 9.14 for the decision node D2. To obtain the costs for each of the three arcs, we must average out the chance nodes C3 and C4. Thus, we obtain the following values for the cost and MTTF and cost consequences:

Chance node C3:

$$\begin{bmatrix}\text{MTTF}\\ \text{Cost}\end{bmatrix} = \begin{bmatrix}(6.6667)(0.25)+(6.6667)(0.5)\\+(6.6667)(0.25)\\(0.20)(0.25)+(0.30)(0.5)\\+(0.40)(0.25)\end{bmatrix}$$

$$= \begin{bmatrix}6.6667\\0.30\end{bmatrix}$$

Chance node C4:

$$= \begin{bmatrix}5.7738\\0.00\end{bmatrix}$$

For the arc *replace*:

$$= \begin{bmatrix}10.0000\\0.00\end{bmatrix}$$

Neither of these three solutions is dominated by any other solution. Because we are maximizing MTTF and minimizing cost, the Pareto-optimal solutions for decision node D2 are

$$[10.0000, 1.00]$$
$$[6.6667, 0.30]$$
$$[5.7738, 0.00]$$

The solutions for the other decision nodes can be similarly obtained.

9.5.2 Virginia Pharmaceuticals

Virginia Pharmaceuticals is a small manufacturer of drugs based in Virginia. Preliminary results from an independent study have shown that its most popular drug causes sudden death. The drug brings in about $3 million/year. The company has three options: It can do nothing and hope that the results of the study are false and the drug is safe, or it can run an advertising campaign warning its customers of the danger of taking the drug, or it can recall the drug. The company estimates that before the final results come out in 9 months, doing nothing will cost no money, running an advertising campaign will cost $2 million, and recalling the drug will cost $10 million in lost research and development (R&D) costs. If the company waits to recall the drug until the final results come out, the cost in lost R&D money will amount to only $9 million since it expects to make $1 million selling the drug in the next 9 months even with the bad press brought about by this study. The results for expected lives lost are given in the decision tree. The company has hired a systems engineer to evaluate the company's options and to recommend a course of action now and in 9 months, when the final results of the study are completed. The systems engineer's results for the expected values of costs and lives are shown in Figure 9.15.

EXAMPLE PROBLEMS 285

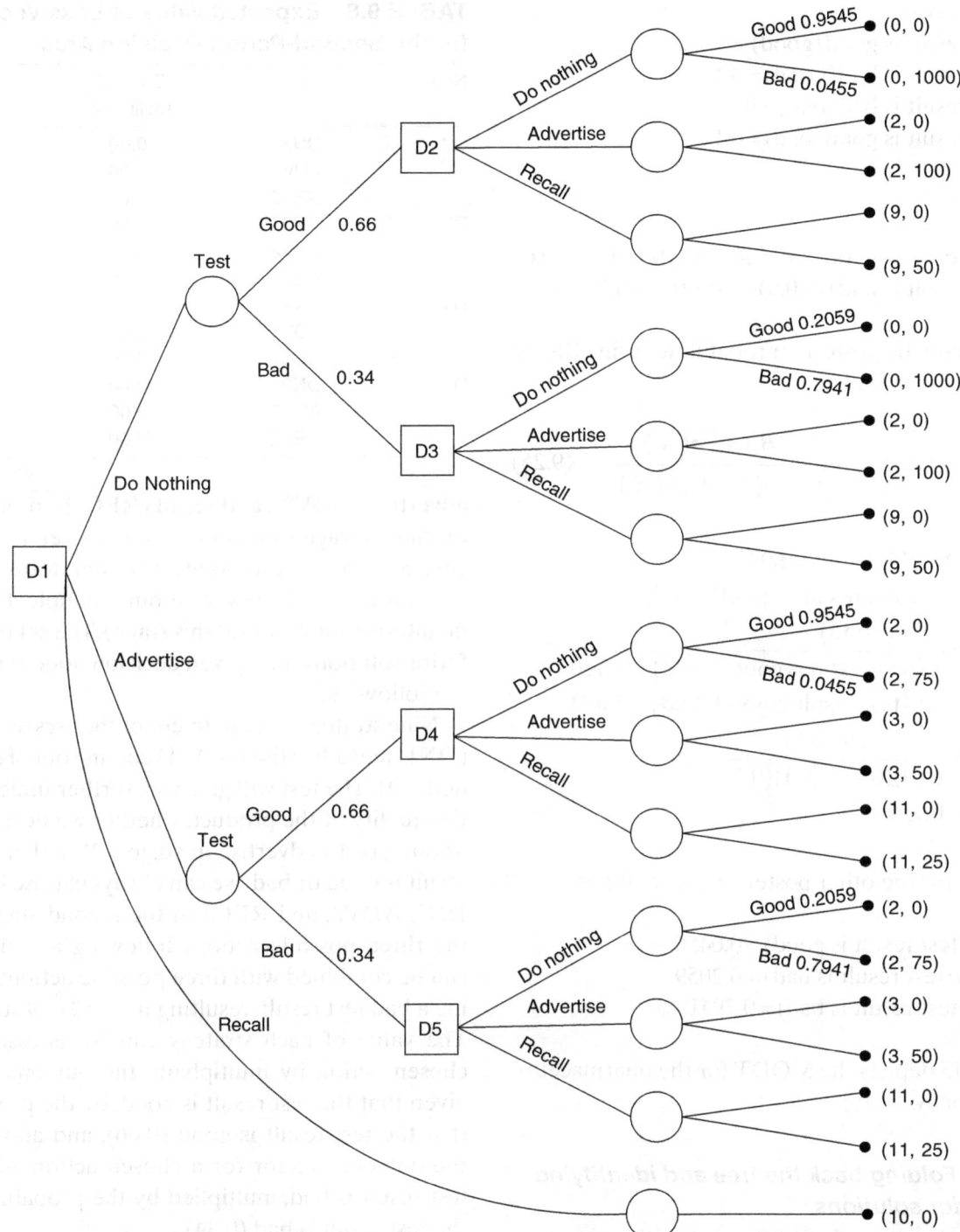

Figure 9.15 Multiobjective decision tree for the pharmaceuticals company.

9.5.2.1 Assumptions and notation

$p(\text{good})$ = probability that product is inherently of good quality

$p(\text{test result is good})$ = probability that product is classified to be of good quality by a certain test procedure

$p(\text{bad})$ = probability that product is inherently of bad quality

$p(\text{test result is bad})$ = probability that product is classified to be of bad quality by a certain test procedure

$p(\text{good}) = 0.7$

$p(\text{bad}) = 0.3$
$p(\text{test result is good} \mid \text{good}) = 0.9$
$p(\text{test result is bad} \mid \text{good}) = 0.1$
$p(\text{test result is bad} \mid \text{bad}) = 0.9$
$p(\text{test result is good} \mid \text{bad}) = 0.1$

Therefore,

$p(\text{test result is good}) = (0.9)(0.7) + (0.1)(0.3) = 0.66$
$p(\text{test result is bad}) = (0.9)(0.3) + (0.1)(0.7) = 0.34$

We calculate the posterior probabilities using Bayes' theorem (Eq. 9.25):

$$p(A_i \mid B_j) = \frac{p(B_j \mid A_i) p(A_i)}{\sum_{i=1}^{2} p(B_j \mid A_i) p(A_i)} \quad (9.25)$$

$p(\text{good} \mid \text{test result is good})$

$= \dfrac{\dfrac{p(\text{test result is good} \mid \text{good})}{p(\text{good})}}{p(\text{test result is good} \mid \text{good}) p(\text{good}) + p(\text{test result is good} \mid \text{bad}) p(\text{bad})}$

$= \dfrac{(0.9)(0.7)}{(0.9)(0.7) + (0.1)(0.3)}$

$= 0.9545$

Similarly, for the other posterior probabilities,

$p(\text{bad} \mid \text{test result is good}) = 0.0455$
$p(\text{good} \mid \text{test result is bad}) = 0.2059$
$p(\text{bad} \mid \text{test result is bad}) = 0.7941$

Figure 9.15 depicts the MODT for the pharmaceuticals company.

9.5.2.2 Folding back the tree and identifying noninferior solutions

For each chance node of type *reality*, the outcomes for the two possible states of nature (good–bad) are averaged according to the conditional probabilities for each. (The probabilities are conditioned on the test results.) For pairs of arcs without probabilities noted, the same probabilities apply as for the pair directly above. The results of this averaging are shown in Table 9.8. For each decision node D2 to D5, three choices are available (do nothing (DN2),

TABLE 9.8 Expected Value of Loss Vectors for the Second-Period Decision Arcs

Node	Arc	Cost ($) (millions)	Lives
D2	DN2	0.00	45.4545
	ADV2	2.00	4.5455
	REC2	9.00	2.2727
D3	DN2	0.00	794.1176
	ADV2	2.00	79.4118
	REC2	9.00	39.7059
D4	DN2	2.00	3.4091
	ADV2	3.00	2.2727
	REC2	11.00	1.1364
D5	DN2	2.00	59.5588
	ADV2	3.00	39.7059
	REC2	11.00	19.8529

advertise (ADV2), and recall (REC2)), represented by the averaged outcome vectors over the reality chance nodes. At this point, only noninferior choices are considered; however, in our example, there are no inferior solutions (at this stage). The set of noninferior solutions for a given decision node is noted in the following.

Nine strategies are defined for the arcs do nothing (DN1) and advertise (ADV1) coming out of decision node D1. The test will give us a further indication of the quality of the product, whether we decide to do nothing or to advertise in stage 1. Whether the test result is good or bad, we can always choose between DN2, ADV2, and REC2 in the second stage. Thus, the three possible actions following a good result can be combined with three possible actions following a bad test result, resulting in $(3 \times 3) = 9$ strategies. The value of each strategy can be assessed for a chosen action by multiplying the outcome vector, given that the test result is good, by the probability that the test result is good (0.66), and adding it to the outcome vector for a chosen action, given the test result is bad, multiplied by the probability that the test result is bad (0.34).

For example, suppose we have decided to do nothing (DN1) in stage 1 and want to evaluate the recall strategy (given that the test result is bad) and the advertise strategy (given that the test result is good). For this strategy, we find an outcome vector:

$$[4.38, 16.50]^T = (0.66)[2.00, 4.5455]^T + (0.34)[9.00, 39.7059]^T$$

Note that here and in the following, the vectors have the format [$ million, lives lost]. For the branch DN1, ADV1, and REC1 (stage 1), the strategies and their associated values are listed in Table 9.9. Overall noninferior strategies or solutions are marked with a superscript a.

TABLE 9.9 Decisions for the First-Period Node

	Test Result		C($)	Lives
	Good	Bad	(millions)	
REC1[a]	—	—	10.00	0.0000
DN1[a]	DN2	DN2	0.00	300.0000
DN1[a]	DN2	ADV2	0.68	57.0000
DN1	DN2	REC2	3.06	43.5000
DN1	ADV2	DN2	1.32	273.0000
DN1	ADV2	ADV2	2.00	30.0000
DN1	ADV2	REC2	4.38	16.5000
DN1	REC2	DN2	5.94	271.5000
DN1	REC2	ADV2	6.62	28.5000
DN1	REC2	REC2	9.00	15.0000
ADV1[a]	DN2	DN2	2.00	22.5000
ADV1[a]	DN2	ADV2	2.34	15.7500
ADV1[a]	DN2	REC2	5.06	9.0000
ADV1	ADV2	DN2	2.66	21.7500
ADV1[a]	ADV2	ADV2	3.00	15.0000
ADV1[a]	ADV2	REC2	5.72	8.2500
ADV1	REC2	DN2	7.94	21.0000
ADV1	REC2	ADV2	8.28	14.2500
ADV1	REC2	REC2	11.00	7.5000

[a] Overall noninferior strategies or solutions.

9.5.2.3 Conclusion

The eight strategies or solutions in Table 9.9 represent the set of noninferior solutions to the sample problem. Note that none of the three alternative courses of action in stage 1 can be excluded from consideration; depending on the decisionmaker's preferences, any one of them could be optimal, in combination with the appropriate actions in stage 2 (as shown in the earlier table). A solution that is actually *preferred* to all others could be found by using, for example, the surrogate worth trade-off method.

Figure 9.16 depicts the Pareto-optimal frontier for the pharmaceutical problem.

9.5.3 Highway Traffic

9.5.3.1 Problem description

This example problem concerns alleviating highway traffic by means of an alternating routing system. Two possible actions—(1) alternate routing and (2) doing nothing—are under consideration. There are cost factors associated with the first option. The decision tree covers two time periods, and the cost associated with each option is a function of the period in which the action will be taken. The complete decision tree is shown in Figure 9.17. The assumptions made in solving the problem are the following:

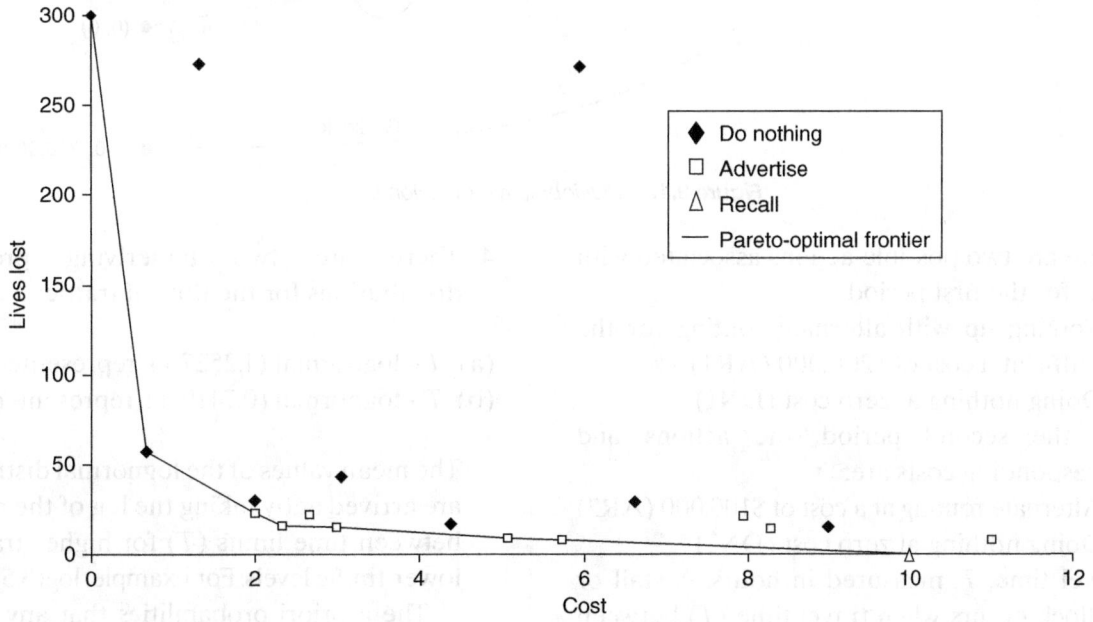

Figure 9.16 Pareto-optimal frontier.

288 MULTIOBJECTIVE DECISION-TREE ANALYSIS

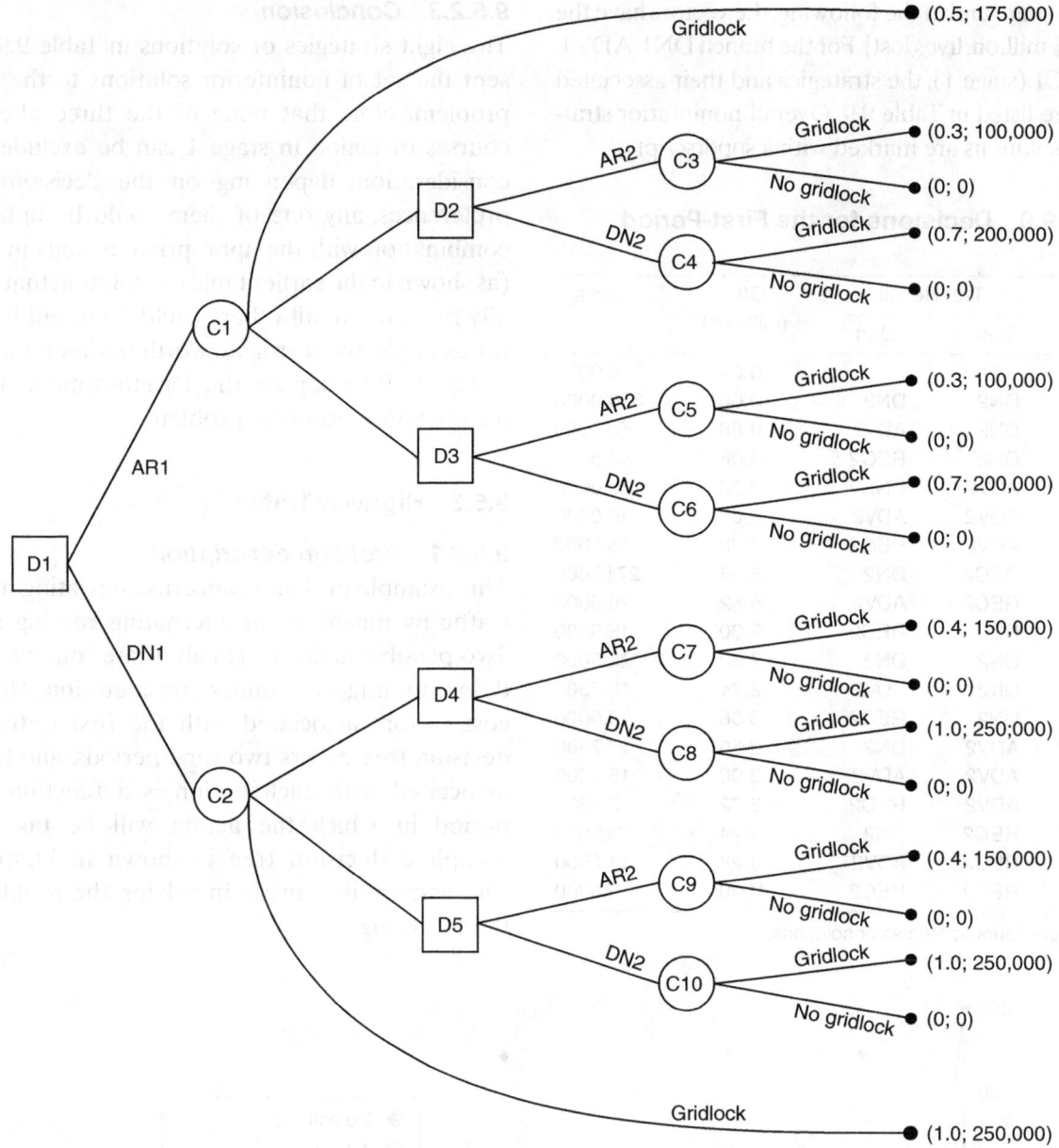

Figure 9.17 Multiobjective decision tree.

1. There are two possible actions associated with costs for the first period:
   a. Coming up with alternate routing for the traffic at a cost of $200,000 (AR1)
   b. Doing nothing at zero cost (DN1)
2. For the second period, the actions and corresponding costs are:
   a. Alternate routing at a cost of $100,000 (AR2)
   b. Doing nothing at zero cost (DN2)
3. Travel time, T, measured in hours. A stall or gridlock occurs when travel time (T) between two points, A and B, is 4 hours or more.
4. There are two underlying probability distributions for the flow of traffic:

   (a) $T \sim$ lognormal (1.2527, 1), represented as LN1
   (b) $T \sim$ lognormal (0.7419, 1), represented as LN2

The mean values of the lognormal distributions are arrived at by taking the log of the midpoint between time limits (T) for higher traffic and lower traffic levels. For example, $\log(3.5) = 1.2527$.

The a priori probabilities that any of these pdf's is the actual pdf are the same (equal).

5. There are three possible events at the end of the first period:
   a. A stall or gridlock ($T \geq 4$ hours)
   b. Higher traffic than current levels ($3 \leq T \leq 4$ hours)
   c. Same or lower traffic levels ($T < 3$)
6. L is the maximum possible loss of lives due to fatal accidents, and C represents money lost due to legal action ensuing from the accident, given no alternate routing.

Calculating a priori probabilities for the first period
To calculate the probabilities of a stall, higher traffic levels, and the same or lower traffic levels at the end of the first period, we use the facts that the possible pdf for traffic level is LN_i with probability 1/2 for $i = 1, 2$ and that a stall, higher, and same or lower traffic levels occur at the values of T described in assumption 5. The a priori probabilities for each of the above outcomes (stall, higher traffic, etc.) are shown in Table 9.10.

9.5.3.2 Calculating the probabilities for the second period

Regardless of whether alternate routing or doing nothing was chosen in the first period, two possible actions must be considered for the second period: alternate routing or do nothing. Depending on the actions taken in the first and the second periods and on the traffic level at the second period, different values of the expected losses for each of the terminal nodes are calculated. At the end of the first period, after measuring the traffic level, the a posteriori probabilities are calculated using Bayes' formula. See Table 9.10.

The required values of the loss vector-valued functions are then computed by multiplying the stall or gridlock probability by the damage vector. For example, $L(DN2|D4) = (0.3591)(1.0) = 0.3591$ (see Table 9.11). When folding back at each decision node takes place, the vector-valued functions are compared, and all dominated inferior solutions are eliminated (in this case, none are inferior).

When averaging out is performed at the chance nodes of the first period, each noninferior decision corresponding to each arc is multiplied by the probability for that arc, yielding a single decision rule for the first-period node. For example, we have four different combinations at AR1, one of which is (AR2|higher, AR2|same or lower).

Table 9.12 presents the values of the vector of objectives for the first-period decision node where none of the decisions are dominated.

TABLE 9.10 A Priori and a Posteriori Probabilities

p(higher|ln 1)	0.1144
p(higher|ln 2)	0.1010
p(stall|ln 1)	0.4469
p(stall|ln 2)	0.2597
p(lower|ln 1)	0.4388
p(lower|ln 2)	0.6393
p(stall)	0.3533
p(higher)	0.1077
p(lower)	0.539
p(ln 1|higher)	0.5311
p(ln 1|lower)	0.4070
p(ln 1|stall)	0.6325
p(ln 2|higher)	0.4689
p(ln 2|lower)	0.5930
p(ln 2|stall)	0.3675
p(stall|higher)	0.3591
p(stall|lower)	0.3358

TABLE 9.11 Stage 2: Expected Value of Loss Vectors for Second-Stage Decision Arcs

Node	Arc	L	C($)
D2	AR2	0.1077	135,900
	DN2	0.2514	71,800
D3	AR2	0.1008	133,600
	DN2	0.2351	67,200
D4	AR2	0.1436	153,900
	DN2	0.3591	89,800
D5	AR2	0.1343	150,400
	DN2	0.3358	84,000

TABLE 9.12 Stage 1: Expected Value of Loss Vectors for First-Stage Decision Arcs

	Higher	Lower	L	C($)
AR1	AR2	AR2	0.2425	348,500
	AR2	DN2	0.3150	312,700
	DN2	AR2	0.2580	341,600
	DN2	DN2	0.3304	305,800
DN1	AR2	AR2	0.4411	185,900
	AR2	DN2	0.5498	150,100
	DN2	AR2	0.4643	179,000
	DN2	DN2	0.5730	143,200

Cost of AR1, $200,000; cost of AR2, $100,000; P(higher), 0.1077; P(lower), 0.5391; P(stall), 0.3533.

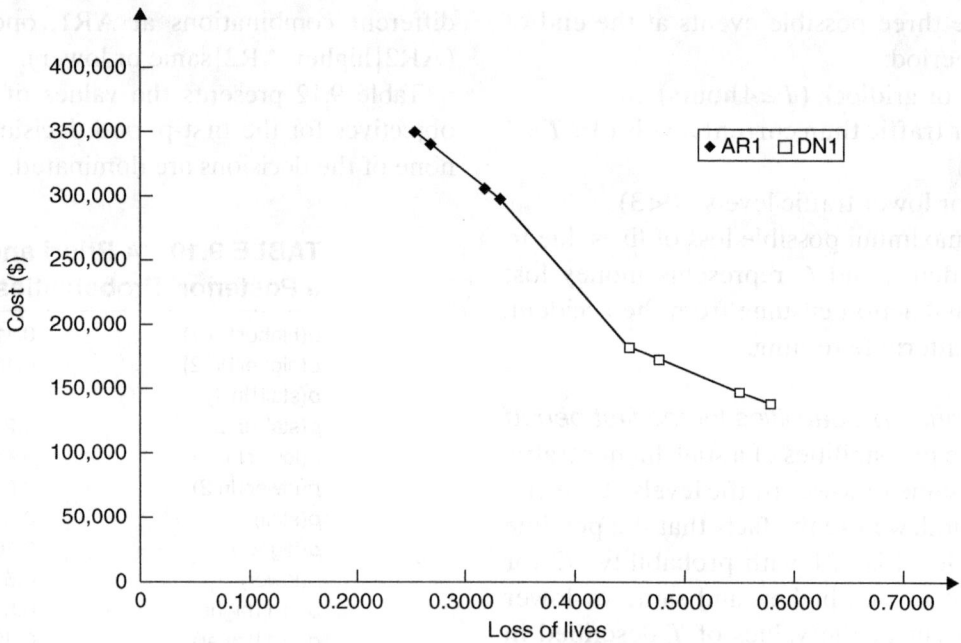

Figure 9.18 Pareto frontier for highway traffic problem.

Figure 9.18 depicts the Pareto-optimal frontier for the highway traffic problem.

9.5.4 Infrastructure Problem

Consider the need to make a decision at the beginning of a planning about a physical infrastructure that has been operating for a long period but can fail. The objective is to determine the best maintenance policy—repair, replace, or do nothing—through the use of MODT analysis [Haimes and Li, 1990].

Assume that the pdf of the failure of a new system is of a known Weibull distribution:

$$p_{NEW}(t) = \lambda \alpha t^{\alpha-1} \exp[-\lambda t^\alpha], \quad t \geq 0, \lambda > 0, \alpha > 0 \quad (9.26)$$

and that a new system costs $1000. At the beginning of this planning period, the system under investigation has been operating for many years, and the pdf of its failure is of the form

$$p_{OLD}(t) = (\lambda + s)\alpha t^{\alpha-1} \exp[-(\lambda + s)t^\alpha], \quad t \geq 0 \quad (9.27)$$

where the parameter s represents a declining factor of an aging system. The exact value of s is unknown. The value of s can be best described by an a priori distribution $p(s)$, which is of a uniform $u[0.05, 0.1]$. A repair action that may be taken at the beginning of the planning period can recover the system's operational capability by updating its failure pdf to

$$p_{REP}(t) = (\lambda + 0.05)\alpha t^{\alpha-1} \exp[-(\lambda + 0.05)t^\alpha], \quad t \geq 0 \quad (9.28)$$

The cost of the repair action is a function of the declining factor s:

$$C_{REP} = 200 + 4000(s - 0.05), \quad 0.05 \leq s \leq 0.1 \quad (9.29)$$

To reduce the uncertainty of the value of s, assume that a test can be performed by an experiment that costs $100. There are three outcomes from the experiment—x_1, x_2, and x_3—and their conditional probabilities are given as

$$P(x_1 \mid s) = 10(0.1 - s) \quad 0.05 \leq s \leq 0.1 \quad (9.30a)$$

$$P(x_2 \mid s) = \begin{cases} 40(s - 0.05) & 0.05 \leq s \leq 0.075 \\ 40(0.1 - s) & 0.075 \leq s \leq 0.1 \end{cases} \quad (9.30b)$$

$$P(x_3 \mid s) = 10(s - 0.05) \quad 0.05 \leq s \leq 0.1 \quad (9.30c)$$

Thus, the posterior distribution of s can be obtained by the Bayesian formula

$$P(s|x_i) = \frac{P(x_i|s)P(s)}{P(x_i)} \quad (9.31)$$

Specifically, we have

$$P(s|x_1) = 800(0.1-s) \quad 0.05 \leq s \leq 0.1 \quad (9.32a)$$

$$P(s|x_2) = \begin{cases} 1600(s-0.05) & 0.05 \leq s \leq 0.075 \\ 1600(0.1-s) & 0.075 \leq s \leq 0.1 \end{cases} \quad (9.32b)$$

$$P(s|x_3) = 800(s-0.05) \quad 0.05 \leq s \leq 0.1 \quad (9.32c)$$

Figure 9.19 presents the corresponding MODT for this example problem, where λ is equal to 0.1 and α is equal to 2.

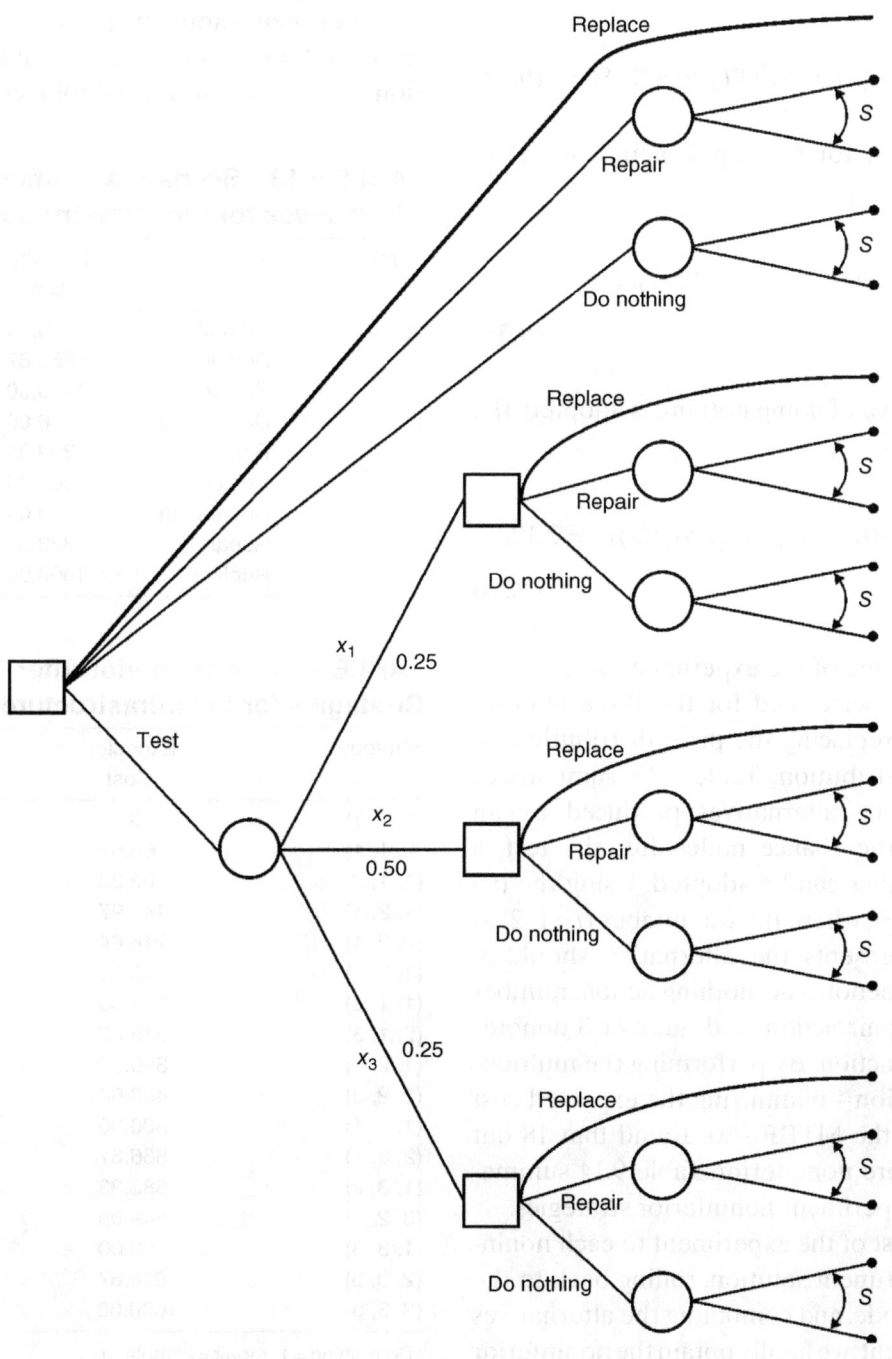

Figure 9.19 Decision tree for the infrastructure problem.

If the alternative to preventive replacement is adopted, the system's mean time before failure (MTBF) is

$$E_{\text{NEW}}(T) = (1/0.1)^{0.5}\Gamma(1.5) = 2.8025 \quad (9.33)$$

where $\Gamma(t)$ is the gamma function defined as $\Gamma(t) = \int_0^\infty u^{t-1}e^{-u}du, t > 0$.

If the repair alternative action is adopted, the system's MTBF is

$$E_{\text{REP}}(T) = (1/0.15)^{0.5}\Gamma(1.5) = 2.2882 \quad (9.34)$$

The expected cost for the repair action is calculated by

$$E(C_{\text{REP}}) = \int_{0.05}^{0.1} [200 + 4000(s - 0.05)]\,p(s)\,ds = 300 \quad (9.35)$$

If the alternative of doing nothing is adopted, the system's MTBF is

$$E_{\text{OLD}}(T) = \int_{0.05}^{0.1} [1/(0.1+s)]^{0.5}\Gamma(1.5)\,p(s)\,ds = 2.1239 \quad (9.36)$$

For each outcome of the experiment, similar calculations can be performed for the three alternatives, except for replacing the prior distribution by the posterior distribution. Table 9.13 summarizes the results for the alternatives produced by an experiment. At the chance node after the test, a total of 27 strategies can be adopted. Using the triple notation $(\cdot, \cdot, \cdot)$, where the ith number ($i = 1, 2, 3$) in the triad represents the alternative should x_i occur, number 1 denotes do-nothing action, number 2 denotes the repair action, and number 3 denotes the replacement action. By performing the multiobjective optimization—minimizing the expected cost and maximizing the MTBF—we found that 18 out of 27 strategies are noninferior. Table 9.14 summarizes the after-experiment noninferior strategies.

Adding the cost of the experiment to each noninferior after-experiment solution, rolling back to the initial decision node, and combining the alternatives without experiment, we finally obtain the noninferior solutions for the planning problem, which are given in Table 9.15. Figure 9.20 depicts the resulting noninferior solutions of the infrastructure problem in the functional space. By evaluating the trade-offs between the additional cost necessary to yield a unit improvement of the MTBF, the decisionmaker can find the preferred solution from among the set of noninferior solutions.

One interesting phenomenon can be observed in this example problem. In SODT analysis, whether an experiment should be performed depends on the EVE. In MODT analysis, the value of experimentation is judged in a multiobjective way. In this

TABLE 9.13 Scenarios of After-Experiment Alternatives for the Infrastructure Problem

Outcome	Alternative	Expected Cost	Mean Time before Failure
x_1	Do nothing	0.00	2.1748
	Repair	266.67	2.2882
	Replace	1000.00	2.8025
x_2	Do nothing	0.00	2.1211
	Repair	299.95	2.2882
	Replace	1000.00	2.8025
x_3	Do nothing	0.00	2.0731
	Repair	333.33	2.2882
	Replace	1000.00	2.8025

TABLE 9.14 Noninferior After-Experiment Strategies for the Infrastructure Problem

Strategy[a]	Expected Cost	Mean Time before Failure
(1, 1, 1)	0	2.1225
(2, 1, 1)	66.67	2.1509
(1, 1, 2)	83.33	2.1763
(1, 2, 1)	149.97	2.2061
(2, 2, 1)	216.64	2.2344
(1, 2, 2)	233.31	2.2599
(1, 1, 3)	250.00	2.3049
(2, 1, 3)	316.67	2.3333
(1, 2, 3)	399.97	2.3884
(2, 2, 3)	466.64	2.4168
(1, 3, 1)	500.00	2.4632
(2, 3, 1)	566.67	2.4916
(1, 3, 2)	583.33	2.5170
(3, 2, 3)	649.99	2.5454
(1, 3, 3)	750.00	2.6456
(2, 3, 3)	816.67	2.6739
(3, 3, 3)	1000.00	2.8025

[a] Do nothing = 1; repair = 2; replace = 3.

TABLE 9.15 Noninferior Solutions for the Infrastructure Problem

Strategy[a]	Expected Cost	Mean Time Before Failure
Doing nothing	0	2.1239
(2, 1, 1)	166.67	2.1509
(1, 1, 2)	183.33	2.1763
(1, 2, 1)	249.97	2.2061
Repair	300.00	2.2882
(1, 1, 3)	350.00	2.3049
(2, 1, 3)	416.67	2.3333
(1, 2, 3)	499.97	2.3884
(2, 2, 3)	566.64	2.4168
(1, 3, 1)	600.00	2.4632
(2, 3, 1)	666.67	2.4916
(1, 3, 2)	683.33	2.5170
(2, 3, 2) (3, 2, 3)	749.99	2.5454
(1, 3, 3)	850.00	2.6456
(2, 3, 3)	916.67	2.6739
Preventive replacement	1000.00	2.8025

[a] Do nothing = 1; repair = 2; replace = 3.

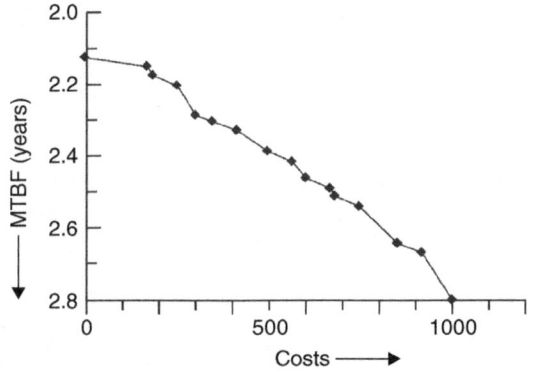

Figure 9.20 Noninferior solutions for the infrastructure problem.

example, the noninferior frontiers generated with and without experimentation do not dominate each other; they supplement each other to generate more solution possibilities for the decisionmaker(s).

REFERENCES

Adams, A.J., 1960, Bernoullian utility theory, In *Mathematical Thinking*, H. Solomon (Ed.), Free Press, New York.

Arrow, K.J., 1963, Utility and expectations in economic behavior, In *Psychology: A Study of Science*, S. Koch (Ed.), McGraw-Hill, New York.

Chankong, V., and Y.Y. Haimes, 1983, *Multiobjective Decision Making: Theory and Methodology*, Elsevier–North Holland, New York.

Edwards, W., 1954, The theory of decision making, *Psychology Bulletin* **51**, 380–417.

Frohwein, H.I., and J.H. Lambert, 2000, Risk of extreme events in multiobjective decision trees: Part 1. Severe events, *Risk Analysis* **20**(1): 113–123.

Frohwein, H.I., J.H. Lambert, and Y.Y. Haimes, 1999, Alternative measures of risk of extreme events in decision trees, *Reliability Engineering and System Safety* **66**(1): 69–84.

Frohwein, H.I., Y.Y. Haimes, and J.H. Lambert, 2000, Risk of extreme events in multiobjective decision trees: Part 2. Rare events, *Risk Analysis* **20**(1): 125–134.

Haimes, Y.Y., and D. Li, 1990, Multiobjective decision tree analysis in industrial systems, In *Control and Dynamic Systems*, Vol. **36**, Academic Press, New York, pp. 1–17.

Haimes, Y.Y., D. Li, and V. Tulsiani, 1990, Multiobjective decision tree method, *Risk Analysis* **10**(1): 111–129.

Haimes, Y.Y., D. Li, V. Tulsiani, and J. Lambert, 1996, *Risk-Based Evaluation of Flood Warning and Preparedness Systems*, Vol. 1: *Overview;* and Vol. 2: *Technical*, IWR Report 96-R-25, U.S. Army Corps of Engineers, Institute for Water Resources, Fort Belvoir, VA.

Luce, R.D., and P. Suppes, 1965, Preference, utility and subjective probability, In *Handbook of Mathematical Psychology*, Vol. **3**, R.D. Luce, R.R. Bush, and E.E. Galanter (Eds.), Wiley, New York.

Pratt, J., H. Raiffa, and R. Schlaifer, 1995, *Introduction to Statistical Decision Theory*, MIT Press, Cambridge, MA.

Raiffa, H., 1968, *Decision Analysis: Introductory Lectures on Choice Under Uncertainty*, Addison-Wesley, Reading, MA.

Santos, J.R., K. Barker, and P.J. Zelinke IV, 2008, Sequential decision-making in interdependent sectors with multiobjective inoperability decision trees: application to biofuel subsidy analysis, *Economic Systems Research* **20**(1): 29–56.

Savage, L.J., 1954, *The Foundation of Statistics*, Wiley, New York.

Schlaifer, R., 1969, *Analysis of Decisions Under Uncertainty*, McGraw-Hill, New York.

Shubik, M., 1964, *Game Theory and Related Approaches to Social Behavior*, Wiley, New York.

von Neumann, J., and O. Morgenstern, 1953, *Theory of Games and Economic Behavior*, Third edition, Princeton University Press, Princeton, NJ.

10

Multiobjective Risk Impact Analysis Method

10.1 INTRODUCTION

The main purpose of systems analysis and optimization is to improve decisionmaking by providing a rational means to obtain a better understanding of a system and its components, generate alternatives for the system's management and control, provide more precise information about its components, and improve communication among the system's managers and controllers.

Ultimately, decisionmaking problems involve, either formally or informally, an evaluation of alternatives and a presentation of the possible consequences of each of these alternatives. These consequences may indicate, for instance, the degree to which the objectives associated with the system can be achieved. Selecting which of several alternatives to adopt will then depend on the decisionmakers' (DMs) preferences for those consequences.

A most common outcome of the use of system modeling and optimization is the derivation of an optimal solution (in a single-optimization model) or a set of Pareto-optimal solutions and their corresponding trade-offs (in a multiobjective optimization model) that may ultimately lead to a preferred, compromise solution. This preferred solution is only part of what DMs hope to accomplish; often, they must also focus on avoiding bad decisions (i.e., it is important to know what not to do). Although this situation may be encountered in most systems, it is particularly relevant in public systems, such as water resources systems, where the preferred or compromise solution reached by the various stakeholders and DMs may not even be Pareto-optimal for multiobjective models or may not be close to an optimum in single-objective models. This fact can be explained from the perspectives of the DMs who are involved in formulating public policy. In evaluating the incremental benefits that might accrue in their quest for the best solution (policy), as compared with their desire to minimize the adverse and irreversible consequences that might result, these DMs often opt to focus on the latter and in a decisive way. This pattern often includes risk-aversive, conservative-minded corporate executives, whose performances may more often be evaluated on the basis of the number of bad decisions they have made.

There has been increasing concern about the effects or consequences of introducing, modifying, or improving large-scale systems. This concern has

Risk Modeling, Assessment, and Management, Fourth Edition. Yacov Y. Haimes.
© 2016 John Wiley & Sons, Inc. Published 2016 by John Wiley & Sons, Inc.

manifested in a variety of fields. In technical communities, a substantial interest exists in guiding the development of technological systems by "looking before you leap" [Porter and Rossini, 1980]; such an approach urges the study of the impact on society that might result from introducing or modifying a particular technological system. Technology has brought enormous social benefits, but these benefits have not been cost-free.

In environmental systems, impact studies have gained great attention through the required preparation of environmental impact statements, while in socioeconomic systems such impact assessment studies have included inflation, arms control, and urban living. Numerous other examples can be found that give evidence, either explicitly or implicitly, of the need for impact analysis as a constituent of informed decisionmaking.

Many systemic methods—such as economic modeling, cost–benefit analysis, mathematical programming, optimum systems control, and system estimation and identification theory—are regarded as potentially useful for impact analysis. Such techniques as expert opinion and interpretative structural modeling are considered useful. Forecasting is considered to be a first step toward developing better guidelines.

The fundamental characteristic of large-scale systems is their inescapably multivarious nature—with multiple and often noncommensurable objectives, multiple DMs, multiple transcending aspects, elements of risk and uncertainty, multiple stages, and multiple overlapping subsystems. Mathematical models that aim at representing real physical systems have become important tools in the synthesis, analysis, planning, operation, and control of complex systems. The nature of the physical system under consideration determines which class of mathematical models will most closely represent it. The models may be finite-or infinite-dimensional and linear or nonlinear; the functional relationships may not be well behaved. Unknown parameters and stochastic disturbances as well as uncertain objectives may also be present.

To provide an impact analysis of various decisions or policy options, the analyst is often asked to develop several scenarios with corresponding assumptions and potential consequences. While this step generally proves to be very valuable, so far it has not been integrated into the modeling and optimization process, and it therefore lacks a formal theoretical and methodological basis. Providing alternate scenarios is often conducted in an ad hoc fashion.

This chapter introduces a methodology, termed the *multiobjective risk impact analysis method (MRIAM)*, that incorporates risk and impact analysis within a multiobjective decisionmaking framework. It thus provides a formal structure for comparing different scenarios that may vary widely, for instance, in the span of time under consideration and the severity of the impact of any adverse consequences.

10.2   IMPACT ANALYSIS

Consider a time-invariant, linear, stochastic, multistage process described by the following:

$$x(k+1) = Ax(k) + Bu(k) + w(k), \qquad k = 0, \ldots, T-1 \tag{10.1}$$

where $x(k)$ represents the state of the system at stage k, $u(k)$ represents the control to be implemented, and A and B are system parameters. The time horizon to be considered is T stages ($k=0,\ldots, T-1$). The variable $w(k)$ is the element of randomness introduced into the model.

Mathematically, the meaning of the problem outlined in Eq. (10.1) is the following: Solve a sequence of static or single-stage multiobjective optimization problems where decisions made at stage k affect stages $k+1, k+2,\ldots, T-1$. In contrast to the strictly static (single-stage) optimization case, we must take into account the consequences that the decisions made will have on future policy options, which is the essence of any dynamic optimization problem. Hence, it is not enough to obtain noninferior solutions at each stage. We must also determine the impact of these decisions in order to obtain a solution that is also noninferior for the time horizon of interest. In this type of multiobjective, multistage decisionmaking problem, we seek to determine for each stage a preferred noninferior solution such that it represents a desirable trade-off among the objectives at that stage as well as among the objectives at different and succeeding stages. To clarify the meaning of Eq. (10.1) as a model for impact analysis, we offer the following example.

Consider an impact assessment problem for a technological system. The central question is how to introduce a new technology in such a way that its benefits to humankind will more than offset its hazards. Suppose that to answer this question, three main objective functions have been quantified. They represent the cost of introducing the technology, its potential benefits, and the expected occurrence of a hazard (e.g., radiation release). The cost and benefit objective functions depend on the current state of the system and on the decisions at that stage. The expected occurrence of a hazard depends on the state of the system, reached as a result of an earlier-stage decision, and on the decision made at the present stage. This situation may be represented by a model, such as the one in Eq. (10.1). The impacts can be assessed by indicating the degree to which the objectives are achieved. They can then be furnished to DMs for policy analysis and decisionmaking.

The multiobjective, multistage impact analysis class of problems is characterized by the fact that decisions must be made at different stages. Hence, since a decision chosen at a particular stage may have consequences at later stages, the effects of decisions must somehow be evaluated in order to improve not only current decisions but also later ones.

Given the limitations of the mathematical models, some sort of postoptimality analysis must be carried out in order to evaluate the adopted decision (the preferred solution) that has been obtained. For example, alternative preferred decisions could be determined, and their stability could be analyzed as the model topology or system parameters change. This is referred to in the literature as *sensitivity analysis*. However, sensitivity analysis is not always sufficient. For instance, the DM, knowing that he will face other decision problems at different stages, may also be interested in determining how the adopted preferred decision will affect his decisionmaking problems at later stages. This means that trade-offs must be made not only among the various objectives at the current stage but also among the various objectives at different stages, thus adding another dimension to decisionmaking problems. We refer to these generalized trade-offs as dynamic multiobjective trade-offs or simply as stage trade-offs.

10.3 THE MULTIOBJECTIVE, MULTISTAGE IMPACT ANALYSIS METHOD: AN OVERVIEW

In the most general sense, impact assessment denotes the study of prospective impacts resulting from current decisions. Among a number of categories of impact assessment are technology assessment, environmental impact assessment, and social impact assessment [Porter and Rossini, 1980]. Gomide and Haimes [1984] more specifically define impact analysis as the study of the effect that decisions have on the decisionmaking problem.

Gomide [1983] developed a theoretical basis for impact analysis in a multiobjective framework. He formulated a multistage multiobjective optimization model and presented a methodology for decisionmaking. This methodology is known as the *multiobjective, multistage impact analysis method (MMIAM)*. For an excellent introduction to multistage optimization and associated multipliers, see Intriligator [1971]. Some of the basic notions and concepts of the MMIAM are used later when the partitioned multiobjective risk method (PMRM) is applied to a dynamic model. A detailed theoretical formulation and examples of applications of the MMIAM may be found in Gomide [1983] and Gomide and Haimes [1984].

Gomide and Haimes introduced the *stage trade-off*, a generalization of the trade-off concept as defined in Haimes and Chankong [1979]. The stage trade-off, given by λ_{ij}^{kl}, represents the marginal rate of change of $f_i(x^*, u^*, k)$ at time k per unit of change in $f_j(x^*, u^*, l)$ at time l. Furthermore, if all other objective functions remain unchanged in value, λ_{ij}^{kl} is a partial trade-off and can be written as

$$\lambda_{ij}^{kl} = -\frac{\partial f_i(x^*, u^*, k)}{\partial f_j(x^*, u^*, l)} = -\frac{\partial f_i^k(x^*, u^*)}{\partial f_j^l(x^*, u^*)} \quad (10.2)$$

All of the theorems concerning partial and total trade-offs given in Haimes and Chankong [1979] and Chankong and Haimes [1983] still apply to stage trade-offs. Gomide [1983] and Gomide and Haimes [1984] present the modifications of these theorems.

A formal definition of *impact* can now be stated. *The impact at stage k means the variations in the levels of the objective function at stage k, due to changes made in the level of the objective function at stage l at the noninferior policies.* The stage trade-offs (total or partial) provide a measure of the impacts on the levels of the objective functions at various stages. Using the impact information provided by the stage trade-offs, it is possible to proceed in the decisionmaking process as defined by the *surrogate worth trade-off (SWT)* method or the interactive SWT method (see Chankong and Haimes [1983]).

10.4 COMBINING THE PMRM AND THE MMIAM

Some multiobjective optimization techniques, such as the SWT method, operate by converting the multiobjective problem into a single-objective one. The ε-constraint approach is a technique for doing this. However, for a problem with many decisions at many stages, the single-objective, single-stage problem that results may be too large to solve efficiently. In addition, the physical nature of the problem is lost by creating a static optimization problem. The integration of the MMIAM with the PMRM—introduced in Chapter 8—constitutes the *MRIAM*.

Gomide [1983] derives a method of solving multiobjective, multistage optimization problems using the ε-constraint approach, augmented Lagrange multiplier functions, and hierarchical optimization concepts. Thus, the dynamic nature of the problem is preserved and exploited, and a decentralized approach to solving the ε-constraint problems is possible. The result is greater efficiency in solving the overall problem.

Equation (10.1) represents the probabilistic nature of the system, which is what makes risk analysis desirable. The disturbance $w(k)$ is assumed to be a normally distributed, purely random sequence. In addition, the initial state, $x(0)$, is also assumed to be normally distributed. This system is assumed to have the following statistical properties (for all $0 \le k \le T-1$, $0 \le l \le T-1$) [Leach, 1984; Leach and Haimes, 1987]:

1. $E[w(k)] = 0$ (10.3)
2. $E[w^2(k)] = P(k) = P$ (10.4)
3. $E[w(k)w(l)] = 0$ for $k \ne l$ (10.5)
4. $E[x(0)] = x_0$ (10.6)
5. $E[(x(0) - x_0)^2] = X_0$ (10.7)
6. $E[(x(0) - x_0)w(k)] = 0$ (10.8)

Suppose now that $y(k)$ represents the measured output of the system at stage k (which may possibly be damage or loss) and is linearly related to the state $x(k)$ by

$$y(k) = Cx(k) + v(k) \quad (10.9)$$

where C is a parameter and $v(k)$ represents another normally distributed, purely random, stationary sequence. Some additional statistical properties are assumed (for all $0 \le k \le T-1, 0 \le l \le T-1$):

7. $E[v(k)] = 0$ (10.10)
8. $E[v^2(k)] = R(k) = R$ (10.11)
9. $E[v(k)v(l)] = 0$ for $k \ne l$ (10.12)
10. $E[(x(0) - x_0)v(k)] = 0$ (10.13)
11. $E[w(k)v(l)] = 0$ (10.14)

From the theory of random variables, it is known that any linear combination of normal random variables is also a normal random variable. Thus, $x(k)$ is a normally distributed random variable and therefore so is $y(k)$. The mean of $y(k)$ is denoted by $m(k)$, and the variance of $y(k)$ is denoted by $s^2(k)$.

To solve Eq. (10.1), to determine the mean of $y(k)$ for any $x(k)$, and to determine the variance of $y(k)$, the following derivations are obtained using mathematical induction.

Given that

$$x(k+1) = Ax(k) + Bu(k) + w(k), k = 0, 1, \ldots, T-1 \quad (10.15)$$

$$x(0) = x_0$$

and

$$y(k) = Cx(k) + v(k) \quad (10.16)$$

where $x(k)$ is normally distributed and $w(k)$ is a normally distributed purely random sequence, prove by induction that

$$x(k) = A^k x(0) + \sum_{i=0}^{k-1} A^i Bu(k-1-i) + \sum_{i=0}^{k-1} A^i w(k-1-i)$$
(10.17)

Proof:
Prove for $k=0$:

$$x(0) = A^0 x(0) + \sum_{i=0}^{-1} A^i Bu(0-1-i) + \sum_{i=0}^{-1} A^i w(0-1-i)$$
(10.18)

$$= x(0)$$

$$x(0) = x(0)$$

Prove for $k=1$:

$$x(1) = A^1 x(0) + \sum_{i=0}^{0} A^i Bu(1-1-i) + \sum_{i=0}^{0} A^i w(1-1-i)$$
(10.19)

$$= A^1 x(0) + A^0 Bu(1-1-0) + A^0 w(1-1-0)$$

$$= Ax(0) + A^0 Bu(0) + A^0 w(0)$$

$$x(1) = Ax(0) + Bu(0) + w(0)$$

Assume for $k > 1$ that

$$x(k) = A^k x(0) + \sum_{i=0}^{k-1} A^i Bu(k-1-i) + \sum_{i=0}^{k-1} A^i w(k-1-i)$$
(10.20)

Therefore, prove that the relationship above exists for $k+1$:

$$x(k+1) = Ax(k) + Bu(k) + w(k) \quad (10.21)$$

$$= A\left[A^k x(0) + \sum_{i=0}^{k-1} A^i Bu(k-1-i) + \sum_{i=0}^{k-1} A^i w(k-1-i) \right]$$
$$+ Bu(k) + w(k)$$

$$= A^{k+1} x(0) + \sum_{i=0}^{k-1} A^{i+1} Bu(k-1-i) + \sum_{i=0}^{k-1} A^{i+1} w(k-1-i)$$
$$+ Bu(k) + w(k)$$

Now, let's change the index of the summation whereby $z = i+1$. Now, we have the following:

$$x(k+1) = A^{k+1} x(0) + \sum_{z=1}^{k} A^z Bu(k-z) + \sum_{z=1}^{k} A^z w(k-z)$$
$$+ Bu(k) + w(k)$$

Now, let's combine terms, and we have the following:

$$x(k+1) = A^{k+1} x(0) + \sum_{z=0}^{k} A^z Bu(k-z) + \sum_{z=0}^{k} A^z w(k-z)$$
(10.22)

Now, let's change the index of the summation whereby $(k+1)-1=k$, which gives the following:

$$x(k+1) = A^{k+1} x(0) + \sum_{z=0}^{(k+1)-1} A^z Bu((k+1)-1-z)$$
$$+ \sum_{z=0}^{(k+1)-1} A^z w((k+1)-1-z)$$
(10.23)

Thus, by mathematical induction, it has been shown that

$$x(k) = A^k x(0) + \sum_{i=0}^{k-1} A^i Bu(k-1-i) + \sum_{i=0}^{k-1} A^i w(k-1-i)$$
(10.24)

for $k=0$, $k=1$, and $k+1$.

Thus, the equation for $x(k)$ holds for all $k \geq 0$.

Prove that

$$m(k) = E[y(k)] = CE[x(k)] \quad (10.25)$$

Proof:

$$m(k) = E[y(k)] = E[Cx(k) + v(k)]$$
$$= CE[x(k)] + E[v(k)]$$
$$= CE[x(k)] + 0$$
$$= CE[x(k)]$$

Prove that

$$m(k) = E[y(k)] = CA^k x_0 + \sum_{i=0}^{k-1} CA^i Bu(k-1-i)$$
(10.26)

Proof:
We start with

$$x(k+1) = Ax(k) + Bu(k) + w(k)$$

and derive the expected value for $x(k+1)$.

For $k=0$,

$$E[x(1)] = E[Ax(0) + Bu(0) + w(0)]$$

Since $u(0)$ = constant and $E[w(0)] = 0$, we obtain

$$E[x(1)] = Ax_0 + Bu(0) + 0$$
$$E[x(2)] = E[Ax(1) + Bu(1) + w(1)]$$
$$= A(Ax_0 + Bu(0)) + Bu(1) + 0$$
$$= A^2 x_0 + ABu(0) + Bu(1)$$

We are going to prove the general equation by induction:

Assume for k: $E[x(k)] = A^k x_0 + \sum_{i=0}^{k-1} A^i Bu(k-1-i)$

We prove for $(k+1)$:

$$E[x(k+1)] = E[Ax(k) + Bu(k) + w(k)]$$
$$= A\left(A^k x_0 + \sum_{i=0}^{k-1} A^i Bu(k-1-i)\right) + BE[u(k)] + E[w(k)]$$
$$= A^{k+1} x_0 + \sum_{i=0}^{k-1} A^{i+1} Bu(k-1-i) + Bu(k) + 0$$

$$\left(\text{Let } s = i+1, \text{ therefore, } \sum_{i=0}^{k-1} \to \sum_{s=1}^{k}\right)$$

$$= A^{k+1} x_0 + \sum_{s=1}^{k} A^s Bu(k-s) + Bu(k)$$

$$E[x(k+1)] = A^{k+1} x_0 + \sum_{s=0}^{k} A^s Bu(k-s)$$

Note that for $s=0$, $A^0 Bu(k-0) = Bu(k)$. Add and subtract 1 and let $s=i$; therefore,

$$E[x(k+1)] = A^{k+1} x_0 + \sum_{i=0}^{(k+1)-1} A^i Bu((k+1)-1-i)$$

Thus, we have so far proved that

$$E[x(k)] = A^k x_0 + \sum_{i=0}^{k-1} A^i Bu(k-1-i) \quad (10.27)$$

Now, we can derive the equation for $m(k)$:

$$m(k) = E[y(k)] = E[Cx(k) + v(k)]$$
$$= CE[x(k)] + 0$$

Thus,

$$m(k) = E[y(k)] = CA^k x_0 + \sum_{i=0}^{k-1} CA^i Bu(k-1-i)$$

Prove that

$$s^2(k) = C^2 A^2 X_0 + \sum_{i=0}^{k-1} C^2 A^{2i} P + R$$

Proof:
We start with the variance of $y(k)$:

$$\text{Var}[y(k)] = \text{Var}[Cx(k)] = C^2 \text{Var}[x(k)]$$
$$\text{Var}[x(k+1)] = \text{Var}[Ax(k) + Bu(k) + w(k)]$$

For $k=0$,

$$\text{Var}[x(1)] = \text{Var}[Ax(0) + Bu(0) + w(0)]$$
$$= \text{Var}[Ax(0)] + 0 + \text{Var}[w(0)]$$
$$= A^2 X_0 + P$$

Assume the following recursive relation is true for k, and we will prove it for $k+1$:

$$\text{Assume: Var}[x(k)] = A^{2k} X_0 + \sum_{i=0}^{k-1} A^{2i} P$$

Start with: $\text{Var}[x(k+1)] = \text{Var}[Ax(k) + Bu(k) + w(k)]$

$$= \text{Var}[Ax(k)] + \text{Var}[Bu(k)] + \text{Var}[w(k)]$$
$$= A^2 \left(A^{2k} X_0 + \sum_{i=0}^{k-1} A^{2i} P\right) + P$$
$$= A^{2(k+1)} X_0 + \sum_{i=0}^{k-1} A^{2(i+1)} P + P$$

Let $s = i+1$:

$$\therefore \sum_{i=0}^{k-1} \to \sum_{s=1}^{k}$$

$$\therefore \text{Var}[x(k+1)] = A^{2(k+1)} X_0 + \sum_{s=1}^{k} A^{2s} P + P$$

$$= A^{2(k+1)} X_0 + \sum_{s=0}^{k} A^{2s} P$$

(Note that for $s=0$, $A^{2s}P = A^0 P = P$.) Let $i = s$:

$$\therefore \mathrm{Var}[x(k+1)] = A^{2(k+1)} X_0 + \sum_{i=0}^{(k+1)-1} A^{2i} P$$

Thus,

$$\mathrm{Var}[x(k+1)] = A^{2(k+1)} X_0 + \sum_{i=0}^{k} A^{2i} P$$

Consequently,

$$s^2(k) = \mathrm{Var}[y(k)] = \mathrm{Var}[Cx(k)] + \mathrm{Var}[v(k)]$$
$$= C^2 \mathrm{Var}[x(k)] + R$$
$$= C^2 \left[A^{2k} X_0 + \sum_{i=0}^{k-1} A^{2i} P \right] + R$$
$$s^2(k) = C^2 A^{2k} X_0 + \sum_{i=0}^{k-1} C^2 A^{2i} P + R$$

Note that the mean of $y(k)$, $m(k)$, is a function of the stage and the control sequence $u(k)$:

$$E[y(k)] = m(k) = CA^k x_0 + \sum_{i=0}^{k-1} CA^i B u(k-1-i)$$

However, the variance of the output $y(k)$, $s^2(k)$ is a function of k alone:

$$s^2(k) = C^2 A^{2k} X_0 + \sum_{i=0}^{k-1} C^2 A^{2i} P + R \quad (10.28)$$

Note that all terms are constants.
Therefore,

$$m(\cdot) = m(k; u) \quad (10.29)$$
$$s^2(\cdot) = s^2(k) \quad (10.30)$$

Note that $y(k)$ is normally distributed, with mean $m(k)$, given by Eq. (10.25), and variance $s^2(k)$, given by Eq. (10.28). An important result is that $m(k)$ is a function of the stage k and the control sequence $u(k)$. The variance is a function of k alone. Intuitively, this is an expected result. Because of the linear dynamics of the system, the control sequence exerts no influence on the amount of uncertainty at any stage.

The PMRM may now be applied to form the risk objective functions at each stage. Leach [1984] and Leach and Haimes [1987] present results of generalized partitioning of the normal distribution, where it is assumed that the mean and standard deviations are known functions of the control variable. These results may be directly applied to partitioning the distribution of $y(k)$, since $y(k)$ is normally distributed and $m(k; u)$ and $s(k)$ are known functions. The results are still applicable even though $s(k)$ is independent of the control. The standard deviation $s(k)$ is simply the positive square root of $s^2(k)$. Leach [1984] shows that the ith objective function is given by the mean plus a constant b_i times the standard deviation. The constant b_i depends explicitly on the partitioning point chosen. Here, the same idea is used except it is assumed that, in general, the partitioning point may vary depending on the stage k. Thus, the constant denoted by β_i^k, which is dependent on the ith partitioning at stage k, is subsequently determined by Eq. (10.41), where the limits on the integrals (s_i and t_i) depend on the stage k. If partitioning is identical for all stages, then the constant β_i^k becomes simply β_i.

Let N_k be the number of partitions for the probability distribution at the kth stage. Denote the ith conditional expected value of $y(k)$ at the kth stage by $f_i^k(u)$, where, again, the u indicates a dependence on the control from stages 0 to $k-1$.

We assume that for any time k, $y(k)$ follows a normal distribution with mean $m(k)$ and variance $s^2(k)$, where $m(k)$ and $s^2(k)$ satisfy

$$m(k) = C E[x(k)] \quad (10.31)$$
$$E[x(k+1)] = A E[x(k)] + B u(k) \quad (10.32)$$
$$s^2(k) = C^2 \mathrm{Var}[x(k)] + R \quad (10.33)$$
$$\mathrm{Var}[x(k+1)] = A^2 \mathrm{Var}[x(k)] + P \quad (10.34)$$

The conditional expectation on the region $[s_k, t_k]$ is by definition

$$f_i^k(\cdot) = \frac{\int_{s_k}^{t_k} \frac{x}{\sqrt{2\pi}\sigma} e^{-\frac{(x-u)^2}{2\sigma^2}} dx}{\int_{s_k}^{t_k} \frac{1}{\sqrt{2\pi}\sigma} e^{-\frac{(x-u)^2}{2\sigma^2}} dx}$$

(10.35a)

$$f_i^k(u) = \frac{\int_{s_k}^{t_k} \frac{y(k)}{\sqrt{2\pi} s(k)} \exp\left[\frac{-(y(k)-m(k))^2}{2 s^2(k)}\right] dy(k)}{\int_{s_k}^{t_k} \frac{1}{\sqrt{2\pi} s(k)} \exp\left[\frac{-(y(k)-m(k))^2}{2 s^2(k)}\right] dy(k)}$$

(10.35b)

For notational simplicity, let

$$\tau = \frac{y(k) - m(k)}{s(k)} \quad (10.36)$$

$$d\tau = \frac{dy(k)}{s(k)}$$

Also, adding $m(k)$ and subtracting it from the numerator yields

$$f_i^k(u) = \frac{\int_{s_k}^{t_k} \frac{(y(k) - m(k) + m(k))}{\sqrt{2\pi} s(k)} e^{-\tau^2/2} dy(k)}{\int_{s_k}^{t_k} \frac{1}{\sqrt{2\pi} s(k)} e^{-\tau^2/2} dy(k)}$$

$$= m(k) \frac{\int_{s_k}^{t_k} \frac{e^{-\tau^2/2}}{\sqrt{2\pi} s(k)} dy(k)}{\int_{s_k}^{t_k} \frac{e^{-\tau^2/2}}{\sqrt{2\pi} s(k)} dy(k)} + \frac{\int_{s_k}^{t_k} \frac{(y(k) - m(k))}{\sqrt{2\pi} s(k)} e^{-\tau^2/2} dy(k)}{\int_{s_k}^{t_k} \frac{e^{-\tau^2/2}}{\sqrt{2\pi} s(k)} dy(k)}$$

$$(10.37)$$

Since $m(k)$ and $s(k)$ are not variables in the integration, we can manipulate them as follows:
Taking the derivative of Eq. (10.36) yields

$$d\left(\frac{y(k) - m(k)}{s(k)}\right) = \frac{dy(k) - 0}{s(k)}$$

Thus,

$$dy(k) = s(k) \left[d\left(\frac{y(k) - m(k)}{s(k)}\right) \right]$$

Note that $dm(k) = 0$, since $m(k)$ is a constant.
Also, since $dy(k)$ is now $d([y(k) - m(k)]/s(k))$, we must change the lower and upper limits of the integrals:

$$t_k \to \frac{t_k - m(k)}{s(k)} = t'_k$$

$$s_k \to \frac{s_k - m(k)}{s(k)} = s'_k$$

Thus,

$$f_i^k(u) = m(k) + \frac{s(k) \int_{s'_k}^{t'_k} \frac{(y(k) - m(k))}{\sqrt{2\pi} s(k)} e^{-\tau^2/2} d\left(\frac{y(k) - m(k)}{s(k)}\right)}{\int_{s'_k}^{t'_k} \frac{1}{\sqrt{2\pi}} e^{-\tau^2/2} d\left(\frac{y(k) - m(k)}{s(k)}\right)}$$

$$(10.38)$$

This finally yields

$$f_i^k(u) = m(k) + s(k) \frac{\int_{s'_k}^{t'_k} \frac{\tau}{\sqrt{2\pi}} e^{-\tau^2/2} d\tau}{\int_{s'_k}^{t'_k} \frac{1}{\sqrt{2\pi}} e^{-\tau^2/2} d\tau} \quad (10.39)$$

where

$$\tau = \frac{y(k) - m(k)}{s(k)}$$

Therefore,

$$f_i^k(u) = m(k; u) + \beta_i^k s(k) \quad (10.40)$$

for $i = 4, f_4^k(u) = m(k; u) + \beta_4^k s(k) \quad (10.40a)$

where

$$\beta_i^\kappa = \frac{\int_{s'_k}^{t'_k} \frac{\tau}{(2\pi)^{1/2}} e^{-\tau^2/2} d\tau}{\int_{s'_k}^{t'_k} \frac{1}{(2\pi)^{1/2}} e^{-\tau^2/2} d\tau} \quad (10.41)$$

and where

$$t'_k = \frac{t_k - m(k)}{s(k)}, \quad s'_k = \frac{s_k - m(k)}{s(k)}$$

$k = 0, \ldots, T-1; i = 1, \ldots, N_i$, and β_i^k is independent of u.

If $y(k)$ represents damage or loss, then $f_i^k(u)$ is an objective function to be minimized. The value of this objective function represents the ith conditional expected value of damage at stage k. In addition to the conditional expected values, the unconditional expected value of damage, given simply by $m(k; u)$, can also be used as an objective function. In fact, the minimization of any one of the conditional expected-value functions is reduced to minimizing the unconditional expected value:

$$\min_u f_i^k(u) = \min_u [\beta_i^k s(k) + m(k; u)]$$

$$= \beta_i^k s(k) + \min_u [m(k; u)] \quad (10.42)$$

This is a direct result of the fact that $s(k)$ is independent of the control. Because of this, the trade-offs associated with the risk functions (for a given k) will be equal. Only the levels of the objectives will be different. In other words, for normal distributions, the risk functions for damage at stage k are parallel curves. This greatly simplifies the multiobjective optimization.

Note that in calculating β_i^k, the numerator can be integrated, but the integral in the denominator has no closed form. It can be evaluated using the cumulative standard normal distribution function $\Phi(0,1)$. Also note we assume that $\mu = 0$ and $\sigma = 1$:

$$\beta_2 = \frac{\int_{-\infty}^{\mu-\sigma} \frac{\tau e^{-\tau^2/2}}{(2\pi)^{\frac{1}{2}}} d\tau}{\int_{-\infty}^{\mu-\sigma} \frac{e^{-\tau^2/2}}{(2\pi)^{\frac{1}{2}}} d\tau} = \frac{-\frac{1}{(2\pi)^{\frac{1}{2}}} e^{-\tau^2/2} \Big|_{-\infty}^{\mu-\sigma}}{\Phi(\mu-\sigma) - \Phi(-\infty)} = \frac{-\frac{1}{(2\pi)^{\frac{1}{2}}} e^{-(\mu-\sigma)^2/2}}{\Phi(\mu-\sigma)}$$

Thus,

$$\beta_2 = \frac{-\frac{1}{(2\pi)^{\frac{1}{2}}} e^{-(-1)^2/2}}{\Phi(-1)} = \frac{-0.24197}{0.158655} = -1.52514 \quad (10.42a)$$

(Note that $\Phi(-1) = 1 - \Phi(1) = 1 - 0.841345 = 0.158655$.)

Similarly,

$$\beta_3 = \frac{\int_{\mu-\sigma}^{\mu+\sigma} \frac{\tau e^{-\tau^2/2}}{(2\pi)^{\frac{1}{2}}} d\tau}{\int_{\mu-\sigma}^{\mu+\sigma} \frac{e^{-\tau^2/2}}{(2\pi)^{\frac{1}{2}}} d\tau} = \frac{-\frac{1}{(2\pi)^{\frac{1}{2}}} e^{-\tau^2/2} \Big|_{\mu-\sigma}^{\mu+\sigma}}{\Phi(\mu+\sigma) - \Phi(\mu-\sigma)}$$

$$= \frac{0}{0.841345 - 0.158655} = 0 \quad (10.42b)$$

$$\beta_4 = \frac{\int_{\mu+\sigma}^{\infty} \frac{\tau e^{-\tau^2/2}}{(2\pi)^{\frac{1}{2}}} d\tau}{\int_{\mu+\sigma}^{\infty} \frac{e^{-\tau^2/2}}{(2\pi)^{\frac{1}{2}}} d\tau} = \frac{-\frac{1}{(2\pi)^{\frac{1}{2}}} e^{-\tau^2/2} \Big|_{\mu+\sigma}^{\infty}}{\Phi(\infty) - \Phi(\mu+\sigma)}$$

$$= \frac{0 + \frac{1}{(2\pi)^{\frac{1}{2}}} e^{-(\mu+\sigma)^2/2}}{1 - 0.841345} = +1.52514 \quad (10.42c)$$

Recall the positioning of the probability axis in Eq. (8.30):

(a) $-\infty \leq P \leq \mu - \sigma$
(b) $\mu - \sigma < P \leq \mu + \sigma$
(c) $P > \mu + \sigma$

The following values of β_2, β_3, and β_4 corresponding to the above partitioning were given in Eqs. (8.34–8.36):

$$\beta_2 = \frac{\int_{-\infty}^{\mu-\sigma} \frac{\tau}{(2\pi)^{1/2}} e^{-\tau^2/2} d\tau}{\int_{-\infty}^{\mu-\sigma} \frac{1}{(2\pi)^{1/2}} e^{-\tau^2/2} d\tau} = -1.5247, \quad 0 \leq P < \mu - \sigma$$

$$\beta_3 = \frac{\int_{\mu-\sigma}^{\mu+\sigma} \frac{\tau}{(2\pi)^{1/2}} e^{-\tau^2/2} d\tau}{\int_{\mu-\sigma}^{\mu+\sigma} \frac{1}{(2\pi)^{1/2}} e^{-\tau^2/2} d\tau} = 0, \quad \mu - \sigma \leq P \leq \mu + \sigma$$

$$\beta_4 = \frac{\int_{\mu+\sigma}^{\infty} \frac{\tau}{(2\pi)^{1/2}} e^{-\tau^2/2} d\tau}{\int_{\mu+\sigma}^{\infty} \frac{1}{(2\pi)^{1/2}} e^{-\tau^2/2} d\tau} = 1.5247, \quad P > \mu + \sigma$$

Since the variance, $s^2(k)$, is independent of the control $u(k)$, we can represent the stochastic system, described by Eqs. (10.15) and (10.16), in terms of the expected values of its state and output variables; namely, the random variables are assigned the values of their means. Denote the variables of the new systems (in terms of their expected values) by (^). The equations for the *equivalent system* become

$$\hat{x}(k+1) = A\hat{x}(k) + B\hat{u}(k)$$

$$\hat{x}(0) = x_o$$

$$\hat{y}(k) = C\hat{x}(k) \quad (10.43)$$

Solving for $\hat{x}(k)$ and $\hat{y}(k)$ yields

$$\hat{x}(k) = A^k x_0 + \sum_{i=0}^{k-1} A^i Bu(k-1-i) \quad (10.44)$$

$$\hat{y}(k) = CA^k x_0 + \sum_{i=0}^{k-1} CA^i Bu(k-1-i) \quad (10.45a)$$

which is the same as $m(k)$ in Eq. (10.25); that is,

$$\hat{y}(k) = m(k) \quad (10.45b)$$

This is an important result because $m(k; u)$ is exactly $\hat{y}(k)$ in the new equivalent system. The conditional expected values are found directly from

$f_i^k(u) = m(k;u) + \beta_i^k s(k)$. The term $m(k; u)$ is determined by the system of Eq. (10.43). The constants β_i^k are determined by partitioning the probability axis used in the PMRM (see Chapter 8), where the partitioning is chosen at each stage (as in Eq. (10.41)), and $s(k)$ is given in Eq. (10.28).

Since the stochastic problem can be formulated using an equivalent system, as given in Eq. (10.43), the concepts and techniques of the MRIAM can be applied in a straightforward manner. The integrated risk and impact analysis can be outlined as follows:

1. Determine the partitioning points at each stage, and calculate the values of all β_i^k.
2. Calculate the variance $s^2(k)$ for each stage according to Eq. (10.28).
3. Formulate the equivalent system of Eq. (10.43).
4. Treating $\hat{y}(k)$ as an objective function along with other nonrisk objective functions (such as cost), find noninferior solutions using the MRIAM. The information generated should include the value of $\hat{y}(k)$, the values of the other objective functions, and the stage trade-offs.
5. The value of $\hat{y}(k)$ is equal to the unconditional expected value. The conditional expected values are determined using Eq. (10.40). The trade-offs, for a given stage k, are the same for all conditional expected values and are equal to the stage trade-offs calculated for $\hat{y}(k)$.
6. Using a multiobjective decisionmaking method (such as the SWT or the interactive SWT methods discussed in Chapter 5), find the preferred solution.

The risk functions may be interpreted the same as before, except for the temporal interpretation suggested by Asbeck [1982] (short-, intermediate-, and long-term risks). This view is no longer valid since the aspect of time has been explicitly incorporated into the system model. The conditional and unconditional expected values taken together form a characterization of risk at each stage, which provides more information than the expected value alone.

One final remark should be made about the combined risk and impact analysis. The solution to the multiobjective, multistage problem consists of the optimal policy choices over the entire planning horizon. As the system progresses through the stages, however, it will be possible to update the model and the information available to the DM. It may be desirable to repeat the risk impact analysis and decisionmaking procedure at several stages of the planning horizon, especially because of the uncertainties involved. Thus, the decisionmaking process itself can be made more dynamic, and control that is implemented at later stages will not be based on outdated information.

10.5 RELATING MULTIOBJECTIVE DECISION TREES TO THE MRIAM*

10.5.1 Introduction

The realization of a successful system entails detailed planning and implementation of every phase of the system life cycle—conception, development, design synthesis, and validation. In each step of the process, problems are encountered, alternatives are investigated, and solutions are implemented. Decisionmaking involves evaluating alternative courses of action. The DM evaluates various alternatives based on a set of preferences, criteria, and objectives, often conflicting in nature. When faced with several alternatives, a DM is assumed to be rational and will select the alternative, which is efficient, noninferior, and optimum. As discussed in Chapter 5, a noninferior or Pareto-optimal solution is defined as follows [Chankong and Haimes, 1983, 2008]:

$\mathbf{x}^*$ is said to be a *noninferior solution* of a vector optimization problem if there exists no other feasible $\mathbf{x}$ (i.e., $\mathbf{x} \in X$) such that $\mathbf{f}(\mathbf{x}) \leq \mathbf{f}(\mathbf{x}^*)$, meaning that $f_j(\mathbf{x}) \leq f_j(\mathbf{x}^*)$ for all $j = 1, \ldots, n$ with strict inequality for at least one j.

When decisions are made in an uncertain environment, select methodologies that incorporate uncertainty are utilized. Maximizing the minimum *(maximin)* gain, maximizing the maximum *(maximax)* gain, and minimizing the maximum *(minimax)* loss are examples of decisionmaking criteria that are used when the risk involved cannot be analyzed explicitly [Chankong and Haimes, 1983, 2008]. Utility theory can also be used in decisionmaking

---

*This section is based on Dicdican [2004] and on Dicdican and Haimes [2005].

under uncertainty (see Keeney and Raiffa [1976]). According to von Neumann and Morgenstern [1980], an individual choosing among alternatives with probabilistic outcomes will select the one with the largest expected subjective value or utility. The influence diagram is another useful technique under uncertainty (see Howard and Matheson [1984]; Shachter [1986, 1990]; Smith [1989]). It offers an intuitive method of showing the decisions, uncertainties, and objectives, along with their relationships. A technique similar to the influence diagram is the valuation network [Shenoy, 1993, 2000], which consists of a set of variables and a set of valuations defined on the subsets of variables. It allows for probability models to be represented by probability valuations, and its solution technique is slightly more efficient than that of the influence diagram [Shenoy, 1994]. The sequential decision diagram [Covaliu and Oliver, 1995] allows for a compact graphical representation for decision problems under uncertainty. It can be used in modeling the asymmetrical and sequential aspects of a decision problem. A commonly used tool in decisionmaking under uncertainty is the decision tree [Raiffa, 1968; von Winterfeldt and Edwards, 1986]. It is a graphical approach that allows for decision analysis over different points in time and for the decomposition of a complex problem into several smaller problems. Call and Miller [1990] presented an approach that combines decision trees with influence diagrams. Studies on the relationship between influence diagrams and decision trees include Howard and Matheson [1984] and Diehl and Haimes [2004].

Kirkwood [1993] developed an algebraic approach to address the combinatorial explosion of decision-tree scenarios. It is a compact representation involving the decision variables, random variables, and the functions relating these variables. The analytic hierarchy process (AHP) [Saaty, 1980] can also be used to model risk and uncertainty [Millet and Wedley, 2002]. According to Millet and Wedley [2002], four cases where AHP could be used involve situations wherein (i) outcome values are known while probabilities are unknown; (ii) the value of an alternative is the expected value of the combination of multiple criteria; (iii) an adjustment is made for variance, regret, and risk aversion; and (iv) risk is modeled as a criterion. Kujawski [2002] developed a project management approach for generating the set of risk response actions that achieves the lowest total cost for a given probability of success, while meeting schedule and technical performance criteria. The approach combines Markowitz's [1952] portfolio selection principles, Monte Carlo simulation, decision trees, and cumulative risk profiles.

The DM also considers the consequences of a decision where the time element must be addressed over a prespecified period. Both short- and long-term effects of decisions are analyzed, including future options and their associated costs, benefits, and risks. In dynamic problems, a sequence of decisions is made at different periods and the effect of each decision is realized at subsequent stages. Prior decisions may affect the range of feasible choices or alternatives that are available at future periods. It becomes important to consider the impact of current decisions on future options. The DM is tasked not only with determining the present optimal course of action but also with projecting into the future and avoiding catastrophic events. Impact analysis is made more important and is a significant part of the DM's considerations. (Impact analysis has been defined by Gomide and Haimes [1984] as *the study of the effect of decisions upon the decisionmaking problem* and by Haimes [1998, 2004] as *the study and investigation of the consequences of present decisions on future policy options.*)

This section develops the theoretical and methodological relationship between *multiobjective decision trees (MODT)* introduced in Chapter 9 and *the MRIAM* introduced earlier in this chapter. Decision trees have been extensively used in decision problems with great success. The MODT includes multiple noncommensurate objective functions over a given period. On the other hand, the MRIAM analyzes risk and decision impacts in a dynamic multiobjective framework. Both methods are used to perform sequential decisionmaking by analyzing the impacts of current decisions on future options. Understanding the advantages and limitations of these two distinct methods and appreciating how they supplement and complement each other contributes synergy and adds depth to an analysis of a dynamic system.

Use of decision trees is widespread, and its applications include decisionmaking [Magee, 1964a;

Frohwein, 1999; Frohwein and Lambert, 2000] and capital investment [Magee, 1964b]. The single-objective decision tree is the more prevalent type. It is "a way of displaying the anatomy of a business investment decision and of showing the interplay among a present decision, chance events, competitors' moves, and possible future decisions and their consequences" [Magee, 1964b]. When decisions involve multiple objectives, the most common approach involves assigning weights to the objectives to arrive at a composite score. The composite score is then folded back to arrive at a decision. Lootsma [1997] used multiplicative AHP [Lootsma, 1993] and the simple multiattribute rating technique [von Winterfeldt and Edwards, 1986] to aggregate the multidimensional consequences in a chance node. Monte Carlo simulation can be used to determine the expected utility of an alternative in a decision tree [Buffett and Spencer, 2003]. Haimes et al. [1990] introduced the MODT where the objective function is expressed as a vector consisting of different objectives. The units of the objectives do not have to be the same; the objectives can be kept in their original metrics. An MODT application in telecommunication showing extreme-event analysis is found in Dillon and Haimes [1996]. Other MODT applications can be found in Haimes [1998, 2004].

Gomide [1983] presented a theoretical method that can be used to address the dynamic nature of decisionmaking. It is known as the MMIAM. Applications are found in Gomide [1983] and Gomide and Haimes [1984]. In the MMIAM, a sequence of multiobjective problems that occur at different stages is analyzed to determine how previous decisions affect subsequent stages. Pareto-optimal solutions for the entire planning horizon, rather than for a single stage only, are evaluated using the ε-constraint method and augmented Lagrange multipliers [Haimes, 1998, 2004]. Extreme-event analysis is an important consideration in impact analysis. The PMRM developed by Asbeck and Haimes [1984] is used to generate conditional expected-value functions. The conditional expected value provides a measure of the expected risk value, given that this value is found within a prespecified range. The MRIAM [Leach and Haimes, 1987] was developed by combining the PMRM and MMIAM. The approach can be applied to time-invariant, linear, stochastic, and multistage processes. This method is used to determine which solutions are superior over the entire time period, rather than for one period at a time.

From a historical perspective, MODT and MRIAM were prompted by different reasons and requirements. MODT came about because of the need to incorporate multiple objectives into a sequential decisionmaking process. MRIAM was developed in order to evaluate discrete options, which are noninferior over a given time horizon, and to incorporate differing risk measures into a decisionmaking problem. Linking these two distinct methods brings a more holistic approach to decisionmaking—leading to a more closely unified risk analysis and dynamic multiobjective decisionmaking. Building on the relationship between the two methods should inspire the development of robust and encompassing software packages that address additional perspectives of the decision problem. Hopefully, scholars will be encouraged to further investigate the commonalities between MODT and MRIAM.

10.5.2 Generalized MRIAM

In Section 10.2, the means of the two random variables $w(k)$ and $v(k)$ are assumed to be zero. In this section, the MRIAM is generalized and relaxes these two assumptions. For pedagogical purposes, the overall time-invariant, linear, stochastic, and multistage processes are reformulated as follows:

$$x(k+1) = Ax(k) + Bu(k) + w(k), k = 0, \ldots, T-1$$
$$x(0) = x_0 \quad (10.46)$$

where $x(k)$ is the state of the system at stage k, $u(k)$ is the control to be implemented at stage k, A and B are system parameters, $w(k)$ is the element of randomness in the model (assumed to be a normally distributed purely random sequence), and k is the time horizon considered with T stages ($k=0, \ldots, T-1$). The measured output of the system at stage k is assumed to be linearly related to the state $x(k)$. This output $y(k)$ is described by the following:

$$y(k) = Cx(k) + v(k) \quad (10.47)$$

where C is a system parameter and $v(k)$ is a normally distributed, purely random, stationary sequence. An equivalent deterministic model is discussed in Haimes [1998, 2004].

According to Leach and Haimes [1987] and Haimes [1998, 2004], the system and its output are assumed to possess the following statistical properties. The system has disturbance $w(k)$, which is assumed to be normally distributed with expected value μ_w and variance σ_w^2. There is no relationship between disturbances at different stages. The initial state of the system is x_0 and has mean x_0 and variance X_0. In addition, there is no relationship between the initial state of the system and the system randomness. For the system output, the disturbance represented by $v(k)$ has mean μ_v and variance σ_v^2, and there is no relation between disturbances at different stages. The initial state of the system and the output randomness are not related, and the disturbances in the system and the output are not related.

In the MRIAM, randomness is assumed to be continuously distributed. The mean and variances of the system state and its output at stage k are given as

$$E[x(k)] = A^k x_0 + \sum_{i=0}^{k-1} A_i Bu(k-1-i) + \sum_{i=0}^{k-1} A^i \mu_w \quad (10.48)$$

$$\mathrm{Var}[x(k)] = A^{2k} X_0 + \sum_{i=0}^{k-1} A^{2i}(\sigma_w^2) \quad (10.49)$$

$$m(k) = E[y(k)] = CA^k x_0 + \sum_{i=0}^{k-1} CA^i Bu(k-1-i)$$
$$+ \sum_{i=0}^{k-1} CA^i \mu_w + \mu_v \quad (10.50)$$

$$s^2(k) = \mathrm{Var}[y(k)] = C^2 A^{2k} X_0 + \sum_{i=0}^{k-1} C^2 A^{2i}(\sigma_w^2) + (\sigma_v^2) \quad (10.51)$$

10.5.3 Relationship between MODT and MRIAM

Determining the relationship between the MODT and the MRIAM is an important factor toward their integration. The question is whether an MODT can be converted into a multiobjective risk impact analysis model and vice versa. It can be shown that the methods are related given that the following requirements of the state equation and system randomness are met. For the state equation requirement, the system is assumed time invariant, linear, and stochastic, and the state equation $x(k+1) = Ax(k) + Bu(k) + w(k)$ must exist for both models. The initial state of the system is $x(0) = x_0$. For the system randomness requirement, randomness follows a normal distribution with parameters μ and σ^2. In each period, there is only one chance event, and it is assumed to have a probability of occurrence equal to 1. The random events are assumed to be independent.

The state equation requirement ensures that a linear relationship is used to model the process. In the MRIAM, a repeated dynamic system is modeled, with the process recursive for all stages or time periods (see Figure 10.1). This requirement is not necessarily followed in all MODT problems. The state equation requirement provides that the system being studied under MODT follows a linear relationship; this problem characteristic is the same in the MRIAM. The system randomness requirement assures that the randomness present is the same in either representation. A necessary condition for applying the MRIAM is that randomness is normally distributed. For an MODT, the system randomness does not necessarily need to follow a normal distribution. Having this

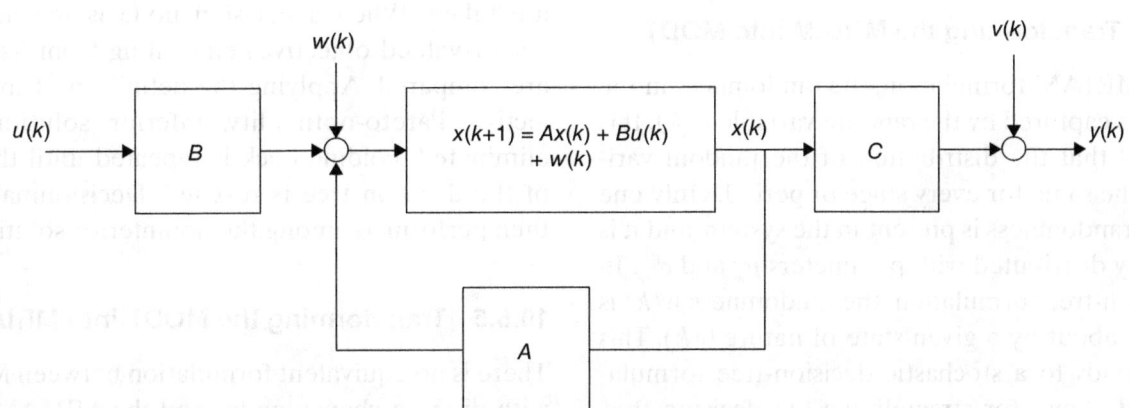

Figure 10.1 Block diagram for decisionmaking process.

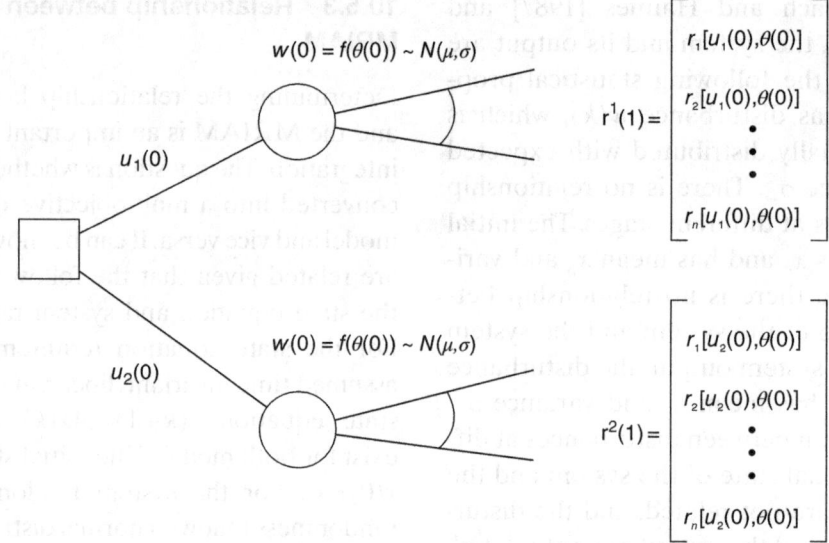

Figure 10.2 Structure of MODT with stochastic chance node.

requirement ensures that both methods can be used to solve the problem. The MODT and MRIAM are both grounded in multiobjective trade-off analysis and Pareto-optimality. The MODT forward calculations are performed using the state equation. With foldback, the expected value is computed for all branches that come from a chance node. Because of the system randomness requirement, the MODT foldback in chance nodes consists of calculating the expected value of the objectives. This is equivalent to computing the expected value in the MRIAM. When both MODT and MRIAM can be used in a problem, MODT can graphically present the problem and its corresponding decision path to the DM, while the MRIAM equations enable the automatic computation of the means and variances.

10.5.4 Transforming the MRIAM into MODT

In the MRIAM formulation, the randomness in the system is captured by the random variable $w(k)$. It is assumed that the distribution of the random variable is the same for every stage or period. Only one type of randomness is present in the system, and it is normally distributed with parameters μ_w and σ^2_w. In a decision-tree formulation, the randomness $w(k)$ is brought about by a given state of nature $\theta(k)$. This corresponds to a stochastic decision-tree formulation and allows for streamlining the decision tree. When the chance node is reached, there is only one chance event, $\theta(k)$, represented by a normal distribution and, consequently, one system randomness $w(k)$ (see Figure 10.2).

A decision policy, which consists of a series of actions for each period or stage $[u(0), u(1), ..., u(k)]$, is investigated in terms of its performance on the objective functions. With the MRIAM, the system state is represented by $x(k)$, and the system output is $y(k)$. The decision tree can be constructed with the different control variables in period k representing the decisions at that period. The system randomness $w(k)$ can be obtained for the given state of nature $\theta(k)$, with a probability of occurrence equal to 1. For each path in the decision tree, the system state and the objective values at each k period are computed until the end of the path is reached. Folding back commences where the means of the objective values are taken. When a decision node is reached, the vector-valued objectives emanating from its nodes are compared. Applying the definition of multiobjective Pareto-optimality, inferior solutions are eliminated. Folding back is repeated until the root of the decision tree is reached. Decisionmaking is then performed among the noninferior solutions.

10.5.5 Transforming the MODT into MRIAM

There is no equivalent formulation between MODT with discrete chance nodes and the MRIAM. With the discrete chance node, it is assumed that different states of nature prevail, and consequently different system randomness occurs. Assuming that the system state is normally distributed, when folding back and averaging occur at the chance node, the normal probability distributions with different means and variances are combined. This forms a mixed distribution, $g(x)$, which is a combination of two or more probability distributions $(f_i s)$, each with a probability of occurrence p_i:

$$g(x) = p_1 f_1(x) + p_2 f_2(x) + \ldots + p_n f_n(x) \quad (10.52)$$

where $0 \le p_i \le 1$ and $\sum_{i=1}^{n} P_i = 1$. With a mixed distribution, the resulting expected system state is not necessarily normally distributed. This is a violation of the MRIAM assumption that the state of the system should be normally distributed. The MODT with discrete chance nodes is not equivalent to the MRIAM [Dicdican, 2004; Dicdican and Haimes, 2005].

An MODT with stochastic chance nodes that exhibits the requirements discussed can be transformed into an MRIAM formulation. The state of the system is assumed to follow Eq. (10.46). The decision tree captures the randomness by the chance event $\theta(k)$, which has normally distributed effects, $w(k)$, on the state of the system $N(\mu_w, \sigma^2_w)$. The tree can be constructed with the corresponding decision nodes and chance nodes. The decision tree shown in Figure 10.3 has two possible paths: Path 1 ($1 \Rightarrow 2 \Rightarrow 3 \Rightarrow 4$) and Path 2 ($1 \Rightarrow 2 \Rightarrow 3 \Rightarrow 5$). For each path, the state of the system can be computed for Periods 1 and 2 (see Table 10.1). The expected values of the system state and output are also computed using Eqs. (10.48) and (10.50).

Starting from the terminal point of the decision tree, folding back is performed. As the infinite events are represented by a normal distribution, the mean of the distribution is taken in the folding-back process. The vector-valued objectives from the choices emanating from the decision node are compared, and inferior solutions are eliminated. The folding-back process is repeated until the root of the decision tree is reached. In the decision tree found in Figure 10.3, folding back is performed on Paths 1 and 2. On Path 1, Chance Node 4 is reached first and the mean of the distribution is taken. On Path 2, Chance Node 5 is reached and again the mean is taken. Decision Node 3 is then encountered in the folding back of both paths. The values of the objective functions of Paths 1 and 2 are compared. Assuming that both paths are Pareto-optimal, the folding back continues. The stochastic Chance Node 2 is reached, and once again, the mean is taken. Decision Node 1, which is the root of the tree, is encountered, and the two paths are compared once again. Given constraints and preferences, the DM must now determine which path is better. It can be seen that an MODT with stochastic chance nodes that are normally distributed is related to the MRIAM. The chance events, decision variables, and paths in the MODT correspond respectively to the random events, control variables, and various

TABLE 10.1 Decision-Tree Paths and State of System

Path	x(1)	x(2)
1	$Ax(0) + Bu_1(0) + w(0)$	$Ax(1) + Bu_1(1) + w(1)$
2	$Ax(0) + Bu_1(0) + w(0)$	$Ax(1) + Bu_2(1) + w(1)$

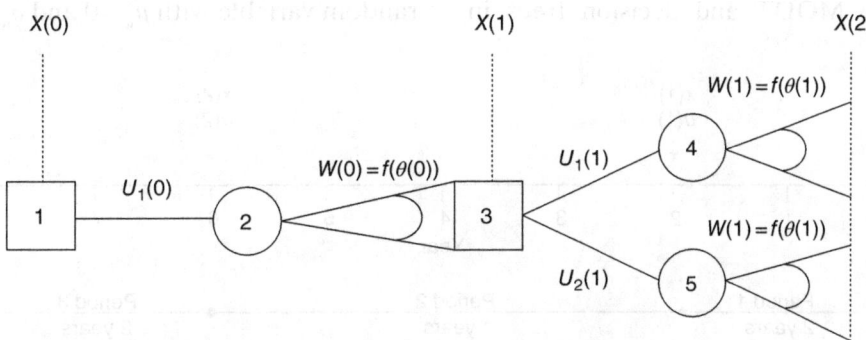

Figure 10.3 Example of a decision tree with two paths.

policies investigated in the MRIAM. Given that the requirements in Section 10.5.3 are followed, the equations for system output can be used for both methods.

10.5.6 Discussion

In the MRIAM, there are explicit functional relationships between the state of the system $x(k)$ and the output $y(k)$ with the random variable (see Eqs. (10.46) and (10.47)). For practical purposes, the stochastic nodes in the MODT are assumed to be normally distributed, even though the chance events can be infinite. The state equation (10.46) is used to calculate the system state at period k for each possible path through the decision tree. In folding back the end values toward the root of the tree, the expected value is calculated at the chance nodes. Both methods lead to the expected value of multiple objectives at each stage. In both the MRIAM and MODT, the analyst must filter non-Pareto-optimal solutions. This illustrates the relationship between the steps and outputs of the MODT and MRIAM, given that the requirements in Section 10.5.3 hold.

However, the conditions for relating both methods are restrictive. MRIAM, by its very nature, is constricted to time-invariant, linear, stochastic, and multistage processes, which exhibit Gaussian randomness. Not many processes exhibit such properties, and this prohibits its extensive use. Nevertheless, it provides general formulas for calculating the means and variances of the system state and output at any given stage and eases computational complexity. MODT can be used in a wider variety of decision problems with no restrictions on the distribution of chance events. It is a graphical tool for displaying the problem and folded-back decisions. Because MODT and decision trees in general allow for the consideration of multiple states of nature, the decision tree and its solution may grow substantially large and difficult to handle. The problem with the MODT solution procedure becoming exponentially large can be resolved when MRIAM equations can be used in its place. MODT and MRIAM can be used together, with MODT utilized to display the problem visually and MRIAM equations aiding the calculations.

10.5.7 Illustrative Example

A one-mile pavement section is assumed to have a current remaining life, $x(0)$, of 5 years. The variance of the initial remaining life is assumed equal to 0. It is also assumed that remaining life is normally distributed. Budget preparation occurs every 2 years. The DM would like to determine (i) what action is best for the next budget, (ii) what the effects of the action will be in two budget periods, and (iii) what options must be adopted at that time. As shown in Figure 10.4, three decision points are identified: *now, 2 years from now,* and *6 years from now.* The interval between decisions made at stage k and the succeeding stage is represented by $l(k)$. Thus, $l(0)=2$, $l(1)=4$, $l(2)=2$. The three policies shown in Table 10.2 are being considered. The objectives are to maximize remaining service life at the end of 8 years, minimize total cost for 8 years, and maximize conditional expected value of remaining service life.

The state equation for remaining service life is assumed to be

$$x(k+1) = x(k) + 0.000146\, u(k) - l(k) - w(k) \quad (10.53)$$

It is assumed that $w(k)$ is a normally distributed random variable with $\mu_w = 0$ and $\sigma_w^2 = 1$.

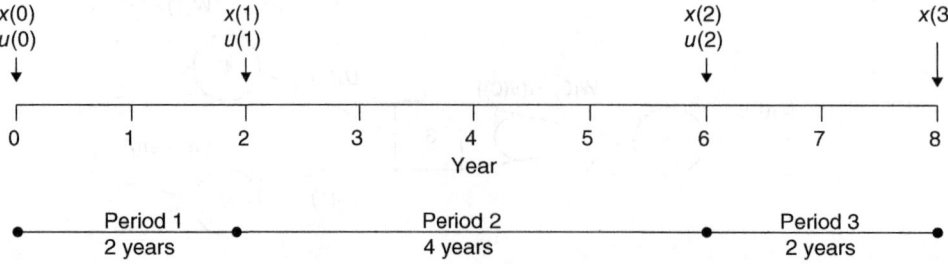

Figure 10.4 Decisionmaking timeline for pavement section.

TABLE 10.2 Policies for Illustrative Example

Policy	Period 1 (Now)	Period 2 (2 Years from Now)	Period 3 (6 Years from Now)
1	Corrective	Corrective	Preventive (crack seal)
2	Corrective	Corrective	Preventive (slurry seal)
3	Restorative	Preventive (crack seal)	Preventive (crack seal)

TABLE 10.3 Values of Control Variable u(k) for Illustrative Example

Policy	u(0)	u(1)	u(2)
1	36,667	36,667	1,500
2	36,667	36,667	6,000
3	80,000	1,500	1,500

The objective functions are:

r_1 = maximize expected remaining service life

$$= \text{maximize } E[y(k)] = E[cx(k)] = cE[x(k)]$$
$$\text{where } c = 1 \quad (10.54)$$

r_2 = minimize total cost

$$= \text{minimize } \Sigma_k u(k) \text{ where } k = 0, 1, \text{ and } 2$$
$$= \text{minimize } u(0) + u(1) + u(2) \quad (10.55)$$

r_3 = maximize conditional expected value of remaining service life with partition made at one standard deviation below the mean

$$= \text{maximize } f_4(k) = \mu(k) - \beta_4 \sigma(k) \quad (10.56)$$

where $\beta_4 = 1.525, \mu(k) = E[y(k)]$, and $\sigma^2(k) = \text{Var}[y(k)]$

The values of the control variable assumed for this example are given in Table 10.3.

10.5.7.1 Solution using MRIAM

The problem is solved through MRIAM utilizing Eq. (10.53). The objective functions are then obtained for each policy. Table 10.4 shows the values for the expected remaining service life, $E[y(k)]$, and the cumulative cost, $\Sigma_k u(k)$, over the study periods.

The high-range conditional expected value for remaining life is also computed where the partition is made at a minus one standard deviation from the mean. The results for the variance of $y(k)$ and for the $f_4(k)$ are found in Table 10.5. Table 10.6 depicts

TABLE 10.4 MRIAM Results for Illustrative Example

Policy	Period	E[y(k)] (Years)	Σ_k u(k) ($)
1	k=0	8.35	36,667
	k=1	9.71	73,334
	k=2	7.93	74,834
2	k=0	8.35	36,667
	k=1	9.71	73,334
	k=2	8.58	79,334
3	k=0	14.68	80,000
	k=1	10.90	81,500
	k=2	9.12	83,000

TABLE 10.5 MRIAM Results for Variance and Conditional Expected Values for Illustrative Example

Policy	Period	Var[y(k)]	$f_4(kn)$
1	k=0	1	6.83
	k=1	2	7.55
	k=2	3	5.28
2	k=0	1	6.83
	k=1	2	7.55
	k=2	3	5.94
3	k=0	1	13.16
	k=1	2	8.74
	k=2	3	6.48

TABLE 10.6 Objective Function Values for Illustrative Example

Objective	Policy 1	Policy 2	Policy 3
r_1 (years)	7.93	8.58	9.12
r_2 ($)	74,834	79,334	83,000
r_3 (years)	5.28	5.94	6.48

the values of the objective functions for the three policies explored in this problem.

10.5.7.2 Solution using MODT

The numeric example is graphically represented in the MODT shown in Figure 10.5. To solve the decision tree, the values at the terminal node or end of the decision path are obtained. For Policy 1 at Period 1, corrective action is performed. Applying Eq. (10.54), the improvement in remaining service life is $E[x(1)] = 5 + 0.000146(36667) - 2 - 0 = 8.35$. At Period 2, corrective action is again performed. Applying Eq. (10.54), the improvement in remaining

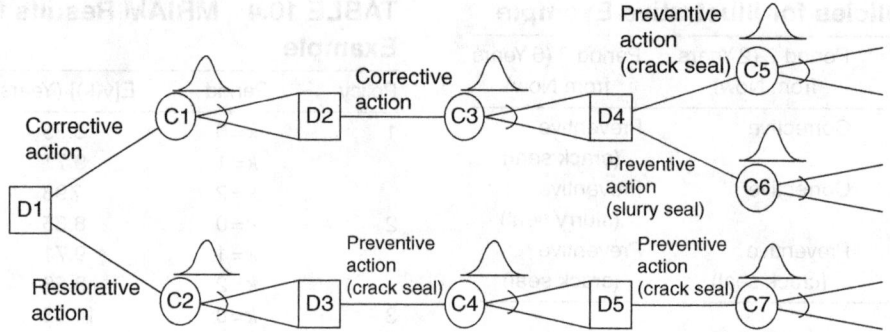

Figure 10.5 Multiobjective decision tree (MODT) for illustrative example.

TABLE 10.7 MODT Results for Illustrative Example

Policy	Period	$r_1 = E[y(k)]$ (years)	$r_2 = \Sigma_k u(k)$ ($)	$r_3 = f_4(k)$ (years)
1	k=0	8.35	36,667	6.83
	k=1	9.71	73,334	7.55
	k=2	7.93	74,834	5.28
2	k=0	8.35	36,667	6.83
	k=1	9.71	73,334	7.55
	k=2	8.58	79,334	5.94
3	k=0	14.68	80,000	13.16
	k=1	10.90	81,500	8.74
	k=2	9.12	83,000	6.48

service life is $E[x(2)] = 8.35 + 0.000146(36667) - 4 - 0 = 9.71$. At Period 3, crack seal is applied. Applying Eq. (10.54), the improvement in remaining service life is $E[x(3)] = 9.71 + 0.000146(1500) - 2 - 0 = 7.93$. The same forward calculations are applied for Policies 2 and 3. The terminal values are folded back until the starting decision node is reached. The results are found in Table 10.7. These are the same results yielded using the MRIAM.

10.5.7.3 Discussion of example results

From the definition of Pareto-optimality, all three policies are Pareto-optimal because gaining in one objective results in a loss in another objective. In Policy 1, the cost is the same for the first two periods and decreases in the third period (see Table 10.3). The total expenditure is $74,834 and 7.93 years of service life remain after 8 years (see Table 10.6). It can be seen from Table 10.4 or 10.7 that the remaining service life does not vary much over the study period. This means that the pavement will be maintained in the same condition during that time. Table 10.5 or 10.7 shows that the high-range conditional expected value increases from 6.83 to 7.55 between the first two periods and decreases from 7.55 to 5.28 between the second and third periods. Policy 2 has the same actions in the first two periods as Policy 1 (see Table 10.2). It differs from Policy 1 in the last period, where slurry seal is used instead of crack seal. The total cost for this policy at the end of the entire study period is $79,334, and the service life remaining is 8.58 years (see Table 10.6). The service life over the years does not vary much, as seen in Table 10.4. The f_4 value in the end is 5.94 years (see Table 10.5 or 10.7). In Policy 3, the total cost is $83,000 (see Table 10.6 or 10.7), and the bulk of the funds is spent immediately (see Table 10.3). The remaining service life after 8 years is 9.12 years (see Table 10.6 or 10.7). Because the amount spent decreases over the study period, the remaining service life also decreases (see Table 10.4 or 10.7). This means that the pavement condition will deteriorate accordingly. Policies 1 and 2 result in more stable pavement conditions compared to those resulting from Policy 3. If the DM prefers to have more consistent pavement conditions, then he may choose between Policies 1 and 2. However, depending on the funds available to the DM in the early and later stages, Policy 3 could be preferred.

10.5.7.4 Summary

This section has compared two methods that are useful to DMs in systems analysis: MODT and the MRIAM. Both methods can handle multiple objectives and sequential decisionmaking. The two tools also facilitate incorporating two types of risk into the analysis. The first is the risk brought about by any random disturbance; system randomness is represented in the state equations for both MODT and the MRIAM. Second, the use of the expected value for some of the objective functions is an acknowledgment that uncertainty occurs. This section has demonstrated that MODT and MRIAM are related when the state equation for a time-invariant, linear, and stochastic system is used and the system random variable is normally distributed. In cases wherein the requirements are not met (i.e., randomness is not normally distributed), MODT is the more robust and general tool and can be used to aid the decisionmaking process. MRIAM can be used for a time-invariant, linear, stochastic, multistage process. Further study can be performed to determine the applicability of both MODT and MRIAM to problems when the system randomness is not necessarily Gaussian, but the consequences of assuming Gaussian randomness have minimal effect.

10.6 EXAMPLE PROBLEMS

10.6.1 Pollution Emission [Leach and Haimes, 1987]

The following is an example of how the PMRM may be applied to a multistage problem as described in the previous section. The model is a stochastic, first-order, linear differential equation representing the relationship between resource damage and pollutant emissions in an environmental system. Although this is a simplistic and hypothetical model, the analysis presented here serves two purposes. First, it demonstrates how the PMRM is applied to multistage models, how the results can be interpreted, and the usefulness of the MRIAM. Second, this problem can be viewed as a first step toward a more realistic model of environmental systems.

Four stages are considered here. There is 5 years between stages, and thus the planning horizon is 15 years long. Associated with each of the first three stages is one control variable that represents the level of pollutant emissions at that stage. This level is assumed to remain constant over the 5-year period. A cost function is formed by summing the present-value costs of emission reductions at each of these stages. Associated with each of the last three stages are the risk functions generated by the PMRM. The levels of risk are affected directly by the levels of emissions. No decision variables or costs are considered at the last stage since the effects of decisions at that stage will not be observed until later. The multiobjective problem is to choose a control sequence $\{u(0), u(1), u(2)\}$ so as to minimize the cost function and the risk functions.

Let $x(k)$ be the state variable at stage k representing environmental damage, expressed as a ratio to the initial environmental damage. The initial damage is assumed to be known exactly, $x(0) = 1$. Let $u(k)$ be the control variable (level of emissions) at stage k, also expressed as a ratio to the latest level of emissions, just before the beginning of the planning horizon. Let $d(k)$ represent a random disturbance at stage k, which has a normal distribution with mean zero and variance s_d^2. The variance is assumed to be constant for all k. Four stages are considered ($k = 0, 1, 2, 3$), as shown in Figure 10.6.

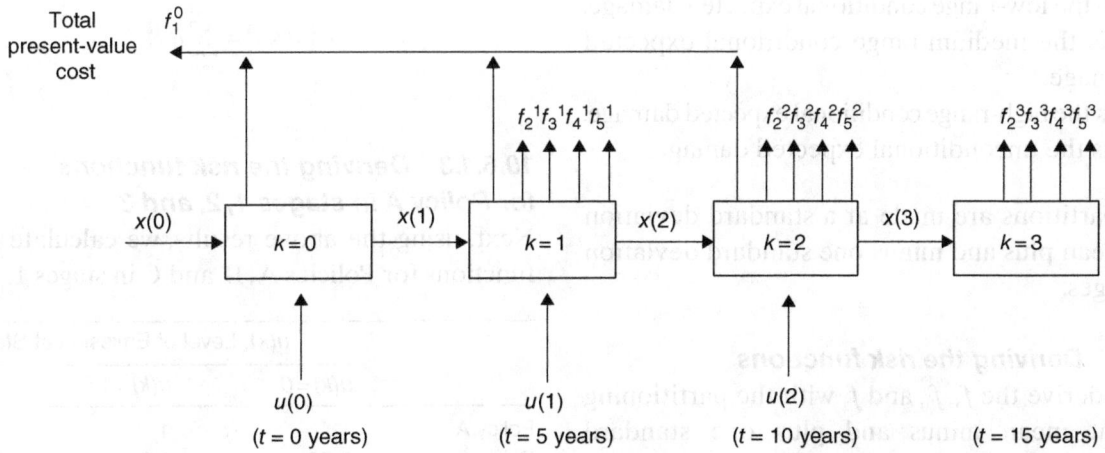

Figure 10.6 The model for environmental damage.

The following equations are used to describe the dynamics of the system:

$$x(k+1) = ax(k) + bu(k) + d(k), \quad x(0) = 1$$
$$0 \leq u(k) \leq 1, \quad k = 0, 1, 2 \quad (10.57)$$

The constraints on $u(k)$ simply imply that emissions are not to be increased from their current levels and that they cannot become negative (physically impossible). Leach [1984] uses the following values for the system's model parameters:

$$a = 0.85, \quad b = 0.25, \quad s_d^2 = 0.04$$

Thus, the system's representation is

$$x(k+1) = 0.85x(k) + 0.25u(k) + d(k), \quad x(0) = 1,$$
$$s_d^2 = 0.04$$
$$0 \leq u(k) \leq 1, \quad k = 0, 1, 2 \quad (10.58)$$

Leach and Haimes [1987] also use the following present-value cost function f_1^0:

$$f_1^0 = K(1-u(0))^2 + K(1-u(1))^2(1+r)^{-5}$$
$$+ K(1-u(2))^2(1+r)^{-10}$$

where $K = \$100 \times 10^6$ and $r = 10\%$, so that the cost function becomes

$$f_1^0 = [100(1-u(0))^2 + 62.1(1-u(1))^2$$
$$+ 38.6(1-u(2))^2](10^6) \quad (10.59)$$

For stages $k = 1, 2$, and 3, the following risk functions are generated by the PMRM:

1. f_2^k is the low-range conditional expected damage.
2. f_3^k is the medium-range conditional expected damage.
3. f_4^k is the high-range conditional expected damage.
4. f_5^k is the unconditional expected damage.

The partitions are made at a standard deviation of the mean plus and minus one standard deviation at all stages.

10.6.1.1 Deriving the risk functions

We first derive the f_2, f_3, and f_4 with the partitioning points at mean minus and plus one standard deviation for a random variable X having normal distribution. Namely,

$$f_2(\cdot): -\infty \leq P \leq \mu - \sigma; \quad f_3(\cdot): \mu - \sigma < P \leq \mu + \sigma;$$
$$f_4(\cdot): P > \mu + \sigma$$

Let $X \sim N(\mu, \sigma^2)$. Let $Y \sim N(0, 1)$; then,

$$f_2 = E[X \mid X \leq \mu - \sigma] = \mu + \sigma E[(X-\mu)/$$
$$\sigma \mid (X-\mu)/\sigma \leq -1] = \mu + \sigma E[Y \mid Y \leq -1] = \mu - 1.52\sigma$$
$$f_3 = E[X \mid \mu - \sigma < X \leq \mu + \sigma] = \mu + \sigma E[Y \mid -1 < Y \leq 1]$$
$$= \mu + 0 = \mu$$
$$f_4 = E[X \mid X > \mu + \sigma] = \mu + \sigma E[Y \mid Y > 1] = \mu + 1.525\sigma$$

The above is true since $E[Y \mid Y \leq -1] = -E[Y \mid Y \geq 1] = -1.525$ and $E[Y \mid -1 \leq Y \leq 1] = 0$. See Eqs. (10.42a)–(10.42c).

10.6.1.2 Deriving the mean and variance of x(k)

Next, we derive the mean and variance of $x(k)$ in Eq. (10.57).

Since $x(0) = 1$, we have

$$x(1) = ax(0) + bu(0) + d(0) = a + bu(0) + d(0)$$
$$x(2) = ax(1) + bu(1) + d(1) = a^2 + abu(0) + bu(1)$$
$$+ ad(0) + d(1)$$

Thus, by induction, we have

$$x(k) = a^k + b\sum_{i=0}^{k-1} a^{k-1-i} u(i) + \sum_{i=0}^{k-1} a^{k-1-i} d(i)$$

$$E[x(k)] = a^k + b\sum_{i=0}^{k-1} a^{k-1-i} u(i)$$

$$\text{Var}[x(k)] = \sum_{i=0}^{k-1} a^{2(k-1-i)} s_d^2$$

10.6.1.3 Deriving the risk functions for Policy A in stages 1, 2, and 3

Next, using the above results, we calculate the risk functions for Policies A, B, and C in stages 1, 2, and 3.

	u(k), Level of Emission at Stage		
	u(k)=0	u(k)=1	u(k)=2
Policy A	1	1	1
Policy B	0.75	0.60	0.50
Policy C	0.50	0.50	0.50

Policy A, stage 1 (k=1):

$$E[x(1)] = a + b = 0.85 + 0.25 = 1.1$$
$$\text{var}(x(1)) = s_d^2 = 0.04$$

Thus,

$$\mu = 1.1, \sigma = 0.04^{1/2} = 0.2, \text{ and}$$
$$f_2^1 = \mu - 1.525\sigma = 1.1 - 1.525 \times 0.2 = 0.795$$
$$f_3^1 = \mu = 1.1$$
$$f_4^1 = \mu + 1.525\sigma = 1.1 + 1.525 \times 0.2 = 1.405$$

Policy A, stage 2 (k=2):

$$E[x(2)] = a^2 + b(au(0) + u(1)) = a^2 + ab + b$$
$$= 0.85^2 + 0.85 \times 0.25 + 0.25 = 1.185$$
$$\text{var}(x(2)) = (a^2 + 1)s_d^2 = (0.85^2 + 1) \times 0.04 = 0.0689$$

Thus,

$$\mu = 1.185, \sigma = 0.0689^{1/2} = 0.262, \text{ and}$$
$$f_2^2 = \mu - 1.525\sigma = 1.185 - 1.525 \times 0.262 = 0.785$$
$$f_3^2 = \mu = 1.185$$
$$f_4^2 = \mu + 1.525\sigma = 1.185 + 1.525 \times 0.262 = 1.585$$

Policy A, stage 3 (k=3):

$$E[x(3)] = a^3 + b(a^2u(0) + au(1) + u(2)) = a^3 + b(a^2 + a + 1)$$
$$= 0.85^3 + 0.25(0.85^2 + 0.85 + 1) = 1.257$$
$$\text{var}(x(3)) = (a^4 + a^2 + 1)s_d^2 = (0.85^4 + 0.85^2 + 1) \times 0.04$$
$$= 0.0898$$

Thus,

$$\mu = 1.257, \sigma = 0.0898^{1/2} = 0.2996, \text{ and}$$
$$f_2^3 = \mu - 1.525\sigma = 1.257 - 1.525 \times 0.2996 = 0.800$$
$$f_3^3 = \mu = 1.257$$
$$f_4^3 = \mu + 1.525\sigma = 1.257 + 1.525 \times 0.2996 = 1.714$$

10.6.1.4 Deriving the risk functions for Policies B and C in stages 1, 2, and 3

Similar to the above derivation, we can calculate the risk functions for Policies B and C in stages 1, 2, and 3. The results are listed in Table 10.8.

10.6.1.5 Deriving the trade-off functions

The values of the trade-offs can be calculated as follows:

$$\min_{u(k)} \{f_1^0(\cdot), f_i^1(\cdot), f_i^2(\cdot), f_i^3(\cdot)\}, \quad i = 2, 3, 4, 5$$

where the superscript denotes the period, the subscript 1 denotes the cost function, the subscripts $i = 2, 3, 4$ denote the conditional expected value of environmental damage, and the subscript $i = 5$ denotes the unconditional expected value of environmental damage.

Using the ε-constraint approach discussed in Chapter 5, we have

$$\min_{u(k)} f_1^0(\cdot)$$

TABLE 10.8 Noninferior Policies for Environmental Model

Stage	Risk Functions	Trade-Offs ($10^6/U$ of Damage)
	Policy A[a]	
k=1	$f_2^1 = 0.795$ $f_3^1 = 1.100$ $f_4^1 = 1.405$ $f_5^1 = 1.100$	$\lambda_{11}^{01} = \lambda_{12}^{01} = \lambda_{13}^{01} = \lambda_{14}^{01} = 0$
k=2	$f_2^2 = 0.785$ $f_3^2 = 1.185$ $f_4^2 = 1.585$ $f_5^2 = 1.185$	$\lambda_{11}^{02} = \lambda_{12}^{02} = \lambda_{13}^{02} = \lambda_{14}^{02} = 0$
k=3	$f_2^3 = 0.800$ $f_3^3 = 1.257$ $f_4^3 = 1.714$ $f_5^3 = 1.257$	$\lambda_{11}^{03} = \lambda_{12}^{03} = \lambda_{13}^{03} = \lambda_{14}^{03} = 0$
	Policy B[b]	
k=1	$f_2^1 = 0.732$ $f_3^1 = 1.038$ $f_4^1 = 1.343$ $f_5^1 = 1.038$	$\lambda_{11}^{01} = \lambda_{12}^{01} = \lambda_{13}^{01} = \lambda_{14}^{01} = 31.109$
k=2	$f_2^2 = 0.632$ $f_3^2 = 1.032$ $f_4^2 = 1.432$ $f_5^2 = 1.032$	$\lambda_{11}^{02} = \lambda_{12}^{02} = \lambda_{13}^{02} = \lambda_{14}^{02} = 67.610$
k=3	$f_2^3 = 0.545$ $f_3^3 = 1.002$ $f_4^3 = 1.459$ $f_5^3 = 1.002$	$\lambda_{11}^{03} = \lambda_{12}^{03} = \lambda_{13}^{03} = \lambda_{14}^{03} = 154.217$
	Policy C[c]	
k=1	$f_2^1 = 0.670$ $f_3^1 = 0.975$ $f_4^1 = 1.280$ $f_5^1 = 0.975$	$\lambda_{11}^{01} = \lambda_{12}^{01} = \lambda_{13}^{01} = \lambda_{14}^{01} = 188.870$
k=2	$f_2^2 = 0.553$ $f_3^2 = 0.954$ $f_4^2 = 1.354$ $f_5^2 = 1.185$	$\lambda_{11}^{02} = \lambda_{12}^{02} = \lambda_{13}^{02} = \lambda_{14}^{02} = 117.284$
k=3	$f_2^3 = 0.479$ $f_3^3 = 0.963$ $f_4^3 = 1.393$ $f_5^3 = 0.936$	$\lambda_{11}^{03} = \lambda_{12}^{03} = \lambda_{13}^{03} = \lambda_{14}^{03} = 154.217$

[a] Control variables: $u(0)=1$, $u(1)=1$, $u(2)=1$; cost: $f_1^0 = (\$10^6)$.
[b] Control variables: $u(0)=0.75$, $u(1)=0.6$, $u(2)=0.5$; cost: $f_1^0 = 25.823(\$10^6)$.
[c] Control variables: $u(0)=0.5$, $u(1)=0.5$, $u(2)=0.5$; cost: $f_1^0 = 50.162(\$10^6)$.

subject to

$$f_i^1(\cdot) \leq \varepsilon_i^1, f_i^2(\cdot) \leq \varepsilon_i^2, f_i^3(\cdot) \leq \varepsilon_i^3$$

Form the Lagrangian function L, and use the following notation:

$$\lambda_1 = -\frac{\partial f_1^0(\cdot)}{\partial f_i^1(\cdot)} = \lambda_{1i}^{01}$$

$$\lambda_2 = -\frac{\partial f_1^0(\cdot)}{\partial f_i^2(\cdot)} = \lambda_{1i}^{02}$$

$$\lambda_3 = -\frac{\partial f_1^0(\cdot)}{\partial f_i^3(\cdot)} = \lambda_{1i}^{03}$$

$$\lambda_{ij}^{kl} = -\frac{\partial f_i^k}{\partial f_j^l}$$

where λ_{ij}^{kl} denotes the trade-off between objective function i at period k and objective function j at period l.

Taking the derivatives with respect to the controls at $u(2)$, $u(1)$, and $u(0)$ yields the following results for Policy B:

$$\frac{\partial L}{\partial u(2)} = (-2)(38.6)(1-u(2)) + \lambda_3(0.25) = 0$$

For Policy B, $u(2) = 0.5$ (see Table 10.8). Thus,

$$\lambda_3 \stackrel{\Delta}{=} \lambda_{1i}^{03} = 154.4.$$

$$\frac{\partial L}{\partial u(1)} = (-2)(62.1)(1-u(1)) + \lambda_2(0.25)$$
$$+ \lambda_3(0.85)(0.25) = 0$$

For Policy B, $u(1) = 0.6$ (see Table 10.8). Thus,

$$\lambda_2 \stackrel{\Delta}{=} \lambda_{1i}^{02} = 67.48$$

$$\frac{\partial L}{\partial u(0)} = (-200)(1-u(0)) + \lambda_1(0.25) + \lambda_2(0.85)(0.25)$$
$$+ \lambda_3(0.25)(0.85)^2 = 0$$

For Policy B, $u(0) = 0.75$ (see Table 10.8). Thus,

$$\lambda_1 \stackrel{\Delta}{=} \lambda_{1i}^{01} = 31.088.$$

Note that all λ_{ij}^{kl} for the same kl are equal. Thus,

$$\lambda_1 \stackrel{\Delta}{=} \lambda_{1i}^{01} = \lambda_{12}^{01} = \lambda_{13}^{01} = \lambda_{14}^{01} = \lambda_{15}^{01} = 31.088$$

$$\lambda_2 \stackrel{\Delta}{=} \lambda_{1i}^{02} = \lambda_{12}^{02} = \lambda_{13}^{02} = \lambda_{14}^{02} = \lambda_{15}^{02} = 67.48$$

$$\lambda_3 \stackrel{\Delta}{=} \lambda_{1i}^{03} = \lambda_{12}^{03} = \lambda_{13}^{03} = \lambda_{14}^{03} = \lambda_{15}^{03} = 154.4$$

Table 10.8 gives three possible noninferior solutions. For each solution, the table gives the values of the control variables, the present-value cost, the levels of the risk functions, and the trade-offs between the cost function and the risk functions, where λ_{1i}^{0j} is the trade-off between the present-value cost function (f_1^0) and the ith objective function at the jth stage (f_i^j).

Policy A represents no change in emissions from the entire planning horizon. Because there is no reduction in emissions, no additional pollution control costs are incurred. However, the conditional and unconditional expected values of damage become increasingly worse over time. By the third stage, the expected value of resource damage has become 1.257, with the high-range conditional expected value at 1.714 and the low-range expected value at 0.800. The trade-offs between all of the risk functions and costs are zero, so that small improvements in the risk functions can be made at little additional cost. Because the trade-offs are zero, this solution is actually an improper noninferior one [Chankong and Haimes, 1983]. This is true because the value of the cost function is at its lowest possible amount (zero) for this policy and cannot be improved by any other choice of controls. The control variables for this policy are also at their upper bounds and cannot be increased.

Policy B is a policy of gradual cutback. Emissions are cut in half over a 10-year period. The expected damage remains approximately the same over the entire time horizon, while the low-range conditional expected value decreases to 0.545, and the high-range conditional expected value increases to 1.459. These are lower than the corresponding values for Policy A, so Policy B leads to a situation of less risk. The cost associated with Policy B, however, is about 25.8×10^6. The trade-offs are also nonzero and increase with time. This is to be expected, since long-range improvements will cost more than short-range ones. As an example of interpreting the trade-offs, consider those at $k=2$ for Policy B, which are equal to $\$67.48 \times 10^6$ per unit of damage. This means that each of the risk objective functions can be improved (by a small amount) at this marginal rate. Because one unit of damage is equal to the initial amount of resource damage and thus represents a large amount, it may be helpful to express the trade-offs in terms of smaller amounts of damage, especially since the trade-offs are only marginal rates. For example, the trade-off mentioned above could be expressed as $\$674.8 \times 10^3$ per 1/100 unit of damage. Thus, to improve the risk objective functions at $k=2$ by 0.01, the marginal cost at their present levels is approximately $675,000.

Policy C represents an immediate cutback to one-half the present level of emissions. The result is a substantial improvement in expected value equal to 0.936 by the third stage. The low-range and high-range expected values at this stage are 0.479 and 1.393, respectively. Of the three solutions in Table 10.1, Policy C has the lowest risk, but it is also the most expensive, at a cost of $\$50.175 \times 10^6$. The trade-offs are also much larger for this policy, indicating that it becomes increasingly expensive to gain additional improvement in the risk functions. Interestingly, the trade-offs at the third stage of Policy C are the same as those at the third stage of Policy B. Since the risk objective functions at stages $k=1$ and $k=2$ are held constant for trade-offs between the cost function and the risk functions at stage $k=3$, the controls at $k=0$ and $k=1$ must also remain fixed. This means that the trade-offs at the third stage depend only on $u(2)$. Since $u(2)=0.5$ for both Policy B and Policy C, and any changes in $u(2)$ will lead to the same changes in the cost function and the risk functions for $k=3$ for either policy, the trade-offs must be equal.

Note that for all three policies, the differences between the conditional expected values increase with time. This reflects growing uncertainty as the effects of policy decisions are projected further into the future. For this reason, these objective functions taken as a set can be viewed as a characterization of risk, since they not only indicate the expected outcome but also provide some measure of the uncertainty of the outcome. This is precisely the goal of applying the PMRM in the first place.

To demonstrate why impact analysis is so useful in a problem such as this one, suppose the multiobjective problem was solved only one stage at a time. (For the purpose of this discussion, only the unconditional expected values are considered.) The cost associated with pollution control in the first stage alone, denoted by C_1, is given by

$$C_1 = 100(1-u(0))^2 \qquad (10.60)$$

Figure 10.7 shows the set of noninferior solutions when the first-stage cost (C_1) and the expected damage at the first stage $\left(f_5^1\right)$ are the only objectives considered. The points corresponding to Policies A, B, and C are indicated on the curve. Consider next the second stage, with the cumulative cost of pollution control, denoted by C_2, given by

$$C_2 = 100(1-u(0))^2 + 62.1(1-u(1))^2 \qquad (10.61)$$

Depending on which policy was followed in the first stage, three different noninferior solution sets are possible in the second stage, as shown in Figure 10.8. Each curve in this figure is labeled with its associated first-stage policy. In fact, there is a

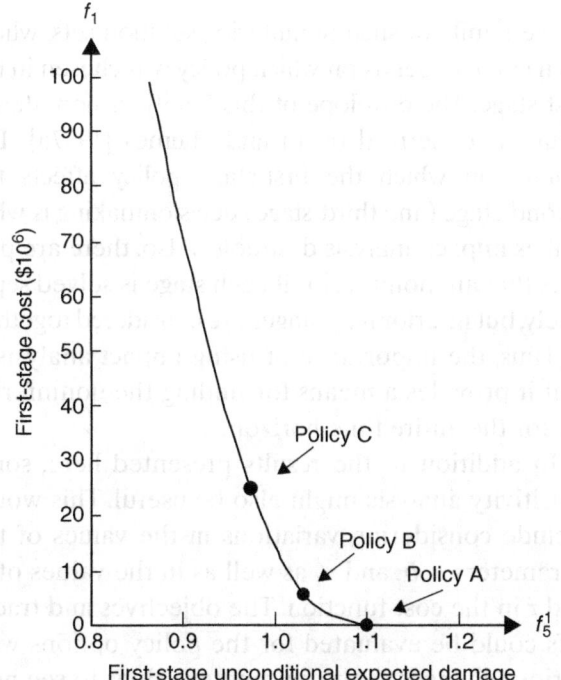

Figure 10.7 *The noninferior solution set when only first-stage objectives are considered.*

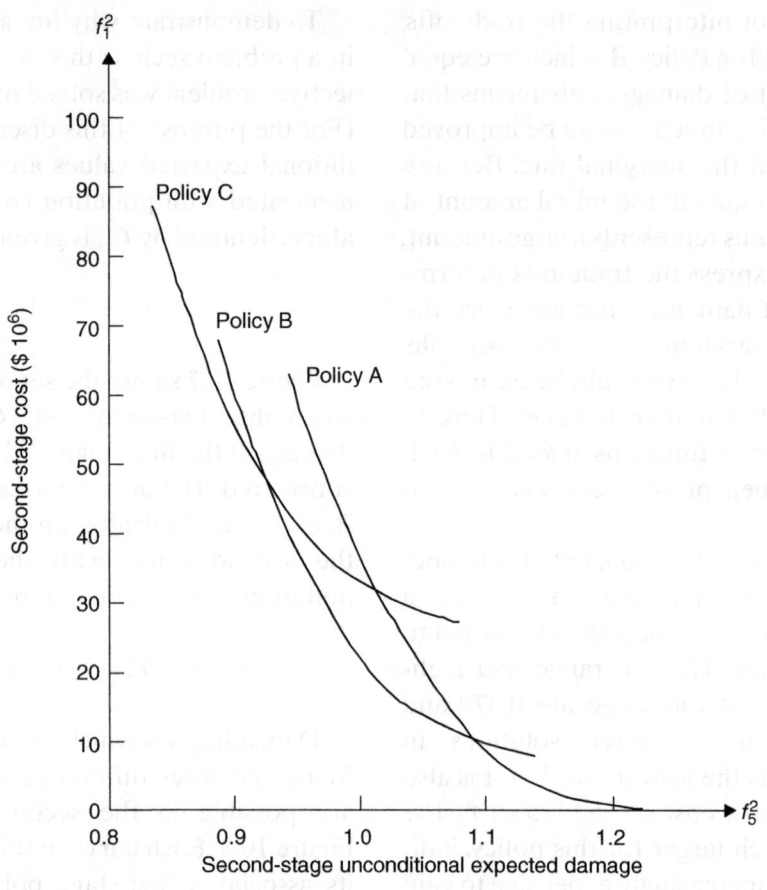

Figure 10.8 Various noninferior solution sets at the second stage.

whole family of such noninferior solution sets, where each curve depends on which policy was chosen in the first stage. The envelope of this family of noninferior solutions is derived by Li and Haimes [1987a]. The manner in which the first-stage policy affects the second-stage (and third-stage) decisionmaking is what makes impact analysis desirable. Also, there are policies that are noninferior if each stage is solved separately, but inferior if all stages are considered together.

Thus, the importance of using impact analysis is that it provides a means for finding the noninferior set for the entire time horizon.

In addition to the results presented here, some sensitivity analysis might also be useful. This would include considering variations in the values of the parameters a, b, and s_d^2 as well as in the values of K and r in the cost function. The objectives and trade-offs could be evaluated for the policy options with various changes in the parameter values to see how sensitive the outcomes are to these changes.

10.6.2 Modified Heroin Addiction Problem

The following example problem is a modified version of a dynamic system presented by Athans et al. [1974]. The original version of the heroin addiction problem assumes that an average heroin addict must steal to support his needs, and in the process, he converts someone from the general population into an addict. An addict who is arrested and convicted would spend 1 year in jail and upon his release would return to the addict population. An addict may undergo methadone treatment to block his heroin craving, but a dropout from the methadone program will return to the addicted population. The Athans et al. [1974] problem had the following five state variables and five state equations:

State variables

$x_1(k)$ = general population at year k
$x_2(k)$ = number of heroin addicts at year k
$x_3(k)$ = number of heroin addicts undergoing methadone treatment in year k

$x_4(k)$ = number of heroin addicts arrested, convicted, and jailed in year k

$x_5(k)$ = number of heroin addicts released from jail in year k

Parameter definitions

b_1 = birthrate of normal population
d_1 = death rate of normal population
a = rate at which a heroin addict converts someone from general population into heroin addiction
d_2 = death rate of heroin addicts (in general higher than d_1)
c = percentage rate of jailed heroin addicts (depends on number of police, tips, judicial, and other factors)
e = percentage rate of heroin addicts attracted to methadone program (depends on advertising budget)
f = rate at which a methadone patient converts a heroin addict to methadone
g = rate of methadone dropouts

The five state equations are as follows:

General population: $x_1(k+1) = x_1(k) + (b_1 - d_1)$
$$x_1(k) - ax_1(k)x_2(k) \quad (10.62)$$

Heroin population:
$$x_2(k+1) = x_2(k) + ax_1(k)x_2(k) - (d_2 + c + e)x_2(k)$$
$$- fx_2(k)x_3(k) + gx_3(k) + x_5(k) \quad (10.63)$$

Methadone population:
$$x_3(k+1) = x_3(k) + fx_3(k)x_2(k) + ex_2(k)$$
$$- d_1 x_3(k) - gx_3(k) \quad (10.64)$$

Jailed population: $x_4(k+1) = cx_2(k) \quad (10.65)$

Released from jail: $x_5(k+1) = x_4(k) \quad (10.66)$

Modified formulation

Assume:
1. $x_1(k+1) = x_1(k)$ general population is constant.
2. f (the rate at which a methadone patient converts a heroin addict to methadone) = 0.
3. Neglect death rate ($d_2 = 0, d_1 = 0$).
4. Rate of methadone dropouts = 0.
5. $x_5(k) = cx_2(k-1)$.

With these assumptions,

$$x_2(k+1) = (1 + ax_1(k) - c - e)x_2(k) + x_5(k) \quad (10.67)$$

$$x_3(k+1) = x_3(k) + ex_2(k) \quad (10.68)$$

The canonical form of one state equation is

$$x(k+1) = ax(k) + u(k) + w(k)$$

where

a = conversion rate of general population = 5×10^{-7}
c = percentage of failed addicts = 20
e = percentage of addicts attached to methadone program = 10
$x_2(0) = 1000$
$x_5(0) = 0$
$x_1(0) = 10^6$
$w(k)$ = a normally distributed random variable with $\mu = 0, s_2^2 = 625$
In other words, $E[w(k)] = 0$, and $[w^2(k)] = 625$.

Consider two objective functions:

$f_4(\cdot)$ = high-range conditional expected value of annual societal cost
$f_5(\cdot)$ = unconditional expected value of annual societal cost

$$\text{Cost} = \gamma_1 x_2(k) + \gamma_2 cx_2(k) + \gamma_3 ex_2(k)$$

γ_1 = cost due to heroin addiction = \$75,000/addict
γ_2 = cost of jail = \$30,000/inmate
γ_3 = cost of methadone = \$10,000/patient

Also,

$E[w(k)] = 0$ for $f_5(\cdot)$
$f_4(\cdot) = m + \beta_4 s_2 = 0 + (1.525)(25) = 38.125$ (based on normal distribution at a partitioning of one sigma)

Consider three possible policy decisions:
A. Linearly increase the percentage of jailed addicts up to 50 over the next 3 years:
$c(1) = 20; c(2) = 30; c(3) = 40; c(4) = 50$
All other variables remain constant.
B. Linearly increase the methadone program up to 40 over the next 3 years:
$e(1) = 10; e(2) = 20; e(3) = 30; e(4) = 40$
All other variables remain constant.
C. Combine methods A and B over the next 3 years. The following results are obtained over 3 years:

Decision A: $x_2(k+1) = (1 + ax_1(k) - c - e)x_2(k) + x_5(k) + w(k)$

Note that $E[x_2(k)] = x_2(k)$ for $E[w(0)] = 0$ (see Eq. 10.27).

Assume $w(k) = 0$, for which we will calculate the unconditional expected value, $f_5(\cdot)$, of the number of heroin addicts at year k:

$f_5(x_2(1)) = (1 + 0.5 - 0.2 - 0.1)1000 + 0 = 1200.0$
$f_5(x_2(2)) = (1 + 0.5 - 0.3 - 0.1)1200 + 0 = 1320.0$
$f_5(x_2(3)) = (1 + 0.5 - 0.4 - 0.1)1320 + 0 = 1320$
$f_5(x_2(4)) = (1 + 0.5 - 0.5 - 0.1)1320 = 1188$

To calculate the conditional expected value of annual societal cost, $f_4(\cdot)$, using Eq. (10.40a) or (8.40), we assume that $w(k) = N(0, 25^2)$; that is, $w(k)$ is a normally distributed random variable with zero mean and a standard deviation of 25. Thus, $\beta_4 s(k) = (1.5247)(25) = 38.125$ (Table 10.9):

$f_4(x_2(1)) = 1200 + 38.125 = 1238.125$
$f_4(x_2(2)) = 1320 + 38.125 = 1358.125$
$f_4(x_2(3)) = 1320 + 38.125 = 1358.125$
$f_4(x_2(4)) = 1188 + 38.126 = 1226.125$

By inspection, it is clear that only policy decision C, which is a combination of Policies A and B, is noninferior. Figure 10.9 depicts the trajectory over four periods of the unconditional and conditional annual societal cost.

TABLE 10.9 Unconditional and Conditional Expected Value of Societal Cost for Four Periods

Decision	Period	$f_5(\cdot)$ ($M/yr) $f_5(x_2(k))$	Cost	$f_4(\cdot)$ ($M/yr) $f_4(x_2(k))$	Cost
A	1	1200	98.4	1238.1	101.5
	2	1320	112.2	1358.1	119.0
	3	1320	116.16	1358.1	126.6
	4	1188	108.1	1226.1	121.2
B	1	1200	98.4	1238.1	101.5
	2	1320	109.6	1358.1	116.2
	3	1320	110.9	1358.1	120.8
	4	1188	101.0	1226.1	113.3
C	1	1200	98.4	1238.1	101.5
	2	1200	103.2	1276.2	109.8
	3	960	86.4	1021.0	91.9
	4	576	54.1	612.6	57.6

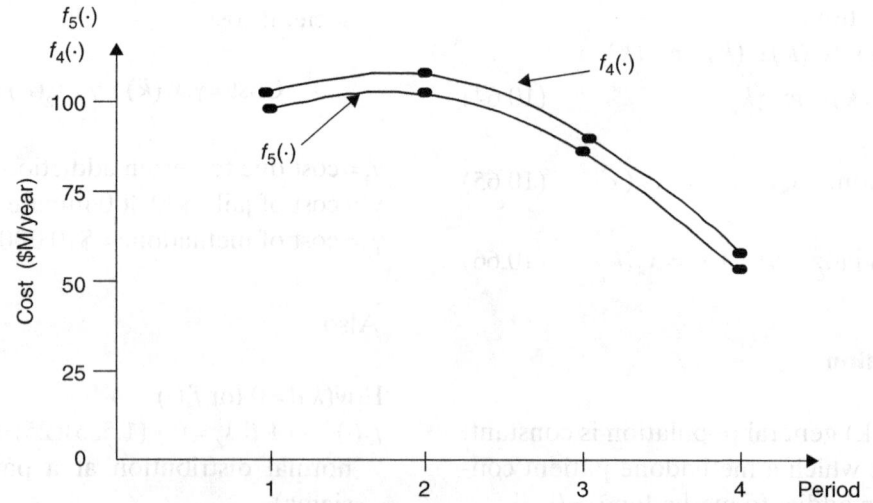

Figure 10.9 Annual societal cost over four periods.

10.6.3 Heroin Addiction Problem (Continued)

10.6.3.1 Introduction
This example further simplifies the heroin addiction model. The DM is the mayor, who can influence the outcome of the state by assigning funding for advertising the methadone program.

10.6.3.2 Modeling
The simplification was performed by redefining the state variable, the control (decision) variable, and random variables. The state of the system is defined as the number of heroin addicts in the city. The control variable is the amount of money spent on advertising to attract addicts to the methadone program. The random variable is a lump-sum noise of conversion, dropouts, and so forth.

The form of the simplified model is

$$x(k+1) = Ax(k) + Bu(k) + w(k),$$

$$y(k) = Cx(k)$$

where

$x(k)$ is the number of heroin addicts at period k
$u(k)$ is the amount of money spent on methadone program advertising at period k
$A\,(=1+a)$ is the growth rate of heroin addicts $= 1.05$
B is the number of addicts per $ spent $=1$ addict/$1000 $=-1/1000$
$C = 1$
$w(k)$ is the lump-sum noise of death, conversion, dropouts, and so forth
$x(0) = 1750$ heroin addicts
$P = E[w^2(k)] = P(k) = 10{,}000$
$f_k^i(\cdot) =$ the conditional expected value of heroin addicts (risk) at period k
$i = 2,3,4$

The simplifications were made under the following assumptions:

1. The general population of the city would stay the same for the modeling period.
2. The most prominent cause of the increase in the number of heroin addicts is due to their converting the nonaddicts to addicts.
3. The effects of jail, deaths, and conversion of addicts into methadone patients are rather significant, and the number of dropouts, as well as other probabilistic sources of changes in output, are lumped to form the white noise part of the above dynamic linear model. This assumption can be justified using the central limit theorem if there are enough variables lumped together to form an approximately normal distribution.
4. In this problem, the control variable $u(k)$ (advertising spending) will result in the conversion of addicts to the methadone program proportional to the amount spent. This will effectively decrease the number of addicts by the value B, where $B = -0.001$. Thus, the state equation for this problem is effectively

$$x(k+1) = Ax(k) - Bu(k) + w(k)$$

10.6.3.3 Implementing MRIAM
The risk functions (number of heroin addicts) for three policies, shown in the table below, are calculated for three stages. All three policies budget the same amount of present-value money during the three periods ($750,000).

	$k=1$	$k=2$	$k=3$
Policy A	250,000	250,000	250,000
Policy B	500,000	150,000	100,000
Policy C	100,000	150,000	500,000

The risk partitions are made at $\mu-\sigma$, μ, and $\mu+\sigma$. The corresponding β_i values are $-1.525, 0$, and 1.525, respectively. The mean and variance of the expected values of the number of heroin addicts at the three periods are calculated using

$$m(k) = CA^k x_0 + \sum_{i=0}^{k-1} CA^i Bu(k-1-i) \quad (10.69)$$

$$\sigma^2(k) = C^2 A^{2k} X_0 + \sum_{i=0}^{k-1} C^2 A^{2i} P \quad (10.70)$$

where x_0 is the initial value of the state variable, the number of heroin addicts, which is equal to 1750 people, and P is the variance of the random variable (equal to 10,000), and $C = 1$ (in Eqs. (10.69) and (10.70)).

The conditional expected values of risk are then calculated for each period k by Eqs. (10.71) to (10.73):

$$f_2(k) = \mu(k) - 1.525\sigma(k) \quad (10.71)$$

$$f_3(k) = f_5(k) = \mu(k) \quad (10.72)$$

$$f_4(k) = \mu(k) + 1.525\sigma(k) \quad (10.73)$$

Note that because β_3 equals 0, the medium-range conditional expected damage, $f_3(\cdot)$, is the same as the unconditional expected value, f_5.

Numerical results

For Policy A: $u(1) = 250{,}000$, $u(2) = 250{,}000$, and $u(3) = 250{,}000$.
   For all policies: $A = 1.05$, $B = -0.001$, $P = 10{,}000$, and $x_0 = 1750$.
The following are the calculations for Policy A:

To synchronize the notation of the mean and standard deviation between Chapters 8 and 10, we define $\mu = m(k)$ and $\sigma = s(k)$.
From Eq. (10.59), we get for $k = 1$
$m(1) = (1.05)^1(1750) - (1.05)^0(0.001)(250{,}000) = 1587.5$
From Eq. (10.60), we get for $k = 1$
$s^2(1) = (1.05)^2(1750) + (1.05)^0(10{,}000) = 11{,}929.375$
Thus, $s(1) = 108.22$

$f_2(1) = m(1) - 1.525 s(1) = 1587.5 - (1.525)(108.22)$
$\quad = 1420.9$
$f_3(1) = m(1) = 1587.5$
$f_4(1) = m(1) + 1.525 s(1) = 1587.5 + (1.525)(108.22)$
$\quad = 1754.1$

Similar calculations can be done for $k = 2$ and $k = 3$.

The tables below show the conditional expected values of the number of heroin addicts for all three periods for Policies A, B, and C, using Eqs. (10.69) and (10.70) to calculate the values of $\mu(k)$ and $\sigma(k)$ for the three periods respectively.

Policy A	$k=1$	$k=2$	$k=3$
f_2	1420.9	1184.8	950.3
f_3	1587.5	1416.9	1237.7
f_4	1754.1	1648.9	1525.2

Policy B	$k=1$	$k=2$	$k=3$
f_2	1170.9	1022.3	929.7
f_3	1337.5	1254.4	1217.1
f_4	1504.1	1486.4	1504.5

Policy C	$k=1$	$k=2$	$k=3$
f_2	1570.9	1442.3	970.7
f_3	1737.5	1674.4	1258.1
f_4	1904.1	1906.4	1545.5

The Pareto-optimal frontiers for Periods 1 and 2 are shown in Figures 10.10 to 10.12.

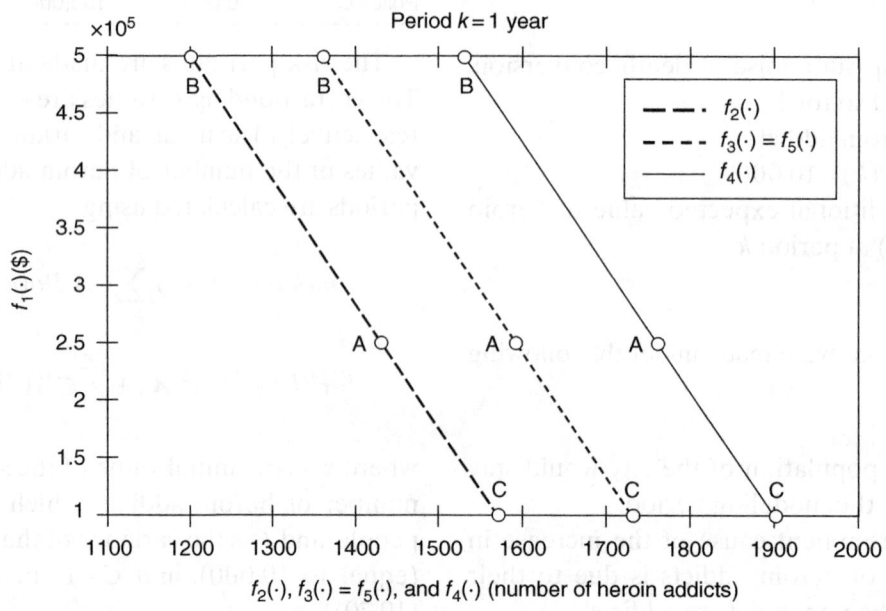

Figure 10.10 Pareto-optimal frontiers for Period 1.

EXAMPLE PROBLEMS 323

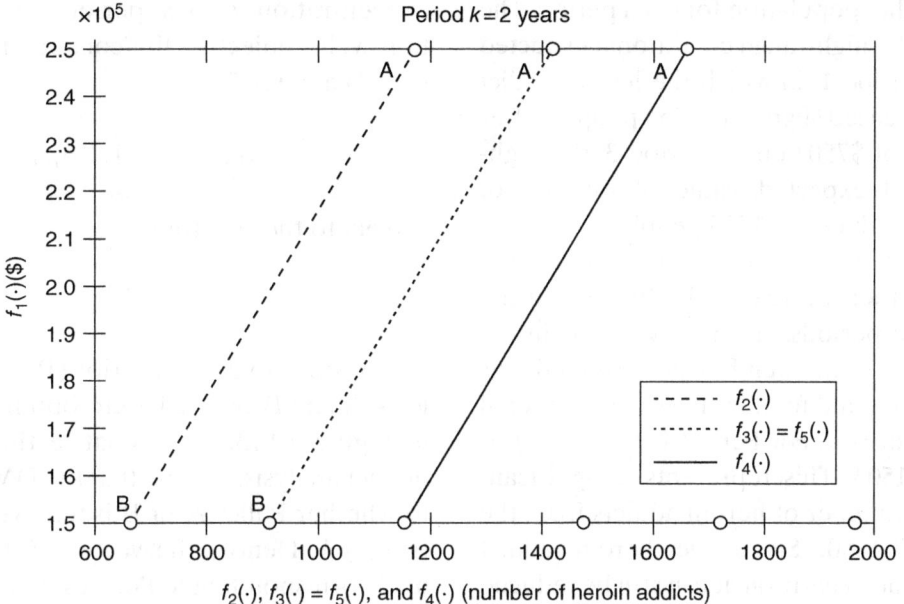

Figure 10.11 Pareto-optimal frontiers for Period 2.

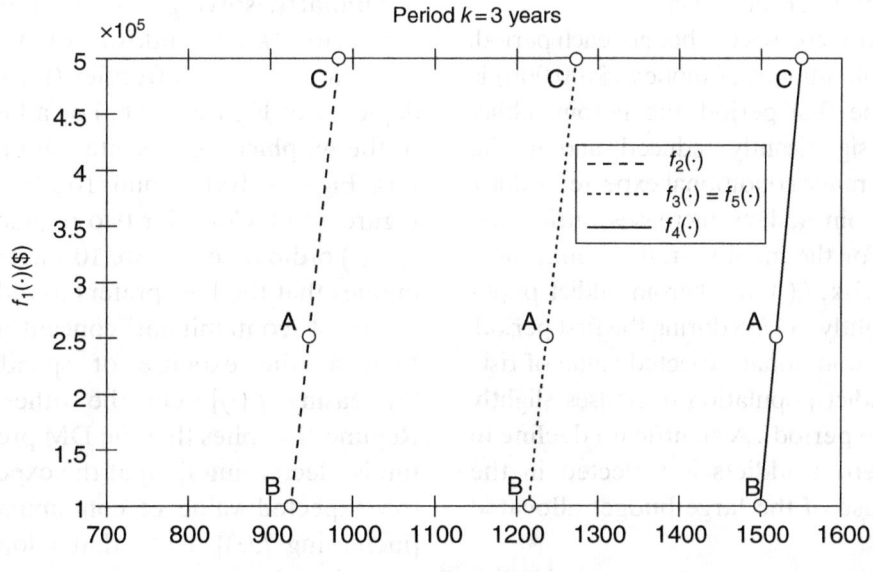

Figure 10.12 Pareto-optimal frontiers for Period 3.

The SWT method (see Chapter 5) can be used to determine the preferred solution from the noninferior solutions for each period. One preferred solution at Period 1 might be to reduce the number of heroin addicts to 1550. At Periods 2 and 3, a preferred solution might be to reduce the number of addicts to 1560.

10.6.3.4 Discussion

Policy A maintains the same budget for heroin addiction prevention and rehabilitation for all three periods. For all three ranges of conditional expected values of risk, the number of heroin addicts is reduced each period from the previous period. This is because the $250,000 spent each year is sufficient to reduce the addict population for each period. The exception is for the high-range conditional expected value ($f_4(\cdot)$) at Period 1, in which the heroin addict population is increased slightly to 1754 people. After spending a total of $750,000 by Period 3, the high-range conditional expected value of number of heroin addicts is reduced to 1525 people.

Policy B spends most of the money during the first period and then decreases the budget during the following two periods. This policy results in the greatest reduction in the number of heroin addicts for all three periods and for all ranges of conditional expected values of risk. For Period 1, f_2 is 1171, f_3 is 1338, and f_4 is 1504. This represents a significant reduction in the number of heroin addicts from the initial value of 1750. Since the heroin addict population is reduced early on, it can still be reduced during the following two periods even though a smaller budget is allocated during that time. During Period 3, the low-range conditional expected value of risk is 930 people, the medium range is 1217, and the high range is 1505 heroin addicts.

Policy C gradually increases the budget each period. Because only a small amount of money ($100,000) is allocated during the first period, the heroin addict population is not significantly reduced; and in the extreme case (high-range conditional expected value), the number of heroin addicts increases during the first two periods. For the medium-range conditional expected value of risk, $f_3(\cdot)$, the heroin addict population decreases slightly to 1738 during the first period. For the low-range conditional expected value of risk, $f_2(\cdot)$, the heroin addict population decreases slightly during the first two periods. A significant decline in the number of heroin addicts is reflected in the third period because of the large budget allocated ($500,000).

It is seen that Policy B is the best policy and Policy C is the worst.

10.6.4 Groundwater Contamination

To understand the essence of the MRIAM, consider a problem of correcting groundwater contamination in which there are two stages with two objective functions that must be evaluated at each stage.

The problem is to minimize the cost of correction, $f_1^k(\cdot)$, and minimize the expected value of contaminant concentration, $f_2^k(\cdot)$, at period k, at a given observation well, subject to the constraints $g_k(\mathbf{u})$ (see Eqs. (10.74) and (10.75)):

$$\min_{\mathbf{u}} f(\mathbf{u}) = [f_1(\mathbf{u}), f_2(\mathbf{u})] \quad (10.74)$$

subject to the constraint

$$g_k(\mathbf{u}) \in U \quad (10.75)$$

Consider two noninferior (Pareto-optimal) policies, A and B, on the Pareto-optimal frontier shown in Figure 10.13. To dramatize the efficacy of the impact analysis, assume that the DM is indifferent as to whether Policy A or B is followed at time period (stage) k. Denote the vector of decision variables $\mathbf{u}(k)$ corresponding to Policies A and B by $\mathbf{u}_A(k)$ and $\mathbf{u}_B(k)$, respectively. Solving systems (10.74) and (10.75) for period $(k+1)$ and for Policy A, $\mathbf{u}_A(k)$, yields a new Pareto-optimal frontier for period $(k+1)$, as depicted in Figure 10.8a.

Similarly, solving systems (10.74) and (10.75) for period $(k+1)$ and for Policy B, $\mathbf{u}_B(k)$, yields a new Pareto-optimal frontier (for period $k+1$), as is depicted in Figure 10.8b. For a better appreciation of the graphical representation of the impact analysis, Figures 10.13a and 10.13b are combined in Figure 10.14. Consider two regimes, I and II, in the $f_1^{k+1}(\cdot)$ ordinate of Figure 10.14. Adopting Regime I implies that the DM prefers to reduce the expected value of contaminant concentration [decreasing $f_2(\cdot)$] at the expense of spending more funds [increasing $f_1(\cdot)$]. On the other hand, adopting Regime II implies that the DM prefers to spend less funds [decreasing $f_1(\cdot)$] at the expense of increasing the expected value of contaminant concentration [increasing $f_2(\cdot)$]. Note that adopting Policy B at stage k yields a set of options that are inferior to Policy A at stage $(k+1)$ in Regime I. Conversely, adopting Policy A at stage k yields a set of options that are inferior to Policy B at stage $(k+1)$ in Regime II. More specifically, assume that the DM is indifferent at stage k to the choice between Policies A and B. If at stage $(k+1)$, however, the DM is more likely to operate in Regime I (i.e., to spend more funds in order to reduce contaminant concentration), then option A at stage k is superior to option B (in terms of its impact on the policy options at stage

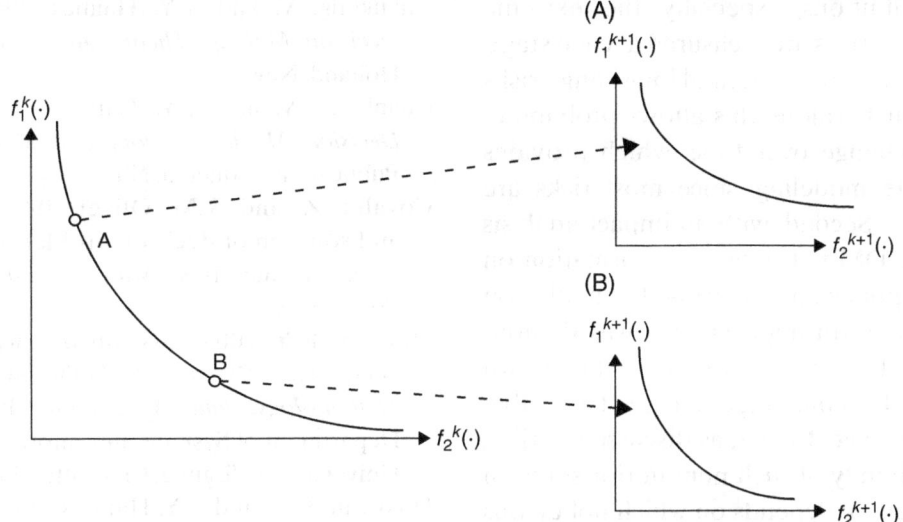

Figure 10.13 Impact of policies at time k on future policies at time k + 1.

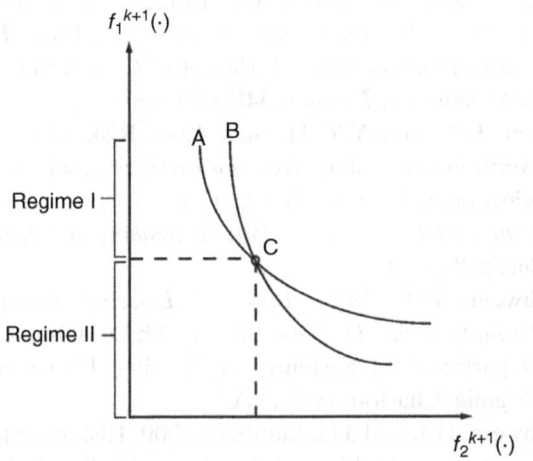

Figure 10.14 Impact analysis with two regimes.

$k+1$). On the other hand, if at stage $(k+1)$ the DM is more likely to operate in Regime II, then obviously, at stage k, Policy B is superior to Policy A. This is because in this regime, the Pareto-optimal frontier at stage $(k+1)$ corresponding to $\mathbf{u}_A(k)$ is inferior to that corresponding to $\mathbf{u}_B(k)$. If he or she is most likely to operate at Point C or in its immediate neighborhood, then further analysis will be required. Refer to Li [1986] and Li and Haimes [1987a, b] and Haimes and Anderegg [2015] for multiple periods using the envelope approach. This is because in this regime, the Pareto-optimal frontier at stage $(k+1)$ corresponding to $\mathbf{u}_A(k)$ is inferior to that corresponding to $\mathbf{u}_B(k)$. If he or she is most likely to operate at Point C or in its immediate neighborhood, then further analysis will be required. Refer to Li [1986]; Li and Haimes [1987a, 1987b] and Haimes and Anderegg [2015] for multiple periods using the envelope approach.

10.7 EPILOGUE

The ultimate purpose of any risk analysis and decisionmaking process should be to answer the fundamental questions posed by Lowrance [1976]: Who should decide on acceptability of what risks, for whom, in what terms, and why? Although the risk and impact analysis methodologies presented here do not specifically address all aspects of this question, the primary motivation for their development has been to provide a more comprehensive analysis of risk. This includes a total decisionmaking process that allows the DM to address the issues raised in Lowrance's question when making value judgments.

The most important aspect of an integrated risk and impact analysis is that time is explicitly built into the modeling and analysis. This has several important implications. First, the representation of risk is more detailed and more comprehensive. In the MRIAM, the probability distributions are represented by conditional expected values rather than by the unconditional expected value alone. This provides the DM with more information about the probability distributions, especially the extreme events. Because the risks are measured at each stage, the short-range, medium-range, and long-range risks are separated. Furthermore, this allows probability distributions to change over time, which provides for more accurate modeling, since most risks are dynamic in nature. Second, with an impact analysis incorporated, the DM not only has information on the risks of each policy option available but also on the potential long-term impacts. This was demonstrated in Figures 10.12 and 10.13, where it is shown that decisions made at one stage can affect available options at a later stage. In fact, as discussed earlier, there is a whole family of such noninferior solution sets, where each curve depends on which policy was chosen in the first stage.

Impact analysis provides solutions that are noninferior over the entire time horizon, whereas solving each stage independently may result in solutions that are inferior. Finally, by using a multiobjective framework, several objectives may be considered simultaneously, and the values of the stage trade-offs can be generated. The stage trade-offs help to quantify the relationships between the objective functions across different stages. The use of trade-offs is an integral part of the SWT method used to solve multiobjective problems.

REFERENCES

Asbeck, E., 1982, *The Partitioned Multiobjective Risk Method*, M.S. thesis, Department of Systems Engineering, Case Western Reserve University, Cleveland, OH.

Asbeck, E.L. and Y.Y. Haimes, 1984, The partitioned multiobjective risk method (PMRM), *Large Scale Systems* **6**: 13–38.

Athans, M., M.L. Dertouzos, R.N. Spann, and S.J. Mason, 1974, *Systems, Networks, and Computation: Multivariable Methods*, McGraw-Hill, New York.

Buffett, S., and B. Spencer, 2003, Efficient Monte Carlo decision tree solution in dynamic purchasing environments, *Proceedings of the International Conference on Electronic Commerce*, Pittsburgh, PA, pp.31–39.

Call, H.J., and W.A. Miller, 1990, A comparison of approaches and implementations for automating decision analysis, *Reliability Engineering and System Safety* **30**(1–3): 115–162.

Chankong, V. and Y.Y. Haimes, 1983, *Multiobjective Decision Making: Theory and Methodology*, North-Holland, New York.

Chankong, V. and Y.Y. Haimes, 2008, *Multiobjective Decision Making: Theory and Methodology*, Dover Publications, Mineola, NY.

Covaliu, Z. and R.M. Oliver, 1995, Representation and solution of decision problems using sequential decision diagrams, *Management Science* **41**(12): 1860–1881.

Dicdican, R.Y., 2004, *Risk-Based Asset Management for Hierarchical Dynamic Multiobjective Systems: Theory, Methodology, and Application*, Ph.D. dissertation, Department of Systems and Information Engineering, University of Virginia, Charlottesville, VA.

Dicdican, R.Y. and Y.Y. Haimes, 2005, Relating multiobjective decision trees to the multiobjective risk impact analysis method, *Systems Engineering* **8**(2):95–108.

Diehl, M. and Y.Y. Haimes, 2004, Influence diagrams with multiple objectives and tradeoff analysis, *IEEE Transactions on Systems, Man, And Cybernetics—Part A: Systems and Humans* **34**(3): 293–304.

Dillon, R.Y. and Y.Y. Haimes, 1996, Risk of extreme events via multiobjective decision trees: Application to telecommunications, *IEEE Transactions on Systems, Man, and Cybernetics—Part A: Systems and Humans* **26**(2): 262–271.

Frohwein, H.I., 1999, *Risk of Extreme Events in Multiobjective Decision Trees*, Ph.D. Dissertation, Department of Systems Engineering, University of Virginia, Charlottesville, VA.

Frohwein, H.I. and J.H. Lambert, 2000, Risk of extreme events in multiobjective decision trees: Part 1. Severe events, *Risk Analysis* **20**(1): 113–123.

Gomide, F., 1983, *Hierarchical Multistage, Multiobjective Impact Analysis*, Ph.D. dissertation, Department of Systems Engineering, Case Western Reserve University, Cleveland, OH.

Gomide, F. and Y.Y. Haimes, 1984, The multiobjective, multistage impact analysis method: theoretical basis, *IEEE Transactions on Systems, Man, and Cybernetics* **14**: 89–98.

Haimes, Y.Y., 1998, *Risk Modeling, Assessment, and Management*, John Wiley and Sons. Inc., New York.

Haimes, Y.Y., 2004, *Risk Modeling, Assessment, and Management*, second edition, John Wiley and Sons. Inc., New York.

Haimes, Y.Y and A. Anderegg, 2015, Sequential Pareto-optimal decisions made during emergent complex systems of systems: an application to the FAA NextGen, *Systems Engineering*, **18** (1): 28–44.

Haimes, Y.Y. and V. Chankong, 1979, Kuhn–Tucker multipliers as trade-offs in multiobjective decision-making analysis, *Automatica* **15**(1): 59–72.

Haimes, Y.Y., D. Li, and V. Tulsiani, 1990, Multi-objective decision-tree analysis, *Risk Analysis* **10**(1): 111–129.

Howard, R.A., and J.E. Matheson, 1984, Influence diagrams, In *The Principles and Applications of Decision Analysis*, Vol. **II**, R.A. Howard and J.E. Matheson (Eds.), Strategic Decisions Group, Menlo Park, CA, pp. 719–762.

Intriligator, M., 1971, *Mathematical Optimization and Economic Theory*, Prentice-Hall, Englewood Cliffs, NJ.

Keeney, R.L. and H. Raiffa, 1976, *Decisions with Multiple Objectives: Preferences and Value Tradeoffs*, John Wiley& Sons. Inc., New York.

Kirkwood, C.W., 1993, An algebraic approach to formulating and solving large models for sequential decisions under uncertainty, *Management Science* **39**(7): 900–913.

Kujawski, E., 2002, Selection of technical risk responses for efficient contingencies, *Systems Engineering* **5**(3): 194–212.

Leach, M.R., 1984, *Risk and Impact Analysis in a Multiobjective Framework*, M.S. thesis, Systems Engineering Department, Case Western Reserve University, Cleveland, OH.

Leach, M.R. and Y.Y. Haimes, 1987, Multiobjective risk-impact analysis method, *Risk Analysis* **7**(2): 225–241.

Li, D. 1986, *Optimization of Large-Scale Hierarchical Multiobjective Systems: The Envelope Approach*, Ph.D. dissertation, Systems Engineering Department, Case Western Reserve University, Cleveland, OH.

Li, D. and Y. Haimes, 1987a, A hierarchical generating method for large scale multiobjective systems, *Journal of Optimization, Theory and Applications* **54**(2): 303–333.

Li, D. and Y. Haimes, 1987b, The envelope approach for multiobjective optimization problems, *IEEE Transactions on Systems, Man, and Cybernetics* **17**(6): 1026–1038.

Lootsma, F.A., 1993, Scale sensitivity in a multiplicative variant of the AHP and SMART, *Journal of MultiCriteria Decision Analysis* **2**: 87–110.

Lootsma, F.A., 1997, Multicriteria decision analysis in a decision tree, *European Journal of Operational Research* **101**: 442–451.

Lowrance, W., 1976, *Of Acceptable Risk*, William Kaufmann, Los Altos, CA.

Magee, J.F., 1964a, Decision trees for decision making, *Harvard Business Review*, July–August: 126–138.

Magee, J.F., 1964b, How to use decision trees in capital investment, *Harvard Business Review*, September–October: 79–96.

Markowitz, H. 1952, Portfolio selection, *Journal of Finance* **7**: 77–91.

Millet, I. and W.C. Wedley, 2002, Modeling risk and uncertainty with the analytic hierarchy process, *Journal of Multi-Criteria Decision Analysis* **11**: 97–107.

von Neumann, J. and O. Morgenstern, 1980, *Theory of Games and Economic Behavior*, fourth edition, Princeton University Press, Princeton, NJ.

Porter, A. and F. Rossini, 1980, Technology assessment/ environmental impact assessment: Toward integrated impact assessment, *IEEE Transactions on Systems, Man, and Cybernetics* **10**(8): 417–424.

Raiffa, H., 1968, *Decision Analysis: Introductory Lectures on Choice under Uncertainty*, Addison-Wesley, Reading, MA.

Saaty, T.L., 1980, *The Analytic Hierarchy Process*, McGraw-Hill, New York.

Shachter, R.D., 1986, Evaluating influence diagrams, *Operations Research* **34**(6): 871–882.

Shachter, R.D., 1990, An ordered examination of influence diagrams, *Networks* **20**: 535–563.

Shenoy, P.P., 1993, Valuation network representation and solution of asymmetric decision problems, *European Journal of Operational Research* **121**(3): 146–174.

Shenoy, P.P., 1994, A comparison of graphical techniques for decision analysis, *European Journal of Operational Research* **78**(1): 1–21.

Shenoy, P.P., 2000, Valuation network representation and solution of asymmetric decision problems, *European Journal of Operational Research* **121**: 579–608.

Smith, J.Q., 1989, Influence diagrams for Bayesian decision analysis, *European Journal of Operational Research* **40**: 363–376.

von Winterfeldt, D. and W. Edwards, 1986, *Decision Analysis and Behavioral Research* Cambridge University Press, Cambridge, UK.

11

Statistics of Extremes: Extension of the PMRM

11.1 A REVIEW OF THE PARTITIONED MULTIOBJECTIVE RISK METHOD

To streamline the discussion on the statistics of extremes and its role in the extension of the partitioned multiobjective risk method (PMRM), a brief review of the method is presented here (see Chapter 8 and Asbeck and Haimes [1984]). The PMRM is a risk analysis method developed for solving multiobjective problems of a probabilistic nature. Instead of using the rational mathematical expectation, the PMRM generates a number of conditional risk functions (or damage functions) that represent the loss, given that the damage falls within specific probability ranges (or damage ranges). Combining any one of the generated conditional expected risk functions with the other objective functions creates a new multiobjective optimization problem. This new optimization problem contains more information about the problem's probabilistic behavior and is therefore superior to the initial one.

Let X be a continuous random variable that represents the amount of damage or loss. To use the PMRM, the marginal probability density function (pdf) $p_X(x; s_j)$ must be known. The pdf can relate the probability of loss to the magnitude of loss, for any policy s_j, $j=1, \ldots, q$. Furthermore, the pdf is assumed to satisfy the following properties: $p_X(x; s_j)$ is nonnegative and piecewise continuous, and $\int_{-\infty}^{\infty} p_X(x; s_j)\, dx = 1$ and $\Pr(a < X \leq b) = \int_a^b p_X(x; s_j)\, dx$. The cumulative distribution function (cdf) is

$$P_X(x; s_j) = \int_{-\infty}^{x} p_X(y; s_j)\, dy, \quad j = 1, \ldots, q \quad (11.1)$$

The assumption about $p_X(x; s_j)$ guarantees the existence of a unique inverse $P_X^{-1}(x; s_j)$ for all nonempty intervals: $x \in [x_1, x_2]$.

The PMRM partitions the probability axis into a set of n ranges, where the selection of n is dependent on the characteristics of the decisionmaking problem and process. However, there are typically three such ranges, and in this chapter, we will focus on the extreme range. Denote the partitioning points on the probability axis α_i. For each α_i and each policy s_j, there exists a unique damage (loss) β_{ij} (see Figure 11.1) such that

---

Risk Modeling, Assessment, and Management, Fourth Edition. Yacov Y. Haimes.
© 2016 John Wiley & Sons, Inc. Published 2016 by John Wiley & Sons, Inc.

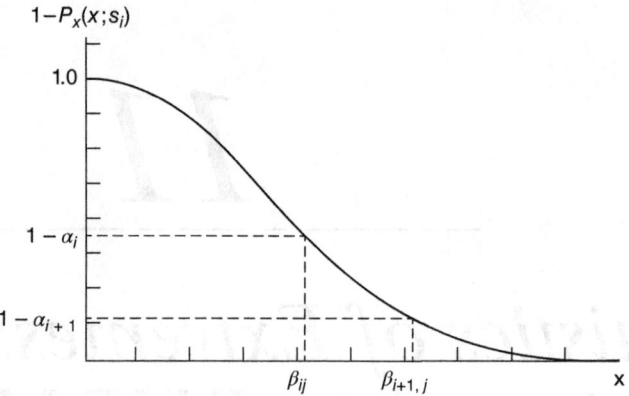

Figure 11.1 Mapping of the partition of the probability axis onto the damage axis.

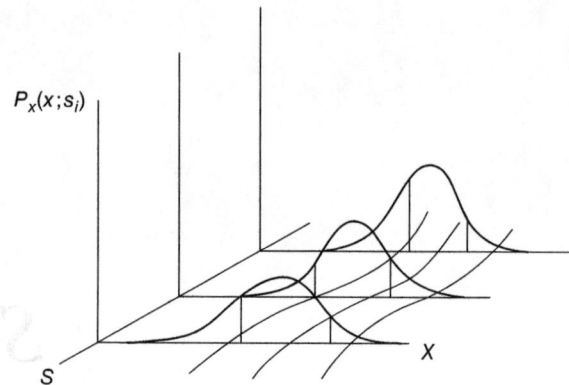

Figure 11.2 Generation of conditional expected risk functions.

$$P_X(\beta_{ij};s_j) = \alpha_i \quad (11.2)$$

Since the inverse $P_X^{-1}(x;s_j)$ exists, we have

$$\beta_{ij} = P_X^{-1}(\alpha_i;s_j) \quad (11.3)$$

This β_{ij} is used in the definition of the conditional expectations:

$$f_i(s_j) = E[X \mid X \in (\beta_{ij}, \beta_{i+1,j})] \quad (11.4)$$

or

$$f_i(s_j) = \frac{\int_{\beta_{ij}}^{\beta_{i+1,j}} x p_X(x;s_j)\, dx}{\int_{\beta_{ij}}^{\beta_{i+1,j}} p_X(x;s_j)\, dx} = \frac{\int_{\beta_{ij}}^{\beta_{i+1,j}} x p_X(x;s_j)\, dx}{\alpha_{i+1} - \alpha_i} \quad (11.5)$$

$i = 2,3,4; j = 1,\ldots,q$

Let θ_i denote the denominator above:

$$\theta_i = P_X(\beta_{i+1,j};s_j) - P_X(\beta_{ij};s_j) = \alpha_{i+1} - \alpha_i \quad (11.6)$$

The θ_i are simply the probabilities that the random variable X falls within range i. Note that the θ_i are not dependent on the policies s_j. If a range on the probability axis, say, $R_i = \{\alpha \mid \alpha \in [\alpha_i, \alpha_{i+1}]\}$, is fixed but the policies s_j are varied, Eq. (11.5) will give a set of conditional expected risks for the range R_i.

There are two approaches that can be taken in order to formulate the conditional expected risk functions [Leach and Haimes, 1987]. The first method is an analytic approach and the second uses simulation and regression. The formulation of the risk functions is depicted in Figure 11.2.

The analytic approach can be used when the probability distribution of damage is a tractable mathematical function in which s appears explicitly. If Eq. (11.5) is solvable, an expression for $f_i(s)$, in terms of s, can be found. These functions (one for each range) will form the different conditional expected risk functions.

In general, however, it is not possible to find an analytic relationship between $f_i(s_j)$ and s_j. Solving Eq. (11.5) numerically for each policy $s_j, j = 1, \ldots, q$ will yield a set of q points $\{f_i(s_j): j = 1, 2, \ldots, q\}$. From this set, an explicit functional relationship between each $f_i(s_j)$ and s_j can be developed by using tabulation, regression, or some other curve-fitting technique. As before, the functions that have been generated are now used as conditional risk functions.

In many applications, one is given a database that represents random values of a probabilistic process. Applying the PMRM requires fitting a distribution function to these observations, using, for example, the method of moments or the method of maximum likelihood. Clearly, in general, the database may not be describable by any particular distribution function, because of the inherent complexity of the randomness of the process. A number of statistical tests have been developed for determining the most appropriate distribution function for a particular database. Among these are the chi-square and the Kolmogorov–Smirnov tests. However, determining the type of distribution that should be chosen is a difficult task and is beyond the scope of this book.

Any of these three generated conditional expected risk functions together with the original cost function $f_1(s)$ constitute the new multiobjective optimization problem [Haimes and Hall, 1974; Chankong and Haimes, 1983]. There are three conditional risk functions of the decision variable s corresponding to the three ranges: $f_2(s)$ represents the high-probability, low-damage conditional expectation; $f_3(s)$ represents the intermediate-probability, intermediate-damage conditional expectation; $f_4(s)$ represents the low-probability, high-damage conditional expectation; and $f_5(s)$ represents the common (unconditional) expectation. Note that the notation $f_1(s)$ is reserved for the cost function and that the term *high-probability, low-damage* for $f_2(s)$ originates from the fact that events belonging to this range have a high probability of being exceeded and causing low damage.

In this chapter, we will mainly consider three different distribution functions: the normal (N), the log-normal (LN), and the Weibull (W). The superscript k ($k = $ N, LN, W) is introduced to facilitate the generalization of the result.

$$p_X^N(x) = \frac{1}{\sqrt{2\pi}\sigma} \exp\left[-\frac{1}{2}\left(\frac{x-\mu}{\sigma}\right)^2\right],$$
$$x \in R, \mu \in R, \sigma > 0 \quad (11.7)$$

$$p_X^{LN}(x) = \frac{1}{\sqrt{2\pi}\tau x} \exp\left[-\frac{1}{2}\left(\frac{\ln x - \eta}{\tau}\right)^2\right],$$
$$x > 0, \eta \in R, \tau > 0 \quad (11.8)$$

$$p_X^W(x) = \frac{c}{a}\left(\frac{x}{a}\right)^{c-1} \exp\left[-\left(\frac{x}{a}\right)^c\right],$$
$$x \geq 0, c > 0, a > 0 \quad (11.9)$$

These three distributions do not represent the majority of probabilistic processes, nor were they chosen with that intention. Rather, they can collectively characterize most important aspects of distributional tail behavior and consequently will make our intuitive derivation more general. To focus on the underlying methodologies, we will include only the deduction of normal distribution in the chapter and give results of the log-normal and Weibull directly. Interested readers may refer to Karlsson and Haimes [1988a, b] for details.

For use in later sections, we will further develop Eq. (11.5). Unlike the exponential distribution, it is very hard to find an explicit equation relating $f_i(\cdot)$ to α_i for these three distributions. For the normal and log-normal, however, it is only possible to find near-closed-form expressions. The following expressions, known as incomplete first moments, can be derived (see, e.g., Raiffa and Schlaifer [1961]). For the normal and partitioning the probability axis between α_i and α_{i+1}

$$f_i^N(s_j) = f_i^N(\cdot) = \frac{\int_{\beta_i}^{\beta_{i+1}} x p_X(x)\, dx}{\int_{\beta_i}^{\beta_{i+1}} p_X(x)\, dx} = \frac{\int_{\beta_i}^{\beta_{i+1}} x \frac{1}{\sqrt{2\pi}\sigma} e^{-\frac{(x-\mu)^2}{2\sigma^2}}\, dx}{\alpha_{i+1} - \alpha_i}$$

Let $y = \dfrac{x - \mu}{\sigma}$

$$f_i^N(\cdot) = \frac{\int_{\frac{\beta_i - \mu}{\sigma}}^{\frac{\beta_{i+1}-\mu}{\sigma}} (\mu + \sigma y)\frac{1}{\sqrt{2\pi}\sigma} e^{-\frac{y^2}{2}}\sigma\, dy}{\alpha_{i+1} - \alpha_i}$$

$$= \mu + \frac{\sigma}{\alpha_{i+1} - \alpha_i}\left(-\frac{1}{\sqrt{2\pi}}\left(-e^{-\frac{y^2}{2}}\right)\Big|_{\frac{\beta_i-\mu}{\sigma}}^{\frac{\beta_{i+1}-\mu}{\sigma}}\right)$$

$$= \mu + \frac{\sigma}{\sqrt{2\pi}(\alpha_{i+1} - \alpha_i)}\left(e^{-\frac{(\beta_i - \mu)^2}{2\sigma^2}} - e^{-\frac{(\beta_{i+1}-\mu)^2}{2\sigma^2}}\right)$$

Let $\Phi\left(\dfrac{\beta_i - \mu}{\sigma}\right) = \alpha_i$; thus, $\left(\dfrac{\beta_i - \mu}{\sigma}\right) = \Phi^{-1}(\alpha_i)$

$$f_i^N(\cdot) = \mu + \frac{\sigma}{\sqrt{2\pi}(\alpha_{i+1} - \alpha_i)}$$
$$\left\{\exp\left[-\frac{1}{2}(\Phi^{-1}(\alpha_i))^2\right] - \exp\left[-\frac{1}{2}(\Phi^{-1}(\alpha_{i+1}))^2\right]\right\}$$
$$(11.10)$$

in which

$$\Phi(x) = \frac{1}{\sqrt{2\pi}} \int_{-\infty}^{x} e^{-\frac{1}{2y^2}} dy \qquad (11.11)$$

is the cdf of the standard normal variate. The unsolvable integrals have hereby been transformed to a standard table search procedure. For log-normal,

$$f_i^{LN}(\cdot) = \frac{e^{(\eta+\tau^2/2)}}{\alpha_{i+1}-\alpha_i}[\Phi(\Phi^{-1}(\alpha_{i+1})-\tau) - \Phi(\Phi^{-1}(\alpha_i)-\tau)] \qquad (11.12)$$

and for Weibull,

$$f_i^{W}(\cdot) = \frac{a}{\alpha_{i+1}-\alpha_i} \int_{\ln 1/(1-\alpha_i)}^{\ln 1/(1-\alpha_{i+1})} t^{1/c} e^{-t} dt \qquad (11.13)$$

Obviously, the expressions above should yield their unconditional expected risk if one lets $\alpha_i = 0$ and $\alpha_{i+1} = 1$; this is also easy to verify.

Karlsson and Haimes [1988a, b] have shown that the conditional expected risk functions $f_i^k(\cdot)$ can be written in the form $f_i^k(\cdot) = \mu g_i^k(\sigma/\mu)$, where k represents different distributions and μ and σ are the mean and standard deviations, respectively, of the initial distribution. For example, Eq. (11.10) can be rewritten as

$$f_i^N(\cdot) = \mu \cdot g_i^N(\sigma/\mu)$$

where

$$g_i^N(\cdot) = 1 + \frac{\sigma/\mu}{\sqrt{2\pi}(\alpha_{i+1}-\alpha_i)} \left\{ \exp\left[-\frac{1}{2}(\Phi^{-1}(\alpha_i))^2\right] - \exp\left[-\frac{1}{2}(\Phi^{-1}(\alpha_{i+1}))^2\right] \right\} \qquad (11.14)$$

The ratio σ/μ is known as the coefficient of variation. If σ/μ is kept constant, the mean value μ will assume the role of a scaling factor. This result is of significant practical importance and will be thoroughly examined in subsequent sections.

Let α denote the lower partitioning point for this low probability of exceedance range, that is, let $\alpha_i = \alpha$ and $\alpha_{i+1} = 1$; then the $f_i^k(\cdot)$ becomes $f_4^k(\cdot)$, $k = N, LN, W$, respectively. The resulting expressions are rather complex, and it is difficult to appreciate their sensitivity to the partitioning point α. However, it is possible to find asymptotic expressions for them as α approaches 1. The low-probability conditional risk function for normal distribution is derived from Eq. (11.10) as

$$f_4^N(\cdot) = \mu + \frac{\sigma}{\sqrt{2\pi}(1-\alpha_i)} \left\{ \exp\left[-\frac{1}{2}(\Phi^{-1}(\alpha_i))^2\right] - 0 \right\}$$

$$= \mu + \frac{\sigma}{\sqrt{2\pi}(1-\alpha)} \left\{ \exp\left[-\frac{1}{2}(\Phi^{-1}(\alpha))^2\right] \right\} \qquad (11.15)$$

The function $\Phi^{-1}(\alpha)$ cannot be found explicitly, but it may be approximated. According to Cramer [1946], $\Phi^{-1}(\alpha)$ may be approximated with

$$\Phi^{-1}(\alpha) = \sqrt{2\ln\frac{1}{1-\alpha}} - \frac{\ln\left(4\pi\ln\frac{1}{1-\alpha}\right)}{2\sqrt{2\ln\frac{1}{1-\alpha}}} + O\left(\frac{1}{\sqrt{\ln\frac{1}{1-\alpha}}}\right) \qquad (11.16)$$

for α sufficiently close to 1. Note that the residue $O(\cdot)$ converges extremely slowly, so that α has to be very close to 1. Thus,

$$f_4^N(\cdot) \cong \mu + \frac{\sigma}{\sqrt{2\pi}(1-\alpha)}$$

$$\left\{ \exp\left[-\frac{1}{2}\left(\sqrt{2\ln\frac{1}{1-\alpha}} - \frac{\ln\left(4\pi\ln\frac{1}{1-\alpha}\right)}{2\sqrt{2\ln\frac{1}{1-\alpha}}}\right)^2\right] \right\}$$

$$= \mu + \frac{\sigma}{\sqrt{2\pi}(1-\alpha)}$$

$$\left\{ \exp\left[\ln\frac{\sqrt{4\pi\ln\frac{1}{1-\alpha}}}{\frac{1}{1-\alpha}} - \frac{\left(\ln\left(4\pi\ln\frac{1}{1-\alpha}\right)\right)^2}{16\ln\frac{1}{1-\alpha}}\right] \right\}$$

$$= \mu + \sigma\sqrt{2\ln\frac{1}{1-\alpha}} \exp\left[-\frac{\left(\ln\left(4\pi\ln\frac{1}{1-\alpha}\right)\right)^2}{16\ln\frac{1}{1-\alpha}}\right]$$

Expanding the exponential term into MacLaurin series yields

$$f_4^N(\cdot) \cong \mu + \sigma\sqrt{2\ln\frac{1}{1-\alpha}}\left(1 - \frac{\left(\ln\left(4\pi\ln\frac{1}{1-\alpha}\right)\right)^2}{16\ln\frac{1}{1-\alpha}} + \cdots\right)$$

Since the second- and higher-order terms of the MacLaurin expansion approach zero, the conditional expectation $f_4^N(\cdot)$ can be written asymptotically as

$$f_4^N(\cdot) \cong \mu + \sigma\sqrt{2\ln\frac{1}{1-\alpha}} \qquad (11.17)$$

Equation (11.17) indicates that $f_4^N(\cdot)$ is very sensitive to variations in α. For α close to 1, the slightest change in α is magnified by taking the reciprocal $1-\alpha$. For log-normal and Weibull, the low-probability risk functions are approximated as [Karlsson, 1986]

$$f_4^{LN}(\cdot) \cong \exp\left[\eta + \tau\left(\sqrt{2\ln\frac{1}{1-\alpha}} - \frac{\ln\left(4\pi\ln\frac{1}{1-\alpha}\right)}{2\sqrt{2\ln\frac{1}{1-\alpha}}}\right)\right]$$

(11.18)

$$f_4^W(\cdot) \cong a\left[\ln\frac{1}{1-\alpha}\right]^{1/c} \qquad (11.19)$$

11.2 STATISTICS OF EXTREMES

An important class of probability problems is those involving the extreme values of random variables. Gumbel [1954, 1958] has made a comprehensive study of how the largest and smallest values from an independent sample of size n are distributed. The statistics of extremes is the study of the largest or smallest values of random variables and is specifically concerned with the maximum or minimum values from sets of independent observations. Galambos [1978] and Castillo [1988] show that for most distributions, as the number of observations approaches infinity, the distributions of these extreme values approach one of three asymptotic forms. These asymptotic forms are influenced by the characteristics of the initial distribution's tail and are independent of the central portion of the initial distribution [Ang and Tang, 1984]. Given an initial random variable X with known initial distribution function $P_X(x)$, the largest and smallest values of a sample of size n taken from the sample space of X will also be random variables. Each sample observation is denoted by $(x_1, x_2, ..., x_n)$. Since every observation is independent of the others, it may be assumed that each observation is a realization of a random variable and the set of observations $(x_1, x_2, ..., x_n)$ represents a realization of sample random variables $(X_1, X_2, ..., X_n)$. The X_i, $i=1, ..., n$, are assumed to be statistically independent and identically distributed, with cdf $P_X(x)$. The maximum and minimum of the sample set $(X_1, X_2, ..., X_n)$ are denoted by Y_n and Y_1, where

$$Y_n = \max(X_1, X_2, ..., X_n) \qquad (11.20a)$$

$$Y_1 = \min(X_1, X_2, ..., X_n) \qquad (11.20b)$$

Note that Y_n and Y_1 are random variables. From now on, only the largest value will be discussed. All the properties for the largest values have their analogous results for the smallest value. (See Ang and Tang [1984] for a more detailed description of the statistics of extremes). Y_n, the largest value among $(X_1, X_2, ..., X_n)$, is less than some y, if and only if all other sample random variables are less than y:

$$P_{Y_n}(y) = \Pr(Y_n \leq y) = \Pr(X_1 \leq y, X_2 \leq y, ..., X_n \leq y)$$
$$= [P_X(y)]^n$$

(11.21)

The corresponding density function for Y_n therefore is

$$p_{Y_n}(y) = \frac{dP_{Y_n}(y)}{dy} = n[P_X(y)]^{n-1} p_X(y) \qquad (11.22)$$

In sum, for a given y, the probability $[P_X(y)]^n$ decreases as n increases; that is, the functions $p_{Y_n}(y)$ and $P_{Y_n}(y)$ will shift to the right with increasing values of n.

Example Problem 11.1 Given the initial variate X with the following probability density function (pdf),

$$p_X(x) = \begin{cases} \dfrac{1}{3}x, & 0 \leq x \leq 2 \\ 2 - \dfrac{2}{3}x, & 2 \leq x \leq 3 \end{cases}$$

derive the largest value from sample of size n.
Solution: Given the above initial variate and pdf, it is necessary to find the cdf, $P_X(x)$. Since the pdf is divided into different parts, the cdf will be divided into the same parts. The general formula for the cdf is defined as follows:

$$P_X(x) = \int_0^x p(y)\, dy$$

Thus, the cdf of the initial variate is

$$P_X(x) = \begin{cases} \dfrac{1}{6}x^2, & 0 \leq x \leq 2 \\ 1 - \dfrac{(3-x)^2}{3}, & 2 \leq x \leq 3 \end{cases}$$

The largest value Y_n is defined as $\max Y_n = \max(X_1, X_2, \ldots, X_n)$
cdf, $P_{Y_n}(y) = [P_X(y)]^n$, and thus,

$$P_{Y_n}(y) = \begin{cases} \left(\dfrac{1}{6}y^2\right)^n, & 0 \leq y \leq 2 \\ \left[1 - \dfrac{(3-y)^2}{3}\right]^n, & 2 \leq y \leq 3 \end{cases}$$

pdf, $p_{Y_n}(y) = \dfrac{dP_{Y_n}(y)}{dy} = n[P_X(y)]^{n-1} P_X(y)$, so

$$p_{Y_n}(y) = \begin{cases} n\left(\dfrac{y^2}{6}\right)^{n-1}\left(\dfrac{1}{3}y\right), & 0 \leq y \leq 2 \\ n\left[1 - \dfrac{(3-y)^2}{3}\right]^{n-1}\left(2 - \dfrac{2}{3}y\right), & 2 \leq y \leq 3 \end{cases}$$

Example Problem 11.2 Given the initial variate X represented by the standard normal pdf

$$p_X(x) = \dfrac{1}{\sqrt{2\pi}} e^{-x^2/2}$$

derive the largest value from samples of size n.
Solution: The cdf of the initial variate is

$$P_X(x) = \dfrac{1}{\sqrt{2\pi}} \int_{-\infty}^{x} e^{-z^2/2}\, dz = \Phi(x)$$

Thus, the cdf of the largest value from samples of size n is

$$P_{Y_n}(y) = \left[\dfrac{1}{\sqrt{2\pi}} \int_{-\infty}^{y} e^{-1/2 z^2}\, dz\right]^n = [\Phi(y)]^n$$

The pdf is

$$p_{Y_n}(y) = \dfrac{n}{\sqrt{2\pi}} [\Phi(y)]^{n-1} e^{-y^2/2}$$

This is the exact expression for the distribution function of the largest value from a sample of size n taken from a population X.

Unfortunately, Eqs. (11.21) and (11.22) are difficult to use in practice and are primarily of theoretical interest. In addition, it seems there exists an asymptotic form of the largest sample distribution when $n \to \infty$.

The asymptotic theory of statistical extremes was developed early in the twentieth century by a number of statisticians, including Fischer and Tippett [1928] and Gnedenko [1943]. This is the analytic theory concerned with the limiting forms of $P_{Y_n}(y)$ and $p_{Y_n}(y)$, which may converge (in distribution) to a particular form as $n \to \infty$. There are typically three such forms, and Gumbel [1958] defined them as types I, II, and III. It can be shown that the behavior of the initial variate's tail determines which type (I, II, and III) of extremal distribution it converges to. If the tail of the initial variate decays exponentially, then the largest value is of type I (also known as Gumbel distribution). Furthermore, if the tail decays polynomially, the extremal distribution is of type II (also known as Frechet distribution). The third type (also known as Weibull distribution) is only for initial variates that have a finite upper bound. These three forms are not exhaustive; however, the most common distributions do converge to either type I, type II, or type III [Lambert et al., 1994].

The cdf of the type I asymptotic form (double exponential) is

$$P_{Y_n}(y) = \exp\left[-e^{\delta_n(y-u_n)}\right] \quad (11.23)$$

where u_n is the characteristic largest value of the initial variate (location parameter) and δ_n is an inverse measure of dispersion (scale parameter). The characteristic largest value, u_n, is a convenient measure of the central location of the possible largest values. The characteristic largest value is defined as the particular value of X such that in a sample of size n from the initial population $(X_1, X_2, \ldots, X_n)$, the expected number of sample values larger than u_n is one [Ang and Tang, 1984]. That is,

$$n[1 - P_X(u_n)] = 1 \quad (11.24a)$$

or

$$P_X(u_n) = 1 - \frac{1}{n} \quad (11.24b)$$

$$u_n = P_X^{-1}\left(1 - \frac{1}{n}\right) \quad (11.24c)$$

In other words, u_n is the value of X with an exceedance probability $[1 - P_X(u_n)]$ of $1/n$ (see Figure 11.3). If we make n observations of a given random variable X, what value of X can we expect to exceed only once? The answer is u_n.

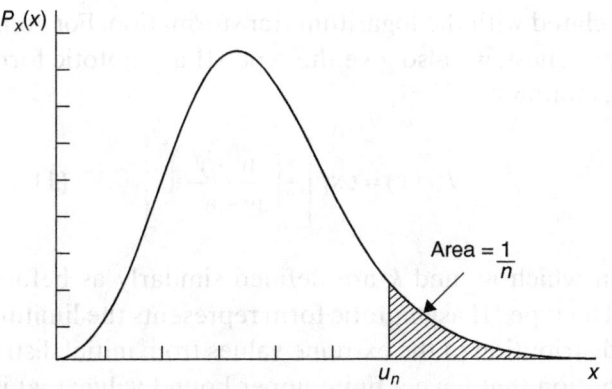

Figure 11.3 Definition of the characteristic largest value u_n.

The characteristic largest value of the initial variate X of the type I asymptotic form, u_n, represents a measure of the most probable largest value of a sample of size n. We now substitute Eq. (11.24b) into Eq. (11.21):

$$P_{Y_n}(u_n) = \left(1 - \frac{1}{n}\right)^n$$

Therefore,

$$\lim_{n\to\infty} P_{Y_n}(u_n) = \lim_{n\to\infty}\left(1 - \frac{1}{n}\right)^n = e^{-1} \quad (11.25)$$

Thus, for large n,

$$P_{Y_n}(u_n) = e^{-1} = 0.368$$

and

$$\Pr(Y_n > u_n) = 0.632$$

This means that among a population of possible largest values from very large samples of size n, about 36.8% are less than or equal to u_n. The characteristic largest value is also the modal value—that is, the most probable value of Y_n [Ang and Tang, 1984; Lambert et al., 1994].

The parameter δ_n is equivalent to the hazard function at the characteristic extreme value [Gumbel, 1958]. That is,

$$\delta_n = \frac{p_X(u_n)}{1 - P_X(u_n)} \quad (11.26a)$$

Equation (11.26a) can be further developed as follows:

$$P_X(u_n) = 1 - \frac{1}{n}$$

Thus,

$$1 - P_X(u_n) = \frac{1}{n}$$

Substitute Eq. (11.24b) into Eq. (11.26):

$$\delta_n = \frac{p_X(u_n)}{1 - P_X(u_n)} = \frac{p_X(u_n)}{1/n}$$

Thus,
$$\delta_n = n p_X(u_n) \quad (11.26b)$$

δ_n is also a measure of the change of u_n with respect to the logarithm of n. Take the derivative of Eq. (11.24b) with respect to n:

$$p_X(u_n) \cdot \frac{du_n}{dn} = \frac{1}{n^2}$$

On the one hand,
$$\frac{dP_X(u_n)}{dn} = p_X(u_n) \frac{du_n}{dn}$$

On the other hand,
$$\frac{dP_X(u_n)}{dn} = \frac{d\left(1 - \frac{1}{n}\right)}{dn} = \frac{1}{n^2}$$

Therefore,
$$p_X(u_n) \cdot \frac{du_n}{dn} = \frac{1}{n^2}$$
$$n^2 p_X(u_n) \cdot \frac{du_n}{dn} = 1$$
$$p_X(u_n) = \left(\frac{1}{n^2}\right) \cdot \frac{dn}{du_n}$$

Thus,
$$\frac{1}{\delta_n} = \frac{1}{n p_X(u_n)} = \frac{1}{n \cdot \left(\frac{1}{n^2}\right) \cdot \frac{dn}{du_n}} = n \cdot \frac{du_n}{dn}$$

But, $\dfrac{d \ln n}{dn} = \dfrac{1}{n}$

Thus,
$$\frac{1}{\delta_n} = \frac{du_n}{d \ln n} \quad (11.27)$$

Though the asymptotic form is independent of the initial variate's distribution, the parameters u_n and δ_n are, as we can observe, dependent on the initial variate's distribution.

Extreme values from an initial distribution with a polynomial tail will converge (in distribution) to the type II asymptotic form [Gumbel, 1958]. The largest value of the type II takes the following form:

$$P_{Y_n}(y) = \exp[-(v_n/y)^k] \quad (11.28)$$

where v_n is the characteristic largest value of the initial variate X of the type II asymptotic form and k is the shape parameter, an inverse measure of dispersion. The characteristic largest value v_n is defined identically as u_n. In fact, the type I and type II forms are related by a logarithmic transformation:

$$u_n = \ln v_n \quad (11.29)$$

and
$$\delta_n = k \quad (11.30)$$

Suppose X_n has the type I asymptotic distribution with parameters u_n and δ_n; Y_n has the type II asymptotic distribution with parameters v_n and k. Let X_n be the logarithmic transformation of Y_n:

$$X_n = \ln Y_n$$

or
$$Y_n = e^{X_n}$$

the distribution of X_n is

$$P_{X_n}(x) = P_{Y_n}(e^x) = \exp[-(v_n/e^x)^k] = \exp[-e^{-k(x - \ln v_n)}]$$

If we define Eqs. (11.29) and (11.30), we have

$$P_{X_n}(x) = \exp[-e^{-\delta_n(x - u_n)}]$$

This is the exact form of type I asymptotic form. Thus, we show how the type I and type II forms are related with the logarithm transformation. For completeness, we also give the type III asymptotic form as follows:

$$P_{Y_n}(y) = \exp\left[-\left(\frac{w-y}{w-w_n}\right)^k\right] \quad (11.31)$$

in which w_n and k are defined similarly as before. The type III asymptotic form represents the limiting distribution of the extreme values from initial distribution that have a finite upper bound value; that is, $P_X(\omega) = 1$.

Gumbel has shown that the normal and log-normal distributions converge to the type I and type II asymptotic forms, respectively. In addition, the Weibull distribution has properties that permit it to cover three subclasses of the type I asymptotic form, adding more generality (see Gumbel [1954]).

For some distributions, it is simple to find an expression for u_n. For others, such as the normal and log-normal distributions, it is hard to find the explicit function for u_n. Consider the normal distribution Φ:

$$P_X^N(u_n) = 1 - \frac{1}{n} = \Phi\left(\frac{u_n - \mu}{\sigma}\right)$$

Thus,

$$u_n^N = \mu + \sigma \Phi^{-1}\left(1 - \frac{1}{n}\right) \quad (11.32a)$$

$$\Phi^{-1}\left(1 - \frac{1}{n}\right) = \frac{u_n^N - \mu}{\sigma} \quad (11.32b)$$

Note that the partitioning point α in Eq. (11.16) is now denoted by p. It will be shown in Section 11.3.1 that $n = \frac{1}{1-p}$ and $p = 1 - \frac{1}{n}$.

Substituting the approximation for Φ^{-1} from Eq. (11.16) into Eq. (11.32b), we will get the following result:

$$u_n^N \cong \mu + \sigma\left[\sqrt{2\ln n} - \frac{\ln(4\pi \ln n)}{2\sqrt{2\ln n}}\right] \quad (11.33a)$$

Let

$$b_n = \left[\sqrt{2\ln n} - \frac{\ln(4\pi \ln n)}{2\sqrt{2\ln n}}\right] \quad (11.33b)$$

Thus,

$$u_n \cong \mu + \sigma b_n \quad (11.33c)$$

The inverse of dispersion can be derived from Eq. (11.26) but first by substituting the approximation of $\Phi^{-1}(1-1/n)$ from Eq. (11.32b) into the value of $p_X(u_n)$ from Eq. (11.7):

$$\delta_n^N = n p_X(u_n)$$
$$= \frac{n}{\sqrt{2\pi}\sigma} \exp\left[-\frac{1}{2}\left(\Phi^{-1}\left(1 - \frac{1}{n}\right)\right)^2\right] \quad (11.34)$$
$$\cong \frac{\sqrt{2\ln n}}{\sigma}$$

The characteristic largest value u_n and inverse measure of dispersion δ_n for normal distribution is only an approximation, and consequently, so are the ones for log-normal distribution:

$$u_n^{LN} \cong \exp\left[\eta + \tau\left(\sqrt{2\ln n} - \frac{\ln(4\pi \ln n)}{2\sqrt{2\ln n}}\right)\right] \quad (11.35)$$

$$\delta_n^{LN} \cong \frac{\sqrt{2\ln n}}{\tau} \quad (11.36a)$$

where

$$\tau = \sqrt{\ln\left[1 + \left(\frac{\sigma}{\mu}\right)^2\right]} \quad (11.36b)$$

As to Weibull distribution, we can get the exact form of u_n and δ_n as

$$u_n^W = a(\ln n)^{1/c} \quad (11.37)$$

$$\delta_n^W = \frac{c}{a}(\ln n)^{1-1/c} \quad (11.38)$$

All the commonly used distributions in environmental analysis belong to the domain of attraction of one of the above distributions. For example, the Cauchy-/Pareto-type distributions belong to the Frechet family, the uniform and triangular distributions belong to the Weibull family, and the exponential and normal distributions belong to the Gumbel family. Thus, it is reasonable to assume that any distribution of practical concern is likely to belong to one of the three asymptotic forms. A powerful and useful aspect of the extreme value theory is that the degrees of freedom in choice of a distribution to model tail data/behavior are limited to the three types. Furthermore, the importance of the above results is that in approximating the extreme values, the choice with infinite degrees of freedom in selecting a parent distribution is reduced to a selection among the three asymptotic distribution families: Gumbel, Frechet, and Weibull. The reduction to three basic families better addresses the uncertainty associated with extreme events; that is, there is no need for fine-tuning among potentially equivalent tail distributions. The use of asymptotic distributions may provide a sound method to characterize extreme data when the distribution model is uncertain. The application of statistics of extremes to approximating uncertain distributions improves on traditional techniques by shifting more focus to the characteristics of a distribution's tail and dealing effectively with the common situation of limited data and information [Lambert et al., 1994]. As the understanding of a problem increases, the use of statistics of extremes may be inadequate. For example, the three types of asymptotic distributions are not exhaustive of the range of choices. They cover a large majority of known distributions, but it must be kept in mind that some distributions do not converge to one of the three forms. New information may prove that the underlying distribution of concern does not converge and another method instead of statistics of extremes should be used. In addition, the statistics of extremes characterizes only the tail of a distribution, meaning that a separate distribution must be used to represent the rest of the distribution. Decisionmakers (DMs) may not be comfortable with using two separate distributions, and application of methods that address tail characteristics and central characteristics with a single distribution may be warranted. As more information becomes available, the use of statistics of extremes may be replaced by more accurate descriptions, but given the current knowledge base, the statistics of extremes provides an initial approach.

11.3 INCORPORATING THE STATISTICS OF EXTREMES INTO THE PMRM

In the PMRM, the conditional expected risk function $f_4(\cdot)$ is a measure of the largest value of damage, given that the damage is of extreme and catastrophic proportions. In a way, $f_4(\cdot)$ is a measure of the same element studied by the statistics of extremes.

11.3.1 Approximation for $f_4(\cdot)$

Since the concept of the return period is very important in water resources as well as in other fields, we will relate the return period to the conditional expected value $f_4(\cdot)$. For this purpose, we will replace from here on the letter n with the letter t to connote the return period of the mapped damage β and replace the letter α with p to connote the partitioning point on the probability axis. Let

$$t = n \qquad (11.39a)$$
$$p = \alpha \qquad (11.39b)$$

As we noted earlier, there exists a relationship between $f_4^k(\cdot)$ and the statistics of extremes, in particular, between $f_4^k(\cdot)$ and u_t^k, where u_t^k is the characteristic largest value of distribution k and for sample size t. Assume that a functional relationship exists between $f_4^k(\cdot)$ and u_t^k. The conditional expectation $f_4^k(\cdot)$ can be viewed as a function of the mean value μ, the standard deviation σ, and the partitioning point p, while u_t^k is a function of μ, σ, and the sample size t. Our goal is to find an H such that

$$f_4^k(\mu,\sigma,p) = H[u_t^k(\mu,\sigma,t)] \qquad (11.40)$$

If there exists such an H, there must also exist an explicit (or implicit) relationship between p and t, say,

$$t = h(p) \qquad (11.41)$$

Assuming that the probability axis is partitioned at p, this partitioning point may be mapped onto the damage axis, yielding a point β such that $P_X^k(\beta) = p$. In other words, β is the value of X (damage) corresponding to an exceedance probability of $1-p$. One could say that by sampling random variables $1/(1-p)$ times from X, one would expect that on the average, one of them would exceed $P_X^{-1}(p) = \beta$. Intuitively, the sample size t can be related to the exceedance probability $(1-p)$ by

$$t = \frac{1}{1-p} \qquad (11.42a)$$

or

$$p = 1 - \frac{1}{t} \qquad (11.42b)$$

For β satisfying $p = P_X(\beta)$, we recognize t in Eq. (11.42b) as being, by definition, the return period of the mapped damage β. Since the inverse $P_X^{-1}(p)$ exists, combining Eqs. (11.24b) and (11.42b), we get

$$P_X(u_t) = 1 - \frac{1}{t} = p \qquad (11.43a)$$

or

$$u_t = P_X^{-1}(p) = \beta \quad (11.43b)$$

This last equation defines the mapping of the probability axis onto the damage axis. In other words, since $P_X(\beta)=p$, then by our definition, $\beta=u_t$.

Using this relationship between p and t enables us to represent $f_4^k(\cdot)$ in terms of u_t^k. Fixing μ and σ, both $f_4^k(\cdot)$ and u_t^k can be considered as functions only of t (or p): $f_4^k(1-1/t)$ and u_t^k. In subsequent sections, it will be shown that by comparing the asymptotic expressions for the expectation $f_4^k(\cdot)$ with the corresponding characteristic largest value as t approaches infinity, $f_4^k(\cdot)$ and u_t^k converge to each other, as shown in Figure 11.4.

Without exception, all the simulations indicated that $f_4^k(\cdot)$ is larger than u_t^k. There seemed to be an almost *constant* difference between them, in the sense that the difference tended to decrease slightly with increasing t.

The simulation indicated

$$\lim_{t\to\infty}\left[f_4^k(\cdot)-u_t^k\right]=0,\quad k=N,LN,W \quad (11.44)$$

We can also verify the relationship between t and p from the above equation with respect to the normal distribution. Recall the asymptotic form $f_4^k(\cdot)$ (Eq. (11.17)) and characteristic extreme value u_t^k (Eq. (11.33a)):

$$f_4^k(\cdot) \cong \mu + \sigma\sqrt{2\ln\frac{1}{1-\alpha}}$$

$$u_n^k \cong \mu + \sigma\left[\sqrt{2\ln n} - \frac{\ln(4\pi\ln n)}{2\sqrt{2\ln n}}\right]$$

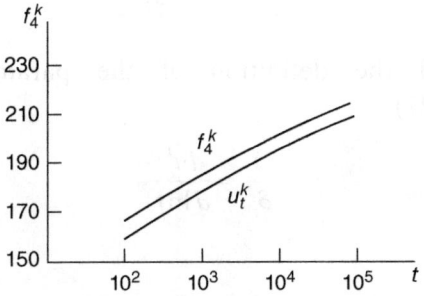

Figure 11.4 $f_4^k(\cdot)$ and u_t^k converge as $t\to\infty$ (normal distribution with $\mu=100$ and $\sigma=25$).

Note that $\dfrac{1}{1-\alpha}=\dfrac{1}{1-p}=n=t$.

$$\lim_{t\to\infty}\left[f_4^N(\cdot)-u_t^N\right]$$
$$=\lim_{t\to\infty}\left[\mu+\sigma\sqrt{2\ln\frac{1}{1-\alpha}}-\mu-\sigma\left(\sqrt{2\ln t}-\frac{\ln(4\pi\ln t)}{2\sqrt{2\ln t}}\right)\right]$$
$$=\lim_{t\to\infty}\sigma\left[\sqrt{2\ln\frac{1}{1-\alpha}}-\sqrt{2\ln t}\right]=0$$

$$(11.45)$$

Obviously, $t=1/(1-p)$ is one of the solutions that satisfy the above equation. Other solutions exist, but asymptotically, they all have to behave as $t=1/(1-p)$. This strengthens the relationship between t and p.

The integral determining $f_4^k(\cdot)$ may be rewritten as

$$f_4^k\left(1-\frac{1}{t}\right)=f_4^k(p)=\frac{\int_\beta^\infty xp_X^k(x)\,dx}{1-p}=t\int_{u_t^k}^\infty xp_X^k(x)\,dx$$

$$(11.46)$$

and it is now dependent on t instead of p for ($k=N$, LN, W).

Equation (11.46) indicates that $f_4^k(\cdot)$ is a weighted sum of values between u_t^k and infinity. Consequently, $f_4^k(\cdot)$ will always be greater than u_t^k. This result can be easily verified as follows:

$$f_4^k(\cdot)=t\int_{u_t^k}^\infty xp_X^k(x)\,dx \geq tu_t^k\int_{u_t^k}^\infty p_X^k(x)\,dx = tu_t^k(1-p)=u_t^k$$

$$(11.47)$$

(Note that $\int_{u_t^k}^\infty p_X^k(x)dx=1-p=\dfrac{1}{t}$.)

Moreover, it can now be explained why $f_4^k(\cdot)$ converges to u_t^k, as $t\to\infty$ for some distributions but not for others. If the decay rate (derivative) of the initial distribution's tail increases with increasing X, then the values of X close to u_t^k will be of more and more significance. In the limit, Eq. (11.46) becomes

$$f_4^k(\cdot)=t\int_{u_t^k}^\infty xp_X^k(x)dx \approx tu_t^k\int_{u_t^k}^\infty p_X^k(x)dx = tu_t^k(1-p)=u_t^k$$

$$(11.48)$$

However, there are distributions for which this assertion might not be true. The exponential distribution, for instance, decays with constant velocity everywhere. Karlsson and Haimes [1988a, b] have also shown that $f_4^{EXP}(\cdot)$ and u_t^{EXP} will always be separated by a constant and therefore do not converge.

To develop a formal expression that relates $f_4^k(\cdot)$ to the characteristic extremes, we take the derivative of Eq. (11.46) with respect to t. We have

$$\frac{df_4^k(\cdot)}{dt} = \int_{u_t^k}^{\infty} xp_X^k(x)\,dx + t\frac{d}{dt}\int_{u_t^k}^{\infty} xp_X^k(x)\cdots dx$$

Observe that the lower bound is dependent on t, while neither the upper bound nor the integrand is; because of this, Leibniz's rule for differentiation of integrals yields

$$\frac{df_4^k(\cdot)}{dt} = \int_{u_t^k}^{\infty} xp_X^k(x)\,dx - tu_t^k p_X^k(u_t^k)\frac{du_t^k}{dt}$$

The first of these terms is almost $f_4^k(\cdot)/t$, but the second term can be further simplified. Taking the derivative of Eq. (11.24b) with respect to t,

$$\frac{dP_X^k(u_t^k)}{dt} = \frac{d}{dt}(1 - 1/t)$$

that is,

$$p_X^k(u_t^k)\frac{du_t^k}{dt} = \frac{1}{t^2}$$

thus generating the following result:

$$\frac{df_4^k(\cdot)}{dt} = \frac{1}{t}f_4^k(\cdot) - t\cdot u_t^k \cdot \left(\frac{1}{t^2}\right)$$

$$\frac{df_4^k(\cdot)}{dt} = \frac{1}{t}f_4^k(\cdot) - \frac{1}{t}u_t^k \quad (11.49)$$

There is indeed a close relationship between the conditional expectation $f_4^k(\cdot)$ and the characteristic largest value u_t^k. This relationship is simply a differential equation. It may seem that solving this differential equation, given an initial variate's characteristic largest value u_t^k, would give an explicit expression for $f_4^k(\cdot)$. Unfortunately, Eq. (11.49) is analytically solvable only for some very special distributions, such as the exponential distribution. In general, integrating the differential equation above yields the same unsolvable integrals that we have been attempting to approximate.

The differential equation relating $f_4^k(\cdot)$ to u_t^k is of theoretical interest, but not of major practical importance. It can be advantageous if the derivative is taken on u_t^k instead of on $f_4^k(\cdot)$. Fortunately, this can be accomplished.

Given our assumptions, it can be shown that the difference between $f_4^k(\cdot)$ and u_t^k converges to zero as $t \to \infty$, at least for the three distributions discussed here and with some restrictions for the Weibull distribution [Karlsson and Haimes, 1988a, b]. The difference between the derivatives converges much faster than the initial difference between $f_4^k(\cdot)$ and u_t^k. The derivatives are almost identical for sufficiently large t. For the time being, we will assume that it is valid to replace df_4^k/dt with du_t^k/dt. Start with Eq. (11.49):

Replace $\frac{df_4^k}{dt}$ with $\frac{du_t^k}{dt}$ and substitute into Eq. (11.49):

Thus,

$$\frac{du_t^k}{dt} = \frac{1}{t}f_4^k(\cdot) - \frac{1}{t}u_t^k(\cdot)$$

$$t\cdot\frac{du_t^k}{dt} = f_4^k(\cdot) - u_t^k(\cdot)$$

$$f_4^k(\cdot) = u_t^k(\cdot) + t\cdot\frac{du_t^k}{dt}$$

But $t\cdot\dfrac{du_t^k}{dt} = \dfrac{du_t^k}{d\ln t}$

$$f_4^k(\cdot) \cong u_t^k + \frac{du_t^k}{d\ln t} \quad (11.50)$$

Recall the definition of the parameter δ_t^k (Eq. 11.27):

$$\frac{1}{\delta_t^k} = \frac{du_t^k}{d\ln t}$$

Thus,

$$f_4^k(\cdot) \cong u_t^k + \frac{1}{\delta_t^k} \quad (11.51)$$

The low-probability expectation $f_4^k(\cdot)$ of an initial distribution X is now expressed as the sum of that distribution's characteristic largest value u_t^k and the inverse of the characteristic dispersion parameter δ_t^k. The characteristic parameters $(u_t$ and $\delta_t)$ of the extremal distribution (corresponding to an initial variate X) are sufficient to determine the conditional expectation $f_4^k(\cdot)$ associated with X, at least for some distributions (N, LN, W) and for very large t.

Note that although the partitioning point p does not appear explicitly in Eq. (11.51), it does implicitly determine the value of $f_4^k(\cdot)$. The sample size t, which is needed to compute the values of the characteristic parameters, is obtained directly from p via Eq. (11.42a); thus, the value of $f_4^k(\cdot)$ computed with Eq. (11.51) will correspond to a certain pre-specified p.

Recall Eq. (11.33a) and (11.34):

$$u_n^N \cong \mu + \sigma\left[\sqrt{2\ln n} - \frac{\ln(4\pi \ln n)}{2\sqrt{2\ln n}}\right]$$

$$\delta_n^N = \frac{\sqrt{2\ln n}}{\sigma}$$

Substituting Eq. (11.33a) and Eq. (11.34) into Eq. (11.51) yields Eq. (11.52):

$$f_4^N(\cdot) \cong u_t^N + \frac{1}{\delta_t^N} = \mu + \sigma\left(\sqrt{2\ln t} - \frac{\ln(4\pi \ln t)}{2\sqrt{2\ln t}} + \frac{1}{\sqrt{2\ln t}}\right)$$
(11.52)

For log-normal and Weibull,

$$f_4^{LN}(\cdot) \cong \left(1 + \frac{\tau}{\sqrt{2\ln t}}\right)\exp\left[\eta + \tau\left(\sqrt{2\ln t} - \frac{\ln(4\pi \ln t)}{2\sqrt{2\ln t}}\right)\right]$$
(11.53)

$$f_4^W(\cdot) \cong a(\ln t)^{1/c}\left[1 + \frac{1}{c\ln t}\right]$$
(11.54)

Example Problem 11.3 Derive an exact form of $f_4^N(\cdot)$ in terms of u_t for a normal distribution function:

$$f_4(\cdot) = t\int_{u_t}^{\infty} x p_X(x)\, dx$$

Note that $p_X^N(x) = \frac{1}{\sqrt{2\pi}\sigma}e^{\frac{-(x-\mu)^2}{2\sigma^2}}$. Thus,

$$f_4^N(\cdot) = t\int_{u_t}^{\infty} \frac{x}{\sqrt{2\pi}\sigma} e^{\frac{-(x-\mu)^2}{2\sigma^2}}\, dx$$

Let $z = \frac{x-\mu}{\sigma}$

$$f_4^N(\cdot) = t\int_{\frac{u_t-\mu}{\sigma}}^{\infty} \frac{\mu + \sigma z}{\sqrt{2\pi}\sigma} e^{-z^2/2}\sigma\, dz$$

$$= t\mu \int_{\frac{u_t-\mu}{\sigma}}^{\infty} \frac{e^{-z^2/2}}{\sqrt{2\pi}}\, dz + \frac{t\sigma}{\sqrt{2\pi}} \int_{\frac{u_t-\mu}{\sigma}}^{\infty} z e^{-z^2/2}\, dz$$

$$= t\mu\left[\Phi(\infty) - \Phi\left(\frac{u_t-\mu}{\sigma}\right)\right] + \frac{t\sigma}{\sqrt{2\pi}}(-e^{-z^2/2})\Big|_{\frac{u_t-\mu}{\sigma}}^{\infty}$$

$$= t\mu\left[1 - \Phi\left(\frac{u_t-\mu}{\sigma}\right)\right] + \frac{t\sigma}{\sqrt{2\pi}}\left[0 - \left(-e^{\frac{-(u_t-\mu)^2}{2\sigma^2}}\right)\right]$$

$$= t\mu\left[1 - \left(1 - \frac{1}{t}\right)\right] + \frac{t\sigma}{\sqrt{2\pi}} e^{\frac{-(u_t-\mu)^2}{2\sigma^2}}$$

Multiply and divide the second term by $\frac{\sigma}{\sigma}$; the exact value of $f_4(\cdot)$ in terms of u_t is

$$f_4^N(\cdot) = \mu + t\sigma^2\left(\frac{1}{\sqrt{2\pi}\sigma}e^{\frac{-(u_t-\mu)^2}{2\sigma^2}}\right) = \mu + t\sigma^2 p_X(u_t)$$

Also, note that since $1 - P_X(u_t) = \frac{1}{t}$, we have

$$f_4^N(\cdot) = \mu + \frac{\sigma^2 P_X(u_n)}{1/t} = \mu + \frac{p_X(u_t)}{1 - P_X(u_t)}\sigma^2$$

From the definition of hazard function, δ_t, $\delta_t = \frac{p_X(u_t)}{1 - P_X(u_t)}$, we can also write $f_4(\cdot)$ as

$$f_4^N(\cdot) = \mu + \delta_t\sigma^2$$

Readers may find that the above equations of $f_4(\cdot)$ are mainly of theoretical interest, since the exact form of u_t and δ_t is not possible to derive for normal distribution.

Example Problem 11.4 Consider an initial normal distribution with mean value $\mu = m = 100$ and standard deviation $\sigma = s = 25$. We wish to compute $f_4^N(\cdot)$ for a given partitioning point $p = 0.9999$. The corresponding value of t is

$$t = 1/(1 - 0.9999) = 10{,}000$$

(see Eq. (11.42a)). Using Eq. (11.52), we immediately get

$$f_4^N(0.9999) = 199.29$$

For comparison, the exact value of $f_4^N(0.9999)$ using Eq. (11.15) is 198.93.

The following are some examples of the exact and approximate values of $f_4^k(\cdot)$, $k = $ N, LN, W, for an initial distribution with mean $m = 100$ and standard deviation $s = 25$.

According to Table 11.1, the approximate expressions for $f_4^k(\cdot)$ give very accurate results with errors of only fractions of a percent. For other combinations of m and s, the situation is quite different. Actually, the accuracy of the approximate expressions is dependent on the coefficient of variation, s/m [Karlsson, 1986; Karlsson and Haimes, 1988a, b]. Large quotients s/m yield relatively large errors. However, the errors do decrease with increasing t.

Table 11.2 indicates that the approximated values of $f_4^k(\cdot)$ for the log-normal and Weibull distributions are inaccurate for $\mu = 100$ and $\sigma = 500$, while the approximations of the normal remain accurate. The approximate expression for $f_4^k(\cdot)$ (Eq. (11.51)) is based on the assumption that df_4^k/dt may be substituted for du_t^k/dt. The accuracy of this approximation depends on the speed of convergence of the difference between $df_4^k(\cdot)/dt$ and du_t^k/dt. In subsequent sections, we will discuss in detail the sensitivity of the approximation for $f_4^k(\cdot)$.

11.3.2 Recursive Method for Obtaining $f_4(\cdot)$

We have previously shown that $f_4(\cdot) = u_t + 1/\delta_t$ is an asymptotically accurate approximation for some distributions under certain conditions. This equation will henceforth be referred to as the first-order approximation of $f_4(\cdot)$. The superscript k, which has so far been used to denote a distribution, will be deleted in this subsection. Let $f_4^{(i)}$ denote the ith order approximation of the low-probability expectation $f_4(\cdot)$ for any initial distribution. A recursive equation for $f_4^{(i)}$ may be derived from Eq. (11.49). We have that

$$f_4^{(i+1)}(\cdot) = u_t + \frac{df_4^{(i)}(\cdot)}{d \ln t} \qquad (11.55)$$

This equation may now be used to develop the exact relationship that exists between the conditional expectation $f_4(\cdot)$ in the PMRM and the statistics of extremes. By substituting the first-order approximation of $f_4(\cdot)$ (Eq. (11.51)) into the recursive equation (Eq. (11.55)), we have

$$f_4^{(2)}(\cdot) = u_t + \frac{d}{d \ln t}\left(u_t + \frac{1}{\delta_t}\right)$$

TABLE 11.1 Comparison of Exact and Approximate $f_4^k(\cdot)$ when $m = 100$ and $s = 25$

Partitioning Probability	Exact	Approximate	Error (%)
Normal			
0.99	166.63	167.39	0.46
0.999	184.17	184.64	0.25
0.9999	198.93	199.29	0.18
0.99999	211.90	212.21	0.15
Log-normal			
0.99	187.57	187.82	0.13
0.999	222.74	222.81	0.03
0.9999	257.51	257.52	0.01
0.99999	292.59	292.59	0.00
Weibull			
0.99	159.69	160.62	0.58
0.999	172.46	172.95	0.28
0.9999	182.52	182.83	0.17
0.99999	190.93	191.14	0.11

TABLE 11.2 Comparison of Exact and Approximate $f_4^k(\cdot)$ when $m = 100$ and $s = 500$

Partitioning Probability	Exact	Approximate	Error (%)
Normal			
0.99	1432.61	1447.88	1.07
0.999	1783.44	1792.75	0.52
0.9999	2078.70	2085.70	0.34
0.99999	2337.91	2344.29	0.27
Log-normal			
0.99	3010.69	2239.45	−25.62
0.999	9935.55	8080.79	−18.67
0.9999	27,805.53	23,743.87	−14.61
0.99999	69,464.08	61,159.96	−11.95
Weibull			
0.99	3639.19	2908.16	−20.09
0.999	10,338.08	9230.77	−10.71
0.9999	22,918.61	21,410.46	−6.58
0.99999	43,503.96	41,575.06	−4.43

Using the definition of δ_t (Eq. 11.27), this expression can be simplified as

$$f_4^{(2)}(\cdot) = u_t + \frac{1}{\delta_t} + \frac{d}{d\ln t}\left(\frac{1}{\delta_t}\right)$$

By repeatedly using the recursive equation, the $(i+1)$th-order approximation of $f_4(\cdot)$ may be written as

$$f_4^{(i+1)}(\cdot) = u_t + \frac{1}{\delta_t} + \sum_{j=1}^{i+1} \frac{d^j}{d(\ln t)^j}\left(\frac{1}{\delta_t}\right) \quad (11.56)$$

Substitute Eq. (11.27) again into Eq. (11.56), we can write it in u_t alone:

$$f_4^{(i+1)}(\cdot) = u_t + \sum_{j=1}^{i+1} \frac{d^j u_t}{d(\ln t)^j} \quad (11.57)$$

For each iteration, a better approximation of $f_4(\cdot)$ is obtained (see Table 11.3). When i approaches infinity, the approximation of $f_4(\cdot)$ converges to its exact value.

$$f_4(\cdot) = u_t + \sum_{j=1}^{\infty} \frac{d^j u_t}{d(\ln t)^j} = u_t + \frac{1}{\delta_t} + \sum_{j=1}^{\infty} \frac{d^j}{d(\ln t)^j}\left(\frac{1}{\delta_t}\right)$$
(11.58)

TABLE 11.3 Comparison of the Exact and Approximate $f_4(\cdot)$ Using Both the First and Second Approximation when $m=100$ and $s=200$

Partitioning Probability	Exact	First Order	Second Order
Normal			
0.99	633.04	639.15	632.00
0.999	773.37	777.10	773.21
0.9999	891.48	894.28	891.75
0.99999	995.17	997.72	995.91
Log-normal			
0.99	1450.94	1276.19	1392.60
0.999	3425.65	3126.81	3340.78
0.9999	7133.82	6648.18	7014.17
0.99999	13,659.78	12,900.71	13,496.73
Weibull			
0.99	1411.54	1343.27	1413.51
0.999	2630.05	2565.45	2631.36
0.9999	4191.51	4129.50	4192.48
0.99999	6082.07	6022.02	6082.83

We can also develop Eq. (11.58) analytically [Mitsiopoulos, 1987]. Note that $\beta = P_X^{-1}(p)$ and from Eq. (11.46) we have

$$f_4(\cdot) = \frac{1}{1-p} \int_{P_X^{-1}(p)}^{\infty} x p_X(x)\, dx$$

Let $P_X(x) = y$; thus, $x = P_X^{-1}(y)$ and $p_X(x)\, dx = dy$

$$f_4(\cdot) = \frac{1}{1-p} \int_p^1 P_X^{-1}(y)\, dy$$

Let $y = 1 - 1/t$; then

$$f_4(\cdot) = \frac{1}{1-p} \int_{\frac{1}{1-p}}^{\infty} \frac{1}{t^2} P_X^{-1}\left(1 - \frac{1}{t}\right) dt$$

Integrating by part and note $P_X^{-1}\left(1 - \frac{1}{t}\right) = u_t$,

$$f_4(\cdot) = \frac{1}{1-p}\left\{-\frac{1}{t}u_t\bigg|_{\frac{1}{1-p}}^{\infty} + \int_{\frac{1}{1-p}}^{\infty} \frac{1}{t}\, du_t\right\}$$

$$= \frac{1}{1-p}\left\{-\frac{1}{t}u_t\bigg|_{\frac{1}{1-p}}^{\infty} + \int_{\frac{1}{1-p}}^{\infty} \frac{1}{t^2} \frac{du_t}{d\ln t}\, dt\right\}$$

$$= \frac{1}{1-p}\left\{-\frac{1}{t}u_t\bigg|_{\frac{1}{1-p}}^{\infty} - \frac{1}{t}\frac{du_t}{d\ln t}\bigg|_{\frac{1}{1-p}}^{\infty} + \int_{\frac{1}{1-p}}^{\infty} \frac{1}{t^2}\frac{d^2 u_t}{d(\ln t)^2}\, dt\right\}$$

$$= \frac{1}{1-p}\left\{-\frac{1}{t}\left[u_t + \sum_{j=1}^{\infty} \frac{d^j u_t}{d(\ln t)^j}\right]\bigg|_{\frac{1}{1-p}}^{\infty}\right\}$$
(11.59)

The upper bound of the integral is unlimited, and we must investigate what happens to Eq. (11.59) as t approaches infinity. A corollary of the Chebyshev inequality states that

$$\Pr(X > m + l) \leq \frac{s^2}{s^2 + l^2} \quad \text{for } l > 0 \quad (11.60)$$

Let $l = u_t - m$:

$$\frac{1}{t} = \Pr(X > u_t) \leq \frac{s^2}{s^2 + (u_t - m)^2}$$

That is,

$$u_t \le m + s\sqrt{t-1} < m + s\sqrt{t} \quad \text{for } u_t > m \quad (11.61)$$

Thus,

$$\lim_{t\to\infty}\frac{1}{t}\left[u_t + \sum_{j=1}^{\infty}\frac{d^j u_t}{d(\ln t)^j}\right] \le \lim_{t\to\infty}\frac{1}{t}\left[m + s\sqrt{t} + s\sum_{j=1}^{\infty}\frac{\sqrt{t}}{2^j}\right] = 0 \quad (11.62)$$

So

$$f_4(\cdot) = \frac{1}{1-p}\left\{-\frac{1}{t}\left[u_t + \sum_{j=1}^{\infty}\frac{d^j u_t}{d(\ln t)^j}\right]\bigg|_{\frac{1}{1-p}}\right\} = u_t + \sum_{j=1}^{\infty}\frac{d^j u_t}{d(\ln t)^j} \quad (11.63)$$

The convergence rate varies with the initial distribution and the value of the quotient s/m, but this convergence is generally rather fast [Karlsson, 1986]. For example, with an initial normal distribution, one iteration is sufficient to obtain approximate values of $f_4(\cdot)$ with errors generally less than 0.1%. For an initial Weibull distribution, the errors are almost always less than 1% after only one iteration. The log-normal distribution, which has a more slowly decaying tail, has the slowest convergence rate. More than one iteration is needed to handle all possible values of s/m.

By comparing Eqs. (11.51) and (11.56), it is clear that Eq. (11.51) can be an asymptotically correct approximation of $f_4(\cdot)$ only, if

$$\lim_{t\to\infty}\sum_{j=1}^{\infty}\frac{d^j}{d(\ln t)^j}\left(\frac{1}{\delta_t}\right) = 0 \quad (11.64)$$

There may exist distributions for which this is not the case. In fact, Mitsiopoulos [1987] showed that such a distribution function of unusual structure actually exists. Consider the [0, 1] uniform distribution, where $u_t = 1 - 1/t$. Then

$$\frac{d^j u_t}{d(\ln t)^j} = (-1)^{j+1}\frac{1}{t}$$

Equation (11.62) will no longer converge under uniform distribution. However, it is asymptotically true for all practical distributions with decaying tails.

11.4 SENSITIVITY ANALYSIS OF THE APPROXIMATION OF $F_4(\cdot)$

Readers may recall from the first section that the conditional risk functions can be written as products of mean m and some function of the coefficient of variation s/m:

$$f_i(x) = m g_i\left(\frac{s}{m}\right) \quad (11.65)$$

Equation (11.65) implies that the values of the conditional risk functions are determined by the quotient s/m. The parameter m plays the role of a scaling factor. Thus, instead of having to choose pairs (m, s), we may now choose values of the coefficient of variation.

By fixing s/m, the relative error that stems from using an approximate formula for $f_4(\cdot)$, instead of the exact formula, will be constant for all values of m. The difficult task of choosing pairs (m, s) is now replaced by a much easier one of choosing quotients s/m. Since the actual value of the scaling factor m is irrelevant, we may fix it without loss of generality. Thus, the mean value of m will be set at $m = 100$ in our discussion.

For the normal distribution, the quotient s/m may take any value from negative to positive infinity. As the mean value of m approaches zero, the quotient s/m will approach either negative or positive infinity depending on the sign of m (since s is always positive). If the dispersion is moderate (small value of s) and the mean value m is large, s/m will be a very small number. In fact, two normal distributions that are identical except for a small variation of the mean may have totally different values of s/m (e.g., N(0,1) and N(1,1)). In conclusion, for the normal distribution, we must consider the entire range of value of s/m.

For the log-normal distribution, the parameters τ and η determine the shape and scale of the pdf, and they are related to the quotient s/m through Eqs. (11.66a) and (11.66b):

$$\tau = \sqrt{\ln\left[1 + \left(\frac{s}{m}\right)^2\right]} \quad (11.66a)$$

$$\eta = \ln\left[\frac{m}{1 + (s/m)^2}\right] \quad (11.66b)$$

Clearly, the parameter τ increases with s/m, and as τ increases, the pdf becomes more and more skewed. Even for the relatively small value $s/m = 5$, the density function is so deformed that it mostly resembles a hyperbole (i.e., a curve of the form $y = 1/x$). However, the log-normal distribution resembles the normal for small quotients s/m. The most interesting values of s/m to consider for the log-normal distribution are values between zero and five.

The Weibull distribution is special because it is highly sensitive to changes in one of its parameters. This parameter, c, is basically the shape parameter for a Weibull distribution. The parameter c is related to m and s through

$$\frac{\Gamma\left(1+\frac{2}{c}\right)}{\Gamma^2\left(1+\frac{1}{c}\right)} = 1 + \left(\frac{s}{m}\right)^2 \qquad (11.67)$$

where $\Gamma(x)$ is the gamma function

$$\Gamma(x) = \int_0^\infty t^{x-1} e^{-t} dt \qquad (11.68)$$

Although not obvious, it can be shown that $c < 1$ whenever $s/m > 1$ and vice versa. The Weibull distribution is unbounded at the origin when $c < 1$. In the special case when $c = 1$ (i.e., the exponential distribution), the quotient $s/m = 1$. Thus, the interesting values of s/m are between zero and one, that is, $c > 1$. The pdf is bounded at the origin and is bell shaped although occasionally skewed.

11.4.1 Dependence of $f_4(\cdot)$ on s/m

The expression for $f_4(\cdot)$ (Eq. (11.58)) is most useful when a very small number of terms are needed for a desired accuracy. The complexity of the expressions increases with the order of the approximation. Karlsson and Haimes [1988a] have shown that for some distributions, the number of items required is dependent on the coefficient of variation s/m. The normal distribution is an exception. This may be seen in Table 11.4, where the first-order approximation of $f_4(\cdot)$ is used. Even for p as low as 0.99, the relative errors are no more than about 1%. Table 11.5 shows that the relative errors of the second-order

approximation are on the order of a tenth of a percent. It seems that the first-order approximation of $f_4(\cdot)$ is sufficient for the normal distribution.

The log-normal distribution does not have these properties. For large quotients of s/m, the first approximation of $f_4(\cdot)$ does not give acceptable values, as can be seen in Table 11.6. The relative errors are below 1% for small values of s/m (below 0.5) but increase rapidly with s/m. These errors decrease as p approaches 1, but they are still appreciable for all s/m greater than one.

TABLE 11.4 Normal Distribution: Comparison of the Exact Values and First-Order Approximations of $f_4(\cdot)$ when $m = 100$ and $p = 0.99$

s/m	Exact	First Order	Error (%)
0.01	102.67	102.70	0.03
0.10	126.65	126.96	0.24
0.25	166.63	167.39	0.46
0.50	233.26	234.79	0.65
1.00	366.52	369.58	0.83
2.00	633.04	639.15	0.96
5.00	1432.61	1447.88	1.07

TABLE 11.5 Normal Distribution: Comparison of the Exact Values and Second-Order Approximations of $f_4(\cdot)$ when $m = 100$ and $p = 0.99$

s/m	Exact	Second Order	Error (%)
0.01	102.67	102.66	−0.01
0.10	126.65	126.60	−0.04
0.25	166.63	166.50	−0.08
0.50	233.26	233.00	−0.11
1.00	366.52	366.00	−0.14
2.00	633.04	632.00	−0.17
5.00	1432.61	1429.99	−0.18

TABLE 11.6 Log-Normal Distribution: Comparison of the Exact Values and First-Order Approximations of $f_4(\cdot)$ when $m = 100$ and $p = 0.99$

s/m	Exact	First Order	Error (%)
0.01	102.70	102.73	0.03
0.10	129.87	130.13	0.20
0.25	187.57	187.82	0.13
0.50	318.72	316.09	−0.82
1.00	676.15	646.16	−4.43
2.00	1450.94	1276.19	−12.04
5.00	3010.69	2239.45	−25.62

If one instead uses the second approximation of $f_4(\cdot)$ (see Table 11.7), the resulting errors are under 1% for all quotients s/m less than 1. The convergence of these approximations to the exact value is rather slow. The first approximation exhibits errors below 1% for all quotients $s/m < 0.5$; the second, for all $s/m < 1$; the third, for s/m less than about 1.5; and the fourth, for s/m less than about 2. Clearly, there is a relationship between the values of t, s/m, and the relative error. It is possible to construct a graph that indicates which combinations of t and s/m produce a particular relative error (see Figure 11.5).

For the Weibull distribution, each higher-order approximation for $f_4(\cdot)$ decreases the error manifold. The first-order approximation is extremely accurate for all cases where s/m is less than 1. These values of s/m are the most interesting since they give a bell-shaped pdf form, albeit skewed. The relative errors are at most 1% (see Tables 11.8 and 11.9). The decrease of the relative error with each higher-order approximation is extraordinary.

When using second-order approximations, all combinations of t and s/m (s/m not more than five) yield errors less than 1%. It seems that each increase in the order of approximation decreases the relative error between 3 and 20 times [Karlsson, 1986]. Figure 11.6 represents combinations of t and s/m yielding a relative error of less than 1% for an initial Weibull distribution.

From now on, the first-order approximation of $f_4(\cdot)$ will be used whenever s/m is less than 1. Otherwise, the second-order approximation will be used for the normal and the Weibull, and the third-order approximation will be used for the log-normal.

11.4.2 Dependence of $f_4(\cdot)$ on the Partitioning Points p

The choice of the partitioning point p has a major impact on the low-probability, high-damage expectation $f_4(\cdot)$. It has hitherto been difficult to

TABLE 11.7 Log-Normal Distribution: Comparison of the Exact Values and Second-Order Approximations of $f_4(\cdot)$ when $m=100$ and $p=0.99$

s/m	Exact	Second Order	Error (%)
0.01	102.70	102.69	−0.01
0.10	129.87	129.82	−0.04
0.25	187.57	187.43	−0.07
0.50	318.72	318.10	−0.20
1.00	676.15	669.22	−1.02
2.00	1450.94	1392.60	−4.02
5.00	3010.69	2645.51	−12.13

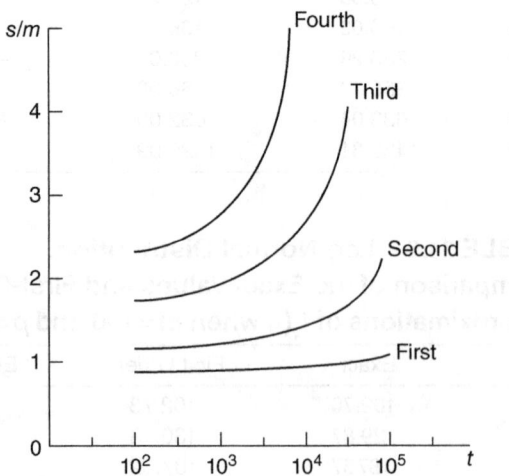

Figure 11.5 All combinations of t and s/m to the right of (below) the 1% error line, corresponding to the ith approximation of $f_4(\cdot)$ for an initial log-normal distribution, yield relative errors less than 1%.

TABLE 11.8 Weibull Distribution: Comparison of the Exact Values and First-Order Approximations of $f_4(\cdot)$ when $m=100$ and $p=0.99$

s/m	Exact	First Order	Error (%)
0.01	101.80	101.82	0.03
0.10	120.07	120.38	0.26
0.25	159.69	160.62	0.58
0.50	255.53	257.66	0.83
1.00	560.52	560.52	0.00
2.00	1411.54	1343.27	−4.84
5.00	3639.19	2908.16	−20.09

TABLE 11.9 Weibull Distribution: Comparison of the Exact Values and Second-Order Approximations of $f_4(\cdot)$ when $m=100$ and $p=0.99$

s/m	Exact	Second Order	Error (%)
0.01	101.80	101.79	−0.01
0.10	120.07	119.96	−0.09
0.25	159.69	159.38	−0.20
0.50	255.53	254.92	−0.24
1.00	560.52	560.52	0.00
2.00	1411.54	1413.51	0.14
5.00	3639.19	3482.09	−4.32

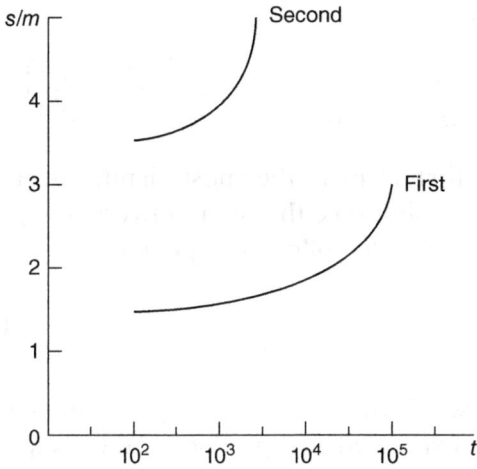

Figure 11.6 All combinations of t and s/m to the right of (below) the 1% error line, corresponding to the ith approximation of $f_4(\cdot)$ for an initial Weibull distribution, yield relative errors less than 1%.

quantify this sensitivity because of the complexity of the integral expressions that determine $f_4(\cdot)$. Combining $f_4(\cdot)$ in the PMRM with the statistics of extremes enables us to study this sensitivity in depth. For a fixed initial distribution, there is a definite relationship between $f_4(\cdot)$ and the distribution's characteristic extremes. In fact, as we have seen in previous sections, there is a formal analytical relationship with which one can determine approximations of $f_4(\cdot)$ with different degrees of accuracy. The conditional expectation may be written as a function of the sample size t. As is pointed out, this t is related to the partitioning point p through $t = 1/(1-p)$. Clearly, t increases very rapidly as p approaches 1. A small variation in p yields an enormous change in t.

Figure 11.7 shows the relation between $f_4(\cdot)$ and ln t (or, equivalently, $\ln[1/(1-p)]$). It would be virtually impossible to use a linear ordinate, since we are interested in the characteristics of the graphs for a very broad spectrum of t values. A linear axis would compress the low t values so that it would not be possible to distinguish among them. In Figure 11.7, the three distributions have the same first two moments, $m = 100$ and $s = 25$.

The curves in the graph are almost linear. Actually, they are slightly concave for both the normal and the Weibull but convex for the log-normal. For a different value of s/m, the graph of $f_4(\cdot)$ is depicted in Figure 11.8.

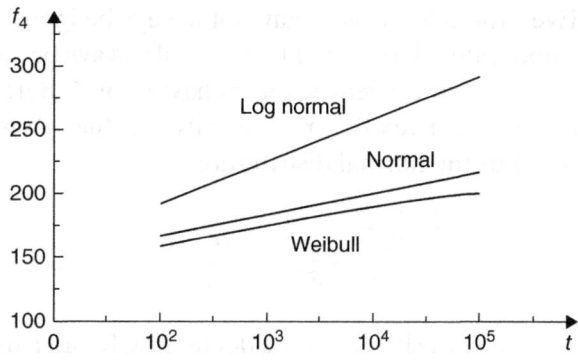

Figure 11.7 Conditional expectation $f_4(\cdot)$ versus ln t for the normal, log-normal, and Weibull distributions ($m = 100$ and $s = 25$).

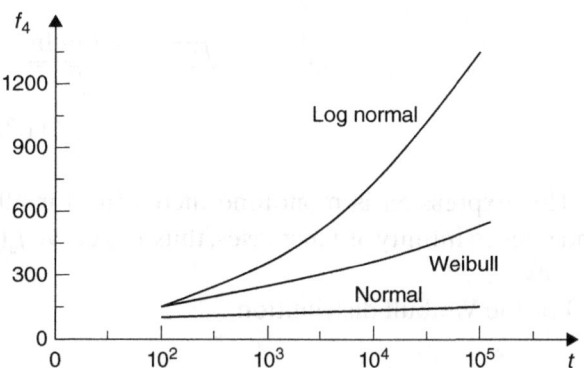

Figure 11.8 Conditional expectation $f_4(\cdot)$ versus ln$t$ for the normal, log-normal, and Weibull distributions ($m = 100$ and $s = 200$).

Here, the normal yields an almost straight but slightly concave curve, while the log-normal and the Weibull both yield convex curves. There is an explanation for this behavior.

From the earlier figures, we have observed that $df_4/d \ln t$ increases for some distributions and decreases for others. This derivative may be developed by taking the derivative with regard to ln t on Eq. (11.58):

$$\frac{df_4}{d\ln t} = \frac{1}{\delta_t} + \sum_{j=1}^{\infty} \frac{d^j}{d(\ln t)^j}\left(\frac{1}{\delta_t}\right) \quad (11.69)$$

For most well-known distributions and large t, the sum converges to zero (Eq. (11.64)), and we are left with the approximation:

$$\frac{df_4}{d\ln t} = \frac{1}{\delta_t} \quad (11.70)$$

Even though the sum may not always be ignored for moderately large t, the term δ_t will always be the most significant. Clearly, the behavior of δ_t determines the convexity or concavity of the above curves. For the normal distribution,

$$\frac{df_4^N}{d\ln t} = \frac{1}{\delta_t^N} = \frac{\sigma}{\sqrt{2\ln t}} \qquad (11.71)$$

which is obviously monotone decreasing for all t and converges to zero. This explains why the curve corresponding to the normal distribution will always be concave.

The log-normal distribution has

$$\frac{df_4^{LN}}{d\ln t} = \frac{1}{\delta_t^{LN}} = \frac{\tau}{\sqrt{2\ln t}} \exp\left[\eta + \tau\left(\sqrt{2\ln t} - \frac{\ln(4\pi \ln t)}{2\sqrt{2\ln t}}\right)\right] \qquad (11.72)$$

This expression is monotone increasing for all t and goes to infinity as t increases; thus, the curve $f_4(\cdot)$ is convex.

For the Weibull distribution,

$$\frac{df_4^W}{d\ln t} = \frac{1}{\delta_t^W} = \frac{a}{c}(\ln t)^{1/c-1} \qquad (11.73)$$

The expression decreases with t whenever the parameter c is greater than 1, increases with t whenever c is less than 1, and is constant for $c=1$. Consequently, the curves corresponding to the Weibull distribution will be concave whenever c is greater than 1, straight for $c=1$, and otherwise convex. It should be noted that the shape of the Weibull distribution's density function changes dramatically with c. The concave curves correspond to a bell-shaped pdf, while the convex corresponds to a hyperbolic-shaped pdf [Karlsson, 1986].

We would like to quantify the sensitivity of $f_4(\cdot)$ to variations of p, $df_4(\cdot)/dp$. Since there exists a relationship between t and p, there must also exist a relationship between d/dt and d/dp. In fact,

$$\frac{d}{dp} = \frac{d}{d\left(1 - \frac{1}{t}\right)} = \frac{d}{\frac{1}{t^2}dt} = t\frac{d}{d\ln t}$$

Thus,

$$\frac{df_4}{dp} = t\frac{df_4}{d\ln t} = t\left[\frac{1}{\delta_t} + \sum_{j=1}^{\infty} \frac{d^j}{d(\ln t)^j}\left(\frac{1}{\delta_t}\right)\right] \qquad (11.74)$$

The first term is the most significant, and for many distributions, the sum converges rapidly to zero, yielding the following equation:

$$\frac{df_4}{dp} \cong \frac{t}{\delta_t} \qquad (11.75)$$

The sensitivity of $f_4(\cdot)$ to the partitioning point p may be expressed as a product of the sample size t and the inverse of the shape parameter δ_t. The inverse of the shape parameter δ_t is also a measure of the dispersion for the extreme distribution associated with the largest value from a sample of t observations identically and independently distributed as the random damage [Gumbel, 1954].

For moderately large t, more of the terms in the sum should be included. Actually, the same number of terms should be used in the approximation of $df_4(\cdot)/dp$ as for the approximation of $f_4(\cdot)$.

Figures 11.9 and 11.10 depict the relationship between $df_4(\cdot)/dp$ and $1/(1-p)$. Although the lines in Figure 11.9 appear to be straight, they are all slightly curved. This curvature may be explained by the same arguments used for $f_4(\cdot)$ as a function on $\ln t$.

Since changes in p correspond to large changes in t, it is obvious that $f_4(\cdot)$ is extremely sensitive to the partitioning point p. This sensitivity will always be greatest for the log-normal distribution because the

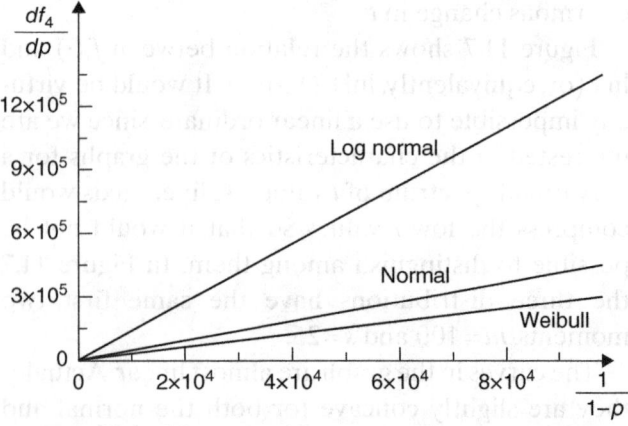

Figure 11.9 Derivative $df_4(\cdot)/dp$ versus $1/(1-p)$ for the normal, log-normal, and Weibull distributions ($m=100$ and $s=25$).

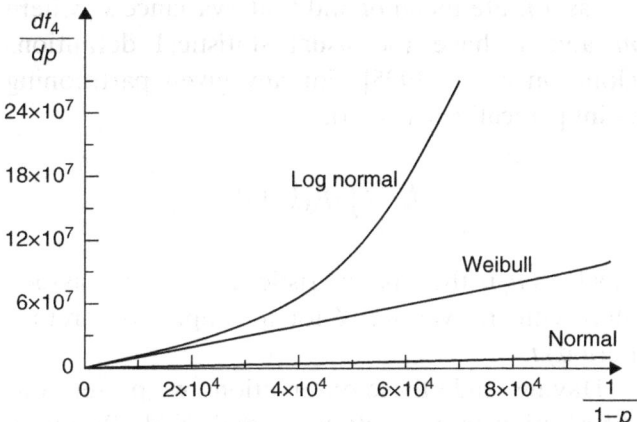

Figure 11.10 Derivative $df_4(\cdot)/dp$ versus $1/(1-p)$ for the normal, log-normal, and Weibull distributions ($m=100$ and $s=200$).

log-normal is of the type II asymptotic form. Its density function's tail decays polynomially, in contrast to the much faster exponential decays for type I distributions. For large c values, the sensitivity of the Weibull distribution will be less than that for the normal. Although they both belong to the type I asymptotic form, for large c, the Weibull distribution's tail decays faster than the normal's. The normal distribution's tail decays approximately as $\exp(-x^2)$, while the Weibull's decays as $x^{c-1}\exp(-x^c)$. The order of the exponential term is greater for the Weibull than for the normal when c is large. For small values of c, the converse is true. Clearly, it is the behavior of the initial distribution's tails that determines not only $f_4(\cdot)$ but also its sensitivity to the partitioning point p.

11.4.3 Sensitivity to the Choice of Distribution

When using the PMRM, it is obvious that the choice of the partitioning points will affect the magnitudes of the conditional risk functions. By studying three distributions in some depth, we realize that distributions belonging to type II class have low-probability expectations that are sensitive to variations in the partitioning point p.

The function $f_4(\cdot)$ is determined primarily by the shape of the distribution's tail. A polynomially decaying tail yields greater probabilities of occurrence for extreme events than an exponentially decaying one. This implies that truly extreme events are given more weight during the integration of the conditional expectations for the type II asymptotic form than for the type I. Consequently, the values of $f_4(\cdot)$ are greater for a type II distribution than for a type I for the same partitioning. Of the three distributions, the log-normal belongs to the type II form, while both the normal and Weibull are of type I. Therefore, we expect the values of $f_4(\cdot)$ that correspond to the log-normal distribution to be greater than the values of $f_4(\cdot)$ that correspond to the other distributions.

We will now examine how partitioning of particular distribution types of type I affect the values of $f_4(\cdot)$. In Figure 11.7, the normal distributions yield greater values of $f_4(\cdot)$ than the Weibull, but as the value of s/m is altered, this is no longer true (see Figure 11.8). Clearly, one cannot generalize as to which distributions will yield greater values of $f_4(\cdot)$. Consider, for example, the normal and the Weibull distributions, whose pdfs are defined in Eqs. (11.7) and (11.9). The normal distribution's tail will always decay as $\exp(-x^2)$, while the tail of the Weibull decays at different rates for different values of the parameter c. For large c, the decay of $x^{c-1}\exp(-x^c)$ will be faster than that for the normal, and consequently, the values of $f_4^W(\cdot)$ will be less than those of $f_4^N(\cdot)$. The converse is true for small values of c. Moreover, small values of s/m correspond to large values of c, and conversely large values of s/m correspond to small values of c.

In Figure 11.7, $s/m=0.25$, a relatively small value, corresponds to a c value of about 4.542. Here, one would expect that the tail of the Weibull decays faster than that of the normal. Consequently, the values of $f_4^N(\cdot)$ are greater than those of $f_4^W(\cdot)$. In Figure 11.8, however, $s/m=2$ corresponds to a c value of 0.543. Clearly, in this case, the tail of the normal decays much faster than that for the Weibull, and we expect that $f_4^N(\cdot)$ is less than $f_4^W(\cdot)$, just as suggested by the simulations.

The values of the expectations not only depend on the choice of the partitioning points but also on the choice of initial distribution. In particular, $f_4(\cdot)$ has shown itself to be very sensitive to the choice of distribution. Although the sensitivity of $f_4(\cdot)$ to the partitioning point can be quantified analytically, our analysis of the sensitivity of $f_4(\cdot)$ to the choice of distribution is solely based on simulations and empirical evidence.

The conservative DM who wishes to consider catastrophic events should use a type II distribution rather than a type I. This is because the slower the decay of the initial distribution's tail, the greater the value of $f_4(\cdot)$.

Combining the statistics of extremes with the PMRM enables us to derive approximate expressions for the low-probability expectation $f_4(\cdot)$. We have also been able to evaluate the sensitivity of $f_4(\cdot)$ to the partitioning points. This latter issue is possibly the more important of the two. The approximate expressions for $f_4(\cdot)$ have also enabled us to assess the impact that the choice of initial distribution has on the values of the conditional expectation.

11.5 GENERALIZED QUANTIFICATION OF RISK OF EXTREME EVENTS

Technical difficulties with the use of the PMRM arise when the behavior of the tail of the risk curve of the underlying frequency of damages is uncertain. This type of problem is particularly evident in the analysis of flood frequencies where the lack of *rare flood* observations makes it difficult to determine the behavior of extreme flood events. When the number of physical observations is small, the analyst is forced to make assumptions about the density of extreme damages (or floods). Each different assumption will generate a different value of $f_4(\cdot)$. Also, there exists an added dimension of difficulty created by the sensitivity of $f_4(\cdot)$ to the choice of the probability partitioning point and distribution-specific approximations. The overall purposes of this section are (i) to present distribution-free results for the magnitude of $f_4(\cdot)$ and (ii) to use these results to obtain a distribution-free estimate of the sensitivity of $f_4(\cdot)$ to the choice of the partitioning point. This section is based on Mitsiopoulos and Haimes [1989].

11.5.1 Distribution-Free Results

Let X be a random variable with density $p_X(x)$ and cdf $P_X(x)$, and let the domain of X be the interval $[L, w]$, where L is the lower bound of the variate (not necessarily finite) and w is the upper bound of the variate (also not necessarily finite). Assume that X has a finite mean m and finite variance s^2, where m and s^2 have the usual statistical definitions [Johnson et al., 1995]. For any given partitioning point p, recall Eq. (11.46),

$$f_4 = t \int_{u_t}^{w} x p_X(x) \, dx$$

in which u_t is the characteristic largest value associated with the variate X for a sample size (return period) t.

Having made these observations, we present our distribution-free results for f_4 and df_4/dt. Proofs of the results can be found in Mitsiopoulos [1987] and Mitsiopoulos and Haimes [1989]. The first major result gives an upper bound for u_t, which is already shown in Eq. (11.61):

$$u_t < m + s\sqrt{t} \quad \text{for } u_t > m$$

where m is the mean and s is the standard deviation of the variate X. The above result is independent of the underlying probability distribution function. The relationship holds for all values of t that satisfy $u_t > m$.

Since u_t is monotone increasing [Mitsiopoulos, 1987], we have $u_t > m$ if and only if $t > t_m$, where t_m satisfies

$$u_{t_m} = m \Leftrightarrow P_X(u_{t_m}) = P_X(m)$$

where $\Leftrightarrow$ denotes *if and only if*. Stated in words, t_m is the return period of the mean of the variate X. Calculating t_m is straightforward. First, we calculate

$$p_m = \Pr(X \leq m)$$

where m is the mean of the variate X. Then, Eq. (11.42a) yields

$$t_m = \frac{1}{1 - p_m} \quad (11.76)$$

The details of computing t_m for the distributions used in this chapter are derived from Mitsiopoulos [1987] and Mitsiopoulos and Haimes [1989]. The normal, exponential, and Gumbel distributions have t_m values that are independent of the parameters of these distributions (Table 11.10).

However, for the log-normal, Weibull, and Pareto distributions, t_m depends on the coefficient of variation, β (where $\beta = s/m$), of the distribution. Table 11.11 summarizes the dependence of t_m on β for these distributions.

We see from the earlier tables that t_m never exceeds 10 when β varies between 0.1 and 5 for all six distributions. In fact, for the normal, Gumbel, and exponential distributions, β can safely range over all positive real numbers. Also, since β is the ratio of the standard deviation to the mean, a value of β of 5 implies that the underlying distribution has a standard deviation that is five times the mean. In theory, we could have a variate whose standard deviation is five times its mean, but such situations seldom arise in practice. In any case, the constraint that $\beta \leq 5$ is, in its own way, too restrictive. As we have seen, such values of β produce values of t_m that are less than 10. In typical applications of the PMRM, the return period of partitioning for f_4 is usually greater than 100.

The point of the preceding discussion is that the results for f_4 and df_4/dt are restricted by the requirement that $t > t_m$, because they are derived from Eq. (11.61). Since we have confirmed that this requirement will be satisfied in most practical PMRM applications, we will now present upper and lower bounds for f_4 and df_4/dt.

TABLE 11.10 Calculation of t_m for the Normal, Exponential, and Gumbel Distributions

Distributions	t_m
Normal	2
Exponential	2.72
Gumbel	2.33

TABLE 11.11 Calculation of the Dependence of t_m on β for the Log-Normal, Weibull, and Pareto Distributions

β	Log-Normal (t_m)	Weibull (t_m)	Pareto (t_m)
0.1	2.08	1.82	2.85
0.25	2.22	1.94	3.04
0.5	2.46	2.17	3.31
1	2.95	2.72	3.64
2	3.80	3.86	3.87
5	5.45	6.70	3.98

Get the derivative of f_4/t. We have

$$\frac{d}{dt}\left(\frac{f_4}{t}\right) = \frac{1}{t}\frac{df_4}{dt} - \frac{1}{t^2}f_4$$

Substituting Eq. (11.49) into the above equation yields

$$\frac{d}{dt}\left(\frac{f_4}{t}\right) = \frac{1}{t^2}f_4 - \frac{1}{t^2}u_t - \frac{1}{t^2}f_4 = -\frac{u_t}{t^2}$$

that is,

$$\frac{d}{dt}\left(\frac{f_4}{t}\right) = -\frac{u_t}{t^2} \qquad (11.77)$$

Integrating both sides of Eq. (11.77) from some point t_0 to some point t, we obtain

$$\frac{f_4}{t} = \frac{f_4}{t}\bigg|_{t_0} + \int_t^{t_0}\frac{u_\tau}{\tau^2}d\tau \qquad (11.78)$$

From Mitsiopoulos [1987], we know that

$$\lim_{t \to \infty}\frac{f_4}{t} = 0 \qquad (11.79)$$

Letting $t_0 \to \infty$ and substituting Eq. (11.79) into Eq. (11.78), we will get

$$\frac{f_4}{t} = \int_t^\infty \frac{u_\tau}{\tau^2}d\tau \qquad (11.80)$$

From Eq. (11.61), we can further develop the above equation:

$$\frac{f_4}{t} = \int_t^\infty \frac{u_\tau}{\tau^2}d\tau < \int_t^\infty \frac{m+s\sqrt{\tau}}{\tau^2}d\tau = \frac{m+2s\sqrt{t}}{t}$$

That is,

$$f_4 < m + 2s\sqrt{t} \quad \text{for } t > t_m \qquad (11.81)$$

Once again, the above inequality is independent of the underlying distribution. If the variate representing the damage is also known to have a finite upper bound, w, then it can be shown that

$$f_4 \leq \min[w, m+2s\sqrt{t}] \quad \text{for } t > t_m \qquad (11.82)$$

Combining inequality in Eq. (11.47), the complete bounds for f_4 are

$$u_t \leq f_4 \leq \min[w, m+2s\sqrt{t}] \quad \forall t > t_m \quad (11.83)$$

The above bounds, together with Eq. (11.49), also yield bounds for df_4/dt:

$$0 \leq \frac{f_4 - u_t}{t} = \frac{df_4}{dt} < \frac{f_4}{t} = \frac{m+2s\sqrt{t}}{t} \quad \forall t > t_m \quad (11.84)$$

From Eq. (11.74), we know

$$\frac{df_4}{dp} = t^2 \frac{df_4}{dt}$$

Thus, the sensitivity bound in terms of the partitioning p becomes

$$0 \leq \frac{df_4}{dp} \leq t(m+2s\sqrt{t}) = \frac{m+2s\sqrt{\frac{1}{1-p}}}{1-p} \quad (11.85)$$

Once again, the preceding inequalities are independent of the underlying distribution function. They are important from a theoretical standpoint because they represent heretofore unknown bounds on conditional expectations (means) of rare events and of the sensitivities of these expectations (means) to the choice of the conditioning (partitioning) probability. From this perspective, these bounds will be valuable to future theoretical analyses of conditional expectations. From a practical standpoint, these bounds also give valuable insight into the conditional risk functions in the PMRM. This insight is especially important when the analyst is completely uncertain about the underlying distribution function. From this perspective, the upper and lower bounds on f_4 and df_4/dp represent, respectively, the most conservative estimates of these risk parameters (i.e., f_4 and df_4/dp). This makes the bounds valuable in practical decisionmaking applications.

11.5.2 Continuation of Bounds

From the previous discussion, we can summarize that for any continuous density function having finite mean m and finite variance s^2 (with $t > t_m$), we obtain

$$f_4 \leq m(1+2\beta\sqrt{t}) \stackrel{D}{=} \overline{f_4} \quad (11.86a)$$

and with $m > 0$, we have

$$\frac{df_4}{dt} \leq \frac{m(1+2\beta\sqrt{t})}{t} \stackrel{D}{=} \frac{\overline{df_4}}{dt}, \quad \text{where } \beta = \left(\frac{s}{m}\right) \quad (11.86b)$$

Because our analysis requires closed-form expressions for f_4, we will concentrate on two distributions for which f_4 can be expressed as a closed-form (or semiclosed-form) function of the variable t, namely, the normal and Pareto distributions. The conformation of the previous sections' results for these two distributions should provide sufficient evidence that the results hold in general, because the normal distribution is representative of most exponential distribution types and the Pareto is representative of most polynomial distribution types. For the normal distribution with parameters $\mu = m$ and $\sigma = s$, we can rewrite Eq. (11.15) as

$$f_4^N(\cdot) = m + \frac{st}{\sqrt{2\pi}} \exp\left[-\frac{1}{2}\left(\Phi^{-1}\left(1-\frac{1}{t}\right)\right)^2\right] \quad (11.87)$$

Combining Eqs. (11.49) and (11.32), we obtain

$$\frac{df_4^N}{dt} = \frac{f_4^N - u_t^N}{t}$$
$$= s\left\{\frac{1}{\sqrt{2\pi}} \exp\left[-\frac{1}{2}\left(\Phi^{-1}\left(1-\frac{1}{t}\right)\right)^2\right] - \frac{1}{t}\Phi^{-1}\left(1-\frac{1}{t}\right)\right\} \quad (11.88)$$

The Pareto distribution has the following density function:

$$p_X(x) = k\frac{b^k}{x^{k+1}} \quad \text{for } x \geq b, k > 2, b > 0 \quad (11.89)$$

The corresponding expression for f_4 is

$$f_4^P(\cdot) = mt^{1/k} \quad \text{for } k > 2 \quad (11.90)$$

where m is the mean of the Pareto distribution ($m = kb/(k-1)$) and k is related to the coefficient of variation β through

$$k = 1 + \sqrt{1 + \left(\frac{1}{\beta}\right)^2} \quad (11.91)$$

When Eq. (11.90) is differentiated with respect to t, we obtain

$$\frac{df_4^P}{dt} = m\left(\frac{1}{k}t^{1/k-1}\right) \quad \text{for } k > 2 \quad (11.92)$$

Having obtained expressions for f_4 and df_4/dt for the normal and Pareto distributions, we compare these expressions with their corresponding upper bounds $\overline{f}_4$ and $\overline{df_4}/dt$ for $t=100$ (p=0.99) and $t=1000$ (p=0.999), with β varying from 0.1 to 5 (see Tables 11.12–11.15).

TABLE 11.12 Comparison of f_4 for the Normal and Pareto Distributions for $t=100$

β	f_4^N (m)	f_4^P (m)	$\overline{f}_4$ (m)
0.1	1.267	1.517	3
0.5	2.333	4.150	11
1	3.665	6.736	21
2	6.330	8.796	41
5	14.326	9.777	101

TABLE 11.13 Comparison of f_4 for the Normal and Pareto Distributions for $t=1000$

β	f_4^N (m)	f_4^P (m)	$\overline{f}_4$ (m)
0.1	1.337	1.869	7.32
0.5	2.683	8.454	32.62
1	4.367	17.484	64.25
2	7.734	26.086	127.49
5	17.834	30.570	317.23

TABLE 11.14 Comparison of df_4/dt for the Normal and Pareto Distributions for $t=100$

β	df_4^N/dt (m)	df_4^P/dt (m)	$\overline{df_4}/dt$ (m)
0.1	0.00034	0.00137	0.03
0.5	0.00169	0.01282	0.11
1	0.00339	0.02790	0.21
2	0.00678	0.04153	0.41
5	0.01694	0.04840	1.01

TABLE 11.15 Comparison of df_4/dt for the Normal and Pareto Distributions for $t=1000$

β	df_4^N/dt (m)	df_4^P/dt (m)	$\overline{df_4}/dt$ (m)
0.1	0.00003	0.00017	0.007
0.5	0.00014	0.00261	0.033
1	0.00028	0.00724	0.064
2	0.00055	0.01232	0.127
5	0.00138	0.01514	0.317

Tables 11.12–11.15 serve two purposes. The primary purpose is that they confirm the upper bounds for both the normal and the Pareto distributions. The secondary purpose is that they also allow comparisons to be made across the two distributions. Specifically, the tables confirm a hypothesis posed by Karlsson [1986]. In simple terms, these results state that among a family of right-side-infinite distributions having identical mean and standard deviations, those distributions with the thickest tails will yield the largest values of f_4 (for large t).

To see why this hypothesis is reasonable, one need only consider the following definition of f_4:

$$f_4 = t \int_{u_t}^{w} x p_X(x)\, dx$$

If we are integrating over a nonnegative region for which t is large (i.e., α, the partitioning probability, is very close to 1), then u_t will be large (recall that u_t always approaches the upper bounds of the variate for large t) and the *largeness* of f_4 depends on the integrand $xp_X(x)$. Thus, we see that the larger $p_X(x)$ is, the larger f_4 will be. This is exactly equivalent to saying that the magnitude of f_4 depends on the tail thickness of the underlying pdf.

Of course, these generalizations are true only when β is of moderate size (when β is somewhere between 0.1 and 5). To see why this restriction is necessary, examine Figure 11.11, which presents normal densities, all with mean $m=100$ but with various values of β. The graph shows quite clearly

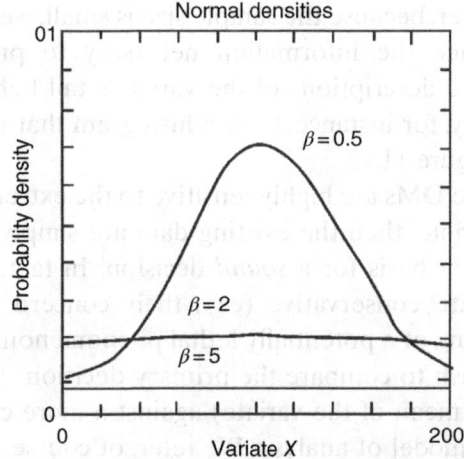

Figure 11.11 Graph of the pdf of normal distribution with $m=100$ and $\beta=0.5, 2,$ and 5.

that the tail of the normal density gets *thicker* as β increases, thus making f_4 larger as β increases. This should also be evident from Eq. (11.87):

$$f_4^N = m\left\{1 + \frac{\beta t}{\sqrt{2\pi}}\exp\left[-\frac{1}{2}\left(\Phi^{-1}\left(1-\frac{1}{t}\right)\right)^2\right]\right\}$$

which is clearly monotone increasing in the parameter β. The underlying point of this discussion is that for variates that go to infinity on the right, large values of β completely determine the shape of the tail. In other words, when β is large and t is moderate, it is impossible to tell whether a distribution with a polynomial tail will produce larger values of f_4 than a distribution with an exponential tail. However, it is true that for large t (p near 1) and β between 0 and 2, the magnitude of f_4 depends largely on the shape of the initial distribution's tail (e.g., polynomial, exponential).

11.5.3 Applications to Decisionmaking

Although the results presented in this section are of a fundamental and theoretical nature, they also have direct potential for applications. These results can be useful in situations in which the tail behavior of a distribution of damages is completely uncertain. Suppose we have a variate X and that we have sampled, say, 20 values of this variate. The central limit theorem implies that we can obtain a fairly accurate estimate for the mean of the variate, given the 20 sampled values. We may also be able to obtain a fairly robust estimate of the standard deviation. However, because the sample size is small, we probably lack the information necessary to produce accurate descriptions of the variate's tail behavior. We may, for instance, have a histogram that resembles Figure 11.12.

If the DMs are highly sensitive to the extremes of the variate, then the existing data are simply not a sufficient basis for a *sound* decision. In fact, if the DMs are conservative (e.g., their concern is the modeling of a potentially lethal phenomenon), they may wish to compare the primary decision (based on the mean of the variate) against a more conservative model of analysis. We refer, of course, to the use of the inequalities derived in the previous discussions.

In order to illustrate the use of the inequalities in a decisionmaking situation, we will create and solve a decisionmaking problem under a condition of complete uncertainty concerning the behavior of *extreme* events.

Consider the following example. We want to modify a reservoir so as to minimize some measure of expected downstream flooding while minimizing the cost of modification. We may choose from five different modification options, each with its specific cost of implementation and each having a distinct impact on downstream flooding. Because of limited data concerning these downstream impacts, we can obtain only reasonable estimates of the mean downstream damage (in millions of dollars) and the standard deviation of the damage of each policy s_j, $j = 1$, ..., 5. Note that we are using a monetary estimate of the damage here and not considering the possibility of human death. The data for the problem are summarized in Table 11.16.

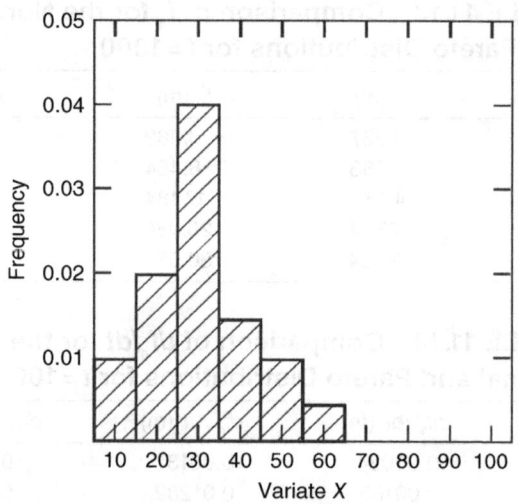

Figure 11.12 The histogram of a variate for which few observations are available.

TABLE 11.16 Cost, Damage, and Deviation Data (in million $)

Policy	Cost of Implementation	Expected Damage (m)	Standard Deviation (s)
s_1	0	345	170
s_2	4	335	140
s_3	7	303	135
s_4	15	298	123
s_5	20	287	100

We can assume that the damage can become effectively infinite if the dam does indeed overflow (or, worse yet, burst). Thus, we can solve this problem by considering trade-offs between:

1. Mean damage and cost
2. Partitioned damage (assuming normality of extreme events) and cost
3. Partitioned damage (using the upper bound) and cost

The advantage in using the inequalities is that we do not have to make any distributional assumptions. For the purposes of this analysis, we will use a partitioning probability of 0.99. This corresponds to a return period of

$$t = \frac{1}{1-p} = \frac{1}{1-0.99} = 100$$

In this case, part (i) is the easiest one to carry out. Table 11.17 yields the cost versus expected damage for the problem.

Assuming that the DMs are indifferent with respect to the outcome of policies s_2 and s_3, then from the graph in Figure 11.13 (and Table 11.17), we can see that consideration of mean damage alone could lead to choosing s_3 as the best decision, since trade-offs leading to s_3 are the largest ($10.7 million saved/$1 million spent). In order to solve part (ii) of the problem, we make use of Eq. (11.87). When $t = 100$, it becomes

$$f_4^N = m + 2.665s$$

We then generate Table 11.18, which is graphed in Figure 11.14.

Note that under the normality assumption and assuming that the DMs are indifferent with respect to policies s_2 and s_3, the best policy would be s_2 since the trade-off leading to it ($22.75 million saved/$1 million spent) is the largest. Thus, we see that

TABLE 11.17 Mean Damage versus Cost for Each Policy

	Expected Damage ($×10⁶)	Cost ($×10⁶)	Trade-Offs
s_1	345	0	—
s_2	335	4	2.5
s_3	303	7	10.7
s_4	298	15	0.63
s_5	287	20	2.20

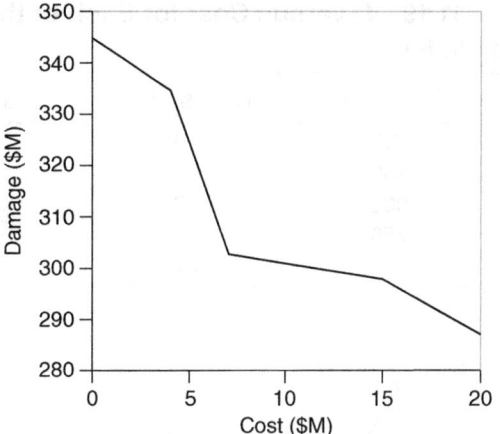

Figure 11.13 Graph of mean damage versus cost.

TABLE 11.18 f_4 versus Cost, Assuming Normality of Damage

	f_4 ($×10⁶)	Cost ($×10⁶)	Trade-Off
s_1	798	0	—
s_2	708	4	22.5
s_3	663	7	15
s_4	626	15	4.6
s_5	554	20	14.4

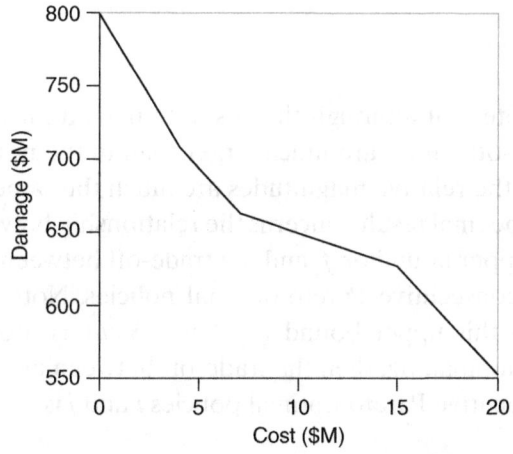

Figure 11.14 Graph of f_4 versus cost assuming normally distributed damage.

applying the PMRM has the potential of changing the DMs' choice of policy. Finally, we can solve part (iii) of the problem using the upper bound $\overline{f}_4$ from Eq. (11.86a). When $t = 100$, it becomes

$$\overline{f}_4 = m + 20s$$

The above equation is used to generate Table 11.19 and Figure 11.15.

TABLE 11.19 f_4 versus Cost for Each of the Five Policies

	$\overline{f}_4$ ($\times 10^6$)	Cost ($\times 10^6$)	Trade-Off
s_1	3745	0	—
s_2	3135	4	152.5
s_3	3003	7	44
s_4	2758	15	30.6
s_5	2287	20	94.2

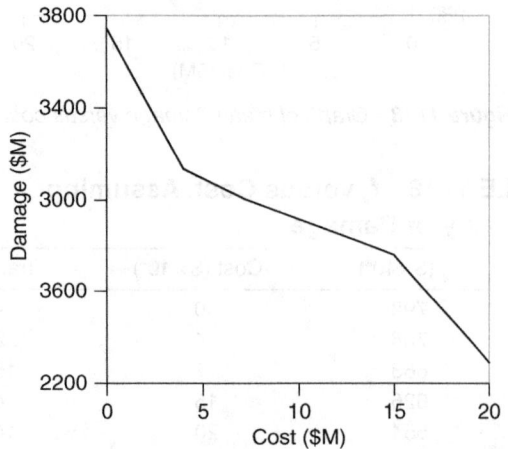

Figure 11.15 Graph of $\overline{f}_4$ versus cost—the worst possible case.

Note that although the absolute magnitudes and trade-off values are much larger than in the normal case, the relative magnitudes are much the same.

One final result concerns the relationship between the upper bound on f_4 and the trade-off between any two consecutive Pareto-optimal policies. Note that when this upper bound $\left(f_4 \leq m + 2S\sqrt{t}\right)$ is used in decisionmaking, then the trade-off between any two consecutive Pareto-optimal policies i and j is

$$\lambda_{ij} = -\frac{(m_j - m_i) + 2(s_j - s_i)\sqrt{t}}{C_j - C_i} \quad (11.93)$$

where (m_j, S_j, C_j) are, respectively, the mean, the standard deviation, and the cost of policy j and (m_i, S_i, C_i) are, respectively, the mean, the standard deviation, and the cost for policy i. We see immediately from Eq. (11.93) that when m_j and m_j are close, that is, when

$$(m_j - m_i) \approx 0$$

then

$$\lambda_{ij} \approx -\frac{2(s_j - s_i)\sqrt{t}}{C_j - C_i} \quad (11.94)$$

The point here is that when the mean damages of two successive policies are not significantly different, using the upper bound for f_4 in decisionmaking is equivalent to basing a decision on the difference between the standard deviations of damages resulting from the two policies. Also, in the normal case, the *form* of f_4 for policy i is

$$f_{4i} = m_i + s_i g(t)$$

where

$$g(t) = \frac{t}{\sqrt{2\pi}} \exp\left[-\frac{1}{2}\left(\Phi^{-1}\left(1 - \frac{1}{t}\right)\right)^2\right]$$

Thus, under the assumption of normality, if the mean damages of two policies i and j are *close*, then the trade-off between policy i and policy j is

$$\lambda_{ij} = -\frac{2(s_j - s_i)g(t)}{C_j - C_i} \quad (11.95)$$

which, up to the constant factor $g(t)$, is the same as the trade-off expression that results from using the upper bound f_4. Thus, if normality is assumed and the means of the two policies i and j are close, then the same *type* of trade-offs results as when using the general upper bound. This is a most promising intuitive result. It states that using f_4 in the decisionmaking process when the means are *close* is actually the same as considering the corresponding standard deviations of damages resulting from the policies.

11.6 SUMMARY

When the PMRM is applied, probabilistic information is preserved through the generalization of a number of conditional expected risk functions. One of these risk functions, the low-probability expectation $f_4(\cdot)$, is of particular interest. The traditional method for solving probabilistic decisionmaking problems has been the use of expected-value functions. By including the low-probability expectation $f_4(\cdot)$ as an objective, the DM is given a means for assessing and incorporating extreme and catastrophic events into the decisionmaking process.

Some elements of the statistics of extremes strongly resemble the properties of the low-probability expectation $f_4(\cdot)$. There exists a relationship between the sample size t and the partitioning point p that relates the two theories. This relationship implies that the sample size t may be viewed as a return period, giving $f_4(\cdot)$ a new interpretation as the expected risk, given that an event with a return period that equals or exceeds t occurs.

The functional relationship between the statistics of extremes and the low-probability expectation may be derived in two ways. A recursive differential equation suggests a way of expressing $f_4(\cdot)$ as an infinite series. Including a sufficient number of terms from that series ensures a very good approximation of $f_4(\cdot)$. Of course, the exact value is obtained when all the terms are considered. However, this more intuitive derivation does not rule out the possible existence of some distribution functions for which the recursive equation might not be correct. An analytically based approach also yields a representation for $f_4(\cdot)$ in terms of an infinite series, that representation being valid only when a certain desirable condition of the initial variate is satisfied. This condition is in the form of a limit; whenever the limit converges to zero, the series representation of $f_4(\cdot)$ is valid.

By comparing the exact values and approximate values of the conditional expected low-probability function, we further found that a different number of terms from the infinite series in the expression of $f_4(\cdot)$ is needed with regard to different quotient values, s/m. Our calculation shows that the first-order approximation is sufficient to achieve the desired approximation error (1%) for all three distributions with $s/m < 1$. Otherwise, the second-order approximation will be used for normal and Weibull and the third-order approximation for log-normal.

The sensitivity of $f_4(\cdot)$ to variations of the partitioning point p is much larger than variations in the quotient. Small changes in p will result in large changes in t, which in turn will affect u_t and δ_t, the key terms in the approximation expression of $f_4(\cdot)$. In fact, we can write the sensitivity of $f_4(\cdot)$ to p as the product of t and the inverse of shape parameter δ_t. Thus, the larger the t is, the more sensitive is $f_4(\cdot)$.

The values of the expectations depend not only on the choice of the partitioning points but also on the choice of initial distributions. Although we cannot quantify the sensitivity of $f_4(\cdot)$ to the choice of distributions, simulations show that the tails of the type II polynomially decaying distributions will be given more weight during the integration than the type I exponentially decaying ones. The conservative DMs should choose type II distribution, as it will have greater $f_4(\cdot)$.

We also derived and explored the distribution-free results for the PMRM's low-probability and high-damage risk function $f_4(\cdot)$. We deducted the lower bounds and upper bounds for both $f_4(\cdot)$ and df_4/dt and then confirmed these inequalities with several distributions. These results can be applied to decision applications under distribution uncertainty.

11.7 EXAMPLE PROBLEMS

11.7.1 Example Problem 1

Calculate u_t and δ_t for standard normal distribution when $t = 50, 100$. Compare the exact value and approximation of $f_4(\cdot)$ and verify the figures given in Table 11.20.

Solution: The characteristic largest value, u_t, is defined by Eq. (11.24c):

$$u_t = P_X^{-1}\left(1 - \frac{1}{t}\right)$$

The second parameter, the inverse measure of dispersion, δ_t, is defined in Eq. (11.26):

$$\delta_t = t p_X(u_t).$$

And the approximation of f_4 is derived in Eq. (11.51):

$$f_{4approx} \cong u_t + \frac{1}{\delta_t}$$

We now proceed to solve u_t, δ_t and f_4 for $t = 50$:

$$t = 50: \quad u_{50} = P_X^{-1}\left(1 - \frac{1}{50}\right) = P_X^{-1}(0.98)$$
$$= \Phi^{-1}(0.98) = 2.054$$

TABLE 11.20 Relative Errors of Approximate f_4 of the Standard Normal Distribution

t	u_t	δ_t	Approximation of f_4	Exact Value of f_4	Error (%)
50	2.054	2.420	2.467	2.420	1.968
100	2.326	2.667	2.701	2.667	1.257

Thus,

$$u_{50} = 2.054.$$

Looking up $P_X^{-1}(0.98)$ in the standard normal probability tables makes it necessary to interpolate in order to find the exact value. From interpolation, it is found that $P_X^{-1}(0.98) = 2.054$. Therefore, $u_{50} = 2.054$:

$$\delta_{50} = \frac{50}{\sqrt{2\pi}} \exp\left(-\frac{2.054^2}{2}\right) = 2.420$$

$$f_{4\text{approx}} \cong u_{50} + \frac{1}{\delta_{50}} = 2.054 + \frac{1}{2.420} = 2.467$$

Similarly, we can calculate

$$t = 100: \quad u_{100} = P_X^{-1}\left(1 - \frac{1}{100}\right) = P_X^{-1}(0.99) = 2.326$$

$$\delta_{100} = \frac{100}{\sqrt{2\pi}} \exp\left(-\frac{2.326^2}{2}\right) = 2.667$$

$$f_{4\text{approx}} \cong u_{100} + \frac{1}{\delta_{100}} = 2.326 + \frac{1}{2.667} = 2.701$$

The exact value of f_4 is defined in Eq. (11.87):

$$f_{4\text{exact}} = \mu + \frac{\sigma t}{\sqrt{2\pi}} \exp\left[-\frac{1}{2}\left(\Phi^{-1}\left(1 - \frac{1}{t}\right)\right)^2\right]$$

Thus,

$$t = 50: \quad f_{4\text{exact}} = \frac{50}{\sqrt{2\pi}} \exp\left[-\frac{1}{2}(\Phi^{-1}(0.98))^2\right] = 2.420$$

$$t = 100: \quad f_{4\text{exact}} = \frac{100}{\sqrt{2\pi}} \exp\left[-\frac{1}{2}(\Phi^{-1}(0.99))^2\right] = 2.667$$

The error is defined as the difference between the approximate value of f_4 and the exact value of $f_4(\cdot)$ divided by the exact value of f_4 multiplied by 100%:

$$\text{Error}(\%) = \left(\frac{f_{4\text{approx}} - f_{4\text{exact}}}{f_{4\text{exact}}}\right) \times 100\%$$

$$t = 50: \quad \text{Error}(\%) = \left(\frac{2.467 - 2.420}{2.420}\right) \times 100\% = 1.968\%$$

$$t = 100: \quad \text{Error}(\%) = \left(\frac{2.701 - 2.667}{2.667}\right) \times 100\% = 1.257\%$$

11.7.2 Example Problem 2

The daily level of dissolved oxygen (DO) concentration for a system is assumed to be of a normal distribution with a mean of 3.5 mg/L and a standard deviation of 0.8 mg/L. Assume that the DO concentration between days is statistically independent.

(a) Determine the most probable one-month maximum DO level.
(b) Determine the probability that the maximum DO level will exceed 4.5 mg/L in a month. Determine the corresponding return period.

Solution: Daily DO levels are $N(\mu = 3.5 \text{ mg/L}, \sigma = 0.8 \text{ mg/L})$.

(a) *Determine the most probable one-month maximum DO level.* Ang and Tang [1984] show that for type I initial variates, the characteristic largest value is the modal (most probable) value of the largest sample value. The characteristic largest value for normal distribution is derived in Eq. (11.32):

$$u_t^N = \mu + \sigma \Phi^{-1}\left(1 - \frac{1}{t}\right)$$

Hence, for the 30-day most likely value, we obtain

$$\text{modal value} = u_{30} = 3.5 + (0.8)\Phi^{-1}\left(1 - \frac{1}{30}\right)$$

$$= 4.967 \text{ mg/L}$$

(b) *Probability that the maximum DO level exceeds 4.5 mg/L in a month.* Let $Y = X_{30}$, the largest value of the month:

$$P_{Y_n}(y) = [P_X(y)]^n = \Phi\left(\frac{y-\mu}{\sigma}\right)^n$$

Evaluate this expression at 4.5 mg/L:

$$\Pr(\max. \text{DO level} > 4.5) = 1 - P_{Y_n}(4.5)$$
$$= 1 - \Phi\left(\frac{4.5 - 3.5}{0.8}\right)^{30}$$
$$= 0.965$$

Return period of maximum DO 4.5 mg/L:

$$= \frac{1}{0.965} = 1.036 \text{ months}$$

That is, 4.5 mg/L is exceeded just about every month on average.

11.7.3 Example Problem 3

Consider the following pdf:

$$p_X(x) = \lambda e^{-\lambda(x-\theta)}, \quad x \geq \theta$$

This is the shifted exponential distribution.

(a) Determine the characteristic largest value, u_t.
(b) Determine the dispersion, δ_t.
(c) Determine the approximation of $f_4(\cdot)$ for a probability partitioning point p.

Solution:

(a) *Determine the characteristic largest value u_t.*
The cdf is

$$P_X(x) = \int_\theta^x \lambda e^{-\lambda(y-\theta)} dy = 1 - e^{-\lambda(x-\theta)}$$

Characteristic largest value (Eq. 11.24b):

$$P_X(u_t) = 1 - \frac{1}{t} = 1 - e^{-\lambda(u_t - \theta)}$$

Thus,

$$1 = te^{-\lambda(u_t - \theta)}$$

$$\ln 1 = \ln t - \lambda(u_t - \theta) = 0$$
$$\ln t = \lambda(u_t - \theta)$$

Solve u_t, we obtain

$$u_t = \theta + \frac{\ln t}{\lambda}$$

(b) *Determine the dispersion δ_t.* Based on its definition (Eq. 11.27),

$$\frac{1}{\delta_t} = \frac{du_t}{d\ln t} = \frac{d}{d\ln t}\left[\theta + \frac{\ln t}{\lambda}\right] = 0 + \frac{d\ln t}{d\ln t}\left(\frac{1}{\lambda}\right) = \frac{1}{\lambda}$$

Thus, $\delta_t = \lambda$.

(c) *Determine the approximation of $f_4(\cdot)$ for partitioning point p.* The approximate $f_4(\cdot)$ is derived in Eq. (11.51):

$$f_4 \cong u_t + \frac{1}{\delta_t}$$
$$= \theta + \frac{\ln t}{\lambda} + \frac{1}{\lambda}$$

Since $t = 1/(1-p)$, thus,

$$f_4 = \theta + \frac{1}{\lambda}\left[1 + \ln\frac{1}{(1-p)}\right]$$
$$= \theta + \frac{1}{\lambda}[1 - \ln(1-p)]$$

11.7.4 Example Problem 4

Given X is of a Weibull distribution with pdf,

$$p_X(x) = \left(\frac{c}{a}\right)\left(\frac{x}{a}\right)^{c-1} e^{-(x/a)^c}, \quad x > 0, a > 0, c \geq 1$$

(a) Using Cramer's method, derive the asymptotical extremal distribution for the largest value and identify its Gumbel type.
(b) Use von Mises' criteria to identify Gumbel type.
(c) Derive expressions for u_n and δ_n.
(d) Obtain the mean and variance for Y_n.

Solution:
Derive the asymptotical extremal distribution for the largest value. The cdf is

$$P_X(x) = \int_0^\infty \left(\frac{c}{a}\right)\left(\frac{x}{a}\right)^{c-1} e^{-(x/a)^c} dx = 1 - e^{-(x/a)^c}$$

Cramer's method: For the largest value Y_n from $(X_1, \ldots, X_n)$, we define

$$g(y) = \xi_n = n[1 - P_X(y)]$$

thus,

$$\Pr(\xi_n \leq \xi) = \Pr(n[1 - P_X(y)])$$
$$= \Pr\left(P_X(Y_n) \geq 1 - \frac{\xi}{n}\right)$$
$$= \Pr\left(Y_n \geq P_X^{-1}\left(1 - \frac{\xi}{n}\right)\right)$$
$$= 1 - \Pr\left(Y_n \leq P_X^{-1}\left(1 - \frac{\xi}{n}\right)\right)$$
$$= 1 - P_{Y_n}\left(P_X^{-1}\left(1 - \frac{\xi}{n}\right)\right)$$
$$= 1 - \left[P_X\left(P_X^{-1}\left(1 - \frac{\xi}{n}\right)\right)\right]^n$$
$$= 1 - \left(1 - \frac{\xi}{n}\right)^n$$

Now take the limit as n approaches infinity:

$$\lim_{n \to \infty} \Pr(\xi_n \leq \xi) = \lim_{n \to \infty}\left[1 - \left(1 - \frac{\xi}{n}\right)^n\right] = 1 - e^{-\xi}$$

We observe that ζ_n decreases as Y_n increases; therefore,

$$P_{Y_n}(y) = \Pr(Y_n \leq y) = \Pr(\xi_n > g(y))$$
$$= 1 - \Pr(\xi_n \leq g(y))$$
$$= \exp[-g(y)]$$

For our problem, we have

$$g(y) = n[1 - P_X(y)] = ne^{-(y/a)^c}$$

So the asymptotic extremal distribution is

$$P_{Y_n}(y) = \exp[-g(y)] = \exp\left[-ne^{-(y/a)^c}\right]$$

This double exponential form implies type I Gumbel distribution.

Use von Mises' criteria to identify Gumbel type. The convergence criteria for the three types of statistics of extremes are stated in the *von Mises' convergence criteria*:

The extreme value of X will converge to the type I asymptotic form if

$$\lim_{x \to \infty} \frac{d}{dx}\left[\frac{1}{h(x)}\right] = \lim_{x \to \infty} \frac{d}{dx}\left[\frac{1 - P_X(x)}{p_X(x)}\right] = 0$$

where $h(x)$ is the hazard function.

The extreme value of X will converge to the type II asymptotic form if

$$\lim_{x \to \infty} xh(x) = k, \quad k > 0 \quad \text{constant}$$

The extreme value of X will converge to the type III asymptotic form if

$$\lim_{x \to w}(w - x)h(x) = k, \quad k > 0 \quad \text{constant}$$

Back to our problem,

$$\frac{1}{h(x)} = \frac{1 - P_X(x)}{p_X(x)} = \frac{e^{-(x/a)^c}}{\left(\frac{c}{a}\right)\left(\frac{x}{a}\right)^{c-1} e^{-(x/a)^c}} = \left(\frac{a}{c}\right)\left(\frac{x}{a}\right)^{1-c}$$

Thus,

$$\lim_{x \to \infty} \frac{d}{dx}\left[\frac{1}{h(x)}\right] = \lim_{x \to \infty} \frac{d}{dx}\left[\left(\frac{a}{c}\right)\left(\frac{x}{a}\right)^{1-c}\right]$$
$$= \lim_{x \to \infty}\left[\left(\frac{1-c}{c}\right)\left(\frac{x}{a}\right)^{-c}\right]$$
$$= 0$$

So the distribution of the largest value from this Weibull-distributed initial variate will converge to Gumbel type I asymptotic form.

Derive expressions for u_n and δ_n. The characteristic largest value is defined as (Eq. (11.24b))

$$1 - \frac{1}{n} = P_X(u_n) = 1 - e^{-(u_n/a)^c}$$

Solving the above equation, we obtain

$$u_n = a(\ln n)^{1/c}$$

With Eq. (11.26), δ_n is defined as

$$\delta_n = h(u_n) = \left(\frac{c}{a}\right)\left(\frac{u_n}{a}\right)^{c-1} = \frac{c}{a}(\ln n)^{1-1/c}$$

We observe that u_n and δ_n are exactly the same as in Eqs. (11.37) and (11.38).

Obtain the mean and variance for Y_n. Define $S = \delta_n(Y_n - u_n)$. Then

$$P_S(s) = \exp[-e^{-s}]$$
$$p_S(s) = e^{-s}\exp[-e^{-s}]$$

We have the moment-generating function

$$G_S(t) = E(e^{ts})$$
$$= \int_{-\infty}^{\infty} e^{ts} e^{-s} e^{-e^{-s}} ds$$

Let $r = e^{-s}$, then $ds = -dr/r$, and

$$G_S(t) = \int_0^{\infty} e^{ts} e^{-r} dr = \int_0^{\infty} r^{-t} e^{-r} dr = \Gamma(1-t)$$

in which Γ is the gamma function defined in Eq. (11.68). The derivatives of $G_S(t)$, evaluated at $t = 0$, will yield respective moments of S [Ang and Tang, 1984]:

$$E[S] = \frac{d\Gamma(1)}{dt} = \gamma = 0.577216\ldots\text{(the Euler number)}$$

$$E[S^2] = \frac{d^2\Gamma(1)}{dt^2} = \gamma^2 + \frac{\pi^2}{6}$$

$$\text{Var}[S] = E[S^2] - E[S]^2 = \frac{\pi^2}{6}$$

Since $Y_n = u_n + S/\delta_n$, thus,

$$E[Y_n] = u_n + \frac{E[S]}{\delta_n}$$
$$= a(\ln n)^{1/c} + \gamma \frac{a}{c}(\ln n)^{1/c-1}$$
$$= a(\ln n)^{1/c}\left(1 + \frac{\gamma}{c\ln n}\right)$$

$$\text{Var}[Y_n] = \frac{\pi^2}{6\delta_n^2}$$
$$= \frac{\pi^2 a^2}{6c^2}(\ln n)^{2/c-2}$$

11.7.5 Example Problem 5

Given

$$p_X(x) = 4x^3 e^{-x^4}, \quad x \geq 0$$

(a) Derive exact formulas of cdf and pdf of the largest value of flood peak from an observation set of n years.
(b) Derive the asymptotic distribution of the largest value of flood peak from an observation set of n years.
(c) What is the return period of a flood peak that exceeds 2 feet in a year?
(d) Calculate the characteristic largest value, u_n, and explain its meaning for $n = 10$.

Solution:

(a) *Derive exact formulas of cdf and pdf of the largest value.* The initial variate's cdf is
Let

$$u = y^4$$
$$du = 4y^3 dy$$

$$P_X(x) = \int_0^x e^{-u} du = -e^{-u}\Big|_0^x = -e^{-x^4} + e^0$$

Thus,

$$P_X(x) = \int_0^x 4y^3 e^{-y^4} dy = 1 - e^{-x^4}$$

From Eq. (11.21), we know that the largest value takes the form of

$$P_{Y_n}(y) = [P_X(y)]^n$$

Thus, the cdf for the largest value is

$$P_{Y_n}(y) = [P_X(y)]^n = (1 - e^{-y^4})^n$$

and the pdf for the largest value is (Eq. (11.22)):

$$p_{Y_n} = n[P_X(y)]^{n-1} p_X(y)$$
$$= 4ny^3 e^{-y^4} (1-e^{-y^4})^{n-1}$$

(b) *Derive the asymptotic distribution of the largest value.* Using Cramer's method, we get

$$g(y) = n[1 - P_X(Y_n)] = ne^{-y^4}$$

so

$$P_{Y_n}(y) = \exp[-g(y)] = \exp[-ne^{-y^4}]$$

Double exponential form $\Rightarrow$ Gumbel type I.

(c) *What is the return period of a flood peak that exceeds 2 feet in a year?* Because $P_X(x)$ is the cdf of the flood peak (maximum elevation), we are interested in the flood peak exceeding 2 feet in a year:

$$P_X(x) = 1 - e^{-x^4}; x = 2$$
$$\Pr(X \geq 2) = 1 - P_X(2) = e^{-2^4} = 1.125 \times 10^{-7}$$

Return period of the flood peak to exceed $x = 2$ is given by

$$t = \frac{1}{1 - P_X(x)} = \frac{1}{e^{-2^4}} = 8.886 \times 10^6$$

This means that a flood peak over 2 feet will occur every 8.886 million years on average. So building a flood protection system that will handle a 2-foot flood peak would seem to do the job for a long time.

(d) *Calculate u_n and explain its meaning for $n = 10$.* u_n is the characteristic largest value, suggesting that it could be used as the probable maximum value over some period. So we want to solve for u_n when $n = 10$, the 10-year most probable maximum flood peak level. We interpret this to mean that we want the most probable central location of possible largest values. This is precisely the definition of u_n. From Eq. (11.24b), we have

$$1 - \frac{1}{n} = P_X(u_n) = 1 - e^{-u_n^4}$$

$$\frac{1}{n} = e^{-u_n^4}$$

$$\ln\left(\frac{1}{n}\right) = -u_n^4$$

$$-\ln n = -u_n^4$$

Solve for u_n:

$$u_n = (\ln n)^{1/4}$$

When $n = 10$, $u_{10} = (\ln 10)^{1/4} = 1.232$ is the characteristic largest value.

11.7.6 Example Problem 6

Consider the pdf of a Rayleigh distribution:

$$p_X(x) = \frac{x}{\mu^2} e^{-x^2/2\mu^2}, \quad x \geq 0, \mu > 0$$

(a) Derive the expression for u_t and δ_t.
(b) Derive the asymptotic form for Rayleigh distribution, and decide the Gumbel type.
(c) Write the approximation of $f_4(\cdot)$. Compare the result with the exact value of $f_4(\cdot)$ for $t = 1000$ and $\mu = 2$.
(d) Derive the approximation and the upper bound of df_4/dt for $t = 1000$ and $\mu = 2$.

Solution:

(a) *Derive the expression for u_t and δ_t.* The cdf of the Rayleigh distribution is as follows:
Let

$$u = y^2$$
$$du = 2y\, dy$$

$$P_X(x) = \int_0^x \frac{y}{\mu^2} e^{-y^2/2\mu^2} dy$$

$$= \int_0^x \frac{e^{-y^2/2\mu^2}}{2\mu^2} du = -\frac{e^{-u^2/2\mu^2} \cdot (2\mu^2)}{2\mu^2}\bigg|_0^x$$

$$= -e^{-x^2/2\mu^2} - (-1) \cdot e^{-0} = 1 - e^{-x^2/2\mu^2}$$

The characteristic largest value u_t is derived from Eq. (11.24b):

$$1 - \frac{1}{t} = P_X(u_t) = 1 - e^{-u_t^2/2\mu^2}$$

Thus,
$$u_t = \mu\sqrt{2\ln t}$$

The dispersion parameter δ_t can be derived from Eq. (11.27):
$$\frac{1}{\delta_t} = \frac{du_t}{d\ln t}$$

From the above, let $y = \ln t$:
$$u_t = \mu\sqrt{2y} = \mu(y)^{1/2}$$
$$dy = d\ln t$$
$$\frac{du}{d\ln t} = \mu\frac{d(2y)^{1/2}}{dy} = \mu\left(\tfrac{1}{2}\right)(2y)^{-1/2}(2)$$

Thus,
$$\frac{1}{\delta_t} = \frac{\mu}{\sqrt{2\ln t}}$$

and
$$\delta_t = \frac{\sqrt{2\ln t}}{\mu}$$

(b) *Derive the asymptotic form and decide the Gumbel type.* Using Cramer's method, we define
$$g(y) = n[1 - P_X(y)] = ne^{-y^2/2\mu^2}$$

So the asymptotic form is
$$P_{Y_n} = \exp[-g(y)] = \exp[-ne^{-y^2/2\mu^2}]$$

Since it takes the double exponential form, it belongs to Gumbel type I. We can also verify it with the von Mises' method:

$$\lim_{x\to\infty}\frac{d}{dx}\left[\frac{1}{h(x)}\right] = \lim_{x\to\infty}\frac{d}{dx}\left[\frac{e^{-x^2/2\mu^2}}{\frac{x}{\mu^2}e^{-x^2/2\mu^2}}\right] = \lim_{x\to\infty}\frac{d}{dx}\left[\frac{\mu^2}{x}\right] = 0$$

(c) *Approximation of $f_4(\cdot)$ when $t = 1000$, $\mu = 2$:*
$$u_{1000} = \mu\sqrt{2\ln t} = 2\sqrt{\ln 1000} = 7.434$$

$$\delta_{1000} = \frac{\sqrt{2\ln t}}{\mu} = \frac{\sqrt{2\ln 1000}}{2} = 1.858$$

The approximation for $f_4(\cdot)$ can be obtained from Eq. (11.51):
$$f_{4\text{approx}} \cong u_t + \frac{1}{\delta_t} = 7.434 + \frac{1}{1.858} = 7.792$$

while the exact value of $f_4(\cdot)$ is derived from Eq. (11.46) as follows:
$$f_{4\text{exact}} = t\int_{u_t}^{\infty}\frac{x^2}{\mu^2}e^{-x^2/2\mu^2}dx$$
$$= t\left\{-xe^{-x^2/2\mu^2}\Big|_{u_t}^{\infty} + \int_{u_t}^{\infty}e^{-x^2/2\mu^2}dx\right\}$$
$$= t\left\{\frac{\mu\sqrt{2\ln t}}{t} + \mu\sqrt{2\pi}\left[1 - \Phi\left(\frac{u_t}{\mu}\right)\right]\right\}$$
$$= 2\sqrt{2\ln 1000} + 2(1000)\sqrt{2\pi}\left[1 - \Phi\left(\sqrt{2\ln 1000}\right)\right]$$
$$= 7.939$$

Therefore, the approximation error is
$$\text{Error}(\%) = \frac{7.972 - 7.939}{7.939} \times 100\% = 0.409\%$$

(d) *Approximation and upper bound of df_4/dt when $t = 1000$, $\mu = 2$.* Combining Eqs. (11.49) and (11.51), we obtain the approximation for df_4/dt:
$$\frac{df_4}{dt} = \frac{1}{t}(f_4 - u_t) \cong \frac{1}{t\delta_t} = \frac{1}{1000 \cdot 1.858} = 5.381 \times 10^{-4}$$

The upper bound of df_4/dt is obtained from inequality Eq. (11.84):
$$\frac{df_4}{dt} < \frac{m + 2s\sqrt{t}}{t}$$

where
$$m = \sqrt{\frac{\pi}{2}}\mu = 2.507$$

and
$$s = \sqrt{\frac{4 - \pi}{2}}\mu = 1.310$$

thus,

$$\frac{df_4}{dt} < \frac{2.507 + 2(1.310)\sqrt{1000}}{1000} = 0.0854$$

11.7.7 Example Problem 7

Reconsider the initial triangular distribution introduced in Example Problem 11.1:

$$p_X(x) = \begin{cases} \frac{1}{3}x, & 0 \le x \le 2 \\ 2 - \frac{2}{3}x, & 2 \le x \le 3 \\ 0, & \text{otherwise} \end{cases}$$

(a) Derive the exact formulas of cdf and pdf for the largest value.
(b) Derive the asymptotic distribution of the largest value from a set of size n.
(c) Derive the expressions for u_n and δ_n.

Solution:

(a) *Derive the exact formulas of cdf and pdf for the largest value.* The initial random variable defined for this problem is X. Given above is the pdf. Thus, it is necessary to find the cdf, $P_X(x)$. Since the pdf is divided into different parts, the cdf will be divided into the same parts. The general formula for the cdf is defined as follows:

$$P_X(x) = \int_0^x p_X(y)\, dy$$

Thus, let us find the cdf for this problem.

For $0 \le x \le 2$, $P_X(x) = \int_0^x \frac{1}{3} y\, dy = \frac{1}{6} x^2$

For $2 \le x \le 3$, $P_X(x) = \int_0^2 \frac{1}{3} y\, dy + \int_2^x \left(2 - \frac{2}{3}y\right) dy$

$$= -\frac{1}{3}x^2 + 2x - 2$$

Thus, the cdf for the initial random variable X is as follows:

$$P_X(x) = \begin{cases} 0, & x < 0 \\ \frac{1}{6}x^2, & 0 \le x \le 2 \\ -\frac{1}{3}x^2 + 2x - 2, & 2 \le x \le 3 \\ 1, & x > 3 \end{cases}$$

Since the cdf has been found, now, the exact formulas of the cdf and pdf for the largest value will be found. The largest value Y_n is defined in Eq. (11.20a):

$$Y_n = \max(X_1, X_2, \ldots, X_n)$$

Assume that $X_1, X_2, \ldots, X_n$ are independent and identically distributed as the initial random variable X. Thus, the general formulas for the cdf and pdf of Y_n are (Eqs. (11.21) and (11.22))

$$P_{Y_n}(y) = [P_X(y)]^n, \text{cdf}$$

$$p_{Y_n}(y) = n[P_X(y)]^{n-1} p_X(y), \text{pdf}$$

The cdf and pdf of Y_n using $P_X(x)$ derived above and $p_X(x)$ given are as follows:

$$P_{Y_n}(y) = \begin{cases} 0, & y < 0 \\ \left[\frac{1}{6}y^2\right]^n, & 0 \le y \le 2 \\ \left[-\frac{1}{3}y^2 + 2y - 2\right]^n, & 2 \le y \le 3 \\ 1, & y > 3 \end{cases}$$

$$p_{Y_n}(y) = \begin{cases} 0, & y < 0 \\ n\left[\frac{1}{6}y^2\right]^{n-1} \frac{1}{3}y, & 0 \le y \le 2 \\ n\left[-\frac{1}{3}y^2 + 2y - 2\right]^{n-1}\left(2 - \frac{2}{3}y\right), & 2 \le y \le 3 \\ 1, & y > 3 \end{cases}$$

(b) *Derive the asymptotic distribution of the largest value.* Use Cramer's method and define:

$$\xi_n = n[1 - P_X(y)] = g(y)$$

Then ζ_n equals

$$\xi_n = \begin{cases} n, & y < 0 \\ n\left(1 - \frac{1}{6}y^2\right), & 0 \le y \le 2 \\ n\left(\frac{1}{3}y^2 - 2y + 3\right), & 2 \le y \le 3 \\ 0, & y > 3 \end{cases}$$

Thus, the cdf of Y_n is obtained from that of ξ_n as

$$P_{Y_n} = \begin{cases} \exp[-n], & y < 0 \\ \exp\left[-n\left(1 - \frac{1}{6}y^2\right)\right], & 0 \le y \le 2 \\ \exp\left[-n\left(\frac{1}{3}y^2 - 2y + 3\right)\right], & 2 \le y \le 3 \\ 1, & y > 3 \end{cases}$$

From observing the asymptotic extremal distribution, it is obvious that there are bounds on the distribution. Thus, the asymptotic extremal distribution is of the Gumbel type III asymptotic form.

Although the asymptotic extremal distribution has been identified as that of the Gumbel type III asymptotic form, it is necessary to verify this with von Mises' criteria:

$$\lim_{x \to w}(w-x)h(x) = \lim_{x \to 3}(3-x)\frac{2 - \frac{2}{3}x}{1 - \left(-\frac{1}{3}x^2 + 2x - 2\right)} = 2$$

Thus, it is obvious that the largest value from the initial variate X converges in distribution to type III form.

(c) *Derive the expressions for u_n and δ_n.* In order to find u_n and δ_n, it is necessary to find what range of n is acceptable for these two measures. In order to solve for u_n, it is necessary to define $P_X(u_n)$, which is as follows:

$$P_X(u_n) = \begin{cases} \frac{1}{6}u_n^2, & 0 \le u_n \le 2 \\ -\frac{1}{3}u_n^2 + 2u_n - 2, & 2 \le u_n \le 3 \end{cases}$$

Let us solve u_n:

$0 \le u_n \le 2$: $\frac{1}{6}u_n^2 = 1 - \frac{1}{n}$, so $u_n = \sqrt{6 - \frac{6}{n}}$.

Since $0 \le u_n \le 2$, that is, $0 \le \sqrt{6 - \frac{6}{n}} \le 2$, thus, we get $1 \le n \le 3$.

$2 \le u_n \le 3$: $-\frac{1}{3}u_n^2 + 2u_n - 2 = 1 - \frac{1}{n}$, so $u_n = 3 - \sqrt{\frac{3}{n}}$ (note that $u_n \le 3$).

Since $2 \le u_n \le 3$, that is, $2 \le 3 - \sqrt{\frac{3}{n}} \le 3$, thus, we get $n \ge 3$.

In summary,

$$u_n = \begin{cases} \sqrt{6 - \frac{6}{n}}, & 1 \le n \le 3 \\ 3 - \sqrt{\frac{3}{n}}, & n \ge 3 \end{cases}$$

Plugging in the values found for u_n above yields the following results:

$1 \le n \le 3$: $\delta_n = n\left(\frac{1}{3}u_n\right) = \frac{n}{3}\sqrt{6 - \frac{6}{n}}$

$n \ge 3$: $\delta_n = n\left(2 - \frac{2}{3}u_n\right) = 2\sqrt{\frac{n}{3}}$

11.7.8 Example Problem 8

Consider the following initial distribution:

$$p_X = \frac{\pi}{6}\sin\frac{\pi x}{3}, \quad 0 \le x \le 3$$

(a) Derive the exact formulas of cdf and pdf for the largest value.

(b) Derive the asymptotic distribution of the largest value from a set of size n.

(c) Derive the expressions for u_n and δ_n.

(d) Assume that the partitioning point on the probability axis is 0.99. Calculate the exact value of f_4.

(e) Write the approximation of f_4 in terms of u_t and δ_t. Determine the value of t for $p=0.99$. Calculate the value of f_4 using the approximation formula. Compare this approximation with the result you obtained from (d).

Solution:

(a) *Derive the exact formulas of cdf and pdf for the largest value.* From the initial variate's pdf, we get the cdf as follows:

$$P_X(x) = \frac{1}{2}\left(1 - \cos\frac{\pi x}{3}\right), \quad 0 \leq x \leq 3$$

Thus, from Eqs. (11.21) and (11.22), we have the cdf and pdf for the largest value:

$$P_{Y_n}(x) = [P_X(x)]^n = \frac{1}{2^n}\left(1 - \cos\frac{\pi x}{3}\right)^n, \quad 0 \leq x \leq 3$$

$$p_{Y_n}(x) = n[P_X(x)]^{n-1} p_X(x)$$

$$= \frac{n\pi}{3 \cdot 2^n} \sin\frac{\pi x}{3}\left(1 - \cos\frac{\pi x}{3}\right)^{n-1}, \quad 0 \leq x \leq 3$$

(b) *Derive the asymptotic distribution of the largest value.* Using Cramer's method, we define

$$g(y) = n[1 - P_X(y)] = \frac{n}{2}\left(1 + \cos\frac{\pi y}{3}\right)$$

so

$$P_{Y_n}(y) = \exp[-g(y)]$$

$$= \exp\left[-\frac{n}{2}\left(1 + \cos\frac{\pi y}{3}\right)\right], \quad 0 \leq y \leq 3$$

(c) *Derive the expressions for u_n, δ_n.* By the definition of u_n, we have

$$1 - \frac{1}{n} = P_X(u_n) = \frac{1}{2}\left(1 - \cos\frac{\pi u_n}{3}\right)$$

Solving for u_n gives

$$u_n = \frac{3}{\pi}\arccos\left(\frac{2-n}{n}\right)$$

For the measure of dispersion, using the definition in Eq. (11.26) gives

$$\delta_n = n p_X(u_n) = \frac{n\pi}{6}\sin\left(\arccos\frac{2-n}{n}\right)$$

$$= \frac{n\pi}{6}\sqrt{1 - \left(\frac{2-n}{n}\right)^2}$$

Note $\sin^2 x + \cos^2 x = 1$.

(d) *Calculate the exact value of f_4 at partitioning point $p=0.99$.* The return period t is

$$t = \frac{1}{1-p} = 100$$

Hence, u_t is computed as follows:

$$u_t = \frac{3}{\pi}\arccos\left(\frac{2-t}{t}\right) = \frac{3}{\pi}\arccos(0.98) = 2.8087$$

There, the exact value of f_4 can be derived from Eq. (11.46):

$$f_{4\text{exact}} = 100\int_{2.8087}^{3} \frac{\pi x}{6}\sin\frac{\pi x}{3}\,dx$$

$$= 50\left(\frac{3}{\pi}\sin\frac{\pi x}{3} - x\cos\frac{\pi x}{3}\right)_{2.8087}^{3}$$

$$= 2.8725$$

(e) *Calculate the approximate value of f_4 at partitioning point $p=0.99$.* Since the initial variate has an upper bound, it belongs to the type III distribution. Thus,

$$f_{4\text{approx}} = u_t + \frac{1}{\delta_t}\left[1 - \frac{1}{(w - u_t)\delta_t + 1}\right]$$

where $w = 3$, $u_t = 2.8087$, and δ_t is given as follows:

$$\delta_t = \frac{t\pi}{6}\sqrt{1 - \left(\frac{2-t}{t}\right)^2} = 10.419$$

Hence, the approximation of f_4 is

$$f_{4approx} = 2.8087 + \frac{1}{10.419}\left[1 - \frac{1}{(3-2.8087)10.419 + 1}\right]$$
$$= 2.8726$$

The approximation error (%) is

$$\text{Error}(\%) = \frac{2.8726 - 2.8725}{2.8725} \times 100\% = 0.002\%$$

It verifies that the approximation of type III distribution is a good approximation.

11.7.9 Example Problem 9

Consider the initial fractile distribution:

$$p_X(x) = \begin{cases} 0.125, & 0 < x \le 2 \\ 0.250, & 2 < x \le 4 \\ 0.125, & 4 < x \le 6 \\ 0, & \text{otherwise} \end{cases}$$

(a) Derive the exact formulas of cdf and pdf for the largest value.
(b) Derive the asymptotic distribution of the largest value from a set of size n.
(c) Derive the expressions for u_n and δ_n.
(d) Assume that the partitioning point on the probability axis is 0.99. Calculate the exact value of f_4.
(e) Write the approximation of f_4 in terms of u_t and δ_t. Determine the value of t for $p=0.99$. Calculate the value of f_4 using the approximation formula. Compare this approximation with the result you obtained from (d).

Solution:

(a) *Derive the exact formulas of cdf and pdf for the largest value.* The initial variate's cdf is

$$P_X(x) = \begin{cases} 0, & x \le 0 \\ 0.125x, & 0 < x \le 2 \\ 0.25(x-1), & 2 < x \le 4 \\ 0.25 + 0.125x, & 4 < x \le 6 \\ 1, & \text{otherwise} \end{cases}$$

Thus, the cdf of the largest value is

$$P_{Y_n}(y) = \begin{cases} 0, & y \le 0 \\ (0.125y)^n, & 0 < y \le 2 \\ (0.25y - 0.25)^n, & 2 < y \le 4 \\ (0.25 + 0.125y)^n, & 4 < y \le 6 \\ 1, & \text{otherwise} \end{cases}$$

And the pdf is

$$p_{Y_n}(y) = \begin{cases} 0.125n(0.125y)^{n-1}, & 0 < x \le 2 \\ 0.25n(0.25y - 0.25)^{n-1}, & 2 < x \le 4 \\ 0.125n(0.25 + 0.125y)^{n-1}, & 4 < x \le 6 \\ 0, & \text{otherwise} \end{cases}$$

(b) *Derive the asymptotic distribution of the largest value.* Use Cramer's method and define

$$g(y) = n[1 - P_X(y)] = \begin{cases} n, & y \le 0 \\ n(1 - 0.125y), & 0 < y \le 2 \\ n(1.25 - 0.25y), & 2 < y \le 4 \\ n(0.75 - 0.125y), & 4 < y \le 6 \\ 0, & \text{otherwise} \end{cases}$$

Therefore, the asymptotic distribution is

$$P_{Y_n}(y) = \exp[-g(y)] = \begin{cases} e^{-n}, & y \le 0 \\ e^{-n(1-0.125y)}, & 0 < y \le 2 \\ e^{-n(1.25-0.25y)}, & 2 < y \le 4 \\ e^{-n(0.75-0.125y)}, & 4 < y \le 6 \\ 1, & \text{otherwise} \end{cases}$$

From this, we can see the asymptotic distribution is of type III with upper bound 6.

(c) *Derive the expressions for u_n and δ_n.*
For $0 < u_n \le 2$,

$$P_X(u_n) = 0.125 u_n = 1 - \frac{1}{n}$$

So,

$$u_n = 8 - \frac{8}{n}$$

As $0 < u_n \le 2$, thus, $0 < 8 - \frac{8}{n} \le 2$, that is, $1 < n \le \frac{4}{3}$

Similarly, we can solve for other ranges, and get the following results:

$$u_n = \begin{cases} 8 - \dfrac{8}{n}, & 1 < n \le \dfrac{4}{3} \\ 5 - \dfrac{4}{n}, & \dfrac{4}{3} < n \le 4 \\ 6 - \dfrac{8}{n}, & n \ge 4 \end{cases}$$

Accordingly,

$$\delta_n = np_X(u_n) = \begin{cases} 0.125n, & 1 < n \le \dfrac{4}{3} \\ 0.25n, & \dfrac{4}{3} < n \le 4 \\ 0.125n, & n \ge 4 \end{cases}$$

(d) *Calculate the exact value of f_4 when $p = 0.99$.* Given the partitioning point $p = 0.99$, we can easily verify that it is within the extreme fractile, $x \in [4,6]$. The return period is given by

$$t = \dfrac{1}{1-p} = 100$$

So the characteristic largest value u_t will be

$$u_{100} = 6 - \dfrac{8}{100} = 5.92$$

By Eq. (11.46), the exact value of f_4 is

$$f_{4\text{exact}} = \int_{5.92}^{6} 0.125x \, dx = 5.96$$

(e) *Calculate the approximate value of f_4 at partitioning point $p = 0.99$.* The initial variate belongs to the type III distribution. So,

$$f_{4\text{approx}} = u_t + \dfrac{1}{\delta_t}\left[1 - \dfrac{1}{(w - u_t)\delta_t + 1}\right]$$

where $w = 6$, $u_t = 5.92$, and δ_t is given as follows:

$$\delta_t = 0.125t = 12.5$$

Hence, the approximation of f_4 is

$$f_{4\text{approx}} = 5.92 + \dfrac{1}{12.5}\left[1 - \dfrac{1}{(6 - 5.92)12.5 + 1}\right] = 5.96$$

The approximation error (%) is

$$\text{Error}(\%) = \dfrac{5.96 - 5.95}{5.96} \times 100\% = 0.0\%$$

This shows that the approximation is exactly the same as the exact value of f_4.

REFERENCES

Ang, A.H.-S. and W.H. Tang, 1984, *Probability Concepts in Engineering Planning and Design*, Vol. **2**, Wiley, New York.

Asbeck, E.L., and Y.Y. Haimes, 1984, The partitioned multiobjective risk method (PMRM), *Large Scale Systems* **6**(1): 13–38.

Castillo, E., 1988, *Extreme Value Theory in Engineering*, Academic Press, Boston, MA.

Chankong, V. and Y.Y. Haimes, 1983, *Multiobjective Decision Making: Theory and Methodology*, Elsevier, New York.

Cramer, H., 1946, *Mathematical Methods of Statistics*, Princeton University Press, Princeton, NJ.

Fischer, R.A. and L.H.C. Tippett, 1928, Limiting forms of the frequency distribution of the largest or smallest member of a sample, *Proceedings of the Cambridge Philosophical Society* **24**, 180–190.

Galambos, J., 1978, *Characterizations of Probability Distributions: A Unified Approach with an Emphasis on Exponential and Related Models*, Springer-Verlag, Berlin.

Gnedenko, B.V., 1943, Sur la distribution limite du terme maxium d'une série aleatoire, *Annals of Mathematics* **44**: 423–453.

Gumbel, E.J., 1954, *Statistical Theory of Extreme Values and Some Practical Applications*, Applied Mathematical Series, **3**, National Bureau of Standards, Washington, DC.

Gumbel, E.J., 1958, *Statistics of Extremes*, Columbia University Press, New York.

Haimes, Y.Y., and W. Hall, 1974, Multiobjectives in water resources systems analysis: The surrogate worth trade-off method, *Water Resources Research* **10**(4), 615–624.

Johnson, L., S. Kotz, and B. Balakrishnan, 1995, *Continuous Univariate Distributions*, Vol. 1 and 2, second edition, John Wiley & Sons, Inc., New York.

Karlsson, P.O., 1986, *Theoretical Foundations for Risk Assessment of Extreme Events: Extensions of the PMRM*, M.S. thesis, Systems Engineering Department, Case Western Reserve University, Cleveland, OH.

Karlsson, P.O. and Y.Y. Haimes, 1988a, Risk-based analysis of extreme events, *Water Resources Research* **24**(1): 9–20.

Karlsson, P.O. and Y.Y. Haimes, 1988b, Probability distributions and their partitioning, *Water Resources Research* **24**(1): 21–29.

Lambert, J.H., N.C. Matalas, C.W. Ling, Y.Y. Haimes, and D. Li, 1994, Selection of probability distributions in characterizing risk of extreme events, *Risk Analysis* **149**(5): 731–742.

Leach, M.R., and Y.Y. Haimes, 1987, The multiobjective risk-impact analysis method, *Risk Analysis* **7**(2): 225–241.

Mitsiopoulos, J., 1987, *Risk of Extreme Events: An Asymptotic Analysis*, M.S. thesis, Systems Engineering Department, Case Western Reserve University, Cleveland, OH.

Mitsiopoulos, J. and Y.Y. Haimes, 1989, Generalized quantification of risk associated with extreme events, *Risk Analysis* **9**(1), 243–254.

Raiffa, H., and R. Schlaifer, 1961, *Applied Statistical Decision Theory*, Harvard University, Cambridge, MA.

12

Systems-Based Guiding Principles for Risk Modeling, Planning, Assessment, Management, and Communication

12.1 INTRODUCTION

In less than half a century, risk analysis has emerged as a requisite area of expertise in almost every professional discipline, ranging from cybersecurity to health care and spanning government, public, and private organizations. By its nature, risk analysis is an intricate, dynamic process—an amalgamation of the arts and sciences—tailored to each specific set of sources of the risks to specific systems. It follows, then, that for any system the balance between quantitative and qualitative risk analysis will be problem and domain specific. Furthermore, meeting the challenges associated with defining, modeling, and quantifying the multidimensional risk function of a system will likely be guided by the specific discipline in which these tasks are being performed and influenced by the specific experiences and expertise of its risk modelers and decisionmakers. For example, although the disciplines of systems engineering and risk analysis share a wide common denominator of philosophy, theory, methodology, and practice, each discipline has historically evolved through separate paths and thus has distinctive followers.

The 10 principles set forth in this chapter are intended to provide a broad framework for understanding and practicing risk analysis—regardless of the specific domain, problem, system, or discipline. These fundamental systems-based principles build on and encapsulate the theory and methodology presented throughout this book. They are designed to guide both quantitative- and qualitative-centered risk analyses. Although these principles may be applied to a range of disciplines, to retain focus, this chapter draws from and is guided by both risk analysis and systems engineering theory, methodology, and practice.

It is constructive first to define the relationship between systems engineering and risk analysis. As a starting point, one may ask: Are risk analysis and systems engineering/analysis grounded on similar principles? Do they represent two distinct fields or disciplines? Or, in reality, do they reinforce and add synergy to each other and thus constitute a unified approach to problem solving and an effective decisionmaking process? On the one hand, the two fields have a common philosophical approach to decisionmaking; on the other hand, they differ in their historical evolution and philosophical and

Risk Modeling, Assessment, and Management, Fourth Edition. Yacov Y. Haimes.
© 2016 John Wiley & Sons, Inc. Published 2016 by John Wiley & Sons, Inc.

technical maturity. Yet both groups aspire to a gestalt–holistic philosophy in their problem-solving approaches. In particular, system modeling, which builds on a plethora of theory, methods, tools, and techniques, offers the instruments with which problems are studied, assessed, understood, managed, and solved.

Systems engineering is distinguished by a practical philosophy that advocates holism in cognition and in decisionmaking. This philosophy is grounded on the arts, natural and behavioral sciences, and engineering, and it is supported by a complement of modeling methodologies, optimization and simulation techniques, data management procedures, and decisionmaking approaches. The ultimate purpose of systems engineering is to (i) build an understanding of a given system's nature, functional behavior, and interaction with its environment; (ii) ensure a holistic, multiperspective representation and systems integration; (iii) improve the decisionmaking process (e.g., in planning, design, development, operation, and management); and (iv) identify, quantify, and evaluate risks, uncertainties, and variability within the decisionmaking process.

The entire process of risk assessment, management, and communication is essentially a synthesis and amalgamation of the empirical and the normative, of the quantitative and the qualitative, and of objective and subjective evidence. It is a process, in fact, that has been built over the years on the contributions of individuals from diverse disciplines. Many of the theories, methods, and quantitative and qualitative tools employed by risk analysts today have been developed by mathematicians, statisticians, biostatisticians, social and behavioral scientists, health scientists, economists, and engineers, among others. In particular, social, behavioral, and organizational scientists have markedly contributed to our understanding and appreciation of the human and social dimensions of risk analysis, such as human perception, organizational and institutional barriers, communication, trust, and conflict resolution. In recognition of the evolving field of risk analysis and its growing constituencies, the Society for Risk Analysis (SRA) was founded in 1980. According to its charter, the SRA "brings together individuals from diverse disciplines and from different countries and provides them opportunities to exchange information, ideas, and methodologies for risk analysis and risk problem-solving."

A reader might ask: *Why do we need a set of principles to perform risk analysis?* Although no one answer can do justice to this encompassing question, one may posit that any decisionmaking process addressing important probable, including dire, consequences resulting from emergent forced changes ought not to be performed on an ad hoc basis. Beyond this, however, because risk analysis has grown to embrace a wide range of specialists from myriad fields, a governing set of principles will provide a conceptual and theoretical framework—a road map—that will enable these individuals to work in concert.

12.2 THE *JOURNEY*: THE GUIDING PRINCIPLES IN THE BROADER CONTEXT OF THE EMERGING NEXT GENERATION DEVELOPED BY THE FEDERAL AVIATION ADMINISTRATION

To demonstrate the efficacy of the ten principles in the risk modeling, planning, assessment, management, and communication process, this chapter offers a conceptual journey, outlining each individual principle and then in turn applying it to a new initiative of the National Airspace System (NAS). At present, the US Federal Aviation Administration (FAA) is planning the most significant overhaul of the NAS since its establishment in 1982. This is an ambitious multiyear and multibillion-dollar system of systems enterprise, which the FAA has termed the Next Generation, or simply NextGen [FAA, 2010]. Aviation safety is vital to the successful implementation of NextGen; furthermore, efficiency and capacity benefits will not occur without the successful integration of new technologies into the existing operational structure. Indeed, safety, aircraft-centric operations, and aircraft equipage are key to NextGen's success. (Aircraft-centric operations rely on a system of autonomous, interconnected aircraft movements, with minimal reliance on external control.) In other words, NextGen is a collection of programs and initiatives designed to improve the capacity, efficiency, safety, security, and environmental impact of aviation in the United States. Indeed, NextGen epitomizes an interconnected and interdependent complex system of systems, where risk modeling, assessment, management, and communication are central to its ultimate successful planning, development, and deployment. The planning of a large and complex project is commonly fraught with myriad sources of risk, termed in this chapter *emergent forced changes*, which connote *trends in external or internal sources of risk to a system that may adversely affect specific states of that system.*

The conceptual journey (see Figure 12.1) starts at part one, "The Evolving Base"; is followed by part two, "The Journey"; proceeds with part three, "The Compass" (the modeling process that is carried out under the guidance of the ten principles); continues with part four, "Interim Destination" (the process of risk assessment, management, and communication); then returns to part one, "The Evolving Base"; and continues. This iterative process is at the heart of the complex risk analysis process [Lowrance, 1976].

12.2.1 Part One: The Evolving Base

Risk analysts and decisionmakers alike must be cognizant of and responsive to the following sample of dynamic shifting rules and realities: (i) goals and objectives; (ii) stakeholders, decisionmakers, and interest groups; (iii) organizational, political, and budgetary baselines; (iv) reorganization and reallocation of key personnel; (v) requirements, specifications, delivery, users, and clients; and (vi) technology and know how.

12.2.2 Part Two: The Journey

The *Journey* provided by the roadmap of the ten principles implies adherence to the following guidelines:

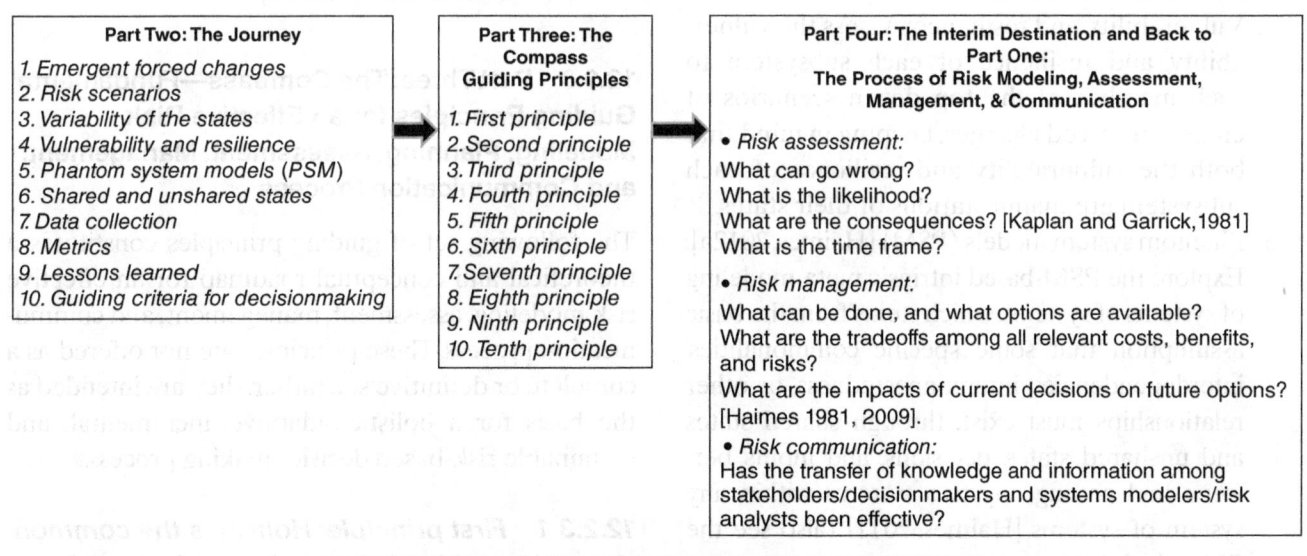

Figure 12.1 Dynamic roadmap for risk modeling, planning, assessment, management, and communication, From Haimes [2012]; © 2012 Wiley.

1. Emergent forced changes: Develop a comprehensive and complete list of emergent forced changes, embracing the modified theory of scenario structuring (TSS) [Haimes, 1981, 2009, 2011, 2012b; Kaplan and Garrick, 1981].
2. Top dozen: Identify the most pertinent emergent forced change scenarios. When properly performed, the TSS may generate hundreds of emergent forced change scenarios. Systemic methods exist to reduce this large number into a manageable one (e.g., a dozen) without missing critically important emergent forced changes [Haimes, 2009, 2012b].
3. Variability of the states: Account for variance of the states of a system over time. The states of a system determine its entire functionality, including its outputs and objective functions. The vulnerability and resilience of and the risk to a system, given specific emergent forced changes and time, are functions (manifestations) of the states; therefore, it is imperative that we focus on the behavior of the states of a system in terms of their variance over time as functions of inputs, decisions, and exogenous and other random variables.
4. Vulnerability and resilience: Assess the vulnerability and resilience of each subsystem to each member of the top dozen scenarios of emergent forced changes, keeping in mind that both the vulnerability and resilience of each subsystem are manifestations of their states.
5. Phantom system models (PSM) [Haimes, 2012a]: Explore the PSM-based intrinsic meta-modeling of systems of systems, which stems from the basic assumption that some specific commonalities, interdependencies, interconnectedness, or other relationships must exist, through shared states and unshared states, decisions, and inputs between and among any two systems within any system of systems [Haimes, 2011] (also see the fifth principle).
6. Shared and unshared states: Develop a comprehensive list of shared and unshared states, decision variables, and inputs among the subsystems of the system of systems under consideration. Shared states represent the interconnectedness and interdependencies among the subsystems, which provide an invaluable mechanism for enhanced system modeling [Haimes, 2011, 2012b].
7. Data collection: Develop a data collection plan and deployment process for each member of the top dozen scenarios of emergent forced changes.
8. Metrics: Building on the collected database, develop metrics with which to:
   a. Evaluate the progress being made
   b. Track precursors to emerging risks
   c. Assess the impacts of current progress on succeeding phases of the planning process
   d. Initiate a comprehensive risk assessment, management, and communication process (see "Part Three: The Compass")
9. Lessons learned: Evaluate the experiences, results, and lessons learned as a prelude to an effective decisionmaking process.
10. Guiding criteria for decisionmaking: Develop guiding criteria for required actions. Keep in mind Lowrance's epistemological statement: "Who should decide on the acceptability of what risk, for whom, in what terms, and why?" [1976].

12.2.3 Part Three: The Compass—Fundamental Guiding Principles for an Effective Risk Modeling, Planning, Assessment, Management, and Communication Process

The following set of guiding principles constitutes a theoretical and conceptual roadmap for an effective risk modeling, assessment, management, and communication process. These principles are not offered as a complete or definitive set; rather, they are intended as the basis for a holistic, adaptive, incremental, and sustainable risk-based decisionmaking process.

12.2.3.1 *First principle: Holism is the common denominator that bridges risk analysis and systems engineering*

The first principle posits that holism is the hallmark of risk analysis. Risk, by its definition as a measure of the probability and severity of adverse effects [Lowrance, 1976], connotes future probable adverse events and probable adverse consequences.

This commonly adopted definition has some implicit ambiguity to it. The term *probability and severity of adverse effects* can simultaneously be interpreted in two ways: (i) in terms of the *probability of the occurrence* of the adverse event and (ii) in terms of the *probability of the severity* of the adverse event, given its occurrence. The latter may be further interpreted in terms of a conditional probability distribution function representing the severity of the adverse effects. Both interpretations are valid; however, each represents significantly varied conceptual and theoretical challenges.

The definition of risk as the likelihood of future adverse consequences resulting from adverse events implies the need for an encompassing holistic identification of all perceived and conceived adverse consequences, as advocated by TSS. Moreover, the holistic characterization of the risk analysis process—the modeling, assessment, management, and communication of risk—does not stop at the TSS level, because each of these four elements of risk analysis, alone and collectively, calls for a systemic, comprehensive, inclusive, and encompassing process. For example, quantifying the risk function requires determining the causal relationship between the timing and severity of the initiating event and the probability of each consequence (as an element of the multidimensional consequences) and thus the resulting risk function. This conceptually, philosophically, and analytically intensive risk analysis task can be effectively achieved through further reliance on and support of the practical philosophy of systems engineering, on its supporting tools and methodologies, and on the guiding principles introduced in this chapter.

In many ways, the gestalt–holistic philosophy can be viewed as the common denominator, unifier, and unique integrator that bridges risk analysis with systems engineering, because both espouse a systemic and comprehensive process. For example, the quest, and indeed the requisite, for identifying all conceivable risk scenarios is harmonious with holism; otherwise, the risk assessment process, and its subsequent management and communication, would be partial, incomplete, biased, and at best ineffective. To ensure such a comprehensive identification of all conceivable risk scenarios, a systems-based methodology is required.

In 1981, both the TSS [Kaplan and Garrick, 1981] and hierarchical holographic modeling (HHM) [Haimes, 1981, 2011] were independently published to address the multiple perspectives of a system and to promote the consideration of every conceivable emergent forced change. (The term *emergent forced changes* is defined in Section 12.2.) Subsequently, a joint paper streamlined the TSS and the HHM, and the revised definition of risk made explicit what was only implicit before [Kaplan et al., 2001]. The TSS and its extension constitute a cornerstone of risk analysis, which requires the generation of all conceivable risk scenarios. This comprehensive process necessarily requires subsequent analysis, filtering, and prioritization of risk scenarios [Haimes, 2009]. Other methods that serve the purpose of analyzing risk scenarios include failure modes and effects analysis (FMEA), event trees, and fault trees, among others [McCormick, 1981].

Relevance to NextGen: *Analyzing a system of systems cannot be a selective process driven by limited perspectives; rather, a holistic systems-based approach must be embraced to account for emergent forced changes; multiple databases and numerous technological and organizational subsystems; multiple objectives, agencies, stakeholders, and decision-makers; and multiple time horizons associated with the phased planning and development process. These considerations, among others, characterize the modeling challenges associated with systems of systems such as the NextGen enterprise.*

12.2.3.2 Second principle: The process of risk modeling, assessment, management, and communication must be methodical, disciplined, systemic, integrated, and commensurate in its comprehensiveness with the criticality of the systems being addressed and of their associated risks

Given the importance of the second principle, it will be useful to establish some key definitions: *risk management* is commonly distinguished from *risk assessment*, even though some may use the term *risk management* to connote the entire process of risk assessment, management, and communication. The term *management* may vary in meaning according to the discipline involved and/or the context.

Risk assessment

Planning for emergent forced changes is central to strategic risk analysis. In risk assessment, the analyst often attempts to answer the following set of three questions [Kaplan and Garrick, 1981]: What can go wrong? What is the likelihood that it would go wrong? What are the consequences? Here, we add a fourth question: What is the time frame? We also amend paraphrase two of the above questions: What can go wrong and when? What are the consequences and for how long? Answers to these questions help risk analysts to identify, measure, quantify, and evaluate risks and their consequences and impacts.

In order to prevent, mitigate, or prepare for undesirable future occurrences, all government agencies as well as public and private sectors seek to understand the trends of risks associated with emergent forced changes. If such trends and changes are unanticipated, undetected, misunderstood, or ignored, they are likely to affect a multitude of states of that system with adverse consequences. It is imperative to be able—through scenario structuring, modeling, and risk analysis—to envision, discover, and track emergent forced changes and their crossovers. In this context, crossovers connote situations where a projected emergent forced change, for example, an impending failure or a terrorist threat, can cross from a state of uncertainty to a forecasted state of high likelihood of occurrence. Emergent forced changes may be caused by natural phenomena or malevolent human actions or accidents. These precursors may be identified based on tangible or intangible information; they are most useful to the risk analysis process when they are examined in conjunction with other sources of knowledge to understand their broader implications and when they are followed with vigilance. Often, a crossover is ignored during the tracking of emergent forced changes and is recognized too late—only after it becomes a disaster [Perrow, 1984/1999; Phimister et al., 2004; Taleb, 2007]. In the context of modeling, crossovers may also connote situations where during simulated scenarios of a dynamic system, the projected emergent forced change crosses the state of knowledge from uncertainty to a more likely eventuality of its occurrence. Duderstadt et al. emphasize that "[c]hange is coming, and the biggest mistake could be underestimating how extensive it will be" [2005]. Of course, positive emergent forced changes may bring with them opportunities in terms of new technological innovations and discoveries.

Bayesian analysis constitutes an important element in modeling, estimating, and tracking over time precursors to the emergence of forced changes. Indeed, Bayes's theorem incorporates probabilistically newly acquired knowledge, whether gained by direct observations or through expert evidence obtained through prior probabilistic assumptions. Scenario development constitutes a basis upon which to track and estimate precursors, and in scenario tracking, both the probabilities and consequences are considered when evaluating the importance of any specific scenario. The uncertain nature of emergent forced changes requires that both aleatory and epistemic sources of uncertainty be considered in the risk analysis process (see the seventh principle). Without such considerations, the representativeness, effectiveness, and credibility of the risk-based decisionmaking process would be questionable.

One important demarcation line that characterizes a sound risk-based decisionmaking process is the extent to which the known and known-unknown future developments and emergent forced changes affecting the system enterprise are addressed and evaluated. Anticipating and planning for emergent forced changes, through preparedness and other risk management actions, will necessarily improve the resilience of the system. Inherently, challenging trade-offs must be made among (i) certain (nonprobabilistic) investments in preparedness in the quest to address probable emergent forced changes and (ii) the added resilience to the system in anticipation of the uncertain nature of the realization of the forced changes.

The often injudicious allocation of precious limited resources (today) in the quest to protect against (future) probabilistic adverse consequences has commonly marred the deployment of risk management options. Ample examples can be cited for the lack of investment in cybersecurity. Indeed, responding to each and every credible scenario of critical emergent forced changes is unrealistic and could lead to bankruptcy. A holistic, adaptive, incremental, and sustainable risk management process, introduced in the tenth principle, alleviates the need for such massive and unsustainable investments in risk management.

Risk management

Risk management builds on the risk assessment process by seeking answers to a second set of three questions [Haimes, 1991]: What can be done, and what options are available? What are the associated trade-offs among all relevant costs, benefits, and risks (both mitigated and residual)? What are the impacts of current management decisions on future options? Note that the last question is the most critical one for any sound and effective managerial decisionmaking. This is so because unless the negative and positive impacts of current decisions on future options are assessed and evaluated (to the extent possible), these policy decisions cannot be deemed to be either effective or indeed desirable.

Risk management can be termed effective only when the three questions are addressed in the broader context of management, where all options and their associated trade-offs are considered within the common hierarchical–organizational structures. Of course, this task cannot be done seriously and meaningfully in isolation from modeling the system and the allocation of appropriate resources. In other words, risk management must be an integral part of the overall organization's policies and procedures. Furthermore, an effective and sustainable risk management approach that harmonizes risk management policies with the overall system's management policies must address the following four categories of major sources of failure [Haimes, 1991, 2009]: hardware, software, human, and organizational.

These four sources of a system's failure are not necessarily independent of each other; for example, the distinction between software and hardware is not always straightforward, and separating human and organizational failure is often difficult. Nevertheless, these four categories provide a meaningful foundation upon which to build a comprehensive risk management framework.

Organizational errors are often at the root of failures of critical engineering systems. Yet when searching for risk management strategies, engineers often tend to focus on technical solutions, in part because of the way risks and failures have been analyzed in the past. For example, in her study of offshore drilling rigs, Paté-Cornell [1990] found that over 90% of documented failures were caused by organizational errors. Hence, the risk assessment and management process must be comprehensive and must, for example, involve all aspects of the system's life cycle: requirements, specifications, planning, design, integration, construction and deployment, operation, and management. This process requires the total involvement of all principal participants: blue- and white-collar workers and managers at all levels of the organizational hierarchy.

Risk communication

The risk assessment and management process aims at answering the previous specific questions in order to allow all participants to make better decisions under uncertain conditions. Similar to system modeling, where a model must be as simple as possible and as complex as desired and required, the risk assessment and management process must follow this same basic maxim. These seemingly conflicting simultaneous attributes—simplicity and complexity—can be harmonized through an effective risk communication process.

Invariably, questions posed to risk analysts during the risk assessment and management process originate from decisionmakers at various levels of responsibility, including managers, designers, stakeholders, journalists and other media professionals, politicians, proprietors, and government or other officials. These decisionmakers must be able to communicate the multidimensional perspectives of the risks and the associated challenges facing them—risks and challenges that they wish to understand better and for which they seek possible answers. In turn, risk analysts must be able to translate complex technical analysis and results into answers that decisionmakers can understand and incorporate into actionable decisions. The questions raised are likely to be complex and may require the translation of similarly complex sets of answers into a language that can be understood by intelligent laypersons.

Relevance to NextGen: *A systemic approach to risk modeling, planning, assessment, management, and communication is an absolute necessary condition for the credibility and effectiveness of the NextGen planning and development process and for the ultimate quantification of the complex multidimensional risk functions associated with its multiple subsystems [Haimes, 2011, 2012b]. As an example, consider five categories of emergent forced changes as risk* *factors associated with the unmanned aircraft system (UAS), which constitutes one subsystem of NextGen: human factors (personnel, culture, and collaboration), technological factors (aircraft type, manufacturer, and communication), infrastructure factors (launch and recovery site, control facilities, and cyber infrastructure), operational factors (aircraft operations, organizational operations, support, and schedule), and policy factors (regulations, security, and liability). Since safety is the level of acceptable risk [Lowrance, 1976] and no risk to a threatened NextGen subsystem can be completely eliminated, trade-offs are made between the cost of reducing one or more components of the risk vector and of the remaining residual risk. Furthermore, risk analysts must be capable of communicating the above reality and must have the tools and training to be capable of addressing the questions raised in the risk assessment and management process.*

12.2.3.3 Third principle: Models and state variables are central to quantitative risk analysis

Consider the following simple problem: An engineer or an environmental scientist is asked to determine the safety of a municipality's drinking water. Can the task be performed without determining the *state* of acidity of the water and the states of turbidity, dissolved oxygen, bacteria, and other pathogens? This example and almost all other challenges facing risk analysts share the following systems-based fact: All systems are characterized at any moment in time by their respective *state variables*. These states are affected and shaped by the other building blocks of mathematical models (e.g., decision, random, and exogenous variables and inputs and outputs). In reality, all state variables are under continuous natural, desired, or forced changes (positive and negative). For example, a patient's high body temperature and high blood pressure—both are state variables—vary over time and can be adjusted with medication. This does not mean that all models must be built with time-dependent state variables, despite the fact that in reality the states of the system are commonly dynamic, and that all risk-based problems are time dependent.

To further highlight the centrality of the states of the system in modeling and in risk analysis, we pose and answer the following seemingly simple question: *Why do farmers irrigate their crops in nonrainy seasons?* Indeed, the answer is fundamental to understanding both the role of state variables in modeling and the definitions of vulnerability, resilience, and thus risk. The farmer irrigates the crops to maintain appropriate soil moisture during the crops' specific growth period. Furthermore, the farmer adds fertilizer to maintain appropriate nutrients in the soil. In other words, the farmer and all decisionmakers endeavor to control or maintain the states of the system under their management at acceptable levels (by deploying appropriate risk management policy options in the case of risk-based decisionmaking).

Chen [1999] offers the following succinct definition of *state variable*: "The state $x(t_0)$ of a system at time t_0 is the information at time t_0 that, together with the input $u(t)$, for $t \geq t_0$, determines uniquely the output $y(t)$ for all $t \geq t_0$." The states of a system constitute the main building blocks of both analytical and simulation models. Models are built to answer specific questions and must be made as simple as possible and as complex as required. Without models, risk analysts would be traveling blindly in unknown and often perilous terrains with potentially dire consequences to those whom they serve. The centrality of state variables in decisionmaking, and particularly in risk analysis, has not been sufficiently emphasized. Indeed, a representative model can effectively provide answers to the questions for which it is built only by first identifying and incorporating all relevant and critical states of the system. Statistical models, which may not suggest a direct cause–effect relationship among the states of the system and the other building blocks of the mathematical model(s), can also play a valuable role in quantitative risk analysis. For example, insurance policies based on statistical distributions of accidents distribute high-cost events across the population through which they occur, thereby spreading the cost through insurance.

Vulnerability and resilience are key concepts in risk analysis. Their systems-based definitions improve our understanding of risk and help make them operational for modeling. *Vulnerability is the manifestation of the inherent states of the system (e.g., physical, technical, organizational, and cultural) that* can be subjected to a natural hazard or be exploited to adversely affect (cause harm or damage to) that system [Haimes, 1981]. The vulnerability of a system is multidimensional, a *vector* in mathematical terms. Consider the risk of terrorism to a military base; the states of the base (as a system) that represent vulnerabilities (to natural hazards or to malevolent acts) are the functionality/availability of telecommunications, electric power, water supply, soldiers' quarters, officers' quarters, perimeter security, and others, all of which are critical to the overall functionality of the base. Furthermore, these state variables are not static; they change and evolve continuously.

The resilience of a system is also *a manifestation of the states of the system, and it is a vector that is time and threat (initiating event) dependent* [Haimes, 2009, 2011, 2012b]. More specifically, *resilience* represents *the ability of the system to withstand a major disruption within acceptable degradation parameters and to recover within an acceptable cost and time.* In other words, resilience is a vector state of the system that is neither abstract or static nor deterministic. Moreover, resilience is similar to vulnerability in that it cannot simply be measured in a single unit metric; its importance lies in the ultimate multidimensional outputs of the system (the consequences) for any specific inputs (threats).

Relevance to NextGen: *Consider the four major initiatives and objectives of the NextGen enterprise: improving the capacity, safety, security, and environmental impacts of aviation in the United States. Each one of these objectives can be addressed and ultimately realized if and only if the multiple states of each subsystem of the NextGen system of systems can be identified, modeled, evaluated, and ultimately controlled to meet these objectives. These states include the quality of trained professionals, the quality and reliability of the supporting physical and cyber infrastructures, and the effectiveness of the states of communication, to cite a few. For example, consider the safety and security objectives. Since the vulnerability and resilience of each subsystem are manifestations of its states at the time of a specific initiating event (e.g., an emergent forced change), then the complex multidimensional risk function can be assessed only when sufficient knowledge of the states of each subsystem is available.*

12.2.3.4 Fourth principle: Multiple models are required to represent the essence of the multiple perspectives of complex systems of systems

When modeling large-scale physical infrastructures, such as dams and levees, more than one mathematical or conceptual model is likely to emerge, each of which may focus on a specific aspect of the system; only when these multiples are integrated do they constitute an effective representation of the system [Haimes, 2009, 2011]. This phenomenon is particularly common in organizational and infrastructure systems, where more than one model may be not only desirable but also essential. For example, a river basin may be modeled from at least four different perspectives: that is, from its (i) hydrology (e.g., controlling floods and soil erosion for different watersheds), (ii) geographical–political boundaries (e.g., state or county boundaries), (iii) temporal decomposition (e.g., different planning horizons), and (iv) functional decomposition (e.g., flood control and hydropower generation). Hence, it is impracticable to represent within a single model all the aspects of complex systems of systems, such as the NextGen enterprise, because they are commonly composed of interconnected and intra- and interdependent subsystems, with multiple functions, operations, and configurations [Haimes, 2009, 2011]. Furthermore, no single model can capture the essence of such systems, that is, their multiple dimensions and perspectives, nor would a risk analysis process derived from such a single model be complete or effective.

Indeed, most physical infrastructures and cyber infrastructures and most government, private, and other public organizations are complex systems composed of myriad subsystems, which in their essence constitute systems of systems. Each subsystem, characterized by its own states (and thus vulnerability and resilience), functionality, and organizational structure, will necessarily respond differently to different emergent forced changes. Hence, when faced with an adverse event, the system as a whole is likely to experience multidimensional consequences, which commensurately requires multiple models to represent the multiple perspectives of the system of systems.

Maier [1998] offers a comprehensive and representative characterization of systems of systems. However, the precise definition of systems of systems remains elusive. In response to the question "What is a system of systems?," Sage and Biemer [2007] provide the following answer: "No universally accepted definition of a system of systems is available at this time." On the other hand, the interdependent and interconnected physical and economic infrastructure systems and sectors of the world introduce significant modeling challenges for systems analysts, requiring that they quantify these causal intra- and interconnected and interdependent relationships. On the other hand, as will be explained in the discussion of the fifth principle, the shared states among interconnected and interdependent subsystems provide a powerful and efficacious instrument for their modeling and understanding. The growing recognition that cyber and physical infrastructures, health care, the military, the environment, and many other entities are *systems of systems* has inspired a commensurate awareness of the need for their modeling—which in turn will enable both better understanding and management of them.

Relevance to NextGen: *Consider a sample of the dozens of subsystems that constitute NextGen: the national airspace UAS; airports and their subsystems, such as runways, airliners, and air traffic controllers; and communications. Each requires one or more specific models to represent the essence of each subsystem. The next principle addresses the metamodeling and integration of these subsystems within a system of systems.*

12.2.3.5 Fifth principle: Meta-modeling and subsystems integration must derive from the intrinsic states of the system of systems

Modeling a simple system, or a complex system of systems, necessarily implies determining its properties, constructing the relationships among its inputs and outputs through its state variables and other variables and parameters (e.g., random, decision, and exogenous variables), quantifying intra- and interdependencies within and among its various components and subsystems, and determining the appropriate model topology (structure) and parameters that best represent its essence and functionality. Models enable us to experiment and test hypotheses, to explore different design options, or to generate responses to or impacts on varied scenarios and policy options. Inversely, by their nature, complex systems constitute, in many respects, black holes to modelers that can be penetrated by acknowledging our inability to directly uncover, understand, or predict their behaviors under different scenarios of emergent forced changes. We commonly lack sufficient knowledge to assess the causal relationships among the subsystems, and to compensate for this shortfall, we revert to multiperspective experimentation aided by the ingenuity, creativity, and domain knowledge of experts, supported by the availability of databases. There is no assurance that modelers will be able to explain the reasons behind variability among submodels; nevertheless, the very process of modeling such variability may highlight limited databases, inconsistent assumptions, unrecognized epistemic and aleatory uncertainties, and a host of other technical or perceptual reasons that ought not to be dismissed.

The PSM [Haimes, 2012a], which builds on the centrality of the states of the system in modeling and in risk analysis, is designed to model inter- and intradependencies between and among the subsystems of a complex system of systems by exploiting vital knowledge and information embedded in the intrinsic and extrinsic shared and unshared state variables among the subsystems.

The coordination and integration of the results of the multiple models of systems of systems are achieved at the meta-modeling phase within the PSM, thereby yielding a better understanding of the system as a whole. The essence of meta-model coordination and integration is to build on all relevant direct and indirect sources of information to gain insight into the interconnectedness and intra- and interdependencies among the submodels and on the basis of this insight to develop representative models of the system of systems under consideration. Indeed, the selection of the trial inputs to the model and the inquisitive process of making sense of the corresponding outputs are at the heart of system identification and parameter estimation. This is not a one-shot process; rather, it can be best characterized by tireless experimentation, trial and error, parameter estimation, and adjustments, as well as by questioning whether the assumed model's topology (structure) is representative of the system being modeled.

In its essence, the PSM is a modeling methodology inspired by philosophical and conceptual thinking from the arts and is driven and supported by systems engineering theory, methodology, and practice [Ehrgott, 2005]. The PSM is designed to model inter- and intradependencies between and among the subsystems of systems of systems by exploiting vital knowledge and information embedded in the intrinsic and extrinsic shared and unshared state variables among the subsystems. The PSM-based intrinsic meta-modeling of systems of systems stems from the basic assumption that *some specific commonalities, interdependencies, interconnectedness, or other relationships must exist, through shared decisions and inputs between and among any two systems within any system of systems.*

Relevance to NextGen: *To plan, develop, and ultimately deploy the NextGen system of systems, all its subsystems must function as harmoniously as possible. This challenge can be achieved if and only if there is a clear understanding of the interdependencies and interconnectedness among the intrinsic states of the entire system of systems. Thus, the metamodeling and integration of the sample of such subsystem models cited in the fifth principle must be performed judiciously and with diligence and patience through a tedious process of learn as you go and trial and error.*

12.2.3.6 Sixth principle: Multiple conflicting and competing objectives are inherent in risk management

All risk-based decisionmaking problems are characterized by multiple, noncommensurate, and conflicting objectives [Pareto, 1896; Keeney and Raiffa, 1976; Chankong and Haimes, 2008]. For example, risk management actions, such as preparedness, reducing vulnerability, or enhancing the resilience of a system, require the expenditure of monetary and human resources to protect lives, property, and the well-being of the public. Furthermore, for most such systems, there exists a hierarchy of objectives, subobjectives, sub-subobjectives, and so on. In risk management and in decisionmaking in general, it is important to identify this hierarchy of objectives, to trade off objectives within the same hierarchical level of importance, and to avoid comparing and trading off objectives that belong to different hierarchical levels. Consider, for example, an electric power generation system of systems that includes power generation turbines, transformers, supervisory control and data acquisition (SCADA), and transmission and electric power distribution systems. Multiple objectives of various levels of importance and criticality are associated with each subsystem of this complex system of systems—a fact that would make meaningless the trade-offs among the objectives associated with the entire SCADA system and those associated only with one turbine. As another example, ensuring the immediate continuity of government operations following an attack is of a different level of importance as compared to repairing a minor bridge that remains only partially functional. This principle is significant because risk to a system is commonly a multifaceted and time-dependent vector that is represented in noncommensurate units.

Pareto-optimality and trade-offs

The concept of Pareto-optimality or identifying a noninferior solution to a multiobjective optimization problem was introduced in 1896 by Pareto [1896], a prominent economist, but it has only been since 1950, when Kuhn and Tucker [1951] published necessary and sufficient conditions for (proper) noninferiority, that considerable effort has been devoted to developing procedures for generating noninferior solutions to multiobjective problems. A solution to a two-objective function is termed Pareto-optimal if improving one objective function can be achieved only by degrading the other. Indeed, Pareto-optimality is central to determining the trade-offs between the additional expenditure required for a further risk reduction. Or, paraphrasing William Lowrance [1976], "[S]afety is the level of acceptable risk"; thus, an acceptable level of safety is a Pareto-optimal level of the associated cost and residual risk.

Building on the Kuhn–Tucker necessary and sufficiency conditions for Pareto-optimality, the epsilon-constraint method [Haimes and Lasdon, 1971; Haimes, 1973, 2009] and the surrogate worth trade-off (SWT) method [Haimes and Hall, 1974] (see Chapter 5) can generate all required Pareto-optimal policies and their associated trade-offs for any multiple-objective optimization problem, without the need for any commensuration among the objective functions. The epsilon-constraint method maintains one objective function and converts the remaining objectives as constraints bounded by variable epsilons. Then, the new constrained optimization problem is further converted into an unconstrained problem through the Lagrangian function. This enables risk analysts and other modelers to generate risk management options for two or more objective functions, for example, monetary cost, human lives, and continuity of operation following a failure (time in hours, days, etc.). Furthermore, the Lagrange multiplier(s) λ_{ij} as function(s) of ε_j, namely, $\lambda_{ij}(\varepsilon_j)$, represents the trade-offs among the ith and jth objective functions, and they can be generated between any two objective functions $f_i(\cdot)$ and $f_j(\cdot)$ at the Pareto-optimal frontier, where

$$\lambda_{1j}(\varepsilon_j) = -\frac{\partial f_1(\cdot)}{\partial \varepsilon_j}, \quad j=2,3,\ldots,n \text{ for } \lambda_{1j}(\varepsilon_j) > 0.$$

Trade-off analyses are central to all decision-making processes, because ultimate decisions among multiple noncommensurate objectives are rarely made solely on the absolute values of each (whether cost, risk, or benefit); rather, they are also determined on the marginal increase or decrease in the corresponding objectives. This fact is particularly salient when two Pareto-optimal policies are closely related but have significantly different associated trade-offs.

In sum, the ultimate efficacy of risk analysis is to develop sound risk management policy options with an acceptable cost, residual risk, and benefit; and this important task can be best accomplished, when possible and practical, through quantitative trade-offs among all relevant competing and commonly noncommensurate objectives. Note, however, that the number of selected objectives in risk management introduces an implicit bias in the ultimate subsequent Pareto-optimal policies generated. For example, answers to the second question in risk management, *What are the trade-offs among all relevant costs, benefits, and risks?* (see the second principle), would squarely depend on the preselection of the specific objective functions. Furthermore, adding or deleting one objective function from consideration in the risk management process would modify the set of Pareto-optimal frontiers. To address this reality, a concerted effort must be devoted to assessing the sensitivity of the resulting policy options to the addition or deletion of one or more objective functions associated with risk management. Note that removing one of the aforementioned three objective functions would eliminate a significant number of Pareto-optimal solutions, and vice versa.

Relevance to NextGen: *The NextGen enterprise epitomizes the multiple, noncommensurate, and competing objectives associated with most, if not all, real systems. The objectives cited earlier to which NextGen aspires—improving the capacity, safety, security, and environmental impacts of aviation in the United States—are a case in point. Furthermore, diverse state, federal, and international organizations as well as the private sector have an interest in and an impact on these objectives. Thus, understanding the dynamics and interplay among these objectives and the stakeholders who are advancing them and ultimately achieving acceptable trade-offs among them through effective multiobjective risk modeling, assessment, management, and communication constitute an imperative and challenging, albeit tedious, task.*

12.2.3.7 Seventh principle: Risk analysis must account for epistemic and aleatory uncertainties

Uncertainty, commonly viewed as the inability to determine the true state of a system, is characterized by two sources that affect system's modeling and thus risk analysis [Morgan and Henrion, 1990; Paté-Cornell, 1996; Apostolakis, 1999]: (i) incomplete knowledge (epistemic) uncertainty, which manifests itself in the selection of model topology (structure) and model parameters, among other sources of ignorance (e.g., lack of knowledge of important interdependencies within the system and among other systems), and (ii) stochastic variability (aleatory) uncertainty, which includes all relevant and important random processes and events, as well as emergent forced changes. Uncertainty dominates most decisionmaking processes and is the Achilles' heel for all deterministic and most probabilistic models. Clearly, both categories of uncertainty markedly affect the quality and effectiveness of the risk analysis efforts and, ultimately, the decisionmaking process.

Significant sources of uncertainty, all too often ignored in the quest for quantitative predictive models, stem from how well the developed models will actually represent the future behavior of a system.

In addition, model uncertainties will commonly be introduced through human errors of both commission and omission. A sample of sources of uncertainties and errors in modeling includes the following six major characteristics: model topology, model parameters, model scope or focus, data, optimization technique, and human subjectivity.

Modelers' subjectivity strongly influences the outcome of the system's model(s) and thus the risk assessment, management, and communication process. This factor includes the background, training, competence, and experience of the analyst(s), as well as personal preference, self-interest, and proficiency. Human subjectivity also plays a very important role in selecting (by omission or commission) the specific building blocks of the model(s). Uncertainty stemming from stochastic variability is associated with the inherent fluctuations or differences in the quantity of concern. In other words, variability occurs when the quantity of concern is not of a specific value, but rather of a population of values. Also, access to sufficiently representative data for model construction, calibration, identification, testing, *validation*, and, ultimately, implementation, which could reduce epistemic uncertainty, is very important in modeling and in risk analysis. Finally, an inaccurate or insufficient database (due to deficiencies in collection, acquisition, processing, and analysis) may cause substantial errors in modeling and thus in the risk analysis process.

Relevance to NextGen: *Epistemic and aleatory uncertainties influence and impact all other principles. This is particularly the case with NextGen due to the numerous sources of uncertainties associated with planning for and deploying new technology; radical utilization of the restructured multilayer airspace for different flights and aircrafts; heavier reliance on GPS and diminished reliance on radar with which to navigate airplanes through narrower airspace corridors; training of professionals at almost all levels, as well as their adjustment to new responsibilities and functionality; and new rules and procedures that must be adapted internationally for all carriers that fly to the United States. These myriad sources of aleatory and epistemic uncertainty are only a scant sample among a host of other new organizational and budgetary environments in which multiple players must confront uncounted unknowns.*

12.2.3.8 Eighth principle: Risk analysis must account for risks of low probability with extreme consequences

The translation of the definition of risk into quantitative terms has been a major source of misuse by many disciplines, often leading to erroneous conclusions [Haimes, 2009, 2011, 2012b]. The most common quantification of risk is the use of the mathematical construct known as the expected-value-of-risk metric, which commensurates low probability of high-consequence events with high probability of low-consequence events. This has played a decisive role in masking the criticality of extreme and catastrophic events. It is important to recognize the misuse of the expected value—the averaging of risk—when it is used as the sole criterion for risk in decisionmaking. The partitioned multiobjective risk method (PMRM) [Haimes, 2009] supplements the expected value measure of risk with a conditional expected value of risk of extreme events (see Chapter 8). A conditional expectation is defined as the expected value of a random variable, given that this value lies within some prespecified probability or consequences range. Furthermore, the expected value of a risk-averse nonlinear utility function provides a greater weight to extreme consequences and also supplements the expected value of risk [Haimes, 2009]. During the last three decades, the interest in risk of extreme events has been the subject of many researchers and practitioners [Perrow, 1984/1999; Paté-Cornell, 1990; Phimister et al., 2004; Taleb, 2007]. The perception of risk plays an important role in the extent to which individuals, the public, and policymakers are willing to invest resources in preparedness, response, and recovery or in the mitigation of low-probability risks of extreme events deemed to exist far in the future [Slovic et al., 1982; Slovic, 2000, 2002].

Statistics of extremes

The theory of the statistics of extremes is concerned with studying the largest, for example, for flood (or the smallest, e.g., for drought), value realized from a sample of n-independent, identically distributed random variables [Ang and Tang, 1984]. The asymptotic theory of statistical extremes was developed early in the twentieth century by a number of statisticians, including Fischer and Tippet [1927] and Gnedenko [1963]. This is an analytic theory concerned with the forms of probability functions with n observations, and as $n \to \infty$, such probabilities may converge to one of three particular forms, which Gumbel [1954] termed as types I, II, and III. It can be shown that the behavior of the tail of the initial probability distribution (variate) determines to which type the extreme distribution corresponds. If the tail of the initial distribution decays exponentially, then the largest value is of type I; if the tail decays polynomially, then the largest value is of type II. The third form, type III, is only for initial variates that have a finite upper bound. The importance of extreme and catastrophic events in risk analysis makes the theory of statistics of extremes an invaluable tool. In particular, the incorporation during the last three decades of the PMRM and the statistics of extremes in risk analysis has proven to be effective and informative [Haimes, 2009].

Relevance to NextGen: Civilian air travel is probably the most scrutinized and protected transportation mode both in the United States and around the world, and historically, it has had an impressive record of safety. Thus, the expected value metric of risk cannot be applied to air transportation. Consider the over 35,000 annual fatalities from car accidents in the United States. This staggering and tragic number of deaths has become the "expected value" in surface transportation; rarely does the media report on such calamities on a regular basis. However, a collision of two airliners would be considered catastrophic by the media and the public. Thus, the risk of low probability and extreme consequences must be modeled and quantified using proper mathematics and statistics as discussed in this, the eighth principle.

12.2.3.9 Ninth principle: The time frame is central to quantitative risk analysis

Can risk analysis be performed effectively without considering future emergent forced changes within the TSS and within an appropriate time frame? The fact that the risk function is composed of the likelihood of adverse initiating events and of their resulting consequences, both of which are time dependent, implies the centrality of the time frame in risk analysis. Thus, a systemic risk analysis approach should (i) embrace continuous and timely assessment and evaluation of precursors to potential forced changes,

(ii) balance present resource allocation for preparedness (to reduce the states of vulnerability and enhance the states of resilience of the system to specific adverse initiating events or threats), (iii) update the likelihood of the tracked initiating events and of the resulting consequences (given that investments in preparedness affect the states of the system, as may be represented by changes in its vulnerability and resilience), and (iv) update, based on the aforementioned, the entire process of risk assessment, management, and communication. In its essence, Bayesian theory and analysis are predicated on the premise that new information can update and enhance prior information. In particular, flexibility in the formulation of risk management policy, with a focus on its future impacts, would add agility to the system and provide a buffer against possible catastrophic or irreversible consequences. Furthermore, a systemic risk analysis should stress the importance of impact analysis due to the dynamics that affect all systems—societal, organizational, and technological. Examples are the natural deterioration over time of all physical infrastructures, demographic changes, and advancements in technology. Furthermore, as discussed in the sixth principle, each stage in this evolving dynamic world is characterized by multiple objectives, and the trade-offs among the attainment of present objectives and future flexibility ought to be incorporated within the overall risk-informed decisionmaking process.

Dynamic systems, where the time domain is of paramount importance, introduce another challenge to the multiple objectives presented in the sixth principle. For example, ignoring the changes in the states of a dynamic safety-critical system as time progresses would render static-based models and decisions misleading, if not disastrous. Recognizing this fact, event trees, decision trees, and process control, among other methods, account for the impacts of current decisions on future options. Figure 12.2 presents the impact of policies at time $t = k$ on future options at time $k + 1$, where the dotted line represents a new Pareto-optimal frontier generated at time $k + 1$. Note that if the decisionmaker were to prefer to operate during period $k + 1$ in regime I, then policy B would be inferior to A. Conversely, if the decisionmaker were to prefer to operate during period $k + 1$ in regime II, then policy A would be inferior to B. The

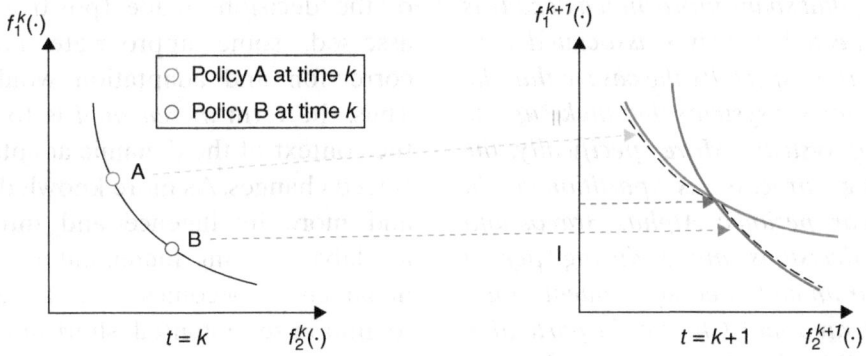

Figure 12.2 *Impact of policies at time t=k on future options at time k+1. From Haimes [2012]; © 2012 Wiley.*

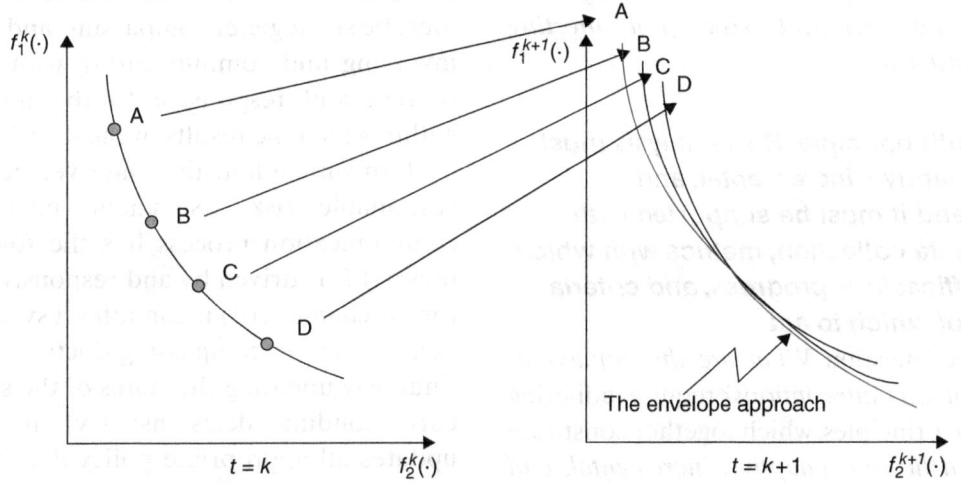

Figure 12.3 *The envelope of the combined Pareto-optimal frontier of policies A, B, C, and D for k+1st period. From Haimes [2012]; © 2012 Wiley.*

envelope approach was developed to account for the propagation of Pareto-optimal solutions from one period to another [Li and Haimes, 1987]. Figure 12.3 demonstrates this concept of the envelope approach, where the Pareto-optimal frontier can be generated for any n periods. Note that the solid-line outer curve represents the envelope of the combined Pareto-optimal frontier of policies A, B, C, and D for the $k+1$st period. Note that, analytically, the envelope approach is applicable to multiple objectives; however, graphical displays are more restricted.

Finally, the dynamic trade-offs among multiple objectives must complement and supplement the determination of Pareto-optimality in risk management, given the dynamic nature of evolving events.

Similarly, the process of eliciting the decisionmakers' preferences must be systemic and harmonious with the realization that conditions, and thus policies, are dynamic and subject to being updated and modified.

Relevance to NextGen: *The time frame has been commonly implicit in risk analysis; yet its centrality in risk and decisionmaking is so dominant that it must be addressed explicitly by decisionmakers, modelers, and other practitioners. The two sets of questions in risk assessment and risk management introduced in the second principle would lose their impetus without adherence to the time frame. Reconsider, in the context of NextGen, the third question in risk management: What are the impacts of current management decisions on future options?*

Multiyear planning that spans more than a decade is fraught with unexpected surprises associated with every subsystem; this is especially the case within the over two dozen major subsystems that make up the NextGen system of systems. More specifically, the NextGen planning process is partitioned in sequences of 3-year periods: Alpha, Bravo, and Charlie. Clearly, decisions made during period Alpha would constrain and affect subsequent future decisions during Bravo and Charlie. In particular, "wrong" or inflexible decisions at any planning period (e.g., period Alpha) could result (during periods Bravo or Charlie) in programmatic risk (the cost overrun of a project and time delay in its completion) and technical risk (not meeting performance criteria).

12.2.3.10 Tenth principle: Risk analysis must be holistic, adaptive, incremental, and sustainable; and it must be supported with appropriate data collection, metrics with which to measure efficacious progress, and criteria on the basis of which to act

Addressing the question *What are the impacts of current decisions on future options?* requires adhering to the first nine principles, which together constitute *a requisite to a holistic, adaptive, incremental, and sustainable risk analysis process.* A holistic risk assessment, management, and communication process of a system of systems must identify all conceivable sources of risk by adhering to the TSS, focusing on and tracking the most critical emergent forced changes. At a minimum, this process must recognize that a system of systems encompasses many of the following attributes: It is dynamic; it is fraught with myriad sources of risk and uncertainty, driven by emergent forced changes; it involves a hierarchy of multiple decisionmakers; it often must address multiple conflicting, competing, and noncommensurate objectives and aspirations; it is driven by ever-evolving requirements, demands, and targets; and, not least of all, it protects, adheres to, and promotes organizational core values.

Fundamental to risk analysis is the consideration of present and future risks associated with any decision or trade-off. Since critical decisions are commonly made under uncertain conditions and as events are tracked over time and the consequences of the decisions made (positive or negative) are assessed, some appropriate *incremental* course correction and adaptation would be imperative. Thus, the term *incremental* is to be understood in the context of the dynamic adaptation to emergent forced changes. As more knowledge is accumulated and more intelligence and information become available, making incremental investments in risk management becomes an effective policy with which to minimize potential shortfalls within budgeted resources.

A sustainable risk-based decisionmaking process must share many of the traits of a holistic process, because the latter implies that the process must be strategic, encompassing, and well balanced, involving and communicating with all concerned parties, and responsive to the uncertain world within which the results of the decision will be realized. In sum, a holistic, adaptive, incremental, and sustainable risk assessment, management, and communication process has the following attributes: (i) It is driven by and responsive to emergent forced changes; (ii) it connotes a systemic, comprehensive, and encompassing doctrine; (iii) it is continuously updating the states of the system and the corresponding decisions; (iv) it incrementally updates all appropriate policy decisions in a step-by-step manner that builds on new knowledge, ensuring a viable and repeatable process; and (v) it is sustainable—that is, it is "maintained at length without interruption, weakening, or losing power or quality" [Webster, 1976].

Finally, risk analysis must be supported with:

- ***Data collection***: Develop a data collection plan and deployment process for each component of the tracked top emergent forced changes and precursors thereto.
- ***Metrics:*** Develop metrics with which to (i) measure and evaluate, on the basis of the collected database, the progress being made in the risk analysis process; (ii) track and evaluate precursors to emerging risks; (iii) assess the impacts of current decisions and progress on future options and on succeeding phases in the risk analysis process; and (iv) evaluate the experiences, results, and lessons learned as a prelude to an effective decisionmaking process.

Guiding criteria for decisionmaking: Develop guiding criteria for required actions when a tracked risk scenario or other risks emerge as imminent and thus must be confronted and acted upon.

Relevance to NextGen: *The necessity for adherence to the tenth principle is grounded on the complexity of all systems of systems (including NextGen) and on all nine principles and other premises discussed in this chapter. The long-term planning horizon and complexity of most such systems require a holistic–comprehensive approach; the myriad emergent forced changes and the dynamic changes within and outside these systems call for an adaptive process; the subsequent responses to these internal and external forces must be incremental (unless a catastrophic failure is discovered); and all decisions must be sustainable. Furthermore, to ensure an effective risk assessment, management, and communication process, a continuous adaptive feedback mechanism must be put in place. This mechanism must be supported with appropriate data collection, metrics with which to measure efficacious progress, and criteria on the basis of which to act when actionable decisions must be made.*

12.2.4 Part Four: The Interim Destination and Back to Part One—The Process of Risk Modeling, Assessment, Management, and Communication

Part four constitutes the ultimate incorporation and deployment of the ten guiding principles within the complex process of risk assessment and management (answering the two sets of questions, respectively) and risk communication.

The multidisciplinary nature of the field of risk analysis necessarily implies that each discipline has over time developed, adopted, and adapted appropriate theories, methodologies, and practices from other disciplines. This cross-fertilization has been the hallmark of this field as is manifested through the diverse membership of the SRA and by its flagship journal *Risk Analysis: An International Journal.* In this context, this chapter is written from the perspectives of systems engineering with the hope and expectation that the principles it introduces will be supplemented and complemented by others.

REFERENCES

Ang, A.H., and W.H. Tang, 1984, *Probability Concepts in Engineering Planning and Design*, Vol. **2**, Wiley Publications, New York.

Apostolakis, G., 1999, The Distinction between aleatory and epistemic uncertainties is important: an example from the inclusion of aging effects into probabilistic safety assessment, *Proceedings of PSA'99*, August 22–25, American Nuclear Society, Washington, DC.

Chankong, V., and Y.Y. Haimes, 2008, *Multiobjective Decision Making: Theory and Methodology*, Dover, New York.

Chen, C., 1999, *Linear System Theory and Design*, Third edition, Oxford University Press, New York.

Duderstadt, J., W. Wulf, and R. Zemsky, 2005, Envisioning a transformed university, *Issues in Science and Technology* **22**(1): 35–41. National Academy Press.

Ehrgott, M., 2005, *Multicriteria Optimization*, Second edition, Springer, New York.

Federal Aviation Administration (FAA), 2010, *FAA's NextGen: Implementation Plan*, NextGen Integration and Implementation Office, Washington, DC.

Fischer, R.A., and L.H.C. Tippett, 1927, Limiting forms of the frequency distribution of the largest of smallest numbers of a sample, *Proceedings of the Cambridge Philosophical Society* **24**: 180–190.

Gnedenko, B., 1963, *The Theory of Probability*. Translated from Russian by Seckler, B., Chelsea Publishing Company, New York.

Gumbel, E.J., 1954, *Statistical Theory of Extreme Values and Some Practical Applications*, Applied Mathematical Series, 3, National Bureau of Standards, Washington, DC.

Haimes, Y.Y., 1973, Integrated system identification and optimization, *Advances in Control System Theory and Applications*, C.T. Leondes, Ed. Vol. 10, Academic Press, New York, pp. 435–518.

Haimes, Y.Y., 1981, Hierarchical holographic modeling, *IEEE Transaction on Systems, Man, and Cybernetics Part A: Systems and Humans* **11**(9): 606–617.

Haimes, Y.Y., 1991, Total risk management, *Risk Analysis* **11**(2): 169–171.

Haimes, Y.Y., 2009, *Risk Modeling, Assessment, and Management*, Third edition, John Wiley & Sons, Inc., Hoboken, NJ.

Haimes, Y.Y., 2011, On the complex quantification of risk: systems-based perspective on terrorism, *Risk Analysis* **31**(8): 1175–1186.

Haimes, Y.Y., 2012a, Systems-based guiding principles for risk modeling, planning, assessment, management, and communication, *Risk Analysis*, **32**(9): 1451–1467.

Haimes, Y.Y., 2012b, Modeling complex systems of systems with phantom system models, *Systems Engineering*, **15**(3): 333–346.

Haimes, Y.Y., and W.A. Hall, 1974, Multiobjectives in water resources systems analysis: The surrogate worth trade-off method, *Water Resources Research* **10**(4): 615–624.

Haimes, Y.Y., L.S. Lasdon, and D.A. Wismer, 1971, On the bicriterion formulation of the integrated system identification and systems optimization, *IEEE Transactions on Systems, Man, and Cybernetics* **SMC-1**: 296–297.

Kaplan, S., and B. Garrick, 1981, On the quantitative definition of risk, *Risk Analysis* **1**(1): 11–27.

Kaplan, S., Y.Y. Haimes, and B. Garrick, 2001, Fitting hierarchical holographic modeling into the theory of scenario structuring and a resulting refinement of the quantitative definition of risk, *Risk Analysis* **21**(5): 807–815.

Keeney, R., and H. Raiffa, 1976, *Decisions with Multiple Objectives*, Wiley, New York.

Kuhn, H.W., and A.W. Tucker, 1951, Nonlinear programming, *Proceedings, 2nd Berkeley Symposium on Mathematical Statistics and Probability*, University of California Press, Berkeley, CA, pp. 481–492.

Li, D., and Y.Y. Haimes, 1987, The envelope approach for multiobjective optimization problems, *IEEE Transactions on Systems, Man and Cybernetics* **17**(6): 1026–1038.

Lowrance, W., 1976, *Of Acceptable Risk*, William Kaufmann, Los Altos, CA.

Maier, M.W., 1998, Architecting principle for systems-of-systems, *Systems Engineering* **1**(4): 267–284.

McCormick, N., 1981, *Reliability and Risk Analysis*, Academic Press, New York.

Morgan, G., and M. Henrion, 1990, *Uncertainty*, Cambridge University Press, Cambridge, MA.

Pareto, V., 1896, Cours d'économie politique professé a l'université de Lausanne, Vol. I, in-8 de v-430 pages. Rouge, editeur a Lausanne; Pichon, libraire a Paris, 23.

Paté-Cornell, E., 1990, Organizational aspects of engineering system safety: the case of offshore platforms, *Science* **250**: 1210–1217.

Paté-Cornell, E., 1996, Uncertainties in risk analysis: six levels of treatment, *Reliability Engineering and System Safety* **54**: 95–111.

Perrow, C., 1984/1999, *Normal Accidents: Living with High-Risk Technologies*, Princeton University Press, Princeton, NJ.

Phimister, J., V. Bier, and H. Kunreuther (Eds.), 2004, *Accident Precursors Analysis and Management: Reducing Technological Risk through Diligence*, National Academy of Engineering, The National Academies Press, Washington, DC.

Sage, A.P., and S.M. Biemer, 2007, Processes for system family architecting, design and integration, *IEEE Systems Journal* **1**(1): 5–16.

Slovic, P., 2000, *The Perception of Risk*, Earthscan Publications Ltd., Sterling, VA.

Slovic, P., 2002, Terrorism as hazard: a new species of trouble, *Risk Analysis* **22**(3): 425–426.

Slovic, P., B. Fischhoff, and S. Lichtenstein, 1982, Why study risk perception? *Risk Analysis* **2**(2): 83–93.

Taleb, N., 2007, *The Black Swan: The Impact of the Highly Improbable*, Random House, New York.

Webster, 1976, *Webster's Third International Unabridged Dictionary*, Merriam-Webster, Springfield, MA.

13

Fault Trees

13.1 INTRODUCTION

One common denominator—unreliability—unifies all of the following undesired events: Two cars collide due to the malfunction of one car's brakes; pollutants are discharged from a wastewater treatment plant due to the failure of a pump; a nuclear reactor power plant is shut down due to the failure of a relay. Safe operation in these three examples depends on the proper functioning, namely, reliability, of all critical components that constitute these systems. Dual brakes installed in parallel in the disabled car, two pumps installed in parallel in the wastewater treatment plant, and two relays installed in parallel in the nuclear reactor could have prevented the above failures.

Myriad components constitute a technologically based system, each with a given reliability and configuration. We define reliability as the conditional probability that the system (or a component thereof) will perform its intended function(s) throughout an interval, given that it was functioning correctly at time t_0. Evaluating the composite reliability of the overall system without a systematic process is a daunting task. Fault-tree analysis is a systematic and quantitative process that takes into account the unreliability contributions of the various components to the overall system [see Apostolakis, 1991; Henley and Kumamoto, 1992]. Recent additions to the literature on fault trees include works from Ebeling [2005], Limnios [2007], and Birolini [2007].

Fault-tree analysis was first conceived in 1961 by H.A. Watson of Bell Telephone Laboratories in connection with a US Air Force contract to study the Minuteman launch control system. At a safety symposium held in 1965 at the University of Washington, cosponsored by the Boeing Company, several papers expounded the virtues of fault-tree analysis. These presentations marked the beginning of a widespread interest in using fault-tree analysis as a safety and reliability tool for complex dynamic systems such as nuclear reactors. Since then, fault-tree analysis has been widely used for evaluating the safety and reliability of complex engineering systems. Thus far, the most widespread use of fault trees has been in the nuclear industry beginning with the *Reactor Safety Study* [U.S. Nuclear Regulatory Commission, 1975] conducted over a 2-year period.

Risk Modeling, Assessment, and Management, Fourth Edition. Yacov Y. Haimes.
© 2016 John Wiley & Sons, Inc. Published 2016 by John Wiley & Sons, Inc.

One of the leading documents on fault-tree analysis is the *Fault Tree Handbook* written by the US Nuclear Regulatory Commission (NRC) [1981]. This handbook remains the primer for all students of fault trees. The following is a succinct description of the fault-tree model [U.S. Nuclear Regulatory Commission, 1981]:

> A fault tree analysis can be simply described as an analytical technique, whereby an undesired state of the system is specified (usually a state that is critical from a safety standpoint), and the system is then analyzed in the context of its environment and operation to find all credible ways in which the undesired event can occur. The fault tree itself is a graphic model of the various parallel and sequential combinations of faults that will result in the occurrence of the predefined undesired event. The faults can be events that are associated with component hardware failures, human errors, or any other pertinent events which can lead to the undesired event. A fault tree thus depicts the logical interrelationships of basic events that lead to the undesired event—which is the top event of the fault tree. It is important to understand that a fault tree is not a model of all possible system failures or all possible causes for system failure. A fault tree is tailored to its top event which corresponds to some particular system failure mode, and the fault tree thus includes only those faults that contribute to this top event. Moreover, these faults are not exhaustive—they cover only the most credible faults as assessed by the analyst.

Fault-tree analysis, which is one of the principal methods for analyzing systems safety, can be used to identify potential weaknesses in a system or the most likely causes of a system's failure. The method is a detailed deductive analysis that requires considerable system information and can also be a valuable design or diagnostic tool.

In fault-tree analysis, the sequence of events leading to the probable occurrence of a predetermined event is systematically divided into primary events whose failure probabilities can be estimated. Several methods have been suggested for handling uncertainty in the failure probabilities of the primary event of interest; however, most of them develop merely an interval of uncertainty. Also, most available research results in fault-tree analysis are applicable only to cases with *point probability distributions*.

Most current methods for fault-tree analysis do not provide the means to use probability distributions for the primary components. When these methods do use probability distributions, at best they develop an interval of uncertainty for the probability of the undesired event of interest. Also, most current methods use the unconditional expected value as a measure of risk. This chapter introduces a relatively new method that incorporates conditional expectations and multiple-objective analysis with fault-tree analysis. It provides managers and decisionmakers with more information about the system rather than merely providing a single point probability for the undesired event.

The conventional approach to fault-tree analysis has been the use of point probabilities for the analysis of the system. The approach is valid when we have accurate data on the component failure rate along with a point distribution. This is, however, practically never the case in most applications. In most cases, the database available for component failure rate is sketchy or has a wide uncertainty interval associated with it. Also, since fault trees deal with rare events, often the failure of some components of the system may not have occurred in the past and thus would not be included in the database.

To overcome the limitations imposed by the unavailability of data, it is common practice to approximate the available data and/or the subjective estimates of the failure rates by a probability distribution.

When different probability distributions are used for basic component failure, existing analytical methods are not very useful because there are no closed-form solutions available for the products and the sums of these distributions. Methods based on analytical techniques (e.g., variance decomposition, variance partitioning, and system moments) develop, at best, an interval of uncertainty or confidence intervals for the overall system failure rate. This is accomplished by approximating the basic component distributions to normal or lognormal distributions and then using known relationships for adding normal or multiplying lognormal distributions. These methods tend to be computationally complex and are difficult to adopt for large systems.

In such cases, methods based on a combination of random variables through numerical simulation are very useful. Numerical methods, when used for fault-tree analysis, can handle most well-known probability distributions, such as normal, lognormal, exponential, and Weibull. System components having these failure rate distributions may be connected in series or in parallel.

Numerical methods are based on the generation of pseudorandom numbers to approximate known or assumed probability distributions for system components. Random numbers generated to approximate a probability distribution can be augmented to obtain the required information about the top event of the system.

The use of simulation methods has grown rapidly with the increased use of high-speed digital computers, since they overcome the one major limitation of numerical methods—the requirement of a large amount of computer time. Personal computers, which can run dedicated programs without having to share processor time with other applications, have also contributed to the widespread use of numerical methods based on simulation.

The main limitation of many current methods stems from the fact that it is not possible to obtain the complete probability distribution for the top event. Among the exception is the Integrated Reliability and Risk Analysis System (IRRAS), which is an integrated computer software for performing probabilistic risk assessment using fault trees [Russell et al., 1987]. These methods can develop the moments of the distribution only for the top event, which can then be used to approximate the top-event distribution using empirical distributions. Alternatively, these methods can develop measures for the components at the basic level. Cox [1982] uses variance as the primary measure in his approach and ranks the input variables according to their contributions to the output uncertainty. Also, all analytical methods use approximations at one stage or another in order to simplify the analytical expressions obtained. This naturally affects the results. The moments of the distributions are represented instead of the distributions themselves. This factor is an approximation in itself, in that two distributions that may have the same moments are treated in the same way even though they may be completely different. Most analytical methods cannot be used to model dependencies among components.

13.2 BASIC FAULT-TREE ANALYSIS

13.2.1 Fault Trees and Extreme Events

The theory, methodology, and utilization of fault trees have become so extensive over the last two decades that no one chapter can do justice to the subject. This chapter is intended to serve two main goals. The first is to provide introductory material on fault trees to readers who are interested in the broader subject of risk analysis. They can then consult any of several references on fault-tree analysis, such as Apostolakis [1991], Henley and Kumamoto [1992], Hoyland and Rausand [1994], Johnson [1989], Martensen and Butler [1987], Rao [1992], Storey [1996], and US NRC [1981]. The second goal is to introduce the distribution analyzer and risk evaluator (DARE) method for fault-tree analysis [Tulsiani, 1989], which incorporates extreme events, focusing on the partitioned multiobjective risk method (PMRM) discussed in Chapters 8 and 11. Theoretical foundations for the incorporation of the statistics of extremes may be found in Pannullo [1992] and Pannullo et al. [1993]. For example, Pannullo et al. [1993] developed an analytical method to determine the parameters of extreme value distributions—the characteristic largest value and the inverse measure of dispersion (which are widely discussed in Chapter 11)—in the fault trees for the overall series system. This methodology also determines the Gumbel type of a series system, given that the Gumbel types of the components are known.

13.2.2 Procedure for Fault-Tree Analysis

To analyze a system using fault trees, we first specify the undesired state of the system whose occurrence probability we are interested in determining. This state may be the failure of the system or of a subsystem. Once this undesired state has been

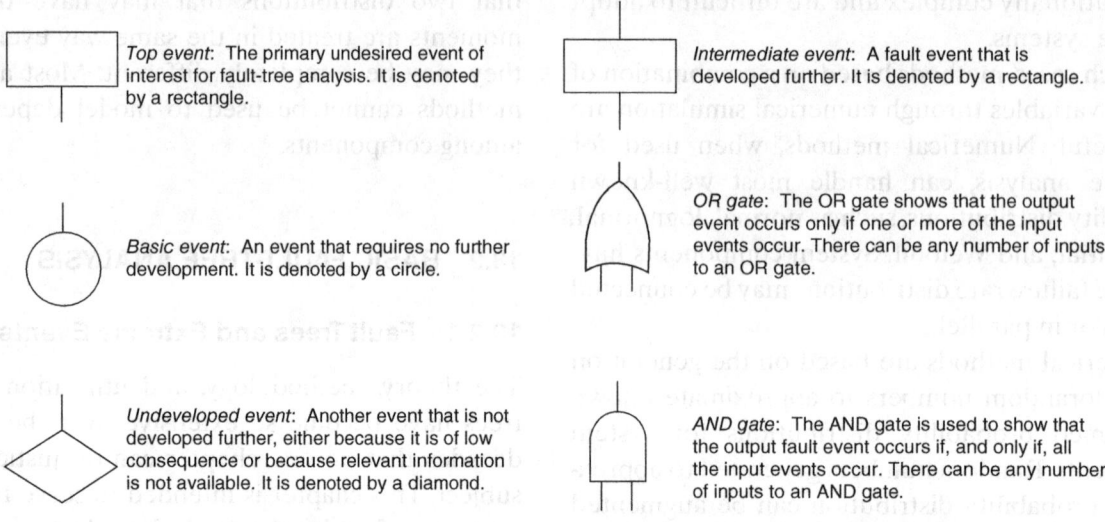

Figure 13.1 Basic components of a fault tree.

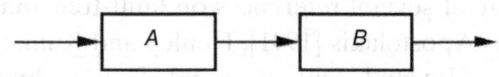

Figure 13.2 Components in series.

specified, a list is made of all the possible ways in which this event can occur. Each of the possible ways is then examined independently to find out how it can occur, until it is no longer feasible or cost-effective to carry out the analysis further.

The lowest-level events are called primary events. All the events are laid out in a *tree* form connected by *gates* that show the relationships between successive levels of the tree. A few of the most common symbols used for fault-tree construction and analysis are shown in Figure 13.1.

A fault tree is a graphic model of the various sequential and parallel combinations of faults (see Figures 13.2 and 13.6a) that will result in the occurrence of the predefined undesired event. The faults can be associated with component hardware failures, human errors, or any other pertinent events that can lead to the undesired outcome. A fault tree thus depicts the logical interrelationships of the basic events that lead to the undesired top event.

13.2.3 Limitations of Fault-Tree Analysis

One major limitation of fault-tree analysis concerns the qualitative aspects of fault-tree construction. It is possible that significant failure modes may be overlooked during the analysis. It is thus very important that the analyst thoroughly understands the system before the fault tree is constructed.

Another limitation is the difficulty in applying Boolean logic to describe the failure modes of some components when their operation can be partially successful. Techniques exist to address this problem, but they increase the complexity of the analysis.

Also, there is the lack of appropriate data on failure modes; even though data might be available, they may not be applicable to the system under consideration. Data on human reliability is very sketchy if at all available.

13.3 RELIABILITY AND FAULT-TREE ANALYSIS

13.3.1 Risk versus Reliability Analysis

The distinction between reliability and risk is not merely a semantic issue; rather, it is a major element in resource allocation throughout the life cycle of a product (whether in design, construction, operation, maintenance, or replacement). The distinction between risk and safety, well articulated over two decades ago by Lowrance [1976], is vital when addressing the design, construction, and maintenance of physical systems, since by their nature such systems are built of materials that are susceptible to failure. The probability of such a failure and its associated consequences constitutes the measure of risk. Safety manifests itself in the level of risk that is acceptable to those in charge of the system. For instance, the selected strength of chosen materials, and their resistance to the loads and demands placed on them, is a manifestation of the level of acceptable safety. The ability of materials to sustain loads and avoid failures is best viewed as a random process—a process characterized by two random variables: (i) the load (demand) and (ii) the resistance (supply or capacity).

Unreliability, as a measure of the probability that the system does not meet its intended functions, does not include the consequences of failures. On the other hand, risk as a measure of the probability (i.e., unreliability) and severity (consequences) of the adverse effects is inclusive and thus more representative.

Clearly, not all failures can justifiably be prevented at all costs. Thus, system reliability cannot constitute a viable metric for resource allocation unless an a priori level of reliability has been determined. This brings us to the duality between risk and reliability on the one hand and multiple-objective and single-objective optimization on the other.

In the multiple-objective model, the level of acceptable reliability is associated with the corresponding consequences (i.e., constituting a risk measure) and is thus traded off with the associated cost that would reduce the risk (i.e., improve the reliability). In the single-objective model, on the other hand, the level of acceptable reliability is not explicitly associated with the corresponding consequences; rather, it is predetermined (or parametrically evaluated) and thus is considered as a constraint in the model.

There are, of course, both historical and evolutionary reasons for the more common use of reliability analysis rather than risk analysis as well as substantive and functional justifications. Historically, engineers have always been concerned with strength of materials, durability of product, safety, surety, and operability of various systems. The concept of risk as a quantitative measure of both the probability and consequences (or an adverse effect) of a failure has evolved relatively recently. From the substantive–functional perspective, however, many engineers or decisionmakers cannot relate to the amalgamation of two diverse concepts with different units—probabilities and consequences—into one concept termed risk. Nor do they accept the metric with which risk is commonly measured. The common metric for risk— the expected value of adverse outcome—essentially commensurates events of low probability and high consequences with those of high probability and low consequences. In this sense, one may find basic philosophical justifications for engineers to avoid using the risk metric and instead work with reliability. Furthermore and most important, dealing with reliability does not require the engineer to make explicit trade-offs between cost and the outcome resulting from product failure. Thus, design engineers isolate themselves from the social consequences that are by-products of the trade-offs between reliability and cost. The design of levees for flood protection may clarify this point.

Designating a *one-hundred-year return period* means that the engineer will design a flood protection levee for a predetermined water level that on the average is not expected to be exceeded more than once every hundred years. Here, ignoring the socioeconomic consequences, such as loss of lives and property damage due to a high water level that would most likely exceed the one-hundred-year return period, the design engineers shield themselves from the broader issues of consequences, that is, risk to the population's social well-being. On the other hand, addressing the multiobjective dimension that the risk metric brings requires much closer interaction and coordination between the design engineers and the decisionmakers. In this case, an interactive process is required to reach acceptable levels of risks, costs, and benefits. In a nutshell, complex issues, especially those involving public policy with health and socioeconomic dimensions, should not be addressed through overly simplified models and tools. As the demarcation line between hardware and software slowly but surely fades away and with the ever-evolving and increasing role of design engineers and systems analysts in technology-based decisionmaking, a new paradigm shift is emerging. This shift is characterized by a strong overlapping of the responsibilities of engineers, executives, and less technically trained managers.

The likelihood of multiple or compound failure modes in infrastructure systems (as well as in other physical systems) adds another dimension to the limitations of a single reliability metric for such infrastructures [Schneiter et al., 1996; Park et al., 1998]. Indeed, because one must address multiple reliabilities of a system, the need for explicit trade-offs among risks and costs becomes more critical. Compound failure modes are defined as two or more paths to failure with consequences that depend on the occurrence of combinations of failure paths. Consider the following examples: (i) a water distribution system, which can fail to provide adequate pressure, flow volume, water quality, and other needs; (ii) the navigation channel of an inland waterway, which can fail by exceeding the dredge capacity and by closing to barge traffic; and (iii) highway bridges, where failure can occur from deterioration of the bridge deck, corrosion or fatigue of structural elements, or an external loading such as flood. Water quality could be used as another basis for the reliability of the water distribution system. None of these failure modes is independent of the others in probability or consequence. For example, deck cracking can contribute to structural corrosion. Structural deterioration in turn can increase the vulnerability of the bridge to floods; nevertheless, the individual failure modes of bridges are typically analyzed in isolation of one another. Acknowledging the need for multiple metrics of reliability of capacity, pressure, hydraulic capacity (joint requirements for flow volume and pressure in the system), or quality could markedly improve decisions regarding maintenance and rehabilitation, especially when these multiple reliabilities are augmented with risk metrics.

Over time, most, if not all, man-made products and structures ultimately fail. Reliability is commonly used to quantify this time-dependent failure of a system. Indeed, the concept of reliability plays a major role in engineering planning, design, development, construction, operation, maintenance, and replacement.

To streamline our discussion on fault-tree analysis, we define the following terms associated with reliability and its modeling:

- Reliability $R(t)$: The probability that the system operates correctly (or performs its intended function) throughout the interval $(0, t)$ given that it was operating correctly at $t=0$.
- Unreliability $Q(t)$: The probability that the system fails during interval $(0, t)$, given that it was operating correctly at $t=0$.
- Failure density $f(t)$: The term $f(t)\,dt$ is the probability that the system fails in time dt about t.
- Failure rate $\lambda(t)$: The term $\lambda(t)\,dt$ is the conditional probability of system failure in time dt about t, given that no failure occurs up to time t:

$$Q(t) = 1 - R(t) \tag{13.1}$$

$$f(t) = \frac{dQ(t)}{dt} = -\frac{dR(t)}{dt} \tag{13.2}$$

$$\lambda(t) = \frac{f(t)}{R(t)} = -\frac{1}{R(t)}\frac{dR(t)}{dt} \tag{13.3}$$

$$R(t) = \exp\left[-\int_0^t \lambda(\tau)d\tau\right] \tag{13.4}$$

13.3.2 Series System

When subsystems are connected in series (see Figure 13.2), the system fails when at least one of its components fails:

$$R(t) = R_A(t)R_B(t) \tag{13.5}$$

$$\begin{aligned}Q(t) &= 1 - [1 - Q_A(t)][1 - Q_B(t)] \\ &= Q_A(t) + Q_B(t) - Q_A(t)Q_B(t)\end{aligned} \tag{13.6}$$

To generalize Eq. (13.5), let $R_i(t)$ represent the reliability of the ith subsystem and let $R_s(t)$ represent the reliability of the entire system:

$$R_s t = \prod_{i=1}^{n} R_i(t) \tag{13.7}$$

$$\begin{aligned}Q_s(t) &= 1 - R_s(t) \\ &= 1 - \prod_i R_i(t) \\ &= 1 - \prod_i (1 - Q_i(t))\end{aligned} \tag{13.8}$$

$$R_s(t) < \min_i \{R_i(t)\} \tag{13.9}$$

Note that Eq. (13.9) is correct for subsystems in series, unless all components have $R_i(t)=1$; then the inequality sign should be modified.

Quantitative fault-tree analysis is based on Boolean algebra, where the events either occur or do not occur. The two basic gates used in fault-tree analysis are the OR gate and the AND gate.

13.3.2.1 The OR gate

The OR gate represents the union of the events attached to the gate. Any one or more of the input events must occur to cause the event above the gate to occur. The OR gate is equivalent to the Boolean symbol +. For example, the OR gate with two input events (as shown in Figure 13.3a) is equivalent to the Boolean expression

$$S = A + B = A \cup B \tag{13.10}$$

In terms of probability,

$$\begin{aligned} P(S) &= P(A) + P(B) - P(AB) \\ &= P(A) + P(B) - P(A)P(B|A) \\ &= P(A) + P(B) - P(B)P(A|B) \end{aligned} \tag{13.11}$$

If A and B are independent events, then $P(B|A) = P(B)$ or $P(A|B) = P(A)$; therefore,

$$P(S) = P(A) + P(B) - P(A)P(B) \tag{13.12}$$

The NRC uses rare-event approximation in its *Fault Tree Handbook* [U.S. Nuclear Regulatory Commission, 1981]. In this case, we have

$$P(S) \approx P(A) + P(B) \tag{13.13}$$

Consider a simple water pumping system [U.S. Nuclear Regulatory Commission, 1981] consisting of a water source, two pumps in parallel, a valve, and a reactor (see Figure 13.4). A no flow of water to the reactor constitutes the undesired event—that is, a failure of the system.

Denote the failure of the system as the top event, T. Then we can represent this simple water pumping system as shown in Figure 13.3b.

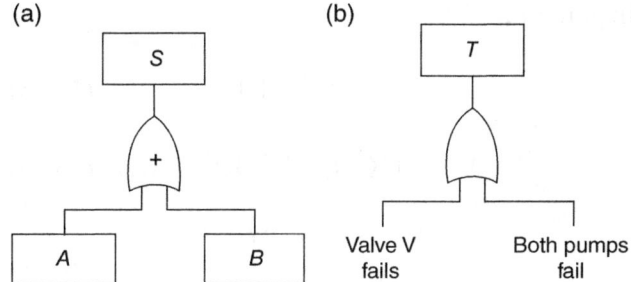

Figure 13.3 (a) OR gate (components in series) and (b) OR gate for pumping system.

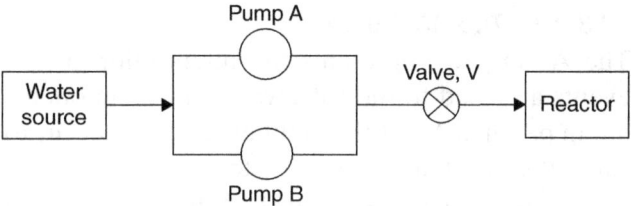

Figure 13.4 Water pumping system. From US Nuclear Regulatory Commission [1981]; © 1981.

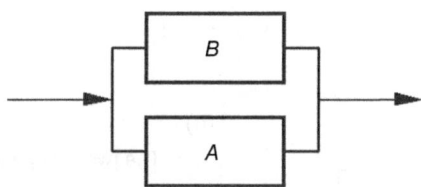

Figure 13.5 Schematic diagram for the two pumps in parallel.

If either valve V or both pumps fail, the top event will occur—failure of the system. The two pumps are designed in parallel as discussed next.

13.3.3 Parallel System

When subsystems are connected in parallel (see Figure 13.5), the system fails only when all of its components fail.

For the system in Figure 13.5, the unreliability of the pumps in parallel is

$$\begin{aligned} Q(t) &= Q_A(t)Q_B(t) \\ R(t) &= 1 - Q(t) = 1 - [1 - R_A(t)][1 - R_B(t)] \\ &= R_A(t) + R_B(t) - R_A(t)R_B(t) \end{aligned} \tag{13.14}$$

In general,

$$Q_s(t) \cong \prod_i Q_i(t) \quad (13.15)$$

$$R_s(t) = 1 - \prod_i Q_i(t) = 1 - \prod_i (1 - R_i(t)) \quad (13.16)$$

$$R_s(t) > \max_i \{R_i(t)\} \quad (13.17)$$

Note that Eq. (13.17) is correct for parallel subsystems only.

13.3.3.1 The AND gate

The AND gate represents the intersection of the events attached to the gate, where the components are in parallel. All of the input events must occur to cause the event above the gate to occur.

The AND gate is equivalent to the Boolean symbol ·. For example, the AND gate with two input events (as shown in Figure 13.6a) is equivalent to the Boolean expression

$$S = A \bullet B \quad (13.18)$$

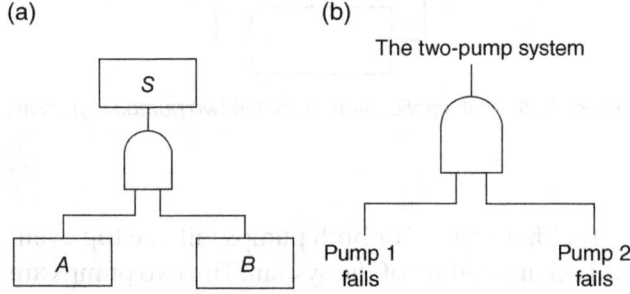

Figure 13.6 (a) AND gate (components in parallel) and (b) AND gate for water pumping system.

If A and B are independent events, then $P(B|A) = P(B)$ or $P(A|B) = P(A)$; therefore,

$$\begin{aligned} P(S) &= P(A)P(B|A) \\ &= P(B)P(A|B) = P(AB) \end{aligned} \quad (13.19)$$

$$P(S) = P(AB) = P(A)P(B) \quad (13.20)$$

The AND gate is used to demonstrate that the output fault occurs only if all the input faults occur, as Figure 13.6b illustrates.

13.3.4 Venn Diagram Representation of Sets

The operational rules of set theory and their graphical representation through Venn diagrams markedly simplify the complexity of fault trees. As will be demonstrated in a subsequent discussion, a system with a large number of components (subsystems) that are connected in series and parallel (through OR gates and AND gates) can be reduced to a simple connection through the use of operational rules of set theory. A brief review of the notation and laws of the algebra of sets is presented in Figure 13.7.

13.3.5 Boolean Algebra

Boolean algebra is the algebra of events; it is especially important in situations where events either occur or do not occur. Understanding the rules of Boolean algebra contributes toward the construction and simplification of fault trees (Table 13.1).

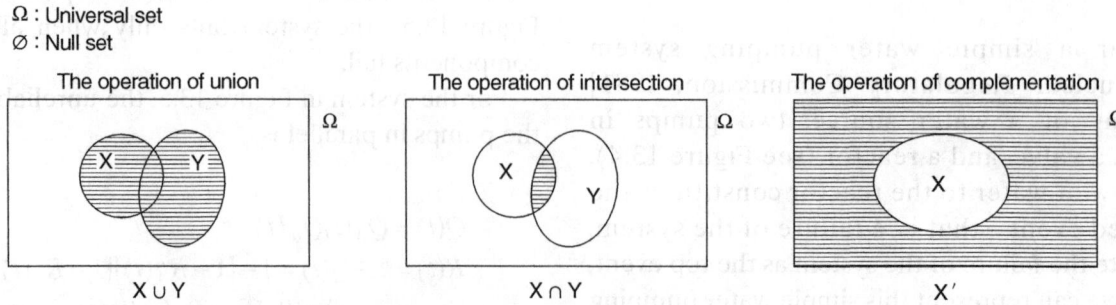

Figure 13.7 Venn diagram representation.

TABLE 13.1 Laws of the Algebra of Sets

Absorption Law
1a. $A \cup A = A$ 1b. $A \cap A = A$

Associative Law
2a. $(A \cup B) \cup C = A \cup (B \cup C)$ 2b. $(A \cap B) \cap C = A \cap (B \cap C)$

Commutative Law
3a. $A \cup B = B \cup A$ 3b. $A \cap B = B \cap A$

Distributive Law
4a. $A \cup (B \cap C) = (A \cup B) \cap (A \cup C)]$ 4b. $A \cap (B \cup C) = (A \cap B) \cup (A \cap C)$

Operations with $\varnothing$ and Ω
5a. $A \cup \varnothing = A$ 5b. $A \cap \Omega = A$
6a. $A \cup \Omega = \Omega$ 6b. $A \cap \varnothing = \varnothing$

Complementation Law
7a. $A \cup A' = \Omega$ 7b. $A \cap A' = \varnothing$
8a. $(A')' = A$ 8b. $\Omega' = \varnothing, \varnothing' = \Omega$

de Morgan's Theorem
9a. $(A \cup B)' = A' \cap B'$ 9b. $(A \cap B)' = A' \cup B'$

From US Nuclear Regulatory Commission [1981].

Operation	Probability	Mathematics	Engineering	Symbol	Structure
Union of A and B	A or B	$A \cup B$	$A + B$		Series
Intersection of A and B	A and B	$A \cap B$	$A \cdot B$ or AB		Parallel
Complement of A	Not A	A' or $\bar{A}$	A' or $\bar{A}$		

Example:

Show that $[(A \cdot B) + (A \cdot B') + (A' \cdot B')]' = A' \cdot B$

$= (A \cdot B)' \cdot (A \cdot B')' \cdot (A' \cdot B')'$ de Morgan's theorem
$= (A' + B') \cdot (A' + B) \cdot (A + B)$ de Morgan's theorem
$= (A' + (B' \cdot B)) \cdot (A + B)$ Distributive law
$= (A' + \varnothing) \cdot (A + B)$ Complementation law
$= A' \cdot (A + B)$
$= (A' \cdot A) + (A' \cdot B)$ Distributive law
$= \varnothing + (A' \cdot B)$
$= A' \cdot B$

13.4 MINIMAL CUT SETS

A minimal cut set is defined as the smallest combination of component failures, which, if they all occur, will cause the top event to occur [U.S. Nuclear Regulatory Commission, 1981].

By definition, a minimal cut set is a combination of intersections of primary events in parallel sufficient for the top event to occur (if all parallel components fail). This combination is the *smallest* combination in that all the failures in the minimal cut set are needed to occur for the top event (system failure) to occur. If any one component in the parallel combination does not occur, then the top event will not occur (by this combination).

A fault tree will consist of a finite number of minimal cut sets, all of which are in series, which are unique for the top event to occur. Since the combination of all minimal cut sets is in series, then the failure of any cut set will cause the failure of the entire system. *In other words, once the minimal cut sets are known, then any system can be written as the series arrangements of its cut sets, and the components of each minimal cut set are arranged in parallel.* Figures 13.8 and 13.9 are a representation of a two-component minimal cut set.

398 FAULT TREES

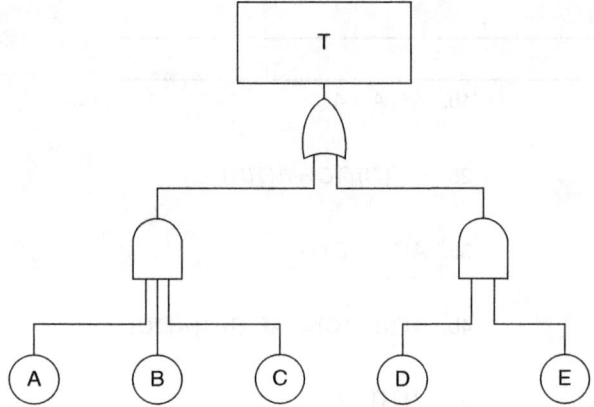

Figure 13.8 A five-component fault tree.

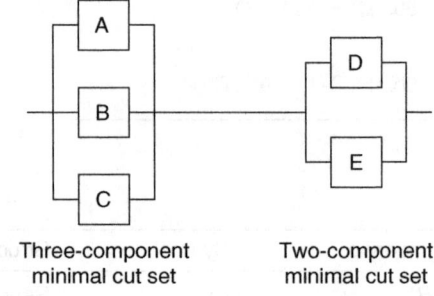

Figure 13.9 Minimal cut sets.

In sum, the one-component minimal cut set represents a single failure that will cause the top event to occur. The two-component minimal cut set represents double failures that together will cause the top event to occur. For an n-component minimal cut set, all n components in the cut set must fail in order for the top event to occur.

The general expression of the minimal cut set for the top event can be written as a combination of OR gates (elements in series):

$$T = M_1 + M_2 + \cdots + M_k \quad (13.21)$$

where T is the top event and each M_i, $i = 1, 2, \ldots, k$, is a minimal cut set and where

$$M_i = X_1 \bullet X_2 \bullet \cdots \bullet X_{n_i} \quad (13.22)$$

and X_i are basic events that can be written as a combination of AND gates (elements in parallel). For the fault tree in Figure 13.3a (OR gate), the minimal cut set expression is

$$T = A + B \quad (13.23)$$

with A and B as the two minimal cut sets. Similarly, for the fault tree in Figure 13.6a (AND gate), the minimal cut set expression is

$$T = A \bullet B \quad (13.24)$$

with $A \bullet B$ as the only minimal cut set.

13.4.1 Fault-Tree Evaluation

Denote the unreliability of the basic event (component) by $q_j(t)$. Then the unreliability of the minimal cut set i, $Q_i(t)$, with n_i components, is given by Eq. (13.25):

$$Q_i(t) = q_1(t) q_2(t) \cdots q_{n_i}(t) \quad (13.25)$$

The unreliability of the system (top event), $Q_s(t)$, is given by Eq. (13.26):

$$Q_s(t) \cong \sum_{i=1}^{n} Q_i(t) \quad (13.26)$$

The fraction of system unreliability contributed by minimal cut set i, $E_i(t)$, is given by Eq. (13.27):

$$E_i(t) = \frac{Q_i(t)}{Q_s(t)} \quad (13.27)$$

The fraction of system unreliability that is contributed by the failure of component k, $e_k(t)$, which represents the importance of component k at time t, is given by Eq. (13.28):

$$e_k(t) = \frac{\sum_{k \text{ in } i} Q_i(t)}{Q_s(t)} \quad (13.28)$$

The importance of the minimal cut sets and of Eqs. (13.25)–(13.28) will become more evident in the specific example problems.

Example [U.S. Nuclear Regulatory Commission, 1981]

Consider the fault tree given in Figure 13.10. The fault tree can be constructed by following either the top-down or bottom-up approaches:

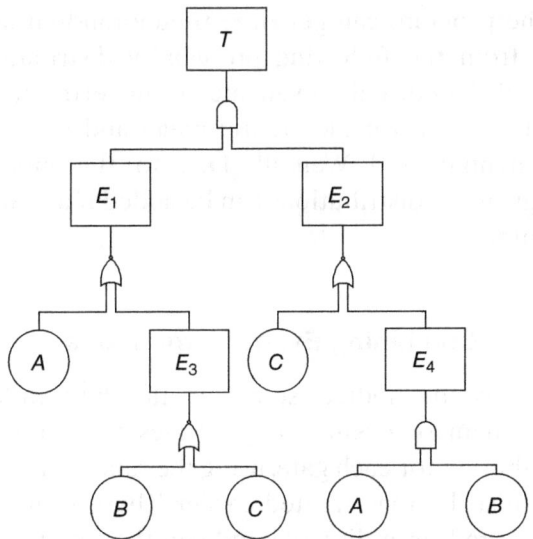

Figure 13.10 Example fault tree.

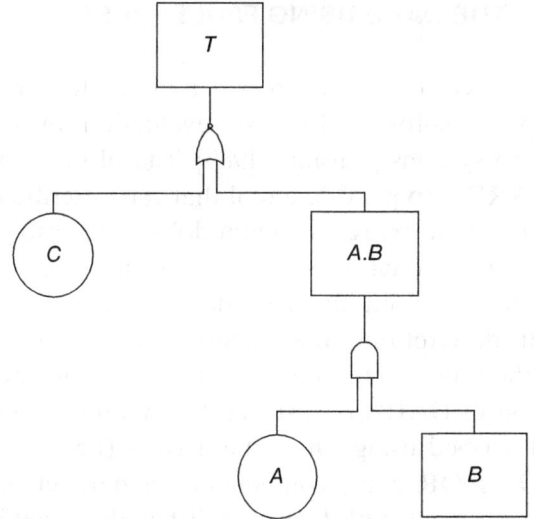

Figure 13.11 Fault tree equivalent of Figure 13.10.

$T = E_1 \cdot E_2$; $E_1 = A + E_3$; $E_3 = B + C$;
$E_2 = C + E_4$; and $E_4 = A \cdot B$

13.4.2 Top-Down Approach

$$T = E_1 \cdot E_2 = (A + E_3) \cdot (C + E_4)$$
$$= A \cdot C + (E_3 \cdot C) + (E_4 \cdot A) + (E_3 \cdot E_4)$$

Substituting for E_3,

$$T = (A \cdot C) + (B + C) \cdot C + E_4 \cdot A + (B + C) \cdot E_4$$
$$= A \cdot C + B \cdot C + C \cdot C + E_4 \cdot A + E_4 \cdot B + E_4 \cdot C$$

By the idempotent law $C \cdot C = C$,

$$\therefore T = A \cdot C + B \cdot C + C + E_4 \cdot A + E_4 \cdot B + E_4 \cdot C$$

But $A \cdot C + B \cdot C + C + E_4 \cdot C = C$ by the law of absorption:

$$\therefore T = C + E_4 \cdot A + E_4 \cdot B.$$

By substituting for E_4 and applying the law of absorption twice,

$$T = C + (A \cdot B) \cdot A + (A \cdot B) \cdot B$$
$$= C + A \cdot B + A \cdot B, \text{ note that } A \cdot B + A \cdot B = A \cdot B$$
$$= C + A \cdot B$$

The minimal cut sets of the top event are thus

$$C \quad \text{and} \quad A \cdot B$$

that is, one simple-component minimal cut set and one double-component minimal cut set. The equivalent final tree is shown in Figure 13.11.

13.4.3 Bottom-Up Approach

$T = E_1 \cdot E_2$; $E_1 = A + E_3$; $E_3 = B + C$; $E_2 = C + E_4$; and $E_4 = A \cdot B$

Because E_4 has only $A \cdot B$ basic failures, we substitute into E_2 to obtain

$$E_2 = C + E_4 = C + A \cdot B$$
$$E_1 = A + E_3 = A + B + C$$
$$T = E_1 \cdot E_2 = (A + B + C) \cdot (C + A \cdot B)$$
$$= A \cdot C + A \cdot A \cdot B + B \cdot C + B \cdot A \cdot B + C \cdot C$$
$$+ C \cdot A \cdot B$$
$$= A \cdot C + A \cdot B + B \cdot C + A \cdot B + C + A \cdot B \cdot C$$

Note that $A \cdot C + B \cdot C + C + A \cdot B \cdot C = C$ by law of absorption. Thus,

$$T = C + A \cdot B$$

The minimal cut sets are of two components: (i) C one-component cut set and (ii) $A \cdot B$ two-component cut set. Indeed, both the top-down and bottom-up approaches led to the same cut sets.

13.5 THE DARE USING FAULT TREES

The DARE using fault trees is a methodology and computer software for risk evaluation of engineering systems [Tulsiani, 1989]. The ultimate goal of DARE is to provide a tool that can describe and evaluate fault trees. The methodology can also generate conditional expectations (as introduced in Chapter 8) given the tree description, the basic event descriptions, the uncertainty distributions, and the time-to-failure distributions. The prototype version of DARE is designed for systems that can be described using simple fault trees (having only AND or OR gates components in parallel or in series, respectively). However, it has the capability for expansion and can be easily enhanced for analysis of complex systems.

13.5.1 Generating the Top-Event Distribution

The DARE program is based on a Monte Carlo simulation. Numerical simulation is used instead of analytical techniques because it can obtain the complete distribution for the top-event probability of occurrence, as opposed to obtaining only the system's moments through analytic methods. This is particularly important when the risk of extreme events is the main object of the study. Numerical simulation methods generate conditional expectations within the overall approximation of the simulation. If analytic methods are used to obtain the conditional expectations, the results are approximated twice: first when the moments of the system are calculated and second when those moments are used to fit a distribution for the top event.

13.5.2 Generating the Basic Event Distribution

The DARE method, which adopts a modification of the normal sampling procedure used in Monte Carlo methods for fault-tree analysis, generates the top-event distribution through two independent modules. The first module is the distribution generator, which generates the random values for each basic component (according to the input statistical distribution for each component) and stores it on a disk file.

The program can generate pseudorandom numbers from the following probability distributions: normal, lognormal (given mean and error factor, median and error factor, and mean and variance), exponential, and Weibull. Due to the modular design, more distributions can be added if and when required.

13.5.3 Combining Event Distributions

The second module starts at the bottom-level component or event and generates the simulated distribution for each gate, using the random component distributions created earlier. These values are also stored on a disk file and can then be used as inputs for higher-level gates. The flowchart for this module is shown in Figure 13.12.

The advantage of using this modified sampling procedure is that the values for all distributions are stored. Thus, if a number of different designs have to be evaluated, only the modified sections of the tree need to be regenerated—a fact that saves considerable time during the simulation procedure. The second advantage of this procedure is that the distributions for all subsystems are generated at the same time and stored. Thus, different subsystems can be studied at the same time as the overall system without having to be programmed separately or generating their corresponding distributions.

The program uses Boolean algebra for computing the point values for the gates; presently, it can handle AND and OR gates with up to 20 components. For OR gates with more than five components, first the mean values of the distributions for all the components are calculated. These distributions are then arranged in increasing order. Equation (13.29) is used for the distributions with the five highest means, and rare-event approximation is used for the rest. For example, if there are five components in series, the point value for all the events through an OR gate would be given by Eq. (13.29), where $\Pr(i), i = 1, \ldots, 5$, are the sampled values of the point probabilities for components $1, \ldots, 5$, respectively. Thus, if there are six components, they are reordered, and the component with the smallest mean is given the number 6. The probability for the gate in this case is given by

$$\Pr(\text{Gate6}) = \Pr(\text{Gate5}) + \Pr(6)$$

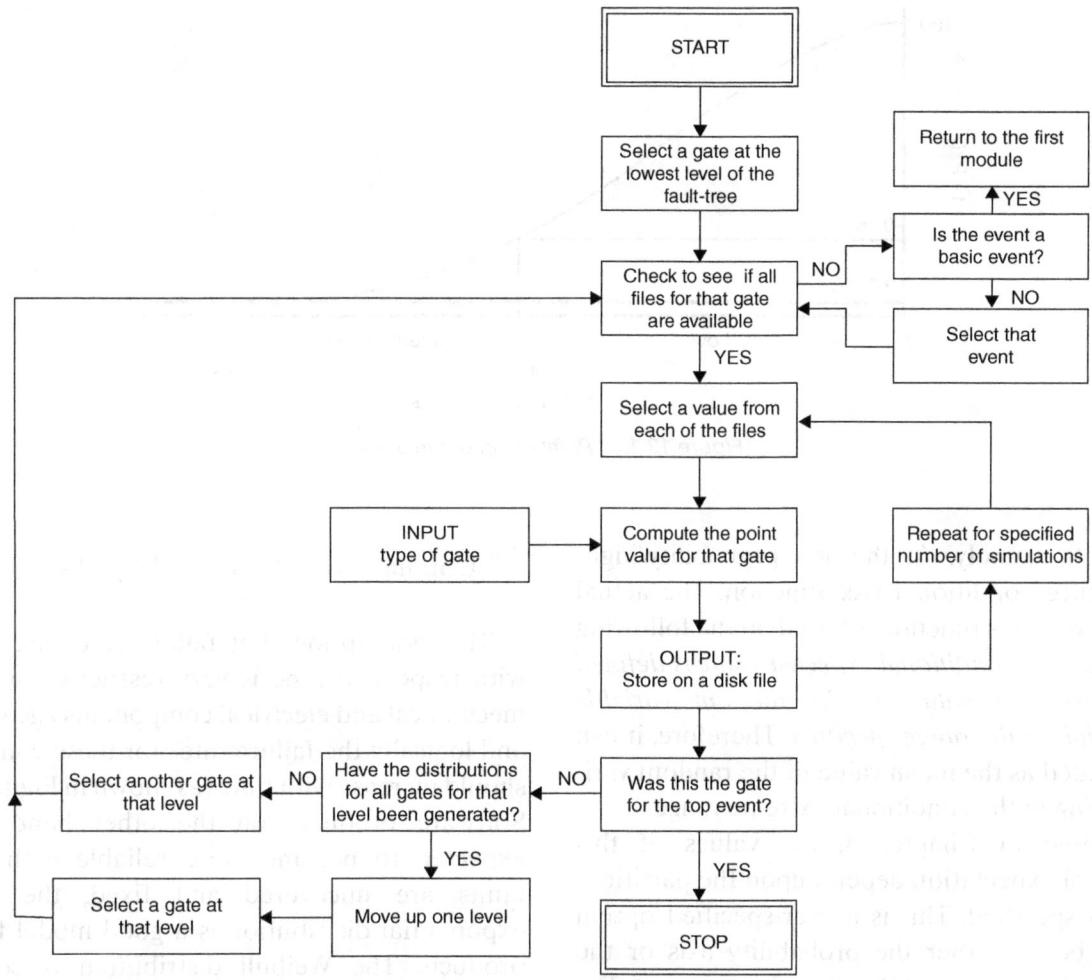

Figure 13.12 Flowchart of DARE simulation procedure, generating top-event distribution.

$$\begin{aligned}
\Pr(\text{Gate}5) = & \Pr(1) + \Pr(2) + \Pr(3) + \Pr(4) \\
& + \Pr(5) - \Pr(1)[\Pr(2) \\
& + \Pr(3) + \Pr(4) + \Pr(5)] \\
& - \Pr(2)[\Pr(3) + \Pr(4) + \Pr(5)] \\
& - \Pr(3)[\Pr(4) + \Pr(5)] - \Pr(4)\Pr(5) \\
& + \Pr(1)\Pr(2)[\Pr(3) + \Pr(4) + \Pr(5)] \\
& + \Pr(2)\Pr(3)[\Pr(4) + \Pr(5)] \\
& + \Pr(1)\Pr(3)[\Pr(4) + \Pr(5)] \\
& + \Pr(1)\Pr(4)\Pr(5) + \Pr(2)\Pr(4)\Pr(5) \\
& + \Pr(3)\Pr(4)\Pr(5) - \Pr(1)\Pr(2)\Pr(3) \\
& \quad [\Pr(4) + \Pr(5)] - \Pr(1)[\Pr(2)\Pr(4)\Pr(5) \\
& + \Pr(3)\Pr(4)\Pr(5)] \\
& - \Pr(2)\Pr(3)\Pr(4)\Pr(5) + \Pr(1)\Pr(2) \\
& \quad \Pr(3)\Pr(4)\Pr(5)
\end{aligned} \quad (13.29)$$

An alternative approximation reduces the computation time by taking two components at a time. For such a case, the approximate probability (ignoring higher-order terms) for an OR gate having six components is given by

$$\begin{aligned}
\Pr(\text{Gate }6) = & \Pr(1) + \Pr(2) - \Pr(1)\Pr(2) + \Pr(3) \\
& + \Pr(4) - \Pr(3)\Pr(4) \\
& + \Pr(5) + \Pr(6) - \Pr(5)\Pr(6)
\end{aligned} \quad (13.30)$$

The method can handle only AND and OR gates, but since it is modular, more gates can be added as required.

13.5.4 Computing the Conditional Extreme Expectation

The DARE program computes the conditional extreme expectation for the top-event probability distribution based on the PMRM, as discussed in

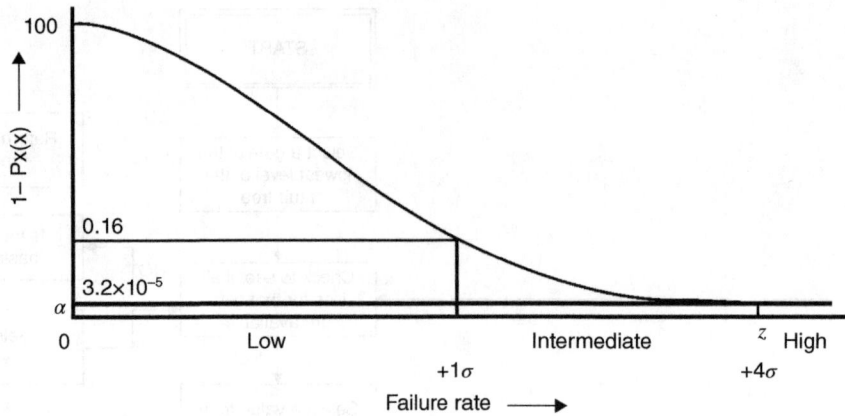

Figure 13.13 Partitioning of the axes.

Chapter 8, namely, f_4, the low-probability/high-consequence conditional risk function. The actual calculation of this function is based on the following definition: *The conditional expected value is defined as the expected value of the random variable conditional on the range specified.* Therefore, it can be computed as the mean value of the random variables falling in the conditional extreme range.

As noted in Chapter 8, the values of this conditional expectation depend upon the partitioning point specified. This is a user-specified option and can be on either the probability axis or the damage axis (in terms of the *exceedance* curve). The exceedance curve is given as the *1 − cdf* curve, where cdf is the cumulative distribution function of the probability distribution. The partitioning is commonly made on the damage axis and is then converted into the partitioning on the probability axis. Figure 13.13 illustrates the partitioning.

13.5.5 Time-to-Failure Distributions

The concept of time-to-failure distributions is very important in reliability analysis, especially in fault-tolerance analysis, where the failure rates are considered as functions of time [Johnson, 1989; Henley and Kumamoto, 1992]. The exponential distribution has been widely used to model system reliability as a function of time. The reliability function for an exponential distribution with parameter λ (for $\lambda \geq 0$) is given in Eqs. (13.31) and (13.32):

$$\text{Reliability:} \quad R(t) = e^{-\lambda t} \quad (13.31)$$

$$\text{Unreliability:} \quad Q(t) = 1 - R(t) = 1 - e^{-\lambda t} \quad (13.32)$$

The assumption that failure rates are constant with respect to time is very restrictive, since most mechanical and electrical components age with time and logically the failure rates for these components should increase with time (as shown in Figure 13.14). Software products, on the other hand, can be expected to become more reliable with time as faults are uncovered and fixed; the negative exponential distribution is a good model for these products. The Weibull distribution is commonly used in reliability analysis to model these components.

The reliability function for a Weibull distribution with shape parameter α and scale parameter λ is given by

$$\text{Reliability:} \quad R(t) = e^{(-\lambda t)^\alpha} \quad (13.33)$$

$$\text{Unreliability:} \quad Q(t) = 1 - R(t) = 1 - e^{(-\lambda t)^\alpha} \quad (13.34)$$

The Weibull distribution is reduced to the exponential distribution for $\alpha = 1$. For $\alpha < 1$, the failure rate decreases with time, and for $\alpha > 1$, the failure rate increases with time.

To analyze the change in the reliability of a system with respect to time, the exponential and Weibull distributions can be used in the DARE method. When performing this analysis, the first module in the distribution generator is changed. The flowchart of this modified module is shown in

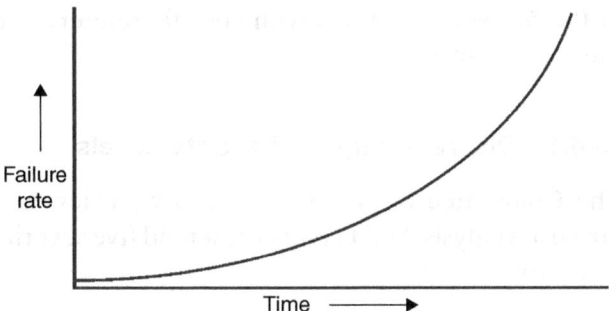

Figure 13.14 Component failure rate as a function of time. Courtesy of the US Nuclear Regulatory Commission [1981].

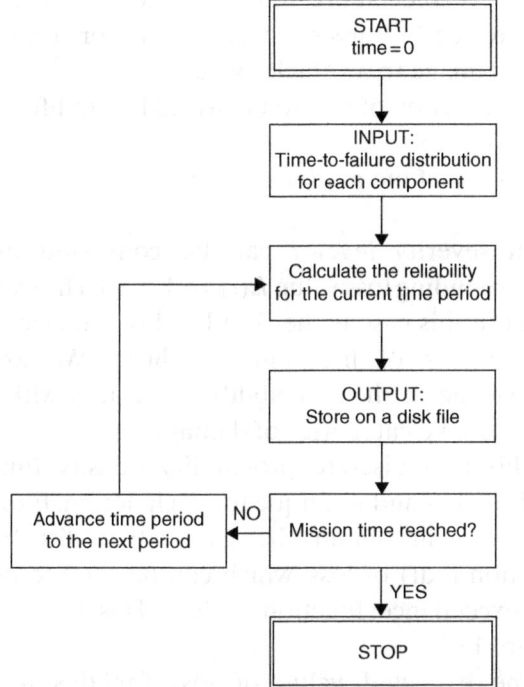

Figure 13.15 Flowchart of DARE module and computation of component time-to-failure distributions.

Figure 13.15. The user specifies the distribution for each component, the parameters for each distribution, the total mission time, and the time increments. The reliability or unreliability of each component (consequently for the entire system) as a function of time is obtained from DARE. This distribution is then used as the input in the second module of the program. Using this module, the reliability or unreliability of the system can be obtained as a function of time.

13.5.6 Combining the Uncertainty and Time-to-Failure Distributions

The earlier sections have discussed DARE's ability to model the component probability distributions and the time-to-failure distributions. This section examines the possibility of combining these concepts.

When the reliability analysis of a system as a function of time is performed, the traditional procedure has been to assume that the failure rates are known perfectly. As discussed earlier, this is practically never the case because of the lack of sufficient information about the component failure rates. Therefore, when the failure rates are uncertain and have an associated probability distribution, the reliability function has a probability distribution as well.

DARE allows the user to evaluate the reliability or unreliability distribution for basic components and for components acting through AND gates. For basic components, both the exponential and the Weibull distributions can be used. For components acting through AND gates, only the exponential distribution can be used. This is based on the fact that the product of two exponential distributions is exponential. Similar results cannot be obtained for systems having OR gates.

The user specifies the total mission time to be considered and the time intervals at which to take the measurement. The data for the distribution are then generated and stored. These data can then be analyzed to find the expected and conditional expected values of reliability or unreliability.

13.6 EXTREME EVENTS IN FAULT TREE ANALYSIS

An extreme catastrophic event is one that has a high value of damage associated with it. The normal concept of an extreme event is one that occurs infrequently, for example, of the order of 1 in 10,000 years or more when dealing with the probable maximum flood (PMF). The magnitude of the flood event determines its associated damage level.

When dealing with fault trees, this concept of an extreme event must be modified, because fault trees are based on Boolean algebra and logic; an event either occurs or does not occur, and no in-between states can be considered (unless one uses fault-tolerance analysis). The top event of the fault tree, usually the failure of the system, has a fixed level of damage associated with it. The damage is thus either zero or a high value, depending on whether the undesired event occurs.

Depending on the subject, the result obtained from fault-tree analysis is the failure rate for the system or the system reliability as a function of time. When the system is considered independent of time, we obtain a probability distribution for the failure rate of the top event. The extreme event in such a case is a high value for the failure rate of the system. If the data for the primary events of the fault tree are given in terms of the number of failures per thousand trials, then the extreme event would be a larger number of failures per thousand trials. Concepts from the theory of the statistics of extremes are invaluable [Pannullo, 1992; Pannullo et al., 1993].

For the space shuttle solid rocket booster (SRB), for example, the damage is given in terms of the failure rate of the system or the number of failures per 10,000 launches (some other suitable unit could have been chosen). Let us say that the extreme event in this case is a large number of failures per 10,000 launches or actual tests. For example, 1 failure per 10,000 launches or actual tests would fall in the high-probability/low-consequence region, while 100 failures in 10,000 launches would fall in the low-probability/high-consequence region.

These failures can be defined in two different ways depending upon the severity levels associated with the failures. These severity levels are explained in the following sections with specific reference to the space shuttle.

13.6.1 Discrete Damage Severity Levels

The Committee on Shuttle Criticality Review and Hazard Analysis Audit [1988] presented five severity levels for damage:

1. No failures
2. Loss of some operational capability of vehicle, but does not affect mission duration
3. Degraded operational capability or unacceptable failure tolerance mode, which leads to early mission termination or results in damage to a vehicle system
4. Abortion of mission to avoid loss of life and/or vehicle
5. Loss of life and/or vehicle

Each severity level i can be converted into a corresponding loss value $L(i), i = 1, \ldots, 5$. The extreme event in this case is the fifth level of damage—that is, the loss of life and/or vehicle. We assume knowledge of the probability associated with each of the above categories of damage or loss.

This is a discrete probability density function (pdf) of loss and is unique for each design. From the pdf, we can obtain the cumulative distribution function (cdf) of loss, which can be used to obtain the exceedance function of loss. This is shown in Figure 13.16.

The expected value of loss for this discrete damage function is given by

$$\sum p(i) L(i) \tag{13.35}$$

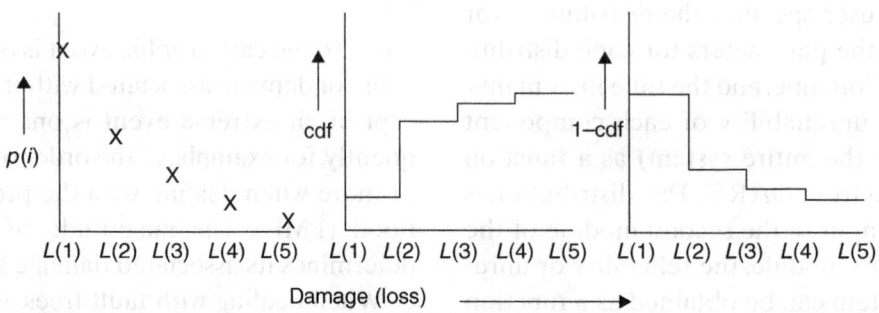

Figure 13.16 Probability versus loss in terms of discrete damage.

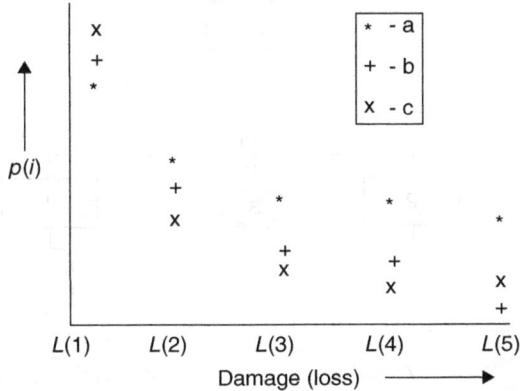

Figure 13.17 Probability versus loss for various design options.

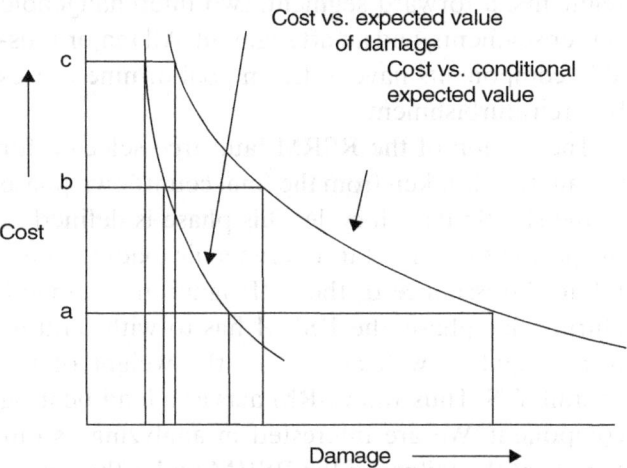

Figure 13.18 Cost versus loss for various design options.

The conditional expected value of loss in the low-probability/high-damage region, f_4 (extreme event), is given by

$$f_4 = \frac{\sum_{i \in \Omega} p(i) L(i)}{\sum_{i \in \Omega} p(i)} \qquad (13.36)$$

where

$\Omega = \{i : L(i) > L(M), p[L(M)] = \alpha\}$
α = partitioning point on the probability axis
$L(M)$ = partitioning point on the damage axis
$p[\cdot]$ = cdf of the loss

Note: If there is only one event in the *extreme* region, the conditional expected value reduces to $L(i)$. When there is only one value of damage, it cannot be minimized. Thus, the better way is to minimize the probability of that loss.

Consider three possible designs (a, b, and c), each with an associated cost whose pdf's for the loss function are as shown in Figure 13.17. Then Figure 13.18 graphs cost versus the risk of the extreme event and cost versus the expected value of risk.

The main disadvantage of this approach is that a fault tree is tailor-made for its top event; and in order to obtain the probability for each of these discrete states of partial failures, a fault tree having the top event as a partial failure would have to be constructed for each partial failure mode. This can make the analysis very cumbersome. Alternatively, one may concentrate on the most extreme event and consider its probability of occurrence as a continuous random variable; this approach is described in the following section. However, it ignores failure modes other than those defined as the extreme event.

13.7 AN EXAMPLE PROBLEM BASED ON A CASE STUDY

The following example problem applies a methodology that integrates multiobjective analysis, the conditional expectation of rare and catastrophic events, and fault-tree analysis. It also demonstrates the interrelationships among the different subsystems and between the subsystems and the overall system. The fault-tree example identifies all possible events and their combinations that could cause the top event to occur.

13.7.1 Problem Description

The problem is based on a case study [Morton Thiokol, Inc., 1988] entitled "Fault-Tree Analysis (FTA) for the Space Shuttle Redesigned Solid Rocket Motor (RSRM)."

The RSRM is the major subsystem of the SRB for the space shuttle. Two SRBs are attached to the space shuttle external tank (ET) and operate in parallel with the shuttle's main engines to provide thrust during lift-off and ascent of the space transportation system (STS). The RSRM is a solid-propellant rocket motor consisting of four motor

406 FAULT TREES

segments: a forward segment, two interchangeable center segments, and an aft segment. All major reusable components have a design goal of nine reuses before refurbishment.

The section of the RSRM fault tree selected for the analysis is taken from the final countdown phase of the shuttle launch cycle. This phase is defined as the period from the start of cryogenic loading of the ET to the issuance of the SRB ignition command. During this phase, the RSRM has to withstand its own weight as well as part of the weight of the overall STS. Thus, the RSRM acts as a load-bearing component. We are interested in analyzing as our top event the failure of the RSRM under this load.

The selected section of the fault tree contains a total of 3 levels and 11 events. The top-level event is at the seventh level of the fault tree for the complete RSRM and is linked to the top event of the RSRM (loss of life/loss of STS) through five OR gates and one AND gate. This means that if the top event of the selected section occurs together with one other event, then the top event of the RSRM fault tree occurs, which is the loss of life/loss of STS.

In the problem structure considered for this example, the three original intermediate events are considered as intermediate events, while all the other eight events are considered as basic events. The undeveloped events are assumed to be that way for the purpose of the example even though some of them are developed further in the complete fault tree for the other elements of the STS. This particular section of the fault tree is selected for pedagogical purposes, because the top event can be easily related to the risk of extreme events and also because the basic events of the tree can be easily understood in terms of their relation to the top event. Furthermore, some of the basic events are mechanical or software failures, and the modifications in design and/or information are easily understandable.

There are a total of three gates in the selected section—two OR gates and one AND gate. The top event has three antecedent events connected through an OR gate. One of the antecedent events is a basic event, while the other two are intermediate events. One of the intermediate events has five antecedent events through an OR gate, while the other event has two antecedent events through an AND gate. The events in this case study have been numbered from the top down, and events on the same level are numbered from left to right (see Figure 13.19).

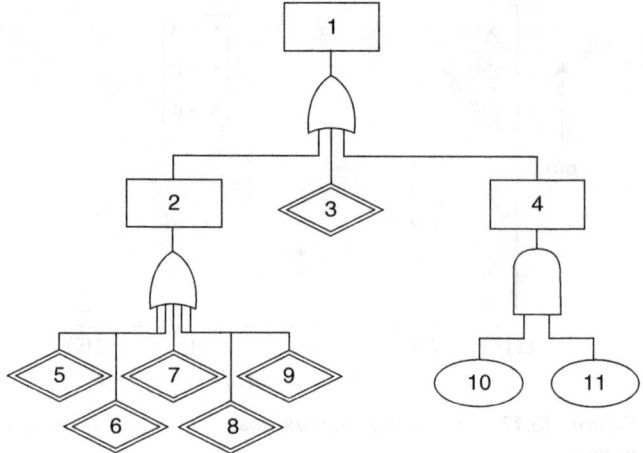

Figure 13.19 Fault tree for case study.

13.7.2 Alternative Designs

Five design options (termed as scenario 1–5) are considered for the case study. Design option 1 corresponds to the basic scenario for the fault tree, shown in Figure 13.19. For the basic scenario (1), it is assumed that the probability of failure for the top event has a realistic value ranging from 1/100 to 1/500. Note that the objective here is to focus on the applicability of the methodology, not on the data collection effort.

Another factor considered in generating the various scenarios is that they should demonstrate the value of information—that is, cases where acquiring additional data at some cost must reduce the uncertainty at the top event while assuming that the mean value remains the same. Note that adding knowledge by itself does not change the functionality of the system, unless the new knowledge is used to modify the system. For these cases, the mean of the component distribution does not change, but the standard deviation does change. If two such scenarios were compared using point probabilities, there would be no change in the value of the expected risk for the top event. If conditional expectations were used instead, however, the value of the conditional expectation would be smaller and thus reflect a reduction in the uncertainty.

TABLE 13.2 Data for the Case Study

Event	Scenario 1 M	EF	Scenario 2 μ	EF	Scenario 3 μ	EF	Scenario 4 μ	EF	Scenario 5 μ	EF
3	0.0005	5	0.0005	5	0.0005	5	0.0005	5	0.0005	5
5	0.0015	5	0.0010	5	0.0015	3	0.0015	5	0.0015	3
6	0.0005	5	0.0005	5	0.0005	5	0.0005	5	0.0005	5
7	0.0020	5	0.0010	5	0.0020	3	0.0020	5	0.0020	3
8	0.0015	5	0.0015	5	0.0015	5	0.0005	5	0.0015	3
9	0.0020	5	0.0020	5	0.0020	5	0.0010	5	0.0020	3
10	0.0300	5	0.0300	5	0.0300	5	0.0300	5	0.0300	5
11	0.0300	5	0.0300	5	0.0300	5	0.0300	5	0.0300	5
Cost ($)	0		150,000		200,000		350,000		450,000	

EF stands for error factor; μ is the mean; μ and EF are the parameters for the lognormal distribution.

The costs for the different scenarios are computed by designating the cost for the basic scenario as the reference and assigning it a value of zero. In scenario 2, the mean value of the failure rate for event 5 (premature separation signal to attach points) is assumed to be reduced by 33%. This is a software error; it can be reduced by duplicating or reconfiguring the hardware or the software, at an assumed cost of $100,000. Also, the mean value of the failure for event 7 (failure of aft bracket ET attach point) is assumed to be reduced by 50%. This is a mechanical failure and is assumed to add $50,000 to the cost. Thus, the additional cost incurred in scenario 2 is $150,000.

In scenario 3, the uncertainties in the failure rates for events 5 and 7 are assumed to be reduced while assuming that the mean failure rate remains the same. This can be accomplished by subjecting the components to various tests at an assumed cost of $100,000 each. Thus, the total additional cost for scenario 3 is $200,000.

In scenario 4, as compared with scenario 2, the mean value of the failure rate for event 8 (aft skirt interface structural failures) is assumed to be reduced by 66%. Also, the mean value for event 9 (failure of forward skirt attach point) is assumed to be reduced by 50%. These are mechanical failures, and their reduction is assumed to cost $175,000 each. Thus, the total cost of scenario 4, as compared to the basic design, is $350,000.

In scenario 5, as compared with scenario 4, the uncertainties in events 5, 7, 8, and 9 are assumed to be reduced by performing tests at a cost of $112,500 each. Thus, the total cost for scenario 5 is $450,000. Table 13.2 summarizes the data for the case study.

13.7.3 Generating Failure Rate Distributions

For each fault tree analyzed, we obtain the probability distribution of the failure rate or the failure probability depending upon the input data. The distribution is generated using Monte Carlo simulation, as per the flowcharts for the DARE program shown in Figures 13.12 and 13.15. The basic output of the simulation procedure is a sequence of random failure rates for the top event and each of the intermediate events. The entire database is generated and stored so that the analysis can be carried out separately for each distribution.

13.7.4 Analyzing the Distribution

The parameters required to analyze the top-event distribution can be specified by the user. Some of the different types of output that can be obtained are as follows:

1. A histogram of failure rates
2. A user-specified number of points on the cdf of the distribution
3. The conditional extreme expectation
4. The mean, median, variance, point probability, and so on

13.7.5 Using Conditional Extreme Expectations

One of the main disadvantages of the conventional methods of fault-tree analysis is the use of expected value. The expected value provides only limited information regarding the probability distribution for the top event. The use of conditional extreme expectation alleviates this limitation. DARE can compute the conditional expected value for the top event in addition to the conventional expected value.

Table 13.3 gives the conditional extreme expected values obtained for the different scenarios.

13.7.6 Using Multiple Objectives

The conditional and unconditional expected values obtained for each design option formulate a multi-objective problem that can then be solved using a preference modeling technique, such as the surrogate worth trade-off (SWT) method, discussed in Chapter 5.

Since we have a discrete number of options, the trade-off functions are not continuous. Table 13.4 summarizes these trade-off values (the change in system failure rate per $1 million spent). The trade-offs are performed between f_1 and f_5 and f_1 and f_4.

It can be seen from Table 13.4 that an expense of $1 million leads to an improvement of 0.00976 in the system failure rate if the expected value is used to analyze the trade-off between the basic design and scenario 2. This leads, however, to an improvement of 0.02898 if the conditional extreme expected value is used. These trade-offs are illustrated in Figure 13.20.

13.7.7 Summary

When large systems are modeled using fault trees, analyzing the impact of the subsystems on the overall system is very important. This impact can have two major components—one corresponding to the mean parameter of interest (the failure rate or

TABLE 13.3 Conditional Extreme Expected Values for the Case Study

Scenario	Mean Value $f_5(\cdot)$	Conditional Extreme Expected Value $f_4(\cdot)$
1	0.00871013	0.03103880
2	0.00724680	0.02699129
3	0.00727024	0.02655157
4	0.00529434	0.02143055
5	0.00533624	0.01977687

TABLE 13.4 Trade-Off Values for $1 Million Expense

	Trade-Off Value for $f_5(\cdot)$	$f_4(\cdot)$
Option 1 versus option 2	0.00976	0.02898
Option 2 versus option 3	−0.00047	0.00279
Option 3 versus option 4	0.01317	0.03414
Option 4 versus option 5	−0.00042	0.01654

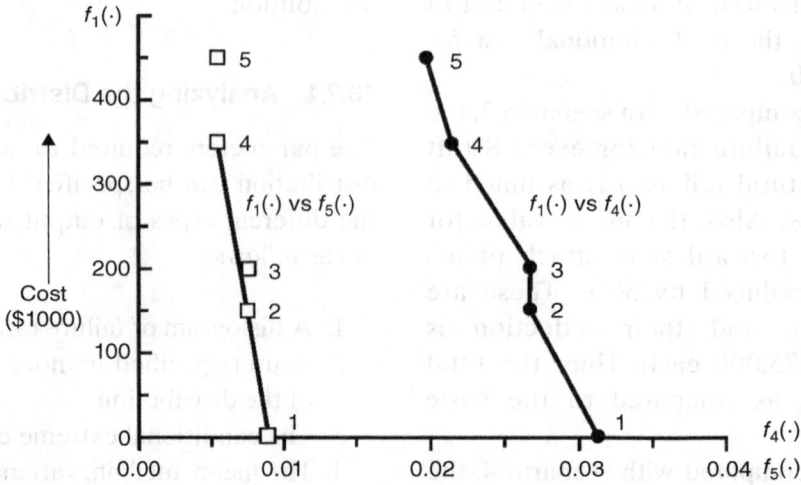

Figure 13.20 Plot of unconditional and conditional extreme expected values versus change in the system failure rate.

the reliability) of the overall system and the other corresponding to the uncertainty in that mean parameter.

Conventionally, this analysis assesses the relative contribution of the subsystems to the overall system in terms of the mean failure rate or the mean reliability. It involves the computation of *importance measures*, which can be called conditional importance measures. Since conditional expectations are functions of the mean value as well as the uncertainty, conditional importance measures should give us an approximation of the percentage contribution of any subsystem's uncertainties to the uncertainty in the overall system.

Often, the database available for the basic components is very sparse and inaccurate. The use of conventional importance measures may not justify further expense for acquiring more accurate data. The use of conditional importance measures would be especially important for this purpose, as additional data would reduce the uncertainties in the parameters of the system and thus in the conditional importance measures.

Alternatively, certain events that had not been developed further (because they were not considered to highly affect the overall system) may actually affect the system's uncertainty due to large uncertainties in their input data. These events, however, do not affect the conventional importance measures. The use of conditional importance measures thus justifies further development of these events to explore their more basic causes.

13.8 FAILURE MODE AND EFFECTS ANALYSIS AND FAILURE MODE, EFFECTS, AND CRITICALITY ANALYSIS

13.8.1 Overview

Failure mode and effects analysis (FMEA) and failure mode, effects, and criticality analysis (FMECA) are reliability-based methods that are widely used for reliability analysis of systems, subsystems, and individual components of systems. They constitute an enabling mechanism with which to identify the multiple paths of system failures.

Indeed, a requisite for an effective risk assessment process is to identify all conceivable failure modes of a system (through all of its subsystems and components as well as the interaction with its environment). In this regard, hierarchical holographic modeling (HHM) (introduced in Chapter 3) constitutes a critical building block in a modified FMEA/FMECA. Just as a team of cross-disciplinary experts is required to construct an effective HHM, such a team is needed for the initial steps of FMEA/FMECA.

Although the quantitative components of the FMEA/FMECA are more simplistic than the quantitative risk analysis methods discussed in this book, the qualitative parts of FMEA/FMECA are quite effective and powerful. They force engineers and other quality control professionals to follow a methodical systemic process with which to track, collect, and analyze critical information that ultimately leads to effectively assessing and managing risks of failure modes of large and small systems. The fact that the military (until 1998) as well as professional associations have embraced FMEA/FMECA and developed standards to guide the users of these methods attests to their value. The MIL-STD-1629A (1980) was the standard for the US military from November 1980 to 1998 (superseding MIL-STD-1629 (SHIPS) dated November 1, 1974, and MIL-STD-2070 (AS) dated June 12, 1977). Other FMEA standards include SAE J1739, which was developed by the Society of Automotive Engineers [2002], and the potential FMEA reference manual first published in 1993 by the DaimlerChrysler, Ford Motor Company, and General Motors Corporation [2001]. In addition, the handbook by Benbow et al. [2002] is another valuable source on these two methods.

13.8.2 The Methodology

The close similarity between FMEA and FMECA, and the fact that the latter is essentially an extension and improvement of the former, justifies a joint treatment in this chapter. Indeed, one may consider FMEA as the first phase of FMECA. Those interested in the details of these methods are encouraged to consult the standard manuals cited in this chapter.

The basic premise of FMEA/FMECA is that the design of engineering and other complex systems is often marred with uncertainty and variability (see Chapter 6). Furthermore, all man-made products are subject to failure sooner or later; thus, prudence calls for comprehensive analyses of such systems at every step throughout their lifecycle. In particular, FMEA and FMECA serve as instruments for identifying and tracking all conceivable failure modes of such systems.

Two major approaches are being followed: the top down and the bottom up or a combination of the two. The focus of the top-down approach is on hardware analysis, while the bottom-up approach focuses on functional analysis. Severity classifications are assigned to each failure mode and each item to provide a basis for establishing corrective action priorities [MIL-STD-1629A]. First priority is given to eliminating catastrophic and critical failure modes. When such failure modes cannot be eliminated or controlled to acceptable levels, alternative controls and other designs or options should be considered. The following is a brief discussion of the severity classifications and the setting of priorities in the control of failure modes.

13.8.2.1 Two- and three-attribute approaches

In the risk filtering, ranking, and management (RFRM) method introduced in Chapter 7, we filtered and ranked the myriad sources of risks (failures in terms of FMEA/FMECA) through the first five phases of the RFRM. In the FMEA/FMECA methods, there are a variety of ways to assign priorities among the sources of failures. Two of the more common approaches differ in the number of attributes used: two or three. The two-attribute approach utilizes "probability" and severity, where both are measured on an ordinal scale between 1 and 10. The term "probability" used in the FMEA/FMECA is in quotation marks to distinguish it from the probability measured on a cardinal scale used throughout this book. The criticality of a failure mode is calculated by multiplying the scores of the "probability" and severity. There are at least two major, fundamental shortcomings associated with this approach:

1. The product of two ordinal numbers is meaningless, because on a scale of 1–10, a "probability" of 10 is not necessarily twice as likely to occur as a "probability" of 5. This is similar with respect to measuring severity.
2. The product of "probability" and severity masks the criticality of the extremes in either the "probability" or the severity. This shortcoming is similar to that of the expected value of risk discussed in Chapter 8.

The three-attribute approach is also based on the use of the ordinal scale, varying between 1 and 10 for three attributes: likelihood of occurrence, severity, and likelihood of detection. The product of these three attributes is commonly known as the *risk priority number*, with a maximum level of $10 \times 10 \times 10 = 1000$ and a minimum of $1 \times 1 \times 1 = 1$. This three-attribute approach for assigning priorities to the identified failure modes suffers from the same shortcomings as the two-attribute approach mentioned above. A third, more elaborate approach is criticality analysis.

13.8.2.2 Criticality analysis

The purpose of criticality analysis is to rank each potential failure mode according to the combined influence of severity classification and its probability of occurrence based on the best available data [MIL-STD-1629A]. The failure mode criticality number, C_m, is the portion of the criticality number for the item due to one of its failure modes under a particular severity classification. The C_m is composed of the failure effect probability, β, also known as the conditional probability of mission loss; the failure mode ratio, α; the part failure rate, λ; and the duration of the applicable mission phase, t:

$$C_m = \{\beta, \alpha, \lambda, t\}$$

The failure effect probability, β, is the conditional probability that the failure effect will result in the identified criticality classification, given that the failure mode occurs. The β values represent the analyst's judgment as to the conditional probability that the loss will occur. It should be quantified in general according to the following [MIL-STD-1629A]:

Failure Effect	β Values
Actual loss	1.00
Probable loss	$0.10 < \beta < 1.00$
Possible loss	$0 < \beta = 0.10$
No effect	0

The failure mode ratio, α, is the probability expressed in a decimal fraction that the part or item will fail in the identified mode. If all potential modes of a particular part or item are listed, the sum of the values for that part or item will equal one (1). The part failure rate, λ, is commonly obtained from its manufacturer. The operating time, t, is commonly measured in hours or the number of operating cycles of the item per mission.

13.8.3 Summary

Risk assessment and risk management are essential in design from concept through development—in other words, through the entire life cycle of the system. No design can foresee all future needs and system evolution; therefore, an iterative process is an integral part of a viable risk analysis. Furthermore, system modeling, which constitutes the sine qua non for any quantitative risk analysis, is also a requisite for a successful deployment of FMECA. An effective team must not only adhere to the above requisites but also know and understand the intricacy of the system being studied.

13.9 EVENT TREES

Event trees offer a graphical methodology for identifying and analyzing both the probabilities and the consequences of failure in a multicomponent system; they provide a unique and compelling mechanism for evaluating the risks associated with any given initiating event. All constructed systems are subject to failure. At the limits of any system, when the time approaches infinity, then the corresponding reliability of that system will approach zero. Moreover, systems with multiple components are likely to have multiple paths to failure. In our examination of fault trees, we saw that it is possible to reduce a large combination of components, connected in series and/or in parallel, into a single set comprised of minimal cut sets in which each individual minimal cut set is composed of one or more components in parallel and in which all minimal cut sets are connected in series. The fault tree provides a graphical depiction of these connections and enables one to understand the complex configurations of the various components within that system. Event-tree analysis serves a similar purpose: It graphically presents multiple paths to failure and shows the probability of failure associated with each path.

In this sense, event trees might be thought of as a graphic illustration of the domino effect: Given multiple paths of standing dominoes that are subjected to an initiating event (the fall of one or more dominoes), some but not all paths will lead to the collapse of other paths of dominoes (i.e., will suffer adverse consequences). Furthermore, similar to mathematical models that are constructed to answer specific questions, event trees provide an excellent mechanism with which to answer questions for a variety of scenarios in terms of *what if?* To identify all plausible paths to failure scenarios and their associated consequences, effective, comprehensive, and complete event trees are best built using HHM (see Chapter 3). Indeed, HHM serves as an excellent medium with which to answer the basic questions in risk assessment discussed in Chapter 1.

The probabilities used in the event trees are commonly estimated from factory reliabilities of each component, from repeated experimentation, or from fault-tree analyses. Furthermore, the consequences of a system's failure may be estimated based on historical as well as on expert evidence. Figure 13.21a depicts how event trees can be used to combine probabilities of events in estimating the probability of a sequence. To avoid abstraction in presenting the event-tree methodology, a study performed by the NRC [1975] will be used as a vehicle with which to demonstrate the methodology. The study provides a basic tool for relating the probabilities of radioactivity releases from a nuclear

412 FAULT TREES

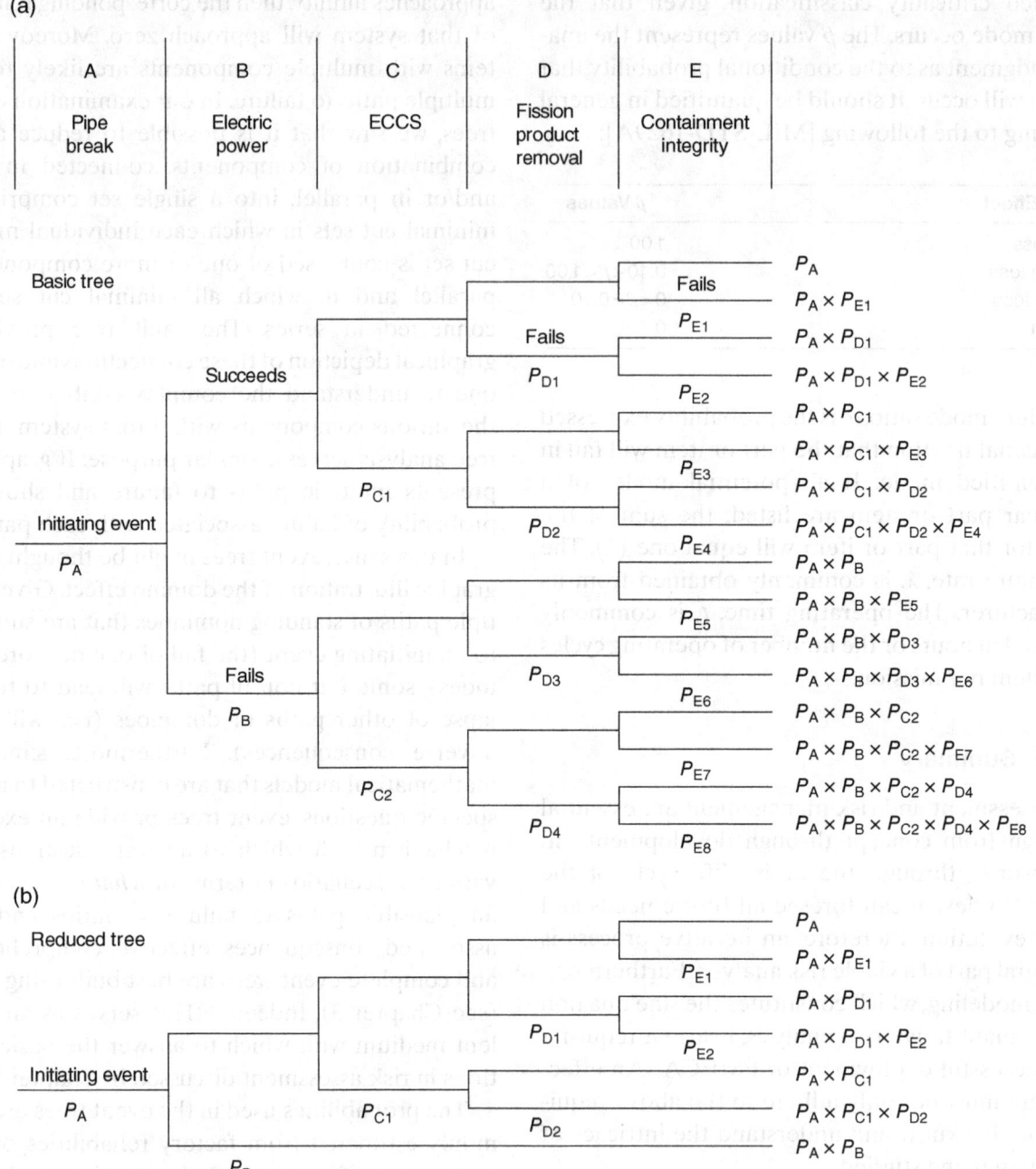

Figure 13.21 (a and b) Contaminant event tree.

contaminant into the environment. Probabilities for the events shown on the trees have been estimated by a number of special analyses: Fault-tree analyses were conducted to identify system elements that might contribute to failures of systems and functions in order to quantify the probability of these failures under accidental conditions; probabilities were estimated for the various modes of contaminant failures; and analyses were done to estimate the probabilities of the occurrence of accident-initiating events. In addition, the mechanisms for radioactivity release and for transporting the fuel into the contaminant atmosphere were analyzed for each accident sequence in the loss-of-coolant accident (LOCA) tree. Modes of contaminant failure were analyzed in order to determine the magnitude of the release from the contaminant into the environment for each sequence. In this sense, the event trees provide a framework and an organizing principle for linking together the results of all these analyses.

Figure 13.21a illustrates how event trees can be used to combine probabilities of events in estimating the probability of a sequence. The simplified LOCA tree in Figure 13.21b shows the probability of a functional failure for each failure branch. Functional failure probabilities are derived from probabilities of failure modes for the systems performing the functions. The same functional failure has different probabilities in different sequences. Assignment of a failure probability to a system requires precise definition of what constitutes its failure, that is, a criterion, and consideration of the conditions under which the system is called upon to perform, that is, a context. Both context and criterion may vary not only with initiating events but also for different paths on the same event tree. For example, for the emergency coolant injection (ECI) function, the criteria for success or failure depend on whether the LOCA is initiated by a small or a large pipe break. Five major critical systems are presented in Figure 13.21a along with their associated probabilities of failure, with a pipe break as the initiating event designated as (A) and with probability P_A; electric power designated as (B) with failing probability P_B; emergency core cooling system (ECCS) designated as (C) and failing with probability P_{Ci} ($i=1,2$); fission product removal designated as (D) and failing with probability P_{Di} ($i=1,2,\ldots,4$); and contaminant integrity designated as (E) and failing with probability P_{Ei} ($i=1,2,\ldots,8$). The product of the probabilities for each branch of the event tree represents the probability of success (or failure) for that branch. The reduced event tree depicted in Figure 13.21b is generated on the assumption that since the probability of failure P is generally small and less than 0.1, then the probability of success $(1-P)$ is always close to 1. Thus, the probability associated with the upper (success) branches of the event tree is assumed to be 1.

Each specific system performing engineering safety feature (ESF) functions may have various failure modes, some of which may be inconsistent with the success of other related systems. For example, for the pressurized water reactor (PWR), since both postaccident heat removal (PAHR) and postaccident radioactivity removal (PARR) are automatically initiated by a single control system, failure modes for PAHR on a success branch for PARR do not include failures. It has been shown in the above NRC study [1975] that the event trees used were an essential component of the overall risk assessment methodology.

The initial requirement for the construction of an event tree is to define the functions to be performed after an initiating event (failure) as well as the interrelationships among those functions. Next, it is necessary to identify the systems provided to perform the functions and then to analyze the interrelationships among the functions to be performed and the operability states of the system itself. Finally, the interrelationships among the operability states of the various systems need to be determined. At each step, dependencies are considered and illogical or meaningless sequence combinations are eliminated. Thus, the event tree can be regarded as a filter into which is fed all pertinent system information affecting the course of events following an initial failure and out of which come only logical and relevant functional and system relationships. The trees are deceptively simple in appearance. Many interrelationships exist that are difficult to represent in a manageable two-dimensional tree. The trees must therefore be split into manageable parts such as a LOCA tree and a contaminant tree, and the sequences on them supplemented with descriptions to assure that all meaningful information about each sequence is used in a quantitative assessment of the trees.

In summary, the following points can be made about event trees as used in the NRC study [1975]:

1. Event trees have provided the overall guidance needed to quantify the risks involved in nuclear power plant accidents because they are well suited for use in combining the probabilities and consequences of accident sequences and in displaying the logic employed.
2. They have assisted in identifying a spectrum of meaningful accident sequences to be quantitatively analyzed.
3. They have assisted in the definition of interrelationships among postaccident functions,

among these functions and the relevant ESF systems provided to perform the functions, and among ESF systems themselves.

4. Their use in eliminating illogical and meaningless relationships has helped to simplify the number of analyses required. This has resulted in an efficient approach to the assessment of potential common mode failures by directing the search for common mode mechanisms only to those systems whose interrelationships are important to risk.

5. They have helped to define which physical processes affected the release and transport of radioactivity from fuel into contaminant and which modes of contaminant failure required analysis for completion of the quantitative risk assessment.

6. They have helped to define how ESF can affect and be affected by the physical processes that can occur in various accident sequences.

7. They have helped in the utilization of fault-tree techniques in the quantification of risk. Fault trees are difficult to use in defining system interrelationships; event trees help to indicate which systems require fault-tree analysis, the conditions of failure, and the ways in which individual system fault trees have to be combined in order to estimate the probabilities of occurrence of applicable accident sequences.

8. They have helped to provide the consequence model with the fundamental inputs regarding the probabilities and magnitudes of radioactivity release from nuclear power plant accidents.

13.10 EXAMPLE PROBLEMS

13.10.1 Water Distribution System

To evaluate the reliability of its water distribution system to a local hospital, a major city in Virginia commissioned a study that applied fault-tree analysis to the distribution of water to a hospital. The study sought to determine the weakest connections where the water valves might fail and completely shut the hospital off from the water distribution system. Pipes to the hospital can collect water from

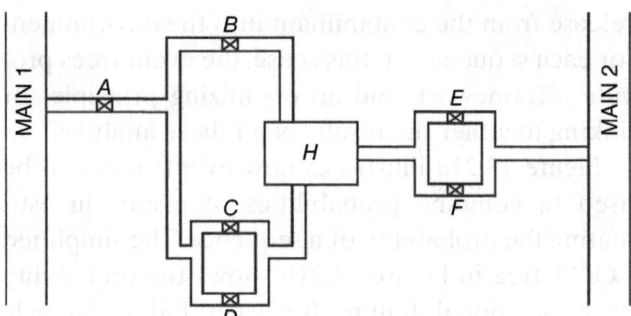

Figure 13.22 Water distribution system to a city hospital.

two mains, 1 and 2, at two distinct connections (points). When a valve fails, it closes, and the water flow stops. Figure 13.22 represents a schematic diagram of this water distribution system.

In Figure 13.22, the hospital is denoted by the letter H, the valves are denoted by |X|, and the two mains are identified explicitly. Note that valves C and D and valves E and F are in parallel. All valves have an equal probability of failure, which is 1/5000. It is assumed that water is flowing through both mains with no failure expected.

Five policy options are considered to improve the reliability of the water distribution system:

1. Adding another pipe from Main 2 with a single valve (cost: $3,000,000)
2. Replacing the valves with new equipment with probability of failure = 1/10,000 (cost: $2 million)
3. Adding another valve in parallel with A (cost: $3.6 million)
4. Removing valve A (cost: $4 million)
5. Adding a small gadget to each valve that decreases the probability of failure by 33% (cost: $1 million)

Solution: The fault tree for the above problem is presented in Figure 13.24. Table 13.5 presents the probabilities and costs associated with the five policy options.

The minimal cut set for the fault tree depicted in Figure 13.23 can be found as follows:

$$N = I \cdot J; \quad I = A + H;$$
$$J = E \cdot F;$$
$$H = B \cdot G; \quad G = C \cdot D$$

EXAMPLE PROBLEMS

TABLE 13.5 Probabilities for the Fault Tree

Policy	Present	1	34.167 pt	3	4	5
A	2.00E−04	2.00E−04	1.00E−04	2.00E−04	—	1.33E−04
B	2.00E−04	2.00E−04	1.00E−04	2.00E−04	2.00E−04	1.33E−04
C	2.00E−04	2.00E−04	1.00E−04	2.00E−04	2.00E−04	1.33E−04
D	2.00E−04	2.00E−04	1.00E−04	2.00E−04	2.00E−04	1.33E−04
E	2.00E−04	2.00E−04	1.00E−04	2.00E−04	2.00E−04	1.33E−04
F	2.00E−04	2.00E−04	1.00E−04	2.00E−04	2.00E−04	1.33E−04
G	4.00E−08	4.00E−08	1.00E−08	4.00E−08	4.00E−08	1.78E−08
H	8.00E−12	8.00E−12	1.00E−12	8.00E−12	8.00E−12	2.37E−12
I	2.00E−04	2.00E−04	1.00E−04	4.00E−08	—	1.33E−04
J	4.00E−08	4.00E−08	1.00E−08	4.00E−08	4.00E−08	1.78E−08
P (fail)	7.99E−12	1.60E−15	1.00E−12	1.60E−15	3.19E−19	2.37E−12
Cost ($ million)	0	3	2	3.6	4	1

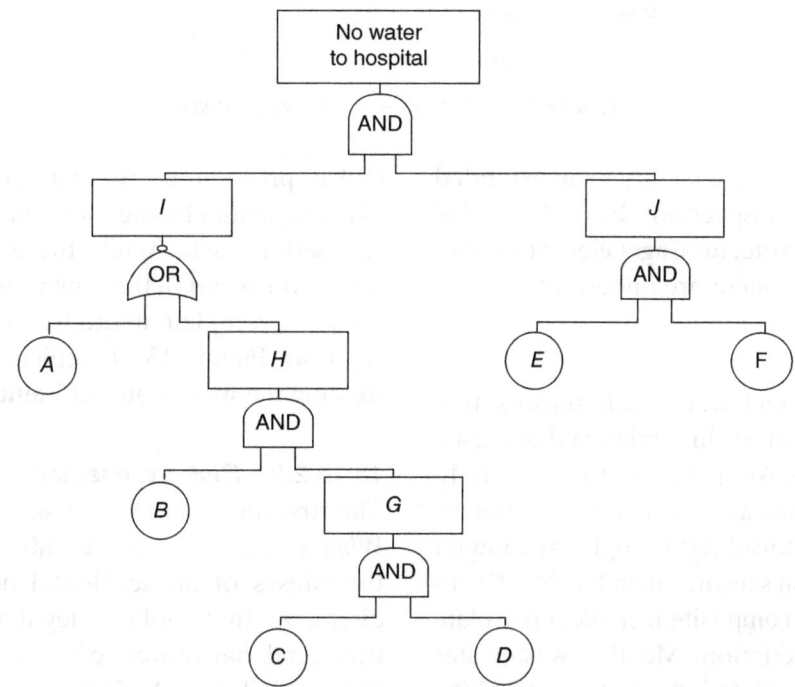

Figure 13.23 Fault tree for the water distribution system.

$$\therefore N = (A + H) \cdot (E \cdot F)$$

Substituting for the value of H,

$$N = (A + B \cdot G) \cdot (E \cdot F)$$

Substituting for the value of G,

$$N = (A + B \cdot (C \cdot D)) \cdot (E \cdot F)$$
$$= (A + B \cdot C \cdot D) \cdot (E \cdot F)$$
$$= A \cdot E \cdot F + B \cdot C \cdot D \cdot E \cdot F$$

13.10.2 Noncompliance with Regulations

This case is concerned with four types of noncompliance with federal, state, and local wastewater treatment regulations in an industrial plant. The treatment system was modeled to demonstrate (i) how a concentration of nickel or copper might exceed regulatory standards and (ii) the possibility of a low or high pH at the sampling point outside the facility. An accident in this case would be a discharge of heavy metals into the public system (stream). This carries a significant financial and

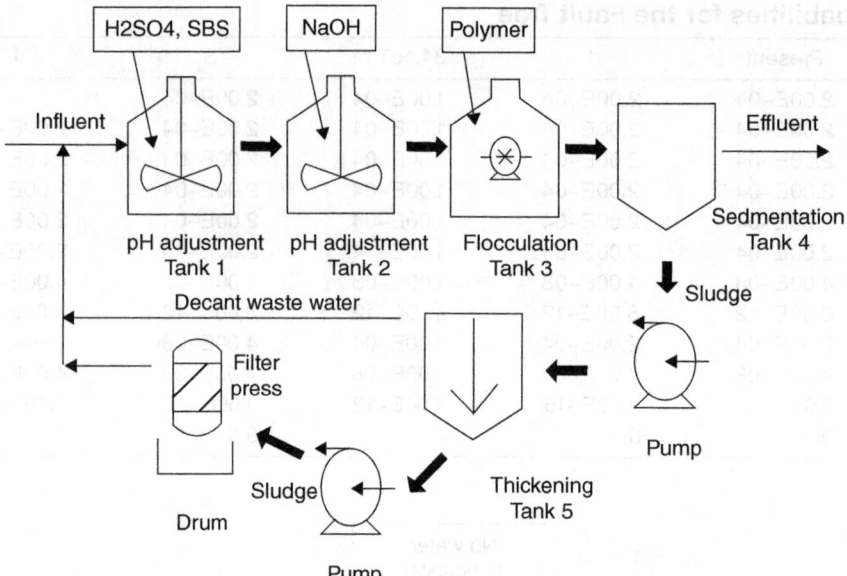

Figure 13.24 Wastewater treatment process.

legal penalty and subjects the facility to an extended period of scrutinized inspections by federal and state environmental protection agencies. Thus, risk assessment and management are imperative.

13.10.2.1 Facility X

Facility X assembles and tests crash sensors that trigger the release of airbags in foreign and domestic passenger automobiles. At the time of this case study, the facility manufactured approximately 12 different crash sensors and employed 1600 people. An integral component of the crash sensor, which Facility X produces on-site, is a thin composite nickel/copper plate with a gold-band inscription. Metallic wastewater from the etching of the plate is treated on-site, along with the wastewater from sulfuric rinsing of the composite nickel/copper film, a swamp cooler (i.e., air conditioner), and the janitorial cleaning room. All wastewater streams lead to a room outside the facility that monitors and adjusts pH levels. Effluent from the treatment process is discharged into the public sewer and is subject to compliance with federal, state, and local pretreatment regulations. Treatment facility permits require quarterly time-composite sampling of the effluent for a variety of metals and toxics or organics, with nickel and copper concentrations the primary pollutants of interest. Approximately three times per shift, or when full, treatment chemicals clarify the water in the storage tank by precipitating the metal contents. Furthermore, pH is regulated by the addition of sodium hydroxide and sulfuric acid. While these chemicals are being added, the water in the tank is mixed for accurate pH readings, complete neutralization, and uniform flocculation. Figure 13.24 depicts the setup used for treating the wastewater effluent at the plant.

13.10.2.2 Risk assessment

The first question in the risk assessment process is, *What can go wrong?* In this case, it is clear that the causes of an accidental heavy metal effluent discharge that violates legal minimum standards transcend hardware, software, organizational, and human failures. A fault tree was developed to capture all conceivable states of failure that might be the cause of the top event—the illegal discharge.

The number of shifts and the number and level of trained operators at the treatment plant were found to constitute minimal cut sets (i.e., their failure would cause the top-level event to occur).

In this example, noncompliance with safety regulations is modeled as the top event. Noncompliance could occur because of excess amounts of heavy metal (Cu or Ni) in the effluent. Another way in which noncompliance can take place is when the pH of the waste stream is not in the desired range, leading to inefficient reduction of the heavy metal concentration. The basic events in this model

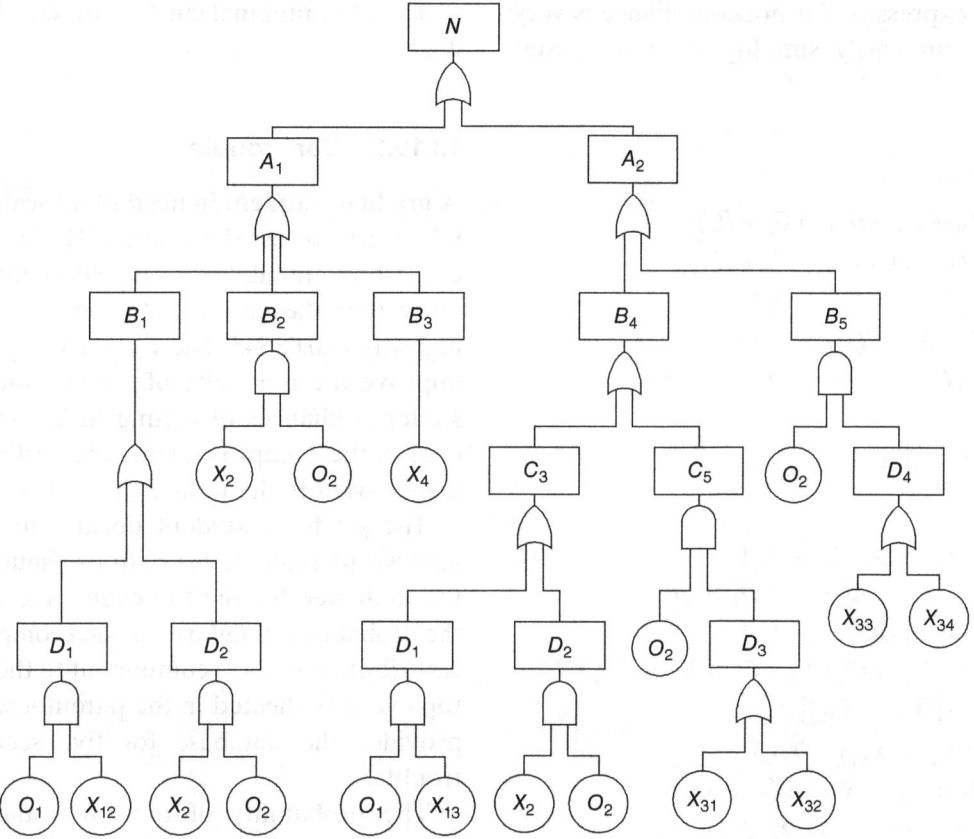

Figure 13.25 Fault tree of the wastewater treatment system.

include operator error, absence of operator, pump failure, and probe failure. The four minimal cut sets for this fault tree, O_1X_1, O_2X_2, O_2X_3, and X_4, will be generated subsequently. Thus, the absence of the operator can cause an unregulated effluent discharge, even if all other components of the system are operating properly. Similarly, failure of the pumps can cause noncompliance at the highest level. Figure 13.25 represents the fault tree of the wastewater treatment system.

The following notation is used to characterize the events in the fault tree:

Top event: N = noncompliance
Intermediate and basic initiating events

A_1 = inefficient removal of reduced colloidal metals
A_2 = inefficient reduction of metal species
B_1 = flocculation problem due to improper dosage of polymer
B_2 = inadequate retention time
B_3 = improper mixing
B_4 = problem in the first reaction tank
B_5 = problem in the second reaction tank due to improper pH
C_3 = improper dosage of metal concentrate
C_5 = improper pH in the first reaction tank
D_1 = improper setting of the feeding device due to operator's error
D_2 = improper setting resulting from flow change
D_3 = problem caused by failure of pH meter 1 or pump 1
D_4 = problem caused by failure of pH meter 2 or pump 2
O_1 = operator present
O_2 = operator absent
X_{12} = operator's error in adding polymer
X_{13} = operator's error in adding metal concentrate
$X_1 = X_{12} + X_{13}$ = operator's error
X_2 = high inflow rate
X_{31} = failure of pH meter 1
X_{32} = failure of pump 1
X_{33} = failure of pH meter 2
X_{34} = failure of pump 2
$X_3 = X_{31} + X_{32} + X_{33} + X_{34}$ = probe or pump failure
X_4 = mixer failure

Although the expression for noncompliance is very complex, we can easily simplify it to an easier equation:

$$N = A_1 + A_2$$
$$= (B_1 + B_2 + B_3) + (B_4 + B_5)$$
$$= [(D_1 + D_2) + O_2 \cdot X_2 + X_4]$$
$$\quad + [C_3 + C_5 + O_2 \cdot D_4]$$
$$= [O_1 \cdot X_{12} + O_2 \cdot X_2 + O_2 \cdot X_2 + X_4]$$
$$\quad + [D_1 + D_2 + O_2 \cdot D_3 + O_2 \cdot D_4]$$

Using idempotent law,

$$N = [O_1 \cdot X_{12} + O_2 \cdot X_2 + X_4]$$
$$\quad + [D_1 + D_2 + O_2 \cdot D_3 + O_2 \cdot D_4]$$
$$= [O_1 \cdot X_{12} + O_2 \cdot X_2 + X_4]$$
$$\quad + [O_1 \cdot X_{13} + O_2 \cdot X_2 + O_2 \cdot (X_{31} + X_{32})$$
$$\quad + O_2 \cdot (X_{33} + X_{34})]$$
$$= O_1 \cdot (X_{12} + X_{13}) + O_2 \cdot X_2 + X_4$$
$$\quad + O_2 \cdot (X_{31} + X_{32} + X_{33} + X_{34})$$
$$= O_1 \cdot X_1 + O_2 \cdot X_2 + O_2 \cdot X_3 + X_4$$

Thus, on simplification of the fault tree, we get

$$N = O_1 X_1 + O_2 X_2 + O_2 X_3 + X_4$$

Thus, the minimal cut sets are $O_1 \cdot X_1, O_2 \cdot X_2, O_2 \cdot X_3, X_4$.

13.10.3 Car Trouble

A graduate student in need of a vacation decides to take a trip across the country. He has a 17-year-old car, whose reliability is questionable. The probability that the car will stall in the middle of the highway is 0.03545. There are five options that will improve the reliability of the car and increase the student's chances of getting to his destinations. Of course, these improvements are costly. He needs to decide what to do in the light of these scenarios.

The graduate student decides to use fault-tree analysis to evaluate his options. Figure 13.26 shows the fault tree for the top event (the car stalls) and the probability of failure for each component and the contribution of each component to the failure of the top event (indicated in the parentheses). Table 13.6 provides the database for five scenarios of car trouble.

The probability of the car's stalling for each scenario and the cost of remedial actions (policy options) associated with each scenario are shown in Table 13.7. Additionally, the Pareto-optimal frontier is shown in Figure 13.27. Clearly, policy options 4 and 5 are inferior.

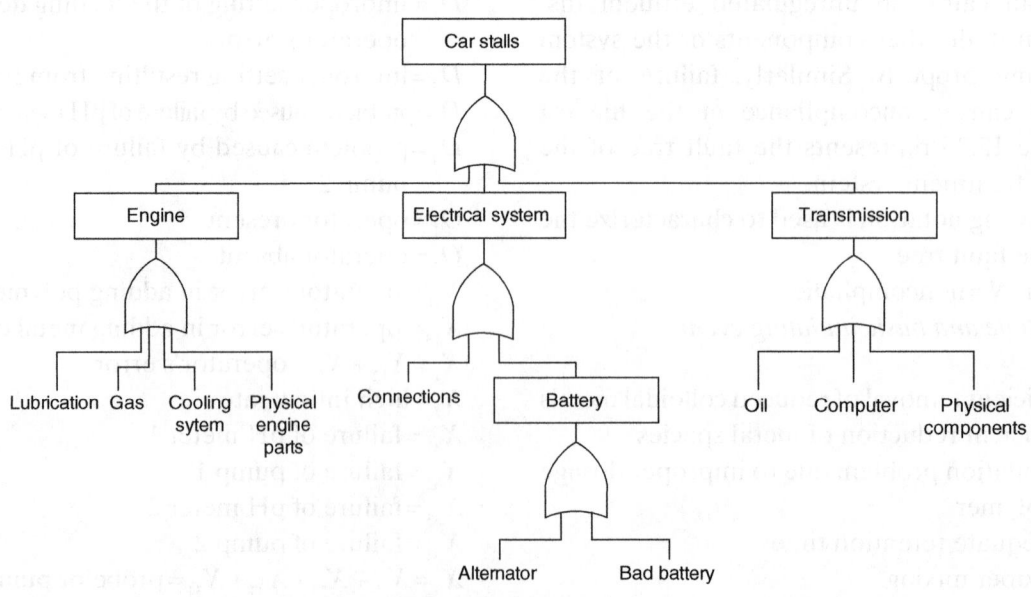

Figure 13.26 Fault tree for the car.

TABLE 13.6 Database for the Car Trouble Problem

Original problem	P (lubricant failure) = 0.001
	P (gas-related failure) = 0.001
	P (cooling system failure) = 0.005
	P (physical engine parts failure) = 0.001
	P (electrical connection) = 0.005
	P (alternator failure) = 0.008
	P (battery failure) = 0.003
	P (transmission oil-related failure) = 0.005
	P (computer failure) = 0.005
	P (physical transmission parts failure) = 0.002
Scenario 1	P (car stalls) = 0.00001
Scenario 2	P (gas-related failure) = 0.00001
	P (cooling system failure) = 0.003
	P (transmission oil-related failure) = 0.00001
Scenario 3	P (gas-related failure) = 0.00001
	P (cooling system failure) = 0.003
	P (transmission oil-related failure) = 0.00001
	P (battery failure) = 0.00001
	P (electrical connection failure) = 0.001
Scenario 4	P (alternator failure) = 0.0002
	P (battery failure) = 0.00001
Scenario 5	P (transmission failure) = 0.002

Original problem
P (engine fails) = $1 - [(1 - 0.001)(1 - 0.001)(1 - 0.005)(1 - 0.001)] = 0.00798$
P (battery fails) = $1 - [(1 - 0.008)(1 - 0.003)] = 0.01098$
P (electrical fails) = $1 - [(1 - 0.005)(1 - 0.01098)] = 0.01592$
P (transmission fails) = $1 - [(1 - 0.005)(1 - 0.005)(1 - 0.002)] = 0.01196$
P (car stalls) = $1 - [(1 - 0.00798)(1 - 0.01592)(1 - 0.01196)] = 0.03545$

Scenario 2
P (engine fails) = $1 - [(1 - 0.001)(1 - 0.00001)(1 - 0.003)(1 - 0.001)] = 0.00500$
P (electrical fails) = 0.01592
P (transmission fails) = $1 - [(1 - 0.00001)(1 - 0.005)(1 - 0.002)] = 0.00700$
P (car stalls) = $1 - [(1 - 0.00500)(1 - 0.01592)(1 - 0.00700)] = 0.02770$

Scenario 3
P (engine fails) = 0.00500
P (electrical fails) = $1 - [(1 - 0.001)(1 - 0.008)(1 - 0.00001)] = 0.00900$
P (transmission fails) = 0.00700
P (car stalls) = $1 - [(1 - 0.00500)(1 - 0.00900)(1 - 0.00700)] = 0.02086$

Scenario 4
P (engine fails) = 0.00798
P (battery fails) = $1 - [(1 - 0.0002)(1 - 0.00001)] = 0.00021$
P (electrical fails) = $1 - [(1 - 0.005)(1 - 0.00021)] = 0.00521$
P (transmission fails) = 0.01196
P (car stalls) = $1 - [(1 - 0.00798)(1 - 0.00521)(1 - 0.01196)] = 0.02495$

Scenario 5
P (engine fails) = 0.00798
P (electrical fails) = 0.01592
P (transmission fails) = 0.002
P (car stalls) = $1 - [(1 - 0.00798)(1 - 0.01592)(1 - 0.002)] = 0.02573$

TABLE 13.7 Probability and Cost Associated with Each Scenario and Policy Option

Scenario 0 (Do nothing)	P (car stalls) = 0.03545	Cost = $0
Scenario 1	P (car stalls) = 0.00001	Cost = $400
Scenario 2	P (car stalls) = 0.02770	Cost = $40
Scenario 3	P (car stalls) = 0.02086	Cost = $100
Scenario 4	P (car stalls) = 0.02495	Cost = $250
Scenario 5	P (car stalls) = 0.02573	Cost = $700

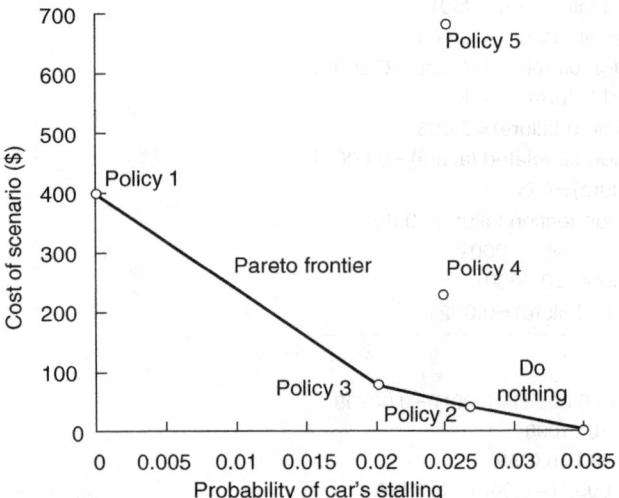

Figure 13.27 Pareto frontiers: cost versus probability of car's stalling.

Among the five policy options examined (excluding the do-nothing policy), three (1, 2, and 3) are Pareto-optimal. The effectiveness of the policy options is directly related to the cost of improving the reliability and the contribution of the components. Inferior solutions, such as rebuilding the transmission (option 5) or replacing the alternator and battery (option 4), improve reliability, but the cost figures are not attractive when compared to other cheaper and better alternatives (1–3).

Overall, the fault tree is a useful tool to analyze system components that contribute to the ultimate undesired system outcome. It provides information about the components that require attention and the cost-effectiveness of solutions.

However, the fault tree has a drawback. One of the critical assumptions made in fault-tree analysis is independence. A system is defined as components working together toward a common goal. Many systems such as automobiles are highly coupled, and the component failures are highly correlated with each other. Thus, it is possible that improper use of fault tree may result in making decisions based on misleading information.

REFERENCES

Apostolakis, G., 1991, *Probabilistic Safety Assessment and Management*, Vol. **1 and 2**, Elsevier, New York.

Benbow, D.W., R.W. Berger, A.K. Elshennaway, and H.F. Walker, 2002, *The Certified Quality Engineer Handbook*, ASQ Quality Press, Milwaukee, WI.

Birolini, A., 2007, *Reliability Engineering: Theory and Practice*, Fifth edition, Springer-Verlag, New York.

Committee on the Shuttle Criticality Review and Hazard Analysis Audit, 1988, *Post Challenger Evaluation of Space Shuttle Risk Assessment and Management*, National Academy Press, Washington, DC.

Cox, D.C., 1982, An analytic method for uncertainty analysis of nonlinear output functions, with applications to fault-tree analysis, *IEEE Transactions of Reliability* **31**(5): 465–462.

DaimlerChrysler Corporation, Ford Motor Company, and General Motors Corporation, 2001, *Potential Failure Mode and Effects Analysis (FMEA): Reference Manual*, Third edition, DaimlerChrysler Corp., Ford Motor Co., General Motors Corp., Detroit, MI.

Ebeling, C.E., 2005, *Introduction to Reliability and Maintainability Engineering*, Waveland Press, Long Grove, IL.

Henley, E. and H. Kumamoto, 1992, *Probabilistic Risk Assessment: Reliability Engineering, Design, and Analysis*, IEEE Press, New York.

Hoyland, A. and M. Rausand, 1994, *System Reliability Theory, Models and Statistical Methods*, John Wiley & Sons, Inc., New York.

Johnson, B.W., 1989, *Design and Analysis of Fault-Tolerance Digital Systems*, Addison-Wesley, Reading, MA.

Limnios, N., 2007, *Fault Trees*, ISTE Publishing Company, London.

Lowrance, W.W., 1976, *Of Acceptable Risk*, William Kaufmann, Los Altos, CA.

Martensen, A.L. and R.W. Butler, 1987, The Fault-Tree Compiler, NASA Technical Memorandum 89098, Langley Research Center, Hampton, VA.

Morton Thiokol, Inc., 1988, Fault Tree Analysis (FTA) for the Space Shuttle Redesigned Solid Rocket Motor (RSRM), Space Operations, UT, Document Number TWR-17262.

Pannullo, J.E., 1992, Risk of Extreme Events: Reliability and the Value of Information, Ph.D. dissertation, Systems Engineering Department, University of Virginia, Charlottesville, VA.

Pannullo, J.E., D. Li, and Y.Y. Haimes, 1993, On the characteristics of extreme value for series systems, *Reliability Engineering and Systems Safety* **40**(2): 101–110.

Park, J.I., J.H. Lambert, and Y.Y. Haimes, 1998, Hydraulic power capacity of water distribution networks in uncertain conditions of deterioration, *Water Resources Research* **34**(2): 3605–3614.

Rao, S.S., 1992, *Reliability-Based Design*, McGraw-Hill, New York.

Russell, K.D., D.M. Snider, M.B. Sattison, H.D. Stewart, S.D. Matthews, and K.L. Wagner, 1987, *Integrated Reliability and Risk Analysis System (IRRAS), User's Guide*, Version 1.0. (Draft), NUREG/CR-4844, EGG-2495, U.S. Nuclear Regulatory Commission, Washington, D.C.

Schneiter, C., Y.Y. Haimes, D. Li, and J.H. Lambert, 1996, Capacity reliability of water distribution networks and optimum rehabilitation decision making, *Water Resources Research* **32**(7): 2271–2278.

Society of Automotive Engineers, 2002, Potential Failure Mode and Effects Analysis in design (Design FMEA), Potential Failure Mode and Effects Analysis in Manufacturing and Assembly Processes (Process FMEA), and Potential Failure Mode and Effects Analysis for Machinery (Machinery FMEA), SAE J1739, Society of Automotive Engineers, Warrendale, PA.

Storey, N., 1996, *Safety Critical Systems*, Addison Wesley, New York.

Tulsiani, V., 1989, Risk, Multiple Objectives, and Fault-Tree Analyses—A Unified Framework, M.S. thesis, Systems Engineering Department, University of Virginia, Charlottesville, VA.

U.S. Nuclear Regulatory Commission, 1975, Reactor Safety Study—An Assessment of Accident Risk in U.S. Commercial Nuclear Power Plants, WASH-1400, NUREG-75/014, U.S. Nuclear Regulatory Commission, Washington, DC.

U.S. Nuclear Regulatory Commission, 1981, *Fault Tree Handbook,* NUREG-81/0492, U.S. Nuclear Regulatory Commission, Washington, DC.

14

Multiobjective Statistical Method*

14.1 INTRODUCTION

Systems modeling is a critically important phase in the quantitative risk assessment and management process. In Chapter 2, we emphasized the central and dominant role that state variables play in the modeling effort. In this chapter, we build on that modeling discussion, on the multiobjective trade-off analysis and the surrogate worth trade-off (SWT) method introduced in Chapter 5, and on the mutual importance of optimization and simulation, benefiting from their complementary and supplementary attributes. A case study on an interior drainage system will illustrate these applications.

Often, the most convenient way to construct risk as well as other objective functions is in terms of state variables rather than in terms of decision variables. For example, a risk function associated with health hazards can be more easily constructed in terms of the level of contaminant concentrations (state vector, $\mathbf{s}$) than in terms of the measures taken to prevent such a contamination (decision vector, $\mathbf{x}$). On the other hand, in the multiobjective optimization and trade-off analysis phase of risk management, it is much more convenient to have these functions expressed explicitly in terms of the decision vector, $\mathbf{x}$, rather than the state vector, $\mathbf{s}$. The multiobjective statistical method (MSM) resolves this dilemma by constructing these risks and other functions in terms of the state variables; then through simulation, regression analysis, or other tools, the MSM regenerates these functions in terms of the decision variables [Haimes et al., 1980].

Essentially, the MSM integrates a multiobjective optimization scheme (the SWT method) and a statistical procedure to assess the different combinations of possible system configurations. The MSM is described here through a levee drainage system design. Given a certain levee height, there are many different possible configurations and capacities for system components to handle interior runoff and provide a certain level of protection for a given area. The level of protection depends explicitly on the river stage and the intensity and duration of rainfall—the two random variables that are considered in the analysis. Given a set of objectives for system

---

*This chapter is based on Haimes et al. [1980].

Risk Modeling, Assessment, and Management, Fourth Edition. Yacov Y. Haimes.
© 2016 John Wiley & Sons, Inc. Published 2016 by John Wiley & Sons, Inc.

performance and a finite set of alternative strategies, there exists some configuration that will be *optimal* in a Pareto-optimal sense in relation to other possible configurations.

Although the methodology presented in this section is generic, its development was aimed at improving the interior drainage system in Moline, Illinois, which is located on the Mississippi River and has a floodplain area of about 475 acres. Much of the floodplain is heavily developed by industry, especially most of the riverfront sites; 85% of industrial acreage in Moline is in the floodplain. Generally when snow melts, heavy rains and frozen ground combine to give a high runoff upstream of Moline, leading to flood stages at Moline itself. In general, the largest floods on the Mississippi at Moline occur between March and late June.

Commonly, the flood control system to reduce damage is a combination of different types of structures and land use policies. The structural modifications include levees, flood walls, pumping stations, and gravity drains; ponding areas also play an integral role in the overall operation. Gravity outlets are openings in the levees that permit discharge of interior drainage flows into the river by gravity when river stages are low. They are equipped with gates to prevent river flows from entering the floodplain area during floods. Pumping stations discharge interior drainage flows over the levees or flood walls or through pressure lines when gravity outlets are blocked by high river stages. Ponding areas consist of any low areas near the inlets to gravity drains or pumping stations that are intended for temporarily storing excess interior drainage flows. They may be storm drains, areas expressly set aside, or even streets and parking lots, if temporary ponding of interior runoff would not cause unacceptable damage.

14.2 MATHEMATICAL FORMULATION OF THE INTERIOR DRAINAGE PROBLEM

Define $E(\mathbf{x}; \eta_i, r_m)$ as the maximum pond elevation given a decision vector, $\mathbf{x} = [x_1, ..., x_k, ..., x_K]$, a river stage η_i, $i = 1, 2, ..., I$, and a storm event r_m, $m \in \{1, 2, ..., M\}$; I and M are to be specified later. A set of decisions $\mathbf{x}$ includes (i) pump capacity, (ii) pump operating sequence, (iii) gravity drain configurations, and (iv) others. The stochastic nature of the problem is reflected in the statistical behavior of the random variables η_i and r_m.

Likewise, define $D(\mathbf{x}; \eta_i, r_m)$ as the duration for which the pond elevation exceeds a specified threshold level, given a sequence of decisions x_k, $k = 1, 2, ..., K$, a river stage η_i, and a rainfall event r_m.

Define the parameters $\hat{\eta}$ and $\bar{\eta}$ as the minimum and maximum attainable river elevations, respectively. Similarly, define $\hat{r}$ and $\bar{r}$ as the minimum and maximum rainfall events, respectively. It is important to observe that not every river elevation is contained in $[\hat{\eta}, \bar{\eta}]$ nor is every rainfall event r_m contained in $[\hat{r}, \bar{r}]$. However, in any given study, we assume that the parameters $\hat{\eta}$, $\bar{\eta}$, $\hat{r}$, and $\bar{r}$ are defined so as to include all possible events of significance, and it is in this context that we can refer to the above parameters as defining a complete system of river and rainfall events within a time period $[0, T]$.

Define the integers I, J, and N, which represent a discretization of the intervals $[\hat{\eta}, \bar{\eta}]$, $[\hat{r}, \bar{r}]$, and $[0, T]$ into a sequence of subintervals of length

$$\Delta\eta \equiv \frac{\bar{\eta} - \hat{\eta}}{I} \quad (14.1)$$

$$\Delta r \equiv \frac{\bar{r} - \hat{r}}{J} \quad (14.2)$$

$$\Delta t \equiv \frac{T}{N} \quad (14.3)$$

respectively.

Define a river stage event $\eta_i = i\Delta\eta$, $i \in \{1, 2, ..., I\}$ and similarly a component rainfall event $r_j = j\Delta r$, $\{i \in 1, 2, ..., J\}$.

Given a sequence of decision variables x_k, a river stage η_i, and a rainfall event r_m, $E(\mathbf{x}; \eta_i, r_m)$ and $D(\mathbf{x}; \eta_i, r_m)$ are computed via simulation.

14.3 FORMULATION OF THE OPTIMIZATION PROBLEM

All objectives are assumed to be a function of the two state variables $E(\cdot)$ and $D(\cdot)$, elevation and duration, respectively. More specifically, consider the problem

$$\min_{x \in X}\left[\tilde{f}(\mathbf{x}) = \begin{Bmatrix} \tilde{f}_1(\mathbf{x}) \\ \vdots \\ \tilde{f}_p(\mathbf{x}) \end{Bmatrix}\right] \quad (14.4)$$

where X is the set of feasible decisions; $f_p(E(\mathbf{x};\eta_i,r_m), D(\mathbf{x};\eta_i,r_m))$ is the value of the pth objective function for each $\mathbf{x}$, η_i, and r_m. Since f_p depends explicitly on the random variables η_i and r_m (the river stage and rainfall events) via the values of the state variables (elevation and duration functions $E(\cdot)$ and $D(\cdot)$, respectively), the optimization problem must be posed to account for this feature.

Let $\tilde{f}_p(\mathbf{x})$ denote the expected value of the function

$$f_p(E(\mathbf{x};\eta_i,r_m), D(\mathbf{x};\eta_i,r_m))$$

Mathematically, $\tilde{f}_p(\mathbf{x}) = E\left[f_p(E(\mathbf{x};\eta_i,r_m), D(\mathbf{x};\eta_i,r_m))\right]$, where $E[\cdot]$ denotes the expected value relative to the joint probability distributions of the random variables η_i and r_m. The set of feasible decisions X is generally characterized by constraints of the form

$$g(\mathbf{x}) \leq 0 \quad (14.5)$$
$$h(\mathbf{x}) = 0 \quad (14.6)$$

Each objective function $\tilde{f}_p(\cdot), p = 1, 2, \ldots, P$, depends implicitly on the decision vector $\mathbf{x}$ since the elevation and duration relationships $E(\cdot)$ and $D(\cdot)$ depend explicitly on $\mathbf{x}$ and on the functional relationship generated via the simulation programs.

In order to proceed, the functions $\tilde{f}_p(\mathbf{x})$, $p = 1, 2, \ldots, P$, must be computed. The random variables involved are η_i and r_m, and they are introduced into the problem formulation by means of the peak ponding elevation and ponding duration relationships.

From the available data, the conditional probabilities $P(r_m|\eta_i)$, $m = 1, 2, \ldots, M$, $i = 1, 2, \ldots, I$ (i.e., the probability of attaining a rainfall event r_m, given that the event η_i has occurred) can be computed. These can be used to compute the conditional expectation of $\tilde{f}_p(\cdot)$ by means of the formula

$$\tilde{f}_p^{(i)}(x;\eta_i) = \sum_{m=1}^{M} f_p(E(x;\eta_i;r_m), D(x;\eta_i,r_m)) P(r_m|\eta_i)$$

$$(14.7)$$

(The conditional expectation $\tilde{f}_p^{(i)}(x;\eta_i)$ can be thought of as the expected value of the function $f_p(\cdot)$, given that the river elevation is in the range (η_{i-1}, η_i).)

Let $P(\eta_i), i = 1, 2, \ldots, I$, denote the probability of the occurrence of the river elevation event η_i, computed from the available data. The expected value of $f_p(\cdot)$ denoted by $\tilde{f}_p(\cdot)$ relative to the joint probability distribution of η_i and r_m is given by

$$\tilde{f}_p(\mathbf{x}) \equiv \sum_{i=1}^{I} \tilde{f}_p^{(i)}(\mathbf{x};\eta_i) P(\eta_i)$$
$$= \sum_{i=1}^{I}\sum_{m=1}^{M} f_p(E(\mathbf{x};\eta_i,r_m), D(\mathbf{x};\eta_i,r_m)) P(r_m|\eta_i) P(\eta_i)$$
$$= \sum_{i=1}^{I}\sum_{m=1}^{M} f_p(E(\mathbf{x};\eta_i,r_m), D(\mathbf{x};\eta_i,r_m)) P(r_m,\eta_i)$$

$$(14.8)$$

where the last equality follows from the multiplication formula of probability for any fixed $\mathbf{x} \in X$ and $p = 1, 2, \ldots, P$. The methodology developed here is valid in a general sense, since it does not require any a priori assumption regarding the statistical dependence or independence of rainfall and river stage events. For example, for the case of statistical independence, we have

$$P(r_m\eta_i) = P(r_m)P(\eta_i) \quad (14.9)$$

and similarly,

$$P(r_m|\eta_i) = P(r_m) \quad (14.10)$$

14.4 THE MSM: STEP-BY-STEP

Figure 14.1 shows a schematic diagram of the following major steps that constitute the MSM:

Step 1. Determine the feasible set of decision/measures, X, for optimizing the objective functions. Also, determine the set of verbal multiple objectives that characterize the interior drainage system.

Step 2. Determine relevant historical records associated with the random variables $\mathbf{r} = \mathbf{r}(\eta_i, r_m)$, and from these data, determine the probability

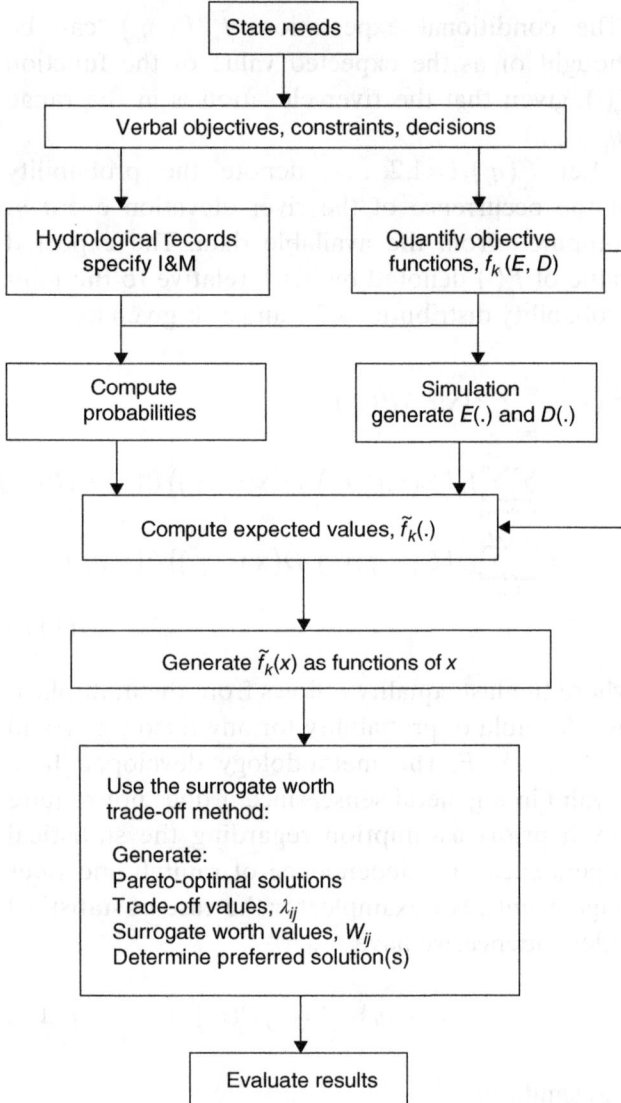

Figure 14.1 *Schematic diagram of the multiobjective statistical method (MSM).*

Step 5. Given a fixed set of feasible decisions, $\mathbf{x} \in X$, for each $m \in \{1,2,\ldots,M\}$ and $i \in [1,2,\ldots,I]$, determine the values of the elevation and duration $E(\cdot)$ and $D(\cdot)$ using hydraulic simulation programs such as Indran [US Army Corps of Engineers, 1975] (a total of $M \cdot I$ values for each).

The Indran simulation model generates ponding durations and peak ponding elevations above three index elevations for various sets of decisions, river stages, and storm events. Elevations and durations are computed for all combinations of river stages, storm events, and decision variables to be used in the study. Output from this module gives the results of the routings, and these values are used as inputs to the regression analysis.

Step 6. For each set of decisions $\mathbf{x}^k$, $k = 1, \ldots, K$, determine the values of the state vector $\mathbf{s}$, and then substitute these values in the risk and other objective functions. Now each $f_p(\cdot)$ is a function of E and D.

Step 7. Let $f_p(E, D)$ denote the pth objective function, which depends explicitly on the pond elevation E and the pond duration D specified numerically in steps 5 and 6. Using the results given by the previous steps, compute (relative to the joint probability distribution of the random variables r_m (rainfall) and η_i (river stage)) the expected value $\tilde{f}_p$ of the function f_p, $p = 1, 2, \ldots, P$.

More specifically,

$$\tilde{f}_p(x) = \sum_{i=1}^{I}\sum_{m=1}^{M} f_p\left(E(x;\eta_i,r_m), D(x;\eta_i,r_m)\right) P(r_m,\eta_i) \quad (14.11)$$

or, equivalently,

$$\tilde{f}_p(x) = \sum_{i=1}^{I}\sum_{m=1}^{M} f_p\left(E(x;\eta_i,r_m), D(x;\eta_i,r_m)\right) P(r_m|\eta_i) P(\eta_i) \quad (14.12)$$

Step 8. Select a different decision vector $\mathbf{x} \in X$, increment q by 1, that is, $q \rightarrow q+1$; if $q \leq Q$ (note that $Q = M \times I$. In the case study, $M = 10$ and $I = 10$; thus, $Q = 100$ simulations run for each decision option), go back to step 5; otherwise, go to step 9.

distribution functions. Compute the required probabilities $P(r_m|\eta_i)$ and $P(\eta_i)$, $m = 1, 2, \ldots, M$, $i = 1, 2, \ldots, I$. Perform the simulation model (set $q = 1$).

Step 3. Construct the risk and other objective functions in terms of pond elevation and duration levels as $f_j(\mathbf{s})$, $j = 1, 2, \ldots, n$, where $\mathbf{s} = (E, D)$ is the state vector.

Step 4. Construct the state variable vector $\mathbf{s}$ (pond duration and pond elevation) in terms of the input vector $\mathbf{u}$, decision vector $\mathbf{x}$, and random variable vector $\mathbf{r}$. In general, $\mathbf{s}$ is dependent on $\mathbf{u}$ and $\mathbf{r}$ (i.e., $\mathbf{s} = \mathbf{s}(\mathbf{x}, \mathbf{u}, \mathbf{r})$).

Step 9. Given the set of ordered pairs $\{\mathbf{x}^{(q)}, \tilde{f}_p^{(q)}\}$, $q=1, 2, \ldots, Q$, $p=1, 2, \ldots, P$, a curve-fitting technique, such as least squares, can be used to determine the functional relationship $\tilde{f}_p : x \to \tilde{f}_p(x)$. Note that the curve-fitting step is not essential and is often not deployed when a tabulation is sufficient.

Two different curve-fitting routines might be used here. If the linear option is chosen, then the pth objective function is assumed to have the form

$$\tilde{f}_p(x) = b_0^{(p)} + b_1^{(p)} x_1 + b_2^{(p)} x_2 + \cdots + b_n^{(p)} x_n \quad (14.13)$$

and the routine uses a least-squares technique to determine the coefficients b_i^p, which give the best fit to the available data.

If the quadratic option is chosen, then the pth objective function is assumed to have the form

$$\begin{aligned}\tilde{f}_p(x) &= b_0^{(p)} + b_1^{(p)} x_1 + b_2^{(p)} x_2 \\ &\quad + \cdots + b_n^{(p)} x_n + b_{n+1}^{(p)} (x_1)^2 + b_{n+2}^{(p)} (x_2)^2 \\ &\quad + \cdots + b_{2n}^{(p)} (x_n)^2\end{aligned} \quad (14.14)$$

The higher-order terms give a closer fit to the inherently nonlinear objective functions, so the quadratic fit should give more accurate results.

The use of expected value, while a sound approximation of the frequency-versus-damage risk distribution in many circumstances, falters when extreme events are considered. High-damage/low-frequency events and low-damage/high-frequency events appear mathematically equivalent in the expected-value context. Here, the partitioned multiobjective risk method (PMRM), discussed in Chapters 8 and 11, through a partitioning scheme, circumvents the drawback of the expected-value approach by constructing risk functions that can be evaluated in a multiobjective framework.

14.5 THE SWT METHOD

Since all associated damages will increase with increasing peak ponding elevation and ponding duration, improving one objective function will improve all of the other objectives at the same time—except for the cost. There is a cost attached to increasing the flood protection level as measured by any of the multiple objectives. Thus, in the end, the problem is essentially reduced to a bicriterion (two-objective) optimization problem. All of the multiple objectives have trade-offs associated with the overall cost, but trade-offs among other multiple objectives would be meaningless (i.e., a trade-off between man-hours lost and flood damage could not be made because their behavior with respect to flooding level and duration is similar). The SWT analysis is most useful when objectives are in conflict (i.e., when the level of one objective can be improved only at the expense of others). The trade-off analysis can then be conducted between the cost and each of the other objectives.

In the ε-constraint formulation (see Chapter 5 and Haimes et al., 1971), one or more of the objectives in the first iteration will be binding, while the others remain nonbinding. A trade-off value for the binding objectives will be generated in the process, but in order to obtain trade-offs for the other objectives, it is necessary to change the right-hand-side ε levels for the nonbinding constraints in such a way as to make them binding. For example, if one of the ε-constraints is $f_3(x) \le 10$ and if, after an initial optimization, $f_3(\mathbf{x})$ has a value of 9, then reformulating the constraint as $f_3(x) \le 9$ (i.e., $\varepsilon_3 = 9$) would make this constraint binding in the neighborhood of $\varepsilon_3 = 9$ at the next iteration. The values for the trade-offs give the marginal costs associated with varying the constraint levels by one unit. These trade-off values can be used to vary the original ε-constraint levels systematically until a preferred Pareto-optimal solution is reached via the use of the surrogate worth functions.

A stepwise procedure corresponding to an algorithm outlined by Haimes et al. [1975], but specialized for this problem, is outlined below.

Step 1. Find minimum and maximum values for each of the multiobjectives by examining the output from the regression module of the program. This is done to find the approximate range of each objective as a function of the decisions to be examined.

Step 2. Set initial right-hand-side values (i.e., ε-values) for each of the constraints corresponding to the multiobjectives. Each

$\varepsilon_j > f_{j\min}$ for $j = 2, 3, \ldots, n$, where n is the number of objectives and $f_{j\min}$ is minimum value of the jth objective function, respectively.

Step 3. Solve

$$\begin{aligned}\min\ &f_1(\mathbf{x})\\ \text{subject to}\ &\mathbf{f}(\mathbf{x}) \leq \varepsilon,\ \mathbf{x} \in X\end{aligned} \quad (14.15)$$

where $X = \{\mathbf{x} : \mathbf{x} \leq 0,$ and all constraints are satisfied$\}$ and

$$\mathbf{f}(\mathbf{x}) = [f_2(\mathbf{x}), \ldots, f_n(\mathbf{x})]$$
$$\mathbf{x} = [x_1, \ldots, x_n]$$

Each solution also gives the trade-off vector, Λ_1, where $\Lambda_1 = [\lambda_{12}, \lambda_{13}, \ldots, \lambda_{1n}]$. If all of the ε-constraints are binding, then $f_1^*(\mathbf{f})$ and $\Lambda_1^*((\mathbf{f})$ are evaluated at $\mathbf{f}(\mathbf{x})) = \varepsilon$. The trade-off values λ_{ij} corresponding to nonbinding constraints should be ignored, that is, $\lambda_{ij} = 0$.

Step 4. If enough information has been generated from previous iterations, then proceed to step 5. Otherwise, select new values for ε and return to step 3. If a constraint j is not binding, then set $\varepsilon_j = f_j(\mathbf{x}^*) - \delta$, where $\delta > 0$ is a very small number—for example, $10^{-3}\varepsilon_j$.

Step 5. Develop the surrogate worth functions $W_{12}(\mathbf{f}), W_{13}(\mathbf{f}), \ldots, W_{1n}(\mathbf{f})$ by generating the decisionmaker's (DM) input as follows. For each set of values $\mathbf{f}$, $\Lambda_1(\mathbf{f})$, and $f_1^*(\mathbf{f})$ at which the value of the worth is desired, ask the DM for his or her assessment of how much $\lambda_{ij}(\mathbf{f})$ additional units of objective f_1 are worth in relation to one additional unit of objective f_j. The DM's assessment then provides the value of $W_{1j}(\mathbf{f})$ for each $j = 1, 2, \ldots, n$. On an ordinal scale of -5 to $+5$, $W_{1j} > 0$ means the DM prefers such a trade, $W_{1j} < 0$ means the DM does not, and $W_{1j} = 0$ implies indifference.

Step 6. Repeat step 5 until a value $\mathbf{f}^*$, the preferred solution, is found, which corresponds to a point at which all of the $W_{1j}(\mathbf{f}^*), j = 1, 2, \ldots, n$, equal zero. Once this is achieved, other values near $\mathbf{f}^*$ can be tried to determine the extent of the indifference band.

Step 7. The preferred decision vector can be found by solving the same problem, with the ε-constraints set equal to the $\mathbf{f}^*$ values.

Step 8. Stop.

This process gives marginal costs (trade-offs) associated with Pareto-optimal solutions for improving any of the objectives at given levels for all of the functions. DMs' preferences are examined to see whether further change is desirable.

14.6 MULTIPLE OBJECTIVES

Several objectives are suggested in the following text without implying that they constitute a complete list. Possible functional forms for the multiple objectives in terms of ponding elevation and duration are also considered in this section.

1. *Business interruption*, $f_1(E, D)$, can be measured in terms of either man-hours lost or monetary loss. For any ponding area, elevation of the ponded water can be related to its inundated area, so that it is always possible to find which businesses and other structures will be affected by a certain ponding level. Each business will probably have a flood-damage prevention plan, which may involve using the employees in flood-prevention efforts or in moving company inventory to safer locations rather than having them work at their normal jobs. This will lead to a certain loss of production for the business. If the flooding becomes serious enough, total evacuation of the business may become necessary, and both workers' wages and normal production will be lost. In this case, duration seems to be a likely candidate for a strictly linear type of measure, since a business that closes when flooding reaches a certain threshold level is likely to remain closed throughout the entire time that flooding is above that level. The business interruption function could then have the form

$$f_1(E, D) = \begin{cases} b_1 D & 0 \leq E \leq 1 \\ a_2 + b_2 D & 1 \leq E \leq 1.5 \\ a_3 + b_3 D & 1.5 \leq E \leq 2.0 \\ \vdots & \vdots \\ a_n + b_n D & E_n \leq E \end{cases} \quad (14.16)$$

The function $f_1(\cdot)$ is the measure (in man-hours lost) of the business interruption objective. The coefficients *a* and *b* are functions of the elevation, and they are determined from a detailed examination of the businesses affected, based on answers from a questionnaire to the business community.

2. *Drowning* as a function of elevation and duration has the same form as business interruption. Once again, the elevation determines some degree of hazard, in this case, the number of drownings per unit of time. Duration enters as a multiplicative factor, which gives the number of drownings for the total time during which the elevation exceeds the threshold level. The drowning versus elevation curve could be in the form of a discontinuous function, since the number of drownings for the low ponding levels anticipated would certainly be small, and the number of deaths is restricted to integer values. The general form of the drowning function $f_2(E, D)$ is given in Eq. (14.17), where $\hat{E}_i$ and $\bar{E}_i$ are the lower and upper limits of E_i, respectively.

$$f_2(E, D) = \{a_i + b_i D, \quad \hat{E}_i \leq E \leq \bar{E}_i, \quad i = 1, 2, \ldots, n\}$$
(14.17)

3. The *general aesthetics* function includes visual, olfactory, and other considerations. An ordinal scale is established with a range from zero (worst) to 10 (best). The form of the objective function determines a value in this range, depending primarily on personal opinion about the aesthetic appeal of a certain ponding level or ponding duration. For a certain elevation range, coefficients can be established to match the ordinal best/worst scale for all the durations that are possible. The constants (d_i and b_j) are such as to increase the value of the function (higher is better) for low elevations and short durations.

   This function is defined by means of inputs that give a rating for different ponding elevations and different durations as measured at any of three possible damage elevations. These inputs are used as grid points for linear interpolation (but linearizing only between durations measured relative to the same index height and not between more than one).

4. The *health hazards* function includes mosquito breeding, water contamination, and similar considerations. Mosquito breeding becomes important only if ponded water is expected to remain for about 2 weeks, which is about the minimum time needed to complete the cycle from egg to adult mosquito. In that case, the number of mosquitoes bred is the most relevant measure. This quantity can be found fairly accurately for any particular area, given the amount of ponded water and the types of mosquitoes.

   Contamination of a municipal water supply can be important in certain locations. If ponded runoff reaches a point where it can leach into groundwater supplies or perhaps enter through the tops of wells, a health hazard for the entire community can be created. The seriousness of this depends on whether an alternate water supply is available and on what percentage of normal water needs can be handled by an alternate system.

5. There seem to be very few *ecological* considerations that are relevant to the ponding scenario examined in this case study. Environmental considerations must be investigated in detail at each ponding site. A measure of this objective might be acres of grass destroyed or number of trees killed by different ponding conditions.

6. *Land use losses* specifically include playgrounds, recreational areas, and parks. Ponded water tends to restrict the use of these areas, and the amount of curtailment will be a function of ponding elevation and ponding duration. The curtailment can be measured in the form of user days lost. This requires a review of the normal use of these recreational areas and a determination of user curtailment versus ponding elevation and duration.

14.7 APPLYING THE MSM

The following case study uses hydrological data for the Moline project area in Illinois. It has three decision variables (pump size, pump-on elevation, and gate closure elevation) and three objectives (pumping cost, job man-hours lost, and aesthetics).

In a normal project evaluation, inputs from affected persons in the project area would be important in determining the forms of the multiobjective functions. Public officials and project engineers would use these inputs to develop objectives that depend upon ponding elevation and duration. In this case study, however, this process is bypassed, and functional forms that approximate the functional form of the objectives are used.

Four different pump sizes (0, 65,000, 90,000, and 150,000 gallons per minute (gpm)), three different pump-on elevations (564, 565, and 566 ft above mean sea level (msl)), and three different gate closure elevations (565, 566, and 567 ft above msl) are examined. Costs are associated with each pump size, and these values are used as grid points for a piecewise linear fit, so that intermediate pump sizes can have costs associated with them by linear interpolation between two of the grid points. The primary cost objective function $f_1(\cdot)$ is of this piecewise linear form.

The overall problem is

$$\min f_1(\mathbf{x}) = \text{cost}$$
$$\min f_2(\mathbf{x}) = \text{man-hours lost}$$
$$\max f_3(\mathbf{x}) = \text{aesthetics}$$

subject to a set of constraints.

The man-hours-lost objective, $f_2(\cdot)$, is formulated as a function of ponding elevation E and ponding duration D_i (i.e., the duration during which the ponded water exceeds an index elevation $E_T(i)$), as follows:

$$\begin{aligned} f_2(\cdot) &= 0 && \text{for } 0 \leq E \leq E_T(1) \\ f_2(\cdot) &= b_1 D_1 E && \text{for } E_T(1) \leq E \leq E_T(2) \\ f_2(\cdot) &= b_2 D_2 E && \text{for } E_T(2) \leq E \leq E_T(3) \\ f_2(\cdot) &= b_3 D_3 E && \text{for } E > E_T(3) \end{aligned} \quad (14.18)$$

$$b_1 = 1, \quad b_2 = 10, \quad b_3 = 100$$
$$E_T(1) = 564, \quad E_T(2) = 565, \quad E_T(3) = 566$$

all in feet above msl.

The aesthetics objective $f_3(\cdot)$ is formulated on a scale from 0 to 10, with 10 being most satisfactory and 0 being least satisfactory. All durations here are taken relative to the lowest index elevation $E_T(1)$.

Separable objective functions and constraints in the form of $f_j(x_j)$ are common in the literature. In particular, overall objective functions that are the sum of the subobjectives constitute a class of problems often referred to as separable problems. The general form is

$$\sum_{j=1}^{n} f_j(x_j)$$

where the jth subobjective function depends only on x_j, the jth decision variable. This format is assumed for the aesthetics function, with two components $f_{3a}(\cdot)$ and $f_{3b}(\cdot)$, where

$$\begin{aligned} f_{3a}(\cdot) &= 10 && \text{for } D = 0 \text{ hours} \\ f_{3a}(\cdot) &= 7 && \text{for } D = 12 \text{ hours} \\ f_{3a}(\cdot) &= 4 && \text{for } D = 24 \text{ hours} \\ f_{3a}(\cdot) &= 0 && \text{for } D \geq 36 \text{ hours} \\ f_{3b}(\cdot) &= 10 && \text{for } E = 0 \\ f_{3b}(\cdot) &= 9 && \text{for } E = E_T(1) \\ f_{3b}(\cdot) &= 6 && \text{for } E = E_T(2) \\ f_{3b}(\cdot) &= 0 && \text{for } E \geq E_T(3) \end{aligned} \quad (14.19)$$

These point values are used as grid points for a piecewise linear fit, so that elevations and durations between the values given above can have a functional value associated with them.

Finally, $f_3(\cdot)$ is defined as $f_3(\cdot) = f_{3a}(\cdot) + f_{3b}(\cdot)$, with 20 as its maximum possible value and zero as its lowest possible value.

14.7.1 Formulation of Linear and Quadratic Forms

Define the following three decision variables:

- x_1 pump size in 10^4 gpm
- x_2 pump-on elevation in feet above 563 ft above msl
- x_3 gate closure elevation in feet above 564 ft above msl

Pump-on elevation is the water elevation at which the pump will turn on. Gate closure elevation is the river height at which gravity drains must be closed to prevent backflow from the river. Four different pump sizes are examined: 0, 65,000, 90,000, and 150,000 gpm, with corresponding costs of $0, $280,000, $350,000, and $520,000, respectively. The three pump-on elevations used are 564, 565, and 566 ft above msl, while the three gate closure elevations are 565, 566, and 567 ft above msl.

Since we have already expressed the cost function $f_1(x_1)$ in terms of the pump size alone, it is not necessary to perform a regression on $f_1(x_1)$ in terms of the other two variables. That is, the cost is assumed here to depend on the pump size x_1 alone and not on the pump-on elevation or gate closure elevation. The four pump sizes and associated pump costs are used as grid points for determining the cost function $f_1(x_1)$ through a piecewise linear fit. Sample pump sizes other than the four given above can then have a cost associated with them by linearizing between the grid points immediately above and below the sample pump size. The equation for determining this cost is

$$\text{Sample pump cost} = \frac{\text{Sample pump size}}{\text{Higher pump} - \text{lower pump}} \times (\text{higher pump cost} - \text{lower pump cost}) \tag{14.20}$$

For example, given a sample pump size of 32,500 gpm, the associated pump cost here is given by

$$\left(\frac{32,500}{65,000-0}\right)(\$280,000 - \$0) = \$140,000 \tag{14.21}$$

The cost function $f_1(x_1)$ is constructed on the basis of the four provided grid points. In order to retain the linear characteristics of the model but still reflect the nonlinearity in pump cost versus pump size, separable programming is used.

The form of the piecewise linear cost function is then given by

$$f_1(x_1) = 0x_{10} + 2.8x_{11} + 3.5x_{12} + 5.2x_{13} \tag{14.22}$$

where the cost coefficients 0, 2.8, and so on, have been divided by 10^5 and where the x_{1i} are special variables. The pump size is defined in terms of these special variables by the equation

$$x_1 = 0x_{10} + 6.5x_{11} + 9.0x_{12} + 15.0x_{13} \tag{14.23}$$

It is also required that

$$x_{10} + x_{11} + x_{12} + x_{13} = 1, \quad 0 \leq x_{1j} \leq 1 \tag{14.24}$$

ensuring that only two adjacent special variables can be nonzero at once.

Expected values for man-hours lost $f_2(\cdot)$ and aesthetics $f_3(\cdot)$ are determined using the approach discussed earlier. These expected values are then regressed in terms of the three decision variables x_1, x_2, and x_3, determining the regression coefficients. Both linear and quadratic fits are performed on the same data, allowing a comparison to be made between the results for each. The output from the regression module then gives the following forms of the expected values for $\tilde{f}_2(\cdot)$ and $\tilde{f}_3(\cdot)$ in terms of the decisions $x_1, x_2,$ and x_3.

Linear regression:

$$\tilde{f}_2(\cdot) = 10,100 - 610x_1 + 279x_2 + 15.2x_3 \tag{14.25}$$

$$\tilde{f}_3(\cdot) = 0.00606 + 0.0341x_1 - 0.0473x_2 - 0.0126x_3 \tag{14.26}$$

Quadratic regression:

$$\tilde{f}_2(\cdot) = 15,700 + \left(43.7x_1^2 - 1570x_1\right) + \left(243x_2^2 - 573x_2\right) + \left(135x_3^2 - 527x_3\right) \tag{14.27}$$

$$\tilde{f}_3(\cdot) = -0.857 + \left(0.126x_1 - 0.00418x_1^2\right) + \left(0.0525x_2 - 0.0313x_2^2\right) + \left(0.425x_3 - 0.11x_3^2\right) \tag{14.28}$$

A formulation of the overall multiobjective problem is carried out using the ε-constraint approach, with the cost objective considered as the primary objective and the other two objectives entering the problem as ε-constraints (see Chapter 5). The form of the overall problem then becomes

$$\min \tilde{f}_1(\mathbf{x})$$

subject to
$$\tilde{f}_2(\mathbf{x}) \leq \varepsilon_2$$
$$\tilde{f}_3(\mathbf{x}) \geq \varepsilon_3$$
$$g(\mathbf{x}) \leq b$$
$$\mathbf{x} \geq 0$$
(14.29)

A separable programming routine is used to perform the optimization, so the problem must be manipulated slightly to satisfy its input requirements. Levels for ε_2 and ε_3 must also be specified. These ε values are manipulated as part of the overall SWT analysis to give the overall preferred solution.

Reformulating both problem forms according to standard procedures for using separable programming (using a grid of five points for x_2 and x_3 in the quadratic case) gives the following results (setting $\varepsilon_2 = 4000, \varepsilon_3 = 0.35$):

$$\min(0x_{10} + 2.8x_{11} + 3.5x_{12} + 5.2x_{13}) \quad (14.30)$$

subject to the constraints

$$10,100 - (0x_{10} + 3,965x_{11} + 5,490x_{12} + 9,150x_{13}) + 279x_2 + 15.2x_3 \leq 4,000 \quad (14.31)$$

$$0.00606 + (0x_{10} + 0.2216x_{11} + 0.3069x_{12} + 0.5115x_{13}) - 0.0473x_2 - 0.0126x_3 \geq 0.35 \quad (14.32)$$

$$1 \leq x_2 \leq 3, \quad 1 \leq x_3 \leq 3 \quad (14.33)$$

$$x_{10} + x_{11} + x_{12} + x_{13} = 1 \quad (14.34)$$

$$\text{Pump size} = 0x_{10} + 6.5x_{11} + 9.0x_{12} + 15.0x_{13} \quad (14.35)$$

Equation (14.30) is f_1, the piecewise linear form of the cost objective. Equation (14.31) is f_2, the man-hours-lost objective. Equation (14.32) is f_3, the aesthetics objective. The inequalities Eq. (14.33) restrict the pump-on elevation and gate closure elevation to values that have been examined in the regression and simulation. Equation (14.34) is the restriction on the special variable associated with the pump size x_1.

Using the optimization routine to generate a Pareto-optimal solution yields the following results:

TABLE 14.1 Sample of Pareto-Optimal Solutions with Their Associated Trade-Off Values

	Run 1	Run 2	Run 3
x_1 pump size, 10^4 gpm	11.84	13.4	14.95
x_2 pump-on elevation, feet above 563 msl	1	1	1
x_3 gate closure elevation, feet above 564 msl	1	1	1
$\tilde{f}_1(x_1)$ pump cost, 10^3	431	475	518.6
$\tilde{f}_2(x_1, x_2, x_3)$ business interruption, man-hours	3170	2213	1274
$\tilde{f}_3(x_1, x_2, x_3)$ aesthetics, units	0.35	0.39	0.39
$\lambda_{12}(\tilde{f}_1, \tilde{f}_2, \tilde{f}_3)$ dollars per man-hour	0.0046	0.0046	0.0046
$\lambda_{13}(\tilde{f}_1, \tilde{f}_2, \tilde{f}_3)$ dollars per unit of aesthetics	8.7	8.7	8.7

Pump size $\quad x_1 = 11.7 = 117,000$ gpm
Pump-on elevation $\quad x_2 = 1 = 564$ ft above msl
Gate closure elevation $\quad x_3 = 1 = 565$ ft above sml
Man-hours lost $\quad \tilde{f}_2 = 3268$
Aesthetics index $\quad \tilde{f}_3 = 0.35$
Pump cost $\quad \tilde{f}_1 = \$426,000$
Trade-offs $\quad \lambda_{12} = \$0.0046$ / man-hour
$\quad \lambda_{13} = \$8.70$ / units of aesthetics

The trade-off value of 0.0046 man-hours lost means that an expenditure of $1000 would reduce the number of man-hours lost by 4.6. Other Pareto-optimal solutions can be obtained for different levels of man-hours lost. Table 14.1 summarizes a sample of Pareto-optimal solutions and their associated trade-off values. The constant values of λ_{12} and λ_{13} are characteristic of the linear form of the objective functions, indicating that the trade-offs remain in the same segment of the Pareto-optimal hyperplane. For a nonlinear formulation, the trade-offs would vary, depending on which linear segment of the nonlinear approximation was being examined.

14.8 EXAMPLE PROBLEMS

The most distinctive feature of the MSM is its focus on the centrality of modeling in quantitative risk analysis and on the dominant role that state variables play in the systems modeling process. Furthermore, by incorporating random variables into the modeling effort, the MSM facilitates generating the expected value or the conditional expected values of various risk functions. For real-world problems, however, it is extremely difficult, if not practically impossible, to analytically quantify the functional relationships of the decision, random, and exogenous variables with the state variables and in turn with the objective functions. However, this task can be achieved via simulation (e.g., Monte Carlo simulation).

To avoid oversimplification and the introduction of trivial analytical derivations in this section, the primary focus will be on model formulation rather than on listing pages of computer-simulated results. Therefore, most example problems will be only introduced along with the building blocks of their mathematical model; readers may complete the modeling effort and generate their own numerical results.

14.8.1 The Farmer's Dilemma Revisited

In this example problem, we revisit the farmer's dilemma of how many acres of corn and sorghum to grow introduced in Chapter 1. Here, we modify it to incorporate one random variable in the model.

14.8.1.1 Assumptions

1. The only constraint introduced is the requirement that more than 0 acres of land be used for raising corn or sorghum.
2. No new exogenous variables are introduced.
3. The same decision variables remain:

   x_1 = number of acres of corn

   x_2 = number of acres of sorghum

4. One random variable is added: r = amount of nutrients incoming as a result of flooding. r has a log-normal distribution; that is, $r \sim$ log-normal $(3, 1)$.
5. One state variable is considered: s = level of nutrients in the soil. We assume that s is a linear function of random variable r. The relationship describing s and r is

$$s(r) = 3.1 + 3.9r \qquad (14.36)$$

Note that even when there are no incoming nutrients through flooding, the soil contains at least 3.1 units of nutrients.

6. Two objective functions are being considered: $f_1(x_1, x_2, s(r))$ = profit function (\$). The profit is a function of the decision variables and the crop yield (bushels), which itself is a function of the level nutrients in the soil. The relationship is summarized as follows:

Crop yield for corn $= 47 + s(r)$

Crop yield for sorghum $= 22 + s(r)$

$$\begin{aligned}f_1(x_1, x_2) = &[(\$2.8 / \text{bushel})(47 + s(r))\text{bushel}/\text{acre} \\ &- (\$40/\text{acre-ft})(3.9 \text{acre-ft}/\text{acre}) \\ &- (\$0.25/\text{lb})(200\text{lb}/\text{acre})](x_1) \\ &+ [(\$2.7/\text{bushel})(22 + s(r))\text{bushel}/\text{acre} \\ &- (\$40/\text{acre-ft})(3 \text{acre-ft}/\text{acre}) \\ &- (\$0.25/\text{lb})(150\text{lb}/\text{acre})](x_2)\end{aligned}$$

$f_2(x_1, x_2)$ = soil erosion function (tons)

We assume that the amount of soil erosion is solely dependent on the amount of acreage planted. The soil erosion function is deterministic:

$$f_2(\cdot) = 2.2(x_1^{1.3}) + 2(x_2^{1.1}) \qquad (14.37)$$

14.8.1.2 Implementing the MSM

The expected value of the profit function $f_1(\cdot)$ is determined by running a simulation using @Risk. This is done for a set of discrete values of r and $s(r)$. Over 1000 iterations were performed in order to compute the expected value. Table 14.2 summarizes the database for the farmer's problem. The results of the @Risk software simulation package can be presented in terms of the expected value of the objective function $f_1(\cdot)$. A regression of $E[f_1|x_1, x_2]$ against x_1 and x_2 produced the linear equation:

TABLE 14.2 Database

	Corn	Sorghum
Soil erosion (tons/acre)	2.2	2.0
Price per bushel	2.8	2.7
Crop yield (S) per bushel	128.1	103.1
Fertilizer cost per lb	0.25	0.25
Fertilizer cost per acre	200.0	150.0
Water cost per acre-ft	40.0	40.0
Water acre-ft per acre	3.9	3.0

$$f_1(\cdot) = \text{Profit} = 12{,}102 + 31.65x_1 - 0.15x_2 \quad (14.38)$$

The *explicit* expression of $f_1(\cdot)$ can now be used to solve the multiobjective optimization problem.

Note that it is desired to maximize profit and minimize soil erosion. In this case, it is convenient to convert maximizing profit to minimizing $(-f_1(\cdot))$ so that we may minimize both objectives:

$$\min f_1(x_1, x_2, s(r)) = -12{,}102 - 31.65x_1 + 0.15x_2 \quad (14.39)$$

$$\min f_2(x_1, x_2) = 2.2(x_1^{1.3}) + 2(x_2^{1.1}) \quad (14.40)$$

subject to
$$x_1 + x_2 \leq 100 \quad (14.41)$$

$$x_1, x_2 > 0 \quad (14.42)$$

14.8.2 Highway Construction

The state has mandated a highway construction road improvement effort along a major commuter artery in Metropolitan Washington, DC. The construction is to be performed on a 1-mile eastbound stretch affecting three lanes of traffic in that direction. The Department of Transportation (DoT) is interested in minimizing labor costs and minimizing delays to commuters along that road. More specifically, the decision is how many lanes to close during construction and how many work crews to schedule for the project.

One crew would require 3000 work hours to complete the project; that estimate is halved by the addition of a second crew and proportionally reduced by a third crew. Each work crew is made up of 10 workers and costs an average rate of $300 per hour. The DoT has options for closing lanes and assigning work crews. In closing lanes, DoT may opt to close off one, two, or all three lanes: Closing one or two lanes would force traffic to the lane(s) not being worked on at the time, while closing all three lanes would force traffic to an alternate route and slow commute time even more. Furthermore, traffic delays are a concern only during the morning rush hour. DoT may assign one, two, or three work crews to the project, but no more because of the need for other crews to work on similar ongoing projects.

The weather, especially rain, affects the schedule for construction. Rain forces the work to be stopped

TABLE 14.3 Probability Distribution

b	p(b)
1	0.125
2	0.125
3	0.100
4	0.100
5	0.150
6	0.150
7	0.125
8	0.125

and not continued until the rain ceases. Rain data are available from a local weather bureau that provides rain frequency data for the relevant season as well as the duration of the rainstorm in hourly intervals. Let the event of rainfall be denoted by Q, and let its frequency be denoted by q, which is derived from weather bureau data and the duration intervals in hours by b, where $b = 1, 2, \ldots, 8$.

14.8.2.1 Variables

Decision variables:

Number of lanes to close $= x$, where $x = 1,2,3$
Number of work crews $= y$, where $y = 1,2,3$

Random variable:

Average duration of rainfall $= b$,

where b is distributed in Table 14.3. Exogenous variables are stated in the problem statement and will not be repeated here.

State variable:

$$\text{Construction time} = C(x, y; b)$$
$$= 3000/y + 0.4(3000)bp(b)$$

Objective functions:

$$\min \{\text{labor costs} = f_1 = 300y \cdot C(x, y; b)\}$$
$$\min \{\text{commuter delays} = f_2 = 0.1x \cdot C(x, y; b)\}$$

Calculating the value of the state variable for possible values of b and the values of the objective functions can be conducted through the use of Excel. The expected values of the objective functions can then be calculated for each combination of x and y.

14.8.3 Manufacturing Problem

14.8.3.1 Introduction
This example addresses a manufacturing problem. A factory makes plastic mold-injected milk crates. The factory manager wants to minimize the cost of producing the crates ($f_1(\cdot)$) while minimizing the time it takes to produce 10,000 units ($f_2(\cdot)$). She may affect these two objectives by changing the number of machines to use and the amount of raw material (pounds of liquid plastic). However, she is uncertain about the length of time the machines may be down and the number of defective units that may be produced during the production period. Below is the mathematical statement of the problem:

$$\min f_1(N, M, D) = 1000N + 57M + 140D \quad (14.43)$$

$$\min f_2(N, D, T_d) = \left(\frac{100}{N} + T_d\right)\left(1 + \frac{D}{10,000}\right) \quad (14.44)$$

where the objective functions are

$$f_1(N, M, D) = \text{production cost}$$
$$f_2(N, D, T_d) = \text{production time}$$

The decision variables are

N = number of machines
M = amount of raw material, pounds

And the random variables are

D = number of defective units manufactured
 = binomial(10,000, 0.05)

T_d = machines downtime (days)
 = exponential (10)

14.8.3.2 Implementing the MSM
The expected values of the two objective functions, $f_1(N, M, D)$ and $f_2(N, D, T_d)$, are determined by using @Risk. This is done for combinations of a set of discrete values of N and M. The results of the simulation show that the output expected values of the objective functions are very similar to objective function values using the expected values of the random variables, D and T_d. As a result, the objective functions (Eqs. 14.43 and 14.44) with the expected values of the random variables are used in implementing the Kuhn–Tucker conditions in the SWT method. The mean of the binomial distribution is (10,000)(0.05) = 500. The mean of the exponential distribution is 10.

$$\min f_1(N, M, D) = 1000N + 57M + 140(500) \quad (14.45)$$

$$\min f_2(N, D, T_d) = \left(\frac{100}{N} + 10\right)\left(1 + \frac{500}{10,000}\right) \quad (14.46)$$

Form the Lagrangian:

$$L = 1000N + 57M + 140(500)$$
$$+ \lambda_{12}\left\{\left(\frac{100}{N} + 10\right)\left(1 + \frac{500}{10,000}\right) - \varepsilon_2\right\} \quad (14.47)$$

Assuming $M > 0$ and $N > 0$ simplifies the Kuhn–Tucker conditions:

$$\frac{\partial L}{\partial N} = 1000 - 100\lambda_{12}N^{-2}\left(1 + \frac{500}{10,000}\right) = 0 \quad (14.48)$$

$$\lambda_{12} = \frac{10N^2}{\left(1 + \frac{500}{10,000}\right)} \quad (14.49)$$

$$= \frac{10N^2}{1.05} \quad (14.50)$$

Pareto-optimum exists since $\lambda_{12} > 0$. Additionally, $M, N > 0$.

14.8.3.3 Discussion
The objective of this example problem is to gain a better insight into the MSM. The model was greatly simplified, so it may not reflect the true production system.

A regression of the conditional expectations of f_1 and f_2 on variable N (number of machines) and M (amount of materials) is not necessary because the model is already simple enough to generate the trade-off (λ_{12}) analytically. By substituting in the mean values of the random variables D (number of defective products) and T_d (machine downtime) together with the decision variables N and M, the conditional expectation of f_1 and f_2 is obtained.

Decreasing M reduces the cost of production without causing a delay. Therefore, the production

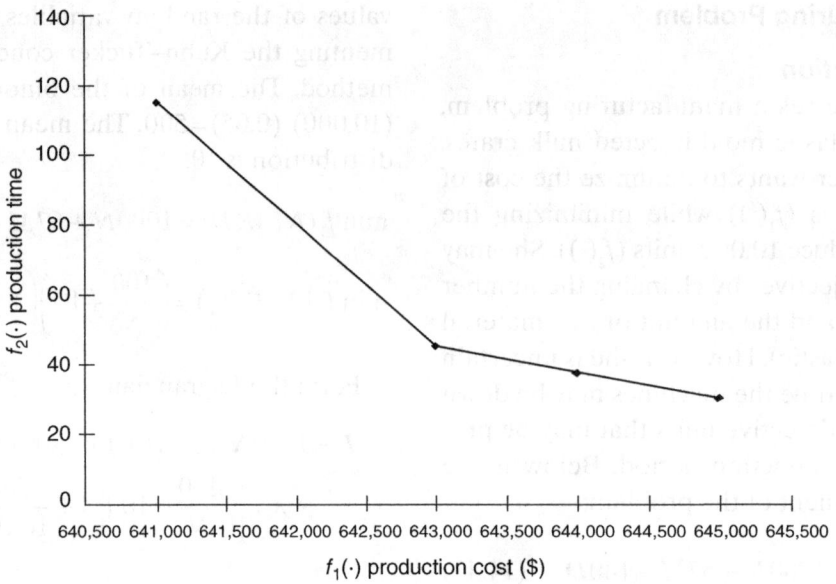

Figure 14.2 Pareto-optimal curve.

time is completely insensitive to changes in M as indicated in the objective functions (production time), and we are able to verify this through the analysis.

The value of the trade-off, λ_{12}, is $10\ N^2/1.05$. As N increases, the magnitude of the trade-off increases quadratically with N. Figure 14.2 depicts the Pareto-optimal frontier for production cost and production time.

14.8.4 Hurricane Problem

This example involves the construction of a beach resort development that could be subject to damage by hurricanes off the southern Atlantic coast. The problem can be formulated as a multiobjective optimization problem within a probabilistic framework. The statistical component stems from the probabilistic behavior of hurricanes and the ensuing property damage. The multiobjective approach takes into account revenues generated by property taxes as well as property damages due to hurricanes.

The MSM allows for integrating a multiobjective optimization scheme such as the SWT method and a statistical procedure to assess the different types of possible beachfront developments relative to economic and property damage objectives. While homes developed and constructed close to the beach generate the greatest amount of revenues, the amount of property damage due to hurricanes is also greatly increased. Thus, using MSM, it is possible that for a given set of system objectives and a finite set of alternative strategies, there will exist some configuration that will be optimal (in a Pareto-optimal sense) in relation to other possibilities.

14.8.4.1 Model formulation

A city on the southern Atlantic coast has approved the development of a 2-mile stretch of beachfront land. While the city hopes to maximize the revenues generated from the sale of beach homes, it also wishes to minimize property damage due to hurricanes, which have been known to ravage the southern coastline. The city planning commission has parceled the 2- by 3-mile stretch of land into four separate zones. The first zone extends inland 1/4 of a mile, the second zone from 1/4 to 3/4 of a mile, the third from 3/4 to $1\frac{1}{2}$, and the fourth from $1\frac{1}{2}$ to 3 miles. Due to the threat of hurricanes, the planning commission will allow only a certain number of homes to be developed within each zone.

This example is based on two objectives: maximizing revenues generated from the sale of homes developed and minimizing property damage due to hurricanes. Four decision variables are given as the number of homes allowed to be built per zones 1, 2, 3, and 4. The variables take on the values A (0–20 homes), B (20–50 homes), C (50–100 homes), or D

(100–200 homes). The state variable of the system is determined as the total number of homes built, which is distributed exponentially with respect to the zoning assignment. The random variables (which follow a log-normal distribution LN ~ (10,1)) include the number of square miles that could be damaged per zone by a hurricane and the possible occurrence of a hurricane along the coastline. Four exogenous variables are defined as the tax revenues generated per home with the homes built in zone 1 generating $2000 each, zone 2 $1500, zone 3 $1000, and zone 4 $500.

Thus, 256 possible zoning combinations for the city planning board are given. The trade-off is that the likelihood of hurricane damage decreases with inland zones and thus with lower tax revenue. Using this information, 100 years is simulated and the value of the objective functions (maximizing expected revenue and minimizing expected homes lost) is generated for all 256 combinations. Of these combinations, approximately 40 prove to be nondominated, and they form a Pareto-optimal frontier in the objective space. With this information and a methodology to evaluate the DM's preference, such as the SWT method, the desired combination for the zoning board can be determined.

14.8.5 Supermarket Checkout Problem

In this example, DMs must evaluate the trade-offs between the number of cashiers of various experience levels and the number of customers waiting in the checkout line. Both decisions will have an impact on costs, either directly through cashiers' salaries or indirectly through lost revenue.

14.8.5.1 Problem formulation

State variable: Average number of people waiting in the checkout queue.

Random variables: Customer arrivals and service times.

Decision variables: Number of slow cashier attendants to employ (s); number of medium-speed cashier attendants to employ (m); and number of experienced cashier attendants to employ (e).

Constraints: Minimum four cashiers at all times; at least one experienced or two medium-speed cashiers to ensure proper assistance at the checkout counters; when three people are in a checkout line, the fourth person will complain, which will result in a loss of goodwill and can lead to lost sales.

Objectives: Minimize checkout cost (represented as the sum of the salaries of the various types of cashiers) and minimize average queue size (the number of people waiting to be served). Minimizing queue size minimizes the number of people who will complain and maximizes the number served efficiently.

Assumptions:

- Customers' arrival at the checkout lines is represented as a Poisson distribution with a mean of 1 customer/minute.
- The service times are represented as uniform distributions with the minimum and the maximum checkout times depending on the type of cashier: slow = 5–8 minutes; medium = 3–6 minutes; and experienced = 2–5 minutes
- Customers complain (will not return to that supermarket) if there are three or more shoppers in every queue.
- Capital and operating costs are independent of the cashiers, and therefore, they are ignored.
- Hourly wages for the three types of cashiers are experienced = $12 per hour; medium = $9 per hour; and slow = $7 per hour.
- Eight possible alternative combinations of cashier types are considered: eemm, eems, emms, eess, mmms, emss, mmss, esss
- No more than two experienced cashiers can be on any 8-hour shift due to cost.
- No more than three slow or medium-speed cashiers can be on any 8-hour shift.

Based on Table 14.4, it can be determined that scenarios eess, esss, and emss are dominated (inferior solutions).

Simulation: The Siman simulation/modeling package was utilized for this example. The customer arrivals are distributed as a Poisson (with mean = 1 customer/minute). The customer gets into the shortest queue available (with queue length of <3), and ties are broken randomly. The delay time in the queue depends on the time of arrival and how

TABLE 14.4 Database for the Supermarket Checkout

	eemm	eems	emms	eess	mmms	emss	mmss	esss
Avg. service time	1.32	1.71	1.98	2.168	2.23	2.28	2.44	2.61
Avg. # served	457	443	418	416	389	386	359	324
Avg. # balked	10	37	51	69	78	89	107	148
Cost $/day	336	320	296	304	272	280	256	264

Note: e = experienced, m = medium speed, s = slow.

long it takes to serve the customers already in line. The average results obtained for different alternatives are given in Table 14.4.

REFERENCES

Haimes, Y.Y., D.A. Wismer, and L.S. Lasdon, 1971, On bicriterion formulation of the integrated system identification and system optimization, *IEEE Transactions on Systems, Man, and Cybernetics* **1**: 296–297.

Haimes, Y.Y., W.A. Hall, and H.T. Freedman, 1975, *Multiobjective Optimization in Water Resources Systems: The Surrogate Worth Trade-Off Method*, Elsevier, New York, NY.

Haimes, Y.Y., K.A. Loparo, S.C. Olenik, and S.K. Nanda, 1980, Multiobjective statistical method (MSM) for interior drainage systems, *Water Resources Research* **16**(3): 467–475.

U.S. Army Corps of Engineers, 1975, *Rock Island District Draft of Material for Moline, Illinois, General Design Memorandum, Hydrology Appendix*, Rock Island, IL, pp. 6–21.

15

*Principles and Guidelines for Project Risk Management**

15.1 INTRODUCTION

The increasing size and complexity of acquisition and development projects in both the public and private sectors have begun to exceed the capabilities of traditional management techniques to control them. With every new technological development or engineering feat, human endeavors inevitably increase in their complexity and ambition. This trend has led to an explosion in the size and sophistication of projects by government and private industry to develop and acquire technology-based systems. These systems are characterized by the often unpredictable interaction of people, organizations, and hardware. This complexity has imposed a complementary rise in the level of adverse events, particularly in acquisition projects, that is often difficult to identify, analyze, and manage.

Indeed, managing the risks associated with the acquisition of large-scale technology-based systems has become a challenging task. Such risks include cost overrun, time delay in project completion, and not meeting a project's performance criteria. For example, several reports issued by the Government Accounting Office (GAO) and papers published in archival journals document the many major acquisition projects by both government and the private sector that were completed behind schedule, well over budget, and not up to promised performance standards [GAO, 1992a, b, 2000; Lam, 1999; Reichelt and Lyneis, 1999]. This untenable mismanagement situation has provided an impetus for the development of new risk management methodologies to combat risk in major development and acquisition projects.

There is a growing body of literature on project risk management composed of a myriad of different approaches and methodologies. Sage [1992, 1995], for example, presents risk from a life cycle perspective, emphasizing the importance of considering the entire project life cycle when implementing project risk management. He discusses the types of risks that can occur and the impact they have on one another. Furthermore, he presents a set of tools and methods for managing risks to a project. R. Chapman [2001] emphasizes the importance of the risk identification process and advocates that a team-based approach is

---

*This chapter is based on Pennock and Haimes [2002], and Schoof and Haimes [1999].

Risk Modeling, Assessment, and Management, Fourth Edition. Yacov Y. Haimes.
© 2016 John Wiley & Sons, Inc. Published 2016 by John Wiley & Sons, Inc.

necessary to properly identify project risks. He presents several methods for eliciting sources of risk and discusses how the risk identification process affects the subsequent management of project risks.

Two other methods worthy of note are *continuous risk management* and *team risk management*, both developed by the Software Engineering Institute (SEI) at Carnegie Mellon University [Dorofee et al., 1996]. These methods aim to engage the entire organization in the risk management process by continuously monitoring the risks to the project in order to manage them before they become major problems.

In general, there is no one right way to conduct project risk management. Often, the best approach for any given project is driven by the unique characteristics of that project. Regardless, there are certain principles that apply universally to all risk management projects. This chapter presents principles and guidelines necessary to conduct risk management in an adaptable and repeatable framework. The first half presents an overview of the tools and methods developed in the course of conducting risk management on many different projects, while the second half focuses specifically on software risk management, developing mathematical models to address this particularly important application.

15.2 DEFINITIONS AND PRINCIPLES OF PROJECT RISK MANAGEMENT

15.2.1 Types of Risk That Threaten a Project

Two basic types of risk—technical risk and programmatic risk—characterize all projects. *Technical risk* denotes the risk that a project will fail to meet its performance criteria. This encompasses the realm of hardware and software failures, requirements shortfalls, and the like. *Programmatic risk* has two major subcomponents: cost overrun (the project exceeds its budget or operating costs) and delay in schedule (the project exceeds its projected completion schedule). In all cases, risk is defined as the probability and severity of adverse effects [Lowrance, 1976]. For example, cost risk includes

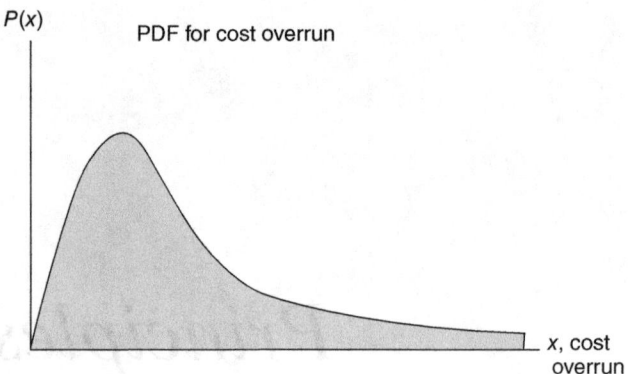

Figure 15.1 Probability density function for cost overrun.

probability and the associated level of cost overrun (see Figure 15.1).

The two-dimensional components of risk capture its complex nature, but they also make risk a more difficult entity with which to work. To that end, the risk assessment and management process can be represented by the six questions discussed in Chapter 1. These two triplets of questions constitute the guiding principles for an effective risk assessment and management process.

15.2.2 The Participating Parties

Numerous parties have stakes in the outcome of large-scale acquisition projects, including, but not limited to, the contractor, the customer, the user, and the analyst. These parties must meet collectively and periodically to ensure that proper risk management is being conducted. No party may be ignored because each brings different perspectives and background knowledge. For example, to develop a new aircraft for an airline, a contractor, such as Boeing, may possess expertise about the design and production of aircraft. The customer may have expert knowledge about past acquisition endeavors and possess knowledge of the airline's financial situation and what it needs to run a successful business. The users, such as the pilots, may bring operational knowledge of aircraft and what is most likely to go wrong while flying and what actions to take to correct failures. In a parallel fashion, the ground crews may know what is needed to minimize maintenance errors. Finally, the analyst may know how to bring together the disparate information of these groups to form a coherent picture of the risk situation and develop a plan to manage it.

15.2.3 Project Life Cycle

An often-neglected concept in project risk management is the consideration of the entire project life cycle; in the past, risk management has been conducted only on the final product. Manufacturing firms, for example, commonly conduct a failure mode and effects analysis (FMEA) and failure mode, effects, and criticality analysis (FMECA) on the product and the assembly line, but ignore the product development and design process (see Chapter 13 for discussion on FMEA and FMECA). Doing so neglects the risks inherent in requirements definition, development, acquisition, and phaseout or upgrade. Sage [1992, 1995] discusses the different types of risk inherent to various stages of the life cycle, such as acquisition schedule risk and fielded system supportability risk, and notes that these risks to life cycle stages are often interdependent and can arise from the design of the life cycle process itself. While concepts such as value engineering and life cycle cost analysis are often employed to assess the intended functionality and cost over the life cycle of the acquisition, analyzing risk over the project life cycle can also yield substantial benefits. Ignoring important stages of the life cycle can lead to substantial problems in terms of programmatic risk for both product development at the beginning of the life cycle and for product upgrade or replacement at the end. If major risks are not handled sufficiently early, they may magnify their effects later in the project. For example, in information technology acquisitions, errors in the requirements definition phase can lead to costly cascading problems later, when the information system fails to meet the customer's needs. As a result, costly modifications may be necessary, causing schedule slips and cost overrun.

15.2.4 Continuous Risk Management

Once the technical and programmatic risks have been identified (e.g., using hierarchical holographic modeling (HHM) introduced in Chapter 3 and in Haimes [1981]) and prioritized (e.g., using risk filtering, ranking, and management (RFRM) introduced in Chapter 7), the process of risk management can commence in earnest. The sources and consequences of emerging problems continue to evolve and change as the project progresses. As more information is obtained about a particular risk, the priority might change; therefore, it is necessary to constantly monitor all risks associated with the project. However, since it is prohibitively expensive, and often impractical, to assess and monitor all possible risks, only those most critical to the project are commonly monitored and managed. In sum, the entire set of risks should be reexamined periodically to ensure that the set of critical risks is still a valid set.

15.2.5 Team Risk Management

Managing the risks inherent in any system is contingent upon having sufficient knowledge of the system's structure and operations. Indeed, this knowledge is imperative in order to comprehensively identify the risks to an acquisition project, accurately estimate the probabilities of failure, and correctly predict the consequences of those failures. While the tendency to collect data and information on the project is important, databases are useful only with an understanding of the way the system they describe operates. Knowledge of a system provides a means to understand and to benefit effectively from the information about the system. Obtaining this knowledge is often difficult enough for a single system; the problem is compounded with the system of systems present in a development or acquisition project. Not only is knowledge of the many component systems required, but also it is critical to understand the boundaries where these systems interact and generate new sources of risk. These interactions include a project's requirements and specifications, design and construction, finance and management, development of new technology, and response to a myriad of changes and conflicting signals from the many participating organizations (among others). Thus, the sheer amount of system knowledge requisite for the risk analysis of even an *average-sized* project imposes some difficulties in the collection, dissemination, and integration of this knowledge.

In their book *Working Knowledge*, Davenport and Prusak [1998] suggest that knowledge moves through an organization via markets just as any other scarce resource does. There are buyers, sellers, and brokers of knowledge. Those who possess it will sell their knowledge if properly compensated with money, reciprocity, repute, bonuses, promotions, or other expected gain. If there is no sufficient compensation for those who sell their knowledge, the transfer will not take place. This market for knowledge has some important implications for risk management. The knowledge necessary to assess the risks to an entire project is spread over many individuals in multiple organizations and at multiple levels in the management hierarchy. For this knowledge to be transferred and collected for the purposes of risk management, an efficient knowledge market must exist.

To this end, management and corporate culture are key influences that must facilitate rather than hinder the operation of knowledge markets. First and foremost, trust is required for the exchange of knowledge [Davenport and Prusak, 1998]. Knowledge markets are informal and lack the security of legal contracts and a system of courts with which to maintain the integrity of exchanges. Therefore, trust is required so that sellers believe that they will receive appropriate compensation and buyers believe that the knowledge they receive is accurate. Management must create an environment that fosters trust. When the factor of concern is risk, knowledge of failures and mistakes is usually the most useful knowledge of all. Incidentally, knowledge of failures and mistakes is also the least likely to be divulged by an organization's members. Consequently, creating a culture of trust is imperative to obtaining the knowledge that is critical for risk management. Punishing personnel for reporting mistakes and failures is certain to short-circuit the entire risk management process. Unfortunately, the large number of participants complicates building trust in a development or acquisition project. System knowledge must be obtained from all of the participating organizations. This means that trust must exist both within each organization and between organizations. A failure in the atmosphere of trust anywhere along the lines can spill over into the rest of project.

Establishing trust is not sufficient for an efficient knowledge market, however. Sellers of knowledge must feel that they are being compensated for the knowledge they are providing to the risk management effort. To that end, project management must take the lead in portraying risk management as crucial to the success of the project. If project managers provide open support for the risk management process and present the success of the project as contingent upon the success of the risk management effort, then it is more likely that those further down the managerial ladder will actively participate and share their knowledge. This will occur because they are being compensated for the knowledge they bring to the table. The more knowledge is shared, the more likely it is that the risk management effort will succeed. When the link has been established that successful risk management leads to a successful project, participants will be compensated by such benefits as perceived value by their organization, potential promotions, bonuses, pay increases, and other rewards.

When trust and compensation are evident on the project, the issue of search costs for knowledge remains [Davenport and Prusak, 1998]. In other words, it is often difficult to ascertain who knows what and with whom relevant knowledge lies. Search costs for knowledge are often imposed by an organization's boundaries. There are four basic types of boundaries that restrict the flow of knowledge through and between organizations. These are horizontal, vertical, external, and geographic [Ashkenas et al., 1995]. *Horizontal* boundaries exist between the subdivisions or specialties within an organization and impose a *stovepipe* structure. The problem with such a structure is that it "fosters a sense of private ownership and discourages communication and cooperation" [FHA, 1999]. When horizontal boundaries are overcome, critical knowledge is transferred about the risks inherent in different sectors of a project. *Vertical* boundaries are those that separate the various levels of the organizational hierarchy. More specifically, they are the boundaries that exist between upper management, middle management, and the operational level. Vertical boundaries prevent understanding of the strategic goals of upper management from reaching the operational-level workers, and they prevent the tactical considerations and constraints of the operational level from reaching upper management [FHA, 1999]. When vertical boundaries are surmounted, lower-level employees will understand the strategic importance of the risk management process, and project management will be able to obtain valuable knowledge of risks to the low-level subsystems. *External* boundaries are the boundaries between organizations. In the case of risk management, these are similar to horizontal boundaries in terms of the difficulties they create. For a major project, the contractor, the client, the customer, and the user will all have specialized knowledge of the systems that they control. Overcoming external boundaries is necessary so that all risks can be identified and assessed. Finally, *geographic* boundaries impose search costs simply by means of distance. The further apart elements of an organization are, the less likely they are to communicate. One way to reduce the search costs inherent in obtaining relevant system knowledge is to foster trust between parties and provide a sense of joint responsibility through team risk management.

Team risk management brings together all of the disparate parties in the risk management effort. "A team is a small number of people with complementary skills who are committed to a common purpose, performance goals, and approach for which they hold themselves mutually accountable" [Katzenbach and Smith, 1999]. When conducting the risk management process in teams, participants are imbued with a common purpose. Risk management is not externally enforced; rather, it is a process within which everyone participates. When all participants have personal stakes in the process, they are much more likely to share their system knowledge as they can see the potential benefits from doing so. Overcoming organizational boundaries means bringing people together in face-to-face meetings: individuals from the various participating organizations, from subdivisions within organizations, and from different levels in the management hierarchy. Each participant brings a personal set of knowledge about the project that is crucial when analyzing and managing risks. While it is obviously not practical to have everyone in all participating organizations involved in every risk management team meeting, it is important to have a representative set. Furthermore, absence from the team meetings is not absence from participation. Everyone should be encouraged to submit potential sources of risk to the risk management team for review. An approach of total organizational involvement will yield a more comprehensive and useful risk management plan than if risk management were simply tasked to a small group of risk experts.

15.3 PROJECT RISK MANAGEMENT METHODS[1]

Changing and evolving risks over the project life cycle are major inhibitors to effectively managing risks. These changes occur due to the predictive nature of risk management and to changes in the priorities and requirements of a project. Without a stationary set of critical risks, risk management becomes more challenging and the risks to a project require constant monitoring to ensure that they remain under control. To that end, risk tracking is critical to effective project risk management.

15.3.1 Risk Tracking

Given a set of risks to a project and set of strategies to manage them, it is necessary to track the status of critical systems to monitor the effectiveness of those strategies. To that end, it is important to identify meaningful risk metrics. Metrics can be any measure of the state of a subsystem or component of the project that is relevant to the risks identified in it. For the collection of the metric data, it is critical to have total organizational participation. Metric levels are only accurate when those actually designing and building the product are fully and accurately disclosing system state information. Risk metrics provide a means of translating the abstract concept of risk into a measurable quantity that can be analyzed.

Once appropriate risk metrics are identified, the question is how to combine metric data with other available information about a risk in order to effectively track it. The *Continuous Risk Management Guidebook* [Dorofee et al., 1996] offers some suggestions. One method is the risk milestone chart, which displays the level of risk exposure over time with respect to the milestones in the risk management plan. The US Navy made use of risk milestone charts during its upgrade of the E-6 fleet [U.S. Navy, 1997]. The Navy risk management team constructed one chart for each risk tracked and included each chart in a monthly risk report. This allowed the project management to quickly identify trouble spots. Figure 15.2 depicts an excerpt from one of those monthly reports.

---

[1]This section is based on Haimes and Chittister [1996], Schooff et al. [1997], and Schooff and Haimes [1999].

444 PRINCIPLES AND GUIDELINES FOR PROJECT RISK MANAGEMENT

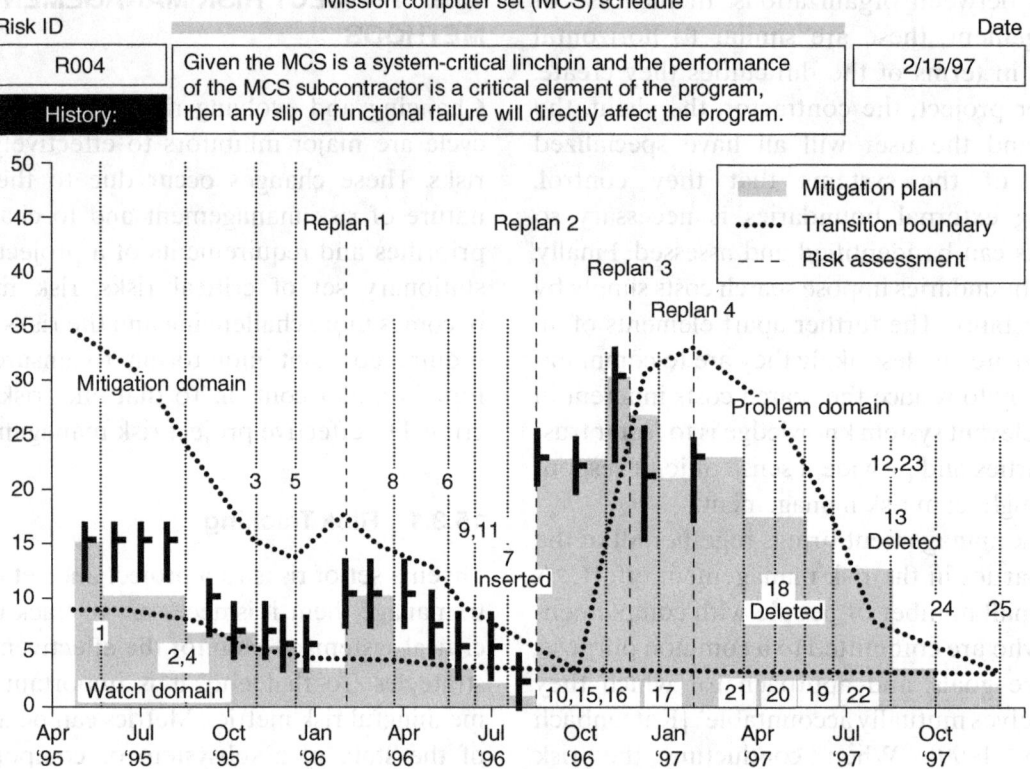

Figure 15.2 Navy milestone chart for risk tracking taken from the E-6 project.

The benefit of this type of display is that it conveys a great deal of information in one place, including the following important features:

- There is one chart for each risk, where the vertical axis indicates the level of that risk and the horizontal axis represents the date.
- The vertical dotted lines represent the milestones for the risk management plan, where each represents a task that, when completed, should lower the level of the risk. In conjunction with the milestones, the shaded areas, determined during the planning phase, represent the anticipated risk level following each milestone.
- The jagged dashed lines divide the chart up into three regions: the problem domain, the mitigation domain, and the watch domain [Dorofee et al., 1996]. When a specific risk is in the *problem* domain of the tracking chart, immediate action is necessary to rectify the situation. When a risk is in the *mitigation* domain, a set of steps must be created to mitigate that risk. When a risk is in the *watch* domain, no action is necessary, but tracking should continue.
- The black markers indicate the measured risk values. The top part of the black line indicates the pessimistic case, the bottom represents the optimistic case, and the horizontal dash represents the most likely case. The measured risk values compare the actual risk level with the predicted risk level. When the measured risk markers are higher than expected, a replan of the risk mitigation milestones may be necessary. This leads to the final feature of the chart.
- The vertical, dashed lines indicate a replan.

During a replan, the milestones in the risk management plan are revised and a new set of anticipated risk levels are developed, which is apparent upon examination of the chart. Note that in the example chart (Figure 15.2) in a period of less than two years, four replans were necessary because of unexpected rises in the risk level. This chart, taken from an actual acquisition project (the E-6 project), clearly demonstrates how continuous risk management can bring technical and programmatic risks under control before they become significant problems.

One important aspect of the risk milestone chart that is of concern is the abstract notion of risk level or risk exposure. This dimensionless measure conveys little meaning to decisionmakers other than that a large number is undesirable. Furthermore, the method of deriving risk exposure deserves improvement. The method used by the Navy team [U.S. Navy, 1997] and demonstrated in the *Continuous Risk Management Guidebook* [Dorofee et al., 1996] is to assign one ordinal number to the probability and one ordinal number to the severity of the adverse event. The two numbers are multiplied to obtain the risk exposure or the level of risk. This procedure violates a basic premise in measurement theory since one cannot multiply ordinal numbers because the ratios between ordinal numbers are meaningless. The result is that when two values for risk exposure are compared, the risk with the higher exposure value may not necessarily be the higher risk. This presents a significant problem when risks are prioritized for the allocation of limited resources. The problems with risk exposure are complicated further by the loss of information incumbent in using an ordinal scale system. If the risk is tracked using some actual measure of the system, then the value of the metric is lost when it is converted into an ordinal value. Consequently, valuable information about the system is likely to be lost. For example, suppose that the risk of concern is that of flooding. The metric of concern might be the current water level. Since the exact level of the water can be measured, that information would be lost if it were subsequently converted to an ordinal value, for example, a discrete number between one and ten. Therefore, using the risk exposure for ranking and tracking risks is fraught with problems.

In order to counteract the problems associated with risk exposure but retain the useful features of the risk milestone chart, the vertical axis of the chart can be replaced with the damage metric for the risk being tracked. For example, if the risk being tracked is cost overrun, *risk level* could be replaced by *amount of cost overrun*. The plotted values on the chart could be the expected value of cost overrun, the conditional expected value of cost overrun (see Chapters 8 and 11), or any other value of interest associated with cost overrun. This method is beneficial in several ways. First, the values on the chart are meaningful to decisionmakers. An expected cost overrun of $10 million is much more meaningful than a risk level of 38. Second, a logical avenue is opened for obtaining both the predicted values on the chart as well as the measured values. The expected value and conditional expected value are both calculated from probability distributions. If historical information is available, a distribution can be developed, and the expected and conditional expected values can be calculated. When historical information is limited, expert evidence can be utilized in terms of developing fractile or triangular distributions. The expected and conditional expected values can be calculated from these distributions as well. As information is collected during the progression of the project, the distributions can be updated. For distributions that are built on historical records, Bayesian updating may be employed, and for distributions that are constructed on the basis of expert evidence, the experts may update their assessments on the basis of new information (see Appendix A.2). This allows for a comparison between the predicted and measured values of risk. Furthermore, the actual values of the risk metric may be compared with the corresponding expected values of risk since they are of the same units. Ultimately, the updating improves the situational awareness of the decisionmakers and remedies the mathematical shortfalls of the risk exposure method.

15.3.2 Risk Identification

Several systemic methods and approaches are available for identifying and tracking risks. While techniques such as FMEA [DoD, 1980] and fault trees [NRC, 1981] (see Chapter 13) work well for mechanical devices, large sociotechnological systems, including acquisition projects, require a more broad-based, multifaceted approach. Indeed, performing a failure analysis on a mechanical device is relatively straightforward because there are a finite number of parts. Each part can be examined individually to determine its failure modes and how those failures will affect other parts and the overall effectiveness of the device. However, in large sociotechnological systems, there are at least four major categories of sources of risk: (i) hardware failure, (ii) software failure, (iii) human failure, and (iv) organizational failure.

446 PRINCIPLES AND GUIDELINES FOR PROJECT RISK MANAGEMENT

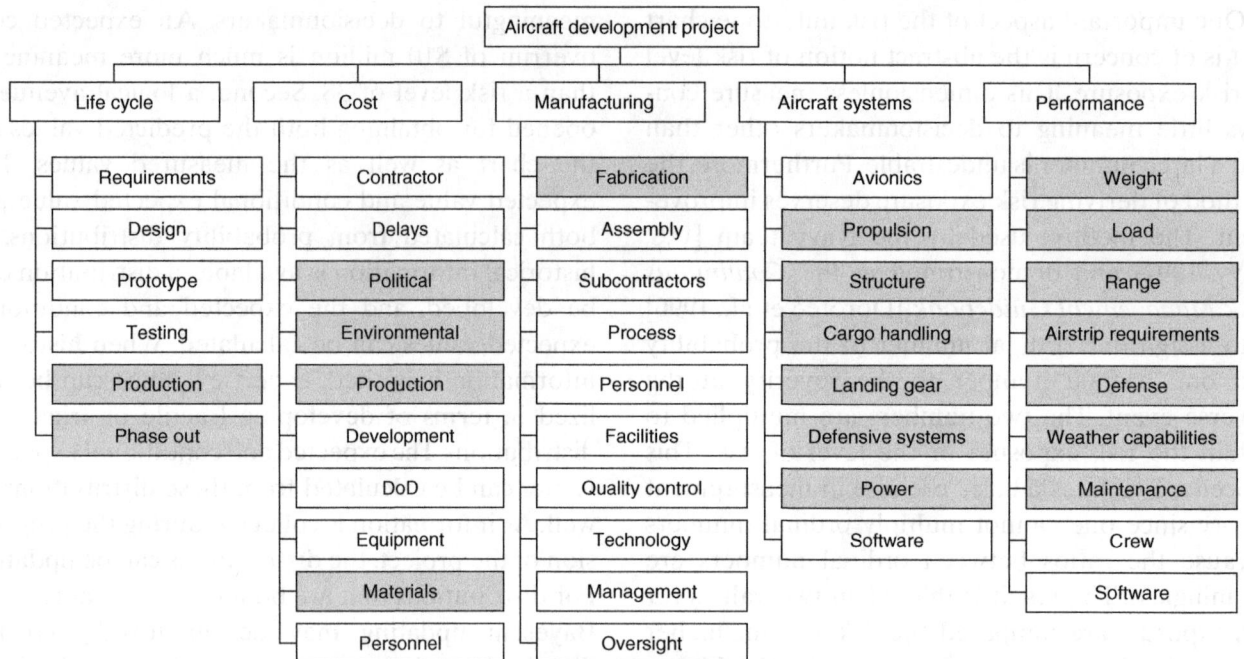

Figure 15.3 HHM for aircraft development with filtered subtopics.

These four categories of risk are highly interactive and complex, and it is very difficult to capture all of them with a single model. It is necessary to use multiple models, each presenting a different perspective of the system. To accomplish this, HHM is employed (see Chapter 3). A sample HHM is provided for an aircraft development project in Figure 15.3.

As is apparent from the example, an HHM can be used to represent most, if not all, of the critical and important facets that make up a system to achieve a more comprehensive assessment. Each subtopic or set of subtopics can be represented by a different model, quantitative or qualitative [Haimes et al., 1998]. This is where the power of HHM lies, as well as its usefulness as a risk identification tool. If each subtopic represents a subsystem or component of the overall system, then each subtopic is a potential source for failure. Therefore, if the HHM represents a global model for the entire system, then modeling the failure of each subtopic will identify the entire set of risk scenarios for the system. In practice, the number of subtopics in an HHM can be very easily in the hundreds.

For constructing the HHM for project risk management, any number of methods is acceptable. For example, brainstorming by domain experts is often an effective means of developing an HHM for a system. This is one area where eliciting contributions from all stakeholders is important. Each stakeholder brings a different perspective and domain expertise that are useful for constructing a comprehensive catalog of risk. Ultimately, other methods may be used as well. Techniques such as anticipatory failure determination (AFD) [Kaplan et al., 1999] or hazard and operability analysis (HAZOP) [AICHE, 1999] may also prove useful in identifying risk scenarios for the HHM.

15.3.3 Risk Filtration

Since it would be prohibitively expensive in terms of both time and resources to model and track every source of risk to a complex system, a method capable of filtering out less critical risks and prioritizing the remainder is necessary. Indeed, a process that discriminates between critical and mundane risks allows for the best allocation of resources for their management. The RFRM method discussed in Chapter 7 offers an eight-step process designed to filter down a large set of risks into those that are most important to decisionmakers.

It is important to note that this method avoids some of the pitfalls of mixing ordinal and cardinal scales and of using ordinal numbers for risk severity measures. Many filtering techniques allow a quick pass filtration by using ordinal scores for both probability and failure effects, where the probability score multiplied by the failure effect score yields the risk severity. The risks are then ordered by severity scores, and those with the highest scores are selected for more analysis and management [Williams, 1996]. The RFRM, however, makes use of both ordinal and cardinal scales, albeit separately and without mixing the two.

The use of this method is advantageous because often when there is insufficient evidence or resources to quantitatively determine the probability and severity of a risk scenario, a more qualitative ordinal classification is the best resort. This procedure is acceptable as long as the ordinal numbers are treated as such and without subjecting them to the algebraic rules of multiplication and addition. The tendency in some tools and methods to multiply the ordinal ratings of probability and severity and add them together to get a risk level should be avoided altogether. As previously noted, this procedure is not mathematically valid since the ratio between two ordinal rankings is meaningless. Thus, when ordinal numbers are multiplied together, an implicit assumption is made in terms of the relative importance of each rating.

Since the RFRM is a methodological framework, it is meant to be adaptable to suit particular situations; for example, it has been adapted in this chapter to suit the needs of project risk management. Several of the steps of the RFRM method that are already accounted for in the proposed project risk management methodology will not be discussed further, keeping the focus on the filtration process. With that in mind, a modification was made to the filtration process discussed in Section 7.3.2 to better suit project risk management, yielding the following five phases:

1. Scenario filtering
2. Bicriteria filtering and ranking
3. Multicriteria filtering
4. Quantitative ranking
5. Interdependency analysis

Each of these phases, and their application to project risk management, is explained in the following text.

15.3.3.1 Phase 1: Scenario filtering

As mentioned earlier, the number of subtopics in an HHM can easily reach the hundreds, and it is unwieldy to work with such a large number of risk scenarios. Therefore, the filtration is accomplished by removing those subtopics not relevant to the current decisionmaker in terms of level, scope, and temporal domain. For example, a midlevel manager in charge of transportation over the next year may be concerned with a different set of risks from those of a CEO whose focus is on corporate strategy over the next five years. This filtering is achieved through expert experience and knowledge of the system in question.

15.3.3.2 Phase 2: Bicriteria filtering and ranking

This phase introduces both probability and consequences to the filtering process. It is based upon an ordinal matrix developed by the US Air Force and the McDonnell Douglas Corporation. Probability and consequence are combined to produce risk severity. For this phase, probability and consequence are each divided into five ordinal categories, with each of the remaining subtopics falling into one block of the matrix. A threshold is set as far as risk severity, and only those subtopics meeting or exceeding the threshold will survive the filtering process.

When placing the risks into the matrix, there are several key considerations. First, one must consider the level of resolution of identified risks. Given the hierarchical nature of HHM, it is possible to break down every risk scenario into individual failure modes and list them as lower levels in the hierarchy. This would provide the risk management team with a comprehensive understanding of the possible failure scenarios and allow for better assessments of probability and severity for the purposes of filtration. Doing so, however, would be a labor-intensive process and defeat the purpose of risk filtration. At the other extreme, if the identified risks are too high level, the risk management team will not have sufficient understanding of the associated failure scenarios to assign the risks realistically to probability and severity categories. Therefore, the risk management team must ascertain the appropriate level of detail with which to filter risk scenarios.

Another critical consideration is the variety of probability and severity combinations that may exist for each identified risk. For example, if the risk of concern is a late delivery of raw materials, there may be a high probability of a short delay and a low probability of a long delay. As it is often the case that a risk will have a range of possible consequences, the best strategy is to score the risk based on the probability/severity combination that presents the highest risk level. To continue with the previous example, if one considers the most likely case of a short delay that is *likely* and *moderate*, the risk of a supplier delay is considered *moderate risk*. If one considers the more extreme case of a long delay that is *seldom* and *critical*, however, the risk of a supplier delay is classified as *high risk*. In order to avoid filtering out a serious risk, the risk of a supplier delay should be scored as high risk.

15.3.3.3 Phase 3: Multicriteria filtering

In this phase, the remaining subtopics must defeat the defensive properties of the system defined in Chapter 7: robustness, redundancy, and resiliency. Since these system attributes are rather vague notions, they have been decomposed into subattributes (Figure 15.4).

The basic idea behind this phase is that a subtopic that lacks redundancy, resiliency, or robustness is more risky than one that does not. For example, a subtopic with multiple paths to failure is more of a problem than a subtopic with only one path to failure. Thus, the analyst must identify how each of the remaining subtopics performs with respect to these defensive properties. The method of accomplishing this is somewhat subject to the preference of the analyst. Methods such as weighting schemes and the analytic hierarchy process [Saaty, 1980] could be used to directly filter subtopics by applying scores for each attribute to each subtopic and then obtaining aggregate scores. Table 7.1, which defines each of the defensive properties, is helpful in understanding this phase.

By juxtaposing the risk scenarios against these attributes, analysts and decisionmakers may revise how they view both likelihood and consequences.

15.3.3.4 Phase 4: Quantitative ranking

In this phase, the probability of each remaining scenario is quantified using all available evidence. The purpose of this process is to replace opinion with evidence and avoid the linguistic confusion of labels such as *high* and *very high*.

15.3.3.5 Phase 5: Interdependency analysis

Most existing systems are complex and highly interdependent, and systems involving development and acquisition are no exception. The role and importance of humans, organizations, hardware, and software in such systems create a highly interactive environment that adds complexity to the situation. Therefore, it is

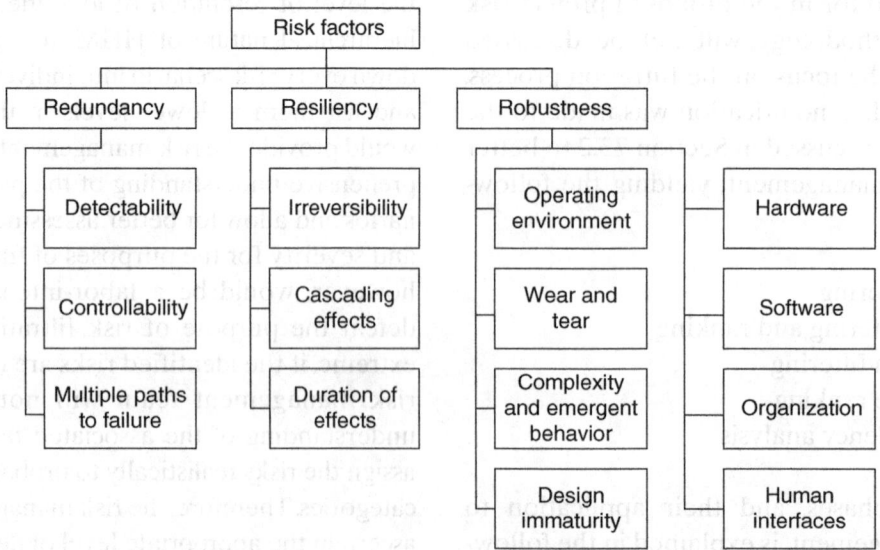

Figure 15.4 Defensive properties of a system.

important to address the interdependencies among the various subsystems or risk scenarios in the project acquisition system. Doing so helps to avert overlooking seemingly innocuous subsystems of the project that are actually critical due to their interconnections with other, more important subsystems. To accomplish this, a simplified dependency analysis has been developed.

Clearly, a comprehensive dependency analysis of every subsystem or risk scenario would be impractical. Therefore, this dependency analysis begins with the remaining set of critical subtopics within the filtration process. For each critical subtopic, the analyst should return to the HHM and identify all the interconnected subtopics. It is important to note that interconnections are directed. While a given pair of subsystems may be interdependent, it might be the case that for another pair of subsystems, one may be dependent upon the other but not vice versa. An example is shown in Figure 15.5 for four fictitious subtopics.

Three critical factors determine the importance of an interconnection. These are the degree of failure transmission, the degree of criticality of the failure, and the duration of the failure:

- *Transmission* is used in the sense that the failure in one subsystem is transmitted to another. The strength of the interdependencies among the subsystems dictates the likelihood that a failure in one subsystem would cause a failure in its dependent subsystem.
- *Criticality* reflects the severity or importance of the derivative failure.
- *Duration* indicates the length of the failure in the initiating subsystem.

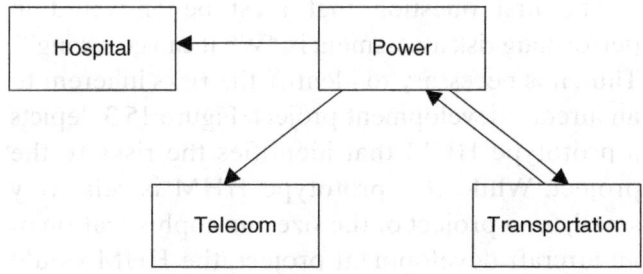

Figure 15.5 Dependency diagram.

These concepts can best be explained with an example. In Figure 15.5, the subtopic hospital is dependent upon the subtopic power. There is a high degree of transmission because most of the equipment at the hospital needs electrical power.

A power failure means that the external source of power to the hospital is lost, and the criticality in this case is a function of duration. A power failure of short duration will have little or no effect because the hospital has backup generators that will maintain power to critical systems. If the failure is long in duration, however, the backup generators may be insufficient and patients' lives might be in jeopardy. Consequently, for a long duration of failure, the connection is critical. From this example, it is clear that the three properties of transmission, criticality, and duration are interrelated. There is also an intuitive notion that because of that interrelation, if power was not considered critical before, it should be now. It is necessary, however, to address this more systematically.

Given a means for assessing the risk imposed by interconnectedness, the question remains as to how to incorporate this into the filtration process. To that end, the rule is that if the receiver is critical, then the transmitter is critical as well. In other words, if a critical subtopic is dependent upon a noncritical subtopic and the interconnection is assessed as critical, then the noncritical subtopic is upgraded to critical. This creates a recursive process for identifying overlooked critical risk scenarios through dependencies. To return to the hospital/power example, assume that hospital is deemed critical but power is not. If the dependency of hospitals on power is identified as critical, then power becomes critical. Since power is now critical, subtopics on which it is dependent have the potential to become critical (e.g., transportation). In highly interactive systems, one may find that the subtopics are so coupled that many subtopics may be reassessed as critical. In that case, the analyst may want to consider a full dependency analysis of the system. One such method is the Leontief-based inoperability input–output model (IIM) to analyze interdependencies [Haimes and Jiang, 2001; Santos and Haimes, 2004; see also Chapter 17].

Ultimately, the filtration process is designed to yield a manageable set of risk scenarios that comprise the most critical risks to the system. The filtration process may not apply uniformly to all systems and may have to be modified to fit special circumstances. As a general rule, it is better to err on the side of being conservative and retain questionable risks rather than filter them out.

15.3.4 Risk Assessment

Once the set of critical risks has been identified, a more in-depth analysis of those risks is required in order to properly manage them. In the case of simple point failures, this may require testing a sample of the failed component to determine a probability distribution. In the case of complex subsystems, it may require building a model of the subsystem and using Monte Carlo simulation. If the problem cannot be quantified, it may require analysis based on expert evidence using fractile or triangular distributions. For situations in which there is no prior experience, scenario analysis may be the best option. Regardless of the adopted approach, all seek to answer the three fundamental risk assessment questions previously mentioned in Chapter 1.

Risk management builds on the art and science of modeling, where the selection of the appropriate risk modeling method for each critical risk is highly dependent on the expertise of the risk analyst. Clearly, each critical risk ought to be modeled effectively, since models provide the means of quantifying the critical risks to a project [Kaplan et al., 2001].

15.3.5 Risk Management

Once the analysts and decisionmakers have thoroughly analyzed the critical set of risks, they are in a better position to determine the best course of action to mitigate those risks. The participating parties identify all viable options available to manage (prevent, mitigate, transfer, or accept) the identified risks. The options must then be traded off against one another in terms of cost, risk, benefit, and any other relevant criteria. Once the best options have been determined, their impact on the rest of the system must be considered. It is possible that some decisions may eliminate sets of options that could resolve future problems or that some decisions may change the risk levels of other scenarios. For example, some risks that were previously deemed noncritical may become critical because of changes made in the system. Therefore, it is imperative that impact analyses be conducted to assess potential changes in the state of the system. The result of the risk management process should be a set of plans to mitigate the critical risks of a project. It is the effectiveness of these plans that risk tracking monitors.

15.3.6 Iteration

As with any other type of analysis, it is unlikely that the analysts, decisionmakers, and other participants did everything right the first time. Therefore, it is necessary to repeat periodically the entire risk assessment and management process with the engagement of the project's participants. In this manner, new critical risks can be identified and reprioritized. To a large extent, this may be due to new information acquired as the project progresses, to changes that take place in the state of the system, or to factors beyond or within the control of the project managers. In other words, a specific risk that originally was deemed noncritical might turn out to be a major bottleneck in the project development and may require proper attention. Risk tracking is pivotal in identifying these unexpected changes. The contractors, clients, users, customers, analysts, and other stakeholders should meet regularly to discuss progress and reassess the risk management strategy.

15.4 AIRCRAFT DEVELOPMENT EXAMPLE

To better explain the risk management methods described in this chapter, the following is a simplified analysis of an aircraft development project [Nordean et al., 1997].

The first question that must be answered in performing risk assessment is, "What can go wrong?" Thus, it is necessary to identify the risks inherent to an aircraft development project. Figure 15.3 depicts a prototype HHM that identifies the risks to the project. While this prototype HHM is relatively small, for a project of the size and sophistication of an aircraft development project, the HHM would probably contain hundreds, if not thousands, of elements. Thus, the next step is to filter down the set of risks to a manageable level.

The first phase of risk filtration is scenario filtering on the basis of the decisionmaker's scope and domain of interest. For the purpose of this retrospective case study, it is assumed that the decisionmaker is a manager in charge of software flight control systems for the duration of the development project. Consequently, subtopics in the HHM not related to the decisionmaker's responsibilities and scope are filtered out. The subtopics filtered out are shown in gray in Figure 15.3.

With the set of risks reduced to the scope of the decisionmaker, bicriteria filtering and ranking can be applied. Table 15.1 lists each subtopic with its corresponding ordinal likelihood of occurrence and effects and thus the relative expected risk. In this case, only those subtopics with high or extremely high risk are retained. The subtopics filtered out are shown in gray.

The remaining subtopics are examined in the multicriteria filtering phase of the RFRM method. The purpose of this phase is to revisit the effects of failures in these subtopics (i.e., the risks) by examining the defensive properties of the project against each risk. For the purpose of consistency, each attribute is scored such that a higher value indicates a higher risk. For example, the criterion *controllability* is scored in terms of *uncontrollability*. Table 15.2 lists the results of this analysis for each subtopic in terms of high, medium, low, and not applicable. The criterion numbering in the table corresponds to the numbered list of attributes in Section 15.3.3.

The results from the criterion scoring are used in the quantitative ranking phase. Each subtopic is placed in the quantitative matrix based on a measured probability and the effect of a failure considering the analysis of the system's criteria, which are summarized in Table 15.2.

Although the probabilities in this example are fictitious, the actual probabilities could be obtained from historical records or through expert evidence. As can be seen from Table 15.3, the effect of two subtopics has been downgraded to *moderate* consequence due to a review of the defensive properties of the system. This reduces the ranking of these two subtopics to *moderate risk*, and they are consequently filtered out.

TABLE 15.1 Subtopics Ordinal Scores

Subtopic	Likelihood	Effect	Risk
Requirements	Likely	Critical	High
Design	Likely	Catastrophic	Extremely high
Prototype	Occasional	Serious	Moderate
Testing	Likely	Critical	High
Contractor	Occasional	Serious	Moderate
Delays	Frequent	Critical	Extremely high
Development	Likely	Serious	High
DoD	Seldom	Marginal	Low
Equipment	Unlikely	Marginal	Low
Personnel—cost	Likely	Critical	High
Assembly	Likely	Serious	High
Subcontractors	Likely	Critical	High
Process	Occasional	Catastrophic	Extremely high
Personnel—manufacturing	Occasional	Moderate	Moderate
Facilities	Unlikely	Marginal	Low
Quality control	Occasional	Critical	High
Technology	Seldom	Moderate	Low
Management	Occasional	Moderate	Moderate
Oversight	Occasional	Moderate	Moderate
Avionics	Likely	Catastrophic	Extremely high
Software—aircraft systems	Likely	Catastrophic	Extremely high
Crew	Occasional	Moderate	Moderate
Software—performance	Likely	Catastrophic	Extremely high

TABLE 15.2 Subtopic Criteria Risk Scoring

Subtopic	1	2	3	4	5	6	7	8	9	10	11
Requirements	L	H	N/A	M	H	H	N/A	N/A	M	N/A	N/A
Design	M	M	N/A	M	H	H	N/A	N/A	M	N/A	N/A
Testing	H	L	M	L	H	L	L	L	M	N/A	L
Delays	N/A	M	H	H	H	H	N/A	N/A	N/A	N/A	N/A
Development	M	M	N/A	M	H	H	N/A	N/A	M	N/A	N/A
Personnel—cost	N/A	M	N/A	H	M	L	N/A	N/A	N/A	N/A	N/A
Assembly	M	M	N/A	H	M	H	N/A	N/A	M	N/A	N/A
Subcontractors	M	M	N/A	M	M	H	N/A	N/A	N/A	N/A	N/A
Process	H	M	N/A	M	H	H	N/A	N/A	M	N/A	H
Quality control	M	M	N/A	M	H	M	N/A	N/A	M	N/A	N/A
Avionics	M	M	H	M	M	H	H	M	H	H	H
Software—aircraft systems	H	H	L	H	H	H	H	N/A	H	H	H
Software—performance	H	H	H	L	H	H	H	N/A	H	H	H

Note: Criterion number corresponds to defensive properties defined in Table 7.1.

TABLE 15.3 Subtopics Quantitative Scores

Subtopic	Likelihood	Effect	Risk
Requirements	0.1–0.5	Critical	High
Design	0.1–0.5	Catastrophic	Extremely high
Testing	0.1–0.5	Moderate	Moderate
Delays	0.5–1	Critical	Extremely high
Development	0.1–0.5	Serious	High
Personnel—cost	0.1–0.5	Moderate	Moderate
Assembly	0.1–0.5	Serious	High
Subcontractors	0.1–0.5	Critical	High
Process	0.02–0.1	Catastrophic	Extremely high
Quality control	0.02–0.1	Critical	High
Avionics	0.1–0.5	Catastrophic	Extremely high
Software—aircraft systems	0.1–0.5	Catastrophic	Extremely high
Software—performance	0.1–0.5	Catastrophic	Extremely high

With the initial filtration complete, it is necessary to check for subtopics that may be critical by association with other subtopics rather than by direct effect. This check is accomplished via interdependency analysis. Each of the remaining subtopics is examined to determine whether it is dependent upon any filtered subtopics. *Software* under the *performance* head topic was found to be dependent upon *personnel* under *manufacturing*. This dependency is then scored using a risk chart. The connection is rated high in transmission and high in criticality over any duration because any failure by the personnel creating the software will most likely lead to a serious error or failure in the software. Therefore, the connection is categorized as extremely high risk. This means that *personnel* is now considered a critical risk. Consequently, any subtopics that are dependent upon it could be critical. A review of the subtopics leads to the conclusion that *personnel* is dependent upon *management*. The connection is scored as medium in transmission and medium in criticality. Therefore, the connection is rated as moderate risk and does not meet the high-risk threshold.

At the completion of the filtration process, the critical set of risks is composed of *requirements, design, delays, development, assembly, subcontractors, process, quality control, avionics, software (aircraft systems), software (performance),* and *personnel (manufacturing).* An initial set of 43 risks has been reduced to 12. In a complete risk assessment and management process, each remaining risk would be extensively studied and modeled, and a risk management plan would be developed. Since that is beyond the scope of this chapter, the risk tracking process is demonstrated here through a single risk.

Suppose that the subtopic of concern is *software* under the *performance* head topic, and the risk management team decides that the most appropriate metric to use is the number of bugs reported per thousand lines of code. This risk metric is tracked throughout the duration of the project. The knowledge of an expert in software development, who has experience with developing software for aircraft, is used to develop probability distributions using the fractile method. To assess the expected performance of the risk management plan, a fractile distribution is developed for each phase of the plan, and the expected value of the bugs can be calculated from each distribution. Figure 15.6 depicts a probability density function (pdf) generated through the fractile method.

A risk tracking chart can be developed on the basis of the constructed probability distributions, as depicted in Figure 15.7. The shaded area represents the forecast expected number of bugs per thousand lines of code in each phase of the project. Each vertical line represents a milestone in the risk mitigation plan. Since the purpose of a risk mitigation plan is to reduce the risk to the project, the probability distribution of the risk metric is expected to change as each milestone is implemented. Consequently, a different probability distribution should be developed for each milestone. To determine the height of the shaded area at each milestone, the expected value of risk is calculated for each corresponding probability distribution. An example of a milestone may be the completion of a testing and debugging task. It is expected that the completion of each task will lower the number of bugs in the software—hence the drop in the expected number of bugs at each milestone. The black tick marks indicate the actual reported number of bugs. In this case, the number of bugs per thousand lines of code is measured through testing, and the error bars indicate the uncertainty of the measurement.

The dashed lines divide the chart into the problem domain, the mitigation domain, and the watch domain. Determining these domains is somewhat arbitrary and subject to the judgment of the risk management team. As a general rule, the problem domain can be delineated by identifying the level of risk that is unacceptable and demands immediate attention. A risk that has entered the problem domain requires immediate mitigation measures or revision of the risk management plan to prevent the risk from jeopardizing the success of the project. Of course, the level of risk that is unacceptable will change as the project progresses, hence the variation in the boundary of the problem domain. The watch domain, on the other hand, can be demarcated by determining the level at which a risk does not merit the expenditure of resources to mitigate it. A risk in this domain is still tracked, however, in case the risk

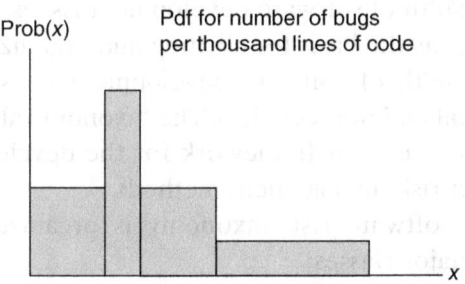

Figure 15.6 Sample fractile probability density function.

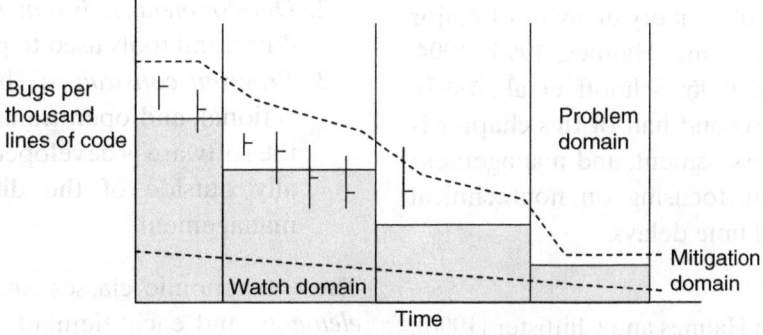

Figure 15.7 Risk tracking chart for bugs.

level begins to exceed the watch domain. Between the problem domain and the watch domain lies the mitigation domain. When a risk is within the mitigation domain, it is within the expected range, and risk management should continue according to plan.

In an actual implementation of this methodology, a tracking chart would be developed for each remaining identified subtopic (risk scenario). Note that each chart may be created using a different method depending on the nature of the risk, allowing each risk to be represented in the most meaningful way possible to the decisionmakers. It may be difficult, however, to measure the actual value of some risk metrics. In such cases, empirical distributions or expert evidence could be used to track changes in the expected value of the risk metric rather than in the actual value of that metric. For example, if the metric of concern is the market demand for a product, it may not be possible to measure the actual value. Instead, the expected market demand may be tracked as marketing surveys and advertising campaigns are implemented. Information gathered can be used to update the expected market demand. Using such a procedure, the changes in the expected market demand can be compared with the demand anticipated at the beginning of the project. Regardless, the key point is that risk metrics must be meaningful and evaluated using sound probabilistic methods.

15.5 QUANTITATIVE RISK ASSESSMENT AND MANAGEMENT OF SOFTWARE ACQUISITION[2]

15.5.1 Taxonomy of Software Development

The more central the role that software plays in overall system integration and coordination, the more likely the impact of delivery delay or of major cost overruns [Chittister and Haimes, 1993, 1994; Haimes and Chittister, 1996; Schooff et al., 1997]. Thus, the focus of the second half of this chapter is on the quantification, assessment, and management of software acquisition, focusing on nontechnical risks: cost overruns and time delays.

The SEI developed a methodology known as software capability evaluation (SCE) and the capability maturity model (CMM) used to assess the software engineering capability of contractors (see, e.g., Humphrey and Sweet [1987]). The SCE asks the question, *Can the organization build the product correctly?* The answer considers three separate aspects of the contractor's expertise:

- Organization and resource management
- The software engineering process and its management
- Available tools and technology

Another tool developed at SEI is a software risk taxonomy (see Table 15.4). It addresses the sources of software technical risk and attempts to answer the question, Is the organization building the right product? [Carr et al., 1993]. Thus, these two processes—the SCE and the taxonomy—offer methods of *assessing* organizational processes and software technical risks. This section presents a process for *quantifying* the risks of project cost and schedule overruns.

Central to the risk identification method is the software risk taxonomy. The taxonomy, which has similar features to the HHM discussed in Chapter 3, provides a framework for organizing and studying the breadth of software development issues. Hence, it serves as the basis for eliciting and organizing the full breadth of software development risks—both technical and nontechnical. The taxonomy also provides a consistent framework for the development of other risk management methods.

The software risk taxonomy is organized into three major classes:

1. *Product engineering*: The technical aspects of the work to be accomplished
2. *Development environment*: The methods, procedures, and tools used to produce the product
3. *Program constraints*: The contractual, organizational, and operational factors within which the software is developed, but which are generally outside of the direct control of local management

These taxonomic classes are further divided into *elements*, and each element is characterized by its *attributes*.

---

[2]This section is based on Haimes and Chittister [1996], Schooff et al. [1997], and Schoof and Haimes [1999].

TABLE 15.4 Taxonomy of Software Development Risks

Product Engineering	Development Environment	Program Constraints

1. Requirements
  a. Stability
  b. Completeness
  c. Clarity
  d. Validity
  e. Feasibility
  f. Precedent
  g. Scale
	1. Development process
	  a. Formality
	  b. Suitability
	  c. Process control
	  d. Familiarity
	  e. Product control
		1. Resources
		  a. Schedule
		  b. Staff
		  c. Budget
		  d. Facilities

2. Design
  a. Functionality
  b. Difficulty
  c. Interfaces
  d. Performance
  e. Testability
  f. Hardware constraints
  g. Nondevelopmental
	2. Development system
	  a. Capacity
	  b. Suitability
	  c. Usability
	  d. Familiarity
	  e. Reliability
	  f. System support
	  g. Deliverability
		2. Contract
		  a. Type of contract
		  b. Restrictions
		  c. Dependencies

3. Code and unit test
  a. Feasibility
  b. Testing
  c. Coding/implementation
	3. Management process
	  a. Planning
	  b. Project organization
	  c. Management experience
	  d. Prime contractor
		3. Program interfaces
		  a. Customer
		  b. Associate contractors
		  c. Subcontractors
		  d. Program interfaces

4. Integration and test
  a. Environment
  b. Product
  c. System
	4. Management methods
	  a. Monitoring
	  b. Personnel management
	  c. Quality assurance
	  d. Configuration management
		e. Corporate management
		f. Vendors
		g. Politics

5. Engineering specialties
  a. Maintainability
  b. Reliability
  c. Safety
  d. Security
  e. Human factors
  f. Specifications
	5. Work environment
	  a. Quality attitude
	  b. Cooperation
	  c. Communication
	  d. Morale	

An overview of the taxonomy groups and their hierarchical organization is provided in Table 15.4. Figure 15.8 depicts the hierarchy of the taxonomy structure.

15.5.2 Overview of the Quantitative Framework

The process of selecting contractors is in itself quite complex; it is driven by legal, organizational, technical, financial, and other considerations—all of which serve as sources of risk. Although factors other than the selection of contractors may decisively affect both technical and nontechnical software risks, they are treated here only as a general background. (See Chittister and Haimes [1993, 1994] and Haimes and Chittister [1996] for a more in-depth discussion of these factors.)

Because the world within which software engineering developed is nondeterministic and because the central tendency measure of random events (i.e., the expected value of software nontechnical risk) conceals vital and critical information about these random events, special attention must be focused on the variance of these events and on their extremes.

Figure 15.9 represents the conceptualization of the quantitative framework, which can be viewed in terms of four major phases. The purpose of Phase I is to quantify the variances in the contractor's cost and schedule estimates by constructing pdfs through

456 PRINCIPLES AND GUIDELINES FOR PROJECT RISK MANAGEMENT

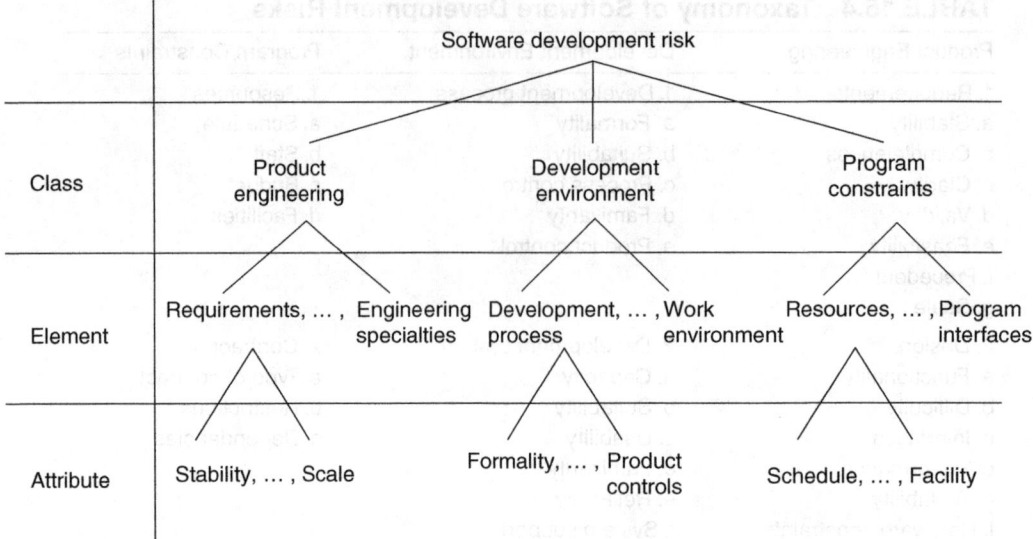

Figure 15.8 Taxonomy structure.

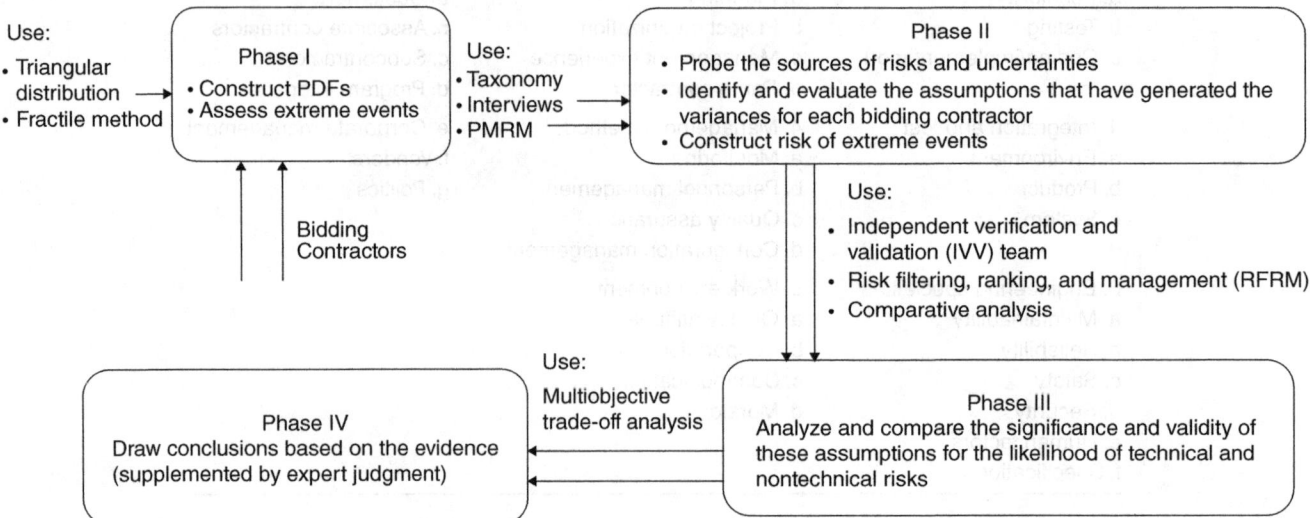

Figure 15.9 Conceptualization of the quantitative framework.

triangular distributions, the fractile method, or any other methods that seem suitable to the contractor (see Chapter 4). Extreme events are also assessed from these pdfs. In Phase II, using the SEI taxonomy, HMM, interviews, and the PMRM (see Chapters 8 and 11), the sources of risks and uncertainties associated with each contractor are probed and evaluated; the assumptions and premises, which provide the basis for generating the variances in the contractor's estimates, are identified and evaluated; and the conditional expected value of risk of extreme cost overruns and time delays is constructed and evaluated. In Phase III, the significance, interpretation, and validity of each contractor's assumptions and premises are analyzed, ranked, filtered, and compared, and the probability of technical and nontechnical risks are assessed. In executing Phase III, three tools and methodologies are used: (i) an independent verification and validation team, (ii) the RFRM method discussed in Chapter 7, and (iii) comparative analysis. In the final phase, Phase IV, conclusions are drawn on the basis of all the previously generated evidence, including the opinions of expert judgment. The ultimate objective of the quantitative framework is to minimize the following three objectives or indices of performance:

$$\min \begin{Bmatrix} \text{risk of project cost overrun} \\ \text{risk of project completion time delay} \\ \text{risk of not meeting performance criteria} \end{Bmatrix}$$

Clearly, multiobjective trade-off analysis, using, for example, the surrogate worth trade-off (SWT) method, should be conducted where all costs and risks are kept and traded off in their own units (see Chapter 5). Good risk analysis must be based on scientifically sound and pragmatic answers to some of the lingering problems and questions concerning the assessment and management of risks of those cost overruns and time delays associated with software engineering development.

It is constructive to discuss the four-phase acquisition process in more detail.

15.5.2.1 Phase I

Phase I will be demonstrated through the construction of the pdfs (using the fractile method and triangular distribution) and through the assessment of extreme events (using the PMRM) by calculating the conditional expected value of extreme events to supplement the common unconditional expected value of cost overrun.

15.5.2.2 Phase II

Through the use of the taxonomy-based questionnaire, interviews, and the quantification of risk of extreme events, Phase II provides a mechanism to (i) probe the sources of risks and uncertainties, (ii) identify and evaluate the assumptions that have generated the variances for each bidding contractor, and (iii) construct the conditional expected value of risk of extreme events, $f_4(\cdot)$.

The taxonomy-based questionnaire, along with the measurements of risk of cost overruns and time delays through $f_4(\cdot)$ and $f_5(\cdot)$, should explain not only the contractor's technical, financial, and other managerial assumptions and premises but also the contractor's attitude toward risk. When a contractor's projection of lowest, most likely, and highest project costs falls, for example, within a close range, there are several possible explanations: (i) The contractor is a risk seeker (a risk-averse contractor would have projected a much wider spread, (ii) the contractor is very knowledgeable and thus has confidence in the tight projections, and (iii) the contractor is ignorant as to the major technical details and complexity of the project's specifications; thus, major inherent uncertainties and variabilities associated with the project have been overlooked. Otherwise, the contractor would have projected a wider spread between the most likely and highest cost projections.

The taxonomy not only constitutes an important instrument with which to discover the reasons for the uncertainties and variabilities associated with the contractor's projections, it also provides a mechanism that allows the customer to assess the validity and soundness of the contractor's assumptions. Indeed, the taxonomy-based questionnaire, which is systematic, structured, and repeatable, is a valuable tool with which the customer can find out the reasons for the contractors' variabilities. The accumulated assumptions of each contractor must then be compared and analyzed.

15.5.2.3 Phase III

In Phase III, an analysis and comparison are conducted on the significance and validity of the contractor's assumptions for the likelihood of technical and nontechnical risks. This is accomplished through the use of an independent verification and validation team, the RFRM method discussed in Chapter 7, and other comparative analysis methods. In comparing assumptions, a number of issues may be addressed:

- Stability of the requirements
- Precedence of the requirements
- Need for research about solutions
- Politics and stability of funding
- Overall knowledge and the lack thereof
- Level of experience of key personnel
- Maturity of technology
- Maturity of the organization

In making these comparisons, the customer could ascertain the reasons for the assumptions and determine whether they are based on knowledge or naiveté and whether the contractor's attitude is conservative/risk averse or liberal/risk seeking.

15.5.2.4 Phase IV

Phase IV is the completion step where conclusions are drawn based on the accumulated evidence. Expert judgment is used in this phase in conjunction with multiobjective trade-off analysis methods, such as the SWT method (see Chapter 5). Adopting the systemic proposed acquisition process should markedly reduce the likelihood of major and catastrophic technical and nontechnical risks.

15.6 CRITICAL FACTORS THAT AFFECT SOFTWARE NONTECHNICAL RISK

The quantitative framework for managing software programmatic risk—the risk of cost overrun and time delay associated with software development—is grounded on the premise that such management must be holistically based. A holistic approach requires complete accounting of all important and relevant forces. Although a holistic view is advocated and discussed here, only limited aspects are ultimately quantified. Intrinsically, the quantification and management of software nontechnical risk (and to a large extent software technical risk) embody (i) the customer, (ii) the contractor(s), (iii) the organizational interface between the customer and the contractor(s), (iv) the state of technology and know-how, (v) the complexity of the specification requirements, (vi) the add-on modifications and refinements, (vii) the availability of appropriate resources, and (viii) the models used for project cost estimation and schedule projection.

Since each element is in itself a complex entity with diverse dimensions, it is essential to recognize which characteristics of each component contribute to programmatic risk. Only by understanding the sources of risk can it ultimately be prevented and managed.

15.6.1 The Customer

The term *customer* is a misnomer because it connotes a singular entity. Yet, in most large-scale software engineering systems, such as Department of Defense's (DoD) systems, projects are initiated, advocated, nourished, and supported by multiple constituencies with some common, but often different, goals and objectives. Furthermore, for DoD projects, there is also the shadow customer—the US Congress, which itself is influenced by various lobbyists, power brokers, and stakeholders. The influence of this multiplicity of clients on the ultimate resources made available for the development of software engineering constitutes a critical source of software risk. It is not uncommon for a pressure group to affect the design specifications and/or the resources allocated for a specific DoD project and, thus, have an impact on its final cost and completion time.

The *organizational maturity* level of the client is another factor that influences software programmatic risk. A client that possesses internal capabilities to communicate with the contractor(s) on both technical and nontechnical levels is more likely to have a better understanding and thus management of software programmatic risk. This attribute will become more evident later in this chapter as specific quantitative information on the variances of cost and schedule is solicited for the proposed methodological framework.

15.6.2 The Contractor(s)

Elaborate procedures and protocols describing contractor selection for the development of software engineering are being employed by government agencies and corporations. A commonly accepted axiomatic premise is that the organizational maturity of the contractor and the experience, expertise, and qualifications of its staff have a marked impact on the management of both software technical and programmatic risks.

15.6.3 The Interface between the Customer and the Contractor(s)

One of the dominant factors in initiating both technical and programmatic risks can be traced to the organizational interface between the customer and the contractor(s). Adequate and appropriate communication between the two parties, along with an understanding and appreciation of each other's role throughout the life cycle of the software development process, is imperative in preventing and/or controlling potential risks.

15.6.4 The State of Technology and Know-How

The contractor's access to know-how and to appropriate technology is a major factor in controlling software technical and programmatic risk. In particular, the lack of such access is likely to cause cost overrun as well as a measurable time delay in project completion.

15.6.5 The Complexity of the Specification Requirements

The more unprecedented the client's specifications in terms of advanced and emerging technology, the higher the risk of time delay in a project's completion and of its cost overrun. Most systems developed by the DoD are advancing the state of the art in some field of technology—for example, software development, stealth, propulsion, or satellites. The requirements in these fields are necessarily complex since the parameters are constrained by the task and are frequently subject to modifications because of changing technology.

15.6.6 The Add-On Modifications and Refinements

Although add-on modifications are often associated with software programmatic risk, they also constitute a critical source of software technical risk. This is because not all modifications are appropriately related to and checked against the original design to ensure ultimate compatibility and harmony. Very large and complex systems are difficult to manage. Systems are now developed by multiple companies (through outsourcing), each having its own area of expertise, and changes often ripple through the entire system. A wide range of factors may cause midcourse modifications; however, the causes that emerge from this range are in three categories:

1. Threat or need change: When a new threat is projected or a new need is contemplated
2. Improved new technology: When a new technology provides improved performance or quality, such as a new sensor
3. Replacing obsolete technology: When the preselected technology becomes obsolete before the project is completed or has even begun

15.6.7 The Availability of Appropriate Resources

One open secret in government procurement and occasionally in the private sector is the level of preallocated funds for a specific project. The competitive zeal of contractors often outweighs the technical judgment of their professional staff; the outcome is a bid that is close to the funds preallocated by the client even though it is clear to the bidder that the job with its specification requirements cannot be delivered at that level of funding. This not-uncommon phenomenon is dramatically illustrated in numerous documented examples by Hedrick Smith [1988] in his book *The Power Game: How Washington Works*:

> The standard technique is to get a project started by having the prime contractor give a low initial cost estimate to make it seem affordable and wait to add fancy electronics and other gadgets much later through engineering "change orders," which jack up the price and the profits. Anyone who has been through building or remodeling a house knows the problem. "This is called the buy-in game," an experienced Senate defense staff specialist confided.

15.6.8 The Models Used for Project Cost Estimation and Schedule Projection

A number of models are used to estimate project cost and completion schedule. Constructive Cost Model (COCOMO) [Boehm, 1981] and Program Evaluation and Review Technique (PERT) are representative examples. Models can be potent tools when they are well understood and supported by an appropriate database and adhere to their operating assumptions. The complexities of such models, however, often result in their misuse or invalid interpretations of their results. They thus ironically become a source of software nontechnical risk. The successful application of our proposed methodological framework, however, does not depend on the specific model used by either the contractor or the customer to estimate the cost or the schedule.

From the text mentioned earlier, it seems that the sources that contribute to software nontechnical risk are organizational and technical in nature; they stem from failures associated with the contractor as well as the customer. In terms of the contractor, these failures primarily originate from, and are functions of, such elements as:

1. The organizational maturity level
2. The process and procedures followed in assessing the project's cost and schedule
3. Management's honesty in communicating the real cost and schedule to the customer (and, of course, vice versa)
4. The extent and level of new and unprecedented technology imposed on the project

5. The level of software engineering experience and expertise of the staff engineers, both in general and in the application domain in particular
6. The level of software engineering experience and expertise of the management team
7. The overall competence of the team developing the software
8. Financial and competitive considerations
9. Immature technology, methods, and tools
10. Using technology in new domains
11. Combining methods and tools in new ways and using them in a new software development environment
12. Requirement modifications causing changes in the system's architecture

In terms of the customer, the nature of organizational failures partially overlaps those of the contractor's but also has distinctive characteristics:

1. The process and procedures followed in assessing the project cost and schedule.
2. How specifically the system and software requirements are detailed.
3. The number of changes and modifications requested by the customer during the software development process; these changes (which generally introduce many new errors) are often not harmonious with earlier specification requirements.
4. The commitment of the customer's project management to closely monitor and oversee the software development process.
5. The specific requirements for technology—for example, specific compilers and database management systems (DBMS).
6. Management's honesty in communicating the real cost to the *real client* (e.g., the DoD as a client and the US Congress as the *real client*).

15.7 BASIS FOR VARIANCES IN COST ESTIMATION

Most, if not all, developers of large, complex software systems use cost models to estimate their costs. These models are structured on a set of relationships based on such parameters as the size and complexity of the software, the experience level of the software developer, and the type of application within which the software will be used. Different models generate different weights or levels of importance for these parameters, and not all models use the same parameters. Radically different cost estimates can result merely on the basis of which parameters are used in the models and how they are implemented. Even when the parameters are consistent, different developers will probably not agree on the value or weight of the parameter in the first place. In fact, many organizations consider their interpretations of these parameters to contribute to their competitive edge because the definition affects their ability to determine costs accurately. For example, an organization that has little experience in developing space system software may not have the same perception of difficulty when developing a complex avionic software system as would an organization that has significant experience in that area. Their understanding of space systems, however, will alter their definition of the avionic system parameters. Do developers with little experience overestimate or underestimate the complexity of the task because of how they define these parameters? The central questions are: What are the sources of risk associated with project cost estimation? How can such risk be quantified?

Although creating, maintaining, and updating project cost estimation metrics and parameters are extremely important for an organization, it is nevertheless unlikely that a future project will be similar enough to previous projects to merit directly importing these metrics or parameters; such metrics and parameters may not be directly applicable without appropriate modifications. Indeed, cost estimators must use judgment when applying these parameters to a new project requirement. Furthermore, cost estimation constitutes a critical area with regard to the sources of risk for software development, which is without parallel to other fields. An analogy would be a contractor estimating the cost to construct a 50-story building. If the contractor had previously built only structures with a maximum of 10 stories, he would not just increase the estimate fivefold. In fact, the contractor would probably question the basic foundations and relevance of extending the 10-story model to the new structure parameters. In software, however, it is not uncommon to increase estimates for new projects by a factor of five from previous projects of one-fifth the size and complexity. Many new systems have size estimates of over 1 million lines of code even though the developers have little experience with systems of this size.

Another example is in the use of commercial off-the-shelf (COTS) software. The original assumption that a commercial DBMS can be used to meet customer requirements may change if the customer requires features not supported by DBMS suppliers. Such changes may have serious ramifications for the cost estimate, depending on how the developer plans to solve the problem. If the developer chooses to deal with a subcontractor in a way similar to dealing with the DBMS vendor, there will be risk associated with the subcontractor—an important subject that will be discussed later. The alternative is for the developer to undertake the development of his or her own DBMS. This requires an additional set of assumptions, design parameters, and judgments regarding the architecture, size, experience level, domain knowledge, software engineering knowledge, and the support environment needed to develop the DBMS. Each of these assumptions, parameters, and judgments has some uncertainty associated with it, which contributes to the overall risk in the cost estimate. If the developer chooses to subcontract the DBMS development to an outside vendor, then the issue for the contractor is understanding and accounting for the set of assumptions that are made by the subcontractors on the DBMS and on the system architecture.

The ability of the developer to make valid assumptions and design decisions is usually based on a set of metrics; these metrics can be based on current measurements or on past performance. Either way, however, there has to be an agreed-upon set of measures that is being evaluated (such as the number of lines of code needed to accomplish specified tasks or productivity rates in terms of lines of code per hour). The difficulty with software development is that the community has not agreed upon basic measures, such as how to count lines of code or how to measure productivity. Using performance history is difficult because the systems under development are sufficiently different such that history may not adequately reflect the new parameters accurately.

In the remainder of this chapter, we will focus on the dynamic nature of software acquisition because it should not be considered a static decision activity. Rather, as captured in the spiral model of software development [Boehm, 1988], the process consists of multiple repetitions of primary stages and often extends over a great length of time. Lederer and Prasad [1993] report that in practice, software estimation is most often prepared at the initial project proposal stage; then, with declining frequency, it is prepared at the requirements, systems analysis, design, and development stages. However, as the software development community continues to move away from the traditional waterfall development process model to the spiral-type models, demand has increased for cost estimation models that account for the dynamics of changing software requirements and design (and the always-present uncertainty) over multiple time periods [Schooff et al., 1997]. Bell's survey of software development and software acquisition professionals indicates that a vast majority believe a dynamic software estimation model would be most applicable for their estimation requirements [Bell, 1995].

At each stage of the acquisition process, decisions are made that affect the events and decision opportunities of subsequent phases. Software estimation is a required activity in every stage of the process. Applying the probabilistic cost estimation method with multiple objective risk functions described in Schooff and Haimes [1999] constitutes a multiple objective decision problem that is solved over the multiple stages of the acquisition life cycle.

15.8 DISCRETE DYNAMIC MODELING

Discrete dynamic modeling is concerned with sequential decision problems involving dynamic systems, where an input–output description and system inputs are selected sequentially after observing past outputs (see Chapter 10). The formulation of optimal control of a dynamic system is very general since the state space, control space, and uncertainty space can be arbitrary and may vary from one state to the next. The system may be defined over a finite or infinite state space. The problem is characterized by the facts that the number of stages of the system is finite and fixed, and the control law is a function of the current state. (Problems where the termination time is not fixed or where termination is allowed prior to the final time can be reduced to the case of fixed termination time [Bertsekas, 1976].)

The discrete-time dynamic system is given by

$$x(k+1) = f(x(k), u(k), w(k)) \quad (15.1)$$

where $x(k)$ is the state of the system at stage k, $u(k)$ represents the control or policy implemented at that stage, and $w(k)$ accounts for the random *disturbance* not otherwise captured in the model. The system output associated with each stage is given by

$$y(k) = g(x(k), v(k)) \quad (15.2)$$

where $y(k)$ is a cost or other output metric associated with the state of the system, $x(k)$ is the state of the system, and $v(k)$ is another purely random sequence accounting for randomness in the observation process.

Given an initial state $x(0)$, the problem is to find a control policy sequence that minimizes both the sum of all output costs $y(k), k=1,\ldots,N$ and the cost associated with the implementation of the control policies $u(k), k=1,\ldots,N$.

Figure 15.10 depicts the dynamic model that has been described. The input to each stage includes the state value from the previous stage $x(k)$, a policy input $u(k)$, and the effect of random process disturbances $w(k)$. These are used in Eqs. (15.1) and (15.2) to produce the cost estimate output $y(k)$ and to update the state variable $x(k+1)$ [Schooff, 1996].

Because the objective of the dynamic model is to find a cost-minimizing control policy sequence, the trade-off among project cost versus policy costs must be examined. Gomide and Haimes [1984] developed a theoretical basis for impact analysis in a multiobjective framework. In Chapter 10, we introduced the multiobjective multistage impact analysis method (MMIAM), where the trade-off decision metric is the marginal rate of change of one objective function f_i per unit change in another objective function f_j. Applying the concepts of the MMIAM along with that of the PMRM in a dynamic model introduces the concept of the stage trade-off given by λ_{ij}^{kl}, which represents the marginal rate of change of $f_i^k(x,u,k)$ per unit change in $f_j^l(x,u,l)$. Stage trade-offs provide a measure of the impacts upon levels of the risk objective functions at various stages. Additional discussion concerning full and partial trade-offs is given in Haimes and Chankong [1979], Chankong and Haimes [1983], and Gomide and Haimes [1984].

15.8.1 A Linear Dynamic Software Estimation Model

As the acquisition process progresses through its several stages, the knowledge regarding the project is updated and the uncertainty is (hopefully) reduced. More specifically, the greater the understanding of the software project as a whole, the better one can estimate key systems characteristics. From this information, appropriate project management policies regarding resource allocation and systems requirements can be made.

Each stage k of the model represents a decision point in the software acquisition process. These include such milestones as the formal decision points of the federal government acquisition process [DoD, 1991] and the less formal, yet more frequent, intermediary review points: preliminary design review (PDR), software specification review (SSR), critical design review (CDR), and others.

For the software cost estimation problem, we define the state variable $x(k)$ to be the estimated thousands of lines of code (KLOC) required for the intended system. As a state variable, KLOC characterizes the overall complexity and feasibility of the desired software system. The system output at each stage of the acquisition process, $y(k)$, is the development effort or cost of the software project. The functional form of $y(k)$ may be that of one of the software cost estimation models described earlier.

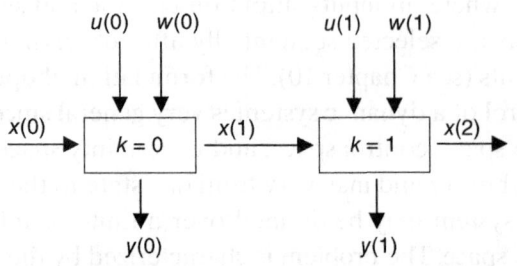

Figure 15.10 Discrete-time dynamic model.

The estimated KLOC requirement of a software system can be affected in several ways, most notably from (i) the characteristics or attributes imposed on the system, (ii) the resource allocation and acquisition strategy policies, and (iii) external factors. Each of these factors is accounted for in the state equation. The performance threshold levels imposed on a system are those metrics required to meet the operational requirements of the user community. Some of these factors are system reliability requirements, software purpose (functionality), execution or turnaround time, and computational throughput [Boehm, 1981; Sage, 1995]. For example, requiring a high degree of system reliability may require increased KLOC for the system. System constraints often increase the complexity of the intended system, further contributing to more KLOC requirements.

The control policy, $u(k)$, represents the acquisition control strategy and project control decisions that are selected. It includes the type and amount of nonbudgetary resources expended for software development. The control policies affect the KLOC requirement for the project and also influence the overall cost. The resource allocation policies considered in this model concern two principal nonbudgetary resources: personnel and technology. Personnel policy decisions relate to the selection and utilization of personnel with suitable experience and qualifications (highly skilled, skilled, limited knowledge, and others). Technological resources include the availability and allocation of specific programming languages and programming tools, the employment of certain programming practices, and database and storage resources.

While there are numerous external factors that have an impact on a software system's characteristics, one common external factor is the user community's changing operational requirements. The dynamic world of the user often results in modifications to the originally specified requirements and functionality of the system. Other external influences that affect the KLOC requirement for the system include political factors, technological advances, and the current status of the software development industry. All these external factors have a possible effect on the system complexity, the estimated KLOC requirements, and the resource allocation policies.

Having introduced the general form of the state and output equations and defined the model elements for a software cost estimation context, we develop a dynamic model for software cost estimation. While this initial model assumes a linear relationship among the parameters, it is anticipated that reality will often dictate a more complex formulation. The intent of this initial model, however, is to describe the general dynamics of the estimated size of the intended software system (measured in KLOC), the control policy and system constraints, and the resultant cost output associated with these elements. The initial model also serves as a vehicle for describing the application of dynamic modeling to software acquisition. Having used a linear model to accomplish these purposes, we will relax the linearity requirement in the following extensions.

In addition to the model parameters described previously, we consider the output of each stage, $y(k)$, to be a vector output as we consider the unconditional as well as the conditional expectation functions associated with the output function. We also introduce a cost function, f_1^k, that accounts for the cost of implementing the chosen control policy at each stage. The problem is to choose a control sequence $\{u(1), u(2), u(3),...,u(n)\}$ so as to minimize the policy implementation cost as well as the development cost vector.

The dynamics of the system are described by

$$x(k+1) = cx(k) + du(k) + w(k), \; x(0) = x_0 \quad (15.3)$$

and the output equation for each stage, representing the cost of project development, is given as

$$y(k) = ax(k) + v(k) \quad (15.4)$$

The multiobjective cost estimation problem *for each stage* is stated as follows:

$$\begin{aligned}
&\text{Minimize:} \quad \mathbf{f}^k = \langle f_1^k, f_4^k, f_5^k \rangle \\
&\text{Subject to:} \quad x(k+1) = cx(k) + du(k) + w(k) \\
&\qquad\qquad\quad\; y(k) = ax(k) + v(k)
\end{aligned} \quad (15.5)$$

where k represents the discrete stages (decision points) of the system; $x(k)$ is the state of the system, the estimated KLOC input to stage k; $y(k)$ is the calculated effort (cost) output of stage k; $u(k)$ is the resource allocation and acquisition strategy control policy of stage k; $w(k)$ is a random variable accounting for process noise; $v(k)$ is a random variable accounting for observation noise; a is a cost-per-KLOC multiplier measured in equivalent terms as $y(k)$; c is the KLOC-adjustment multiplier reflecting system and environment attributes; d is a KLOC requirements-per-selected policy multiplier; f_4^k is the conditional expectation of the output variable $y(k)$ at stage k; f_5^k is the unconditional expectation of the output variable $y(k)$ at stage k; and f_1^k is the cost of implementing control policy $u(k)$.

15.8.2 Solution Approach for the Linear Dynamic Problem

The solution to a deterministic formulation of the problem given by Eq. (15.5), in which the values of all model parameters are known with certainty and the preferred control policy is ascertained, is a straightforward application of multiobjective mathematical programming methods (see Chapter 5). In order to introduce the considerations of uncertainty and variance in the model parameters, we apply the probabilistic approach discussed in Chapter 10 to describe the model parameters where the disturbances $v(k)$ and $w(k)$ are permitted to be normally distributed, purely random sequences with mean zero and variance σ_v^2 and σ_w^2, respectively (constant for all k).

The selection of normal random variants is based on the knowledge that any linear combination of normal random variables is also a normal random variable [Ross, 1989]. Thus, examining Eq. (15.3), we conclude $x(k+1)$ is a normally distributed random variable and, therefore, so is $y(k)$ by Eq. (15.4). In Chapter 10, we derived exact-form solutions for the unconditional and conditional expected values of normally distributed functions with the form of Eqs. (15.3) and (15.4). The conditional expectation objective function σ_4^k of the normally distributed $y(k)$ is defined in terms of the mean $m(k)$ and variance $\sigma^2(k)$ of the cost distribution. The conditional expectation on the region $[s_k, t_k]$, $s_k < t_k$ is (see Chapter 10)

$$f_4^k(u) = \mu(k;u) + \beta_4^k \sigma(k) \quad (15.6)$$

where

$$\beta_4^k = \frac{\int_{s_k'}^{t_k'} \frac{t}{\sqrt{2\pi}} e^{-t^2/2} dt}{\int_{s_k'}^{t_k'} \frac{1}{\sqrt{2\pi}} e^{-t^2/2} dt} \quad (15.7)$$

$$t_k' = \frac{t_k - \mu(k)}{\sigma(k)}, \quad s_k' = \frac{s_k - \mu(k)}{\sigma(k)} \quad (15.8)$$

$$\mu(k;u) = E[y(k)] = E[ax(k) + v(k)] \\ = aE[x(k)] + 0 = aE[x(k)] \quad (15.9)$$

$$\sigma^2(k) = \text{Var}[y(k)] \quad (15.10)$$

Because $y(k)$ represents cost, the conditional expectation shown in Eq. (15.6) is an objective function to be minimized.

The unconditional expected cost, f_5^k, is the expected value of the output cost function which, using Eq. (15.9), can be represented as

$$f_5^k = E[y(k)] = aE[x(k)] \quad (15.11)$$

The general solution to Eq. (15.11) can be proven by induction (see Chapter 10), resulting in

$$f_5^k = E[y(k)] = ac^k x_0 + \sum_{i=0}^{k-1} ac^i du(k-1-i) \quad (15.12)$$

Observe from Eqs. (15.6), (15.7), and (15.10) that the term $\beta_4^k \sigma(k)$ is a function of k only, and not of the control $u(k)$. Therefore, minimizing the conditional expected value function Eq. (15.6) is reduced to minimizing the unconditional expected value:

$$\min_u f_4^k(u) = \min_u \{\mu(k;u) + \mu_4^k \sigma(k)\} \\ = \beta_4^k \sigma(k) + \min_u \mu(k;u) \quad (15.13)$$

This implies that minimizing the mean of $y(k)$—that is, minimizing $\mu(k;u)$—should yield the same controls as minimizing f_4^k. Because of this, the trade-offs associated with the conditional and unconditional expectation functions for any given k will be equal. Only the levels of the objectives will be different. In other words, the expectation functions f_4^k and f_5^k at stage k are parallel lines.

Using the results of Eq. (15.13) and the fact that the variance is independent of the control, we can consider the equivalent formulation described by Eqs. (15.3) and (15.4), where all random variables are assigned the value of their mean.

Let ^ denote the equivalent variables to the stochastic system, and Eqs. (15.3) and (15.4) become

$$\hat{x}(k+1) = c\hat{x}(k) + d\hat{u}(k)$$
$$\hat{y}(k) = a\hat{x}(k) \qquad (15.14)$$
$$\hat{x}(0) = x_0$$

Solving Eq. (15.14) for $\hat{y}(k)$ yields the same solution as Eq. (15.12). Hence, the important result is

$$\hat{y}(k) = f_5^k = E[y(k)] = \mu(k;u) \qquad (15.15)$$

An outline of a methodology for solving the multiobjective, multistage problem Eq. (15.5) is given below (see Chapter 10 for more details):

1. Determine the partitioning scheme for each component of damage (cost) for each stage and calculate the values of all β_4^k.
2. Calculate the variance $\sigma^2(k)$ for each stage.
3. Formulate the equivalent deterministic system Eq. (15.14).
4. Include the deterministic cost equation $\hat{y}(k)$ with the other objective functions in finding noninferior solutions.
5. The value of $\hat{y}(k)$ is equal to the unconditional expected value. Determine the conditional expected values by Eq. (15.6). Trade-offs for a given stage are the same since all conditional expected values are equal to the stage trade-offs calculated for $\hat{y}(k)$.
6. Use a multiobjective decisionmaking method such as the SWT method to find the preferred solution.

15.8.3 Example 1: Policy Evaluation Using the Linear Dynamic Software Estimation Model

The following is an example of how the multistage model described in the previous section may be applied. The model is a stochastic, time-invariant, linear-difference equation representing the relationship between software development management control policies, estimated model size, and project cost. Three stages are considered here, representing original cost and system requirement estimates obtained through a prebid conference, which are then updated at decision points early in the requirements determination and design phases of the software acquisition process.

Let $x(k)$, the state variable at stage k representing estimated KLOC, be expressed as a ratio to the initial estimate. The initial state is known with certainty, hence $x(0) = 1$. The control policy $u(k)$ (level of resource allocation) is expressed as a ratio to the normal level of allocations, just before the beginning of the planning horizon. Implementation of a particular policy is selected as a risk prevention measure, reducing the risk of excessive project cost overruns. This value can be considered as incorporating the personnel and product elements of the intermediate COCOMO [Boehm, 1981], along with acquisition management options such as additional review and study, the hiring of external consultants, and requirements for the development of prototype systems, among others.

The cost-per-KLOC constant, a, is fixed at \$1 million, an oft-quoted figure for mission-critical flight control software [Rifkin, 1995]. Let the performance characteristics constant, c, be fixed at $c = 1.44$, representing increasing complexity due to operational demands imposed on the system. This value is obtained by considering the product and computer attributes of the intermediate COCOMO [Boehm, 1981]. The parameter d, the KLOC adjustment due to policy selected, is fixed at $d = -0.25$. This value is negative, assuming a modest moderating effect of the application of resources on the otherwise increasing system complexity. Finally, let $w(k)$ represent an external random disturbance with mean zero and variance $\sigma_w^2 = 0.04$. The system's representation is then

$$x(k+1) = 1.44x(k) - 0.25u(k) + w(k)$$
$$y(k) = x(k); \quad x(0) = 1; \quad \mu_w = 0; \qquad (15.16)$$
$$\sigma_w^2 = 0.04; \quad u(k) \geq 0; \quad k = 0,1,2,3$$

For this example, the present-value cost function associated with the implementation of a particular policy is given by

$$f_1 = \sum_{k=0}^{n-1} K[u(k)-1]^2 \left(1+\frac{r}{2}\right)^{-2k} \qquad (15.17)$$

where $K=\$(100)10^3$, $r=10\%$, the annual discount rate, and the time period between stages is 6 months. Note that the cost function does not change with time—the dynamics are incorporated through the present value.

Following the procedure outlined earlier, we now formulate the equivalent system. The multiobjective optimization that includes the unconditional and conditional expected project costs, Eqs. (15.6) and (15.11), and the control policy implementation cost, Eq. (15.17), can now be stated as

$$\min_{u(k)} \begin{bmatrix} f_1^0(u(k)) \\ f_4^1(u(k)) \\ f_4^2(u(k)) \\ f_4^3(u(k)) \end{bmatrix} \text{ and } \min_{u(k)} \begin{bmatrix} f_1^0(u(k)) \\ f_5^1(u(k)) \\ f_5^2(u(k)) \\ f_5^3(u(k)) \end{bmatrix} \quad (15.18)$$

When we use the ε-constraint approach (see Chapter 5) to generate the needed Pareto-optimal solutions, the problem formulation is given as

$$\min_{u(k)} f_1^0(u(k)) \quad (15.19)$$

subject to the constraints

$$f_i^1(u(k)) \le \varepsilon_1$$
$$f_i^2(u(k)) \le \varepsilon_2 \quad (15.20)$$
$$f_i^3(u(k)) \le \varepsilon_3, \quad i=4,5$$

This leads to forming the Lagrangian function

$$L(\cdot) = f_1^0 + \lambda_1\left(f_i^1 - \varepsilon_1\right) + \lambda_2\left(f_i^2 - \varepsilon_2\right) + \lambda_3\left(f_i^3 - \varepsilon_3\right),$$
$$i=4,5$$
$$(15.21)$$

where the Lagrange multipliers describing the trade-offs between the cost function and risk functions are represented by

$$\lambda_k = \lambda_{1i}^{0k} = -\frac{\partial f_1^0}{\partial f_i^k} \quad (15.22)$$

To solve the multiobjective optimization problem, we need only to generate the unconditional expected cost function, f_5^k, for each stage $k=1,2,3$ (due to Eqs. (15.13) and (15.15)). Applying Eqs. (15.11), (15.12), and (15.16) produces the following unconditional expectation functions for each stage:

$$f_5^1 = E[y(1)] = aE[x(1)]$$
$$= aE[cx(0)] = [du(0) + w(0)] = acE[x(0)]$$
$$+ adE[u(0)] + E[w(0)]$$
$$= (1)(1.44)(1) + (1)(-0.25)u(0) + 0$$
$$= 1.44 - 0.25u(0)$$
$$f_5^2 = E[y(2)]$$
$$= (1.44)^2 - 0.25u(1) - (1.44)(0.25)u(0)$$
$$= 2.074 - 0.36u(0) - 0.25u(1)$$
$$f_5^3 = E[y(3)]$$
$$= (1.44)^3 - 0.25u(2) - (1.44)(0.25)u(1)$$
$$- (1.44)^2(0.25)u(0)$$
$$= 2.986 - 0.25u(2) - 0.36u(1) - 0.518u(0)$$

When we substitute the earlier three results and that of Eq. (15.17) into Eq. (15.21), the Lagrangian becomes now

$$L(\cdot) = \left[100(u(0)-1)^2 + 90.70(u(1)-1)^2 \right.$$
$$\left. + 82.27(u(2)-1)^2\right]$$
$$+ \lambda_1[1.44 - 0.25u(0)]$$
$$+ \lambda_2[2.074 - 0.36u(0) - 0.25u(1)]$$
$$+ \lambda_3[2.986 - 0.25u(2) - 0.36u(1) - 0.5184u(0)]$$
$$(15.23)$$

Taking the derivatives of Eq. (15.23) with respect to the controls at $u(2)$, $u(1)$, and $u(0)$ and applying first-order stationary conditions, we determine the trade-off values of Eq. (15.22). Table 15.5 gives three possible noninferior solutions. For each solution, the table gives the values of the control variables, the value and the levels of the risk functions, and the trade-off values between the cost f_1^0 and risk functions.

Policy A represents no change in resource allocation over the planning horizon. Because there is no additional application of resources, no policy implementation costs are incurred. However, the conditional and unconditional expected project costs become increasingly higher over time. By the third stage, the expected value of project cost has become 1.858, with the extreme-event conditional expected value at 2.135. The trade-offs between all risk functions and cost are zero (due to no implementation costs), so that small improvements in the

TABLE 15.5 Noninferior Policies for Software Acquisition

Stage	Risk Function	Trade-Offs
	Policy A[a]	
$k=1$	$f_4^1 = 1.467$	
	$f_5^1 = 1.190$	$\lambda_{14}^{01} = \lambda_{15}^{01} = 0$
$k=2$	$f_4^2 = 1.741$	
	$f_5^2 = 1.464$	$\lambda_{14}^{02} = \lambda_{15}^{02} = 0$
$k=3$	$f_4^3 = 2.135$	
	$f_5^3 = 1.858$	$\lambda_{14}^{03} = \lambda_{15}^{03} = 0$
	Policy B[b]	
$k=1$	$f_4^1 = 1.442$	
	$f_5^1 = 1.165$	$\lambda_{14}^{01} = \lambda_{15}^{01} = 863.62$
$k=2$	$f_4^2 = 1.642$	
	$f_5^2 = 1.365$	$\lambda_{14}^{02} = \lambda_{15}^{02} = 181.40$
$k=3$	$f_4^3 = 1.868$	
	$f_5^3 = 1.591$	$\lambda_{14}^{03} = \lambda_{15}^{03} = 329.08$
	Policy C[c]	
$k=1$	$f_4^1 = 1.342$	
	$f_5^1 = 1.065$	$\lambda_{14}^{01} = \lambda_{15}^{01} = 804.84$
$k=2$	$f_4^2 = 1.436$	
	$f_5^2 = 1.159$	$\lambda_{14}^{02} = \lambda_{15}^{02} = 362.80$
$k=3$	$f_4^3 = 1.570$	
	$f_5^3 = 1.293$	$\lambda_{14}^{03} = \lambda_{15}^{03} = 329.08$

[a]Control variables: $u(0) = 1$, $u(1) = 1$, $u(2) = 1$; cost, $f_1^0 = 0$ (10^3).
[b]Control variables: $u(0) = 1.10$, $u(1) = 1.25$, $u(2) = 1.50$; cost, $f_1^0 = 27.24$ (10^3).
[c]Control variables: $u(0) = 1.5$, $u(1) = 1.5$, $u(2) = 1.5$; cost, $f_1^0 = 68.24$ (10^3).

risk functions can be made at little additional cost. Because the trade-offs are zero, this is an improper noninferior solution (see Chapter 5 and Chankong and Haimes [1983]).

Policy B is a policy of gradual increase in personnel and technological resources allocated for project development. The expected project cost increases less dramatically over the time period, and the conditional and unconditional expected values indicate less risk than Policy A. The lower project costs over those of Policy A are achieved with relatively low policy implementation costs. This will be demonstrated graphically.

Policy C represents an immediate increase in resource allocation. The result is a significant decrease in expected project cost, with the expected value rising to only 1.293 by the third stage. Of the three solutions in Table 15.5, Policy C is the one of lowest risk, but it is also the most expensive. The trade-offs are also much larger for this policy, indicating that it becomes increasingly expensive to gain additional improvement in the risk functions.

The decisionmaker selects the most preferred of the three noninferior policies, taking into account his or her personal preferences in the trade-offs between the cost function and the risk functions. Formal methods such as the SWT method are appropriate. The impact that control policies have on later-stage decisionmaking options must also be taken into account and analyzed.

To demonstrate why impact analysis is so useful in a problem such as this, suppose the multiobjective problem was solved only one stage at a time. The cost associated with resource allocation policy (Eq. (15.17)) in the first stage, denoted by f_1^1, is

$$f_1^1 = 100[u(0) - 1]^2$$

Figure 15.11 shows the set of noninferior solutions when the first-stage costs f_1^1 and the expected damage at the first stage f_5^1 are the only objectives considered. The points corresponding to the Policies A, B, and C are indicated on the curve. Considering only the first-stage objectives, the selection of Policy C over the other alternative policies would appear desirable; the initial $25,000 policy implementation cost produces an expected $125,000 project cost reduction over the project's life cycle.

Now, consider the second stage, with the cumulative control costs, denoted by f_5^2, given by

$$f_1^2 = 100[u(0) - 1]^2 + 90.70[u(1) - 1]^2$$

Depending on which policy was implemented in the first stage, three different noninferior solution sets are possible in the second stage, as shown in Figure 15.12. Each curve is labeled with its associated first-stage policy. It is desirable to analyze the way in which the first-stage policy affects the second-stage (and subsequent-stage) decisionmaking. Li and Haimes [1987, 1988] show that there is a

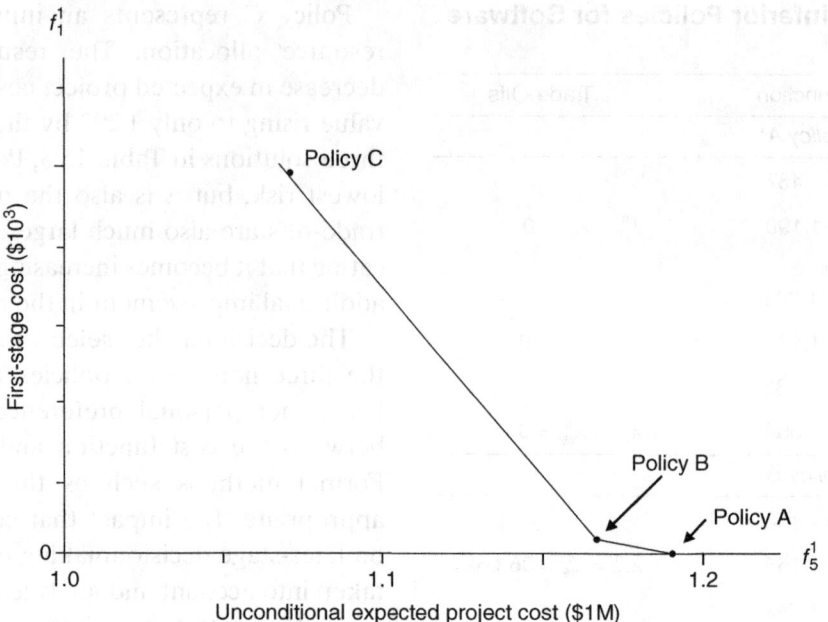

Figure 15.11 Noninferior solution set considering only first-stage objectives.

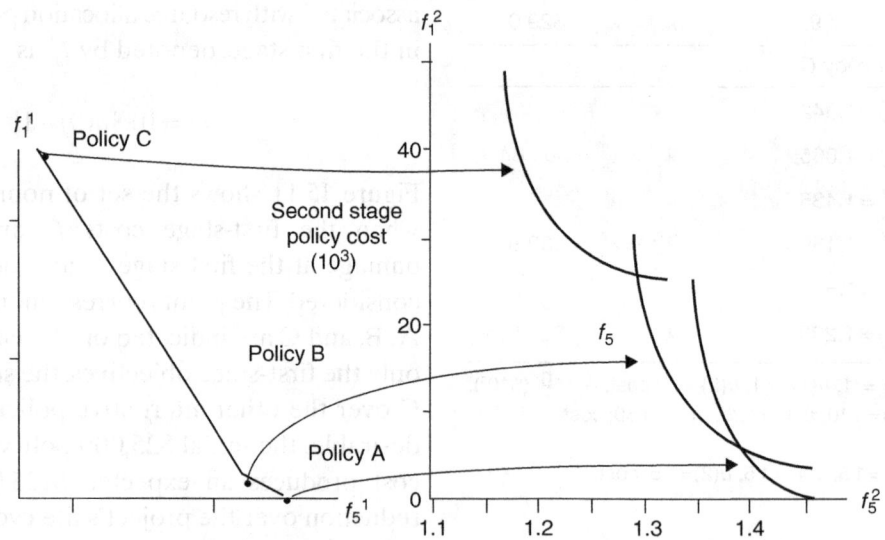

Figure 15.12 Impact analysis at the second stage.

family of such noninferior solution sets, where each curve depends on the chosen policy of the previous stage. The envelope of this family of curves engulfs all the noninferior solutions of each stage, thereby defining the noninferior frontier for the multistage problem. Additional decisionmaking information can be provided by plotting the conditional expectation curves for each alternative policy. Trade-offs are then made in terms of both expectation values (see Chapter 10).

15.8.4 Observations

The linear, multistage software estimation model has provided a framework for understanding and analyzing the software cost estimation parameters. The closed-form solution enabled an analytical description of the dynamics of the model parameters. The example problem demonstrated the benefits to decisionmaking by using this approach—in terms of the importance of both impact analysis and multiobjective trade-off analysis. This model opens the door to the development of a multistage software cost estimation model that is more closely associated with existing methods.

15.9 SUMMARY

Project risk management, when implemented correctly, can reap tremendous benefits in terms of reducing programmatic and technical risks. The US Navy's E-6 program is one example of how project risk management can work. Any reductions achieved in cost and schedule overrun as well as improvements in technical performance will go a long way in improving the perceived value of any organization. This means more interest on the part of stockholders and clients for corporations as well as the public and policymakers for government agencies. Thus, project risk management has a definite place in the knowledge base of any major organization.

The quantitative risk management framework described in this chapter is designed to provide a systematic means of addressing risk to acquisition or development projects. By providing methods for identifying, filtering, and tracking risks, it provides a means of addressing the complex and often qualitative nature of the large-scale sociotechnological systems that characterize major development and acquisition projects. While much research remains to be done in this field, this methodology provides a small step forward in managing the risks to these expensive and complicated projects.

Controlling the cost and time schedule of major projects has been and continues to be a major problem facing governmental and nongovernmental acquisition managers. Because of the close influence and interaction between software technical and non-technical risks and the diverse sources and causes that constitute the driving force behind these risks, the acquisition managers' job is complicated. One of the major premises of this chapter is that a careful, systemic, and analytically based process for contractor selection is imperative to prevent major risks of cost overruns and time delays. The four-phase process can be best viewed as a framework rather than as a rigid step-by-step procedure. Space limitations prevent a full demonstration here of each of the four phases of the proposed acquisition process. Those who are more familiar with the SEI taxonomy-based questionnaire will be able to relate more easily to its use in Phases II and III. Similar statements can be made on familiarity with the risk ranking and filtering method discussed in Chapter 4, the independent verification and validation team, and other methods used in the proposed acquisition process.

In the second part of this chapter, we extended the traditional application of software cost estimation methods by developing multistage, dynamic models. As the software development community continues to move away from the traditional waterfall development models to repetitive, spiral-type models, software cost estimation methods must be responsive to this new development paradigm. One overriding characteristic of this environment, particularly in the early stages of the development life cycle, is the uncertainty regarding the desired software system. To this end, a probabilistic approach that explicitly accounts for parameter variability is required.

The dynamic models developed in this chapter account for the need to update software cost estimates due to changes in requirements, improved system design information, and various resource allocation policies associated with the early stages of the software development life cycle. Incorporating a probabilistic extension of traditional software cost estimation methods, the models utilize the conditional expected value as an additional decisionmaking metric. Stagewise updating of software cost estimates gives the decisionmaker greater understanding of anticipated project costs and development effort requirements, as well as information concerning the expected impact of various control policy options in reducing project risk.

REFERENCES

American Institute of Chemical Engineers (AICHE), 1999, *HAZOP: Guide to Best Practice*, U-66.

Ashkenas, R., D. Ulrich, T. Jick, and S. Kerr, 1995, *The Boundaryless Organization*, Jossey-Bass, San Francisco, CA.

Bell, G.A., 1995, Applying the system design dynamics technique to the software cost problem: A rationale, in

Proceedings of the Tenth Annual COCOMO User's Group Meeting, Software Engineering Institute, Carnegie Mellon University, Pittsburgh, PA.

Bertsekas, D.P., 1976, *Dynamic Programming and Stochastic Control*, Academic Press, New York, NY.

Boehm, B.W., 1981, *Software Engineering Economics*, Prentice-Hall, Englewood Cliffs, NJ.

Boehm, B.W., 1988, A spiral model of software development and enhancement, *Computer* **21**(5): 61–72.

Carr, M.J., S.L. Konda, I. Monarch, F.C. Ulrich, and C.F. Walker, 1993, *Taxonomy-based risk identification*, CMU/SEI-93-TR-6, ESC-TR-93-183, Software Engineering Institute, Carnegie Mellon University, Pittsburgh, PA.

Chankong, V., and Y.Y. Haimes, 1983, *Multiobjective Decision Making*, North-Holland, New York, NY.

Chapman, R.J., 2001, The controlling influences on effective risk identification and assessment for construction design management, *International Journal of Project Management* **19**(3): 147–160.

Chittister, C., and Y.Y. Haimes, 1993, Risk associated with software development: A holistic framework for assessment and management, *IEEE Transactions on Systems, Man, and Cybernetics* **23**(3): 710–723.

Chittister, C., and Y.Y. Haimes, 1994, Assessment and management of software technical risk, *IEEE Transactions of Systems, Man, and Cybernetics* **24**(2): 187–202.

Davenport, T.H., and L. Prusak, 1998, *Working Knowledge*, Harvard Business School Press, Boston, MA.

Department of Defense (DoD), 1980, *Procedures for Performing a Failure Mode, Effects, and Criticality Analysis*, MILSTD-1629A, Department of Defense, Washington, DC.

Department of Defense (DoD), 1991, *Directive 5000.1, Defense Acquisition*, Office of the Undersecretary of Defense (Acquisition), U.S. Department of Defense, Washington, DC.

Dorofee, A.J., J.A. Walker, C.J. Alberts, R.P. Higuera, R.L. Murphy, and R.C. Williams, 1996, *Continuous Risk Management Guidebook*, Software Engineering Institute, Carnegie Mellon University, Pittsburgh, PA.

Federal Highway Administration (FHA), 1999, *Asset Management Primer*, Office of Asset Management, U.S. Department of Transportation, Washington, DC.

General Accounting Office (GAO), 1992a, *Embedded Computer Systems: Significant Software Problems on C-17 Must be Addressed*, GAO/IMTEC-92-48, Government Printing Office, Washington, DC.

General Accounting Office (GAO), 1992b, *Information Technology: An Audit Guide for Assessing Acquisition Risks*, GAO/IMTEC-8.1.4, Government Printing Office, Washington, DC.

General Accounting Office (GAO), 2000, *Defense Acquisitions: Recent F-22 Production Cost Estimates Exceeded Congressional Limitations*, GAO/NSIAD-00-178, Government Printing Office, Washington, DC.

Gomide, F., and Y.Y. Haimes, 1984, The multiobjective multistage impact analysis method: Theoretical basis, *IEEE Transactions on Systems, Man, and Cybernetics* **SMC-14**(1): 88–98.

Haimes, Y.Y., 1981, Hierarchical holographic modeling, *IEEE Transactions on Systems, Man, and Cybernetics* **SMC-11**(9): 606–617.

Haimes, Y.Y., 1991, Total risk management, *Risk Analysis* **11**(2): 169–171.

Haimes, Y.Y., and V. Chankong, 1979, Kuhn–Tucker multipliers as trade-offs in multiobjective decision-making analysis, *Automatica* **15**(1): 59–72.

Haimes, Y.Y., and C. Chittister, 1996, Systems integration via software risk management, *IEEE Transactions on Systems, Man, and Cybernetics* **26**(9): 521–532.

Haimes, Y.Y., N.C. Matalas, J.H. Lambert, B.A. Jackson, and J.F.R. Fellows, 1998, Reducing the vulnerability of water supply systems to attack, *Journal of Infrastructure Systems*, **4**(4): 164–177.

Haimes, Y.Y., and P. Jiang, 2001, Leontief-based model of risk in complex interconnected infrastructures, *Journal of Infrastructure Systems* **7**(1): 1–12.

Humphrey, W.S., and W.L. Sweet, 1987, *A method for assessing the software engineering capability of contractors*, CMU/SEI-87-TR-23, Software Engineering Institute, Carnegie Mellon University, Pittsburgh, PA.

Kaplan, S., S. Visnepolschi, B. Zlotin, and A. Zusman, 1999, *New Tools for Failure and Risk Analysis, Anticipatory Failure Determination (AFD) and the Theory of Scenario Structuring*, Ideation International, Southfield, MI.

Kaplan, S., Y.Y. Haimes, and B.J. Garrick, 2001, Fitting hierarchical holographic modeling into the theory of scenario structuring and a resulting refinement to the quantitative definition of risk, *Risk Analysis* **21**(5): 807–819.

Katzenbach, J.R., and D.K. Smith, 1999, *The Wisdom of Teams*, Harper Collins, New York.

Lam, P.T.I., 1999, A sectoral review of risks associated with major infrastructure projects, *International Journal of Project Management* **17**(2): 77–87.

Lederer, A.L., and J. Prasad, 1993, Information systems software cost estimating: A current assessment, *Journal of Information Technology* **8**: 22–33.

Li, D., and Y.Y. Haimes, 1987, The envelope approach for multiobjective optimization problems, *IEEE Systems, Man, and Cybernetics* **17**(6): 1026–1038.

Li, D., and Y.Y. Haimes, 1988, Decomposition technique in multiobjective discrete-time dynamic problems, *Control and Dynamic Systems: Advances in Theory and Applications* **28**: 109–180.

Lowrance, W.W., 1976, *Of Acceptable Risk*, William Kaufmann, Los Altos, CA.

Nordean, D.L., R.L. Murphy, R.P. Higuera, and Y.Y. Haimes, 1997, *Risk Management in Accordance with DODD 5000 series: An Executive Perspective*, Software Engineering Institute, Carnegie Mellon University, Pittsburgh, PA.

Nuclear Regulatory Commission (NRC), 1981, *The Fault Tree Handbook, NUREG 0492*, Government Printing Office, Washington, DC.

Pennock, M.J., and Y.Y. Haimes, 2002, Principles and guidelines for project risk management, *Systems Engineering* **5**(2): 98–108.

Reichelt, K., and J. Lyneis, 1999, The dynamics of project performance: Benchmarking the drivers of cost and schedule overrun, *European Management Journal* **17**(2): 135–150.

Rifkin, S., 1995, *Level 5 CMM Companies, electronic mail communication*, Master Systems Inc., George Washington University, Washington, DC.

Ross, S., 1989, *Introduction to Probability Models*, Academic Press, Boston, MA.

Sage, A., 1992, *Systems Engineering*, John Wiley & Sons, Inc., New York.

Sage, A., 1995, *Software Systems Engineering*, John Wiley & Sons, Inc., New York.

Saaty, T. L., 1980, *The Analytic Hierarchy Process*, McGraw-Hill, New York.

Santos, J., and Y.Y. Haimes, 2004, Modeling the demand–reduction input-output (I-O) inoperability due to terrorism of interconnected infrastructures, *Risk Analysis* **24**(6): 1437–1451.

Schooff, R.M., 1996, *Hierarchical holographic modeling for software acquisition risk assessment and management*, Ph.D. dissertation, Systems Engineering Department, University of Virginia, Charlottesville, VA.

Schooff, R.M., Y.Y. Haimes, and C.G. Chittister, 1997, A holistic management framework for software acquisition, *Acquisition Review Quarterly* **4**(1): 55–85.

Schooff, R.M., and Y.Y. Haimes, 1999, Dynamic multistage software estimation, *IEEE Transactions on Systems, Man, and Cybernetics* **SMC-29**(2): 272–284.

Smith, H., 1988, *The Power Game: How Washington Works*, Random House, New York.

U.S. Navy, 1997, *E-6B Monthly Risk Report*, U.S. Navy Integrated Project Team.

Williams, T.M., 1996, The two-dimensionality of project risk, *International Journal of Project Risk Management* **14**(3): 185–186.

16

Modeling Complex Systems of Systems with Phantom System Models

16.1 INTRODUCTION

There is an unfortunate imbalance in the curricula of most undergraduate and graduate programs in systems and industrial engineering and in operations research, which typically are devoted to *system optimization* rather than system *modeling*—whether modeling systems with single or with multiple objectives. Such imbalance in education and experience, and possibly in knowledge as well, could lead to optimizing a system with a poorly constructed or misrepresentative model. This reality of system modeling—often termed the *inverse problem* for system identification—was recognized and gained the interest and contributions of many researchers in the 1960s and 1970s in books, technical reports, and archival papers on system identification (see, e.g., Haimes, 1970; Graupe, 1972; Eykhoff, 1974). In system optimization, we assume knowledge of the system model, under specific assumptions, where, for each set of inputs, we can generate, or probabilistically estimate, the outputs. For example, in the context of risk management, no effective risk management policy options can be developed; neither can the associated trade-offs among all critical costs, benefits, and risks be evaluated; and nor can the impacts of current decisions on future options be assessed, without having constructed a model or a set of interdependent models that represent the multiple perspectives of the system and thus the essence of the system.

The fact that modeling is as much an art as a science—a tedious investigative trial-and-error, learn-as-you-go process—means that an equally imaginative approach is necessary to discover the inner functionality of complex systems through modeling. In this context, this chapter (i) addresses system modeling, and the inverse problem, or the system identification problem, through the phantom system models (PSM) [Haimes 2007, 2008, 2012a]; (ii) analyzes the contributions of PSM as a modeling mechanism through which to experiment with creative approaches to modeling complex systems of systems (S-o-S); and (iii) relates (at the meta-modeling level) the intrinsic common/shared state variables among the subsystems of the S-o-S, thereby offering more insight into the intra- and interdependencies among the subsystems.

Risk Modeling, Assessment, and Management, Fourth Edition. Yacov Y. Haimes.
© 2016 John Wiley & Sons, Inc. Published 2016 by John Wiley & Sons, Inc.

16.2 WHAT HAVE WE LEARNED FROM OTHER CONTRIBUTORS?

Reflecting on the history of modern systems theory and its close ties to the Gestalt psychology first introduced in 1912, we cannot underestimate the intellectual power of this multidisciplinary field and the holistic philosophy that has sustained it, allowing it to transcend the arts, the humanities, and the natural, social, and physical sciences, as well as engineering, medicine, and law. The fact that systems engineering and systems analysis have continued to grow and infiltrate other fields of study over the years can be attributed to the fundamental premise that a system can be understood only if all the intra- and interdependencies among its parts and its environment are also understood and accounted for. For more than a century, particular mathematical models, upon which systems-based theory and methodologies were developed, have been deployed in myriad large-scale projects in the natural and constructed environments. Moreover, if we were to identify a single concept that has dominated systems thinking and modeling, it would be the *state space*. (See Chapter 2 for more details on the building blocks of mathematical models.) Indeed, the centrality of state variables in this context is so dominant that no meaningful mathematical model of a real system can be built without identifying the states of that system and relating all other building blocks of the model to them (including decision, random, and exogenous variables and inputs and outputs). (More will be discussed on the centrality of state variables in modeling as it relates to the entire theme of this chapter.) In this respect, the art and science of system modeling has served, in many ways, as the medium through which the holistic systems philosophy has informed the practice not only of engineering but of a broad range of other fields. As the discipline of systems engineering continues to develop and expand its domains of application, the need for new organizational and modeling paradigms to represent complex systems has emerged and has ultimately led to the study of S-o-S.

Complex systems are commonly composed of myriad subsystems, which in their essence constitute S-o-S. Each complex system is characterized by a hierarchy of interacting components, with multiple functions, operations, efficiencies, costs, and stakeholders. Clearly, no single model can ever attempt to capture the essence of such systems—their multiple dimensions and perspectives. Indeed, almost every living entity, all infrastructures, and both the natural and constructed environments are S-o-S [Haimes, 2008, 2012a]. For example, different organs and parts of the human body, as an S-o-S, are continuously bombarded by a variety of bacteria, viruses, and other pathogens; however, only a subset of the (states of the) human body is vulnerable to the threats from yet another subset of the would-be attackers, and due to our immune system, only a smaller subset of the human body would experience adverse effects. Thus, composites of low-level, measurable states integrate to define higher-level fundamental state variables that characterize the system. Indeed, the vulnerability of a system is a manifestation of the inherent *states* of that system (and to specific threats), and each state of a system can be dynamic and change in response to inputs, other random variables, and the building blocks of mathematical models (as discussed in the next section).

The precise definition of S-o-S, however, is more elusive. In a seminal paper, Sage and Cuppan [2001] directly ask, "What is a system of systems?" They conclude, "Unfortunately, there is no universally accepted definition of these 'super systems.' What distinguishes a system of systems from other systems does not at this point have a definitive answer." In a relatively more recent paper, Sage and Biemer [2007] provide the following answer to the same question: "No universally accepted definition of an S-o-S is available at this time." To address this problem, Sage and Cuppan [2001] build on the following five properties of S-o-S suggested by Maier [1998]:

(i) *Operational independence of the individual systems.* An S-o-S is composed of systems that are independent and useful in their own right.
(ii) *Managerial independence of the systems.* The component systems not only can operate independently, but they generally are operated independently to achieve an intended purpose.
(iii) *Geographic distribution.* Geographic dispersion of component systems is often large.

Often, these systems can readily exchange only information and knowledge with one another and not substantial quantities of physical mass or energy.

(iv) *Emergent behavior.* The S-o-S performs functions and carries out purposes that do not reside in any component system.

(v) *Evolutionary development.* An S-o-S is never fully formed or complete. Development of these systems is evolutionary over time and with structure, function, and purpose added, removed, and modified as experience with the system grows and evolves over time.

Building on the above five principles, this chapter attempts to improve our understanding of S-o-S by extending the multiperspective modeling schema, through hierarchical holographic modeling (HHM) [Haimes, 1981, 2012a], into the PSM.

Several modeling philosophies and methods have been developed over the last five decades to address the complexity of modeling complex large-scale systems and to offer various modeling schema. They are included in the following volumes: *New Directions in General Theory of Systems* [Mesarović, 1965], *General Systems Theory* [Macko, 1967], *Systems Theory and Biology* [Mesarović, 1968], *Advances in Control Systems* [Leondes, 1969], *Theory of Hierarchical Multilevel Systems* [Mesarović et al., 1970], *Methodology for Large-Scale Systems* [Sage, 1977], *Systems Theory: Philosophical and Methodological Problems* [Blauberg et al., 1977], *Hierarchical Analyses of Water Resources Systems: Modeling and Optimization of Large-Scale Systems* [Haimes, 1977], and *Multifaceted Modeling and Discrete Event Simulation* [Zigler, 1984]. *Synectics: The Development of Creative Capacity* [Gordon, 1968] introduced an approach that uses metaphoric thinking as a means to solve complex problems. Gheorghe [1982] presented the philosophy of systems engineering as it is applied to real-world systems. Hall [1989] developed a theoretical framework to capture the multiple dimensions and perspectives of a system. Other works include Sage [1977, 1992, 1995], Shenhar [1994], and Sage and Rouse [1999]. Eisner [1993], Maier [1998], and Sage and Cuppan [2001] provide valuable insight into S-o-S and definitions of emergent behavior of complex systems in the context of S-o-S.

Most of the works on S-o-S have been devoted to their organizational, functional, and structural nature; on the other hand, there has been comparatively less inquiry into the problem of modeling S-o-S, and much of it has emerged within the last decade. For example, Ottino [2003] reviews three major tools for quantitative modeling and studying complex systems: nonlinear dynamics, agent-based models, and network theory. Shalizi [2006] also reviews the main methods and techniques of complex systems, which include tools for analyzing data, constructing and evaluating models, and measuring complexity. Chang and Harrington [2005] provide a comprehensive description of agent-based models of organizations. Amaral and Ottino [2004] describe network theory and its importance in augmenting the framework for the quantitative study of complex systems. Lloyd and Lloyd [2003] present a general method for modeling complex systems in terms of flows of information. Page [1999] discusses robust computational models. In an analysis of the challenges associated with complex systems engineering, Johnson [2006] provides a comprehensive review of emergent properties and how they affect the engineering of complex systems. Bar-Yam [2003a] reviews past lessons learned from problems with systems engineering and suggests adopting an evolutionary paradigm for complex systems engineering. Within the application of complex systems theory, in a multiscale analysis of military littoral warfare, Bar-Yam [2003b] suggests the necessity of considering the specific organizational and technological requirements needed to perform effectively in a high-complexity environment. In health care, Funderburk [2004] presents a brief survey of several formal dynamic and/or network-based models that are relevant for health-care policy development and evaluation. Tivnan [2007] describes the formulation, successful replication, and critical analysis of Levinthal's model of emergent order for economic firms.

Jamshidi [2009a, b] edited two volumes on S-o-S engineering. In the preface of the first volume [2009a], he writes: "The SoS [Systems of Systems] concept presents a high-level viewpoint and explains the interactions between each of the independent systems. However, when it comes to engineering and engineering tools of SoS, we have a long way to go. This is the main goal of this volume." Indeed, Jamshidi confirms the need for concerted efforts in modeling complex S-o-S.

16.3 THE CENTRALITY OF THE STATES OF THE SYSTEM IN MODELING AND IN RISK ANALYSIS

Chen [1999] offers the following succinct definition of *state variable*: "The state $x(t_0)$ of a system at time t_0 is the information at time t_0 that, together with the input $u(t)$, for $t \geq t_0$, determines uniquely the output $y(t)$ for all $t \geq t_0$." The states of a system, commonly a multidimensional vector, characterize the system as a whole and play a major role in estimating its future behavior for any given inputs. Thus, the behavior of the states of the system, as a function of time, enables modelers to determine, under certain conditions, its future behavior for any given *inputs* or *initiating events*. For example, to determine the reliability and functionality of a car, one must know the *states* of the fuel, oil, tire pressure, and other mechanical and electrical components. In other words, all systems are characterized at any moment by their respective state variables and the conditions thereof, and these conditions are subject to continuous change. In addition, a modeler who has determined to select only those state variables that represent the *essence* of a system must decide whether its state variables should be modeled as static (constant) or dynamic (time dependent), deterministic or stochastic, and so forth.

Given that all systems large and small can be characterized by their state variables, recognizing the hierarchy of states, substates, and sub-substates is crucial to system modeling. For example, a simplified water resources system that supplies water to a large community can be characterized by the states of the water distribution (groundwater and surface water) storage, purification, and sewer systems. The data for each of the states can be further presented by substates. For example, the states of the water distribution system may be represented by the status of the main carriers, local pipes, pumps, and storage tanks. Similarly, the status of each organ of the human body may be represented by state and substate variables. For instance, the state of the heart of the human body may be represented by its components, muscles, compartments, and so forth. With any complex system, the most critical fact to note is the intra- and interdependencies that exist among the states of the system, which necessarily overlap the multiple perspectives of the system represented by the multiple models. In other words, a central role of modeling S-o-S is to coordinate, to integrate, or to *make a whole* of the various systems perspectives represented by the multiple models through the states of the systems. This important task cannot be achieved without carefully identifying and discovering those states that characterize the most important perspectives of the system. The fact that all state variables are uncertain functions of uncertain initiating events requires that modeling efforts take into account both epistemic and aleatory uncertainties [Paté-Cornell, 1996].

Consider the following definitions of the vulnerability and resilience of a system [Haimes, 2007, 2009, 2011, 2012b]:

Vulnerability is the manifestation of the inherent states of the system (e.g., physical, technical, organizational, and cultural) that if exploited by an adversary, or affected by a harmful initiating event, can result in adverse consequences to that system. The vulnerability of a system is a multidimensional, *vector that is time- and threat (initiating event)-dependent.*

The resilience of a system is also *a manifestation of the states of the system and it is a vector that is time- and threat (initiating event)-dependent. More specifically, resilience represents the ability of the system to withstand a major disruption within acceptable degradation parameters and to recover within an acceptable cost and time.* In other words, resilience is a vector state of the system that is neither abstract or static, nor deterministic. Moreover, resilience is similar to vulnerability in that it cannot simply be measured in a single unit metric; its importance lies in the ultimate multidimensional outputs of the system (the consequences) for any specific inputs (threats).

The question "What is the resilience of a specific system X?" is unanswerable. This question cannot be answered without reverting to the states of the system and to the specific threat and its timing. Thus, such a question can be answerable only when the threat (initiating event) scenario (or a set of scenarios) is specifically identified and the essential states of the system at the initiating event (threat) are known. Resilience is not merely an abstract concept; it is a state of the system (composed of a vector of substates) that may have different responses to different inputs (threat scenarios) from any specific substate within the hardware, software, policies and procedures, or connections to the Internet.

This discussion of the centrality of states of the system in modeling will be further explored and will be related to the intrinsic meta-modeling coordination and integration of the multiperspective models and the necessity of relying on the states of the system. This approach is in contrast to relying solely on the *extrinsic* output-to-input model coordination and integration, which does not build explicitly on the common and overlapping *intrinsic* states among the submodels.

16.4 THE CENTRALITY OF TIME IN MODELING MULTIDIMENSIONAL RISK, UNCERTAINTY, AND BENEFITS

The time frame is central to all decisions, whether implicitly or explicitly. For a pilot, the time frame may be measured in mere seconds; for a planner, it may be years or decades. For example, all real-world systems are characterized by multiple objectives (often noncommensurate, competing, and in conflict with each other); thus, Pareto-optimal policies associated with such system models are achieved through the manipulation of the appropriate states of the system; and since the latter are a function of time, the time frame is thus critical for modeling all systems. Models, which are built to answer specific questions, must also be constructed to address the following basic question: What are the impacts of current decisions on future options, given the inevitable occurrence of emergent forced changes? (The term *emergent forced changes* connotes external or internal trends in sources of risk and uncertainty to a system that may adversely affect or enhance specific states of that subsystem and consequently affect the entire S-o-S.) Unanticipated, undetected, misunderstood, or ignored emergent forced changes, whether they originate from within or from outside a subsystem, are likely to affect a multitude of states of that system with potentially adverse consequences to the entire S-o-S. Therefore, it is imperative to be able—through scenario structuring, modeling, and risk analysis—to envision, discover, and track emergent forced changes. Consider, again, the US Federal Aviation Administration's Next Generation (FAA NextGen) project, with its multiple goals and objectives, agencies, functionality, geographic dispersion, and stakeholders. This multibillion-dollar, decade-effort S-o-S enterprise will be subjected to emergent changes in technology spanning satellite communication, airspace congestion, trends in air traffic, and pollution emission, among myriad other changes. (For additional discussion of NextGen, see Chapters 12 and 19.)

These emergent forced changes may be characterized, as appropriate, through uncertainty and through risk analysis. Uncertainty, commonly viewed as *the inability to determine the true state of a system*, can be caused by *incomplete knowledge* and/or by *stochastic variability*. Two major sources of uncertainty in modeling affect risk analysis [Paté-Cornell, 1990, 1996; Apostolakis, 1999]. *Knowledge (epistemic) uncertainty* manifests itself in the selection of model topology (structure) and model parameters, among other sources of ignorance (e.g., lack of knowledge of important interdependencies within the states of the system and among other systems). *Variability (aleatory) uncertainty* includes all relevant and important random processes and other random events. Uncertainty dominates most decisionmaking processes and is the Achilles' heel for all deterministic and most probabilistic models. This uncertainty is commonly introduced through the selection of incorrect model topology (structure) (e.g., linear for a highly nonlinear system, its parameters, data collection, and the employed processing techniques). Model uncertainties will often be introduced through human errors of both commission and omission.

The multidimensional probabilistic consequences resulting from an initiating event yield a multidimensional risk function whose modeling and quantification complexity present considerable challenges. The selection of appropriate models to represent the essence of the system's multiple perspectives determines the effectiveness of the entire risk assessment, management, and ultimately communication process. In particular, the scope and effectiveness of strategic risk management options are implicitly and explicitly dependent on the perspectives of the system that are included in or excluded from the ultimate modeling efforts. In particular, a probable initiating event would necessarily affect only substates of a subsystem, but not necessarily the entire S-o-S. Thus, one must model the different probability distribution functions of consequences affecting each subsystem resulting from the same initiating event. Each perspective of a system—manifested through its structure, functionality, the services it provides, the customers it supports, and the other systems on which it depends—will experience specific and, likely, unique consequences resulting from the same initiating event.

16.5 EXTENSION OF HHM TO PSM

HHM is a holistic philosophy/methodology aimed at capturing and representing the essence of the inherent diverse characteristics and attributes of a system—its multiple aspects, perspectives, facets, views, dimensions, and hierarchies [Haimes, 1981, 2007, 2012a]. In the abstract, a mathematical model may be viewed as a one-sided image of the real system that it portrays. With single-model analysis and interpretation, it is virtually impossible to represent the multiple perspectives of the system.

The term *holographic* refers to the desire to have a multiview image of a system. A hologram captures the multiple features of an object through multiple scattered light fields. In our attempt to model a system, each model represents either one or limited aspects, dimensions, or perspectives of the system. The term *hierarchical* refers to the desire to understand the intricacy that characterizes the many different levels of the system's organizational, temporal, functional, and decisionmaking hierarchy.

HHM has turned out to be particularly useful in modeling large-scale, complex, and hierarchical systems, such as defense and civilian infrastructure systems. The multiple visions and perspectives of HHM add strength to risk analysis. It has been extensively and successfully deployed to study risks for government agencies such as the President's Commission on Critical Infrastructure Protection (PCCIP), the TRW, the FBI, the NASA, the US Army, the US Army Corps of Engineers, the US Department of Homeland Security, the FAA, the Virginia Governor's Office for Preparedness, the Virginia Department of Transportation (VDOT), and the National Ground Intelligence Center, among others. Details of several of these case studies are shared with the reader throughout this book. To present a holistic view of the elements that must be included in the model, the HHM approach involves organizing a team of experts with widely varied experiences and knowledge bases (technologists, psychologists, political scientists, criminologists, and others). The broader the base of expertise that goes into identifying potential risk scenarios, the more comprehensive is the ensuing HHM.

These multiple perspectives of complex systems have been often characterized and represented through the hierarchical nature of the system. Indeed, many organizational as well as technology-based systems are hierarchical in nature, and most states of a system (state variables) are hierarchical with substates and sub-substates (e.g., any organ of the human body and any physical or cyber infrastructure); thus, the modeling of such systems has been driven by and responsive to this hierarchical structure. This hierarchical structure of the subsystems and sub-subsystems, when it is understood and taken advantage of, can simplify the modeling process and the ultimate management of the system as a whole [Haimes et al., 1990]. Hierarchical modeling makes it possible to decompose an overall system into smaller subsystems, which are easier to model, analyze, and subsequently integrate with other subsystem models. The decomposition can be based on functional, technical, geographical, organizational, political, social, and myriad other perspectives of a system and especially of S-o-S. For example, an economic system may be decomposed into, or represented through, geographic regions or activity sectors. An electric power management system may be decomposed according to, or represented through, the various functions of the system (e.g., power generation units, power transformer units, and transmission units) or along geographic or political boundaries. Another decomposition might be a time-wise decomposition into various planning periods. If several aspects of the system are to be dealt with, such as the geographic regions and activity sectors of an economic system, it could be advantageous to consider several decompositions or to model representations of the multiple perspectives and functionalities of the system.

Haimes and Macko [1973] have identified four major decomposition structures in water resources systems on the basis of hydrological, political–geographical, functional, and temporal considerations. The decomposition of a regional area into subregions depends on the viewpoint and aims of the analyst. One decomposition may be performed with respect to the region's hydrology. The region would be decomposed into subregions, such as river basins and subbasins, having topographical divisions as their boundaries. A second decomposition might be with respect to political boundaries. The regional area would be decomposed into political subregions such as townships, municipalities, counties, and so forth. A third decomposition might be with respect to regional goals and functions. A fourth decomposition might address the time frame and resource allocation that would affect the planning for irrigation, navigation, hydroelectric power generation, recreation, and so forth. In regional water resource management, the major aspects of the regional area cannot be divorced from each other. The decompositions just cited overlap one another. Hydrological subregions can easily overlap or span political boundaries; and hydroelectric generating stations may be dispersed through a region and not be confined to any one political or hydrological subregion. Indeed, the subregional boundaries in hydrological decomposition generally do not coincide with the subregional boundaries in geographical decomposition. Since multiple models are required when modeling complex S-o-S, which are common in hierarchical multilevel modeling, hierarchical overlapping coordination between two or more hierarchical structures has been proven to serve as an effective schema to supplement and complement the knowledge and information provided by each structure separately [Haimes and Macko, 1973; Macko and Haimes, 1978; Haimes et al., 1990; Yan and Haimes, 2010].

PSM builds on and takes advantage of hierarchical overlapping coordination.

The principal advantage of hierarchical multilevel modeling is that it breaks down a large complex system into its component subsystems. It allows each subsystem to be studied, analyzed, understood, and possibly managed at a lower level of the hierarchy independently of the other levels and coordinated at a higher level of the hierarchy. It might be argued that decomposition is fairly easy; the real challenge is resolving the conflicts and interactions between and among the subsystems and ensuring that the submodels account for all critical states of the system, as well as for the specified system's overall objectives and constraints. The hierarchical approach meets these requirements via higher-level coordination. For example, general coordination methodologies [Lasdon, 1964, 1970, 2002; Haimes, 1977; Singh, 1987; Haimes et al., 1990] distribute the total planning and management task among the component subsystems.

Hierarchical modeling also has significant implications for risk modeling, assessment, and management [Tarvainen and Haimes, 1981]. For example, the risks associated with each subsystem within the hierarchical structure may contribute to and ultimately determine the risks of the overall system. Furthermore, the distribution of risks within critical subsystems often plays a dominant role in the allocation of resources for the entire system. This is manifested in the quest to achieve a level of risk that is deemed acceptable when the tradeoffs among all the costs, benefits, and risks are considered. By virtue of the existence of multiple subsystems, hierarchical systems commonly have multiple noncommensurate and often competing and conflicting objectives and multiple decisionmakers and stakeholders (e.g., departments in a factory or subregions in a regional planning problem).

PSM builds on and extends the basic theory and philosophy of HHM by offering operational guidelines and principles on the basis of which to model S-o-S; one of its most salient features is that it offers modelers a four-decade-old tested approach to learning the inherent characteristics and interdependencies of S-o-S. In his book *Metasystems Methodology*, Hall [1989] states: "In this way, history becomes one model needed to give a rounded view of our subject within the philosophy of hierarchical holographic modeling [Haimes, 1981] being used throughout this book, defined as using a family of models at several levels to seek understanding of diverse aspects of a subject, and thus to comprehend the whole."

16.6 PSM AND META-MODELING

16.6.1 Philosophical–Conceptual Foundations

Architects, painters, and music composers share similar challenges with analysts who are involved in the art and science of system modeling. The similarities are manifested in a seemingly endless process of discovery and creativity and in continuous learning through experimentation, measurement, assessment, and trial and error. Creative artists invariably start with a visionary theme through which they deliver one or multiple messages. Through their creative artistic talent and capability and by intuitive inquiries and exploration of a variety of motifs, artists and composers strive to express their visionary themes by answering imaginary or invisible questions (at least to the layperson).

Artists, as the quintessential modelers, represent through their artwork the influence of the culture and social environment within which they live. In an analogous way, system modelers attempt to represent the multiple perspectives and facets of the system under study so that they may gain a better understanding of the composition of its inherent intra- and interconnectedness and interdependencies and thus be able to answer specific questions relevant to the system. Thus, both artists and system modelers assume a similar creative, systemic, and challenging task of representation. Finally, not dissimilar to an artistic composition, a system model ought to be as simple as possible but as complex as required—resulting in a model that offers an acceptable representation of the system and is capable of providing answers and clarifications to the important questions that the model was designed to address. It is in the context of the modeler's scientific knowledge and artistic creativity, the quest of the modeler to represent complex reality with models that are guided by the intricate basic knowledge of the system under consideration, and the experience and creativity embedded within the modeler's artistic vision that the term *phantom* is used to epitomize the representation, through multiple models, of the inherent multiple perspectives of a complex system.

Indeed, models must represent broad perspectives, and modelers must possess matching capabilities, wisdom, and foresight for futuristic and out-of-the-box thinking. Emergent forced changes, the need for agile and flexible multiplicity of models, and building on the human systems engineering experience, expertise, and capabilities together contribute to the need for the PSM. In this sense, the PSM constitutes a real-to-virtual laboratory for experimentation, a learn-as-you-go facility, and a process "for exploring existing or emergent systems that are not yet completely designed and developed" [Horowitz and Lambert, 2006]. The Human Genome Project may be considered an audacious complex S-o-S, fraught with uncertainties and involving participants from multiple disciplines with varied perspectives, experience, skills, and backgrounds.

It is not unrealistic to compare the evolving process of the PSM to the *modeling* experience of children at play. They experiment and explore their uncorrupted imaginative emergent world with Play-Doh® and LEGO® while patiently embracing construction and reconstruction in an endless trial-and-error process with great enjoyment and some success. The innovation, imagination, and initiatives of modelers experimenting with the PSM on S-o-S can be instrumental in creating a learning process that can benefit decisionmakers.

Modeling emerging unprecedented and complex systems (e.g., a new national electric power grid system, a new and safe generation of cars fueled by hydrogen, or a human space mission to Mars and back), which are inherently elusive and visionary, as well as modeling existing large-scale S-o-S, by and large involve phantom entities of multiple perspectives. This modeling effort is driven and constrained by a mix of evolving future needs and available resources, technology, emergent forced changes and developments, and myriad other unforeseen events.

Consider the trade-offs between (i) the relatively low cost of modeling a complex S-o-S—and the inherently invaluable, often unrecognized and unappreciated efficacy that such modeling generates or offers—and (ii) the cost (higher by many orders of magnitude) associated with the conception, development, construction, and planning for operation of a new generation of physical infrastructures (e.g., water and sewers, electric power grids, transportation systems, communications, public support buildings, etc.). Indeed, the cost associated with bringing to life complex infrastructure systems could be in the billions of dollars, while the associated modeling cost would be in the millions of dollars. Thus, a ratio of three orders of magnitude ought to encourage and justify essential investments in modeling.

Models enable us to experiment and test hypotheses, to explore different designs options, or to generate responses to or impacts on varied policy options. Inversely, by their nature, complex systems constitute, in many respects, black holes to modelers that can be penetrated only by acknowledging our inability to directly uncover, understand, or predict their behaviors under different scenarios of disturbances (inputs). We commonly lack sufficient knowledge to assess the causal relationships among the subsystems, and to compensate for this shortfall, we revert to multiperspective experimentation aided by the ingenuity, creativity, and domain knowledge of experts, supported by the availability of databases. There is no assurance that modelers would be able to explain the reasons behind any variability among submodels; nevertheless, the very process of modeling such variability may highlight limited databases, inconsistent assumptions, unrecognized epistemic and aleatory uncertainties, and a host of other technical or perceptual reasons that ought not to be dismissed. For example, in a closed-loop process control of a system in operation, the automatic controller adjusts the parameters of the system in response to internal or external disturbances or initiating events. In contrast, the adjustment of the parameters in an open-loop process (in response to the initiating events) is made by the system's operator or engineer.

In the meta-model coordination and integration of the multiple submodels (to be discussed subsequently), the task is exceedingly more complicated, because the modeler assumes the roles of both the closed-loop controller and the open-loop controller. More specifically, the modeler at the meta-modeling level makes extensive use of the knowledge generated through lessons learned from (i) the subsystems' coordination, (ii) interdependencies within and among the states of the subsystems, (iii) innovation and creativity in model experimentation, and (iv) intrinsic overlapping and mutual characteristics, functionality, objectives, and states that combine to make all the subsystems an S-o-S.

16.6.2 Meta-modeling Coordination and Integration

16.6.2.1 Methodological approach

The essence of meta-model coordination and integration is to build on all relevant direct and indirect sources of information to gain insight into the interconnectedness and intra- and interdependencies among the submodels and on the basis of this insight to develop representative models of the S-o-S under consideration. The coordination and integration of the results of the multiple models are achieved at the meta-modeling phase within the PSM, thereby yielding a better understanding of the system as a whole. More specifically, modeling the intra- and interdependencies within and among the subsystems of complex S-o-S requires an understanding of the intricate relationships that characterize the dynamics within and among the states of the subsystems. This very important task is achieved at the meta-modeling level of the PSM by observing, estimating, and assessing the outputs for given inputs and by building on the intrinsic common states within and among the subsystems. Note that although the intrinsic common states constitute a key element of the PSM, the extrinsic (input–output) relationships are also very important and support the intrinsic one. Indeed, the selection of the trial inputs to the model and the inquisitive process of making sense of the corresponding outputs are at the heart of system identification and parameter estimation. This is not a one-shot process; rather, it can be best characterized by tireless experimentation, trial and error, and parameter estimation and adjustments, as well as by questioning whether the assumed model's topology is representative of the system being modeled.

The PSM-based intrinsic meta-modeling of S-o-S stems from the basic assumption that *some specific commonalities, interdependencies, interconnectedness, or other relationships must exist between and among any two systems within any S-o-S*. More specifically:

(i) An S-o-S connotes a specific group of subsystems. A subsystem will denote any system member of the S-o-S. A model of a subsystem will be denoted as a submodel.

(ii) A meta-model represents the overall coordinated and integrated submodels of the S-o-S. We define a meta-model as a family of submodels, each representing specific aspects of the subsystem for the purpose of gaining knowledge and understanding of the multiple interdependencies among the submodels and thus allowing us to comprehend the S-o-S as a whole.

(iii) *The essence of each subsystem can be represented by a finite number of essential state variables.* (The term *essence* of a system connotes the quintessence of the system, the heart of the system, i.e., everything critical about the system.) Given that a system may have a large number of state variables, the term *essential states of a system* connotes the minimal number of state variables in a model with which to represent the system in a manner that permits the questions at hand to be effectively answered. Thus, these state variables become fundamental for an acceptable model representation.

(iv) For a properly defined S-o-S, any interconnected subsystem will have at least one (typically more) essential state variable(s) and objective(s) shared with at least one other subsystem. This requirement constitutes a necessary and sufficient condition for modeling interdependencies among the subsystems (and thus interdependencies across an S-o-S). This ensures an overlapping of state variables within the subsystems. Of course, the more we can identify and model joint (overlapping) state variables among the subsystems, the greater is the representativeness of the submodels and the meta-model of the S-o-S.

(v) The importance of the availability of multiple, albeit overlapping, databases can be effectively utilized by multiple submodels, each of which is built to answer the specific questions for which it is built. Furthermore, each submodel's characterization, whether modeled separately or in groups, is likely to share common state variables—a fact that facilitates the ultimate coordination and integration of the modeled multiple submodels at the meta-modeling level. Thus, a common database that supports the family of S-o-S must be available.

(vi) The fusion of multiple submodels via the intrinsic meta-modeling coordination and integration enhances our understanding of the inherent behavior and interdependencies of existing and emergent complex systems.

16.6.2.2 PSM-based modeling of a prototype S-o-S

This subsection, which focuses on saltwater intrusion into groundwater systems and seawater rise due to climate change, explores and highlights some concepts associated with modeling a real S-o-S with PSM, albeit not sufficiently developed to generate results. Figure 16.1 graphically depicts the commonly used extrinsic nonreliance on state variables in systems integration (by using inputs from submodels as inputs to others). In contrast to Figure 16.1, Figure 16.2 depicts the intrinsic reliance on shared and unshared state variables for meta-modeling coordination and integration.

Freshwater has been and continues to be a scarce resource, and groundwater plays a major role in the overall water supply of the United States and around the world. Many models predict a significant seawater rise due to climate change [US DOT Report, 2008], which would cause saltwater intrusion into coastal groundwater aquifer systems. We consider three subsystem models: hydrologic, agricultural–social, and regional economic, where the only inputs are provided from external climatological models.

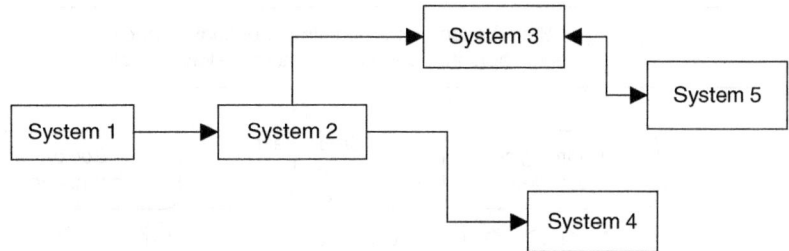

Figure 16.1 Extrinsic input–output submodel coordination and integration. From Haimes [2012]; © 2012 Wiley.

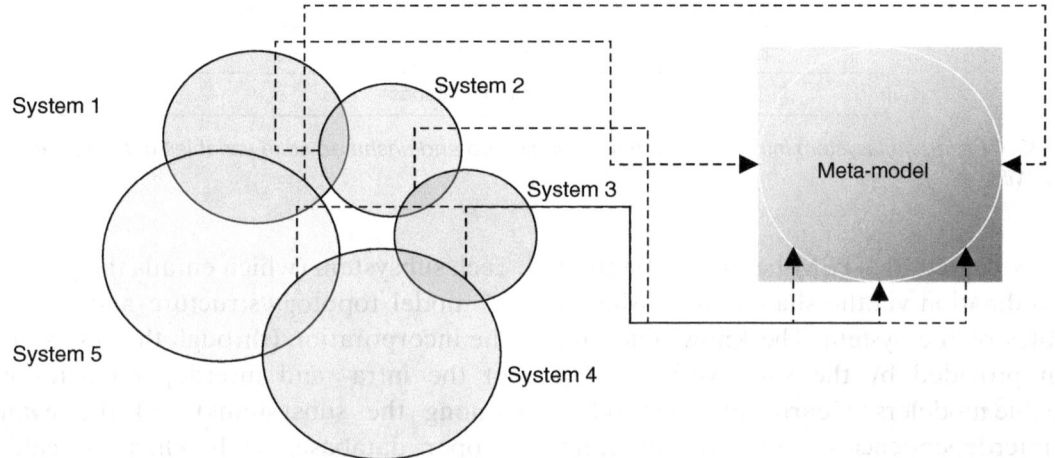

Figure 16.2 Intrinsic submodel, coordination, and integration via system state variables. From Haimes [2012]; © 2012 Wiley.

The role of the meta-model, which is composed of the aforementioned submodels, is to explore and learn about the intra- and interdependencies among the submodels and to evaluate the information necessary to assess the ultimate multiple impacts of the rise of groundwater salinity (due to the expected seawater rise resulting from climate change) on crop yield and variety, drinking water quality, farmers' economic well-being, and the regional economy. We envision the following multiple models:

(i) Hydrologic modeling effort can focus on a representative set of scenarios of climate change and seawater rise and can address the questions regarding the resulting consequences of seawater intrusion into the groundwater system.

(ii) Agricultural–social model can focus on the impacts of increased groundwater salinity on (a) agriculture, affecting the quality and yield of crops that are heavily dependent on groundwater, and (b) domestic water supply.

(iii) Regional economic model can focus on the regional economic impacts of the above on the agricultural and domestic use of groundwater.

(iv) Meta-modeling of the groundwater system serves as the coordinator and integrator of the multiple models, building on the shared and unshared state variables.

Let $c(t)$ represent an initiating event of climatological input that impacts seawater level and temperature; let $s_1(t)$ represent seawater level at time t; and let $s_2(t)$ represent the temperature at time t. Note the common and uncommon state variables in the following functional relationships:

(a) Groundwater salinity level $s_3(t) = s_3(t, c(t), s_1(t), s_2(t))$
(b) Groundwater yield $s_4(t) = s_4(t, c(t), s_1(t), s_2(t))$
(c) Crop quality and variety $s_5(t) = s_5(t, c(t), s_3(t), s_4(t))$
(d) Income to farmers $s_6(t) = s_6(t, c(t), s_5(t))$
(e) Regional viability of farms $s_7(t) = s_7(t, c(t), s_5(t), s_6(t))$

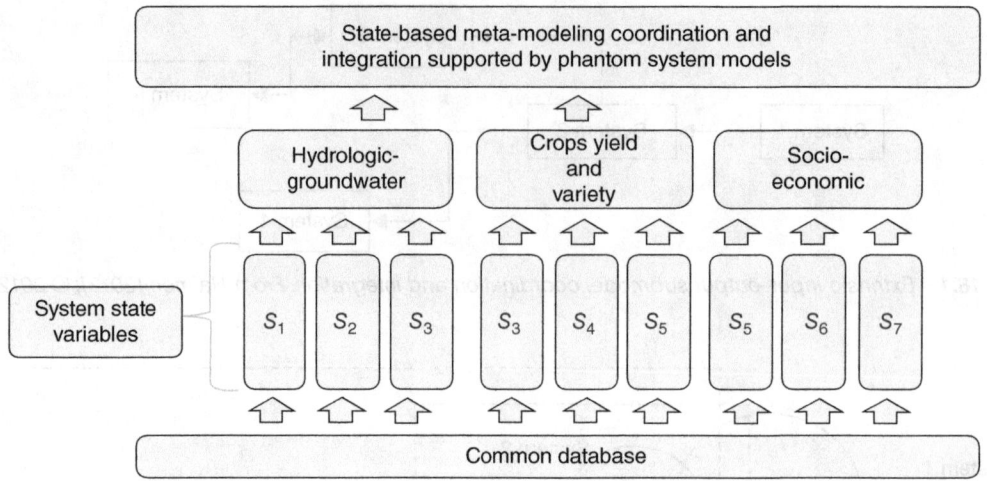

Figure 16.3 *PSM-based metasystem intrinsic coordination via shared and unshared state variables of the system. From Haimes [2012]; © 2012 Wiley.*

Figure 16.3 depicts the PSM-based metasystem intrinsic coordination via the shared and unshared state variables of the system. The knowledge and information provided by the state variables (s_1, s_2, ..., s_5) enable modelers to learn and better understand the interdependencies among the different submodels. For instance, the following set of intersections of state variables $s_1 \cap s_2, s_2 \cap s_3, s_3 \cap s_4, s_4 \cap s_5$, and $s_5 \cap s_1$ can help modelers to identify causal relationships among the multiple perspectives of the groundwater system.

The effectiveness of the PSM-based meta-model intrinsic coordination and integration is grounded on (i) the number of common state variables shared between two or more subsystems' models (a minimum of one shared state is required; otherwise, modelers can reasonably assume that a subsystem without any shared state variable is completely independent of the other subsystems); (ii) the domain knowledge of each of the subsystems' perspectives to ensure proper and effective modeling of the corresponding subsystems; (iii) the appropriate modeling efforts, skills, and expertise invested in modeling each subsystem, including, most importantly, the skill and ability of modelers to learn through the *mixing bowl* of infused knowledge, information, and learning generated through the meta-model at higher-level model coordination and knowledge integration; (iv) the appropriate modeling methodologies and tools (e.g., analytical, simulation) devoted in modeling each subsystem (which entails the proper selection of model topology/structure and parameters and the incorporation (through the states of the S-o-S) of the intra- and interdependencies within and among the subsystems); (v) the availability of proper databases with which to calibrate, test, validate, and verify the model of each subsystem (submodel) under varied conditions; and (vi) the availability of an appropriate computing laboratory that supports all of the aforementioned modeling efforts.

The intrinsic shared states provide a powerful mechanism for understanding and exploiting the strong interdependencies among the subsystems of S-o-S. The efficacy of shared states among subsystems may be manifested through (i) sharing databases, (ii) realizing that decisions made by the stakeholders of subsystem *A* can have a direct impact on subsystem *B*, and (iii) encouraging and enticing stakeholders of different subsystems to collaborate on inputs, constraints, and decisions that can affect the shared states, in order to obtain win–win outcomes. By the same token, unshared states can result in adverse, organizationally induced consequences due to competitiveness among subsystems. Thus, understanding and exploiting unshared states could, for example, allow decisionmakers (i) to defuse potential conflicts among the subsystems and (ii) to generate collaboration in the face of joint budgetary constraints or unfavorable policies affecting the subsystems.

16.6.3 Systems-Based Theoretical and Methodological Foundations

The following is a sample of systems-based methodologies that support the PSM.

16.6.3.1 Decomposition and hierarchical coordination

Hierarchical decomposition of complex large-scale systems enables modelers and systems analysts to use the decentralized approach to analyze and comprehend the behavior of subsystems at the lower level of the hierarchy and to transmit the information gained to fewer subsystems at the higher level. More specifically, the system's model is decomposed at the lower level of the hierarchy into *independent* subsystems (using pseudovariables), and the interdependencies are coordinated at a higher level. This system's decomposition and hierarchical coordination methodology, which is well documented in copious books and archival papers, has been successfully deployed for modeling and optimizing hierarchical complex systems, and it constitutes one of the methodologies that supports PSM [Dantzig and Wolf, 1961; Bauman, 1966; Lasdon and Scheffler, 1966; Lasdon, 1970; Wismer, 1971; Haimes, 1977; Haimes et al., 1990]. For example, consider a system composed of two subsystems that are coupled by one state variable (s_1). The system can be decomposed into two independent subsystems by assuming at the lower level of decomposition a pseudovariable (σ_1) as a surrogate for the state variable (s_1) of one subsystem and keeping (s_1) for the state variable for the second subsystem. Then, the sources of the difference $[(s_1)-(\sigma_1)]$ must be investigated, understood, and if possible minimized at the second level of the hierarchy. Several higher-level coordination methods for different types of decomposition, such as feasible and nonfeasible decomposition, have been developed and successfully deployed. This approach is applicable to any number of shared state variables with complex interdependencies among the subsystems. (See, e.g., Lasdon, 1970; Haimes, 1975) When observing (measuring) different values of shared states or outputs between two subsystems and when there are sufficient reasons to believe that the outputs associated with the two subsystems (corresponding to the common states) ought to be the same or with an acceptable difference, then the use of pseudovariables can become a useful instrument in the system identification and parameter estimation process within the PSM. On the other hand, differences between state variables representing a common perspective of two subsystems could also be due to a lack of understanding of the interdependencies between the two subsystems. More specifically, in intrinsic meta-modeling, we aim to reconcile the differences between common state variables to compensate for our ignorance. The availability of sufficient database is a requisite for an effective PSM modeling effort, given that most states of the system are time variant, and comparing the differences of $[s_1(t+1)-\sigma_1(t+1)]$ over time can shed more light on the system's behavior.

16.6.3.2 Coordinated hierarchical Bayesian model

Reliance on both direct and indirect information and database is common in systems modeling with sparse database and when empirical data are either sparse or lacking, in particular in risk of extreme events [Yan, 2007; Yan and Haimes, 2010]. Furthermore, with sparse data, important model parameters may not be estimated and tested within an acceptable level of significance. When a large database is available, standard statistical techniques can be applied to estimate the parameters and create a fairly accurate and well-parameterized model. Researchers and practitioners in systems engineering and risk analysis are commonly plagued by the data scarceness problem, which can be prevalent in modeling complex S-o-S. On the other hand, it is well known that when estimating the parameters of a model by traditional statistical methods using relatively small datasets, those methods generate *unstable* results with large estimation variance [Farrell et al., 1997; Assuncao and Castro, 2004]. Consequently, important model parameters cannot be estimated and tested within an acceptable level of significance. For example, Ferson [1997] argues that "problems in risk analysis often involve extreme events, which rarely happen, or are even hypothetical at the time of the assessment."

In this chapter, we adopt an alternative approach to address this problem at the meta-modeling level, by borrowing strength from indirect but relevant data from one subsystem and applying it to another. Strength-borrowing methods aim to borrow strength from indirect data to compensate for the sparseness of direct data. Subjective methods include expert evidence solicitation and Bayesian analysis; the latter provides a natural way to combine expert evidence with limited direct data.

We decompose the term *data* into three parts—direct data, indirect data, and expert evidence: (i) direct data may represent testing, experimentation, measurements, and observations from a system (or a subsystem) with unknown parameters; (ii) indirect data represent observations from different but related (or similar) subsystems; and (iii) expert evidence is information received by soliciting evidence from one or multiple experts. Hierarchical Bayesian models (HBMs), which have been applied in the reliability, risk, and system safety fields, offer an objective method suitable for addressing the data sparseness problem [Ghosh and Rao, 1994; Ghosh and Meeden, 1997; Carlin and Louis, 2000; Gelman et al., 2004]. Coordinated hierarchical Bayesian models (CHBMs), which borrow strength from indirect data or expert evidence to compensate for the sparseness of direct data [Yan, 2007; Yan and Haimes, 2010], can provide valuable support to the meta-modeling process.

The structures of HBM and CHBM are described in Figure 16.4a and b, respectively. In HBM, y_i represents the dataset observed from subsystem i, θ_i represents the parameter for the subsystem i, and η represents the hyperparameter. In CHBM, y_{ij} represents the dataset observed from scenario (i,j), u_i represents the fixed effect of perspective i, α_{ij} represents the cross-classified random effects from scenario (i,j), and $\tau\alpha_i$ represents the variance of the hyperdistribution of the random effects in perspective i.

As opposed to HBM, where there is only one dimension and a single hierarchy, the CHBM has two cross-hierarchies, each corresponding to one dimension of strength borrowing. Note that the bidimensional model can be easily extended to accommodate multiple-dimensional cross-classified random effects as multiple dimensions presenting in a system.

16.6.3.3 Influence diagrams

The combined art and science of system modeling builds on diverse philosophies, theories, tools, and methodologies. Probably the most basic, logical, and intuitive of all are influence diagrams [Oliver and Smith, 1990]. They are effective because they enable systems engineers and decisionmakers alike to represent the causal relationships among the large number of variables affecting and characterizing the system. Furthermore, through the use of conventional symbols, such as decision nodes and chance nodes, influence diagrams capture the probabilistic nature of the randomness associated with the system. Consequently, the quantification of risks and benefits can be performed on sound foundations.

The most effective deployment of influence diagrams is through brainstorming sessions with all principal parties involved with the system. In this setting, the varied expertise of the study team members produces a deeper understanding of the interactions between and among the subsystems. Similar to an engineering design project, the initial phase of constructing an influence diagram may result in an unwieldy *mess chart* that includes trivial as well as critical components. Through an open and constructive dialogue among the analyst(s) and decisionmaker(s), the mess chart becomes more coherent and includes what are deemed to be only essential variables and building blocks of the system's model.

The systems-based approaches presented here constitute only a sample of methodologies that support the modeling of complex S-o-S through the PSM. The challenges associated with modeling S-o-S necessarily require reliance on every applicable theory and methodology that can support this effort.

16.7 PSM LABORATORY

There is a need for a PSM laboratory (PSML) to *support, coordinate, and integrate results from a plurality of computer-based analytical (and simulation models)*, each providing a unique system perspective, with the outlook that the combination of such results can improve our learning and ability to gain knowledge. A PSML configuration offers a group of modelers

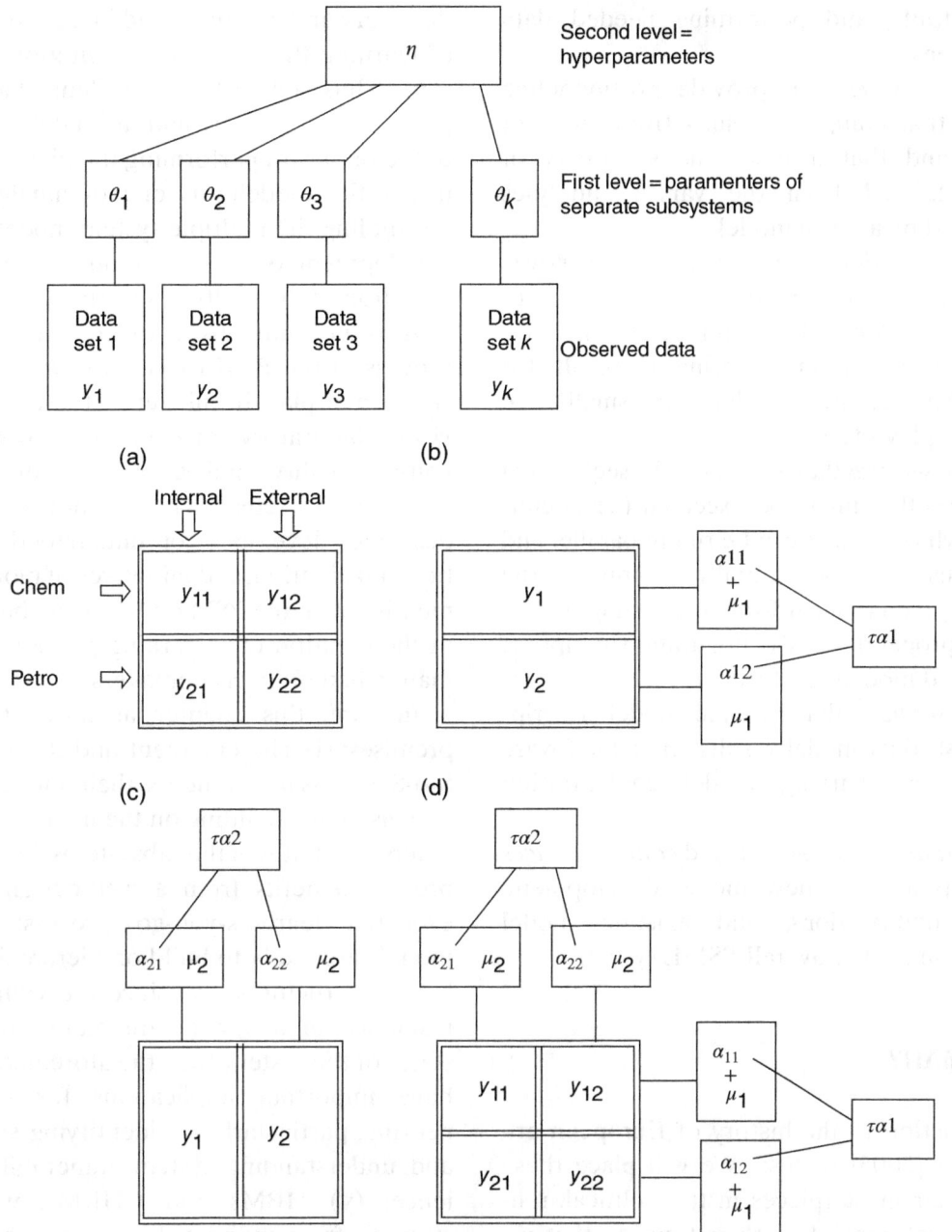

Figure 16.4 (a) Structure of HBM; (b) structure of CHBM. From Haimes [2012]; © 2012 Wiley.

the following benefits: (i) desired software-based models for a particular analysis, (ii) an array of data sources to support the desired modeling activities, (iii) tools for organizing the components of the modeling system so as to achieve the desired model relationships, and (iv) support for the intrinsic meta-modeling coordination and integration.

To perform these functions, the PSML ought to provide a structure based upon service-oriented architecture that will enable the user to perform desired modeling efforts. The following constitute representative services that a PSML ought to provide modelers in support of their modeling efforts:

- *Modeling services* that determine which models will be executed
- *Data services* that include organizing the needed data inputs, collecting the desired

data outputs, and performing needed data conversions
- *Analysis services* that provide postmodeling analysis that compares results from different models and that assesses the sensitivity of results derived from the varying analyses performed by a given model
- *Data presentation* and visualization services that include composing the analytical results to aid the modeler and decisionmakers in interpreting results and in preparing the results for different presentation media (print, small/large screen display, etc.)
- *Workflow services* that determine the sequencing of services that might be executed (e.g., determining which models can be run in parallel and which must be run sequentially) and provide the necessary data to a hardware organizing service to allow proper physical configuration to support the desired modeling effort
- *Library services* that provide model descriptions, historical model results, model software designs, and varying model configuration descriptions
- *Configuration management* and *control services* that help manage new model development, model modifications, and existing model integration for the overall PSML system

16.8 SUMMARY

In an introduction to the history of European art, William Kloss [2005] writes: "We will place these artists and their masterpieces in the political, religious, and social context of their time, so that we have a profound understanding of both why an artwork was created and how it responded to a particular set of historical circumstances." The creative work of system's modelers is not dissimilar from that of artists. Just as no single model is capable of representing the multiple perspectives of a complex system, whether in harmony or in juxtaposition, no symphony by Beethoven could have been composed using one instrument or one motif or theme. The similarities among artists and analysts involved in the art and science of system modeling are manifested in a seemingly endless process of discovery and creativity and in a continuous process of learning through experimentation and trial and error. Thus, similar to the challenge facing the composer of a symphony and ultimately the conductor of the orchestra performing the piece, the challenge facing the modeler is one of amalgamating and assembling the multiple system models to yield the development of a harmonious multimodel representation of the multiperspective system.

It is not unrealistic to compare the evolving process of the PSM to the *modeling* experience of children at play. In this sense, the PSM is a real-to-virtual laboratory for experimentation, a learn-as-you-go facility, and a process for existing and emergent systems that are not yet completely designed, developed, or understood. The innovation, imagination, and initiatives of modelers experimenting with the PSM on S-o-S can be instrumental in the creation of a learning process that can ultimately benefit decisionmakers.

In sum, this chapter advances the following premises: (i) The emergent and dynamic nature of S-o-S necessarily renders their models ephemeral and visionary, building on the intrinsic relationships among the states of the subsystems; (ii) this modeling process benefits from a well-designed and well-executed learn-as-you-go process; (iii) system models are likely to build on hierarchical and overlapping structures; (iv) since the vulnerability and resilience of a system are manifestations of the states of the system, then the aforementioned points have important implications for systems engineering, particularly for identifying sources of risk and understanding system vulnerability and resilience; (v) HBMs and CHBMs, which borrow strength from indirect data or expert evidence to compensate for the sparseness of direct data, can provide valuable support to the meta-modeling process; (vi) other systems engineering methods, such as hierarchical decomposition and higher-level coordination, influence diagrams, and others can be instrumental in the meta-modeling process; and (vii) building on the intrinsic interplay among the shared and unshared state variables among the subsystems, the philosophy and theory of the PSM provide a modeling paradigm that complements and supplements the commonly used extrinsic (input–output-based) modeling approach.

REFERENCES

Amaral, L., and J.M. Ottino, 2004, Complex networks: augmenting the framework for the study of complex systems, *The European Physical Journal B* **38**(2): 147–162.

Apostolakis, G., 1999, The distinction between aleatory and epistemic uncertainties is important: an example from the inclusion of aging effects into probabilistic safety assessment, Proceedings of PSA'99, American Nuclear Society, Washington, DC, August 22–25.

Assuncao, R.M., and M.S. Castro, 2004, Multiple cancer sites incidence rates estimation using a multivariate Bayesian model, *International Journal of Epidemiology* **33**: 508–516.

Bar-Yam, Y., 2003a, When systems engineering fails—toward complex systems engineering, *International Conference on Systems, Man & Cybernetics* **2**: 2021–2028.

Bar-Yam, Y., 2003b, Complexity of Military Conflict: Multiscale Complex Systems Analysis of Littoral Warfare, Chief of Naval Operations, Report for Contract: F30602-02-C-0158, Multiscale Representations Phase II: Task 2: Multiscale Analysis of Littoral Warfare, U.S. Department of the Navy, Washington, DC, 30p.

Bauman, E.J. "Multilevel optimization techniques with application to trajectory decomposition." C.T. Leondes (Editor) *Advances in Control Systems*, 4th edition, Vol. 6, Academic Press, New York, 1968.

Blauberg, I.V., V.N. Sadovsky, and E.G. Yudin, 1977, *Systems Theory: Philosophical and Methodological Problems*, Progress Publishers, New York, p. 132.

Carlin, B.P., and T.A. Louis, 2000, Empirical Bayes: past, present, and future, *Journal of the American Statistical Association* **95**: 1286–1289.

Chang, M., and J. Harrington, 2005, Agent-based models of organizations, In *Handbook of Computational Economics* II: Agent-Based Computational Economics, K. Judd and L. Tesfatsion (Eds.), Elsevier, Amsterdam, pp. 1–66.

Chen, C., 1999, *Linear System Theory and Design*, Third edition, Oxford University Press, New York.

Dantzig, G.B., and P. Wolfe, The decomposition algorithm for linear programs, *Econometrica* **29**(4)(1961), 767–778.

Eisner, H., 1993, RCASSE: rapid computer-aided systems of systems engineering, Proceedings of the 3rd International Symposium of the National Council of System Engineering, *INCOSE* **1**: 267–273.

Eykhoff, P., 1974, *System Identification: Parameter and State Estimation*, John Wiley & Sons, New York.

Farrell, P., B. MacGibbon, and T. Tomberlin, 1997, Empirical Bayes estimators of small area proportions in multistage designs, *Statistica Sinica* **7**: 1065–1083.

Ferson, S., 1997, *Bayesian Methods in Risk Assessment*, Technical Report, Applied Biomathematics, Setauket, NY. Available online: http://citeseerx.ist.psu.edu/viewdoc/download?doi=10.1.1.87.4577&rep=rep1&type=pdf (Accessed January 20, 2015).

Funderburk, F., 2004, Organizational culture from a complex dynamic systems perspective: moving from metaphor to action in healthcare, *System Models of Organizational Behavior*. Available online: http://www.necsi.edu/events/iccs/openconf/author/s/f183.pdf (Accessed February 14, 2011).

Gelman, A., J. Carlin, H. Stern, and D. Rubin, 2004, *Bayesian Data Analysis*, Chapman & Hall, New York.

Gheorghe, A.V., 1982, *Applied Systems Engineering*, John Wiley & Sons, New York.

Ghosh, M., and G. Meeden, 1997, *Bayesian Methods for Finite Population Sampling*, Chapman & Hall, London.

Ghosh, M., and J. Rao, 1994, Small area estimation: an appraisal, *Statistical Science* **9**: 55–76.

Gordon, W.J., 1968, *Synectics: The Development of Creative Capacity*, Collier Books, New York.

Graupe, D., 1972, *Identification of Systems*, Van Nostrand Reinhold Company, New York.

Haimes, Y.Y., Hierarchical modeling for the planning and management of a total regional water resource system, *Automatica*, (January)(1975) pp. 25–36.

Haimes, Y.Y., 1970, *The Integration of System Identification and System Optimization*, Report No. UCLA-ENG-7029, University of California, Los Angeles, CA.

Haimes, Y.Y., 1977, *Hierarchical Analyses of Water Resources Systems: Modeling and Optimization of Large-Scale Systems*, McGraw-Hill, New York.

Haimes, Y.Y., 1981, Hierarchical holographic modeling, *IEEE Transactions on Systems, Man, and Cybernetics* **11**(9): 606–617.

Haimes, Y.Y., 2007, Phantom system models for emergent multiscale systems, *Journal of Infrastructure Systems* **13**(2): 81–87.

Haimes, Y.Y., 2008, Models for risk management of systems of systems, *International Journal of Systems of Systems Engineering* **1**(1/2): 222–236.

Haimes, Y.Y., 2009, On the complex definition of risk: a systems-based approach, *Risk Analysis* **29**(12): 1647–1654.

Haimes, Y.Y., 2011 On the complex quantification of risk: systems-based perspective on terrorism, *Risk Analysis*, **31**(8): 1175–1186.

Haimes, Y.Y., 2012a, Modeling complex systems of systems with phantom system models, *Systems Engineering* **15**(3): 333–346.

Haimes, Y.Y., 2012b, Systems-based guiding principles for risk modeling, planning, assessment, management, and communication, *Risk Analysis*, **32**(9): 1451–1467.

Haimes, Y.Y., and D. Macko, 1973, Hierarchical structures in water resources systems management, *IEEE Transactions on Systems, Man, and Cybernetics* **3**(4): 396–402.

Haimes, Y.Y., K. Tarvainen, T. Shima, and J. Thadathil, 1990, *Hierarchical Multiobjective Analysis of Large-Scale Systems*, Hemisphere, New York.

Hall III, A.F., 1989, *Metasystems Methodology: A New Synthesis and Unification*, Pergamon Press, New York.

Horowitz, B.M., and J.H. Lambert, 2006, Learn as you go systems engineering, *IEEE Transactions on Systems, Man, and Cybernetics* **36**(2): 286–297.

Jamshidi, M. (Ed.), 2009a, *System of Systems Engineering: Innovations for the 21st Century*, John Wiley & Sons, Inc., Hoboken, NJ.

Jamshidi, M. (Ed.), 2009b *System of Systems Engineering: Principle and Applications*, CRC Press/Taylor & Francis Group, New York.

Johnson, C., 2006, What are emergent properties and how do they affect the engineering of complex systems? *Reliability Engineering and System Safety* **91**(12): 1475–1481.

Kloss, W., 2005, *History of European art, 48 lectures for The Great Courses*, The Teaching Company, Chantilly, VA.

Kossiakoff, A., and W.N. Sweet, 2002, *Systems Engineering Principles and Practice*, John Wiley & Sons, Inc., New York.

Lasdon, L.S., 1964, *A Multi-Level Technique for Optimization*, Ph.D. thesis, Case Institute of Technology, Cleveland, OH.

Lasdon, L.S., 1970, *Optimization Theory for Large Systems*, Macmillan, New York.

Lasdon, L.S., 2002, *Optimization Theory for Large Systems*, Second edition, Dover Publications, New York.

Lasdon, L.S., and J.D. Scheffler, 1966, Decentralized plant control, *ISA Transactions* **5**(2): 175–183.

Leondes, C.T. (Ed.), 1969, *Advances in Control Systems*, Vol. **6**, Academic Press, New York.

Lloyd, S., and T. Lloyd, 2003, *Bits and Bucks: Modeling Complex Systems by Information Flow*, Working Paper Series ESD-WP-2003-01.17, MIT Engineering Systems Division, Cambridge, MA.

Macko, D., 1967, General System Theory Approach to Multilevel Systems, Report SRC 106-A-67-44, Systems Research Center, Case Western Reserve University, Cleveland, OH.

Macko, D., and Y.Y. Haimes, 1978, Overlapping coordination of hierarchical structures, *IEEE Transactions on Systems, Man, and Cybernetics* **8**(10): 745–751.

Maier, M.W., 1998, Architecting principle for systems-of-systems, *Systems Engineering* **1**(4): 267–284.

Mesarović, M.D., 1965, Multilevel concept of systems engineering, *Proceedings of the Systems Engineering Conference*, Chicago, IL.

Mesarović, M.D. (Ed.), 1968, *Systems Theory and Biology*, Springer-Verlag, New York.

Mesarović, M.D., D. Macko, and Y. Takahara, 1970, *Theory of Hierarchical, Multilevel Systems*, Academic Press, New York.

Oliver, R.M., and J.Q. Smith (Eds.), 1990, *Influence Diagrams, Belief Nets and Decision Analysis*, John Wiley & Sons, New York.

Ottino, J.M., 2003, Complex systems, *AIChE Journal* **49**(2): 292–299.

Page, S.E., 1999, Computational models from A to Z, *Complexity* **5**(1): 35–41.

Paté-Cornell, E., 1990, Organizational aspects of engineering system safety: the case of offshore platforms, *Science* **250**: 1210–1217.

Paté-Cornell, E., 1996, Uncertainties in risk analysis: six levels of treatment, *Reliability Engineering and System Safety* **54**(2–3): 95–111.

Sage, A.P., 1977, *Methodology for Large Scale Systems*, McGraw-Hill, New York.

Sage, A.P., 1992, *Systems Engineering*, John Wiley & Sons, New York.

Sage, A.P., 1995, *Systems Management for Information Technology and Software Engineering*, Wiley, New York.

Sage, A.P., and S.M. Biemer, 2007, Processes for system family architecting, design and integration, *IEEE Systems Journal* **1**(1): 5–16.

Sage, A.P., and C.D. Cuppan, 2001, On the systems engineering and management of systems of systems and federation of systems, Information, Knowledge, *Systems Management* **2**(4): 325–345.

Sage, A.P., and W.B. Rouse (Eds.), 1999, *Handbook on Systems Engineering and Management*, Second edition, John Wiley & Sons, New York.

Shalizi, C., 2006, Methods and Techniques of Complex Systems Science: An Overview, 96p. Available online: http://www.scribd.com/complex-overview/d/28107 (Accessed January 20, 2015).

Shenhar, A., 1994, A new systems engineering taxonomy, Proceedings of the 4th International Symposium of the National Council of System Engineering, *INCOSE* **2**: 261–276.

Singh, M.G., 1987, *Systems and Control Encyclopedia: Theory, Technology*, Applications, Pergamon Press, New York.

Tarvainen, K., and Y.Y. Haimes, 1981, Hierarchical-multi-objective framework for energy storage systems, In *Organizations: Multiple Agents with Multiple Criteria*, Vol. **190**, J. Morse (Ed.), Springer-Verlag, Berlin, pp. 424–446.

Tivnan, B., 2007, Modeling organizational adaptation: a replication of Levinthal's model of emergent order, In *Proceedings of the 2007 Winter Simulation Conference*, S.G. Henderson, B. Biller, M.H. Hsieh, J. Shortle, J.D. Tew, and R.R. Barton (Eds.), ACM, New York, pp. 1241–1246.

United States Department of Transportation (USDOT), 2008, *Impacts of Climate Change and Variability on Transportation Systems and Infrastructure: Gulf Coast Study, Phase I*, Washington, D.C.

Wismer, D.A. (Editor). *Optimization methods for large-scale systems: with applications*. McGraw-Hill Companies, New York, NY, 1971.

Yan, Z., 2007, *Risk Assessment and Management of Complex Systems with Hierarchical Analysis Methodologies*, Ph.D. thesis, University of Virginia, Charlottesville, VA.

Yan, Z., and Y.Y. Haimes, 2010, Cross-classified hierarchical Bayesian models for risk-based analysis of complex systems under sparse data, *Reliability Engineering and System Safety* **95**: 764–776.

Zigler, B.P., 1984, *Multifaceted Modeling and Discrete Event Simulation*, Academic Press, New York.

17

Adaptive Two-Player Hierarchical Holographic Modeling Game for Counterterrorism Intelligence Analysis*

17.1 INTRODUCTION

Intelligence gathering and analysis for countering terrorism are a vital and costly venture; therefore, approaches need to be explored that can help determine the scope of collection and improve the efficacy of analysis efforts. The adaptive two-player hierarchical holographic modeling (HHM) game introduced in this chapter is a repeatable, adaptive, and systemic process for tracking terrorism scenarios. It builds on fundamental principles of systems engineering, system modeling, and risk analysis. The game creates two opposing views of terrorism: One is developed by a Blue Team defending against acts of terrorism, and the other by a Red Team planning to carry out a terrorist act. The HHM process identifies the vulnerabilities of potential targets that could be exploited in attack plans. These vulnerabilities can be used by the Blue Team to identify corresponding surveillance capabilities that can help to provide warning of a possible attack. Vulnerability-based scenario structuring, comprehensive risk identification, and the identification of surveillance capabilities that can support preemption are all achieved through the deployment of HHM.

State variables, which represent the essence of the system, play a pivotal role in the adaptive two-player HHM game, providing an enabling road map to intelligence analysts. Indeed, vulnerabilities are defined in terms of the system's state variables: Vulnerability is the manifestation of the inherent states of a system (e.g., physical, technical, organizational, cultural) that can be exploited by an adversary to cause harm or damage. Threat is a potential adversarial intent to cause harm or damage by adversely changing the states of the system. Threat to a vulnerable system may lead to risk, which is a measure of the probability and severity of adverse effects.

Each player in the adaptive two-player HHM game deploys the same modeling tools. This ensures that the results from different models can be compared and integrated. If the membership of different teams is drawn from groups with different value systems, skills, and experience, it can be expected

---

*Chapter 17 is based on and draws from Haimes and Horowitz [2004].

Risk Modeling, Assessment, and Management, Fourth Edition. Yacov Y. Haimes.
© 2016 John Wiley & Sons, Inc. Published 2016 by John Wiley & Sons, Inc.

that modeling results will differ. This should help to identify the appropriate mix of skills for a modeling team to develop a robust model. In addition, Bayesian analysis is central to determining the adaptive characteristics of the proposed methodology. Not only do new samples of evidence serve as likelihood functions to generate additional probabilities for given scenarios, but the probabilities associated with one scenario can be used as likelihood functions for other scenarios. This cross-updating process is further exploited by the construction of multiple decompositions, each representing a different perspective, for example, geographical, functional, or temporal.

To demonstrate our approach, in this chapter, we present a food-poisoning scenario with Red and Blue Teams.

17.2 BAYES' THEOREM

Intelligence gathering and analysis for combating terrorism constitute a complex process that must deal with large-scale systems of systems (S-o-S) with numerous components. These S-o-S may be characterized as dynamic and nonlinear and/or as spatially distributed. They may involve multiple government and nongovernment agencies, agents, and decisionmakers; agencies with different missions, resources, timetables, agendas, and cultures; and multiple constituencies. Furthermore, within any S-o-S, the risk of extreme and catastrophic events is of paramount importance; organizational and human errors/failures are common; and operations at all levels are fraught with multiple conflicting and competing objectives.

Clearly, no silver-bullet approach can address this complexity; neither can a single model do justice to the inherent difficulties associated with the intelligence process. Furthermore, no single technical study, including this chapter, can be expected to provide a unified and comprehensive scientific basis for intelligence gathering, analysis, and decisionmaking. The scope and objective of this chapter are to advance our scientific analytical capabilities and thus contribute an important building block to the overall gigantic modeling, analysis, and decisionmaking efforts facing the myriad intelligence agencies both in the United States and around the world.

The plethora of data, information, and other intelligence reports that security agencies receive daily on threats to the homeland cry for a search for connectedness, motives, patterns, hidden terrorist plans, and ultimately the discovery of a road map of the terrorist networks. There is a crucial need for quantitative and systemic intelligence analyses; thus, we have extended in this chapter the application of the HHM methodology [Haimes 1981]. The following are three focus areas that intelligence collection must encompass:

1. Intelligence gathering focused on the terrorists *and* their organizations. This area includes gaining information about such elements as membership, sympathizers, recruiting, money gathering, training, and strategizing.
2. Intelligence gathering focused on the various *methods* that terrorists might employ to carry out a specific mission. This area includes weapons as well as processes for committing terrorist acts.
3. Intelligence gathering about potential *targets*. This includes developing lists of the most likely points of attack, vulnerabilities surrounding these targets that can be exploited to form the basis for attack, and target-specific observables that can serve as a basis for surveillance and protection.

All three focus areas must be addressed if we are to develop a successful intelligence-collection system for homeland defense. This is due to the fact that the initial likelihood of a specific attack is very low, so that evidence suggesting an attack must be convincing enough to overcome the initial belief of low likelihood. Unless a conclusive piece of evidence emerges, the sum of many suggestive pieces of evidence must result in a persuasive outcome. This assertion is supported by the following analysis. Assume that a set of evidence, e, is collected about a specific terrorist attack, T. In general, we expect that the likelihood of T will be quite small. Assuming, for example, that an attack T has a probability of 0.001 of actually occurring within a given evidence **e**, the probability $p(T|e)$ of the attack T occurring is

$$p(T|e) = p(e|T)p(T)/[p(e|T)p(T) + p(e|\bar{T})p(\bar{T})] \quad (17.1)$$

where $p(\bar{T})$ is the likelihood of no terrorist attack occurring. Note that $p(\bar{T})$ is equal to 0.9999 for the situation being presented.

If we divide the numerator and denominator of Equation (17.1) for $p(T|e)$ by the numerator, we can transform it into the following form:

$$p(T|e) = 1/\left[1 + \{p(e|\bar{T})/p(e|T)\}\{p(\bar{T})/p(T)\}\right] \quad (17.2)$$

For the example numerical values of $p(T) = 0.0001$ and $p(\bar{T}) = 0.9999$, thus, Equation (17.2) becomes

$$p(T|e) = 1/\left[1 + \{p(e|\bar{T})/p(e|T)\}9{,}999\right] \quad (17.3)$$

It can be observed that unless $p(e|T)/p(e|\bar{T})$ — which we will call the *evidence ratio*—is a large enough number to offset the ratio of initial value of $p(\bar{T})/p(T)$ in a significant way, the value of $p(T|e)$ will remain small. That is, unless the evidence that has been collected is much more likely to have come from a potential terrorist attack as opposed to other possibilities, the likelihood of the attack will remain small. For illustration purposes, sample values are presented below.

Evidence Ratio	p(T|e)
1.0	0.00010
10.0	0.000999
100.0	0.0099
1000.0	0.0909
10,000.0	0.50

This example illustrates the point that the evidence ratio must be very large in order to offset an initial estimate that a specific terrorist attack is unlikely. Unless evidence that obviously points to a terrorist attack is discovered (i.e., a smoking gun), substantial evidence-collection efforts related to each of the three focus areas identified previously will need to be integrated to achieve useful evidence ratios.

Intelligence collection does not start and end with government agencies—civilian or military. Within the protection of civil liberty, the private sector can and should be an important source of intelligence that supplements and complements the others, since the ability to conduct surveillance at potential target locations most frequently must take place at the local level. For example, in the case of cyber attacks, without the sharing of historical data, organizations must operate on their limited experiences combined with sketchy information about what has happened elsewhere. This does not serve to provide a sound basis for broad intelligence analysis and decisionmaking.

The ultimate efficacy of the adaptive two-player HHM game is predicated on the assumption that the agencies that are collecting and analyzing intelligence will overcome vertical, horizontal, and organizational barriers, share the vast amount of collected information with each other, and disseminate it in a timely fashion [Ashkenas et al. 1995].

17.3 MODELING THE MULTIPLE PERSPECTIVES OF COMPLEX SYSTEMS

Evaluating intelligence cannot be subject to the single perspective of an analyst who is responsible for deciphering the maze of disparate data that may or may not be related to a scenario under consideration [Haimes, 2012b]. Rather, a holistic approach encompasses the multiple visions and perspectives inherent in the vast pool of intelligence data. Such a systemic process is imperative in order to successfully understand and address the complexity of the terrorism networks' S-o-S [NRC 2002]. Thus, appropriate modeling efforts are needed to support the adaptive two-player HHM game.

Several modeling philosophies and methods have been developed over the years to address the complexity of modeling large-scale systems and to offer various modeling schema. In his book *Methodology for Large-Scale Systems*, Sage [1977] addresses the "need for value systems which are structurally repeatable and capable of articulation across interdisciplinary fields" with which to model the multiple dimensions of societal problems. Blauberg et al. [1977] point out that, to understand and analyze a large-scale system, the fundamental principles of *wholeness* (representing the integrity of the system) and *hierarchy* (representing the internal structure of the system) must be supplemented by the principle of "the multiplicity of description for any system." To capture the multiple dimensions and perspectives of a system, HHM was introduced by Haimes [1981]: "To clarify and document not only the multiple components, objectives and constraints of a system but also its welter of societal aspects (functional, temporal geographical, economic, political, legal, environmental, sectoral, institutional, etc.) is quite impossible with a single model analysis and interpretation" (see also Chapter 3). Recognizing that a system "may be subject to a multiplicity of management, control and design objectives," Zigler [1984] addressed such modeling complexity in *Multifaceted Modeling and Discrete Event Simulation*. Zigler [1984] introduced the term *multifaceted* "to denote an approach to modeling which recognizes the existence of multiplicities of objectives and models as a fact of life." In *Synectics: The Development of Creative Capacity*, Gordon [1968] introduced an approach that uses metaphoric thinking as a means to solve complex problems. Hall [1989] developed a theoretical framework, which he termed *metasystems methodology*, to capture the multiple dimensions and perspectives of a system. Other early seminal works in this area include the book on societal systems and complexity by Warfield [1976] and *Systems Engineering* by Sage [1992]. In his book, Sage identifies several phases of the systems engineering life cycle; embedded in his and other such analyses is the notion of multiple perspectives—which necessarily entail structural, functional, and purposeful definitions. Finally, the multiple volumes of the *Systems & Control Encyclopedia: Theory, Technology, Applications* [Singh, Editor-in-Chief 1987] offer a plethora of theory and methodology on modeling large-scale and complex systems. In this sense, multifaceted modeling, metasystems, HHM, and other contributions to the field of large-scale systems constitute the fundamental philosophy upon which the adaptive two-player HHM game is grounded.

HHM is based on the premise that an inescapably multifarious nature is the fundamental attribute of large-scale systems such as intelligence systems. These systems include hierarchical noncommensurate objectives, multiple decisionmakers, multiple transcending aspects, and elements of risk and uncertainty. Thus, for purposes of intelligence gathering and analysis, it is impracticable to represent within a single model all the aspects and perspectives of either the terrorist networks or homeland protection organizations. Recognizing this, two separate HHMs are developed in this chapter, one to capture the terrorist networks' multiple perspectives and the other to capture the multiple dimensions of the homeland. The terrorist perspective does not include the same level of detailed knowledge about the vulnerabilities of potential targets as does the homeland perspective, but it does include a rich appreciation of the goals, objectives, and methodologies of terrorist groups [Arquilla and Ronfeldt, 2001]. HHM can be also regarded as a general method for identifying a set of risk scenarios [Kaplan et al. 2001]. Its multiple visions and perspectives have been used extensively and successfully for identifying the risk scenarios in numerous projects (see Haimes [1981]; Lambert et al. [2001]; Dombroski et al. [2002]).

17.3.1 Applying HHM to Tracking Terrorism

HHM is a structured approach to organizing a team effort for performing a risk analysis. The three questions that a risk analysis must address [Kaplan and Garrick 1981] are: *What can go wrong? What are the consequences? What is the likelihood?* Here, we add: *What is the time frame?* These questions form the basis for an HHM methodology directed toward tracking and preempting terrorist attacks. The premise of our efforts to apply HHM to counterterrorism intelligence systems is that potential attacks will consist of a set of activities that exploit particular vulnerabilities at the planned target. In order to develop approaches to preempt an attack, it is necessary to identify and characterize these vulnerabilities.

The methodology consists of the following steps:

1. Select classes of potential terrorist attacks to be tracked (e.g., meat poisoning, water poisoning, nuclear power plant attacks).
2. For each class, conduct an HMM analysis as described in the following sections. The results are sets of attack elements; when combined in various ways, these can be the basis for

coherent attack scenarios. For example, some elements could be the following: (i) gain employment at target location, (ii) steal weapon(s) for an attack, and (iii) bribe an employee at the target location.
3. Combine elements into packages of potential attacks. For each package, evaluate the consequences and likelihood, as described in the following.
4. Rank the attacks and attack elements in order of concern, as described in the following.
5. For the highest-ranking attacks, evaluate the potential observables that could result if a terrorist were to undertake such a plan of action.
6. For the attack elements and combinations that provide the most unusual observations, as defined later, consider setting up an intelligence-collection capability; then evaluate actual collections based on observing these elements in isolation and in combination.
7. When defined thresholds of observation are exceeded, raise the level of likelihood for the corresponding terrorist attack.

Based on numerous prior applications of HHM for risk assessment, a critical step in the methodology is the integration of a team of interested parties and experts to identify *everything* that could possibly go wrong and the corresponding system vulnerabilities that make these scenarios possible. While in the case of complex systems it may not be possible to develop a complete result, the team is nonetheless inspired to be as complete as possible.

17.3.2 What Can Go Wrong?

The term *holographic* refers to a multiview image of a system to identify vulnerabilities (as opposed to a single view or flat image). Views of terrorism risk can include, but are not limited to, (i) economic, (ii) health, (iii) technical, (iv) political, and (v) social. In addition, the risks can be identified as geography related, time related, and so on. In order to capture a holographic outcome, the team that performs the analysis must possess a broad array of experience and knowledge.

The term *hierarchical* refers to the importance of understanding what can go wrong at many different levels of the system hierarchy. A complete HHM recognizes that the macroscopic risks that are understood at the upper-management level of an organization are very different from the microscopic risks observed at lower levels. In a particular situation, a microscopic risk can become a critical factor in making things go wrong. For example, the security guard who does not do his job properly can add significant risk to the overall security of a facility, as can a poorly performing sensor for detecting and supporting response to a system failure. In order to perform a complete HHM analysis, the team must include individuals with knowledge up and down the hierarchy.

By its nature, the systemic HHM process yields a very large number of target vulnerabilities that provide the foundation for identifying risk scenarios. These vulnerabilities are hierarchically organized into major sets (known as *head topics*) and subsets (*subtopics*). When done well, the set of vulnerabilities and corresponding set of scenarios at any level of the hierarchy will converge to a *complete set* [Kaplan et al. 2001]. Figure 17.1 represents a subset of an HHM for a meat-poisoning scenario at a slaughterhouse. The figure contains a variety of potential attack elements, such as avoiding the security process at the slaughterhouse, gaining employment there, and bribing the owners or key employees.

Once the vulnerabilities of a target are identified, an analysis of consequences is required. The head topics of the HHM model help to pinpoint the many dimensions that consequences can have (e.g., dollars, lost lives, increased levels of fear, and possible corresponding losses of freedom). The team of experts is usually able to directly apply its knowledge to estimating consequences, but additional team members may be needed for this. Predicting likelihoods is more difficult. To do this, it is necessary to understand the terrorists' processes of selecting the attack target and method. Contributing factors might logically include (i) consequences, (ii) expertise required versus available expertise, (iii) cost versus available resources, (iv) logistical complexity, (v) availability of information about the target, and (vi) likelihood of getting intercepted and caught. Many of these factors go beyond the expertise and knowledge of the HHM team as

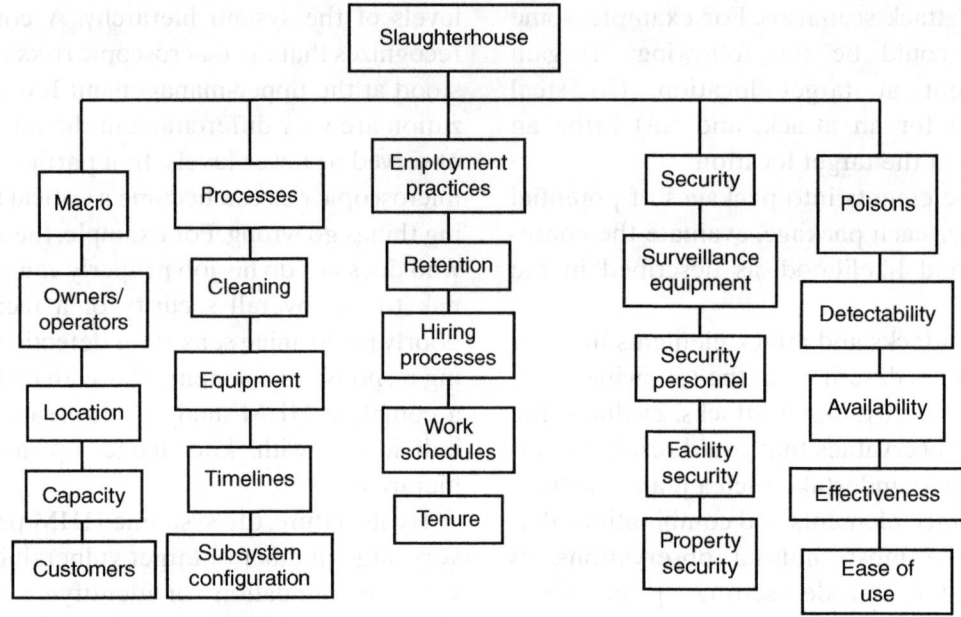

Figure 17.1 Food-poisoning scenarios. From Haimes and Horowitz [2004]; © 2004 The Walter de Gruyter Publishing Group.

discussed to this point. Obviously, experts with specialized knowledge about terrorist organizations are required.

17.3.3 Consequences and Likelihood of Attacks

The risk filtering, ranking, and management (RFRM) method [Haimes et al. 2002], introduced in Chapter 7, reduces to a manageable size the large number of scenarios developed by the HHM. It filters and ranks them by giving strong preference to those elements that are considered most important from several different perspectives, such as likelihoods and consequences. When HHM and RFRM are applied to terrorist scenario tracking, the result is a filtered scenario model that is commensurate with the workload of the intelligence-collection and analysis system [Horowitz and Haimes 2003]. Assuming that the consequences of various attacks can be divided into sets (e.g., worst case, bad case, ..., least-critical case) and that the same thing can be done for likelihoods (e.g., highest, high, ..., lowest), the multiple of the size of these two sets is the total number of combinations of consequences and likelihoods. For example, if there are five consequence levels and four likelihood levels, there would be 20 possible combinations across all elements of potential attacks. The HHM analysis team can position its identified vulnerabilities according to an agreed-upon ranking of the possible combinations of consequences and likelihoods.

The HHM-derived model would add or delete elements systemically as a function of intelligence system workload by referring to a master model that incorporates all of the HHM-identified elements. This systemic feedback process requires that practical subsystem workload measures be defined and monitored (see Figure 17.2) [Horowitz and Haimes 2003].

17.3.4 Observation and Decision Thresholds

Observables are defined *as the possible observations related to a particular event that has been determined to have the potential to be a part of an attack plan.* For example, observing employment records at a target location along with lists of known terrorist sympathizers can potentially identify an attack plan that includes using an insider for access. Similarly, combining information from crime reports about a stolen poison and from an HHM analysis that identifies potent food poisons can potentially identify a food-poisoning plan. If the employment situation such as that given previously is at a particular meat-packing plant, then the

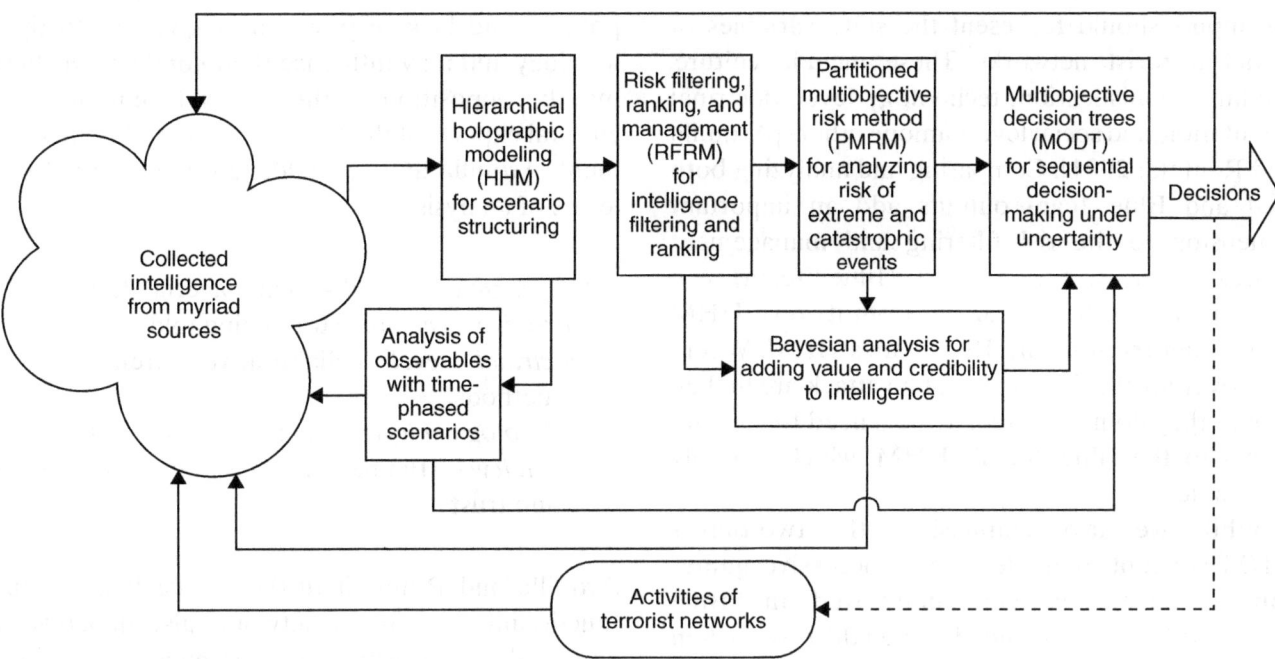

Figure 17.2 Risk-based methodological framework for scenario tracking and intelligence gathering and analysis for countering terrorism. From Horowitz and Haimes [2003]; © 2003 Wiley.

combination of the two elements can potentially provide stronger evidence about a specific attack in progress.

17.4 ADAPTIVE TWO-PLAYER HHM GAME: TERRORIST NETWORKS VERSUS HOMELAND PROTECTION

17.4.1 Overview

The adaptive two-player HHM game provides a methodology for intelligence collection and analysis. (The relationship between this game and the classical game theory as introduced by von Neumann and Morgenstern [1972] and extended by others, e.g., Kuhn [1997], is discussed subsequently.) For this application, intelligence analysts are divided into two teams: *offense* and *defense*. The objectives of each player are as follows:

1. For the *Blue Team—homeland defenders*: Develop a comprehensive HHM of its own system as a way of evaluating vulnerabilities and opportunities for adversaries to exploit such vulnerabilities. The results will be used to develop a set of surveillance efforts that could provide attack warning and assessment information to support attack preemption efforts. This team has access to all available information about the system it is defending and a set of risk specifications to consider in its analysis (e.g., level of protection against financial loss).

2. For the *Red Team—terrorist networks*: Develop a comprehensive HHM of the defender's system by collecting intelligence on potential targets and focusing on the opponent's vulnerabilities and strengths, that is, their *state variables*. This would be used as a basis for selecting possible attack scenarios.

It is imperative that two independent HHMs be developed—one from the homeland perspective and one from the terrorist perspective. Note that having the defensive Blue Team consider the opponent's HHM perspectives results in (i) the consolidation of both perspectives, thus yielding a more complete HHM, and (ii) valuable benchmark information on the depth and breadth of the assessment. Additional benefits are greater self-understanding and knowledge of the opponent. To maximize the effectiveness of the Red Team's HHM, the inputs should represent the state variables of actual terrorist networks. These include culture, funding, sophistication, technology level, doctrinal orientation, and social levels, among others [Arquilla and Ronfeldt 2001]. Comparing and analyzing both Red and Blue Team outputs add an important dimension to the risk-filtering and management process. Clearly, the defense (Blue Team) can temper the conclusions drawn from its own HHM by relating them to the Red Team's HHM. Where they overlap, the likelihoods of an attack are higher. Where they do not, there may be a need to add elements to the Blue Team's HHM, which is easily adaptable.

While we have emphasized the two-player HHM concept, it is clear that successive games can be played involving many Red and Blue Teams. Two questions need to be addressed when conducting multiple games. First, *how many game iterations involving the same situations are needed to achieve a comprehensive and relatively stable set of intelligence-collection observables?* The answer is that measures of convergence can potentially be developed based on Bayesian and decision-tree analyses. Thus, when the observables and their corresponding probabilistic results converge using the decision trees that emerge from successive HHM analyses, the utility of the new changes to the stable HHM models will have little, if any, value. Experiments involving Blue and Red Teams and using measures of convergence can establish the characteristics of HHM convergence.

The second question is: *How do results vary as the basic characteristics of the teams' players are varied?* To address this question, both teams need to possess a variety of skills, experience, and interests. Results can be compared, again using Bayesian and decision-tree analyses to determine the importance of the variations (see Monahan [2000]; Monahan et al. [2001]; Slovic et al. [2000]). Ultimately, the choice of Red Team and Blue Team participants is critical for the intelligence community.

17.4.2 Red Team Perspectives

Effective Red Teams must be cognizant of the cultural and societal environments within which terrorist networks live and are nourished. For example, poverty and lack of power may give rise to their ideology and may influence their conduct. Or there may be opposition to the values, technology, and cultural exports of the West. To explore this environment, Arquilla and Ronfeldt [2001] identified five levels of analysis:

Organizational level—its managerial design
Narrative level—the story being told
Doctrinal level—collaborative strengths and methods
Technological level—the information system
Social level—the personal ties that ensure loyalty and trust

Arquilla and Ronfeldt further argue that the full functioning of terrorist networks also depends on how well, and in what ways, the members are personally known and connected to each other.

Wulf et al. [2003] identify the following eight, not necessarily independent, state variables that may serve as an initial representation of the environments that nourish and sustain the terrorist networks (see Figure 17.3):

1. *Nationalism*: This worldwide movement has led to the creation of a large number of new independent countries during the last four decades. This trend in turn continues to inspire nationalism within and beyond developing countries.
2. *Globalization*: Information communications and technology have virtually removed many international barriers in commerce and communications, as well as in the arts, movies, television, and other cultural activities. This facilitates the free movement and activities of terrorist networks.
3. *Extremism*: Extremism has hijacked not only religions but also political discourse around the world.
4. *Oppression*: The worldwide oppression from which many populations suffer breeds extremism and unhappiness.
5. *Autocratic regimes*: Many developing countries remain governed by autocratic regimes that often seek personal gratification and financial gain at the expense of the populace.

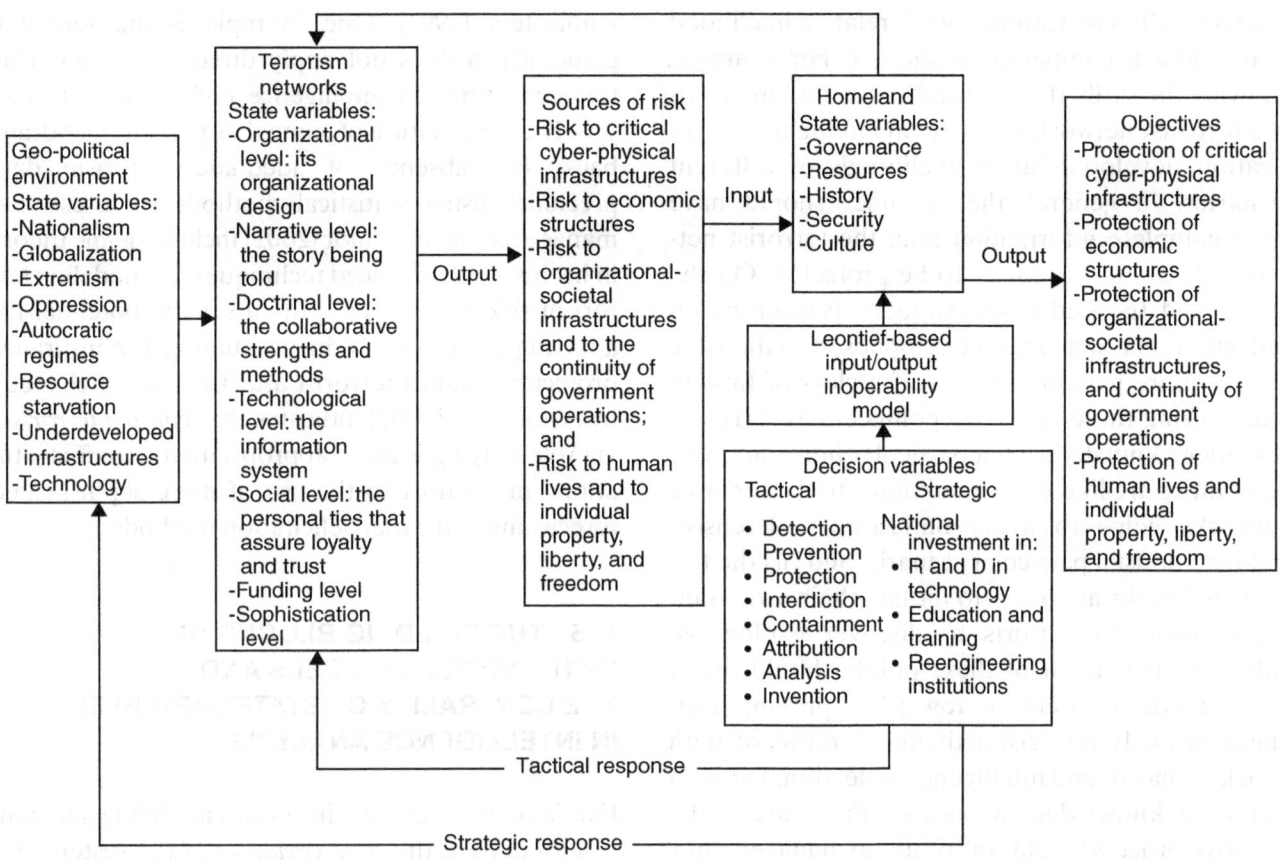

Figure 17.3 A road map of risks of terrorism to the homeland. From Haimes and Horowitz [2004]; © 2004 The Walter de Gruyter Publishing Group.

Such regimes sow the seeds of poverty, oppression, and terrorism.

6. *Resource starvation*: The exploitation of natural and human resources by autocratic regimes is a central cause of the prevalence of poverty and poor health in many developing countries.

7. *Underdeveloped infrastructures*: The lack of adequate investments, especially in critical physical infrastructures, has markedly contributed to the low standard of living and poor quality of life in many developing countries.

8. *Technology*: Developing countries that lack the deployment of technology are struggling in their quests to pull out of poverty.

An authentic Red Team cannot be ignorant either of these variables or of the cultural and societal environments that produce terrorist networks. In particular, an HHM generated by a Red Team might include the following elements as sources for deriving attack scenarios: *psychology, emotions and jealousy, hatred and revenge, resentment and anger, pride and honor, religion, symbols*, and *power*. Taken as a whole or in part, these characteristics may be viewed as a strong driving force of the terrorist networks, with threats and attacks providing outlets for emotion and frustration.

17.4.3 Procedures for the Adaptive Two-Player HHM Game

To start, the defender develops an HHM to consider the range of possible scenarios that a terrorist might choose to initiate. To do this, the Blue Team must gather all information related to a class of attacks (e.g., food poisoning) and assess all the vulnerabilities in related systems that can be exploited in targets of concern. Potential attack scenarios can then be evaluated for their consequences, likelihoods of success, and likelihoods of occurrence. Since terrorist attacks have been relatively rare, there is little information available for directly estimating the likelihood of an attack. However, the intelligence we do have about terrorist networks can provide a basis for indirectly estimating the relative likelihood of one attack compared to another. For example, knowing the skills, the financial status, and the goals of a terrorist network can help an intelligence organization develop relative likelihoods of different scenarios. In general, the defender should have more complete information than the terrorist networks do about the assets to be protected. On the other hand, the Red Team can focus its information collection on a single target, as opposed to the Blue Team's more general analysis of a class of targets. Recognizing these facts, two points emerge: (i) The defender's knowledge that particular homeland vulnerabilities are likely to be unknown to the terrorist networks helps to avoid unnecessary defensive actions against a particular scenario, and (ii) the terrorist networks are likely to initiate their own intelligence-collection efforts to discover exploitable vulnerabilities. Each of these points should direct the defender's attention toward improving estimates of likely terrorist activities in terms of both attack scenarios and intelligence collection. Through increased knowledge in both of these areas, the adaptive process would contribute to management's decisions related to its own intelligence collection as well as to improved defense.

17.4.4 Classical Game Theory and the Adaptive Two-Player HHM Game

Classical game theory was originally developed by John von Neumann as early as 1928, was formalized by von Neumann and Morgenstern [1972] in 1944, and subsequently was extended by many other scholars. Because of its value in intelligence analysis for countering terrorism, this theory has provided the impetus for the methodology presented in this chapter. However, the purpose of the adaptive two-player HHM game is to provide a quantitative road map for intelligence collection and analysis, *not* to defeat the other player(s) through a game-generated *optimal strategy* per se. This use of a two-player game to enhance intelligence collection and analysis accentuates the limitations of classical game theory and also restricts the utilization of its extended richness by Nash [1997], Shapley [1997], Kuhn [1997], and others. Other books on game theory include Luce and Raiffa [1989], Davis [1983], and Osborne and Rubinstein [1994]. One example is the zero-sum game, which does not apply directly to countering terrorism through intelligence collection and analysis. The static nature of game theory is another drawback. The absence of adequate terrorism data precludes using statistical methods to assess and manage terrorism. Major [2002] includes game theory in his roster of advanced techniques for modeling terrorism risk. Oster [2002] discusses the potential for applying game theory in accounting for insurance protection against terrorist attacks. A research study by Armstrong [2002] indicates that *gaming terrorism via role-playing* is more appropriate for conflict situations, in contrast to the use of more sophisticated forecasting and expert-elicitation methods.

17.5 THE BUILDING BLOCKS OF MATHEMATICAL MODELS AND THE CENTRALITY OF STATE VARIABLES IN INTELLIGENCE ANALYSIS

Fundamental elements in system modeling and control theory are the *state variables* of the system. For example, to control the production of steel requires an understanding of the states of the steel at any instant—its *temperature*, *viscosity*, and other physical and chemical properties. To know when to irrigate and fertilize a farm to maximize crop yield, a farmer must assess the *soil moisture* and the level of *nutrients* in the soil. To treat a patient, a physician first must know the *temperature*, *blood pressure*, and other states of the patient's physical health. State variables also constitute the building blocks of the adaptive two-player HHM game.

To relate the centrality of state variables in intelligence analysis to countering terrorism, it is important to define two major terms—vulnerability and threat:

- *Vulnerability* is the manifestation of the inherent states of the system (e.g., physical, technical, organizational, cultural) that can be exploited by an adversary to harm or damage the system.
- *Threat* is a potential intent to cause harm or damage to the system by adversely changing its states. A *threat to a vulnerable* system with adverse effects results in *risk*.

Lowrance [1976] defines *risk* as a measure of the probability and severity of adverse effects.

Threats to our vulnerability include the terrorist networks that aim to change the states of our homeland—for example, from a stable to an unstable government, from operable to inoperable infrastructures, and from a trustworthy to an untrustworthy cyber system. In other words, these terrorist networks that threaten our homeland have the same goal as those commissioned to protect its safety, albeit in opposite directions—*both want to control the states of the systems in order to achieve their objectives.* However, the objectives of the terrorist network will determine the selection of an attack plan, and these objectives may be different from those used by a Blue Team in an analysis aimed at anticipating possible attacks. Therefore, for homeland protection, the ultimate objective of an effective intelligence analysis is to:

1. Identify the states of the system being defended
2. Associate the vast set of intelligence data with the way terrorist networks might select a target and thus attempt to transform the homeland's states of safety and security into states of risk

17.5.1 Role of State Variables in Adaptive Two-Player HHM Game

An important input to the analysis of specific scenarios and targets is the a priori likelihood of an attack, which is dependent on the outlook and current situation faced by terrorist organizations. As shown in the example in Section 17.2, the a priori values selected for an attack greatly influence the evidence required for warning. For an improved understanding of the potential behavior of terrorist organizations, the adaptive two-player HHM game can be used to model the a priori values as well as the target-specific aspects. A unique feature of the adaptive two-player HHM game is the direct relationship that exists between the *head topics* of the HHM and the *state variables* of the system (e.g., the homeland or the terrorist networks). In general, each head topic represents an important state variable. This would be the focus of an effective attack as well as of the system's defense, hence the importance of identifying these states when developing a complete HHM. Consider, for example, *resources* as a head topic. Effective counterterrorism requires credible knowledge of the levels, sources, and channels operating within the terrorist networks. To this end, reliable intelligence must address a critical state variable—namely, the level of resources available to the terrorist networks and their allocations for different functions.

It is important to note that not all state variables (head topics) in the HHM are necessarily directly observable or controllable. Many can be represented only by several surrogates. For example, addressing quality, Garvin [1988] argues that there are eight dimensions or categories of product or service quality: performance, features, reliability, conformance, durability, serviceability, aesthetics, and perceived quality. Each category may be viewed as distinct and self-contained, yet they define quality only when integrated. The same can be argued for the state variables representing the terrorism networks as depicted in Figure 17.3: *organization*, *narrative*, *doctrine*, *technology*, *resources*, and *sophistication*. Over the years, systems engineers have used surrogates due to the lack of controllability or observability of state variables in process control. Similarly, counterterrorism analysts can resort to surrogate variables to represent the unobservable and/or uncontrollable state variables listed in Figure 17.3.

17.5.2 Synergy between the Adaptive Two-Player HHM Game and State Variables

In classical game theory [von Neumann and Morgenstern 1972], the actions of the players and their consequences as well as the anticipated or perceived reactions and countermeasures are explicit in the ensuing game. The adaptive two-player HHM game is based not only on the actions of the players and their consequences but also on an explicit understanding of the inherent characteristics of the players that necessarily lead to the observed actions and consequences. In other words, the strategies and actions of the homeland Blue Team in the HHM game respond to the states of their own system as well as to those of the terrorist Red Team. Therefore, intelligence analyses for countering terrorism will be far more effective if they are driven not only by the symptoms (i.e., the actions of the terrorist networks) but also by the *root causes* (i.e., the states that characterize the terrorist networks). To this end, the adaptive two-player HHM game also offers a road map for scenario tracking that accounts for the characteristics of both the root causes and the target (see Haimes [2002]; Horowitz and Haimes [2003]).

Because of the organizational complexity of the terrorist networks, intelligence gathering and analysis associated with them require an adaptive process. For example, the organizational infrastructure of al-Qa'ida is composed of a large number of hierarchically decentralized yet interconnected cells. Useful predictions about the selection of targets and weapons for attack would be difficult to make without accounting for this structure. Therefore, the information/intelligence-gathering system for counterterrorism must be inherently responsive to this reality in terms of its own hierarchical–organizational and functional decisionmaking infrastructure. This realignment of the organizational infrastructure implies reengineering the process through which intelligence is collected, shared, and analyzed—resulting in a revolutionary process that necessarily affects the roles of federal, state, county, and city agencies in sharing and disseminating information. Managing this change is a daunting task because it affects the culture of myriad decisionmakers, constituencies, power brokers, stakeholders, and users of the information/intelligence system, as well as a host of institutional, legal, and other societal aspects (e.g., profiling) that require consideration. Through the convergence of the HHM processes, the adaptive two-player HHM game is attuned to the needs of this adaptive, malleable process.

17.6 HIERARCHICAL ADAPTIVE TWO-PLAYER HHM GAME

So far, the focus of the adaptive two-player HHM game has been on helping to understand terrorist organizations, the likelihoods of various classes of attack, and the details that might accompany an attack scenario. Since it is unreasonable to assume that terrorists would select only one class of attack scenarios, the following steps are suggested:

1. Organizing multiple Red and Blue Teams
2. Collecting and analyzing intelligence on many classes of attack scenarios (e.g., food poisoning, agricultural attacks, destruction of critical infrastructures, nuclear weapon attacks, bioattacks)
3. Structuring multiple hierarchies of related scenarios to represent the multifarious dimensions and perspectives of the risk to the homeland

The following summarizes the rationale for extending the single-scenario adaptive two-player HHM game to the hierarchical model:

1. When a Red Team reaches a convergence on a single scenario through the HHM game, select members can join another Red Team that is working on another scenario, thereby adding valuable knowledge, experience, and synergy to the augmented team.
2. Infusing new intelligence into the Red Teams improves their performances and insights into the terrorist networks' thinking. Integrated, the multiple scenarios studied by the various Red Teams can be reevaluated, distilled, and generalized to reflect the collective accumulated knowledge gained through the continually iterative process. This by-product of the hierarchical model provides the opportunity for a major gain from the adaptive two-player HHM game because:
   - It enhances the ability of the intelligence agencies to better understand the a priori threats posed by the terrorist networks. It does this through a process of systemic feedback, integration, and synthesis that is grounded on sound risk and systems analyses, thus strengthening and improving the intelligence-gathering process itself.
   - It enables the Blue Teams (intelligence agencies and the homeland defenders) to identify commonalities among various scenarios, thus improving the protective measures undertaken by federal, state, and local security agencies.

3. Each of the multiple hierarchies, within which the various scenarios can be organized and structured, can be associated with a decomposition that represents a specific vision or perspective of terrorism risks. For example, decomposition and its corresponding hierarchy of scenarios may address a geographical, temporal, or infrastructure-functional interest or perspective.

17.7 COLLABORATIVE COMPUTING SUPPORT FOR ADAPTIVE TWO-PLAYER HHM GAMES

This section presents the detailed features and technical aspects of a collaborative computing tool for the HHM, elements of which are still being developed, as noted in the following. There are several practical issues that must be considered in order to put the two-player HHM game into practice. First, the availability of appropriate team members could create limiting logistics issues. Second, the results of successive games would need to be archived in a fashion that would help integrate results over a potentially long period of time. These requirements have inspired the development of an HHM computer tool to support the Red and Blue Teams' efforts. It is an application that runs on a commercially available, general-purpose collaborative computing tool; among other capabilities, it provides a vehicle for Internet-based team interaction and for archiving results. Initial applications of this tool have served to demonstrate the potential that commercially available collaborative computing capabilities offer in support of the two-player HHM game concept.

The purpose of the software is to enable the HHM process for teams of geographically separated participants. In addition, it permits an ongoing, continuous collaboration, so that inputs can be modified as real-world events change. Each participant runs the software on his or her own computer to access an interface through which the HHM can be viewed and manipulated. The model is consistent across all computers, and changes made by any user appear in real time. The software recognizes a particular user as the manager and provides him or her with tools for managing the process. The software was built on the Groove platform in Microsoft's C# programming language. While the tool is immature, a 14-person, two-player game has been conducted, providing the first use-based feedback to help direct improvements.

17.7.1 Features

17.7.1.1 HHM interface

The main function of the tool is to allow users to collaboratively develop an HHM on their computers. An HHM includes a set of hierarchies of topics, each one coming from a different perspective. The root of each hierarchy is the head topic. The user can add a head topic by clicking an *Add Head Topic* button. The name and description of each head topic are entered when the head topic is created. Below the head topics are subtopics, which represent more specific areas of risk. Subtopics can branch off from head topics or other subtopics, and there is no limit to the depth of the hierarchy. The user can add a subtopic by first clicking on an existing head topic or subtopic to select it and then clicking an *Add Subtopic* button. As with head topics, each subtopic has a name and description that are entered upon creation. Both head topics and subtopics can be removed by clicking first on the topic and then on the *Remove Topic* button. Aside from removing a topic or adding subtopics to it, selecting a topic (by clicking on its box) can also be used to view more information about it.

All additions, deletions, and modifications of topics will update the model immediately for all users who are running the software. Users who are off-line will have their models updated the next time they log on.

17.7.1.2 Anonymity

When a user first logs on to the software, he or she is assigned a random identification (Bill, Mary, Bob, etc.) and will be known only by that name throughout the collaborative HHM process. This prevents biases and allows people to interact without concern about others' personal opinions of them.

17.7.1.3 Commenting

Checking a *Comments* checkbox will bring up a window that allows the user to view, add, and remove comments related to the risk identified in the box. Each comment has a title and text field. A comment can be edited or removed by the user who made it. When a comment is added to a topic, the comment counter for that topic box is incremented immediately for all active users.

17.7.1.4 Risk assessment

Each user can assign a rating to the consequences and likelihood of each topic. This allows more attention to be given to topics that are considered more serious. To make or change an assessment, the user simply clicks on the topic he or she wants to assess and then selects one of a set of choices for consequence and one for likelihood from the assessment window. The data is updated immediately for all active users.

Individual assessments for topics can be indicated by a colored bar across the bottom of each topic box. A color code can be used to denote the seriousness of a particular risk (e.g., red is the most concerning and blue is the least). An integrated risk assessment can also be presented through the use of a multicolored bar. If, for example, four people have assessed a certain topic and three selected a combination of consequence and likelihood with red priority while one selected a combination with orange priority, three-quarters of the length of the bar would be red, while one-quarter would be orange. If a topic has not been assessed by anyone, the bar will not be present.

17.7.1.5 Groups

A designated manager can restrict access to parts of the HHM by creating groups and assigning privileges to them. A manager can view a list of groups and of users for each group and add and remove groups and users. Changes made to group membership cause each active user's view to be updated immediately. If a user is added to a group with exclusive access to part of the HHM hierarchy, that section, in addition to the generally accessible portion of the HHM, will appear immediately on the user's display or the next time the user logs on. By default, all users are members of a group called *Everyone* that has access to the lowest-common-denominator portion of the HHM.

17.7.1.6 Private discussion

Users can communicate with each other by checking a *Chat Window* checkbox. This window can contain multiple tabs, one for each group in which the user is a member. Each tab has a list of members and a chat area where text messages appear. Users can create their own groups and give others access, which allows private discussion to take place independent of the manager. Unlike manager-created groups, user-created chat groups cannot be used for access control (discussed in the next section).

17.7.1.7 Access control

The manager can restrict access to parts of the HHM hierarchy to protect sensitive information. Access control is based on group membership and access lists for individual topics. The manager can expose the access control window by checking an *Access Control* checkbox. This window displays a list of the groups with access to the selected topic. Groups can be added and removed from this list by the manager. Access settings for a topic in the hierarchy automatically apply to all descendants of the topic. By default, the *Everyone* group is in the access list for new topics.

17.7.1.8 Risk filtering and ranking

The manager can rank the combinations of consequence and likelihood and can also sort the topics by assessment or date. There are 25 combinations of consequence and likelihood, each of which can be moved up or down in priority relative to the others. One of five colors (red, orange, yellow, green, or blue) can be assigned to each combination. The colors chosen will show how each topic has been assessed.

17.7.1.9 Procedures

As indicated earlier, an initial two-player game was conducted using the tools described in this section. Initial results included refining the procedures for conducting the game. Based on this, it is recommended that the defending team (Blue) be provided with initial information about the target, while the team planning an attack (Red) should not be provided with any information. In addition, the Blue Team should be provided with a risk management specification of the boundaries to consider for the target in question. These might include, for example, considering defenses only for those attacks that can cause more than 100 deaths or serious injuries or more than $1 million of loss or that can be replicated at many similar targets to achieve an equivalent integrated result. On the other hand, the Red Team should start with a range of attack types to initiate its members' thinking, but should not be restricted to this set. For example, for the case of attacking an airport, the Red Team might be asked to include in its considerations hijacking an aircraft, as well as taking hostages in the terminal building. Finally, the manager of the two-player game should provide an initial bare-bones template for each team to help them get started. The template should be based on some of the critical state variables, as discussed earlier. The start-up template should be different for each team and consistent with the other initial information provided to them.

17.8 SUMMARY

The adaptive two-player HHM game has several attributes that support intelligence gathering and analysis:

- It represents a repeatable, adaptive, and systemic process that builds on fundamental principles of systems engineering, system modeling, and risk analysis. Scenario structuring and comprehensive risk identification from multiple perspectives are achieved through the deployment of HHM. State variables, which represent the essence of a system, play a pivotal role in the game, providing an enabling road map to intelligence analysts. The following sample of questions can be answered through this road map:

    What intelligence should be collected, and why?

    How can diverse intelligence reports be related to specific scenarios?

    What intelligence is of interest to the terrorist networks?

    How can priorities be introduced in intelligence collection and analysis?

    How can we corroborate and add credibility to intelligence reports?

- Answers to such questions can potentially be generated through a number of well-tested methodologies and methodological frameworks; these form the basis for the game. They include:

    HHM—for scenario structuring and risk identification (see Chapter 3).

    RFRM—for adding priorities to the generated scenarios and intelligence database (see Chapter 7).

    Bayesian analysis—for corroboration and adding credibility to intelligence.

    Building blocks of mathematical models and the centrality of state variables—for identifying, in conjunction with HHM, the critical elements that are of interest to the terrorist networks. These form the basis for collecting intelligence. Such knowledge can result in a priori likelihoods of attacks using specific classes of weapons.

- Each player in the adaptive two-player HHM game deploys the same modeling tools. This ensures the maximum reliability of the process. It also constitutes a learning-oriented approach in the sense that both teams can benefit from the same multiple-perspective procedures.

- At the end of the first round of the game, the Blue Team's HHM can be augmented with new elements (head topics and subtopics) from the Red Team's HHM. When this process is repeated with new Red Teams, the Blue Team's HHM converges to a *complete set* of risk scenarios (head topics and subtopics). As a result, intelligence analysts can be assured that most, if not all, important and critical risk scenarios have been explored.

The breadth and complexities involved in evaluating possible terrorist attacks demand the involvement of an enormous group of intelligence sources. Organizations working on countering terrorism need to integrate the knowledge and experience of large communities of experts who can create families of models. The two-player game process provides an opportunity for these organizations to establish a basis for decisionmaking through interaction. A structured, cooperative modeling approach would provide significant benefits beyond the models themselves; inevitably, it would initiate other valuable collaborations. A vehicle for collaboration in cyberspace could lead to the creation of even more effective tools. The two-player game provides an excellent start for such teamwork and should provide the impetus for increased opportunities for communities of experts to work together.

REFERENCES

Armstrong, J.S., 2002, *How to Avoid Surprises in the War on Terrorism: What to Do When You Are between Iraq and a Hard Place*. The Wharton School, University of Pennsylvania, Philadelphia, PA. Online at http://www.chforum.org/library/gaming.shtml (Accessed January 6, 2003).

Arquilla, J. and D. Ronfeldt, 2001, *Networks and Netwars*, National Defense Research Institute, RAND, Pittsburgh, PA.

Ashkenas, R., D. Ulrich, T. Jick, and S. Kerr, 1995, *The Boundaryless Organization: Breaking the Chains of Organizational Structure*, Jossey-Bass Publishers, San Francisco, CA.

Blauberg, I.V., V.N. Sadovsky, and E.G. Yudin, 1977, *Systems Theory: Philosophical and Methodological Problems*, Progress Publishers, New York.

Davis, M.D., 1983, *Game Theory: A Nontechnical Introduction*, Dover Publications, Inc., Mineola, NY.

Dombroski, M., Y.Y. Haimes, J.H. Lambert, K. Schlussel, and M. Sulkoski, 2002, Risk-based methodology for support of operations other than war, *Military Operations Research* 7(1): 19–38.

Garvin, D.A., 1988, *Managing Quality: The Strategic and Competitive Edge*, The Free Press, New York.

Gordon, W.J.J., 1968, *Synectics: The Development of Creative Capacity*, Collier Books, New York.

Haimes, Y.Y., 1981, Hierarchical holographic modeling, *IEEE Transactions on Systems, Man, and Cybernetics*, 11(9): 606–617.

Haimes, Y.Y., 2002, Roadmap for modeling risks of terrorism to the homeland, *Journal of Infrastructure Systems*, 8(2): 35–41.

Haimes, Y.Y., 2012a, Modeling complex systems of systems with phantom system models, *Systems Engineering*, 15(3): 333–346.

Haimes, Y.Y., 2012b, Systems-based guiding principles for risk modeling, planning, assessment, management, and communication, *Risk Analysis*, 32(9): 1451–1467.

Haimes, Y.Y., S. Kaplan, and J.H. Lambert, 2002, Risk filtering, ranking, and management framework using hierarchical holographic modeling, *Risk Analysis*, 22(2): 383–397.

Haimes, Y.Y. and B.H. Horowitz, 2004, Adaptive two-player hierarchical holographic modeling game for counterterrorism intelligence analysis, *Journal of Homeland Security and Emergency Management*, 1(3): 1–21.

Hall, A.D. III, 1989, *Metasystems Methodology: A New Synthesis and Unification*, Pergamon Press, New York.

Horowitz, B. and Y.Y. Haimes, 2003, Risk-based methodology for scenario tracking for terrorism: a possible new approach for intelligence collection and analysis, *Systems Engineering*, 6(3), 152–169.

Kaplan, S. and B.J. Garrick, 1981, On the quantitative definition of risk, *Risk Analysis*, 1(1): 11–27.

Kaplan S., Y.Y. Haimes, and B.J. Garrick, 2001, Fitting hierarchical holographic modeling (HHM) into the theory of scenario structuring and a refinement to the quantitative definition of risk, *Risk Analysis*, 21(5): 807–819.

Kuhn, H.W. (Ed), 1997, *Classics in Game Theory*, Princeton University Press, Princeton, NJ.

Lambert, J.H., Y.Y. Haimes, D. Li, R. Schooff, and V. Tulsiani, 2001, Identification, ranking, and management of risks in a major system acquisition, *Reliability Engineering and System Safety*, 72(3): 315–325.

Lowrance, W.W., 1976, *Of Acceptable Risk: Science and the Determination of Safety*, William Kaufmann, Inc., Los Altos, CA.

Luce, D.R. and H. Raiffa, 1989, *Games and Decisions: Introduction and Critical Survey*, Dover Publications, Inc., New York.

Major, J.A., 2002, SVP, The Guy Carpenter & Company, Inc., *Advanced Techniques for Modeling Terrorism Risk*. National Bureau of Economic Research Insurance Group Conference, February 1, 2002. http://www.emeraldinsight.com/doi/abs/10.1108/eb022950 (Accessed: January 6, 2003).

Monahan, J., 2000, The scientific status of research on clinical and actuarial predictions of violence, In *Modern Scientific Evidence: The Law and Science of Expert Testimony*, 1, Second edition, D. Faigman, D. Kaye, M. Saks, and J. Sanders (Eds.), West Publishing Company, St. Paul, MN, pp. 423–445.

Monahan, J., H. Steadman, E. Silver, P. Appelbaum, P. Robbins, E. Mulvey, L. Roth, T. Grisso, and S. Banks, *Rethinking Risk Assessment: The MacArthur Study of Mental Disorder and Violence*. New York: Oxford University Press, 2001.

Nash, J.F. Jr., 1997, Non-cooperative games, In *Classics in Game Theory*, H.W Kuhn (Ed.), Princeton University Press, Princeton, NJ.

NRC (National Research Council), 2002, *Making the Nation Safer: The Role of Science and Technology in Countering Terrorism*, The National Academies Press, Washington, DC.

Osborne, M.J. and A. Rubinstein, 1994, *A Course in Game Theory*, The MIT Press, Cambridge, MA.

Oster, C., 2002, Can the risk of terrorism be calculated by insurers? game theory might do it, The Wall Street Journal Online http://www.wsj.com/articles/SB1018213759791798240, April 8, 2002. Accessed: January 6, 2003.

Sage, A.P., 1977, *Methodology for Large-Scale Systems*, McGraw-Hill, New York.

Sage, A.P., 1992, *Systems Engineering*, John Wiley & Sons, Inc., New York.

Shapley, L.S., 1997, A value for n-person games, In *Classics in Game Theory*, H.W. Kuhn (Ed.), Princeton University Press, Princeton, NJ.

Singh (Editor-in-Chief), M.G., 1987, *Systems & Control Encyclopedia: Theory, Technology, Applications*, Pergamon Press, New York.

Slovic, P., J. Monahan, and D. MacGregor, 2000, Violence risk assessment and risk communication: the effects of using actual cases, providing instruction, and employing probability versus frequency formats, *Law and Human Behavior*, **24**: 271–296.

von Neumann, J. and O. Morgenstern, 1972, *Theory of Games and Economic Behavior*, Princeton University Press, Princeton, NJ.

Warfield, J.N., 1976, *Social Systems—Planning and Complexity*, John Wiley & Sons, New York.

Wulf, W.A., Y.Y. Haimes, and T.A. Longstaff, 2003, Strategic alternative responses to risks of terrorism, *Risk Analysis*, **23**(3): 429–444.

Zigler, B.P., 1984, *Multifaceted Modeling and Discrete Simulation*, Academic Press, New York.

18

Inoperability Input–Output Model and Its Derivatives for Interdependent Infrastructure Sectors

18.1 OVERVIEW

The advancement in information technology has markedly increased the interconnectedness and interdependencies of our critical infrastructures, such as telecommunications, electrical power systems, gas and oil storage and transportation, banking and finance, transportation, water supply systems, emergency services, and continuity of government. Due to the vulnerability of these infrastructures to the threats of terrorism, there is an emerging need to better understand and advance the art and science of modeling their complexity.

To illustrate this complexity, let us consider the US electric power utility, which is a large-scale, hierarchical, and interconnected system. At the national level, it consists of three interconnected networks: (i) the Eastern Interconnected System, covering the eastern two-thirds of the United States; (ii) the Western Interconnected System, covering the Southwest and areas west of the Rocky Mountains; and (iii) the Texas Interconnected System, consisting mainly of Texas.

At the network level, each network, as its name implies, is an interconnected system in itself, comprising numerous generators, distribution and control centers, transmission lines, converters, and other elements. Proper functioning of these interacting components is crucial to the continuous operation of the entire power system. In addition to its essential internal dependency, the US power system is externally dependent upon other infrastructure systems, notably telecommunications, fuel supply, and transportation. For example, its operation is heavily dependent upon voice and data communications. Data communications provide real-time updates (i.e., every few seconds) of electrical system status to supervisory control and data acquisition (SCADA) systems in distribution and bulk electric control centers. Data communications are also used for the remote control of devices in the field, such as circuit breakers, switches, transformer taps, and capacitors. Moreover, data communications allow generating units to follow the real-time signals from the control center that are necessary to balance electricity generation with consumer demand instantaneously. Although the power industry owns and operates the majority of its communications equipment, a substantial portion is dependent upon local telephone carriers, long-distance carriers, satellites,

---

Risk Modeling, Assessment, and Management, Fourth Edition. Yacov Y. Haimes.
© 2016 John Wiley & Sons, Inc. Published 2016 by John Wiley & Sons, Inc.

cellular systems, paging systems, networking service providers, Internet service providers, and others.

Historically, many critical infrastructures around the world were physically and logically separate systems with little interdependence. This situation is rapidly changing, and close relationships among infrastructures can now take many forms. For example, telecommunications, power, transportation, banking, and others are marked by immense complexity, characterized predominantly by strong intra- and interdependencies as well as hierarchies. These interconnections take many forms, including flows of information, shared security, and physical flows of commodities, among others. There is a need for a high-level, overarching modeling framework capable of describing the risks to our nation's critical infrastructures and industry sectors—focusing on risks arising from interdependencies.

In assessing a system's vulnerability, it is important to analyze both the intraconnectedness of the subsystems that compose it and its interconnectedness with other external systems. Addressing the importance of interconnectedness can be achieved by modeling the way *inoperability* propagates throughout our critical infrastructure systems or industry sectors. The inoperability caused by willful attacks, accidental events, or natural causes can set off a complex chain of cascading impacts on other interconnected systems. For example, similar to other critical infrastructures, water resource systems—surface water and groundwater sources, water transport, treatment, distribution, storage, and wastewater collection and treatment—heretofore have been designed, built, and operated without a threat to their integrity. Today, the interdependencies and interconnectedness among infrastructures pose a threat to our water systems. This section addresses modeling these interdependencies.

18.2 BACKGROUND: THE ORIGINAL LEONTIEF INPUT–OUTPUT MODEL

Wassily Leontief received the 1973 Nobel Prize in Economics for developing what came to be known as the Leontief input–output (I–O) model of the economy [Leontief, 1951a, b; 1986]. The economy (and thus the model) consists of a number of subsystems, or individual economic sectors or industries, and is a framework for studying the equilibrium behavior of an economy. The model enables understanding and evaluating the interconnectedness among the various sectors of an economy and forecasting the effect on one segment of a change in another. The Leontief I–O model describes the equilibrium behavior of both regional and national economies [Isard, 1960; Liew, 2000; Lahr and Stevens, 2002], and the I–O model is a useful tool in the economic decisionmaking processes used in many countries [Miller et al., 1989].

The Leontief model enables accounting for the intraconnectedness within each critical infrastructure as well as the interconnectedness among them. Miller and Blair [1985] provide a comprehensive overview of I–O analysis with deep insights into the Leontief economic model and its applications. Recent literature in the area of cascading failures through interconnected sectors can be found in the US Department of Commerce (DOC) [2003] and Embrechts et al. [1997]. Many notable extensions were later created based on the original Leontief model, including the nonlinear Leontief model [Krause, 1992], energy I–O analysis [Griffin, 1976; Proops, 1984], and environmental I–O analysis [Converse, 1971; Lee, 1982]. Haimes and Nainis [1974] and Haimes [1977] developed an I–O model of supply and demand in a regional water resources system. Olsen et al. [1997] developed an I–O model for risk analysis of distributed flood protection. Extensions of I–O analysis were described by Lahr and Dietzenbacher [2001].

The brief outline below is based on Intriligator [1971], Haimes [1977], and Haimes et al. [2005a and 2005b]. It provides a simplified version of the Leontief [1951a] I–O model to trace resources and products within an economy. The economy (system) is assumed to consist of a group of n interacting sectors or industries, where each *industry* produces one product (commodity). A given industry requires labor, input from the outside, and also goods from interacting industries. Each industry must produce enough goods to meet both interacting demands (from other industries in the group) plus external demands (e.g., foreign trade and industries outside the group). A static (equilibrium-competitive) economy, with constant coefficients for a fixed unit of time (1 year), is assumed. Define the following notation:

x_j	is the output (for the total economy) of jth goods, $j=1, 2, \ldots, n$
r_k	is the input (for the total economy) of kth resource, $k=1, 2, \ldots, m$
x_{ij}	is the amount of the ith goods used in the production of the jth goods
r_{kj}	is the amount of the kth resource input used in the production of the jth goods

The Leontief model assumes that the inputs of both goods and resources required to produce any commodity are proportional to the output of that commodity:

$$x_{kj} = a_{kj} x_j, \quad j, k, = 1, 2, \ldots, n \quad (18.1)$$

$$r_{ij} = b_{ij} x_j, \quad k = 1, 2, \ldots, m, j = 1, 2, \ldots, n \quad (18.2)$$

Furthermore, the output of any commodity is used either as input for the production of other commodities or as final demands, c_k. The balance equation (18.1) is a key to the subsequent development of the Leontief-based equation (18.3):

$$x_k = \sum x_{kj} + c_k, \quad k = 1, 2, \ldots, n \quad (18.3)$$

Combining Eqs. (18.1) and (18.3) yields the Leontief equation

$$x_k = \sum_j a_{kj} x_j + c_k, \quad k = 1, 2, \ldots, n \quad (18.4)$$

Similarly, the proportionality assumption applies to the resources

$$r_{ij} = b_{ij} x_j \quad (18.5)$$

$$\sum_i r_{ij} = \sum_i b_{ij} x_j \quad (18.6)$$

Since the demand for the ith resource cannot exceed its supply, then

$$\sum_j b_{ij} x_j \leq r_i, r_i \geq 0, \quad i = 1, 2, \ldots, m \quad (18.7)$$

The above basic model of the economy is written in a compact matrix notation in Eq. (18.8):

$$\mathbf{x} = \mathbf{Ax} + \mathbf{c} \Leftrightarrow \{ x_i = \sum_j a_{ij} x_j + c_i \} \forall i \quad (18.8)$$

18.3 INOPERABILITY INPUT–OUTPUT MODEL

Grounded on Leontief's work, a first-generation Inoperability Input–Output Model (IIM) of interconnected systems was developed by Haimes and Jiang [2001]. This *physical-based* model considers multiple intra- and interconnected systems. The primary purpose of the model is to improve understanding of the impact of complexity on the continued and sustained operability of these systems under adverse conditions. Other related works on infrastructure interdependencies and risks of terrorism are presented in Haimes [2002, 2004], Haimes and Horowitz [2004], Santos and Haimes [2004], Crowther and Haimes [2005], and Jiang and Haimes [2004].

Note that the *supply* and *demand* concepts in the Leontief economic model assume a different interpretation and have been inverted to some extent in the IIM risk model. Although the mathematical construct of the two models is similar, the interpretation of the model parameters is fundamentally different. *Dollars* are the units used in the Leontief I–O model for the economy. The infrastructure model uses units of *risk of inoperability* [0, 1], defined above as a measure of the probability (likelihood) and degree (percentage) of the inoperability (dysfunctionality) of a system. An inoperability of 1 would mean that an infrastructure is totally out of commission. As stated earlier, inoperability may take various forms according to the nature of the system. When the model is applied to study any infrastructure system, one of the very first tasks is to define the specific inoperability and the associated risks.

This model addresses the equilibrium state of the system in the event of an attack, provided that the interdependency matrix is known. The *input* to the system is an initial perturbation triggered by an attack of terrorism, an accidental event, or a natural disaster.

The *outputs* of the system are the resulting risks of inoperability of different infrastructures due to their connections to one another. The output can be triggered by one or multiple failures due to their inherent complexity or to external perturbations (e.g., natural hazards, accidents, or acts of terrorism).

In his basic I–O model, Leontief considered an economy that produces n goods as output and that uses m primary resources as input. For the IIM, we consider a system consisting of n critical complex intra- and interconnected infrastructures [Haimes and Jiang, 2001]. Although the equations are similar, there is a major difference in the interpretation of the variables. In other words, the basic Leontief equations (18.1) to (18.8) are similar to the IIM equations (18.9) to (18.12) that will be introduced subsequently; however, they connote different meanings.

In the IIM, the output is the infrastructure's *risk of inoperability* or simply *inoperability* that can be triggered by one or multiple failures due to complexity, accidents, or acts of terror. *Inoperability* is defined as *the inability of the system to perform its intended natural or engineered functions*. In the model, the term *inoperability* can denote the level of the system's dysfunction, expressed as a percentage of the system's *as-planned* level of operation. Alternatively, inoperability can be interpreted as a degradation of a system's capacity to deliver its intended output (or supply). Although inoperability in its current scope applies to physical and economic losses, it can be extended to assess impacts due to information failure. In addition, other factors for assessing failures, such as loss of lives, environmental quality, and others, can supplement the economic factors used in the context of inoperability.

Inoperability is assumed to be a continuous variable evaluated between 0 and 1, with 0 corresponding to a flawlessly operable system state and 1 corresponding to the system being completely inoperable. Inoperability may take different forms, depending upon the nature of the problem and the type of the system. When the production level is of major concern, inoperability may well be defined as the unrealized production (i.e., the actual production level subtracted from the desired production level). For instance, if the system under consideration is a power plant, then the inoperability may be defined as the ratio of the actual amount of power produced (in appropriate units) to the desired amount. Furthermore, the notion of inoperability also attempts to capture the quality of a system's function. Assuming that quality can be measured numerically, a defective system whose performance is of degenerate quality is considered partially operable and thus has inoperability greater than zero. For instance, a television set that has a picture but no sound is only partially operable and thus has inoperability greater than zero. By the same token, a water supply system producing slightly contaminated water is also considered partially operable and thus has inoperability greater than 0. Finally, inoperability of a system is not necessarily a continuous variable. Under certain circumstances, it may take discrete values such as binary values. Here, we focus our discussion on the continuous case.

Risk of inoperability can also be viewed as an extension of the concept of *unreliability*. Unreliability is the conditional probability that a system will fail during a specified period of time t, given that it operates perfectly at $t=0$. In fact, the system may not fail completely during this time span; it may fail partially with certain probability. For instance, during this period of time, it may fail 100% with probability 0.1, it may lose 50% of its functionality with probability 0.4, it may lose 10% of its functionality with probability 0.8, and so forth (provided that the functionality is quantifiable). Thus, a natural extension of the notion of unreliability is to average out all these possibilities by considering both the failure level and the likelihood. In so doing, we end up with a quantity that represents the expected value of the failure level during a certain period of time. In other words, if the expected-value metric is adopted in the definition of risk, then the risk of inoperability can be viewed as the expected inoperability. A conditional expected-value metric, to supplement the expected-value metric, will be introduced later. Hence, for the sake of brevity, in the following discussion, we sometimes use *inoperability* in lieu of *risk of inoperability*.

The inoperability of an infrastructure may be manifested in several dimensions, for example, geographical, functional, temporal, or political. On the one hand, these and other perspectives markedly influence the values assigned to the probability (coefficient) of inoperability in the model. On the other hand, each may justify the construction of a different inoperability model addressing a specific dimension. An example would be inoperability that spans regional or statewide, short-term or long-term, one-function failure or multiple failures of an infrastructure. In such cases, each model will require specific and different probabilities of inoperability. In addition, one such inoperability model might evaluate, and measure in monetary terms, the risk of inoperability or damage to property, production, service, or injury under extreme natural and accidental conditions or due to acts of terrorism.

In the following discussions, we assume that each infrastructure system performs a uniquely defined function, that is, no two systems perform the same function. In other words, in this preliminary model, we do not consider the issue of redundancy. The systems that we consider here fall into the category of *unparallel* systems.

Let x_j, $j = 1, 2, \ldots, n$, be the overall risk of inoperability of the jth intra- and interconnected infrastructure that can be triggered by one or multiple failures caused by accidents or acts of terrorism.

Let x_{kj} be the degree of inoperability triggered by one or multiple failures that the jth infrastructure can contribute to the kth infrastructure due to their complex intra- and interconnectedness.

Let a_{kj} be the probability of inoperability that the jth infrastructure contributes to the kth infrastructure. In our model, a_{kj} describes the degree of dependence of the kth infrastructure on the jth infrastructure. For example, if $a_{kj} = 1$, then this means a complete failure of the jth infrastructure will lead to a complete failure of the kth infrastructure. A value of $a_{kj} = 0$, on the other hand, indicates that the failure of the jth infrastructure has no effect on kth infrastructure.

Let c_k be the natural or man-made perturbation into the kth critical infrastructure.

At this stage, the proportionality assumption that underpins the Leontief economic model is assumed to hold for the Inoperability Input–Output risk model as well; then we have

$$x_{kj} = a_{kj} x_j \qquad j, k = 1, 2, \ldots, n \qquad (18.9a)$$

The following balance equation is a key to the subsequent development of the linear model:

$$x_k = \sum_j x_{kj} + c_k, \qquad j, k = 1, 2, \ldots, n \qquad (18.9b)$$

Combining the balance equation with the proportionality equation yields the inoperability equation for the infrastructure model:

$$x_k = \sum_j a_{kj} x_j + c_k, \qquad j, k = 1, 2, \ldots, n \qquad (18.10)$$

The above equation can be written in matrix notation as follows:

$$\mathbf{x} = A\mathbf{x} + \mathbf{c} \qquad (18.11)$$

where $\mathbf{x} = [x_1, x_2, \ldots, x_n]^T$, $\mathbf{c} = [c_1, c_2, \ldots, c_n]^T$, $\mathbf{r} = [r_1, r_2, \ldots, r_m]^T$, $[.]^T =$ a column vector, and $A = [a_{kj}]$ $n \times n$ matrix.

Defining $I = n \times n$ identity matrix and assuming that $(I - A)$ is nonsingular, the vector of inoperability $\mathbf{x}$ in Eq. (18.11) can be solved using the following matrix operation:

$$\mathbf{x} = (I - A)^{-1} \mathbf{c} \qquad (18.12)$$

Determining the values of the **A** matrix during the modeling process is a very challenging undertaking, and extensive data collection and data mining may be required to complete this step. The following are general guiding principles for determining the **A** matrix:

- Explore the potential use of publicly available I–O tables to enable understanding the transactions among various sectors in the economy (e.g., see US DOC [1998] and Kuhbach and Planting [2001]). Prior to conducting actual field surveys and interviews, these I–O tables can provide valuable insights into the interdependencies among various infrastructures.
- Define the level of resolution and the boundary conditions of each infrastructure, because a system may be analyzed at different levels of resolution. Note, however, that the level of resolution adopted in the analysis must be harmonious with the accuracy of the data

and the analytical tractability (including the determination of the I–O relationships). The realism that exists at high granularity should not be sacrificed in the process of aggregation.

- Identify physical connections among the infrastructures. In general, if there are no physical connections between infrastructures i and j, then $a_{ij} = a_{ji} = 0$. Physical boundary conditions are very critical in identifying the physical connections among different infrastructures.
- If there are any deterministic correlations among any infrastructures, then these relationships should be singled out first. For instance, if the failure of infrastructure i will definitely lead to failure of infrastructure j, then $a_{ji} = 1$. By the same token, if the failure of infrastructure i will definitely lead to failure of one of the two subsystems of infrastructure j, which performs 50% of that infrastructure's functions, then $a_{ji} = 0.5$.
- If the correlation between two infrastructures (e.g., infrastructures i and j) is of a stochastic nature, then all conceivable scenarios must be analyzed and a statistical average has to be taken to obtain a_{ij} and a_{ji}. (For example, if the failure of infrastructure i leads, with probability 0.3, to complete failure of infrastructure j, and with probability 0.7, leads infrastructure j to be 50% inoperable, then $a_{ji} = (0.3)(1) + (0.7)(0.5) = 0.65$. If the real data are not sufficient, a simulation may be helpful in order to obtain data for the probability distributions.

18.4 REGIMES OF RECOVERY

Several time frames, or regimes, exhibit different features of interdependencies following an attack or other extreme events affecting infrastructure. The nature and extent of sector interactions will vary from one time frame to the next. Moreover, the metrics of outcomes will be allowed to vary from time frame to time frame [Lambert and Patterson, 2002; Tsang et al., 2002]. Within each time frame, the Inoperability Input–Output risk model can describe a conceptual situation of equilibrium. Before equilibrium is reached, the system will have evolved to a distinct and new frame of interactions. A sample of several time frames that will be addressed by the IIM is presented in Figure 18.1. Further uses of the regimes include comparing the physical versus psychological effects of an attack. While the physical-based Inoperability Input–Output risk model analyzes the physical losses caused by either natural or human-caused disasters, it is important to consider psychological factors as well. A comprehensive survey of the psychological effects of various types of disasters is documented in Norris et al. [2002]. Specific empirical studies such as those by Susser et al. [2002] and Galea et al. [2002] show the significance of the *fear factor* induced by the September 11, 2001 (9/11) terrorist attacks. Fear can cause the public to reduce their demand for the goods and services produced by an attacked industry.

For example, public apprehension after 9/11 about the safety of air transportation caused a

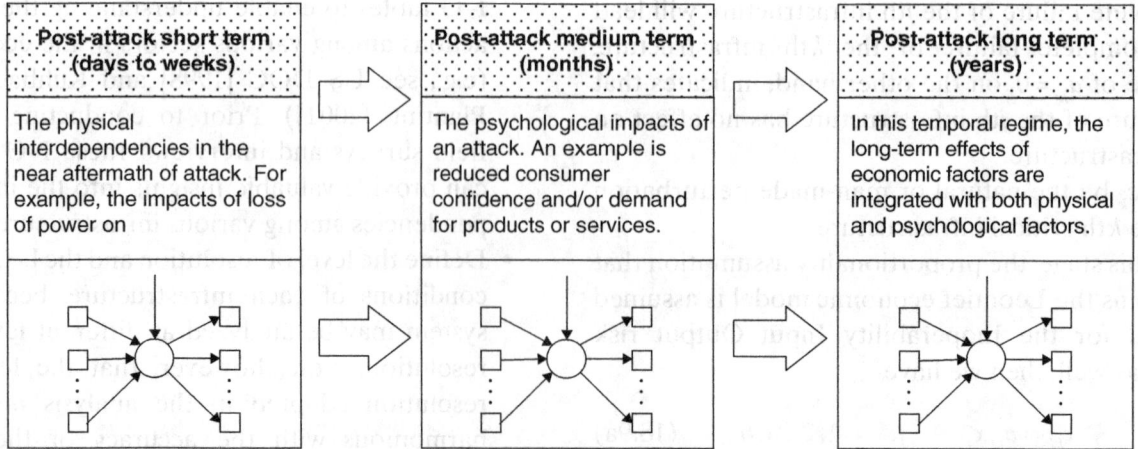

Figure 18.1 Three temporal regimes of recovery that are considered in IIM analysis of attack impacts.

drastic reduction in the operations of the airlines and other airline-dependent industries. These retrenchments and changes in demand can have large economic repercussions that compound the physical losses (e.g., degraded production capacity). Both physical and psychological considerations ought to be accounted for in analyzing the long-term adverse economic impacts on the *as-planned* operation levels of interconnected sectors.

18.5 SUPPORTING DATABASES FOR IIM ANALYSIS

An advantage of building on the Leontief I–O model is that it is supported by major ongoing data collection efforts. These available databases of interdependency statistics provide an essential foundation for applying the IIM to model a terrorist attack. In this section, we review two main data resources: (i) the Bureau of Economic Analysis (BEA) database of national I–O accounts and (ii) the Regional Input–Output Multiplier System (RIMS II) accounts. The BEA database provides an overview of the national economic I–O accounts; this is a series of tables depicting the production and consumption of commodities (i.e., goods and services) by various sectors in the US economy. The BEA consumption and production tables are combined to calculate the Leontief technical coefficient matrix for nearly 500 industry sectors of the US economy and their corresponding interdependencies with the workforce sector. RIMS II is a set of regional data maintained by the BEA, Regional Economic Analysis Division. Empirical tests suggest that regional multipliers can be used as surrogates for time-consuming and expensive surveys without compromising accuracy.

Utilizing the BEA database [US DOC, 1998], the demand reduction Inoperability Input–Output Model (demand reduction IIM) complements and supplements the physical-based model developed by Haimes and Jiang [2001]. While this new model quantifies inoperability in terms of degraded capacity to deliver the intended outputs, the demand-based model addresses the demand reductions that can potentially stem from perturbations [Santos, 2003; Santos and Haimes, 2004]. Logically, the demand reduction of a perturbed sector produces further adverse impacts on the operations of other dependent sectors. For example, the demand reduction of the airline industry—an industry primarily affected by the 9/11 terrorism—caused the demand for other dependent industries to decline as well (e.g., travel and hotel industries). Specifically, a 33.2% reduction in passenger enplanements [FAA, 2002] and a 19.2% reduction in hotel occupancy [Ernst and Young, 2002] were realized in the aftermath of 9/11, relative to 2000 figures. Integrating the concept of inoperability into the Leontief economic I–O model makes it possible to analyze how demand reduction inoperability affects other interdependent infrastructures.

Two motivations have driven the use of an economic model to study physical interactions. One deals with the general issue of translating physical to economic values, while the other accounts for the effects of perturbations (e.g., terrorist attacks) on power sources as well as on equipment operated by the using sectors (e.g., computers, control systems). An assumption made when applying the IIM is that the level of economic dependency constitutes a surrogate measure of the level of physical dependency. That is, it is assumed that two companies with a large amount of economic interaction will have an approximately similar high level of physical interdependency. However crude this assumption may be, it is founded on BEA data that reflects real physical interactions between economic sectors. These are commensurated into dollar units by multiplying interactions of physical quantities by producers' prices. In turn, these prices indicate how a sector values the physical interdependencies. However, when compared with the availability of economic data from the BEA, the corresponding lack of data on physical interdependencies, and the extraordinary cost required to collect such information on the scale of the economic data collections, the degree of inaccuracy in IIM results becomes a question. A case study discussed in Haimes et al. [2005b] determines the rank order of interdependent sectors, the loss in their production capacities, and the corresponding economic impact. This can be used to determine the size of the risk and where to invest to reduce it. One possible way to add confidence in the results is to carry out a study of the top sectors resulting from an IIM analysis to determine how close the physical ties are relative to economic ties. Such a study might be bounded enough to be carried out at an acceptable cost when compared with costs of poor risk management and could result in modifying prioritizations.

When applying the IIM to a potential terrorist attack, the BEA's data can be used to determine the expenditures of all economic sectors on items that use electricity (i.e., how much a sector spends on computers and other electrical equipment). Using the percentage of each sector's total resources that are spent on electrical equipment to estimate the production-focused level of dependence on electric power, we can estimate the percentage loss in production level that each sector would suffer due to its own electrical devices failing. This permits us to create an input vector for inoperability that includes not only the unavailability of power sources but the production losses of power-dependent sectors even with power restored (e.g., due to dysfunctional equipment). However limited it may be by substituting economic for physical data, this use of the IIM provides a direct approach for understanding interdependencies.

18.6 NATIONAL AND REGIONAL DATABASES FOR IIM ANALYSIS

18.6.1 BEA Database

The US BEA publishes the national economic I–O accounts, which are a series of tables depicting the production and consumption of commodities (i.e., goods and services) of various sectors in the US economy. The detailed national tables are composed of hundreds of industries, organized according to the North American Industry Classification System (NAICS) codes.

In the original Leontief model formulation, each industry is assumed to produce a distinct commodity. The term *commodity* here refers to the output of an industry, which can be in the form of goods or services. Realistically, however, it is possible that a given industry can produce more than one commodity. On the other hand, a given commodity may not be a unique output of a given industry. The BEA recognizes that the one-to-one correspondence assumption between an industry and commodity is generally not true. The BEA makes a distinction between an industry and a commodity in its published I–O data via the *industry-by-commodity* and *commodity-by-industry* matrices. Figure 18.2, adapted from Miller and Blair [1985], shows a summary of the types of national I–O accounts maintained by the BEA.

The *make* matrix in Figure 18.2, denoted by **V**, would show the monetary values of the different column commodities *produced* by the different row industries. A sample of *make* matrix data is shown in Table 18.1. The *use* matrix on the other hand, denoted by **U**, would show the monetary values of the different row commodities *consumed* by the different column industries. A sample of *use* matrix data is shown in Table 18.2. Note that Figure 18.2 does not directly specify the I–O matrix representing the industry-by-industry transactions. This matrix, denoted by **A** in the Leontief formulation, is called the industry-by-industry technical coefficient matrix in Leontief parlance. It would show the input of industry i to j, expressed as a proportion of the total production inputs to industry j. The BEA does not publish the elements of the **A** matrix because this task is left to the analyst. Typically, the **A** matrix is established from the *make* and *use* matrices using various assumptions (e.g., commodity-technology assumption (CTA) and industry-technology assumption (ITA); see Guo et al. [2002]). One approach is carried out by first normalizing the values of the *make* and *use* matrices. The following sections discuss the operations for deriving the *normalized make* matrix ($\hat{\mathbf{V}}$) from the *make* matrix (**V**) and the *normalized use* matrix ($\hat{\mathbf{U}}$) from the *use* matrix (**U**).

18.6.2 Coefficients of Production in National and Regional Economies

The *make* matrix (**V**) in BEA I–O reports shows the itemized production of commodities by various industries. Each element of the *make* matrix (v_{ij}) shows industry i's production of commodity j (typically measured in millions of dollars). If there are m commodities and n industries, then the total

	Commodity	Industry		
Commodity		Use matrix (U)	Exogenous Demand (e)	Total commodity output (y)
Industry	Make matrix (V)			Total industry output (x)
		Value added (w)		
	Total commodity input (y^T)	Total industry input (x^T)		

Figure 18.2 Summary of economic input–output accounts.

TABLE 18.1 Sample *Make* Matrix for 1992 US Economy

Industry (SIC Code)	Commodity Output (SIC Code)	Value (in Million $)
1.0100		20,285
	1.0100	19,646
	4.0001	86
	14.0600	365
	76.0206	188

Excerpt from US Department of Commerce [1998, p. 47].

TABLE 18.2 Sample *Use* Matrix for 1992 US Economy

Commodity	Using Industry	Value (in Million $)
2.0502		2,162
	2.0502	55
	14.1900	2,099
	93.0000	4
	94.0000	4

Excerpt from US Department of Commerce [1998, p. 83].

industry output for the ith industry (x_i) must follow the balance equation below (see Figure 18.2):

$$x_i = v_{i1} + v_{i2} + \cdots + v_{im} = \sum_{j \leq m} v_{ij}; \quad \forall i = 1,2,\ldots,n \quad (18.13)$$

Denoting **x** as the vector of total industry outputs, Σ as a unity vector (i.e., a vector whose elements are all 1's, also known as a summation vector), and **V** as the make matrix, it can be shown that Eq. (18.13) can be written in the following matrix form:

$$\mathbf{x} = \mathbf{V}\Sigma \quad (18.14)$$

Due to the volume of data, the BEA does not present the *make* matrix in the format of v_{ij} (i.e., with the industries arranged along the rows and commodities along the columns). Rather, referring to Table 18.1, one industry is given at a time (see first column), the second column enumerates the commodities produced by that industry, and the third column gives the value of those commodities. For example, the dairy farm products *industry* in Table 18.1 (1.0100) produces $19,646 million worth of the dairy farm products *commodity* (1.0100); $86M worth of the agricultural, forestry, and fishery services *commodity* (4.0001); $365M worth of the fluid milk *commodity* (14.0600); and $188M worth of the other amusement and recreation services *commodity* (76.0206).

Equation (18.19) shows the formulation for the *normalized make* matrix $\hat{\mathbf{V}} = [\hat{v}_{ij}]$. It is an industry-by-commodity matrix because it shows the industries along the rows and the commodities along the columns. To better understand how Eq. (18.19) is derived, we dissect the elements of the underlying *make* matrix (**V**) and the total commodity output vector ($\mathbf{y}^T$) as follows (see Figure 18.2):

$$V = \begin{vmatrix} \vdots & \vdots & \vdots \\ v_{i1} & \cdots & v_{ij} & \cdots & v_{im} \\ \vdots & \vdots & \vdots \end{vmatrix} \quad (18.15)$$

$$y^T = \left[y_1 = \sum_i v_{i1} \quad \cdots \quad y_j = \sum_i v_{ij} \quad \cdots \quad y_m = \sum_i v_{im} \right] \quad (18.16)$$

The *normalized make* matrix, whose elements are denoted by $\hat{v}_{ij}$, can be obtained by dividing each element of the make matrix (v_{ij}) by the respective column sum (y_j) as follows:

$$\hat{V} = \begin{vmatrix} \vdots & \vdots & \vdots \\ v_{i1}/y_1 & v_{ij}/y_j & v_{im}/y_m \\ \vdots & \vdots & \vdots \end{vmatrix} \quad (18.17)$$

As Eq. (18.18) shows, Eq. (18.17) can be written in a compact matrix notation by first denoting the operator diag (θ) as the resulting diagonal matrix constructed from a given vector θ. (Note that this notation will also be used later.)

$$\text{diag}(\boldsymbol{\theta}) = \text{diag} \begin{vmatrix} \theta_1 \\ \theta_2 \\ \vdots \\ \theta_m \end{vmatrix} = \begin{vmatrix} \theta_1 & 0 & \cdots & 0 \\ 0 & \theta_2 & \ddots & \vdots \\ \vdots & \ddots & \ddots & 0 \\ 0 & \cdots & 0 & \theta_m \end{vmatrix} \quad (18.18)$$

Thus, from Eqs. (18.17) and (18.18),

$$\hat{V} = V[\text{diag}(y)]^{-1} \Leftrightarrow \left\{ \hat{v}_{ij} = \frac{v_{ij}}{y_j} \right\} \forall i, j \quad (18.19)$$

18.6.3 Coefficients of Consumption in National and Regional Economies

The *use* matrix (**U**) in BEA I–O reports shows the itemized consumption of commodities by various industries. Each element of the *use* matrix (u_{ij}) shows industry *j*'s consumption of the *i*th commodity (typically measured in millions of dollars). Suppose there are *m* commodities and *n* industries. Denoting e_i as exogenous consumptions for commodity *i* (or final commodity demands), the total commodity output for the *i*th commodity (y_i) must follow balance Eq. (18.20). (The notation **c** or c_i throughout this chapter refers to final *industry* demand. It should be distinguished from **e** or e_i, which refers to the exogenous or final *commodity* demand.) (See Figure 18.2.)

$$y_i = u_{i1} + u_{i2} + \cdots + u_{im} + e_i = \sum_{j \le n} u_{ij} + e_i; \quad \forall i = 1, 2, \ldots, m \quad (18.20)$$

Denoting the total commodity output vector by **y**, a summation vector by Σ, and the *use* matrix by **U**, Eq. (18.20) can be written in the following matrix notation. (The notation **x** or x_i throughout this chapter refers to total *industry* output. It should be distinguished from **y** or y_i, which refers to the total *commodity* output.)

$$\mathbf{y} = \mathbf{U}\Sigma + \mathbf{e} \quad (18.21)$$

Sample data from the *use* matrix is depicted in Table 18.2. Due to the volume of data, the BEA does not present the *use* matrix in the format of u_{ij} (i.e., the commodities arranged along the rows and industries along the columns). Rather, one commodity is listed at a time (see the first column of Table 18.2), the second column enumerates the industries that use that commodity, and the third column gives the amount of that commodity used by the industries. For example, the usage of the sugar crops *commodity* in Table 18.2 (2.0502) is as follows: $55 million by the sugar crops *industry* (2.0502), $2,099M by the sugar *industry* (14.1900), $4M as change in business inventories (93.0000), and $4M as exports of goods and services (94.0000). Note that the last two codes, 93.0000 and 94.0000, are not industries per se. Rather, they are the final commodity consumptions (e_i) in the balance equation 18.20.

Equation 18.25 shows the formulation for the normalized *use* matrix $\hat{\mathbf{U}} = [\hat{u}_{ij}]$. This is a commodity-by-industry matrix because it shows the commodities along the rows and the industries along the columns. To better understand how Eq. 18.25 is derived, we dissect the elements of the underlying *use* matrix (**U**) and the total industry output vector (**x**) as follows (see Figure 18.2):

$$\mathbf{U} = \begin{bmatrix} u_{11} & \cdots & u_{1j} & \cdots & u_{1n} \\ \vdots & & \vdots & & \vdots \\ u_{i1} & \cdots & u_{ij} & \cdots & u_{in} \\ \vdots & & \vdots & & \vdots \\ u_{m1} & \cdots & u_{mj} & \cdots & u_{mn} \end{bmatrix} \quad (18.22)$$

$$\mathbf{x} = \begin{bmatrix} x_1 = \sum_j v_{1j} \\ \vdots \\ x_i = \sum_j v_{ij} \\ \vdots \\ x_n = \sum_j v_{nj} \end{bmatrix} \Leftrightarrow \mathbf{x}^T = \begin{bmatrix} x_1 & \cdots & x_j & \cdots & x_n \end{bmatrix} \quad (18.23)$$

The *normalized use* matrix, whose elements are denoted by $\hat{u}_{ij}$, can be obtained by dividing each element of the *use* matrix (u_{ij}) by its respective column sum, which happens to be x_j (see Figure 18.2). The *normalized use* matrix is represented by the following matrix notations:

$$\hat{\mathbf{U}} = \begin{bmatrix} u_{11}/x_1 & \cdots & u_{1j}/x_j & \cdots & u_{1n}/x_n \\ \vdots & & \vdots & & \vdots \\ u_{i1}/x_1 & \cdots & u_{ij}/x_j & \cdots & u_{in}/x_n \\ \vdots & & \vdots & & \vdots \\ u_{m1}/x_1 & \cdots & u_{mj}/x_j & \cdots & u_{mn}/x_n \end{bmatrix}$$

$$(18.24)$$

Thus,

$$\hat{\mathbf{U}} = \mathbf{U}[\text{diag}(\mathbf{x})]^{-1} \Leftrightarrow \left\{ \hat{u}_{ij} = \frac{u_{ij}}{x_j} \right\} \forall i,j \quad (18.25)$$

18.6.4 Technical Coefficient Matrix

The *technical coefficient* matrix, denoted by $\mathbf{A}$, has industries along the rows as well as the columns. It can be shown that $\mathbf{A}$ is the product of the *normalized make* and the *normalized use* matrices:

$$\mathbf{A} = \hat{\mathbf{V}}\hat{\mathbf{U}} \Leftrightarrow \left\{ a_{ij} = \sum_k \hat{v}_{ik}\hat{u}_{kj} \right\} \forall i,j \quad (18.26)$$

On the other hand, the vector of industry final demands (**c**) can be shown to be the product of the *normalized make* matrix and the *exogenous commodity demand* vector:

$$\mathbf{c} = \hat{\mathbf{V}}\mathbf{e} \Leftrightarrow \left\{ c_i = \sum_k \hat{v}_{ik} e_k \right\} \forall i \quad (18.27)$$

Deriving Eqs. (18.26) and (18.27) involves the following steps:
Substituting Eq. (18.19) for Eq. $\mathbf{x} = \mathbf{V}\Sigma$ (18.14), we have

$$\mathbf{x} = \mathbf{V}\Sigma = (\hat{\mathbf{V}}\text{diag}(\mathbf{y}))\Sigma \quad (18.28)$$

Equation (18.28) can be simplified further by using the fact that $\text{diag}(\mathbf{y})\Sigma = \mathbf{y}$

$$\mathbf{x} = \hat{\mathbf{V}}\mathbf{y} \quad (18.29)$$

Similarly, we substitute Eq. (18.25) for Eq. (18.21) to form the following equation:

$$\mathbf{y} = (\hat{\mathbf{U}}\text{diag}(\mathbf{x}))\Sigma + \mathbf{e} \quad (18.30)$$

Equation (18.30) can be simplified further by using the fact that $\text{diag}(\mathbf{x})\Sigma = \mathbf{x}$

$$\mathbf{y} = \hat{\mathbf{U}}\mathbf{x} + \mathbf{e} \quad (18.31)$$

Premultiply $\hat{\mathbf{V}}$ to Eq. (18.31):

$$\hat{\mathbf{V}}\mathbf{y} = \hat{\mathbf{V}}\hat{\mathbf{U}}\mathbf{x} + \hat{\mathbf{V}}\mathbf{e} \quad (18.32)$$

Substitute Eq. (18.29) for Eq. (18.32):

$$\mathbf{x} = \hat{\mathbf{V}}\hat{\mathbf{U}}\mathbf{x} + \hat{\mathbf{V}}\mathbf{e} \quad (18.33)$$

For Eq. (18.33) to become equivalent to the usual Leontief balance equation (18.8), then Eqs. (18.26) and (18.27) must be true. Thus, we have shown that the Leontief industry-by-industry coefficient matrix (**A**) can be calculated on the bases of the *normalized make* and *normalized use* matrices as described in Eq. (18.26). In addition, the industry final demand can be constructed from the exogenous commodity demand by premultiplying it by the *normalized make* matrix, as described in Eq. (18.27).

18.6.5 Relevant Data for Workforce Sector Vulnerability Analysis

We have added a new *Workforce* row and column to the original national technical coefficient matrix (**A**). By extracting the household portion of the exogenous demand (measured in terms of personal consumption expenditures) and the household portion of the value added (measured in terms of personnel compensations), we were able to generate an updated **A** matrix. This integrates information on additional interdependency impacts contributed by the household sector. The extraction of household portions from exogenous demand and value-added vectors is described in Figure 18.3. The household sector, a standard BEA sector classification, is the source of labor inputs in various sectors of the economy. Thus, from here on, we refer to it as the *workforce sector*.

18.7 RIMS II

Regional decomposition enables a more focused and thus more accurate analysis of interdependencies for regions of interest in the United States. Miller et al. [1989] and Lahr and Dietzenbacher [2001] discuss the validity of *closing* the I–O analysis to a particular region (i.e., a single regional I–O framework as opposed to multiregional) since interregional feedbacks empirically are found to be *small*. The RIMS II division of the US DOC is responsible for releasing multipliers for various regions in the United States. Empirical tests suggest that regional multipliers can be used as surrogates for time-consuming and expensive surveys without compromising accuracy [Brucker et al., 1990]. With the availability of national I–O tables and location quotients [US DOC, 1997, 1998], analysts can convert and customize the national data according to the region of interest.

The RIMS II utilizes location quotients derived from *personal income data* and *wage-and-salary data* to regionalize the national Leontief technical coefficient matrix (i.e., the **A** matrix). A location quotient indicates how well an industry's production capacity satisfies the regional local demand. In addition, as the value of an industry's location quotient tends to 1, its relative concentration in the region approaches that of the national level:

$$l_i = \frac{\hat{x}_i^R / \hat{x}_s^R}{\hat{x}_i / \hat{x}_s} \qquad (18.34)$$

where:

- $\hat{x}_i^R$ is the regional output for the ith industry
- $\hat{x}_s^R$ is the total regional output for all region-level industries
- $\hat{x}_i$ is the national output for the ith industry
- $\hat{x}_s$ is the total national output for all national-level industries

	Commodity	Sector			
Commodity		Use matrix (**U**)	Workforce (**e₁**)	Exogenous demand (**e₂**)	Total commodity output (**y**)
Sector	Make matrix (**V**)				Total sector output (**x**)
		Workforce ($\mathbf{z_1}^T$)			
		Value added ($\mathbf{z_2}^T$)			
	Total commodity input ($\mathbf{y}^T$)	Total sector input ($\mathbf{x}^T$)			

Figure 18.3 Economic input–output accounts reconfigured for workforce analysis.

The regional industry-by-industry technical coefficient matrix $\mathbf{A}^R$, whose elements are denoted by a_{ij}^R, is then established as follows:

$$a_{ij}^R = \begin{cases} (a_{ij})(l_i) & l_i < 1 \\ a_{ij} & l_i \geq 1 \end{cases} \quad (18.35)$$

When $\mathbf{l}$ is used to denote a vector of location quotients and Σ a unity vector, Eq. (18.35) can be written in the following matrix notation:

$$\mathbf{A}^R = \text{diag}[\text{Min}(\mathbf{l},\Sigma)]\mathbf{A} \Leftrightarrow \left\{ a_{ij}^R = \text{Min}(l_i,1)a_{ij} \right\} \quad \forall i,j \quad (18.36)$$

RIMS II issues a series of multipliers for various sectors of a specified region, generated via the region's location quotients (see Eq. (18.34)). Some examples are as follows:

- *Output multiplier* gives the change in the *production output* of a sector resulting from a $1 change in the demand for another sector's output.
- *Earnings multiplier* gives the change in the *workforce earnings* of a sector resulting from a $1 change in the demand for another sector's output.
- *Employment multiplier* gives the change in the *number of workers* of a sector resulting from a $1M change in the demand for another sector's output.

RIMS II multipliers are presented in the form of 38×490 matrices. The columns in Figure 18.4 represent detailed sectors (e.g., column 420 (C420), electric services/utilities). On the other hand, the rows in the matrix of RIMS II multipliers represent an aggregation of several column sectors (e.g., R26 (electric, gas, and sanitary services)). Thus, this specific row corresponds to the aggregated version of C420–C424, which includes C421, natural gas transportation; C422, natural gas distribution; and so on.

An extreme event such as a terrorist attack degrades the capability of a sector to supply its *as-planned* level of output. A sector's supply reduction necessarily leads to demand reduction (e.g., consumption adjusts when available supply is below the *as-planned* demand level). The RIMS II multipliers can be utilized for predicting the impact of reduced demand or supply on various interconnected sectors of a region due to extreme events.

18.8 DEVELOPMENT OF THE IIM AND ITS EXTENSIONS

18.8.1 Physical-Based IIM

A first-generation physical-based Inoperability Input–Output Model (or physical IIM, for simplicity) was developed by Haimes and Jiang [2001], Jiang [2003], and Jiang and Haimes [2004] to

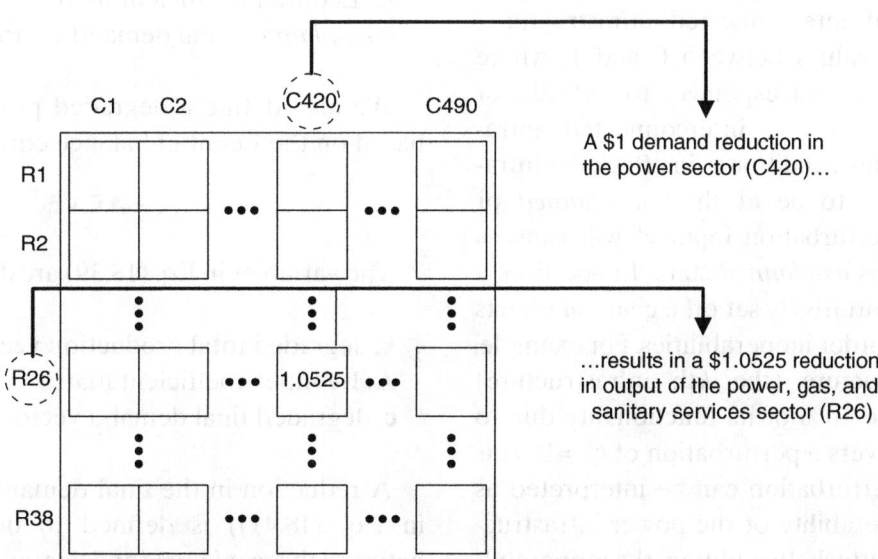

Figure 18.4 Sample interpretation of RIMS II multipliers.

describe how the impact of willful attacks can cascade through a system of interconnected infrastructures. Inoperability connotes degradation in the system's functionality (expressed as a percentage relative to the intended state of the system). The formulation of the physical-based model is as follows:

$$x_i^P = \sum_j a_{ij}^P x_j^P + c_i^P \Leftrightarrow \mathbf{x}^P = \mathbf{A}^P \mathbf{x}^P + \mathbf{c}^P \quad (18.37)$$

Haimes and Jiang [2001] added the superscript P in Eq. (18.37) to the original formulation to distinguish it from the Leontief model. Although the mathematical construct of the two models is similar, the interpretation of the model parameters is fundamentally different. The *supply* and *demand* concepts in the Leontief economic model now assume different interpretations and have been inverted to some extent in the physical IIM. In the Leontief model, **c** and **x** represent commodities typically measured in production or monetary units. In the physical-based model, the vector $\mathbf{c}^P$ represents the input to the interconnected infrastructures—perturbations in the form of natural events, accidents, or willful attacks. The output is defined as the resulting vector of inoperability of the different infrastructures, denoted by $\mathbf{x}^P$, due to their connections to the perturbed infrastructure and to one another. The long-run inoperabilities of the interconnected infrastructures following an attack can be calculated using Eq. (18.37).

The inoperability vector $\mathbf{x}^P$ describes the degree of functionality of interconnected infrastructures. Thus, it takes on values between 0 and 1, where flawless operation corresponds to $\mathbf{x}^P = 0$ or $x_1^P = x_2^P = \cdots = x_n^P = 0$ for n interconnected infrastructures. When this condition is in effect, the infrastructures are said to be at their *as-planned* or *ground state*. A perturbation input $\mathbf{c}^P$ will cause a departure from this *as-planned* state. In addition, a perturbation can intuitively set off a chain of effects leading to higher-order inoperabilities. For example, a power infrastructure (the kth infrastructure) would initially lose 10% of its functionality due to an attack that delivers a perturbation of $c_k^P = 0$. This means that the perturbation can be interpreted as the resulting inoperability of the power infrastructure *right after* an attack. In addition, the inoperability propagated by the power infrastructure to other power-dependent infrastructures will in turn cause more inoperabilities and ultimately may cause additional inoperability in the power infrastructure itself. In general, we expect the long-run inoperability of an attacked infrastructure to increase from its postattack value (i.e., the perturbation).

18.8.2 Demand Reduction IIM

The demand reduction Inoperability Input–Output Model is derived by combining the insight and intuition gained from the physical I–O model with the rigor and proven BEA databases that accompany the original Leontief model. The BEA data is a record of the physical exchange of commodities between various interconnected industrial sectors of the economy that have been scaled by producers' prices into one common unit of dollars. Therefore, this will be the foundation for our measure of interdependency.

Using the definition of normalized production loss, we derive the demand-based model on the basis of the Leontief model. We first define an *as-planned* production scenario based on the Leontief balance equation:

$$\hat{\mathbf{x}} = \mathbf{A}\hat{\mathbf{x}} + \hat{\mathbf{c}} \quad (18.38)$$

The variables in Eq. (18.38) are defined as follows:

$\hat{\mathbf{x}}$: *as-planned* total production vector
$\mathbf{A}$: Leontief coefficient matrix
$\hat{\mathbf{c}}$: *as-planned* final demand vector

We also define a degraded production scenario based on the Leontief balance equation:

$$\tilde{\mathbf{x}} = \mathbf{A}\tilde{\mathbf{x}} + \tilde{\mathbf{c}} \quad (18.39)$$

The variables in Eq. (18.39) are defined as follows:

$\tilde{\mathbf{x}}$: degraded total production vector
$\mathbf{A}$: Leontief coefficient matrix
$\tilde{\mathbf{c}}$: degraded final demand vector

A reduction in the final demand (denoted by $\delta\mathbf{c}$ in Eq. (18.41)) is defined to be the difference between the *as-planned* and degraded final demands. This reduction in final demand consequently triggers a reduction in production (denoted by $\delta \mathbf{x}$ in Eq. (18.40)), which is defined to be the difference between the *as-planned* and degraded productions:

$$\delta \mathbf{x} = \hat{\mathbf{x}} - \tilde{\mathbf{x}} \quad (18.40)$$

$$\delta \mathbf{c} = \hat{\mathbf{c}} - \tilde{\mathbf{c}} \quad (18.41)$$

Subtracting Eq. (18.39) from Eq. (18.38) will result in the following relationship between $\delta \mathbf{x}$ and $\delta \mathbf{c}$:

$$(\hat{\mathbf{x}} - \tilde{\mathbf{x}}) = \mathbf{A}(\hat{\mathbf{x}} - \tilde{\mathbf{x}}) + (\hat{\mathbf{c}} - \tilde{\mathbf{c}}) \Leftrightarrow \delta \mathbf{x} = \mathbf{A}\delta \mathbf{x} + \delta \mathbf{c} \quad (18.42)$$

The transformations in Eqs. (18.43), (18.44), and (18.45) are needed to derive the demand-based model in a form analogous to the balance equation of the Leontief model:

$$\mathbf{c}^* = [(\text{diag}(\mathbf{x}))^{-1} \delta \mathbf{c}] \quad (18.43)$$

$$\mathbf{A}^* = [(\text{diag}(\hat{\mathbf{x}}))^{-1} \mathbf{A}(\text{diag}(\hat{\mathbf{x}}))] \quad (18.44)$$

$$\mathbf{q} = [(\text{diag}(\hat{\mathbf{x}}))^{-1} \delta \mathbf{x}] \quad (18.45)$$

Define the transformation matrix:

$$\mathbf{P} = [\text{diag}(\hat{\mathbf{x}})]^{-1} \quad (18.46)$$

Using the transformation matrix in Eq. (18.46), Eq. (18.42) becomes Eq. (18.48) by the transformation defined in Eq. (18.47):

$$[\mathbf{P}\delta \mathbf{x}] = [\mathbf{P}\mathbf{A}\mathbf{P}^{-1}][\mathbf{P}\delta \mathbf{x}] + [\mathbf{P}\delta \mathbf{c}] \quad (18.47)$$

$$\mathbf{q} = \mathbf{A}^*\mathbf{q} + \mathbf{c}^* \quad (18.48)$$

Assuming that the demand-based interdependency matrix $\mathbf{A}^*$ is nonsingular and stable, the demand-based inoperability q can be calculated as follows:

$$\mathbf{q} = [\mathbf{I} - \mathbf{A}^*]^{-1} \mathbf{c}^* \quad (18.49)$$

18.8.3 Regional IIM

At the national level, the derived form of the demand reduction IIM is $\mathbf{q} = \mathbf{A}^*\mathbf{q} + \mathbf{c}^*$. The regional model takes a similar form:

$$\mathbf{q}^R = \mathbf{A}^{*R}\mathbf{q}^R + \mathbf{c}^{*R} \quad (18.50)$$

The system of equations corresponding to Eq. (18.50) is as follows:

$$q_1^R = a_{11}^{*R} q_1^R + a_{12}^{*R} q_2^R + \cdots + a_{1n}^{*R} q_n^R + c_1^{*R}$$
$$q_2^R = a_{21}^{*R} q_1^R + a_{22}^{*R} q_2^R + \cdots + a_{2n}^{*R} q_n^R + c_2^{*R}$$
$$\vdots$$
$$q_n^R = a_{n1}^{*R} q_1^R + a_{n2}^{*R} q_2^R + \cdots + a_{nn}^{*R} q_n^R + c_n^{*R} \quad (18.51)$$

The term a_{ij}^{*R} in Eq. (18.51) can be expressed in terms of the regional technical a_{ij}^R coefficient using the identity shown in Eq. (18.52). This identity is analogous to the corresponding national-level formula $\mathbf{A}^* = [(\text{diag}(\hat{\mathbf{x}}))^{-1} \mathbf{A}(\text{diag}(\hat{\mathbf{x}}))]$:

$$\mathbf{A}^{*R} = [\text{diag}(\hat{\mathbf{x}}^R)]^{-1} \mathbf{A}^R [\text{diag}(\hat{\mathbf{x}}^R)]$$
$$\Leftrightarrow \left\{ a_{ij}^{*R} = a_{ij}^R \left(\frac{\hat{x}_j^R}{\hat{x}_i^R} \right) \right\} \forall i,j \quad (18.52)$$

Now, we express the regional industry-by-industry technical coefficient matrix ($\mathbf{A}^R$) in terms of the counterpart national matrix ($\mathbf{A}$). The resulting $\mathbf{A}^R$ matrix in Eq. (18.53) is obtained by substituting Eq. (18.36) for Eq. (18.52). Thus, the regional interdependency matrix $\mathbf{A}^{*R}$ can be established on the bases of the location quotients, the national industry-by-industry technical coefficients, and the *as-planned* production outputs of the regional industries:

$$\mathbf{A}^{*R} = [\text{diag}(\hat{\mathbf{x}}^R)]^{-1} [\text{diag}[\text{Min}(\mathbf{l})] \mathbf{A}] [\text{diag}(\hat{\mathbf{x}}^R)]$$
$$\Leftrightarrow \left\{ a_{ij}^{*R} = \text{Min}(l_i, 1) a_{ij} \left(\frac{\hat{x}_j^R}{\hat{x}_i^R} \right) \right\} \forall i,j \quad (18.53)$$

18.8.4 Multiregional IIM

The Regional IIM can be interconnected to develop a multiregional version that improves spatial explicitness, model flexibility, and analysis coverage. The construction of the Multiregional IIM (MRIIM) builds on the regionalized IIM from the previous section by accounting for cross-regional flows of goods and services that interconnect regions [Crowther, 2007]. Accounting for cross-regional flows enables calculating multiregional coefficients, which in turn adjust the intraregional interdependency matrices $\mathbf{A}^*$. A spatially explicit interdependency matrix can be formed as a block diagonal matrix in Eq. (18.54), where $\mathbf{A}^s$ is a matrix containing all intraregional technical coefficients for region s calculated above:

$$\mathbf{A} = \begin{bmatrix} \mathbf{A}^1 & & & \\ & \mathbf{A}^2 & & \\ & & \ddots & \\ & & & \mathbf{A}^p \end{bmatrix} \quad (18.54)$$

Multiregional coefficients are calculated using commodity and service flow data. These coefficients describe the way that multiple regions are interconnected as larger regional systems, due to their economic transactions of goods and services across geographical areas. To decisionmakers in *large regional*, these coefficients provide a measure of economic intraconnections across smaller (sub) regions that can result in either cascades of impacts or sources of resilience following a disaster scenario. To the decisionmakers in smaller (sub)regions, they systemically provide (i) a demand *footprint* describing other regions from which they purchase goods and services and (ii) a supply *footprint* describing other regions to which they deliver goods and services. Such decisionmakers can adapt strategic preparedness to mitigate risks against disaster scenarios that produce (i) supply perturbations in their demand footprint and (ii) demand perturbations in their supply footprint. We will henceforth refer to commodities and services as commodities. Let z_i^{rs} be the value of commodity i produced in region r and consumed in region s. For each commodity, we form an origin–destination matrix similar to the matrix in Table 18.3.

TABLE 18.3 Multiregion *Origin–Destination* Table for Commodity *i*

		\multicolumn{4}{c}{Region of Destination}
		1	2	…	p
Region of origin	1	z_i^{11}	z_i^{12}	…	z_i^{1p}
	2	z_i^{21}	z_i^{22}		
	⋮	⋮	⋮	⋱	
	p	z_i^{p1}			z_i^{pp}
Column sums		s_i^1	s_i^2	…	s_i^p

The ratio of commodity flow z_i^{rs} to the total consumed commodities at the final destination s_i^s represents the portion of commodities consumed in region s that arrived from region r.

Equation (18.55) estimates the interregional technical coefficient, given the demand pooling assumption

$$a_{ij}^{rs} = \left(\frac{z_i^{rs}}{s_i^s}\right) z_{ij}^{\bullet s} / x_j^s = t_i^{rs} z_{ij}^{\bullet s} / x_j^s = t_i^{rs} a_{ij}^{\bullet s} \quad (18.55)$$

where $t_i^{rs} = z_i^{rs} / s_i^s$ is the proportion of commodity i consumed by region s that originated in region r.

Equation (18.56) defines the spatially explicit interregional flow matrix $\mathbf{T}$, and Eqs. (18.57) and (18.58) define $\mathbf{x}$ and $\mathbf{f}$, respectively, for a p-region economy. Note that each block matrix $\mathbf{T}^s$ in $\mathbf{T}$ is a diagonal matrix by construction:

$$\mathbf{T} = \begin{bmatrix} \mathbf{T}^{11} & \cdots & \mathbf{T}^{1p} \\ \vdots & \ddots & \vdots \\ \mathbf{T}^{p1} & \cdots & \mathbf{T}^{pp} \end{bmatrix} \quad (18.56)$$

$$\mathbf{x} = \begin{bmatrix} \mathbf{x}^1 \\ \vdots \\ \mathbf{x}^p \end{bmatrix} \quad (18.57)$$

$$\mathbf{f} = \begin{bmatrix} \mathbf{f}^1 \\ \vdots \\ \mathbf{f}^p \end{bmatrix} \quad (18.58)$$

This composition of the various components results in a multiregional Leontief-based model for p regions with n sectors per region. Equation (18.59) shows the multiregional Leontief-based model used to construct the MRIIM. Constructions similar to Eq. (18.59) can be found in Miller and Blair [1985] and Isard et al. [1998]:

$$\mathbf{x} = \mathbf{TAx} + \mathbf{Tf} \iff \left\{ x_i^r = \sum_{js} t_i^{rs} a_{ij}^{\bullet s} x_j^s + \sum_s t_i^{rs} f_i^s \right\}, \forall i, r \quad (18.59)$$

Each component of the multiregional Leontief-based model can be transformed according to the equations above. Following the same derivation, **T** is transformed similarly as shown in Eq. (18.60):

$$\mathbf{T}^* = [\hat{\mathbf{x}}^{-1}\mathbf{T}\hat{\mathbf{x}}] = \begin{bmatrix} [\hat{\mathbf{x}}^1]^{-1}\mathbf{T}^{11}\hat{\mathbf{x}}^1 & \cdots & [\hat{\mathbf{x}}^1]^{-1}\mathbf{T}^{1p}\hat{\mathbf{x}}^p \\ \vdots & \ddots & \vdots \\ [\hat{\mathbf{x}}^p]^{-1}\mathbf{T}^{p1}\hat{\mathbf{x}}^1 & \cdots & [\hat{\mathbf{x}}^p]^{-1}\mathbf{T}^{pp}\hat{\mathbf{x}}^p \end{bmatrix}$$

(18.60)

18.9 THE DYNAMIC IIM

To address more effectively the temporal dynamic behavior of industry recoveries in the static IIM, a Dynamic IIM (DIIM) is proposed and formulated. In this section, the concept of an *industry resilience coefficient* is introduced as a key element in the dynamic extension that supplements and complements the static IIM. Fundamentals on how to define a resilience coefficient and its connection to parameters of recovery are also discussed. A comparison of dynamic and static models at the end of this section shows the consistency of the two models.

18.9.1 Introduction to the DIIM

In I–O literature, the classic Leontief dynamic I–O model takes the following form (see Miller and Blair [1985]):

$$\mathbf{x}(t) = \mathbf{A}\mathbf{x}(t) + \mathbf{c}(t) + \mathbf{B}\dot{\mathbf{x}}(t) \quad (18.61)$$

Matrix **B** in Eq. (18.61) is a square matrix of capital coefficients. It represents the willingness of the economy to invest in capital resources. Blanc and Ramos [2002] argue that the elements of **B** must either be zero or negative for an economic system to be stable. Such a condition will produce an economic behavior consistent with the static model, independent of initial conditions and final demand. Therefore, the capital coefficient matrix **B** can be interpreted as an expression of short-term countercyclical policy instead of long-term growth. For intuition about **B**, consider the case investigated by Blanc and Ramos [2002] where $\mathbf{B} = -\mathbf{I}$, which represents an economy that quickly adjusts its production levels following information about mismatches in supply and demand:

$$\dot{\mathbf{x}}(t) = \mathbf{A}\mathbf{x}(t) + \mathbf{c}(t) - \mathbf{x}(t) \quad (18.62)$$

Using the classic Leontief I–O model and the results above, we can extend IIM to model the industry sectors' dynamic recovery behaviors and dynamic interactions caused by demand reduction or terrorist attacks on industry sectors. Consider a diagonal matrix form of the capital coefficient matrix **B**:

$$\mathbf{B} = \mathrm{diag}(b_i) \quad \forall i = 1, 2, \ldots, n \quad (18.63)$$

Furthermore, we define a **K** matrix as follows:

$$\mathbf{K} = \mathrm{diag}(k_i) \quad \forall i = 1, 2, \ldots, n \quad (18.64a)$$

We relate Eqs. (18.64a) and (18.64b) as follows:

$$\mathbf{K} = -\mathbf{B}^{-1} \Leftrightarrow k_i = \frac{1}{b_i} \quad \forall i = 1, 2, \ldots, n \quad (18.64b)$$

Substituting Eq. (18.64b) for Eq. (18.61) and rearranging the terms will yield the following equation:

$$\dot{\mathbf{x}}(t) = \mathbf{K}[\mathbf{A}\mathbf{x}(t) + \mathbf{c}(t) - \mathbf{x}(t)] \quad (18.65)$$

Or in discrete form,

$$\mathbf{x}(k+1) - \mathbf{x}(k) = \mathbf{K}[\mathbf{A}\mathbf{x}(k) + \mathbf{c}(k) - \mathbf{x}(k)] \quad (18.66)$$

Transforming Eqs. (18.65) and (18.66) into normalized inoperability form will yield the following equations:

$$\dot{\mathbf{q}}(t) = \mathbf{K}[\mathbf{A}^*\mathbf{q}(t) + \mathbf{c}^*(t) - \mathbf{q}(t)] \quad (18.67)$$

$$\mathbf{q}(k+1) - \mathbf{q}(k) = \mathbf{K}[\mathbf{A}^*\mathbf{q}(k) + \mathbf{c}^*(k) - \mathbf{q}(k)] \quad (18.68)$$

In Eqs. (18.65) and (18.66), matrix **A** is the Leontief technical coefficient matrix; vector $\mathbf{c}(t)$ is the final demand vector at time t; vector $\mathbf{x}(t)$ represents the total output of sectors at time t. In Eqs. (18.67) and (18.68), matrix $\mathbf{A}^*$ is the normalized interdependency matrix; vector $\mathbf{c}^*(t)$ is the normalized final demand vector at time t; $\mathbf{q}(t)$ is the inoperability vector at time t. Collectively, Eqs. (18.65), (18.66), (18.67), and (18.68) give the formulation for the DIIM.

Matrix $\mathbf{K}$ will be referred to as the *industry resilience coefficient matrix*; each element k_i in the matrix measures the resilience of sector i, given an imbalance between supply and demand. In the case of a terrorist attack or other catastrophic events, it measures the recovery rate of the industry sectors. In the case of demand reduction, k_i measures the production adjustment rate of the sector.

The resilience coefficient k_i can be controlled and managed. Each resilience coefficient k_i in the matrix $\mathbf{K}$ is determined by the nature of the individual sector itself as well as by the controls on it via risk management policies. Hardening and other risk mitigation efforts in the industry sectors increase k_i during the recovery. Consequently, economic losses and other adverse impacts are minimized with shorter recovery times. This would enable policymakers to assess the return on investments associated with candidate risk management actions for expediting recovery. A general solution to Eq. (18.61) is

$$\mathbf{q}(t) = e^{-\mathbf{K}(\mathbf{I}-\mathbf{A}^*)t}\mathbf{q}(0) + \int_0^t \mathbf{K}e^{-\mathbf{K}(\mathbf{I}-\mathbf{A}^*)(t-z)}\mathbf{c}^*(z)\,dz \quad (18.69)$$

If the final demand $\mathbf{c}^*(t)$ is stationary, Eq. (18.69) can be further simplified to

$$\mathbf{q}(t) = (\mathbf{I}-\mathbf{A}^*)^{-1}\mathbf{c}^* + e^{-\mathbf{K}(\mathbf{I}-\mathbf{A}^*)t}\left[\mathbf{q}(0) - (\mathbf{I}-\mathbf{A}^*)^{-1}\mathbf{c}^*\right] \quad (18.70)$$

or

$$\mathbf{q}(t) = \mathbf{q}_\infty + e^{-\mathbf{K}(\mathbf{I}-\mathbf{A}^*)t}[\mathbf{q}(0) - \mathbf{q}_\infty] \quad (18.71)$$

In the equation above, $\mathbf{q}_\infty$ stands for the equilibrium inoperability, determined by the final demand vector. The exponential term $e^{-\mathbf{K}(\mathbf{I}-\mathbf{A}^*)t}[\mathbf{q}(0) - \mathbf{q}_\infty]$ is the temporal term that is decaying with time. When Eq. (18.70) reaches its equilibrium, it becomes

$$\mathbf{q}(t) = (\mathbf{I}-\mathbf{A}^*)^{-1}\mathbf{c}^* \quad (18.72)$$

In the equilibrium state, the DIIM reduces to the form of the static IIM. It can be viewed as a more general extension of the static IIM, and/or the static model can be viewed as a description of the dynamic model at its equilibrium condition.

18.9.2 Assessing the Industry Resilience Coefficient

As discussed in the previous section, the industry resilience coefficient is the key to modeling the DIIM. The resilience coefficient reflects the output response of each individual industry sector to an imbalance of supply and demand. For a detailed assessment of the industry resilience coefficients, consider an economy consisting of n sectors. It is assumed that initially sector i is attacked by terrorists. Based on the postattack economic response, two sets of sectors should be analyzed.

The first is sector i. After an attack, sector i will start the recovery process (e.g., rebuild the factories, machines, and so on) with a recovery rate k_i, $0 \le k_i < 1$. Depending on the risk mitigation efforts and the damage, the faster sector i recovers, the larger the value of k_i will be. The second set of sectors encompasses all the others in the economy that are affected by the attack due to their dependence on sector i. To be able to respond efficiently to the attack scenario, the production outputs of these sectors must be immediately adjusted relative to the new level of demand. Such immediate adjustments to mismatches in supply and demand correspond to the maximum recovery rates $k_j = 1$, $j \ne i$.

For a special case where $k_i = 0$, and momentarily neglecting the dependence of i on j, $a_{ij}^* = 0$ $\forall j \ne i$, and if final demand stays constant, the ith row in Eq. (18.68) will read as follows:

$$q_i(k+1) - q_i(k) = 0 \quad (18.73)$$

from which follows:

$$q_i(k+1) = q_i(0) \quad \forall k = 0, 1, 2, \ldots, T \quad (18.74)$$

In other words, during the period of time under consideration, sector i has a constant inoperability equal to the initial perturbation.

In the following discussion, the assessment of the recovery rate of the attacked sector, corresponding to k_i, $0 < k_i < 1$, is addressed and formulated in greater detail.

In Eq. (18.69), if $k_i > 0$, $a_{ij}^* = 0$ $\forall j \neq i$, and if final demand stays constant, then the inoperability equation for sector i becomes

$$1 - q_i(t) = 1 - e^{-k_i(1-a_{ii}^*)t} q_i(0) \quad (18.75)$$

Equation (18.75) is called an *individual sector recovery trajectory*. Similar to the concept of inoperability, $q_i(t)$, the term $1 - q_i(t)$ is defined as the operability of sector i at time t. From this, we conclude that a recovery trajectory that follows an exponential curve in temporal space will have a recovery parameter $k_i(1 - a_{ii}^*)$.

The recovery trajectory of a sector can also be written in the following form typically found in reliability literature (note: the ratio λ/τ will be clarified in the forthcoming example):

$$1 - q_i(t) = 1 - e^{-(\lambda/\tau)t} q_i(0) \quad (18.76)$$

Comparing Eqs. (18.75) and (18.76) generates the following formula that can be used to estimate the resilience coefficient of sector i:

$$k_i = \frac{\lambda}{\tau(1 - a_{ii}^*)} \quad (18.77)$$

When $a_{ii}^* \ll 1$, Eq. (18.77) can be approximated further as follows:

$$k_i \approx \frac{\lambda}{\tau} \quad (18.78)$$

This equation provides the connection between the resilience coefficient (recovery rate) and recovery parameter. It justifies the definition of k_i in the DIIM as a *sector resilience coefficient* or *recovery rate* (λ/τ). As an example, the derivation of the recovery rate for the electric power generation and supply sector is shown to illustrate the process. Consider a power blackout scenario that follows an exponential recovery such that 99% recovery is achieved in 60 days. The resilience coefficient of the power sector (denoted by the subscript p) can be derived as shown below. According to Eq. (18.76), the recovery parameter can be calculated as follows:

$$\frac{\lambda}{\tau} = \frac{-\ln\left[\frac{q_p(60)}{q_p(0)}\right]}{60} = 0.0768 \,/\, \text{day}$$

Through the BEA data, we determined for the power sector that $a_{pp}^* = 1.217 \times 10^{-4}$. From Eq. (18.77), we can calculate the recovery rate to be $k_p \approx 0.0768\,/\,\text{day}$. Therefore, the individual recovery trajectory for the power sector has the following function, which is depicted in Figure 18.5:

$$1 - q_p(t) = 1 - e^{-k_p(1-a_{pp}^*)t} q_p(0) = 1 - e^{-0.0768\,t} q_p(0)$$

18.9.3 Assessing Economic Loss during the Recovery through the DIIM

To understand better what the impacts of the attack will be and facilitate the trade-off analysis in the risk management decisionmaking, it is imperative that the economic loss during the recovery from each individual industry sector be estimated in quantitative dollar amounts for all kinds of possible scenarios. During the recovery process, it is important to know not only a sector's own loss compared to the *as-planned* level; all the indirect losses from its interdependent industry sectors should be quantified and taken into account as well. The national and regional case studies both consider the two measures in dollar amounts: (i) economic losses of the attacked sector and (ii) economic losses (direct and indirect) from all sectors. According to the dynamic model, in the continuous form, the cumulative economic loss for each individual industry i is given by

$$Q_i(t) = \hat{x}_i \int_{t=0}^{T} q_i(t)\, dt \quad (18.79)$$

$\hat{x}_i$: the *as-planned* output rate of industry i ($\$/$time unit)
$q_i(t)$: the inoperability of industry i at time t
$Q_i(t)$: the cumulative economic loss of industry i by time t
$q_i(t)$: is subject to $\mathbf{q}(t) = (\mathbf{I} - \mathbf{A}^*)^{-1}\mathbf{c}^* + e^{-\mathbf{K}(\mathbf{I}-\mathbf{A}^*)t} [\mathbf{q}(0) - (\mathbf{I} - \mathbf{A}^*)^{-1}\mathbf{c}^*]$

Therefore, $Q_i(t)$ will also be exponential due to the exponential recovery trajectory of sector i.

Similarly, the total economic loss from all n sectors by time t (denoted by $Q(t)$) is assessed as

$$Q(T) = \sum_{i=1}^{n}\left(\hat{x}_i \int_0^t q_i(t)dt\right) \quad (18.80)$$

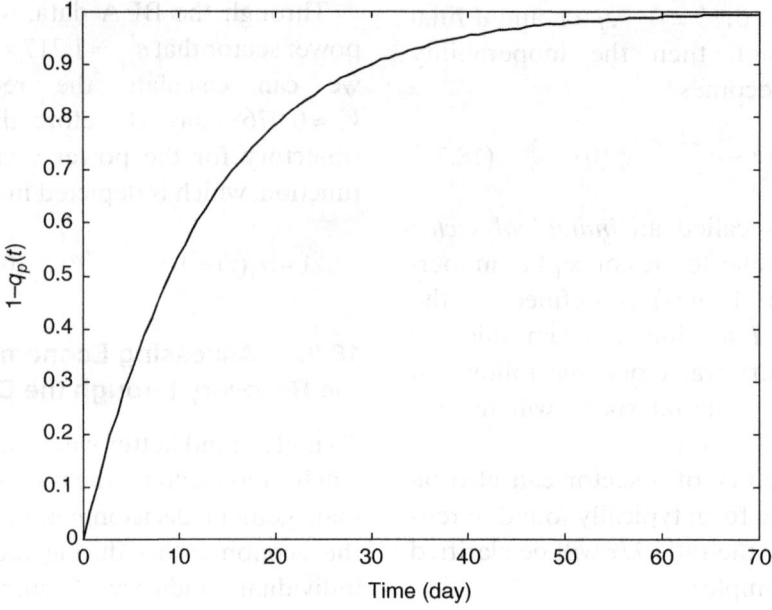

Figure 18.5 Individual recovery trajectory of power sector.

18.9.4 Comparing the Static IIM and DIIM

Static and dynamic models are consistent under equilibrium conditions, and the dynamic model can be transformed into a static model through the concept of *equivalent static inoperability*.

As noted in the previous section, the DIIM takes the form of $\dot{\mathbf{q}}(t) = \mathbf{K}[\mathbf{A}^*\mathbf{q}(t) + \mathbf{c}^*(t) - \mathbf{q}(t)]$. When equilibrium is reached, $\dot{\mathbf{q}}(t) = 0$. It follows that $\mathbf{A}^*\mathbf{q}(t) + \mathbf{c}^*(t) - \mathbf{q}(t) = 0$ or $\mathbf{q}(t) = [\mathbf{I} - \mathbf{A}^*\mathbf{q}(t)]\mathbf{c}^*(t)$. Therefore, under equilibrium conditions, the dynamic model becomes the static model.

18.9.5 Specializing the Static IIM to the DIIM

In the dynamic model, suppose that sector i follows a dynamic inoperability function $q_i(t)$ from time $t = 0$ to T. A static inoperability $\overline{q}_i$ exists, defined as follows:

$$\overline{q}_i = \frac{1}{T}\int_{t=0}^{T} q_i(t)\,dt \qquad (18.81)$$

$\overline{q}_i$ is called equivalent static inoperability during $[0, T]$. Through the equivalent static inoperability, the economic loss accumulated during the dynamic recovery can be estimated statically using the following equation derived from Eq. (18.79):

$$Q_i(t) = \hat{x}_i \int_{t=0}^{T} q_i(t)\,dt = (\hat{x}_i)(\overline{q}_i T) \qquad (18.82)$$

As depicted in Figure 18.6, the scenario where the power sector recovers from 100% inoperability to 1% in 60 days has an equivalent constant static inoperability of 22% for the 60-day period.

18.10 PRACTICAL USES OF THE IIM

The IIM provides a computation base for risk impact analysis that, as noted earlier, utilizes I–O data from the BEA—the agency responsible for documenting the transactions of approximately 500 producing and consuming sectors within the US economy. Through our direct use of the detailed national I–O tables published by the BEA, we benefit from their intensive data collection efforts and resource base. In addition, we utilize data available through the RIMS II for conducting region-level analysis. This provides a solid foundation for any analysis, especially one as sensitive to the unknown as the analysis of a terrorist attack.

Given that BEA data provides each producing sector's requirements or support from other sectors (i.e., production inputs such as products and services), the IIM is capable of:

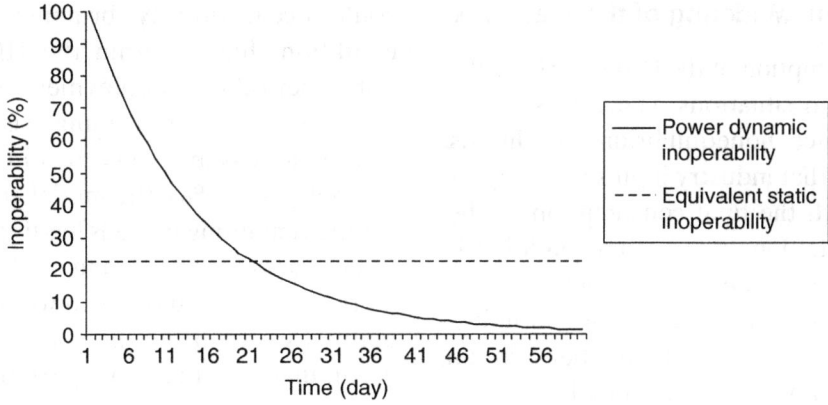

Figure 18.6 Dynamic inoperability and equivalent static inoperability.

- Computing the propagating impacts of diverse perturbation scenarios for various regions.
- Computing the impact of varying recovery rates for interdependent sectors.
- Computing various perspectives of impact, including *inoperability* and *economic loss*. This yields insight into societal consequences and provides a quantitative method for resource allocation.

As part of using economic-based data for analyzing a terrorist attack situation, the IIM application is based upon the assumption that the level of economic interdependencies between sectors is also representative of physical interconnectedness (i.e., in general, two sectors that have a large number of economic transactions similarly have a large degree of physical linkage). Therefore, utilizing economic interdependencies made accessible to us through the BEA and RIMS II is an efficient and cost-effective alternative for comprehensively accounting for physical linkages between national sectors. (Otherwise, a similar or even greater special data collection effort would be required.) By allowing a holistic integration of sectors, the IIM provides analysts with a tool for systemically prioritizing sectors deemed to be economically and physically critical, in addition to identifying those sectors whose products are critical during recovery operations. The IIM's prioritization capability also serves to avoid erroneous assumptions that might otherwise occur in preselecting *most vulnerable* sectors or commodities.

Specifically for a power sector analysis, for example, the IIM could provide the following information essential for assessing and managing the propagating impacts of a terrorist attack:

- Direct economic and power-production impacts of a terrorist attack on the power generation and power supply sectors
- Economic and production capacity impacts to electrical power users (manufacturing, commerce, household, and others) due to terrorist destruction of vulnerable electronic equipment
- Trade-offs between possible reductions in economic losses and the corresponding cost of investment required for carrying out various equipment recovery/resource allocation options
- Labor requirements to support production, delivery, and use of *as-planned* power outputs
- Economic and production impacts due to the possible psychological effects of a terrorist attack

18.10.1 Assumptions and Limitations of the IIM

Several assumptions from the original Leontief economic structure are retained in the IIM formulation. Many of these remain unchanged because of the need to capitalize on the vast BEA databases, which were designed specifically for the Leontief linear, deterministic, equilibrium model. It is important to address the underlying model assumptions for optimal understanding and interpretation of IIM analysis results.

18.10.2 Equilibrium Modeling of the Static IIM

The equilibrium assumption of the IIM is perhaps the hardest to manage in situations where it is highly possible to experience nonequilibrium conditions. Equilibrium implies that industry inputs and outputs will find balance with the final consumption of the sectors' outputs. In the long run, such a condition is evidently true. Moreover, during a recovery process, equilibrium conditions will also dominate, as industries are constantly improving their states in an interdependent fashion, as illustrated by the DIIM. However, in the short time immediately following scenarios that impose large, widespread perturbations, nonequilibrium conditions could dominate, and the IIM results would not exactly reflect real recovery production rates or economic losses.

A terrorist attack would most likely impact only a defined region of the country while fortunately leaving surrounding regions intact. The specific attributes of the attack scenario determine the size and location of the impacted region and which regional economies are categorized by equilibrium economic data. Where the impacted region is relatively small, the consequences, while large within that region, can potentially be dealt with either by importing resources from the rest of the country or exporting resources or problems out (e.g., hospital patients or unusable inventory to support increased production in other regions). These transfers would complement other activities to restore normal operations. When applying the IIM, we anticipate that the national impact on the economy and production capacity due to a terrorist attack is important, but not approaching anything like 100%, even during the time period immediately following an attack. The smaller the fraction of inoperability (e.g., less than 10%), the more applicable are the results obtained from the IIM. This is because the overall capacity of the country to produce goods plays a significant role. It does this through redistributions that could be feasible without drastic economic adjustments that would go beyond the IIM's assumptions of constant technology and overall economic structure. At the regional level, the results from using the IIM would be more suspect, since the region under attack would likely have very large disruptions. However, the redistribution of national resources to the region and the recovery process could occur quickly, bringing the region into a condition that is within the IIM boundaries. The initial period of redeployment and recovery would not be suitable for IIM analysis of inoperability or economic loss, but once within a close fraction of normal (e.g., 10%), the model could provide results for the remaining periods leading up to full recovery.

For cases encompassing a relatively small region, a brief time period (days to weeks), or slight inoperability (less than 10%), the bulk of economic losses accumulate in the long period of final recovery, where companies operate normally, quarter by quarter, with constantly improving states dominated by equilibrium conditions. Since the bulk of economic losses are accumulated later, the costs in data and time required for building highly accurate transient models are prohibitively large. Therefore, although the initial period of redeployment and recovery is not suitable for IIM analysis of inoperability or economic loss, once within a close fraction of normal (e.g., 10%), the model could provide results for the remaining periods leading up to full recovery. Additionally, it can provide an optimal prioritization strategy for recovery during the uncertain transient periods to minimize the overall losses when the equilibrium condition is reached.

18.10.3 Stability of the Technical Coefficient Matrix

At the core of the Leontief model and the IIM is the technical coefficient matrix A, derived from the BEA's databases. These equilibrium data define deterministic guidelines for sector interactions based on the assumptions of constant technology and economic structure. The properties of these data provide the remaining assumptions that lead to model limitations.

Again, the BEA decomposes the US economy into about 500 sectors whose outputs contribute to the input of other sectors. Each element in this matrix gives a constant, deterministic value that represents the contributions to one sector, say, j, from any other sector, say, i, which is proportional to the output of Infrastructure j. There are obvious examples where this assumption is exact and valid. Moreover, as the number of industry subdivisions increases, a first-order approximation becomes increasingly accurate. (500 sectors create a matrix defining 250,000 interdependencies.) For example, if Infrastructure *i* is tire production and Infrastructure *j* is automobile production, then the value of tires used by Infrastructure *j* increases linearly with the value of automobiles produced. On the other hand, there are also cases where the linearity assumption may not be valid. For example, consider any process where human innovation is involved. As production increases proportionally, producers seek to increase resource sharing, among other measures, to consume fewer manufacturing materials. For example, an increase in the production of cars may not necessarily lead to a proportional increase in the requirement for alloy wheels (i.e., basic car models often use simple lug-bolt rims with hubcaps). Nonlinearity in the use of input requirements may also be observed in the case of shared resources (e.g., multiple computers sharing external drives or peripheral devices). Thus, in addition to the extreme cost of such added modeling, the database would become overwhelming and probably less meaningful to the analyst.

It is also important to note that the constants that define the relationships between industries are time invariant and deterministic. This stems from the fact that the BEA generates data every 5 years directly from the US Census data, because of economic momentum that causes values to change very little from year to year. Therefore, the variance of commodity flows changes very little from year to year and in most industry sectors. (This is intuitive because production procedures change very little over short periods of time.) Returning to the car example, four tires for every one car will certainly vary negligibly over time. However, this obviously will not always be the case. In order to protect against major flaws, studies have compared various years of economic data. Analysis results have shown only negligible changes between two recent 5-year periods.

18.11 UNCERTAINTY IIM

Chapter 6 discussed two major sources of uncertainty in modeling a system: uncertainty in the system itself and uncertainty in the ability of the modeler to capture the behavior of the system. This second source of uncertainty, also referred to as epistemic uncertainty [Paté-Cornell, 1996], is of interest in this discussion and is characterized by several types of errors, including those in model topology, model scope, data, and, most important for this IIM discussion, model parameters [Haimes and Hall, 1977]. The results of the IIM and its derivatives are particularly susceptible to errors in model parameters that describe interdependencies among infrastructure sectors due to the time-invariant and deterministic nature of $\mathbf{A}^*$ as defined in Eq. (18.44). These limitations on $\mathbf{A}^*$, discussed in Section 18.10.3, may hinder the IIM and its derivatives from accurately modeling the behavior of interdependent infrastructures, particularly following disruptive events when forced substitution and other effects on $\mathbf{A}^*$ may occur. This section extends the DIIM by measuring the sensitivity of its output to changes in $\mathbf{A}^*$. It uses the uncertainty sensitivity index method (USIM) discussed in Chapter 6 and is adapted from Barker [2008] and Barker and Haimes, [2008a, b]. The approach for evaluating uncertainty in infrastructure interdependencies specifically when modeling recovery, titled the Uncertainty DIIM (U-DIIM), enhances robust risk-based decision-making by incorporating sensitivity into infrastructure interdependency parameters when comparing multiple risk management strategies.

Violating the assumption of the time-invariant and deterministic $\mathbf{A}^*$ is typically due, in the economics literature, to structural change or temporal variability in the interdependency matrix. A number of approaches have been used to loosen the deterministic assumption of the input–output model, including works by Quandt [1958, 1959], Sebald and Bullard [1976], and Percoco et al. [2006], among others. The U-DIIM integrates the DIIM with the USIM [Haimes and Hall, 1977; Li and Haimes, 1988] to account for potential changes in $\mathbf{A}^*$ due to substitution for the purpose of comparing risk management strategies. That is, the DIIM is used as a means to measure the efficacy of risk management [Crowther and Haimes, 2005], where strategies can affect sector *i* in various ways. These include reducing the initial effects experienced after a disruptive event (lower $q_i(0)$), reducing the time to recover from the event (lower T_i), and reducing the linger demand reductions (lower $c_i^*(t)$). The U-DIIM aids in determining what risk management strategies better *absorb* the changes that could occur in the interdependent relationships between sectors, that is, add more resilience to the system.

Recall x_{ij} from Eq. (18.9), the flow of commodities from the *i*th sector to the *j*th sector. Define **X** as an $n \times n$ matrix whose *i*, *j*th element is x_{ij} and that describes the commodities flows between all sectors. The relationship between **A** and **X** is defined as

$$a_{ij} = \frac{x_{ij}}{x_j} \Rightarrow \mathbf{A} = \mathbf{X}[\text{diag}(\mathbf{x})]^{-1} \quad (18.83)$$

The $\mathbf{A}^*$ matrix used in IIM and DIIM calculations, as a function of **X**, would then be

$$\mathbf{A}^* = [\text{diag}(\mathbf{x})]^{-1}\mathbf{A}[\text{diag}(\mathbf{x})] = [\text{diag}(\mathbf{x})]^{-1}\mathbf{X} \quad (18.84)$$

A discrete-time version of the recursive DIIM from Eq. (18.62) is introduced in a nonrecursive form:

$$\mathbf{q}(t) = [\mathbf{I} + \mathbf{K}\mathbf{A}^* - \mathbf{K}]^t \mathbf{q}(0)$$
$$+ \sum_{i=0}^{t-1}[\mathbf{I} + \mathbf{K}\mathbf{A}^* - \mathbf{K}]^i \mathbf{K}\mathbf{c}^*(t-1-i) \quad (18.85)$$

Also, an equation for total economic loss, similar to Eq. (18.74), is introduced here for the discrete form of the DIIM. Economic losses for each time period are summed over τ time periods, assuming that total output, **x**, is time invariant:

$$Q = \mathbf{x}' \sum_{j=1}^{\tau} \mathbf{q}(j) \quad (18.86)$$

Recall from Chapter 6 that the USIM measures sensitivity of an objective function with respect to multiple uncertain parameters by summing the squared partial derivatives of that objective function with respect to each of the uncertain parameters. If a decisionmaker is interested in minimizing the total economic loss of a set of interconnected infrastructure sectors following a disruptive event, the sensitivity of economic loss can be measured. Based on the potential substitution effects forced by the disruptive event, the elements of **X**, the commodity flows between sectors, become uncertain parameters. That is, a disruption in sector *i* may require sector *j* to look to alternative sources of inputs,

thereby altering the value of x_{ij}. The sensitivity of Q with respect to changes in individual x_{ij} can be found with the following sensitivity index, ψ:

$$\psi = \sum_{i=1}^{n}\sum_{j=1}^{n}\left[\frac{\partial Q}{\partial x_{ij}}\right]^2 \quad (18.87)$$

Combining Eqs. (18.84)–(18.87) provides the specific calculation of the sensitivity index:

$$\psi = \sum_{i=1}^{n}\sum_{j=1}^{n}\left[\frac{\partial}{\partial x_{ij}}\mathbf{x}'\sum_{t=1}^{\tau}\begin{bmatrix}[\mathbf{I}+\mathbf{K}[\text{diag}(\mathbf{x})]^{-1}\mathbf{X}-\mathbf{K}]^t\mathbf{q}(0)+\\ \sum_{l=0}^{t-1}[\mathbf{I}+\mathbf{K}[\text{diag}(\mathbf{x})]^{-1}\mathbf{X}-\mathbf{K}]^l\\ \mathbf{K}\mathbf{c}^*(t-1-l)\end{bmatrix}\right]^2$$
$$(18.88)$$

Note that, given the relationship between $\mathbf{A}^*$ and **X**, calculating the resilience coefficient in **K** involves commodity flow. The following calculation of the resilience coefficient is derived from Eq. (18.69):

$$k_i = \frac{\ln[q_i(0)/q_i(T_i)]}{T_i}\left(\frac{1}{1-x_{ij}/x_i}\right) \quad (18.89)$$

Assume that the cost function for implementing a particular strategy is shown below. Reflected in the cost function, $h(\cdot)$, are the various DIIM parameters that are affected by the strategy. That is, implementation cost is a function of the risk-based planning required to lower $\mathbf{q}(0)$, **T**, and $\mathbf{c}^*(t)$:

$$\text{cost} = h\big(\mathbf{q}(0),\mathbf{T},\mathbf{c}^*(0),\mathbf{c}^*(1),\ldots,\mathbf{c}^*(\tau-1)\big) \quad (18.90)$$

With the U-DIIM, we desire a means to compare strategies on the basis of economic loss and implementation cost while also minimizing the effect of changes in x_{ij} on our strategies. In minimizing ψ, we seek a strategy that will be effective in the face of unforeseen substitution strategies. Therefore, the following multiobjective formulation is provided, minimizing economic loss, the sensitivity of economic loss to changes in x_{ij}, and implementation cost. Recall from Chapter 6 that the optimization problem is solved for nominal values of the uncertain parameters; here, the matrix of nominal values is represented by $\hat{\mathbf{x}}$. The decision variables in the following formulation are the quantifiable strategy-specific parameters of $\mathbf{q}(0)$, $\mathbf{T}$, and $\mathbf{c}^*(t)$.

Barker [2008] demonstrates that the sensitivity index is computable using the product rule in matrix calculus (see, e.g., Turkington [2002]):

$$\min_{\mathbf{q}(0),\mathbf{T},\mathbf{c}^*(\cdot)} \left\{ \begin{array}{l} \mathbf{x}' \sum_{t=1}^{\tau} \left[\begin{array}{l} [\mathbf{I}+\mathbf{K}[\mathrm{diag}(\hat{\mathbf{x}})]^{-1}\mathbf{X}-\mathbf{K}]^t \mathbf{q}(0) + \\ \sum_{l=0}^{t-1} [\mathbf{I}+\mathbf{K}[\mathrm{diag}(\hat{\mathbf{x}})]^{-1}\mathbf{X}-\mathbf{K}]^l \mathbf{K}\mathbf{c}^*(t-1-l) \end{array} \right] \Bigg|_{\mathbf{x}=\hat{\mathbf{x}}} \\ \sum_{i=1}^{n}\sum_{j=1}^{n} \left[\frac{\partial}{\partial x_{ij}} \mathbf{x}' \sum_{t=1}^{\tau} \left[\begin{array}{l} [\mathbf{I}+\mathbf{K}[\mathrm{diag}(\mathbf{x})]^{-1}\mathbf{X}-\mathbf{K}]^t \mathbf{q}(0) + \\ \sum_{l=0}^{t-1} [\mathbf{I}+\mathbf{K}[\mathrm{diag}(\mathbf{x})]^{-1}\mathbf{X}-\mathbf{K}]^l \mathbf{K}\mathbf{c}^*(t-1-l) \end{array} \right] \right]^2 \Bigg|_{\mathbf{x}=\hat{\mathbf{x}}} \\ h(\mathbf{q}(0),\mathbf{T},\mathbf{c}^*(0),\mathbf{c}^*(1),\ldots,\mathbf{c}^*(\tau-1)) \end{array} \right\} \quad (18.91)$$

The analysis will surely become computationally intensive when a large number of sectors are studied, for example, the nearly 500 sectors of commodity flow data published by the BEA every 5 years. It is likely that only a handful of these elements will be meaningful in an uncertainty analysis. One may focus the sensitivity analysis on a subset of sector interdependencies, calculating partial derivatives for a subset of the X matrix elements. The choice of the subset could be based on the set of 17 critical infrastructure and key resources identified by the Department of Homeland Security [DHS, 2006] or on the interests of the decisionmaker. A particularly interesting method of choosing the elements of set S could come from the *fields of influence* approach discussed by Sonis and Hewings [1992], Percoco et al. [2006], and Percoco [2006], among others, wherein the elements of the interdependency matrix with greatest impact on the rest of the economy are found.

Models are built to answer specific questions. For example, we make use of the DIIM to answer the following question: what are the impacts (e.g., proportion of inoperability/dysfunctionality, economic loss) on all infrastructure sectors, given a natural or man-made hazard affecting multiple sectors? However, to answer such specific question, models must be built with sufficient complexity to capture the essence of the system. The quality and effectiveness of the IIM and its derivatives, as a modeling enterprise, are indeed subject to model assumptions, topology, and selection of parameters, among others, whose associated uncertainties hinder the complexity required to deal realistically with the specific questions that we seek to answer. Therefore, short of modifying the basic, widely accepted Leontief-based inoperability model for which considerable data are collected, the U-DIIM attempts to compensate for the inherent uncertainties in the model and its assumption that $\mathbf{A}^*$ is time invariant and deterministic. The U-DIIM is capable of evaluating and quantifying parameter uncertainties in interdependent models to improve the risk management policymaking process by more accurately measuring the efficacy of risk management strategies. It also promotes using a multiobjective framework for comparing strategies and calculating trade-offs among competing objectives for each strategy. The U-DIIM minimizes the sensitivity of DIIM metrics (about the nominal values of the elements of $\mathbf{A}^*$) with respect to unforeseeable substitution strategies. It is advantageous because it does not limit the modeler to a predefined substitution policy.

18.12 EXAMPLE PROBLEMS

18.12.1 Example Problem 1

The following example demonstrates the workings of the Inoperability Input–Output risk model. Twelve infrastructure sectors have been selected [Santos, 2003]: (i) coal, (ii) petroleum refining, (iii) railroads and related services, (iv) trucking and courier, (v) water transportation, (vi) air transportation, (vii) telephone and telegraph, communication services, (viii) electric services, (ix) water supply and sewerage systems, (x) banking, (xi) eating and drinking places, and (xii) hospitals. For illustration purposes, the interdependency matrix (A) for these sectors shown in Table 18.4 has been generated through a transformation of the I–O matrices published by the BEA [US DOC, 1998].

Assume that the electric services infrastructure ($k=8$) is attacked, causing its operability to be reduced by 20% (i.e., $c_8 = 0.20$). Also, assume that the rest of the elements of the c vector are zeroes (i.e., electric services is the only infrastructure that was attacked). Applying Eq. (18.11), $x = Ax + c$, Table 18.5 presents the inoperability vector (x) after the attack, arising from infrastructure interconnectedness.

The framework embedded in our infrastructure Inoperability Input–Output risk model is represented by selected infrastructures as depicted in Figure 18.7, where $\{x_{kj}\}$ and c_k are the inputs to Infrastructure k and $\{x_{jk}\}$ is the output of Infrastructure k. The matrix $A = \{a_{kj}\}$ plays a central role in the problem definition. When the system is in a perfect condition—a condition where all components are operating flawlessly (i.e., $x_k = 0$ for all k and $\Sigma_k x_k = 0$)—then it is said to be in *ground state*.

In risk assessment, we ask the triplet questions posed in Chapter 1: (i) What can go wrong? (ii) What is the likelihood that it would go wrong? (iii) What are the consequences? [Kaplan and Garrick, 1981] Applying these questions to the analysis of risks inherent in a system of interconnected infrastructures requires in-depth definition and understanding of S_0—the system's *as-planned scenario*. Any deviations from S_0 are *risk scenarios* (S_i's), which can have adverse impacts on the functionality of a system. The hierarchical holographic modeling (HHM) [Haimes, 1981, 1998] and the theory of scenario structuring (TSS) [Kaplan, 1996] are risk assessment methods appropriate for defining the S_0, and the corresponding S_i's, of a particular system. Furthermore, the fusion of HHM and TSS, as described by Kaplan et al. [2001], offers a risk

TABLE 18.4 Interdependency Matrix (A)

	$j=1$	$j=2$	$j=3$	$j=4$	$j=5$	$j=6$
$i=1$	0.1130	0.0826	0.2236	0.0667	0.0060	0.0118
$i=2$	0.0000	0.0618	0.0026	0.0050	0.0010	0.0003
$i=3$	0.0000	0.0308	0.0617	0.0020	0.0002	0.0019
$i=4$	0.0000	0.0385	0.0025	0.1569	0.0003	0.0021
$i=5$	0.0002	0.0848	0.0020	0.0169	0.1247	0.0066
$i=6$	0.0000	0.1307	0.0014	0.0047	0.0006	0.0614
$i=7$	0.0000	0.0006	0.0001	0.0012	0.0000	0.0016
$i=8$	0.0144	0.0093	0.0338	0.0035	0.0008	0.0012
$i=9$	0.0000	0.1635	0.0313	0.4487	0.0000	0.0455
$i=10$	0.0000	0.0005	0.0001	0.0064	0.0000	0.0013
$i=11$	0.0000	0.0011	0.0008	0.0054	0.0000	0.0010
$i=12$	0.0000	0.0015	0.0009	0.0040	0.0000	0.0014

	$j=7$	$j=8$	$j=9$	$j=10$	$j=11$	$j=12$
$i=1$	0.0090	0.1204	0.0000	0.0644	0.0370	0.0000
$i=2$	0.0015	0.0120	0.0000	0.0134	0.0036	0.0000
$i=3$	0.0010	0.0013	0.0000	0.0252	0.0067	0.0000
$i=4$	0.0148	0.0050	0.0000	0.0112	0.0080	0.0000
$i=5$	0.0046	0.0170	0.0000	0.1203	0.0151	0.0020
$i=6$	0.0245	0.0050	0.0000	0.0185	0.0576	0.0000
$i=7$	0.1236	0.0030	0.0000	0.0137	0.0055	0.0000
$i=8$	0.0018	0.0001	0.0000	0.0194	0.0046	0.0000
$i=9$	1.0060	0.7701	0.0000	0.8386	0.2021	0.0006
$i=10$	0.0062	0.0033	0.0000	0.0474	0.0038	0.0000
$i=11$	0.0023	0.0118	0.0000	0.0074	0.0150	0.0000
$i=12$	0.0047	0.0058	0.0000	0.0040	0.0168	0.0000

TABLE 18.5 Inoperabilities Resulting from 20% Degraded Functionality of the Electric Infrastructure

Infrastructure (i)	$i=1$	$i=2$	$i=3$	$i=4$	$i=5$	$i=6$
Inoperability (x_i)	0.0279	0.0026	0.0004	0.0014	0.0044	0.0016
Infrastructure (i)	$i=7$	$i=8$	$i=9$	$i=10$	$i=11$	$i=12$
Inoperability (x_i)	0.0007	0.2005	0.1574	0.0007	0.0024	0.0012

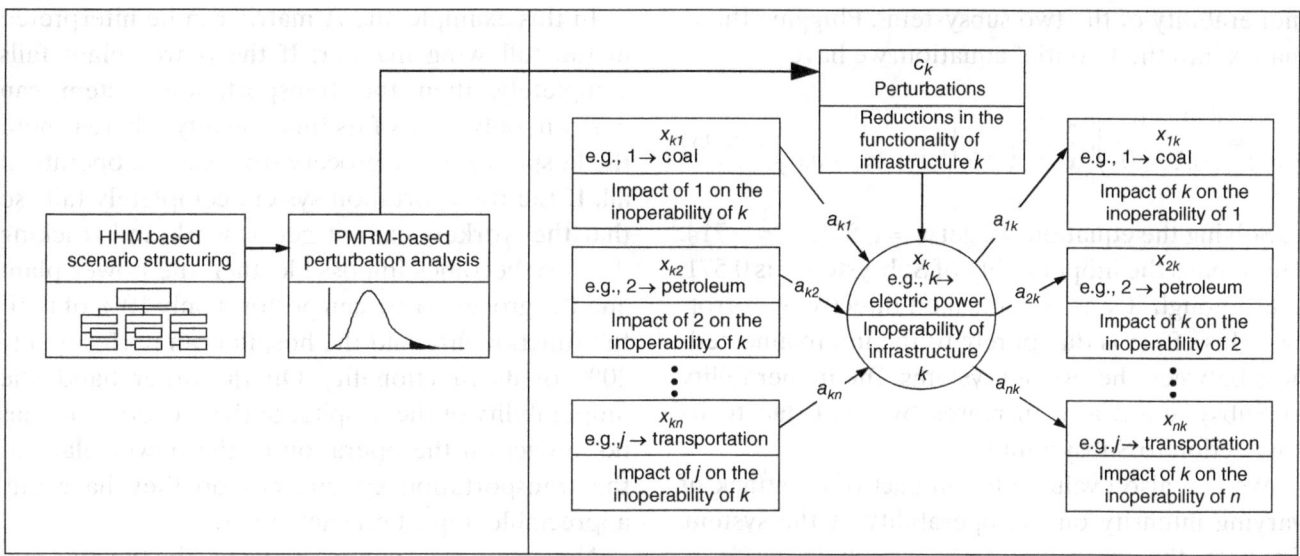

Figure 18.7 *Framework for the infrastructure Inoperability I–O model.*

assessment framework for systematically identifying the myriad S_i's that can perturb a system's S_0.

So far, the treatment of the term *perturbation* (i.e., events that trigger a reduction in the functionality of a system) is compatible with the above risk assessment framework. In the 12-infrastructure example, however, we specifically analyze a scenario corresponding to a reduction of 20% in the functionality of the electric power infrastructure (i.e., $c_k = 0.2$). In doing so, we skip the important process underlying the quantification of this perturbation (c_k). To adhere to the risk assessment framework in the infrastructure Inoperability Input–Output risk model, identifying the system's risk scenarios should be a prerequisite to generating specific c_k's. For instance, a bombing incident could manifest as a 0.2 (or 20%) reduction in the operability of the kth infrastructure. Through the use of an elaborate HHM-based scenario structuring process, we posit c_k to be a preliminary *output* (or *throughput*), which is rooted in a risk scenario. The perturbation c_k resulting from a specific risk scenario then becomes the *input* to our IIM. Additionally, the partitioned multiobjective risk method (PMRM) [Asbeck and Haimes, 1984; Haimes, 1998] will be utilized to generate the magnitude of the perturbation. Since perturbations are generally nondeterministic, a distribution may be more appropriate to use than point estimates. For example, a specific risk scenario may degrade at least 1% of a system's functionality or worse 50%. The PMRM enables analyzing various *what-ifs* based on a risk scenario's perceived distribution. The conditional expectations used in the PMRM can effectively distinguish low-consequence/high-probability events from high-consequence/low-probability events (i.e., extreme events). Figure 18.7 shows a framework that integrates HHM and PMRM into our infrastructure Inoperability Input–Output risk model.

18.12.2 Example Problem 2

To show how to apply the Leontief equation, we solve the following example. Suppose we have a system with two subsystems [Jiang, 2003]. The inoperability of these two subsystems is represented as x_1, x_2, respectively. Now, suppose a failure at Subsystem 2 will lead Subsystem 1 to be 80% inoperable and a failure at Subsystem 1 will lead Subsystem 2 to be 20% inoperable.

Thus, the A matrix reads

$$A = \begin{pmatrix} 0 & 0.8 \\ 0.2 & 0 \end{pmatrix} \quad (18.92)$$

Now, suppose that Subsystem 2 loses 60% of its functionality due to an external perturbation (such as an attack by terrorists). We want to know the inoperability of the two subsystems. Plugging the A matrix into the Leontief equation, we have

$$\begin{pmatrix} x_1 \\ x_2 \end{pmatrix} = \begin{pmatrix} 0 & 0.8 \\ 0.2 & 0 \end{pmatrix} \begin{pmatrix} x_1 \\ x_2 \end{pmatrix} + \begin{pmatrix} 0 \\ 0.6 \end{pmatrix} = \begin{pmatrix} 0.8x_2 \\ 0.2x_1 + 0.6 \end{pmatrix} \quad (18.93)$$

Solving the equation, we get $x_1 = 0.571$, $x_2 = 0.714$. This means the inoperability of Subsystem 1 is 0.571, even though it was *not* attacked directly by terrorists. This effect is due purely to the interconnectedness between the two subsystems. The inoperability of Subsystem 2 also increases by 0.114 due to its connection to Subsystem 1.

We can also evaluate the impact of an attack of varying intensity on the operability of the system. Suppose the intensity of the attack is $h \times 100\%$, meaning that $h \times 100\%$ of the operability of Subsystem 2 is lost due to the attack alone. The Leontief equation reads

$$\begin{pmatrix} x_1 \\ x_2 \end{pmatrix} = \begin{pmatrix} 0 & 0.8 \\ 0.2 & 0 \end{pmatrix} \begin{pmatrix} x_1 \\ x_2 \end{pmatrix} + \begin{pmatrix} 0 \\ h \end{pmatrix} = \begin{pmatrix} 0.8x_2 \\ 0.2x_1 + h \end{pmatrix} \quad (18.94)$$

The solution is $x_1 = 0.952h$, $x_2 = 1.190h$, for $0 \le h \le 0.84$ and $x_1 = 0.8$, $x_2 = 1.0$, for $0.84 < h \le 1$. Note that due to the constraint $0 \le x_1, x_2 \le 1$, Subsystem 2 fails completely when the external attack brings down 84% of its operability. The remaining 16% is taken away by its dependency on Subsystem 1.

18.12.3 Example Problem 3

Consider a system consisting of four subsystems:

Subsystem 1: a power plant
Subsystem 2: a transportation system (roads, signs, signaling facilities, etc.)
Subsystem 3: a hospital
Subsystem 4: a grocery store

Suppose the A matrix for this system is as simple as

$$A = \begin{pmatrix} 0 & 0.9 & 0 & 0 \\ 0.4 & 0 & 0 & 0 \\ 1 & 0.8 & 0 & 0 \\ 1 & 0.9 & 0 & 0 \end{pmatrix} \quad (18.95)$$

In this example, the A matrix can be interpreted in the following manner: If the power plant fails completely, then the transportation system can perform only 60% of its functionality, whereas both the hospital and the grocery store cannot operate at all. If the transportation system completely fails so that the workers cannot get to work and trucking delivery becomes impossible, then the power plant and the grocery store can perform only 10% of their full functionality, and the hospital can perform only 20% of its functionality. On the other hand, the inoperability of the hospital or the grocery store has no impact on the operation of the power plant or the transportation system, nor do they have any appreciable impact on each other.

Now, suppose a major hurricane hits the area and destroys 50% of the functionality of the transportation system. Due to this disaster, many workers are not able to get to work, and the delivery trucks are not able to arrive as scheduled. Using the given A matrix, we have the following Leontief equation:

$$\begin{pmatrix} x_1 \\ x_2 \\ x_3 \\ x_4 \end{pmatrix} = \begin{pmatrix} 0 & 0.9 & 0 & 0 \\ 0.4 & 0 & 0 & 0 \\ 1 & 0.8 & 0 & 0 \\ 1 & 0.9 & 0 & 0 \end{pmatrix} \begin{pmatrix} x_1 \\ x_2 \\ x_3 \\ x_4 \end{pmatrix} + \begin{pmatrix} 0 \\ 0.5 \\ 0 \\ 0 \end{pmatrix} \quad (18.96)$$

The solution to this equation is $(x_1, x_2, x_3, x_4) = (0.78, 0.70, 1, 1)$. Thus, due to the ravages of the hurricane, the power plant loses 78% of its functionality, the transportation system loses 70% of its operability, and the hospital and the grocery store are not able to operate at all.

As with the previous example, we can also evaluate the impact of a hurricane attack of varying intensity on the operability of the system. Suppose the intensity of the attack is $h \times 100\%$ with respect to the transportation system, meaning that $h \times 100\%$ of the operability of the transportation system is lost due to the hurricane alone. Solving the Leontief equation yields

$$(x_1, x_2, x_3, x_4) = (1.41h, 1.56h, 2.66h, 2.80h),$$
$$\text{for } 0 \le h \le 0.357 \quad (18.97)$$

When $h = 0.357$, the value of x_4 reaches 1. The value of x_3 reaches 1 when h increases to 0.376. This means that when the magnitude of the attack on the transportation system is 0.376, the actual inoperability of the power plant is 0.53, and the actual inoperability of the transportation system is 0.59, due to the compound effect of the direct attack and the inoperability of the power plant. Both the grocery store and the hospital are out of commission at this point, due to the inoperability of both the transportation system and the power plant. When h reaches the level of $0.641, x_2$ becomes 1, which means the transportation system completely fails.

18.13 SUMMARY

The Input–Output model is based on the Leontief input–output model to study the effects that intra- and interconnectedness can have on the inoperability of a system composed of a large number of components (subsystems). We proposed viewing the generic Leontief-based model as a first-order approximation of a general model, and we discussed the approximation of linearization under various circumstances.

The underlying economic data utilized in the IIM provide each sector's requirements of support from other sectors (i.e., production inputs such as products and services). The IIM is capable of calculating the propagating impacts of diverse perturbation scenarios for various regions. In using economic-based data for analyzing a terrorist attack, the application of the IIM is based upon the observation that the level of economic interdependency between sectors is often representative of physical interconnectedness (i.e., in general, two sectors that have a large volume of economic transactions have a similarly large degree of physical linkages). Utilizing interdependencies based on economic data gives IIM the capability to comprehensively assess the vulnerability of approximately 500 nationwide sectors given an attack to one or multiple sectors. By allowing the holistic integration of sectors, the IIM (i) provides analysts with a tool for systemically prioritizing sectors deemed to be economically critical and (ii) identifies those sectors whose continued operability is critical during recovery operations.

The following features and capabilities of the IIM make it particularly useful for conducting an impact analysis of a terrorist attack:

- The IIM considers numerous sectors of the economy. This is to avoid misleading results that can stem from studying only those sectors that are selected because they are related to direct vulnerabilities without also accounting for indirect ripple effects.
- The IIM provides a comprehensive ranking of approximately 500 BEA sectors according to inoperability and economic loss impact metrics. It does this in a graphic format (e.g., histograms) that is relevant to risk management of terrorist attacks.
- The IIM is capable of modeling workforce recovery. This enables the modeling of critical and electronically vulnerable sectors such as the power and health services sectors and identification of the most essential personnel for response to a terrorist attack.
- The IIM provides various geographic resolutions (e.g., counties, states, and economic regions) that can be customized via RIMS II data to closely approximate different attack strengths and intensities.
- The DIIM is capable of modeling different temporal frames of recovery (e.g., recovery rates for different sectors) and establishing various interdependent adjustments to levels of equilibrium that may occur after a terrorist attack.

Developing a meaningful model capable of capturing the complex essence of the intra- and interconnectedness of our critical infrastructures is by any account a daunting task that requires the contributions of many individuals from several disciplines. The IIM contributes to the gigantic effort needed to better our understanding of these dependencies and subsequently to manage more cost-effectively the threats and risks that critical infrastructures encounter today. Thus, this model is another important tool in a comprehensive risk assessment and management framework for ensuring the integrity and continued operability of our complex critical infrastructures.

However, before the model is brought to bear on any complex problems, there are three issues meriting special attention. First and foremost, we must define inoperability for each of the subsystems.

This is important to capture the essence of the problem and to be sure that the characteristics of all subsystems pertinent to the objectives of the problem are appropriately and effectively represented. Second, we must make sure that the assumption of linearization is justified. If the circumstance suggests that nonlinearity reigns, then starting from Eq. (18.9a), we must unearth the underlying relationships (f functions) by looking into the detailed structure and coupling of the system. Last but not least, the matrices A, B, and C play overarching roles in both formulating and solving the problem. To determine the elements of any of these matrices could become an overwhelming task and may entail very extensive data collecting and data mining. The extent to which we can project the principles of the Leontief input–output model into analyzing the interdependencies and interconnectedness among critical infrastructures constitutes an opportunity to those in charge of ensuring their continued operation, as well as a challenge to the research community. In this sense, the Leontief-based Input–Output Infrastructure Model (IIM) is intended to be used for at least two purposes:

- The primary and dominant purpose is to improve our understanding of the effects of impacts on the continued and sustained operability of our critical infrastructures under all plausible conditions.
- The secondary purpose is to serve as a tool to allocate resources for an effective process of risk assessment and risk management. In particular, the added security, survivability, assurance, and integrity of our critical infrastructures can be best accomplished through a multiobjective-based risk management framework, where all important costs, benefits, and risks are traded off in a systemic way.

The challenge in realizing both purposes undoubtedly lies in further developing the needed theoretical foundations, methodological instruments, and essential vast appropriate database. Countries worldwide, including the United States, have successfully developed and deployed Leontief input–output models for their economies. Their accomplishments in meeting similar data collection challenges (for thousands of coefficients in the Leontief model) should be a source of encouragement and optimism in our quest to ensure the operability of our complex critical infrastructures.

REFERENCES

Asbeck, E.L., and Y.Y. Haimes, 1984, The partitioned multiobjective risk method (PMRM), *Large Scale Systems* **6**(1): 13–38.

Barker, K., 2008, *Extensions of inoperability input-output modeling for preparedness decisionmaking: Uncertainty and inventory*, Ph.D. Dissertation, University of Virginia, Charlottesville, VA.

Barker, K., and Y.Y. Haimes, 2008a, *Uncertainty Analysis of Interdependencies in Dynamic Infrastructure Recovery: Applications in Risk-Based Decisionmaking*, Technical Report-08-2008, Center for Risk Management of Engineering Systems, University of Virginia, Charlottesville, VA.

Barker, K., and Y.Y. Haimes, 2008b, *Assessing Uncertainty in Extreme Events: Applications to Risk-Based Decisionmaking in Interdependent Infrastructure Sectors*, Technical Report 21-08, Center for Risk Management of Engineering Systems, University of Virginia, Charlottesville, VA.

Blanc, M., and C. Ramos, 2002, *The Foundations of Dynamic Input-Output Revisited: Does Dynamic Input-Output Belong to Growth Theory?* http://www.researchgate.net/publication/28092604_The_foundations_of_dynamic_input-output_revisited_Does_dynamic_input-output_belong_to_growth_theory. Accessed online April 6, 2006.

Blanc, M., and C. Ramos, Dynamic equilibrium in input-output models: theory and empirical applications, in Proceedings of the 11th ASEPUMA Conference, Orviedo, Spain, 2003, accessed online April 6, 2004, www.iioa.org/conferences/intermediate-2004/pdf/blanc.pdf.

Brucker, S.M., S.E. Hastings, and W.R. Latham III, 1990, The variation of estimated impacts from five regional input–output models, *International Regional Science Review* **13**: 113–139.

Converse, A.O., 1971, On the extension of input–output analysis to account for environmental externalities, *American Economic Review* **61**: 197–198.

Crowther. K.G., 2007, *Development and Deployment of the Multiregional Inoperability Input-Output Model (MRIIM) for Strategic Preparedness of Interdependent*

Regions, Ph.D. Dissertation, Systems and Information Engineering Department, University of Virginia, Charlottesville, VA.

Crowther, K.G., and Y.Y. Haimes, 2005, Application of the Inoperability Input–Output Model (IIM) for systemic risk assessment and management of interdependent infrastructures, *Systems Engineering* **8**(4): 323–341.

DHS, Department of Homeland Security, 2006, *National Infrastructure Protection Plan*, Office of the Secretary of Homeland Security, Washington, DC.

Embrechts, P., C. Kluppelberg, and T. Mikosch, 1997, *Modeling External Events for Insurance and Finance*, Springer, New York.

Ernst and Young, 2002, *Manhattan Lodging Forecast*, Ernst and Young Real Estate Advisory Group, New York.

FAA, 2002, *Aviation Industry Overview Fiscal Year 2001*, Federal Aviation Administration, Office of Aviation Policy and Plans, Washington, DC.

Galea, S., J. Ahern, H. Resnick, D. Kilpatrick, M. Bucuvalas, J. Gold, and D. Vlahov, 2002, Psychological sequelae of the September 11 terrorist attacks in New York City, *The New England Journal of Medicine* **346**(13): 982–987.

Griffin, J., 1976, *Energy Input–Output Modeling*, Electric Power Research Institute, Palo Alto, CA.

Guo, J., A.M. Lawson, and M.A. Planting, 2002, From make-use to symmetric I-O tables: an assessment of alternative technology assumptions, *14th International Conference on Input-Output Techniques*, Montreal, CA.

Haimes, Y.Y., 1977, *Hierarchical Analyses of Water Resources Systems: Modeling and Optimization of Large-Scale Systems*, McGraw-Hill, New York.

Haimes, Y.Y., 1981, Hierarchical holographic modeling, *IEEE Transactions on Systems, Man and Cybernetics* **11**(9): 606–617.

Haimes, Y.Y., 1998, *Risk Modeling, Assessment, and Management*, first edition, John Wiley & Sons, New York.

Haimes, Y.Y., 2002, Roadmap for modeling risks of terrorism to the homeland, *Journal of Infrastructure Systems* **8**(2): 35–41.

Haimes, Y.Y., 2004, *Risk Modeling, Assessment, and Management*, second edition, John Wiley & Sons, New York.

Haimes, Y.Y., and W.S. Nainis, 1974, Coordination of regional water resource supply and demand planning models, *Water Resources Research* **10**(6): 1051–1059.

Haimes, Y.Y., and W.A. Hall, 1977, Sensitivity, responsivity, stability, and irreversibility as multiobjectives in civil systems, *Advances in Water Resources* **1**: 71–81.

Haimes, Y.Y. and B.M. Horowitz, 2004, Adaptive two-player hierarchical holographic modeling game for counterterrorism intelligence analysis, *Journal of Homeland Security and Emergency Management* **1**(3): Article 302.

Haimes, Y.Y., B.M. Horowitz, J.H. Lambert, J.R. Santos, C. Lian, and K.G. Crowther, 2005a, Inoperability input-output model (IIM) for interdependent infrastructure sectors, I: Theory and methodology, *Journal of Infrastructure Systems* **11**(2): 67–79.

Haimes, Y.Y., and P. Jiang, 2001, Leontief-based model of risk in complex interconnected infrastructures, *ASCE Journal of Infrastructure Systems* **7**(1): 1–12.

Haimes, Y.Y., B.M. Horowitz, J.H. Lambert, J.R. Santos, K.G. Crowther, and C. Lian, 2005b, Inoperability Input–Output Model (IIM) for interdependent infrastructure sectors, II: Case study, *Journal of Infrastructure Systems* **11**(2): 80–92.

Intriligator, M.D., 1971, *Mathematical Optimization and Economic Theory*, Prentice-Hall, Englewood Cliffs, NJ.

Isard, W., 1960, *Methods of Regional Analysis: An Introduction to Regional Science*, MIT Press, Cambridge, MA.

Isard, W., I.J. Azis, M.P. Drennan, R.E. Miller, S. Saltzman, and E. Thorbecke, 1998, *Methods of Interregional and Regional Analysis*, Ashgate, Brookfield, VT.

Jiang, P., 2003, *Input–Output Inoperability Risk Model and Beyond: A Holistic Approach*, Ph.D. Dissertation, University of Virginia, Charlottesville, VA.

Jiang, P., and Y.Y. Haimes, 2004, Risk management for Leontief-based interdependent systems, *Risk Analysis* **24**(5): 1215–1229.

Kaplan, S., 1996, *An Introduction to TRIZ, the Russian Theory of Inventive Problem Solving*, Ideation International, Inc., Southfield, MI.

Kaplan, S., and B.J. Garrick, 1981, On the quantitative definition of risk, *Risk Analysis* **1**(1): 11–27.

Kaplan, S., Y.Y. Haimes, and B.J. Garrick, 2001, Fitting hierarchical holographic modeling (HHM) into the theory of scenario structuring and a refinement to the quantitative definition of risk, *Risk Analysis* **21**(5): 807–819.

Krause, U., 1992, Path stability of prices in a nonlinear Leontief model, *Annals of Operations Research* **37**: 141–148.

Kuhbach, P.D. and M.A. Planting, 2001, Annual input–output accounts of the US economy, *Survey of Current Business* **9**: 42.

Lahr, M.L. and E. Dietzenbacher (Eds.), 2001, *Input–Output Analysis: Frontiers and Extensions*, Palgrave, New York.

Lahr, M.L. and B.H. Stevens, 2002, A study of regionalization in the generation of aggregation error in regional input–output models, *Journal of Regional Science* **42**(3): 477–507.

Lambert, J.H. and C.E. Patterson, 2002, Prioritization of schedule dependencies in hurricane recovery of a transportation agency, *Journal of Infrastructure Systems* **8**(3):103–111.

Lee, K.S., 1982, A generalized input–output model of an economy with environmental protection, *Review of Economics and Statistics* **64:** 466–473.

Leontief, W.W., 1951a, Input/output economics, *Scientific American* **185**(4).

Leontief, W.W., 1951b, *The Structure of the American Economy, 1919–1939*, Second Edition, Oxford University Press, New York.

Leontief, W.W., 1986, *Input–Output Economics*, Second Edition, Oxford University Press, New York.

Li, D. and Y.Y. Haimes, 1988, The uncertainty sensitivity index method (USIM) and its extension, *Naval Research Logistics* **35**(6): 655–672.

Liew, C.J., 2000, The dynamic variable input-output model: an advancement from the Leontief dynamic input–output, *The Annals of Regional Science* **34:** 591–614.

Miller, R.E., and P.D. Blair, 1985, *Input-Output Analysis: Foundations and Extensions*, Prentice-Hall, Englewood Cliffs, NJ.

Miller, R.E., K.R. Polenske, and A.Z. Rose, 1989, *Frontiers of Input–Output Analysis*, Oxford University Press, New York.

Norris, F.H., C.M. Byrne, E. Diaz, and K. Kaniasty, 2002, *The Range, Magnitude, and Duration of Effects of Natural and Human-Caused Disasters: A Review of Empirical Literature*, A National Center for Post-Traumatic Stress Disorder (PTSD) Fact Sheet, NCPTSD, White River Junction, VT.

Olsen, J.R., P.A. Beling, J.H. Lambert, and Y.Y. Haimes, 1997, Leontief input-output model applied to optimal deployment of flood protection, *Journal of Water Resources Planning and Management* **124**(5): 237–245.

Paté-Cornell, M.E., 1996, Uncertainties in risk analysis: six levels of treatment, *Reliability Engineering and System Safety* **54**(2): 95–111.

Percoco, M., 2006, A note on the inoperability input-output model, *Risk Analysis* **26**(3): 589–594.

Percoco, M., G.J.D. Hewings, and L. Senn, 2006, Structural change decomposition through a global sensitivity analysis of input–output models, *Economic Systems Research* **18**(2): 115–131.

Proops, J.L., 1984, Modeling the energy-output ratio, *Energy Economics* **6:** 47-51.

Quandt, R.E., 1958, Probabilistic errors in the Leontief system, *Naval Research Logistics Quarterly* **5**: 155–170.

Quandt, R.E., 1959, On the solution of probabilistic Leontief systems, *Naval Research Logistics Quarterly* **6**: 295–305.

Santos, J.R., 2003, *Interdependency Analysis: Extensions to Demand Reduction Input–Output Inoperability Modeling and Portfolio Selection*, Ph.D. Dissertation, Systems and Information Engineering Department, University of Virginia, Charlottesville, VA.

Santos, J.R., and Y.Y. Haimes, 2004, Modeling the demand reduction input–output (I–O) inoperability due to terrorism of interconnected infrastructures, *Risk Analysis* **24**(6): 1437–1451.

Sebald, A.V., and C.W. Bullard III, 1976, Simulation of effects of uncertainty in large linear models, *Proceedings of the 76 Bicentennial Conference on Winter Simulation*, National Bureau of Standards, Gaithersburg, MD, 135–143.

Sonis, M. and G.J.D. Hewings, 1992, Coefficient change in input-output models: theory and applications, *Economic Systems Research* **4**(2): 143–157.

Susser, E.S., D.B. Herman, and B. Aaron, 2002, Combating the terror of terrorism, *Scientific American* **287**(2): 70–77.

Tsang, J.L., J.H. Lambert, and R.C. Patev, 2002, Extreme event scenarios for planning of infrastructure projects, *Journal of Infrastructure Systems* **8**(2): 42–48.

Turkington, D.A., 2002, *Matrix Calculus and Zero-One Matrices*, Cambridge University Press, Cambridge, UK.

US DOC, United States Department of Commerce, Bureau of Economic Analysis, 1997, *Regional Multipliers: A User Handbook for the Regional Input–Output Modeling System (RIMS II)*, US Government Printing Office, Washington, DC.

US DOC, US Department of Commerce, Bureau of Economic Analysis, 1998, *Benchmark Input–Output Accounts of the United States, 1992*, US Government Printing Office, Washington, DC.

US DOC, US Department of Commerce, 2003, *Digital Economy*, US Government Printing Office, Washington, DC (Hard copy can be requested at http://www.esa.doc.gov/DigitalEconomy2003.cfm)

19

Case Studies

This chapter presents six case studies that make use of the risk methodologies discussed throughout this book. The first three cases deploy the Inoperability Input–Output Model (IIM) and its derivatives.

19.1 A RISK-BASED INPUT–OUTPUT METHODOLOGY FOR MEASURING THE EFFECTS OF THE AUGUST 2003 NORTHEAST BLACKOUT[1]

This section demonstrates the IIM to measure the financial and inoperability effects of the Northeast Blackout. The case study uses information from sources such as the US input–output (I–O) tables and sector-specific reports to quantify losses for specific inoperability levels. The IIM estimates losses of the same magnitude as other published reports; however, with a detailed accounting of all affected economic sectors (see Chapter 18). Finally, a risk management framework is proposed to extend the IIM's capability for evaluating investment options in terms of their implementation costs and loss reduction potentials.

19.1.1 Introduction

Exploiting an aging power grid and faulty transmission lines, an unforeseen surge of electricity hit large portions of the Midwest and Northeast United States and Ontario, Canada, on August 14, 2003, at 4:09 P.M. Eastern Daylight Time (EDT). Within a matter of seconds, an area of approximately 50 million people experienced the largest electric power blackout ever to hit North America, and 61,800 MW of electric load disappeared into quiet darkness [UCPSOTF, 2004]. With society so dependent on electric power, the blackout caused major disruptions to many facets of life. Commuters stood trapped in subways for hours, restaurants and grocery stores dispensed masses of unprotected perishable food, cars waiting for gas backed up in lines stretching around city blocks, and households had to manage without water service, as power was not restored for 4 days in some parts of the United States.

The resulting estimates of the 2003 Northeast Blackout put the total cost to the US in the range

---

[1]This case study is based on Anderson et al. [2007].

Risk Modeling, Assessment, and Management, Fourth Edition. Yacov Y. Haimes.
© 2016 John Wiley & Sons, Inc. Published 2016 by John Wiley & Sons, Inc.

between $4 and $10 billion, according to the Anderson Economic Group (AEG) paper [Anderson and Geckil, 2003] and the US–Canada Power System Outage Task Force (UCPSOTF) Final Report [UCPSOTF, 2004]. These loss estimates are typically in aggregate values; hence, they lack specificity in regard to the sector-by-sector distribution of economic losses. The complex, interconnected network of today's economy and infrastructure makes it difficult to discern the exact effect of such crippling events. The challenge to policymakers is to allocate resources in the most effective manner so as to mitigate the current damage and prepare for future scenarios. Who exactly bears the burden of these economic losses? How is the overall economic loss distributed among different sectors within the impacted region? And what sectors are the most susceptible to electric power outages? The IIM captures the essence of the foregoing questions because it is capable of (i) itemizing the regional economic loss estimates, (ii) describing the economic impact and percentage of inoperability incurred from a large-scale electric power disruption on a sector-by-sector basis, and (iii) assessing the efficacy of risk management of the electric power system through a multiobjective cost–benefit–risk trade-off analysis. The IIM uses published economic data from the Bureau of Economic Analysis (BEA), US Department of Commerce (USDOC), as well as the electric power service recovery profiles from Anderson and Geckil [2003]. This model is capable of quantifying itemized economic loss estimates for each of the affected sectors, in addition to the aggregate economic impacts suggested in the literature.

19.1.2 A Brief Recap of the IIM

For completeness and convenience, the basic equations introduced earlier in the derivation of the IIM are repeated below. As defined in Chapter 18, the normalized production loss is

$$\text{Normalized production loss} = \frac{\text{"As planned" production} - \text{degraded production}}{\text{Nominal production}} \quad (19.1)$$

Recall that to derive the IIM, we start with the traditional I–O equation as follows, where $\mathbf{x}$ is the production vector, $\mathbf{A}$ is the Leontief technical coefficient matrix, and $\mathbf{c}$ is the final demand vector:

$$\mathbf{x} = \mathbf{A}\mathbf{x} + \mathbf{c} \quad (19.2)$$

Introducing a new term, $\tilde{x}_i$, for degraded production of sector i, we can use Eq. (19.1) to define the inoperability q_i as the normalized production loss between the *as-planned* production x_i and $\tilde{x}$ as shown below:

$$q_i = \frac{x_i - \tilde{x}_i}{x_i} \quad (19.3)$$

We can build upon this relationship to reestablish the original Leontief formulation in Eq. (19.2) in terms of inoperability as shown in the series of matrix operations below:

$$(\mathbf{x} - \tilde{\mathbf{x}}) = \mathbf{A}(\mathbf{x} - \tilde{\mathbf{x}}) + (\mathbf{c} - \tilde{\mathbf{c}}) \quad (19.4)$$

Next, let $\hat{\mathbf{x}}$ be the diagonal matrix derived from vector $\mathbf{x}$ and introduce it into Eq. (19.4):

$$\hat{\mathbf{x}}^{-1}(\mathbf{x} - \tilde{\mathbf{x}}) = \hat{\mathbf{x}}^{-1}\mathbf{A}(\mathbf{x} - \tilde{\mathbf{x}}) + \hat{\mathbf{x}}^{-1}(\mathbf{c} - \tilde{\mathbf{c}}) \quad (19.5)$$

It can be shown that Eq. (19.5) is equivalent to the following IIM equation:

$$\mathbf{q} = \mathbf{A}^*\mathbf{q} + \mathbf{c}^* \quad (19.6)$$

where

$$\mathbf{q} = \hat{\mathbf{x}}^{-1}(\mathbf{x} - \tilde{\mathbf{x}}) \quad (19.7)$$

$$\mathbf{A}^* = \hat{\mathbf{x}}^{-1}\mathbf{A}(\hat{\mathbf{x}}) \quad (19.8)$$

$$\mathbf{c}^* = \hat{\mathbf{x}}^{-1}(\mathbf{c} - \tilde{\mathbf{c}}) \quad (19.9)$$

Hence, through matrix manipulations, we have introduced the concept of inoperability into the traditional I–O model, which augments typical economic loss analysis.

19.1.2.1 Importance of the Regional IIM

An advantage of building on the I–O model is that it is supported by the reliable data collection methodology of the BEA, USDOC. Perhaps most important for the blackout study, which involves many different regions, is the fact that the data are gathered in a comparable fashion across states and industries [Bezdek and Wendling, 2005]. The I–O data provide an available basis for the IIM modeling of such large-scale perturbations as the 2003 Northeast power outage. The Regional IIM utilizes the BEA data [USDOC, 1998] as well as the Regional Input–Output Multiplier System II (RIMS II) database [BEA, 1997]. Thus, it is capable of pinpointing the *top n* sectors with the greatest sensitivity to a given perturbation input. The sector rankings can provide guidance for policymaking in addressing resource allocation issues. A recent *American Scientist* article [Bezdek and Wendling, 2005] showed through such a quantitative model how fuel efficiency and the economy may be related. Their study makes use of I–O analysis to evaluate the impacts of proposed fuel efficiency standards. Certain tools, such as the Federal Emergency Management Agency's Hazard Loss Estimation Methodology (HAZUS) software [FEMA, 2006], provide decisionmakers with estimated losses. However, as this tool only analyzes the impacts of earthquakes, floods, and hurricane winds, structuring production loss scenarios as an input to the IIM allows decisionmakers to perform uncertainty analysis and sector sensitivity analysis in terms of both economic loss and inoperability perspectives.

While the IIM is a linear transformation of the Leontief I–O model [Leontief, 1951a, b], it is especially helpful for decisionmakers who are concerned with inoperability of a particular sector. The inoperability complements the economic loss metric that can be directly obtained from straightforward I–O analysis. An interesting observation is that the sectors that suffer the largest financial losses due to a disruptive economic event are not always the same set of sectors that suffer the highest inoperabilities. For example, a $1 million loss to a sector whose total production is $10 M (i.e., a 10% inoperability) has a higher inoperability in contrast to a $10 million loss for a sector with a total production of $1 billion (i.e., a 1% inoperability). In the case of the August 2003 Northeast Blackout, our model showed that the retail trade sector had the second largest economic loss at $140 million but finished 15th in the sector inoperability rankings, with inoperability of less than 1%. Hence, risk management actions in the event of disruptive economic events must address mitigating both the magnitude of monetary loss (i.e., economic loss) and the relative impact on the size of the sector (i.e., inoperability). Moreover, it is also important to place some level of priority on either relatively low-value sectors with high frequency of interconnectedness with other sectors or low-value sectors that provide essential support to critical sectors (e.g., the electric power sector depends upon the coal sector).

Due to the equilibrium assumption of the Leontief model, the economic losses are typically estimated on an annual basis. Hence, for smaller time resolutions, we may assume that the losses are evenly distributed throughout the applicable year corresponding to the horizon of interest (e.g., dividing the annual loss by 365 produces an estimate of the daily economic loss). Such uniform-loss assumption has been extended by Lian and Haimes [2005] to take into account nonlinear recovery processes wherein economic losses are adjusted to reflect the resilience of the impacted sectors. An application of the nonlinear recovery process has been applied by Haimes et al. [2005a, b] to a high-altitude electromagnetic pulse case study. An alternative approach to modeling the nonlinear behavior of sector recovery pursuant to a disruptive event has been proposed by Rose and Liao [2005] using the concept of resilience in the context of computable general equilibrium.

19.1.3 Applying the IIM to the Risk Analysis of the 2003 Blackout

19.1.3.1 Scenario description and sources of data

To apply the IIM to the specific occurrence of the August 2003 Northeast Blackout, an ex post analysis of the event can be used to structure a perturbation input into the model. Three main sources of information have been used to calculate the blackout losses: (i) electric power outage data from the AEG [Anderson and Geckil, 2003] (e.g., *fraction affected* or unserved electric power demand), (ii) the IIM data matrices developed by the Center for

Risk Management of Engineering Systems at the University of Virginia, and (iii) gross state production (GSP) and local area personal income (LAPI) information from the BEA.

To estimate with the IIM the economic losses resulting from the 2003 Northeast Blackout, we first define specific electric power perturbations experienced in the eight affected states (Connecticut, Massachusetts, Michigan, New Jersey, New York, Ohio, Pennsylvania, and Vermont). In their 2003 blackout study, the AEG provided *fraction affected* data for the eight states over the course of 3 days. Although the magnitude and length of the power outage varied widely across the region, each state's perturbation was adjusted according to its contribution to the region's total economy, in terms of GSP. Table 19.1 displays the final results of our calculations, combining both AEG's data on the fraction affected and regional economic data from the BEA. The table also defines power outages for each day in terms of an adjusted average, which incorporates each state's economic contribution to the region. The adjusted-average results are used to characterize the blackout's input to the IIM. Note that the recovery trajectories supplied by the AEG (see Table 19.1) are already based on actual sector recoveries; hence, the *factor* of temporal resilience is integrated into the analysis.

The first step in deploying the IIM is to define **A**, the Leontief technical coefficient matrix described by Santos and Haimes [2004]. Transformations are needed to convert the **A** matrix to the interdependency matrix, **A***, found in Eq. (19.6). Since the affected states covered a wide range of geographic and economic perspectives, the study assumed that the blackout region was a microcosm of the country; hence, we assumed that the interdependency matrix for the Greater Northeast region of the United States is similar to that of the entire country. This matrix represents the magnitude of interdependencies between the 37 economic sectors defined in Table 19.2. The economic structure of a region, as described by **A***, is usually established by using location quotients [Miller and Blair, 1985]. Location quotients typically describe the similarity of the technical coefficients of a region relative to the nation. Empirically, as the size of the region increases, the economic structure converges to that of the overall national economic structure. The **A** matrix, partially shown in Figure 19.1, illustrates how each sector relies on another for its own production output.

19.1.3.2 Model description

With sector interdependencies defined, the direct impacts of a blackout event on the electric power and the workforce sectors can be structured as inputs into the IIM. However, after generating this **A** matrix, a variety of calculations are still needed to create the Regional IIM. In order to do so, the parameters for the Leontief model in Eq. (19.2) must be established for the blackout region (i.e., the regional production **x** and the regional demand **c**). These parameters were used to make the necessary transformations for the IIM, described by Eq. (19.6). To calculate the inoperability resulting from a given

TABLE 19.1 Fraction Affected Power Outage Data

	Day 1 (%)	Day 2 (%)	Day 3 (%)	Economic Contribution (%)
Michigan	40	40	5	11
New York	40	40	20	30
New Jersey	25	5	0	13
Ohio	25	15	0	13
Pennsylvania	10	5	0	15
Connecticut	10	5	0	6
Vermont	5	0	0	7
Massachusetts	0.5	0	0	11
Average loss	20	14	3	
Adjusted average	26	20	7	

From Anderson Economic Group [2003].

TABLE 19.2 Breakdown of Economic Sectors

#	Sector Name	Abbreviations
1	Farm products and agriculture, forestry, and fishing services	FARM
2	Forestry and fishing products	FRST
3	Coal mining	COAL
4	Oil and gas extraction	O&G
5	Metal mining and nonmetallic minerals, except fuels	MIN
6	Construction	CONS
7	Food and kindred products and tobacco products	FOOD
8	Textile mill products	TEXT
9	Apparel and other textile products	APPR
10	Paper and allied products	PAPR
11	Printing and publishing	PRNT
12	Chemicals and allied products and petroleum and coal products	CHEM
13	Rubber and miscellaneous plastic products and leather and leather products	RUBR
14	Lumber and wood products and furniture and fixtures	LMBR
15	Stone, clay, and glass products	STNE
16	Primary metal industries	PMET
17	Fabricated metal products	FMET
18	Industrial machinery and equipment	MACH
19	Electronic and other electric equipment	ELEQ
20	Motor vehicles and equipment	MOTR
21	Other transportation equipment	TREQ
22	Instruments and related products	INST
23	Miscellaneous manufacturing industries	MSMG
24	Transportation	TRNS
25	Communications	COMM
26	Electric, gas, and sanitary services	UTIL
27	Wholesale trade	WTRD
28	Retail trade	RTRD
29	Depository and nondepository institutions and security and commodity brokers	DEP
30	Insurance	INSC
31	Real estate	REAL
32	Hotels and other lodging places, amusement and recreation services, and motion pictures	HTL
33	Personal services	PSRV
34	Business services	BSRV
35	Eating and drinking places	ETNG
36	Health services	HLTH
37	Miscellaneous services	MSRV

Sector	MACH	ELEQ	MOTR	TREQ	INST	MSMG	TRNS
MACH	0.1353	0.0157	0.0447	0.0371	0.0163	0.0113	0.0053
ELEQ	0.0838	0.1740	0.0488	0.0355	0.0973	0.0165	0.0028
MOTR	0.0009	0.0000	0.2100	0.0087	0.0001	0.0001	0.0045
TREQ	0.0000	0.0000	0.0000	0.1883	0.0000	0.0000	0.0101
INST	0.0020	0.0051	0.0040	0.0390	0.0342	0.0003	0.0004
MSMQ	0.0001	0.0004	0.0000	0.0001	0.0001	0.0678	0.0005
TRNS	0.0159	0.0135	0.0263	0.0154	0.0114	0.0210	0.1465

Figure 19.1 Cross section of the interdependency matrix A.

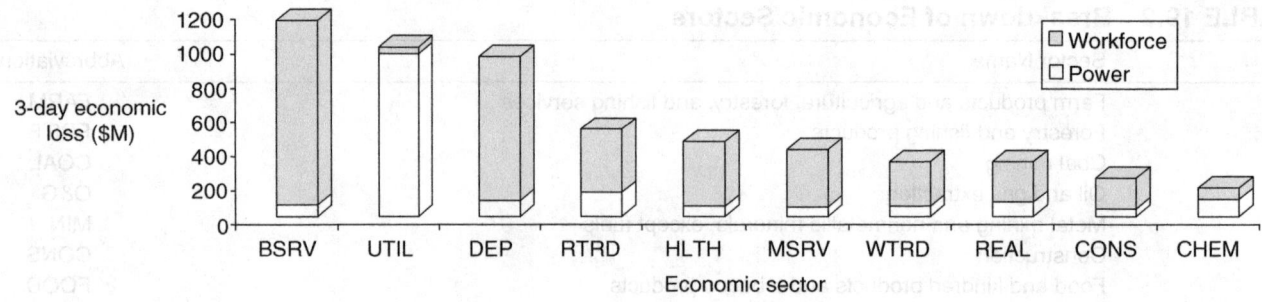

Figure 19.2 Total economic losses from the August 2003 Northeast Blackout.

Rank	Sector	Total ($)	Rank	Sector	Total($)
1	Business services	1,083	6	Misc. services	392
2	Electric, gas and sanitary	993	7	Wholesale trade	320
3	Depository institutions and brokers	936	8	Real estate	320
4	Retail trade	511	9	Construction	224
5	Health services	439	10	Chemical, allied, petroleum, and coal products	167

disaster, **q** is represented in Eq. (19.10) as a function of a prespecified **c*** perturbation vector, a conformable identity matrix **I**, and the interdependency matrix **A***. (Note that **A*** and **c*** were defined in Eqs. (19.8) and (19.9), respectively, and can be obtained using transformations of the BEA-published data.) The solution for the inoperability metric can then be written as follows:

$$\mathbf{q} = (\mathbf{I} - \mathbf{A}^*)^{-1}\mathbf{c}^* \quad (19.10)$$

Using this relationship, the necessary matrix multiplication was performed and generated the inoperability **q** resulting from a particular perturbation **c*** as a function of the interdependencies described by **A***. The final step, converting this inoperability into economic loss, is achieved by multiplying inoperability by the output vector for final demand, thus producing the dollars lost for a given perturbation. Summing all of these losses gives results for the entire regional economy.

Using this methodology, the perturbations of the blackout were used to create an ex post case study of the 3-day outage in the Greater Northeastern US region. The IIM output for this particular scenario resulted in a total 3-day loss of $6.53—$2.12 billion lost from the electric power perturbation and $4.41 billion lost from the workforce perturbation. While these estimates are consistent with the previously mentioned studies [Anderson and Geckil, 2003; UCPSOTF, 2004], an important additional value derived from IIM analysis was the sector-by-sector decompositions of economic loss and inoperability impacts. These are not otherwise available in the published economic loss estimates for the 3-day blackout. The top 10 affected sectors, in terms of 3-day economic loss from the August 2003 Northeast Blackout, are shown in Figure 19.2.

19.1.3.3 Losses resulting from unfulfilled electric power demand

For the case study, we decomposed the initial sector perturbations (**c***) into (i) direct effects on the electric power production and (ii) direct effects caused by reduced workforce productivity. Whereas the electric power perturbation is applied directly to one sector (i.e., the electric power sector), the workforce productivity effects are distributed to practically all sectors of the region. In this section, the focus is on analyzing the electric power perturbation. The losses resulting from workforce impacts are discussed in Section 19.1.3.4.

In particular, the amount of economic loss due to electric power perturbation was generated by using the percentage of unfulfilled electric power demand in each of the eight affected states. For day 1, a 26% perturbation generated a loss of $1.03 billion; for day 2, a 20% perturbation generated a loss of $0.82 billion; and for day 3, a 7% perturbation generated a loss of $0.27 billion.

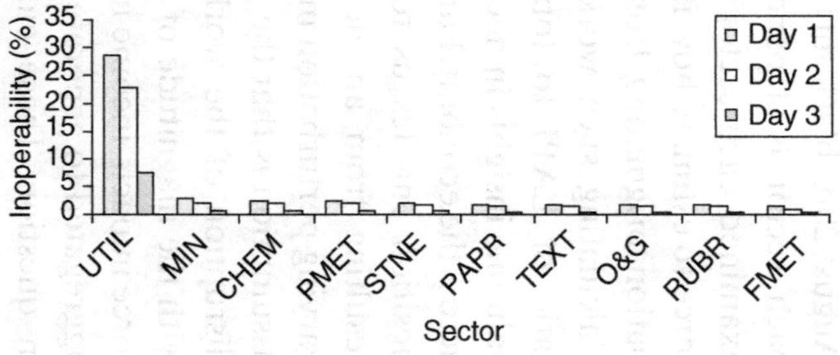

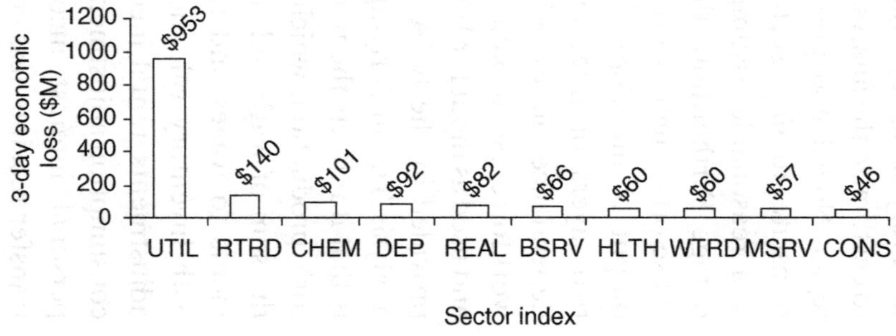

Top-10 sectors with highest production inoperability

Sector	Day 1 (%)	Day 2 (%)	Day 3 (%)
Electric, gas and sanitary	28.72	22.739	7.41
Metal and minerals except fuels	2.85	2.253	0.73
Chemical, allied, petroleum and coal products	2.62	2.073	0.68
Primary metal industries	2.58	2.043	0.67
Stone, clay and glass products	2.13	1.684	0.55
Paper and allied products	1.90	1.502	0.49
Textile mill products	1.80	1.428	0.47
Wholesale trade	1.79	1.415	0.46
Oil and gas extraction	1.64	1.297	0.42
Rubber, leather and misc. plastic products	1.53	1.209	0.39
Fabricated metal products			

Top-10 sectors with highest production output losses

Loss ($M)	Sector	Index
$953	Electric, gas and sanitary	26
$140	Retail trade	28
$101	Chemical, allied, petroleum and coal products	12
$92	Depository institutions and brokers	29
$82	Real estate	31
$66	Business services	34
$60	Health services	36
$60	Wholesale trade	27
$57	Miscellaneous services	37
$46	Construction	6

Figure 19.3 Loss and inoperability from the power perturbation.

The top 10 affected sectors from the electric power perturbation, both in terms of inoperability and economic loss, are shown in Figure 19.3. These values placed the 3-day total loss due to the blackout at $2.12 billion (excluding foregone earnings, which will be discussed subsequently).

Inspecting the top sectors shown in Figure 19.2 provides useful information for risk assessment and management. As expected, the *utilities* sector took the biggest hit both in terms of economic loss and inoperability; however, the additional production effects were not as intuitively obvious. *Metal and minerals except fuels* had the second highest inoperability from the loss of electric power, while *retail trade* had the second highest economic loss from that specific perturbation. These findings can assist policymakers when determining best practices for electric power distribution and responsibility across different industries.

The multidimensional metrics used to describe the impacts also add significant merit to their results. In complex situations such as the Northeast Blackout, no single metric can adequately measure what went wrong; however, by describing not only economic loss but also the inoperability of the various affected sectors of the economy, the IIM provides complementary views for identifying critical sectors. For example, while the *metal and minerals except fuels* sector had the second highest inoperability, it ranked 34th in economic loss. If not for the multiple perspectives generated by the IIM, the significant disruption to this sector may have been easily overlooked. The bicriteria evaluation process using the metrics of economic loss and percentage of inoperability allows users to identify criticality beyond the sole focus on a financial-based impact; it also incorporates a more holistic view of the problem by factoring the analysis into the criticality of sectors with relatively low value but high interconnectedness.

19.1.3.4 Losses resulting from workforce impacts

Another feature that distinguishes the IIM is its ability to incorporate the workforce into the scenarios under consideration. Since the IIM can use workforce statistics as perturbation inputs, several approaches were considered to forecast the *inoperability* of the workforce across all sectors. Our approach is to determine the workforce requirements for every sector and make subsequent adjustments to customize the analysis for the Northeast Blackout region. Note that we posit that workforce unavailability translates to *direct* sector productivity effects, which is a reasonable assumption for an electric power outage. Furthermore, these direct workforce effects cause other higher-order effects in the productivity/output of interdependent sectors. By quantifying two layers of interdependencies (viz., a sector's dependence on *workforce* and *electric power*), a workforce perturbation vector can be structured and used as input for IIM analysis. The data on LAPI provided by the BEA allows us to accomplish this analysis. The underlying assumption in using LAPI is that it reflects the workforce component of a sector's production, which is defined by the BEA "as the sum of wage and salary disbursements, supplements to wages and salaries, proprietors' income with inventory valuation and capital consumption adjustments, rental income of persons with capital consumption adjustment, personal dividend income, personal interest income, and personal current transfer receipts, less contributions for government social insurance" [BEA, 2004].

To determine how much LAPI would be affected by the inability to work (as occurred during the August 2003 blackout), we considered how much of each sector is dependent on the workforce. We examined each sector's contribution to the workforce to estimate how it will be affected by a perturbation originating from the electric power sector. Calculating such workforce percentages (i.e., the ratio of LAPI to total sector production) gives valuable insights in decomposing a workforce *shock* across the economy. Furthermore, workforce decomposition enables us to translate the inoperability resulting from an electric power disruption into varying perturbation inputs to multiple sectors. Our assumption is that the way a sector is affected by a disruption of the workforce is directly correlated with the magnitude of its LAPI. To calculate workforce impacts, income information from each state is aggregated to generate a holistic view of the region in question. These earnings are then divided by the total LAPI for the region to determine the workforce sector's relative weight. A more sophisticated approach for decomposing workforce income effects uses Miyazawa multiplier analysis wherein economic–demographic coefficients are augmented to the basic Leontief technical coefficient matrix (see Okuyama et al. [1999] for implementation of this method in the context of unscheduled economic disruptions). Here, we have simplified the analysis of workforce unavailability and its direct impact on the productivity of each sector by converting the given output (supply) constraints into equivalent demand reductions. This process is consistent with the mixed I–O models discussed in Miller and Blair [1985].

The results generated from the IIM workforce analysis show a 3-day earnings loss of $4.41 billion. Figure 19.4 shows the top 10 sectors affected by workforce loss, both in terms of inoperability and economic loss. When combined with the previously calculated $2.12 billion lost as a result of perturbing only the electric power sector, the IIM results raise the total blackout cost to $6.53 billion. The fact that approximately 2/3 of the total economic loss came from workforce inoperability plays an extremely important role in risk management. With such a large impact from workforce delays and productivity losses, initial mitigation strategies could focus on restoring the workforce to normalcy as a top priority (e.g., providing backup power sources and ensuring workforce mobility). Note that an electric power outage may significantly affect mobility in terms of unavailable power-dependent transportation modes, and the absence of traffic control lights can also prolong the workforce commute.

19.1.3.5 Comparing the IIM results with published loss estimates

The $6.53 billion IIM results correlated closely with the AEG [Anderson and Geckil, 2003] and UCPSOTF [2004] reports. For example, the AEG concluded that the Northeast Blackout was likely to reduce US earnings by $6.4 billion, of which $4.2 billion was suffered by workers and investors in terms of income losses (i.e., reductions in wage and salary earnings and profits). The difference between AEG's $6.4 and $4.2 billion figures is $2.2 billion—remarkably close (within 4%) to the $2.12 billion in electric power-related losses calculated from the IIM. These findings demonstrate that, both for each day individually and for the perturbation time as a whole, about two-thirds of the economic loss was a result of the workforce disruption and one-third was due to the disturbance of the electric power sector.

19.1.4 Risk Management Considerations[2]

The ultimate efficacy of the risk assessment efforts described in the previous section is a prelude to risk management, which asks the triplet questions introduced in Chapter 1: (i) What can be done and what options are available? (ii) What are the trade-offs in terms of all costs, benefits, and risks? (iii) What are the impacts of current decisions on future options? In this section, we employ a variety of tools to consider these questions from multiple perspectives and present a methodology to evaluate the efficacy of risk management.

19.1.4.1 Hierarchical holographic modeling: A precursor to risk management

Because the electric power problem crosses many disciplines (e.g., economic, social, political, and cultural concerns), its complexity demands a holistic risk analysis. With different stakeholders, considerations, and interdependent systems all laying claim to the electric power infrastructure, more than one mathematical or conceptual model is likely to emerge. To better capture the many perspectives from which to view large-scale systems such as the US Northeast Grid, hierarchical holographic modeling (HHM) (see Chapter 3) can be employed to supplement the process of generating IIM scenario inputs.

HHM, which is a holistic philosophy and methodology, captures and represents the essence of the inherent diverse characteristics and attributes of a system—its multiple aspects, perspectives, facets, views, dimensions, and hierarchies. HHM can be viewed as a master chart that depicts the different perspectives governing a system of interest [Leung et al., 2003]. These perspectives are then further decomposed into subtopics and specific risk scenarios. In particular, potential perspectives relating to the August 2003 blackout might include the following: (i) *geographic*, examines the physical,

---

[2]This section is based on Haimes and Chittester [2005].

Sector	Day 1 (%)	Day 2 (%)	Day 3 (%)
Business services	5.01	3.97	1.29
Depository institutions and brokers	3.87	3.06	1.00
Health services	3.80	3.01	0.98
Retail trade	2.89	2.29	0.75
Construction	2.78	2.20	0.72
Wholesale trade	2.50	1.98	0.64
Communications	2.32	1.84	0.60
Insurance	2.24	1.77	0.58
Motor vehicles and equipment	2.05	1.62	0.53
Personal services	1.88	1.49	0.49

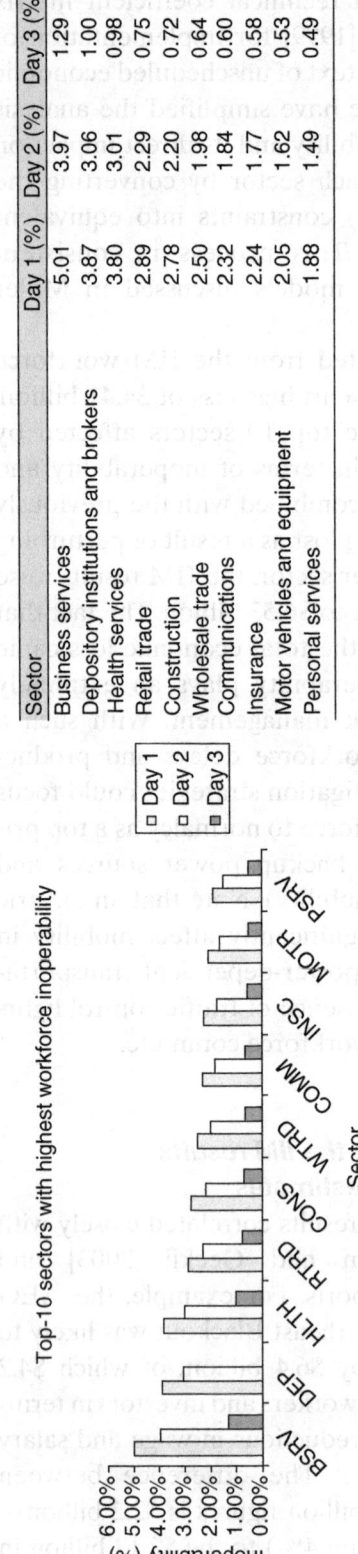

Figure 19.4 *Loss and inoperability from the workforce perturbation.*

Loss ($M)	Sector description	Index
1018	Business services	34
845	Depository institutions	29
379	Health services	36
372	Retail trade	28
335	Miscellaneous services	37
261	Wholesale trade	27
238	Real estate	31
179	Construction	6
99	Insurance	30
97	Communications	25

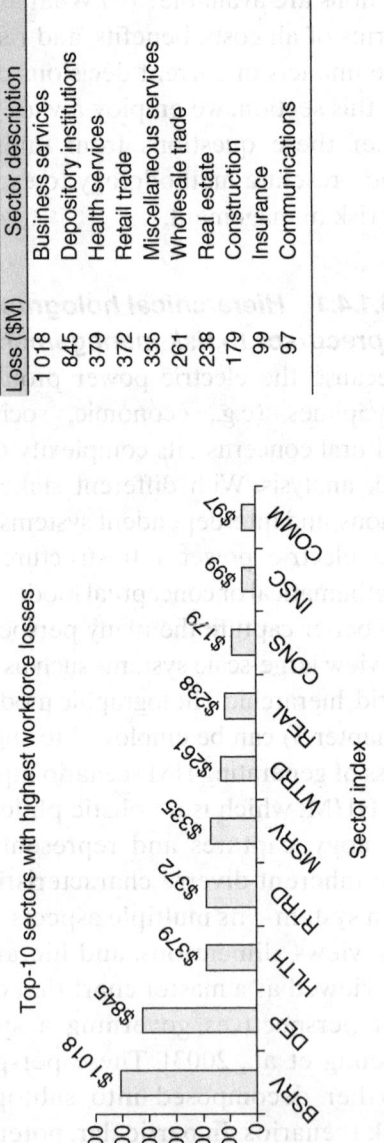

political, and industrial compositions of the various components of the electric power infrastructure; (ii) *sectoral*, relates the impacts of power perturbations to other sectors (e.g., transportation, communication, health services, food supply, etc.); (iii) *temporal*, considers the dynamic behavior of risk over varying periods of time; (iv) *workforce*, observes various outcomes that major disruptions will have on human resources, such as transportation difficulties or psychological hysteria; (v) *social*, reflects the work breakdown structure associated with different areas of social response to a widespread utility failure (as inspired by New York City's drastically different responses to the 1977 and 2003 blackouts); (vi) *emergency response*, examines the interplay between different aspects of emergency response activities and management of *responsibilities*; (vii) *political*, looks at government policies and regulations associated with power grid operation and the role of state officials in expediting recovery from emergency events such as large-scale power outages; (viii) *economy*, reviews the influence of power loss on production output, capabilities, sales, and substitution effects across the economy; and (ix) *security*, incorporates *lessons learned* and best practices to minimize power infrastructure vulnerabilities for better protection in the future.

The above perspectives provide a starting point that policymakers could use to identify a wide array of possible risk scenarios. After developing a comprehensive set of risks, the systemic process of risk filtering and ranking (see Chapter 7) can help prioritize the scenarios for risk management.

19.1.4.2 Relating the IIM to risk management
The task force recommendations from UCPSOTF [2004] underscore the importance of identifying and developing risk management solutions to prevent another mass power outage. Citing compromised independence and repeated conflicts of interest, the task force recommended that a regulator-approved mechanism be developed to objectively fund the North American Electricity Reliability Council (NERC) and the regional reliability councils. Instead of the dues currently paid by market participants to fund NERC's $13 million annual budget, such a proposed mechanism would derive funding from a surcharge on transmission rates and free the councils from responsibility to the parties they oversee. The final report notes that this change would increase NERC's budget but concludes that the additional costs are relatively small compared to costs of another major blackout.

The financial feasibility of implementing risk management solutions can be assessed using the IIM. Assuming that the aforementioned UCPSOTF recommendation would reduce the geographical scope and the recovery time of potential blackouts, the IIM can demonstrate the economic viability of this recommendation relative to the cost of another major power failure. With the IIM for the Northeast Blackout region presented earlier, we conducted a parametric analysis to find the trade-off point at which economic and workforce loss from electric power inoperability would equal the $13 million NERC budget. At this level, the additional investment from the transmission rate surcharge would essentially pay for the savings gained by avoiding additional blackout losses. The IIM shows that only a 0.33% outage to the UTIL sector—a significantly smaller value compared to the actual outage percent experienced during the 2003 Greater Northeast Blackout region—would create a 1-day loss of approximately $13 million. Hence, if implementing the new system (i.e., the proposed funding system that will allow more freedom for the NERC and more accountability from market participants) would prevent at least a 0.33% outage during the lifespan of the proposed system, one can say that the decision is economically feasible.

19.1.4.3 Evaluating the efficacy of risk management
This section discusses a framework for evaluating the efficacy of risk management options that can reduce the economic loss effects of regional blackouts. We adapt the framework proposed by Haimes and Chittester [2005] in quantifying the net benefit of available risk management options. Let the notation $i = 1, 2, \ldots, n$ represent each of the sector categories included in the analysis (Table 19.2 shows the $n = 37$ BEA sectors that have been utilized in the case study). Subsequently, $\mathbf{x}$ is defined as the as-planned production vector, while $\tilde{\mathbf{x}}$ represents

a degraded production vector. The economic loss for each individual sector i can be calculated as the difference between the as-planned production and the degraded production of the ith sector:

$$\Delta x_i = x_i - \tilde{x}_i \tag{19.11}$$

Santos and Haimes [2004] show that the economic loss can be represented as a function of the inoperability of sector i:

$$\Delta x_i = q_i x_i \tag{19.12}$$

where q_i is the resulting inoperability to sector i derived from Eq. (19.6), which when multiplied by the ideal production (x_i) will yield an estimate of the economic loss (Δx_i). Summing all the individual sectors' economic losses will yield the cumulative economic loss. For a baseline scenario that assumes no explicit application of risk management, the cumulative economic loss to the economy, denoted by $\Gamma_{w[0]}$, can be calculated as follows:

$$\Gamma_{w[0]} = \sum_{i=1}^{n} \Delta x_{w[0],i} = \sum_{i=1}^{n} q_{w[0],i} x_i \tag{19.13a}$$

where $\Delta x_{w[0],i}$ is the economic loss and $q_{w[0],i}$ is the resulting inoperability for sector i. For simplicity, we drop the subscript $w[0]$ because $\Delta x_i = \Delta x_{w[0],i}$ and $q_i = q_{w[0],i}$ from Eq. (19.13a):

$$\Gamma_{w[0]} = \sum_{i=1}^{n} \Delta x_i = \sum_{i=1}^{n} q_i x_i \tag{19.13b}$$

To consider the effectiveness of risk management, one must take into account many factors. These may include the availability and capability of manpower in charge of the deployment and operations of the selected policy options, as well as the extent to which risk managers respond and adapt to the continuously evolving nature of disasters [Haimes and Chittester, 2005]. Applying such factors to the formulation shown in Eq. (19.6), the IIM equation can be customized to take into account the application of risk management policies. For a specific risk management policy option j, the IIM formulation is defined as follows:

$$\mathbf{q}_{w[j]} = \mathbf{A}^* \mathbf{q}_{w[j]} + \mathbf{c}^*_{w[j]} \tag{19.14}$$

where $\mathbf{q}_{w[j]}$ is a new level of inoperability and $\mathbf{c}^*_{w[j]}$ is a new level of perturbation resulting from a scenario with risk management policy j in place. On the other hand, the equation for the cumulative economic loss to the economy *with* risk management policy option j is

$$\Gamma_{w[j]} = \sum_{i=1}^{n} \Delta x_{w[j],i} = \sum_{i=1}^{n} q_{w[j],i} x_i \tag{19.15}$$

To evaluate the efficacy of risk management, we first assess its effect on the perturbation input to the IIM (i.e., for risk management j, the corresponding perturbation $\mathbf{c}^*_{w[j]}$ in Eq. (19.14) is first established). The resulting sector inoperabilities are then calculated using Eq. (19.14). When entered in Eq. (19.15), this will yield an estimate of the cumulative economic loss that reflects implementing risk management j. The difference in the magnitude of cumulative economic loss calculated using the baseline scenario $\Gamma_{w[0]}$ and the cumulative economic loss when risk management j is applied ($\Gamma_{w[j]}$) represents the effectiveness of risk management j in reducing the adverse effects of a disruptive event (such as the regional blackout considered here). In addition, each risk management option has associated implementation costs (e.g., capital, maintenance, and other operating costs). Thus, we must consider γ_j, the costs that are associated with implementing risk management policy option j, to determine the resulting net benefit (δ_j):

$$\delta_j = \Gamma_{w[0]} - \Gamma_{w[j]} - \gamma_j \tag{19.16}$$

The δ_j can also be viewed as a potential loss reduction associated with applying the jth risk management policy option. When $\delta_j > 0$, the risk management option is economically viable; hence, its implementation is justified (unless a more detailed feasibility study—considering other dimensions such as political, legal, and ethical—indicates otherwise). The decisionmaker would logically be interested in increasing the value of δ_j, as a larger value reflects a more cost-effective risk management option. In addition to the formulation in Eq. (19.9), we may also represent the efficacy of risk management by the ratio (φ_j)

$$\varphi_j = \frac{\Gamma_{w[0]} - \Gamma_{w[j]}}{\gamma_j} \tag{19.17}$$

where the difference between the magnitude of cumulative economic losses and risk management policy option *j* relative to a baseline scenario is divided by the costs associated with implementing that *j*th option. The notation φ_j represents a benefit–cost ratio where the numerator term is the risk reduction potential (benefit), while the denominator term is the investment required (cost).

To demonstrate this analysis for the blackout study, we conducted a case study of three potential risk management policy options. The economic cost estimates for the policy options enumerated in the following were derived from the American Society of Civil Engineers (ASCE) *Report Card for America's Infrastructure: US Electric Power Grid* [ASCE, 2005]. The ASCE gathers sources from across the public and private sectors to evaluate current conditions, trends, and policy options in the power grid area. As recent growth in electricity demand has not been matched by investment or maintenance expenditures, the report card specifically mentions operations, maintenance, and investment costs to spur action. Examples of these are applied here to risk management of the August 2003 Northeast Blackout to provide a framework for evaluating potential options for future events. In the following are the three risk management policy options considered in the case study and their expected benefits:

- Policy option *j* = 0 refers to the actual recovery process that occurred in the aftermath of the 2003 Northeast Blackout (i.e., the scenario depicted in Table 19.1, which resulted in a cumulative economic loss of $6.4 billion based on IIM calculations). We designate *option j = 0* to be the baseline scenario; hence, in reference to other risk management options, it would require $0 in additional risk management investment. Thus, the net benefit δ_0 for policy option *j* = 0 is $0, as follows:

$$\delta_0 = \Gamma_{w[0]} - \Gamma_{w[0]} - \gamma_0 = \$0 \qquad (19.18)$$

where $\gamma_0 = \$0$ because there are no additional investment costs for this scenario. The default case (i.e., baseline scenario) represents what actually occurred in the aftermath of the blackout.

- Policy option *j* = 1 describes increasing overall investment in the transmission grid to $1 billion; this investment stayed at $286 million annually between 1999 and 2003 [ASCE, 2005]. For the purposes of this case study, we assumed that the Northeast Blackout region would receive 1/3 of that national $1 billion investment. Therefore, the cost γ_1 associated with implementing policy *j* = 1 is $0.33 billion. We then evaluated a scenario for which this investment would reduce the electric power outage by 5% each day, relative to the data shown in Table 19.1. After running the IIM with these new perturbation inputs, we found the cumulative economic loss with risk management policy option *j* = 1 as $\Gamma_{w[1]} = \$5.9$ billion, as compared with the $6.4 billion. Thus, the net benefit $\delta_1 = \$0.17$ billion and the benefit–cost ratio $\varphi_1 = 1.5$ for this option are calculated as follows:

$$\begin{aligned}\delta_1 &= \Gamma_{w[0]} - \Gamma_{w[1]} - \gamma_1 \\ &= \$6.4 - \$5.9 - \$0.33 \\ &= \$0.17 \text{ billion}\end{aligned} \qquad (19.19)$$

$$\varphi_1 = \frac{\Gamma_{w[0]} - \Gamma_{w[1]}}{\gamma_1} = \frac{\$6.4 - \$5.9}{\$0.33} = 1.5 \qquad (19.20)$$

- Policy option *j* = 2 considers increasing national maintenance and operation spending for electric utilities by adding $2 billion to the $3 billion reported in 1999 [ASCE, 2005], for a total of $5 billion. We again assumed that 1/3 of that $2 billion additional investment would occur in the Northeast Blackout region. Thus, this would increase spending on maintenance and operation by $0.67 billion. We then evaluated the effect of that additional $\gamma_2 = \$0.67$ billion investment, assuming it would provide enough additional power to reduce the period of the August 2003 Northeast Blackout from 3 to 2 days (i.e., there are no more outages during the *day 3* column of Table 19.2). If this increase provided the resources necessary to recover to normalcy after 2 days, the total economic loss would decrease from $6.4 billion to $\Gamma_{w[2]} = \$5.6$ billion. Therefore, the net benefit $\delta_2 = \$0.13$ billion and the benefit–cost ratio $\varphi_2 = 1.2$ for this option are calculated as follows:

$$\delta_2 = \Gamma_{w[0]} - \Gamma_{w[2]} - \gamma_2$$
$$= \$6.4 - \$5.6 - \$0.67 \quad (19.21)$$
$$= \$0.13 \text{ billion}$$

$$\varphi_2 = \frac{\Gamma_{w[0]} - \Gamma_{w[2]}}{\gamma_2} = \frac{\$6.4 - \$5.6}{\$0.67} = 1.2 \quad (19.22)$$

19.1.4.4 Multiobjective trade-off analysis

The surrogate worth trade-off (SWT) method [Haimes and Hall, 1974] (see Chapter 5) provides a methodology to address these multiple-objective problems. For the three policy options described in Figure 19.4, we consider two-objective functions, f_1 and f_2, denoting the cumulative economic loss and policy cost, respectively. Policymakers would aim to minimize both of these objective functions in order to minimize economic loss while investing the minimal amount of resources necessary. The first objective is to minimize the cumulative economic loss (f_1):

$$\min f_1 = \Gamma_{w[j]} = \sum_{i=1}^{n} \Delta x_{w[j],i} \text{ from Eq. (19.15)} \quad (19.23)$$

where $\Delta x_{w[j],i} = \Delta x_i$ when $j = 0$, the baseline scenario from Eqs. (19.13a) and (19.13b). The second objective is to minimize the investment cost (f_2):

$$\min f_2 = \gamma_j \quad (19.24)$$

Recall that the notation $\Delta x_{w[j],i}$ refers to the economic loss for each sector i, which is a function of the policy option j, while γ_j represents the investment (or implementation) cost. An effective policy option is capable of driving down the value of the expected economic loss to the minimum level possible subject to the constraint of allowable implementation cost. To solve the two-objective optimization problem via the SWT method, we convert Eqs. (19.23) and (19.24) into the following ε-constraint formulation:

$$\min f_1$$
$$\text{subject to } f_2 \leq \varepsilon_2 \quad (19.25)$$

We can reformulate Eq. (19.25) in terms of a Lagrangian function using the ε-constraint approach (see Chapter 5). The problem becomes

$$L(\cdot) = f_1 + \lambda_{12}(f_2 - \varepsilon_2) \quad (19.26)$$

From this equation, a necessary condition for optimality states that

$$\lambda_{12} = -\frac{\partial f_1}{\partial f_2} > 0 \quad (19.27)$$

where λ_{12} represents the trade-off, or slope, between the two objective functions. Once these relationships are defined, the decisionmaker may interact with the model and evaluate different policy options, subject to preferences regarding constraints such as cost and acceptable risk.

To visualize the trade-offs between different policy options, a decisionmaker may plot each option's investment cost, γ_j, against the corresponding cumulative economic loss associated with each risk management policy option, $\Gamma_{w[j]}$. The resulting graph gives a sense of the potential *returns* associated with each level of investment. The trade-offs at each point between policy costs and economic losses are represented by the slope λ_{12}. The results from the three risk management policy options previously computed in Section 19.1.4.3 are graphically shown in Figure 19.5. For this scenario, we find that $\lambda_{12} = 0.66$ at the location where policy option $j = 1$. This value of λ_{12} simply shows us the ratio of investment $j = 1$ with cost of $\$33.3$ million (f_1) with respect to its economic loss reduction of $\$5.9–\$6.4$ billion. (Note that $\lambda_{12} = 0.66$ is the reciprocal of $\varphi_1 = 1.5$ calculated from Eq. (19.20).) Such trade-off analysis helps policymakers who face multiple, noncommensurate, and often conflicting objectives in most (if not all) real-world decisionmaking problems.

Evaluating the efficacy of risk management as depicted in Figure 19.5 is a repeatable framework that can be generalized to other risk management options. The analysis can improve as decisionmakers incorporate more detailed estimates of implementation costs and their corresponding reductions in terms of recovery time, number of affected sectors, and geographical scope of major blackouts. For example, policy options $j = 1, 2$ can be combined to synergistically form another option (say, $j = 3$),

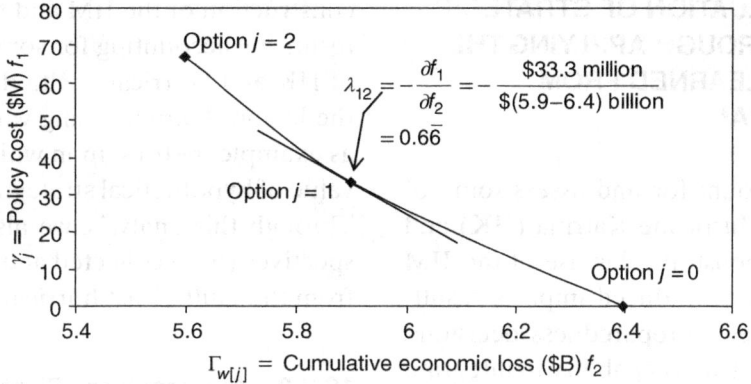

Figure 19.5 Sample risk management trade-off analysis.

which could be more cost-effective than any of the original scenarios. The IIM's capability to evaluate the efficacy of available risk management options can provide policymakers with important insights by comparing the economic cost of future blackouts relative to the costs (capital, operation, and maintenance) of those options. Risk management can enhance both the infrastructural and organizational integrity of the electric power systems. In conducting cost–benefit–risk analyses, policymakers often face the following difficult questions: (i) What constitutes an acceptable level of risk (i.e., safety)? (ii) How robust are the current safety factors for preventing future disasters? Just as these questions are applicable to the electric power infrastructure, they also serve as guiding principles for assessing the risks to other critical infrastructures and evaluating the potential risk management options.

19.1.5 Summary and Conclusions

This section presented a framework for modeling, assessing, and managing the risks associated with the August 2003 Northeast Blackout. While the blackout sparked a renewed commitment to reduce the likelihood of similar future events, this study provides a framework from which future extreme events may be forecasted and evaluated. An accurate risk assessment of possible complex system failures is important in order to effectively manage potential solutions. Although the IIM has been deployed in part for planning purposes [Haimes et al., 2005a, b], it is a valuable tool to conduct an ex post analysis of the blackout, identifying critical sectors and calculating impacts in terms of both inoperability and economic loss. The loss of $2.12 billion from the electric power perturbation and $4.41 billion from the workforce disruption matched well with published findings by the AEG and the USCPOTF. In addition to the total loss estimates, however, the study also showed how the IIM can be applied to all aspects of the US economy to identify the specific effects of an event on a sector-by-sector basis.

To effectively evaluate decisionmaking policies, one must be cognizant of the trade-offs between risk and resource allocation issues. For example, to harden the electric power infrastructure, one may consider how the risk of power outages may be reduced with an increased level of investment in electric power security. Resource allocation and other risk management options to restore the sectors rendered inoperable by an outage can be addressed via sensitivity and uncertainty analyses. The strength of the IIM results in our study is also evident from the intuitive perspectives provided by the numerical analysis. Calculations of lost productivity can be easily communicated and understood by most decisionmakers. A danger of using computer-generated metrics for risk management scenarios, however, is that some users may instinctively trust the computer's results. It is important to note that the results only provide initial findings for this particular scenario; other external factors, such as market trends and human variability, can lead to important changes.

19.2 SYSTEMIC VALUATION OF STRATEGIC PREPAREDNESS THROUGH APPLYING THE IIM WITH LESSONS LEARNED FROM HURRICANE KATRINA[3]

In this section, we account for and assess some of the major impacts of Hurricane Katrina (HK) and Hurricane Rita to demonstrate this use of the IIM and illustrate hypothetical reduced impacts resulting from various strategic preparedness decisions. The results indicate the IIM's capability to integrate various data sources into singular results and to guide the decisionmaking processes involved in developing a preparedness strategy.

19.2.1 Introduction

More than three decades ago, White and Haas [1975] collaborated to publish a classic assessment of natural disaster research wherein they reported escalating costs to lives, property, and the economy of disasters that extended the current research. They explained that technical disaster research had failed to result in a significant impact on regional preparedness due to a lack of social, economic, and political factors that would lead to adapting such research. Interestingly, they illustrated these points with familiar anecdotes of decisions made over 30 years ago, such as permitting "millions of people [to reside] in coastal areas where one day they will be hit by hurricane wind and storm surge, ... without also providing adequate means of evacuating the area when [a] storm warning is issued" [p. 3]. Other assessments such as Lindell [1997], Mileti [1999], Tierney et al. [2001], and Zimmerman [2005] continue to make similar recommendations for tighter integration of research and decisionmaking beyond the progress in disaster research.

This section addresses the need to develop risk management methodology and *risk-based formulas* that enhance the ability of decisionmakers to understand trade-offs and prioritize preparedness activities across multiple regions and sectors of the economy. Building on the IIM and its derivatives enables governing regions to valuate and prioritize strategic preparedness efforts. We explore the construction of the IIM and the interpretation of its results by accounting for some of the major impacts of HK and Hurricane Rita during August 2005 and the lessons learned, using various open-source data as example metrics upon which we demonstrate the value of hypothetical strategic preparedness options. Through this analysis, we also present several perspectives (both collected and original) of the impact from the Gulf Coast hurricanes of 2005.

19.2.2 Approach for Strategic Preparedness Valuation

Strategic preparedness connotes a decision process and its resulting actions, implemented in advance of a natural or man-made disaster. It is aimed at reducing disaster consequences (e.g., recovery time and cost) and/or likelihood to levels considered acceptable (through the decisionmakers' implicit and explicit acceptance of various risks and trade-offs). It is harmonious with the risk assessment and management process guided by the two sets of triplet questions presented in Chapter 1. In risk assessment, we ask: What can go wrong? What is the likelihood? What are the consequences? [Kaplan and Garrick, 1981] These fundamental questions have resulted in various methodologies that capture sources of risk and estimate their likelihoods and consequences. In risk management, we ask: What can be done and what options are available? What are the associated cost–benefit–risk trade-offs? What are the impacts of current decisions on future options? [Haimes, 1991] Answers to these questions critically enhance the capacity to plan strategic decisions about acceptable levels of risk and preparedness.

Solving strategic preparedness problems is not straightforward due to the complex and multifaceted nature of preparedness—it is a multiobjective, multilevel decision process requiring many trade-offs in the allocation of scarce resources (see Chapter 5). Moreover, it is infeasible to eliminate all risk to largescale systems. Some *after action* and retrospective evaluations during and post recovery often ignore this complexity by criticizing preparedness efforts without recognizing the inherent required trade-offs (and their impacts) that had been made at the preparedness stage. Indeed, a systemic process is required

---

[3]This case study is based on the paper by Crowther et al. [2007].

to evaluate investments in preparedness and resilience to account for competing objectives. Such a process should account for economic interdependencies and the distribution of impacts; define impact groups rigorously in terms of regions, economic activity, and time; and be tractable, cost-effective, and holistic to produce reasonable estimates of trade-offs between preparedness options. Most importantly, this process must be able to integrate distributed results generated from various analyses into a single picture of strategic preparedness options and associated trade-offs at an aggregate level.

19.2.2.1 Preparedness valuation approach

In the previous case study (see Section 19.1 on the Northeast Power Blackout), we introduced a method of estimating the economic value of risk management by calculating the difference between loss estimates with and without specific risk management options for cybersecurity [Haimes and Chittester, 2005]. Burton et al. [1978] present a multistep philosophy for making decisions about proper preparedness against natural hazards, which includes assessing regional vulnerabilities, perceptions, and decision processes and comparing them against possible solutions to indicate the best regional adjustments for preparedness. These steps can be generalized and adapted into a risk analysis framework for strategic preparedness, where the IIM plays a major role. Figure 19.6 illustrates the framework.

The standard V-shaped iterative process depicted in Figure 19.6 illustrates the various levels of analysis. The three left descending steps depict the decomposition and distribution of analysis work, while the right ascending steps depict collecting and integrating assessments and analyses. The stacked boxes indicate a level of distribution, and the width of the V indicates the level of decision coverage. At the top of Figure 19.6, a small number of decisionmakers make decisions that cover large regions, multiple scenarios, and many sectors. In contrast, at the bottom, highly distributed analyses are detailed at the intersection of specific regions, scenarios, and economic sectors. Proper integration of these analyses is critical for effective decisionmaking at the highest level (e.g., the national government). Such V-shaped decision processes are common in systems engineering design literature, such as Buede [2000] or Sage [1992]. This particular process is used to illustrate the value of utilizing the IIM for strategic preparedness decisions.

The IIM framework provides several benefits that are compatible with the decomposition and integration steps. It uses the North American Industry Classification Systems (NAICS) taxonomy, which defines industry groups that have related production activity and relies on the US Census Bureau definition of regions. A decomposition guided by such a taxonomy enables a structured way to define sectors systemically and benefit from the Census Bureau's major data collection efforts. Given the definition that *vulnerability* is *a manifestation of the inherent states (characteristics) of the system*, the IIM provides several metrics that can be used to structure distributed vulnerability analyses of step 4 to enable the smooth integration of independent results. These metrics include inoperability, reduced production output, demand reduction, and workforce reduction.

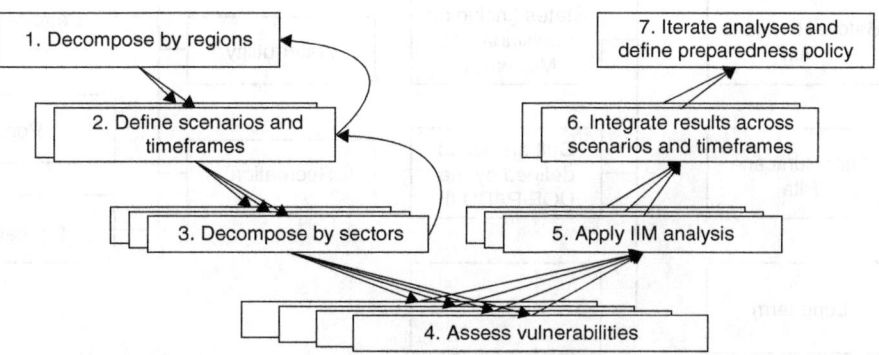

Figure 19.6 A process for strategic preparedness policy development.

19.2.3 Industry and Infrastructure Impacts of HK on the Gulf Coast Region

This section provides several perspectives of the impacts of HK on the Gulf Coast region. We identify major sectors according to the IIM-compatible data about the region and assess the aftermath of the hurricane based on available data. Similar analyses could be performed by assessing the major impacted sectors and their vulnerabilities to specific scenarios. Impact on a sector output can be used as a measure of its vulnerability and lack of preparedness. Other important perspectives, such as governing organizations, response organizations, and physical infrastructure, are beyond the scope of this section. Figure 19.7 depicts the scope of risk management analysis with respect to steps 1–3 in the strategic preparedness process presented in Figure 19.6.

The data and models developed across the three perspectives in Figure 19.7 are integrated through deploying the IIM to enable analyzing preparedness among connected economic and infrastructure sectors spanning the various regions under consideration in the next section to illustrate step 5.

19.2.3.1 Background on 2005 Gulf Coast hurricanes

HK is by many metrics the worst natural disaster in US history and had a more adverse effect on a region and on the entire United States than any other recorded catastrophe [EAS, 2005]. Yet, the lessons that can be learned from the effects of the disaster are broadly applicable and can enhance strategic efforts toward hardening, preparedness, recovery, and resilience across a large range of future potential catastrophes. HK hit land as a Category 3 on August 29, 2005 [Knabb et al., 2005]. The results were catastrophic, but the risks were the result of trade-off decisions made in the presence of scarce resources and estimates of the known risks. (See, e.g., US Army Corps of Engineers [USACE, 2005] and Fischetti [2001].)

A question explored in this chapter is whether a complete accounting for economic cascading effects

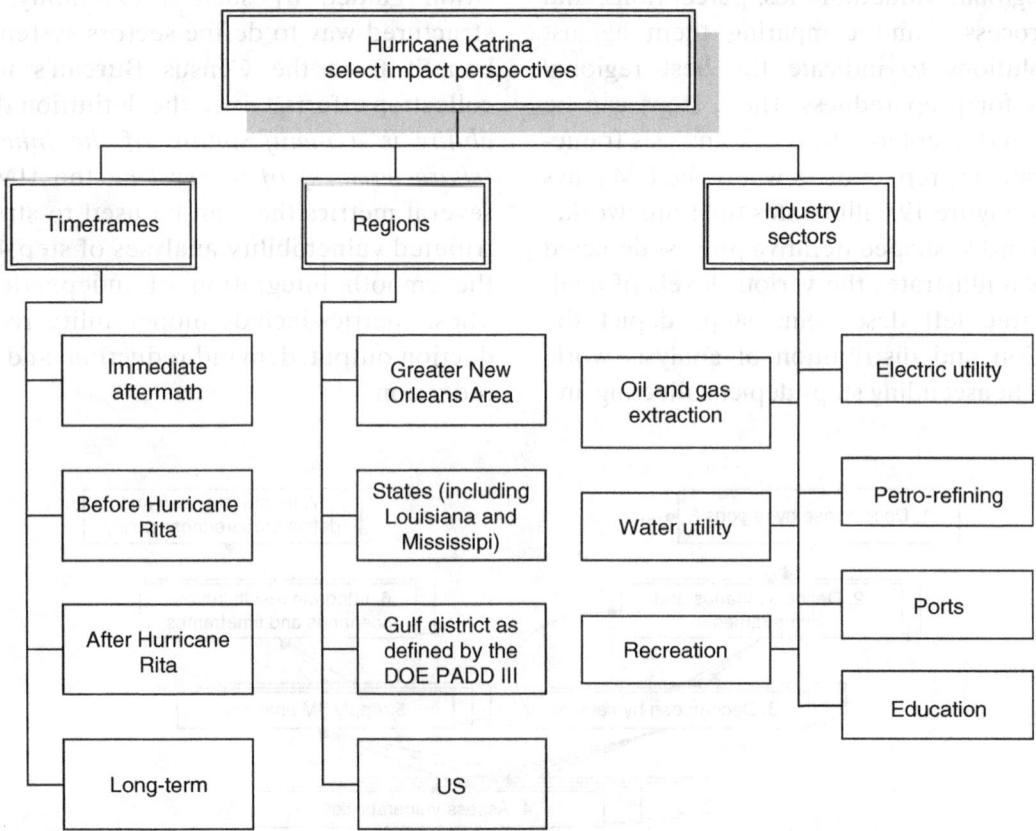

Figure 19.7 Select analysis perspectives to study the impact of Hurricane Katrina.

would have resulted in different decision trade-offs. Louisiana and especially the city of New Orleans were hit extremely hard. The levee system—which protects the city from both the Mississippi River and Lake Pontchartrain—failed, flooding the city.

The entire Gulf region was economically devastated. Close to 1 million people evacuated the region prior to the storm; many of them did not or were not able to return once the storm had cleared. Large numbers of houses and buildings had been destroyed by either the storm or the flooding in the days following. Electricity was out in many areas for extended periods of time. Drinking water treatment plants were unable to fully function in the aftermath of the storm due to damage, loss of electricity, and extreme amounts of debris in water sources [EPA, 2005]. Major industries, which rely directly on the Gulf of Mexico and the waterways, could not be accessed because of the storm [PNO, 2005]. Oil and gas refining, fishing, coffee importation, and gambling are some of the major industries that relied directly on access to water. The high adverse consequences in lives and property loss and the long recovery time for the region and the country seem unacceptable to many due to the lack of preparedness for such a not-unlikely event. Techniques for evaluating preparedness measures can help to illustrate the value of strategic preparedness compared to the cascading impacts of risks in an interdependent economy.

19.2.3.2 Oil and gas sector

The oil and gas sector represents a critical infrastructure of the United States and an important consideration in strategic preparedness for the Gulf Coast region. Over 60% of the US output from the oil and gas extraction industry comes from the Gulf Coast region, and over 40% of our crude oil is refined there [EIA, 2006]. There are two NAICS economic sector classifications of interest in the study of oil and gas. First is the so-called oil and gas extraction industry, and the second is the so-called petroleum and coal products manufacturing industry. These each represent a standard decomposition economic activity for measuring regional economic production and growth. This section reviews data that quantify the impact on these sectors as defined in NAICS, and it deploys the IIM to estimate indirect impacts that would result from the pre-HK structure of the regional economies. Using the prehurricane structure enables us to illustrate what would have been envisioned in the planning process. The result still provides an impact perspective that illustrates the cascading effect of HK. The Energy Information Administration (EIA) reports monthly the actual amounts of refined crude for regions, according to voluntary reports from the refineries themselves.

Deploying the IIM for the oil and gas sector

The IIM can be deployed through the use of interdependency data available before the Gulf Coast hurricanes. This enables estimates of sector inoperability and economic losses that may cascade from direct impacts to the *oil and gas extraction* (OILG) and the *petroleum and coal products manufacturing* (PETR) sectors during the first months after HK. It also serves as a framework to integrate the results and will begin to shape decisionmakers' understanding of preparedness trade-offs. Figure 19.8 presents results from the IIM displaying the top 12 sectors that report regular transactions resulting from activity with the OILG and PETR sectors. Utilizing this information about intersector transactions, we obtain the approximate reductions in manufacturing that result from output constraints to the OILG and PETR sectors. Figure 19.8 shows the estimated cascading economic impacts to Louisiana from the first month after HK, given the direct reduction discussed in the previous sections. The left bars represent the cascading impact from the first month reduction to the OILG sector and the right bars to the PETR sector.

Using the IIM, we estimate that the total loss for the month-long reduction of average 40% in output of the OILG sector is $320 million, representing the sum of all the left-hand bars in Figure 19.8. We estimate that higher-order impacts cascading from reduced refinery output result in approximately $800 million in losses to the State of Louisiana for the 1-month period directly after the hurricane. The sectors that experience cascading losses are those that most highly utilize the supply of refined crude and as a result are impacted due to a reduction in its output. Such knowledge of cascading impact enables decisionmakers to begin to scope proper funding

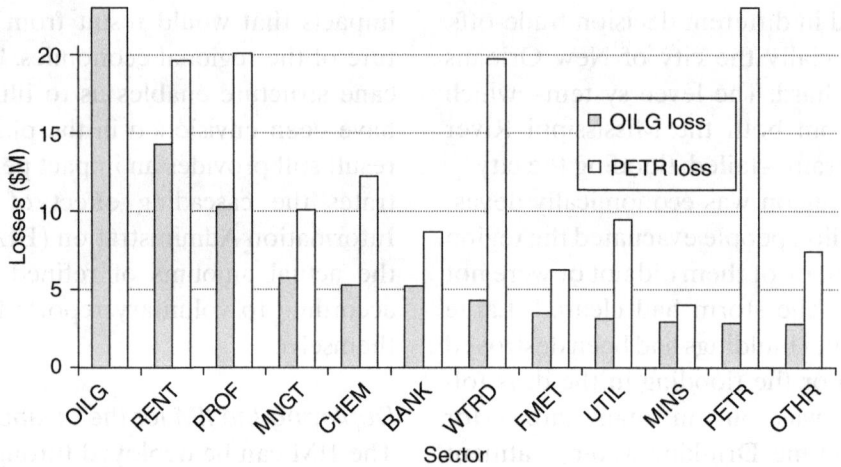

Figure 19.8 IIM results: top 12 cascading economic losses from PETR and OILG sectors.

for preparedness activities and to focus preparedness decisions on economic sectors that impose large indirect impacts to the community.

In addition to approximate the magnitude of the total impact compared to the direct impact, the IIM framework enables us to estimate which sectors are receiving the heaviest indirect impacts. One might expect a reduction in the impact on the *chemical manufacturing* (CHEM) sector. However, Figure 19.8 points to several service sectors that received large indirect impacts due to the output constraints on the OILG and PETR sectors. These include the *rental and leasing services and lessors of intangible assets* (RENT) and *professional, scientific, and technical services* (PROF). The large magnitude of indirect impact calls for further investigation, during future iterations of the preparedness decision process, to discover the specific connection among these sectors. Such up-front knowledge of large indirect impacts enables decisionmakers to better allocate research efforts during preparedness activities.

Finally, as a preview of step 5 in the example preparedness decision process, the IIM enables us to integrate all the impacts from the aforementioned vulnerability assessments. IIM results presented in Figure 19.9 show the top 10 sectors that receive strictly indirect effects from direct constraints to the outputs of the OILG and PETR sectors. The approximate portion of indirect impact that initiates from a particular sector is illustrated by the color of the bar.

The total combined impact calculation generated by the IIM results from constraining OILG and PETR simultaneously yields a total impact that is only slightly larger than the single impact on the PETR sector. The total approximate economic loss resulting from disruption of the OILG and PETR sectors in Louisiana is $870 million for the first month following HK. Note that this is smaller than the sum of direct impacts calculated previously. The reason for the small increase when the total impacts are combined is that impact to the PETR sector gives rise to some of the reductions that would have resulted from the impact to OILG individually. This is explained by the fact that when the PETR sector is unable to refine fuel, then it no longer needs a certain amount of extracted crude. Therefore, all or some of the independent reduction occurring from OILG reduction would happen anyway from a different initiating source. Unifying and integrating views of impact are critical to making broad decisions about strategic preparedness and will be discussed in more detail later. It is important to note that individual analyses that disregard higher-order impacts could be misleadingly large or small when making decisions about a regional system.

This type of preparedness analysis through the IIM enables decisionmakers to begin to understand the range of trade-offs that are accepted when not preparing specific industry sectors against natural and man-made disasters. The total impact, including the higher-order cascades of losses, becomes an important factor for the governing agent responsible

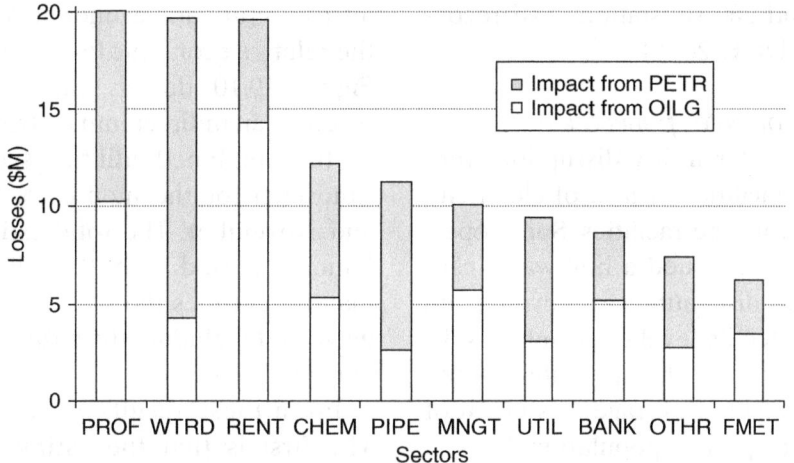

Figure 19.9 IIM results: top 10 sectors that receive indirect impacts from combined supply constraints of OILG and PETR, along with the approximate sources of the indirect impacts.

for preparing the regional community for optimal resilience and recovery. These results are still industry specific, but the IIM provides a framework where many such direct and indirect impacts can be accumulated to provide impact summaries for various disaster scenarios and preparedness efforts, as will be seen in Section 19.2.4.

If we expand this analysis to the Gulf Coast region Petroleum Administration for Defense Districts (PADD III), the initiating perturbation becomes smaller because the impact was concentrated in a small portion of the region. However, the total impact from OILG and PETR sectors accumulates to approximately $5.1 billion for the first month after HK. This amount is large compared to various options available for preparedness. However, if this cost is figured in with the probability of 1 month out of 100 years, then it is not sufficiently large to justify any preparation. This is why a critical part of preparedness must be an analysis of extreme consequences, rather than expectations only. Indeed, the expected-value-of-risk measure, when used as the sole metric for risk, can lead to erroneous and misleading results (see Chapter 8).

19.2.3.3 Public utilities (electric power and water)

Public utilities are essential to the economy and to infrastructure sector operations and represent a core resource for any region's economic prosperity. Their vitality seems to be included in most regional strategic planning activities, and together, they are considered a critical infrastructure.

The major electric utility company in the Gulf Coast region is Entergy (electricity and natural gas), while local public works departments offer water and waste treatment services. The following sections review the availability of electric power and water purification operations in both Louisiana and the Gulf Coast region following HK and Hurricane Rita. Though devastated at first, the electric utilities industry was able to recover steadily and relatively quickly following the hurricanes, which behavior we also see in the water infrastructures. The resilience of these infrastructures is likely a result of preparedness investments that had been made due to the criticality of the infrastructures. On the US national level, electric power generation, transmission, and distribution comprise nearly 70% of the utilities sector output and as such are the focus of this section.

Vulnerability analysis of the electric power utility
The Department of Energy (DOE) reported electrical outages in its daily situational reports that varied significantly throughout the Gulf region. Florida was hit first by HK on August 26, 2005, resulting in approximately 1.1 million customers without power, all of whom had full power by September [DOE, 2005]. On August 29, 2005, HK hit Alabama, Mississippi, and Louisiana, resulting in 181 power lines and 283 substations out of service,

of which 152 lines and 260 substations had recovered within a month [DOE, 2005].

Vulnerability analysis of water purification

The major causes of water utility disruption were direct damage to the facilities or lack of electricity or fuel supply to operate the facilities. Some operable purification facilities issued a boil-water caution so that they could distribute water even when they could not guarantee its level of cleanliness. To measure the inoperability of the water utility through the IIM, a weighted average was taken of the three states with respect to population. For the month of September, an average of 23.7% of residents of the Gulf Coast region did not have access to pure water, which decreased to 8.42% for October and to 3.8% for November. These data will be averaged with the electric utility data with respect to the value of the sector output in order to combine them into the IIM framework. This will enable estimating an integrated impact for the region according to step 5 of the preparedness process shown in Figure 19.6.

Deploying the IIM for public utilities

Table 19.3 summarizes the percentage of inoperability reported in the previous sections. Natural gas numbers can be included based on an independent business report [Allbusiness, 2005], which stated that approximately 230,000 natural gas customers were without service in Louisiana and Mississippi and that impacts to natural gas distribution across other areas were more manageable.

The IIM is used as in the previous sections to estimate the cascade of inoperability to other infrastructure and economic sectors and to estimate the relative economic losses to each of these sectors. Figure 19.10 depicts the top nine sectors that received an indirect impact from constrained output of the combined utilities sector in the State of Louisiana for the months of September, October, and November. The total estimated losses for the 3-month period are $201 million, which is approximately 1.5 times the size of the losses to the utility itself (not including the costs of physically replacing damaged assets).

From these results, we derive several insights. The first is that the estimated economic losses from inactive services are much smaller than the impact from other sectors. This is in part due to already enacted preparedness activities that resulted in increased resilience in the utility sectors. A second note is that the affected sectors are much different than those that received impacts from OILG and PETR. For example, *rail transportation* (RAIL) and *construction* (CNST) sectors now appear in the top 10, but *chemical manufacturing* (CHEM) and *truck transportation* (TRCK) no longer receive as significant an indirect impact. When these insights and data are integrated, they enable preparedness decision-makers to begin to understand the magnitude of cascading impacts resulting from sector vulnerabilities. Thus, they would have a framework with which to quantitatively estimate the value of investment needed to provide incentives or to mandate preparedness practices that would ensure a better sector resilience. These results will be integrated to produce a systemic picture of impact for the region in Section 19.2.4.

TABLE 19.3 Summary of Inoperability Percentages for Utilities Infrastructures Presented in This Section

		Electric Utility (67%)	Water Utility (10%)	Natural Gas (23%)	Total
Louisiana	Sept.	0.3338	0.2081	0.25	0.3020
	Oct.	0.0695	0.1537	0.25	0.1194
	Nov.	0.016	0.0565	0.25	0.0739
Mississippi	Sept.	0.1177	0.2474	0	0.1036
	Oct.	0	0.0683	0	0.0068
	Nov.	0	0.0455	0	0.0046
Alabama	Sept.	0.0090	0.2681	0	0.0328
	Oct.	0	0	0	0
	Nov.	0	0	0	0

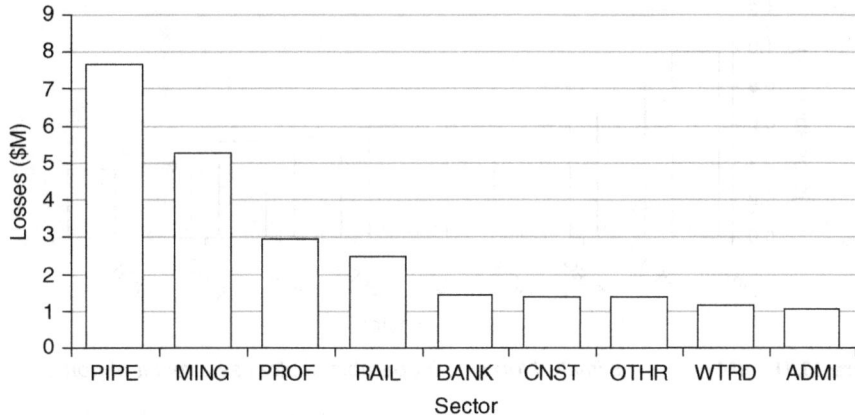

Figure 19.10 IIM results: estimated distribution of economic losses across Louisiana economic sectors.

TABLE 19.4   Rank and Volume of the Ports of Louisiana

Port	US Rank	Millions of Short Tons		
		Total	Foreign	Domestic
Port of South Louisiana	1	217.8	98.6	119.1
New Orleans	4	90.8	52.5	38.3
Baton Rouge	9	65.6	23.1	42.5
Port of Plaquemines	11	59.9	21.0	38.9
Lake Charles	12	55.5	35.0	20.5

From: BTS [2002].

19.2.3.4   Ports and water transportation

All Louisiana ports were ranked in the top 15 highest volume ports by a Bureau of Transportation Statistics report [BTS, 2002]. Table 19.4 is reproduced from that publicly available report.

Port operations and related services in Louisiana comprise a large portion of the *other transportation and support activities* (TRNM) sector. After HK, nearly 50% of employees of the Port of New Orleans (PNO) were without homes. In order to recover the port and its operation, the Maritime Administration (MARAD) provided temporary housing for over 1000 employees [PNO, 2005]. After just over a week, the port was able to resume operations to 40% [PNO, 2005] and quickly recovered until it was operating at full normal capacity by February 2006, ahead of recovery schedule [PNO, 2006]. The Port of South Louisiana is located just a bit further up the Mississippi, not many miles from the PNO. Assuming that both were affected similarly for the 3-month time span, we estimated the approximate impact to the State of Louisiana in terms of inoperability and economic losses of reduced port availability. Results from the IIM are presented in Figure 19.11, which shows the top nine sectors that are unable to operate fully in the absence of the ports. Figure 19.11 represents another metric that is significant in preparedness operations, where sectors with small economic value may incur larger inoperability due to their reliance on other specific infrastructures. The inoperability metric provides a means of understanding the fraction of operation that is impacted.

The indirect impact from the ports is mainly in industries whose operations perform transactions with them, such as the *warehousing and storage* (WRHS) sector. Note again that this would lead preparedness decisionmakers to direct attention to these sectors during future iterations of the preparedness decision process. Section 19.2.4 integrates these results with those from the other economic sectors reviewed in this case study.

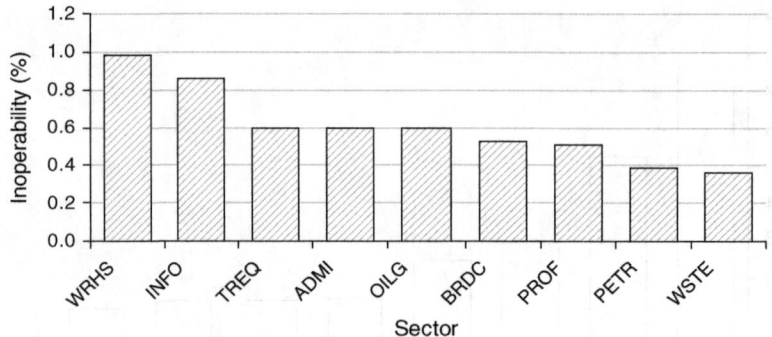

Figure 19.11 IIM results: estimated top nine inoperable sectors as a result of port closures.

TABLE 19.5 IIM Results: Estimated Cascade to Louisiana Economy Due to Closure of Major New Orleans Universities [Tulane University, 2004]

Sector	Inoperability	Sector Description
MPIC	0.0155	Motion picture and sound recording industries
ADMI	0.0147	Administrative and support services
REAL	0.0100	Real estate
PRNT	0.0091	Printing and related support activities
NMET	0.0089	Nonmetallic mineral product manufacturing
INFO	0.0080	Information and data processing services
PUBL	0.0069	Publishing, including software
AIRT	0.0065	Air transportation
GRND	0.0065	Transit and ground passenger transportation

19.2.3.5 Education, recreation, and others

Tulane University is the largest university in New Orleans, employing approximately 8000 people, spending over $650 million/year in the local economy, and attracting almost $500 million/year in local spending by students, parents of students, and professional collaborators [Tulane University, 2004].

HK and the resulting floods resulted in relocating most of the school's operations, which impact cascaded through the local economy of New Orleans and Louisiana. Table 19.5 lists the indirect impact in terms of inoperability, or the approximate percentage of reduced normal operation, to the local economy due to closure of this major university's operations.

The recreation and gambling industry supplied approximately 30,000 jobs and supported a large tourism industry. Because 8 of the 12 floating casinos and the only land casino in the region were destroyed, the vast majority of employees in these sectors were forced to look elsewhere to find jobs. In Biloxi, the casinos were 7 of the 10 largest taxpayers to the city. The industry combined pays a total of $11.6 million toward the city's general fund and another $11.6 million toward the local school and public safety departments. Repairs were expected to take between 2 and 3 years, and the region would continue to suffer financially from the impact of the storm on this industry [Robertson, 2005].

This process can continue through all major sectors that are either critical or vulnerable. The IIM can be used to decompose economic functional sectors and regions, and industry-specific reports can either be gathered or funded concerning the state of preparedness and impacts of well-defined scenarios. In a planning situation, funding should be allocated to industry-knowledgeable teams to make reasonable estimates of probable impacts that can be integrated through the IIM framework. The degree of trust to put into these analyses and create a market for them is an important topic beyond the scope of this chapter.

19.2.4 Efficacy of Preparedness

The impact of any catastrophe cannot be summarized in a single number, just as any picture or painting cannot adequately be summarized in a single paragraph. Similarly, preparedness planning cannot hinge on any single metric such as the avoidance of total expected economic loss. At the beginning of this section, we presented in Figure 19.7 a decomposition of measures for viewing the impact of HK. These include three broad categories of *industry sectors*, *regions*, and *time*. The results in the previous sections attempted to provide a preview of the various impacts to *industries* across various *regions* during various *time* frames after the hurricane. They were estimated through a decomposition directed by the structure of the IIM and sector-specific analyses. The inputs were common metrics that are reported after incidents and could serve as consistent metrics to begin an analysis of impact. Whether or not their totality results in a meaningful estimate of total economic loss is difficult to determine, but multiple such analyses from various perspectives can form a credible and effective process for preparedness planning to prioritize resource allocation, understand preparedness trade-offs, and direct more thorough analyses in a very systematic and systemic way. The IIM becomes an inexpensive and effective tool for integrating multiple analyses into a strategic decisionmaking process, such as that framework presented in Figure 19.6. Using results from the IIM, Figure 19.12 summarizes the impacts from Section 19.2.3 across the population that is supporting and being supported by the Louisiana economy; it also illustrates the distribution of direct and indirect per capita impacts across the entire working population. This metric enables decisionmakers to estimate how the total impact to each sector might be distributed across a working population. The integration was performed similarly to that described in Section 19.2.3.2, wherein the IIM generates the total combined impact by constraining multiple directly impacted sectors simultaneously.

In Figure 19.12, sectors were aggregated into larger economic groups to better enable demonstrating the IIM results in this section. These results show that the petroleum refining, utilities, and oil and gas extraction sectors received large impacts per employee. This impact then cascaded to other industries whose smaller direct impacts from the storm are coupled with the cascading indirect impacts that result from economic interdependencies. To understand the value of strategic preparedness options, the framework in Figure 19.6 would need to be repeated across various preparedness options. Such a second iteration would result in another chart that would incorporate both (i) the increased impact from policy funding (e.g., taxes)

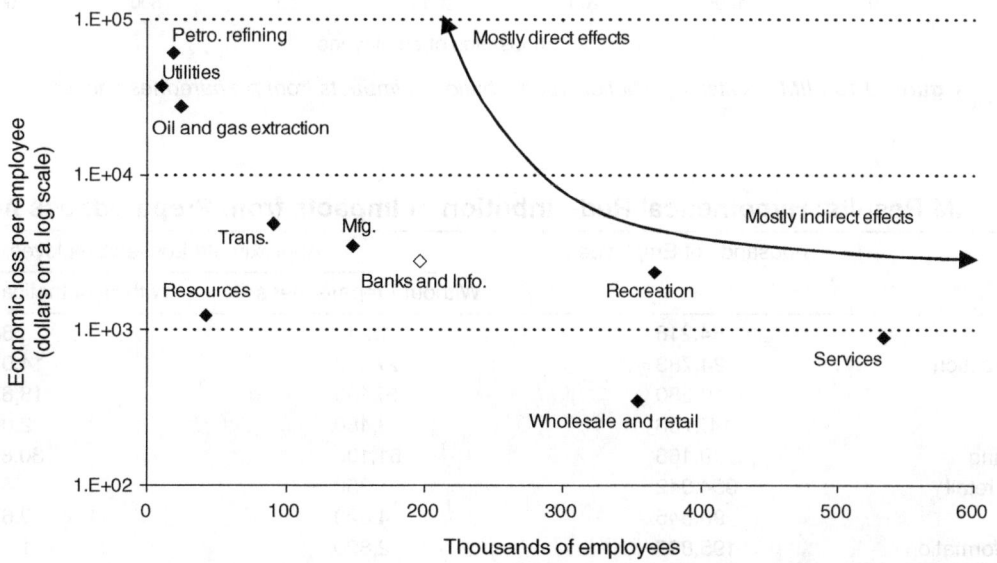

Figure 19.12 IIM results: approximate distribution of direct and indirect impacts across Louisiana economic sectors during the month after Katrina.

and (ii) the decreased impact from the strategy's ability to reduce the direct impacts to sectors of the economy through regulation, audits, and/or other incentives (e.g., preparedness grants). For example, suppose that the State of Louisiana decided to generate $500 million dollars through taxes and to allocate these funds for preparedness activities. Furthermore, suppose that with these funds the Louisiana government supplemented activities that would reduce the direct impacts from Gulf Coast storms by approximately half. Figure 19.13 illustrates the redistribution of impact from such preparedness activities, and Table 19.6 presents the results.

Clearly, the overall impact is reduced due to a lower direct impact on the sectors of concern. However, we see from the chart that if taxes are evenly distributed across employees, the *wholesale and retail trade* sector does not benefit from such a distribution, even in the face of the hurricane scenario. This is due to the fact that the reduced indirect impacts do not exceed the cost of taxes across the number of employees that work in that industry. This insight is possible due to the ability to decompose sectors of the economy in a structured way. Thus, they can be analyzed independently, and the results of independent analyses can be integrated

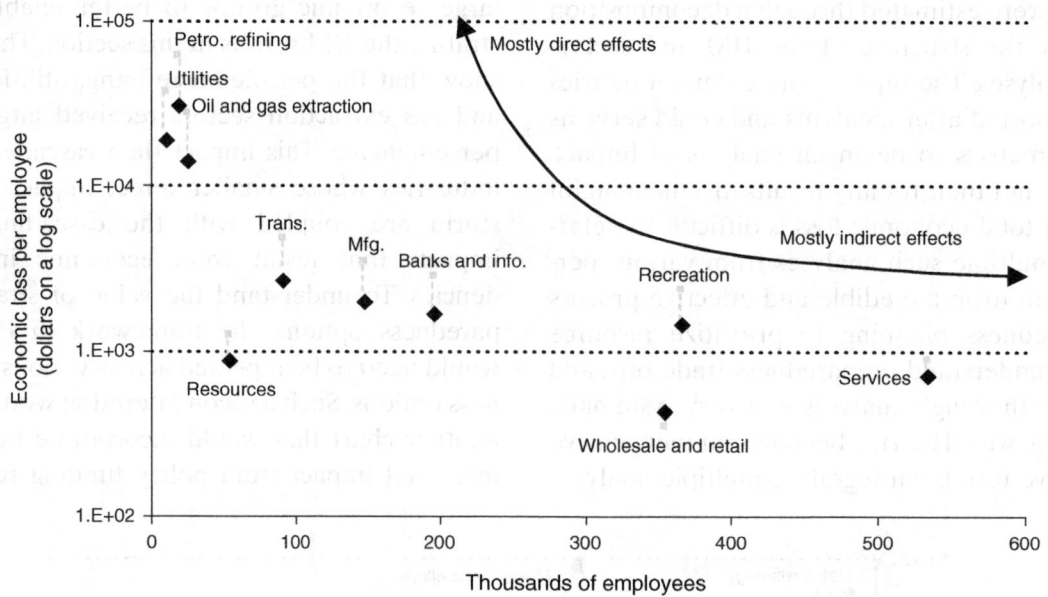

Figure 19.13 IIM results: hypothetical redistribution of impacts from preparedness activity.

TABLE 19.6   IIM Results: Hypothetical Redistribution of Impacts from Preparedness Activity

Sector	Thousands of Employees	Approximate Losses/Employee	
		Without Preparedness	With Hypothetical Preparedness
Resources	54,216	1,240	880
Oil and gas extraction	24,783	27,560	14,030
Utilities	10,280	37,190	18,850
Mfg.	147,753	3,480	2,000
Petroleum refining	19,166	61,100	30,810
Wholesale and retail	354,342	350	430
Transportation	90,545	4,820	2,670
Banking and information	195,807	2,820	1,670
Services	533,822	890	700
Recreation	366,076	2,370	1,440

back into a single picture that incorporates interdependencies represented by economic transactions. Such insights would enable decisionmakers to valuate preparedness actions and understand the trade-offs that are being made by accepting such a strategic preparedness policy. Variations of the policy (such as redistributing the preparedness tax according to risk) could be implemented to make trade-offs more *acceptable*. The IIM provides an integrator to provide well-defined decompositions and vulnerability metrics as well as a methodology to integrate analysis results.

19.2.5 Conclusions

This case study illustrates the use of the IIM for the valuation of strategic preparedness. It is hierarchical and consistent with organizational decisionmaking structures. It thus enables a method by which a region can be decomposed into several operational sectors, the sectors can be assessed independently, and the results can be integrated in a fashion that accounts for economic interdependencies. This process provides a systemic view of a region and enables decisionmakers to focus on the most critical sectors necessary to illuminate the trade-offs between preparedness options. Currently, most methods of assessing impact to large-scale systems for preparedness activities lack details at the broad decisionmaking level or do not effectively integrate multiple high-resolution analyses. The example method depicted in Figure 19.6 naturally emerges from the structure of the IIM and data structures that are standard through the US Census Bureau and collaborating government agencies. The definitions of regions and sectors, coupled with regional jurisdictions' definitions of scenarios of concern, could lead to a decomposition and distribution of vulnerability analyses that can be integrated for a holistic view of the state of preparedness. The IIM and other equilibrium approaches that are grounded on standard datasets allow for such standardization to occur in a framework that is simple and inexpensive and enables findings to be powerfully integrated.

This work has illustrated the usefulness of the IIM through the framework presented in Figure 19.6. It is based on a scenario that has already happened so that the large-scale task of distributed vulnerability analysis could be accomplished in a brief time frame through the gathering of various agency reports. In practice, this process would require defining scenarios that have not happened and several analyses directed from an understanding of the scenario impacts. In the process of illustrating this framework, we have also presented an accounting of the impact from HK on New Orleans, Louisiana, and the Gulf Coast that presents more than single metrics but paints a picture of the impact and approximates how it was distributed across the workforce.

19.3 *EX POST* ANALYSIS USING THE IIM OF THE SEPTEMBER 11, 2001, ATTACK ON THE UNITED STATES[4]

19.3.1 Scenario Description

The case study described in this section is an ex post analysis of the September 11, 2001 (9/11), attack on the United States. Although many sectors of the economy were affected initially by this disruptive event, this economic loss estimation focuses on those sectors that suffered the largest demand reductions. The study uses integrated information derived from I–O matrices (make, use, and capital flow), supported with data published by the Federal Aviation Administration (FAA) [2002] and Ernst and Young [2002]. It considers a 33.2% reduction in passenger enplanements and a 19.2% reduction in hotel occupancy. These demand reduction percentages are then used as inputs to the IIM. The 59-sector NAICS classification scheme is utilized, which can be found in Table 19.7. (Note that aside from the NAICS sector definitions, this table contains other information that will be discussed further in the section on "dynamic multipliers.")

19.3.2 Inoperability and Economic Loss Rankings

Demand-side inoperability (or *inoperability*, for brevity) is one type of metric that results from the IIM analysis. It represents the percentage gap between a sector's *business-as-usual* and current

---

[4]This case study is based on a paper by Santos [2006].

TABLE 19.7 Dynamic Multipliers for the 59-Sector Classification Scheme for Three Lags

Codes	Sector Description	d(0)	p(0) (%)	d(1)	p(1) (%)	d(2)	p(2) (%)
CROP	Crop and animal production	2.42	75	0.70	22	0.10	3
FRST	Forestry, fishing, and related activities	1.96	85	0.29	13	0.04	2
OILG	Oil and gas extraction	2.00	61	1.01	31	0.24	7
MINE	Mining, except oil and gas	2.00	72	0.66	24	0.11	4
SUPM	Support activities for mining	2.01	72	0.66	24	0.10	4
UTIL	Utilities	1.97	69	0.75	26	0.13	5
CONS	Construction	1.95	83	0.34	15	0.05	2
WOOD	Wood products	2.50	88	0.30	10	0.04	2
NMET	Nonmetallic mineral products	2.04	83	0.35	14	0.05	2
PMET	Primary metal products	2.42	85	0.35	12	0.05	2
FMET	Fabricated metal products	2.11	87	0.27	11	0.04	2
MACH	Machinery manufacturing	2.26	87	0.28	11	0.04	2
COMP	Computer and electronic products	2.20	85	0.34	13	0.05	2
ELEC	Electrical equipment and appliances	2.29	87	0.28	11	0.04	2
MOTR	Motor vehicle, body, trailer, and parts	2.79	87	0.35	11	0.05	2
TREQ	Other transportation equipment	2.36	85	0.36	13	0.05	2
FURN	Furniture and related products	2.21	88	0.25	10	0.04	1
MSMN	Miscellaneous manufacturing	2.07	87	0.26	11	0.04	2
FOOD	Food, beverage, and tobacco products	2.52	85	0.38	13	0.06	2
TXTL	Textile and textile product mills	2.59	85	0.38	13	0.06	2
APRL	Apparel, leather, and allied products	2.44	89	0.25	9	0.04	1
PAPR	Paper manufacturing	2.39	84	0.38	13	0.06	2
PRNT	Printing and related support activities	2.17	85	0.32	13	0.05	2
PETR	Petroleum and coal products	2.81	74	0.78	21	0.17	4
CHEM	Chemical manufacturing	2.33	82	0.44	16	0.07	2
PLST	Plastics and rubber products	2.30	89	0.23	9	0.04	1
WTRD	Wholesale trade	1.61	87	0.21	11	0.03	2
RTRD	Retail trade	1.76	81	0.34	16	0.05	2
AIRT	Air transportation	2.03	68	0.80	27	0.13	4
RAIL	Rail transportation	1.90	66	0.84	29	0.13	5
WATR	Water transportation	2.10	75	0.60	21	0.09	3
TRCK	Truck transportation	1.97	80	0.41	17	0.06	2
GRND	Transit and ground passenger	2.37	78	0.56	18	0.09	3
PIPE	Pipeline transportation	2.34	70	0.82	25	0.14	4
OTRN	Other transportation	1.58	84	0.26	14	0.04	2
WRHS	Warehousing and storage	1.50	84	0.25	14	0.04	2
PUBL	Publishing including software	1.72	87	0.21	11	0.03	2
MPIC	Motion picture and sound recording	2.12	86	0.29	12	0.04	2
BRDC	Broadcasting and telecommunications	1.83	69	0.70	26	0.11	4
INFO	Information and data processing	1.60	83	0.28	14	0.04	2
BANK	Federal banks, credit intermediation	1.47	84	0.23	13	0.04	2
SECU	Securities, commodity contracts	1.68	87	0.21	11	0.03	2
INSR	Insurance carriers and related activities	1.92	88	0.23	11	0.03	2
FUND	Funds, trusts, and other financial vehicles	2.51	89	0.26	9	0.04	1
REAL	Real estate	1.45	66	0.62	28	0.11	5
RENT	Rental and leasing services	1.21	59	0.72	35	0.10	5
PROF	Professional, scientific, and technical services	1.48	86	0.21	12	0.03	2
MNGT	Management of companies and enterprises	1.48	89	0.16	9	0.02	1
ADMI	Administrative and support services	1.46	86	0.21	12	0.03	2
WSTE	Waste management and remediation services	1.91	80	0.39	17	0.06	3
EDUC	Educational services	1.81	74	0.54	22	0.09	4
HLTH	Ambulatory health care services	1.53	88	0.18	10	0.03	2
HOSP	Hospitals, nursing and residential care	1.79	78	0.42	18	0.06	3

TABLE 19.7 Continued

Codes	Sector Description	d(0)	p(0) (%)	d(1)	p(1) (%)	d(2)	p(2) (%)
SOCL	Social assistance	1.84	86	0.26	12	0.04	2
PERF	Performing arts, museums, and related activities	1.80	84	0.29	14	0.05	2
AMUS	Amusements, gambling, and recreation	1.58	82	0.30	15	0.05	2
ACCO	Accommodation	1.84	69	0.69	26	0.11	4
FDSV	Food services and drinking places	1.90	85	0.28	13	0.04	2
MISC	Other services	1.67	86	0.23	12	0.04	2

Note: $d(k)$ is the multiplier for lag k and $p(k)$ is the corresponding % relative to the total dynamic multiplier $(I-A-B)^{-1}$.

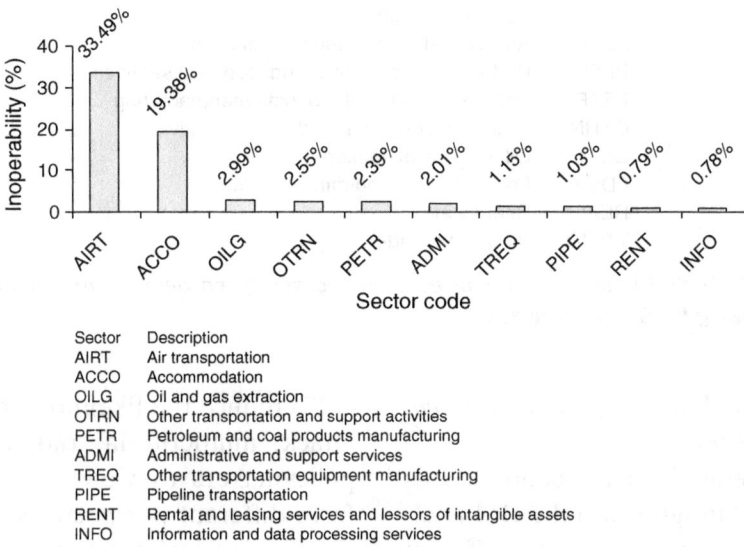

Figure 19.14 *Sectors most affected in terms of inoperability given demand reductions in* air transportation *and* accommodation *sectors following the 9/11/2001 attacks.*

levels of production due to demand reductions caused by a disruptive event. Using the initial demand reductions of 33.2 and 19.2% to the air transportation and accommodation sectors from the 9/11 attacks, respectively, the resulting ripple effects throughout the entire set of US economic sectors are calculated.

The sectors are shown in Figure 19.14: (i) *air transportation*, (ii) *accommodation*, (iii) *oil and gas extraction*, (iv) *other transportation and support activities*, (v) *petroleum and coal products manufacturing*, (vi) *administrative and support services*, (vii) *other transportation equipment manufacturing*, (viii) *pipeline transportation*, (ix) *rental and leasing services and lessors of intangible assets*, and (x) *information and data processing services*.

While the inoperability metric quantifies the nonachievement of a target production level, it is also meaningful to express the resulting impact of a demand reduction in terms of monetary values. Examples of questions that can be raised from Figure 19.14 are: How much economic loss is associated with a 33.49% inoperability of the air transportation sector? 19.38% inoperability of the accommodation sector? 2.99% inoperability of the oil and gas extraction sector? Such questions can be answered by ranking the economic losses as depicted in Figure 19.15. The top 10 sectors with the highest *economic losses* resulting from demand reductions in the air transportation and accommodation sectors are as follows: (i) *air transportation*; (ii) *accommodation*; (iii) *administrative and support services*; (iv) *professional, scientific, and technical services*; (v) *petroleum and coal products manufacturing*; (vi) *other transportation and support activities*; (vii) *oil and gas extraction*;

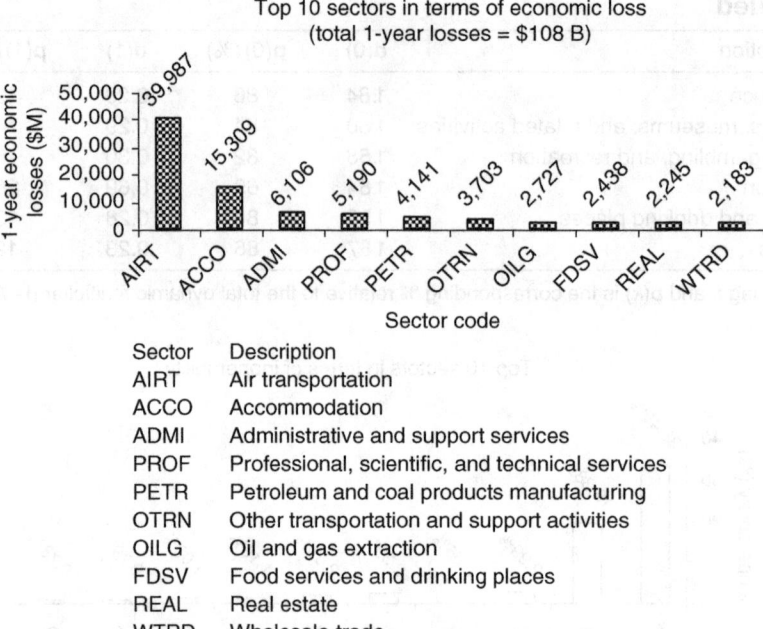

Figure 19.15 *Sectors most affected in terms of 1-year economic losses given demand reductions in* air transportation *and* accommodation *sectors following the 9/11/2001 attacks.*

(viii) *food services and drinking places*; (ix) *real estate*; and (x) *wholesale trade*.

Integrating the inoperability and economic loss metrics offers additional insights into the IIM analysis. Specifically, these metrics generate different sector rankings, which may be attributable to the sectors' different production scales. For example, a $1 million economic loss in one sector (*sector X*) is lower compared to a $10 million economic loss in another sector (*sector Y*). However, *sector X* can have a larger inoperability value than *sector Y* if the ideal production levels are $5 million and $1 billion, respectively. This would lead to a 20% inoperability for *sector X* ($1 million/$5 million) versus a 1% inoperability for *sector Y* ($10 million/$1 billion). Therefore, both IIM metrics, inoperability and economic loss, need to be considered when conducting sector risk assessments because they yield different criticality rankings of sector effects. The effects can be prioritized either in terms of the magnitude of monetary loss or of the *normalized* loss relative to a sector's total production. Logically, different sets of priority sectors are generated depending upon the type of objective being considered (i.e., minimizing inoperability vs. minimizing economic loss).

To reduce the likelihood of a successful terrorist attack, multicriteria trade-off analysis can allow policymakers to view the effects of such attacks from different perspectives. This can provide decisionmaking insights (e.g., hardening of vulnerable sectors and providing redundancies to tightly coupled sectors). When multiple scenarios are considered (in addition to the *air transportation* and *accommodation* sector scenarios considered in the current case study), a more holistic comparison of the resulting rankings of critical sectors can be performed. To some extent, this can provide guidance for generating courses of action to reduce the likelihood and adverse effects of a successful attack.

19.3.3 Dynamic I–O Analysis

A dynamic form of the I–O model is discussed in Miller and Blair [1985]. In discrete form, the mathematical formulation of the dynamic model with time-invariant technical and capital coefficient matrices is as follows:

$$\mathbf{x}(k) = \mathbf{A}\mathbf{x}(k) + \mathbf{c}(k) + \mathbf{B}[\mathbf{x}(k+1) - \mathbf{x}(k)] \quad (19.28)$$

where k is a time index, $\mathbf{x}$ is the total production, $\mathbf{A}$ is the technical coefficient matrix, $\mathbf{c}$ is the final demand, and **B** is the capital coefficient matrix. This dynamic model converges to the static equation when the difference between the outputs approaches zero for two successive periods.

The economic loss estimates provided in Figure 19.15 are underestimated because they capture the losses only for the year following the 9/11 catastrophe. In addition, these estimates do not include other costs such as emergency response, costs for repairing the physical damage and cleanup, and equity losses (e.g., from stock market drops), among others (see related discussions in Center on Contemporary Conflict [2002]). The losses for subsequent years can be estimated via the concept of dynamic multipliers. Liew [2000] derives a total multipliers formula (**t**) that includes a capital coefficient component to reflect losses that are accumulated over several production lags (measured in years). Denoting a vector of ones by $\mathbf{i}' = [1\ 1\ \ldots\ 1]$ and a conformable identity matrix by **I**, we have

$$\mathbf{t} = \mathbf{i}'(\mathbf{I} - \mathbf{A} - \mathbf{B})^{-1} \quad (19.29)$$

The above vector of total multipliers can be decomposed into a series of dynamic multipliers ($\mathbf{d}(k)$) for various production lags k as follows:

$$\mathbf{d}(0) = \mathbf{i}'(\mathbf{I} - \mathbf{A})^{-1} \quad (19.30a)$$

$$\begin{aligned}\mathbf{d}(1) &= \mathbf{i}'(\mathbf{I} - \mathbf{A})^{-1}\mathbf{B}(\mathbf{I} - \mathbf{A})^{-1} \\ &= \mathbf{d}(0)\mathbf{B}(\mathbf{I} - \mathbf{A})^{-1}\end{aligned} \quad (19.30b)$$

$$\mathbf{d}(k) = \mathbf{d}(k-1)\mathbf{B}(\mathbf{I} - \mathbf{A})^{-1} \quad \forall k > 0 \quad (19.30c)$$

The first-ever NAICS-based capital flow data for the United States (which is useful for generating the **B** matrix) was released by the BEA in September 2003. The capital flow data can indicate to some extent the degree of sector dependence on capital commodities (investment in structures, equipment, or software on which other sectors are dependent).

To generate the **B** matrix, it is necessary to first conduct a mapping of the sector classifications used in the capital flow data that is compatible with those used in the **A** matrix. This mapping process enables the formation of an adjusted capital flow data, which involves aggregating the lower-resolution capital sectors to match any of the 59 industry sectors utilized in the **A** matrix. (As one example, uranium, radium, and vanadium ore-mining capital sectors are subsets of the mining except oil and gas industry sector.) The resulting **B** matrix is then established by normalizing the entries along the jth column of the adjusted capital flow data (for all j) with the total production output of the corresponding column sector (i.e., the notation x_j in Eq. (19.1)).

Using Eqs. (19.30a)–(19.30c), Table 19.7 shows the temporal decomposition of the 59-sector dynamic multipliers over five periods. On the average, roughly 80% of the demand reduction impacts are expected to be realized during the first year, while the remaining 20% of the impacts are spread over the remaining years following a disruptive event. The demand reduction scenarios in the 9/11 case study are those that initially render direct disruptions to the air transportation and accommodation sectors.

Applying these same demand reduction scenarios, Figure 19.16 shows the top 10 sectors most affected in terms of 5-year economic losses. When compared to the first-year losses in Figure 19.15, several observations can be made. First, additional economic losses amounting to $50 billion ($158B 5-year loss–$108B 1-year loss) are expected to be realized after the first year. Second, the rankings of the most affected sectors are different (e.g., the wholesale trade sector that was initially ranked #10 in Figure 19.15 is raised to #8 in Figure 19.16). Third, three sectors that were not in the first year's top 10 highest economic losses (computer and electronic product manufacturing, other transportation equipment manufacturing, and construction) are now part of the 5-year top 10 highest economic losses. This result is not surprising because these three sectors produce capital outputs that have relatively low replacement frequencies; this explains the somewhat delayed economic losses.

19.3.4 Discussion

The 9/11 case study was intended as an ex post analysis for model validation purposes. It is worth noting that the economic loss estimates are in the same ballpark as a previously published estimate by the Government Accountability Office [GAO, 2002]: "…while all the metropolitan areas in the country sustained losses of about $191 billion."

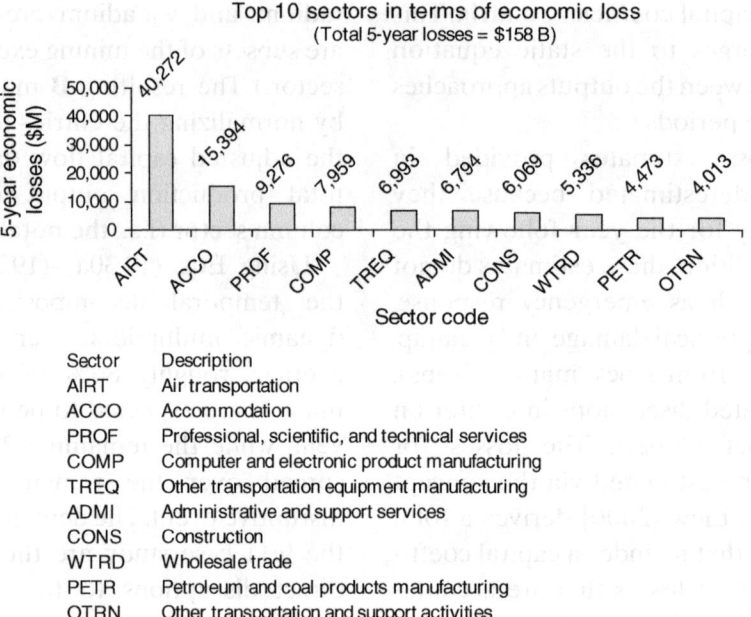

Figure 19.16 *Sectors most affected in terms of 5-year economic losses, given demand reductions in* air transportation *and* accommodation *sectors.*

Although economic loss estimates of the 9/11 attacks have been published previously, the IIM analysis offers additional perspectives on the event and results beyond the scope of other studies. An additional feature offered by the IIM is its capability of showing the distributions of economic losses according to different sectors, using the 59-sector NAICS classification scheme. Typically, economic loss estimates are published in highly aggregated values that would comprise only broad sector categories. Due to the insufficient level of sector details in other published estimates, it is usually difficult to validate the sector rankings obtained here.

Nevertheless, an ex post analysis can be conducted by comparing the actual total output data of each sector after year 2001 with the corresponding data before 2001. (Note that the case study performed here assumed that 9/11 had just occurred.) For example, one can compare the difference in the total output of each sector for years 2002 and 2000 using the BEA data [http://www.bea.gov/bea/dn2/i-o_annual.htm, date accessed July 29, 2005]. Such analysis will reveal the sectors with the greatest losses post-9/11 as follows: computer and electronic products; securities, commodity contracts, and investments; machinery; motor vehicles, bodies and trailers, and parts; petroleum and coal products; chemical products; oil and gas extraction; electrical equipment, appliances, and components; primary metals; and air transportation. When compared to Figure 19.16 results, the majority of these sectors have been included in the ranking of the most affected sectors in terms of economic losses. In addition, one of the equipment-producing sectors, computer and electronic products, which was not ranked in the BEA data, has been captured using the dynamic IIM approach (see Figure 19.16).

19.3.5 Summary and Conclusions

The study highlights the importance of assessing economic interdependencies to identify the sectors that are most sensitive to the adverse effects of a disruptive event. Through the IIM, an I–O-based framework analyzed the negative demand effects of the 9/11 catastrophe on the air transportation and accommodation sectors and also the ripple effects to other sectors in the economy. IIM analysis reveals the most affected sectors through inoperability and economic loss metrics, which can be used to prioritize sectors for planning and evaluating potential policy actions to manage the adverse effects of disruptive events. The resulting rankings produced by the IIM metrics can differ, thus motivating the development of a multicriteria visualization tool to introduce the importance of trade-off analysis.

Several useful databases can be fused to form input scenarios for the I–O computations. These include passenger enplanement data published by the FAA and hotel occupancy data from research agencies such as Ernst and Young. Additionally, posttraumatic stress disorder (PTSD) data and consumer confidence surveys can be used to create perturbation inputs to the IIM, especially for disruptive events that are impending or have just recently taken place. The BEA datasets (e.g., make, use, and capital flow tables) are primarily utilized to generate the coefficient matrices that serve as engines for the inoperability and economic loss calculations. Although this case study considers only one particular IIM input scenario, it is possible to incorporate parametric analysis to determine the sensitivity of sector rankings to different values of demand reductions. For example, one can tweak the air transportation demand reduction value while fixing that of the accommodation sector (and vice versa) to determine the *tipping points* where changes in rank start to occur (i.e., each sector has a different set of linkages to other sectors).

The IIM can serve as a tool for forecasting any future impacts of a disastrous event. For example, the US economy could suffer from some remaining effects of 9/11 for years afterward. This can be potentially quantified via a dynamic I–O framework. The 9/11 case study presented here serves as a demonstration of the IIM. A similar approach can be customized for modeling other disruptive events that can potentially cause prolonged demand reductions (e.g., the effect of the severe acute respiratory syndrome (SARS) epidemic on global tourism).

While we have identified useful features of an I–O-based framework for analyzing disruptive events, it is also necessary to carry out supplementary analyses that deal with other important modeling aspects and dimensions. A holistic 9/11 impact analysis study requires estimating losses other than those resulting from demand reductions. For example, I–O analysis is not appropriate for estimating the physical losses that reduce the production capacity of sectors (e.g., destroyed structures and production equipment—see discussions in Oosterhaven [1988]). Also, not all disruptive events result entirely in demand reductions. Clearly, 9/11 has triggered increased spending on defense, intelligence, and other activities related to homeland security.

In conclusion, the IIM is a useful tool for identifying and managing the sectors that are critically affected by disruptive events. When used in combination with other tools, a more powerful and robust analysis can be performed to address other modeling issues beyond its current capabilities. Therefore, a more detailed effort to integrate I–O models with other tools deserves continued research attention.

19.4 RISK MODELING, ASSESSMENT, AND MANAGEMENT OF LAHAR FLOW THREAT

19.4.1 Introduction[5]

Mount (Mt.) Pinatubo is a volcano that stands 5770 ft, located in the Philippines along the coordinates 15°N, 120°E. After 500 years of dormancy, its eruption on June 15, 1991, was the second most violent volcanic activity in the twentieth century and the largest eruption ever in terms of affected population [USGS, 1997a, b]. In just a few months posteruption, it released nearly 20 million tons of pyroclastic debris, destroyed more than 200,000 acres of land [USGS, 1997c], and caused major casualties and damage. These included the death of more than 700 people and the destruction of more than 200,000 homes [USDOC, 1992]. The huge amount of volcanic materials deposited on the slopes of the volcano, about 1 mi^3 in volume, posed continued threats years after the eruption [Pierson et al., 1992]. Heavy rains visiting the Mt. Pinatubo region during the months of June and October mixed easily with the volcanic deposits to create mudflows called *lahar* (the Indonesian term for volcanic ash). Lahar flows are dangerous because their high volume and speed can easily obstruct natural water channels such as rivers or result in massive bank erosion. The residential communities along the paths of these flows were buried in several feet of hardened lahar—destroying lives, agricultural products, properties, and infrastructures.

---
[5] This case study is adopted from Leung et al. [2003].

Risks of lahar flow were immediately recognized after the eruption. A systematic process to observe lahar was organized within days, and vital scientific data were provided to national and local officials by the Philippine Institute of Volcanology and Seismology (PHIVOLCS), the University of Illinois, the US Geological Survey (USGS), and other research institutions [Janda et al., 1994]. A series of hazard maps were prepared to highlight the areas vulnerable to lahar flows; the first map was made available in August 1991. In addition to this, impact scenarios were developed for three types of rainfall representative of the rainfall pattern in the region [Punongbayan et al., 1992]. These scenarios proved to be accurate; almost all of the predicted events occurred in 1992 and 1993.

The benefits resulting from research efforts geared toward monitoring the lahar flow risks were perceived to substantially outweigh the costs incurred [USGS, 1997b]. With the wealth of scientific information available on lahar threat, it seemed that developing mitigation options would be guided considerably by the information available. However, competing political, economic, and social agendas subordinated the importance of scientific information in policymaking [Janda et al., 1994]. This chapter addresses the issues surrounding the failure to use scientific information in the disaster mitigation policymaking process for the Mt. Pinatubo lahar threat, which resulted in costly, yet futile, mitigation efforts in the initial stages of the lahar flow period.

This case study is organized as follows: Section 19.4.2 documents several disaster mitigation methodologies employed in the aftermath of the Mt. Pinatubo incident, along with their associated benefits and shortcomings. Section 19.4.3 explains how the six questions of risk assessment and management can serve as a road map in the study of lahar flow threats. Section 19.4.4 describes how HHM can be deployed in identifying risks such as those surrounding the Mt. Pinatubo problem. Section 19.4.5 discusses various statistical tools, highlighting the importance of graphic forms of statistical data as aids in the policymaking process. Section 19.4.6 presents two other minicase studies to demonstrate risk analysis tools that capture multiple objectives and extreme events inherent in the Mt. Pinatubo incident. Finally, Section 19.4.7 offers conclusions.

19.4.2 Methodologies for Disaster Management of Lahar Threat

Janda et al. [1994] describe the initial mitigation actions implemented by the Philippine government. The first engineering measures, mostly involving construction of small *sabo* dams, were perceived to be little more than an employment program for affected residents and a show of action by the government. These *sabo* dams proved to be hopelessly undersized and were promptly overrun. Succeeding larger dikes also were no match for the magnitude and erosive power of the lahar flows. However, dike construction persisted, because it presented a more acceptable alternative than resettlement for both local leaders and residents. Therefore, throughout the first 3 years after the eruption, bigger and better dikes were constructed, but inevitably, they were overtopped by lahar flows. In 1993, policymakers were forced to reexamine long-range mitigation plans for the lahar hazard. But at this point, roughly half of the funds available for risk management already had been spent.

Formal methodologies were employed to evaluate the long-term mitigation plans for lahar hazard. Cost–benefit (C/B) models developed by different organizations, including the US Army Corps of Engineers (USACE) and the Japan International Cooperation Agency (JICA), were applied in these long-term planning studies. USACE applied this model in evaluating long-term structural prevention measures for eight river basins around Mt. Pinatubo [USACE, 1994]. JICA and the Philippine Department of Public Works and Highways used the model in a master plan study on floods and mudflow control in the Pinatubo hazard region [DPWH/JICA, 1996]. However, the C/B models proved to be useful only within the context of certain limitations [Dedeurwaerdere, 1998]. First, limited data availability poses serious impediments to the use of the model. Second, the C/B modeling framework requires monetary terms for valuation of cost and benefit. This raises questions on the valuation of nonmonetary factors such as human lives and geological and environmental damages, among others.

Haimes [2001] raises the issue of trade-off analysis in C/B modeling. In essence, the C/B framework involves trade-offs between two conflicting objectives—minimize costs and maximize benefits (or minimize risk/damage). However, C/B analysis converts these multiple objectives into a single-objective problem, precommensurating objectives with multiple dimensions into a single monetary value. Trading off risk with other objectives (usually with cost) inherently involves judging safety—the level of acceptable risk [Lowrance, 1976]. Haimes [2001] suggests assessing risk within a multiobjective framework. This type of framework has been applied to a number of problems, including risk-based management of hurricane preparedness and recovery [CRMES, 2001].

19.4.3 Risk Assessment and Management

The following triplet questions in risk assessment posed by Kaplan and Garrick [1981] were discussed in Chapter 1: (i) What can go wrong? (ii) What is the likelihood that it would go wrong? (iii) What are the consequences? Similarly, the following triplet questions posed by Haimes [1991] were discussed in Chapter 1: (iv) What can be done and what options are available? (v) What are their associated trade-offs in terms of all costs, benefits, and risks? (vi) What are the impacts of current management decisions on future options? Modeling, an integral part of risk analysis, is central to this study and is applied to the ex post data on the Mt. Pinatubo eruption. The three risk assessment questions helped focus on the following issues: (1) the threats posed by the massive lahar deposits on the volcano's slopes during rainy seasons, (2) the predicted frequency and amount of rainfall based on records by weather bureaus, and (3) the quantification of potential damages such as loss of property and lives, among others. In addition, the three questions of risk management allowed us to pinpoint the following issues: (4) identifying specific alternatives to control lahar flow, such as dikes and the excavation of artificial channels; (5) the resource requirements, cost, and completion time for each of the identified alternatives; and (6) the short- and long-run effectiveness of such alternatives.

The two sets of triplet questions of risk assessment and management defined the scope of the framework used in the study, as illustrated in Figure 19.17. An important step in modeling the various aspects

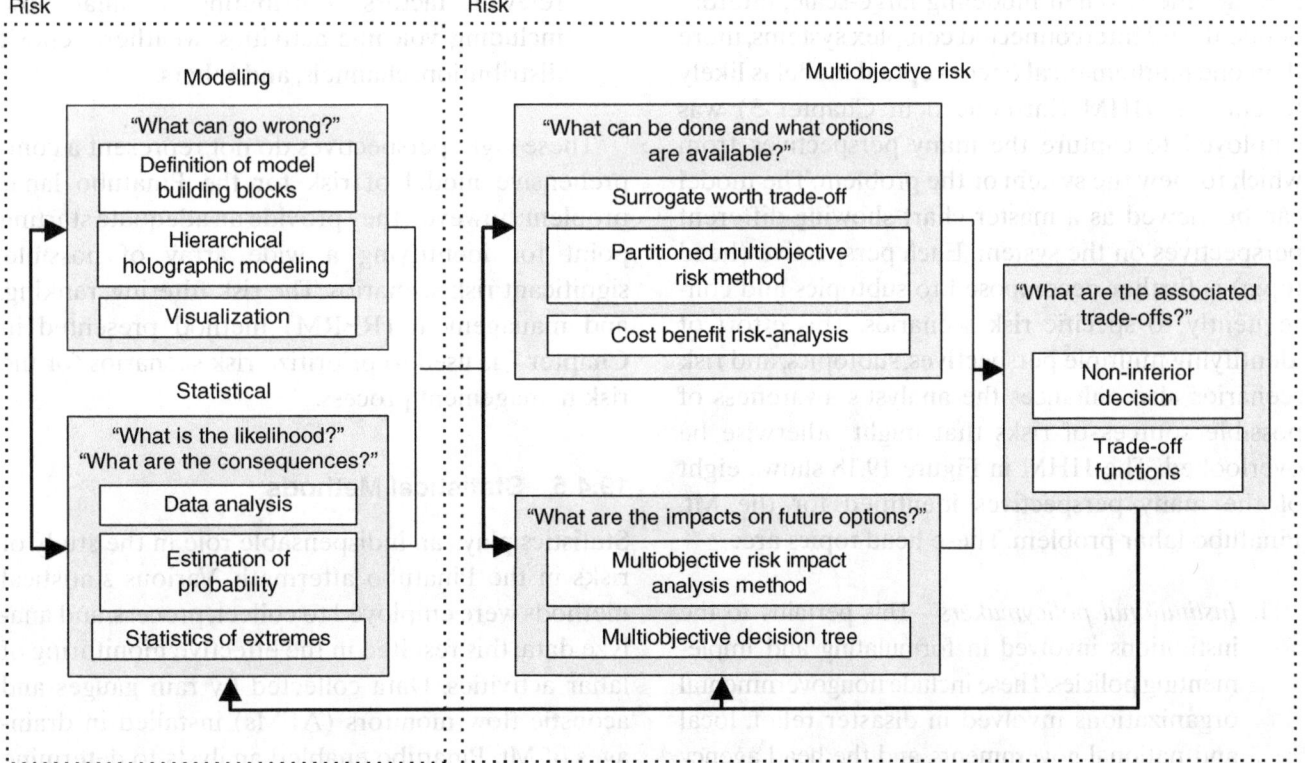

Figure 19.17 Study framework for lahar risk assessment and management.

of the lahar system was accomplished using HHM, introduced in Chapter 3. Then, from the mathematical modeling of the problem, the system variables were identified. The relevant variables and their relationships determined the type of data to be collected and analyzed for this study. Statistical analysis enabled generating suitable probability density functions (pdfs), as well as the functional relationships among the critical variables. A number of multiobjective risk methods can be employed in risk management depending on the nature of the problem. Here, two of these methods were used to illustrate different types of decision problems for lahar hazard mitigation. Finally, noninferior decision policies were generated. There are feedback loops in the system, showing that the process is iterative.

19.4.4 HHM: A Holistic Approach

The Mt. Pinatubo lahar problem is very complex. It crosses scientific boundaries to include social, political, and economic concerns. There are various aspects of risk involved—different stakeholders, issues of engineering capabilities, meteorological considerations, and a government bureaucracy, among others. When modeling large-scale, interdependent, and interconnected complex systems, more than one mathematical or conceptual model is likely to emerge. HHM (introduced in Chapter 3) was employed to capture the many perspectives from which to view the system or the problem. The model can be viewed as a master chart showing different perspectives on the system. Each perspective (head topic) is further decomposed to subtopics and consequently to specific risk scenarios. This effort of identifying multiple perspectives, subtopics, and risk scenarios also enhances the analyst's awareness of possible sources of risks that might otherwise be overlooked. The HHM in Figure 19.18 shows eight of the many perspectives identified for the Mt. Pinatubo lahar problem. These head topics are:

1. *Institutional policymakers*—This pertains to the institutions involved in formulating and implementing policies. These include nongovernmental organizations involved in disaster relief, local and national governments, and the head agency investigating the lahar hazard, the PHIVOLCS.

2. *Sources of failure*—Failures can be attributed to four elements of the system: human, organizational, hardware, and software.
3. *Spatial*—The spatial perspective addresses geographical elements of the system, such as terrain and location of channels, communities, lahar deposits, and others.
4. *Temporal*—This provides for integrating a time dimension in modeling, including seasonal variations, depletion of lahar deposits, population dynamics, turnover of experts, and planning horizons.
5. *Stakeholders*—This lists all the stakeholders in the system in order to incorporate their expectations into the modeling.
6. *Disaster assessment*—This pertains to the elements affected by the risk and damage to the system. It involves human casualties and damages to property, the infrastructure, and the environment, among others.
7. *Disaster management*—This provides a view of all possible risk prevention and mitigation measures available to decisionmakers.
8. *Lahar flow*—Lahar is the primary focus of the investigation. This perspective gives all relevant factors contributing to lahar flow, including volcanic activities, weather, deposit distribution, channels, and others.

These eight perspectives do not represent a comprehensive model of risk for the Pinatubo lahar problem; however, they provide an adequate starting point for identifying a wide array of possible, significant risk scenarios. The risk filtering, ranking, and management (RFRM) method presented in Chapter 7 is used to prioritize risk scenarios for the risk management process.

19.4.5 Statistical Methods

Statistics plays an indispensable role in the study of risks in the Pinatubo aftermath. Various statistical methods were employed to collect, process, and analyze data; this resulted in the effective monitoring of lahar activities. Data collected by rain gauges and acoustic flow monitors (AFMs) installed in drainages of Mt. Pinatubo enabled analysts to determine the categories of lahar activities that are triggered

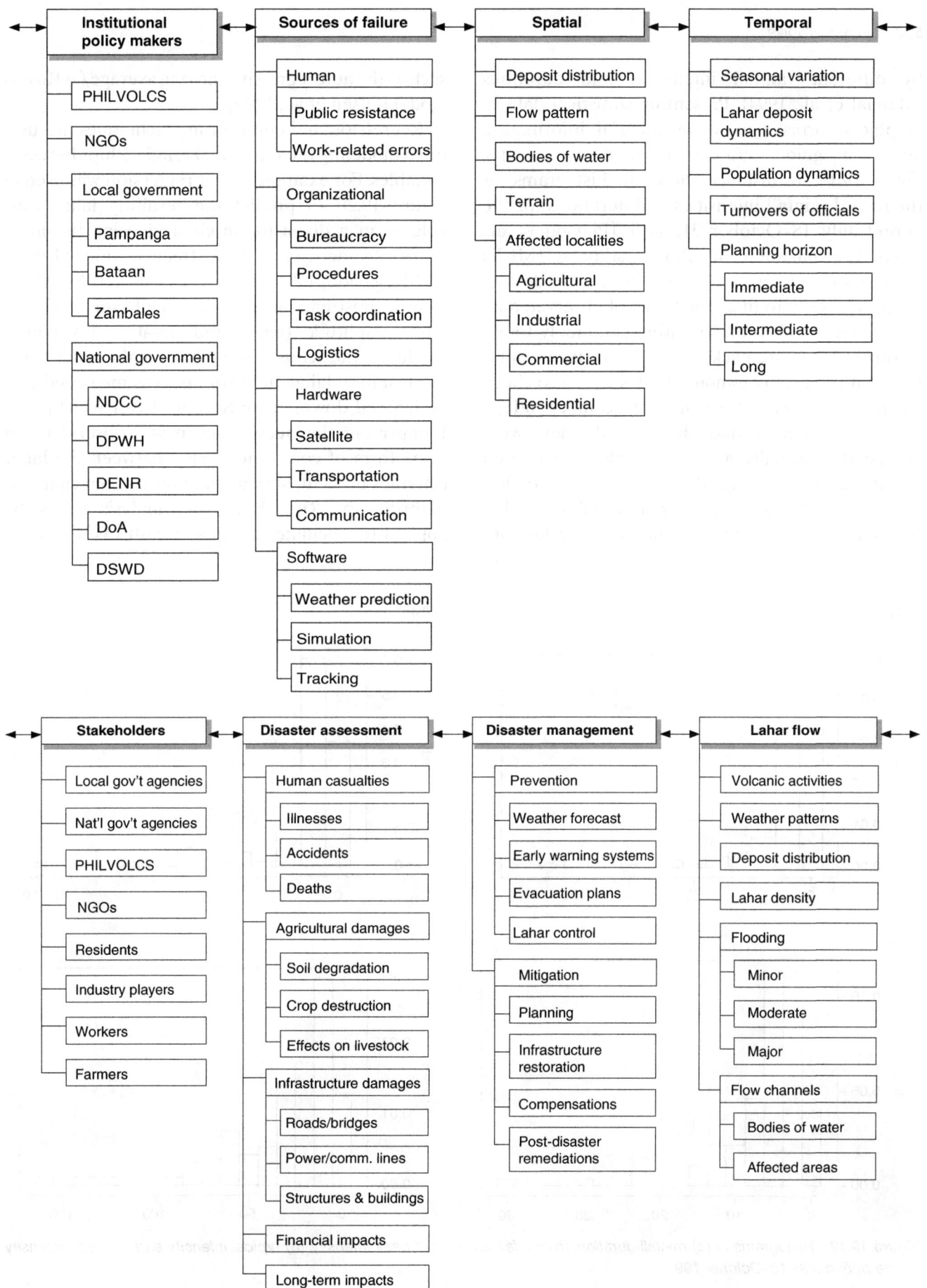

Figure 19.18 Partial HHM (multiple perspectives) of Pinatubo lahar problem.

by different rainfall intensities and durations [Marcial et al., 1994]. Presenting statistical data in graphical forms conveys meaningful information and can guide subsequent modeling efforts. Figure 19.19, for instance, shows the histograms for the record rainfall intensities and durations for the period July 18–October 31, 1991 [Pierson et al., 1994]. The maximum likelihood estimator (MLE) can be used to determine the parameters of a density function chosen to fit a specific set of observed data (i.e., a sample). It is an algorithm commonly found in various software packages. When fitting samples to distributions, a common pitfall is failing to check whether the observations are indeed independent. Treating autocorrelated data as if they were independent usually generates faulty distribution functions. In studying floods, the USACE has embarked on the use of more sophisticated probability functions that incorporate time series techniques such as the autoregressive moving average (ARMA) model [Olsen et al., 1999].

Regression and curve-fitting techniques are useful in modeling the functional relationships between variables. For example, linear regression was used in Figure 19.20 to predict the resulting lahar flow volume in a downstream channel using the measured lahar activity on an upstream channel [Tungol and Regalado, 1994]. The data measured by AFMs located upstream in the Sacobia River were the lahar amplitude (cm/s) and duration (s), whose product is the acoustic flux (cm). On the other hand, the resulting lahar flow volume was measured in a designated downstream Sacobia River watchpoint. Using regression techniques, it was observed that some form of correlation exists between the lahar activities in the upstream and downstream channels of the Sacobia River. Regression analysis can also be applied to calculate the lahar runoff (as measured

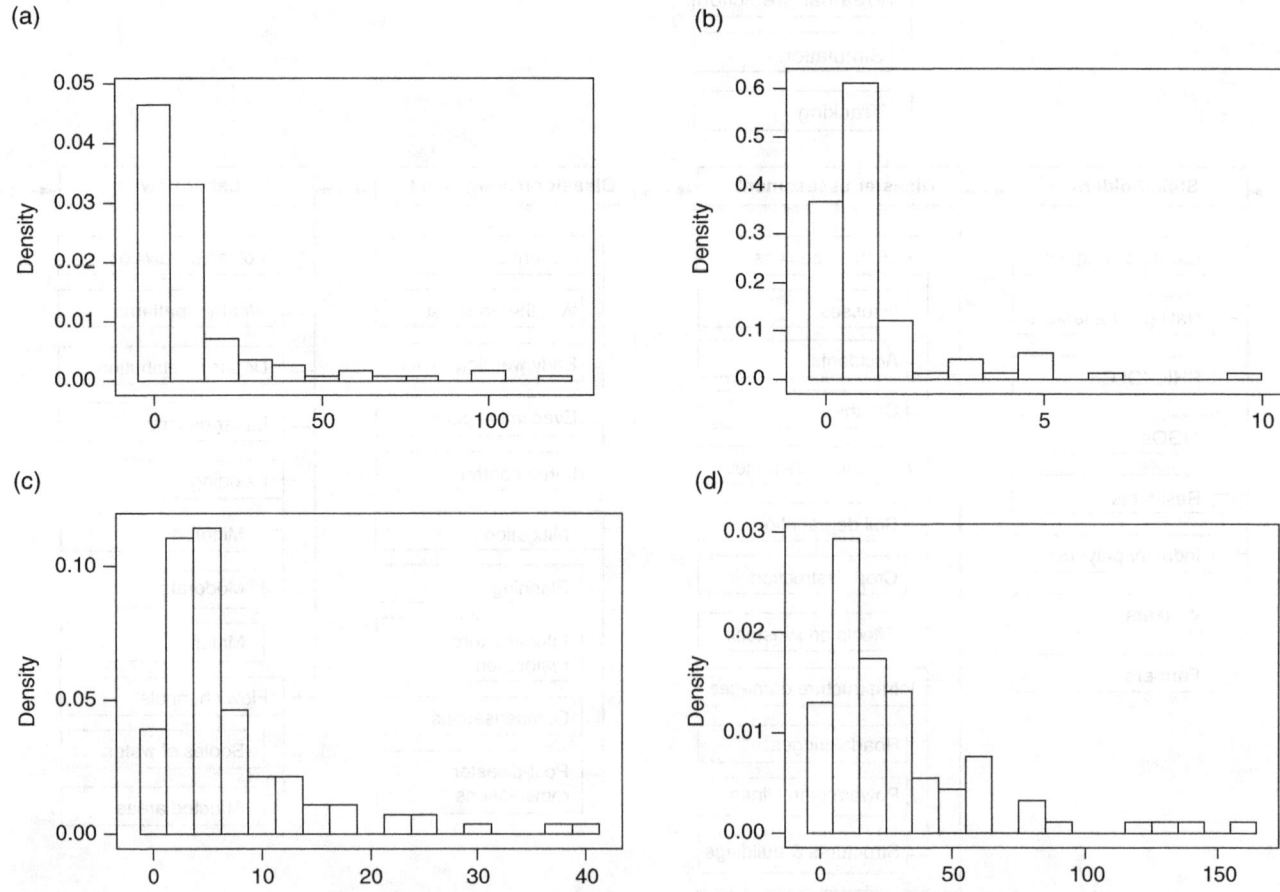

Figure 19.19 Histograms of (a) rainfall duration, (b) rainfall duration at peak intensity, (c) typical intensity, and (d) peak intensity for the period July 18–October 1991.

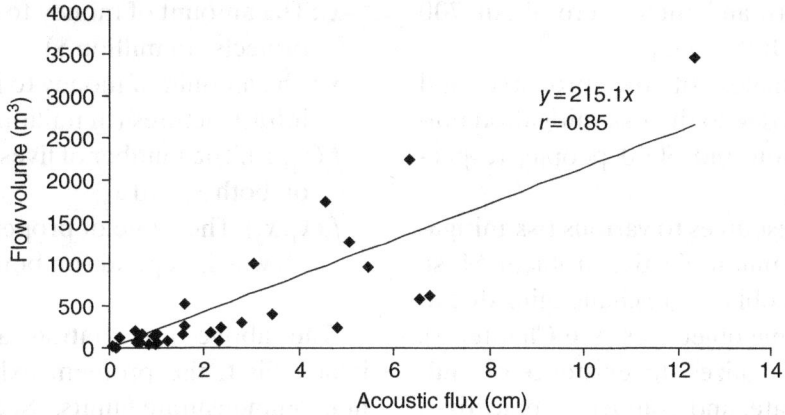

Figure 19.20 Regression for lahar acoustic flux versus flow volume in Sacobia River (July 1–August 29, 1992).

by acoustic flux) resulting from different rainfall magnitudes. For functional relationships that do not appear to be linear, curve-fitting techniques (e.g., maximum likelihood estimation, least squares) can be used in lieu of linear regression. For example, curve-fitting routines have been implemented to determine threshold curves for lahar-triggering rainfall events, which are assumed to follow the form of a power function [Ang and Tang, 1984]. Analyzing data from monitoring instruments using regression and curve-fitting models enables relaying prompt warnings to the residents to prevent casualties.

Extreme-event analysis, presented in Chapters 8 and 11, is another statistical area appropriate for studying rare and catastrophic phenomena such as the Pinatubo incident. In a set of repeated observations, maximum (or minimum) values can be obtained to form a random variable of extreme values. These extreme values comprise a population of their own and can be modeled using probability distributions [Castillo, 1987]. There are three asymptotic types (i.e., limiting forms) of the extreme value distributions: the Gumbel, Frechet, and Weibull. Of these three, the Gumbel type is the relevant distribution to use in modeling the maximum rainfalls and floods (see Chapter 11 and Castillo [1987]). The value of incorporating statistics of extremes into the analysis of risks is that it allows the forecasting of catastrophic events and their return periods (RP). Clearly, there are advantages to predicting when catastrophic events can strike (e.g., rainfall intensity >150 mm/h as shown in Figure 19.19d). One advantage is being able to incorporate reasonable safety factors when designing and constructing infrastructures.

19.4.6 Multiobjective Risk Methods: Applying Risk Trade-Off and Impact Analyses to Lahar Management

19.4.6.1 Multiobjective trade-off analysis using the SWT method

Various stakeholders have different and often opposing views of how to approach the lahar mitigation issue. The eruption of Mt. Pinatubo is a force *majeure* for which little prior scientific data are available; thus, geoscientists believe that the most rational policy action in response to the threats of lahar flow would be resettlement [Janda et al., 1994]. On the other hand, engineers believe that it is possible to control and rechannel the lahar flows by constructing structures such as dikes and dams. Government officials view natural hazards as a political and policymaking problem in the sense that limited resources must be used to everyone's advantage and decisions have to be made on conflicting issues. The Philippine Republic Act 7437 enacted in September 1992 allocated a total budget of PhP10 billion or $370 million for mitigation actions relating to the Pinatubo incident (note that the conversion rate in 1992 was $1 ≈ PhP27). Of this total amount, about $92 million was utilized for resettlement and about $156 million for lahar-control infrastructures [Mercado et al., 1994]. The value of the property damage and production losses in 1991–1992 was estimated to be $463 million

[Mercado et al., 1994], and there were about 700 human fatalities [USDOC, 1992].

Conservative estimates of the property and number of lives saved due to the risk mitigation policies were $250 million and 5000 people, respectively [USGS, 1997b].

The allocation of resources to various risk mitigation policies typifies a multiobjective problem. Most, if not all, real-world problems are challenging due to the presence of multiple objectives (see Chapter 5). This methodology recognizes the existence of multiple, noncommensurate, and conflicting objectives and has the ability to keep the units of the objectives in their natural forms. It avoids the need to precommensurate all objectives, say, in monetary units (e.g., human lives). As a result, analysts and decisionmakers are able to determine Pareto-optimal solutions based upon the values of the objective functions and their associated trade-offs. (In Pareto-optimal solutions, an increase in the value of one objective will lead to a decrease in the value of another objective.)

This case study employs the SWT method (see Chapter 5 and Haimes and Hall, 1974), which makes use of an ε-constrained approach to express objectives in their original, tractable measurement units. In doing so, it avoids the weaknesses of aggregation and normalization techniques in addressing noncommensurate units of measurements. By invoking the Lagrangian formulation and Kuhn–Tucker conditions, the SWT method is able to show the trade-offs between competing objectives. The trade-off functions enable decisionmakers to base a choice on the set of Pareto-optimal policy options. To apply the SWT method to the risk mitigation policies for the Pinatubo incident, let us define the following:

x_1: The amount of money to invest in resettlement projects (in million $)

x_2: The amount of money to invest in lahar-control infrastructures (in million $)

$f_1(x_1,x_2)$: The number of lives saved, which depends on both x_1 and x_2

$f_2(x_1,x_2)$: The value of properties saved (in million $), which depends on both x_1 and x_2.

The above formulation addresses two major issues. First, the problem exhibits objectives with noncommensurate units. Second, the objectives exhibit some form of conflict because not all residents are willing to resettle at the expense of giving up their homes and properties. To conduct the analysis, we need information such as the population size/density and the value of properties in the affected localities. In addition, expert resources can be invoked to assess the sensitivity of the objective functions to changes in the decision variables. For this problem, we are specifically interested in determining the impacts of various government investment policies (e.g., resettlement and lahar-control infrastructures) on the objective functions (i.e., maximizing both the amount of property and the number of lives saved). To demonstrate the SWT method, we used the data shown in Table 19.8, which were derived and inferred from various references (see USGS [1997b], USDOC [1992], Mercado et al. [1994], and various unpublished data by the Philippine Department of Budget and Management). Fitting a quadratic model for the data in Table 19.8 will yield Eqs. (19.31) and (19.32). The adjusted R^2 for both equations were found to be greater than 90% (R^2 is a measure of how well a set of data fits a specified model):

TABLE 19.8 Different Investment Scenarios for Resettlement (x_1) and Lahar-Control Infrastructures (x_2) and Resulting Effects to the Number of Lives (f_1) and Valuation of Properties (f_2) Saved

x_1 (Million $)	x_2 (Million $)	f_1 (Number)	f_2 (Million $)
0[a]	0[a]	200	0.05
50	50	2000	50
50	75	3500	80
75	100	4500	150
75	150	4800	220
100	150	5000	250

[a]With the *do-nothing* policy, properties and lives can be saved by the residents' initiatives.

$$f_1(x_1, x_2) = 170 - 89.5x_1 + 0.588x_1^2 \\ + 120x_2 - 0.442x_2^2 \quad (19.31)$$

$$f_2(x_1, x_2) = -1.3 - 0.91x_1 + 0.0134x_1^2 \\ + 1.24x_2 + 0.00118x_2^2 \quad (19.32)$$

Equations (19.31) and (19.32) represent two objective functions: the number of lives saved ($f_1(x_1, x_2)$) and the value of properties saved ($f_2(x_1, x_2)$), respectively. Since both objectives are to be maximized, the corresponding SWT problem formulation is as follows:

Maximize $f_1(x_1, x_2) = 170 - 89.5x_1 + 0.588x_1^2$
$\qquad\qquad + 120x_2 - 0.442x_2^2$

subject to $f_2(x_1, x_2) = -1.3 - 0.91x_1 + 0.0134x_1^2$
$\qquad\qquad + 1.24x_2 + 0.00118x_2^2 \geq \varepsilon_2$

$\qquad\qquad\qquad\qquad\qquad\qquad (19.33)$

Figure 19.21 shows the Pareto frontier in the space of the decision variables, which means that any combinations of x_1 and x_2 that lie along the Pareto frontier yield optimal values of the objective functions. Several iso-curves for the two objective functions are also shown in Figure 19.21. Moving along a particular iso-curve changes the values of the decision variables but does not change the value of the objective function.

The Pareto frontier in the space of the objective functions is shown in Figure 19.22. This graphically depicts the trade-offs in terms of the natural units of the two objectives: namely, the number of lives (in terms of head count) and the value of properties saved (in million $). It is clear from the graph that the two objectives compete—the more emphasis given to saving properties, the worse the impact will be on the number of lives saved, and vice versa. This result is intuitive, since the residents who opt to stay in lahar-prone areas,

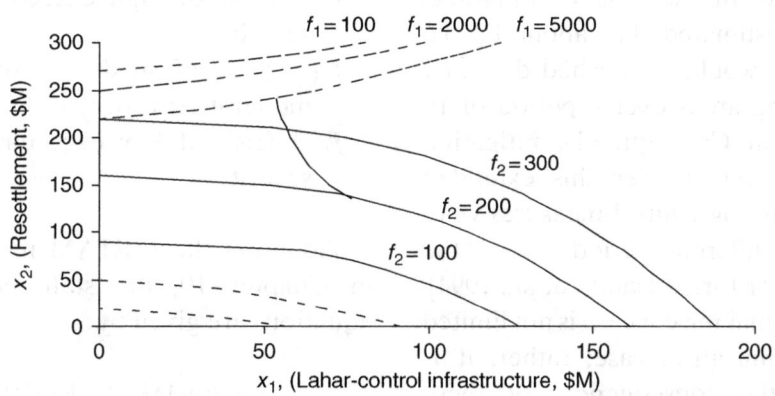

Figure 19.21 Pareto frontier in decision space (x_1, x_2), iso-f_1 curves (---), and iso-f_2 curves (—).

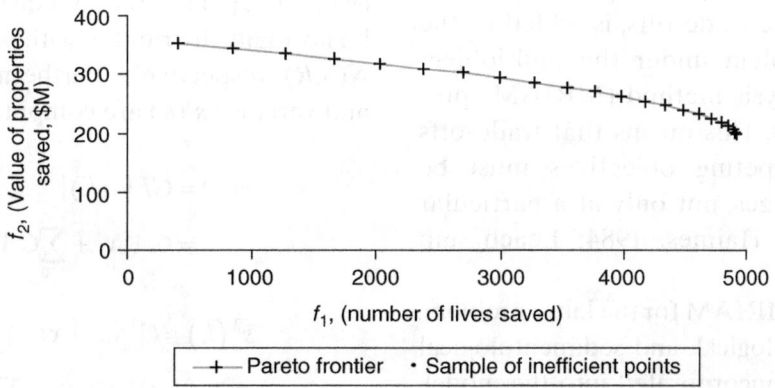

Figure 19.22 Pareto frontier in objective function space.

whether because of emotional attachment to their homes or their trust in the lahar-control infrastructures, face more risk than those who evacuate to the resettlement sites. Taking into account the ethical and legal issues involved in this problem, we can argue that human lives cannot be compromised to any other objectives. This would mean that the rational choice would be to save as many lives as possible—approximately 5000 people. This number can be increased when other policy options are considered in the analysis, such as ensuring that residents in lahar-hazard zones follow evacuation advisories and increasing the effectiveness of rescue operations, to cite a few.

19.4.6.2 Multiobjective risk impact analysis method

Another aspect of the lahar problem that needed attention was its duration. Lahar fallout begins immediately after a major eruption and persists over a period of time. In the case of Pinatubo, Pierson et al. [1994] estimated that about 1.2–3.6 billion m³ of sediment would be washed down by lahar flows to low-lying areas over a period of 10 years after the eruption. Consequently, mitigation measures were programmed over this extended period of time. Engineering control measures were also implemented at different periods, given the changing nature of lahar threat [Janda et al., 1994]. Therefore, the decisionmaking concern is not limited to a single-stage optimization case; rather, it is extended to assess the consequences of these decisions on future policy options.

In Section 19.4.6.1, we studied two issues of noncommensurate and conflicting objectives. A third dimension, stage trade-offs, is added to the decisionmaking problem under the multiobjective risk impact analysis method (MRIAM) presented in Chapter 10. This means that trade-offs among various competing objectives must be made at different stages, not only at a particular stage [Gomide and Haimes, 1984; Leach and Haimes, 1987].

To implement the MRIAM for the lahar problem, meteorological, hydrological, and sedimentological perspectives must be incorporated into the model. Although the following example is simplified by the limited available information, it serves to present the application of the method to the multistage lahar problem. Specifically, we are interested in determining the impacts of various government fund-release schedules on the objective function of minimizing risk. We define the following:

$x(k)$: The state of the system at stage k, defined as the uncontrolled volume of erodible sediment (in millions of m³)

$y(k)$: The output of the system at stage k, in terms of damage

$u(k)$: The control implemented at stage k (amount of budget released in million $)

$f_i^k(\cdot)$: The conditional expected-value-of-risk functions calculated for each stage k

Recall the conditional expected-value-of-risk functions f_i, $i=2,3,4$, generated by the partitioned multiobjective risk method (PMRM) presented in Chapter 8:

$f_2(\cdot)$ (risk of high exceedance probability, low severity)

$f_3(\cdot)$ (risk of medium exceedance probability, moderate severity)

$f_4(\cdot)$ (risk of low exceedance probability, high severity)

Adapting the MRIAM methodology presented in Chapter 10, the state equation and output equations are given by

$$x(k+1) = Ax(k) + Bu(k) + \omega(k) \quad (19.34)$$

$$y(k) = Cx(k) + v(k) \quad (19.35)$$

$\omega(k)$ and $v(k)$ are random disturbances, assumed to be normally distributed with parameters $N(0,P)$ and $N(0,R)$, respectively. Furthermore, the mean $m(k)$ and variance $s^2(k)$ are computed as

$$\begin{aligned} m(k) &= CE[x(k)] \\ &= CA^k x_0 + \sum_{i=0}^{k-1} CA^i Bu(k-1-i) \end{aligned} \quad (19.36)$$

$$\begin{aligned} s^2(k) &= C^2 \text{Var}[x(k)] + R \\ &= C^2 A^{2k} X_0 + \sum_{i=0}^{k-1} C^2 A^{2i} P + R \end{aligned} \quad (19.37)$$

The policies are evaluated by minimizing the conditional expected values of risk (f_i^k) given by the expression in Eq. (19.38):

$$f_i^k = m(k) + \beta_i^k s(k) \qquad (19.38)$$

where β_i^k is a constant associated with each conditional value of risk f_i^k. Partitioning at one standard deviation from the mean, β_i^k, results in the following: $\beta_2^k = -1.525, \beta_3^k = 0,$ and $\beta_4^k = 1.525$. A sample problem follows.

19.4.6.3 Sample problem: Implementing MRIAM to lahar fund release

Table 19.9 summarizes three policies that define the schedule of budget releases to fund lahar risk mitigation measures. It is assumed that the funding released at a previous stage would have a realizable effect at period k in terms of $y(k)$. A total budget amount of $800 million is used.

For this illustration, we set the following parameters, corresponding to Eqs. (19.34)–(19.38), to calculate $m(k)$ and $s(k)$ using Eqs. (19.36) and (19.37):

- $A = \exp(-0.5)$: Effective decay rate of sediment yield per stage k (computed from estimated annual volumes of sediment provided in Table III, page 2 of Pierson et al. [1992]). The sediment deposit on the slope was expected to be depleted due to continuous lahar erosion and the compaction factor over time. The sediment delivery rate was projected to decay exponentially [Pierson et al., 1992].
- $B = (-)0.25$: Controlled volume of sediment per unit investment (in million m³/million $ spent). In this problem, a controlled volume of sediment can be achieved by investing in different control measures, both structural (e.g., dike construction) and nonstructural (e.g., evacuation, resettlement). Note that, realistically, each measure would have its corresponding value of B. However, this illustration is simplified by assuming a uniform impact on the control of erodible sediment per dollar invested in each risk mitigation measure.
- $C = 1$: The volume of uncontrolled erodible sediment ($x(k)$) used as the surrogate measure of risk. Naturally, risk is higher with a larger volume.
- $x_0 = 1000$ millions m³: Erodible sediment at stage $k = 0$. (First-year volume estimated by Pierson et al. [1992] is based on 10–15% of sediment eroded by September 10, 1991.)
- $X_0 = 250$ millions m⁶: The assumed value of variance of x_0.
- $P = 0.001$: The variance of the random variable $\omega(k)$.
- $R = 0$.

The conditional expected values are directly calculated from computed values of $m(k), s(k)$ using Eqs. (19.39)–(19.41). A summary for all options is given in Table 19.10. The solution to the multiobjective, multistage problem consists of the optimal policy choices over the entire planning horizon. Figure 19.23 shows the optimal frontiers for stages 1–3. All policies are Pareto-optimal at stages 1 and 3 (see Figure 19.23a and c). However, at stage 2 (see

TABLE 19.9 Three Scheduling Alternatives for Releasing Lahar Risk Mitigation Funds

	$u(k-1)$ = Amount Spent on Lahar Risk Mitigation at Stage $k-1$ (in Fraction of Total Fund[a])		
Policy	$k=1$	$k=2$	$k=3$
A	0.7	0.2	0.1
B	0.4	0.4	0.2
C	0.2	0.3	0.5

[a] A total budget amount of $800 million is used in the example.

TABLE 19.10 Summary of Conditional Expected Values of Uncontrolled Erodible Sediment (Millions of m³) for the Different Policies

Policy A	$k=1$	$k=2$	$k=3$
f_2	452	234	122
f_3	467	243	127
f_4	481	252	133
Policy B			
f_2	512	230	100
f_3	527	239	105
f_4	541	248	111
Policy C			
f_2	552	275	67
f_3	567	284	72
f_4	581	292	77

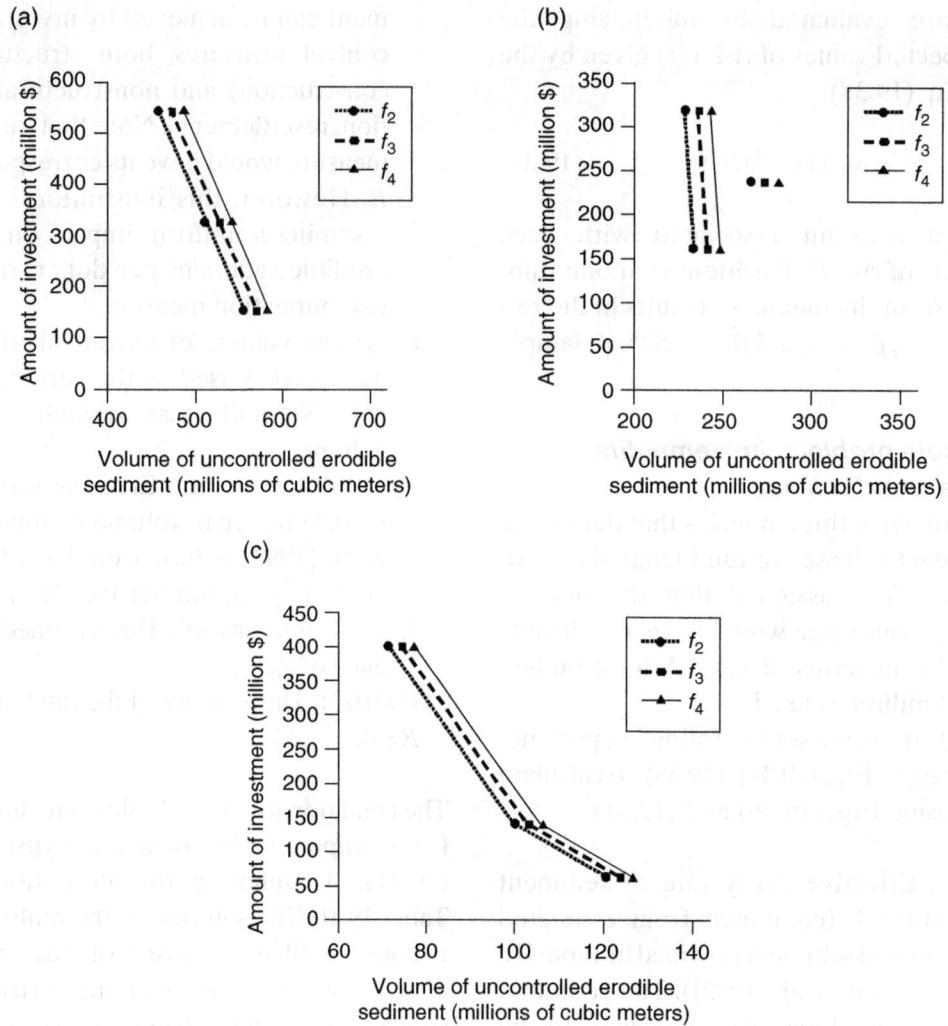

Figure 19.23 Pareto frontiers at various k stages: (a) stage 1, (b) stage 2, and (c) stage 3.

Figure 19.23b), Policy C is dominated by Policy A. This is so despite the fact that funds released for Policy A are lower than for Policy C at stage 2. The impact of the larger invested fund at stage 1 for Policy A resulted in greater control of erodible sediment at stage 2. Therefore, Policy C is no longer a candidate for the preferred solution. Policies A and B are Pareto-optimal for all stages. Policy A allocated most of the budget at the start of the period. As expected, this generated the most impact in controlling the volume of erodible sediment at stage 1. Although the small amount of investment in the succeeding periods had a marginal effect, the impact of the initial investment kept the uncontrolled volume small overall. Policy A generated the smallest total volume of uncontrolled erodible sediment for all risk functions ($f_1, f_2,$ and f_3) in the entire period for the given mitigation budget. Therefore, it is the preferred fund-release policy.

19.4.7 Conclusions

This case study presented a framework for modeling, assessing, and managing the risks associated with Mt. Pinatubo's lahar flow. The two sets of triplet questions in risk analysis and management, along with HHM, provided a systemic road map for identifying, prioritizing, and evaluating policies for risk management. Specifically, this study highlights the following issues: (i) integrating scientific data and techniques into the decisionmaking process, (ii) recognizing the presence of multiple objectives and their trade-offs, (iii) evaluating various policy options for mitigating extreme risks, and (iv) incorporating temporal issues in resource allocation projects. Although ex post data were utilized in this case study, the SWT and MRIAM methods can be tailor-made to analyze and manage future disaster mitigation problems.

19.5 THE STATISTICS OF EXTREME EVENTS AND 6-SIGMA CAPABILITY[6]

19.5.1 Introduction

During the 1970s and 1980s, the Japanese demonstrated that high quality is achievable at low cost and with greater customer satisfaction. This movement was the result of total quality management (TQM) [Cartin, 1993]. The goal of TQM is to focus on satisfying the needs and expectations of the customer and operating to continuously improve the quality of all company processes.

However, organizations find it extremely difficult and expensive to provide customers with flawless products. The difficulty stems from the variability within any process. In cases where variability is very small, there may be no effect on customer satisfaction. However, if the variability is too large, the customer may perceive the unit to be undesirable and unacceptable, and thus, it is considered defective [Montgomery, 1991]. This variability is often described in terms of its sigma [Pande et al., 2000].

The purpose of this study is to show how sigma limit capabilities, popularized by Motorola engineers in the 1980s, can be put in the perspective of extreme-event analysis. Specifically, we apply the statistics of extremes analysis (discussed in Chapter 11) to the normal distribution for the various levels of sigma capability to develop a relationship between TQM and risk analysis.

By using zone control charts (see Figure 19.24), management is able to gain a better understanding of the variability within a process. A fundamental assumption of control charts is that the underlying distribution of the quality characteristic is normal.

---

[6]This case study builds and expands on a term project inspired by Fischer and Oelrich [1994].

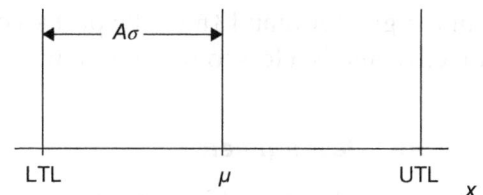

Figure 19.24 Process capability and tolerance limits.

To identify where variability occurs within a process, tolerance limits (TL), which are specifications set by management, are developed for an acceptable level of variability within a given process. An item produced inside the TL is acceptable; otherwise, it is considered defective.

If the specification limits are set at $\pm A\sigma$ away from the nominal specification (μ), as in Figure 19.24, and production follows a normal distribution, it can be approximated by $N(\mu, \sigma^2)$ and the process is said to have an A-sigma capability.

During the 1980s and part of the 1990s, most companies considered a 3-sigma limit capability acceptable, while some were satisfied with a 1- or 2-sigma limit capability.

19.5.2 Problem Definition

In high-quality manufacturing, a defect can be considered an extreme event. While strides have been made to achieve acceptable levels of variability, examining the extreme events and their corresponding probabilities shows that this will not fully be accomplished until 6-sigma capability limits are adopted universally. The following analysis demonstrates how various sigma capability levels have different variances in their statistics of extremes. The purpose of increasing the capability level is to decrease the number of extreme events.

The following sample problem illustrates the differences between capability levels. An industrial machine uses gaskets with a 10 mm diameter. This machine can use gaskets between 7 and 13 mm; a gasket diameter that does not fall in this range cannot be used and will need to be replaced. A company manufactures the gasket for these machines with a diameter that is normally distributed with a mean of 10 and a variance of σ^2. If this process produces a gasket with a diameter smaller than 7 mm or greater than 13 mm, the part is considered defective and is a loss to the company.

19.5.3 Model Development

There is no closed-form solution to the integral of the cumulative normal distribution function:

$$\Pr(X \le x) = F(x) = \frac{1}{\sigma\sqrt{2\pi}} \int_{-\infty}^{x} \exp\left[-(\tau - \mu)^2 / 2\sigma^2\right] d\tau \quad (19.39)$$

Therefore, to obtain a solution to Eq. (19.39), we transform a normally distributed random variable, X, with mean, μ, and standard deviation, σ, into a standardized random variable, Z, with a mean of zero and a standard deviation of one, using the formula:

$$z = \frac{x - \mu}{\sigma} \quad (19.40)$$

The corresponding z-values and probabilities, $\Phi(z) = \Phi\left(\frac{x - \mu}{\sigma}\right)$, can be obtained by using either standard tables or appropriate software packages. For the 6-sigma case, we define the term $A\sigma$ – capability as the *distance to the TL*. To use the standard normal tables for any $A\sigma$ – capability, we use the same transformation except that the value of σ will be calculated from A. For example, for a 6-sigma capability, that is, $A = 6$, then

$$A = \frac{x - \mu}{\sigma} \quad \text{or} \quad \sigma = \frac{x - \mu}{A} \quad (19.41)$$

Assume that the gasket manufacturing follows a normal distribution with a mean of 10. The distance from the mean to both the upper and lower TL is 3:

$$x - \mu = \text{UTL} - \mu = 13 - 10 = 3,$$
$$\text{then } \sigma = \frac{3}{A} = \frac{3}{6} = 0.5$$

Figure 19.25 plots the different probability distribution functions for the various capability levels. The difference in the number of defects or extreme events can be seen by looking at how much the tails of the distributions exceed the TL. Six-sigma capability produces a curve that is much narrower and is less likely to be outside the limits. Throughout the following analysis, examples will be given for 3- and 6-sigma limit capabilities.

The first quantitative calculations examined are the actual probabilities of having defective parts at each capability level. Because the normal distribution is symmetric and the TL are an equal distance from the mean, the probability that a part will lie below the lower tolerance limit (LTL) is the same as the probability that a part will lie above the upper tolerance limit (UTL). Thus, the probability that a part is defective is just $2[1 - \Phi \text{(capability)}]$ where Φ (capability) can be obtained from the standard normal distribution table or software packages. These probabilities are multiplied by 1 million to get the estimated number of defects per 1 million gaskets produced.

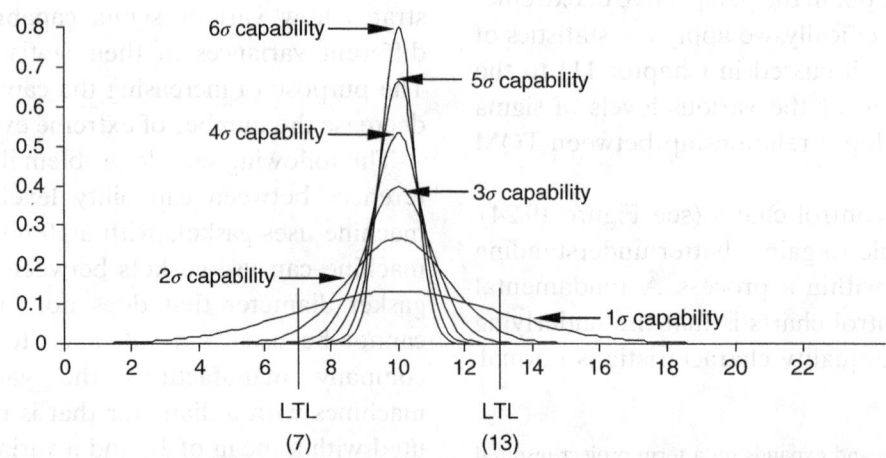

Figure 19.25 Normal probability distribution functions at different capabilities.

TABLE 19.11 Possible Values for σ Depending on Capability

σ-Capability	σ	σ^2
1	3	9
2	1.5	2.25
3	1	1
4	0.75	0.5625
5	0.6	0.36
6	0.5	0.25

Percent defective

$$\Pr[x < \text{LTL}] + \Pr[x > \text{UTL}] = 2[1 - F(\text{UTL})]$$
$$= 2\left[1 - \Phi\left(\frac{\text{UTL} - \mu}{\sigma}\right)\right]$$
$$= 2[1 - \Phi(\text{Capability})] \quad (19.42)$$

For 3-sigma capability:

$$2[1 - \Phi(3)] = 2[1 - 0.99865]$$
$$= 0.0027$$

For 6-sigma capability:

$$2[1 - \Phi(6)] = 2[1 - 0.999999999]$$
$$= 0.000000002$$

Defect rate in parts per million (ppm) (= percentage defective × 1,000,000)

For 3-sigma capability:

$$(0.0027)(1,000,000) = 2700 \text{ ppm}$$

For 6-sigma capability:

$$(0.000000002)(1,000,000) = 0.002 \text{ ppm}$$
$$= 2 \text{ ppb}$$

Note that the foregoing calculations assume no *drift* in the process mean value. Such a drift is attributable to variations in manufacturing batches, especially more pronounced in cases of multipart products. Six-sigma literature assumes that the process mean can drift 1.5 standard deviations for a more pessimistic estimation of defect rates [Pande et al., 2000]. For example, one would find in the literature that a manufacturing system with 6-sigma capability typically considers a defect rate of 3.4 ppm (instead of the 2 ppb calculated previously).

For the remaining discussion, focus will be on defects associated with exceeding the UTL. Similar analysis can be conducted for defects associated with nonconformities below the LTL. Having seen the contrast in the probabilities and the defects per million, the next step is to investigate the differences in the conditional expected values (given that a part is defective) for each capability level, since each one had the same unconditional expected value (f_5) or mean = 10. Conditional expectations, denoted by f_4, can be calculated using the following formula presented in Chapter 8:

$$f_4 = \frac{\int_{\text{UTL}}^{\infty} \frac{x}{\sigma\sqrt{2\pi}} e^{\frac{-(x-\mu)^2}{2\sigma^2}} dx}{\int_{\text{UTL}}^{\infty} \frac{1}{\sigma\sqrt{2\pi}} e^{\frac{-(x-\mu)^2}{2\sigma^2}} dx} \quad (19.43)$$

Thus, for 3-sigma capability,

$$f_4 = \frac{\int_{13}^{\infty} \frac{x}{1\sqrt{2\pi}} e^{\frac{-(x-10)^2}{2(1)^2}} dx}{\int_{13}^{\infty} \frac{1}{1\sqrt{2\pi}} e^{\frac{-(x-10)^2}{2(1)^2}} dx} = 13.283 \quad (19.44)$$

and for 6-sigma capability,

$$f_4 = \frac{\int_{13}^{\infty} \frac{x}{0.5\sqrt{2\pi}} e^{\frac{-(x-10)^2}{2(0.5)^2}} dx}{\int_{13}^{\infty} \frac{1}{0.5\sqrt{2\pi}} e^{\frac{-(x-10)^2}{2(0.5)^2}} dx} = 13.068 \quad (19.45)$$

The value of f_5 is equal to 10 mm (the specified mean) regardless of the capability level, since the normal distribution is symmetric about the mean:

$$f_5 = \int_{-\infty}^{\infty} \frac{x}{\sigma\sqrt{2\pi}} e^{\frac{-(x-\mu)^2}{2\sigma^2}} dx = 10 \quad (19.46)$$

Following the statistics of extremes analysis, the cumulative distribution function (cdf) for the maximum value $\left(F_{Y_n}\right)$ was determined from the formula (see Chapter 11)

$$F_{Y_n}(y) = [F_X(y)]^n = \left[\Phi\left(\frac{y-\mu}{\sigma}\right)\right]^n \quad (19.47)$$
$$= [\Phi(\text{capability})]^n$$

Thus, for 3-sigma capability,

$$F_{Y_n}(13) = [\Phi(3)]^{1000} = [0.99865]^{1000}$$
$$= 0.259 \quad \text{for } n = 1000$$
$$F_{Y_n}(13) = [\Phi(3)]^{2000} = [0.99865]^{2000}$$
$$= 0.067 \quad \text{for } n = 2000$$

and for 6-sigma capability,

$$F_{Y_n}(13) = [\Phi(6)]^{1000} = [0.999999999]^{1000}$$
$$= 0.999999 \quad \text{for } n = 1000$$
$$F_{Y_n}(13) = [\Phi(6)]^{2000} = [0.999999999]^{2000}$$
$$= 0.999998 \quad \text{for } n = 2000$$

This number is important because if the largest value of the process capability is much larger than the UTL, it is a concern for the manufacturer. The corresponding RP is the amount of time for the maximum value of the random variable to exceed the value y and is determined by

$$\frac{1}{1-F_{Y_n}(y)} \quad (19.48)$$

Thus, for 3-sigma capability,

$$\frac{1}{1-0.259} = 1.350 \quad \text{for } n = 1000$$
$$\frac{1}{1-0.067} = 1.072 \quad \text{for } n = 2000$$

and for 6-sigma capability,

$$\frac{1}{1-0.999999} \cong 1 \times 10^6 \quad \text{for } n = 1000$$
$$\frac{1}{1-0.999998} \cong 5 \times 10^6 \quad \text{for } n = 2000$$

Note that the RP of 6-sigma capability is several orders of magnitude higher than that of 3-sigma capability. For example, a 3-sigma capability could yield an RP of 1.072 for a batch run of 2000 gaskets. In other words, we may expect to observe 1 defect per a batch of 2000 gaskets. A 6-sigma system, on the other hand, will observe virtually no defect in a production run of 2000 gaskets. Likewise, the characteristic largest or smallest values can be determined as follows.

Thus, for 3-sigma capability,

$$\frac{u_{1000}-10}{1} \approx 3.09 \quad u_{1000} = 13.09$$
$$\frac{u_{2000}-10}{1} \approx 3.29 \quad u_{2000} = 13.29$$

and for 6-sigma capability,

$$\frac{u_{1000}-10}{0.5} \approx 3.09 \quad u_{1000} = 11.545$$
$$\frac{u_{2000}-10}{0.5} \approx 3.29 \quad u_{2000} = 11.645$$

These results can be interpreted as follows: If the manufacturing company produces 1000 gadgets at 3-sigma capability, what gadget size can we expect to be exceeded only once? The answer is 13.09 mm (clearly exceeding the UTL of 13 mm). Similarly, if the manufacturing company produces 1000 gadgets at 6-sigma capability, what gadget size can we expect to be exceeded only once? The answer is 11.545 mm (clearly well within the TL).

Another interesting statistic to study is the most probable maximum/minimum value in n observations (u_n), along with the inverse measure of dispersion (δ_n):

$$F_X(u_n) = 1 - \frac{1}{n} \quad (19.49)$$

$$\delta_n = nf_X(u_n) = \frac{n}{\sigma\sqrt{2\pi}} e^{\frac{-(u_n-\mu)^2}{2\sigma^2}} \quad (19.50)$$

Thus, for 3-sigma capability,

$$\delta_{1000} = \frac{1000}{1\sqrt{2\pi}} e^{\frac{-(13.09-10)^2}{2(1)^2}} = 3.37,$$

$$\delta_{2000} = \frac{2000}{1\sqrt{2\pi}} e^{\frac{-(13.29-10)^2}{2(1)^2}} = 3.55$$

and for 6-sigma capability,

$$\delta_{1000} = \frac{1000}{0.5\sqrt{2\pi}} e^{\frac{-(11.545-10)^2}{2(0.5)^2}} = 6.73,$$

$$\delta_{2000} = \frac{2000}{0.5\sqrt{2\pi}} e^{\frac{-(11.645-10)^2}{2(0.5)^2}} = 7.11$$

These figures were determined for $n = 1000$ and 2000. The best capability will have the most probable maximum and minimum values falling inside the TL. Another computation explored was the difference between the exact value of f_4 following the formula above and the approximation done by

$$f_4 = u_n + \frac{1}{\delta_n} \tag{19.51}$$

$$u_n = \mu + \sigma \left[(2\ln n)^{1/2} - \frac{\ln(4\pi \ln n)}{2(2\ln n)^{1/2}} \right] \tag{19.52}$$

$$\delta_n = \frac{(2\ln n)^{1/2}}{\sigma} \tag{19.53}$$

Thus, for 3-sigma capability,

$$f_4 = 13.12 + \frac{1}{3.72} = 13.39 \quad \text{for } n = 1000$$

$$f_4 = 13.31 + \frac{1}{3.90} = 13.57 \quad \text{for } n = 2000$$

and for 6-sigma capability,

$$f_4 = 11.56 + \frac{1}{7.43} = 11.69 \quad \text{for } n = 1000$$

$$f_4 = 11.66 + \frac{1}{7.80} = 11.79 \quad \text{for } n = 2000$$

Note that the exact values calculated earlier for f_4 are 13.283 and 13.068 for 3- and 6-sigma capabilities, respectively. The calculated approximate values of f_4 depend on the specified value of n. The larger the value of n, the closer the approximate f_4 values will be to the corresponding exact f_4 values for each capability level.

19.5.4 Analysis of Computational Results

The difference is great between the diverse probabilities of defect at each capability level. This can be seen in Figure 19.26, which displays the number of defects per million for each capability level. Going from 3 to 6 sigma, or from 2700 defects per million to approximately 2 per billion (note that considering a 1.5-sigma drift will cause this value to become more conservative—3.4 ppm), results in great savings for a company.

Figure 19.27 stresses the importance of examining the probability of extreme events. It demonstrates that the expected value for the diameter of the gasket at each capability level is the same. The fallacy of the expected value is that it distorts the relative importance of the extreme events by not accentuating them and their consequences, misrepresenting what otherwise would be considered an unacceptable risk. As capability increases, this conditional expected value approaches the UTL from its position outside the TL. Likewise, the conditional value, given that the diameter is smaller than the LTL, approaches the LTL as capability is increased. This shows that at lower capability levels, there are more observations that are further outside

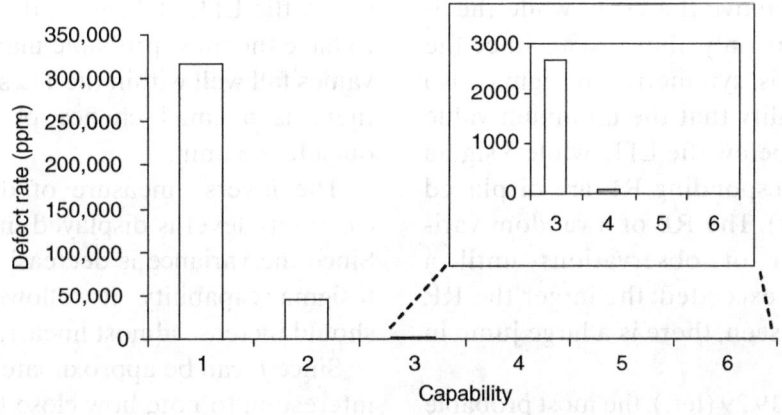

Figure 19.26 *Capability versus defective parts per million (no 1.5-sigma drift).*

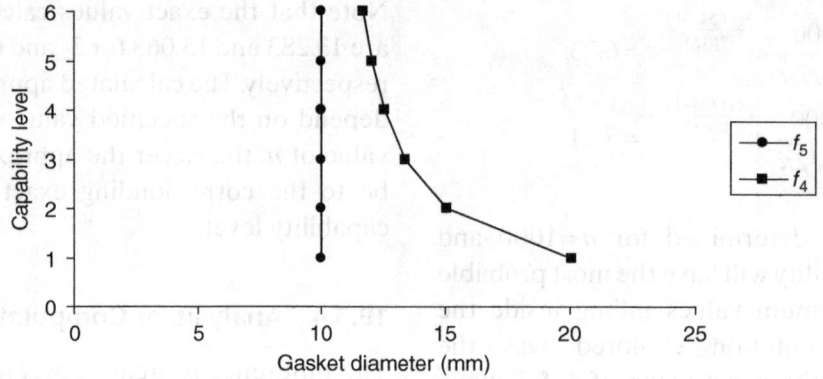

Figure 19.27 Exact f_4 and f_5 values versus capability levels.

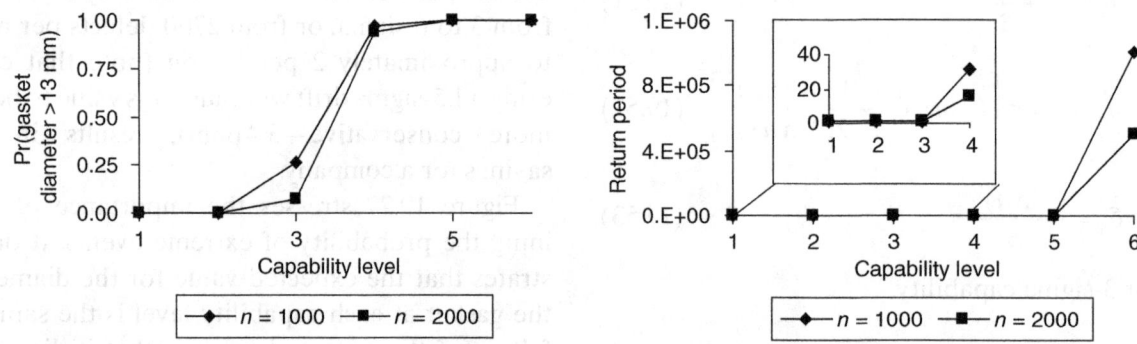

Figure 19.28 Left: capability versus $F_{Y_n}(13)$ for n = 1000 and 2000. Right: corresponding return period for $F_{Y_n}(13)$ for n = 1000 and 2000.

the TL. It is important that while reducing the variance of the risk does not contribute to changing the expected value, it has a large impact on the conditional expected value of the extreme events.

From Figure 19.28 (left), it is clear that there is a big jump in the cdf of the largest value between 3 sigma and 6 sigma. The 6-sigma capability offers about a 99% probability that the maximum value observed will not be above the UTL, while the 3-sigma capability offers only about 26%. Since the normal distribution is symmetric, 6 sigma also offers a 99% probability that the minimum value observed will not be below the LTL, while 3 sigma offers 26%. The corresponding RP are displayed in Figure 19.28 (right). The RP of a random variable is the number of observations until a maximum value, y, is exceeded; the larger the RP, the better. As can be seen, there is a large jump in the RP at 6 sigma.

Note that in Figure 19.29 (left), the most probable maximum value for 6-sigma capability falls well below the UTL, which is very desirable, while the most probable maximum value for the 3-sigma capability was about 13.09, slightly above the UTL. Since the normal distribution is symmetric, the most probable minimum value would fall above the LTL for 6-sigma capability.

Likewise, as u_n was right above the UTL at 3 sigma, the most probable minimum would fall right below the LTL at 3 sigma. It is definitely beneficial to have the most probable maximum and minimum values fall well within the TL, since that implies that there is a small chance of having observations outside the limits.

The inverse measure of dispersion δ_n at each capability level is displayed in Figure 19.29 (right). Since the variance is decreasing linearly from 1- to 6-sigma capability, it follows intuitively that δ_n should increase almost linearly.

Since f_4 can be approximated using u_n and δ_n, it is interesting to note how close these approximations are to the exact value. As n approaches infinity, the

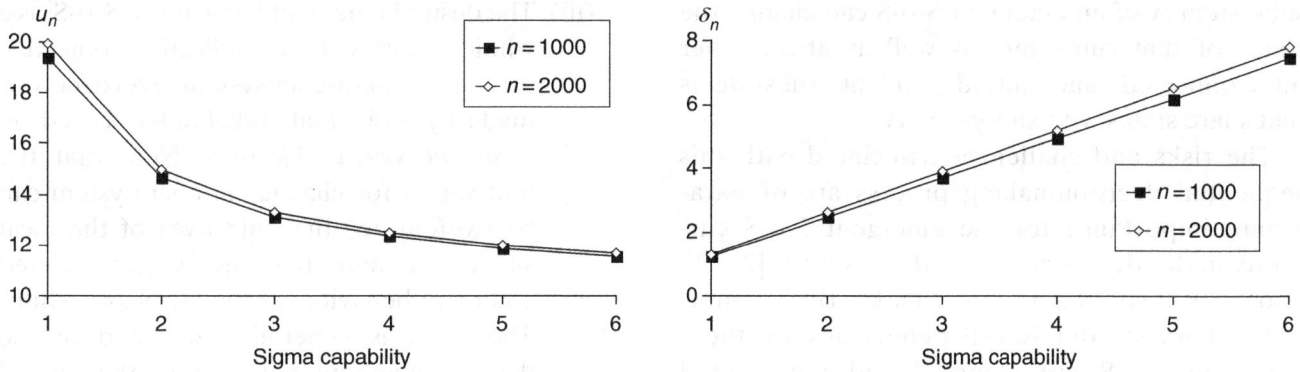

Figure 19.29 Left: capability versus u_n. Right: capability versus δ_n.

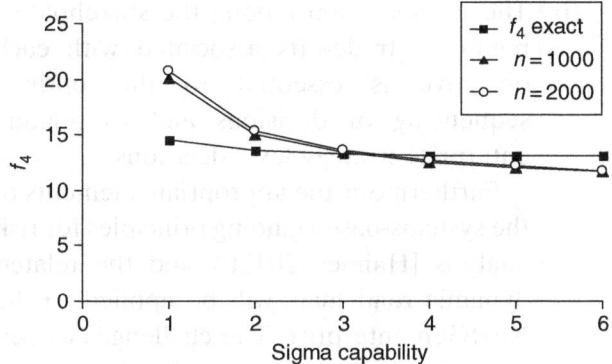

Figure 19.30 Capability versus f_4 approximation.

approximation should approach the exact value. As can be seen in Figure 19.30, n equal to 1000 and 2000 are not good approximations for f_4, resulting in errors ranging from 0.8 to 12%. The results are shown in Table 19.12.

19.5.5 Conclusions

This study has shown how sigma capabilities can be defined through the analysis of extreme events. It has established how f_4 can be approximated through extreme-event analysis, provided insight about defect RP, and shown the limitations of depending on only the mean for quality control. These findings aid TQM through the ease of computing f_4 and its insights about a likely maximum value in a given number of trials (as demonstrated when $n=1000$ and 2000). Moreover, this case study establishes the close relationship between the worlds of risk analysis and TQM.

TABLE 19.12 Summary of u_n and δ_n Values for Approximating f_4

		\multicolumn{6}{c}{Capacity}
		1	2	3	4	5	6
σ		3	1.5	1	0.75	0.6	0.5
u_n	$N=1000$	19.35	14.68	13.12	12.34	11.87	11.56
	$N=2000$	19.94	14.97	13.31	12.49	11.99	11.66
δ_n	$N=1000$	1.24	2.48	3.72	4.96	6.20	7.43
	$N=2000$	1.30	2.60	3.90	5.20	6.50	7.80

19.6 SEQUENTIAL PARETO-OPTIMAL DECISIONS MADE DURING EMERGENT COMPLEX SYSTEMS OF SYSTEMS: AN APPLICATION TO THE FAA NEXTGEN[7]

Addressing the third question in risk management—"What are the impacts of current decisions on future options?"—is critically important for emergent complex systems of systems (S-o-S), because their conception, the evolution of their requirements and specifications, their design and development, and their ultimate operation can span several years. Furthermore, the sequential decisions made during the development of each subsystem of these complex S-o-S will most likely affect the development of other subsystems in the future with the expectation that, in their totality, all subsystems are ultimately expected to operate as an integrated harmonious S-o-S. For example, decisions to achieve specific outcomes made on

---
[7]This case study is adopted from Haimes and Anderegg [2015].

subsystem A of an emergent S-o-S can change the states of that subsystem as well as affect other interconnected and interdependent subsystems that share states with subsystem A.

The risks and challenges associated with this sequential decisionmaking process are of paramount importance for the emergent S-o-S currently under development by the US FAA [2012], known as Next Generation (NextGen). The interconnectedness and interdependencies of these emergent S-o-S are modeled and represented through shared (common) states and decision variables existing among the subsystems, as advanced by the phantom system models [Haimes, 2012a]. To avoid abstraction and to more clearly communicate the concepts and methodologies introduced in this article, appropriate examples of the shared states and decisions are linked to NextGen to demonstrate their relevance and practicality.

In this sense, relating the systems-based methodological approach and its application to the NextGen enterprise becomes one integral theme. The technical presentation is described through the current and ongoing modernization of the National Airspace System (NAS), one of the most all-encompassing projects ever undertaken by the FAA.

The following premises guide the technical presentation in this article, given that the time frame for the development of NextGen spans two decades:

(i) Sequential decisions made during the emergence of complex S-o-S must be dynamically Pareto-optimal and can be generated using the *Envelope Approach* [Li and Haimes, 1987; Haimes, 2009b]. The criticality of the time frame, within which the ultimate decisions are most likely to accomplish their intended objectives, and the multiple decisionmakers and stakeholders will require agreed-upon policies commensurate with the satisficing principle introduced by Simon [1956].

(ii) The emergent behavior of S-o-S arises from the various trade-offs made by multiple stakeholders who have multiple goals and objectives affecting future decisions and outcomes.

(iii) The desired emergent behavior of S-o-S, as a whole, represents a collective objective driven by separate subsystem-level decisions made by individual stakeholders based on *local subsystem objectives*. Note that the motivation for changes in a subsystem can be twofold: (a) the objectives of the local subsystem and (b) the larger desired emergent behavior for the S-o-S as a whole. The FAA is generally interested in the desired emergent behavior of the overall enterprise in addition to the local effect. DeLaurentis [2005] was a pioneer in addressing the air transportation as S-o-S.

(iv) The collaboration among the stakeholders regarding trade-offs associated with each objective is essential to the optimal sequencing of decisions and to making enterprise strategy-level decisions.

Furthermore, the appropriate elements of the systems-based guiding principles for risk analysis [Haimes, 2012b], and the related dynamic road map, will be applied to the NextGen enterprise. The challenges associated with the modeling and development of emergent S-o-S are highlighted, with a focus on the centrality of the state variables and time frame. One theme that transcends the entire discussion in this article is the critical role that shared (common) states, decisions, and the time frame play in modeling the interconnectedness and interdependencies among the subsystems that constitute emergent S-o-S. The theoretical and methodological concepts are harmonized through their relevance to ongoing emergent complex S-o-S in the FAA's NextGen program.

19.6.1 Methodological Approach

19.6.1.1 Introduction

The increasing complexities of shared infrastructure projects that span the public and private sectors have begun to exceed the capabilities of traditional management techniques to plan, construct, and control them. Every new generation of technology and operational practice increases in complexity and ambition. This trend has led to an explosion in the sophistication of government and private industry projects designed to develop and/or acquire technology-based systems. Unpredictable interactions and interdependencies among individual and community users, software–hardware, and organizations commonly characterize these systems. This complexity has imposed a complementary rise in the level of *unpredictable* adverse consequences, in terms of technological risk (not meeting a project's performance criteria, e.g., failing to deliver expected benefits for individual investors or the desired behavior of the emergent S-o-S as a whole) and programmatic risk (cost overrun, time delay in project completion.)

In general, there is no one right way to perform the complex process of risk modeling, assessment, management, and communication associated with small or large projects. Often, the best approach is driven by the unique characteristics of the system under consideration. *Regardless, there are certain principles and guidelines, introduced in this article, that apply universally to emergent complex S-o-S.*

19.6.1.2 The complexity of planned and emergent S-o-S

Emergent complex S-o-S are by definition composed of several subsystems and are commonly in a constant state of flux, modification, and readjustment caused by the fluctuations in one or more subsystems. Most notably, these seemingly random driving forces, denoted by the *Evolving Base*, are dynamically shifting interdependencies and realities based on changing [Haimes, 2012b] (i) goals and objectives; (ii) stakeholders, decisionmakers, and interest groups; (iii) organizational, political, and budgetary considerations; (iv) reorganization and reallocation of key personnel; and (v) requirements, specifications, delivery dates, users, and clients.

The literature is replete with definitions of what constitute S-o-S, albeit not how to model them. Maier [1998] identifies five characteristics of S-o-S: (i) operational independence of the individual systems, (ii) managerial independence of the systems, (iii) geographic distribution, (iv) emergent behavior, and (v) evolutionary development.

Other contributors to the literature on complex S-o-S include Bar-Yam [2003a, b], Blauberg et al. [1977], Eisner [1993], Jamshidi [2009a, b], and Haimes [2007, 2012a]. Sage and Cuppan [2001] provide the following representative definition of emergent behavior in the context of an S-o-S:

The S-o-S performs functions and carries out purposes that do not reside in any component system. These behaviors are emergent properties of the entire S-o-S and not the behavior of any component system. The principal purposes supporting engineering of these systems are fulfilled by these emergent behaviors.

One of the major challenges facing professionals in charge of the multiple stages in the development of complex S-o-S, such as the FAA's NextGen (to be subsequently introduced), is harmonizing the myriad sequential decisions, whether regarding policies and procedures or structural issues, so that decisions made at one point will not conflict with, or adversely affect, future options. By their nature, emergent S-o-S are composed of multiple interconnected and intra- and interdependent subsystems, each with a unique mission, and human, cultural, organizational, geographical, and technological characteristics that span cyber and physical infrastructures. In addition, a hierarchy of diverse stakeholders and decisionmakers, multiple noncommensurate and often conflicting and competing goals and objectives that span multiple time frames, and assorted policies and procedures serving each subsystem are features of these systems. Clearly, it is essential to coordinate the planning, design, specification, and construction of each subsystem; to develop appropriate operational policies and procedures that meet the needs of each subsystem; and to ensure a harmonious integration of the subsystems with each other. This constitutes a challenging process of risk modeling, assessment, management, and communication. The common denominators that both transcend and link the aforementioned activities are the *time frame* and *the shared states* that constrain as well as enable the subsystems to function, to the extent possible, as a harmonious S-o-S.

One of the challenges facing the developers of emergent S-o-S is to navigate the sequential temporal decisionmaking process while being cognizant of the dynamic multiobjective trade-offs leading to a satisficing set of policies that are responsive to the broader goals and objectives of emergent

S-o-S enterprise. This process is also aimed at harmonizing the myriad decisions associated with each emerging subsystem and within emergent S-o-S.

In sum, the development of a sequential multiobjective decisionmaking process and its deployment on the NextGen S-o-S would demonstrate that decisions made, policies formulated, and procedures and regulations promulgated during one period would not cause future detrimental technical and programmatic risks to the subsystems within NextGen, nor would they adversely affect or impact the multiple objectives and stakeholders associated with the NextGen S-o-S.

19.6.1.3 The FAA NextGen S-o-S

Applications presented in this case study are drawn from the FAA NextGen multiyear, multiagency, and multibillion dollar project. NextGen is a project to modernize the NAS. Decisions for this complex enterprise are made by a combination of individual stakeholders and organizations in order to achieve specific objectives and outcomes. At the same time, the FAA has a desired set of emergent behaviors for the evolution of the NAS, which will be the product of these many separate decisions. NextGen epitomizes interconnected and interdependent complex S-o-S.

NextGen is a public–private endeavor involving over 30 major systems and 190 government subsystems, each of which has its own platform and many stakeholders. The private sector consists of hundreds of commercial aircraft operators with hundreds of models of aircraft and thousands of private aircraft operators. These aircrafts operate out of over 5000 airports, all with local management. All of these components are part of the NextGen emergent S-o-S. From the FAA perspective as a regulator and service provider, the decisions for FAA components must serve the diverse population of private sector stakeholders. An aim of NextGen is to create incentives for all participants to invest in capabilities that will contribute to the desired emergent behaviors of safety, flight efficiency, capacity, and the environmental impact of the S-o-S as a whole.

FAA's NextGen is being planned with three major 5-year implementation windows, known as Alpha (2010–2015), Bravo (2015–2020), and Charlie (2020–2025). An inherent characteristic of an emergent S-o-S such as NextGen is the long time frame from its development to its realization. Components, such as aircraft and airport modifications, will be expected to remain in service for 20 years or more. Each decision, therefore, will have long-lasting consequences. Furthermore, the timing of changes must be planned and aligned with changes in other subsystems that share common states. In particular, the coordination of the planned services will require changes in ground systems and synchronizing of aircraft and flight crews.

19.6.1.4 Three major sources of risk that threaten emergent interdependent S-o-S characterized by intrinsic shared states

Technical, programmatic, and enterprise performance risks are all concerns of emergent S-o-S. The first two risks apply to the subsystem level, while enterprise risk is the unexpected consequence in the desired emergent behavior of the S-o-S as a whole. More specifically:

(i) *Technical risk* occurs when a project fails to meet its performance criteria. This includes hardware and software failures and requirement shortfalls. Technical risk can occur when an emergent forced change causes significant changes in the behavior of one or more subsystems, leading to a cascading failure of the S-o-S. The phrase *emergent forced changes* connotes *trends in external or internal sources of system risk that may adversely affect specific states of that system.*

(ii) *Programmatic risk* denotes cost overruns and delays in the schedule. This source of risk characterizes many emergent large-scale systems, especially for subsystems with significant software components.

(iii) *Enterprise performance risk* leads to negative outcomes due to changes in the system. When the performance of an *emergent* S-o-S does not meet the collective desired behavior, whether by unanticipated, undetected, misunderstood, or ignored emergent forced changes from within or from outside a system, the state changes may result in a negative emergent behavior. Therefore, it is imperative to be able, through scenario structuring, modeling, and risk analysis, to envision, discover, and track emergent forced changes [Haimes, 1981, 2009b] and to understand the likely consequences for emergent behavior. The interdependency among subsystems of S-o-S is a manifestation of the intrinsic shared states among them. Consequently, emergent forced changes that affect intrinsic shared states of one subsystem will affect all other subsystems that share the same intrinsic states. In sum, the essence of interdependencies among subsystems of S-o-S is grounded on the shared states among the subsystems.

Consider the case where a decisionmaker acts believing that a subsystem change will create a desired behavior; however, the change is actually insufficient to provide the desired performance outcome. In a hypothetical example from NextGen, consider two capabilities that may each be necessary yet not sufficient to achieve the desired benefit. The first is an *airborne reroute* path-dependent *capability*, which replaces a series of paper note exchanges and phone calls with electronic exchange of information between multiple parties needed to direct an aircraft to an alternate route, as depicted in Figure 19.31. The second capability, *collaborative trajectory options program*, also shown in Figure 19.31, enables the aircraft operator flight operations center to indicate a preference for airborne trajectory options. Each of these capabilities is necessary to create more flexible routing around weather but is individually insufficient to do so. It may be that without including both capabilities, the use of airborne reroutes will fail to achieve the desired performance. This could be anticipated by examining the shared states for the two capabilities. Both capabilities involve knowledge of the airborne flight trajectory, procedures for avoiding weather, and operating environments involving weather. The implication is that all capabilities with shared states must be evaluated in collaboration with the independent decisionmakers for the necessary and sufficient conditions to create a desired emergent behavior. Such decisions become corequisite.

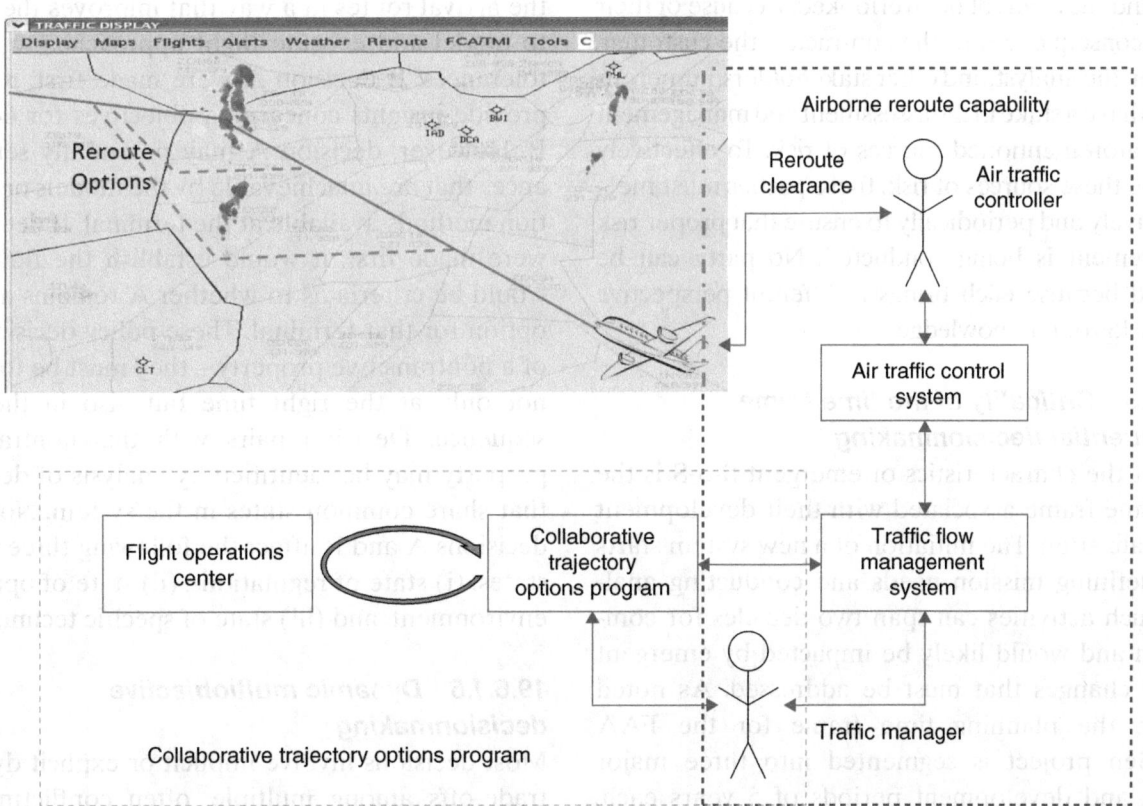

Figure 19.31 *Airborne reroute and airborne trajectory options capabilities.*

In other cases, decisions may transfer program or technical risk among decisionmakers. This would be the case if one of the capabilities in the example earlier, either the airborne reroute capability or the airborne trajectory options capability, were implemented with specifications that reduced the risk of one by adding program risk to the other. If coordinated as part of the overall design of a combined capability, then this risk management should lower the overall risk. If this transfer of risk occurs as a result of separate, uncoordinated program designed activities, then this represents a risk transfer from one decisionmaker to another, which ultimately can degrade both the desired performance of the S-o-S as a whole and the level of trust. A key concept in managing the risk of not achieving the desired emergent behavior is that *the critical decisions that warrant additional collaboration among decisionmakers can be identified through the analysis of the shared states and can be addressed proactively.* Risk mitigation should include coordination of the sequence and scope of the individual decisions with knowledge of the implications on other decisions.

Emergent forced changes are tied to risk management and they cannot be overlooked. Because of their broad consequences to the contractor, the customer, the user, the analyst, and other stakeholders, numerous parties have a stake in the assessment and management of the aforementioned sources of risk. To effectively manage these sources of risk, these parties must meet collectively and periodically to ensure that proper risk management is being conducted. No party can be ignored because each brings a different perspective and background knowledge.

19.6.1.5 Criticality of the time frame in sequential decisionmaking

One of the characteristics of emergent S-o-S is the long time frame associated with their development and realization. The initiation of a new system starts with defining mission needs and conducting analysis. Such activities can span two decades for completion and would likely be impacted by emergent forced changes that must be addressed. As noted earlier, the planning time frame for the FAA NextGen project is segmented into three major design and development periods of 5 years each, known as Alpha, Bravo, and Charlie. Indeed, decisions made at any time period are *path dependence*. Liebowitz and Margolis [2000] eloquently elaborate on the importance of the time frame concept: "Most generally, path dependence means that where we go next depends not only on where we are now, but also upon where we have been." Decisions made during any period can have significant consequences on the state of the system both at the time decisions are made and during subsequent periods; namely, we have path-dependence decisions. By altering the states of the S-o-S, each decision may adversely affect the size, location, functionality, and schedule of future changes. In addition, the large number of major structural and policy decisions exacerbates tracking such forced changes.

Consider, as an example, the following two decisions made in parallel. Decision A will be an FAA decision to introduce schedule-based merges within a terminal area. Referred to as terminal sequencing and scheduling (TSS), this is a flow management capability in which specific times and tolerances are used to smoothly interleave two flows of traffic. Decision B will be a decision to structure the arrival routes in a way that improves the ability to model and execute flight approaches to tighter tolerances. If decision A were made first, it would provide insights concerning objectives for decision B. However, decision A may potentially set tolerances that are unachievable by the models or execution methods available at the terminal. If decision B were made first, it would establish the flows that would be criteria as to whether A remains a viable option for that terminal. These policy decisions are of a nontransitive property—they must be followed not only at the right time but also in the right sequence. Decision pairs with this nontransitive property may be identified by analysis of decisions that share common states in the system. Note that decisions A and B affect the following three shared states: (i) state of regulations, (ii) state of operating environment, and (iii) state of specific technology.

19.6.1.6 Dynamic multiobjective decisionmaking

Most decisions involve implicit or explicit dynamic trade-offs among multiple, often conflicting, and noncommensurate goals and objectives. *Indeed, analyzing and understanding the impacts of decisions on multiple objectives and their associated trade-offs that affect different time frames constitute a foundational maxim in prudent decisionmaking.*

Given the often conflicting, competing, and noncommensurate nature of the multiple objectives associated with NextGen and because the trade-offs among them are complex from multiple perspectives, it is impractical to consider assigning them weights to use in making meaningful decisions. The decisionmaking process must avoid an overly constrained feasible set for which there is no answer. In this case study, we adopt the concept of satisficing originally introduced by Herbert Simon [1956], which builds on his earlier work [Simon, 1947].

Any attempt to *optimize* a set of objectives, even within the Pareto-optimal sense, that are limited to present needs and aspirations or that are not responsive to emergent or future forced changes could lead to severe unintended consequences. (Informally, *a solution to two-objective functions is termed Paretooptimal if improving one objective function can be achieved only by degrading the other* [Intriligator, 1970].) In other words, given the nature of the evolving, adaptive, incremental, and multistage S-o-S project, present decisions and policies must take into consideration that emergent forced changes may produce catastrophic or irreversible consequences on the project in the future. Thus, a systemic approach to addressing the multistage decisions during Alpha, Bravo, and Charlie periods should (i) continuously assess and evaluate precursors to potential emergent forced changes (building on the theory of scenario structuring), (ii) balance present multiple objectives with potential or perceived emergent future needs and objectives, and (iii) add more flexibility to present policy formulation to ensure against adverse emergent or unintended catastrophic consequences.

Robustness stems from evaluating the consequences and future flexibility of two preferred Paretooptimal policies where potential future environments would yield a distinct choice between two options presently perceived as seemingly equivalent. Assessing the temporal impacts on future multiobjective-based options is especially critical for rapidly changing technological systems, such as NextGen. The *Envelope Approach*, which enables dynamic multiobjective trade-off analyses [Li and Haimes, 1987], will be introduced in the following section.

Furthermore, dynamic systems such as NextGen, where the time frame is of paramount importance, introduce another challenge to both modelers and decisionmakers. Namely, ignoring or not accounting for the changes in the states of a safety-critical dynamic system as time progresses (e.g., such as the integration of the unmanned aircraft system (UAS) within the NAS) would render static-based decisions misleading, if not disastrous. Recognizing this fact, methods such as event trees, decision trees, and process control can account for the impact of current decisions on future options.

Consider the following two objectives at time k depicted in Figure 19.32: access, $f_1(\cdot)$, in the Y-axes

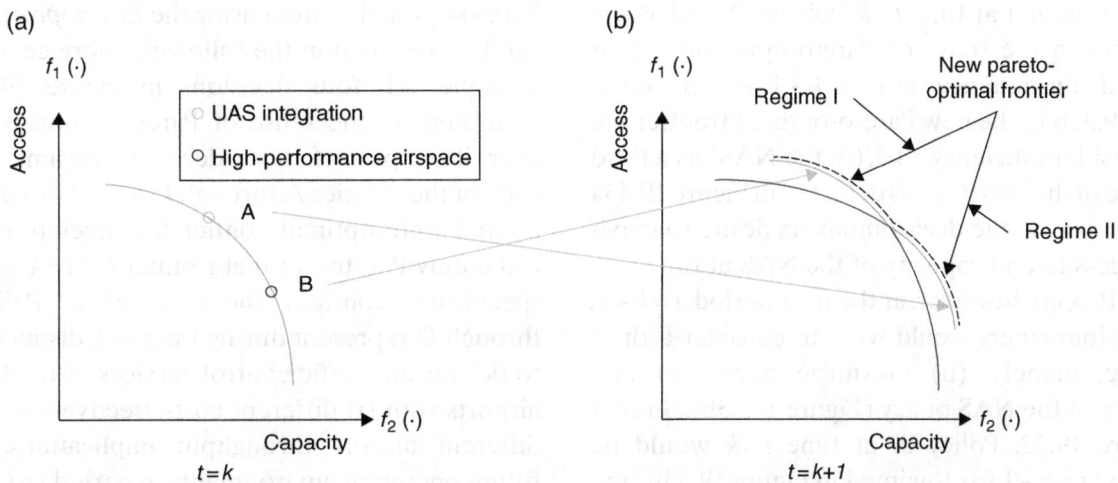

Figure 19.32 A dynamic Pareto-optimal frontier is related for two objective functions: access versus capacity.

600 CASE STUDIES

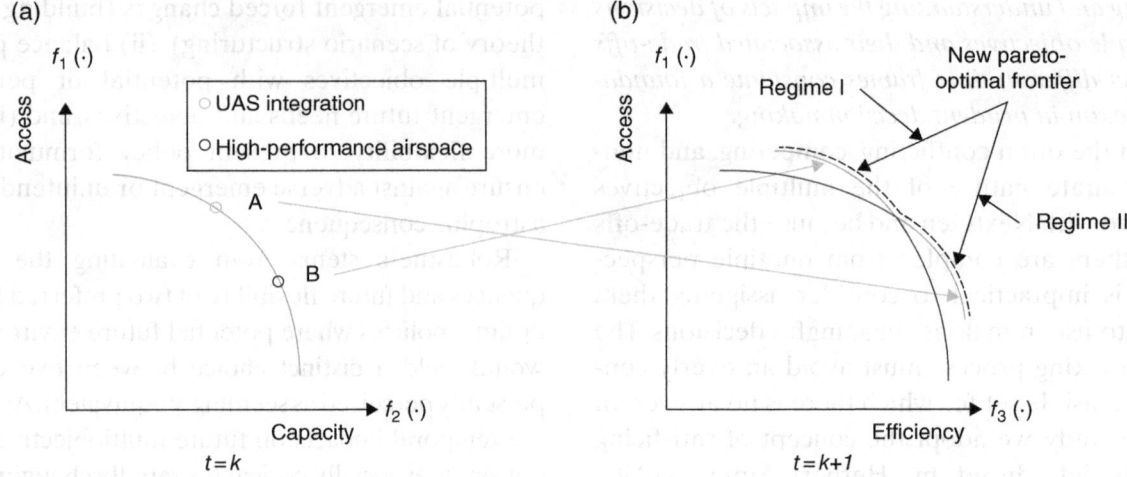

Figure 19.33 A dynamic Pareto-optimal frontier is related for two objective functions: (a) access versus capacity and (b) access versus efficiency.

representing increased flight capacity resulting from NextGen, and capacity, $f_2(\cdot)$, in the X-axes representing improving flight efficiency at time k. Figure 19.32a presents two Pareto-optimal Policies **A** and **B** selected at time $t=k$. When the projections of Policies **A** and **B** are plotted for period $t=k+1$ in Figure 19.32b, they reveal that only the *envelope* depicted in the dotted line of the intersection of the two Pareto-optimal frontiers constitutes the new Pareto-optimal frontier (for $t=k+1$). Thus, given the availability of a dynamic model of the two objectives, each Policy **A** (associated with UAS integration) and **B** (associated with high-performance airspace) at time $t=k$ (see Figure 19.32a) would generate future (new *envelope*) Pareto-optimal frontier at $t=k+1$ (see Figure 19.32b). In sum, although at time $t=k$ Policies A and B are equivalent in the sense of Pareto-optimality, their projected consequences at time $k+1$ as depicted in Figure 19.32b in the new Pareto-optimal frontier are not. Consider efficiency, $f_3(\cdot)$, (of the NAS) as a third objective of the NextGen as depicted in Figure 19.33a and b: (i) at $t=k$, the decisionmakers desire to maximize access to and capacity of the NAS at time $t=k$ (Figure 19.33a); however, at the next period, $t=k+1$, the decisionmakers would want to consider a third objective, namely, (ii) maximize access to and efficiency of the NAS policy (Figure 19.33b). Similar to Figure 19.32, Policy **A** at time $t=k$ would be inferior at $t=k+1$ for Regime I in Figure 19.33b; and Policy **B** at time $t=k$ would be inferior at $t=k+1$ for Regime II in Figure 19.33b. In other words, given the *Evolving Base* and the expected changes in circumstances and priorities, decisionmakers and policy analysts can use the dynamic multiobjective analysis, using the *Envelope Approach* to project scenarios, given the inevitable *Evolving Base*. More will be discussed in the next section. The *Envelope Approach* was developed to account for the propagation of Pareto-optimal solutions from one period to another [Li and Haimes, 1987].

Figure 19.34 presents a final example of a dynamic Pareto-optimal frontier to n periods through the *Envelope Approach*. Note that the solid line's outer curve represents the envelope of the combined Pareto-optimal frontier of Policies A, B, C, and D for the $k+1$st period. In sum, the value of the dynamic Pareto-optimal frontier using the *Envelope Approach* can be explained in the following more generalized example: All four decisions in Figure 19.34 are equivalent in the sense of Pareto-optimality; however, if we project those decisions to time $t=k+1$, each of the Policies A through D would project a different Pareto-optimal frontier, the envelope of which is the only Pareto-optimal frontier for $t=k+1$. More specifically, consider the case where Policies A through D represent during time $t=k$ distinct means to deliver air traffic control services at nontowered airports with (i) different cost-effectiveness and (ii) different airport throughput implications. Given future operating environments in period $t=k+1$, the introduction of different classes of UAVs would

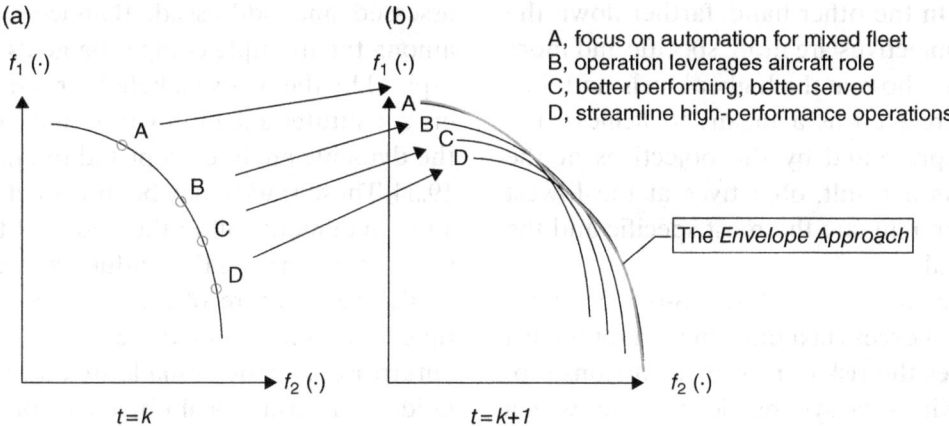

Figure 19.34 The envelope of the combined Pareto-optimal frontiers of policies A, B, C, and D for k+1st period.

produce different Pareto-optimal frontiers associated with cost-effectiveness and throughput for each option.

The future environment would likely be similarly affected by other changes in period $t = k+1$, such as the use of data communications between pilot and controller, the expansion of surveillance capabilities, changes in the high-altitude airspace flows, and allocation of controller workforce across facilities driven by demand.

19.6.1.7 Judgment and values in decisionmaking

In every decision process, there are factual elements and value elements. Factual elements are those that can be verified scientifically and subjected to scientific machinery and are likely to lead to other verifiable elements. Value elements, on the other hand, defy all forms of scientific verification and treatment. A collection of value elements and their sources constitutes a value system. Judgment, which signifies the act of giving an opinion, is the most common value element in any decisionmaking process [Chankong and Haimes, 1983, 2008]. The decisionmaking process is often an amalgamation of real or perceived facts and of value judgment.

The development of the emergent NextGen S-o-S is necessarily driven by both fact-based and judgmental decisions. For the publicly funded and operated NextGen complex S-o-S, analytically harmonizing logical inputs with the political decisionmaking process is often not a smooth process. In such cases, past experiences can skew the sense of where investments are needed to properly assess and manage technological or programmatic risks. Similarly, biases affect the relative value of desired behaviors and outcomes in a multiobjective decisionmaking environment.

In this case study, we posit that the amalgamation of facts and value judgments has an important role to play in the complex dynamic multiobjective decisionmaking process, where judgment plays an integral role. Modeling and managing NextGen as S-o-S enable stakeholders and decisionmakers to analyze, model, and measure desired emergent behaviors' performance of the S-o-S as a whole. This dynamic multiobjective process constitutes a nontrivial challenge to the community that must decide on acceptable trade-offs among dynamic, noncommensurate, competing, and conflicting objectives that characterize the overall performance of NextGen.

Similar to the existence of a hierarchy of substates and sub-substates of a system, so too are there sub-objectives and sub-subobjectives, especially for emergent complex S-o-S. The highest level of this structure generally represents the broad overall objectives that are instrumental in initiating the multiobjective decision problem in the first place. The overall objectives of NextGen, safety, efficiency, capacity, and environment, are a case in point. However, these objectives are often vaguely stated and, hence, not operational. For example, the NextGen project must minimize inefficiencies for flight operators while also holding the accident rate to 10^{-6} and simultaneously increasing the national airspace capacity and ensuring stewardship of the environment. On the other hand, farther down the hierarchy, the objectives are more specific and more operational than those at the higher levels, and they are at least perceived as a means to achieve the higher ends represented by the objectives at the higher level. As a result, objectives at the lowest level of the hierarchy are the most specific and the most operational.

For example, in the NextGen S-o-S, simultaneously reducing excess taxi time on the ground in a way that reduces the risk of runway incursions represents a win–win—two synergistic outcomes with a positive change. For example, Figure 19.33 represents a dynamic Pareto-optimal frontier for two objectives: access versus capacity and efficiency. From a capacity perspective, the two capabilities have a clear trade-off with access, but when considering efficiency the relative trade-off may reverse, then the criteria for the judgment call become important.

Consider the introduction of procedures to integrate UAS within the NAS or alternatively to create performance-based airspace as a means of accommodation of the UAS. The trade-offs could be quantified, but the acceptable policy on the Pareto-optimal frontier would be a judgment call.

These judgment calls are made more complex by the combination of trade-offs among shared states highlighted in Table 19.13. For example, three NextGen decisions that fit in such a cluster of shared state variables include (i) a decision to create high-performance airspace (segregating certain operations), (ii) a decision on automation needs to serve a more diverse fleet in terms of aircraft and navigation performance, and (iii) a decision on the aircraft role in separation and collision avoidance. Each of these decisions has a different effect on the three objectives: access, capacity, and efficiency. The criteria or basis for judgment in balancing between these objectives must be clear to the analyst who must study the three decisions as a set. The resulting dynamic multiobjective trade-offs must be represented in a three dimensional space.

The US Congress mandated that the FAA develop a plan for the integration of UAS within the NAS within the NextGen enterprise. The known and precursors to unknown multiple sources of risk associated with such integration must be carefully assessed and addressed. Balancing the trade-offs among the multiple competing goals and objectives aspired by the many stakeholders and decisionmakers constitutes a daunting task that can benefit from the dynamic analyses depicted in Figures 19.33 and 19.34. These trades may be informed by fact but are still a judgment call in the end, and they constitute some of the drivers that influence such trade-offs.

The very nature of emergent S-o-S is that multiple decisionmakers direct the evolution of the enterprise. Decisions made by an individual stakeholder reflecting local objectives of one subsystem might be perceived as isolated, but it will likely affect—positively or negatively—other public decisions made to incentivize a desired emergent behavior. Consider NextGen's efforts to increase access to runways in marginal metrological conditions. A capability using improved navigation procedures applicable to all trained aircraft operators would raise the total throughput for the airport by mitigating the loss of parallel runway throughput. An alternative solution uses enhanced visual systems to deliver benefit in the form of greater access as reduced levels of visibility directly to flights that are so equipped and trained. The overall capacity of the airport in reduced visibility is not affected by the second solution. Both decisions impact on the local objectives, but they require different investments by aircraft operators and result in different emergent behavior objectives. In the first case, all aircraft operators share in the benefit from the investment in training the flight crews regardless of who invests, while in the second case those who invest in the visual system have more exclusive rights to the benefit as only they will receive improved service (a differentiated value for the investor).

19.6.2 Principles and Guidelines

This section presents definitions, principles, and premises for the planning, deployment, and management of emergent S-o-S.

19.6.2.1 Centrality of the states of a system

Given a system's model, the smallest set of independent system variables (such that the values of the members of the set at time t_0 along with known inputs, decisions, random, and exogenous variables) determine the value of all system variables for all $t>t_0$. The shared states among the subsystems that constitute the NextGen S-o-S play a major role in modeling, understanding, and ultimately harmonizing (to the extent possible) the myriad time-critical structural and nonstructural policy decisions that are made by the multiple stakeholders engaged in this complex process. Table 19.13 presents shared states among the major NextGen objectives. These shared states play a central role in highlighting and appreciating the complexities in the decisionmaking process that are caused by the interconnectedness and interdependencies among the subsystems, by the objectives, and by the corresponding stakeholders and decisionmakers. Furthermore, the shared states streamline the essential harmonization process associated with reaching acceptable trade-offs among the dynamic multiple objectives as well as among the principal stakeholders and decisionmakers and, ultimately, acting upon those trade-offs.

The complexity of the quantification of the multidimensional risk function has been modeled by relying on the states of the system. Indeed, the multidimensional probabilistic consequences, resulting from an initiating event, yield a multidimensional risk function whose modeling and quantification can be achieved most effectively through the states of the affected system. Furthermore, both the vulnerability and the resilience of a system to any specific initiating event are manifestations of the states (e.g., physical, technical, organizational, and cultural) of the affected system. Thus, the consequences resulting from any specific initiating event, which are necessarily functions of the states and of the specific time frame, are also functions of the vulnerability and resilience of the system to the specific initiating event [Haimes, 2009a]. For example, resilience represents the ability of the system to withstand a disruption within acceptable degradation parameters and to recover within acceptable losses and time. Other characterizations of the risk function in terms of robustness, flexibility, and sustainability are discussed in the literature [Jugulum and Frey, 2007; de Neufville and Scholtes, 2011; Cardin et al., 2013; Hu et al., 2013].

The integration of UAS into the NAS, the transition to cloud computing, and the change in pilots' or controllers' roles are often identified as sources of vulnerabilities to the NAS. The projected likelihood of these or other vulnerabilities causing specific emergent changes should not be a roadblock to advancing the NAS; instead, they should be viewed as keys to shaping the decisions needed to manage the emerging sources of risk to NextGen. Scenario structuring is a useful tool with which to refine and assess the vulnerabilities of NextGen in the *Evolving Base* of changing conditions (to be discussed in subsequent sections). For example, in the integration of UAS into the NAS, a decision on the

TABLE 19.13 Shared States among the Major NextGen Objectives

Efficiency	Safety	Capacity	Environment
1. ATM technology	1. ATM technology	1. ATM technology	
2. Operating environment	2. Operating environment	2. Operating environment	2. Operating environment
3. Personnel and operations	3. Personnel and operations	3. Personnel and operations	
4. Aircraft responsiveness	4. Aircraft responsiveness	4. Aircraft responsiveness	
5. Information quality	5. Information quality		
7. Culture	7. Culture	7. Culture	
8. Budget	8. Budget	8. Budget	8. Budget
9. Reliability (6 sigma)	9. Reliability (6 sigma)	9. Reliability (6 sigma)	9. Reliability (6 sigma)
10. Trust	10. Trust	10. Trust	
11. Standardization	11. Standardization		
12. Congestion	12. Congestion		
	13. Product reliability		
21. Runway usage		21. Runway usage	21. Runway usage
23. Fuel usage			23. Fuel usage
			24. Noise

flight rules governing a UAS flight will affect three shared states in the NAS: (i) the controller workload, (ii) the vehicle technology requirements, and (iii) the pilot qualifications. Emergent forced changes to any one of these states will likely alter options available to stakeholders and other decisionmakers. Similar to vulnerability, the resilience of a system is also multidimensional, and in many ways, the vulnerability and resilience of a system are two sides of the same coin.

The current NAS emerged from a series of development efforts that did not explicitly include resilience of the air traffic management capability as a design goal. Individual systems were protected against single point failures, but the collective S-o-S has no established overall plan for resilience. So the state of resilience is not assured. One of the design objectives of NextGen would be to build in resilience of the S-o-S, making it less vulnerable to emergent forced changes. To avoid recreating the kind of crystal-like growth that characterizes the present NAS where capabilities were added on as needed, the future must capture a more complete design of services.

19.6.2.2 Project life cycle and team risk management

An often-neglected aspect of project risk management is attention to the entire project life cycle. Manufacturing firms, for example, commonly conduct a failure mode and effects analysis (FMEA) and failure mode, effects, and criticality analysis (FMECA) on the product and the assembly line, but they ignore the product development and design processes. Doing so ignores the risks inherent in requirements definition, development, acquisition, and phaseout or upgrade. Sage [1992, 1995] discusses the different types of risk inherent in the various stages of the life cycle. Ignoring important stages of the life cycle can lead to substantial problems in terms of programmatic risk for both product development at the beginning of the life cycle and for product upgrade or replacement at the end. If major risks are not handled sufficiently early, they may magnify their effects later in the project. For example, in information technology acquisitions, errors in the requirements definition phase can lead to costly cascading problems later when the information system fails to meet the customer's needs. As a result, costly modifications may be necessary, causing schedule slips and cost overrun.

Considering the shifting operating environment for NextGen tends to alter the requirements. For example, an operational tool designed initially to sequence flights to maximize runway occupancy during capability constraints may be consequently required to sequence and space flights for efficiency as fuel prices rise. Versatility and adaptability to uncertain future demands, fleet mix, and general operating environment (the external forced changes) are characteristics of NextGen. A case in point is the need to be cognizant of and responsive to the *Evolving Base* [Haimes, 2012b]: "Dynamic shifting rules and realities: (i) Goals and objectives; (ii) Stakeholders, decisionmakers, and interest groups; (iii) Organizational, political, and budgetary baselines; (iv) Reorganization and reallocation of key personnel; and (v) Requirements, specifications, and delivery." For example, requirements necessarily evolve in response to external changes. If they are not adequately managed in response to the *Evolving Base*, the resulting consequences can be not only expensive, but they may not serve future needs. Similarly, if the requirements are too fluid, costs and schedules cannot be maintained. Common to all decisions, appropriate trade-offs must be made.

Managing the risks inherent in any system is contingent upon having sufficient knowledge of the system's structure and operations. Indeed, this knowledge is imperative in order to comprehensively identify the risks associated with project acquisition and management, accurately estimate the probabilities of failure, and correctly predict the consequences of those failures. While the tendency to collect data and information on the project is important, databases are useful only when accompanied by an understanding of the way the system they describe operates. Knowledge of a system provides a means to understand and effectively use information from the system. Obtaining this knowledge is often difficult even for a single system, and the problem is compounded with the development or acquisition of S-o-S. Knowledge of the many component systems is required, and it is also critical to understand the boundaries where these systems interact and generate new sources of risk. These interactions include project requirements and specifications, design and construction, finance and management, development of new technology, and response to a myriad of changes and conflicting signals from the many participating organizations. Thus, the sheer amount of system knowledge requisite for the risk analysis imposes some difficulties in its collection, dissemination, and integration.

In their book *Working Knowledge*, Davenport and Prusak [1998] suggest that knowledge moves through an organization via markets just as any other scarce resource does. There are buyers, sellers, and brokers of knowledge. Those who possess it will sell their knowledge if properly compensated with money, reciprocity, repute, bonuses, promotions, or other gains. If there is not sufficient compensation for those who sell their knowledge, the transfer will not take place. This market for knowledge has some important implications for risk management. The knowledge necessary to assess the risks to an entire project is spread over many individuals in multiple organizations and at multiple levels in the management hierarchy. For this knowledge to be transferred and collected for the purposes of risk management, an efficient knowledge market must exist. To this end, management and corporate culture are key influences that must facilitate rather than hinder the operation of knowledge markets. First and foremost, trust is required for the exchange of knowledge [Davenport and Prusak, 1998; Covey, 2008].

Team risk management [Haimes, 2009b] brings together all of the disparate parties in the risk management effort. Katzenbach and Smith [1999] introduce the following definition: "A team is a small number of people with complementary skills who are committed to a common purpose, performance goals, and approach for which they hold themselves mutually accountable." In other words, when conducting risk management in teams, participants are imbued with a common purpose. Risk management is not externally enforced; rather, it is a process in which everyone participates. When all participants have personal stakes in the process, they are much more likely to share their system knowledge since they can see the potential benefits from doing so. To facilitate this process, organizational boundaries must be overcome. An effective means is to bring people together in face-to-face meetings: individuals from the various participating organizations, from subdivisions within organizations, and from different levels in the management hierarchy.

19.6.2.3 FAA NextGen: A Dynamic Roadmap

NextGen is comprised of S-o-S encompassing hardware and software, humans, policies and procedures, national airspace, and vehicles including unmanned vehicles. It depends on other S-o-S, such as GPS, commercial telecommunications, national weather service, passengers, military, and international cooperation and partnership. Furthermore, decisions for this complex enterprise are made by a combination of individual stakeholders and organizations in order to maintain control of the states of the system under consideration to achieve specific local objectives and outcomes. To address the myriad of sources of risk associated with the emergent development of NextGen, an extension of the *Dynamic Roadmap for Risk Modeling, Planning, Assessment, Management, and Communication for Emergent Systems of Systems* [Haimes, 2012b] is applied to NextGen with a focus on the automation portion of NextGen. The major flight data processing subsystems addressed here, which constitute an S-o-S, include: (i) system-wide information management (SWIM), (ii) en route automation modernization (ERAM), (iii) terminal automation modernization and replacement (TAMR), (iv) collaborative air traffic management technologies (CATMT), and (v) terminal flight data manager (TFDM). The motivation for this application is that the current path will not drive automation convergence or alignment without a systemic and principle-guided risk modeling, assessment, management, and communication processes.

The FAA must sequence and manage its decision-making activities under the following constraining realities: (i) adhering to a desired *time frame*; (ii) addressing all competing *objectives* affected by the decisions; (iii) operating in the context of assumed linkage of *alternatives* to outcomes; (iv) functioning in the context of *decisions made by individual power-brokers and stakeholders* who are striving for state changes that are beneficial to their objectives and outcomes; (v) addressing the operating modes explicitly, including *degraded modes*; (vi) addressing all questions in the risk assessment, management, and communication processes to identify the *emergent forced changes* that may alter outcomes; and (vii) recognizing *first-order and indirect impacts on each subsystem*, especially those cases that pose a risk of low-probability yet extreme and dire consequences.

19.6.2.4 Applying the Dynamic Roadmap and its extensions

All emergent S-o-S are subject to emergent forced changes (external or internal sources of risk to a subsystem or to the entire S-o-S that may adversely affect specific states of that system). Thus, the *Journey* (associated with the *Dynamic Roadmap*) starts with the *Evolving Base* (dynamic shifting rules and realities: (i) goals and objectives; (ii) stakeholders, decisionmakers, and interest groups; (iii) organizational, political, and budgetary baselines; (iv) reorganization and reallocation of key personnel; and (v) requirements, specifications, delivery, users, and clients), namely, the dynamic emergent forced changes that dominate the entire life cycle of emergent S-o-S.

First principle: Holism is the common denominator that bridges risk analysis and systems engineering

The imperative adherence to the gestalt–holistic philosophy must be the sine qua non for an effective modeling of the NextGen as emergent S-o-S. Consider the following sample of multiple perspectives that must be addressed: (i) the integration of the navigation, surveillance, and communication systems, which currently operate semi-independently to avoid one path to failure, and (ii) the integration of the UAS with the NAS. To be more specific, this sample of systems integration, among numerous others, requires accounting for the corresponding subsystems; the associated shared states; decisions, decisionmakers, and stakeholders; policies, procedures, and regulations; integration of old with technology, retraining of technical and service personnel, and streamlining operational software with the corresponding hardware systems; harmonizing organizational culture and tradition that characterize each of the current navigation, surveillance, and communication systems (which have been operating semi-independently for close to 40 years);

new privacy, safety, and many heretofore unforeseen sources of risk associated with the integration of the UAS with the NAS; and the impacts of nearly 90 decisions (structural, operational, policy, etc.) that must be made during the Alpha period, which are likely to affect and constrain the progress to be made during the subsequent Bravo and Charlie periods. These and a welter of other considerations must be addressed within the gestalt–holistic philosophy.

Second principle: The process of risk modeling, assessment, management, and communication must be methodical, disciplined, systemic, integrated, and commensurate in its comprehensiveness with the criticality of the systems being addressed and of their associated risks

Being cognizant of the reality of the *Evolving Base* associated with NextGen must guide the risk modeling, assessment, management, and communication processes. Given the importance of risk and uncertainty analyses to NextGen, the development of which spans two decades, it will be useful to distinguish between risk analysis and uncertainty analysis and, within each, its different parts and perspectives. *Risk management* is commonly distinguished from *risk assessment*, even though some may use the term *risk management* to connote the entire process of risk assessment, management, and communication. The term *management* may vary in meaning according to the discipline involved and/or the context.

In risk assessment, the analyst often attempts to answer the following three questions [Kaplan and Garrick, 1981]: What can go wrong? What is the likelihood that it will go wrong? What are the consequences? Here, we add a fourth question: *What is the time frame?* Answers to these questions help risk analysts identify, measure, quantify, and evaluate risks and their consequences and impacts.

Risk management builds on the risk assessment process by seeking answers to a second set of three questions [Haimes, 1991]: What can be done and what options are available? What are the associated trade-offs among all relevant costs, benefits, and risks (both mitigated and residual)? What are the impacts of current management decisions on future options? This third question is the most significant for the emergent NextGen S-o-S because decisions during one time period (e.g., during the Alpha 5-year period) will affect many decisions at the subsequent Bravo and Charlie periods lasting more than 10 years combined. Risk communication must harmonize the risk assessment and management processes, as well as the communication between the risk analysts and decisionmakers and other stakeholders.

Effective risk communication undoubtedly plays a dominant role in harmonizing the responses to the *Evolving Base* (within and among the subsystems of the NextGen S-o-S) by all involved parties. Decisionmakers and other stakeholders benefit when they are able to clearly describe the challenges facing them for which they are seeking a better understanding in order to formulate appropriate policies and possible solutions. In turn, risk analysts, systems engineers, and systems integrators must be able to translate complex technical concepts, analyses, and results into language that decisionmakers and other stakeholders can relate, understand, and incorporate into actionable decisions.

NextGen technical and infrastructure-based decisions, policies, regulations, and procedures are continuously being formulated and implemented. Recognizing that any and all current decisions are likely to have major impacts on future options on the emergent NextGen S-o-S, it becomes imperative to harmonize decisions made during the Alpha period with decisions to be made during the Bravo and Charlie periods to avoid structural foundational mishaps affecting current and future infrastructure systems. Between 2014 and 2017, the FAA will make numerous decisions, ranging from flight protocol standards and procedures to infrastructure design options that will dictate the course by which the US airspace evolves throughout the next decade. Although some of these strategy decisions affect only a single subsystem of the NAS, a number of the decisions have broader implications for the NAS as a whole and have a high likelihood of incurring programmatic risks (e.g., cost overruns and delays in meeting delivery schedules) if they are not made carefully. The NextGen decisions were analyzed to understand potential interactions as examples for the principles presented here.

There is a continuous need to harmonize arising conflicts among the planners and stakeholders, keeping in mind the *Evolving Base* and responsibilities during the three planning periods. In this context, it is reasonable to assume that the NextGen planners and executives, Congressional committees, the US Department of Homeland Security, airline carriers, pilot associations, and other stakeholders have the following emergent issues on their agendas: (i) revisit the scope and objectives of select infrastructure systems and communicate them to the team; (ii) provide a clear insight on what would be an acceptable range of alternatives and their projected impacts on future options; (iii) collect data on each infrastructure during its construction; (iv) develop metrics with which to measure progress, and use the results to populate the *Dynamic Roadmap*; (v) continuously revisit the *Dynamic Roadmap* to ensure an adherence to its principles and guidelines; (vi) be cognizant of the fact that two-thirds of the major decisions made during the Alpha period relate to infrastructure decisions affecting navigation and surveillance systems; and (vii) assess the impact of the above on workforce, budget, schedule, cost, and the system's performance criteria.

In the multidecisionmaker environment that characterizes NextGen, the perspectives and reactions of one decisionmaker to decisions and actions of another may alter the technical and programmatic risks for the enterprise. Consider efforts to define services offered based on better flight performance or better performing—better served. Procedures for certain airport and weather conditions would allow flights with improved navigational performance to continue to their destination airport, while those less equipped would have to be delayed or diverted. The decision for how to implement this procedure may be motivated by either a desire to incentivize aircraft operators to equip aircraft with higher levels of navigational performance or a desire by the operators already equipped with improved performance to get the most from their investment. If the FAA procedures are designed to operate with low percentage of the fleet equipped to start but continue to deliver benefits even as the incentives take affect and the percentage grows, then they satisfy both objectives. If they only give benefits so long as the fraction of the fleet able to participate is small but the benefits diminish with higher levels of participation, then the incentives also fade with time. Each decisionmaker must decide to join or not as others are weighing similar decisions.

Third principle: Models and state variables are central to quantitative risk analysis

The states of a system constitute the main building blocks of both analytical and simulation models. Models are built to answer specific questions and must be made as simple as possible and as complex as required. Without models, risk analysts would find it nearly impossible to do their work. The centrality of state variables in decisionmaking, and particularly in risk analysis, has not been sufficiently emphasized. Indeed, a representative model can effectively provide answers to the questions for which it is built only by first identifying and incorporating all relevant and critical states of the system. As noted earlier, the vulnerability and resilience of a system are *manifestations of the states of the system, and each constitutes a vector that is time and threat (emergent forced changes) dependent.*

Each decision in NextGen can be modeled as one or more state changes that represent standard elements of a concept of operations, including the expected performance, operating environment, policy constraints, operating procedures, roles and responsibilities, technology, and information quality in use by the system. To understand joint susceptibility to emergent forced changes, these state changes can be analyzed for their overlap and the magnitude of their impact on related decisions. A clustering technique can be used to group decisions with significant shared state space for further consideration regarding sequencing of decisions. This clustering was demonstrated in section A for the NextGen decisions on (i) a possible decision to create high-performance airspace (segregating certain operations), (ii) a decision on automation needs to serve a more diverse fleet, and (iii) a decision on the aircraft role in separation and collision avoidance.

Fourth principle: Multiple conflicting and competing objectives are inherent in risk management

This principle, common to all emergent S-o-S, was discussed in Section A.6, where the concept of Pareto-optimality or identifying a noninferior solution to a multiobjective optimization problem was introduced. A solution to a two-objective function is termed Pareto optimal if improving one objective function can be achieved only by degrading the other. Indeed, Pareto optimality is central to determining the trade-offs between the additional expenditure required for further risk reduction.

In the NextGen S-o-S, interest in high-performance airspace is tied to a gain in efficiency for advanced capability aircraft, while automation is designed to integrate a more diverse fleet and would favor access. Both may coexist at some level, based on where performance diversity and similarities are greatest in the NAS. The third decision on avionics' role in separation and collision avoidance could support either paradigm, but it depends on knowledge of the constraints imposed by the first two.

Fifth principle: Risk analysis must account for epistemic and aleatory uncertainties

Uncertainty, commonly viewed as *the inability to determine the true state of a system*, is characterized by two sources that affect a system's modeling and thus its risk analysis [Morgan and Henrion, 1990; Paté-Cornell, 1996; Apostolakis, 1999]: (i) *incomplete knowledge (epistemic) uncertainty*, which manifests itself in the selection of model topology (structure) and model parameters, among other sources of ignorance (e.g., lack of knowledge of important interdependencies within the system and among other systems), and (ii) *stochastic variability (aleatory) uncertainty*, which includes all relevant and important random processes and events, as well as emergent forced changes. Uncertainty dominates most decisionmaking processes and is the Achilles' heel for all deterministic, and most probabilistic, models. Both categories of uncertainty markedly affect the quality and effectiveness of risk analysis efforts and, ultimately, the decisionmaking process.

For the NextGen model, three of the states, the service policy, the operational roles and personnel qualities, and the ATM technologies, are more susceptible to epistemic uncertainties because the planning horizon covers multiple decades. The initial intent of policy constraints; the operating procedures, roles, and responsibilities; and technology changes can be approximated, but there is little practical basis for how to model these changes over time, especially in reaction to external forced changes from the other stakeholders. The other NextGen shared states, operating environment, aircraft responsiveness, and information quality are a function of industry economics and operating practices and so are more conducive to treatment as stochastic variables with external drivers such as fuel price and market demographics. Thus, a range of future operating environments and expected performance can be used to understand the coming 20-year need for high-performance airspace or divergent fleet access. The policy and operating procedures will change with time in ways that cannot be modeled; consequently, the technology must be robust and adaptable enough to serve a wide range of policies and procedures. The epistemic and aleatory uncertainties identified in Figure 19.35 are associated with a select set of states of the NAS, noting that often most of these states are subject to both categories of uncertainties.

Sixth principle: Risk analysis must account for risks of low probability with extreme consequences

The most common quantification of risk is the use of the mathematical construct known as the expected-value-of-risk metric, which equates low probability of high-consequence events with high probability of low-consequence events. This metric has played a decisive role in masking the criticality of extreme and catastrophic events. It is important to recognize the misuse of the expected value—the averaging of risk—when it is used as the sole criterion for risk in decisionmaking. The PMRM [Asbeck and Haimes, 1984; Haimes, 2009b] supplements the expected value measure of risk with a conditional expected value of risk of extreme events. A conditional expectation is defined as the expected value of a random variable, given that this value lies within some prespecified probability or consequences range.

In the NextGen model, risk ratings for the state changes were based on the consequence of decisions made by one stakeholder on other stakeholders. Thus, state changes to the operational roles were rated very high if public/private roles, size of workforce, and harmonization between workforces of

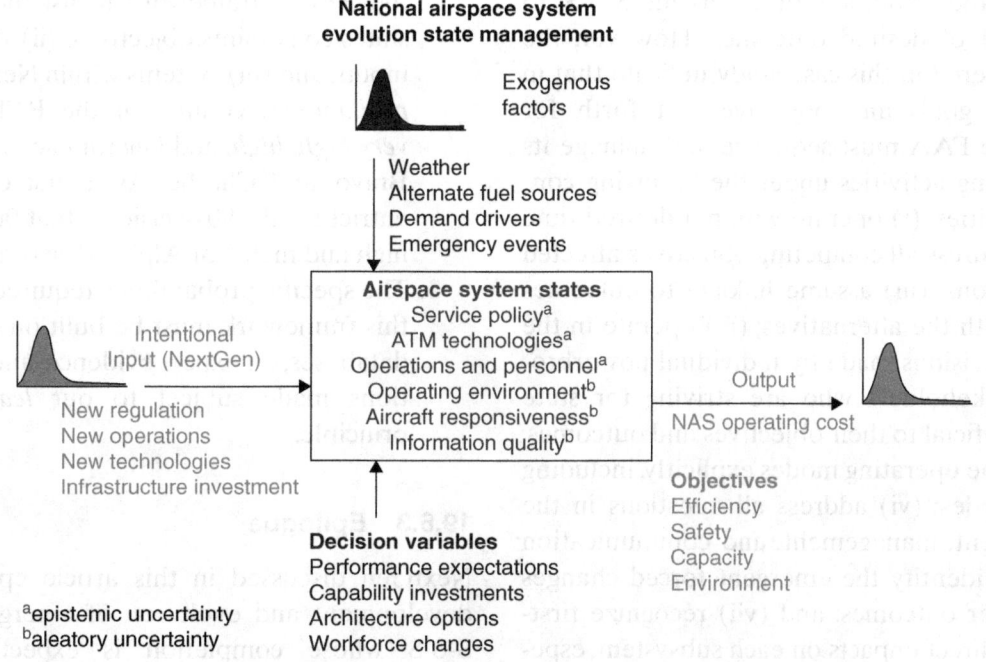

Figure 19.35 State space depiction of the National Airspace System.

different organizations were required, regardless of their probabilities. The assumption was that with multiple stakeholders involved in emergent system planning, it is necessary to assume the most significant consequences will result from the groups' interactions and then work to mitigate them.

Seventh principle: The time frame is central to quantitative risk analysis

System modeling and risk analysis cannot be performed effectively, if at all, without considering the time frame of the analysis and decisionmaking processes. It is important to note that the states of all systems, and thus their risk functions, are dynamic. All are functions of the time frame—even though they may be assumed, in some studies, to be static for simplicity. Furthermore, the dynamic trade-offs among multiple objectives both complement and supplement the determination of Pareto-optimality in risk management, given the dynamic nature of evolving events. Similarly, the process of eliciting the decisionmakers' preferences must be systemic and harmonious with the realization that conditions, and thus policies, are dynamic and subject to being updated and modified.

19.6.2.5 Summary

There is no single method to manage the risks associated with the evolution of emergent S-o-S to achieve a set of desired outcomes. However, the principles offered in this case study indicate that in meeting the goals and objectives set forth for NextGen, the FAA must sequence and manage its decisionmaking activities under the following constraining realities: (i) operate within a desired time frame; (ii) address all competing objectives affected by the decisions; (iii) assume linkage to outcomes associated with the alternatives; (iv) operate in the context of decisions made by individual powerbrokers and stakeholders who are striving for state changes beneficial to their objectives and outcomes; (v) address the operating modes explicitly, including degraded modes; (vi) address all questions in the risk assessment, management, and communication processes to identify the emergent forced changes that may alter outcomes; and (vii) recognize first-order and indirect impacts on each subsystem, especially those cases that pose a risk of low probability with extreme and dire consequences. Many of the critical decisions were modeled at a high level in terms of six states (service policy, ATM technology, operations and personnel, operating environment, aircraft responsiveness, and information quality) to understand how to apply this reality to the practice of continuously monitoring the *Evolving Base* for such complex S-o-S. The insights gained by the FAA show that by focusing on the most significant consequence of each state change, it is possible to cluster decisions and work on mitigation strategies for the sequencing of decisions. In the following, we recommend a more comprehensive and continuous treatment of NextGen evolution:

1. Develop criteria in order to prioritize decisions made during each period (Alpha, Bravo and Charlie) and to assess the impact of such decisions on safety, efficiency, environment, and capacity.
2. Focus on decisions where expected impacts are assessed to be high and very high and the associated consequences are deemed to be significant. To accomplish this, apply the risk filtering and ranking method (RFRM) [Haimes et al., 2002; Haimes, 2009b] to reduce the 90 decisions to a manageable high-priority number of decisions. Use the *consequences* column in the RFRM framework to list the affected (i) states/constraints/objectives, (ii) decisions and inputs, and (iii) systems within NextGen. In the *probabilities* column of the RFRM, indicate *very high*, *high*, and *low* for each period Alpha, Bravo, and Charlie. As a first cut, select all entries (of the 90 decisions) that fall within very high and high (for Alpha, Bravo, and Charlie).
3. The specific probabilities required to populate this framework must be built on the collected databases, on expert evidence, and on assumptions made subject to our *learn-as-you-go* principle.

19.6.3 Epilogue

NextGen discussed in this article epitomizes the development and evolution of emergent complex S-o-S whose completion is expected to span two decades. Simultaneous with its evolving development, it is subject to a tedious process of modeling, assessing, managing, and communicating the myriad corresponding sources of risk. The flowchart depicted in Figure 19.36 associates the S-o-S-based theory and methodology with the FAA's NextGen complex project. Consider the following challenges facing the parties involved in this enterprise:

(i) How can we correctly estimate the life cycle cost and the overall budget for the NextGen enterprise? Which decisions are the major cost drivers? Are they being taken in proper sequence to drive the most beneficial outcome for the objectives, the outputs, or the states of NextGen?

(ii) Given that early decisions made during the Alpha time span may determine or constrain the options in ways that would drive the majority of the total cost, how can planners and decisionmakers ensure that by eliminating structural or policy options they are not adversely impacting or eliminating future options? For example, wrong decisions on operational changes even though they are not acquisitions can be very costly because they have direct impacts on workforce or technology cost drivers for the remaining decisions.

(iii) The FAA is also charged with making a large number of strategy decisions concerning the development of NextGen standards, procedures, and design choices in a short amount of time, many of which will not manifest themselves until several years from now. The challenge in making each of these decisions is being able to evaluate its impact not only on the project that the decision pertains to but also on interdependent subsystems of the NAS.

(iv) Since the FAA is also a regulatory government agency, requiring aircraft operators to comply with service policies can represent a transfer of cost to others. For example, solutions that segregate flows to adjacent airports based on performance-based procedures in the flight deck may require operator investment yet are fundamentally an FAA actions/decisions on its own trade space. The formal rule-making process addresses this type of joint investment decision but may be too onerous to do for each small decision. Packaging multiple decisions into a single dialogue for joint decisionmaking requires understanding the interactions and therefore the sequence of the smaller decisions to construct a fruitful joint investment decision.

(v) The developers of the NextGen enterprise must be continuously cognizant of, and responsive to, the *Evolving Base* introduced earlier. These forced changes also have direct impacts on the dynamic multiple objectives and the associated trade-offs that must be continuously addressed by the concerned decisionmakers and stakeholders.

(vi) The motivation for changes in any subsystem of the NextGen can be twofold: (1) the objectives of the local subsystem and (2) the larger desired emergent behavior for the S-o-S as a whole. In particular, the FAA is commonly interested in the desired emergent behavior of the overall enterprise in addition to the local effect. At any time, there is a desired set of tolerances for the qualities of service performance. As emergent forced changes influence the operation of the national airspace, the desired emergent behavior may correspondingly change. For example capacity has long been the primary objective of modernization programs. The new emergent behavior was manifested by higher load factors, fewer flights, and ultimately more pressure for optimal fuel routes. These emergent behaviors can also be geographical with different desired goals—locally and nationally. Thus, the Pareto-optimal frontier is a function of time and of the *Evolving Base*. One way to model this evolution is to observe and react to the changes in the *Evolving Base* or, alternatively, to use scenarios as precursors to forecast the shifts in frontiers and in desired behaviors.

Clearly, the aforementioned challenges constitute a daunting systems engineering task.

612 CASE STUDIES

Figure 19.36 Flowchart that associates the S-o-S-based theory and methodology with the FAA's NextGen complex project.

REFERENCES

Allbusiness.com (2005) *Utilities Respond to Katrina*. Available at http://www.allbusiness.com/periodicals/article/1060104-1.html.

American Society of Civil Engineers (ASCE), 2005, *Report Card for America's Infrastructure: US Electric Power Grid*. Available online: http://ascelibrary.org/doi/book/10.1061/9780784478851.

Anderson, P., and I. Geckil, 2003, *Northeast Blackout Likely to Reduce US Earnings by $6.4 Billion*, Anderson Economic Group Working Paper AEG 2003-2. Available online: http://www.andersoneconomicgroup.com/Publications/Detail/tabid/125/articleType/ArticleView/articleId/6749/Northeast-Blackout-Likely-to-Reduce-US-Earnings-by-64-Billion-AEG-Working-Paper-20032.aspx.

Anderson, C.W., J.R. Santos, and Y.Y. Haimes, 2007, A risk-based input-output methodology for measuring the effects of the August 2003 Northeast blackout, *Economics Systems Research* **19**(2): 183–204.

Ang, A.H.S., and W.H. Tang, 1984, *Probability Concepts in Engineering Planning and Design, Volume II: Decision, Risk, and Reliability*, Wiley, New York.

Apostolakis, G., 1999, The Distinction between aleatory and epistemic uncertainties is important: an example from the inclusion of aging effects into probabilistic safety assessment, *Proceedings of PSA'99*, August 22–25, American Nuclear Society, Washington, DC.

Asbeck, E.L., and Y.Y. Haimes, 1984, The partitioned multiobjective risk method (PMRM), *Large Scale Systems* **6**(1): 13–38.

Bar-Yam, Y., 2003a, When systems engineering fails—toward complex systems engineering, *International Conference on Systems, Man & Cybernetics* **2**: 2021–2028.

Bar-Yam, Y., 2003b, *Complexity of Military Conflict: Multiscale Complex Systems Analysis of Littoral Warfare*, Chief of Naval Operations, Report for Contract: F30602-02-C-0158, Multiscale Representations Phase II: Task 2: Multiscale Analysis of Littoral Warfare, U.S. Department of the Navy, Washington, DC, 30p.

Bezdek, R., and R. Wendling, 2005, Fuel efficiency and the economy, *American Scientist* **93**: 132–139.

Blauberg, I.V., V.N. Sadovsky, and E.G. Yudin, 1977, *Systems Theory: Philosophical and Methodological Problems*, Progress Publishers, New York, p. 132.

Buede, D.M., 2000, *The Engineering Design of Systems: Models and Methods*, John Wiley and Sons, New York.

Bureau of Economic Analysis (BEA), 1997, *Regional Multipliers: A User Handbook for the Regional Input–Output Modeling System (RIMS II)*, US Department of Commerce, Washington, DC.

Bureau of Economic Analysis (BEA), 1998, *Benchmark Input–Output Accounts of the United States for 1992*, US Department of Commerce, Washington, DC.

Bureau of Economic Analysis (BEA), 2004, *Bureau of Economic Analysis: Regional Economic Accounts: Gross State Product*. Available online: http://www.bea.gov/bea/regional/gsp/ (Accessed January 16, 2015).

Bureau of Transportation Statistics (BTS), 2002, *Louisiana: Transportation Profile*. Available online: http://www.bts.gov/publications/state_transportation_profiles/louisiana/ (Accessed May 5, 2006).

Burton, I., R.W. Kates, and G.F. White, 1978, *The Environment as Hazard*, Oxford University Press, New York.

Cardin, M.-A., G.L. Kolfschoten, D.D. Frey, R. de Neufville, O.L. de Weck, and D.M. Geltner, 2013, Empirical evaluation of procedures to generate flexibility in engineering systems and improve lifecycle performance, *Research in Engineering Design* **24**(3): 277–295.

Cartin, T.J., 1993, *Principles & Practices of TQM*, ASQC Quality Press, Milwaukee, WI.

Castillo, E., 1987, *Extreme Value Theory in Engineering*, Academic Press, Boston, MA.

Center for Contemporary Conflict, 2002, Economic costs to the United States stemming from the 9/11 attacks, *Strategic Insights* **1**: 1–4.

Center for Risk Management of Engineering Systems (CRMES), 2001, *Hurricane Preparedness and Recovery by a Transportation Agency*, Technical Report, University of Virginia, Charlottesville, VA.

Chankong, V., and Y.Y. Haimes, 1983, *Multiobjective Decisionmaking: Theory and Methodology*, North Holland, New York.

Chankong, V., and Y.Y. Haimes, 2008, *Multiobjective Decisionmaking: Theory and Methodology*, Dover, New York.

Covey, S.M.R., 2008, *The Speed of Trust: The One Thing that Changes Everything*, Free Press a Division of Simon & Schuster, Inc., New York.

Crowther, K.G., Y.Y. Haimes, and S.G. Taub, 2007, Systemic valuation of strategic preparedness with illustrations from Hurricane Katrina, *Risk Analysis* **27**(5): 1345–1364.

Davenport, T.H., and L. Prusak, 1998, *Working Knowledge*, Harvard Business School Press, Boston, MA.

Dedeurwaerdere, A., 1998, *Cost–benefit Analysis for Natural Disaster Management—A Case-study in the Philippines*, CRED Working Paper 143, Center for Research on the Epidemiology of Disaster, Brussels, Belgium.

DeLaurentis, D., 2005, Understanding transportation as a system-of-systems design problem, *43rd AIAA Aerospace Sciences Meeting and Exhibit*, AIAA-2005-0123, January 10–13, Reno, NV.

de Neufville, R., and S. Scholtes, 2011, *Flexibility in Engineering Design*, MIT Press, Cambridge, MA.

Department of Energy (DOE), 2005, *Daily Report on Hurricane Impacts on US Energy*, Reports for August 26–October 24, 2005. Available online: http://tonto.eia.doe.gov/oog/special/eia1_katrina.html (Accessed November 1, 2005).

Department of Public Works and Highways, Republic of The Philippines/Japan International Cooperation Agency (DPWH/JICA), 1996, *The Study on Floods, Mudflow Control for Sacobia, Bamban, Abacan River Draining from Mt. Pinatubo*, Office of the Secretary, Department of Public Works and Highways, Manila.

Earth and Atmospheric Sciences (EAS), 2005, *10 Worst Natural Disasters*, Department of Earth and

Atmospheric Sciences, St. Louis University, St. Louis, MO. Available online: http://mnw.eas.slu.edu/hazards.html (Accessed December 22, 2005).

Eisner, H., 1993, RCASSE: rapid computer-aided systems of systems engineering, *Proceedings of the 3rd International Symposium of the National Council of System Engineering, INCOSE* **1**: 267–273.

Energy Information Administration (EIA), 2006, *Monthly Refinery Report*. Available online: http://tonto.eia.doe.gov/dnav/pet/pet_pnp_refp2_aep00_ypy_mbbl_m.htm (Accessed April 4, 2006).

Environmental Protection Agency (EPA), 2005, *Response Activity Reports*, August 29, 2005 through September 26, Washington, D.C.

Ernst and Young, 2002, *Manhattan Lodging Forecast*, Ernst and Young Real Estate Advisory Group, New York, 2005, EPA, Washington, D.C.

Federal Aviation Administration (FAA), 2002, *Aviation Industry Overview Fiscal Year 2001*, FAA Office of Aviation Policy and Plans, Washington, DC.

Federal Aviation Administration (FAA), 2012, *NextGen: Implementation Plan*, NextGen Integration and Implementation Office, Washington, DC.

Federal Emergency Management Agency (FEMA), 2006, *HAZUS: Hazard Loss Estimation Methodology*. Available online: https://www.fema.gov/hazus.

Fischer and Oelrich, 1994, SYS 6050, Risk Analysis Graduate Course, Systems and Information Department, University of Virginia, Charlottesville, Virginia.

Fischetti, M., 2001, Drowning New Orleans, *Scientific American* **285**(4): 76–85.

Gomide, F., and Y.Y. Haimes, 1984, The multiobjective, multistage impact analysis method: theoretical basis, *IEEE Transactions on Systems, Man and Cybernetics* **14**: 89–98.

Government Accountability Office (GAO), 2002, *Review of Studies of the Economic Impact of the September 11, 2001 Terrorist Attacks on the World Trade Center*, GAO, Washington, DC, p. 3.

Haimes, Y.Y., 1981, Hierarchical holographic modeling, *IEEE Transactions on Systems, Man, and Cybernetics* **11**(9): 606–617.

Haimes, Y.Y., 1991, Total risk management, *Risk Analysis* **11**(2): 169–171.

Haimes, Y.Y., 2001, Water system complexity and the misuse of modeling and optimization, In *Risk-Based Decisionmaking in Water Resources IX*, Y.Y. Haimes, D.A. Moser, and E. Z. Stakhiv (Eds.), ASCE, New York.

Haimes, Y.Y., 2007, Phantom system models for emergent multiscale systems, *Journal of Infrastructure Systems* **13**(2): 81–87.

Haimes, Y.Y., 2009a, On the complex definition of risk: a systems-based approach, *Risk Analysis* **29**(12): 1647–1654.

Haimes, Y.Y., 2009b, *Risk Modeling, Assessment, and Management*, Third edition, John Wiley & Sons, Inc., Hoboken, NJ.

Haimes, Y.Y., 2012a, Modeling complex systems of systems with phantom systems models, *Systems Engineering*, **15**(3) 333–346.

Haimes, Y.Y., 2012b, Systems-based guiding principles for risk modeling, planning, assessment, management, and communication, *Risk Analysis* **32**(9): 1451–1467.

Haimes, Y.Y. and A. Anderegg, 2015, Sequential Pareto-optimal decisions Made during emergent complex systems of systems: An application to the FAA NextGen, *Systems Engineering*, **18**(1) 28–44.

Haimes, Y.Y., and C.G. Chittester, 2005, A roadmap for quantifying the efficacy of risk management of information security and interdependent SCADA systems, *Journal of Homeland Security and Emergency Management* **2**(2): 1–21.

Haimes, Y.Y., and W.A. Hall, 1974, Multiobjectives in water resources analysis: the surrogate worth trade-off method, *Water Resources Research* **10**(4): 615–624.

Haimes, Y.Y., S. Kaplan, and J.H. Lambert, 2002, Risk filtering, ranking, and management framework using hierarchical holographic modeling, *Risk Analysis* **22**(2): 383–397.

Haimes, Y.Y., B. Horowitz, J. Lambert, J. Santos, C. Lian, and K. Crowther, 2005a, Inoperability input–output model (IIM) for interdependent infrastructure sectors: theory and methodology, *Journal of Infrastructure Systems* **11**: 67–79.

Haimes, Y.Y., B. Horowitz, J. Lambert, J. Santos, K. Crowther, and C. Lian, 2005b, Inoperability input–output model (IIM) for interdependent infrastructure sectors: case study, *Journal of Infrastructure Systems* **11**: 80–92.

Hu, J., M.-A. Cardin, K.-L. Poh, and E.S. Chia, 2013, An approach to generate flexibility in engineering design of sustainable waste-to energy systems, *The 19th International Conference on Engineering Design*, Seoul, Korea.

Intriligator, M.D., 1970, *Mathematical Optimization and Economic Theory*, Prentice-Hall, Inc., Englewood Cliffs, NJ.

Jamshidi, M. (Ed.), 2009a, *System of Systems Engineering: Innovations for the 21st Century*, John Wiley & Sons, Inc., Hoboken, NJ.

Jamshidi, M. (Ed.), 2009b, *System of Systems Engineering: Principle and Applications*, CRC Press/Taylor & Francis Group, New York.

Janda, R.J., A.S. Daag, P.J. Delos Reyes, C.G. Newhall, T.C. Pierson, R.S. Punongbayan, K.S. Rodolfo, R.U. Solidum, and J.V. Umbal, 1994, Assessment and response to lahar hazard around Mount Pinatubo 1991 to 1993, In *Fire and Mud Lahars of Mount Pinatubo, Philippines*, C.G. Newhall and R.S. Punongbayan (Eds.), University of Washington Press, Seattle, WA.

Jugulum, R., and D.D. Frey, 2007, Toward a taxonomy of concept designs for improved robustness, *Journal of Engineering Design* **18**(2): 139–156.

Kaplan, S., and B.J. Garrick, 1981, On the quantitative definition of risk, *Risk Analysis* **1**(1): 11–27.

Katzenbach, J.R., and D.K. Smith, 1999, *The Wisdom of Teams*, HarperCollins, New York.

Knabb, R.D., J.R. Rhome, and D.P. Brown, 2005, *Tropical Cyclone Report: Hurricane Katrina, 23–30 August 2005*, Publication of the National Hurricane Center 20 December 2005. Available online: http://www.nhc.noaa.gov/pdf/TCR-AL122005_Katrina.pdf (Accessed January 16, 2015).

Leach, M.R., and Y.Y. Haimes, 1987, Multiobjective risk-impact analysis method, *Risk Analysis* **7**(2): 225–241.

Leontief, W.W., 1951a, Input/output economics, *Scientific American* **185**(4): 15–21.

Leontief, W.W., 1951b, *The Structure of the American Economy, 1919–1939*, Second edition, Oxford University Press, New York.

Leung, F., J. Santos, and Y. Haimes, 2003, Risk modeling, assessment, and management of lahar flow threat, *Risk Analysis* **23**(6): 1323–1335.

Li, D., and Y.Y. Haimes, 1987, The *Envelope Approach* for multiobjective optimization problems, *IEEE Transactions on Systems, Man, and Cybernetics* **17**(6): 1026–1038.

Lian, C., and Y. Haimes, 2005, Risk management of terrorism to interdependent infrastructure systems through the dynamic inoperability input–output model, *Systems Engineering* **9**(3): 241–258.

Liebowitz, S., and S. Margolis, 2000, *Encyclopedia of Law and Economics*, Elgar, Cheltenham, p. 981.

Liew, C.J., 2000, The dynamic variable input-output model: an advancement from the Leontief dynamic input–output, *The Annals of Regional Science* **34**: 591–614.

Lindell, M.K., 1997, Adoption and implementation of hazard adjustments, *International Journal of Mass Emergencies and Disasters* **15**(Special Issue): 327–453.

Lowrance, W.W., 1976, *Of Acceptable Risk: Science and the Determination of Safety*, William Kaufman, Los Altos, CA.

Maier, M.W., 1998, Architecting principle for systems-of-systems, *Systems Engineering* **1**(4): 267–284.

Marcial, S., A.A. Melosantos, K.C. Hadley, R.G. LaHusen, and J.N. Marso, 1994, *Instrumental Lahar Monitoring at Mount Pinatubo*. Available online: http://pubs.usgs.gov/pinatubo/marcial/index.html (Accessed January 16, 2015).

Mercado, R., J. Lacsamana, and G. Pineda, 1994, Assessment and response to lahar hazard around Mount Pinatubo 1991 to 1993, In *Fire and Mud Lahars of Mount Pinatubo, Philippines*, C.G. Newhall, and R.S. Punongbayan (Eds.), University of Washington Press, Seattle, WA.

Mileti, D.S., 1999, *Disasters by Design: A Reassessment of Natural Hazards in the United States*, Joseph Henry Press, Washington, DC.

Miller, R.E., and P.D. Blair, 1985, *Input–Output Analysis: Foundations and Extensions*, Prentice-Hall, Englewood Cliffs, NJ.

Montgomery, D.C., 1991, *Introduction to Statistical Quality Control*, Wiley, New York.

Morgan, G., and M. Henrion, 1990, *Uncertainty*, Cambridge University Press, Cambridge, MA.

Okuyama, Y., M. Sonis, and G.J.D. Hewings, 1999, Economic impacts of an unscheduled, disruptive event: a Miyazawa multiplier analysis, In *Understanding and Interpreting Economic Structure*, G.J.D. Hewings, M. Sonis, M. Madden, and Y. Kimura (Eds.), Springer-Verlag, New York, pp. 113–143.

Olsen, J.R., J.R. Stedinger, N.C. Matalas, and E.Z. Stakhiv, 1999, Climate variability and flood frequency estimation for the Upper Mississippi and Lower Missouri rivers, *Journal of American Water Resources Association* **35**(6): 1509–1524.

Oosterhaven, J., 1988, On the plausibility of the supply-driven input–output model, *Journal of Regional Science* **28**: 203–217.

Pande, P.S., R.P. Neuman, and R.R. Cavanagh, 2000, *The Six Sigma Way: How GE, Motorola, and Other Top Companies Are Honing Their Performance*, McGraw-Hill, New York.

Pareto, V., 1896, Cours d'économie politique professé a l'université de Lausanne, Vol. I, in-8 de v-430 pages. Rouge, editeur a Lausanne; Pichon, libraire a Paris, 23.

Paté-Cornell, M.E., 1996, Uncertainties in risk analysis: six levels of treatment, *Reliability Engineering and System Safety* **54**(2–3): 95–111.

Pierson, T.C., R.C. Janda, J.V. Umbal, and A.S. Daag, 1992, *Immediate and Long-Term Hazards from Lahars and Excess Sedimentation in Rivers Draining Mt. Pinatubo, Philippines*, US Geological Survey Water-Resources Investigations Report 92-4039, Vancouver, WA.

Pierson, T.C., A.S. Daag, P.J. Delos Reyes, M.T.M. Regalado, R.U. Solidum, and B.S. Tubianosa, 1994, Flow deposition of posteruption hot lahars on the east side of Mount Pinatubo, July–October 1991, In *Fire and Mud Lahars of Mount Pinatubo, Philippines*, C.G. Newhall and R.S. Punongbayan (Eds.), University of Washington Press, Seattle, WA.

Port of New Orleans (PNO), 2005, Life after Katrina, *Port Record: The Worldwide Publication of the Port of New Orleans*, Winter. Available online: http://www.portno.com/PortRecord.pdf (Accessed May 5, 2006).

Port of New Orleans (PNO), 2006, *Recovery Message*. Available online: http://www.portno.com/message.htm (Accessed May 5, 2006).

Punongbayan, R.S., J. Umbal, R. Torres, A.S. Daag, R. Solidum, P. Delos Reyes, K.S. Rodolfo, and C.G. Newhall, 1992, *Three Scenarios for 1992 Lahars of Pinatubo Volcano*, Unpublished Report, Philippine Institute of Volcanology and Seismology, Philippines.

Robertson, C., 2005, Coastal cities of Mississippi in the shadows, *New York Times*, September 12. Available online: http://www.nytimes.com/2005/09/12/national/nationalspecial/12gulf.html?th&em=th (Accessed September 12, 2005).

Rose, A., and S. Liao, 2005, Modeling regional economic resilience to disasters: a computable general equilibrium analysis of water service disruptions, *Journal of Science* **45**: 75–112.

Sage, A.P., 1992, *Systems Engineering*, John Wiley and Sons, New York.

Sage, A.P., 1995, *Systems Management for Information Technology and Software Engineering*, John Wiley & Sons, New York.

Sage, A.P., and C.D. Cuppan, 2001, On the systems engineering and management of systems of systems and federation of systems, *Information Knowledge Systems Management* **2**(4): 325–345.

Santos, J.R., 2006, Inoperability input–output modeling of disruptions to interdependent economic systems, *Systems Engineering* **9**(1): 20–34.

Santos, J.R., and Y.Y. Haimes, 2004, Modeling the demand reduction input–output (I–O) inoperability due to terrorism of interconnected infrastructures, *Risk Analysis* **24**: 1437–1451.

Simon, H.A., 1947, *Administrative Behavior: A Study of Decisionmaking Processes in Administrative Organization*, First edition, Macmillan, New York.

Simon, H.A., 1956, Rational choice and the structure of the environment, *Psychology Review* **63**(2): 129–138.

Tierney, K.J., M.K. Lindell, and R.W. Perry, 2001, *Facing the Unexpected: Disaster Preparedness and Response in the United States*, Joseph Henry Press, Washington, DC.

Tulane University, 2004, *Strength in Numbers*, Tulane University statistics based on economic impact study for fiscal year 2002–2003, completed in 2004. Available online: http://impact.tulane.edu/numbers.html (Accessed January 16, 2015).

Tungol, N.M., and M.T.M. Regalado, 1994, Rainfall, acoustic flow monitor records, and observed lahars in the Sacobia River in 1992, In *Fire and Mud Lahars of Mount Pinatubo, Philippines*, C.G. Newhall, and R.S. Punongbayan (Eds.), University of Washington Press, Seattle, WA.

US Army Corps of Engineers (USACE), 1994, *Mount Pinatubo Recovery Action Plan Long Term Report: Eight River Basins, Republic of the Philippines*, US Department of Commerce (USDOC), National Oceanic and Atmospheric Administration, National Geophysical Data Center.

US Army Corps of Engineers (USACE), 2005, *Project Fact Sheet: Lake Pontchartrain, LA. and Vicinity Hurricane Protection Project, St. Bernard, Orleans, Jefferson, and St. Charles Parishes, LA*. Available online: http://www.gao.gov/new.items/d06244t.pdf.

US-Canada Power System Outage Task Force (UCPSOTF), 2004, *Final Report on the August 14, 2003 Blackout in the United States and Canada: Causes and Recommendations*. Available online: https://reports.energy.gov/BlackoutFinal-Web.pdf (Accessed January 16, 2015).

US Department of Commerce, Bureau of Economic Analysis (USDOC), 1998, *Benchmark Input–Output Accounts of the United States, 1992*, US Government Printing Office, Washington, DC.

U.S. Department of Commerce, National Oceanic and Atmospheric Administration, National Geographic Data Center. 1992, *Mount Pinatubo: The 1991 Eruptions*. Boulder, Co.

United States Geological Survey (USGS), 1997a, *Lahars of Mount Pinatubo, Philippines*, US Geological Survey Fact Sheet 114-97, Vancouver, WA.

United States Geological Survey (USGS), 1997b, *Benefits of Volcano Monitoring Far Outweigh Costs—The Case of Mount Pinatubo*, US Geological Survey Fact Sheet 115-97, Vancouver, WA.

United States Geological Survey (USGS), 1997c, *The Cataclysmic 1991 Eruption of Mount Pinatubo, Philippines*, US Geological Survey Fact Sheet 113-97, Vancouver, WA.

White, G.F., and J.E. Haas, 1975, *Assessment of Research on Natural Hazards*, MIT Press, Cambridge, MA.

Zimmerman, R., 2005, Critical infrastructure and interdependencies, In *McGraw-Hill Handbook of Homeland Security*, D. Kamien (Ed.), McGraw-Hill, New York.

Appendix: Optimization Techniques

A.1 INTRODUCTION TO MODELING AND OPTIMIZATION[1]

Systems engineering provides systematic methodologies for studying and analyzing the various structural and nonstructural aspects of a system and its environment by using mathematical and/or physical models. It also assists in the decisionmaking process by selecting the best alternative policies subject to all pertinent constraints by using simulation and optimization techniques.

In general, to obtain a way to control or manage a physical system, we introduce a mathematical model that closely represents the physical system. A mathematical model is a set of equations that describes and represents the real system. This set of equations uncovers the various aspects of the problem, identifies the functional relationships between all of the system's components and elements and its environment, establishes measures of effectiveness and constraints, and thus indicates what data should be collected to deal with the problem quantitatively. These equations could be algebraic, differential, or other, depending on the nature of the system being modeled. The mathematical model is solved, and its solution is applied to the physical system.

Figure A.1 depicts a schematic representation of the process of system modeling and optimization. The same input applied to both the real system and the mathematical model yields two different responses, namely, the system's output and the model's output. The closeness of these responses indicates the merit and validity of the mathematical model. Figure A.1 also applies solution strategies, often referred to as optimization and simulation techniques, to the mathematical model. The optimal decision is then implemented on the physical system.

In this book, the vector notation will be adopted wherever it is possible and easier to use.

Optimization is the procedure of selecting that set of decision variables (also known as manipulated variables) that maximizes the objective function (also known as performance function or index of performance) subject to the system's constraints.

The following is a general optimization problem.

Select the set of decision variables, $x_1^*, x_2^*, \ldots, x_n^*$, that maximize (minimize) the objective function $f(x_1, x_2, \ldots, x_n)$:

---

[1] Section A.1 is based on chapter 1 of Yacov Y. Haimes [1977].

Risk Modeling, Assessment, and Management, Fourth Edition. Yacov Y. Haimes.
© 2016 John Wiley & Sons, Inc. Published 2016 by John Wiley & Sons, Inc.

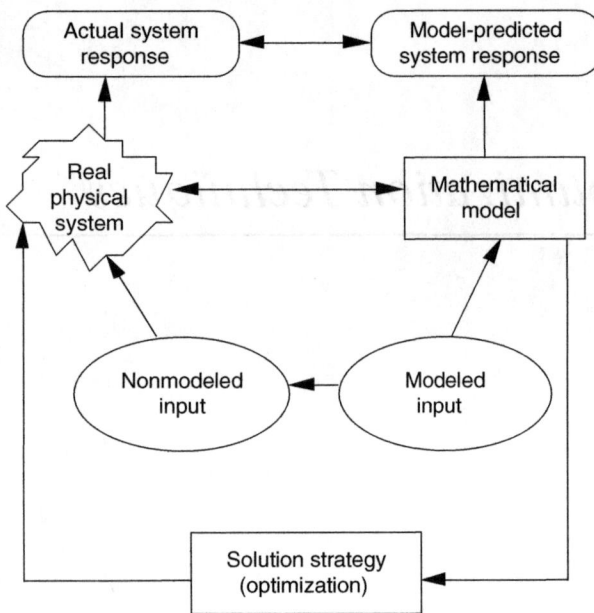

Figure A.1 System modeling and optimization.

$$\max_{x_1,\ldots,x_n} f(x_1, x_2, \ldots, x_n)$$

subject to the constraints

$$\begin{aligned} g_1(x_1, x_2, \ldots, x_n) &\leq b_1 \\ g_2(x_1, x_2, \ldots, x_n) &\leq b_2 \\ &\vdots \\ g_m(x_1, x_2, \ldots, x_n) &\leq b_m \end{aligned} \quad \text{(A.1a)}$$

where $b_1, \ldots, b_m$ are known values.

Let $\mathbf{x}^T = [x_1, x_2, \ldots, x_n]$ denote an n-dimensional row vector. The superscript T denotes the transpose operation. Thus, the system in Equation A.1a can be rewritten as

$$\left.\begin{aligned} &\max_x f(\mathbf{x}) \\ \text{subject to the constraints } &g_j(\mathbf{x}) \leq b_j, j = 1, 2, \ldots, m \end{aligned}\right\} \quad \text{(A.1b)}$$

Depending on the nature of the objective function and the constraints, the general optimization problem posed by Equation A.1b can be classified accordingly. The following are four possible ways that may be considered to classify mathematical models:

1. Linear versus nonlinear
2. Deterministic versus probabilistic (stochastic)
3. Static versus dynamic
4. Lumped parameters versus distributed parameters

Linear versus nonlinear. A *linear* model is one that is represented by linear equations; that is, all constraints and the objective function(s) are linear. A *nonlinear* model is represented by nonlinear equations; that is, part or all of the constraints and/or the objective function are nonlinear.

Examples:

$$\begin{aligned} \text{Linear equations}: y &= 5x_1 + 6x_2 + 7x_3 \\ \text{Nonlinear equations}: y &= 5x_1^2 + 6x_2 x_3 \\ y &= \log x_1 \\ y &= \sin x_1 + \log x_2 \end{aligned}$$

Deterministic versus probabilistic. Deterministic models or elements of models are those in which each variable and parameter can be assigned a definite fixed number or a series of fixed numbers for any given set of conditions.

In *probabilistic* (stochastic) models, the principle of uncertainty is introduced. Neither the variables nor the parameters used to describe the input–output relationships and the structure of the elements (and the constraints) may be precisely known.

Example:
The value of x is in $(a-b, a+b)$ with 90% probability, meaning that in the long run, the value of x will be less than $(a-b)$ or greater than $(a+b)$ in 10% of the cases.

Static versus dynamic. *Static* models are those that do not explicitly take the variable time into account. In general, static models are of the form given by Equation A.1a.

Dynamic models are those involving difference or differential equations. An example is given in Equation A.2:

$$\max_{u_1,\ldots,u_M} \int_{t_0}^{t} F(x_1, \ldots, x_N, u_1, \ldots, u_M, t) dt \quad \text{(A.2)}$$

subject to the constraints

$$\frac{d}{dt}(x_i) = G_i(x_i, \ldots, x_N, u_1, \ldots, u_M, t),$$
$$i = 1, 2, \ldots, N$$

$$x_i(t_0) = x_i^0, \quad i = 1, 2, \ldots, N$$

Static optimization problems are often referred to as mathematical programming, while dynamic optimization problems are often referred to as optimal control problems.

Distributed parameters versus lumped parameters. A *lumped parameter* model ignores variations, and the various parameters and dependent variables can be considered homogeneous throughout the entire system.

A *distributed parameter* model takes into account detailed variations in behavior from point to point throughout the system.

Most physical systems are distributed parameters systems. For example, the diffusion equation

$$T\left[\frac{1}{r}\frac{\partial}{\partial r}\left(r\frac{\partial P}{\partial r}\right)\right] = S\frac{\partial P}{\partial t} \pm Q \qquad (A.3)$$

represents a distributed parameter system.

Several techniques of optimization are available to solve the above optimization problems, such as:

1. Calculus
2. Linear programming (LP)
3. Non-LP
   a. Direct search
   b. Lagrange multipliers (penalty functions)
   c. Gradient methods (e.g., GRE)
   d. Geometric programming
   e. Others
4. Dynamic programming
5. Simulation
6. Decomposition and multilevel approach
7. Others

Related theories and techniques for dynamic systems:

1. Queueing theory
2. Game theory
3. Network theory
4. The calculus of variations
5. The maximum principle
6. Quasilinearization
7. Decomposition and multilevel approach
8. Others

This Appendix briefly introduces some of the above techniques.

Simulation is one of the systems engineering tools that has been used heavily in decisionmaking. Simulation involves setting up a mathematical model of a real situation and then performing experiments on the model by trying to answer the question, what if? Three major simulation techniques can be identified. They are based on the use of a digital computer, an analog computer, or a hybrid computer (combination of digital and analog). Digital computer simulation is more accurate than analog computer simulation.

The model plays an extremely important role in determining the optimal solution to the real physical problem. Thus, it is imperative that the choice of the model topology (structure) and its parameters be carried on scientifically and systematically.

The task of determining the structural parameters on the basis of observations over time and the positions of the inputs and outputs is termed systems identification.

If the response of both the real system and the mathematical model to the same signal input is identical (ideally), then the mathematical simulation is considered *perfect*. In general, however, these two responses are not identical and an error exists. Thus, the purpose is to construct a mathematical model so that such an error is minimized.

The following outline summarizes the various phases of a systems engineering study:

1. Analyzing in detail all components of the system and collecting pertinent data.
2. Formulating a comprehensive mathematical model of the problem, focusing on a chosen technique or techniques; analyzing the subsystem interconnections (i.e., the couplings within a system).
3. Solving the formulated model and testing its validity. This includes:
   a. Developing pertinent algorithms for computing the solution of the problem formulated in step 2 above
   b. Computer programming of the optimization technique used to solve the problem
   c. Parameterizing the system model variables and investigating the solution's stability. Establishing control over the solution
4. Putting the solution to work: implementation.

A.1.1 Classification of Mathematical Programming Problems

Mathematical programming problems, often referred to as static optimization problems, can be classified as follows:

1. *Unconstrained problem*:

$$\min_x f(\mathbf{x})$$

2. *Classical equality constraint problem*:

$$\min_x f(\mathbf{x}) \quad \text{so that} \quad g_j(\mathbf{x}) = b_j, \quad j = 1, 2, \ldots, m$$

(especially when all functions can be differentiated).

3. *Non-LP*:

$$\min_x f(\mathbf{x}) \quad \text{so that} \quad g_j(\mathbf{x}) \geq 0, \quad j = 1, 2, \ldots, m$$

where $f(\mathbf{x})$ and/or $g_j(\mathbf{x})$ are nonlinear functions.

4. *LP*:

$$\min_{x \geq 0} \mathbf{c}^T \mathbf{x} \quad \text{so that} \quad A\mathbf{x} \geq \mathbf{b}$$

where A is a matrix of coefficients and b is a vector of constraints (resources).

5. *Quadratic programming*:

$$\min_x \mathbf{x}^T Q \mathbf{x} + \mathbf{c}^T \mathbf{x} \quad \text{so that} \quad A\mathbf{x} \geq \mathbf{b}$$

where A and Q are matrices of coefficients.

6. *Separable programming*:

$$\min \sum_{i=1}^n f_i(x_i) \quad \text{so that} \quad \mathbf{x} \geq \mathbf{0}$$

7. *Discrete (integer) programming*: Any of the above plus the requirement that certain of the variables be integers.

Some of the above classes of problems, as well as the optimization techniques available for their solution, will be discussed subsequently.

A.1.2 Classical Unconstrained Optimization Problems

The general unconstrained problems can be formulated as

$$\min_x f(\mathbf{x})$$

where $f(\mathbf{x})$ is any linear or nonlinear function. If $f(\mathbf{x})$ is differentiable, then necessary and sufficient conditions for a minimum can be derived via calculus. Stationary conditions are necessary but not sufficient for a local minimum of a function. Stationary points are those where the function assumes its minimum, maximum, or inflection.

A.1.2.1 Necessary conditions for stationarity

A necessary condition for a point $\mathbf{x} = \mathbf{x}^0$ to be a stationary point for the function $f(\mathbf{x})$ is that the gradient of $f(\mathbf{x})$ at $x = x^0$ equals zero:

$$\nabla_x f(\mathbf{x}^0) = 0, \quad \text{where} \quad \nabla_x f(\mathbf{x}) = \begin{bmatrix} \dfrac{\partial f(\mathbf{x})}{\partial x_1} \\ \dfrac{\partial f(\mathbf{x})}{\partial x_2} \\ \vdots \\ \dfrac{\partial f(\mathbf{x})}{\partial x_n} \end{bmatrix}$$

Since this is true at minimum, maximum, and inflection points of the function, the conditions are necessary but not sufficient for a minimum.

A.1.2.2 Sufficient conditions for minimum

Sufficient condition for a point $\mathbf{x} = \mathbf{x}^0$ to be a local minimum of the function $f(\mathbf{x})$ is that the Hessian matrix, $H[f(\mathbf{x})]$, be positive definite where

$$H[f(\mathbf{x})] = \begin{bmatrix} \dfrac{\partial^2 f}{\partial x_1^2} & \dfrac{\partial^2 f}{\partial x_1 \partial x_2} & \cdots & \dfrac{\partial^2 f}{\partial x_1 \partial x_n} \\ \dfrac{\partial^2 f}{\partial x_2 \partial x_1} & \dfrac{\partial^2 f}{\partial x_2^2} & \cdots & \dfrac{\partial^2 f}{\partial x_2 \partial x_n} \\ & & \vdots & \\ \dfrac{\partial^2 f}{\partial x_n \partial x_1} & \dfrac{\partial^2 f}{\partial x_n \partial x_2} & \cdots & \dfrac{\partial^2 f}{\partial x_n^2} \end{bmatrix}$$

Note that the Hessian matrix is symmetric. Necessary and sufficient conditions for the Hessian matrix H to be positive definite are that the principal minors of H be positive (Sylvester's theorem).

Likewise, a sufficient condition for a maximum is that the Hessian matrix $H[f(x)]$ be negative definite. Necessary and sufficient conditions for the Hessian H to be negative definite are that all odd number principal minors be negative and all even number principal minors be positive (Sylvester's theorem).

Example:

$$\min f(\mathbf{x}) = (x_1 - 2)^2 + (x_2 - 1)^2$$

$$\frac{\partial f}{\partial x_1} = 2(x_1 - 2), \quad \frac{\partial f}{\partial x_2} = 2(x_2 - 1)$$

The stationary points are at $\partial f / \partial x_1 = 0$, $\partial f / \partial x_2 = 0$, or $x_1^0 = 2$ and $x_2^0 = 1$:

$$\frac{\partial^2 f}{\partial x_1 \partial x_2} = 0, \quad \frac{\partial^2 f}{\partial x_1^2} = 2, \quad \frac{\partial^2 f}{\partial x_2^2} = 2$$

Then the Hessian matrix H is

$$H = \begin{bmatrix} 2 & 0 \\ 0 & 2 \end{bmatrix}$$

The first principal minor is $2>0$; the second principal minor is $4>0$. Thus, the Hessian is positive definite and the function $f(\mathbf{x})$ has a minimum at $x^0 = (2, 1)$.

The above necessary and sufficient conditions for a minimum or maximum do not hold for constrained optimization problems. For example, given the constraint $x_1 \geq 0$, a negative value of x, obtained from the previous conditions yielding a minimum, would be infeasible. Nonlinear constrained optimization will be discussed in Section A.3.

A.1.3 Classical Equality Constraint Problem

The Lagrangian formulation is the general formulation of the classical equality constraint problem and can be given as follows:

$$\min_{\mathbf{x}} f(\mathbf{x})$$

subject to constraints

$$g_j(\mathbf{x}) \geq 0, \quad j = 1, 2, \ldots, m$$

where $f(\mathbf{x})$ is a continuous function (differentiable) and $g_j(\mathbf{x})$, $j = 1, 2, \ldots, m$, have continuous first derivatives. This is usually stated as

$$f'(\mathbf{x}) \in C' \quad \text{and} \quad g_j(\mathbf{x}) \in C^2, \quad j = 1, 2, \ldots, m$$

where C^1 denotes the set of all continuous functions and C^2 denotes the set of functions with continuous first derivatives.

If the constraint set is linear, namely, if all $g_j(\mathbf{x})$, $j = 1, 2, \ldots, m$, are linear functions, then these functions can be substituted into the objective function $f(\mathbf{x})$, yielding an unconstrained optimization problem.

A.1.3.1 The Lagrangian function

For nonlinear equality constraints, the Lagrangian formulation can be utilized. Consider the following simple optimization problem with two decision variables—x_1 and x_2:

$$\min_{x_1 x_2} f(x_1, x_2)$$

subject to the equality constraint

$$g(x_1, x_2) = b$$

where

$$f(x_1, x_2) \in C^1$$
$$g(x_1, x_2) \in C^2$$

writing the constraint as

$$g(x_1, x_2) - b = 0$$

We define a function L, called the Lagrangian, as follows:

$$L(x_1, x_2, \lambda) = f(x_1, x_2) + \lambda [g(x_1, x_2) - b]$$

where λ is a Lagrange multiplier. Note that if the constraints are satisfied at (x_1^*, x_2^*) then $g(x_1^*, x_2^*) = b$ and $g(x_1^*, x_2^*) - b = 0$ therefore, $L(x_1^*, x_2^*, \lambda^*) = f(x_1^*, x_2^*)$, where λ^* is the optimal Lagrange multiplier; that is, the value of the Lagrangian is the same as the optimal value of the objective function.

Necessary conditions for a stationary point of L are

$$\frac{\partial L}{\partial x_1} = \frac{\partial L}{\partial x_2} = \frac{\partial L}{\partial \lambda} = 0$$

or, by expanding,

$$\frac{\partial L}{\partial x_1} = \frac{\partial f}{\partial x_1} + \lambda \frac{\partial g}{\partial x_1} = 0$$

$$\frac{\partial L}{\partial x_2} = \frac{\partial f}{\partial x_2} + \lambda \frac{\partial g}{\partial x_2} = 0$$

$$\frac{\partial L}{\partial \lambda} = g - b = 0$$

These are three equations with three unknowns x_1, x_2, and λ; their solution yields the stationary points x_1^*, x_2^* and λ^*. If the sufficiency conditions for minimum are satisfied, then $(x_1^*, x_2^*, \lambda^*)$ yields the minimum to $f(x_1, x_2)$.

A.1.3.2 General formulation of the Lagrangian function

Consider the following general, equality-constrained, nonlinear optimization problem:

$$\min_{\mathbf{x}} f(\mathbf{x})$$

subject to the equality constraints

$$g_j(\mathbf{x}) = b_j, \quad j = 1, 2, \ldots, m$$

where

$$f(\mathbf{x}) \in C^1$$
$$g_j(\mathbf{x}) \in C^2, \quad j = 1, 2, \ldots, m$$

Form the Lagrangian L:

$$L(\mathbf{x}, \boldsymbol{\lambda}) = f(\mathbf{x}) + \sum_{j=1}^{m} \lambda_j [g_j(\mathbf{x}) - b_j]$$

where

$$\boldsymbol{\lambda}^T \underline{\underline{\Delta}} (\lambda_1, \lambda_2, \ldots, \lambda_m)$$

Necessary conditions for a stationary point of L are

$$\frac{\partial L}{\partial x_i} = 0 \quad \text{yields} \quad \frac{\partial f(x)}{\partial x_i} + \sum_{j=1}^{m} \lambda_j \frac{\partial g_j}{\partial x_i} = 0,$$
$$i = 1, 2, \ldots, n$$

$$\frac{\partial L}{\partial \lambda_k} = 0 \quad \text{yields} \quad \frac{\partial f(x)}{\partial \lambda_k} + \sum_{j=1}^{m} \left[g_j(x) - b_j\right] = 0$$
$$k = 1, 2, \ldots, m$$

Note that

$$\frac{\partial \lambda_j}{\partial \lambda_k} = \delta_{jk}$$

where

$$\delta_{jk} = \begin{cases} 1 & \text{for } j = k \\ 0 & \text{for } j \neq k \end{cases}$$

The Lagrangian function L has many important properties that will be discussed in Section A.7.

A.1.3.3 Example problem

Find the radius, r, and the length, h, of a water reservoir of a closed cylinder shape and of a given volume V_0, having the minimum surface area.

We know that $V_0 = \pi r^2 h$. Let S = surface area = $2\pi rh$. The objective function is

$$\min_{r,h} \{f(r, h) = 2\pi rh + 2\pi r^2\}$$

subject to the constraint

$$g(r, h) = \pi r^2 h = V_0$$

Form the Lagrangian L:

$$L(r, h, \lambda) = f(r, h) + \lambda[g(r, h) - V_0]$$
$$= 2\pi rh + 2\pi r^2 + \lambda(\pi r^2 h - V_0)$$

Necessary conditions for optimum:

$$\frac{\partial L}{\partial r} = \frac{\partial f}{\partial r} + \frac{\partial g}{\partial r} = 0$$
$$\frac{\partial L}{\partial h} = \frac{\partial f}{\partial h} + \lambda \frac{\partial g}{\partial h} = 0$$
$$\frac{\partial L}{\partial \lambda} = \pi r^2 h - V_0 = 0$$

Substituting the following derivatives:

$$\frac{\partial f}{\partial r} = 2\pi(h + 2r), \quad \frac{\partial f}{\partial h} = 2\pi r,$$
$$\frac{\partial g}{\partial r} = 2\pi rh, \quad \frac{\partial g}{\partial h} = \pi r^2$$

yields the following three equations:

$$2\pi(h + 2r) + 2\lambda \pi rh = 0$$
$$2\pi r + \lambda \pi r^2 = 0$$
$$\pi r^2 h - V_0 = 0$$

or

$$h + 2r + \lambda rh = 0$$
$$2r + \lambda r^2 = 0$$
$$\pi r^2 h - V_0 = 0$$

Solving the last three equations simultaneously yields

$$\lambda^* = -\frac{2}{r}$$

$$r^* = \sqrt[3]{\frac{V_0}{2\pi}}$$

$$h^* = \sqrt[3]{\frac{4}{\pi}V_0}$$

and

$$f(r^*, h^*) = 2\pi \sqrt[3]{\frac{V_0}{2\pi}} \sqrt[3]{\frac{4}{\pi}V_0} + 2\pi \left(\sqrt[3]{\frac{V_0}{2\pi}}\right)^2$$

A.1.3.4 Lagrange multipliers and inequality constraints

The Lagrangian approach given above is suitable for handling a non-LP problem with equality constraints. Generalized Lagrange multipliers (also called the Kuhn–Tucker multipliers) can be introduced to handle nonlinear problems with inequality constraints. In order to facilitate the discussion on the Kuhn–Tucker theory in non-LP, the classical method for inequality constraints will be discussed first. Consider the problem

$$\min_{x_1, x_2} f(x_1, x_2)$$

subject to the constraints

$$g(x_1, x_2) = b$$

and

$$x_1 \geq a$$

We convert this to equality by introducing the variable θ, which is defined by

$$\theta^2 = x_1 - a$$

We require θ to be real; and if $x_1 \geq a$, then $\theta^2 > 0$. (If $x_1 < a$, then $\theta^2 < 0 - \theta$ is imaginary.) Hence, the two constraints can be rewritten as

$$g(x_1, x_2) - b = 0$$
$$\theta^2 - x_1 + a = 0$$

We form the Lagrangian L:

$$L(x_1, x_2, \theta, \lambda_1, \lambda_2) = f(x_1, x_2) + \lambda_1[g(x_1, x_2) - b]$$
$$+ \lambda_2[\theta^2 - x_1 + a]$$

The necessary conditions for stationary points are

1. $\dfrac{\partial L}{\partial x_1} = \dfrac{\partial f}{\partial x_1} + \lambda_1 \dfrac{\partial g}{\partial x_1} - \lambda_2 = 0$

2. $\dfrac{\partial L}{\partial x_2} = \dfrac{\partial f}{\partial x_2} + \lambda_1 \dfrac{\partial g}{\partial x_2} = 0$

3. $\dfrac{\partial L}{\partial \lambda_1} = g - b = 0$

4. $\dfrac{\partial L}{\partial \lambda_2} = \theta^2 - x_1 + a = 0$

5. $\dfrac{\partial L}{\partial \theta} = 2\lambda_2\theta = 0$

In analyzing condition 5, two cases can be distinguished for $2\lambda_2\theta = 0$:

Case 1: $\theta = 0$, then $x_1 = a$. The solution in this case is on the boundary; that is, the constraint is binding. Often, a binding constraint is referred to as an active constraint; then λ_2 is not necessarily equal to zero.

Case 2: $\lambda_2 = 0$, then $\theta \neq 0$. The solution in this case is not on the boundary; that is, the constraint is not binding. Often, a nonbinding constraint is referred to as an inactive one.

Example problem A desalination plant produces freshwater in each of three successive periods. The requirements for freshwater are at least 5 units (acre-ft) at the end of the first period, 10 units at the end of the second period, and 15 units at the end of the third period, for a total of 30. The cost of producing x units in any period is $f(x) = x^2$.

Additional water may be produced in one period and carried over to a subsequent one. A holding cost of $2 per unit is charged for any freshwater carried over from one period to the next. Assuming no initial inventory, how many units should be produced each period?

Formulation: Let $x_1, x_2,$ and x_3 represent production in periods 1, 2, and 3, respectively. Total cost = production cost plus holding cost:

$$f(x_1, x_2, x_3) = x_1^2 + x_2^2 + x_3^2 + 2(x_1 - 5) \\ + 2(x_1 + x_2 - 15)$$

The constraints are

1. $x_1 \geq 5$
2. $x_1 + x_2 \geq 15$
3. $x_1 + x_2 + x_3 = 30$
4. $x_2 \geq 0$
5. $x_3 \geq 0$

The optimization problem is

$$\min_{x_1, x_2, x_3} f(x_1, x_2, x_3)$$

subject to constraints 1 through 5.

One possible approach is to ignore inequality constraints and form the Lagrangian L, and then check the solution for feasibility:

$$L(x_1, x_2, x_3, \lambda) = x_1^2 + x_2^2 + x_3^2 + 2(x_1 - 5) \\ + 2(x_1 + x_2 - 15) \\ + \lambda(x_1 + x_2 + x_3 - 30)$$

The necessary conditions for minimum are

$$\frac{\partial L}{\partial x_1} = 2x_1 + 2 + 2 + \lambda = 0 \quad \text{or} \quad 2x_1 = -4 - \lambda$$

$$\frac{\partial L}{\partial x_2} = 2x_2 + 2 + \lambda = 0 \quad \text{or} \quad 2x_2 = -2 - \lambda$$

$$\frac{\partial L}{\partial x_3} = 2x_3 + \lambda = 0 \quad \text{or} \quad 2x_3 = -\lambda$$

$$\frac{\partial L}{\partial \lambda} = x_1 + x_2 + x_3 - 30 = 0 \quad \text{or} \quad x_1 + x_2 + x_3 = 30$$

Solving the above simultaneously yields

$$-4 - \lambda - 2 - \lambda = 60; \quad \text{thus} \quad \lambda^* = -\frac{66}{3} = -22$$

and

$$x_1^* = 9, \, x_2^* = 10, \, x_3^* = 11, \, f(x_1^*, x_2^*, x_3^*) = \$318$$

The above result should be tested for feasibility. Substituting the values of $x_1 = 9$, $x_2 = 10$, and $x_3 = 11$ into constraints 1 and 2 does not violate them. Thus, the optimal solution is feasible and constraints 1 and 2 are not binding (not active); that is,

1. $x_1 > 5$
2. $x_1 + x_2 > 15$

A.1.4 Newton–Raphson Method

Solving simultaneous n nonlinear equations with n unknowns can be carried out very effectively via the Newton–Raphson method. We will first discuss the one-dimensional case—that is, finding the roots of one nonlinear equation with one unknown variable. Then a generalized solution to the n-dimensional case will be developed.

A.1.4.1 One-dimensional problem

Find a sequence of approximations to the root of the equation

$$f(x) = 0$$

We assume that $f(x)$ is monotone decreasing for all x and strictly convex—that is, the second derivative is positive, $f''(x) > 0$—and that the root, r, is simple:

$$f'(r) \neq 0$$

Let x_0 be the initial approximations to the root r, with $x_0 < r$, $[f(x_0) >]$, and let us approximate $f(x)$ by a linear function of x determined by the value of the slope of the function $f(x)$ at $x = x_0$ (ignoring all nonlinear terms of a Taylor expansion):

$$f(x) \cong f(x_0) + (x - x_0) f'(x_0)$$

A further approximation to r is then obtained by solving the linear equation in x:

$$f(x_0) + (x - x_0) f'(x_0) = 0$$
$$f(x_0) + x f'(x_0) - x_0 f'(x_0) = 0$$

which yields the second approximation,

$$x_1 = x_0 - \frac{f(x_0)}{f'(x_0)}, \quad f'(x_0) \neq 0$$

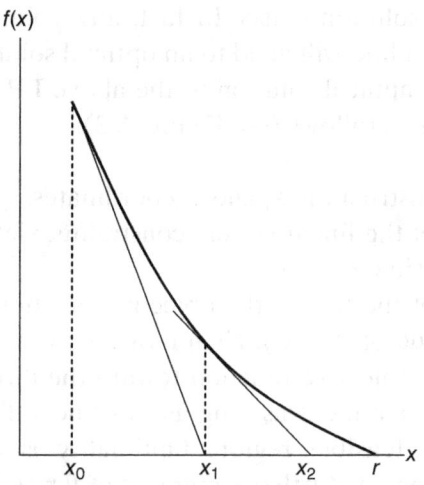

Figure A.2 Newton–Raphson method.

The process is repeated at x_1, leading to a new value x_2, and repeatedly to a new value x_n. The general recurrence equation is given by Equation A.4:

$$x_{n+1} = x_n - \frac{f(x_n)}{f'(x_n)}, \quad f'(x_n) \neq 0 \quad (A.4)$$

where the subscript n denotes the nth iteration. A graphical description of the iterative procedure is depicted in Figure A.2.

Note that if the iterative procedure of using Equation A.4 converges, then it converges quadratically:

$$|x_{n+1} - x_n| \leq k |x_n - x_{n-1}|^2$$

where k is independent of n.

Example problem Find the square root of 2; that is, find the solution to $x^2 = 2$. Let

$$f(x) = x^2 - 2 = 0$$

$$x_{n+1} = x_n - \frac{f(x_n)}{f'(x_n)}$$

$$x_{n+1} = x_n - \frac{x_n^2 - 2}{2x_n} = \frac{x_n}{2} + \frac{2}{2x_n}$$

$$x_{n+1} = \frac{x_n}{2} + \frac{2}{2x_n}$$

In this problem ($\sqrt{2} = 1.4142$) Let $x_0 = 1$:

$$x_1 = \frac{1}{2} + \frac{2}{2(1)} = \frac{3}{2} = 1.50$$

$$x_2 = \frac{3/2}{2} + \frac{1}{3/2} = \frac{17}{12} = 1.4167$$

$$x_3 = \frac{1.4167}{2} + \frac{2}{2(1.4167)} = 1.4142$$

A.1.4.2 Multidimensional problem

Determine a sequence of approximations to the roots of a system of n simultaneous equations with n unknown variables

$$f_i(x_1, x_2, x_3, \ldots, x_n) = 0, \quad i = 1, 2, 3, \ldots, n$$

In vector notation: $\mathbf{f(x)} = \mathbf{0}$. Let x_0 be the initial approximation; then

$$\mathbf{f(x)} \cong \mathbf{f(x_0)} + J(\mathbf{x_0})(\mathbf{x} - \mathbf{x_0})$$

where $J(\mathbf{x_0})$ is the Jacobian matrix evaluated at $\mathbf{x} = \mathbf{x_0}$:

$$J(\mathbf{x_0}) = \left[\frac{\partial f_i}{\partial x_j}\right]_{\mathbf{x}=\mathbf{x_0}} = \begin{bmatrix} \frac{\partial f_1}{\partial x_1} & \frac{\partial f_1}{\partial x_2} & \cdots & \frac{\partial f_1}{\partial x_n} \\ \vdots & & & \\ \frac{\partial f_n}{\partial x_1} & \frac{\partial f_n}{\partial x_2} & \cdots & \frac{\partial f_n}{\partial x_n} \end{bmatrix}_{\mathbf{x}=\mathbf{x_0}}$$

The new approximation is thus obtained as follows:

$$\mathbf{f(x)} = \mathbf{0} \quad \text{yields} \quad \mathbf{f(x_0)} + J(\mathbf{x_0})(\mathbf{x} - \mathbf{x_n}) = \mathbf{0}$$

$$\mathbf{f(x_0)} + J(\mathbf{x_0})\mathbf{x_1} - J(\mathbf{x_0})\mathbf{x_0} = \mathbf{0}$$

or

$$\mathbf{x_1} = \mathbf{x_0} - J(\mathbf{x_0})^{-1}\mathbf{f(x_0)} \quad J(\mathbf{x_0})^{-1} \neq \mathbf{0}$$

The recurrence relationship

$$\mathbf{x_n} = \mathbf{x_{n-1}} - J(\mathbf{x_{n-1}})^{-1}\mathbf{f(x_{n-1})}$$

where

$$J(\mathbf{x_{n-1}}) = \begin{bmatrix} \frac{\partial f_1}{\partial x_1} & \frac{\partial f_1}{\partial x_2} & \cdots & \frac{\partial f_1}{\partial x_n} \\ \frac{\partial f_2}{\partial x_1} & \frac{\partial f_2}{\partial x_2} & \cdots & \frac{\partial f_2}{\partial x_n} \\ \vdots & & & \\ \frac{\partial f_n}{\partial x_1} & \frac{\partial f_n}{\partial x_2} & \cdots & \frac{\partial f_n}{\partial x_n} \end{bmatrix}_{\mathbf{x}=\mathbf{x_{n-1}}}$$

A.1.5 LP

LP is a very effective optimization technique for solving linear static models—namely, a linear objective function and linear constraints—where all functions are in algebraic form.

A.1.5.1 Graphical solution

The following simple two-dimensional example demonstrates the basic concepts and principles of the technique.

Maximize the function $f(x_1, x_2)$, where $f(x_1, x_2) = 2x_1 + 3x_2$, subject to the following constraints:

1. $x_1 + 2x_2 \leq 4$
2. $x_1 \geq 0$
3. $x_2 \geq 0$

$$f(x_1^*, x_2^*) = \max\{f = 2x_1 + 3x_2\}$$
$$= 2x_1^* + x_2^* = (2)(4) + (3)(0) = 8$$

Note that each inequality constraint in x_1 and x_2 represents a half plane and the intersection of all these half planes yields the feasible region, as shown in Figure A.3.

From the theory of LP [Dantzig, 1963], the optimal solution, if it exists, will be obtained on one of the vertices of the feasible region. In special cases when the slope of the objective function coincides with one of the active constraints, more than one optimal solution exists. In fact, any point on that constraint line will yield to an optimal solution.

The graphical solution to the above LP problem proceeds as follows (see Figure A.3):

1. Construct the x_1 and x_2 coordinates.
2. Plot the linear system constraints, yielding the feasible region.
3. Plot the line of the objective function for any value of $f(x_1, x_2)$. Then move with the slope of the objective function toward the direction of its increment as long as this line still touches the feasible region. Optimality is achieved when any further increment of $f(x_1, x_2)$ yields a point outside the feasible region.
4. The coordinates of this vertex are the desired optimal x_1^* and x_2^*, where the function $f(x_1, x_2)$ attains its maximum value at (x_1^*, x_2^*).

A.1.5.2 General problem formulation

Consider the following problem.

Find a vector $\mathbf{x}^T = (x_1, x_2, \ldots, x_n)$ that minimizes the following linear function, $f(\mathbf{x})$:

$$f(\mathbf{x}) = c_1 x_1 + c_2 x_2 + \cdots + c_n x_n$$

or

$$f(\mathbf{x}) = \sum_{j=1}^{n} c_j x_j \quad (A.5a)$$

subject to the restrictions

$$x_j \geq 0, \quad j = 1, 2, \ldots, n \quad (A.6a)$$

and the linear constraint

$$\begin{aligned} a_{11}x_1 + a_{12}x_2 + \cdots + a_{1n}x_n &\leq b_1 \\ a_{21}x_1 + a_{22}x_2 + \cdots + a_{2n}x_n &\leq b_2 \\ &\vdots \\ a_{m1}x_1 + a_{m2}x_2 + \cdots + a_{mn}x_n &\leq b_m \end{aligned} \quad (A.7a)$$

where a_{ij}, b_i, and c_j are given constants for

$$j = 1, 2, \ldots, n$$
$$i = 1, 2, \ldots, m$$

and $f(\mathbf{x})$ is the objective function.

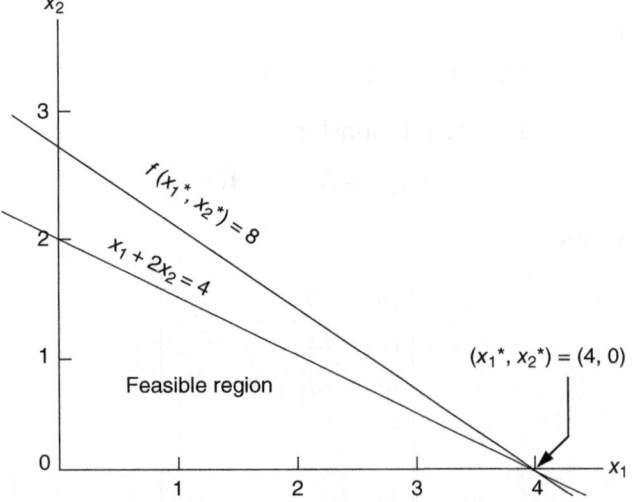

Figure A.3 Geometric solution of a linear programming problem.

In matrix notation, the problem can be formulated as follows:

$$\min_{\mathbf{x}}\{f(\mathbf{x}) = (\mathbf{c}^T\mathbf{x})\} \quad (A.5b)$$

subject to

$$\mathbf{x} \geq 0 \quad (A.6b)$$

$$A\mathbf{x} \geq \mathbf{b} \quad (A.7b)$$

where

$$\mathbf{c}^T = (c_1, c_2, \ldots, c_N)$$

$$\mathbf{x} = \begin{bmatrix} x_1 \\ x_2 \\ \vdots \\ x_n \end{bmatrix}, \quad \mathbf{b} = \begin{bmatrix} b_1 \\ b_2 \\ \vdots \\ b_m \end{bmatrix}, \quad A = \begin{bmatrix} a_{11} \cdots a_{1n} \\ a_{m1} \cdots a_{mn} \end{bmatrix} \quad (A.8)$$

A.1.5.3 Definitions

Solution: Any set x_j that satisfies the constraints (A.7a, b).

Feasible solution: Any solution that satisfies the nonnegative restrictions (A.6a, b) (and the constraints A.7a, b).

Optimal feasible solution: Any feasible solution that optimizes (minimizes or maximizes) the objective function (A.5a, b) is optimal.

In order to solve the above general LP problem, the simplex method is used [Dantzig, 1963; Hillier and Lieberman, 1995].

The inequality constraints are converted to equality constraints by introducing new variables, called *slack variables*, into the model as follows:

$$a_{11}x_1 + a_{12}x_2 + \cdots + a_{1n}x_n \leq b_1$$
$$a_{21}x_1 + a_{22}x_2 + \cdots + a_{2n}x_n \leq b_2$$
$$\vdots$$
$$a_{m1}x_1 + a_{m2}x_2 + \cdots + a_{mn}x_n \leq b_m$$

where all $x_{n+1} \geq 0, \ldots, x_{n+m} \geq 0$.

The problem posed by Equations A.5a, b, A.6a, b, and A.8 is now in its canonical form. LP computer packages are available in almost all digital computer systems.

A.1.5.4 Duality in LP

The duality concept has many important applications in projective geometry, electrical and mechanical systems, LP, and others. Associated with each LP problem, called the *primal*, is another problem, called the *dual*.

Define the following primal problem:

$$\min_{x}\{f(\mathbf{x}) = (\mathbf{c}^T\mathbf{x})\}$$

subject to the constraints

$$\mathbf{x} \geq 0, \quad A\mathbf{x} \geq \mathbf{b} \quad (A.9)$$

Introduce the following transformation of variables:

Primal		Dual
x	↔	y
n	↔	m
c	↔	b
A	↔	A^T
a_{ij}	↔	a_{ji}
≤	↔	≥
max	↔	min

where the vector $\mathbf{y} > 0$ is known as a vector of dual variables, shadow prices, or imputed prices.

Hence, the dual problem for the system shown in Equation A.9 becomes

$$\min_{y} G(\mathbf{y}) = (\mathbf{b}^T\mathbf{y})$$

subject to the constraints

$$\mathbf{y} \geq 0, \quad A^T\mathbf{y} \geq \mathbf{c} \quad (A.10)$$

The dual of the dual is the primal, and the optimal solution of the dual is equal to the optimal solution of the primal.

A.1.5.5 Lagrange multipliers and the dual in LP

Consider the following problem:

$$\min_{x_i}\left\{f = \sum_{i=1}^{n} c_i x_i\right\} \quad (A.11)$$

subject to the constraints

$$\sum_{i=1}^{n} a_{ij} x_i = b_j, \quad j = 1, 2, \ldots, m \quad (A.12)$$

where

$$x_i \geq 0, \quad i = 1, 2, \ldots, n$$

(note that the slack variables are already included in this formulation). Form the Lagrangian L:

$$L = \sum_{i=1}^{n} c_i x_i + \sum_{j=1}^{m} \lambda_j \left(\sum_{i=1}^{n} a_{ij} x_i - b_j \right) \quad (A.13)$$

Necessary conditions for optimal solution are

$$\frac{\partial L}{\partial x_i} = 0, \quad i = 1, 2, \ldots, n \quad (A.14)$$

$$\frac{\partial L}{\partial x_j} = 0, \quad j = 1, 2, \ldots, m \quad (A.15)$$

Solving Equations A.14 and A.15 yields

$$\frac{\partial L}{\partial x_i} = c_i + \sum_{j=1}^{m} \lambda_j a_{ji} = 0$$

Thus,

$$c_i = -\sum_{j=1}^{m} \lambda_j^* a_{ji} \quad (A.16)$$

$$\frac{\partial L}{\partial \lambda_j} = \sum_{i=1}^{n} a_{ji} x_i - b_j = 0 \quad (A.17)$$

Substituting Equation A.17 into Equation A.13 yields

$$L = \sum_{i=1}^{n} c_i x_i^* + \sum_{j=1}^{m} \lambda_j^* (0)$$

$$L = \sum_{i=1}^{n} c_i x_i^* = f(\mathbf{x}^*) \quad (A.18)$$

Substituting Equation A.16 into Equation A.13 yields

$$L = -\sum_{i=1}^{n} \sum_{j=1}^{m} \lambda_j^* a_{ji} x_i^* + \sum_{i=1}^{n} \sum_{j=1}^{m} \lambda_j^* a_{ji} x_i^* - \sum_{j=1}^{m} \lambda_j^* b_j$$

Therefore,

$$L = -\sum_{j=1}^{m} \lambda_j^* b_j$$

But $L = f(\mathbf{x}^*)$, as derived from Equation A.18; therefore,

$$f(\mathbf{x}^*) = -\sum_{j=1}^{m} \lambda_j^* b_j \quad (A.19)$$

Equation A.19 represents the objective function of the dual problem

$$-f(\mathbf{x}^*) = \sum_{j=1}^{m} \lambda_j^* b_j$$

λ_j are the dual variables, also called simplex multipliers. Equation A.19 establishes the fact that the solutions to the primal and dual problems are equal. Since the primal problem is minimized and the dual problem is maximized, the negative sign appears in Equation A.19. An important result is

$$\frac{\partial f(\mathbf{x})}{\partial b_j} = -\lambda_j, \quad j = 1, 2, \ldots, m \quad (A.20)$$

Example problem and economic interpretation of the dual primal problem

$$\max_{x} \{ f(\mathbf{x}) = 8x_1 + 9x_2 + 10x_3 \}$$

subject to the constraints

$$6x_1 + 7x_2 + 8x_3 \leq 1$$
$$5x_1 + 6x_2 + 7x_3 \leq 2$$
$$4x_1 + 5x_2 + 6x_3 \leq 3$$
$$3x_1 + 2x_2 + x_3 \leq 4$$

$$x_1 \geq 0, \quad x_2 \geq 0, \quad x_3 \geq 0$$

In matrix notation, we have

$$\max \left\{ f(\mathbf{x}) = (8 \ 9 \ 10) \begin{bmatrix} x_1 \\ x_2 \\ x_3 \end{bmatrix} \right\}$$

subject to

$$\begin{bmatrix} 6 & 7 & 8 \\ 5 & 6 & 7 \\ 4 & 5 & 6 \\ 3 & 2 & 1 \end{bmatrix} \begin{bmatrix} x_1 \\ x_2 \\ x_3 \end{bmatrix} \leq \begin{bmatrix} 1 \\ 2 \\ 3 \\ 4 \end{bmatrix}$$

$$x_1 \geq 0, \quad x_2 \geq 0, \quad x_3 \geq 0$$

The dual problem becomes

$$\max_{y}\left\{G(\mathbf{y})=\begin{bmatrix}1 & 2 & 3 & 4\end{bmatrix}\begin{bmatrix}y_1\\y_2\\y_3\\y_4\end{bmatrix}\right\}$$

subject to

$$\begin{bmatrix}6 & 5 & 4 & 3\\7 & 6 & 5 & 2\\8 & 7 & 6 & 1\end{bmatrix}\begin{bmatrix}y_1\\y_2\\y_3\\y_4\end{bmatrix}\geq\begin{bmatrix}8\\9\\10\end{bmatrix}$$

$y_1 \geq 0$, $y_2 \geq 0$, $y_3 \geq 0$, $y_4 \geq 0$ where y_i, $i = 1, 2, \ldots, 4$, are the dual variables.

Carrying out the matrix multiplication, the dual problem can be rewritten as follows:

$$\min_{y}\{G(\mathbf{y}) = y_1 + 2y_2 + 3y_3 + 4y_4\}$$

subject to

$$6y_1 + 5y_2 + 4y_3 + 3y_4 \geq 8$$
$$7y_1 + 6y_2 + 5y_3 + 2y_4 \geq 9$$
$$8y_1 + 7y_2 + 6y_3 + y_4 \geq 8$$
$$y_1 \geq 0, \quad y_2 \geq 0, \quad y_3 \geq 0, \quad y_4 \geq 0$$

Note that the primal problem has three variables and four constraints, whereas the dual problem has four variables and three constraints.

To each resource i, there corresponds a dual variable y_i that by its dimensions is a price, or cost, or value to be associated with one unit of resource i. That is, the dimension of the dual variables is $/unit of resource. If it were possible to increase the amount available of resource i by one unit without changing the solution to the dual, the maximum profit would be increased by y_i. This is the basis for the opportunity-cost interpretation.

A.1.5.6 The diet problem

Given five kinds of foods, all containing either calories or vitamins or both, select that combination of foods that costs the minimum and satisfies the minimal standards for nutrients, namely, 700 calorie units and 400 vitamin units. Table A.1 presents the

TABLE A.1 Nutrients and Costs per Unit of Food

	Foods					Requirements
	1	2	3	4	5	
Calories	1	0	1	1	2	700
Vitamins	0	1	0	1	1	400
Price ($/unit)	2	20	3	11	12	

price of each food and the nutrients per unit of food [Dorfman et al., 1958].

Problem formulation:

$$\min_{x}\{f(\mathbf{x}) = (\mathbf{c}^T\mathbf{x})\}$$

subject to the constraints

$$A\mathbf{x} \geq \mathbf{b} \quad \text{and} \quad \mathbf{x} \geq 0$$

where

$$\mathbf{x}^T = (x_1 \quad x_2 \quad x_3 \quad x_4 \quad x_5)$$
$$\mathbf{c}^T = (2 \quad 20 \quad 3 \quad 11 \quad 12)$$
$$\mathbf{b}^T = (700 \quad 400)$$
$$A = \begin{bmatrix}1 & 0 & 1 & 1 & 2\\0 & 1 & 0 & 1 & 1\end{bmatrix}$$

or

$$\min\{f(\mathbf{x}) = 2x_1 + 20x_2 + 3x_3 + 11x_4 + 12x_5\}$$

subject to the constraints

$$x_1 + x_3 + x_4 + 2x_5 \geq 700$$
$$x_2 + x_4 + x_5 \geq 400$$
$$x_i \geq 0, \quad i = 1, 2, \ldots, 5$$

The optimal solution that can be derived from the simplex method (to be discussed subsequently) is

$$\mathbf{x}^* = [0, 0, 0, 100, 300]$$
$$\min_{x}\{f(\mathbf{x})\} = f(\mathbf{x}^*) = f(0, 0, 0, 100, 300) = \$4700$$

The dual of the diet problem. Let $\mathbf{y}$ be the dual variable; hence,

$$\max_y \{G(\mathbf{y}) = (\mathbf{b}^T\mathbf{y})\}$$

subject to the constraints

$$A^T\mathbf{y} \le \mathbf{c} \quad \text{and} \quad \mathbf{y} \ge 0$$

where

$$\mathbf{y} = \begin{bmatrix} y_1 \\ y_2 \end{bmatrix}$$

or

$$\max_{y_1,y_2}\left\{ G(y_1,y_2) = [700,400]\begin{bmatrix} y_1 \\ y_2 \end{bmatrix} \right\}$$
$$= \max_{y_1,y_2}\{700 y_1 + 400 y_2\}$$

subject to

$$\begin{bmatrix} 1 & 0 \\ 0 & 1 \\ 1 & 0 \\ 1 & 1 \\ 2 & 1 \end{bmatrix}[y_1 \; y_2]^T \le \begin{bmatrix} 2 \\ 20 \\ 3 \\ 11 \\ 12 \end{bmatrix} \quad \text{or} \quad \begin{matrix} y_1 \le 2 \\ y_2 \le 20 \\ y_1 \le 3 \\ y_1 + y_2 \le 11 \\ 2y_1 + y_2 \le 12 \end{matrix}$$

$$\mathbf{y} \ge 0 \quad \text{or} \quad y_1 \ge 0, \; y_2 \ge 0$$

The graphical solution of the dual problem is given in Figure A.4.

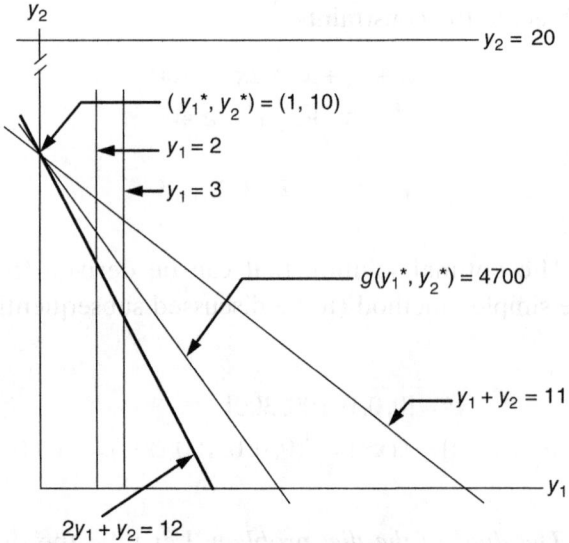

Figure A.4 *Graphical solution of the dual problem.*

Note:

$$\max_x f(\mathbf{x}) = \max_y G(\mathbf{y}) = \$4700$$

where

$$y_1^* = 1(\$/\text{calorie})$$
$$y_2^* = 10(\$/\text{vitamin})$$

Economic interpretation. The dimensions of the dual variables y_1 and y_2 are

$$[y_1] = \$/\text{calorie}, \quad [y_2] = \$/\text{vitamin}$$

These can be interpreted as the prices of calories and vitamins. The function $G(\mathbf{y})$ to be maximized is the imputed value of an adequate diet—that is, the imputed value of 700 calories plus 400 vitamins. The five inequalities, one for each food, state that in every case, the price of a food must be at least as great as the imputed value of the calories and vitamins that it provides.

Table A.2 compares the cost of each food with the value of the nutritive elements that it contains. Consider food 5: It contains 2 calories (each worth $1) and 1 vitamin (worth $10), giving a total value of $12. The two prices $y_1 = \$1$ and $y_2 = \$10$ satisfy the inequality conditions because the value of the nutrients is in no case greater than the price of the food.

The economy-minded shopper will never buy a food unless the value of its nutrients is at least as great as its price. Accordingly, she will not buy the first three foods, but will buy only the fourth and fifth (as we found by solving the primal, $x_1^* = 0, x_1^* = 0, x_3^* = 100, x_5^* = 300$).

TABLE A.2 Cost and Nutrient Value Comparison

Food	Value of Nutrients (1)	Price of Food (2)	Excess of (1) over (2)
1	1	2	−1
2	10	20	−10
3	1	3	−2
4	11	11	0
5	12	12	0

A.1.5.7 The simplex method

$$\max_{x_1,x_2}\{f = x_1 + 2x_2\} \quad (A.21)$$

subject to the constraints

$$x_1 \leq 5$$
$$x_2 \leq 8$$
$$2x_1 + x_2 \leq 12$$
$$x_1 \geq 0, \quad x_2 \geq 0 \quad (A.22)$$

The inequality constraints (Eq. A.22) will be converted to a system of equality constraints by introducing slack variables. Consequently, Equation A.22 becomes

$$x_1 + x_3 = 5$$
$$x_2 + x_4 = 8$$
$$2x_1 + x_2 + x_5 = 12$$
$$x_i \geq 0, \quad i = 1, 2, \ldots, 5$$

The objective function can be written as

$$f - x_1 - 2x_2 = 0$$

In Table A.3, for this problem, two classes of variables are distinguished: basic and nonbasic. The nonbasic variables are set equal to zero, where the basic variables are nonzero variables that are solved in terms of the nonbasic variables. The basic variables in Table A.3 are $x_3, x_4,$ and x_5, where the nonbasic variables are $x_1 = 0$ and $x_2 = 0$. The solution to Table A.3 is $x_3 = 5, x_4 = 8,$ and $x_5 = 12$. This solution is obviously feasible since the constraints in Equation A.22 are satisfied; however, this solution is not optimal since there are negative coefficients in row 0.

In Table A.4, the above solution is improved by selecting a new set of basic variables. The variable corresponding to the smallest coefficient of row 0 on Table A.3 is chosen as the new entering basic variable. This smallest coefficient is –2; thus, x_2 would be chosen as the new entering basic variable. The variable that will leave the basic variable set is chosen as follows: The right-hand column is divided by all positive numbers in the column of new entering basic variables— namely, the column of x_2 (excluding row 0). The variable corresponding to the smallest ratio will be the leaving basic variable. In Table A.3, the smallest ratio is 8 and the corresponding basic variable that leaves is x_4.

The final step in the construction of Table A.4 is the process of pivoting, where the coefficient of x_2 is made to be 1 and all other coefficients in the x_2 column are 0.

This is accomplished, for example, by multiplying row 2 by 2 and adding it to row 0 and by multiplying row 2 by (–1) and adding it to row 3.

The new basic feasible solution is

$$x_3 = 5, \quad x_2 = 8, \quad x_5 = 4, \quad x_1 = 0, \quad x_4 = 0 \text{ with } f = 16$$

Since one of the coefficients in row 0 is negative (–1), this solution is not optimal, and another iteration is required. Table A.5 is constructed similarly. There is only one negative coefficient in row 0; thus, the variable entering the basic variables set is x_1. The positive ratios are 5 and 2; thus, the smallest ratio is 2 and the corresponding variable leaving the basic variables set is x_5 after the pivoting process takes place. Table A.5 is obtained as given.

$$\text{Optimal solution}: f = 18, \text{ where } \mathbf{x}^* = (2, 8, 3, 0, 0)$$

The dual variables can be obtained directly from the final simplex table. In Table A.5, the coefficients associated with x_3, x_4, and x_5 in row 0 are the

TABLE A.3 First Simplex Tableau

Basic Variables	Equation Number	Coefficient of						Right Side of Equation
		f	x_1	x_2	x_3	x_4	x_5	
f	0	1	–1	–2	0	0	0	0
x_3	1	0	1	0	1	0	0	5
x_4	2	0	0	1	0	1	0	8
x_5	3	0	2	1	0	0	1	12

TABLE A.4 Second Simplex Tableau

Basic Variables	Equation Number	Coefficient of						Right Side of Equation
		f	x_1	x_2	x_3	x_4	x_5	
f	0	1	–1	0	0	2	0	16
x_3	1	0	1	0	1	0	0	5
x_2	2	0	0	1	0	1	0	8
x_5	3	0	2	0	0	–1	1	4

TABLE A.5 Third Simplex Tableau

Basic Variables	Equation Number	Coefficient of f	x_1	x_2	x_3	x_4	x_5	Right Side of Equation
f	0	1	0	0	0	1.5	0.5	18
x_3	1	0	0	0	1	0.5	−0.5	5
x_2	2	0	0	1	0	1	0	8
x_1	3	0	1	0	0	−0.5	0.5	2

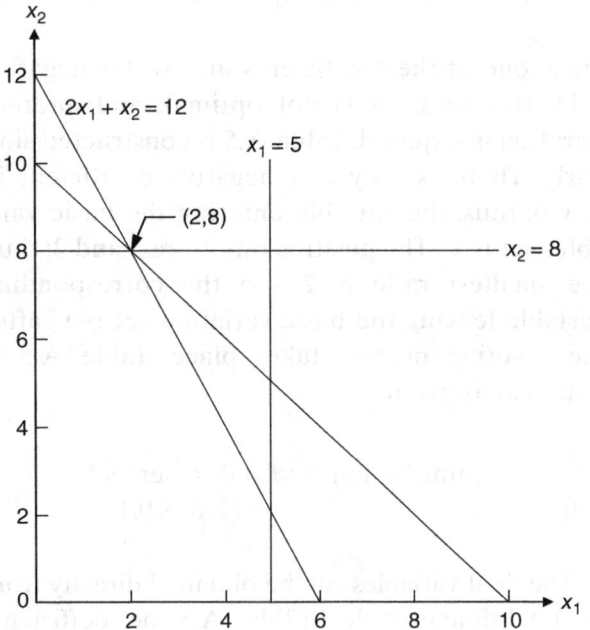

Figure A.5 Graphical solution to the linear programming problem.

corresponding dual variables to the first, second, and third constraints, respectively. A graphical solution to this example LP problem is given in Figure A.5. More will be said on the dual variables and their relationship to the constraints under the discussion on the Kuhn–Tucker theory.

A.1.5.8 The transportation problem

(See, e.g., Dantzig [1963] and Hillier and Lieberman [1995].) An important LP formulation is the transportation problem formulation because of its wide applicability to many problems. Consider M origin points and N destination points. We have the ability to transfer integer number of quantities from each origin to each destination.

Definitions:

a_i = number of units at origin i, $i = 1,...,M$.
b_j = number of units required at destination j, $j = 1,...,N$.
x_{ij} = number of units shipped from origin i to destination j; x_{ij} is a decision variable.
c_{ij} = cost to ship one unit from origin i to destination j, for all i,j.

It is assumed that no origin can ship more units than are available and that all demands are satisfied.
Hence,

$$\sum_{i=1}^{M} a_i = \sum_{j=1}^{N} b_j \qquad (A.23)$$

This is implicit in the problem and does not enter as a constraint.

Objective: Select $M \times N$ decision variables x_{ij} to minimize shipping cost f:

$$f = \sum_{j=1}^{N}\sum_{i=1}^{M} c_{ij} x_{ij} \qquad (A.24)$$

The LP formulation becomes

$$\min_{x_{ij}} \left\{ f = \sum_{j=1}^{N}\sum_{i=1}^{M} c_{ij} x_{ij} \right\} \qquad (A.25)$$

$$\sum_{j=1}^{N} x_{ij} = a_i, \quad i = 1, 2, ..., M \qquad (A.26)$$

$$\sum_{i=1}^{M} x_{ij} = b_j, \quad j = 1, 2, ..., N \qquad (A.27)$$

$$x_{ij} \geq 0$$

There is no loss in generality by assuming that

$$a_i > 0, \quad i = 1,...,M$$
$$b_j > 0, \quad j = 1,...,N$$

Note: If a_i and b_j are integers, then the resultant x_{ij} is also an integer.
Note: It should be carefully noted that the model has feasible solutions only if

$$\sum_{i=1}^{M} a_i = \sum_{j=1}^{N} b_j$$

This may be verified by observing that the restrictions require that both

$$\sum_{i=1}^{M} a_i = \sum_{i=1}^{M}\sum_{j=1}^{N} x_{ij} \quad \text{and} \quad \sum_{j=1}^{N} b_j = \sum_{i=1}^{M}\sum_{j=1}^{N} x_{ij}$$

This condition that the total supply must equal the total demand merely requires that the system be in balance.

If the problem has physical significance and this condition is not met, it usually means that either a_i or b_j actually represents a bound rather than an exact requirement. If this is the case, a fictitious *origin* or *destination* can be introduced to take up the slack in order to convert the inequalities into equalities and satisfy the feasibility condition.

The canonical transportation problem. We have $M+N$ constraints and $M \times N$ decision variables. Define

$$\mathbf{x} = \begin{bmatrix} x_{11} \\ \vdots \\ x_{1N} \\ x_{21} \\ \vdots \\ x_{M1} \\ \vdots \\ x_{MN} \end{bmatrix} N \times M \text{ vector} \quad \mathbf{b} = \begin{bmatrix} a_1 \\ \vdots \\ a_M \\ b_1 \\ \vdots \\ b_N \end{bmatrix} M+N \text{ vector}$$

$$A = \begin{bmatrix} U_N^T & O_N^T & \cdots & O_N^T \\ O_N^T & U_N^T & \cdots & O_N^T \\ \vdots & & & \\ O_N^T & O_N^T & \cdots & U_N^T \\ I_N & I_N & \cdots & I_N \end{bmatrix} \begin{matrix} \\ M \\ \downarrow \\ \\ N \\ \downarrow \end{matrix}$$

where

$$U_N = \begin{bmatrix} 1 \\ \vdots \\ 1 \end{bmatrix} N, \quad O_N = \begin{bmatrix} 0 \\ \vdots \\ 0 \end{bmatrix} N,$$

$$I_N = \begin{bmatrix} 1 & & 0 \\ & \ddots & \\ 0 & & 1 \end{bmatrix} N \Rightarrow A\mathbf{x} = \mathbf{b}$$

Thus, the transportation problem is an LP problem.

Example problem

$$2 \text{ origins}: M = 2$$
$$4 \text{ destinations}: N = 4$$

Thus, the LP problem is

$$\min_x \{\mathbf{c}^T \mathbf{x}\}$$

subject to

$$A\mathbf{x} = \mathbf{b}$$

where

$$\begin{bmatrix} 1 & 1 & 1 & 1 & 0 & 0 & 0 & 0 \\ 0 & 0 & 0 & 0 & 1 & 1 & 1 & 1 \\ 1 & 0 & 0 & 0 & 1 & 0 & 0 & 0 \\ 0 & 1 & 0 & 0 & 0 & 1 & 0 & 0 \\ 0 & 0 & 1 & 0 & 0 & 0 & 1 & 0 \\ 0 & 0 & 0 & 1 & 0 & 0 & 0 & 1 \end{bmatrix} \begin{bmatrix} x_{11} \\ x_{12} \\ x_{13} \\ x_{14} \\ x_{21} \\ x_{22} \\ x_{23} \\ x_{24} \end{bmatrix} = \begin{bmatrix} a_1 \\ a_2 \\ b_1 \\ b_2 \\ b_3 \\ b_4 \end{bmatrix}$$

Now, we can use the simplex method. However, many shorter ways are available because of the form of matrix A.

A.1.6 Dynamic Programming

Dynamic programming is the most used nonlinear optimization technique in water resource systems. This is because the Markovian and sequential nature of the decisions that arise in water resource problems nicely fits into Bellman's principle of optimality on which dynamic programming is based [Bellman and Dreyfus, 1962]. The concept of dynamic programming is relatively simple, as will be shown here; however, formulating the recursive equation of real problems often requires some ingenuity. The following network example will illustrate the principle upon which dynamic programming is based.

A.1.6.1 Network example

In the game described by Figure A.6, one must choose a feasible path that maximizes the total return. The rules of the game are as follows: Start at any point A, B, C, or D and terminate at any destination point a, b, c, or d. The return via a feasible path connecting any two points in the network is shown in Figure A.6. Four stages are designated in the network, and it is possible to proceed from stage i to $i+1$ only where a path is shown.

Clearly, the solution to this combinatorial problem becomes prohibitive for large networks. However, it will be shown that solving this problem via dynamic programming will circumvent the combinatorial calculations and hence effectively reduce the computations needed. The example problem will be used as a vehicle to demonstrate some of the principles upon which dynamic programming is based—in particular, Bellman's principle of optimality.

The complexity of the problem is reduced by decomposing it into smaller subproblems that are sequentially coupled to one another. In Figure A.6, four stages are identified where a limited objective function is associated with each stage. A stage can be viewed as a subproblem or a subsystem.

At the first stage, the following questions are asked: Having reached stage 2 from stage 1, what is the maximum reward (return) that can be achieved at each node (circle in the network), and what is the corresponding path? The answers to these questions are given in Figure A.7, where the maximum rewards for each node in the network (shown in the circles of Figure A.7) are 10, 6, 8, and 7. Often, more than one path may yield the same maximum reward. In that case,

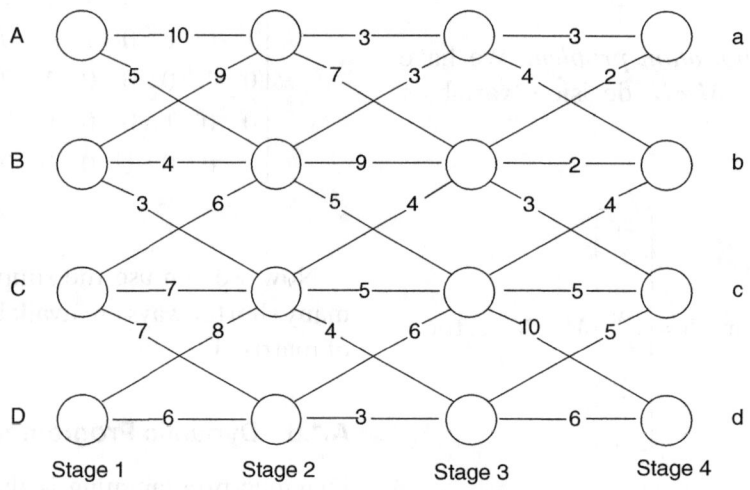

Figure A.6 Network example.

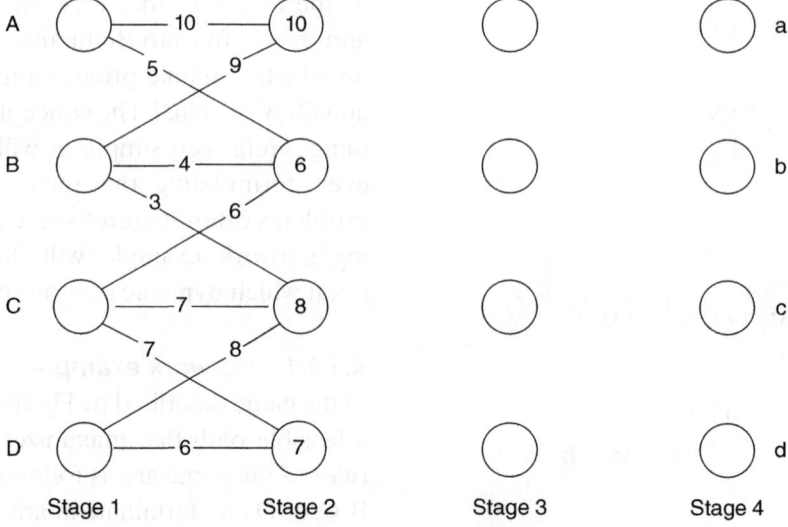

Figure A.7 Stage 1 to stage 2.

more than one solution can be derived. It is important to note that these four maximum reward values replace all 10 information values given at this stage. In other words, proceeding from stage 2 to stage 3, there is no need to be concerned with the 10 reward values given between stages 1 and 2 and only the four new values in the circles suffice to optimally proceed to stage 3.

At the second stage, the following question is asked: Having reached stage 3 from stage 2, what is the maximum reward that can be achieved at each node (circle in the network) and what is the corresponding path, assuming that an optimal path was chosen in progressing from stage 1 to stage 2? The answer to this question is given in Figure A.8, where the maximum cumulative rewards for each node in the network (shown in the circles) are 13, 17, 13, and 12. For example, advancing from node A2, which has a previous maximum reward of 10, to node A3 yields a cumulative sum of 13. Alternatively, advancing from node B2, which has a previous maximum reward of 6, to node A3 yields a cumulative sum of 9. Clearly, the maximum cumulative reward that can be achieved for node A3 is 13.

Similarly, the maximum cumulative rewards that can be achieved for each of the nodes A4, B4, C4, and D4 are 19, 19, 20, and 23, respectively, as given in Figure A.9. Obviously, the overall maximum cumulative reward is 23 achieved at node D4 (also designated as d). In order to find the optimal path, which has resulted in the maximum

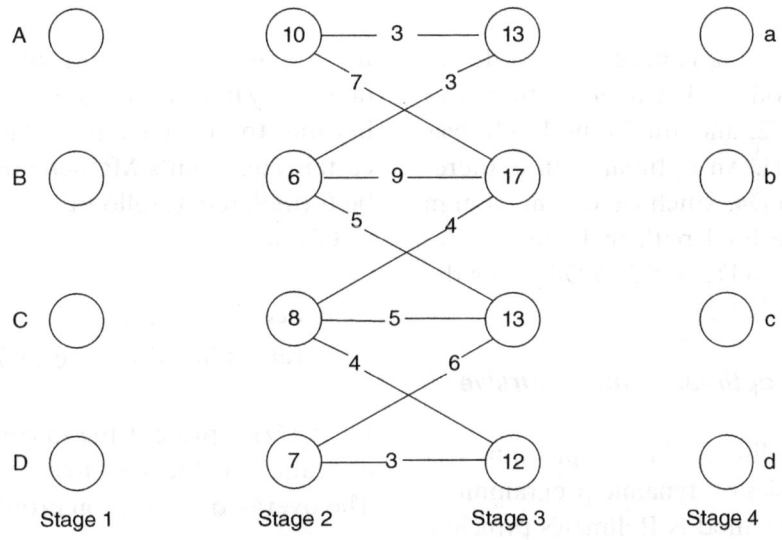

Figure A.8 Stage 2 to stage 3.

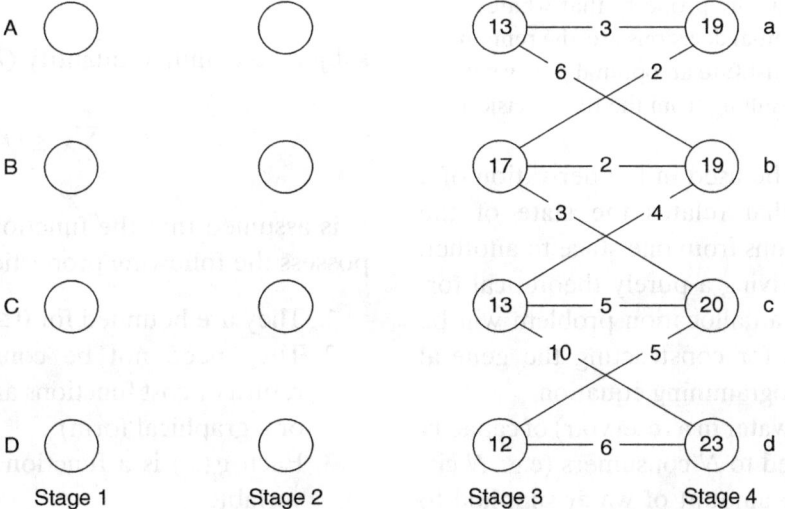

Figure A.9 Stage 3 to stage 4.

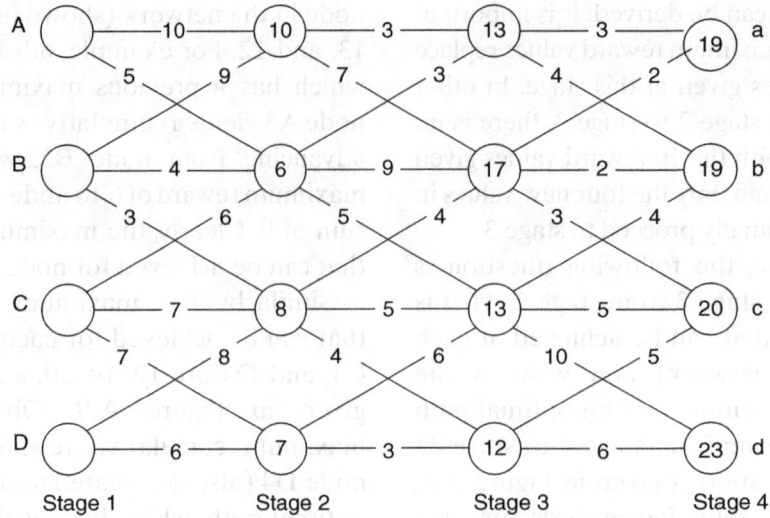

Figure A.10 Optimal path.

reward, a backward tracing is necessary. Node D4 was reached from node C3, which in turn was reached from node C2, and finally node C2 was reached from node D1. An optimal path is therefore D1 to C2 to C3 to D4, which yields a maximum reward of 23 (see the bold path in Figure A.10). Note that the path C1 to D2 to C3 to D4 yields the same maximum reward of 23.

A.1.6.2 Principle of optimality and recursive equation

The network example discussed in the previous section illustrates the basics of dynamic programming. Fundamental to this method is Bellman's principle of optimality, which states that

> An optimal policy has the property that whatever the initial state and initial decisions are, the remaining decisions must constitute an optimal policy with regard to the state resulting from the first decision.

This principle will be used in the derivation of a recursive equation that relates the state of the system and the decisions from one stage to another. In order to avoid deriving a purely theoretical formula in this section, an allocation problem will be utilized as a vehicle for constructing the general recursive dynamic programming equation.

Given a resource (water in a reservoir) of capacity Q that can be supplied to N consumers (e.g., N cities), let x_i denote the amount of water supplied to the ith city with a return of $g_i(x_i)$. The problem is how many units of water, x_i, to allocate to the ith city in order to maximize the total net return subject to certain constraints. Mathematically, the problem can be formulated as follows.

Given:

$$N \text{ decisions}: x_1, x_2, \ldots, x_N$$
$$N \text{ return functions}: g_1(x_1), g_2(x_2), \ldots, g_N(x_N)$$

Let $f_N(Q)$ represent the maximum return from the allocation of the resource Q to the N consumers. The overall optimization problem is

$$\max_{x_i}\{g_1(x_1) + g_2(x_2) + L + g_N(x_N)\}, \quad x_i \geq 0,$$
$$i = 1, 2, \ldots, N$$

subject to a limited quantity Q

$$\sum_{i=1}^{N} x_i \leq Q \qquad (A.28)$$

It is assumed that the functions $g_i(x_i)$, $i = 1, 2, \ldots, N$, possess the following properties:

1. They are bounded for $0 \leq x_i \leq Q$.
2. They need not be continuous (often these return or cost functions are given in a tabulated or a graphical form).
3. Each $g_i(x_i)$ is a function of only one decision variable.

Note that in this problem, there is one state variable only, namely, the water to be allocated to the various consumers. The state variable will be represented by q, indicating the amount of resource available for allocation. The number of decision variables is N and so is the number of stages. Note that the number of decisions and stages is not always the same.

At the first stage, we assume that there is only one potential user of the resource, which will be designated by the subscript 1. Then, since we would still wish to make maximum use of the resource, we define

$$f_1(q) = \max_{\substack{0 \leq x_1 \leq q \\ 0 \leq q \leq Q}} \{g_1(x_1)\} \quad (A.29)$$

At the second stage, we assume that there are two potential users of the resource. The new user is designated by the subscript 2. If we allocate to this user an amount x_2, $0 \leq x_2 \leq q$, there will be a return $g_2(x_2)$, and a remaining quantity of the resource $(q - x_2)$ can be allocated to user 1. Applying the principle of optimality, the optimal return of the resource for two potential users is

$$f_2(q) = \max_{\substack{0 \leq x_2 \leq q \\ 0 \leq q \leq Q}} \{g_2(x_2) + f_1(q - x_2)\} \quad (A.30)$$

The recursive calculation is now established and $f_3, f_4, \ldots, f_N$ can be written and solved in succession for all possible values of q. When this process is completed, $f_N(q)$ represents the return from allocating the resource optimally to N users as a function of the quantity of the resources, whatever it may be.

The general recursive relationship for N stages is

$$f_N(q) = \max_{\substack{0 \leq x_N \leq q \\ 0 \leq q \leq Q}} \{g_N(x_N) + f_{N-1}(q - x_N)\} \quad (A.31)$$

A direct derivation of the above dynamic programming recursive equation is given below, following Bellman and Dreyfus [1962]:

$$\max_{\substack{x_1 + x_2 + \cdots + x_N = q \\ x_j \geq 0}} = \max_{0 \leq x_N \leq q} \left[\max_{\substack{x_1 + x_2 + \cdots + x_{N-1} = q - x_N \\ x_1 \geq 0}} \right] \quad (A.32)$$

We can write

$$f_N(q) = \max_{\substack{x_1 + x_2 + L + x_N = q \\ x_i \geq 0}} [g_N(x_N) + g_{N-1}(x_{N-1}) + \cdots + g_1(x_1)]$$

$$= \max_{0 \leq x_N \leq q} \left[\max_{\substack{x + x_2 + L + x_{N-1} = q - x_N \\ x_1 \geq 0}} [g_N(x_N) + g_{N-1}(x_{N-1}) + \cdots + g_1(x_1)] \right]$$

$$= \max_{0 \leq x_N \leq q} \left[g_N(x_N) + \max_{\substack{x + x_2 + L + x_{N-1} = q - x_N \\ x_1 \geq 0}} [g_{N-1}(x_{N-1}) + \cdots + g_1(x_1)] \right]$$

$$= \max_{0 \leq x_N \leq q} [g_N(x_N) + f_{N-1}(q - x_N)]$$

Example problem: Given three cities, Los Angeles, Long Beach, and San Diego, and a limited amount of water $Q = 8$ available to them, let x_i be the amount of water allocated to the ith city with a return of $g_i(x_i)$, $i = 1, 2, 3$. It is assumed that Q is the excess capacity over all other mandatory supplies, and thus there are no constraints on minimum or maximum water supply. The overall optimization problem can be formulated as follows. The objective is to maximize the total return over all x_i:

$$\max_{x_1, x_2, x_3} \{g_1(x_1) + g_2(x_2) + g_3(x_3)\} \quad (A.33)$$

subject to the constraints

$$x_1 + x_2 + x_3 \leq 8, \quad x_1 \geq 0, \quad x_2 \geq 0, \quad x_3 \geq 0 \quad (A.34)$$

The return functions are given in Table A.6.

Stage 1: Only one city is under consideration. Define $f_1(q)$:

$f_1(q)$ = the maximum return from the allocation of the resource q to one city

$$f_1(q) = \max_{\substack{0 \leq x_1 \leq q \\ 0 \leq q \leq 8}} \{g_1(x_1)\} \quad (A.35)$$

Solving the above optimization problem for all discrete values of q, $0 \leq q \leq 8$ yields Table A.7.

Stage 2: Only two cities are under consideration. Define $f_2(q)$:

$f_2(q) \triangleq$ the maximum return from the allocation of the resource q to (one or) *two cities*

$$f_2(q) = \max \left\{ \begin{array}{l} \text{return from} \\ \text{city 2} \end{array} + \begin{array}{l} \text{optimal return} \\ \text{from city 1} \end{array} \right\} \quad (A.36)$$

TABLE A.6 Return Functions

q	$g_1(q)$	$g_2(q)$	$g_3(q)$
0	0	0	0
1	6	5	7
2	12	14	30
3	35	40	42
4	75	55	50
5	85	65	60
6	90	70	70
7	96	75	72
8	100	80	75

$$f_2(q) = \max_{\substack{0 \leq x_2 \leq q \\ 0 \leq q \leq 8}} \{g_2(x_2) + f_1(q - x_2)\}$$

Solving the above optimization problem for all discrete values of q, $0 \leq q \leq 8$, yields Table A.8.

A computational example of solving the recursive equation for the second stage for $q=7$ is

$$f_2(7) = \max_{0 \leq x_2 \leq 7} \{g_2(x_2) + f_1(7 - x_2)\}$$

$$f_2(7) = \max \begin{bmatrix} g_2(0) + f_1(7) \\ g_2(1) + f_1(6) \\ g_2(2) + f_1(5) \\ g_2(3) + f_1(4) \\ g_2(4) + f_1(3) \\ g_2(5) + f_1(2) \\ g_2(6) + f_1(1) \\ g_2(7) + f_1(0) \end{bmatrix}$$

$$f_2(7) = \max \begin{bmatrix} 0+96 \\ 5+91 \\ 14+85 \\ 40+75 \\ 55+35 \\ 65+12 \\ 70+6 \\ 75+0 \end{bmatrix} = \max \begin{bmatrix} 96 \\ 96 \\ 99 \\ 115 \\ 90 \\ 77 \\ 76 \\ 75 \end{bmatrix} = 115$$

$$f_2(7) = 115$$
$$x_2^* = 3$$

Stage 3: All three cities are under consideration. Define $f_3(q)$:

TABLE A.7 Results for the First Stage

q	x_1	$f_1(q)$
0	0	0
1	1	6
2	2	12
3	3	35
4	4	75
5	5	85
6	6	90
7	7	96
8	8	100

TABLE A.8 Results for the Second Stage

q	x_1	$f_1(q)$	x_2	$f_2(q)$
0	0	0	0	0
1	1	6	0	6
2	2	12	2	14
3	3	35	3	40
4	4	75	0	75
5	5	85	0	85
6	6	90	0	91
7	7	96	3	115
8	8	100	4	130

$f_3(q) \triangleq$ the maximum return from the allocation of the resource q to (one or two or) three cities

$$f_3(q) = \max \left\{ \begin{array}{l} \text{return from} \\ \text{city 3} \end{array} + \begin{array}{l} \text{optimal return} \\ \text{from cities 1 \& 2} \end{array} \right\}$$

$$f_3(q) = \max_{\substack{0 \leq x_3 \leq q \\ 0 \leq q \leq 8}} \{g_3(x_3) + f_2(q - x_3)\}$$

(A.37)

Table A.9 summarizes the solution to the overall optimization problem.

To analyze the result, we noted that 130 was the maximum return that can be expected with 8 available units of water to allocate. How many units should be sold to each city? What is the optimum policy?

We find that selling zero units to San Diego provides a maximum return of 130 with 8 units still remaining to be sold. Thus, $x_3^* = 0$. Looking at x_2 for $q = 8$, the decision is to allocate 4 units to Long Beach; thus, $x_2^* = 4$. Looking at x_1 for $q = 4$, the decision is to allocate the remaining 4 units to Los Angeles; thus, $x_1^* = 4$. The optimum solution is then

TABLE A.9 Summary of Results

q	x_1	$f_1(q)$	x_2	$f_2(q)$	x_3	$f_3(q)$
0	0	0	0	0	0	0
1	1	6	0	6	1	7
2	2	12	2	14	2	30
3	3	35	3	40	3	42
4	4*	75	0	75	0	75
5	5	85	0	85	1	82
6	6	90	0	91	2	105
7	7	96	3	115	3	117
8	8	100	4*	130	0*	130

$$x_1^* = 4, \quad x_2^x = 4, \quad x_3^* = 0$$

and the maximum return is 130.

If in the future only 7 units are available, Table A.9 reveals the following solution: For $q = 7$, the maximum return is 117 with $x_3 = 3$, which leaves 4 remaining units to sell to Long Beach and/or to Los Angeles. For $q = 4, x_2 = 0$, which leaves 4 remaining units to sell to Los Angeles. The optimum solution for $q = 7$ is then

$$x_1^* = 4, \quad x_2^x = 0, \text{ and } x_3^* = 3$$

A.1.6.3 An inventory (procurement) problem

A common problem in systems engineering is the optimal operation and management of storage facilities such as warehouses and reservoirs. This section formulates and solves a prototype procurement problem via dynamic programming, using a simplified reservoir system for illustration.

Assume that a water resources agency is in charge of the water supply for a region. It must make water available in sufficient quantities to meet all demands in N time periods. The agency procures water from various sources and stores it in the reservoir, which has a maximum capacity of Q acre-feet. It is assumed that a procurement order by the agency can be initiated once at the beginning of each period (a period may be a day, a week, a month, etc.) and that the water is made available to the agency without a lead time delay. It is also assumed that the agency delivers water to all its customers at the beginning of each period.

Water may be procured (pumped or imported) by the agency in one period, stored in the reservoir, and delivered at a later period. The associated storage cost is \$$a$ per acre-ft per period, and the procurement cost is \$$b$ per procurement. The water procured by the agency at the beginning of the ith period, x_i, can be ordered in quantities with integer increments, Δ. It is assumed that the initial storage of water in the reservoir at the beginning of the first period and the final storage at the end of the last period are zero.

The objective is to minimize the total cost of supplying water to meet all demands over the entire planning horizon.

In the above statement of the procurement problem, simplified assumptions about renewals were made for pedagogical purposes. Note that these assumptions require that procurement lead time be zero; procurement orders be initiated only at the beginning of each period; inventory at the beginning or at the end of the last period be zero; demand occurs at the beginning of the period; and there be no shortages. All these constraints can be removed with proper modifications of the mathematical model developed here. The price of a unit of water (\$/acre-ft) is not given (since it does not affect the optimization problem). Storage at the beginning and the end of the planning horizon is fixed.

A.1.6.3.1 Model formulation

Let

q_i = the stock level during period i, $i = 1, 2, ..., N$ (state variable)

x_i = the quantity procured at the beginning of period i, $i = 1, 2, ..., N$ (decision variable)

D_i = the demand at the beginning of period i, $i = 1, 2, ..., N$

$g_i(x_i, q_i)$ = the procurement cost and holding cost for period i, $i = 1, 2, ..., N$

Note that the number of stages in this dynamic programming formulation coincides with the number of periods, N, of the planning horizon.

The overall objective function is

$$\min_{x_i} \sum_{i=1}^{N} g_i(x_i, q_i) \quad (A.38)$$

The constraints are

$$x_i \geq 0, \quad i=1,2,\ldots,N \quad (A.39)$$

$$q_i = q_{i-1} + x_i - D_i, \quad i=1,2,\ldots,N \quad (A.40)$$

where

$$q_0 \equiv 0$$

Thus, $q_{i-1} = q_i + D_i - x_i$.

Note that the maximum storage at any period cannot exceed Q. Therefore,

$$0 \leq q_{i-1} \leq Q$$

or

$$0 \leq q_i + D_i - x_i \leq Q$$

Rearranging the above constraint yields a lower and upper bound on x_i:

$$q_i + D_i - Q \leq x_i \leq q_i + D_i, \quad i=1,2,\ldots,N$$

Define a new function $f_1(q_1)$ as follows:

$f_1(q_1)$ = the minimum cost of meeting water demand at the first period with a water storage level at q_1.

Mathematically, the optimization problem for the first stage can be written as

$$f_1(q_1) = \min_{x_1} g_1(x_1, q_1) \quad (A.41)$$

$$q_1 + D_1 - Q \leq x_1 \leq q_1 + D_1 \quad (A.42)$$

Similarly, define the general function $f_n(q_n)$ to be

$f_n(q_n)$ = the minimum cost of meeting all water demands for all n previous periods with a water storage level q_1 during the nth period.

Mathematically, the general recursive equation for the dynamic programming formulation can be written as

$$f_n(q_n) = \min_{x_n}\{g_n(x_n, q_n) + f_{n-1}(q_{n-1})\} \quad (A.43)$$

$$q_n + D_n - Q \leq x_n \leq q_n + D_n, \quad i=1,2,\ldots,N \quad (A.44)$$

Substituting the value of q_{n-1} in $f_{n-1}(q_{n-1})$ yields

$$f_n(q_n) = \min_{x_n}\{g_n(x_n, q_n) + f_{n-1}(q_n + D_n - x_n)\} \quad (A.45)$$

The above recursive equation should be solved for all possible stock levels q_n for all planning periods, $n=1,2,\ldots,N$. Then the optimal procurement policy, x_n^*, for all periods, $n=1,2,\ldots,N$, can be determined using the state equation $q_{n-1} = q_n + D_n - x_n$ that relates the state variable at $(n-1)$st period to the state and decision variables at the nth period. A detailed discussion on the determination of the overall optimal procurement policy is given in the following numerical example.

A.1.6.3.2 Example problem

Given the following numerical values to the general procurement problem discussed above,

$N = 5$ (periods in the planning horizon)
$Q = 40$ (maximum storage capacity)
$\Delta = 10$ (integer units of procurement increments)
$a = \$0.10$ (holding cost per unit period based on the stock level at the end of the period)
$b = \$20$ (procurement cost per procurement)
$D_1 = 10$ (water demand at period 1)
$D_2 = 20$ (water demand at period 2)
$D_3 = 30$ (water demand at period 3)
$D_4 = 30$ (water demand at period 4)
$D_5 = 20$ (water demand at period 5)

Find the optimal procurement policy for all five periods at a minimum total cost.

Solution: *First stage*. The recursive equation for the first stage ($n=1$) is

$$f_1(q_1) = \min_{x_1}\{g_1(x_1, q_1)\}$$

$$q_1 + 10 - 40 \leq x_1 \leq q_1 + 10$$

and

$$x_1 \geq 0$$

This recursive equation should be solved for all feasible incremental values of q_1 ($q_1 = 0, 10, 20, 30,$ and 40):

$$f_1(0) = g_1(10,0) = 20 + 0 = 20$$
$$f_1(10) = g_1(20,10) = 20 + 1 = 21$$
$$f_1(20) = g_1(30,20) = 20 + 2 = 22$$
$$f_1(30) = g_1(40,30) = 20 + 3 = 23$$
$$f_1(40) = g_1(50,40) = 20 + 4 = 24$$

Note that the cost is composed of two parts: the fixed procurement cost of $20 and the corresponding per-unit per-period holding cost.

Second stage. The recursive equation for the second stage ($n=2$) is

$$f_2(q_2) = \min_{x_2}\{g_2(x_2,q_2) + f_1(q_2 + 20 - x_2)\}$$

$$q_2 + 20 - 40 \leq x_2 \leq q_2 + 20$$

$$x_2 \geq 0$$

Again, the last recursive equation should be solved for $q_2 = 0, 10, 20, 30,$ and 40.

1. For $q_2 = 0$:

$$f_2(0) = \min_{x_2}\{g_2(x_2,0) + f_1(0 + 20 - x_2)\}$$

$$0 \leq x_2 \leq 20$$

$f_2(0) =$

$$\min \begin{pmatrix} g_2(0,0) + f_1(0+20-0) = 0+0+22 = 22 \\ g_2(10,0) + f_1(0+20-10) = 20+0+21 = 41 \\ g_2(20,0) + f_1(0+20-10) = 20+0+21 = 40 \end{pmatrix}$$

$$f_2(0) = 22, \quad x_2^* = 0$$

2. For $q_2 = 10$:

$$f_2(10) = \min_{x_2}\{g_2(x_2,10) + f_1(10+20-x_2)\}$$

$$0 \leq x_2 \leq 30$$

$f_2(10) =$

$$\min \begin{pmatrix} g_2(0,10) + f_1(10+20-0) = 0+1+23 = 24 \\ g_2(10,10) + f_1(10+20-10) = 20+1+22 = 43 \\ g_2(20,10) + f_1(10+20-20) = 20+1+21 = 42 \end{pmatrix}$$

$$f_2(0) = 24, \quad x_2^* = 0$$

3. For $q_2 = 20$:

$$f_2(20) = \min_{x_2}\{g_2(x_2,20) + f_1(20+20-x_2)\}$$

$$0 \leq x_2 \leq 40$$

$f_2(20) =$

$$\min \begin{pmatrix} g_2(0,20) + f_1(20+20-0) = 0+2+24 = 26 \\ g_2(10,20) + f_1(20+20-10) = 20+2+23 = 45 \\ g_2(20,20) + f_1(20+20-20) = 20+2+22 = 44 \\ g_2(30,20) + f_1(20+20-30) = 20+2+21 = 43 \\ g_2(40,20) + f_1(20+20-40) = 20+2+20 = 42 \end{pmatrix}$$

$$f_2(20) = 26, \quad x_2^* = 0$$

4. For $q_2 = 30$:

$$f_2(30) = \min_{x_2}\{g_2(x_2,30) + f_1(30+20-x_2)\}$$

$$10 \leq x_2 \leq 50$$

$f_2(30) =$

$$\min \begin{pmatrix} g_2(10,30) + f_1(30+30-10) = 20+3+24 = 47 \\ g_2(10,30) + f_1(30+30-10) = 20+3+24 = 47 \\ g_2(20,30) + f_1(30+20-20) = 20+3+23 = 46 \\ g_2(40,30) + f_1(30+20-40) = 20+3+21 = 44 \\ g_2(50,30) + f_1(30+20-50) = 20+3+20 = 43 \end{pmatrix}$$

$$f_2(30) = 43, \quad x_2^* = 50$$

5. For $q_2 = 40$:

$$f_2(40) = \min_{x_2}\{g_2(x_2,40) + f_1(40+20-x_2)\}$$

$f_2(30) =$

$$20 \leq x_2 \leq 60$$

$$\min \begin{cases} g_2(20,40) + f_1(40+30-20) = 20+4+24 = 48 \\ g_2(30,40) + f_1(40+20-30) = 20+4+23 = 47 \\ g_2(40,40) + f_1(40+20-40) = 20+4+22 = 46 \\ g_2(50,40) + f_1(40+20-50) = 20+4+21 = 45 \\ g_2(60,40) + f_1(40+20-60) = 20+4+20 = 44 \end{cases}$$

$$f_2(40) = 44, \quad x_2^* = 60$$

This concludes the calculations for the second stage.

Similar calculations have been made for the third, fourth, and fifth stages. The results of these calculations are summarized in Table A.10. Note that at the fifth stage, there is no need to solve the recursive equation for all feasible values of $g_5(\cdot)$, since it was assumed that no inventory will be left at the end of that period. Table A.10, however, does give the corresponding values for g_5 at all feasible increments for pedagogical purposes.

The final step in solving the procurement problem is tracing the optimal procurement policies for all five stages, or periods. This is done in a reverse order, starting with the fifth stage. All optimal values in Table A.10 are identified by an asterisk.

The optimal procurement policy and the minimum cost for meeting water demand at all five stages are given in columns 9 and 10, respectively. These values are $x_5 = 0, f_5(0) = 64$. The corresponding optimal inventory level at $N=5$ is $q_5 = 0$. It is now possible to find the optimal inventory level for stage 4:

TABLE A.10 Summary of Results for All Five Stages

	Column Number									
	1	2	3	4	5	6	7	8	9	10
q	x_1	$f_1(q_1)$	x_2	$f_2(q_2)$	x_3	$f_3(q_3)$	x_4	$f_4(q_4)$	x_5	$f_5(q_5)$
0	10	20	0*	22	30*	42	0	45	0*	64
10	20	21	0	24	40	43	0	49	30	66
20	30*	22	0	26	50	44	50*	64	40	67
30	40	23	50	43	60	45	60	65	50	68
40	50	24	60	44	60	46	60	67	60	69

$$q_4 = q_5 + D_5 - x_5$$
$$q_4 = 0 + 20 - 0 = 20$$

The optimal procurement policy at the fourth stage corresponds to $q_4 = 20$ and can be found in column 7 to be $x_4 = 50$.

The optimal inventory level for stage 3 is

$$q_3 = 20 + 30 - 50 = 0$$

The optimal procurement policy at the third stage corresponds to $q_3 = 0$, and it can be found in column 5 to be $x_3 = 30$.

The optimal inventory level for stage 2 is

$$q_2 = 0 + 30 - 30 = 0$$

The optimal procurement policy at the second stage corresponds to $q_2 = 0$, and it can be found in column 3 to be $x_2 = 0$.

Finally, the optimal inventory level for stage 1 is

$$q_1 = 0 + 20 - 0 = 20$$

The optimal procurement policy at the first stage corresponds to $q_1 = 20$, and it can be found in column 1 to be 30.

In summary, the optimal procurement policy vector, $\mathbf{x}^*$, for all five periods is $\mathbf{x}^* = (30, 0, 30, 50, 0)$, and the minimum cost is \$64.

A.1.7 Generalized Non-LP

A general non-LP problem is formulated in this section, and optimality conditions are derived based on the work of Kuhn and Tucker [1951], Lasdon [1968, 1970], and Wismer [1971]. The Kuhn–Tucker conditions provide much of the foundation for non-LP.

Consider the following general mathematical programming problem:

$$\min_x f(\mathbf{x}) \qquad (A.46)$$

subject to the constraints

$$g_k(\mathbf{x}) \leq 0, \quad k = 1, 2, \ldots, K \qquad (A.47)$$

$$\mathbf{x} \geq 0 \qquad (A.48)$$

where

- $\mathbf{x}$ is an n-dimensional vector of decision variables defined over the n-dimensional E^N Euclidean space, S_1.
- $f(\mathbf{x})$ and $g_k(\mathbf{x})$ are real-value differentiable functions defined over S_1.

The general nonlinear optimization problem posed by Equations A.46 to A.48 will be called the *primal problem*. A more compact formulation of the primal problem is often found in the literature and is given below:

$$\min f(\mathbf{x}), \quad \mathbf{x} \in S \qquad (A.49)$$

where

$$S = S_1 \cap S_2 \qquad (A.50)$$
$$S_1 = \{\mathbf{x} \mid \mathbf{x} \geq 0\} \qquad (A.51)$$
$$S_2 = \{\mathbf{x} \mid g_k(\mathbf{x}) \leq 0, k = 1, 2, \ldots, K\} \qquad (A.52)$$

A.1.7.1 The Kuhn–Tucker conditions

The nonlinear mathematical programming problem posed by Equations A.46 to A.48 may be solved by several available nonlinear optimization techniques. The major purpose of the Lagrangian formulation presented by Equation A.53 is for generating conditions for optimality and not necessarily for solving the optimization problem directly:

$$L(\mathbf{x}, \lambda) = f(\mathbf{x}) + \sum_{k=1}^{K} \lambda_k g_k(\mathbf{x}) \qquad (A.53)$$

where $L(\mathbf{x}, \lambda)$ is the generalized Lagrangian function, λ_k are the generalized Lagrange multipliers, and λ is a k-dimensional vector of λ_k.

A.1.7.2 Necessary conditions for stationarity

The point $(\mathbf{x}^0, \lambda^0)$ is a stationary point of the Lagrangian function $L(\mathbf{x}, \lambda)$ if the following necessary conditions are satisfied:

Group 1:

$$\nabla_x L(\mathbf{x}^0, \lambda^0) \geq 0 \quad (A.54)$$

$$(x^0)^T \nabla_x L(\mathbf{x}^0, \lambda^0) = 0 \quad (A.55)$$

$$\mathbf{x}^0 \geq 0 \quad (A.56)$$

Group 2:

$$\nabla_\lambda L(\mathbf{x}^0, \lambda^0) \leq 0 \quad (A.57a)$$

or

$$g_k(\mathbf{x}^0) \leq 0, \quad k-1, 2, \ldots, K \quad (A.57b)$$

$$(\lambda^0)^T \nabla_\lambda L(\mathbf{x}^0, \lambda^0) = 0 \quad (A.58a)$$

or

$$\lambda_k^0 g_k(\mathbf{x}^0) = 0, \quad k = 1, 2, \ldots, K \quad (A.58b)$$

$$\lambda^0 \geq 0 \quad (A.59)$$

The conditions (A.54) to (A.59) can be easily explained as a generalization of the necessary conditions for stationarity for the classical Lagrangian discussed in Section A.3.

The first group of the necessary conditions (A.54) to (A.56) ensures that if a stationary point happened to be on one of the negative ordinates of x (e.g., on x_j), then the stationary point will be forced to be on the boundary of x_j—that is, on $x_j = 0$. Condition (A.56) ensures that the nonnegativity constraints (A.48) are satisfied. The second group of necessary conditions (A.57a, b) to (A.59) ensures that if the kth inequality constraint is not binding (not active) (i.e., $g_k(\mathbf{x}) < 0$), then the corresponding Lagrange multiplier, λ_k, is equal to zero (i.e., $\lambda_k = 0$). This is guaranteed by condition (A.58a, b); since $\nabla_{\lambda_k} L(\mathbf{x}^0, \lambda^0) = g_k(\mathbf{x}^0), g_k(\mathbf{x}^0) \leq 0$, and $\lambda_k^0 \geq 0$, the product $\lambda_k^0 g_k(\mathbf{x}^0)$ is always equal to zero at $(\mathbf{x}^0, \lambda^0)$.

The economic interpretation of condition (A.58a, b) is very useful. A nonbinding kth constraint (i.e., $g_k(\mathbf{x}^0) < 0$) means that there is an unused excess of the kth resource at the optimal point $\mathbf{x}^0$. Consequently, the corresponding Lagrange multiplier, λ_k^0, which is also the marginal benefit or shadow price, should be zero. In other words, when a resource is not utilized to its full capacity at the optimal solution, there should be no further improvement in the optimal solution with the increase of the availability of that resource.

Associated with the above Kuhn–Tucker necessary conditions for stationarity are conditions that safeguard against singularities on the boundaries of the inequality constraints (A.47). These are presented in the literature in many different ways and are termed Kuhn–Tucker constraint qualifications, Kuhn–Tucker regularity conditions, constraint conditions, regularity assumptions, and so on. Two forms are presented here.

A.1.7.3 Kuhn–Tucker regularity conditions

Let $\mathbf{x}^0$ solve the optimization problem posed by Equations A.46 to A.48. There must exist an n-dimensional vector in the Euclidean space E^N, $\mathbf{h} \in E^N$, so that at each equality constraint, the inner product at each binding inequality constraint is $[\nabla_x g_k(\mathbf{x}^0) \bullet \mathbf{h}] < 0$.

The above regularity condition ensures that the Lagrange multipliers associated with the binding constraints are finite or bounded.

A.1.7.4 An alternative condition

The Lagrange multipliers associated with the Lagrangian presented by Equation A.53 are uniquely determined if the rank of the matrix of gradients of all binding constraints is maximal.

A.1.7.5 Saddle point

The concept of a saddle point plays an important role in non-LP and in multiobjective optimization.

Definition. A point $(\mathbf{x}^0, \lambda^0)$ with $\lambda^0 \geq 0$ is said to be a constrained saddle point for $L(\mathbf{x}, > \lambda)$ as defined by Equation A.53 if it satisfies

$$L(\mathbf{x}^0, \lambda^0) \leq L(\mathbf{x}, \lambda^0) \quad \text{for all } \mathbf{x} \geq 0 \quad (A.60)$$

$$L(\mathbf{x}^0, \lambda^0) \geq L(\mathbf{x}^0, \lambda) \quad \text{for all } \lambda \geq 0 \quad (A.61)$$

That is, $\mathbf{x}^0$ minimizes $L(\mathbf{x}, \lambda^0)$ for $\mathbf{x} \geq 0$ and λ^0 maximizes $L(\mathbf{x}^0, \lambda)$ over all $\lambda \geq 0$; that is,

$$L(\mathbf{x}^0, \lambda) = \min_x L(\mathbf{x}, \lambda) \quad \text{for } \mathbf{x} \geq 0 \quad (A.62)$$

$$L(\mathbf{x}, \lambda^0) \leq \max_\lambda L(\mathbf{x}, \lambda) \quad \text{for } \lambda \geq 0 \quad (A.63)$$

The inequalities (A.60) and (A.61) may be combined as follows:

$$L(\mathbf{x}^0, \lambda) \leq L(\mathbf{x}^0, \lambda^0) \leq L(\mathbf{x}, \lambda^0) \quad (A.64)$$

Theorem A.1 Let $\lambda^0 \geq 0$. A point $(\mathbf{x}^0, \lambda^0)$ is a constrained saddle point for $L(\mathbf{x}, \lambda)$ if and only if $\mathbf{x}^0$ minimizes $L(\mathbf{x}, \lambda^0)$ over S_1 and the conditions (A.57a, b)–(A.58a, b) are satisfied. See Lasdon [1968] for the complete proof of this theorem.

If $(\mathbf{x}^0, \lambda^0)$ is a saddle point for $L(\mathbf{x}, \lambda)$, then $\mathbf{x}^0$ solves the primal problem. Thus, if a saddle point exists, then the following equalities hold:

$$\begin{aligned} f(\mathbf{x}^0) &= L(\mathbf{x}^0, \lambda) = \max_{\lambda \geq 0} L(\mathbf{x}^0, \lambda) \\ &= \min_{\mathbf{x} \geq 0} L(\mathbf{x}, \lambda^0) \end{aligned} \quad (A.65)$$

In order to determine the saddle point of the Lagrangian $L(\mathbf{x}, \lambda)$, the concept of duality in non-LP will be introduced in the next section.

A.1.7.6 The dual function

The dual function in non-LP has characteristics similar to the dual function in LP discussed in Section A.1.5.6. The dual function for the Lagrangian given by Equation A.53 will be denoted by $H(\lambda)$, where

$$H(\lambda) = \min_{\mathbf{x} \geq 0} L(\mathbf{x}, \lambda) \quad (A.66)$$

The domain D of the dual function $H(\lambda)$ is given by Equation A.67:

$$D = \{\lambda | \lambda \geq 0, \min_{\mathbf{x} \geq 0} L(\mathbf{x}, \lambda) \text{ exists}\} \quad (A.67)$$

The dual problem is defined by Equations A.68–A.69:

$$\max_\lambda H(\lambda) \quad (A.68)$$

subject to

$$\lambda \in D \quad (A.69)$$

Theorem A.2 A saddle point for $L(\mathbf{x}, \lambda)$ exists if and only if the optimal values of the primal and dual objectives are equal.

The above theorem yields

$$\min_{\mathbf{x} \geq 0} \max_{\lambda \geq 0} L(\mathbf{x}, \lambda) = \max_{\lambda \geq 0} \min_{\mathbf{x} \geq 0} L(\mathbf{x}, \lambda) \quad (A.70)$$

that is, the dual of the dual yields the primal.

Theorem A.3 The following inequality $H(\lambda) \leq f(\mathbf{x})$ holds for all $\mathbf{x}$ satisfying Equations A.47 and A.48 and for $\lambda \in D$.

The proof of this theorem is given below for pedagogic purposes.

Assuming that $\mathbf{x}$ satisfies Equations A.47 and A.48, then

$$H(\lambda) = L(\mathbf{x}^0, \lambda) \leq L(\mathbf{x}, \lambda) \quad (A.71)$$

Expanding the above relation yields

$$H(\lambda) \leq f(\mathbf{x}) + \sum_{k=1}^{K} \lambda_k g_k(\mathbf{x}) \quad (A.72)$$

However, since $\mathbf{x}$ satisfies Equation A.47 (i.e., $g_k(x) \leq 0$) and $\lambda \in D$ (i.e., $\lambda \geq 0$), then

$$\sum_{k=1}^{K} \lambda_k g_k(\mathbf{x}) \leq 0$$

Hence,

$$H(\lambda) \leq f(\mathbf{x}) \quad (A.73)$$

Theorem A.4 The point $(\mathbf{x}^0, \lambda^0)$ is a constrained saddle point to $L(\mathbf{x}, \lambda)$ if and only if:

1. $\mathbf{x}^0$ solves the primal problem (defined by Eqs. A.46–A.48).
2. λ^0 solves the dual problem (defined by Eqs. A.68–A.69).
3. $f(\mathbf{x}^0) = H(\lambda^0)$.

See Schaeffler in Wismer [1971] and Lasdon [1968] for proof of this theorem.

It is very instructive to derive the duality in LP from the duality in non-LP [Intriligator, 1971].

A.1.7.7 Example problem
Given the following nonlinear optimization problem,

$$\min_{x_1, x_2}\{f(x_1, x_2) = x_1^2 + x_2^2\} \quad (A.74)$$

subject to the constraints

$$x_1 + x_2 \geq 4 \quad (A.75a)$$

$$2x_1 + x_2 \geq 5 \quad (A.76a)$$

$$x_1 \geq 0, \quad x_2 \geq 0$$

1. Solve the problem graphically.
2. Solve the problem by using the Kuhn–Tucker necessary conditions.
3. Check the Kuhn–Tucker regularity conditions.
4. Derive and solve the dual problem.
5. Check the saddle point conditions.

For notational convenience, the constraints (A.75a) and (A.76a) are rewritten in the canonical form. Let

$$g_1(x_1, x_2) = 4 - x_1 - x_2 \leq 0 \quad (A.75b)$$

$$g_2(x_1, x_2) = 5 - 2x_1 - x_2 \leq 0 \quad (A.76b)$$

A.1.7.8 Graphical solution
The graphical solution to the optimization problem posed by Equations A.74 to A.76b yields $x_1^* = 2, x_2^* = 2, f(x_1^*, x_2^*) = 8$ (the reader is encouraged to derive this solution). It is evident that the constraint $g_1(x_1, x_2)$ is binding, whereas $g_2(x_1, x_2)$ is not.

A.1.7.9 Kuhn–Tucker necessary conditions
Form the Lagrangian function $L(\mathbf{x}, \lambda)$:

$$L(x, \lambda) = x_1^2 + x_2^2 + \lambda_1(4 - x_1 - x_2) + \lambda_2(5 - 2x_1 - x_2) \quad (A.77)$$

For simplicity in notation, the argument in the Lagrangian function will be dropped unless it is significant. The Kuhn–Tucker necessary conditions for stationarity are

$$\frac{\partial L}{\partial x_1} = 2x_1^* - \lambda_1^* - 2\lambda_2^* \geq 0 \quad (A.78)$$

$$\frac{\partial L}{\partial x_2} = 2x_2^* - \lambda_1^* - \lambda_2^* \geq 0 \quad (A.79)$$

$$x_1^* \frac{\partial L}{\partial x_1} = x_1^*(2x_1^* - \lambda_1^* - 2\lambda_2^*) = 0 \quad (A.80)$$

$$x_2^* \frac{\partial L}{\partial x_2} = x_2^*(2x_2^* - \lambda_1^* - \lambda_2^*) = 0 \quad (A.81)$$

$$x_1^* \geq 0, \quad x_2^* \geq 0 \quad (A.82)$$

$$\frac{\partial L}{\partial \lambda_1} = 4 - x_1^* - x_2^* \leq 0 \quad (A.83)$$

$$\frac{\partial L}{\partial \lambda_2} = 5 - 2x_1^* - x_2^* \leq 0 \quad (A.84)$$

$$\lambda_1^* \frac{\partial L}{\partial \lambda_1} = \lambda_1^*(4 - x_1^* - x_2^*) = 0 \quad (A.85)$$

$$\lambda_2^* \frac{\partial L}{\partial \lambda_2} = \lambda_2^*(5 - 2x_1^* - x_2^*) = 0 \quad (A.86)$$

$$\lambda_1^* \geq 0, \quad \lambda_2^* \geq 0 \quad (A.87)$$

To solve conditions (A.78) to (A.87), certain assumptions must be made on the constraints (each constraint is either binding or not binding). Then the Kuhn–Tucker conditions are solved and a check is made as to whether the assumptions on the constraints were correct. If all constraints are satisfied, the assumptions were correct and a solution has been obtained; otherwise, new assumptions must be made on the constraints. Note that the Kuhn–Tucker conditions are not usually used as a computational procedure for solving non-LP problems. This example problem is presented here for pedagogical purposes.

Assume that one constraint is not binding (e.g., $g_2(x_1^*, x_2^*) < 0$) and that both $x_1^* > 0$ and $x_2^* > 0$ yield Equations A.88–A.91 as follows: Assuming $g_2(x_1^*, x_2^*) < 0$, condition (A.86) yields

$$\lambda_2^* = 0 \quad (A.88)$$

Assuming $x_1^* > 0$, condition (A.80) yields

$$2x_1^* - \lambda_1^* - 2\lambda_2^* = 0 \quad (A.89)$$

Assuming $x_2^* > 0$, condition (A.81) yields

$$2x_2^* - \lambda_1^* - \lambda_2^* = 0 \qquad (A.90)$$

Assuming $g_1(x_1^*, x_2^*)$ is binding, condition (A.85) yields

$$4 - x_1^* - x_2^* = 0 \qquad (A.91)$$

Solving Equations A.89 to A.91 simultaneously yields

$$x_1^* = 2, \quad x_2^* = 2, \quad \lambda_1^* = 4, \quad \lambda_2^* = 0$$

Substituting the above values into Equations A.78 to A.87 indicates that all the Kuhn–Tucker conditions are satisfied. (The reader is encouraged to do so.)

A.1.7.10 Kuhn–Tucker regularity conditions

There is only one binding constraint, namely, $g_1(x_1^*, x_2^*) = 0$. The Kuhn–Tucker regularity conditions require that there exists a vector **h** so that

$$\left[\nabla_x g_1(x_1^*, x_2^*) \cdot \mathbf{h}\right] < 0 \qquad (A.92)$$

$$\frac{\partial g_1(x_1^*, x_2^*)}{\partial x_1} = -1$$

$$\frac{\partial g_1(x_1^*, x_2^*)}{\partial x_2} = -1$$

$$\nabla_x g_1(x_1^*, x_2^*) = [-1, -1] \qquad (A.93)$$

$$\mathbf{h} = \begin{bmatrix} h_1 \\ h_2 \end{bmatrix} \qquad (A.94)$$

Substituting Equations A.93 and A.94 into Equation A.92 yields

$$-h_1 - h_2 < 0 \quad \text{or} \quad h_1 + h_2 > 0 \qquad (A.95)$$

It is easy to show that a vector **h** exists (e.g., [1, 1]), so that Equation A.95 is satisfied.

A.1.7.11 Dual function

The dual function $H(\lambda_1, \lambda_2)$ is defined again by Equation A.96:

$$H(\lambda_1, \lambda_2) = \min_{\substack{x_1 \geq 0 \\ x_2 \geq 0}} L(x_1, x_2, \lambda_1, \lambda_2)$$
$$= L(x_1^*, x_2^*, \lambda_1, \lambda_2) \qquad (A.96)$$

Equations (A.89) and (A.90) yield x_1 and x_2 in terms of λ_1 and λ_2:

$$2x_1^* = \lambda_1 + 2\lambda_2 \qquad (A.97)$$

$$2x_2^* = \lambda_1 + \lambda_2 \qquad (A.98)$$

or

$$x_1^* = \frac{\lambda_1}{2} + \lambda_2$$

$$x_2^* = \frac{\lambda_1 + \lambda_2}{2}$$

Substituting Equations A.97 and A.98 into A.96 yields

$$H(\lambda_1, \lambda_2) = L(x_1^*, x_2^*, \lambda_1, \lambda_2)$$
$$= \left(\frac{\lambda_1}{2} + \lambda_2\right)^2 + \left(\frac{\lambda_1 + \lambda_2}{2}\right)^2$$
$$+ \lambda_1\left[4 - \left(\frac{\lambda_1}{2} + \lambda_2\right) - \left(\frac{\lambda_1}{2} + \frac{\lambda_2}{2}\right)\right]$$
$$+ \lambda_2\left[5 - 2\left(\frac{\lambda_1}{2} + \lambda_2\right) - \left(\frac{\lambda_1}{2} + \frac{\lambda_2}{2}\right)\right]$$
$$H(\lambda_1, \lambda_2) = -\frac{\lambda_1^2}{2} - \frac{3}{2}\lambda_1\lambda_2 - \frac{5}{2}\lambda_2^2 + 4\lambda_1 + 5\lambda_2$$
$$(A.99)$$

The domain D of the dual function was given by Equation A.67:

$$D = \{\lambda_1, \lambda_2 | \lambda_1 \geq 0, \lambda_2 \geq 0, L(x_1^*, x_2^*, \lambda_1, \lambda_2) \text{ exists}\}$$

$$\lambda_1 \geq 0 \qquad (A.100)$$

$$\lambda_2 \geq 0 \qquad (A.101)$$

$$x_1^* = \frac{\lambda_1}{2} + \lambda_2 \geq 0 \qquad (A.102)$$

$$x_2^* = \frac{\lambda_1 + \lambda_2}{2} \geq 0 \qquad (A.103)$$

Therefore,

$D = \{\lambda_1, \lambda_2 \mid$ conditions (A.100) to (A.103) are satisfied$\}$

The dual problem is thus

$$\max_{\lambda_1, \lambda_2} H(\lambda_1, \lambda_2) \qquad (A.104)$$

subject to the constraints

$$\lambda_1 \in D, \quad \lambda_2 \in D \qquad (A.105)$$

The dual problem posed by Equations A.104 and A.105 may be solved using the following Kuhn–Tucker necessary conditions for a maximum:

$$\frac{\partial H(\lambda_1^*, \lambda_2^*)}{\partial \lambda_1} \leq 0 \qquad (A.106)$$

$$\frac{\partial H(\lambda_1^*, \lambda_2^*)}{\partial \lambda_2} \leq 0 \qquad (A.107)$$

$$\lambda_1^* \frac{\partial H(\lambda_1^*, \lambda_2^*)}{\partial \lambda_1} = 0 \qquad (A.108)$$

$$\lambda_2^* \frac{\partial H(\lambda_1^*, \lambda_2^*)}{\partial \lambda_2} = 0 \qquad (A.109)$$

Here again, in order to solve Equations A.106 to A.109, certain assumptions should be made. Assuming that $g_1(x_1^*, x_2^*) = 0$, thus $\lambda_1^* \geq 0$, and $g_2(x_1^*, x_2^*) < 0$, thus $\lambda_2^* = 0$ (as is the case):

$$\frac{\partial H(\lambda_1^*, \lambda_2^*)}{\partial \lambda_1} = 0$$

$$\frac{\partial H}{\partial \lambda_1} = -\lambda_1 - \frac{3}{2}\lambda_2 + 4 = 0 \qquad (A.110)$$

Reducing Equation A.108 into Equation A.110 yields

$$\lambda_1^* = 4 \qquad (A.111)$$

since it was assumed that

$$\lambda_2^* = 0 \qquad (A.112)$$

Equations A.111 and A.112 yield the solution to the dual problem, since they satisfy conditions (A.100) to (A.103) and (A.106) to (A.109). Substituting Equations A.111 and A.112 into A.99 yields

$$H(\lambda_1^*, \lambda_2^*) = 8$$

which is the same solution obtained by solving the primal problem.

A.1.7.12 Saddle point conditions

The saddle point conditions are:

1. (x_1^*, x_2^*) minimize $L(x_1, x_2, \lambda_1^*, \lambda_2^*)$ for $x_1 \geq 0$ and $x_2 \geq 0$
2. $g_1(x_1^*, x_2^*) \leq 0$, $g_2(x_1^*, x_2^*) \leq 0$
3. $\lambda_1^* g_1(x_1^*, x_2^*) \leq 0$, $\lambda_2^* g_2(x_1^*, x_2^*) \leq 0$

Condition 1:

$$\begin{aligned} L(x_1, x_2, \lambda_1^*, \lambda_2^*) &= x_1^2 + x_2^2 + \lambda_1^*(4 - x_1 - x_2) \\ &\quad + \lambda_2^*(5 - 2x_1 - x_2) \\ &= x_1^2 + x_2^2 + 4(4 - x_1 - x_2) \\ &\quad + 0(5 - 2x_1 - x_2) \\ L(x_1, x_2, \lambda_1^*, \lambda_2^*) &= x_1^2 + x_2^2 + 16 - 4x_1 - 4x_2 \end{aligned}$$
$$(A.113)$$

It can easily be shown that $x_1^* = 2$ and $x_2^* = 2$ minimize Equation A.113 at

$$L(x_1, x_2, \lambda_1^*, \lambda_2^*) = 8$$

Condition 2:

Both $g_1(x_1^*, x_2^*) \leq 0$ and $g_2(x_1^*, x_2^*) \leq 0$ are satisfied.

Condition 3:

$$\lambda_1^* g_1(x_1^*, x_2^*) = 4(4 - 2 - 2) = 0$$
$$\lambda_2^* g_2(x_1^*, x_2^*) = 0(5 - 2x_1 - x_2) = 0$$

Thus, condition 3 is also satisfied and a saddle point exists.

A.1.8 Multiobjective Decision Trees

The following calculations supplement the text in Sections 9.2.3.2 and 9.2.3.3:

$$\Pr(\mathbf{flood})$$
$$= \Pr\left(w > 50{,}000\right) = \sum_{i=1}^{3} \Pr(\text{flood}|LN_i)\Pr(LN_i) \quad \text{(A.114)}$$

$$\Pr(\mathbf{flood}) = \sum_{i=1}^{3} \frac{1}{3} \times \Pr\left(z > \frac{\ln w_i - \mu_i}{\sigma_i}\right) \quad \text{(A.115)}$$

For LN_1: $\Pr\left(z > \dfrac{\ln 50{,}000 - \mu_1}{\sigma_1}\right)$
$$= 1 - \Pr\left(z \le \frac{\ln 50{,}000 - 10.4}{1}\right)$$
$$= 1 - \phi(0.42) = 0.3372$$

For LN_2: $\Pr\left(z > \dfrac{\ln 50{,}000 - \mu_2}{\sigma_2}\right)$
$$= 1 - \Pr\left(z \le \frac{\ln 50{,}000 - 9.1}{1}\right)$$
$$= 1 - \phi(1.72) = 0.0427$$

For LN_3: $\Pr\left(z > \dfrac{\ln 50{,}000 - \mu_3}{\sigma_3}\right)$
$$= 1 - \Pr\left(z \le \frac{\ln 50{,}000 - 7.8}{1}\right)$$
$$= 1 - \phi(3.02) = 0.0013$$

$$\Pr(\mathbf{flood}) = (0.3372 + 0.0427 + 0.0013)\frac{1}{3} = 0.1271$$

$$\Pr(\mathbf{Higher}) = \Pr(15{,}000 \le w \le 50{,}000)$$
$$= \sum_{i=1}^{3} \Pr(\text{Higher}|LN_i)\Pr(LN_i) \quad \text{(A.116)}$$

$\Pr(\text{Higher}|LN_1)$
$$= \Pr\left(\frac{\ln 15{,}000 - 10.4}{1} \le z \le \frac{\ln 50{,}000 - 10.4}{1}\right)$$
$$= \phi(0.42) - \phi(-0.78)$$
$$= 0.1628 - (-0.2823) = 0.4451$$

$\Pr(\text{Higher}|LN_2)$
$$= \Pr\left(\frac{\ln 15{,}000 - 9.1}{1} \le z \le \frac{\ln 50{,}000 - 9.1}{1}\right)$$
$$= \phi(1.72) - \phi(0.52) = 0.2588$$

$\Pr(\text{Higher}|LN_3)$
$$= \Pr\left(\frac{\ln 15{,}000 - 7.8}{1} \le z \le \frac{\ln 50{,}000 - 7.8}{1}\right)$$
$$= \phi(3.02) - \phi(1.82) = 0.0331$$

$$\Pr(\mathbf{Higher}) = (0.4451 + 0.2588 + 0.0331)\frac{1}{3} = 0.2457$$

$$\Pr(\mathbf{Same}) = \Pr(5000 \le w \le 15{,}000)$$
$$= \sum_{i=1}^{3} \Pr(\text{Same}|LN_i)\Pr(LN_i)$$
$$\text{(A.117)}$$

$\Pr(\text{Same}|LN_1)$
$$= \Pr\left(\frac{\ln 5000 - 10.4}{1} \le z \le \frac{\ln 15{,}000 - 10.4}{1}\right)$$
$$= \phi(-0.78) - \phi(-1.88) = 0.1876$$

$\Pr(\text{Same}|LN_2)$
$$= \Pr\left(\frac{\ln 5000 - 9.1}{1} \le z \le \frac{\ln 15{,}000 - 9.1}{1}\right)$$
$$= \phi(0.52) - \phi(-0.58) = 0.4175$$

$\Pr(\text{Same}|LN_3)$
$$= \Pr\left(\frac{\ln 5000 - 7.8}{1} \le z \le \frac{\ln 15{,}000 - 7.8}{1}\right)$$
$$= \phi(1.82) - \phi(0.72) = 0.2014$$

$$\Pr(\mathbf{Same}) = (0.1876 + 0.4175 + 0.2014)\frac{1}{3} = 0.2689$$

$$\Pr(\mathbf{Lower}) = \Pr(w \le 5000)$$
$$= \sum_{i=1}^{3} \Pr(\text{Lower}|LN_i)\Pr(LN_i)$$
$$\text{(A.118)}$$

$$\Pr(\text{Lower}|LN_1) = \Pr\left(z \le \frac{\ln 5000 - 10.4}{1}\right)$$
$$= \phi(-1.88) = 0.0301$$

$$\Pr(\text{Lower} \mid LN_2) = \Pr\left(z \le \frac{\ln 5000 - 9.1}{1}\right)$$
$$= \phi(-0.58) = 0.2810$$

$$\Pr(\text{Lower} \mid LN_3) = \Pr\left(z \le \frac{\ln 5000 - 7.8}{1}\right)$$
$$= \phi(0.72) = 0.7642$$

$$\Pr(\textbf{Lower}) = (0.0301 + 0.2810 + 0.7642)\frac{1}{3} = 0.3584$$

To check these calculations, the probabilities of all the events must be equal to 1:

$$\Pr(\text{Total}) = \Pr(\text{Flood}) + \Pr(\text{Higher}) + \Pr(\text{Same}) + \Pr(\text{Lower})$$
$$= 0.1271 + 0.2457 + 0.2689 + 0.3584 \approx 1.000$$

Posterior probabilities

$$Pr(LN_i \mid w_j) = \frac{\Pr(w_j \mid LN_i)\Pr(LN_i)}{\sum_{i=1}^{3}\Pr(w_j \mid LN_i)\Pr(LN_i)} \quad (A.119)$$

$$\Pr(LN_1 \mid w_0) = \Pr(LN_1 \mid \text{Flood})$$
$$= \frac{\Pr(\text{Flood} \mid LN_1)\Pr(LN_1)}{\sum_{i=1}^{3}P(\text{Flood} \mid LN_i)\Pr(LN_i)}$$
$$= \frac{(0.3372)(1/3)}{0.1271} = 0.8843$$

$$\Pr(LN_2 \mid w_0) = \Pr(LN_2 \mid \text{Flood})$$
$$= \frac{\Pr(\text{Flood} \mid LN_2)\Pr(LN_2)}{\sum_{i=1}^{3}P(\text{Flood} \mid LN_i)\Pr(LN_i)}$$
$$= \frac{(0.0427)(1/3)}{0.1271} = 0.1120$$

$$\Pr(LN_3 \mid w_0) = \Pr(LN_3 \mid \text{Flood})$$
$$= \frac{\Pr(\text{Flood} \mid LN_3)\Pr(LN_3)}{\sum_{i=1}^{3}P(\text{Flood} \mid LN_i)\Pr(LN_i)}$$
$$= \frac{(0.0013)(1/3)}{0.1271} = 0.0034$$

$$\Pr(LN_1 \mid w_1) = \Pr(LN_1 \mid \text{Higher})$$
$$= \frac{\Pr(\text{Higher} \mid LN_1)\Pr(LN_1)}{\sum_{i=1}^{3}P(\text{Higher} \mid LN_i)\Pr(LN_i)}$$
$$= \frac{(0.4451)(1/3)}{0.2457} = 0.6039$$

$$\Pr(LN_2 \mid w_1) = \Pr(LN_2 \mid \text{Higher})$$
$$= \frac{\Pr(\text{Higher} \mid LN_2)\Pr(LN_2)}{\sum_{i=1}^{3}P(\text{Higher} \mid LN_i)\Pr(LN_i)}$$
$$= \frac{(0.2588)(1/3)}{0.2457} = 0.3511$$

$$\Pr(LN_3 \mid w_1) = \Pr(LN_3 \mid \text{Higher})$$
$$= \frac{\Pr(\text{Higher} \mid LN_3)\Pr(LN_3)}{\sum_{i=1}^{3}P(\text{Higher} \mid LN_i)\Pr(LN_i)}$$
$$= \frac{(0.0031)(1/3)}{0.2457} = 0.0450$$

$$\Pr(LN_1 \mid w_2) = \Pr(LN_1 \mid \text{Same})$$
$$= \frac{\Pr(\text{Same} \mid LN_1)\Pr(LN_1)}{\sum_{i=1}^{3}P(\text{Same} \mid LN_i)\Pr(LN_i)}$$
$$= \frac{(0.1876)(1/3)}{0.2689} = 0.2326$$

$$\Pr(LN_2 \mid w_2) = \Pr(LN_2 \mid \text{Same})$$
$$= \frac{\Pr(\text{Same} \mid LN_2)\Pr(LN_2)}{\sum_{i=1}^{3}P(\text{Same} \mid LN_i)\Pr(LN_i)}$$
$$= \frac{(0.4175)(1/3)}{0.2689} = 0.5175$$

$$\Pr(LN_3 \mid w_2) = \Pr(LN_3 \mid \text{Same})$$
$$= \frac{\Pr(\text{Same} \mid LN_3)\Pr(LN_3)}{\sum_{i=1}^{3}P(\text{Same} \mid LN_i)\Pr(LN_i)}$$
$$= \frac{(0.2014)(1/3)}{0.2689} = 0.2497$$

$$\Pr(LN_1 \mid w_3) = \Pr(LN_1 / \text{Lower})$$
$$= \frac{\Pr(\text{Lower} \mid LN_1)\Pr(LN_1)}{\sum_{i=1}^{3}P(\text{Lower} \mid LN_i)\Pr(LN_i)}$$
$$= \frac{(0.0301)(1/3)}{0.3584} = 0.0280$$

$$\Pr(LN_2 | w_3) = \Pr(LN_2 / Lower)$$
$$= \frac{\Pr(Lower | LN_2)\Pr(LN_2)}{\sum_{i=1}^{3} P(Lower | LN_i)\Pr(LN_i)}$$
$$= \frac{(0.2810)(1/3)}{0.3584} = 0.2614$$

$$\Pr(LN_3 | w_3) = \Pr(LN_3 / Lower)$$
$$= \frac{\Pr(Lower | LN_3)\Pr(LN_3)}{\sum_{i=1}^{3} P(Lower | LN_i)\Pr(LN_i)}$$
$$= \frac{(0.7642)(1/3)}{0.3584} = 0.7108$$

A.1.9 Derivation of the Expected Value of a Log-Normal Distribution

By definition, a random variable Y is said to follow a log-normal distribution if its logarithm is normally distributed:

$$X = \ln Y \sim N(\mu, \sigma^2) \quad (A.120)$$

The probability density function (pdf) of a normal distribution denoted by the variate X in the above definition is as follows:

$$f(x) = \frac{1}{\sqrt{2\pi}\sigma} e^{-\frac{1}{2}(x-\mu)^2/\sigma^2}, \quad -\infty < x < +\infty \quad (A.121)$$

On the other hand, a log-normal distribution with variate Y has the following pdf:

$$f(y) = \frac{1}{\sqrt{2\pi}\sigma y} e^{-\frac{1}{2}(\ln y - \mu)^2/\sigma^2}, \quad 0 < y < +\infty \quad (A.122)$$

The aim of this section is to derive an expression for the expected value of a log-normal distribution. The expected value for a pdf is denoted by f_5 consistently throughout this book. For a log-normal distribution whose pdf is given in Equation A.122, its f_5 can be established by using the following formula for expectation of a random variable:

$$f_5 = \int_0^\infty y f(y) dy = \int_0^\infty y \cdot \frac{1}{\sqrt{2\pi}\sigma y} e^{-\frac{1}{2}(\ln y - \mu)^2/\sigma^2} dy \quad (A.123)$$

Simplifying Equation A.123 can be made possible by transforming the log-normal variate Y to the corresponding normal variate X. Let

$$y = e^x \quad (A.124)$$
$$x = \ln y \quad (A.125)$$
$$dx = (1/y) dy \quad (A.126)$$

Substituting Equations A.124, A.125, and A.126 into A.123,

$$f_5 = \int_0^\infty y \cdot \frac{1}{\sqrt{2\pi}\sigma y} e^{-\frac{1}{2}(\ln y - \mu)^2/\sigma^2} (dy)$$
$$= \int_{-\infty}^\infty e^x \cdot \frac{1}{\sqrt{2\pi}\sigma} e^{-\frac{1}{2}(x-\mu)^2/\sigma^2} (dx) \quad (A.127)$$

Combining the exponential terms in Equation A.127, we have

$$f_5 = \int_{-\infty}^\infty \frac{1}{\sqrt{2\pi}\sigma} e^{x - \frac{1}{2}(x-\mu)^2/\sigma^2} dx = \int_{-\infty}^\infty \frac{1}{\sqrt{2\pi}\sigma} e^{g(x)} dx \quad (A.128)$$

For the subsequent steps, we implement a process commonly referred to as completing the squares to the terms in the exponential operator e (i.e., denoted by $g(x)$ in Eq. A.128):

$$g(x) = x - \frac{1}{2}(x-\mu)^2/\sigma^2$$
$$= \frac{2x\sigma^2 - (x^2 - 2x\mu + \mu^2)}{2\sigma^2}$$
$$= \frac{2x\sigma^2 - x^2 + 2x\mu - \mu^2}{2\sigma^2}$$
$$= \frac{-x^2 + 2x(\mu + \sigma^2) - \mu^2}{2\sigma^2}$$
$$= \frac{-x^2 + 2x(\mu + \sigma^2) - (\mu^2 + 2\mu\sigma^2 + \sigma^4) + 2\mu\sigma^2 + \sigma^4}{2\sigma^2}$$
$$= \frac{-x^2 + 2x(\mu + \sigma^2) - (\mu + \sigma^2)^2 + 2\mu\sigma^2 + \sigma^4}{2\sigma^2} \quad (A.129)$$

Therefore,

$$g(x) = \frac{-\left[x - (\mu + \sigma^2)\right]^2}{2\sigma^2} + \mu + \frac{1}{2}\sigma^2 \quad (A.130)$$

Substituting Equation A.130 into Equation A.128 will yield

$$f_5 = \int_{-\infty}^{\infty} \frac{1}{\sqrt{2\pi}\sigma} e^{\frac{-[x-(\mu+\sigma^2)]^2}{2\sigma^2} + \mu + \frac{1}{2}\sigma^2} dx \quad (A.131)$$

This can be rearranged to the following form:

$$f_5 = e^{\mu + \frac{1}{2}\sigma^2} \left(\int_{-\infty}^{\infty} \frac{1}{\sqrt{2\pi}\sigma} e^{\frac{-[x-(\mu+\sigma^2)]^2}{2\sigma^2}} dx \right) \quad (A.132)$$

Define a parameter μ' as follows:

$$\mu' = \mu + \sigma^2 \quad (A.133)$$

Substituting Equation A.133 into Equation A.132 will yield the following:

$$f_5 = e^{\mu + \frac{1}{2}\sigma^2} \left[\int_{-\infty}^{\infty} \frac{1}{\sqrt{2\pi}\sigma} e^{\frac{-(x-\mu')^2}{2\sigma^2}} dx \right] \quad (A.134)$$

Observe that the integral in the bracketed quantity in Equation A.134 is a pdf for a normal distribution with parameters $N(\mu', \sigma^2)$. Thus, it must obey the probability law that states that the total probability for all realizations of a random variable is unity. Therefore from Equation A.134, it is easy to see that the expected value (f_5) of a log-normal distribution is as shown in Equation A.135. Note that this f_5 is expressed in terms of the original parameters μ and σ appearing in the definition of a log-normal pdf in Equation A.122:

$$f_5 = e^{\mu + \frac{1}{2}\sigma^2} \quad (A.135)$$

A.1.10 Derivation of the Conditional Expected Value of a Log-Normal Distribution

An upper-tail conditional expectation of a log-normal distribution, denoted by $f_4(\cdot)$ (or f_4 for simplicity), will be developed in this section. The value of f_4 refers to the conditional expected value of a partition of a pdf with high consequence, although with low likelihoods of occurrence. Suppose an upper-tail partition β along the x-axis (i.e., the damage axis) is specified. For a log-normal distribution, f_4 can be established using the following definition:

$$f_4(\cdot) = \frac{\int_{\beta}^{\infty} y f(y) dy}{\int_{\beta}^{\infty} f(y) dy} \quad \beta > 0 \quad (A.136)$$

Regardless of the underlying pdf, the denominator of (A.136) is the exceedance probability $1-\alpha$ (see Figure A.11):

$$\Pr(y > \beta) = 1 - \Pr(y \leq \beta) = 1 - \alpha \quad (A.137)$$

Substituting (A.122) and (A.137) into (A.136), we obtain

$$f_4 = \frac{1}{1-\alpha} \int_{\beta}^{\infty} y \cdot \frac{1}{\sqrt{2\pi}\sigma y} e^{-\frac{1}{2}(\ln y - \mu)^2/\sigma^2} dy \quad (A.138)$$

The steps similar to those employed in the derivation of expected value for a log-normal distribution (see Eqs. A.128–A.132) yield the following transformed version of Equation A.138:

$$f_4 = \frac{e^{\mu + \frac{1}{2}\sigma^2}}{1-\alpha} \left[\int_{\ln \beta}^{\infty} \frac{1}{\sqrt{2\pi}\sigma} e^{\frac{-[x-(\mu+\sigma^2)]^2}{2\sigma^2}} dx \right] \quad (A.139)$$

Note that the bracketed quantity is a normal distribution evaluated in the interval between $\ln \beta$ and $+\infty$, with parameters $N(\mu',\sigma^2)$, where the shifted mean μ' is defined to be $\mu' = \mu + \sigma^2$ (see A.133).

In the meantime, we analyze a normal distribution with the original mean parameter μ, which we can subsequently transform to a normal distribution of interest with mean μ'. Referring to Figure A.11, we can establish the following identities:

$$\Pr(y > \beta) = \Pr(x > \ln \beta) = 1 - \alpha \quad (A.140)$$

Using the standard normal distribution formula and denoting $z_{\ln \beta}$ as the standard normal distribution partition corresponding to $x = \ln \beta$, we have

$$z_{\ln \beta} = \frac{\ln \beta - \mu}{\sigma} \quad (A.141)$$

Taking the cumulative probabilities on both sides of Equation A.141, we obtain

$$\Pr(z \leq z_{\ln \beta}) = \Pr\left(z \leq \frac{\ln \beta - \mu}{\sigma}\right) \quad (A.142)$$

Denote Φ as the cumulative probability function (cdf) of a standard normal distribution. For example,

$$\Pr(z \leq z_{\ln \beta}) = \Phi(z_{\ln \beta}) \quad (A.143)$$

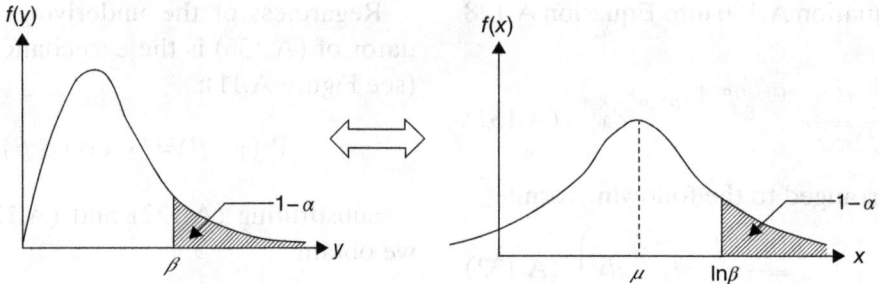

Figure A.11 Transforming a log-normal distribution to a normal distribution.

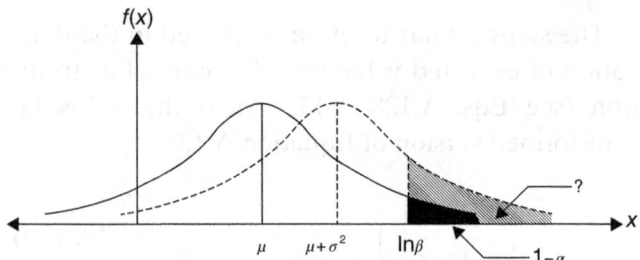

Figure A.12 Shifting the mean of a normal distribution.

Substituting Equation A.143 into A.142 and knowing that the right-hand side of Equation A.142 is simply α (see Figure A.11), we get

$$\Phi(z_{\ln\beta}) = \alpha \qquad (A.144)$$

As with any cdf, Φ is an increasing function—thus, the existence of a corresponding inverse function is guaranteed. Using this property, we can rewrite (A.144) as follows:

$$z_{\ln\beta} = \Phi^{-1}(\alpha) \qquad (A.145)$$

Now, we need to find $z'_{\ln\beta}$ corresponding to a normal distribution with shifted mean μ'. In Figure A.12, we see that this normal distribution (i.e., with shifted mean μ') is only a linear translation of the original normal distribution (i.e., with mean μ) because they have the same variance σ^2. Therefore,

$$z'_{\ln\beta} = \frac{\ln\beta - \mu'}{\sigma} = \frac{\ln\beta - (\mu+\sigma^2)}{\sigma} \qquad (A.146)$$

Simplifying,

$$z'_{\ln\beta} = \left[\frac{\ln\beta - \mu}{\sigma}\right] - \sigma \qquad (A.147)$$

Note that the bracketed quantity in Equation A.147 is $z_{\ln\beta}$ (see Eq. A.141):

$$z'_{\ln\beta} = z_{\ln\beta} - \sigma \qquad (A.148)$$

Substituting Equation A.145 into Equation A.148, we obtain

$$z'_{\ln\beta} = \Phi^{-1}(\alpha) - \sigma \qquad (A.149)$$

Taking the standard normal cumulative probability of Equation A.149, we get

$$\Phi(z'_{\ln\beta}) = \Phi(\Phi^{-1}(\alpha) - \sigma) \qquad (A.150)$$

Thus, the upper-tail probability (i.e., the complement) associated with Equation A.150 will be

$$1 - \Phi(z'_{\ln\beta}) = 1 - \Phi(\Phi^{-1}(\alpha) - \sigma) \qquad (A.151)$$

We revisit the f_4 expression for a log-normal distribution in Equation A.139 and conclude from Figure A.12 that

$$\begin{aligned}f_4 &= \frac{e^{\mu+\frac{1}{2}\sigma^2}}{1-\alpha}\left[\int_{\ln\beta}^{\infty}\frac{1}{\sqrt{2\pi}\sigma}e^{-[x-(\mu+\sigma^2)]^2/2\sigma^2}dx\right]\\ &= \frac{e^{\mu+\frac{1}{2}\sigma^2}}{1-\alpha}\left[1-\Phi(z'_{\ln\beta})\right]\end{aligned} \qquad (A.152)$$

Finally, substituting Equation A.151 into Equation A.152, the f_4 for a log-normal distribution is established:

$$f_4 = \frac{e^{\mu+\frac{1}{2}\sigma^2}}{1-\alpha}[1-\Phi(\Phi^{-1}(\alpha)-\sigma)] \qquad (A.153)$$

Example:
A market hypothesis asserts that the relative change in a stock's value (i.e., ratio of stock's current price vs. a prior price), denoted by Y, is log-normally distributed. An increase in the price of a stock may be considered either a profit or a loss depending on the scenario. For instance, selling a stock today on the belief that its value will depreciate tomorrow will translate to an opportunity loss when its price tomorrow actually increases. For this example, we consider stock price increases as losses. Therefore, assuming that the relative change in a stock's value behaves log-normally as hypothesized, the upper tail will correspond to high-consequence, low-probability events.

Suppose an investor who has faith in this market hypothesis asked you to conduct an analysis for a stock with mean return $\mu = 0.2$ and volatility $\sigma = 0.4$. Calculate: (i) the expected relative change in the stock's value and (ii) the conditional expected relative change in the stock's value for an upper-tail probability partition of $1 - \alpha = 0.1$.

1. The expected relative change in the stock's value refers to the f_5 of the log-normal distribution with parameters $\mu = 0.2$ and volatility $\sigma = 0.4$ as specified in this example. Using (A.135),

$$f_5 = e^{\mu + \frac{1}{2}\sigma^2} = \exp[0.2 + 0.5(0.4)^2] = 1.3231$$

Therefore, the expected relative change in the stock's value is 1.3231 times its current value. Suppose that the parameters μ and σ correspond to annual data. Then, in a course of 1 year, a stock whose unit price now is $20 has an expected price of $20(1.3231) = $26.462.

2. The conditional expected relative change in the stock's value for an upper-tail probability partition of $1 - \alpha = 0.1$ refers to the f_4 as derived in Equation A.153:

$$f_4 = \frac{e^{\mu + \frac{1}{2}\sigma^2}}{1 - \alpha}[1 - \Phi(\Phi^{-1}(\alpha) - \sigma)]$$

Let us progressively calculate the expression in the bracketed quantity:

$$\Phi^{-1}(\alpha) = \Phi^{-1}(0.90) = 1.281552$$

$$\Phi^{-1}(\alpha) - \sigma = 1.281552 - 0.4 = 0.881552$$

$$\Phi(\Phi^{-1}(\alpha) - \sigma) = \Phi(0.881552) = 0.81099$$

$$1 - \Phi(\Phi^{-1}(\alpha) - \sigma) = 1 - 0.81099 = 0.18901$$

Therefore,

$$f_4 = \frac{e^{\mu + \frac{1}{2}\sigma^2}}{1 - \alpha}[1 - \Phi(\Phi^{-1}(\alpha) - \sigma)]$$
$$= \frac{f_5}{1 - \alpha}[1 - \Phi(\Phi^{-1}(\alpha) - \sigma)] = \frac{1.3231}{0.1}[0.18901]$$
$$f_4 = 2.5$$

Using an upper-tail probability partition of $1 - \alpha = 0.1$, the conditional expected relative change in the stock's value is therefore 2.5, which is almost twice compared to the expected value of $f_5 = 1.3231$.

A.1.11 Triangular Distribution: Unconditional and Conditional Expected Values

Using the notation $a, b,$ and c to denote the minimum, maximum, and most likely (mode) values of a triangular distribution, the resulting pdf in terms of the random variable x follows the form as shown in Equation A.154. A triangular distribution is depicted in Figure A.13 representing the general locations of the parameters a, b, and c. The figure also shows an upper-tail partition of $x = \beta$ corresponding to a probability of $P(x = \beta) = \alpha$ required for calculating a desired conditional expected value:

$$f(x) = \begin{cases} \dfrac{2(x-a)}{(b-a)(c-a)}, & a \leq x \leq c \\ \dfrac{2(b-x)}{(b-a)(b-c)}, & c < x \leq b \\ 0, & \text{otherwise} \end{cases} \quad (A.154)$$

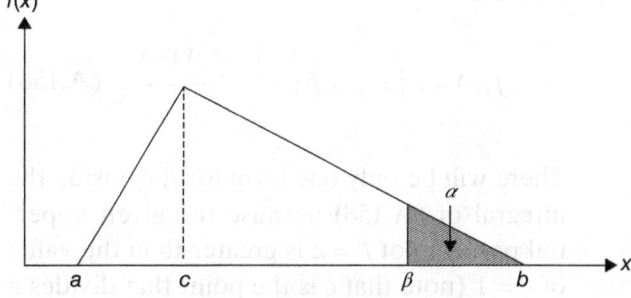

Figure A.13 *Triangular distribution.*

654 APPENDIX: OPTIMIZATION TECHNIQUES

Example:
Consider a triangular distribution with parameters $a = 0, c = 1$, and $b = 3$.

Calculate:

1. The expected value
2. The conditional expected value for $x > \beta$, using $\beta = 2$

Solution:

1. The expected value, which the book denotes by $f_5(\cdot)$, generally takes the following form:

$$f_5(\cdot) = E[x] = \int_x^\infty x f(x)\, dx \quad \text{(A.155)}$$

Substituting the given parameters into (A.154) and (A.155) gives the expected value of the triangular distribution for this example:

$$f_5(\cdot) = \int_0^1 x \left\{ \frac{2(x-0)}{(3-0)(1-0)} \right\} dx + \int_1^3 x \left\{ \frac{2(3-x)}{(3-0)(3-1)} \right\} dx$$

$$= \left. \frac{2x^3}{9} \right|_0^1 + \left. \left(\frac{x^2}{2} - \frac{x^3}{9} \right) \right|_1^3$$

$$= \frac{4}{3}$$

(A.156)

A simpler approach for calculating $f_5(\cdot)$ for a triangular distribution (i.e., no integration required) is achieved through the direct use of the following formula:

$$f_5(\cdot) = \frac{a+c+b}{3} = \frac{0+1+3}{3} = \frac{4}{3} \quad \text{(A.157)}$$

2. The conditional expected value of a triangular distribution corresponding to an upper-tail partition of $x > \beta$ is derived using the following formula:

$$f_4(\cdot) = E[x \mid x > \beta] = \frac{\int_\beta^b x f(x)\, dx}{\int_\beta^b f(x)\, dx} \quad \text{(A.158)}$$

There will be only one term for $f(x)$ inside the integral of (A.158) because the given upper-tail partition of $\beta = 2$ is greater than the value of $c = 1$ (note that c is the point that divides a triangular distribution's pdf into two parts).

Therefore, we only use the second part of the pdf of the given triangular distribution. Referring to (A.154), we obtain

$$f(x) = \frac{2(b-x)}{(b-a)(b-c)} = \frac{6-2x}{6} = 1 - \frac{1}{3}x, \quad c \le x < b$$

(A.159)

Substituting (A.159) into (A.158) and knowing that $\beta = 2$ and $b = 3$,

$$f_4(\cdot) = \int_2^3 x \left\{ 1 - \frac{1}{3}x \right\} dx \div \int_2^3 \left\{ 1 - \frac{1}{3}x \right\} dx$$

(A.160a)

$$f_4(\cdot) = \left. \left\{ \frac{x^2}{2} - \frac{x^3}{9} \right\} \right|_2^3 \div \left. \left\{ x - \frac{x^2}{6} \right\} \right|_2^3$$

(A.160b)

$$f_4(\cdot) = \left\{ \frac{9}{2} - \frac{27}{9} - \frac{4}{2} + \frac{8}{9} \right\} \div \left\{ 3 - \frac{9}{6} - 2 + \frac{4}{6} \right\}$$

$$= \left\{ \frac{5}{2} - \frac{19}{9} \right\} \div \left\{ 1 - \frac{5}{6} \right\}$$

(A.160c)

$$f_4(\cdot) = \left\{ \frac{45}{18} - \frac{38}{18} \right\} \div \left\{ \frac{1}{6} \right\} = \frac{7(6)}{18} = \frac{7}{3} \quad \text{(A.160d)}$$

A more straightforward calculation of $f_4(\cdot)$ for this example is by using a generalized formula for the conditional expected value of a triangular distribution:

$$f_4(\cdot) = E[x \mid x > \beta] = \frac{\int_\beta^b x \left\{ \frac{2(b-x)}{(b-a)(b-c)} \right\} dx}{\int_\beta^b x \left\{ \frac{2(b-x)}{(b-a)(b-c)} \right\} dx}$$

$$= \frac{\int_\beta^b (2bx - 2x^2)\, dx}{\int_\beta^b (2b - 2x)\, dx} \quad \text{(A.161a)}$$

$$f_4(\cdot) = \frac{\left. \left\{ bx^2 - \frac{2}{3}x^3 \right\} \right|_\beta^b}{\left. \left\{ 2bx - x^2 \right\} \right|_\beta^b} = \frac{b^3 - \frac{2}{3}b^3 - b\beta^2 + \frac{2}{3}\beta^3}{2b^2 - b^2 - 2b\beta + \beta^2}$$

$$= \frac{\frac{1}{3}b^3 - b\beta^2 + \frac{2}{3}\beta^3}{b^2 - 2b\beta + \beta^2} \quad \text{(A.161b)}$$

$$f_4(\cdot) = \frac{\frac{1}{3}(b^3 - 3b\beta^2 + 2\beta^3)}{b^2 - 2b\beta + \beta^2} = \frac{\frac{1}{3}(b+2\beta)(b-\beta)^2}{(b-\beta)^2}$$
$$= \frac{b+2\beta}{3}$$

(A.161c)

Substituting $\beta = 2$ and $b = 3$ to the derived formula for $f_4(\cdot)$ of a triangular distribution in (A.161c) will yield the same result as Equation A.160d:

$$f_4(\cdot) = \frac{b+2\beta}{3} = \frac{3+2(2)}{3} = \frac{7}{3} \quad (A.161d)$$

A.2 BAYESIAN ANALYSIS AND THE PREDICTION OF CHEMICAL CARCINOGENICITY[2]

The vast number of new chemicals produced in today's economy creates a risk of exposure to carcinogens. The use of short-term laboratory bioassays (tests) to predict whether a chemical is a carcinogen has continuously been on the rise. Furthermore, epidemiological and occupational exposure studies of human subjects together with experimental results on laboratory animals have shown that certain synthetic and natural chemicals can produce cancer in humans. Each year, new chemicals are introduced into drugs, foods, consumer goods, and the environment. Yet, the human health effects of many of these chemicals are unknown [Kleindorfer and Kunreuther, 1987]. It is very important that a good procedure be developed that will accurately identify chemicals as suspected carcinogens or as noncarcinogens. With such a system in place, regulatory agencies can take appropriate measures to prevent or reduce human exposure to the higher-risk chemicals. Such regulatory actions can result in an overall reduction in the risk of cancer.

One of the greatest weaknesses of the currently available data on the impact of chemicals on human health is that information on animal carcinogenicity

---
[2] Section A.2 is based on Chankong et al. [1985]. Reprinted with permission.

is available on fewer than 1% of chemicals that are known as carcinogens. This is because animal carcinogenicity bioassays are both time-consuming and costly—well over $3 million per chemical. However, we do have short-term *in vitro* bioassay results on over 20,000 chemicals. There is a high prevalence of known carcinogens in the data based on these short-term results. The Gene-Tox database is the major database used in the carcinogenicity prediction and battery selection (CPBS) methodology, the subject of this chapter [see Pet-Edwards, 1986; Pet-Edwards et al., 1985a, b, 1989; Rosenkranz et al., 1984a, b, and Haimes et al., 1987]. The database was first assembled and published under the auspices of the US Environmental Protection Agency (EPA). Even though the Gene-Tox database probably encompasses less than 25% of the published literature, it serves as an appropriate base because of its unbiased (peer-review) character. It is also sufficiently complex to provide a good test for the CPBS methodology.

The ability of the assays to predict carcinogenicity can be characterized by analyzing the test results as to sensitivity, specificity, and accuracy, which are defined as follows:

$$\alpha^+ \equiv \text{Sensitivity} = \frac{\text{number of known carcinogens that test positive in assay}}{\text{number of carcinogens tested}} \times 100$$

(A.162)

$$\alpha^- \equiv \text{Specificity} = \frac{\text{number of known noncarcinogens that test negative in assay}}{\text{number of noncarcinogens tested}} \times 100$$

(A.163)

$$\text{Accuracy} = \frac{\text{number of correct test results}}{\text{number of chemicals tested}} \times 100$$

Notation:

+ = positive result
− = negative result
CA = carcinogen
NC = noncarcinogen
α^+ = sensitivity
α^- = specificity

$\alpha^+ \equiv \Pr(+|CA) =$ the probability that the test result is positive, given that the tested chemical is known to be a carcinogen

$\alpha^- \equiv \Pr(-|NC) =$ the probability that the test result is negative, given that the tested chemical is known to be a noncarcinogen

Accuracy $= \alpha^+ \Pr(CA) + \alpha^- \Pr(NC)$

Two main objectives are associated with the CPBS methodology:

1. Determine the reliability and predictive capability of both individual and batteries of short-term tests.
2. Develop a strategy for formulating and selecting optimally preferred batteries of short-term tests for screening chemicals for further testing.

The CPBS can be used as a means of estimating costs and risks associated with programs that test chemicals of suspected carcinogenicity. The method can assist in risk-based decisionmaking where cost–risk trade-off analysis can be brought about efficiently and effectively for policy and regulatory purposes. The five major components of the CPBS are:

1. Data consolidation
2. Parameter estimation
3. Predictability calculation
4. Battery selection
5. Risk assessment

A.2.1 Calculating Sensitivity and Specificity

Calculation of the sensitivity and specificity of an assay is critical to understanding its predictive capabilities. When the available database is not ideal (containing many gaps), it is difficult to estimate the sensitivity ($\hat{\alpha}^+$) and specificity ($\hat{\alpha}^-$). Our task is then to determine whether $\hat{\alpha}^+$ and $\hat{\alpha}^-$ reflect the *true* sensitivity and specificity of an assay.

Assays that give the same positive and negative responses on a set of chemicals should have the same sensitivities and specificities. Thus, if we are able to group the assays that give similar responses, based on an expanded database, possibly including test results of chemicals of unknown carcinogenicity (as well as known ones), we should be able to assume that assays within such a group have the same sensitivity and the same specificity. If the sensitivity and/or specificity of an assay within the group is known with a high degree of assurance, then the estimates of the other assays within the group are strengthened by this information.

Cluster analysis [Anderberg, 1973, Pet-Edwards et al., 1985a, b] can be used to determine which of the assays are most similar to each other in terms of their responses. To accomplish this, comparisons are made between the responses of each pair of assays to determine their similarity. These pairwise similarities are utilized in several different hierarchical clustering schemes to uncover the natural groupings (clusters) of assays. The clusters uncovered by the analysis are a characteristic of the responses of the assays on a large number of chemicals. The sensitivities and specificities are also response characteristics for the assays. Assays within a cluster that have a high degree of similarity should have similar responses and, in turn, similar sensitivities and specificities.

A.2.2 Predictivity of an Assay

An assay is said to be predictive if, based on the results alone, we can conclude with a reasonable degree of confidence that the tested chemicals are carcinogens or noncarcinogens. We say that the assay is $p\%$ predictive (or reliable) if there is a $p\%$ chance that each prediction given by the test is correct.

The following is used as a composite measure of predictability:

$\theta^+ \equiv \Pr(CA|+) =$ the probability that the tested chemical is a carcinogen, given that the test result is positive

$\theta^- \equiv \Pr(NC|-) =$ the probability that the tested chemical is a noncarcinogen, given that the test result is negative

Bayes' formula. To calculate the predictivity of an assay, we can use the well-known Bayes' formula, the sensitivity and specificity of an assay, and the prior probability ($\Pr(CA)$) that a tested chemical is a carcinogen [Leemis, 1995; Pratt et al., 1995]:

$$\theta^+ = \Pr(CA\,|\,+) = \frac{\Pr(CA)\Pr(+|CA)}{\Pr(CA)\Pr(+|CA)+\Pr(NC)\Pr(+|NC)}$$

$$= \frac{\Pr(CA)\alpha^+}{\Pr(CA)\alpha^+ + \Pr(NC)(1-\alpha^-)} \quad \text{(A.164)}$$

Since $\Pr(-|NC) = \alpha^-$; $\Pr(-|NC) + \Pr(+|NC) = 1$ we have $\Pr(+|NC) = 1 - \alpha^-$

$$\theta^- = \Pr(NC\,|\,-) = \frac{\Pr(NC)\Pr(+|NC)}{\Pr(NC)\Pr(-|NC)+\Pr(CA)\Pr(-|CA)} \quad \text{(A.165)}$$

$$= \frac{\Pr(NC)\alpha^-}{\Pr(NC)\alpha^- + \Pr(CA)(1-\alpha^+)} \quad \text{(A.166)}$$

If we cannot make a good estimate of Pr(CA), we may assume an uninformative prior, namely, Pr(CA) = 0.5. In other words, we have no reason to believe that the probability of carcinogenicity is higher or lower than 50–50.

Example:
Assay A_1 has the following characteristics:

$$\alpha^+ = 0.7$$
$$\alpha^- = 0.9$$
$$\Pr(CA) = 0.5$$

Then the probability that the tested chemical is a carcinogen (positive), given that the test result is positive, is

$$\theta^+ = \Pr(CA\,|\,+) = \frac{(0.5)(0.7)}{(0.5)(0.7)+(0.5)(0.1)} = 0.875$$

(A.167)

Thus, the positive test result of assay A_1 has increased the chance estimate of carcinogenicity in the tested chemical from an initial estimate of 0.5–875. Similarly,

$$\Pr(NC) = 1 - \Pr(CA) = 035$$
$$\Pr(-|CA) = 1 - \Pr(+|CA) = 1 - 0.7 = 0.3$$

Using Equation A.165 yields

$$\theta^- = \Pr(NC\,|\,-) = \frac{(0.5)(0.9)}{(0.5)(0.9)+(0.5)(0.3)} = 0.750$$

(A.168)

The above calculation is intended to demonstrate (i) how the predictivity indices of a test can be computed and (ii) how the test result of an assay, such as A_1, improves our knowledge regarding the carcinogenicity of a chemical. A more complete use of predictivity formulas, such as those in Equations A.167 and A.168, may be ascertained if they are translated into the graphical form shown in Figures A.14 and A.15. Given one's intuitive feeling about the carcinogenicity of a substance so that an initial guess of Pr(CA) or Pr(NC) can be obtained, the new estimate of Pr(CA) or Pr(NC) based on the result of A_1 can be read directly from the graph in Figures A.14 or A.15, depending, respectively, on whether the test result is positive or negative. For example, if the expert's intuitive feeling leads to an initial estimate of Pr(CA) of 0.60, then from Figure A.14, the positive value of the test result of A_1 will enhance the chance, from 0.60 to 0.86, that the substance is a carcinogen.

Should the test result be negative, the probability of its being noncarcinogenic is raised from the initial estimate of 0.40 to 0.73, according to Figure A.15. Thus, the test result can be viewed as an aid that helps us improve our subjective value judgment or enhances the state of our knowledge. Note that Figures A.14 and A.15 are based on $\alpha^+ = 0.8$, $\alpha^- = 0.8$, and Pr(CA) = 0.6.

A similar analysis can be carried out for any other assay of known sensitivity and specificity indices.

A.3 THE FARMER'S DILEMMA: LINEAR MODEL AND DUALITY

To demonstrate the progressive modeling process through the use of the building blocks of models, the statement of the farmer's dilemma, introduced in Chapter 1, is repeated for completeness and modeled using a deterministic linear model.

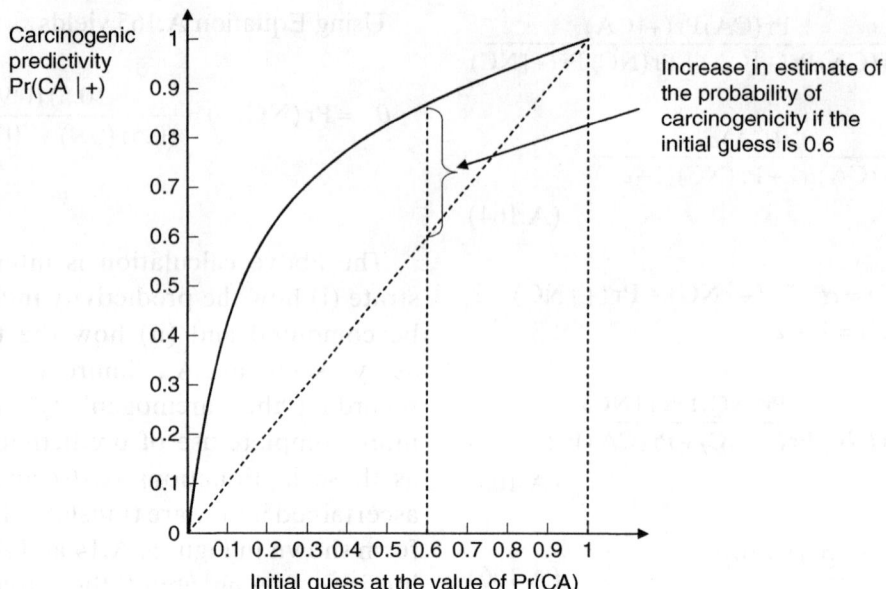

Figure A.14 Carcinogenic predictivity curves for assay A_1 (sensitivity = 0.8 and specificity = 0.8) [Chankong et al., 1985].

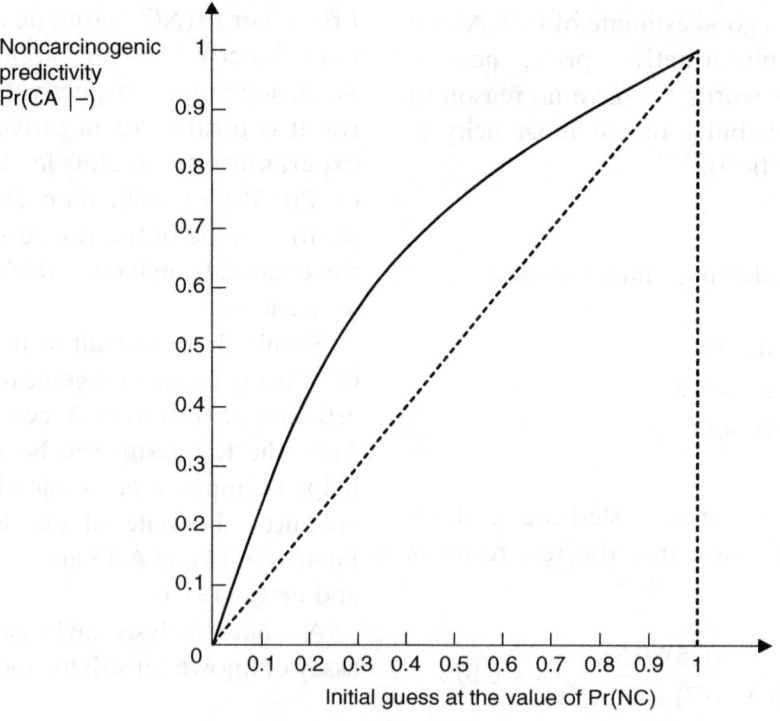

Figure A.15 Noncarcinogenic predictivity curves for assay A_1 (sensitivity = 0.8 and specificity = 0.8) [Chankong et al., 1985].

A.3.1 Problem Definition

Consider, for illustrative purposes, the following oversimplified problem.

A farmer who owns 100 acres of agricultural land is considering two crops for next season—corn and sorghum. Due to a large demand for these crops, he can safely assume that he can sell all his yield (the term *he* is used generically to connote either gender). From his past experience, the farmer has found out that the climate in his region requires (a) an irrigation of 3.9 acre-ft of water per acre of corn and 3 acre-ft of water per acre of sorghum at a subsidized cost of $40 per acre-ft and (b) nitrogen-based fertilizer of 200 lb per acre of corn and 150 lb per

TABLE A.11 Summary of Verbal Information

	Corn	Sorghum	Availability
Land (acres)	x_1	x_2	100
Water (acre-ft)	3.9/acre	3/acre	320 acre-ft
Fertilizer (lb)	200/acre	150/acre	
Fertilizer cost ($)	0.25/lb	0.25/lb	
Water cost ($/acre-ft)	40	40	
Crops yield (bushels)	125	100	
Price of crops ($)	2.80/bushel	2.70/bushel	
Soil erosion (tons)	2.2/acre	2/acre	210 acres

acre of sorghum at a cost of $25 per 100 lb of fertilizer (an acre-ft of water is a measure of 1 acre of area covered with 1 ft of water).

The farmer believes that his land will yield 125 bushels of corn per acre and 100 bushels of sorghum per acre. The farmer expects to sell his crops at $2.80 per bushel of corn and $2.70 per bushel of sorghum.

The farmer has inherited his land and is very concerned about the loss of topsoil due to soil erosion resulting from flood irrigation—the method used in his farm. A local soil conservation service extension expert has determined that the farmer's land loses about 2.2 tons of topsoil per acre of irrigated corn and about 2 tons of topsoil per acre of irrigated sorghum. The farmer is interested in limiting the total topsoil loss from his 100-acre land to no more than 210 tons per season.

The farmer has a limited allocation of 320 acre-ft of water available for the growing season, but he can draw all the credit needed to purchase fertilizer. He would like to determine his optimal planting policy in order to maximize his income. He considers his labor to be equally needed for both crops, and he is not concerned about crop rotation. Note that at this stage of discussion, water quality (e.g., salinity and other contamination), impact on groundwater quality and quantity, and other issues are not addressed.

The results are summarized in Table A.11.

A.3.2 Model Formulation

Let:

x_1 = the number of acres allocated for corn
x_2 = the number of acres allocated for sorghum
s_1 = the level of soil erosion (tons per acre)
s_2 = the level of soil moisture
s_3 = the state of crop growth
s_4 = the state of soil nutrients
c_1 = the market price per one bushel of corn
c_2 = the market price per one bushel of sorghum
a_{11} = the number of tons of soil erosion resulting from growing corn on 1 acre of land
a_{12} = the number of tons of soil erosion resulting from growing sorghum on 1 acre of land
a_{21} = the number of pounds of fertilizer applied per acre for growing corn
a_{22} = the number of pounds of fertilizer applied per acre for growing sorghum
a_{41} = the number of bushels of corn produced from 1 irrigated acre of land
a_{42} = the number of bushels of sorghum produced from 1 irrigated acre of land
c_3 = the cost of 1 lb of fertilizer
c_4 = the cost of 1 acre-ft of water
a_{31} = the amount of water in acre-ft applied for growing corn on 1 acre of land
a_{32} = the amount of water in acre-ft applied for growing sorghum on 1 acre of land
b_1 = the total acres of agricultural land available to the farmer for growing corn and sorghum (assumed fixed; otherwise, it becomes a state variable)
b_2 = the number of tons of soil that the farmer does not want to exceed due to his irrigation practice (assumed fixed; otherwise, it becomes part of an objective function)
b_3 = the total number of acre-ft of water assumed available to the farmer during the growing season (assumed fixed; otherwise, it becomes a random variable or a state variable)
u_1 = the total amount of water in acre-ft applied to growing corn
u_2 = the total amount of water in acre-ft applied to growing sorghum
u_3 = the total amount of fertilizer in pounds applied to growing corn
u_4 = the total amount of fertilizer in pounds applied to growing sorghum
y_1 = the total yield of corn in bushels
y_2 = the total yield of sorghum in bushels
y_3 = the total number of tons of topsoil eroded due to the production of corn
y_4 = the total number of tons of topsoil eroded due to the production of sorghum

Note that for model simplicity, no explicit relationships between the state variables and the other variables are presented. Rather, the level of soil erosion, s_1, is given as a constant (e.g., 2.2 ton of soil erosion/acre for corn). Yet, we know that soil erosion is a function of the level and intensity of precipitation (a random variable), irrigation pattern (a decision variable), cultivation practices, crops selection and crops rotation (decision variables), and so on. Similarly, the level of soil moisture, s_2, which is dependent on precipitation (random variable) and irrigation (decision variable), is assumed constant for each crop, where a fixed amount of irrigation water is assumed (e.g., 3.9 acre-ft of water/acre of corn). The same applies for the state of growth of the crops, s_3, which is dependent on many factors, including fertilizer, irrigation, climatic conditions, and state of soil nutrients. Finally, the state of soil nutrients, s_4, which depends on many factors, including crop rotation and the application of fertilizer, is assumed constant (e.g., 200 lb/acre of fertilizer is required to grow corn). In general, modeling physical relationships among building blocks is determined through experimentation and historical records. For example, the Extension Stations of the US Department of Agriculture provide soil erosion rates under various irrigation or water runoff conditions. A major challenge in the modeling process is quantifying the causal relationships among the state variables and all other relevant variables on which they depend. Exploring these relationships is beyond the scope of this book. The results are summarized in Table A.12.

The objective function of the farmer can be written as

$$f(\cdot) = c_1 y_1 + c_2 y_2 - c_3(u_3 + u_4) - c_4(u_1 + u_2) \quad (A.169)$$

where

$c_1 y_1 + c_2 y_2$ is the income from the sale of his crops
$c_3(u_3 + u_4)$ is the cost of fertilizer
$c_4(u_1 + u_2)$ is the cost of irrigation water

$$y_1[\text{bushels of corn}] = a_{41}\left[\frac{\text{bushels of corn}}{\text{acres of corn}}\right] x_1[\text{acres of corn}]$$
$$= a_{41} x_1 [\text{bushels of corn}]$$

TABLE A.12 Summary of Numerical Values

	Corn	Sorghum	Availability
Land (acres)	x_1	x_2	$b_1 \leq 100$ b_1
Water (acre-ft/acre)	3.9 a_{31}	3 a_{32}	$b_3 \leq 320$
Fertilizer (lb/acre)	200 a_{21}	150 a_{22}	
Fertilizer cost ($/lb)	0.25 c_3	0.25 c_3	
Water cost ($/acre-ft)	40 c_4	40 c_4	
Crop yield (bushel/acre)	125 a_{41}	100 a_{42}	
Price of crops ($/bushel)	2.80 c_1	2.70 c_2	
Soil erosion (ton/acre)	2.2 a_{11}	2 a_{12}	$b_2 \leq 210$

$$y_2[\text{bushels of sorghum}] = a_{42}\left[\frac{\text{bushels}}{\text{acres}}\right] x_1[\text{acres}]$$
$$= a_{42} x_2 [\text{bushels of sorghum}]$$

$$y_3[\text{tons of topsoil eroded to corn}]$$
$$= a_{11}\left[\frac{\text{tons}}{\text{acres}}\right] x_1[\text{acres}]$$
$$= a_{11} x_1 [\text{tons of topsoil eroded due to corn productivity}]$$

$$y_4[\text{tons of topsoil eroded due to sorghum}]$$
$$= a_{12}\left[\frac{\text{tons}}{\text{acres}}\right] x_2[\text{acres}]$$
$$= a_{12} x_2 [\text{tons of topsoil eroded due to sorghum productivity}]$$

$$u_1[\text{water in acre-ft used for corn}]$$
$$= a_{31}\left[\frac{\text{acre-ft}}{\text{acres}}\right] x_1[\text{acres}]$$
$$= a_{31} x_1 [\text{acre-ft of water applied to corn}]$$

$$u_2[\text{water in acre-ft used for sorghum}]$$
$$= a_{32}\left[\frac{\text{acre-ft}}{\text{acres}}\right] x_2[\text{acres}]$$
$$= a_{32} x_2 [\text{acre-ft of water applied to sorghum}]$$

$$u_3[\text{lbs of fertilizer for corn}] = a_{21}\left[\frac{\text{lb}}{\text{acres}}\right]x_1[\text{acres}]$$
$$= a_{21}x_1[\text{lb of fertilizer applied to corn}]$$

$$u_4[\text{lbs of fertilizer for sorghum}] = a_{22}\left[\frac{\text{lb}}{\text{acres}}\right]x_2[\text{acres}]$$
$$= a_{22}x_2[\text{lb of fertilizer applied to sorghum}]$$

Thus, after appropriate substitution, the objective function becomes

$$\begin{aligned}f(\cdot) &= c_1 y_1 + c_2 y_2 - c_4(u_1 + u_2) - c_3(u_3 + u_4) \\ &= c_1 a_{41} x_1 + c_2 a_{42} x_2 - c_4[a_{31}x_1 + a_{32}x_2] - c_3[a_{21}x_1 \\ &\quad + a_{22}x_2] \\ &= [c_1 a_{41} - c_4 a_{31} - c_3 a_{21}]x_1 + [c_2 a_{42} - c_4 a_{32} - c_3 a_{22}]x_2\end{aligned}$$
(A.170)

Simplifying further, we get

$$f(x_1, x_2) = \hat{c}_1 x_1 + \hat{c}_2 x_2 \quad (A.171)$$

where

$$\hat{c}_1 = c_1 a_{41} - c_4 a_{31} - c_3 a_{21} \quad (A.172)$$

$$\hat{c}_2 = c_2 a_{42} - c_4 a_{32} - c_3 a_{22} \quad (A.173)$$

There are constraints on land, soil erosion, and water. Note that most constraints are exchangeable with objective functions and vice versa. For example, instead of limiting soil erosion so as not to exceed 2.0 tons per acre, we may add another objective function, that is, minimize soil erosion [see Chankong and Haimes, 1983, 2008] and Chapter 5. Furthermore, most constraints also are state variables, as is the case in the farmer's problem:

Land: The total available is b_1; thus,

$$x_1 + x_2 \leq b_1 \quad (A.174)$$

Soil erosion: The total allowed eroded soil is not to exceed b_2; thus,

$$a_{11}x_1 + a_{12}x_2 \leq b_2 \quad (A.175)$$

Water: The total water available is b_3; thus, the crops cannot receive more than b_3:

$$a_{31}x_1 + a_{32}x_2 \leq b_3 \quad (A.176)$$

Additional constraints can be applied to the availability of capital to purchase fertilizer and so on. Since the farmer cannot choose to allocate fewer than zero acres to either corn or sorghum, we add nonnegativity constraints:

$$x_1 \geq 0 \text{ and } x_2 \geq 0 \quad (A.177)$$

A.3.3 Model Optimization

The overall mathematical model for the farmer's resource allocation problem (allocation of land, water, fertilizer, etc. to different crops) can be rewritten in Equation A.178. The fact that no state variable (e.g., soil moisture or nutrients) appears explicitly in the objective function does not minimize the centrality of state variables in modeling. For example, the yield coefficient of corn (bushels of corn per acres of corn), a_{41}, is an implicit function of two state variables (soil moisture and nutrients). Multiplying a_{41} by the number of acres of corn, x_1 provides the total yield of corn, y_1:

$$\underset{x_1, x_2}{\text{Maximize}} f(x_1, x_2) = \hat{c}_1 x_1 + \hat{c}_2 x_2 \quad (A.178)$$

subject to the constraints

$$\begin{aligned} x_1 + x_2 &\leq b_1 \\ a_{11}x_1 + a_{12}x_2 &\leq b_2 \\ a_{31}x_1 + a_{32}x_2 &\leq b_3 \\ x_1 \geq 0 \text{ and } x_2 &\geq 0 \end{aligned} \quad (A.179)$$

By substituting for the known values of the variables, the optimization problem becomes

$$\text{Maximize } f(x_1, x_2) = 144x_1 + 112.5x_2 \quad (A.180)$$

subject to the constraints

$$\begin{aligned} x_1 + x_2 &\leq 100 \\ 2.2x_1 + 2x_2 &\leq 210 \\ 3.9x_1 + 3x_2 &\leq 320 \\ x_1 \geq 0 \text{ and } x_2 &\geq 0 \end{aligned} \quad (A.181)$$

where
$$\hat{c}_1 = c_1 a_{41} - c_4 a_{31} - c_3 a_{21}$$
$$= (2.8)(125) - (40)(3.9) - (0.25)(200) \quad (A.182)$$
$$= 350 - 156 - 50 = \$144 / \text{acre-ft of corn}$$

$$\hat{c}_2 = c_2 a_{42} - c_4 a_{32} - c_3 a_{22}$$
$$= (2.7)(100) - 40(3) - (0.25)(150)$$
$$= 270 - 120 - 37.5 = \$112.5/\text{acre-ft of sorghum} \quad (A.183)$$

Definitions:

Solution: Any set of decision variables that satisfies all the constraints is a solution.

Feasible solution: Any solution that also satisfies the nonnegative restrictions (and the constraints) is a feasible solution.

Optimal feasible solution: Any feasible solution that optimizes (minimizes or maximizes) the objective function is an optimal solution.

Solving this problem graphically yields the following results (see Figure A.16).

Note that x_1^* denotes an optimal solution:

$$x_1^* = 22.2 \text{ acres of corn}$$
$$x_2^* = 77.8 \text{ acres of sorghum}$$
$$f(x_1^*, x_2^*) = (144)(22.2) + (112.5)(77.8) = \$11,950$$

The dual problem and its solution provide valuable insight into the system being studied and analyzed. As discussed in the Section A.1, a dual variable associated with a constraint (resource) represents a shadow price, that is, the marginal value added to the objective function by increasing the constrained resource by one unit of that resource. (See Section A.1 for a review of the primal and dual optimization problems.) The following text outlines the primal and dual formulation of the farmer's dilemma problem:

Primal Problem	Dual Problem
Maximize	Minimize
$f(x_1, x_2) = c_1 x_1 + c_2 x_2$	$h(y_1, y_2, y_3) = b_1 y_1 + b_2 y_2 + b_3 y_3$
Subject to the constraints	Subject to the constraints
$a_{11} x_1 + a_{12} x_2 \leq b_1$	$a_{11} y_1 + a_{21} y_2 + a_{31} y_3 \geq c_1$
$a_{21} x_1 + a_{22} x_2 \leq b_2$	$a_{12} y_1 + a_{22} y_2 + a_{32} y_3 \geq c_2$
$a_{31} x_1 + a_{32} x_2 \leq b_3$	
$x_1 \geq 0$ and $x_2 \geq 0$	$y_1 \geq 0, y_2 \geq 0,$ and $y_3 \geq 0$

where $y_1, y_2,$ and y_3 are the three dual decision variables corresponding to the three constraints of the primal problem and $h(y_1, y_2, y_3)$ is the dual objective function.

Note that:

- Maximizing $f(x_1, x_2)$ corresponds and is equal to minimizing $h(y_1, y_2, y_3)$.
- The cost coefficients c_1 and c_2 in the primal become the resource coefficients in the dual and vice versa.
- The resource coefficients in the primal, b_1, b_2, b_3 become the cost coefficients in the dual.
- The inequalities in the constraints reverse themselves.
- However, the nonnegativities remain for both primal and dual decision variables.
- Each technological coefficient, a_{ij}, in the primal problem becomes the a_{ji} in the dual.

To better illustrate the relationship between the primal and dual problems, consider the following problem.

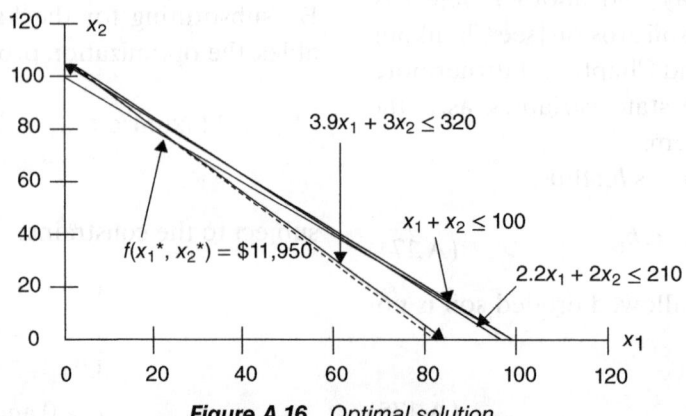

Figure A.16 Optimal solution.

Find a vector $\mathbf{x}^T = (x_1, x_2, \ldots, x_n)$ that maximizes the following linear function $f(x)$:

$$f(x) = c_1 x_1 + c_2 x_2 + \cdots + c_n x_n \quad (A.184)$$

or

$$f(\mathbf{x}) = \sum_{i=1}^{n} c_i x_i \quad (A.185)$$

subject to the restrictions

$$x_i \geq 0, \quad i = 1, 2, \ldots, n \quad (A.186)$$

and the linear constraints

$$\begin{aligned} a_{11} x_1 + a_{12} x_2 + \cdots + a_{1n} x_n &\leq b_1 \\ a_{21} x_1 + a_{22} x_2 + \cdots + a_{2n} x_n &\leq b_2 \\ &\vdots \\ a_{m1} x_1 + a_{m2} x_2 + \cdots + a_{mn} x_n &\leq b_m \end{aligned} \quad (A.187)$$

where a_{ij} are given constants for

$$i = 1, 2, \ldots, n; \quad j = 1, 2, \ldots, m$$

and $f(\mathbf{x})$ is the objective function. For a more detailed discussion on optimization, see Hillier and Lieberman [1990].

In matrix notation, the primal problem can be formulated as follows:

$$\max f(\mathbf{x}) = \mathbf{c}^T \mathbf{x} \quad (A.188)$$

subject to the constraints

$$\mathbf{x} \geq 0 \quad \text{and} \quad A\mathbf{x} \leq \mathbf{b} \quad (A.189)$$

where A represents the matrix of coefficients (see Section A.1).

Introduce the following transformation of variables:

Primal Variables		Dual Variables
x	↔	Y
n	↔	M
c	↔	B
A	↔	A^T
a_{ij}	↔	a_{ji}
≤	↔	≥
Max	↔	Min

where the vector $\mathbf{y} \geq 0$ is known as a vector of dual variables, shadow prices, or imputed prices and the superscript T denotes the transpose operation. Note that both $\mathbf{x} \geq 0$ and $\mathbf{y} \geq 0$ remain unchanged.

Dual problem

$$\min h(\mathbf{y}) = \mathbf{b}^T \mathbf{y} \quad (A.190)$$

subject to the constraints

$$\mathbf{y} \geq 0 \quad \text{and} \quad A^T \mathbf{y} \geq \mathbf{c} \quad (A.191)$$

The dual of the dual problem is the primal problem, and the optimal solution of the dual is equal to the optimal solution of the primal.

Primal Problem	Dual Problem
Maximize	Minimize
$f(x_1, x_2) = 144 x_1 + 112.5 x_2$	$h(y_1, y_2, y_3) = 100 y_1 + 210 y_2 + 320 y_3$
Subject to the constraints	Subject to the constraints
$x_1 + x_2 \leq 100$	$y_1 + 2.2 y_2 + 3.9 y_3 \geq 144$
$2.2 x_1 + 2 x_2 \leq 210$	$y_1 + 2 y_2 + 3 y_3 \geq 112.5$
$3.9 x_1 + 3 x_2 \leq 320$	
$x_1 \geq 0$ and $x_2 \geq 0$	$y_1 \geq 0, y_2 \geq 0,$ and $y_3 \geq 0$

From the graphical solution of the primal problem (see Figure A.16), we note that the optimal solution is

$$(x_1^*, x_2^*) = (22.2, 77.8) \quad (A.192)$$

This optimal solution is obtained at the intersection of the constraint on water availability $3.9 x_1 + 3 x_2 = 320$ (at the equality) with the constraint on land availability $x_1 + x_2 = 100$.

This leaves the soil erosion constraint to be nonbinding—that is, at the strictly inequality sign:

$$2.2(22.2) + 2(77.8) = 204.4 < 210$$
$$f(x_1, x_2) = 114 x_1 + 112.5 x_2$$
$$h(y_1, y_2, y_3) = 100 y_1 + 210 y_2 + 320 y_3$$

Equating $f(x_1, x_2)$ to $h(y_1, y_2, y_3)$ and focusing on the units of each element yields

$$[f(x_1, x_2)] = [\$] = [h(y_1, y_2, y_3)]$$

$$[h(y_1, y_2, y_3)] = [\text{acres}][y_1] + [\text{tons of top soil}][y_2]$$
$$+ [\text{acre-ft of water}][y3] = [\$]$$

Thus, the units of the dual variables are:

$[y_1] = \$/\text{acre of land}$
$[y_2] = \$/\text{ton of eroded top soil}$
$[y_3] = \$/\text{acre} - \text{ft of water}$

Note that each dual variable corresponds to a constraint in the primal and that all dual variables corresponding to nonbinding constraints of the primal problem are equal to zero at the optimal solution. Also, at the optimal solution, a strictly positive dual variable corresponds to a binding constraint:

$$y_1 + 2.2y_2 + 3.9y_3 = 144$$
$$y_1 + 2y_2 + 3y_3 > 112.5$$

We know, however, that the constraint corresponding to y_2^* is nonbinding. Thus

$$y_2^* = \$0/\text{ton of eroded soil}$$

Therefore,

$$y_1 + 3.9y_3 = 144$$
$$y_1 + 3y_3 = 112.5$$
$$y_1^* = 7.5, \quad y_3^* = 35$$

We could also have obtained this result by equating the values of the objective functions of the primal and the dual problems:

$$f(x_1^*, x_2^*) = h(y_1^*, y_2^*, y_3^*) = \$11{,}950 \quad (A.193)$$

and since

$$h(y_1^*, y_2^*, y_3^*) = (100y_1^* + 210y_2^* + 320y_3^*) \quad (A.194a)$$

then

$$11{,}950 = 100y_1 + 210(0) + 320y_3 \quad (A.194b)$$

Thus, given that the units of 11,950, 100, and 320 are dollars, acres of land, and acre-ft of water, respectively, then from dimensionality analysis, Equation A.194b yields

$$y_1^* = \$7.5/\text{acre}, \quad y_2^* = \$35.00/\text{acre-ft}$$

A.4 STANDARD NORMAL PROBABILITY TABLE

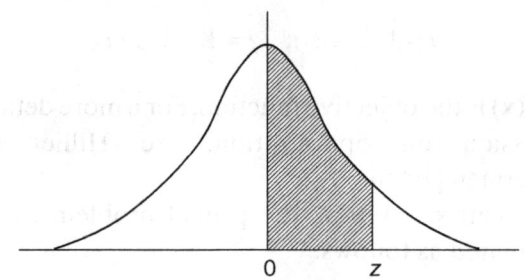

Figure A.17 Standard normal probability density function.

TABLE A.13 Standard Normal Probability Table

z	0.00	0.01	0.02	0.03	0.04	0.05	0.06	0.07	0.08	0.09
0.0	0.0000	0.0040	0.0080	0.0120	0.0160	0.0199	0.0239	0.0279	0.0319	0.0359
0.1	0.0398	0.0438	0.0478	0.0517	0.0557	0.0596	0.0636	0.0675	0.0714	0.0753
0.2	0.0793	0.0832	0.0871	0.0910	0.0948	0.0987	0.1026	0.1064	0.1103	0.1141
0.3	0.1179	0.1217	0.1255	0.1293	0.1331	0.1368	0.1406	0.1443	0.1480	0.1517
0.4	0.1554	0.1591	0.1628	0.1664	0.1700	0.1736	0.1772	0.1808	0.1844	0.1879
0.5	0.1915	0.1950	0.1985	0.2019	0.2054	0.2088	0.2123	0.2157	0.2190	0.2224
0.6	0.2257	0.2291	0.2324	0.2357	0.2389	0.2422	0.2454	0.2486	0.2517	0.2549
0.7	0.2580	0.2611	0.2642	0.2673	0.2704	0.2734	0.2764	0.2794	0.2823	0.2852
0.8	0.2881	0.2910	0.2939	0.2967	0.2995	0.3023	0.3051	0.3078	0.3106	0.3133
0.9	0.3159	0.3186	0.3212	0.3238	0.3264	0.3289	0.3315	0.3340	0.3365	0.3389
1.0	0.3413	0.3438	0.3461	0.3485	0.3508	0.3531	0.3554	0.3577	0.3599	0.3621
1.1	0.3643	0.3665	0.3686	0.3708	0.3729	0.3749	0.3770	0.3790	0.3810	0.3830
1.2	0.3849	0.3869	0.3888	0.3907	0.3925	0.3944	0.3962	0.3980	0.3997	0.4015

TABLE A.13 Continued

z	0.00	0.01	0.02	0.03	0.04	0.05	0.06	0.07	0.08	0.09
1.3	0.4032	0.4049	0.4066	0.4082	0.4099	0.4115	0.4131	0.4147	0.4162	0.4177
1.4	0.4192	0.4207	0.4222	0.4236	0.4251	0.4265	0.4279	0.4292	0.4306	0.4319
1.5	0.4332	0.4345	0.4357	0.4370	0.4382	0.4394	0.4406	0.4418	0.4429	0.4441
1.6	0.4452	0.4463	0.4474	0.4484	0.4495	0.4505	0.4515	0.4525	0.4535	0.4545
1.7	0.4554	0.4564	0.4573	0.4582	0.4591	0.4599	0.4608	0.4616	0.4625	0.4633
1.8	0.4641	0.4649	0.4656	0.4664	0.4671	0.4678	0.4686	0.4693	0.4699	0.4706
1.9	0.4713	0.4719	0.4726	0.4732	0.4738	0.4744	0.4750	0.4756	0.4761	0.4767
2.0	0.4772	0.4778	0.4783	0.4788	0.4793	0.4798	0.4803	0.4808	0.4812	0.4817
2.1	0.4821	0.4826	0.4830	0.4834	0.4838	0.4842	0.4846	0.4850	0.4854	0.4857
2.2	0.4861	0.4864	0.4868	0.4871	0.4875	0.4878	0.4881	0.4884	0.4887	0.4890
2.3	0.4893	0.4896	0.4898	0.4901	0.4904	0.4906	0.4909	0.4911	0.4913	0.4916
2.4	0.4918	0.4920	0.4922	0.4925	0.4927	0.4929	0.4931	0.4932	0.4934	0.4936
2.5	0.4938	0.4940	0.4941	0.4943	0.4945	0.4946	0.4948	0.4949	0.4951	0.4952
2.6	0.4953	0.4955	0.4956	0.4957	0.4959	0.4960	0.4961	0.4962	0.4963	0.4964
2.7	0.4965	0.4966	0.4967	0.4968	0.4969	0.4970	0.4971	0.4972	0.4973	0.4974
2.8	0.4974	0.4975	0.4076	0.4077	0.4977	0.4978	0.4979	0.4979	0.4980	0.4981
2.9	0.4981	0.4982	0.4982	0.4983	0.4984	0.4984	0.4985	0.4985	0.4986	0.4986
3.0	0.4987	0.4987	0.4987	0.4988	0.4988	0.4989	0.4989	0.4989	0.4990	0.4990
3.1	0.4990	0.4991	0.4991	0.4991	0.4992	0.4992	0.4992	0.4992	0.4993	0.4993
3.2	0.4993	0.4993	0.4994	0.4994	0.4994	0.4994	0.4994	0.4995	0.4995	0.4995
3.3	0.4995	0.4995	0.4995	0.4996	0.4996	0.4996	0.4996	0.4996	0.4996	0.4997
3.4	0.4997	0.4997	0.4997	0.4997	0.4997	0.4997	0.4997	0.4997	0.4997	0.4998
3.5	0.4998	0.4998	0.4998	0.4998	0.4998	0.4998	0.4998	0.4998	0.4998	0.4998
3.6	0.4998	0.4998	0.4999	0.4999	0.4999	0.4999	0.4999	0.4999	0.4999	0.4999
3.7	0.4999	0.4999	0.4999	0.4999	0.4999	0.4999	0.4999	0.4999	0.4999	0.4999
3.8	0.4999	0.4999	0.4999	0.4999	0.4999	0.4999	0.4999	0.4999	0.4999	0.4999
3.9	0.5000	0.5000	0.5000	0.5000	0.5000	0.5000	0.5000	0.5000	0.5000	0.5000

REFERENCES

Anderberg, M.R., 1973, *Cluster Analysis with Applications*, Academic Press, New York.

Bellman, R.E., and S.E. Dreyfus, 1962, *Applied Dynamic Programming*, Princeton University Press, Princeton, NJ.

Chankong, V., and Y.Y. Haimes, 1983, *Multiobjective Decision making: Theory and Methodology*, North Holland, New York.

Chankong, V., and Y.Y. Haimes, 2008, *Multiobjective Decisionmaking: Theory and Methodology*, Dover, New York.

Chankong, V., Y.Y. Haimes, H.S. Rosenkranz, and J. Pet-Edwards, 1985, The carcinogenicity prediction and battery selection (CPBS) method: a Bayesian approach, *Mutation Research* **153**: 135–166.

Dantzig, G.B., 1963, *Linear Programming and Extensions*, Princeton University Press, Princeton, NJ.

Dorfman, R., P.A. Samuelson, and R.M. Solow, 1958, *Linear Programming and Economic Analysis*, McGraw-Hill, New York.

Haimes, Y.Y., 1977, *Hierarchical Analyses of Water Resource Systems: Modeling and Optimization of Large-Scale Systems*, McGraw-Hill, New York.

Haimes, Y.Y., V. Chankong, J. Pet-Edwards, and H.S. Rosenkranz, 1987, Carcinogenicity prediction and battery selection procedure: an in depth analysis of cyclamate and its major metabolite cyclohexylamine, *Molecular Toxicology* **1**: 49–60.

Hillier, S.F., and G.J. Lieberman, 1990, *Introduction to Mathematical Programming*, Fifth edition, McGraw-Hill, New York.

Hillier, S.F., and G.J. Lieberman, 1995, *Introduction to Operations Research*, Fifth edition, Holden-Day, San Francisco.

Intrilligator, M.D., 1971, *Mathematical Optimization and Economic Theory*, Prentice Hall, Englewood Cliffs, NJ.

Kleindorfer, P.R., and H.C. Kunreuther (Eds.), 1987, *Insuring and Managing Hazardous Risks: From Seveso to Bhopal and Beyond*, Springer-Verlag, New York.

Kuhn, H.W., and A.W. Tucker, 1951, Nonlinear programming, *Proceedings of Second Berkeley Symposium on*

Mathematical Statistics and Probability, 1950, University of California Press, Berkeley, CA, pp. 481–492.

Lasdon, L.S., 1968, Duality and decomposition in mathematical programming, *IEEE Transactions on Systems Science and Cybernetics* SSC-**4**(2): 86–100.

Lasdon, L.S., 1970, *Optimization Theory for Large Systems*, Macmillan, New York.

Leemis, Lawrence A., 1995, *Reliability: Probabilistic Models and Statistical Methods*, Prentice-Hall, New Jersey.

Pet-Edwards, J., 1986, *Selection and Interpretation of Conditionally Dependent Tests for Binary Predictions*, Ph.D. dissertation, Systems Engineering Department, Case Western Reserve University, Cleveland, OH.

Pet-Edwards, J., H.S. Rosenkranz, V. Chankong, and Y.Y. Haimes, 1985a, Cluster analysis in predicting the carcinogenicity of chemicals using short-term assays, *Mutation Research* **153**: 167–185.

Pet-Edwards, J., V. Chankong, H.S. Rosenkranz, and Y.Y. Haimes, 1985b, Application of the carcinogenicity prediction and battery selection (CPBS) method to the Gene-Tox data base, *Mutation Research* **153**: 187–200.

Pet-Edwards, J., Y.Y. Haimes, V. Chankong, H.S. Rosenkranz, and F.K. Ennever, 1989, *Risk Assessment and Decision Making Using Test Results: The Carcinogenicity Prediction and Battery Selection Approach*, Plenum Press, New York.

Pratt, J., H. Raiffa, and R. Schlaifer, 1995, *Introduction to Statistical Decision Theory*, MIT Press, Cambridge, MA.

Rosenkranz, H.S., G. Klopman, V. Chankong, J. Pet-Edwards, and Y.Y. Haimes, 1984a, Prediction of environmental carcinogens: a strategy of the mid 1980's, *Environmental Mutagenesis* **6**: 231–258.

Rosenkranz, H.S., J. Pet-Edwards, V. Chankong, and Y.Y. Haimes, 1984b, Assembling a battery of assays to predict carcinogenicity: a case study, *Mutation Research* **141**: 65–68.

Wismer, D.A. (Ed.), 1971, *Optimization Methods for Large-Scale Systems*, McGraw-Hill, New York.

Author Index

Adams, A., 259
Adelman, L., 7
Agrawal, A., 79
Amaral, L., 475
Anderberg, M., 656
Anderson, C. (et al.), 34, 224, 543
Anderson, P., 543, 545, 548, 551
Ang, A., 333, 335, 358, 383, 581
Apostolakis, G., 60, 187, 382, 389, 391, 477, 608
Armstrong, J., 502
Arquilla, J., 80, 496, 500
Arrow, K., 259
Asbeck, E., 24, 30, 32, 62, 232, 234, 237, 304, 306, 329, 537, 609
Ashkenas, R. (et al.), 442, 495
Assuncao, R., 485
Athans, M. (et al.), 318

Barker, K., 201, 533, 535
Baron, J. (et al.), 213
Bar-Yam, Y., 475, 595
Bauman, E., 485
Bell, G., 461
Bellman, R., 633, 637
Belton, V., 160
Benbow, D. (et al.), 409
Bernstein, P., 14

Bertalanffy, L., 6, 141
Bertsekas, D., 462
Bezdek, R., 545
Biemer, S., 380, 474
Bier, V., 201
Bier, V. (et al.), 53, 62, 64, 233
Birolini, A., 389
Blair, P., 512, 518, 526, 527, 546, 557, 572
Blanc, M., 527
Blanchard, B., 7, 9
Blauberg, I. (et al.), 7, 74, 475, 495, 595
Blum, B., 92
Boehm, B., 24, 92, 459, 461, 463, 465
Boulding, K., 7
Brooks, C., 12
Brucker, S. (et al.), 522
Buede, D., 7, 559
Bullard, C., 533
Burke, T. (et al.), 14
Burton, I. (et al.), 559
Butler, R., 391

Call, H., 305
Cardin, M. (et al.), 603
Carlin, B., 486
Carr, M. (et al.), 93, 454
Cartin, T., 587

Risk Modeling, Assessment, and Management, Fourth Edition. Yacov Y. Haimes.
© 2016 John Wiley & Sons, Inc. Published 2016 by John Wiley & Sons, Inc.

AUTHOR INDEX

Castillo, E., 333, 581
Castro, M., 485
Chang, M., 475
Chankong, V., 19, 21, 22, 29, 54, 155, 160, 162, 167, 168, 172, 231, 259, 297, 298, 304, 316, 331, 381, 462, 467, 601, 661
Chankong, V. (et al.), 655
Chapman, R., 439
Chen, C., 378, 476
Chittister, C., 24, 252, 443, 454, 455
Chu, C., 70
Cohen, I., 6
Collette, Y., 160
Collins, J., 16
Converse, A., 512
Covaliu, Z., 305
Covey, S., 5, 9, 14, 605
Cox, D., 391
Cramer, H., 332
Crowther, K., 34, 513, 525, 533
Crowther, K. (et al.), 34, 558
Cuppan, C., 7, 474, 475, 595

Dantzig, G., 485, 626, 627, 632
Davenport, T., 12, 16, 441, 442, 605
Davis, M., 502
Dedeurwaerdere, A., 576
DeLaurentis, D., 594
Deming, W., 13
deNeufville, R., 603
Dicdican, R., 31, 304
Diehl, M., 305
Dillon, R., 306
Dombroski, M., 223
Dombroski, M.(et al.), 98, 99, 102, 215, 220, 496
Dorfman, R. (et al.), 629
Dorofee, A. (et al.), 440, 443–445
Douglas, M., 22
Dracup, J., 43
Dreyfus, S., 633, 637
Drucker, P., 17
Duderstadt, J. (et al.), 64, 376
Durango-Cohen, P., 70

Ebeling, C., 389
Edwards, W., 259, 305, 306
Ehrgott, M., 160, 381
Eisner, H., 475, 595
Embrechts, P. (et al.), 512
Everett, H., 165

Eykhoff, P., 473
Ezell, B., 252
Ezell, B. (et al.), 252, 253

Fabrycky, W., 7, 9
Fang, L. (et al.), 7
Farrell, P. (et al.), 485
Feinberg, H., 22
Ferson, S., 485
Feynman, R. (et al.), 6, 35
Fiering, M., 58, 217
Fineberg, H., 22
Finkel, A., 29, 186–189, 191
Fischer, R., 334, 383
Fischetti, M., 560
Fischoff, B.(et al.), 14, 22, 26
Frey, D., 603
Frohwein, H., 282, 306
Frohwein, H. (et al.), 282
Funderburk, F., 475

Galambos, J., 333
Galea, S. (et al.), 516
Gardner, H., 12
Garrick, B., 17, 19, 26, 76, 140, 213, 373–376, 496, 536, 558, 577, 606
Garvin, D., 503
Gass, S., 202
Geckil, I., 543, 545, 548, 551
Gelman, A. (et al.), 486
Geoffrion, A., 166, 167
Gharajedaghi, J., 7
Gheorghe, A., 7, 475
Ghosh, M., 486
Gnedenko, B., 51, 334, 383
Golding, D., 14
Goleman, D., 12, 16
Gomide, F., 31, 297, 298, 306, 462, 584
Gordon, W., 7, 8, 74, 475, 496
Grant, P. (et al.), 51
Graupe, D., 473
Griffin, J., 64, 512
Grygier, J., 240
Gumbel, E., 333–337, 348, 384
Guo, J. (et al.), 518

Haas, J., 139, 558
Haimes, Y., 7–9, 12, 14, 15, 17, 19, 21, 22, 24, 26, 28–32, 34, 42, 44, 50, 51, 53, 54, 56–60, 62, 64, 69–71, 73, 74, 76, 79, 80, 83, 85, 109, 112, 139–142, 155, 160, 162, 167, 168, 172, 175, 180, 188, 201, 213, 231, 232, 234, 237, 238, 252, 259,

290, 297, 298, 301, 304–307, 314, 316, 318, 325, 329–332,
340, 342, 345, 350, 373–375, 377, 379–381, 383–385,
439, 441, 443, 449, 454, 455, 461, 462, 467, 473–476,
478–480, 483–487, 493, 494, 496, 498, 504, 512–514, 517,
523, 524, 533, 536, 537, 545, 551, 553, 554, 556, 558,
559, 576, 577, 582, 584, 593–595, 597, 599–601, 603–606,
609, 610, 617
Haimes, Y. (et al.), 8, 9, 12, 14, 19, 21, 22, 24, 29–32, 34, 57,
 58, 71, 80, 83, 94, 104, 158, 161, 181, 186, 201, 202, 211,
 212, 217, 233, 237, 259, 261, 263, 264, 266, 269, 275–277,
 280, 306, 381, 423, 427, 446, 478, 479, 485, 498, 512, 517,
 545, 557, 610, 655
Hall, A., 7, 8, 74, 475, 479, 480, 496
Hall, W., 21, 24, 29, 30, 43, 180, 188, 232,
 331, 533
Hammonds, J., 186
Harrington, J., 475
Hatch, T., 12
Hatley, D. (et al.), 7
Hegel, G., 6
Henley, E., 389, 391, 402
Henrion, M., 186–190, 382, 608
Hertwich, E. (et al.), 64
Hewings, G., 535
Hillier, S., 627, 632
Hirshleifer, J., 43, 186
Hobson, K., 144
Hoffman, F., 186
Hogg, R., 190
Holling, C., 57
Hollnagel, E. (et al.), 57
Horowitz, B., 79, 80, 109, 112, 142, 480, 493, 498,
 504, 512, 513, 517
Howard, R., 190, 305
Hoyland, A., 391
Hu, J., 603
Humphrey, W., 454
Hutchins, R., 5

Imai, M., 13, 18
Intriligator, M., 297, 512, 599, 644
Isard, W., 512
Isard, W. (et al.), 526

Jackson, D., 24
Jamshidi, M., 475, 476, 595
Janda, R. (et al.), 576, 581, 584
Jiang, P., 34, 449, 513, 514, 517, 523, 524
Johnson, B., 92, 391, 402
Johnson, C., 51, 475
Jugulum, R., 603

Kahneman, D., 190
Kaplan, R., 224
Kaplan, S., 17, 19, 26, 44, 52, 53, 76, 77, 140, 213, 373–376,
 496, 536, 558, 577, 606
Kaplan, S. (et al.), 14, 53, 76, 77, 375, 446, 450, 496,
 497, 536
Karlsson, P., 237, 238, 331–333, 340, 342, 344–346,
 348, 353
Katzenbach, J., 443, 605
Keeney, R., 54, 305, 381
Kelton, W., 127
Kendall, R., 24
Kirkwood, C., 305
Kleindorfer, P., 655
Kloss, W., 488
Knabb, R. (et al.), 560
Kossiakoff, A., 7
Krause, U., 512
Krimsky, S., 14
Kuhbach, P., 515
Kuhn, H., 79, 166, 197, 381, 499, 502, 642
Kujawski, E., 305
Kumamoto, H., 389, 391, 402
Kunreuther, H., 14, 43, 231, 655

Lahr, M., 512
Lam, P., 439
Lamanna, C., 140
Lambert, J., 306, 480, 516
Lambert, J. (et al.), 233, 282, 335, 338, 496
Lamm, G., 223
Larsen, R., 22
Lasdon, L., 479, 485, 642, 644
Law, A., 127
Leach, M., 31, 298, 301, 306, 307, 314,
 330, 584
Lederer, A., 461
Lee, K., 512
Lee, L. (et al.), 64
Leemis, L., 233, 656
Leondes, C., 7, 475
Leontief, W., 34, 141, 512, 545, 584
Leung, F. (et al.), 551, 575
Leung, M. (et al.), 224
Lewis, H., 14
Li, D., 30, 70, 180, 199, 201, 203, 204, 207, 208,
 290, 318, 325, 385, 467, 533, 594, 599, 600
Li, W., 140
Lian, C., 34, 545
Liao, S., 57, 545
Lieberman, G., 627, 632

Liebowitz, S., 598
Liew, C., 512, 573
Limnios, N., 389
Lin, S., 201
Lindell, M., 558
Ling, C., 186, 188
Lloyd, S., 475
Lloyd, T., 475
Longstaff, T., 12
Lootsma, F., 306
Louis, T., 486
Lowrance, W., 3, 4, 7, 19, 43, 51, 54, 60, 139, 140, 192, 230, 231, 325, 373–375, 378, 381, 392, 440, 503, 577
Luce, D., 502
Luce, R., 259
Lyneis, J., 439

Macko, D., 7, 70, 80, 475, 479
Magee, J., 305, 306
Mahoney, B., 223
Maier, M., 7, 380, 474, 475, 595
Major, J., 502
Marcial, S. (et al.), 580
Margolis, S., 598
Markowitz, H., 305
Martensen, A., 391
Matalas, N., 58, 217
Matheson, J., 305
McCormick, N., 375
McGill, W. (et al.), 64
McQuaid, J., 4
Meeden, G., 486
Mercado, R. (et al.), 581, 582
Mesarović, M., 7, 475
Mesarović, M. (et al.), 7, 475
Meyer, W., 28, 132, 136, 137
Mileti, D., 558
Miller, R., 512, 518, 526, 527, 546, 557, 572
Miller, R. (et al.), 512, 522
Miller, W., 305
Millet, I., 305
Mitsiopoulos, J., 350
Monahan, J., 80, 500
Monahan, J. (et al.), 80, 500
Montgomery, D., 587
Morgan, G., 186–190, 382, 608
Morgan, M. (et al.), 212, 213
Morgenstern, O., 79, 80, 259, 305, 499, 502
Moriarty, B., 216
Morrall, J., 22, 23

Nainis, W., 512
Nash, J., 502
Nauta, M. (et al.), 64
Nordean, D. (et al.), 450
Norris, F. (et al.), 516
Norton, D., 204

Okuyama, Y. (et al.), 551
Oliver, R., 128, 305, 486
Olsen, J. (et al.), 512, 580
Oosterhaven, J., 575
Osbourne, M., 502
Oster, C., 502
Ottino, J., 475

Page, S., 475
Pande, P. (et al.), 587, 589
Pannullo, J., 391, 404
Pannullo, J. (et al.), 391, 404
Pareto, V., 381, 593
Park, J. (et al.), 393
Paté-Cornell, M., 18, 60, 187, 377, 382, 476, 477, 533, 608
Patterson, C., 516
Pejtersen, A., 7
Pennock, M., 439
Percoco, M., 533, 535
Percoco, M. (et al.), 535
Perrow, C., 62, 64, 376
Pet-Edwards, J., 655
Pet-Edwards, J. (et al.), 655, 656
Peterson, D., 158
Petrakian, R. (et al.), 240, 242
Phimister, J. (et al.), 53, 64, 376
Pierson, T. (et al.), 575, 580, 584, 585
Planting, M., 515
Porter, A., 296, 297
Porter, M., 140
Post, D. (et al.), 24
Prasad, J., 461
Pratt, J. (et al.), 259, 656
Proops, J., 512
Prusak, L., 12, 16, 441, 442, 605
Punongbayan, R. (et al.), 576

Quandt, R., 533

Raiffa, H., 31, 54, 118, 259, 281, 305, 331, 381, 502
Ramos, C., 527
Rao, J., 486

Rao, S., 391
Rasmussen, J. (et al.), 7
Rausand, M., 391
Rechtin, E., 7
Regalado, M., 580
Reichelt, K., 439
Reid, R., 175
Reiss, R., 64
Rifkin, S., 465
Riley, J., 43, 186
Robertson, C., 566
Roland, H., 216
Ronfeldt, D., 80, 496, 500
Rose, A., 57, 545
Rosenkranz, H. (et al.), 655
Ross, S., 464
Rossini, F., 296, 297
Rouse, W., 7, 475
Rowe, W., 186, 189
Rubinstein, A., 502
Runyon, R., 231
Russell, K. (et al.), 391
Ryan, P., 190

Saaty, T., 123, 202, 212, 305, 448
Sage, A., 7–9, 22, 73, 74, 380, 439, 441, 463, 474, 475, 495, 496, 559, 595, 604
Santos, J., 34, 449, 513, 517, 546, 554, 569
Santos, J. (et al.), 282
Savage, L., 259
Scheffler, J., 485
Schlaifer, R., 259, 331
Schleifstein, M., 4
Schneiter, C., 8, 9, 14
Schneiter, C. (et al.), 70, 393
Scholtes, S., 603
Schoof, R., 439, 443, 454, 461, 462
Schoof, R. (et al.), 90–92, 443, 454, 461
Sebald, A., 533
Senge, P., 8
Shachter, R., 305
Shafer, G., 64
Shalizi, C., 475
Shapley, L., 502
Shenhar, A., 475
Shenoy, P., 305
Shubik, M., 259
Siarry, P., 160
Simon, H., 594, 599
Singh, M., 8, 74, 479, 496
Slovic, P., 7, 14, 43, 54, 55, 62, 80, 231, 383

Slovic, P. (et al.), 383, 500
Smith, D., 443, 605
Smith, H., 459
Smith, J., 128, 305, 486
Sokal, R., 212
Sonis, M., 535
Spedden, S., 190
Spencer, B., 306
Starr, C., 180
Stedinger, J., 240
Stern, P., 22
Stevens, B., 512
Stewart, T., 160
Storey, N., 391
Suppes, P., 259
Susser, E. (et al.), 516
Sweet, W., 7, 454

Taleb, N., 55, 64, 201, 229, 230, 376
Tang, W., 333, 335, 358, 383, 581
Tanis, E., 190
Tarvainen, K., 69, 479
Taylor, A., 186
Teigen, K., 186
Tierney, K. (et al.), 558
Tippett, L., 334, 383
Tivnan, B., 475
Toffler, A., 3, 26
Tsang, J. (et al.), 516
Tucker, A., 166, 197, 381, 642
Tulsiani, V., 129, 391, 400
Tulsiani, V. (et al.), 32
Tungol, N., 580
Turkington, D., 535
Tversky, A., 190

Vemuri, V., 175
von Neumann, J., 79, 80, 259, 305, 499, 502
vonWinterfeldt, D., 305, 306

Warfield, J., 8, 74, 496
Webler, T. (et al.), 212
Wedley, W., 305
Wendling, R., 545
Wernick, I., 14
Westrum, R., 57
White, G., 139, 558
Whitman, A., 6
Wiener, N., 6, 15
Wierzbicki, A., 203
Wiese, I. (et al.), 253

Williams, T., 447
Wismer, D., 485, 642, 644
Wolfe, P., 485
Woods, D., 57
Wulf, W. (et al.), 500

Xu, L., 140

Yan, Z., 71, 479, 485, 486
Yan, Z. (et al.), 71

Zadeh, L., 190, 202
Zeleny, M., 7
Zigler, B., 7, 74, 475, 496
Zimmerman, R., 558

Subject Index

Absorption law, 397, 399
Acquisition
   conditional expected value, 250–252, 401, 403–406, 409
   expected value, 229–230, 250–251
   software, 440, 454–456, 458–460
   unconditional expected value, 250, 252
Adaptive two-player HHM game, 493
   Blue team, 499
   Red team, 499
Adverse effects, 4, 7, 16, 31, 230, 393, 440
Agricultural–social model, 483
Aircraft development example (project management), 445, 450–454
Airplane acquisition problem (fractile method), 90–92, 250–252
Analysis, levels of societal, 500
   doctrinal, 500
   narrative, 500
   organizational, 500
   social, 500
   technological, 500
Analytic hierarchy process (AHP), 123, 212, 306
AND gate (Boolean algebra), 391–392, 396, 398–403
Anticipatory failure determination (AFD), 76, 77, 446
Aristotle, systems engineering, 6
Artificial intelligence (AI), 156
Assays (CPB), 371–387
Associative law, 396

Automated Highway System (AHS), 103–108
   engineered systems, 104, 105
   failure, 104–108
   hierarchical holographic model (HHM), 105–108

Bands of indifference, 162, 163
Bank problem (uncertainty), 192–196
Bayes' formula, 265, 286, 288, 376
   carcinogenic predictivity, 32
   decision tree and, 118
   multiobjective decision-tree analysis (MODT), 265–268
   theorem, 494
Bayesian analysis, terrorism and, 498–499
Bellman's principle of optimality, 633, 636, 637
Benefit–cost concept, 180, 232
Bertalanffy, systems engineering, 6, 7
Boolean (logic) algebra, 395–402
   AND gate, 391, 396, 398–403
   distribution analyzer and risk evaluator (DARE), 401–405
   fault trees, 395–399
   OR gate, 396, 399–403, 408
   rules, 398
Boulding, systems engineering, 7
Brainstorming
   hierarchical holographic modeling (HHM), 53, 99
   influence diagrams, 128–132
Bureau of Economic Analysis (BEA), 518–519

Risk Modeling, Assessment, and Management, Fourth Edition. Yacov Y. Haimes.
© 2016 John Wiley & Sons, Inc. Published 2016 by John Wiley & Sons, Inc.

SUBJECT INDEX

Calculus, optimization and, 618–620, 622
   dynamic systems, 384, 619
Canonical transportation problem (linear programming), 632–633
Capability maturity model (CMM), 454, 455
Car trouble problem (fault tree), 418–420
Cardinal scale, 413
Case studies, 543
   extreme events and 6-sigma capability, 587–593
   Hurricane Katrina, 558–569
   management of lahar flow threat, 575–587
   Northeast blackout, 543–593
   September 11, 2001, 569–575
Catastrophic events
   damage, 15, 24
   decision analysis and, 115
   partitioned multiobjective risk method (PMRM), 346–347
   risk of, 229, 230, 232, 233, 259
Cauchy–Schwarz inequality, 200
Chebyschev inequality, 343
Chi-square test, 330
Classical game theory, 502
COCOMO *see* Constructive cost model (COCOMO)
Commercial off-the-shelf (COTS), 461, 503
Commutative law, 397
Complementation law, 397, 398
Complex systems, 379, 474
Computer manufacturing problem (decision analysis), 149–150
Conceptual framework, 372
Conditional expected loss, 275
Conditional expected value
   crutches problem (uncertainty), 116–122
   extreme events, 230–233, 259
   functions, 234, 263
   log-normal distribution, 651–653
   multiobjective decision-tree analysis (MODT), 274–275, 376–377
   multiobjective multistage impact analysis method (MMIAM), 297–299
   multiobjective risk-impact analysis method (MRIAM), 298
   partitioned multiobjective risk method (PMRM), 297, 301
   project risk management, 444, 455
   triangular distribution, 653–655
ε-constraint approach, 160–161, 167, 298, 315, 431, 466, 556
Constraints, 8, 9, 26, 29, 43, 47, 50, 150
Constructive cost model (COCOMO), 93, 459, 465–466

Continuous risk management, 440, 441
Coordinated hierarchical Bayesian model, 485
Cost function, 464
Cost–benefit (C/B) models, 21, 295, 544, 557, 558, 576–578
COTS *see* Commercial off-the-shelf (COTS)
Covey's *Seven Habits of Highly Effective People,* 5–15
   abundance mentality, 11
   circle of concern, 10
   circle of influence, 10
   iteration, 12
   kaizen, 12
   knowledge management, 12
   state variables, 11
   systems approach, 12–13
   time management, 11
   trade-offs, 12
Cramer's method, 359, 362, 364–366
Critical path method (CPM), 11
Criticality analysis (FMEA, FMECA), 411–413
Crutches problem, 116–122
Cumulative distribution function (cdf), 124, 230, 248, 367, 401–402, 404–405, 409
Cyber attack
   SCADA, 252, 511
   on water utility (extreme events), 252–257
Cyber infrastructure, 379
Cybernetics, 6
Cyberterrorism, 511

Dam failure problem (PMRM), 240–243
   probable maximum flood (PMF), 241
Data collection, 374, 386
Database management system (DBMS), 461
Data-into-intelligence, 45–46
de Morgan's theorem, 396, 397
Decentralized approach, 485
Decision analysis
   channel reliability problem, 129–131
   crutches problem, 116–118
   decision matrix, 122–123
   decision rules under uncertainty, 116–118
   decision trees, 118–122
      chance node, 119
      consequences, 119–120
      crutches problem, 116–118
      decision node, 119–120
   example problems, 144–152
      computer manufacturing, 149–150
      deicing, 147–149
      dingo population, 150–152

expected monetary value, 146–147
    testing, 145–148
expected value of opportunity loss, 119–122
expected value of outcome, 119
fractile method, 110–113, 124–127
Hurwitz
    model, 150
    rule, 116–118
influence diagrams, 128–132
    channel reliability problem, 129–132
    variable impact, 132
Leslie matrix, 135–138
optimistic rule, 116
pessimistic rule, 116
population dynamics, 132–138
    example problems, 132–135
    exponential model, 132
    macro population model, 132–133
    Malthusian parameter, 132
    micro population model, Leslie, 135–138
triangular distribution, 127–128
Decision matrix, 122–123
Decision rules under uncertainty
    conditional expected value of risk, 115
    Hurwitz rule, 116–118, 145, 150
    optimistic rule, 116
    pessimistic rule, 116
Decision tree *see also* Multiobjective decision tree analysis (MODT)
    Bayes' theorem, 118
    chance node, 119
    consequences, 119
    crutches problem, 119–122
    decision node, 119
    defined, 29
    differences, SODT *vs.* MODT, 279–281
    expected opportunity loss (EOL), 119–122
    fold back, 149, 271, 273, 274, 286–287
    methods, 73
    most likely value (MLV), 122–123
    multiple objectives, 50, 259–260
    representation (illustration), 145, 147
    single objective (SODT), 279
Decision uncertainty
    defined, 191–192
    risk measurement, 191, 195–196
    social cost of risk, 192, 195–196
    social values, quantification of, 192, 195–196
Decision variables
    definition, 47, 48
    hierarchical holographic modeling (HHM), 53

    modeling and, 45, 47 48, 56–57
    multiobjective trade-off analysis, 169, 170, 172, 173, 175, 177, 224
    uncertainty and, 180
Decisionmaking
    extreme events, quantification, 354–356
    major categories of, 179
    mathematical models, 30
    multiobjective decision tree (MODT), 259
    process, 372
    risk-based, 3
Decision-support systems (DSS), 156
Decomposition
    hierarchical holographic modeling (HHM), 53, 70, 78, 479, 485
    multilevel approach and, 619
Deicing problem (decision analysis), 147–149
Desalination plant problem (Lagrange), 623–624
Deterministic models
    linear programming, 56
    probabilistic, and, 42, 619
    uncertainty and, 180
Diet problem (Lagrange, linear programming), 629–630
Diffusion equation, 619
Dingo population problem (decision analysis), 150–152
Direct search, 619
Discrete (integer) programming, 620
Discrete-time dynamic system, 449
Dissolved oxygen problem (PMRM), 358–359
Distributed parameter, 42, 619
Distribution analyzer and risk evaluator (DARE)
    Boolean algebra, 401–405
    conditional expectation, 400
    conditional expected value, 401
    conditional extreme expectation, 401
    extreme events, risks of, 391
    failure rate distributions, 419
    fault trees and, generally, 32, 400–403
    flowcharts of, 403
    Monte Carlo simulation, 400
    partitioning, 401
    time-to-failure distributions, 401–403
    Weibull distribution, 400
Distributive law, 396, 397
Dual function in nonlinear programming, 644–645
Dual in linear programming, 627–628
Dual problems, 53–56
Duality in linear programming, 627–630
Dynamic model, 42, 619
Dynamic multiobjective decisionmaking, 598

Dynamic programming
  applications of, generally, 633–642
  Bellman's principle of optimality, 636
  inventory problem, 639–642
  model formulation example, 639–640
  network example, 633–636
  water inventory problem, 639–642
Dynamic Roadmap, 605
  first principle, 606
  second principle, 606
  third principle, 608
  fourth principle, 608
  fifth principle, 609
  sixth principle, 609
  seventh principle, 610
Dynamic systems, 384

Emergent behavior, 475
Emergent forced changes, 64, 374, 376
Engineering-based systems, risk characteristics, 25–26
Envelope approach, 385
Environmental health and safety (PMRM), 248–250
Environmental trade-off, 43
Equality constraint problem, 620–624
Errors, measurement, 43
Essential state, 329
Event trees, 76, 77, 411, 414
Events, 44
Evolutionary development, 475
Evolving base, 595, 600
Exogenous variables
  generally, 8, 32
  hierarchical holographic modeling (HHM), 76
  modeling and, 48–49, 56
Expected monetary value (EMV), 145–147
Expected opportunity loss, 119–121
Expected value
  adverse consequences, of, 61
  decision-tree analysis, 285
  denoted, 236
  fallacy of, 32, 232–233
  log-normal distribution, 650–651
  opportunity loss, 119–121
  outcome, 119
  partitioned multiobjective risk method (PMRM), 55, 233–236, 383
  probability density function (pdf), 651
  project risk management, 445
  random variable, 651
  risk see Expected value of risk
  triangular distribution, 653–655
Expected value of experimentation (EVE), 262–263

Expected value of risk
  extreme events, 230–231
  extreme failures, 234
  failure mode effects analysis (FMEA, FMECA), 413
  fallacy, 232–233
  generally, 22–23
  highway design, 246–247
  metric, 61, 374
  risk reduction, 241
Expected-value function, 262–263
Expert evidence, 115
Extreme event(s)
  dam failure, 240–243
  expected value of risk, 230–231
  fallacy of expected value, 232–233
  fault trees and, 403–406
  lahar flow case study, 575–587
  probability density function (pdf), 455
  quantification, 457
  risk and, 230–231, 260–261
  statistics of, 587–593
  sustainable development and, 84
Extreme failure, value of, 240
Extreme values
  partitioned multiobjective risk method (PMRM), 55, 319–322, 383
  sensitivity analysis, 241
  Weibull, 332–334, 337–338, 340–344

Faculty population problem, 133
Failure
  automated highway system (AHS) example, 104–107
  engineering-based systems, organizational, 26
  generally, 17–19
  risk assessment process and, 45
  software risk management, 25
  sources, 16–17, 41
Failure density, 394
Failure mode effects analysis (FMEA, FMECA)
  criticality analysis, 413–414
  failure effect probability, 413
  failure mode ratio, 412
  fault trees, 33
  hierarchical holographic modeling (HHM), 53, 70, 76, 78, 411
  methodology, 375, 393–395
  MIL-STD, 119, 393–395
  overview, 411–412
  part failure rate, 412
  probability, 413–414
  project risk management, 411

risk filtering, ranking, and management (RFRM), 413
risk priority number, 413
scenarios, 76
SAE J1739, 409
space missions, 476
time management, 11
two- and three-attribute approaches, 413
weaknesses, 413
Failure rate, fault-tree analysis, 402, 407
Fallacy of expected value, 229–233
Farmer's dilemma problem, 26–35, 41, 43, 51–56, 432–434, 657–664
Fault trees
    AND gate, 391, 396, 398–404
    applications of, generally, 32, 52, 78
    approximations, 391
    background, 389–391
    Boolean (logic) algebra, 392, 595
    car trouble problem, 418–420
    conditional expectation, rare and catastrophic events, 409
    conditional expected value, 402, 404
    conditional extreme expectations, 402, 409
    damage severity levels, 400–401, 404
    DARE *see* Distribution analyzer and risk evaluator (DARE)
    definition, 391
    example problems, 412–420
    exponential distributions, 402–403
    extreme events and, 391, 398–401
    Gumbel types, 391
    hierarchical holographic modeling (HHM), 53, 76–77
    idempotent law, 417
    importance measure, 411
    integrated reliability and risk analysis (IRRAS), 391
    intermediate event, 402
    limitations, 392
    minimal cut set, 397–399
    model description, 390
    normal distribution, 354
    OR gate, 391, 393, 395–396, 401–402
    overview, 389–391
    partitioned multiobjective risk method (PMRM), 391
    point probability distribution, 390
    procedure, 391–392
    regulation noncompliance problem, 415–418
    reliability, 389
        analysis, 392–397
    safety and reliability, 389–390
    scenarios, 76–77
    space shuttle redesigned solid rocket motor (RSRM) problem, 406–411
    top-down, 399
    top event, 390, 395
    unconditional expected values, 410
    variance, 391
    water distribution system problem, 412–414
    Weibull distributions, 391, 402–403
Federal Aviation Administration, 372
Fifth principle, 380
First Principle, 374
Filtering
    expert opinion, 214
    risk filtering, ranking and management (RFRM), 211–224
Flood frequency distribution problem, 241
Flood protection levee (fault tree), 387
Flood warning problem (MODT), 263–273, 275–279
FMEA, FMECA *see* Failure mode effects analysis
Fourth principle, 379
Fractile method
    airplane acquisition problem, 124–127
    applications, generally, 28
    cumulative distribution function (cdf), 124–125
    extreme events and, 246–250
    probability density function (pdf), 124–125, 455
Frechet distribution, 334
Furniture company problem, 72–73
Fusion submodels, 329

Gambler's ruin problem, 181
Game theory, 502, 619
Gamma function, 291–292
Gasket quality problem (6-sigma), 587–594
General nonlinear programming problem, 642
General optimization problem, 618–619
Gene-Tox database, 371–372
Geographic distribution, 474
Geometric programming, 620
Gestalt philosophy, systems engineering, 5
    Gestalt psychology, 474
    Gestalt–holistic philosophy, 375
Gradient methods, 619
Grapes, screening of, problem, 56–57
Groundwater problem(s)
    conditional expected value, 245–246
    constraints, 51, 56–57
    contamination problem, 21, 47–51, 56–577
    decision variables, 52, 56–57
    exogenous variables, 47–48, 56–57
    extreme events, 231, 243–245
    input variables, 47
    mathematical models and, 47

Groundwater problem(s) (cont'd)
  model parameters, 183–184
  model scope, 184
  multiobjective risk-impact analysis (MRIAM), 316–317
  objective functions, 56
  optimization techniques, 184–185
  output variables, 48–49
  partitioned multiobjective risk method (PMRM), 239–240, 243–245
  random variables, 48
  risk assessment process, 49
  risk, generally, 21, 56–57
  salinity, 483
  state variables, 48, 56
  unconditional expected value, 229–245
  variables, 48–50
Guiding principles, system based, 371, 372
Gumbel type distribution, generally, 31, 335, 351, 360–361

Hackers, 493
Hazard and operability analysis (HAZOP), 77, 446
Hazard U.S.-multihazard (HAZUS-MH), 143, 502
Hegel, systems engineering, 5
Heisenberg's uncertainty principle, 35
Heroin addiction problem (MRIAM), 311–312
Hessian matrix, 620
Hierarchical aspects of risk management
  engineered systems, 53, 70
  generally, 69, 497
  risk, identifying, quantifying, and evaluating, 60, 69
Hierarchical Bayesian models (HBMs), 486
  coordinated hierarchical Bayesian models (CHBM), 485, 486
Hierarchical holographic modeling (HHM)
  action horizon, 90
  adaptive two-player, 493
  aircraft development example, 450–454
  applications, generally, 28, 70–72
  attributes, 71–72
  automated highway system (AHS), 103–108
  comparison charts, 102–103
  decision variables, 74
  defined, 69–70
  exogenous variables, 72
  failure mode effects analysis (FMEA, FMECA), 76–78, 375, 411–414, 440, 477
  fault-tree analysis, 70
  hardening the water supply infrastructure, 96–98
  head topics, 70, 212
  hierarchical aspects, 69–70
  hierarchical holographic submodels (HHS), 83–85

lahar flow case study, 578–580
multiple-criteria decisionmaking (MCDM), 156
operations other than war (OOTW), 98–103
Pareto-optimal solution, 70, 475, 478
partitioning, 76–77
project risk management, 445
risk
  identification, 28, 74–76, 89, 90
  ranking of system, 86–87
risk filtering, ranking, and management (RFRM), 102
scenario model, 74
scenario structuring, 72–74
software acquisition, 90–95
space missions, 473, 478–480
subtopics, 71, 212–213
sustainable development, 83–87
system acquisition project, 86–90
  program consequences, 94–95
  system risks, 69
system constraints, 74, 80
terrorism, analysis of, 79
trade-offs, 69, 70, 90, 91, 481
triplet questions, 76
water resource system, 80–83
water supply, hardening infrastructure, 95–98
Hierarchical overlapping coordination (HOC)
  decomposition, 70–74, 485
  defined, 70
  furniture company problem, 72–73
  hierarchical holographic modeling (HHM), 53, 72–76
  matrix organization, 70–72
Hierarchical representation, of risk, 85–86
Hierarchical structures, 479, 497
Hierarchical-multiobjective framework, 18–20
Hierarchy of objectives, 381
High consequence-low probability, 55, 230
Highway construction problem (MSM), 434–435
Highway design, conditional expected value, 246–247
Highway traffic problem (MODT), 287–290
Holism, first principle, 374
Holism in cognition, 372
Holistic systems philosophy, 474
Holographic, 53, 478, 497 see also Hierarchical holographic modeling (HHM)
Homeland security
  adding resilience to systems, 499
  protection of assets, 499
  risk management phases, 499–502
  strategic preparedness, 502
Homework optimization problem, 56–57

Hurricane Katrina case study
    background, 560
    efficacy of preparedness, 567–569
    impacts on industry and infrastructure, 560–567
        education, recreation, and others, 566–567
        oil and gas sector, 561–563
        ports and water transportation, 565–566
        public utilities, 563–565
Hurwitz rule
    decision analysis, influence diagram, 145
    model, 150
    rules under uncertainty, 115–117
Hydrologic modeling, 483

Idempotent law, 417
Impact analysis, multiobjective decision-tree analysis (MODT), 260
Imprecision, defined, 43
Independent subsystems, 485
Indifference band, 162, 163, 354
Inequality constraints, 623
Influence diagrams
    channel reliability problem, Mississippi River, 129–132
    exogenous variables
        defined, 48, 132
        examples, 134–135
    generally, 128–132
    Hurwitz rule and, 145
    quantification of risk, 128
Information, intelligence, and models, 45–47
Infrastructure
    defined, 70
    interdependencies, 502, 511–540
Infrastructure problem (MODT), 290–293
Inoperability
    demand-side, 571
    risk of, 513–515
Inoperability input–output model (IIM), 511–540
    background, 512–513
    demand-side inoperability, 571
    development and extensions, 523–527
    dynamic, 527–530
    example problems, 536–539
    multiregional, 525–527
    practical uses, 530–533
    regimes of recovery, 516–517
    regional I/O multiplier system (RIMS II), 522–524
    supporting databases, 517–522
    uncertainty (U-IIM), 533–536
Integer (discrete) programming, 263–309

Integrated reliability and risk analysis system (IRRAS), 390
Intelligence, information, and models, 45–46
Interconnectedness, 380
Interdependencies, and intra, 380
Interdependency matrix (BEA), 517–540
Interstate transportation problem (MODT), 279–284
Intrinsic shared states
    enterprise performance risk, 596
    programmatic risk, 596
    technical risk, 596
Intrinsic states, 380
    shared states, 484
Inventory (water supply) problem (dynamic programming), 637–639
Inverse problem, 473
Irreversibility, uncertainty and, 180–182

Journey, The, 372–373
    compass, the, 373–374
    evolving base, 373
Judgment and values in decisionmaking, 601

Kaizen, 13, 18
KLOC, software risk management, 462–464
Knowledge management, 432–434
    boundaries, 443
    communication, 12–13
    compensation, 442
    defined, 12
    organizational failure, 18
    system goals, 13
    trust, 12, 13, 442
Knowledge uncertainty
    abnormal situations, 187
    approximation, 187
    disagreement, 187
    epistemic, 60, 382, 383, 477
    excluded variables, 187
    incorrect form, 187
    model uncertainty, 187–189, 383
    sources of, 191
    surrogate variables, 187
Kolmogorov, theory of probability, 52, 53
Kolmogorov–Smirnov test, 320
Kuhn–Tucker
    conditions, 381, 642, 645–646
    multipliers, 167, 168, 175–176
        multiobjective statistical method (MSM), 435–436
        simplex method, 630–632
        uncertainty sensitivity index model (USIM), 197

Lagrangian function
    desalinization problem, 623–624
    diet problem, 629–632
    formulation of, 621–624
    generalized nonlinear programming, 642–647
    Kuhn–Tucker multipliers, 622, 632
    Lagrange multipliers, 382, 619, 622, 626, 642–644
    multiobjective statistical method (MSM), 436
    nonlinear equality constraints, 622
    software acquisition, 466
Lahar flow case study
    auto regressive moving average (ARMA), 578
    conditional expected value, 585
    extreme events, 576, 581
    hierarchical holographic modeling (HHM), 576–578
    infrastructure, 581
    linear regression, 580
    mathematical modeling, 576–577
    modeling, 576–577
    multiobjective risk impact analysis method (MRIAM), 584–586
    multiple objective trade-off analysis, 581–584
    multiple objectives, 576
    Pareto frontier, 382, 384, 584
    Pareto-optimal solutions, 265, 582
    partitioned multiobjective risk method (PMRM), 585
    probability distributions, 580
    risk assessment and management, 577–578
    statistics of extremes, 581
    surrogate worth trade-off (SWT), 581–584
    triplet questions, 577, 587
    variables, 577
Leibniz
    rule, 340
    systems engineering, 5
Leontief input–output model, 512–514
Leslie matrix, 137, 150–152
Levee drainage system problem (MSM), 423–425
Life-cycle cost analysis, 440
Linear dynamic software estimation model, 462–464
Linear model, 41, 515, 618
Linear programming (LP), 626–633
    diet problem, 629–632
    dual primal problem, 628–629
    dual variables, 627, 629–631
    duality
        generally, 627–628
        Lagrange multipliers, 626–627
    feasible solution, defined, 626
    formulation, 625–626
    geometric solution of, 626
    model, 41
    objective function, 626
    optimal feasible solution, defined, 625
    primal, 626, 627
    primal problem, 627
    problem see Linear programming problem
    simplex method, 630–632
    simplex tableaus, 631
    slack variables, 627
    solution, defined, 626
    transportation problem, 632–634
Linear programming problem, 626–627
    optimization and, 626–627
Linear regression, 580
Log–Pearson Type III, 498
Low consequence-high probability, 230
Low frequency-high consequence, 241
Lumped parameter, 42, 619

MacLaurin expansion, 487
Maintainability, defined, 69
Malthusian parameter, 132
Management information system, 503–504
Managerial independence of the systems, 474
Manufacturing
    matrix organization, 71
    software risk management and, 25
Manufacturing problem (MSM), 435–436
Marginal rate of substitution, 169–171
Markov process models, 93
Mathematical models
    defined, 42, 618
    deterministic vs. probabilistic, 42
    distributed parameters vs. lumped parameters, 42
    hierarchical holographic modeling (HHM), 53, 69
    linear model, 42, 51
    linear vs. nonlinear, 42
    modeling errors, 182–185
    multiobjective multistage impact analysis method (MMIAM), 297–298
    multiobjective risk-impact analysis method (MRIAM), 295–326
    multiple-criteria decisionmaking (MCDM), 155
    nonlinear model, 42
    optimization techniques, 184–185
    parameters, distributed vs. lumped, 42–43
    probabilistic, stochastic model, 42
    risk function, 41
    static vs. dynamic, 42, 43
    uncertainty sensitivity index method (USIM), 196–199

variables, 47
see also Model(s) and modeling
Mathematical programming
   classical equality constraint problem, 620–624
   discrete (integer) programming, 620
   linear programming, 620
   nonlinear programming *see* Nonlinear programming
   quadratic programming, 620
   separable programming, 620
   static optimization, 619, 620
   unconstrained problem, 620
Maumee River Basin problem (HHM, SWT), 80–83, 165
Maximax criterion, 115–117
Maximin criterion, 115, 116
Maximum likelihood measures, 87
Maximum principle, 619
Mean time to failure (MTTF), 281–284
Medfly problem, (extreme events), 247–250
Meta-model Coordination, 380
   integration, and, 481
Meta-modeling, 479
   system of systems, of, 482
Metasystems methodology, 74, 326
   PSM-based metasystem, 484
Metrics, 386
MIL-STD, 411
Minimal cut sets, 411–414
Minimax criterion, 115–116
Minimum sufficient conditions, 620
Mises' criteria, 359–361, 363
Model(s) and modeling
   categories, 43
   decision variables, 47, 48
   deterministic, 618
   dynamic, 42, 619
   errors *see* Modeling errors
   hierarchical holographic modeling *see* Hierarchical holographic modeling (HHM)
   information (intelligence), 45–47
   input variables, 47
   intelligence, 45–47
   lahar flow case study, 576–577
   Leontief input–output, 503, 512–513
   Leslie, 135–138
   linear, 618
   mathematical, 47
      building blocks of, 47, 50, 149–150
      classifications, 42
      defined, 617
      exogenous variables, 47
      output variables, 47
      random variables, 47
      state variables, 47, 50
   mess chart, 129
   micropopulation, 135–138
   models, 295
   multiobjective statistical method (MSM), 33–34
   national goals, 43
   nonlinear, 618
   objective functions, 56–57
   optimization, 52–54
   optimization techniques, 184–185, 617–620
   parameters, 42, 183, 619
   probabilistic, 42, 618
   random variables, examples, 56–57
   regional economic, 483
   risk assessment process, 27–29, 41–65
   scope, 184
   sources of skepticism, 46–47
   static, 42, 619
   system, 41–42
   topology, 183, 619
   uncertainty, 182–183
   water resource system, 81
Modeling errors
   data, 184
   human subjectivity, 185, 383
   hydrologic, 483
   optimization techniques, 184–185
   parameters, 183–184
   scope, 184
   topology, 183
Monte Carlo simulation
   fault trees, 400, 409
   multiobjective statistical method (MSM), 432
   project risk management, 450
Most likely value (MLV), 122
Multidisciplinary, 387
Multiobjective analysis
   multiobjective decision, tree analysis (MODT), 281
   noninferior solution, 19, 43
   Pareto-optimum concept, 19, 43
Multiobjective decision tree (MODT)
   analysis, 259–324
   Bayes' formula, 265–268
   calculations, 647–650
   chance nodes, 261–262
   continuous case, 275–279
   decision node, 261–263
   defined, 31
   discrete case, flood warning, 263–273
   expected value, 293
   experimentation, 262–263

Multiobjective decision tree (MODT) (cont'd)
   extreme events, 275
   fold-back step, 262, 266, 274, 275
   impact analysis, 260
   loss vectors, 276–278
   mean time to failure (MTTF), 282–284
   multiple-risk measures, 274–275
   sample problems, 263–293
   sequential structure of, 262
   single objective (SODT), differences, 262, 279–281
   solution procedure, 261–262
   terrorism, analysis of, 494
   Weibull distribution, 290
Multiobjective, multistage impact analysis method (MMIAM), 297–326
   equivalent system, 302
   expected values, 302
   impact, defined, 298
   partitioning, 304
   stage trade-off, 297
Multiobjective multistage problem, software risk management, 465–468
Multiobjective problems
   multiobjective risk-impact analysis method (MRIAM), 317
   statistics of extremes, 329
Multiobjective risk-impact analysis method (MRIAM)
   defined, 298
   example problems, 313–325
   groundwater contamination problem, 323–325
   impact defined, 298
      impact analysis, 296–297, 304, 314–316
   lahar flow case study, 584–586
   modified heroin addiction problem, 318–323
   multiobjective models, 295
   pollution emission problem, 313–318
   purpose, 295
   risk functions, 302, 304
   risk-objective functions, 301
   sensitivity analysis, 318
   single-objective models, 295
   time horizon, 296
Multiobjective statistical method (MSM)
   conditional expectation, 432
   decision variables, 423, 427, 428
   exogenous variables, 432, 434, 437
   expected value, 424, 427, 432
   farmer's dilemma problem, 432–434
   highway construction, problem, 434–435
   Indran simulation model, 426
   Kuhn–Tucker conditions, 381, 435
   Lagrangian multipliers, 435
   levee drainage system problem, 423–425
   linear regression, 431
   log-normal distribution, 433
   manufacturing problem, 434–435
   methodology, 426–427
   Monte Carlo simulation, 432
   multiobjective functions, 431
   objective functions, 426
   optimization, 423–425
   Pareto-optimal, 424, 427, 428, 432, 433, 435
   partitioned multiobjective risk method (PMRM), 427
   Poisson distribution, 437, 438
   quadratic regression, 431
   random variables, 424, 432, 435
   regression analysis, 423
   risk functions, 627
   schematic diagram of, 424
   Siman simulation, 438
   simulation, 423, 424, 433
   state variables, 423, 424
   supermarket checkout problem, 437–438
   surrogate worth functions, 427
   surrogate worth trade-off (SWT), 423, 427–428, 431, 437
   trade-off values, 427
Multiobjective trade-off analysis, 155–177, 455
   advanced systems concepts, 155
   artificial intelligence (AI), 156
   ε-constraint method, 160–161, 163, 167
      decision space, 160
      functional space, 160
   decision-support system (DSS), 156
   example problems, 172–177
   hierarchical holographic modeling (HHM), 157
   improper inferior solution, 166–168
   improper noninferior solution, 166–168
   indifference band, 162–164
   Kuhn–Tucker multipliers, 161, 167, 168, 172, 174, 176–177
   lahar flow case study, 581–584
   Lambda theorem, 165
   modeling, 156
   multiobjective statistical method (MSM), 423
   multiple decisionmakers, 155, 165
      ideal case, 165–166
      probable case, 166
   multiple environmental objectives, 157–159
      categories of concern, 157
      client, 157
      constraints, 157
      nature, 157

scope, 157
   time horizon, 157
multiple perspectives, 157
multiple-criteria decisionmaking (MCDM), 155–157
multiple objective (vector) optimization problem (MOP), 159–160, 174
noninferior solution (Pareto-optimal), 156, 161
optimization, 155, 156
   impact analysis, 156
optimum solution, 162
philosophy, 155–156
proper noninferior solution, 166–168
Reid–Vemuri (water resources problem), 175–177
reservoir problem, 157–159
surrogate worth function, 162–164
surrogate worth trade-off (SWT), 159–174
   ε-constraint method, 160–161
   indifference band, 162
   Kuhn–Tucker condition, 161
   Lagrange multiplier, 165
   Lambda theorem, 165
   multiple decisionmakers, 165–166
   multiple objective optimization problem (MOP), 159–160
trade-off function, 160–162
trade-off surface, 169–170
trade-offs, 156, 159, 170
utility function approach, 168–172
water resources planning problem (Reid–Vemuri), 175–177
Weierstrass' theorem, 164
weighting method, 159
Multiobjective-criteria decisionmaking (MCDM), 155–157
Multiple decision makers (MDMs), 156
Multiple environmental objectives
   categories of concern, 157
   examples, 157–159
   reservoir problem, 157–159
Multiple-criteria decisionmaking (MCDM), 155–157, 188
   chemical carcinogenicity, prediction of, 371
   examples, 157–159
   extreme events, 230–232
   multiobjective decision-tree analysis (MODT), 259
   multiobjective trade-off analysis, 155–178

NASA *see* Space missions
National Environmental Policy Act (NEPA), 84
Network example (dynamic programming), 633–636
Network theory, 619
Newton, Isaac, and systems engineering, 5

Newton–Raphson method, 624–625
Next generation, 372, 377, 379, 381–383, 385, 594, 596
   FAA, 477
   National airspace system (NAS), 594, 596, 606
   UAS, 606
NextGen *see* Next generation
   NextGen S-o-S, 607
Ninth Principle, 384
Noncompliance with regulations problem (fault tree), 415–418
Noninferior solution *see* Pareto-optimum
Nonlinear model, 42, 618
Nonlinear programming
   direct search, 619
   dual function, 644, 646
   dual problem, 644
   dynamic programming, 633–642
   general nonlinear optimization problem, 642–647
   geometric programming, 619
   gradient methods, 619
   Kuhn–Tucker, 642–643, 645–647
   Lagrangian multipliers, 619
   optimization and, 619
   primal problem, 642
   problem, 645–647
   saddle point, 643, 644, 647
   theorem II, III, IV, 644–645

Objective functions, in risk assessment process, 56
Operational independence, 474
Operations other than war (OOTW)
   case study, 220–223
   hierarchical holographic model (HHM), 98–103, 214–215
Opportunity loss matrix, 120, 121
Optimality principle *see* Bellman's principle of optimality
Optimistic rule, 116
Optimization
   defined, 618
   linear programming, 626–633
   modeling, 617
   multiobjective decision-tree analysis (MODT), 274
   multiobjective risk-impact analysis method (MRIAM), 295
   multiple-criteria decisionmaking (MCDM), 155
   Newton–Raphson method, 624–625
Optimum solution, 162
OR gate and Boolean algebra, 396, 399–402, 408
Ordinal scale, 411–413
Output variables, in risk assessment process, 49–51

Parameter uncertainty
  identification of, 189–191
  measurement
    error, 189
    generally, 42, 194–195
  random error in direct measurements, 189–190
  sampling
    error, 195
    generally, 190–191
  statistical variation, 189
  systematic
    error, 190
    generally, 194–195
  unpredictability, 190, 195
Pareto distribution, Pearson type IV, 241
Pareto-optimal
  extreme events, 241
  fault trees, 419, 420
  frontier, 242, 268, 279, 287, 290, 382, 384, 440
    extreme events, 241
    fault trees, 419
    multiobjective risk-impact analysis method (MRIAM), 314–316
  hierarchical holographic modeling (HHM), 53, 69
  limitations, 175
  multiobjective decision-tree analysis (MODT), 267–268, 279, 287
  multiobjective risk-impact analysis method (MRIAM), 295, 314–316
  multiobjective trade-off analysis, 156, 158, 175, 382
  policies, 324
  set, 284
Pareto-optimum
  multiobjective decision-tree (MODT), 260
  multiobjective trade-off analysis, 155, 164
  multiple-criteria decisionmaking (MCDM), 156, 158–159
  risk assessment process, 43–45
Partitioned multiobjective risk method (PMRM)
  catastrophic events, 257
  chi-square test, 330
  coefficient of variation, 334, 335
  conditional expectations, 233, 326, 349
  conditional expected-risk function, 329–332, 338, 356
  conditional expected value, 233–235, 651–653
  conditional expected-value functions, 230, 232–233
  cost function, 313
  Cramer's method, 359, 362, 364, 365
  cumulative distribution function (cdf), 230, 232
  damage ranges, 233
  defined, 229

dissolved oxygen problem, 357–358
distribution-free results, 350–352, 356
example problems, 357–368
exceedance probability, 230
expected value, 229–231, 313, 383
exponential distribution, 331, 334–340, 345, 351
extension of, 329–333
extreme events, 55, 228–230, 349–351
extreme failure, 226
extreme values, 236
extremes, 338–344
flood peak problem, 361–362
formulation, 235–237, 259–261
fractile distribution, 367–368
Gumbel Type I distribution, 335, 336, 338, 351, 360–361, 363
Gumbel Type II, 335, 336, 338, 351, 360–361
Gumbel Type III, 241, 335, 336, 351, 360–361, 365
Kolmogorov–Smirnov test, 330
lahar flow case study, 584
log-normal distribution (LN), 241, 288, 331–332, 345–349
MacLaurin series, 333
mean, 234, 298, 301, 332, 350
medfly problem, 247–250
Mises' criteria, 359–361, 363
multiobjective multistage impact analysis method (MMIAM) and, 297, 304
multiobjective optimization, 292, 303
multiobjective statistical method (MSM), 427
natural hazards, 233, 240, 257
nonexceedance probability, 234
normal case, 356
normal (N) distribution, 265, 302, 303, 336, 345, 346, 352
Pareto distribution, 241, 351–353
partitioning points, 302, 347–349
partitioning probability, 235–237, 341–342, 352, 353
probability density function (pdf), 234, 329, 334–335, 345, 354, 404
random damage, 348
ranges, 314
Rayleigh distribution, 362
risk analysis, 242–244
risk modeling, 231–232
risk objective, 241
sensitivity, 237, 241, 242, 332, 341
s/m, 342, 344–347, 349
software risk management, 455
space mission Challenger, 476–477
standard deviations, 321

standard normal distribution, 347
statistics of extremes, 238, 322–335, 340, 350
surrogate worth trade-off (SWT), 237
terrorism, analysis of, 499
triangular distribution, 338, 363–365
Type I, 335, 336, 338, 351, 360–361, 363
Type II, 335, 336, 338, 351, 360–361
Type III, 241, 335, 336, 351, 360–361, 365
unconditional expected damages, 240
Weibull (W) distribution, 290, 331–333, 335–337, 340–349
Partitioning, 76–78, 338–340, 349
Pearson type IV, 240
Pessimistic rule, 116
Phantom system models (PSM), 139–144, 380, 381, 473, 479, 484
   decision-based modeling and simulation, 143
   graphical depiction of the methodological framework, 143
   risk modeling, 140
Pharmaceuticals problem (MODT), 284–287
Physical infrastructure, 379
Plato, systems engineering, 5
Poisson distribution, 437, 438
Pollution emission problem (MRIAM), 304–311
Population dynamics
   example problems, 133–135, 138–147
   exponential model, 133
   macro population model, 132–135
   Malthusian parameter, 132
   micro population model, Leslie, 135–138
   model
      exponential, 133
      macro, 132–133
      micro, 135–138
   problems, 133–135, 139–148
Population testing problem (decision analysis), 150–152
Preparedness, 58, 502
Probabilistic model, 618
Probabilities
   adverse effects, 554
   capability, 554
   cumulative (probability) density function (cdf), 229
   evidence-based, 28–29
   exceedance probability, 229, 233
   extreme event, 227–228, 580–581
      statistics of extremes, 322
   failure mode effects analysis (FMEA, FMECA), 411
   generation of, 124
      fractile method, 124
      triangular distribution, 127

   intent, 495, 498
   objective, 115
   subjective, 115
   terrorism, analysis of, 494, 495, 498
Probability density function (pdf), 48, 455, 498
Probability distributions, 309, 325, 330
Probable maximum flood (PMF), 240–242
Problem-solving steps, 10–13
Program evaluation and review technique (PERT), 11, 93, 459
Programmatic risk, 440
Project risk management
   aircraft development example, 450–452
   continuous risk management, 440, 441
   cost estimation, 459–461
   cost overrun, 440
      failure mode and effects analysis (FMEA, FMECA), 441
      hierarchical holographic modeling (HHM), 444
   iteration, 448
   knowledge management, 440–442
   life cycle, 439–440
   literature review, 437–438
   methods, 442–448
   overview, 437–438
   participating parties, 439
   quantitative framework, 454–456
   quantitative risk assessment, 452–456
   risk
      assessment, 447–448
      defined, 438
      filtration, 445–447
      generally, 439
      identification, 444–445
      management, 448
      tracking, 448
   schedule delays, 440
   software acquisition, 452–456
   software development taxonomy, 452–454
   software estimation model, 461–463
      examples, 464–465
      solution approach, 462–464
   software nontechnical risk, 459
   taxonomy, 452–454
   teams, 440–442
Proper noninferior solution, 166–168
   ε-constraint method, 167, 168
   improper noninferior solution, 166
   Kuhn–Tucker multipliers, 166
PSM laboratory (PSML), 486

Quadratic programming, 620
Quality control, defined, 18
Quantitative framework for software acquisition, 454–455
Quantitative risk analysis, 76, 115, 378, 411, 493
Quantitative risk assessment, 115, 454–458
Quantitative risk index, 180
Quasilinearization, 619
Queuing theory, 619

Random variable, in risk assessment process, 47, 49, 56, 57, 299, 393
Rayleigh distribution, 362
Recursive equation, 636–639
　water supply problem, 637–639
Redundancy, 217, 501–502
Regional economic model, 483
Regional Input–Output Multiplier System (RIMS II), 522–524
Regret matrix *see* Opportunity loss matrix
Reid–Vemuri (water resources problem), 175–178
Reliability, 389–420
　analysis, 392
　defined, 69, 389, 394
　exponential distribution, 402
　fault trees, 391–397
　flood protection levee, 394
　parallel system, 395–396
　random variables, 393
　series system, 394–395
　　Boolean (logic) algebra, 395
　time-to-failure distributions, 401–403
　trade-offs, 392
　unreliability, 515
　Venn diagram, 396–397
Requirements, 604
Reservoir modification problem (PMRM), 344–346
Reservoir problem, 157–159
Resilience
　defined, 52, 56–57, 60, 217, 378–379, 384, 476, 501–502
　in emergent systems, 501–502
Responsivity, uncertainty, 179–180
Restaurant problem, choosing, 122–123
Risk analysis, generally
　criteria for, 26
　automated highway system (AHS), 104
　conditional expectation, 233
　decision rules, 116–118
　extreme events and, 229–230
　holistic approach, 4–5, 43
　modeling, 476

multiobjective risk-impact analysis method (MRIAM), 298
partitioned multiobjective risk method (PMRM), 233–235
sources of skepticism, 46–47
total quality management (TQM), 586
universities, role of, 16–17
Risk assessment
　engineering-based systems, 25–26
　expected value of risk, 61, 222
　extreme events, 22
　hierarchical holographic modeling (HHM), 53, 73–79, 85
　lahar flow case study, 577–578
　modeling, 41, 375
　multiple-criteria decisionmaking (MCDM), 155
　perception of, 21
　phantom system models, 139–141
　process overview *see* Risk assessment process, 44
　software acquisition, 90–94
　sustainable development, 83–86
　triplet questions, 44
Risk assessment process, 44, 46, elements
　defined, 44
　models *see* Model(s) and modeling)
　risk evaluation, 45
　risk identification, 45
　risk management, 45, 375
　risk modeling, 45
　risk quantification and measurement, 45
　steps of, 44
　systems analysis, 47
　triplet questions, 44
Risk communication, 16–17, 55
Risk evaluation, 45
Risk filtering, ranking, and management (RFRM), 211–223
　Bayes' theorem, 218, 222, 494
　failure mode effects analysis (FMEA, FMECA), 411
　hierarchical holographic modeling (HHM), 213–217, 220
　　head topics, 214
　　subtopics, 214
　large-scale systems, 211
　operations other than war (OOTW), 98–103, 219–223
　phases, 214–219
　　bicriteria filtering and ranking, 214, 216–217, 220–221
　　multicriteria evaluation, 214, 221–222
　　operational feedback, 214, 219, 223
　　quantitative ranking, 214, 218, 222–223
　　risk management, 214, 218

SUBJECT INDEX 687

safeguarding against missing critical items, 214, 218–219, 223
scenario filtering, 214–216, 220
scenario identification, 214–215
principles, 213–214
project risk management, 446–450
redundancy, 217
resilience, 52, 56, 60, 217
risk ranking and filtering (RRF), 212–214
robustness, 217
software acquisition quantitative framework, 455–458
terrorism, analysis of, 62, 215
U.S. Air Force matrix, 215
Risk functions, in multiobjective decision-tree analysis (MODT), 260–261, 274
Risk, generally
acceptance and avoidance, 45
analysis *see* Risk analysis
assessment of *see* Risk assessment
catastrophic events, 229–231
conditional expected value, 115
decision analysis, 115
defined, 4, 16, 44, 75, 76, 139, 179, 230
empirical process, 232
evaluation, 45
expected value of, 49, 115, 231–233
extreme events, 229–230
fault trees, 414, 415, 417, 427, 587
feeling of, 43
groundwater contamination, 21–22
identification, 45
irreversibility, 179–180
management *see* Risk management
modeling, 45
multiobjective risk-impact analysis method (MRIAM), 296–297, 304
multiobjective statistical method (MSM), 32–34
perception of, 21
quantification, 60, 222
responsivity, 179–181
sensitivity, 179–181
sources, 45
stability, 179–181
sustainable development, 83–86
systems engineering, 14–15
triplet questions, 76
Risk identification, 45
Risk management
comparison charts, 102–103
defined, 17, 375
engineering-based systems, 25–26

failure, sources of, 16–17
hierarchical aspects, generally, 69
hierarchical holographic modeling (HHM), 29, 551
lahar flow case study, 577–578
linear programming model, 41
multiple-criteria decisionmaking (MCDM), 155
phantom system models, 139–141
projects *see* Project risk management
software acquisition, 81, 82
software risk management, 90–91
sustainable development, 83–86
time frame, 55, 384, 595, 610
total risk management (TRM), 17–18
triplet questions, 44–45, 76–78
Risk modeling, 45
Risk priority number (FMEA, FMECA), 410
Risk ranking and filtering (RRF)
defined, 212–213
Risk scenario, 477, 502, 537
Risk tracking, 443–444
milestone chart, 444
Robustness, 217, 599

Saddle point, 643–644
SAE J1739 (FMEA, FMECA), 409
Satisficing principle, 594, 599
SCADA, 97, 252, 381
Scenario model, 75–76
single-failure, 78
success or as-planned, 76
Scenario structuring, theory of (TSS), 52, 53, 76–77, 374, 375, 384
Scenario tracking, intelligence gathering, 499
Second Principle, 375
Sensitivity
analysis *see* Sensitivity analysis
partitioned multiobjective risk method (PMRM), 335–340, 342
partitioning, 338–339
Sensitivity analysis
applications of, generally, 29–30
uncertainty and, 179–181
Separable programming, 431–432, 620
Sequential Pareto-optimal, 593
Seven Habits *see* Covey's Seven Habits
Seventh principle, 382
Severity, failure mode effects analysis (FMEA, FMECA), 413
Shared states, 374, 484, 595–596
Sigma-limit capabilities, 587–588

Siman simulation, 437
Simplex method, 627–633
Simplex tableaus, 631–632
Single-objective decision tree (SODT), 115, 262, 279–281
Six-sigma case study
   cumulative distribution function (cdf), 589
   gasket quality problem, 587–593
   return period (RP), 590
   sigma, defined, 587
   sigma-limit capabilities, 587
   standard normal distribution table, 588
   statistics of extremes, 587–593
   tolerance limits (TL), 587
   total quality management (TQM), 586
      risk analysis, 587
   zone control charts, 587
Sixth Principle, 381
Society for General Systems Research, systems engineering, 7
Society for risk analysis, 372
Software acquisition, 90–94, 454–458
Software capability evaluation (SCE), 454
Software development
   acquisition, 454–455
   commercial off-the-shelf (COTS), 461
   conditional expected value, 455, 464, 465, 467
   constructive cost model (COCOMO), 92, 459, 465
   database management system (DBMS), 461
   development environment, 456
   discrete dynamic modeling, 461–469
   extreme events
      conditional expected value, 467
      risk of, 456
   nontechnical risk factors, 458–460
   product engineering, 455, 456
   program constraints, 455, 456
   program evaluation and review techniques (PERT), 459
   taxonomy, 455
   trade-off values, 466
   variance, 460–461
Software engineering, 24, 454
Software Engineering Institute (SEI), 59, 440
Space missions
   risk filtering, ranking and management
   solid rocket boosters, 406
   space shuttle redesigned Solid Rocket Motor (RSRM), 405–406
   space transportation system (STS), 406

Space Shuttle Redesigned Solid Rocket Motor (RSRM) problem, 405–406
Specifications, 604
Stability, uncertainty and, 179, 181–182
Standard normal probability, 664–665
State variables, 50, 52, 476
   autocratic regimes, 500
   definition, 47, 49, 378, 476
   extremism, 500
   globalism, 500
   nationalism, 500
   oppression, 500
   resource starvation, 501
   technology, 501
   underdeveloped infrastructure, 501
Static model, 42, 618
Stationary point, 620
Statistics of extremes
   cumulative distribution function (cdf), 332, 334
   extension of partititioned multiobjective risk method (PMRM), 329–331, 384
   generalized quantification of risk, 333, 350–356
   lahar flow case study, 580–581
   Leibniz's rule, 340
   partititioned multiobjective risk method (PMRM), 340–344
   recursive method, 342–344
   risk functions, 330
   sensitivity to partitioning, 344–350
   six-sigma case study, 586–593
   standard deviation, 332
Stochastic variability, uncertainty and, 192–193, 382–383
Student dilemma problem, 19–21
Supermarket checkout problem (MSM), 437–438
Supervisory control and data acquisition (SCADA), 97, 252, 381
Surrogate variables, uncertainty and, 187–188, 194
Surrogate worth function
   ε-constraint method, 163, 171–172
   interactive procedures, 164–166
   Lambda theorem, 165
   Weierstrass' theorem, 164
Surrogate worth trade-off (SWT) method
   ε-constraint method, 160–161, 163, 170–171
   example problems, 172–178
   fault-tree analysis, 420
   interactive procedures, 170–172
   lahar flow case study, 581–584
   marginal rate of substitution, 170–171
   Maumee River Basin problem, 165–166
   multiobjective optimization problems, 166–167

multiobjective risk-impact analysis method (MRIAM) and, 298, 304
partitioned multiobjective risk method (PMRM), 237, 298
probable case, 166
space missions, 476–477
trade-off function, 161–163
utility function approach, 168–172
Sustainable development
　defined, 83
　hierarchical holographic modeling (HHM)
　　framework, 84–85
　　for risk identification, 85
　science and engineering, 84–86
Sustainable future, 84–85
Sylvester's theorem, 620–621
Synergy, defined, 12–13
System acquisition (HHM), 86–90
System optimization, 473
Systems
　linear dynamic, 49
　phantom system models, 131–144
Systems analysis, sources of skepticism, 46
Systems engineering
　circle of concern, 10
　circle of influence, 10
　Covey's Seven Habits, 5, 8–13
　fundamentals, 41
　historical perspective, 5–8
　humans, 16
　iteration, 13, 14
　life cycle, 7
　management, 16–17
　optimization, 617
　organization, 16
　phases, 619
　philosophers, 5–7
Systems modeling, 473
　deterministic
　　defined, 618
　　probabilistic *vs.*, 618
　distributed parameters, lumped parameters *vs.*, 42, 618
　dynamic models, 618
　linear
　　defined, 618
　　nonlinear *vs.*, 618
　lumped parameters, 42, 618
　nonlinear model, 618
　probabilistic model, 618
　schematic, 618
　static

　　defined, 618
　　dynamic *vs.*, 618
Systems of systems, 8, 50, 380, 381, 473, 474, 485, 494
　emergent, 595
Systems theory, 6
Systems-based theory, 474

Target industry markets (TIMs), stochastic variability, 192–193
Taxonomy, 454–456
　uncertainty, 185–196
Taylor series expansion, 200
Tech-Com study, 43, 157
Technical coefficient matrix, 521–522
　stability of, 532–533
Technical risk, 454
Tenth Principle, 386
Terrorism
　Bayesian analysis, 499–500
　capability, defined, 56
　evidence ratio, 495
　hierarchical holographic model (HHM), 493–506
　inoperability, defined, 34
　inoperability input–output model (IIM), 513–516
　intent, defined, 56
　Leontief input–output model, 512–513
　modeling, 495–499
　multiobjective decision tree (MODT), 499
　network system, 499–502
　observables, 498
　　analysis of, 498–499
　probability, 494–495
　risk analysis, five premises on, 62
　risk filtering, ranking, and management (RFRM), 498
　risks of, 497, 501
　SCADA, 252–256
　scenario model, 498
　scenario tracking, intelligence gathering, 494, 496–498
　threat scenario, 59–65
　vulnerability, defined, 56, 502
Terrorist network system, model, 63, 500–501
Theoretical framework, 372
Theory of scenario structuring (TSS), 52, 53, 76–77, 374, 375, 384
Third principle, 378
Time management, 11
Time-to-failure distributions, 402–403
Total quality management (TQM), 587
Total risk management (TRM), 17–18
Transportation problem (linear programming), 632–633
　canonical transportation problem, 633

Triangular distribution
   decision analysis, 115, 127–129
   expected value, 653–655
   performance assessment problem, 127–128
   statistics of extremes, 329, 353–354
Triplet questions, 16–17, 537, 551, 558, 577

Uncertainty
   decision, 191–192
   defined, 43, 179
   fault trees, 60, 389–391, 400
   hardening the water supply infrastructure, 94–98
   hierarchical holographic modeling (HHM), 53, 73–76
   knowledge model, 187–192
   multiobjective multistage impact analysis method (MMIAM), 297–298
   multiobjective risk-impact analysis method (MRIAM), 298
   multiobjective statistical method (MSM), 31–32
   parameter *see* Parameter uncertainty
   responsivity, 181–182
   sensitivity, generally, 180–181
   sensitivity index method (USIM), 29–30
   social cost of risk, 191
   social values, quantification of, 191
   sources, 38–39
   stability, 180–182
   surrogate variables, 192–194
   target industry markets (TIMs), 192–196
   taxonomy, 185–186, 196
   uncertainty-sensitivity index method (USIM), 180–181
   variability, 186–187
Uncertainty-sensitivity index method (USIM), 196–204
   algorithm of, 204–208
   applications to dynamic systems, 201–202
   ε-constraint form, 196–197
   defined, 196
   equality constraints, 203–204
   Kuhn–Tucker, 197
   Lagrangian function, 197
   multiobjective optimization problem, 199–204
   multiple uncertain parameters, 200–204
   optimality, 207–209
   parameter optimization, 199–204
   sensitivity, generally, 180–181
   sensitivity objective function, 198
   surrogate worth trade-off (SWT), 198–201
Unconditional expected value
   triangular distribution, 653–655
Unconstrained problem, 620
Unmanned aircraft system, 378
Unreliability, 514
Unshared state *see* Shared state
U.S. Air Force matrix, 215
Utility function, 168–172

Variability, 44
Variability uncertainty
   aleatory, 60, 382, 383, 477
   component sources of, 186–187
   defined, 186
   temporal, individual, and spatial, 193–194
Variables, defined, 32, 47–51
Variance
   multiobjective risk-impact analysis method (MRIAM), 300–303
   partitioned multiobjective risk method (PMRM), 341–342
   uncertainty, 274
Venn diagram, 396
von Mises' criteria, 359–360, 363, 365
Vulnerability, 52, 56
   resilience, and, 374, 378–379, 384, 476

Water cylinder (tower) problem, 622–623
Water distribution system problem (fault tree), 414–415
Water inventory problem (dynamic programming), 639–642
Water resources problem (Reid–Vemuri), 175–177
Water supply, hardening infrastructure, 94–98
Waverly bank problem (uncertainty), 192–196
Weibull distribution *see specific types* of analysis methods
Weierstrass' theorem, 165
Wertheimer, systems engineering, 6
World [Bruntland] Commission on Environment and Development, 8

WILEY SERIES IN SYSTEMS ENGINEERING AND MANAGEMENT

Andrew P. Sage, Editor

ANDREW P. SAGE and JAMES D. PALMER
Software Systems Engineering

WILLIAM B. ROUSE
Design for Success: A Human-Centered Approach to Designing Successful Products and Systems

LEONARD ADELMAN
Evaluating Decision Support and Expert System Technology

ANDREW P. SAGE
Decision Support Systems Engineering

YEFIM FASSER and DONALD BRETTNER
Process Improvement in the Electronics Industry, Second Edition

WILLIAM B. ROUSE
Strategies for Innovation

ANDREW P. SAGE
Systems Engineering

HORST TEMPELMEIER and HEINRICH KUHN
Flexible Manufacturing Systems: Decision Support for Design and Operation

WILLIAM B. ROUSE
Catalysts for Change: Concepts and Principles for Enabling Innovation

LIPING FANG, KEITH W. HIPEL, and D. MARC KILGOUR
Interactive Decision Making: The Graph Model for Conflict Resolution

DAVID A. SCHUM
Evidential Foundations of Probabilistic Reasoning

JENS RASMUSSEN, ANNELISE MARK PEJTERSEN, and LEONARD P. GOODSTEIN
Cognitive Systems Engineering

ANDREW P. SAGE
Systems Management for Information Technology and Software Engineering

ALPHONSE CHAPANIS
Human Factors in Systems Engineering

YACOV Y. HAIMES
Risk Modeling, Assessment, and Management, Third Edition

DENNIS M. BUEDE
The Engineering Design of Systems: Models and Methods, Second Edition

ANDREW P. SAGE and JAMES E. ARMSTRONG, Jr.
Introduction to Systems Engineering

WILLIAM B. ROUSE
Essential Challenges of Strategic Management

YEFIM FASSER and DONALD BRETTNER
Management for Quality in High-Technology Enterprises

THOMAS B. SHERIDAN
Humans and Automation: System Design and Research Issues

ALEXANDER KOSSIAKOFF and WILLIAM N. SWEET
Systems Engineering Principles and Practice

HAROLD R. BOOHER
Handbook of Human Systems Integration

JEFFREY T. POLLOCK and RALPH HODGSON
Adaptive Information: Improving Business Through Semantic Interoperability, Grid Computing, and Enterprise Integration

ALAN L. PORTER and SCOTT W. CUNNINGHAM
Tech Mining: Exploiting New Technologies for Competitive Advantage

REX BROWN
Rational Choice and Judgment: Decision Analysis for the Decider

WILLIAM B. ROUSE and KENNETH R. BOFF (editors)
Organizational Simulation

HOWARD EISNER
Managing Complex Systems: Thinking Outside the Box

STEVE BELL
Lean Enterprise Systems: Using IT for Continuous Improvement

J. JERRY KAUFMAN and ROY WOODHEAD
Stimulating Innovation in Products and Services: With Function Analysis and Mapping

WILLIAM B. ROUSE
Enterprise Tranformation: Understanding and Enabling Fundamental Change

JOHN E. GIBSON, WILLIAM T. SCHERER, and WILLAM F. GIBSON
How to Do Systems Analysis

WILLIAM F. CHRISTOPHER
Holistic Management: Managing What Matters for Company Success

WILLIAM B. ROUSE
People and Organizations: Explorations of Human-Centered Design

MO JAMSHIDI
System of Systems Engineering: Innovations for the Twenty-First Century

ANDREW P. SAGE and WILLIAM B. ROUSE
Handbook of Systems Engineering and Management, Second Edition

JOHN R. CLYMER
Simulation-Based Engineering of Complex Systems, Second Edition

KRAG BROTBY
Information Security Governance: A Practical Development and Implementation Approach

JULIAN TALBOT and MILES JAKEMAN
Security Risk Management Body of Knowledge

SCOTT JACKSON
Architecting Resilient Systems: Accident Avoidance and Survival and Recovery from Disruptions

JAMES A. GEORGE and JAMES A. RODGER
Smart Data: Enterprise Performance Optimization Strategy

YORAM KOREN
The Global Manufacturing Revolution: Product-Process-Business Integration and Reconfigurable Systems

AVNER ENGEL
Verification, Validation, and Testing of Engineered Systems

WILLIAM B. ROUSE (editor)
The Economics of Human Systems Integration: Valuation of Investments in People's Training and Education, Safety and Health, and Work Productivity

ALEXANDER KOSSIAKOFF, WILLIAM N. SWEET, SAM SEYMOUR, and STEVEN M. BIEMER
Systems Engineering Principles and Practice, Second Edition

GREGORY S. PARNELL, PATRICK J. DRISCOLL, and DALE L. HENDERSON (editors)
Decision Making in Systems Engineering and Management, Second Edition

ANDREW P. SAGE and WILLIAM B. ROUSE
Economic Systems Analysis and Assessment: Intensive Systems, Organizations, and Enterprises

BOHDAN W. OPPENHEIM
Lean for Systems Engineering with Lean Enablers for Systems Engineering

LEV M. KLYATIS
Accelerated Reliability and Durability Testing Technology

BJOERN BARTELS, ULRICH ERMEL, MICHAEL PECHT, and PETER SANDBORN
Strategies to the Prediction, Mitigation, and Management of Product Obsolescence

LEVANT YILMAS and TUNCER ÖREN
Agent-Directed Simulation and Systems Engineering

ELSAYED A. ELSAYED
Reliability Engineering, Second Edition

BEHNAM MALAKOOTI
Operations and Production Systems with Multiple Objectives

MENG-LI SHIU, JUI-CHIN JIANG, and MAO-HSIUNG TU
Quality Strategy for Systems Engineering and Management

ANDREAS OPELT, BORIS GLOGER, WOLFGANG PFARL, and RALF MITTERMAYR
Agile Contracts: Creating and Managing Successful Projects with Scrum

KINJI MORI
Concept-Oriented Research and Development in Information Technology

KAILASH C. KAPUR and MICHAEL PECHT
Reliability Engineering

MICHAEL TORTORELLA
Reliability, Maintainability, and Supportability

YACOV Y. HAIMES
Risk Modeling, Assessment, and Management, Fourth Edition

Printed in the USA/Agawam, MA
November 22, 2023

855431.020